Routledge Revivals

The Philosophical Works of Francis Bacon

First published in 1905, this reissued edition of *The Philosophical Works of Francis Bacon* is an edited collection based upon the definitive seven volume edition of 1857, translated and prefaced by Robert Leslie Ellis and James Spedding.

Of great historical, philosophical and scientific interest, this collection brings together translations of Bacon's most important works, including the *Novum Organum*, the *De Augmentis Scientarium*, the *Parasceve*, and the *De Principiis atque Originibus,* as well as works originally written in English, such as the *Valerius Terminus* and the *Filum Labyrinthi*. The reissue offers a comprehensive and provocative collection of the key writings of the man we now consider to be the father of Empiricism who popularised inductive methodologies for scientific inquiry. All works include prefaces by Robert Leslie Ellis and James Spedding, and the collection includes an introductory note from the editor John M. Robertson.

The Philosophical Works of Francis Bacon

Reprinted from the texts and translations, with the notes and prefaces, of

Ellis & Spedding

Edited with an Introduction by

John M. Robertson

Routledge
Taylor & Francis Group
Routledge Revivals

First published in 1905
by George Routledge & Sons Limited

This edition first published in 2011 by Routledge
2 Park Square, Milton Park, Abingdon, Oxon, OX14 4RN

Simultaneously published in the USA and Canada
by Routledge
711 Third Avenue, New York, NY 10017

Routledge is an imprint of the Taylor & Francis Group, an informa business

Publisher's Note
The publisher has gone to great lengths to ensure the quality of this reprint but points out that some imperfections in the original copies may be apparent.

Disclaimer
The publisher has made every effort to trace copyright holders and welcomes correspondence from those they have been unable to contact.

A Library of Congress record exists under LC Control Number: 70119952

ISBN 13: 978-0-415-67955-8 (hbk)
ISBN 13: 978-0-203-80632-6 (ebk)

THE PHILOSOPHICAL WORKS OF FRANCIS BACON

THE PHILOSOPHICAL WORKS OF FRANCIS BACON

BARON OF VERULAM, VISCOUNT ST. ALBANS, AND LORD HIGH CHANCELLOR OF ENGLAND

REPRINTED FROM THE TEXTS AND TRANSLATIONS, WITH THE NOTES AND PREFACES, OF

ELLIS AND SPEDDING

EDITED WITH AN INTRODUCTION

By JOHN M. ROBERTSON

LONDON
GEORGE ROUTLEDGE AND SONS LIMITED
NEW YORK: E. P. DUTTON & Co.
1905.

EDITOR'S PREFACE

An adequate collection of the chief works of Francis Bacon, at a price within the reach of the mass of readers, has long been wanting, and the present reprint, from the magistral edition of Ellis and Spedding (7 vols., 1857), is an attempt to meet the need. It contains far more than has ever before been included in any popular collection, and is believed to give all of Bacon's philosophical writings that are likely to be read by any save a few special students. Such treatises as those collected in the fifth volume of the Ellis-Spedding edition under the title of *Natural and Experimental History*—the "Histories" of the Winds, of Life and Death, of Dense and Rare, and the rest—have now not only no scientific value, but almost no literary or philosophical interest, the subject matter being such as would give small scope to Bacon's style even had they been written in English, whereas, like so much of the more interesting matter in this volume, they were published in Latin. The *Sylva Sylvarum*, though in Bacon's English, is no less obsolete. Our collection, however, includes translations not only of the *Novum Organum* and the *De Augmentis Scientiarum*, but of the *Parasceve*, the *De Principiis atque Originibus*, the *Descriptio Globi Intellectualis*, the *Thema Coeli*, and the *De Sapientia Veterum*, as well as the original English treatises entitled *Valerius Terminus*, and *Filum Labyrinthi*—the latter a version by Bacon of his Latin *Cogitata et Visa*.

With the aid of the valuable prefaces by Messrs. Ellis and Spedding, which are in every case retained, the student can gather from this mass of matter a thorough knowledge of Bacon's work, in system and in detail, in its strength and its weakness.

Holding that Bacon's didactic works were to be reproduced as far as they had literary value, the editor has included not only the original *Advancement of Learning* and the classic *Essays* and *New Atlantis*, but the *Apophthegms*, though, like the *Wisdom of the Ancients*, they strain somewhat on the title of *Philosophical Works*. The *History of Henry VII* is omitted as being wholly outside that title, and as being easily accessible in cheap editions; and the legal works are omitted as appealing only to a few even among law students.

The large mass of Bacon's work here brought together is presented to the reader in the most accurate texts and the most accurate translations in existence —those of the complete edition of Ellis and Spedding. On the English texts of that edition Mr. Spedding lavished a care which can be partially appreciated from his notes in the following pages. The bulk of the translations, which were made originally by Mr. Francis Headlam, was thoroughly revised by Mr. Spedding, and part of that of the *Novum Organum* by Mr. Ellis also. The present editor will not venture to say that they are absolutely faultless; but after making a number of comparisons he is satisfied that they are quite the most trustworthy that have been published.

In addition to the boon of accurate versions, and the skilled guidance given in the sectional prefaces, the reader of the present edition has the help of the multitude of learned notes appended by Mr. Ellis to the Latin *Novum Organum* and *De Augmentis*, as well as those added by Mr. Spedding to these works, and his annotations and various readings to the English works. With the help of the latter, which note changes made in the translations as well as the variants of the earlier editions, Bacon's thought can be followed with critical closeness.

Mr. Ellis's notes, in so far as they were not philological, and applicable only to the originals, have been transferred to the corresponding passages in the translations in this reprint. A small number of notes, some of them borrowed from other editions, have been added, always in brackets, by the present editor; but beyond adding a few instances to those singled out by Mr. Ellis, he has not attempted either to add to the specification of Bacon's mistakes in physics or to note the advances made in science since Mr. Ellis wrote. The *Novum Organum* and *De Augmentis* are now read not for scientific information, but as the exposition of a great writer's conception of the needs and the methods of the sciences, some three centuries ago.

As Mr. Spedding indicated in his first preface to the seven-volume edition, no chronological or other "order" of Bacon's works has ever been agreed upon, and that followed in the present volume is a compromise on his. After Rawley's "Life" and Mr. Ellis's General Preface, the original *Advancement of Learning*, in two books, is placed first, as the simplest and most attractive, as well as the earliest, of Bacon's fuller expositions of his aims; the pregnant fragments entitled *Valerius Terminus* and *Filum Labyrinthi* coming next. Then follow the translations of the *Novum Organum*, the *Parasceve*, of which the original appeared in the same volume with the *Organum* in 1620; the *De Augmentis Scientiarum*, which appeared in Latin in 1623 (reproducing Book I of the *Advancement*, expanding and modifying Book II, and adding seven books more); the *De Principiis*, which is also late; and the *Descriptio Globi Intellectualis* and *Thema Coeli*, which, though of earlier date (1612), are rather appendices than preparatives to the main scheme of the "Great Instauration". As to the projection and the imperfect fulfilment of that scheme, full information is given in Mr. Ellis' preface to the *Novum Organum*, Mr. Spedding's notes to that preface, and his own prefaces. The other works in the present collection are not placed in order of date, but merely grouped apart from the more strictly philosophical works, the *New Atlantis* coming first because of its relation to these.

In the somewhat anxious task of reading the proofs of this volume, as before in his edition of Buckle in the same series, the editor has had the invaluable assistance of Mr. Ernest Newman. He may therefore hope that this reprint from a standard edition is not less accurate than that—of which, indeed, he has been able to correct some typographical errors.

EDITOR'S INTRODUCTION

I

THE manifold debate which has circled round the name of Bacon for over two hundred years, but especially in the past century, may be divided under two heads—that of his character, and that of his intellectual merit. For many students, happily, the first issue is settled, and the second is perhaps near settlement. But for the general reading public each problem is still somewhat confused by the influence of Macaulay's famous Essay, which seriously mishandled both.

Logically considered, the two questions are quite independent: that is to say, a decision on either leaves the other still open. But for any one in doubt on the first, it must be nearly impossible to read a page of Bacon (who so constantly passes moral judgments) without having the critical faculty either primed or puzzled by the reflection that this moralist is charged by a series of eminent writers with being as base in conduct as he was brilliant in thought and speech. Pope's line—

> "The brightest, wisest, meanest of mankind"—

s still, it is to be feared, the common estimate, as it was in effect Maçaulay's; and the real paradox of great powers in combination with low instincts is common enough in life to permit of Pope's extravagance—which asserts something quite different—passing as a statement of possible psychological fact.

It is best, then, to come straight to the historical facts. The main charges against Bacon as a man are two: treachery to his patron and benefactor, Essex; and corruptness as a judge; and Macaulay presses both with all his force. The argument on the first is that because Essex, for great services rendered him by Bacon, had given him a fairly adequate reward, Bacon's duty, when Essex had not only long ceased to take his counsel but had grossly contravened it, was to refuse to take any action against him as a crown lawyer. When it is remembered that Essex, on his part, had received from the Queen a hundred times the benefits he had bestowed on Bacon, and was thus, on the principles assumed, guilty past all apology, not only in his act of insane sedition but in his previous complots, the attack is seen to break down. Bacon, who held the normal view of his duty to the head of the State, acted on principles of public fealty which then as now were as clearly of plenary force as his obligation to Essex was limited. And his action in the prosecution was that of a man concerned to save an offender who, unwise to the verge of madness, would not let himself be saved. So much has been established for all open-minded readers by the admirable treatise of the late James Spedding, entitled *Evenings with a Reviewer*, a work unique in literature. In that patient and exhaustive discussion Macaulay's case is once for all destroyed; and with it the additional indictments framed by some later and closer investigators. The one fresh contention since set up on the hostile side—that Bacon counselled Essex to accept the Irish command, whereas Spedding declared he had often dissuaded him—also breaks down on scrutiny. The letter in which Bacon acquiesced is quite compatible with previous dissuasion, the tone being that of a lenient friend trying to make the best of an unalterable arrangement.

In regard to the charge of bribery, the case is different. Bacon did take some gifts from suitors; and therein he sinned against his own precepts for the management of judicial affairs. But in not a single case is there any reason to believe that he was corruptly swayed by the gift; and in taking presents from suitors he conformed to the common judicial practice of his day, though he heedlessly took one or two of them at dates which brought him under suspicion, instead of in the annual or otherwise usual way of official routine. The folly of this heedlessness can hardly be exaggerated, inasmuch as it brought on him absolute ruin; and it can be explained only as it has been by Spedding, on the ground of his failure to check the scandalous profusion of his many underlings at home, which kept him in constant embarrassment. But his fault was laxity, never iniquity; and he could truly claim, while admitting the justice of the sentence passed on him by the House of Lords, that he had been the justest judge of his day. Not one of his thirty-six thousand decrees as Lord Chancellor appears to have been overturned on the score of corruption. And his one serious lapse from right conduct has a more profoundly pathetic aspect for us when we realize that, as Mr. Spedding puts the case in his preface to the *De Interpretatione Naturæ Procœmium*, it was an indirect consequence of his devotion to his disinterested intellectual enterprise.

"He began by conceiving that a wiser method of studying nature would give man the key to all her secrets, and therewith the mastery of her powers. . . . But the work would be long and arduous, and the event remote; and in the meantime he was not to neglect the immediate and peculiar service which as an Englishman he owed to his country, and as a Protestant to his religion. He set out with the intention of doing what he could towards the discharge of all three obligations, and planned his course accordingly. With regard to the two last, however, he found as life wore away that the means and opportunities which he hoped for did not present themselves; and fearing that all would fail together if he lost more time in waiting for them, he resolved to fall back upon the first as an enterprise which depended for success on himself alone.

"So his case stood when [between forty and fifty] he drew up this paper. Afterwards, though new exigencies of state gave him an opening for service and drew him again into business and politics, he did not cease to devote his leisure to the prosecution of his main object; and as soon as his fall restored to him the entire command of his time, he again made it his sole occupation.

"So far, therefore, his actual course was quite consistent with his first design; and it is even probable that this very constancy was in some degree answerable for the great error and misfortune of his life. That an absorbing interest in one thing should induce negligence of others not less important, is an accident only too natural and familiar; and if he did not allow the *Novum Organum* to interfere with his attention to the causes which came before him in Chancery, it did probably prevent him from attending as carefully as he should, and otherwise would have done, to the proceedings of his servants and the state of his accounts."

II

If, finally, Bacon be judged in the only fair way, by comparison with his leading contemporaries, he is found to be in essentials a much better man than most of them. His successor in the Chancellorship, Bishop Williams, was convicted of real corruption, and disgraced accordingly. Another of his impeachers, Cranfield, was found guilty of gross and manifold embezzlement as Lord Treasurer, and disgraced likewise; and of most of those active against him it may be said that they were as much morally as intellectually his inferiors. No public man of that age of whose career we have any full knowledge makes after a close examination so strong an impression of general worthiness and fairness. "All that were great and good," says the unromantic Aubrey, "loved and honoured him". Spedding, generally held to be one of the most sagacious men of *his* age in England, has deliberately said of him, on the strength of a quite unrivalled knowledge of his whole career: "I doubt whether there was ever any man whose evidence upon matters of fact may be more absolutely trusted".

The only ground on which that judgment is now likely to be disputed is the occasional semblance of servility in Bacon's relations to King James and his favourites. But in truth it is only a semblance. Bacon in his youth, by the admission of his detractors, bore himself rather haughtily than otherwise to his social equals; and it was only under the discipline of life, after suffering from slow advancement and the sense of wasted powers, that he learned something of the necessary arts of courts. Those who check Macaulay's essay by Spedding's commentary will realize how far he was at all times from self-prostration, despite his own resolve in later life to put a check on his acquired habit of compliment. All the testimony goes to show that the impression he made on most of his contemporaries was one of noble dignity and courtesy; and if he flattered the King in his books, the measure of that weakness is not to be taken without comparing his eulogies with those passed by the bishops in the dedication of their version of the Bible. Of the King, indeed, the King's favoured officer might reasonably be panegyrical. To him, on his accession, Bacon looked with eager hope for help in his scheme for the advancement of science, James being of all monarchs of that day indisputably the most learned; and even when that hope was visibly not to be fulfilled, he could say with truth: "As my good old mistress [Elizabeth] was wont to call me her watch-candle, because it pleased her to say I did continually burn (and yet she suffered me to waste almost to nothing), so I must much more owe the like duty to your Majesty, by whom my fortunes have been settled and raised". From any point of view, he owed immeasurably more to James than he ever did to Essex. And when his entire political career is read in the light of Spedding's consummate knowledge and intimate appreciation, it stands out no less fully redeemed than his personal character from the charges so zealously pressed by Macaulay. The accusation of cherishing monopolies and judicial torture, and the lawless use of the King's prerogative, all fall to the ground on full confrontation with the facts. It is safe to say that had Bacon's life ended in undimmed official lustre, and not in technical disgrace, he would pass without challenge as one of the most sagacious and most upright public men of his day.

He was not indeed morally original in any noteworthy degree. In his readiness to advocate unnecessary war, to the end of national aggrandisement, he falls below Burleigh, and even below James. But such perversion of ethical judgment by the spirit of statecraft has been nearly normal in all ages, and may be noted, in that of Bacon, in so esteemed a spirit as Coligny. In other regards he is at worst over-wary in precept, never unscrupulous. To infer from his inculcations of worldly wisdom in the *Essays* and elsewhere that he was abnormally crafty and self-seeking in his own life, is to misconceive his age in general and his environment in particular. Save for his one fatal lapse, his life comports worthily enough with his pretensions.

III

The case is somewhat otherwise, however, with Bacon's credit as a thinker. In his own day, amid a volume of praise not always authoritative, there was heard the note of not incompetent detraction, when Harvey, "though he esteemed him much for his wit and style," said of him that he wrote philosophy "like a Lord Chancellor". After two centuries in which that challenge was overborne by a chorus of admiration coming from men of science and men of letters alike, it has been revived and amplified with a zeal very different from that of Macaulay, who won the repute of impartiality by following up his expansion of the conventional censure with a reverberation of the conventional eulogy. That summing-up loosely credits "the Baconian philosophy" with having "performed the wonders of subsequent scientific progress", and the *Novum Organum* and *De Augmentis* with having "moved the intellects which have moved the world". It is true that Bacon has greatly impressed many great minds, beginning in his own century with Leibnitz, Comenius, and Vico; but it is also true that such great intellects as Descartes, Gassendi, and Newton, though they appreciated Bacon's work, did their own independently; while Galileo and Kepler and Gilbert, in Bacon's day, did theirs, which he did not assimilate, on the stimulus of Copernicus, whose vital doctrine he never accepted; and Harvey certainly owed him nothing. The problem must be disengaged from the rhetoric which obscures such facts.

No more searching and more judicial analysis of Bacon's scientific and philosophic work has ever been made than that of Mr. R. L. Ellis, in the prefaces and notes reproduced in the following pages; and he, without any lack of sympathy, disallows all the more specific claims made for Bacon as a renovator and reformer of scientific thought. The special Baconian method, he observes, is "nearly useless". And Spedding, the most devoted and the most effectual champion that a libelled reputation ever had, acquiesces in that verdict. In his preface to the third part of his and Ellis' edition—containing those works of Bacon originally designed to be included in the *Instauratio Magna*, but superseded or abandoned—Spedding thus writes of their author's unfinished system:

"We no longer look for the discovery of any great treasure by following in that direction. His peculiar system of philosophy—that is to say, the peculiar method of investigation, the 'organum,' the 'formula,' the 'clavis,' the 'ars ipsa interpretandi naturam,' the 'filum Labyrinthi,' or by whichever of its many names we choose to call that artificial process by which alone he believed that man could attain a knowledge of the laws and a command over the powers of nature—of this philosophy we can make nothing. If we have not tried it, it is because we feel confident that it would not answer. We regard it as a curious piece of machinery, very subtle, elaborate, and ingenious, but not worth constructing, because all the work it could do may be done more easily another way."

It is true that Spedding in his preface to the *Parasceve* repeats, as a kind of forlorn hope, the argument first put by him in the *Evenings with a Reviewer*, that Bacon counted mainly on the sheer collection of data for the attainment of that command over Nature which he desired for men. But that plea too is implicity disallowed by Ellis, and must be disallowed by the critical reader now. Spedding's half-abandoned contention that Bacon would not have dropped his theoretical work to make a great collection of facts if he had not felt the latter to be the vitally important procedure, does not advance the case if it be granted. In point of fact, Bacon's reason for undertaking his collection of data was obvious enough. There was little use in his telling men that his method of inquiry would yield them an unparalleled harvest of truth, unless he showed them some sheaves won by it; and this he sought to do. Spedding admits that he failed, but urges that if only men did systematically and comprehensively what Bacon asked, they might ere this have attained immense results. The answer is that men have always been doing what Bacon urged, to the best of their ability; and their slow progress has been partly due to what made his own success so small—the essential and irremovable difficulty of discovering general truths or natural laws. As Spedding actually remarks, Kepler had had to his hand a great mass of observations; and yet he fumbled long and variously before he hit upon true theories. It is arguable that a still larger collection of facts might have shortened his task; but on the one hand this is very doubtful, and on the other hand the decisive answer is given by Ellis, that beyond a certain point men do not know how to look for facts save in the light of a new hypothesis.

Bacon was really on a truer track when he began by arguing for a better discipline in inference, and a rigorous revision of beliefs. Here, despite the final miscarriage of the method he schemed, which was to have worked as it were mechanically, enabling even ordinary minds to arrive at new and true scientific generalisations, there is perhaps more to be said for him than is in effect allowed either by Macaulay in his summing-up or by Ellis in his. Macaulay's is finally void, resolving itself into his characteristic bracketing of unresolved contradictions. On one page he declares that "if by wit be meant the power of perceiving analogies between things which appear to have nothing in common", the feats of Bacon's wit in the *Wisdom of the Ancients* and the second book of the *De Augmentis* "were not merely admirable, but portentous, almost shocking". "Indeed he possessed this faculty, or rather the faculty possessed him, to a morbid degree." On the next page we are told, of the same man, that "no imagination was ever at once so strong and so thoroughly subjugated. It never stirred but at a signal from good sense. It stopped at the first check from good sense." In the same fashion we are told on the one hand that Bacon "was the person who first turned the minds of speculative men, long occupied in verbal disputes, to the discovery of new and useful truth"; and on the other hand that Bacon's inductive rules " though accurate, are not wanted, because in truth they only tell us to do what we are all doing". That is to say, Macaulay also holds that Bacon had achieved nothing in the matter of method, but claims that he first taught men what kind of truth was best worth seeking for. " He was the person who first called the public attention to an inexhaustible mine of wealth, which had been utterly neglected, and which was accessible by that road alone " [i.e. by the inductive method—the doing of what " we are all doing "]. " By doing so he caused that road, which had previously been trodden only by peasants and higglers, to be frequented by a higher class of travellers."

This claim in turn, which repeats Bacon's most sanguine estimate of his own performance, cannot stand for a moment. The kinds of truth which Bacon sought, were the kinds that many other men before and around him had sought for. The six names of Copernicus, Leonardo da Vinci, Kepler, Galileo, Gilbert, and Harvey, suffice to dispose of Macaulay's generalization. It was indeed one that so erudite a man could hardly have framed, little as he knew of the detail history of the sciences and the useful arts, had he not been bent on making out anyhow a case which should justify him in his endorsement of the conventional admiration for Bacon's works after he had endorsed the conventional blame of Bacon's life. Bacon was really deficient in his appreciation of what had been achieved by his predecessors in the way of " fruits " of right reasoning. From the prefaces of his colleague, Spedding has compiled a formidable list of the oversights and signs of ignorance in the various treatises. The would-be reformer of astronomy " appears to have been utterly ignorant of the discoveries which had just been made by Kepler's calculations. Though he complained in 1623 of the want of compendious methods for facilitating arithmetical computations, especially with regard to the doctrine of Series . . . he does not say a word about Napier's Logarithms, which had been printed only nine years before, and reprinted more than once in the interval. He complained that no considerable advance had been made in geometry beyond Euclid, without taking any notice of what had been done by Archimedes and Apollonius." Seeking to determine specific gravities, he shows ignorance of the better methods previously tried by Archimedes, Ghetaldus, and Porta. Discussing the backwardness of mechanics, he names neither any of these, nor Galileo, nor Stevinus, nor Guldinus. He discusses the rate of fall of weights in ignorance of Galileo's doctrine, published thirty years before, and makes inquiries concerning the lever without knowledge of the theory of it, which was well established in his day. Speaking of the poles of the earth as fixed, he shows inacquaintance with the then familiar fact of the precession of the equinoxes. There is no sign that he sought the acquaintance of able contemporary English astronomers like Harriot ; and though Harvey was court physician, and had been publicly discussing his theory for at least nine years before he published his great treatise (1628), Bacon gives no indication of having heard of it.

And on the side of the advances in mechanics and the useful arts he was equally ill-informed. Not only did he repel Copernicus, ignore Kepler, and disparage Gilbert where Gilbert was substantially right and he wrong—as on the nature and movement of the stellar bodies, and the existence of lightless globes—but, as Dean Kitchin has noted, he was denying progress even in the mechanical arts when fly-clocks, telescopes, and miscroscopes were being newly made around him. Macaulay speaks as if until Bacon's day thinking men had been merely marking time in metaphysics and theology, taking no thought of utilities. To a lamentable extent, certainly, time had been so wasted ; but throughout the sixteenth and in the early years of the seventeenth century, thousands had been toiling at physics, mechanics, astronomy, anatomy, physiology, botany, and medicine. By Macaulay's own showing, they must have been using the inductive method, because we all use it all the time. And as regards the kind of progress that Macaulay seems to regard as alone worth reckoning, that made in the applied sciences or the useful arts, much if not most of it since Bacon's day has been made by men who probably never read one of his treatises. The great inventions in spinning and weaving made in the eighteenth century, like the locomotive, were the work of practical men, who would have been so had Bacon never written. The Marquis of Worcester's steam engine was probably an improvement on a previous contrivance described by Porta ; and it has not been shown that he, or De Caus, or Papin, or Savery, or Newcomen, or Watt, was a Baconian student. And the economic circumstances which delayed till the eighteenth century the commercial exploitation of the idea were not to be overruled by the popularity of Bacon's works among the early members of the Royal Society. Macaulay's praise, in short, is wrong, as his blame was wrong.

How little could be done for men by a mere exhortation to look about them for facts in the fashion of the authorities of Solomon's House in the *New Atlantis*, may be realized after a glance through Sprat's *History of the Royal Society*. The narrative part of that extremely interesting work ends with an account of the Society's procedure in " gathering and dispersing Queries " ; and the first questions tabulated (with the answers to them by Sir Philiberti Vernatti, Resident in Batavia) are these—

" Q. 1. Whether diamonds and other precious stones grow again after three or four years, in the same places where they have been digged out ?

" Q. 2. Whether the quarries of stone in India, near Fetipoca, not far from Agra, may be cleft like logs, and sawn like planks.

" Q. 3. Whether there be a hill in Sumatra which burneth continually, and a fountain which runneth pure balsam ?

" Q. 4. What river is there in Java Major that turns wood into stone ?

" Q. 5. Whether it be true that upon the coast of Achin in Sumatra the sea, though it be calm, groweth very high when no rain falls, but is smooth in rain, though it blows hard ? "

There are many more of the same order, one of which, with the answer, may serve to round our extract—

" Q. 8. What ground there may be for that relation, concerning Horns taking root and growing about Goa ?

" A. Inquiring about this, a Friend laught and told me it was a jeer put upon the Portugees."

It may be said that this was not the sort of questioning that Bacon counselled ; but it was the sort natural to men undisciplined and unguided by scientific study. It is only fair to note that the answers are much more intelligent than the questions, and that in this way some enlightenment was being circuitously gathered ; but it was not on these lines that the useful arts flourished, that Newton reached his conclusions, and that Franklin was led to his experiment and Galileo to his discoveries.

IV

We are on much sounder ground when we come to the finding of Mr. Ellis, that " It is neither to the technical part of his method nor to the details of his view of the nature and progress of science that his fame is justly owing. His merits are of another kind. They belong to the spirit rather than to the positive precepts of his philosophy." The last words, however, I would venture to modify, by way of bringing out the writer's real intention. It is rather the concrete conclusions and the alleged potentialities of the method of investigation in Bacon that have to be disallowed : the precepts, in the proper sense of that term, are for the most part admirable ; and it is in the unique force and insight with which he urged them that the real triumph of Bacon lies. Without fully compassing any important new truths, and without recognizing many of those reached by other men, he yet saw and stated, with a vividness never surpassed, the intellectual vices which incapacitated most men for either discovering or appreciating truth. To Bacon belonged in the very highest degree two faculties —that of utterance or statement, and that of insight into human character. He has truly written of himself, addressing the King in the *De Augmentis* (below, p. 606), that he was " a man naturally fitted rather for literature than for anything else, and borne by some destiny against the inclination of his genius into active life". In rightly recognizing the predominance of his literary gift, he has implicitly undervalued his gift for public life, which was only less great, his moral sagacity being so keen that only his chronic failure to reck his own rede —the disparity between his insight and his force of will—put him at any disadvantage as a man of action. And it is in virtue of his combination of the gifts of speech and of moral insight that he is so memorable and so convincing in his demonstration of the *why* of most men's failure to think rightly. It is between his commanding and irrefutable censure of the vices of normal mental habit, and his thrilling prediction of the great things to be done when those vices are amended, that he holds still the admiration which he had conquered within a generation of his death.

The question of Bacon's "influence" has been somewhat obscured by the inference of some of his later censors that writings so far from reaching right scientific results can never have helped men to be scientific. But long before Macaulay wrote, the fact of the influence had been established by the research of Macvey Napier, published in the *Edinburgh Philosophical Transactions* for 1818. The many proofs there given of the delighted stir set up by Bacon in the English and European mind during the greater part of the seventeenth century have been supplemented somewhat in Thomas Martin's *Character of Lord Bacon* (1835), and in Professor Fowler's praiseworthy edition of the *Novum Organum*; and the general fact cannot be gainsaid. What had happened was not the sudden calling of all hands to the work of useful invention, as imagined by Macaulay, but the pervading thrill of a new critical perception and a new hope. And that impact was not the less momentous because it was felt by many men who did no scientific work. It can hardly be better realized than in a perusal of Sprat's *History of the Royal Society* (1667), unless it be in Cowley's impressive though precariously poetic Ode, prefixed to that work. The Ode is substantially a celebration of Bacon, of whom it sings that

"Autority, which did a Body boast,
Though 'twas but Air condens'd, and stalked about
Like some old Giant's more Gigantic Ghost
To terrifie the Learned Rout,
With the plain Magique of tru Reason's Light
He chac'd out of our sight;
Nor suffered living Men to be misled
By the vain Shadows of the Dead."

No doubt some of Bacon's errors passed muster for truths with pupils of this temper, beginning to swear by the new master who had adjured them so persuasively not to swear by any. But there is no mistaking the sense of liberation, the instinct of new destinies, which pulsates in that generation at Bacon's touch, not only among Englishmen, but among continentals such as Comenius and Leibnitz, and later in Vico. Cowley becomes positively inspired by it:

"Bacon, like Moses, led us forth at last;
The barren Wilderness he past;
Did on the very Border stand
Of the blest promised Land;
And from the Mountain Top of his Exalted Wit,
Saw it himself, and shew'd us it."

And with all his adoration, Cowley takes note, in duly descending diction, of an allowance that is reasonably to be made for the main circumstances of Bacon's life; going on to sigh--

"But Life did never to one Man allow
Time to Discover Worlds, and Conquer too;
Nor can so short a Line sufficient be
To fadome the vast depths of Nature's Sea:

> The work he did we ought t' admire,
> And were unjust if we should more require
> From his few years, divided 'twixt th' Excess
> Of low Affliction and high Happiness:
> For who on things remote can fix his sight,
> That's always in a Triumph, or a Fight?"

The last question is worth remembering. Spedding, despite his devotion, is fain to explain Bacon's practical failure in science as being due to a lack of "the faculty of distinguishing differences". While disposed to qualify some of Spedding's approbations, I hesitate to assent to this criticism; and venture to suggest that a sounder explanation is conveyed by Cowley. Bacon missed success in detail because he was striving to compass nearly the whole field of Nature in a life which was engrossed with work enough of other kinds to keep a strong man busy. No gift of distinguishing differences in things natural could have availed for success under such circumstances. The gift must be allowed, by implication, to Descartes and Galileo, who both made great scientific discoveries, the former in a life of peculiar seclusion, the latter in periods of partial relief from less drudgery than Bacon's. And yet both Descartes and Galileo made notable scientific blunders; the former, in particular, reasoning often in as arbitrary a fashion as Bacon ever did. The enormous range of mere observation shown by Bacon—a range seldom noted, either by friends or foes—is hardly compatible with a positive lack of faculty for discrimination. Rather we are led by the history of scientific discovery to think that new truth is reached in a way ultimately obscure, by the long absorption of a competent mind in a certain subject matter, the recognition of a new meaning coming at last suddenly, rather than by any quasi-mechanical process of reasoning.

Now, Bacon never could have been long enough absorbed in any one problem to attain to this kind of consummation. His field of speculation was too vast to be anywhere minutely and intensely explored. The sheer amount of mental power expended by him is marvellous; but with his fixed determination to find Nature finite, his ardent ambition to make the entire circuit, he never roots himself, never grows tranced in clairvoyant insight. At times, indeed, as in his discussion of the inconceivability of the atom—which extorted from Leibnitz a glowing eulogy at the expense of Descartes, and which in a manner anticipates the latest scientific speculation—he shows original force which challenges us to reconsider all our detractions; but the fact remains that, miscalculating the scope of his problem to begin with, he aimed at the impossible. With splendid powers, he undertook what no powers could achieve; and he did it in the random leisures of a responsible official career. What such powers might have done had they been concentrated for long periods on separate problems, is a question not to be lightly answered. In any case it must be reckoned one of the supreme flights of human perversity to surmise that the man so immeasurably preoccupied with the two orbs of natural and civil lore actually wrought, in addition to what he vainly sought to do, the stupendous imaginative performance of Shakspere.

V

We come back to the recognition of Bacon's enduring success—his magistral indictment of what can best be described as intellectual barbarism—the dogged adhesion to superstitious tradition, the wayward subjection of facts to feelings, the obstinate refusal to part with a predilection, the puerile imposition of a fanciful order on the face of things, the feudal partisanship towards the dicta of authority—all the stiff-necked and self-complacent follies arraigned under the classic nomenclature of Idols of the Cave, the Tribe, the Theatre, and the Forum. Some day that arraignment may be out of date; but the time is not yet. Its force, as put finally in the first book of the *Novum Organum*, is in no way lessened by the fact that Bacon a hundred times flatly transgressed his own precepts. Protesting against the self-will of those who would make their anthropomorphic guess the measure of Nature, he again and again does the thing he denounces, even as Aristotle did before him. After scolding Aristotle, in season and out, for his sins in that sort, the English moralist with perfect simplicity proceeds to solve the problem of sidereal motion by the principle, among others, that there *ought* to be rest in Nature. Hence a grievous sequence of miscarriages. But in the intellectual as in the social life it is happily possible for men to show the way they do not tread. It is with the ethic of opinion as with the Golden Rule of reciprocity, which has been current for ten millenniums, and dubbed divine through two of them, and is yet daily trodden under foot by millions who profess to revere it. And when we note the failure of Bacon's undertaking to frame an automatic organon or mechanical induction of truth, we shall miss half the significance of the matter if we do not realise that the very aspiration was possible only to a great intelligence and a great personality. Till the time of Comte, his was the only current classification of the sciences.

In Bacon's day, every important new idea was arrested in the name of dogma, and as nearly strangled as was possible to those in authority. He himself, in the very books in which he is driven to protest against the procedure, gives it his countenance in the concrete as often as he censures it in the abstract. He could not realise the full truth of his own diagnosis, and he blessed tradition with the left hand, while banning it with the right. Those who have come after him have done the same, even unto the tenth generation. Untaught by the unvarying record of quashed vetoes, exploded rebuttals, and outlived dooms, the majority of men, having accepted all innovations but the last, proceed to treat that exactly as their fathers treated the earlier. Perhaps in our day, at least in the physical sciences, there has emerged a new intellectual self-consciousness, the result of the unceasing percussion of novelty; but to the moral and historical sciences we stand very much as Bacon's generation did to the physical. Hence the enduring significance of what is best in his message and finest in his phrase.. When all is said, we are listening to a man of genius, one of the great masters of English prose, and so great a master of the essentials of all diction that his Latin was to foreigners almost what his English was and is to his countrymen.

The late Professor Bain was not a man likely to be caught by mere rhetoric, or to be lightly enthusiastic about anything; and he could find no word save "wonderful" to express his sense of the sustained power and splendour of the first book of the *Novum Organum.* There Bacon put forth all his power of gnomic concentration and august style; and the result would to-day have been still nobler had he left us his own English. Of how it might have gone, the English reader may get some idea from the earlier English *Filum Labyrinthi*, which approximates to it in purpose.

The resort to Latin was part of the strategy by which he strove to counterbalance his own bias to over-confidence, now partly plain to himself. In youth he had begun with a Latin tractate headed *The Greatest Birth of Time*, the said birth being his proposed reform of investigation; but the walls of Jericho had not fallen before his trumpeting; and ere long he had silently altered the *Maximus* of his title to *Masculus.* Even after that, he had deliberately tried the experiment, in the second chapter of the *Temporis Partus Masculus*, of a style of scornful invective, to find whether haply he might win adherents by hectoring where persuasion had failed. Of that experiment in turn he had duly repented; and for his later works he took nearly every precaution that his ripened worldly wisdom could suggest, short of withdrawing his radical charges against men's average intellectual procedure. Finally, in his *De Augmentis Scientiarum*, published after his fall, he not only recast in Latin his earlier English work, but prudently softened or expunged every allusion to Catholicism that might offend the more liberal Catholics of the Continent, to whom, among others, he began to look for the hearing that he feared his own countrymen, daily more absorbed in theological disputes, might for ever deny him. At the same time to guard himself after his keen thrusts in the *Novum Organum* at theological hindrances to the sciences, he emphasised in the *De Augmentis* his authoritarian theology, which, after all, did not avail to prevent severe imputations on his orthodoxy.

Once again he had miscalculated, by reason of the very over-hopefulness which was at the core of his character, and which gave him so much of his energy. No mere tactic of propitiation could countervail the force of the prejudices against which his work was a protest. If his way of working were so new as he claimed, it followed that it would be slow of adoption. If men could be at once docile to his appeal, his polemic against their intellectual vices would have been proved false. It happened to be true; and they treated his work, on the whole, as he ought to have expected. Foreigners were naturally less unready than his countrymen to listen to a disgraced judge who took a high tone of censure towards the average mind; and his compliments to the Jesuits naturally softened them. Meanwhile, however, as Bacon had with signal prescience predicted in his *De Interpretatione Naturæ Procœmium*, English civil troubles drifted into civil war; and it was at the close of the struggle that, in the new order of things, his teaching began to have the "stupendous" success ascribed to it by Oldenburg, the Secretary of the Royal Society, in the preface to the collected Philosophical Transactions for 1672–77. As Sprat recognizes, the convulsion of the civil war had broken down many old walls and stirred many stagnant waters. Even during the progress of the strife, every species of discussion was quickened and deepened; and under the new regimen of the Restoration, in which Dutch practice and French ideas were thrust upon English usage, especially when London was burnt down and had to be rebuilt, Bacon's better day began, in such sort that scolding old Dr. Henry Stubbe, one of the last of the out-and-out Aristotelians in England, designated the experimental philosophers of his day a "Bacon-faced generation".

In later times the debate has turned as we have seen; but down to our own day Bacon's fame is relatively undiminished, having survived even the attempt of some of his worshippers to prove that he wrote the plays of Shakspere, and a whole library besides. And it is perhaps to-day that he can be best appreciated, seeing that our day knows more fully than any other how true are his charges against men's way of living their mental life—nay, how true they are against himself—and how fundamentally right was his prediction that men's power over Nature would be increased a thousand-fold when they learned to interpret her with the humility of truthseekers, casting aside all prepossessions. That much of the purification of spirit of modern science is due to Bacon will hardly be denied by any one who will make a "Baconian" induction from the records, instead of arguing *a priori*, with Carlyle and Liebig and Lange, that a pioneer who himself went so far astray cannot have helped or stirred men to do otherwise. As Columbus found the New World in seeking for the Asiatic Indies, and while believing he had actually found them, so Bacon, by his teaching, "builded better than he knew". Like Columbus, he was the hero of an Idea; and like so many heroes of fabulous quests, he bore a magic sword, to wit, his unrivalled power of speech. Hence, of all those who in or before his time warred by precept against the tyranny of tradition, he alone retains his spell. Ramus was slain by fanaticism; but even his martyr's death has not moved posterity to cherish his works. Telesius, being dull, is simply buried under the dust of time. Only Descartes, with his masterly *Discours de la Méthode*, written in *his* mother tongue for the next generation, compares with Bacon in his sustained hold upon posterity. And when we are making so many comparisons, it is meet to remember that Descartes in his turn, with all his scientific faculty, showed constant disrespect to the great Galileo, perhaps for a worse reason than that of Bacon's attitude to Copernicus, namely, a mere concern to propitiate the Catholic Church. However that may be, the spectacle of the strength and weakness, the successes and failures, of two such men recalls us to the true and final attitude of retrospective criticism, a recognisant compassion before the mysterious self-frustration of men.

CONTENTS

THE LIFE OF THE HONOURABLE AUTHOR [1]

[BY DR. W. RAWLEY, BACON'S "FIRST AND LAST CHAPLAIN"]

FRANCIS BACON, the glory of his age and nation, the adorner and ornament of learning, was born in York House, or York Place, in the Strand, on the two and twentieth day of January, in the year of our Lord 1560. His father was that famous counsellor to Queen Elizabeth, the second prop of the kingdom in his time, Sir Nicholas Bacon, knight, lord-keeper of the great seal of England; a lord of known prudence, sufficiency, moderation, and integrity. His mother was Anne, one of the daughters of Sir Anthony Cook; unto whom the erudition of King Edward the Sixth had been committed; a choice lady, and eminent for piety, virtue, and learning; being exquisitely skilled, for a woman, in the Greek and Latin tongues. These being the parents, you may easily imagine what the issue was like to be; having had whatsoever nature or breeding could put into him.

His first and childish years were not without some mark of eminency; at which time he was endued with that pregnancy and towardness of wit, as they were presages of that deep and universal apprehension which was manifest in him afterward; and caused him to be taken notice of by several persons of worth and place, and especially by the queen; who (as I have been informed) delighted much then to confer with him, and to prove him with questions; unto whom he delivered himself with that gravity and maturity above his years, that Her Majesty would often term him, *The young Lord-keeper.* Being asked by the queen *how old he was*, he answered with much discretion, being then but a boy, *That he was two years younger than Her Majesty's happy reign*; with which answer the queen was much taken [2].

At the ordinary years of ripeness for the university, or rather something earlier, he was sent by his father to Trinity College, in Cambridge [3], to be educated and bred under the tuition of Doctor John White-gift, then master of the college,

[1] This Life was first published in 1657, as an introduction to the volume entitled "Resuscitatio; or bringing into public light several pieces of the works, civil, historical, philosophical, and theological, hitherto sleeping, of the Right Honourable Francis Bacon, Baron of Verulam, Viscount St. Alban; according to the best corrected copies". Of this volume a second edition, or rather a re-issue with fresh titlepage and dedication, and several sheets of new matter inserted, appeared in 1661; the "Life of the Honourable Author" being prefixed as before, and not altered otherwise than by the introduction of three new sentences; to make room for which two leaves were cancelled.

[Mr. Spedding has "modernized the spelling; altered at discretion the typographical arrangement as to capitals, italics, and punctuation," and added the notes.]

[2] This last sentence was added in the edition of 1661. The substance of it had appeared before in the Latin Life prefixed to the *Opuscula Philosophica* in 1658, which is only a free translation of this, with a few corrections.

[3] He began to reside in April 1573; was absent from the latter end of August 1574 till the beginning of March, while the plague raged; and left the university finally at Christmas 1575, being then on the point of sixteen. See Whitgift's accounts, printed in the *British Magazine*, vol. xxxii. p. 365, and xxxiii. p. 444.

afterwards the renowned Archbishop of Canterbury ; a prelate of the first magnitude for sanctity, learning, patience, and humility ; under whom he was observed to have been more than an ordinary proficient in the several arts and sciences. Whilst he was commorant in the university, about sixteen years of age (as his lordship hath been pleased to impart unto myself), he first fell into the dislike of the philosophy of Aristotle ; not for the worthlessness of the author, to whom he would ever ascribe all high attributes, but for the unfruitfulness of the way ; being a philosophy (as his lordship used to say) only strong for disputations and contentions, but barren of the production of works for the benefit of the life of man ; in which mind he continued to his dying day.

After he had passed the circle of the liberal arts, his father thought fit to frame and mould him for the arts of state ; and for that end sent him over into France with Sir Amyas Paulet then employed ambassador lieger into France [4] ; by whom he was after awhile held fit to be entrusted with some message or advertisement to the queen ; which having performed with great approbation, he returned back into France again, with intention to continue for some years there. In his absence in France his father, the lord-keeper, died [5], having collected (as I have heard of knowing persons) a considerable sum of money, which he had separated, with intention to have made a competent purchase of land for the livelihood of this his youngest son (who alone was unprovided for ; and though he was the youngest in years, yet he was not the lowest in his father's affection) ; but the said purchase being unaccomplished at his father's death, there came no greater share to him than his single part and portion of the money dividable amongst five brethren ; by which means he lived in some straits and necessities in his younger years. For as for that pleasant site and manor of Gorhambury, he came not to it till many years after, by the death of his dearest brother, Mr. Anthony Bacon [6], a gentleman equal to him in height of wit, though inferior to him in the endowments of learning and knowledge ; unto whom he was most nearly conjoined in affection, they two being the sole male issue of a second *venter*.

Being returned from travel, he applied himself to the study of the common law, which he took upon him to be his profession [7] ; in which he obtained to great excellency, though he made that (as himself said) but as an accessary, and not his principal study. He wrote several tractates upon that subject ; wherein, though some great masters of the law did outgo him in bulk, and particularities of cases, yet in the science of the grounds and mysteries of the law he was exceeded by none. In this way he was after awhile sworn of the queen's council learned, extraordinary ; a grace (if I err not) scarce known before [8]. He seated himself, for the commodity of his studies and practice, amongst the Honourable Society of Gray's-Inn, of which house he was a member ; where he erected that

[4] Sir Amyas landed at Calais on the 25th of September 1576, and succeeded Dr. Dale as ambassador in France in the following February. See *Burghley's Diary*, Murdin, pp. 778, 779.

[5] In February 1578–9.

[6] Anthony Bacon died in the spring of 1601. See a letter from Mr. John Chamberlain to Sir Dudley Carlton, in the State Paper Office, dated 27th May, 1601.

[7] He had been admitted *de societate magistrorum* of Gray's Inn on June 27, 1576 ; commenced his regular career as a student in 1579 ; became "utter barrister" on the 27th of June 1582 ; bencher in 1586 ; reader in 1588 ; and double reader in 1600. See Harl. MSS. 1912, and Book of Orders, p. 56.

[8] In the Latin version of this memoir, for "after a while" Rawley substitutes *nondum tyrocinium in lege egressus*, by which he seems to assign a very early period as the date of this appointment. But I suspect he was mistaken, both as to the date and the nature of it. The title he got no doubt from a letter addressed by Bacon to King James, about the end of January 1620–1. "You found me of the Learned Council, Extraordinary, without patent or fee, a kind of *individuum vagum*. You established me and brought me into Ordinary." Coupling this probably with an early but undated letter to Burghley, in which Bacon thanks the queen for "appropriating him to her service", he imagined that the thanks were for the appointment in question. This however is incredible. A copy of this letter in the Landsdowne Collection gives the date,—18 October 1580 ; at which time Bacon had not been even a student of law for more than a year and a half,

elegant pile or structure commonly known by the name of *The Lord Bacon's Lodgings*, which he inhabited by turns the most part of his life, (some few years only excepted) unto his dying day. In which house he carried himself with such sweetness, comity, and generosity, that he was much revered and beloved by the readers and gentlemen of the house.

Notwithstanding that he professed the law for his livelihood and subsistence, yet his heart and affection was more carried after the affairs and places of estate; for which, if the majesty royal then had been pleased, he was most fit. In his younger years he studied the service and fortunes (as they call them) of that noble but unfortunate earl, the Earl of Essex; unto whom he was, in a sort, a private and free counsellor, and gave him safe and honourable advice, till in the end the earl inclined too much to the violent and precipitate counsel of others his adherents and followers; which was his fate and ruin [9].

His birth and other capacities qualified him above others of his profession to have ordinary accesses at court, and to come frequently into the queen's eye, who would often grace him with private and free communication, not only about matters of his profession or business in law, but also about the arduous affairs of estate; from whom she received from time to time great satisfaction. Nevertheless, though she cheered him much with the bounty of her countenance, yet she never cheered him with the bounty of her hand; having never conferred upon him any ordinary place or means of honour or profit, save only one dry reversion of the Register's Office in the Star Chamber, worth about £1,600 *per annum*, for which he waited in expectation either fully or near twenty years [10]; of which his lordship would say in Queen Elizabeth's time, *That it was like another man's ground buttalling upon his house, which might mend his prospect, but it did not fill his barn*; (nevertheless, in the time of King James it fell unto him); which might be imputed, not so much to Her Majesty's averseness and disaffection

and could not therefore have been qualified for such a place; still less could such a distinction have been conferred upon him without being much talked of at the time and continually referred to afterwards. Moreover, we have another letter of Bacon's to King James, written in 1606, in which he speaks of his "*nine* years' service of the crown". This would give 1597 as the year in which he began to serve as one of the learned council; at which time it was no extraordinary favour, seeing that he had been recommended for solicitor-general three or four years before, both by Burghley and Egerton. It appears however to have been no regular or formal appointment. He was not sworn. He had no patent; not even a written warrant. His tenure was only *ratione verbi regii Elizabethæ* (see Rymer, A.D. 1604, p. 121). Elizabeth, who "looked that her word should be a warrant", chose to employ him in the business which belonged properly to her learned council, and he was employed accordingly. His first service of that nature,—the first at least of which I find any record,—was in 1594. In 1597 he had come to be employed regularly, and so continued till the end of the reign, and was familiarly spoken of as "Mr. Bacon of the learned council".

9 The connexion between Bacon and Essex appears to have commenced about the year 1590 or 1591, and furnishes matter for a long story—too long to be discussed in a note. His conduct was much misunderstood at the time by persons who had no means of knowing the truth, and has been much misrepresented since by writers who cannot plead that excuse. The case is not however one in which a unanimous verdict can be expected. Always where choice has to be made between fidelity to the state and fidelity to a party or person, popular sympathy will run in favour of the man who chooses the narrower duty; for the narrower duty is not only easier to comprehend, but, being seen closer, *appears* the larger of the two. But though sentiments will continue to be divided, facts may be agreed upon; and for the correction of all errors in matter of fact, I must refer to the Occasional Works, where the whole story will necessarily come out in full detail. In the mean time I may say for myself that I have no fault to find with Bacon for any part of his conduct towards Essex, and I think many people will agree with me when they see the case fairly stated.

10 The reversion, for which he considered himself indebted to Burghley, was granted to him in October 1589. He succeeded to the office in July 1608. In the Latin version Rawley adds that he administered it by deputy.

towards him, as to the arts and policy of a great statesman then, who laboured by all industrious and secret means to suppress and keep him down ; lest, if he had risen, he might have obscured his glory [11].

But though he stood long at a stay in the days of his mistress Queen Elizabeth, yet after the change, and coming in of his new master King James, he made a great progress ; by whom he was much comforted in places of trust, honour, and revenue. I have seen a letter of his lordship's to King James, wherein he makes acknowledgement, *That he was that master to him, that had raised and advanced him nine times ; thrice in dignity, and six times in office.* His offices (as I conceive) were Counsel Learned Extraordinary [12] to His Majesty, as he had been to Queen Elizabeth ; King's Solicitor-General ; His Majesty's Attorney-General ; Counsellor of Estate, being yet but Attorney ; Lord-Keeper of the Great Seal of England ; lastly, Lord Chancellor ; which two last places, though they be the same in authority and power, yet they differ in patent, height, and favour of the prince ; since whose time none of his successors, until this present honourable lord [13], did ever bear the title of Lord Chancellor. His dignities were first Knight, then Baron of Verulam ; lastly, Viscount St. Alban ; besides other good gifts and bounties of the hand which His Majesty gave him, both out of the Broad Seal and out of the Alienation Office [14], to the value in both of eighteen hundred pounds per annum ; which, with his manor of Gorhambury, and other lands and possessions near thereunto adjoining, amounting to a third part more, he retained to his dying day.

Towards his rising years, not before, he entered into a married estate, and took to wife Alice, one of the daughters and coheirs of Benedict Barnham, Esquire and Alderman of London ; with whom he received a sufficiently ample and liberal portion in marriage [15]. Children he had none ; which, though they be the means to perpetuate our names after our deaths, yet he had other issues to perpetuate his name, the issues of his brain ; in which he was ever happy and admired as Jupiter was in the production of Pallas. Neither did the want of children detract from his good usage of his consort during the intermarriage, whom he prosecuted with much conjugal love and respect, with many rich gifts and endowments, besides a robe of honour which he invested her withal ; which she wore unto her dying day, being twenty years and more after his death [16].

The last five years of his life, being withdrawn from civil affairs [17] and from an

[11] The person here alluded to is probably his cousin Robert Cecil, who, though he always professed an anxiety to serve him, was supposed (apparently not without reason) to have thrown obstacles secretly in the way of his advancement.

[12] See note 8, p. 2. Rawley should rather have said "counsel learned, *no longer extraordinary*". It is true indeed that King James did at his first entrance confirm Bacon by warrant under the sign manual in the same office which he had held under Elizabeth by special commandment. But it was the "establishing him and bringing him into *ordinary*" with a salary of 40*l*., which he reckons as first in the series of advancements. This was in 1604. He was made solicitor in 1607, attorney in 1613, counsellor of state in 1616, lord-keeper in 1617, lord chancellor in 1618. His successive dignities were conferred respectively in 1603, 1618, and 1620–1.

[13] Sir Edward Hyde, made Lord Chancellor June 1, 1660. This clause was added in 1661 ; the leaf having been cancelled for the purpose.

[14] Here the paragraph ended in the first edition. The rest was added in 1661.

[15] It appears, from a manuscript preserved in Tenison's Library, that he had about 220*l*. a-year with his wife, and upon her mother's death was to have about 140*l*. a-year more.

[16] By the "robe of honour" is meant, I presume, the title of viscountess. It appears however that a few months before Bacon's death his wife had given him some cause of grave offence. Special provision is made for her in the body of his will, but revoked in a codicil, "for just and great causes," the nature of which is not specified. Soon after his death she married Sir John Underwood, her gentleman-usher. She was buried at Eyworth in Bedfordshire on the 29th of June 1650.

[17] On the 3rd of May 1621, Bacon was condemned, upon a charge of corruption to which he pleaded guilty, to pay a fine of 40,000*l*. ; to be imprisoned in the Tower during

active life, he employed wholly in contemplation and studies—a thing whereof his lordship would often speak during his active life, as if he affected to die in the shadow and not in the light; which also may be found in several passages of his works. In which time he composed the greatest part of his books and writings, both in English and Latin, which I will enumerate (as near as I can) in the just order wherein they were written [18]: *The History of the Reign of King Henry the Seventh; Abcedarium Naturæ*, or a Metaphysical piece which is lost [19]; *Historia Ventorum; Historia Vitæ et Mortis; Historia Densi et Rari*, not yet printed [20];

the king's pleasure; to be for ever incapable of sitting in parliament or holding office in the state; and to be banished for life from the verge of the court. From that time his only business was to find means of subsistence and of satisfying his creditors, and to pursue his studies.

His offence was the taking of presents from persons who had suits in his court, in some cases while the suit was still pending; an act which undoubtedly amounted to corruption as corruption was defined by the law. The degree of *moral* criminality involved in it is not so easily ascertained. To judge of this, we should know, First, what was the understanding, open or secret, upon which the presents were given and taken,—for a gift, though it be given to a judge, is not necessarily in the nature of a bargain to pervert justice: Secondly, to what extent the practice was prevalent at the time,—for it is a rare virtue in a man to resist temptations to which all his neighbours yield: Thirdly, how far it was *tolerated*,—for a practice may be universally condemned and yet universally tolerated; people may be known to be guilty of it and yet received in society all the same: Fourthly, how it stood with regard to other abuses prevailing at the same time,—for it is hard to reform all at once, and it is one thing for a man to leave a single abuse unreformed while he is labouring to remove or resist greater ones, and another thing to introduce it anew, or to leave all as it was, making no effort to remove any. Now all this is from the nature of the case very difficult to ascertain. But the whole question, as it regards Bacon's character, must be considered in connexion with the rest of his political life, and will be fully discussed in its place in the Occasional works; where all the evidence I can find shall be faithfully exhibited. In this place it may be enough to say that he himself always admitted the taking of presents as he had taken them to be indefensible, the sentence to be just, and the example salutary; and yet always denied that he had been an unjust judge, or "had ever had bribe or reward in his eye or thought when he pronounced any sentence or order"; and, that I cannot find any reason for doubting that this was true. It is stated, indeed, in a manuscript of Sir Matthew Hale's, published by Hargrave, that the censure of Bacon "for many decrees made upon most gross bribery and corruption . . . gave such a discredit and brand to the decrees thus obtained that they were easily set aside"; and it is true that some bills were brought into the House of Commons *for the purpose of* setting aside such decrees; but I cannot find that any one of them reached a third reading; and it is clear from Sir Matthew's own argument that he could not produce an instance of one reversed by the House of Lords; and if any had been reversed by a royal commission appointed for the purpose (which according to his statement was the only remaining way), it must surely have been heard of; yet where is the record of any such commission? Now if of all the decrees so discredited none were reversed, it is difficult to resist the conclusion that they had all been made *bonâ fide* with regard only to the merits of the cases, and were in fact unimpeachably just; and we may believe that Bacon pronounced a true judgment on his own case when he said to his friends (as I find it recorded in a manuscript of Dr. Rawley's in the Lambeth Library), "I was the justest judge that was in England these fifty years; but it was the justest censure in parliament that was these two hundred years."

18 In the Latin version Rawley adds, *quam præsens observavi:* which gives this list a peculiar value.

19 A fragment of this piece was recovered and printed by Tenison in the *Baconiana*; and will appear in this edition after the *Historia Ventorum*, which it was intended to accompany. [Not in the present reprint.]

20 This was true in 1657; but it was printed the next year in the *Opuscula Philosophica;* and, therefore, for "not yet printed", the Latin version substitutes *jam primum typis mandata.* In the edition of 1661 a corresponding alteration ought to have been made in the English, but was not; and as the words occur in one of the cancelled leaves they must have been left by oversight.

Historia Gravis et Levis, which is also lost [21]; *a Discourse of a War with Spain; a Dialogue touching an Holy War; the Fable of the New Atlantis; a Preface to a Digest of the Laws of England; the beginning of the History of the Reign of King Henry the Eighth; De Augmentis Scientiarum*, or the Advancement of Learning, put into Latin [22], with several enrichments and enlargements; *Counsels Civil and Moral*, or his book of *Essays*, likewise enriched and enlarged; *the Conversion of certain Psalms into English Verse; the Translation into Latin of the History of King Henry the Seventh, of the Counsels Civil and Moral* [23], *of the Dialogue of the Holy War, of the Fable of the New Atlantis*, for the benefit of other nations [24]; his revising of his book *De Sapientiâ Veterum; Inquisitio de Magnete; Topica Inquisitionis de Luce et Lumine*; both these not yet printed [25]; lastly, *Sylva Sylvarum*, or *the Natural History*. These were the fruits and productions of his last five years. His lordship also designed, upon the motion and invitation of his late majesty, to have written the reign of King Henry the Eighth; but that work perished in the designation merely, God not lending him life to proceed farther upon it than only in one morning's work; whereof there is extant an *ex ungue leonem*, already printed in his lordship's *Miscellany Works*.

There is a commemoration due as well to his abilities and virtues as to the course of his life. Those abilities which commonly go single in other men, though of prime and observable parts, were all conjoined and met in him. Those are, *sharpness of wit, memory, judgment*, and *elocution*. For the former three his books do abundantly speak them; which [26] with what sufficiency he wrote, let the world judge; but with what celerity he wrote them, I can best testify. But for the fourth, his *elocution*, I will only set down what I heard Sir Walter Raleigh once speak of him by way of comparison (whose judgment may well be trusted), *That the Earl of Salisbury was an excellent speaker, but no good penman; that the Earl of Northampton (the Lord Henry Howard) was an excellent penman, but no good speaker; but that Sir Francis Bacon was eminent in both.*

I have been induced to think, that if there were a beam of knowledge derived from God upon any man in these modern times, it was upon him. For though he was a great reader of books, yet he had not his knowledge from books [27], but from some grounds and notions from within himself; which, notwithstanding, he vented with great caution and circumspection. His book of *Instauratio Magna* [28]

[21] This was probably the tract which Gruter says he once had in his hands, and which he describes as merely a skeleton, exhibiting heads of chapters not filled up. "*De Gravi et Levi in manibus habui integrum et grande volumen, sed quod, præter nudam delineatæ fabricæ compagem ex titulis materiam prout eam conceperat Baconus absolventibus, nihil descriptionis continebat.*" See his letter to Rawley, May 29, 1652, in the *Baconiana*, p. 223.

[22] In this [original] edition I have placed the *De Augmentis* before the *Historia Ventorum*, because, though published after, it was prepared and arranged, and in that sense composed, before. And in this view I am supported by a slight variation which is introduced here in the Latin version, viz. "Intervenerat *opus de Augmentis Scientiarum*", &c.

We learn also from the Latin version that Bacon worked at the translation of the *Advancement of Learning* himself: *in quo e linguâ vernaculâ*, proprio Marte, *in Latinam transferendo honoratissimus auctor plurimum desudavit.*

[23] These were the Essays as they appeared in the third and last edition; but he gave them a weightier title when he had them translated into "the general language": *exinde dicti, sermones fideles, sive interiora rerum.*

[24] The Latin version adds, *apud quos expeti audiverat.*

[25] These words are omitted in the Latin version, and must have been left by oversight in the edition of 1661; for they occur in one of the cancelled leaves; and the works in question had been printed in 1658. The error is the more worth noticing because it shows that wherever the English and the Latin differ, the Latin must be regarded as the later and better authority. [26] The Latin version adds, *ut de Julio Cæsare Hirtius.*

[27] *i.e.* not from books *only*: *Ex libris tamen solis scientiam suam deprompsisse haudquaquam concedere licet.*

[28] For *Instauratio Magna* in this place, and also for *Instauration* a few lines further on, the Latin version substitutes *Novum Organum*. Rawley, when he spoke of the *In-*

(which in his own account was the chiefest of his works) was no slight imagination or fancy of his brain, but a settled and concocted notion, the production of many years' labour and travel. I myself have seen at the least twelve copies of the *Instauration*, revised year by year one after another, and every year altered and amended in the frame thereof, till at last it came to that model in which it was committed to the press; as many living creatures do lick their young ones, till they bring them to their strength of limbs.

In the composing of his books he did rather drive at a masculine and clear expression than at any fineness or affectation of phrases, and would often ask if the meaning were expressed plainly enough, as being one that accounted words to be but subservient or ministerial to matter, and not the principal. And if his style were polite [29], it was because he would do no otherwise. Neither was he given to any light conceits, or descanting upon words, but did ever purposely and industriously avoid them; for he held such things to be but digressions or diversions from the scope intended, and to derogate from the weight and dignity of the style.

He was no plodder upon books; though he read much, and that with great judgment, and rejection of impertinences incident to many authors; for he would ever interlace a moderate relaxation of his mind with his studies, as walking, or taking the air abroad in his coach [30], or some other befitting recreation; and yet he would lose no time, inasmuch as upon his first and immediate return he would fall to reading again, and so suffer no moment of time to slip from him without some present improvement.

His meals were refections of the ear as well as of the stomach, like the *Noctes Atticæ*, or *Convivia Deipno-sophistarum*, wherein a man might be refreshed in his mind and understanding no less than in his body. And I have known some, of no mean parts, that have professed to make use of their note-books when they have risen from his table. In which conversations, and otherwise, he was no dashing man [31], as some men are, but ever a countenancer and fosterer of another man's parts. Neither was he one that would appropriate the speech wholly to himself, or delight to outvie others, but leave a liberty to the co-assessors to take their turns. Wherein he would draw a man on and allure him to speak upon such a subject, as wherein he was peculiarly skilful, and would delight to speak. And for himself, he contemned no man's observations, but would light his torch at every man's candle.

His opinions and assertions were for the most part binding, and not contradicted by any; rather like oracles than discourses; which may be imputed either to the well weighing of his sentence by the scales of truth and reason, or else to the reverence and estimation wherein he was commonly had, that no man would contest with him; so that there was no argumentation, or *pro* and *con* (as they term it), at his table: or if there chanced to be any, it was carried with much submission and moderation.

I have often observed, and so have other men of great account, that if he had occasion to repeat another man's words after him, he had an use and faculty to dress them in better vestments and apparel than they had before; so that the author should find his own speech much amended, and yet the substance of it

stauration, was thinking, no doubt, of the volume in which the *Novum Organum* first appeared, and which contains all the pieces that stand in this edition before the *De Augmentis*.

29 The Latin version adds: *Siquidem apud nostrates eloquii Anglicani artifex habitus est.*

30 In the Latin version Rawley adds gentle exercise on horseback and playing at bowls: *Equitationem non citam sed lentam, globorum lusum, et id genus exercitia.*

31 The word *dash* is used here in the same sense in which Costard uses it in *Love's Labour's Lost*: "There, an't please you; a foolish, mild man; an honest man, look you, and soon *dashed*": Rawley means that Bacon was not a man who used his wit, as some do, to put his neighbours out of countenance: *Convivantium neminem aut alios colloquentium* pudore suffundere *gloriæ sibi duxit, sicut nonnulli gestiunt.*

still retained [32]; as if it had been natural to him to use good forms, as Ovid spake of his faculty of versifying,

"Et quod tentabam scribere, versus erat".

When his office called him, as he was of the king's council learned, to charge any offenders, either in criminals or capitals, he was never of an insulting and domineering nature over them, but always tender-hearted, and carrying himself decently towards the parties (though it was his duty to charge them home), but yet as one that looked upon the *example* with the eye of severity, but upon the *person* with the eye of pity and compassion. And in civil business, as he was counsellor of estate, he had the best way of advising, not engaging his master in any precipitate or grievous courses, but in moderate and fair proceedings: the king whom he served giving him this testimony, *That he ever dealt in business* suavibus modis; *which was the way that was most according to his own heart.*

Neither was he in his time less gracious with the subject than with his sovereign. He was ever acceptable to the House of Commons [33] when he was a member thereof. Being the king's attorney, and chosen to a place in parliament, he was allowed and dispensed with to sit in the House; which was not permitted to other attorneys.

And as he was a good servant to his master, being never in nineteen years' service (as himself averred) rebuked by the king for anything relating to His Majesty, so he was a good master to his servants, and rewarded their long attendance with good places freely [34] when they fell into his power; which was the cause that so many young gentlemen of blood and quality sought to list themselves in his retinue. And if he were abused by any of them in their places, it was not only the error of the goodness of his nature, but the badges of their indiscretions and intemperances.

This lord was religious: for though the world be apt to suspect and prejudge great wits and politics to have somewhat of the atheist, yet he was conversant with God, as appeareth by several passages throughout the whole current of his writings. Otherwise he should have crossed his own principles, which were,

[32] This is probably the true explanation of a habit of Bacon's which seems at first sight a fault, and perhaps sometimes is; and of which a great many instances have been pointed out by Mr. Ellis;—a habit of inaccurate quotation. In quoting an author's words,—especially where he quotes them merely by way of voucher for his own remark, or in acknowledgment of the source whence he derived it, or to suggest an allusion which may give a better effect to it,—he very often quotes inaccurately. Sometimes, no doubt, this was unintentional, the fault of his memory; but more frequently, I suspect, it was done deliberately, for the sake of presenting the substance in a better form, or a form better suited to the particular occasion. In citing the evidence of witnesses, on the contrary, in support of a narrative statement or an argument upon matter of fact, he is always very careful.

[33] The Latin version adds, *in quo sæpe peroravit, non sine magno applausu*; a statement of the truth of which abundant evidence may be found in all the records which remain of the proceedings of the House of Commons. The first parliament in which he sate was that of 1584: after which he sate in every parliament that was summoned up to the time of his fall.

As an edition of Bacon would hardly be complete unless it contained Ben Jonson's famous description of his manner of speaking, I shall insert it here:—"Yet there happened in my time one noble speaker, who was full of gravity in his speaking. His language (where he could spare or pass by a jest) was nobly censorious. No man ever spoke more neatly, more pressly, more weightily, or suffered less emptiness, less idleness, in what he uttered. No member of his speech but consisted of his own graces. His hearers could not cough, or look aside from him, without loss. He commanded where he spoke; and had his judges angry and pleased at his devotion. No man had their affections more in his power. The fear of every man that heard him was, lest he should make an end." —*Discoveries*: under title *Dominus Verulamius.*

[34] *Gratis*, in the Latin version; *i.e.* without taking any money for them, an unusual thing in Bacon's time, when the sale of offices was a principal source of all great men's incomes.

That a little philosophy maketh men apt to forget God, as attributing too much to second causes ; but depth of philosophy bringeth a man back to God again. Now I am sure there is no man that will deny him, or account otherwise of him, but to have him been a deep philosopher. And not only so ; but he was able *to render a reason of the hope which was in him*, which that writing of his of the *Confession of the Faith* doth abundantly testify. He repaired frequently, when his health would permit him, to the service of the church, to hear sermons, to the administration of the sacrament of the blessed Body and Blood of Christ ; and died in the true faith, established in the Church of England.

This is most true—he was free from malice, which (as he said himself) *he never bred nor fed* [35]. He was no revenger of injuries ; which if he had minded, he had both opportunity and place high enough to have done it. He was no heaver of men out of their places, as delighting in their ruin and undoing. He was no defamer of any man to his prince. One day, when a great statesman was newly dead, that had not been his friend, the king asked him, *What he thought of that lord which was gone ?* he answered, *That he would never have made His Majesty's estate better, but he was sure he would have kept it from being worse* ; which was the worst he would say of him : which I reckon not among his moral, but his Christian virtues.

His fame is greater and sounds louder in foreign parts abroad, than at home in his own nation ; thereby verifying that divine sentence, *A prophet is not without honour, save in his own country, and in his own house.* Concerning which I will give you a taste only, out of a letter written from Italy (the storehouse of refined wits) to the late Earl of Devonshire, then the Lord Candish : *I will expect the new essays of my Lord Chancellor Bacon, as also his History, with a great deal of desire, and whatsoever else he shall compose ; but in particular of his History I promise myself a thing perfect and singular, especially in Henry the Seventh, where he may exercise the talent of his divine understanding. This lord is more and more known, and his books here more and more delighted in ; and those men that have more than ordinary knowledge in human affairs, esteem him one of the most capable spirits of this age ; and he is truly such.* Now his fame doth not decrease, with days since, but rather increase. Divers of his works have been anciently and yet lately translated into other tongues, both learned and modern, by foreign pens. Several persons of quality, during his lordship's life, crossed the seas on purpose to gain an opportunity of seeing him and discoursing with him ; whereof one carried his lordship's picture from head to foot [36] over with him into France, as a thing which he foresaw would be much desired there, that so they might enjoy the image of his person as well as the images of his brain, his books. Amongst the rest, Marquis Fiat, a French nobleman, who came ambassador into England, in the beginning of Queen Mary, wife to King Charles, was taken with an extraordinary desire of seeing him ; for which he made way by a friend ; and when he came to him, being then through weakness confined to his bed, the marquis saluted him with this high expression, *That his lordship had been ever to him like*

[35] " He said he had breeding swans and feeding swans ; but for malice, he neither bred it nor fed it." From a commonplace book of Dr. Rawley's in the Lambeth Library. " Et posso dir," says Sir Tobie Matthew, in his dedication to Cosmo de' Medici of an Italian translation of the *Essays* and *Sapientia Veterum*, 1618, " et posso dir con verita (per haver io havuto l'honore di pratticarlo molti anni, et quando era *in minoribus*, et hora quando sta in colmo et fiore della sua grandezza) di non haver mai scoperto in lui animo di vendetta, per qualsivoglia aggravio che se gli fosse fatto ; nè manco sentito uscirgli di bocca parola d'ingiuria contra veruno, che mi paresse venite da passione contra la tal persona ; ma solo (et questo ancora molto scarsamente) per giudicio fattone in sangue freddo. Non è già la sua grandezza quel che io ammiro, ma la sua virtù ; non sono li favori fattimi da lui (per infiniti che siano) che mi hanno posto il cuore in questi ceppi et catene in che mi ritrovo, ma si bene il suo procedere in commune ; che se egli fosse di conditione inferiore, non potrei manco honorarlo, e se mi fosse nemico io dovrei con tutto ciò amar et procurar di servirlo."

[36] This picture was presented to him by Bacon himself, according to the Latin version.

the angels ; of whom he had often heard, and read much of them in books, but he never saw them. After which they contracted an intimate acquaintance, and the marquis did so much revere him, that besides his frequent visits, they wrote letters one to the other, under the titles and appellations of father and son. As for his many salutations by letters from foreign worthies devoted to learning, I forbear to mention them, because that is a thing common to other men of learning or note, together with him.

But yet, in this matter of his fame, I speak in the comparative only, and not in the exclusive. For his reputation is great in his own nation also, especially amongst those that are of a more acute and sharper judgment ; which I will exemplify but with two testimonies, and no more. The former, when his *History of King Henry the Seventh* was to come forth, it was delivered to the old Lord Brook, to be perused by him ; who, when he had dispatched it, returned it to the author with this eulogy, *Commend me to my lord, and bid him take care to get good paper and ink, for the work is incomparable.* The other shall be that of Doctor Samuel Collins, late provost of King's College in Cambridge, a man of no vulgar wit, who affirmed unto me [37], *That when he had read the book of the Advancement of Learning, he found himself in a case to begin his studies anew, and that he had lost all the time of his studying before.*

It hath been desired, that something should be signified touching his diet and the regimen of his health, of which, in regard of his universal insight into nature, he may perhaps be to some an example. For his diet, it was rather a plentiful and liberal diet, as his stomach would bear it, than a restrained ; which he also commended in his book of the *History of Life and Death.* In his younger years he was much given to the finer and lighter sort of meats, as of fowls, and such like ; but afterward, when he grew more judicious [38], he preferred the stronger meats, such as the shambles afforded, as those meats which bred the more firm and substantial juices of the body, and less *dissipable ;* upon which he would often make his meal, though he had other meats upon the table. You may be sure he would not neglect that himself, which he so much extolled in his writings, and that was the use of nitre ; whereof he took in the quantity of about three grains in thin warm broth every morning, for thirty years together next before his death. And for physic, he did indeed live physically, but not miserably ; for he took only a maceration of rhubarb [39], infused into a draught of white wine and beer mingled together for the space of half an hour, once in six or seven days, immediately before his meal (whether dinner or supper), that it might dry the body less ; which (as he said) did carry away frequently the grosser humours of the body, and not diminish or carry away any of the spirits, as sweating doth. And this was no grievous thing to take. As for other physic, in an ordinary way (whatsoever hath been vulgarly spoken) he took not. His receipt for the gout, which did constantly ease him of his pain within two hours, is already set down in the end of the *Natural History.*

It may seem the moon had some principal place in the figure of his nativity : for the moon was never in her passion, or eclipsed [40], but he was surprised with a sudden fit of fainting ; and that, though he observed not nor took any previous knowledge of the eclipse thereof ; and as soon as the eclipse ceased, he was restored to his former strength again.

37 In the Latin version Rawley has thought it worth while to add that this may have been said playfully : *Sive festive sive serio.*

38 More judicious (that is) by experience and observation : *experientiâ edoctus* is the expression in the Latin version,

39 In the Latin version Rawley gives the quantity : *Rhabarbari sesquidrachmam.*

40 Lord Campbell (who appears to have read Rawley's memoir only in the Latin, where the words are *quoties luna* defecit *sive eclipsin passa est*), supposing *defecit* to mean *waned,* discredits this statement, on the ground that " no instance is recorded of Bacon's having fainted in public, or put off the hearing of any cause on account of *the change of the moon,* or of any approaching eclipse, visible or invisible ". And it is true that if *defectus lunæ* meant a change of the moon, or even a dark moon (which it might have meant well enough if the Romans had not chosen to appropriate the word to quite another meaning),

He died on the ninth day of April in the year 1626, in the early morning of the day then celebrated for our Saviour's resurrection, in the sixty-sixth year of his age, at the Earl of Arundel's house in Highgate, near London, to which place he casually repaired about a week before; God so ordaining that he should die there of a gentle fever, accidentally accompanied with a great cold, whereby the defluxion of rheum fell so plentifully upon his breast, that he died by suffocation; and was buried in St. Michael's Church at St. Albans; being the place designed for his burial by his last will and testament, both because the body of his mother was interred there, and because it was the only church then remaining within the precincts of old Verulam: where he hath a monument erected for him in white marble (by the care and gratitude of Sir Thomas Meautys, knight, formerly his lordship's secretary, afterwards clerk of the King's Honourable Privy Council under two kings); representing his full portraiture in the posture of studying, with an inscription composed by that accomplished gentleman and rare wit, Sir Henry Wotton[41].

But howsoever his body was mortal, yet no doubt his memory and works will live, and will in all probability last as long as the world lasteth. In order to which I have endeavoured (after my poor ability) to do this honour to his lordship, by way of conducing to the same.

the accident must have happened in public too often to pass unnoticed. But Rawley was too good a scholar to misapply so common a word in that way. He evidently speaks of eclipses only, and of eclipses visible at the place. Now it is not at all likely that lunar eclipses visible at Westminster would have coincided with important business in which Bacon was conspicuously engaged, often enough (even if he did not faint every time) to establish a connexion between the two phenomena. Of course Rawley's statement is not sufficient to prove the reality of any such connexion; but the fact of the fainting-fits need not be doubted, and may be fairly taken, I think, as evidence of the extreme delicacy of Bacon's temperament, and its sensibility to the skiey influences. That Bacon himself never alluded to this relation between himself and the moon is easily accounted for by supposing that he was not satisfied of the fact. He may have observed the coincidence, and mentioned it to Rawley; and Rawley (whose commonplace book proves that he had a taste for astrology) may have believed in the physical connexion, though Bacon himself did not.

FINIS.

41 FRANCISCUS BACON, BARO DE VERULAM, S^T. ALBANI VIC^mes,
SEU NOTIORIBUS TITULIS
SCIENTIARUM LUMEN FACUNDIÆ LEX
SIC SEDEBAT.

QUI POSTQUAM OMNIA NATURALIS SAPIENTIÆ
ET CIVILIS ARCANA EVOLVISSET
NATURÆ DECRETUM EXPLEVIT
COMPOSITA SOLVANTUR
AN. DÑI. M.DC.XXVI.
ÆTAT^is LXVI.

TANTI VIRI
MEM.
THOMAS MEAUTUS
SUPERSTITIS CULTOR
DEFUNCTI ADMIRATOR
H. P.

GENERAL PREFACE TO BACON'S PHILOSOPHICAL WORKS

By ROBERT LESLIE ELLIS

(1) OUR knowledge of Bacon's method is much less complete than it is commonly supposed to be. Of the *Novum Organum*, which was to contain a complete statement of its nature and principles, we have only the first two books; and although in other parts of Bacon's writings, as for instance in the *Cogitata et Visa de Interpretatione Naturæ*, many of the ideas contained in these books recur in a less systematic form, we yet meet with but few indications of the nature of the subjects which were to have been discussed in the others. It seems not improbable that some parts of Bacon's system were never perfectly developed even in his own mind. However this may be, it is certain that an attempt to determine what his method, taken as a whole, was or would have been, must necessarily involve a conjectural or hypothetical element; and it is, I think, chiefly because this circumstance has not been sufficiently recognized, that the idea of Bacon's philosophy has generally speaking been but imperfectly apprehended.

(2) Of the subjects which were to have occupied the remainder of the *Novum Organum* we learn something from a passage at the end of the second book. "Nunc vero," it is said at the conclusion of the doctrine of prerogative instances, "ad adminicula et rectificationes inductionis, et deinceps ad concreta, et latentes processus, et latentes schematismos, et reliqua quæ aphorismo XXI ordine proposuimus, pergendum".* On referring to the twenty-first aphorism we find a sort of table of contents of the whole work. "Dicemus itaque primo loco, de prærogativis instantiarum; secundo, de adminiculis inductionis; tertio, de rectificatione inductionis; quarto, de variatione inquisitionis pro naturâ subjecti · quinto, de prærogativis naturarum quatenus ad inquisitionem, sive de eo quod inquirendum est prius et posterius; sexto, de terminis inquisitionis, sive de synopsi omnium naturarum in universo; septimo, de deductione ad praxin, sive de eo quod est in ordine ad hominem; octavo, de parascevis ad inquisitionem; postremo autem, de scalâ ascensoriâ et descensoriâ axiomatum."† Of these nine subjects the first is the only one with which we are at all accurately acquainted.

(3) Bacon's method was essentially inductive. He rejected the use of syllogistic or deductive reasoning, except when practical applications were to be made of the conclusions, axiomata, to which the inquirer had been led by a systematic process of induction. "Logica quæ nunc habetur inutilis est ad inventionem scientiarum. . . . . Spes est una in inductione verâ." [1] It is to be observed that wherever Bacon speaks of an "ascending" process, he is to be understood to mean induction, of which it is the character to proceed from that which is *nobis notius* to that which is *notius simpliciter*. Contrariwise when he speaks of a descent, he always refers to the correlative process of deduction. Thus when in the *Partis secundæ Delineatio* he says, . . . "meminerint homines in inquisitione activâ necesse esse rem per scalam decensoriam (cujus usum in contemplativâ sustulimus) confici: omnis enim operatio in individuis versatur quæ

[* **Trans.** below, "But now I must proceed," etc.]
[† **Trans.** below, "I propose to treat them in the first place", etc.]

[1] Nov. Org. i. 11. and 14.

infimo loco sunt,"—we are to understand that in Bacon's system deduction is only admissible in the inquisitio activa; that is, in practical applications of the results of induction. Similiarly in the *Distributio Operis* he says, "Rejicimus syllogismum; neque id solùm quoad principia (ad quæ nec illi eam adhibent) sed etiam quoad propositiones medias"*. Everything was to be established by induction. "In constituendo autem axiomate forma inductionis alia quàm adhuc in usu fuit excogitanda est, eaque non ad principia tantùm (quæ vocant) probanda et invenienda, sed etiam ad axiomata minora, et media, denique omnia."[2]

(4) It is necessary to determine the relation in which Bacon conceived his method to stand to ordinary induction. Both methods set out "a sensu et particularibus," and acquiesce "in maximè generalibus"[3]; but while ordinary induction proceeds "per enumerationem simplicem," by a mere enumeration of particular cases, "et precario concludit et periculo exponitur ab instantiâ contradictoriâ", the new method "naturam separare debet, per rejectiones et exclusiones debitas; et deinde post negativas tot quot sufficiunt super affirmativas concludere"[4]. A form of induction was to to be introduced, "quæ ex aliquibus generaliter concludat ita ut instantiam contradictoriam inveniri non posse demonstretur"[5]. In strong contrast with this method stands "the induction which the logicians speak of", which "is utterly vicious and incompetent". . . . "For to conclude upon an enumeration of particulars, without instance contradictory, is no conclusion, but a conjecture". . . . "And this form, to say truth, is so gross, as it had not been possible for wits so subtile as have managed these things to have offered it to the world, but that they trusted to their theories and dogmaticals, and were imperious and scornful towards particulars[6]." We thus see what is meant by the phrase "quot sufficiunt" in the passage which has been cited from the *Novum Organum*; it means "as many as may suffice in order to the attainment of certainty", it being necessary to have a method of induction, "quæ experientiam solvat et separet, et per exclusiones et rejectiones debitas necessario concludat"[7]. Absolute certainty is therefore one of the distinguishing characters of the Baconian induction. Another is that it renders all men equally capable, or nearly so, of attaining to the truth. "Nostra verò inveniendi scientias ea est ratio ut non multum ingeniorum acumini et robori relinquatur; sed quæ ingenia et intellectus ferè exæquet"[8]; and this is illustrated by the difficulty of describing a circle liberâ manu, whereas every one can do it with a pair of compasses. "Omninò similis est nostra ratio." The cause to which this peculiarity is owing, is sufficiently indicated by the illustration: the method "exæquat ingenia", "cùm omnia per certissimas regulas et demonstrationes transigat".

(5) Absolute certainty, and a mechanical mode of procedure such that all men should be capable of employing it, are thus two great features of the Baconian method. His system can never be rightly understood if they are neglected, and any explanation of it which passes them over in silence leaves unexplained the principal difficulty which that system presents to us. But another difficulty takes the place of the one which is thus set aside. It becomes impossible to justify or to understand Bacon's assertion that his method was essentially new. "Nam nos," he says in the preface to the *Novum Organum*, "si profiteamur nos meliora afferre quam antiqui, eandem quam illi viam ingressi, nullâ verborum arte efficere possimus, quin inducatur quædam ingenii, vel excellentiæ, vel facultatis comparatio, sive contentio. . . . Verùm cùm per nos illud agatur, ut *alia omnino via* intellectui aperiatur illis intentata et incognita, commutata tota

[* Trans. below, "I therefore reject the syllogism," etc.]

[2] Nov. Org. i. 105.

[3] Nov. Org. i. 22. [4] Nov. Org. i. 105. [5] Cogitata et Visa § 18.

[6] Advancement of Learning. The corresponding passage in the De Augm. is in the 2nd chap. of the 5th book. [7] Distrib. Operis, § 10.

[8] Nov. Org. i. 61., and comp. i. 122. Also the Inquisitio legitima de Motu, and Valerius Terminus, c. 19.

jam ratio est ", etc.* He elsewhere speaks of himself as being " in hâc re plane protopirus, et vestigia nullius sequutus " [9]. Surely this language would be out of place, if the difference between him and those who had gone before him related merely to matters of detail; as, for instance, that his way of arranging the facts of observation was more convenient than theirs, and his way of applying an inductive process to them more systematic. And it need not be remarked that induction in itself was no novelty at all. The nature of the act of induction is as clearly stated by Aristotle as by any later writer. Bacon's design was surely much larger than it would thus appear to have been. Whoever considers his writings without reference to their place in the history of philosophy will I think be convinced that he aimed at giving a wholly new method,—a method universally applicable, and in all cases infallible. By this method, all the knowledge which the human mind is capable of receiving might be attained, and attained without unnecessary labour. Men were no longer to wander from the truth in helpless uncertainty. The publication of this new doctrine was the *Temporis Partus Masculus*; it was as the rising of a new sun, before which " the borrowed beams of moon and stars " were to fade away and disappear [10].

(6) That the wide distinction which Bacon conceived to exist between his own method and any which had previously been known has often been but slightly noticed by those who have spoken of his philosophy, arises probably from a wish to recognize in the history of the scientific discoveries of the last two centuries the fulfilment of his hopes and prophecies. One of his early disciples however, who wrote before the scientific movement which commenced about Bacon's time had assumed a definite form and character—I mean Dr. Hooke—has explicitly adopted those portions of Bacon's doctrine which have seemingly been as a stumbling-block to his later followers. In Hooke's *General Scheme or Idea of the Present State of Natural Philosophy* [11], which is in many respects the best commentary on Bacon, we find it asserted that in the pursuit of knowledge, the intellect " is continually to be assisted by some method or engine which shall be as a guide to regulate its actions, so as that it shall not be able to act amiss. Of this engine no man except the incomparable Verulam hath had any thoughts, and he indeed hath promoted it to a very good pitch." Something however still remained to be added to this engine or art of invention, to which Hooke gives the name of philosophical algebra. He goes on to say, " I cannot doubt but that if this art be well prosecuted and made use of, an ordinary capacity with industry will be able to do very much more than has yet been done, and to show that even physical and natural inquiries as well as mathematical and geometrical will be capable also of demonstration; so that henceforward the business of invention will not be so much the effect of acute wit, as of a serious and industrious prosecution " [12]. Here the absolute novelty of Bacon's method, its demonstrative character, and its power of reducing all minds to nearly the same level, are distinctly recognized.

(7) Before we examine the method of which Bacon proposed to make use, it is necessary to determine the nature of the problems to which it was, for the most part at least, to be applied. In other words, we must endeavour to determine the idea which he had formed of the nature of science.

Throughout his writings, science and power are spoken of as correlative— " in idem coincidunt " ; and the reason of this is that Bacon always assumed that the knowledge of the cause would in almost all cases enable us to produce the observed effect. We shall see hereafter how this assumption connected itself with the whole spirit of his philosophy. I mention it now because it presents

[* Trans. below, " For if I should profess that I," etc. " As it is, however," etc.]

9 Nov. Org. i. 113.

10 See, for instance, the *Præfatio Generalis*, where Bacon compares his method to the mariner's compass, until the discovery of which no wide sea could be crossed; an image probably connected with his favourite device of a ship passing through the pillars of Hercules, with the motto " Plus ultra ".

11 Published posthumously in 1705.

12 Present State of Nat. Phil. pp. 6, 7.

itself in the passage in which Bacon's idea of the nature of science is most distinctly stated. "Super datum corpus novam naturam, sive novas naturas, generare et superinducere, opus et intentio est humanæ potentiæ. Datæ autem naturæ formam, sive differentiam veram, sive naturam naturantem, sive fontem emanationis, (ista enim vocabula habemus quæ ad indicationem rei proxime accedunt) invenire, opus et intentio est humanæ scientiæ." This passage, with which the second book of the *Novum Organum* commences, requires to be considered in detail.

In the first place it is to be remarked, that natura signifies "abstract quality" —it is used by Bacon in antithesis with corpus or "concrete body". Thus the passage we have quoted amounts to this, that the scope and end of human power is to give new qualities to bodies, while the scope and end of human knowledge is to ascertain the formal cause of all the qualities of which bodies are possessed.

Throughout Bacon's philosophy, the necessity of making abstract qualities (naturæ) the principal object of our inquiries is frequently insisted on. He who studies the concrete and neglects the abstract cannot be called an interpreter of nature. Such was Bacon's judgment when, apparently at an early period of his life, he wrote the *Temporis Partus Masculus* [13]; and in the *Novum Organum* he has expressed an equivalent opinion: "quòd iste modus operandi, (qui naturas intuetur simplices licet in corpore concreto) procedat ex iis quæ in naturâ sunt constantia et æterna et catholica, et latas præbeat potentiæ humanæ vias [14]". Quite in accordance with this passage is a longer one in the *Advancement of Learning*, which I shall quote in extenso, as it is exceedingly important. "The forms of substances, I say, as they are now by compounding and transplanting multiplied, are so perplexed as they are not to be inquired; no more than it were either possible or to purpose to seek in gross the forms of those sounds which make words, which by composition and transposition of letters are infinite. But on the other side to inquire the form of those sounds or voices which make simple letters is easily comprehensible, and being known induceth and manifesteth the forms of all words which consist and are compounded of them. In the same manner, to inquire the form of a lion, of an oak, of gold—nay of water, or air—is a vain pursuit; but to inquire the forms of sense, of voluntary motion, of vegetation, of colours, of gravity and levity, of density, of tenuity, of heat, of cold, and all other natures and qualities which like an alphabet are not many, and of which the essences upheld by matter of all creatures do consist,—to inquire, I say, the true forms of these, is that part of metaphysique which we now define of." And a little farther on we are told that it is the prerogative of metaphysique to consider "the simple forms or difference of things" (that is to say, the forms of simple natures), "which are few in number, and the degrees and co-ordinations whereof make all this variety".

We see from these passages why the study of simple natures is so important—namely because they are comparatively speaking few in number, and because, notwithstanding this, a knowledge of their essence would enable us, at least in theory, to solve every problem which the universe can present to us.

As an illustration of the doctrine of simple natures, we may take a passage which occurs in the *Silva Silvarum*. "Gold," it is there said, "has these natures: greatness of weight, closeness of parts, fixation, pliantness or softness, immunity from rust, colour or tincture of yellow. Therefore the sure way, though most about, to make gold, is to know the causes of the several natures before rehearsed, and the axioms concerning the same. For if a man can make a metal that hath all these properties, let men dispute whether it be gold or no [15]."

Of these simple natures Bacon has given a list in the third book of the *De*

[13] Mr. Ellis alludes, I think, to the *De Interpretatione Naturæ Sententiæ* XII., which M. Bouillet prints as part of the *Temporis Partus Masculus*. My reasons for differing with M. Bouillet on this point, and placing it by itself, and assigning it a later date, will be found in a note to Mr. Ellis's Preface to the *Novum Organum*.—*J. S.*

[14] Nov. Org. ii. 5. [15] Compare Nov. Org. ii. 5.

Augmentis. They are divided into two classes: schematisms of matter, and simple motions. To the former belong the abstract qualities, dense, rare, heavy, light, &c., of which thirty-nine are enumerated, the list being concluded with a remark that it need not be carried farther, "neque ultra rem extendimus". The simple motions—and it will be observed that the word "motion" is used in a wide and vague sense—are the motus antitypiæ, which secures the impenetrability of matter; the motus nexûs, commonly called the motus ex fugâ vacui, &c.; and of these motions fourteen are mentioned. This list however does not profess to be complete, and accordingly in the *Novum Organum* (ii. 48) another list of simple motions is given, in which nineteen species are recognised.

The view of which we have now been speaking—namely, that it is possible to reduce all the phenomena of the universe to combinations of a limited number of simple elements—is the central point of Bacon's whole system. It serves, as we shall see, to explain the peculiarities of the method which he proposed.

(8) In what sense did Bacon use the word "Form"? This is the next question which, in considering the account which he has given of the nature of science, it is necessary to examine. I am, for reasons which will be hereafter mentioned, much disposed to believe that the doctrine of Forms is in some sort an extraneous part of Bacon's system. His peculiar method may be stated independently of this doctrine, and he has himself so stated it in one of his earlier tracts, namely the *Valerius Terminus.* It is at any rate certain, that in using the word "Form" he did not intend to adopt the scholastic mode of employing it. He was much in the habit of giving to words already in use a new signification. "To me," he remarks in the *Advancement of Learning,* "it seemeth best to keep way with antiquity usque ad aras, and therefore to retain the ancient terms, though I sometimes alter the uses and definitions." And thus though he has spoken of the scholastic forms as figments of the human mind [16], he was nevertheless willing to employ the word "Form" in a modified sense, "præsertim quum hoc vocabulum invaluerit, et familiariter occurrat" [17]. He has however distinctly stated that in speaking of Forms, he is not to be understood to speak of the Forms "quibus hominum contemplationes et cogitationes hactenus assueverunt" [18].

As Bacon uses the word in his own sense, we must endeavour to interpret the passages in which it occurs by means of what he has himself said of it; and this may I think be satisfactorily accomplished.

We may begin by remarking that in Bacon's system, as in those of many others, the relation of substance and attribute is virtually the same as the relation of cause and effect. The substance is conceived of as the causa immanens of its attributes [19], or in other words it is the formal cause of the qualities which are referred to it. As there is a difference between the properties of different substances there must be a corresponding difference between the substances themselves. But in the first state of the views of which we are speaking this latter difference is altogether unimaginable: "distincte quidem intelligi potest, sed non explicari imaginabiliter" [20]. It belongs not to natural philosophy, but to metaphysics.

These views however admit of an essential modification. If we divide the qualities of bodies into two classes, and ascribe those of the former class to substance as its essential attributes, while we look on those of the latter as connected with substance by the relation of cause and effect—that is, if we recognise the distinction of primary and secondary qualities—the state of the question is changed. It now becomes possible to give a definite answer to the question, Wherein does the difference between different substances, corresponding to the difference between their sensible qualities, consist?

The answer to this question of course involves a reference to the qualities which have been recognised as primary; and we are thus led to the principle that in the sciences which relate to the secondary qualities of bodies the primary ones are to be regarded as the causes of the secondary [21].

16 Nov. Org. i. 51. 17 Nov. Org. ii. 2. 18 Nov. Org. ii. 17.

19 See Zimmermann's Essay on the Monadology of Leibnitz, p. 81, (Vienna, 1807).

20 Leibnitz, De ipsâ Naturâ. 21 Whewell, Phil. Ind. Science, [book iv. ch. i.].

This division of the qualities of bodies into two classes is the point of transition from the metaphysical view from which we set out to that of ordinary physical science. And this transition Bacon had made, though not perhaps with a perfect consciousness of having done so. Thus he has repeatedly denied the truth of the scholastic doctrine that Forms are incognoscible because supra-sensible [22]; and the reason of this is clearly that his conception of the nature of Forms relates merely to the primary qualities of bodies. For instance, the Form of heat is a kind of local motion of the particles of which bodies are composed [23], and that of whiteness, a mode of arrangement among those particles [24]. This peculiar motion or arrangement corresponds to and engenders heat or whiteness, and this in every case in which those qualities exist. The statement of the distinguishing character of the motion or arrangement, or of whatever else may be the Form of a given phenomenon, takes the shape of a law; it is the law in fulfilling which any substance determines the existence of the quality in question. It is for this reason that Bacon sometimes calls the Form a law; he has done this particularly in a passage which will be mentioned a little farther on.

With the view which has now been stated, we shall I think be able to understand every passage in which Bacon speaks of Forms;—remembering however that as he has not traced a boundary line between primary and secondary qualities, we can only say in general terms that his doctrine of Forms is founded upon the theory that certain qualities of bodies are merely subjective and phenomenal, and are to be regarded as necessarily resulting from others which belong to substance as its essential attributes. In the passage from which we set out [25], the Form is spoken of as vera differentia, the true or essential difference,—as natura naturans—and as the fons emanationis. The first of these expressions refers to the theory of definition by genus and difference. The difference is that which gives the thing defined its specific character. If it be founded on an accidental circumstance the definition, though not incorrect if the accident be an inseparable one, will nevertheless not express the true and essential character of its subject; contrariwise, if it involve a statement of the formal cause of the thing defined.

The second of these phrases is now scarcely used, except in connexion with the philosophy of Spinoza. It had however been employed by some of the scholastic writers [26]. It is always antithetical to natura naturata, and in the passage before us serves not inaptly to express the relation in which the Form stands to the phenomenal nature which results from it.

The phrase fons emanationis does not seem to require any explanation. It belongs to the kind of philosophical language which attempts, more or less successfully, to give clearness of conception by means of metaphor. It is unnecessary to remark how much this is the case in the later development of scholasticism.

A little farther on in the second book of the *Novum Organum* than the passage we have been considering—namely in the thirteenth aphorism—Bacon asserts that the "forma rei" is "ipsissima res", and that the thing and its Form differ only as "apparens et existens, aut exterius et interius, aut in ordine ad hominem et in ordine ad universum". Here the subjective and phenomenal character of the qualities whose form is to be determined is distinctly and strongly indicated.

The principal passage in which the Form is spoken of as a law occurs in the second aphorism of the same book. It is there said that, although in nature nothing really exists (vere existat) except "corpora individua edentia actus puros individuos ex lege", yet that in doctrine this law is of fundamental importance, and that it and its clauses (paragraphi) are what he means when he speaks of Forms.

In denying the real existence of anything beside individual substances, Bacon opposes himself to the scholastic realism; in speaking of these substances as "edentia actus," he asserts the doctrine of the essential activity of substance;

22 See Scaliger, Exercit. in Cardan. 23 Nov. Org. ii. 20.

24 [Valerius Terminus, II. 1.]

25 [Nov. Org ii. 1.] 26 See Vossius De Vitiis Serm. in voce Naturare; and Castanæus, Distinctiones in voc. Natura.

by adding the epithet "puros" he separates what Aristotle termed ἐντελέχειαι from mere motions or κινήσεις, thereby by implication denying the objective reality of the latter; and, lastly, by using the word "individuos", he implies that though in contemplation and doctrine the form law of the substance (that is, the substantial form) is resoluble into the forms of the simple natures which belong to it, as into clauses, yet that this analysis is conceptual only, and not real.

It will be observed that the two modes in which Bacon speaks of the Form, namely as ipsissima res and as a law, differ only, though they cannot be reconciled, as two aspects of the same object.

Thus much of the character of the Baconian Form. That it is after all only a physical conception appears sufficiently from the examples already mentioned, and from the fact of its being made the most important part of the subject-matter of the natural sciences.

The investigation of the Forms of natures or abstract qualities is the principal object of the Baconian method of induction. It is true that Bacon, although he gives the first place to investigations of this nature, does not altogether omit to mention as a subordinate part of science, the study of concrete substances. The first aphorism of the second book of the *Novum Organum* sufficiently explains the relation in which, as he conceived, the abstract and the concrete, considered as objects of science, ought to stand to one another. This relation corresponds to that which in the *De Augmentis* [iii. 4.], he had sought to establish between Physique and Metaphysique, and which he has there expressed by saying that the latter was to be conversant with the formal and final causes, while the former was to be confined to the efficient cause and to the material. It may be asked, and the question is not easily answered, Of what use the study of concrete bodies was in Bacon's system to be, seeing that the knowledge of the Forms of simple natures would, in effect, include all that can be known of the outward world? I believe that, if Bacon's recognition of physique as a distinct branch of science which was to be studied apart from metaphysique or the doctrine of Forms, can be explained except on historical grounds—that is, except by saying that it was derived from the quadripartite division of causes given by Aristotle [27]—the explanation is merely this, that he believed that the study of concrete bodies would at least at first be pursued more hopefully and more successfully than the abstract investigations to which he gave the first rank [28].

However this may be, it seems certain that Bacon's method, as it is stated in the *Novum Organum*, is primarily applicable to the investigation of Forms, and that when other applications were made of it, it was to be modified in a manner which is nowhere distinctly explained. All in fact that we know of these modifications results from comparing two passages which have been already quoted[29], namely the two lists in which Bacon enumerates the subjects to be treated of in the latter books of the *Novum Organum*.

It will be observed that in one of these lists the subject of concrete bodies corresponds to the "variation of the investigation according to the nature of the subject" in the other, and from this it seems to follow that Bacon looked on his method of investigating Forms as the fundamental type of the inductive process, from which in its other applications it deviated more or less according to the necessity of the case. This being understood, we may proceed to speak of the inductive method itself.

(9) The practical criterium of a Form by means of which it is to be investigated and recognised, reduces itself to this,—that the form nature and the phenomenal nature (so to modify, for the sake of distinctness, Bacon's phraseology) must constantly be either both present or both absent; and moreover that when either increases or decreases, the other must do so too [30]. Setting aside the vagueness of the second condition, it is to be observed that there is nothing in this criterium to decide which of two concomitant natures is the Form of the

[27] For an explanation of which, see note to De Augmentis, iii. 4.—*J. S.*

[28] See, in illustration of this, Nov. Org. ii. 5.

[29] Vide supra, § 2.

[30] Nov. Org. ii. 4, 13, 16.

other. It is true that in one place Bacon requires the form nature, beside being convertible with the given one, to be also a limitation of a more general nature. His words are " natura alia quæ sit cum naturâ datâ convertibilis et tamen sit limitatio naturæ notioris instar generis veri " [31]. Of this the meaning will easily be apprehended if we refer to the case of heat, of which the form is said to be a kind of motion—motion being here the natura notior, the more general natura, of which heat is a specific limitation; for wherever heat is present there also is motion, but not vice versâ. Still the difficulty recurs, that there is nothing in the practical operation of Bacon's method which can serve to determine whether this subsidiary condition is fulfilled; nor is the condition itself altogether free from vagueness.

To each of the three points of that which I have called the practical criterium of the Form corresponds one of the three tables with which the investigation commences. The first is the table " essentiæ et præsentiæ ", and contains all known instances in which the given nature is present. The second is the table of declination or absence in like case (declinationis sive absentiæ in proximo), and contains instances which respectively correspond to those of the first table, but in which, notwithstanding this correspondence, the given nature is absent. The third is the table of degrees or comparison (tabula graduum sive tabula comparativæ), in which the instances of the given nature are arranged according to the degree in which it is manifested in each.

It is easy to see the connexion between these tables, which are collectively called tables of appearance, " comparentiæ," and the criterium. For, let any instance in which the given nature is present (as the sun in the case of heat, or froth in the case of whiteness) be resolved into the natures by the aggregation of which our idea of it is constituted; one of these natures is necessarily the form nature, since this is always to be present when the given nature is. Similarly, the second table corresponds to the condition that the Form and the given nature are to be absent together, and the third to that of their increasing or decreasing together.

After the formation of these tables, how is the process of induction to be carried into effect? By a method of exclusion. This method is the essential point of the whole matter, and it will be well to show how much importance Bacon attached to it.

In the first place, wherever he speaks of ordinary induction and of his own method he always remarks that the former proceeds " per enumerationem simplicem ", that is, by a mere enumeration of particular cases, while the latter makes use of exclusions and rejections. This is the fundamental character of his method, and it is from this that the circumstances which distinguish it from ordinary induction necessarily follow. Moreover we are told that whatever may be the privileges of higher intelligences, man can only in one way advance to a knowledge of Forms: he is absolutely obliged to proceed at first by negatives, and then only can arrive at an affirmative when the process of exclusion has been completed (post omnimodam exclusionem) [32]. The same doctrine is taught in the exposition of the fable of Cupid. For according to some of the mythographi Cupid comes forth from an egg whereon Night had brooded. Now Cupid is the type of the primal nature of things; and what is said of the egg hatched by Night refers, Bacon affirms, most aptly to the demonstrations whereby our knowledge of him is obtained; for knowledge obtained by exclusions and negatives results, so to speak, from darkness and from night. We see, I think, from this allegorical fancy, as clearly as from any single passage in his writings, how firmly fixed in his mind was the idea of the importance, or rather of the necessity, of using a method of exclusion.

It is not difficult, on Bacon's fundamental hypothesis, to perceive why this method is of paramount importance. For assuming that each instance in which the given nature is presented to us can be resolved into (and mentally replaced by) a congeries of elementary natures, and that this analysis is not merely subjective or logical, but deals, so to speak, with the very essence of its subject

[31] Nov. Org. ii. 4. [32] Nov. Org. ii. 15.

matter, it follows that to determine the form nature among the aggregate of simple natures which we thus obtain, nothing more is requisite than the rejection of all foreign and unessential elements. We reject every nature which is not present in every affirmative instance, or which is present in any negative one, or which manifests itself in a greater degree when the given nature manifests itself in a less, or vice versâ. And this process when carried far enough will of necessity lead us to the truth; and meanwhile every step we take is known to be an approximation towards it. Ordinary induction is a tentative process, because we chase our quarry over an open country; here it is confined within definite limits, and these limits become as we advance continually narrower and narrower.

From the point of view at which we have now arrived, we perceive why Bacon ascribed to his method the characters by which, as we have seen, he conceived that it was distinguished from any which had previously been proposed. When the process of exclusion has been completely performed, only the form nature will remain; it will be, so to speak, the sole survivor of all the natures combined with which the given nature was at first presented to us. There can therefore be no doubt as to our result, nor any possibility of confounding the Form with any other of these natures. This is what Bacon expresses, when he says that the first part of the true inductive process is the exclusion of every nature which is not found in each instance where the given one is present, or is found where it is not present, or is found to increase where the given nature decreases, or vice versâ. And then, he goes on to say, when this exclusion has been duly performed, there will in the second part of the process remain, as at the bottom, all mere opinions having been dissipated (abeuntibus in fumum opinionibus volatilibus), the affirmative Form, which will be solid and true and well defined [33]. The exclusion of error will necessarily lead to truth.

Again, this method of exclusion requires only an attentive consideration of each "instantia", in order first to analyse it into its simple natures, and secondly to see which of the latter are to be excluded—processes which require no higher faculties than ordinary acuteness and patient diligence. There is clearly no room in this mechanical procedure for the display of subtlety or of inventive genius.

Bacon's method therefore leads to certainty, and may be employed with nearly equal success by all men who are equally diligent.

In considering the only example which we have of its practical operation, namely the investigation of the form of heat [34], it is well to remark a circumstance which tends to conceal its real nature. After the three tables of Comparentia, Bacon proceeds to the Exclusiva, and concludes by saying that the process of exclusion cannot at the outset (sub initiis) be perfectly performed. He therefore proposes to go on to provide additional assistance for the mind of man. These are manifestly to be subsidiary to the method of exclusions; they are to remove the obstacles which make the Exclusiva defective and inconclusive. But in the meanwhile, and as it were provisionally, the intellect may be permitted to attempt an affirmative determination on the subject before it: "Quod genus tentamenti Permissionem Intellectûs, sive Interpretationem inchoatam, sive Vindemiationem primam, appellare consuevimus". The phrase Permissio Intellectûs sufficiently indicates that in this process the mind is suffered to follow the course most natural to it; it is relieved from the restraints hitherto imposed on it, and reverts to its usual state. In this Vindemiatio we accordingly find no reference to the method of exclusion: it rests immediately on the three tables of Comparentia; and though of course it does not contradict the results of the Exclusiva, yet on the other hand it is not derived from them. If we lose sight of the real nature of this part of the investigation, which is merely introduced by the way "because truth is more easily extricated from error than from confusion", we also lose sight of the scope and purport of the whole method. All that Bacon proposes henceforth to do is to perfect the Exclusiva; the Vindemiatio prima, though it is the closing member of the example which Bacon makes

[33] Nov. Org. ii. 16. [34] Nov. Org. ii. 11–20.

use of, is not to be taken as the type of the final conclusion of any investigation which he would recognise as just and legitimate. It is only a parenthesis in the general method, whereas the Exclusiva, given in the eighteenth aphorism of the second book, is a type or paradigm of the process on which every true induction (inductio vera) must in all cases depend.

It may be well to remark that in this example of the process of exclusion, the table of degrees is not made use of.

Bacon, as we have seen, admits that the Exclusiva must at first be in some measure imperfect; for the Exclusiva, being the rejection of simple natures, cannot be satisfactory unless our notions of these natures are just and accurate, whereas some of those which occur in his example of the process of rejection are ill-defined and vague [35]. In order to the completion of his method, it is necessary to remove this defect. A subsidiary method is required, of which the object is the formation of scientific conceptions. To this method also Bacon gives the name of induction; and it is remarkable that induction is mentioned for the first time in the *Novum Organum* in a passage which relates not to axioms but to conceptions [36]. Bacon's induction therefore is not a mere ἐπαγωγή, it is also a method of definition; but of the manner in which systematic induction is to be employed in the formation of conceptions we learn nothing from any part of his writings. And by this circumstance our knowledge of his method is rendered imperfect and unsatisfactory. We may perhaps be permitted to believe that so far as relates to the subject of which we are now speaking, Bacon never, even in idea, completed the method which he proposed. For of all parts of the process of scientific discovery the formation of conceptions is the one with respect to which it is the most difficult to lay down general rules. The process of establishing axioms Bacon had succeeded, at least apparently, in reducing to the semblance of a mechanical operation; that of the formation of conceptions does not admit of any similar reduction. Yet these two processes are in Bacon's system of co-ordinate importance. All commonly received general scientific conceptions Bacon condemns as utterly worthless [37]. A complete change is therefore required; yet of the way in which induction is to be employed in order to produce this change he has said nothing.

This omission is doubtless connected with the kind of realism which runs through Bacon's system, and which renders it practically useless. For that his method is impracticable cannot I think be denied, if we reflect not only that it never has produced any result, but also that the process by which scientific truths have been established cannot be so presented as even to appear to be in accordance with it. In all cases this process involves an element to which nothing corresponds in the tables of comparence and exclusion; namely the application to the facts of observation of a principle of arrangement, an idea, existing in the mind of the discoverer antecedently to the act of induction. It may be said that this idea is precisely one of the naturæ into which the facts of observation ought in Bacon's system to be analysed. And this is in one sense true; but it must be added that this analysis, if it be thought right so to call it, is of the essence of the discovery which results from it. To take for granted that it has already been effected is simply a petitio principii. In most cases the mere act of induction follows as a matter of course as soon as the appropriate idea has been introduced. If, for instance, we resolve Kepler's discovery that Mars moves in an ellipse into its constituent elements, we perceive that the whole difficulty is antecedent to the act of induction. It consists in bringing the idea of motion in an ellipse into connexion with the facts of observation; that is, in showing that an ellipse may be drawn through all the observed places of the planet. The mere act of induction, the ἐπαγωγή, is perfectly obvious. If all the observed places lie on an ellipse of which the sun is the focus, then every position which the planet successively occupies does so too. This inference, which is so obvious that it must have passed through the mind of the discoverer almost unconsciously, is an instance

[35] Nov. Org. ii. 19; and compare i. 15, which shows the necessity of a complete reform. [36] Nov. Org. i. 14, and comp. i. 18. [37] Nov. Org. i. 15, 16.

of induction "per enumerationem simplicem"; of which kind of induction Bacon, as we have seen, has said that it is utterly vicious and incompetent.

The word realism may perhaps require some explanation. I mean by it the opinion, which Bacon undoubtedly entertained, that for the purpose of investigation, the objects of our thoughts may be regarded as an assemblage of abstract conceptions, so that these conceptions not only correspond to realities, which is of course necessary in order to their having any value, but may also be said adequately to represent them. In his view of the subject, ideas or conceptions (notiones) reside in some sort in the objects from which we derive them; and it is necessary, in order that the work of induction may be successfully accomplished, that the process by which they are derived should be carefully and systematically performed. But he had not perceived that which now at least can scarcely be doubted of, that the progress of science continually requires the formation of new conceptions whereby new principles of arrangement are introduced among the results which had previously been obtained, and that from the necessary imperfection of human knowledge our conceptions never, so to speak, exhaust the essence of the realities by which they are suggested. The notion of an alphabet of the universe, of which Bacon has spoken more than once, must therefore be given up; it could at best be only an alphabet of the present state of knowledge. And similarly of the analysis into abstract natures on which the process of exclusion, as we have seen, depends. No such analysis can be used in the manner which Bacon prescribes to us; for every advance in knowledge presupposes the introduction of a new conception, by which the previously existing analysis is rendered incomplete, and therefore erroneous.

We have now, I think, succeeded in tracing the cause both of the peculiarities of Bacon's method, and of its practical inutility. Some additional information may be derived from an examination of the variations with which it is presented in different parts of his writings;—less however than if we could arrange his smaller works in chronological order. Nevertheless two results, not without their value, may be thus obtained; the one, that it appears probable that Bacon came gradually to see more of the difficulties which beset the practical application of his method; and the other, that the doctrine of Forms is in reality an extraneous part of his philosophy.

(10) In the earliest work in which the new method of induction is proposed, namely, the English tract entitled *Valerius Terminus*, no mention is made of the necessity of correcting commonly received notions of simple natures. The inductive method is therefore presented in its simplest form, unembarrassed with that which constitutes its principal difficulty. But when we advance from *Valerius Terminus* to the *Partis secundæ Delineatio et Argumentum*, which is clearly of a later date, we find that Bacon has become aware of the necessity of having some scientific method for the due construction of abstract conceptions. It is there said that the "pars informans", that is, the description of the new method, will be divided into three parts—the ministration to the senses, the ministration to the memory, and the ministration to the reason. In the first of these, three things are to be taught; and of these three the first is how to construct and elicit from facts a duly formed abstract conception (bona notio); the second is how the sense may be assisted; and the third, how to form a satisfactory collection of facts. He then proposes to go on to the other two ministrations.

Thus the construction of conceptions would have formed the first part of the then designed *Novum Organum*; and it would seem that this arrangement was not followed when the *Novum Organum* was actually written, because in the meantime Bacon had seen that this part of the work involved greater difficulties than he had at first supposed. For the general division into "ministrationes" is preserved in the *Novum Organum*[38], though it has there become less prominent than in the tract of which we have been speaking. In the ministration to the senses, as it is mentioned in the later work, nothing is expressly included but a good and sufficient natural and experimental historia; the theory of the formation of conceptions has altogether disappeared, and both this ministration and

[38] Nov. Org. ii. 10.

that to the memory are postponed to the last of the three, which contains the theory of the inductive process itself. We must set out, Bacon says, from the conclusion, and proceed in a retrograde order to the other parts of the subject. He now seems to have perceived that the theory of the formation of conceptions and that of the establishment of axioms are so intertwined together, that the one cannot be presented independently of the other, although in practice his method absolutely requires these two processes to be carried on separately. His view now is, that at first axioms must be established by means of the commonly received conceptions, and that subsequently these conceptions must themselves be rectified by means of the ulterior aids to the mind, the fortiora auxilia in usum intellectûs, of which he has spoken in the nineteenth aphorism of the second book. But these fortiora auxilia were never given, so that the difficulty which Bacon had once proposed to overcome at the outset of his undertaking remained to the last unconquered. The doctrine of the *Novum Organum* (that we must first employ commonly received notions, and afterwards correct them) is expressly laid down in the *De Interpretatione Naturæ Sententiæ Duodecim* [39]. Of this however the date is uncertain.

It is clear that while any uncertainty remains as to the value of the conceptions (notiones) employed in the process of exclusion, the claim to absolute immunity from error which Bacon has made on behalf of his general method must be more or less modified : and of this he seems to have been aware when he wrote the second book of the *Novum Organum* [40].

(11) Thus much of the theory of the formation of conceptions. With regard to the doctrine of Forms, it is in the first place to be observed that it is not mentioned as a part of Bacon's system, either in *Valerius Terminus* or in the *Partis secundæ Delineatio*, or in the *De Interpretatione Naturæ Sententiæ Duodecim*, although in the two last named tracts the definition of science which is found at the outset of the second book of the *Novum Organum* is in substance repeated. This definition, as we have seen, makes the discovery of Forms the aim and end of science ; but in both cases the word *form* is replaced by *causes*. It is however to be admitted that in the *Advancement of Learning*, published in 1605, Forms are spoken of as one of the subjects of Metaphysique. Their not being mentioned except ex obliquo in *Valerius Terminus* is more remarkable, because Bacon has there given a distinct name to the process which he afterwards called the discovery of the Form. He calls it the freeing of a direction, and remarks that it is not much other matter than that which in the received philosophies is termed the Form or formal cause. Forms are thus mentioned historically, but in the dogmatic statement of his own view they are not introduced at all [41].

The essential character of Bacon's philosophy, namely the analysis of the concrete into the abstract, is nowhere more prominent than in *Valerius Terminus*. It is there said " that every particular that worketh any effect is a thing compounded more or less of diverse single natures, more manifest and more obscure, and that it appeareth not to whether (which) of the natures the effect is to be ascribed " [42]. Of course the great problem is to decide this question, and the method of solving it is called " the freeing of a direction ". In explanation of this name, it is to be observed that in *Valerius Terminus* the practical point of view predominates. Every instance in which a given nature is produced is regarded as a *direction* for its artificial production. If air and water are mingled together, as in snow, foam, &c., whiteness is the result. This then is a direction for the production of whiteness, since we have only to mingle air and water together in order to produce it. But whiteness may be produced in other ways, and the direction is therefore not free. We proceed gradually to free it by rejecting, by means of other instances, the circumstances of this which are unessential : a process which is the exact counterpart of the Exclusiva of the *Novum Organum*. The instance I have given is Bacon's, who developes it at some length.

[39] Vide § viii. of this tract. [40] Nov. Org. ii. 19. [41] I refer to my preface to *Valerius Terminus* for an illustration of some of the difficulties of this very obscure tract.
[42] Val. Ter. c. 17.

Here then we have Bacon's method treated entirely from a practical point of view. This circumstance is worthy of notice because it serves to explain why Bacon always assumes that the knowledge of Forms would greatly increase our command over nature, that it " would enfranchise the power of man unto the greatest possibility of works and effects ". It has been asked what reason Bacon had for this assumption. " Whosoever knoweth any Form," he has said in the *Advancement,* " knoweth the utmost possibility of superinducing that nature upon any variety of nature ". Beyond question, the problem of superinducing the nature is reduced to the problem of superinducing the Form ; but what reason have we for supposing that the one is more easy of solution than the other ? If we knew the Form of malleability, that is, the conditions which the intimate constitution of a body must fulfil in order that it may be malleable, does it follow that we could make glass so ? So far as these questions admit of an answer, *Valerius Terminus* appears to suggest it. Bacon connected the doctrine of Forms with practical operations, because this doctrine, so to speak, represented to him his original notion of the freeing of a direction, which, as the phrase itself implies, had altogether a practical significance.

Even in the *Novum Organum* the definition of the form is made to correspond with the præceptum operandi, or practical direction [43]. The latter is to be " certum, liberum, et disponens sive in ordine ad actionem ". Now a direction to produce the Form as a means of producing the given nature is certain, because the presence of the Form necessarily determines that of the nature. It is free, because it requires only that to be done which is necessary, since the nature can never be present unless its Form is so too. Thus far the agreement between the practical and the scientific view is satisfactory. But to the third property which the practical direction is to possess, namely its being in ordine ad actionem, or such as to facilitate the production of the proposed result, corresponds the condition that the Form is to be " the limitation of a more general nature ; " that is to say, the Form presents itself as a limitation of something more general than the given nature, and as determining, not merely logically but also causatively, the existence of the latter. At this point the divergence between the practical and the scientific view becomes manifest ; practical operations do not, generally speaking, present to us anything analogous to the limitation here spoken of, and there is no reason to suppose that it is easier to see how this limitation is to be introduced than to see how the original problem, the ἐξ ἀρχῆς προκείμενον, may be solved. But this divergence seems to show that the two views are in their origin heterogeneous ; that the one contains the fundamental idea of Bacon's method, while the other represents the historical element of his philosophy. We shall however hereafter have occasion to suggest considerations which may seem to modify this conclusion.

(12) In a survey of Bacon's method it is not necessary to say much of the doctrine of prerogative instances, though it occupies the greater part of the second book of the *Novum Organum.* It belongs to the unfinished part of that work ; at least it is probable that its practical utility would have been explained when Bacon came to speak of the Adminicula Inductionis.

Twenty-seven kinds of instances are enumerated, which are said to excel ordinary instances either in their practical or their theoretical usefulness. To the word *instance* Bacon gives a wide range of signification. It corresponds more nearly to *observation* than to any other which is used in modern scientific language.

Of some classes of these instances collections are to be made for their own sake, and independently of any investigation into particular natures. Such, for instance, are the instantiæ conformes ; Bacon's examples of which are mostly taken from comparative anatomy. One of them is the analogy between the fins of fishes, the feet of quadrupeds, and the feet and wings of birds ; another the analogy of the beak of birds and the teeth of other animals, &c [44].

[43] Nov. Org. ii. 4, which is the best comment on the dictum, Knowledge is power.

[44] Nov Org. ii. 27. It does not seem that Bacon added much to what he found in Aristotle on the subject of these analogies.

The other classes of prerogative instances have especial reference to particular investigation, and are to be collected when individual tables of comparence are formed.

It would seem from this that the theory of prerogative instances is intended to guide us in the formation of these tables. But it is difficult to see how the circumstances which give any instance its prerogative could have been appreciated à priori. An instantia crucis [45], to take the most celebrated of all, has its distinguishing character only in so far as it is viewed with reference to two contending hypotheses. In forming at the outset of an inquiry the appropriate tables, nothing would have led the interpreter to perceive its peculiar value.

This theory, whatever may be its practical utility, may supply us with new illustrations of the importance in Bacon's method of the process of exclusions.

At the head of the list—and placed there, we may presume, from the importance of the end which they promote—stand the instantiæ solitariæ, whose prerogative it is to accelerate the Exclusiva [46]. These are instances which exhibit the given nature in subjects which have nothing in common, except that nature itself, with the other subjects which present it to us. Thus the colours shown by the prism or by crystals are a solitary instance of colour, because they have nothing in common with the fixed colours of flowers, gems, &c. Whatever therefore is not independent of the particular constitution of these bodies must be excluded from the form of colour.

Next to the instantiæ solitariæ are placed the instantiæ migrantes, which show the given nature in the act of appearing or of disappearing ; as when glass, being pounded, becomes white. Of these it is said they not only accelerate and strengthen the Exclusiva, but also confine within narrow limits the Affirmative, or Form itself, by showing that it is something which is given or taken away by the observed change. A little farther on Bacon notices the danger in these cases of confounding the efficient cause with the Form, and concludes by saying " But this is easily remedied by a legitimately performed Exclusiva ".

Other remarks to the same effect might be made with reference to other classes of instances ; but these are probably sufficient.

I shall now endeavour to give an account of Bacon's views on some questions of philosophy, which are not immediately connected with the reforms he proposed to introduce.

(13) It has sometimes, I believe, been supposed that Bacon had adopted the atomic theory of Democritus. This however is by no means true ; but certainly he often speaks much more favourably of the systems of the earlier physicists and especially of that of Democritus, than of the philosophy of Plato and Aristotle. In doing this he may perhaps have been more or less influenced by a wish to find in antiquity something with which the doctrines he condemned might be contrasted. But setting this aside, it is certain that these systems were more akin to his own views than the doctrine of the schools of which Socrates may be called the founder. The problems which they proposed were essentially physical:—given certain material first principles, to determine the origin and causes of all phenomena. They were concerned, for the most part, with that which is accessible to the senses, or which would be so if the senses were sufficiently acute. In this they altogether agree with Bacon, who, though he often speaks of the errors and shortcomings of the senses, yet had never been led to consider the question which stands at the entrance of metaphysical philosophy, namely whether the subjective character of sensation does not necessarily lead to scepticism, if no higher grounds of truth can be discovered. The scepticism of Protagoras, and Plato's refutation of it, seemed to him to be both but idle subtleties. Plato, Aristotle, and their followers, were in his opinion but a better kind of sophists. What Dionysius said to Plato, that his discourse was but dotage, might fitly be applied to them all [47].

It cannot be denied, that to Bacon all sound philosophy seemed to be included in what we now call the natural sciences ; and with this view he was naturally led to prefer the atomic doctrine of Democritus to any metaphysical speculation.

[45] Nov. Org. ii. 36. [46] Nov. Org. ii. 22. [47] Redargut. Phil. et Nov. Org. i. 71.

Every atomic theory is an attempt to explain some of the phenomena of matter by means of others ; to explain secondary qualities by means of the primary. And this was what Bacon himself proposed to do in investigating the Forms of simple natures. Nevertheless he did not adopt the peculiar opinions of Democritus and his followers. In the *Novum Organum* he rejects altogether the notion of a vacuum and that of the unchangeableness of matter [48]. His theory of the intimate constitution of bodies does not, he remarks, relate to atoms properly so called, but only to the actually existing ultimate particles. Bacon cannot therefore be said to be a follower of Democritus, though he has spoken of him as being, of all the Greek philosophers, the one who had the deepest insight into nature [49].

But though Bacon was not an atomist, he was what has been called a mechanical physiologist. Leibnitz's remark that the restorers of philosophy [50] all held the principle that the properties of bodies are to be explained by means of magnitude, figure, and motion (a statement which envelopes every such theory of matter as that of Descartes, together with the old atomic doctrine), is certainly true of Bacon.

(14) The opinion which Bacon had formed as to the class of subjects which ought to be included in Summary Philosophy (the English phrase by which he renders the expression he sometimes uses, namely prima philosophia), is worthy of attention.

In the writings of Aristotle, the first philosophy denotes the science which since his time has been called metaphysic. It is the science of first principles, or as he has himself defined it, the science of that which is, as such. In the first book of the *Metaphysics* we find a proof of the necessity of having such a science, distinct from and in a manner superior to all others.

Bacon, adopting Aristotle's name, applied it differently. With him, the first philosophy is divided into two parts. Of these the first is to be a receptacle of the axioms which do not belong exclusively to particular sciences, but are common to more than one ; while the second is to inquire into the external or adventitious conditions of existences—such as the much and the little, the like and the unlike, the possible and impossible, &c.

In illustration of the contents of the first part, Bacon quotes several axioms which are applicable in more than one science. Of these the first is, " If to unequals are added equals, the sums are unequal," which is a mathematical principle, but which, Bacon says, referring to the distinction laid down by Aristotle between commutative and distributive justices, obtains also in moral science ; inasmuch as it is the rule by which distributive justice must be guided. The next is, " Things which agree with a third, agree with one another,"—which is also a mathematical principle, but yet, differently stated, forms the foundation of the theory of syllogism. Thus far Bacon's doctrine does not materially dissent from Aristotle's, who has taught the necessity of recognising in all sciences two kinds of principles, those which are proper to the subject of each science, and those which, connecting themselves with the doctrine of the categories, are common to all. The last are in his nomenclature axioms, though Bacon, following probably Ramus, who in his turn followed Cicero and the Stoics, gives a much more general sense to this word ; and it is to be remarked that Aristotle has given as an instance of an axiom the first of the two which I have quoted from Bacon, or at any rate another which is in effect equivalent to it. But most of the instances which Bacon goes on to give are of a different nature. They are not derived from the laws of thought, but on the contrary involve an empirical element, and therefore are neither self-evident nor capable of an à priori proof. Thus the axiom that " a discord resolved into a concord improves the harmony", is, Bacon says, not only true in music, but also in ethics and the doctrine of the affections. But this axiom is in its literal sense merely a result of observation, and its application to moral subjects is clearly only analogical or tropical. Again, that " the organs of the senses are

[48] Nov. Org. ii. 8. Compare Cogit. De Nat. Rerum.

[49] Nov. Org. i. 51. ; also Parm. Teles. and Dem. Phil.

[50] Namely, the Cartesians, Verulam, Hobbes, &c. See his letter to Thomasius, p. 48 of the edition of his philosophical works by Erdmann.

analogous to instruments which produce reflection", is, Bacon says, true in perspective, and also in acoustics; being true both of the eye and ear. Here we have a result of observation which is made to enter into two different sciences simply in virtue of the classification employed. For this axiom, if true, properly belongs to physiology, and neither to perspective nor to acoustics; though in a secondary and derivative manner a portion of the truth it includes may be introduced into these sciences. And so on. There is however one of these axioms which is of higher authority: "Quantum naturæ nec minuitur nec augetur"; which, Bacon says, is true not only in physics, but also in natural theology, if it be stated in a modified form; viz. if it be said that it belongs to Omnipotence to make something out of nothing, or vice versâ. Of this axiom it may be remarked, that it is common to physics and natural theology simply because the subjects of these sciences are in some measure common to both; wherein it differs from the Aristotelian conception of an axiom. But it is of more interest to observe that this axiom, of which the truth is derived from our notion of substance, and which can never be established by an empirical demonstration, is constantly quoted by Bacon as a principle of incontestable truth; of which his theory of specific gravities is in some sort only an application.

The question arises both with regard to this axiom and to the others, In what manner Bacon supposed that they ought to be demonstrated; or, if he thought they required no demonstration, in what manner he conceived that the mind apprehended their truth? He has certainly affirmed in express terms that there can be only two ways of arriving at truth, namely syllogism and induction; both of which are manifestly inapplicable to some at least of the principles which he includes in the philosophia prima. But whether he would have admitted that this dictum admits of exception in relation to these cases, or on the other hand had not been led to consider the nature of the difficulty which they present, we have, I think, no means of deciding. It is to be observed that the philosophia prima is spoken of as a collection (receptaculum) of axioms—a phrase which implies that it is not a science in itself, having its own principles and an independent development, but that, contrariwise, it derives from the contributions of other sciences the elements of which it is composed. Of the second part we are unable to speak more definitely than of the first. It is obviously a reflexion of the Aristotelian doctrine of the categories[51], from which, however, Bacon intended to contrast it by requiring that the "conditiones entium," which he has doubtless called transcendent from their applicability to all classes of objects, should be treated not logically but physically[52].

But then what are the questions to be resolved in this mode of treating them? Bacon gives some examples of the discussions which ought to occupy this part of philosophy. The first is, why there is so much of one kind of substance, and so little of another—why, for instance, so much more iron in the world than gold, &c. This belongs to the inquiry "de multo et parvo". Again, in treating "de simili et diverso", it ought to be explained why between dissimilar species are almost always interposed others which partake of the nature of both, and form, as it were, ambiguous species—for instance, bats between birds and quadrupeds, or moss between corruption and plants, &c. The difficulty however which I have already mentioned in speaking of the other part of the philosophia prima recurs with reference to this, namely by what method were the questions here proposed to be answered? If by induction, by induction on what data? and if not, by what other way of arriving at truth?

The illustrations which Bacon has given, and perhaps his way of looking at the whole subject, connect themselves with what has recently been called palætiology. The questions which Bacon proposes are questions as to how that which actually exists, and which in the present order of things will continue to exist, came into being—whether abruptly or by slow transitions, and

[51] Trendelenberg has accordingly quoted the passages in the *De Augmentis* which relate to it, in the historical part of his work on the categories.

[52] De Augmentis iii. 4.

under what agency. He seems to point, though from a distance, to discussions as to the formation of strata and the succession of species. Yet on the other hand the discussion on Like and Unlike was to include at least one portion of a different character, namely why, in despite of the maxim " similia similibus gaudent ", iron does not attract iron but the magnet, nor gold gold, but quicksilver.

(15) Another subject, sufficiently interesting to be here mentioned, though less connected with Bacon's general views, is the doctrine which he entertained touching the nature of the soul. He distinguishes in several parts of his writings between the animal soul, common, at least in kind, to man and to the brutes, and the immortal principle infused by the divine favour into man only [53]. To the latter he gave the name of spiraculum, which was of course suggested by the text, " Spiravit in faciem ejus spiraculum vitæ ". M. Bouillet, in his edition of Bacon's philosophical works [54], condemns this doctrine of man's having two souls, and goes on to remark that Bacon was led to adopt it in deference to the opinions of the schoolmen, and that it is also sanctioned by S. Augustine. In these remarks he is much less accurate than usual; the truth being that the doctrine of the duality of the soul is condemned very strongly by S. Augustine and by the schoolmen, and that there is no doubt as to the source from which Bacon derived it, namely from the writings of Telesius. The notion of a lower soul, distinct in essence from the higher principle of man's nature, is in reality much older than Telesius. We find it for instance among the Manichees—a circumstance which makes it singular that S. Augustine should have been supposed to countenance it. Both in his work *De Ecclesiæ Dogmatibus*, and nearly in the same words in that *De Animâ*, he rejects in the most precise and accurate manner the doctrine of two distinct souls, affirming that there is but one, which is at once the principle of nutrition, of sensation, and of reason. In opposing the tenets of the Manichæans, he has more than once condemned the same doctrine, though less at length than in the works just mentioned. The schoolmen also peremptorily rejected the doctrine which M. Bouillet has affirmed that Bacon derived from them. Thus S. Thomas Aquinas says, "Impossibile est in uno homine esse plures animas per essentiam differentes sed una tantum est anima intellectiva quæ vegetativæ et sensitivæ et intellectivæ officiis fungitur " [55]. And this follows at once from the received opinion, that the soul is joined to the body as its form (ut forma unitur corpori). It would be easy to multiply citations to the same effect; but as no schoolman could venture to contradict an emphatically expressed opinion of S. Augustine, it appears unnecessary to do so [56].

Telesius of Cosenza, whom Bacon has commended as " the best of the novellists ", was one of the Italian reformers of philosophy. Tennemann's remark that the reform which he attempted to introduce was but partial, as having reference only to the natural sciences, is not altogether accurate, but it describes with sufficient correctness the general character of his writings. They contain an attempt to explain all phenomena, including those of animal life, on the hypothesis of the continuous conflict and reciprocal action of two formal principles, heat and cold. His other doctrines are either subordinated to this kind of dualism, or are merely the necessary complements of a system of philosophy. In proposing to inquire into the nature and origin of the soul, he had no other end in view than to arrive at an explanation of the phenomena of sensation, voluntary motion, &c., which should be in accordance with his fundamental hypothesis. He therefore

[53] De Augmentis iv. 3. [54] Œuvres Philosophiques de Bacon, Paris, 1834.—*J. S.*

[55] S. Thom. Prim. Q. 76. a. 3. Concl.

[56] With what bold ignorance the schoolmen are sometimes spoken of is well seen in Dr. Guhrauer's preface to his edition of Leibnitz *De Principio Individui.* The sixth proposition in the *Corollarium* attached to this disputation is as follows :—" Hominis solum una est anima quæ vegetativam et sensitivam virtualiter includat." The learned Doctor declares that in this statement Leibnitz set himself in direct opposition to the schoolmen, and that it contains the germ of Leibnitz's own psychology ; the statement being almost a literal transcript of that of St. Thomas Aquinas, Sum i. Q. 76. a. 3 : to which I have already referred. Leibnitz scarcely thought that in following the Angelic Doctor he was protesting against scholasticism.

sets out from the physiological point of view ; and in order to explain the phenomena of animal and vegetable life, refers them to an indwelling spiritus, or animal soul, which in plants resides in the bark and fibres, and in animals in the white and exsanguine parts of the body, the bones being however excepted [57]. The animal and vegetable souls are in essence alike, but the latter is " paulo quam qui in animalibus inest crassior ". In both cases the origin of this anima is the same ; it is educed from the seed (educta ex semine), and is to all intents as truly material as any other part of the body.

In the application of these views to the soul of man, Telesius was met by considerations of another order. The soul educed ex semine, was (like the body which it animated, and of which it was only the subtlest portion) propagated by generation ; whereas it was decided by orthodox theology that souls are not ex traduce, do not pass from parent to child in the way Telesius must have supposed. The soul is a gift, which after death is to return to Him who gave it. I do not conceive that Telesius's attempt to co-ordinate this doctrine with his own views arose merely from a wish to avoid the imputation of heresy. His writings are, I think, free from that tone of mocking deference to authority by which those of many of his contemporaries are disfigured. They have, on the contrary, much of the melancholy earnestness which characterises those of his disciple Campanella. The difference between the faculties of men and brutes appeared to him to be such that merely a subtler organisation of the spiritus would be insufficient to account for it. Man's higher faculties are to be ascribed to a higher principle, and this can only be conceived of as a divinely formed soul. The question as to the relation between the two souls may be presented under two aspects, namely what are the faculties in man which ought to be ascribed to each of them ? and again are these two souls wholly independent, and if not, how are they connected ? The criterion by which Telesius would decide what ought to be reserved as the peculiar appanage of the divinely created soul, appears to be this—that which in man is analogous to the faculties we recognise in brutes ought to be ascribed to the principle by which they are animated and which we possess in common with them. Whatever, on the contrary, seems peculiar to man, more especially the sense of right and wrong, which is the foundation of all morality, ought to be ascribed to the principle which it is our prerogative to possess [58].

As to the connexion between the two, Telesius decides " both on grounds of human reason and from the authority of Scripture " that they cannot be wholly independent of each other, and he accordingly affirms that the divinely created soul is the Form of the whole body, and especially of the spiritus itself. That the soul is the Form of the body he could not without heresy deny [59], although he condemns Aristotle for saying so ; asserting that Aristotle refers to the spiritus, and not to the true soul, with which probably he was unacquainted [60]. The tendency of these views is towards materialism ; the immaterial principle being annexed to the system, as it were, ab extra. Accordingly Telesius's disciple Donius, whom Bacon has more than once referred to, omits it altogether [61].

Comparing the views of Telesius with those of Bacon, we see that in both the duality of the soul is distinctly asserted, and that in both the animal soul is merely material [62]. Our knowledge of the divinely derived principle must rest principally on revelation. Let this knowledge be drawn, he counsels us, from the same fountain of inspiration from whence the substance of the soul itself proceeded.

Bacon rejects or at least omits Telesius's formula, that this higher soul is the

[57] De Rerum Nat. v. 1. et vi. 26.

[58] De Rerum Naturâ, v. 2.

[59] The collection known as the Clementines contains an authoritative decision on this point. " Ut quisque deinceps asserere defendere aut tenere pertinaciter præsumpserit, quod anima rationalis non sit forma corporis humani per se et essentialiter tanquam hæreticus sit censendus ". I quote from Vulpes on Duns Scotus, 46 a. 5. To this decision Telesius seems to allude, De Rer. Nat. v. 40. Campanella has expressly mentioned it.

[60] De Rer. Nat. v. 3.

[61] See his De Nat. Hominis.

[62] Proceeding e matricibus elementorum, De Augm. iv. 3.

Form of the body—a formula to which either in his system or that of Telesius no definite sense could be attached. He differs from his predecessor in this also, that with him the spiritus is more a physiological and less a psychological hypothesis than with Telesius—it is at least less enwrapped in a psychological system than we find it in the *De Rerum Naturâ*.

On the other hand, he has not, I think, recognised so distinctly as Telesius or Campanella the principle that to the rational soul alone is to be referred the idea of moral responsibility ; and the fine passage on the contrast of public and private good in the seventh book of the *De Augmentis* seems to show (if Bacon meant that the analogy on which it is based should be accepted as anything more than an illustration) that he conceived that something akin to the distinction of right and wrong is to be traced in the workings, conscious or unconscious, of all nature.

(16) We are here led to mention another subject, on which again the views of Telesius appear to have influenced those of Bacon. That all bodies are animated, that a principle of life pervades the whole universe, and that each portion, beside its participation in the life of the world, has also its proper vital principle, are doctrines to which in the time of Bacon the majority of philosophical reformers were at least strongly inclined. The most celebrated work in which they are set forth is perhaps the *De Sensu Rerum* of Campanella. The share which it had in producing the misfortunes of his life is well known, and need not here be noticed.

In one of his letters to Thomasius [63], Leibnitz points out how easy the transition is from the language which the schoolmen held touching substantial forms and the workings of nature to that of Campanella : " Ita reditur ad tot deunculos quot formas substantiales et Gentilem prope polytheismum. Et certe omnes qui de substantiis illis incorporalibus corporum loquuntur non possunt mentem suam explicare nisi translatione a Mentibus sumptâ. Hinc enim attributus illis appetitus vel instinctus ille naturalis ex quo et sequitur cognitio naturalis, hinc illud axioma : Natura nihil facit frustra, omnis res fugit sui destructionem, similia similibus gaudent, materia appetit formam nobiliorem, et alia id genus. Quum tamen reverâ in naturâ nulla sit sapientia, nullus appetitus, ordo vero pulcher ex eo oriatur, quia est horologium Dei ". To the censure implied in these remarks Aristotle is himself in some measure liable, seeing that he ascribed the various changes which go on around us to the half-conscious or unconscious workings of an indwelling power which pervades all things, and to which he gives the name of Nature. Nature does nothing in vain and of things possible realizes the best, but she does not act with conscious prevision. She is, so to speak, the instinct of the universe.

It is on account of these views that Bacon charges Aristotle with having set aside the doctrine of a providence, by putting Nature in the place of God [64]. Nevertheless Bacon himself thought it possible to explain large classes of phenomena by referring them, not certainly to the workings of Nature, but to the instincts and appetites of individual bodies. His whole doctrine of simple motions is full of expressions which it is very difficult to understand without supposing that Bacon had for the time adopted the notion of universally diffused sensation. Thus the " motus nexûs " is that in virtue of which bodies, as delighting in mutual contact, will not suffer themselves to be separated. All bodies, we are told, abhor a solution of continuity, and the rising of cream is to be explained by the desire of homogeneous elements for one another.

The distinction which Bacon has elsewhere taken between sensation and perception, which corresponds to Leibnitz's distinction between apperception and perception, does not appear to accord with these expressions. He there asserts that inanimate bodies have perception without sensation. But such words as *desrie* and *horror* imply not only a change worked in the body to which they are applied in virtue of the presence of another, but also a sense of that presence,—that is, in Bacon's language, not only perception but sensation.

The contrast between the expressions I have quoted and those of which he made use in other parts of his writings, is remarkable. In stating the doctrine of

[63] P. 48. of Erdmann's edition of his philosophical works.

[64] De Aug. iii. 4.

simple motions, he speaks as if all phenomena were to be explained by means of the desires and instincts of matter, every portion of which is more or less consciously sentient. But in other passages we find what at first appears to be a wholly different view, namely that phenomena are to be explained by the site, form, and configuration of atoms or ultimate particles, capable neither of desire nor fear, and in all their motions simply fulfilling the primary law impressed on them by Providence.

Nevertheless there is here no real inconsistency. For Bacon, following Telesius, ascribed all the phenomena of animal life to the spiritus, which, though it is the subtlest portion of the body which it animates, is notwithstanding as truly material as any other part. In every body, whether animated or not, dwells a portion of spirit, and it was natural therefore to ascribe to it some share of the powers which the more finely constituted spirits of animals were supposed to possess. How far however this analogy between animate and inanimate bodies ought to be carried, was a doubtful question ; and we need not be surprised to find that Bacon sometimes denies and sometimes appears to admit that the latter as well as the former are, to a certain extent at least, consciously sentient. But in all cases he proposed to explain the phenomena of animal life by means of the ultimate constitution of matter. Thus such phenomena as the rising of cream, the subsidence of the lees of wine, the clinging of gold leaf round the finger, &c., were to be explained in the first instance by the instincts and appetites of portions of matter, and afterwards to receive a deeper and more fundamental explanation when these instincts and appetites were themselves shown to result from the site, form, and configuration of the ultimate particles of which all bodies are composed.

To the doctrine of universally diffused sensation, so far as he adopted it, Bacon was led by the writings of many of his contemporaries, and in particular by those of Telesius. Brucker has remarked, and with perfect truth, that this doctrine is stated as distinctly, though not so conspicuously, by Telesius as by Campanella. Added to which this doctrine serves to explain phenomena of which, without it, no explanation could readily be given. Thus Bacon is much disposed to ridicule Gilbert for the pains he had bestowed on the subject of electrical attraction, affirming that it is merely the result of the power which friction possesses to excite the appetite of bodies for contact. This appetite " aerem non bene tolerat, sed aliud tangibile mavult ".

(17) Bacon's opinion as to Final Causes has often been discussed. It seems however scarcely necessary to refute the interpretation which on no just grounds has been given to the phrase, "causarum finalium inquisitio tanquam virgo Deo consecrata nihil parit [65] ". Nihil parit, as the context plainly shows, [means simply *non parit opera*] [66]. Bacon is speaking of the classification of physics and metaphysics—the one being the science of the material and efficient cause, and the other containing two parts, namely the doctrine of forms and the doctrine of final causes. To physics corresponds in practical application mechanica or mechanics—to metaphysics, magia or natural magic. But magia corresponds to metaphysique because the latter contains the doctrine of Forms ; that of final causes admitting from its nature of no practical application. It is this idea which Bacon has expressed by saying that the doctrine in question is, as it were, a consecrated virgin.

It is not sufficiently remarked that final causes have often been spoken of without any reference to a benevolent intention. When it is said that the final cause of a stone's falling is " locus deorsum," the remark is at least but remotely connected with the doctrine of an intelligent providence. We are to remember that Bacon has expressly censured Aristotle for having made use of final causes without referring to the fountain from which they flow, namely the providence of the Creator. And in this censure he has found many to concur.

Again, in any case in which the benevolent intention can be perceived, we are at liberty to ask by what means and according to what laws this benevolent in-

[65] De Augm. iii. 5. See note on the place.—*J.S.* [66] I have supplied these words to complete the sentence, which ends abruptly at the bottom of a page, a fresh page having apparently been substituted for that which originally followed—*J. S.*

tention is manifested and made efficient. If this question is not to be asked, there is in the first place an end of physical science, so far as relates to every case in which a benevolent intention has been or can be recognised ; and in the second, the argument à posteriori founded on the contrivance displayed in the works of creation is entirely taken away.

This is, in effect, what Bacon says in the passage of the *De Augmentis* in which he complains of the abuse of final causes. If, he affirms, the physical cause of any phenomenon can be assigned as well as the final, so far is this from derogating from our idea of the divine wisdom, that on the contrary it does but confirm and exalt it. " Dei sapientia effulget mirabilius cum natura aliud agit, providentia aliud elicit, quam si singulis schematibus et motibus naturalibus providentiæ characteres essent impressi[67]." And a little farther on he expresses an opinion which we shall do well always to remember, namely that so far is the study of physical causes from withdrawing men from God and providence, that on the contrary those who have occupied themselves in searching them out have never been able to find the end of the matter without having recourse at length to the doctrine of divine providence.

In one respect Bacon seems to have overlooked the advantage which is to be derived from the study of final causes. In the sciences which relate to animal and vegetable life, the conviction that every part of the organisation has its appropriate function which conduces to the well-being of the whole, serves not only to direct our thoughts to the wisdom of the Creator, but also to guide our investigation into the nature of the organisation itself.

(18) It will now, I think, be well to attempt to arrange the fundamental ideas of Bacon's system in the order in which, as we may conceive, they presented themselves to his mind. To do this will necessarily involve some degree of repetition ; but it will enable us to form a better idea of the scope and spirit of his philosophy.

When, at the outset of his philosophical life, he looked round on the visible universe, it would seem that to him the starry heavens, notwithstanding the grandeur of the spectacle they present to us, were of less interest than things on earth. The stars in their courses declare the glory of God ; but, excepting the great lights which rule the day and night, they exert no conspicuous influence on the welfare of mankind. And on the other hand it is certain that we can in nowise affect the causes by which these phenomena are produced. But on the earth beneath, and in the waters under the earth, Nature is perpetually working in ways which it is conceivable that we may be able to imitate, and in which the beneficence of the Creator, wherein His glory is to us chiefly visible, is everywhere to be traced. Wherever we turn, we see the same spectacle of unceasing and benevolent activity. From the seed of corn Nature develops the stalk, the blade, and the ear, and superinduces on the yet immature produce the qualities which make it fit for the sustenance of man. And so, too, animal life is developed from its first rudiments to all the perfection which it is capable of attaining. And though this perfection is necessarily transitory, yet Nature, though she cannot perpetuate the individual, yet continues the species by unceasing reproduction.

But the contemplation of God's works, glorious as they are, is not the whole of man's business here on earth. For in losing his first estate he lost the dominion over the creatures which was its highest privilege, and ever since has worn out few and evil days, exposed to want, sickness, and death. His works have all been vanity and vexation of spirit, his labour nearly profitless, his knowledge for the most part useless. Is his condition altogether hopeless, or may it not be possible to soften, though not to set aside, the effects of the primal curse ? To this question Bacon unhesitatingly made answer, that of His great mercy God would bless our humble endeavours to restore to suffering humanity some part at least of what it had lost ; and thus he has more than once described the instauration of the sciences as an attempt to regain, so far as may be, that of which the Fall deprived us.

A deep sense of the misery of mankind is visible throughout his writings. The principal speaker in the *Redargutio Philosophiarum*, and the son [father] of Solomon's House in the *New Atlantis*, both express Bacon's idea of what the philos-

[67] De Aug. iii. 4.

opher ought to be ; and of both it is said that their countenance was as the countenance of one who pities men. Herein we see the reason why Bacon has often been called an utilitarian ; not because he loved truth less than others, but because he loved men more.

The philosopher is therefore not merely to contemplate the works of the Creator, but also to employ the knowledge thus obtained for the relief of man's estate. If we ask how this is to be done, we find, Bacon tells us (and here he still seems to recur to the idea that the new philosophy is to be in some sort a restoration to man of his original condition), that as no one can enter into the kingdom of heaven "nisi sub personâ infantis," so, too, in order to obtain a real and fruitful insight into Nature, it is necessary to become as a little child, to abnegate received dogmas and the idols by which the mind is most easily beset, and then to follow with child-like singleness of purpose the indications which Nature gives us as to how her operations are performed. For we can command Nature only by obeying her ; nor can Art avail anything except as Nature's handmaiden. We can affect the conditions under which Nature works ; but things artificial as well as things natural are in reality produced not by Art but Nature. Our power is merely based upon our knowledge of the procedure which Nature follows. She is never really thwarted or controlled by our operations, though she may be induced to depart from her usual course, and under new and artificial conditions to produce new phenomena and new substances.

Natural philosophy, considered from this point of view, is therefore only an answer to the question, How does Nature work in the production of phenomena ? When, to take a trivial instance, she superinduces yellowness on the green leaf, or silently and gradually transforms ice into crystal, we ask how are these changes brought about ?—what conditions are necessary and sufficient in order that the phenomena we observe may be engendered ? If we knew what these conditions are, we might ourselves be able to determine their existence, and then the corresponding phenomena would necessarily follow, since the course of Nature is absolutely uniform.

At this point of the development of Bacon's system, the question of method would naturally present itself to him. Having determined what the object of our inquiries is to be, we must endeavour to find a way of attaining it.

For this end Bacon, as we have seen, proposes to examine all the cases in which the phenomenon to be reproduced has been observed, and to note all the conditions which in each case accompany its production. Of all these those only can be necessary which are universally concomitant. Again he proposes to observe all the cognate cases in which, though certain of the conditions before mentioned are present, they are not accompanied by the required phenomenon. By these two classes of observations all the superfluous conditions may be rejected, and those which remain are what we seek. Wherever we can determine their existence we can produce the phenomenon in question.

This process is what Bacon calls, in *Valerius Terminus*, the freeing of a direction, and in his later writings the investigation of the Form.

His thinking that this process would in all cases, or even generally, be successful, arose from his not having sufficiently appreciated the infinite variety and complexity of Nature. Thus he strongly condemns as most false and pernicious the common opinion that the number of individual phenomena to be observed is sensibly infinite, and commends Democritus (a commendation which seems rather to belong to Lucretius) for having perceived that the appearance of limitless variety which the first aspect of Nature presents to us disappears on a closer inspection.

The transition from this view of Nature to the idea that it was possible to form an alphabet of the universe, and to analyse all phenomena into their real elements, is manifestly easy.

By the new method of induction it would be possible to ascertain the conditions requisite and sufficient for the production of any phenomenon ; and as this determination was meant chiefly to enable us to imitate Nature, or rather to direct her operations, Bacon was naturally led to assume that the conditions in question would be such that it would in all cases be possible to produce them artificially.

Now the power of man is limited to the relations of space. He brings bodies together; he separates them; but Nature must do the rest. On the other hand the conditions of the existence of any phenomenon must be something which inheres more closely in the essence of the substance by which that phenomenon is exhibited than the phenomenon itself. And this something is clearly the inward configuration of the substance; that is, the form and arrangement etc. of its ultimate particles. Whiteness, for instance, depends on an even arrangement of these particles in space; and herein we perceive a perfect analogy between what man can do and what Nature requires to be done. The familiar processes of the arts consist simply in giving particular forms to portions of matter, in arranging them and setting them in motion according to certain rules. Between arranging stones so as to form a house, and arranging particles so as to produce whiteness, there is no difference but that of scale. So in other cases. The difference of scale once set aside, it seemed to follow that the knowledge of the Form would in all cases lead to great practical results.

Thus far of the end which the new philosophy proposes to itself, and of the method which it must employ. The next question relates to the mode of procuring and arranging the materials on which this method is to work. In this part of the subject we again perceive the influence of Bacon's opinion touching the limitedness of Nature. No one acquainted with the history of natural philosophy would think it possible to form a collection of all the facts which are to be the materials on which any science is to operate, antecedently to the formation of the science itself.

In the first place, the observations necessary in order to the recognition of these facts would never have been made except under the guidance of some preconceived idea as to the subject of observation; and in the second, the statement which embodies the result of observation always involves some portion of theory. According to the common use of language, it is a fact and not a theory that in ordinary refraction the sine of the angle of incidence is to the sine of the angle of refraction in a given ratio. But the observations on which this statement is based, and the statement itself, presuppose the recognition of a portion of the theory of light, namely that light is propagated in straight lines—in other words they presuppose the conception of a ray. Nor would these observations have been made but for the idea in the mind of the observers that the magnitude of the angle of refraction depends on that of the angle of incidence.

As we advance farther in any science, what we call facts involve more and more of theory. Thus it is a fact that the tangent of the angle of polarisation is equal to the index of refraction. But no one could have made the observations which prove it, or have stated their result in words, without a distinct conception, first of the law of refraction, and secondly of the distinguishing character of polarised light.

The history of science and the nature of the case concur in showing that observation and theory must go on together;—it is impossible that the one can be completed before the other begins. Now although Bacon did not think that observation and experiments might altogether be laid aside when once the process of interpretation had begun (we see on the contrary that one of the works of Solomon's House was the trying of experiments suggested by previously obtained conclusions), he certainly thought it possible so to sever observation from theory that the process of collecting facts and that of deriving consequences from them might be carried on independently and by different persons. This opinion was based on an imperfect apprehension of the connection between facts and theories; the connection appearing to him to be merely an external one, namely that the former are the materials of the latter. With these views that which has been already noticed touching the finiteness of Nature, namely that there are but a finite and not very large number of things which for scientific purposes require to be observed[68], is altogether in accordance.

The facts on which the new philosophy was to be based being conceivable apart from any portion of theory, and moreover not excessively numerous, they might

68 See the Phænomena Universi, and the Partis secundæ Del., &c.

be observed and recorded within a moderate length of time by persons of ordinary diligence.

If this registering of facts were made a royal work, it might, Bacon seems to have thought, be completed in a few years: he has at least remarked that *unless* this were done, the foundation of the new philosophy could not be laid in the life-time of a single generation. The instauration, he has said in the general preface, is not to be thought of as something infinite and beyond the power of man to accomplish; nor does he believe that its mission can be fully completed (rem omnino perfici posse) within the limits of a single life. Something was therefore left for posterity to do; and probably the more Bacon meditated on the work he had in hand, the more was he convinced of its extent and difficulty. But the *Distributio Operis* sufficiently shows that he believed, when he wrote it, that the instauration of the sciences might speedily become an opus operatum. Of the *Historia Naturalis* on which it was to be based he there speaks, not less than of the *Novum Organum*, as of a work which he had himself accomplished,—" Tertia pars operis complectitur Phænomena Universi",—not " complecti debet". Doubtless the preface was written before the work itself was commenced; still if he had not thought it possible to make good what he here proposes to do, he would have expressly said so [69].

In a letter to Fulgenzio, written probably when Bacon was " dagli anni e da fortuna oppresso", he remarks that " these things " (the instauration of the sciences) require some ages for the ripening of them. But though he despaired of completing his design himself, and even thought that some generations must pass before it received its consummation, yet he always regarded it as a thing which sooner or later would be effectually accomplished, and which would thenceforth remain as a κτῆμα ἐς ἀεί. His instauration of the sciences had a definite end, in which when it was once attained it would finally acquiesce; nor is there anything in his writings to countenance the assumption which has been often made, that in his opinion the onward progress of knowledge was to continue throughout all time. On the contrary, the knowledge which man is capable of might, he thought, be attained, not certainly at once, but within the compass of no very long period. In this doubtless he erred; for knowledge must always continue to be imperfect, and therefore in its best estate progressive.

Bacon has been likened to the prophet who from Mount Pisgah surveyed the Promised Land, but left it for others to take possession of. Of this happy image perhaps part of the felicity was not perceived by its author. For though Pisgah was a place of large prospect, yet still the Promised Land was a land of definite extent and known boundaries, and moreover it was certain that after no long time the chosen people would be in possession of it all. And this agrees with what Bacon promised to himself and to mankind from the instauration of the sciences.

A truer image of the progress of knowledge may be derived from the symbol which, though on other grounds, Bacon himself adopted. Those who strive to increase our knowledge of the outward universe may be said to put out upon an apparently boundless sea; they dedicate themselves

" To unpathed waters—undreamed shores ";

and though they have a good hope of success, yet they know they can subdue but a small part of the new world which lies before them.

(19) In this respect, then, as in others, the hopes of Francis Bacon were not destined to be fulfilled. It is neither to the technical part of his method nor to the details of his view of the nature and progress of science that his great fame is justly owing. His merits are of another kind. They belong to the spirit rather than to the positive precepts of his philosophy.

He did good service when he declared with all the weight of his authority and of his eloquence that the true end of knowledge is the glory of the Creator and the relief of man's estate. The spirit of this declaration runs throughout his writings,

[69] The sixth part containing the new philosophy itself is spoken of at the end of the *Distributio* as at least an inchoate work, which others must finish, but to which he hopes to give " initia non contemnenda ".

and we trust has worked for good upon the generations by which they have been studied. And as he showed his wisdom in coupling together things divine and human, so has he shown it also in tracing the demarcation between them, and in rebuking those who by confounding religion and philosophy were in danger of making the one heretical and the other superstitious.

When, not long before Bacon's time, philosophy freed itself from the tutelage of dogmatic theology, it became a grave question how their respective claims to authority might be most fitly co-ordinated. It was to meet, perhaps rather to evade, this question, that the distinction between that which is true in philosophy and that which is true in religion was proposed and adopted. But it is difficult to believe that the mind of any sincere and truth-loving man was satisfied by this distinction. Bacon has emphatically condemned it. " There is," he affirms, " no such opposition between God's word and his works". Both come from him who is the father of lights, the fountain of all truths, the author of all good ; and both are therefore to be studied with diligence and humility. To those who wish to discourage philosophy in order that ignorance of second causes may lead men to refer all things to the immediate agency of the first, Bacon puts Job's question, " An oportet mentiri pro Deo,"—will you offer to the God of truth the unclean sacrifice of a lie ?

The religious earnestness of Bacon's writings becomes more remarkable when we contrast it with the tone of the most illustrious of his contemporaries. Galileo's works are full of insincere deference to authority and of an affected disbelief in his own discoveries. Surely he who loves truth earnestly will be slow to believe that the cause of truth is to be served by irony. But we must not forget the difference between the circumstances in which the two men were placed.

Next to his determination of the true end of natural philosophy and of the relation in which it stands to natural and to revealed theology, we may place among Bacon's merits his clear view of the essential unity of science. He often insists on the importance of this idea, and has especially commended Plato and Parmenides for affirming " that all things do by scale ascend to unity ". The Creator is holy in the multitude of his works, holy in their disposition, holy in their unity : it is the prerogative of the doctrine of Forms to approach as nearly as possible towards the unity of Nature, and the subordinate science of Physics ought to contain two divisions relating to the same subject. One of these ought to treat of the first principles which govern all phenomena,and the other of the fabric of the universe [70]. All classifications of the science ought to be as veins or markings, and not as sections or divisions ; nor can any object of scientific inquiry be satisfactorily studied apart from the analogies which connect it with other similar objects.

But the greatest of all the services which Bacon rendered to natural philosophy was, that he perpetually enforced the necessity of laying aside all preconceived opinions and learning to be a follower of Nature. These counsels could not to their full extent be followed, nor has he himself attempted to do so. But they contain a great share of truth, and of truth never more needful than in Bacon's age. Before his time doubtless the authority of Aristotle, or rather that of the scholastic interpretation of his philosophy, was shaken, if not overthrown. Nevertheless the systematising spirit of the schoolmen still survived; and of the reformers of philosophy not a few attempted to substitute a dogmatic system of their own for that from which they dissented.

Nor were these attempts unsuccessful. For men still leaned upon authority, and accepted as a test of truth the appearance of completeness and sicentific consistency. This state of things was one of transition ; and probably no one did more towards putting an end to it than Bacon. To the dealers in systems and to their adherents he opposed the solemn declaration, that they only who come in their own name will be received of men. He constantly exhorted the seeker after truth to seek it in intercourse with Nature, and has repeatedly professed that he was no founder of a sect or school. He condemned the arrogance of those who thought it beneath the dignity of the philosopher to dwell on matters of observa-

[70] The latter is in effect what is now called Kosmos.

tion and experiment, and reminded them that the sun " æque palatia et cloacas ingreditur ; nec tamen polluitur". We do not, he continues, erect or dedicate to human pride a capitol or a pyramid ; we lay the foundations in the mind of man of a holy temple, whereof the exemplar is the universe. Throughout his writings the rejection of systems and authority is coupled with the assertion, that it is beyond all things necessary that the philosopher should be an humble follower of Nature. One of the most remarkable parts of the *Novum Organum* is the doctrine of Idola. It is an attempt to classify according to their origin the false and ill-defined notions by which the mind is commonly beset. They come, he tells us, from the nature of the human mind in general, from the peculiarities of each man's individual mind, from his intercourse with other men, from the formal teaching of the received philosophies. All these must be renounced and put away, else no man can enter into the kingdom which is to be founded on the knowledge of Nature [71]. Of the four kinds of idols Mersenne has spoken in his *Vérité des Sciences*, published in 1625, as of the four buttresses of the *Organum* of Verulam. This expression, though certainly inaccurate, serves to show the attention which in Bacon's time was paid to his doctrine of idola [72].

His rejection of syllogistic reasoning, in the proposed process for the establishment of axioms, was not without utility. In the middle ages and at the reform of philosophy the value of the syllogistic method was unduly exalted. Bacon was right in denying that it was possible to establish by a summary process and à priori the first principles of any science, and thence to deduce by syllogism all the propositions which that science could contain ; and though he erred in rejecting deductive reasoning altogether, this error could never have exerted any practical influence on the progress of science, while the truth with which it was associated was a truth of which his contemporaries required at least to be reminded. The reason of his error seems to have been that he formed an incorrect idea of the nature of syllogism, regarding it rather as an entirely artificial process than as merely a formal statement of the steps necessarily involved in every act of reasoning. However this may be, it is certain that whenever men attempted to set aside every process for the discovery of truth except induction, they must always have been led to recognise the impossibility of doing so.

Lastly, the tone in which Bacon spoke of the future destiny of mankind fitted him to be a leader of the age in which he lived. It was an age of change and hope. Men went forth to seek in new-found worlds for the land of gold and for the fountain of youth ; they were told that yet greater wonders lay within their reach. They had burst the bands of old authority ; they were told to go forth from the cave where they had dwelt so long, and look on the light of heaven. It was also for the most part an age of faith ; and the new philosophy upset no creed, and pulled down no altar. It did not put the notion of human perfectibility in the place of religion, nor deprive mankind of hopes beyond the grave. On the contrary, it told its followers that the instauration of the sciences was the free gift of the God in whom their fathers had trusted—that it was only another proof of the mercy of him whose mercy is over all his works.

71 Nov. Org. i. 68. The word idolon is used by Bacon in antithesis to idea. He does not mean by it an idol or false object of worship. 72 Compare Gassendi, Inst. Log.

The two Books of Francis Bacon of the Proficience and Advancement of Learning Divine and Human

PREFACE.

BY JAMES SPEDDING.

THE first edition of the *Advancement of Learning* is dated 1605. In what month it appeared is doubtful; but from certain allusions in a letter sent by Bacon to Tobie Matthew with a presentation copy, I gather (for the letter bears no date) that it was not out before the latter end of October.

Tobie Matthew, eldest son of the Bishop of Durham, was then about 27 years old, and had been intimate with Bacon, certainly for the last three years, and probably for more. Bacon had a high opinion of his abilities and seems to have consulted him about his works. "I have now at last (he says in this letter) taught that child to go, at the swaddling whereof you were. My work touching the *Proficiency and Advancement of Learning* I have put into two books, whereof the former, which you saw, I account but as a Page to the latter. I have now published them both, whereof I thought it a small adventure to send you a copy, who have more right to it than any man, except Bishop Andrews, who was my Inquisitor [1]."

Now Matthew had been abroad since April, 1605; and as he had seen the first book only, it is probable that the second was not then written; a circumstance which may be very naturally accounted for, if I am right in supposing that the *Advancement of Learning* was begun immediately after the accession of James I. From the death of Elizabeth, 24th March, 1602–3, to the meeting of James's first Parliament, 19th March, 1603–4, Bacon had very little to do. He held indeed the same place among the Learned Counsel which he had held under Elizabeth, but his services were little if at all used. On the 3rd of July, 1603, we find him writing to Lord Cecil:—" For my purpose or course, I desire to meddle as little as I can in the King's causes, his Majesty now abounding in counsel. . . . My ambition now I shall only put upon my pen, whereby I shall be able to maintain memory and merit of the times succeeding." And in the trial of Sir Walter Raleigh at Winchester the following November (though it was a complicated case involving many persons and requiring a great number of examinations) he does not appear to have been employed at all. But from the meeting of Parliament in March till the end of 1604 he was incessantly employed; first during the session (which lasted till the 7th of July) in the business of the House of Commons; then during the vacation, in preparation for the Commission of the Union [2] which was to meet in October; and from that time to the beginning of December in the business of the Commission itself;—all matters of extreme urgency and importance, and the "labour whereof, for men of his profession, rested most upon his hand" [3].

[1] Sir Tobie Matthew's collection of English letters, p. xi. Andrews was made a Bishop on the 3rd of November, 1605.

[2] See "Certain Articles or Considerations touching the union of the Kingdoms of England and Scotland; collected and dispersed for His Majesty's better service".

[3] Letter to the King, touching the Solicitor's place.

On the 4th of December the Commissioners signed their report; and on the 24th the next meeting of Parliament, which had been fixed for February, was postponed till October. This prorogation secured Bacon another interval of leisure; an interval longer perhaps, considering the nature of the public services which had now fallen upon him, than he was likely soon again to enjoy; and which it was the more important therefore to use in finishing the great literary work which he had begun. The same consideration may have determined him to be content with a less perfect treatment of the subject than he had originally designed; for certainly the second book, though so much the more important of the two, is in point of execution much less careful and elaborate than the first, and bears many marks of hasty composition. The presumption that an interval occurred between the writing of the two is further confirmed by the fact that they were not printed at the same time. The first ends with a half-sheet, and the second begins upon a fresh one with a new signature; whence I suppose we may infer that the first had been printed off before the second was ready for the press.

Of the motives which induced Bacon to undertake and hurry forward the *Advancement of Learning* at that particular time, and of those which afterwards suggested the incorporation of it into his great work on the Interpretation of Nature, I have explained my own view in my preface to the *De Augmentis*. Upon all matters requiring explanation or illustration the reader is referred to Mr. Ellis's notes upon the corresponding passages in that more finished work; and that the reference may be more easy I have marked the places where the several chapters begin; adding some account, more or less complete, of the principal differences between the two. In many cases these differences are so extensive that no adequate idea of their nature could be given within the limits of a note; and in such cases I have been content with a simple reference to the place. But where the substance of any addition or alteration which seemed to me material could be stated succinctly, — especially if it involved any modification of the opinion expressed in the text, — I have generally endeavoured to state it; sometimes translating Bacon's words, sometimes giving the effect in my own, as I found most convenient.

For the text, I have treated the edition of 1605 as the only original authority; the corrections introduced by later editors, though often unquestionably right, being (as far as I can see) merely conjectural. And therefore, though I have adopted all such corrections into the text whenever I was satisfied that they give the true reading, I have always quoted in a note the reading of the original. Only in the typographical arrangement with respect to capitals, italics, etc. (which in the original was probably left to the printer's taste, and is inconsistent in itself, and would be perplexing to modern eyes) and also in the punctuation, which is extremely confused and inaccurate, I have used the full liberty of my own judgment; altering as much as I pleased, and endeavouring only to make the sense clear to an eye accustomed to modern books, without encumbering the page with any notice of such alterations.

There is one innovation however which I have ventured to introduce and which it is necessary to explain. The *Advancement of Learning* was written for readers who were familiar with Latin, and abounds with Latin quotations. In these days it may be read with profit by many persons of both sexes to whom such quotations are a very perplexing obstruction. Forming as they generally do a part of the context, so that the sentence is not complete without them, those who cannot read Latin are in many cases unable to follow the sense of the English. To give such readers the means of understanding them seemed therefore no less than necessary; and I thought the true effect of them would be conveyed to the mind most perfectly and satisfactorily by presenting the interpretations in such a form that they might be read in their places, just as they would have been had they formed part of the original text, and just as they are in those passages where Bacon has himself furnished the interpretation. Following his example therefore as nearly as I could, I have endeavoured to give the effect of each of these Latin quotations in such a form as seemed to suit best the English idiom and to fall best into the English context; not tying myself to literal translation, but rather preferring to vary the expression, especially where I could by that means give it such a turn as to throw the emphasis more distinctly upon that part of the quotation which was more particularly in point. Thus it will be found, I think, that those who understand the Latin may still read the English without feeling it to be a mere repetition, while those who do not will in reading the English alone find the sense always complete. It was evident however that translations of this kind could not be read in this way conveniently if inserted in notes at the bottom of the page ; and therefore, there being no room in the margin, I have ventured to insert them in the text ; from which however, that they may not be mistaken for a part of it, I have always taken care to distinguish them by brackets. In a few cases where a Latin quotation occurs, not followed by a translation within brackets, it is to be understood that it is introduced merely as a voucher for what has just been said in the English, or for the purpose of suggesting a classical allusion which a translation would not suggest except to a classical reader, and that the sense is complete without it. In a few other cases where a quotation is followed by a translation *not* included within brackets, it is to be understood that it is Bacon's own translation and forms part of the original text.

For all the notes except those signed *R. L. E.*, which are Mr. Ellis's, I am responsible.

J. S.

THE FIRST BOOK

TO THE KING.

THERE were under the Law (excellent King) both daily sacrifices and freewill offerings; the one proceeding upon ordinary observance, the other upon a devout cheerfulness. In like manner there belongeth to kings from their servants both tribute of duty and presents of affection. In the former of these I hope I shall not live to be wanting, according to my most humble duty, and the good pleasure of your Majesty's employments: for the later, I thought it more respective to make choice of some oblation which might rather refer to the propriety and excellency of your individual person, than to the business of your crown and state.

Wherefore representing your Majesty many times unto my mind, and beholding you not with the inquisitive eye of presumption to discover that which the Scripture telleth me is inscrutable, but with the observant eye of duty and admiration; leaving aside the other parts of your virtue and fortune, I have been touched, yea and possessed, with an extreme wonder at those your virtues and faculties which the philosophers call intellectual; the largeness of your capacity, the faithfulness of your memory, the swiftness of your apprehension, the penetration of your judgment, and the facility and order of your elocution: and I have often thought that of all the persons living that I have known, your Majesty were the best instance to make a man of Plato's opinion, that all knowledge is but remembrance,* and that the mind of man by nature knoweth all things, and hath but her own native and original notions [1] (which by the strangeness and darkness of this tabernacle of the body are sequestered) again revived and restored: such a light of nature I have observed in your Majesty, and such a readiness to take flame and blaze from the least occasion presented, or the least spark of another's knowledge delivered. And as the Scripture saith of the wisest king, *That his heart was as the sands of the sea;* which though it be one of the largest bodies yet it consisteth of the smallest and finest portions; so hath God given your Majesty a composition of understanding admirable, being able to compass and comprehend the greatest matters, and nevertheless to touch and apprehend the least; whereas it should seem an impossibility in nature for the same instrument to make itself fit for great and small works. And for your gift of speech, I call to mind what Cornelius Tacitus saith of Augustus Cæsar; *Augusto profluens, et quæ principem deceret, eloquentia fuit*; [that his style of speech was flowing and prince-like [2]:] for if we note it well, speech that is uttered with labour and difficulty, or speech that savoureth of the affectation of art and precepts, or speech that is framed after the imitation of some pattern of eloquence, though never so excellent, — all this has somewhat servile, and holding of the subject. But your Majesty's manner of speech is indeed prince-like, flowing as from a fountain, and yet streaming and branching itself into nature's order, full of facility and felicity, imitating none, and inimitable by any. And as in your civil estate there appeareth to be an emulation and contention of your Majesty's virtue with your fortune; a virtuous disposition with a fortunate regiment; a virtuous expectation (when time was) of your greater fortune, with a prosperous possession thereof in the due time; a virtuous observation of the laws of marriage; a virtuous and most Christian desire of peace

[* *Phædo*, p. 75; *Meno*, ad. init. Cf. Aristot. *Anal. Pri.* ii. 21.]

[1] So edd. 1629 and 1633. Ed. 1605 has *motions.*

[2] Observe that the translations within brackets are not in the original, but inserted by myself. My reasons for adopting this plan, and the principle upon which I have proceeded in translating, are explained in the preface.—J.S.

with a fortunate inclination in your neighbour princes thereunto: so likewise in these intellectual matters, there seemeth to be no less contention between the excellency of your Majesty's gifts of nature and the universality and perfection of your learning. For I am well assured that this which I shall say is no amplification at all, but a positive and measured truth; which is, that there hath not been since Christ's time any king or temporal monarch which hath been so learned in all literature and erudition, divine and human. For let a man seriously and diligently revolve and peruse the succession of the emperors of Rome, of which Cæsar the dictator, who lived some years before Christ, and Marcus Antoninus were the best learned; and so descend to the emperors of Græcia, or of the West, and then to the lines of France, Spain, England, Scotland and the rest; and he shall find this judgment is truly made [3]. For it seemeth much in a king, if by the compendious extractions of other men's wits and labours he can take hold of any superficial ornaments and shews of learning, or if he countenance and prefer learning and learned men: but to drink indeed of the true fountains of learning, nay to have such a fountain of learning in himself, in a king, and in a king born, is almost a miracle. And the more, because there is met in your Majesty a rare conjunction as well of divine and sacred literature as of profane and human; so as your Majesty standeth invested of that triplicity which in great veneration was ascribed to the ancient Hermes; the power and fortune of a King, the knowledge and illumination of a Priest, and the learning and universality of a Philosopher. This propriety inherent and individual attribute in your Majesty deserveth to be expressed not only in the fame and admiration of the present time, nor in the history or tradition of the ages succeeding; but also in some solid work, fixed memorial, and immortal monument bearing a character or signature both of the power of a king and the difference and perfection of such a king.

Therefore I did conclude with myself, that I could not make unto your Majesty a better oblation than of some treatise tending to that end; whereof the sum will consist of these two parts: the former concerning the excellency of learning and knowledge, and the excellency of the merit and true glory in the augmentation and propagation thereof; the latter [4], what the particular acts and works are which have been embraced and undertaken for the advancement of learning, and again what defects and undervalues I find in such particular acts; to the end that though I cannot positively or affirmatively advise your Majesty, or propound unto you framed particulars, yet I may excite your princely cogitation to visit the excellent treasure of your own mind, and thence to extract particulars for this purpose agreeable to your magnanimity and wisdom.

In the entrance to the former of these,—to clear the way, and as it were to make silence to have the true testimonies concerning the dignity of learning to be better heard without the interruption of tacit objections,—I think good to deliver it from the discredits and disgraces which it hath received; all from ignorance; but ignorance severally disguised; appearing sometimes in the zeal and jealousy of divines, sometimes in the severity and arrogancy of politiques, and sometimes in the errors and imperfections of learned men themselves.

I hear the former sort say, that knowledge is of those things which are to be accepted of with great limitation and caution; that the aspiring to over-much knowledge was the original temptation and sin, whereupon ensued the fall of man; that knowledge hath in it somewhat of the serpent, and therefore where it entereth into a man it makes him swell,—*Scientia inflat* [knowledge puffeth up]; that Salomon gives a censure, *That there is no end of making books, and that*

[3] In the translation the reference to the particular dynasties is omitted; he only says, —*Percurrat qui voluerit imperatorum et rerum seriem, et juxta mecum sentiet.*

[4] I have observed elsewhere, that it was only the latter part which entered into the original scheme of the *Instauratio Magna*. And though in adapting the *Advancement of Learning* to it, he retained the former part, yet he marks it in the translation as comparatively unimportant; adding with regard to the first, *quæ levior est, neque tamen ullo modo prætermittenda*, and with regard to the second, *quod caput rei est.*

much reading is weariness of the flesh ; and again in another place, *That in spacious knowledge there is much contristation, and that he that increaseth knowledge increaseth anxiety ;* that St. Paul gives a caveat, *That we be not spoiled through vain philosophy ;* that experience demonstrates how learned men have been archheretics, how learned times have been inclined to atheism, and how the contemplation of second causes doth derogate from our dependence upon God, who is the first cause.

To discover then the ignorance and error of this opinion and the misunderstanding in the grounds thereof, it may well appear these men do not observe or consider that it was not the pure knowledge of nature and universality, a knowledge by the light whereof man did give names unto other creatures in Paradise, as they were brought before him, according unto their proprieties, which gave the occasion to the fall ; but it was the proud knowledge of good and evil, with an intent in man to give law unto himself and to depend no more upon God's commandments, which was the form of the temptation. Neither is it any quantity of knowledge how great soever that can make the mind of man to swell ; for nothing can fill, much less extend, the soul of man, but God and the contemplation of God ; and therefore Salomon speaking of the two principal senses of inquisition, the eye and the ear, affirmeth that the eye is never satisfied with seeing, nor the ear with hearing ; and if there be no fulness, then is the continent greater than the content : so of knowledge itself and the mind of man, whereto the senses are but reporters, he defineth likewise in these words, placed after that calendar or ephemerides which he maketh of the diversities of times and seasons for all actions and purposes ; and concludeth thus : *God hath made all things beautiful, or decent, in the true return of their seasons ; Also he hath placed the world in man's heart, yet cannot man find out the work which God worketh from the beginning to the end ;* declaring not obscurely that God hath framed the mind of man as a mirror or glass capable of the image of the universal world, and joyful to receive the impression thereof, as the eye joyeth to receive light ; and not only delighted in beholding the variety of things and vicissitude of times, but raised also to find out and discern the ordinances and decrees which throughout all those changes are infallibly observed. And although he doth insinuate that the supreme or summary law of nature, which he calleth *the work which God worketh from the beginning to the end,* is not possible to be found out by man ; yet that doth not derogate from the capacity of the mind, but may be referred to the impediments, as of shortness of life, ill conjunction of labours, ill tradition of knowledge over from hand to hand, and many other inconveniences whereunto the condition of man is subject. For that nothing parcel of the world is denied to man's inquiry and invention he doth in another place rule over, when he saith, *The spirit of man is as the lamp of God, wherewith he searcheth the inwardness of all secrets.* If then such be the capacity and receit of the mind of man, it is manifest that there is no danger at all in the proportion or quantity of knowledge, how large soever, lest it should make it swell or out-compass itself ; no, but it is merely the quality of knowledge, which be it in quantity more or less, if it be taken without the true corrective thereof, hath in it some nature of venom or malignity, and some effects of that venom, which is ventosity or swelling. This corrective spice, the mixture whereof maketh knowledge so sovereign, is Charity, which the apostle immediately addeth to the former clause ; for so he saith, *knowledge bloweth up, but charity buildeth up ;* not unlike unto that which he delivereth in another place : *If I spake* (saith he) *with the tongues of men and angels, and had not charity, it were but as a tinkling cymbal ;* not but that it is an excellent thing to speak with the tongues of men and angels, but because if it be severed from charity, and not referred to the good of men and mankind, it hath rather a sounding and unworthy glory than a meriting and substantial virtue. And as for that censure of Salomon concerning the excess of writing and reading books and the anxiety of spirit which redoundeth from knowledge, and that admonition of St. Paul, *That we be not seduced by vain philosophy ;* let those places be rightly understood, and they do indeed excellently set forth the true bounds and limitations whereby human knowledge is confined and circumscribed ; and yet without any such contracting or coarctation, but that it may comprehend all the universal nature of things. For these limitations are three. The first, *that we do not so place our felicity in knowledge, as we forget our mortality.* The second, *that we make application of our knowledge to give ourselves repose and contentment, and not distaste or repining.* The third, *that we do not presume by the contemplation of nature to attain to the mysteries of God.* For as touching the first of these, Salomon doth excellently expound himself in another place of the same book, where he saith; *I saw well that knowledge recedeth as far from ignorance as light doth from darkness, and that the wise man's eyes keep watch in his head, whereas the fool roundeth about in darkness; but withal I learned that the same mortality involveth them both.* And for the second, certain it is, there is no vexation or anxiety of mind which resulteth from knowledge otherwise than merely by accident; for all knowledge and wonder (which is the seed of knowledge) is an impression of pleasure in itself: but when men fall to framing conclusions out of their knowledge, applying it to their particular, and ministering to themselves thereby weak fears or vast desires, there groweth that carefulness and trouble of mind which is spoken of: for then knowledge is no more *Lumen siccum* [a dry light], whereof Heraclitus the profound said, *Lumen siccum optima anima*[5] [the dry light is the best soul]; but it becometh *Lumen madidum* or *maceratum* [a light charged with moisture], being steeped and infused in the humours of the affections. And as for the third point, it deserveth to be a little stood upon and not to be lightly passed over: for if any man shall think by view and inquiry into these sensible and material things to attain that light whereby he may reveal unto himself the nature or will of God, then indeed is he spoiled by vain philosophy: for the contemplation of God's creatures and works produceth (having regard to the works and creatures themselves) knowledge; but having regard to God, no perfect knowledge, but wonder, which is broken knowledge. And therefore it was most aptly said by one of Plato's school, *That the sense of man carrieth a resemblance with the sun, which (as we see) openeth and revealeth all the terrestrial globe; but then again it obscureth and concealeth the stars and celestial globe; so doth the sense discover natural things, but it darkeneth and shutteth up divine.* And hence it is true that it hath proceeded that divers great learned men have been heretical, whilst they have sought to fly up to the secrets of the Deity by the waxen wings of the senses. And as for the conceit that too much knowledge should incline a man to atheism, and that the ignorance of second causes should make a more devout dependence upon God which is the first cause; first, it is good to ask the question which Job asked of his friends, *Will you lie for God, as one man will do for another, to gratify him?* For certain it is that God worketh nothing in nature but by second causes; and if they would have it otherwise believed, it is mere imposture, as it were in favour towards God; and nothing else but to offer to the author of truth the unclean sacrifice of a lie. But farther, it is an assured truth and a conclusion of experience, that a little or superficial knowledge of philosophy may incline the mind of man to atheism, but a farther proceeding therein doth bring the mind back again to religion; for in the entrance of philosophy, when the second causes, which are next unto the senses, do offer themselves to the mind of man, if it dwel' and stay there, it may induce some oblivion of the highest cause; but when a man passeth on farther, and seeth the dependence of causes and the works of Providence; then, according to the allegory of the poets, he will easily believe that the highest link of nature's chain must needs be tied to the foot of Jupiter's chair. To conclude therefore, let no man, upon a weak conceit of sobriety or an ill-applied moderation, think or maintain that a man can search too far or be too well studied in the book of God's word or in the book of God's works; divinity or philosophy; but rather let men endeavour an endless progress or proficience in both; only let men beware that they apply both to charity, and not to swelling; to use,

[5] *αὐγὴ ξηρὴ ψυχὴ σοφωτάτη*: a corruption, according to the conjecture of Professor W. H. Thompson, of *αὔη ψυχὴ σοφωτάτη*; *ξηρὴ* having been first inserted by one commentator, to explain the unusual word *αὔη*, and so passed into the text; *αὔη* having been turned into *αὐγὴ* by another, to make sense. See *Remains of Professor Archer Butler*, vol. i. p. 314.

and not to ostentation ; and again, that they do not unwisely mingle or confound these learnings together.

AND as for the disgraces which learning receiveth from politiques, they be of this nature ; that learning doth soften men's minds, and makes them more unapt for the honour and exercise of arms ; that it doth mar and pervert men's dispositions for matter of government and policy, in making them too curious and irresolute by variety of reading, or too peremptory or positive by strictness of rules and axioms, or too immoderate and overweening by reason of the greatness of examples, or too incompatible and differing from the times by reason of the dissimilitude of examples ; or at least that it doth divert men's travails from action and business, and bringeth them to a love of leisure and privateness ; and that it doth bring into states a relaxation of discipline, whilst every man is more ready to argue than to obey and execute. Out of this conceit Cato surnamed the Censor, one of the wisest men indeed that ever lived, when Carneades the philosopher came in embassage to Rome, and that the young men of Rome began to flock about him, being allured with the sweetness and majesty of his eloquence and learning, gave counsel in open senate that they should give him his despatch with all speed, lest he should infect and inchant the minds and affections of the youth, and at unawares bring in an alteration of the manners and customs of the state. Out of the same conceit or humour did Virgil, turning his pen to the advantage of his country and the disadvantage of his own profession, make a kind of separation between policy and government and between arts and sciences, in the verses so much renowned, attributing and challenging the one to the Romans, and leaving and yielding the other to the Grecians ; *Tu regere imperio populos, Romane, memento, Hæ tibi erunt artes, etc.*

[Be thine, O Rome,
With arts of government to rule the nations.]

So likewise we see that Anytus, the accuser of Socrates, laid it as an article of charge and accusation against him that he did with the variety and power of his discourses and disputations withdraw young men from due reverence to the laws and customs of their country ; and that he did profess a dangerous and pernicious science, which was to make the worse matter seem the better, and to suppress truth by force of eloquence and speech.

But these and the like imputations have rather a countenance of gravity than any ground of justice : for experience doth warrant that both in persons and in times there hath been a meeting and concurrence in learning and arms, flourishing and excelling in the same men and the same ages. For as for men, there cannot be a better nor the like instance, as of that pair, Alexander the Great and Julius Cæsar the dictator ; whereof the one was Aristotle's scholar in philosophy, and the other was Cicero's rival in eloquence ; or if any man had rather call for scholars that were great generals than generals that were great scholars, let him take Epaminondas the Theban, or Xenophon the Athenian ; whereof the one was the first that abated the power of Sparta, and the other was the first that made way to the overthrow of the monarchy of Persia. And this concurrence is yet more visible in times than in persons, by how much an age is greater object than a man. For both in Ægypt, Assyria, Persia, Græcia, and Rome, the same times that are most renowned for arms are likewise most admired for learning ; so that the greatest authors and philosophers and the greatest captains and governors have lived in the same ages. Neither can it otherwise be : for as in man the ripeness of strength of the body and mind cometh much about an age, save that the strength of the body cometh somewhat the more early ; so in states, arms and learning, whereof the one correspondeth to the body, the other to the soul of man, have a concurrence or near sequence in times.

And for matter of policy and government, that learning should rather hurt than enable thereunto, is a thing very improbable. We see it is accounted an error to commit a natural body to empiric physicians, which commonly have a few pleasing receits whereupon they are confident and adventurous, but know neither the causes of diseases, nor the complexions of patients, nor peril of accidents, nor the true method of cures. We see it is a like error to rely upon advocates or lawyers which are only men of practice and not grounded in their books, who are many times easily surprised when matter falleth out besides their experience, to the prejudice of the cause they handle. So by like reason it cannot be but a matter of doubtful consequence, if states be managed by empiric statesmen, not well mingled with men grounded in learning. But contrariwise, it is almost without instance contradictory, that ever[6] any government was disastrous that was in the hands of learned governors. For howsoever it hath been ordinary with politic men to extenuate and disable learned men by the names of *Pedantes* ; yet in the records of time it appeareth in many particulars, that the governments of princes in minority (notwithstanding the infinite disadvantage of that kind of state) have nevertheless excelled the government of princes of mature age, even for that reason which they seek to traduce, which is, that by that occasion the state hath been in the hands of *Pedantes* : for so was the state of Rome for the first five years, which are so much magnified, during the minority of Nero, in the hands of Seneca, a *Pedanti* : so it was again for ten years space or more, during the minority of Gordianus the younger, with great applause and contentation in the hands of Misitheus, a *Pedanti* : so was it before that, in the minority of Alexander Severus, in like happiness, in hands not much unlike, by reason of the rule of the women, who were aided by the teachers and preceptors. Nay let a man look into the government of the bishops of Rome, as by name into the government of Pius Quintus and Sextus Quintus in our times, who were both at their entrance esteemed but as pedantical friars, and he shall find that such popes do greater things, and proceed upon truer principles of estate, than those which have ascended to the papacy from an education and breeding in affairs of estate and courts of princes ; for although men bred in learning are perhaps to seek in points of convenience and accommodating for the present, which the Italians call *ragioni di stato*, whereof the same Pius Quintus could not hear spoken with patience, terming them inventions against religion and the moral virtues; yet on the other side, to recompense that, they are perfect in those same plain grounds of religion, justice, honour, and moral virtue; which if they be well and watchfully pursued, there will be seldom use of those other, no more than of physic in a sound or well-dieted body. Neither can the experience of one man's life furnish examples and precedents for the events of one man's life : for as it happeneth sometimes that the grandchild or other descendant resembleth the ancestor more than the son ; so many times occurrences of present times may sort better with ancient examples than with those of the later or immediate times : and lastly, the wit of one man can no more countervail learning than one man's means can hold way with a common purse.

And as for those particular seducements or indisposition of the mind for policy and government, which learning is pretended to insinuate ; if it be granted that any such thing be, it must be remembered withal, that learning ministereth in every of them greater strength of medicine or remedy, than it offereth cause of indisposition or infirmity. For if by a secret operation it make men perplexed and irresolute, on the other side by plain precept it teacheth them when and upon what ground to resolve ; yea, and how to carry things in suspense without prejudice till they resolve. If it make men positive and regular, it teacheth them what things are in their nature demonstrative, and what are conjectural ; and as well the use of distinctions and exceptions, as the latitude of principles and rules. If it mislead by disproportion or dissimilitude of examples, it teacheth men the force of circumstances, the errors of comparisons, and all the cautions of application ; so that in all these it doth rectify more effectually than it can pervert. And these medicines it conveyeth into men's minds much more forcibly by the quickness and penetration of examples. For let a man look into the errors of Clement the seventh, so lively described by Guicciardine, who served under him, or into the errors of Cicero painted out by his own pencil in his epistles to Atticus, and he will fly apace from being irresolute. Let him look into the errors of Phocion, and he will beware how he be obstinate or inflexible. Let him

[6] So in all the editions.

but read the fable of Ixion, and it will hold him from being vaporous or imaginative. Let him look into the errors of Cato the second, and he will never be one of the Antipodes, to tread opposite to the present world.

And for the conceit that learning should dispose men to leisure and privateness, and make men slothful; it were a strange thing if that which accustometh the mind to a perpetual motion and agitation should induce slothfulness; whereas contrariwise it may be truly affirmed that no kind of men love business for itself but those that are learned; for other persons love it for profit, as an hireling that loves the work for the wages; or for honour, as because it beareth them up in the eyes of men, and refresheth their reputation which otherwise would wear; or because it putteth them in mind of their fortune, and giveth them occasion to pleasure and displeasure; or because it exerciseth some faculty wherein they take pride, and so entertaineth them in good humour and pleasing conceits toward themselves; or because it advanceth any other their ends. So that as it is said of untrue valours that some men's valours are in the eyes of them that look on, so such men's industries are in the eyes of others, or at least in regard of their own designments [7]; only learned men love business as an action according to nature, as agreeable to health of mind as exercise is to health of body, taking pleasure in the action itself, and not in the purchase; so that of all men they are the most indefatigable, if it be towards any business which can hold or detain their mind.

And if any man be laborious in reading and study and yet idle in business and action, it groweth from some weakness of body or softness of spirit, such as Seneca speaketh of; *Quidam tam sunt umbratiles, ut putent in turbido esse quicquid in luce est** [there are some men so fond of the shade, that they think they are in trouble whenever they are in the light]; and not of learning. Well may it be that such a point of a man's nature may make him give himself to learning, but it is not learning that breedeth any such point in his nature.

And that learning should take up too much time or leisure; I answer, the most active or busy man that hath been or can be hath (no question) many vacant times of leisure, while he expecteth the tides and returns of business, (except he be either tedious and of no dispatch, or lightly and unworthily ambitious to meddle in things that may be better done by others); and then the question is but how those spaces and times of leisure shall be filled and spent; whether in pleasures or in studies; as was well answered by Demosthenes to his adversary Æschines [8], that was a man given to pleasure, and told him *that his orations did smell of the lamp*: *Indeed* (said Demosthenes) *there is a great difference between the things that you and I do by lamp-light.* So as no man need doubt that learning will expulse business; but rather it will keep and defend the possession of the mind against idleness and pleasure, which otherwise at unawares may enter to the prejudice of both.

Again, for that other conceit that learning should undermine the reverence of laws and government, it is assuredly a mere depravation and calumny without all shadow of truth. For to say that a blind custom of obedience should be a surer obligation than duty taught and understood, it is to affirm that a blind man may tread surer by a guide than a seeing man can by a light. And it is without all controversy that learning doth make the minds of men gentle, generous, maniable, and pliant to government; whereas ignorance makes them churlish, thwart, and mutinous: and the evidence of time doth clear this assertion, considering that the most barbarous, rude, and unlearned times have been most subject to tumults, seditions, and changes.

And as to the judgment of Cato the Censor, he was well punished for his blasphemy against learning, in the same kind wherein he offended; for when he was past threescore years old, he was taken with an extreme desire to go to school again and to learn the Greek tongue, to the end to peruse the Greek authors; which doth well demonstrate, that his former censure of the Grecian learning

[7] *i.e.* they have for their object either the applause of others or some inward gratification of their own. (*hoc videntur agere, aut ut alii plaudant, aut ut ipsi intra se gestiant.*)

[* Seneca, Ep. 3. Quotation inaccurate.] [8] Pytheas, according to Plutarch.

was rather an affected gravity, than according to the inward sense of his own opinion. And as for Virgil's verses, though it pleased him to brave the world in taking to the Romans the art of empire, and leaving to others the arts of subjects; yet so much is manifest, that the Romans never ascended to that height of empire till the time they had ascended to the height of other arts; for in the time of the two first Cæsars, which had the art of government in greatest perfection, there lived the best poet, Virgilius Maro; the best historiographer, Titus Livius; the best antiquary, Marcus Varro; and the best, or second orator, Marcus Cicero, that to the memory of man are known. As for the accusation of Socrates, the time must be remembered when it was prosecuted; which was under the thirty tyrants, the most base, bloody, and envious persons that have governed; which revolution of state was no sooner over, but Socrates, whom they had made a person criminal, was made a person heroical, and his memory accumulate with honours divine and human; and those discourses of his, which were then termed corrupting of manners, were after acknowledged for sovereign medicines of the mind and manners, and so have been received ever since till this day. Let this therefore serve for answer to politiques, which in their humourous severity or in their feigned gravity have presumed to throw imputations upon learning; which redargution nevertheless (save that we know not whether our labours may extend to other ages) were not needful for the present, in regard of the love and reverence towards learning which the example and countenance of two so learned princes, queen Elizabeth and your Majesty, being as Castor and Pollux, *lucida sidera*, stars of excellent light and most benign influence, hath wrought in all men of place and authority in our nation.

Now therefore we come to that third sort of discredit or diminution of credit that groweth unto learning from learned men themselves, which commonly cleaveth fastest. It is either from their fortune, or from their manners, or from the nature of their studies. For the first, it is not in their power; and the second is accidental; the third only is proper to be handled. But because we are not in hand with true measure, but with popular estimation and conceit, it is not amiss to speak somewhat of the two former. The derogations therefore which grow to learning from the fortune or condition of learned men, are either in respect of scarcity of means, or in respect of privateness of life and meanness of employments.

Concerning want, and that it is the case of learned men usually to begin with little and not to grow rich so fast as other men, by reason they convert not their labours chiefly to lucre and increase; it were good to leave the common place in commendation of poverty to some friar to handle [9], to whom much was attributed by Machiavel in this point, when he said, *That the kingdom of the clergy had been long before at an end, if the reputation and reverence towards the poverty of friars had not borne out the scandal of the superfluities and excesses of bishops and prelates**. So a man might say that the felicity and delicacy of princes and great persons had long since turned to rudeness and barbarism, if the poverty of learning had not kept up civility and honour of life. But without any such advantages, it is worthy the observation what a reverend and honoured thing poverty of fortune was for some ages in the Roman state, which nevertheless was a state without paradoxes. For we see what Titus Livius saith in his introduction: *Cæterum aut me amor negotii suscepti fallit, aut nulla unquam respublica nec major, nec sanctior, nec bonis exemplis ditior fuit; nec in quam tam seræ luxuriaque immigraverint; nec ubi tantus ac tam diu paupertati ac parsimoniæ honos fuerit*: [that if affection for his subject did not deceive him, there was never any state in the world either greater or purer or richer in good examples, never any into which avarice and luxury made their way so late; never any in which poverty and frugality were for so long a time held in so great honour]. We see likewise, after that the state of Rome was not itself but did degenerate, how that person that took upon him to be counsellor to Julius Cæsar after his victory, where to begin his restoration of the state, maketh it of all points the

[9] Patribus mendicantibus (pace eorum dixerim).—*De Aug.* [* *Discorsi*, iii, 1.]

most summary to take away the estimation of wealth: *Verum hæc et omnia mala pariter cum honore pecuniæ desinent; si neque magistratus, neque alia vulgo cupienda, venalia erunt*: [but these and all other evils (he says) will cease as soon as the worship of money ceases; which will come to pass when neither magistracies nor other things that are objects of desire to the vulgar shall be to be had for money]. To conclude this point, as it was truly said that *rubor est virtutis color* [a blush is virtue's colour], though sometime it come from vice; so it may be fitly said that *paupertas est virtutis fortuna* [poverty is virtue's fortune], though sometime it may proceed from misgovernment and accidents. Surely Salomon hath pronounced it, both in censure, *Qui festinat ad divitias non erit insons* [he that maketh haste to be rich shall not be innocent]; and in precept, *Buy the truth, and sell it not; and so of wisdom and knowledge;* judging that means were to be spent upon learning, and not learning to be applied to means. And as for the privateness or obscureness (as it may be in vulgar estimation accounted) of life of contemplative men; it is a theme so common to extol a private life, not taxed with sensuality and sloth, in comparison and to the disadvantage of a civil life, for safety, liberty, pleasure, and dignity, or at least freedom from indignity, as no man handleth it but handleth it well; such a consonancy it hath to men's conceits in the expressing and to men's consents in the allowing. This only I will add, that learned men forgotten in states, and not living in the eyes of men, are like the images of Cassius and Brutus in the funeral of Junia; of which not being represented, as many others were, Tacitus saith, *Eo ipso præfulgebant, quod non visebantur* [they had the pre-eminence over all—in being left out].

And for meanness of employment, that which is most traduced to contempt is that the government of youth is commonly allotted to them; which age, because it is the age of least authority, it is transferred to the disesteeming of those employments wherein youth is conversant, and which are conversant about youth. But how unjust this traducement is (if you will reduce things from popularity of opinion to measure of reason) may appear in that we see men are more curious what they put into a new vessel than into a vessel seasoned, and what mould they lay about a young plant than about a plant corroborate; so as the weakest terms and times of all things use to have the best applications and helps. And will you hearken to the Hebrew Rabbins? *Your young men shall see visions, and your old men shall dream dreams;* say they [10] youth is the worthier age, for that visions are nearer apparitions of God than dreams. And let it be noted, that howsoever the conditions of life of *Pedantes* have [11] been scorned upon theatres, as the ape of tyranny; and that the modern looseness or negligence hath taken no due regard to the choice of schoolmasters and tutors; yet the ancient wisdom of the best times did always make a just complaint that states were too busy with their laws and too negligent in point of education: which excellent part of ancient discipline hath been in some sort revived of late times by the colleges of the Jesuits; of whom, although in regard of their superstition I may say, *quo meliores, eo deteriores* [12] [the better the worse]; yet in regard of this, and some other points concerning human learning and moral matters, I may say, as Agesilaus said to his enemy Pharnabazus, *talis quum sis, utinam noster esses* [they are so good that I wish they were on our side]. And thus much touching the discredits drawn from the fortunes of learned men.

[10] So the original. Edd. 1629 and 1633 have *the*. The meaning is, "upon this text they observe," etc. (*Ex hoc textu colligunt.*)]

[11] So ed. 1633. The original has *hath*.

[12] This parenthesis is omitted in the translation, no doubt as offensive to the Roman Catholics. Several other passages of the same kind occur in the *Advancement*, and they are all treated in the same way. The motive for which is sufficiently explained by Bacon himself in the letter which he sent to the King along with the *De Augmentis*. "I have been also (he says) mine own *Index Expurgatorius*, that it may be read in all places. For since my end of putting it into Latin was to have it read everywhere, it had been an absurd contradiction to free it in the language and to pen it up in the matter." Mr. Ellis made a list of these passages, which will be noticed in their places. The word *enemy* in the next clause is omitted, probably from the same motive.

As touching the manners of learned men, it is a thing personal and individual: and no doubt there be amongst them, as in other professions, of all temperatures; but yet so as it is not without truth which is said, that *abeunt studia in mores*, studies have an influence and operation upon the manners of those that are conversant in them [13].

But upon an attentive and indifferent review, I for my part cannot find any disgrace to learning can proceed from the manners of learned men; not inherent to them as they are learned [14]; except it be a fault (which was the supposed fault of Demosthenes, Cicero, Cato the second, Seneca, and many more, that because the times they read of are commonly better than the times they live in, and the duties taught better than the duties practised, they contend sometimes too far to bring things to perfection, and to reduce the corruption of manners to honesty of precepts or examples of too great height. And yet hereof they have caveats enough in their own walks. For Solon, when he was asked whether he had given his citizens the best laws, answered wisely, *Yea, of such as they would receive*: and Plato, finding that his own heart could not agree with the corrupt manners of his country, refused to bear place or office; saying, *That a man's country was to be used as his parents were, that is, with humble persuasions, and not with contestations*: and Cæsar's counsellor put in the same caveat, *Non ad vetera instituta revocans quæ jampridem corruptis moribus ludibrio sunt* [not to attempt to bring things back to the original institution, now that by reason of the corruption of manners the ancient simplicity and purity had fallen into contempt]: and Cicero noteth this error directly in Cato the second, when he writes to his friend Atticus; *Cato optime sentit, sed nocet interdum reipublicæ; loquitur enim tanquam in republica Platonis, non tanquam in fæce Romuli* [Cato means excellently well; but he does hurt sometimes to the state; for he talks as if it were Plato's republic that we are living in, and not the dregs of Romulus:] and the same Cicero doth excuse and expound the philosophers for going too far and being too exact in their prescripts, when he saith, *Isti ipsi præceptores virtutis et magistri videntur fines officiorum paulo longius quam natura vellet protulisse, ut cum adultimum animo contendissemus, ibi tamen, ubi oportet, consisteremus**: [that they had set the points of duty somewhat higher than nature would well bear; meaning belike to allow for shortcomings, and that our endeavours aiming beyond the mark and falling short, should light at the right place]: and yet himself might have said, *Monitis sum minor ipse meis* [that he fell short of his own precepts]; for it was his own fault, though not in so extreme a degree.

Another fault likewise much of this kind hath been incident to learned men; which is, that they have esteemed the preservation, good, and honour of their countries or master before their own fortunes or safeties. For so saith Demosthenes unto the Athenians: *If it would please you to note it, my counsels unto you are not such whereby I should grow great amongst you, and you become little amongst the Grecians; but they be of that nature, as they are sometimes not good for me to give, but are always good for you to follow.* And so Seneca, after he had consecrated that *Quinquennium Neronis* to the eternal glory of learned governors, held on his honest and loyal course of good and free counsel, after his master grew extremely corrupt in his government. Neither can this point otherwise be; for learning endueth men's minds with a true sense of the frailty of their persons, the casualty of their fortunes, and the dignity of their soul and vocation; so that it is impossible for them to esteem that any greatness of their own fortune can be a true or worthy end of their being and ordainment; and therefore are desirous to give their account to God, and so likewise to their masters under God (as kings and the states that they serve), in these words; *Ecce tibi lucrefeci*, and not *Ecce mihi lucrefeci*, [' Lo, I have gained for thee,' not ' Lo, I have gained for myself ']: whereas the corrupter sort of mere politiques, that have not their thoughts established by learning in the love and apprehension of duty, nor never

[13] And that learning (the translation adds), unless the mind into which it enters be much depraved, corrects the natural disposition and changes it for the better.

[14] *i.e.* not [I mean, from such manners as are] inherent etc.

[**Pro Muraenâ*, c. 31. Quotation inaccurate.]

look abroad into universality, do refer all things to themselves, and thrust themselves into the centre of the world, as if all lines should meet in them and their fortunes; never caring in all tempests what becomes of the ship of estates, so they may save themselves in the cockboat of their own fortune; whereas men that feel the weight of duty, and know the limits of self-love, use to make good their places and duties, though with peril. And if they stand in seditious and violent alterations, it is rather the reverence which many times both adverse parts do give to honesty, than any versatile advantage of their own carriage. But for this point of tender sense and fast obligation of duty which learning doth endue the mind withal, howsoever fortune may tax it and many in the depth of their corrupt principles may despise it, yet it will receive an open allowance, and therefore needs the less disproof or excusation.

Another fault incident commonly to learned men, which may be more probably defended than truly denied, is that they fail sometimes in applying themselves to particular persons; which want of exact application ariseth from two causes; the one, because the largeness of their mind can hardly confine itself to dwell in the exquisite observation or examination of the nature and customs of one person: for it is a speech for a lover and not for a wise man, *Satis magnum alter alteri theatrum sumus* [each is to other a theatre large enough]. Nevertheless I shall yield, that he that cannot contract the sight of his mind as well as disperse and dilate it, wanteth a great faculty. But there is a second cause, which is no inability but a rejection upon choice and judgment. For the honest and just bounds of observation by one person upon another extend no farther but to understand him sufficiently, whereby not to give him offence, or whereby to be able to give him faithful counsel, or whereby to stand upon reasonable guard and caution in respect of a man's self: but to be speculative into another man, to the end to know how to work him or wind him or govern him, proceedeth from a heart that is double and cloven, and not entire and ingenuous; which as in friendship it is want of integrity, so towards princes or superiors is want of duty. For the custom of the Levant, which is, that subjects do forbear to gaze or fix their eyes upon princes, is in the outward ceremony barbarous; but the moral is good: for men ought not by cunning and bent observations to pierce and penetrate into the hearts of kings, which the Scripture hath declared to be inscrutable.

There is yet another fault (with which I will conclude this part) which is often noted in learned men, that they do many times fail to observe decency and discretion in their behaviour and carriage, and commit errors in small and ordinary points of action; so as the vulgar sort of capacities do make a judgment of them in greater matters by that which they find wanting in them in smaller. But this consequence doth oft deceive men; for which I do refer them over to that which was said by Themistocles, arrogantly and uncivilly being applied to himself out of his own mouth, but being applied to the general state of this question pertinently and justly; when being invited to touch a lute, he said *he could not fiddle, but he could make a small town a great state.* So no doubt many may be well seen in the passages of government and policy, which are to seek in little and punctual occasions. I refer them also to that which Plato said of his master Socrates, whom he compared to the gallypots of apothecaries, which on the outside had apes and owls and antiques, but contained within sovereign and precious liquors and confections; acknowledging that to an external report he was not without superficial levities and deformities but was inwardly replenished with excellent virtues and powers. And so much touching the point of manners of learned men.

But in the mean time I have no purpose to give allowance to some conditions and courses base and unworthy, wherein divers professors of learning have wronged themselves and gone too far; such as were those trencher philosophers, which in the later age of the Roman state were usually in the houses of great persons, being little better than solemn parasites; of which kind, Lucian maketh a merry description of the philosopher that the great lady took to ride with her in her coach, and would needs have him carry her little dog, which he doing officiously and yet uncomely, the page scoffed, and said, *That he doubted the philosopher of a Stoic would turn to be a Cynic.* But above all the rest, the gross and palpable flattery whereunto many (not unlearned) have abased and abused their wits and pens, turning (as Du Bartas saith) Hecuba into Helena and Faustina into Lucretia, hath most diminished the price and estimation of learning. Neither is the moral [15] dedications of books and writings, as to patrons, to be commended: for that books (such as are worthy the name of books) ought to have no patrons but truth and reason; and the ancient custom was to dedicate them only to private and equal friends, or to intitle the books with their names; or if to kings and great persons, it was to some such as the argument of the book was fit and proper for. But these and the like courses may deserve rather reprehension than defence.

Not that I can tax or condemn the morigeration or application of learned men to men in fortune. For the answer was good that Diogenes* made to one that asked him in mockery, *How it came to pass that philosophers were the followers of rich men, and not rich men of philosophers?* He answered soberly, and yet sharply, *Because the one sort knew what they had need of, and the other did not.* And of the like nature was the answer which Aristippus made, when having a petition to Dionysius and no ear given to him, he fell down at his feet, whereupon Dionysius staid and gave him the hearing and granted it; and afterward some person tender on the behalf of philosophy, reproved Aristippus that he would offer the profession of philosophy such an indignity, as for a private suit to fall at a tyrant's feet: but he answered, *It was not his fault, but it was the fault of Dionysius, that had his ears in his feet.* Neither was it accounted weakness, but discretion, in him that would not dispute his best with Adrianus Cæsar; excusing himself, *That it was reason to yield to him that commanded thirty legions.* These and the like applications and stooping to points of necessity and convenience cannot be disallowed; for though they may have some outward baseness, yet in a judgment truly made they are to be accounted submissions to the occasion and not to the person.

Now I proceed to those errors and vanities which have intervened amongst the studies themselves of the learned; which is that which is principal and proper to the present argument; wherein my purpose is not to make a justification of the errors, but, by a censure and separation of the errors, to make a justification of that which is good and sound, and to deliver that from the aspersion of the other. For we see that it is the manner of men to scandalize and deprave that which retaineth the state and virtue, by taking advantage upon that which is corrupt and degenerate: as the Heathens in the primitive church used to blemish and taint the Christians with the faults and corruptions of heretics. But nevertheless I have no meaning at this time to make any exact animadversion of the errors and impediments in matters of learning which are more secret and remote from vulgar opinion; but only to speak unto such as do fall under, or near unto, a popular observation.

There be therefore chiefly three vanities in studies, whereby learning hath been most traduced. For those things we do esteem vain, which are either false or frivolous, those which either have no truth or no use: and those persons we esteem vain, which are either credulous or curious; and curiosity is either in matter or words: so that in reason as well as in experience, there fall out to be these three distempers (as I may term them) of learning; the first, fantastical learning; the second, contentious learning; and the last, delicate learning; vain imaginations, vain altercations, and vain affectations; and with the last I will begin [16]. Martin Luther, conducted (no doubt) by an higher Providence

[15] *i.e.* customary. *Morem illum receptum libros patronis nuncupandi.*—De Aug. Ed. 1629 has *moderne.* [* Not Diogenes, but Aristippus. Diog. Laërt. *in Aristip,* c. 69].

[16] The passage which follows is much curtailed in the translation; no doubt for the reason mentioned in note p. 50. All allusion to the "higher Providence", the "degenerate traditions" of the church, the study of the ancient authors, and the "primitive but seeming new opinions" is left out: and we are only told that this distemper of luxuriance of speech (though in former times it had been occasionally in request) began

but in discourse of reason finding what a province he had undertaken against the Bishop of Rome and the degenerate traditions of the church, and finding his own solitude, being no ways aided by the opinions of his own time, was enforced to awake all antiquity, and to call former times to his succors to make a party against the present time; so that the ancient authors, both in divinity and in humanity, which had long time slept in libraries, began generally to be read and revolved. This by consequence did draw on a necessity of a more exquisite travail in the languages original wherein those authors did write, for the better understanding of those authors and the better advantage of pressing and applying their words. And thereof grew again a delight in their manner of style and phrase, and an admiration of that kind of writing; which was much furthered and precipitated by the enmity and opposition that the propounders of those (primitive but seeming new) opinions had against the schoolmen; who were generally of the contrary part, and whose writings were altogether in a differing style and form; taking liberty to coin and frame new terms of art to express their own sense and to avoid circuit of speech, without regard to the pureness, pleasantness, and (as I may call it) lawfulness of the phrase or word. And again, because the great labour then [17] was with the people (of whom the Pharisees were wont to say, *Execrabilis ista turba, quæ non novit legem*) [the wretched crowd that has not known the law], for the winning and persuading of them there grew of necessity in chief price and request eloquence and variety of discourse, as the fittest and forciblest access into the capacity of the vulgar sort. So that these four causes concurring, the admiration of ancient authors, the hate of the schoolmen, the exact study of languages, and the efficacy of preaching, did bring in an affectionate study of eloquence and copie * of speech, which then began to flourish. This grew speedily to an excess; for men began to hunt more after words than matter; and more after the choiceness of the phrase, and the round and clean composition of the sentence, and the sweet falling of the clauses, and the varying and illustration of their works with tropes and figures, than after the weight of matter, worth of subject, soundness of argument, life of invention, or depth of judgment. Then grew the flowing and watery vein of Osorius, the Portugal bishop†, to be in price. Then did Sturmius ‡ spend such infinite and curious pains upon Cicero the orator and Hermogenes the rhetorician, besides his own books of periods and imitation and the like. Then did Car of Cambridge, and Ascham, with their lectures and writings, almost deify Cicero and Demosthenes, and allure all young men that were studious unto that delicate and polished kind of learning. Then did Erasmus take occasion to make the scoffing echo; *Decem annos consumpsi in legendo Cicerone* [I have spent ten years in reading Cicero]: and the echo answered in Greek, *one, Asine.* Then grew the learning of the schoolmen to be utterly despised as barbarous. In sum, the whole inclination and bent of those times was rather towards copie than weight.

Here therefore [is] the first distemper of learning, when men study words and not matter: whereof though I have represented an example of late times, yet it hath been and will be *secundum majus et minus* in all time. And how is it possible but this should have an operation to discredit learning, even with vulgar capacities, when they see learned men's works like the first letter of a patent or limned book; which though it hath large flourishes, yet it is but a letter? It seems to me that Pygmalion's frenzy is a good emblem or portraiture of this vanity: for words are but the images of matter; and except they have life of reason and invention, to fall in love with them is all one as to fall in love with a picture.

But yet notwithstanding it is a thing not hastily to be condemned, to clothe and adorn the obscurity even of philosophy itself with sensible and plausible elocution. For hereof we have great examples in Xenophon, Cicero, Seneca,

to prevail very much about the time of Luther; chiefly on account of the demand for fervour and efficacy of preaching, etc. The remarks on the style of the schoolmen and the hatred which at that time began to be conceived against them are retained.

[17] So edd. 1629 and 1633. The original has *that then*. [* *I.e.* Copiousness.]
[† Author of *De rebus gestis Emamuelis.* D. 1580.] [‡ B. 1507; d. 1589.]

Plutarch, and of Plato also in some degree; and hereof likewise there is great use; for surely to the severe inquisition of truth, and the deep progress into philosophy, it is some hinderance; because it is too early satisfactory to the mind of man, and quencheth the desire of further search, before we come to a just period; but then if a man be to have any use of such knowledge in civil occasions, of conference, counsel, persuasion, discourse, or the like; then shall he find it prepared to his hands in those authors which write in that manner. But the excess of this is so justly contemptible, that as Hercules, when he saw the image of Adonis, Venus' minion, in a temple, said in disdain, *Nil sacri es* [you are no divinity]; so there is none of Hercules' followers in learning, that is, the more severe and laborious sort of inquirers into truth, but will despise those delicacies and affectations, as indeed capable of no divineness [18]. And thus much of the first disease or distemper of learning.

The second, which followeth, is in nature worse than the former; for as substance of matter is better than beauty of words, so contrariwise vain matter is worse than vain words: wherein it seemeth the reprehension of St. Paul was not only proper for those times, but prophetical for the times following; and not only respective to divinity, but extensive to all knowledge: *Devita profanas vocum novitates, et oppositiones falsi nominis scientiæ* [shun profane novelties of terms and oppositions of science falsely so called]. For he assigneth two marks and badges of suspected and falsified science; the one, the novelty of and strangeness of terms; the other, the strictness of positions, which of necessity doth induce oppositions, and so questions and altercations. Surely, like as many substances in nature which are solid do putrefy and corrupt into worms, so it is the property of good and sound knowledge to putrefy and dissolve into a number of subtile, idle, unwholesome, and (as I may term them) vermiculate questions, which have indeed a kind of quickness and life of spirit, but no soundness of matter or goodness of quality. This kind of degenerate learning did chiefly reign amongst the schoolmen; who having sharp and strong wits, and abundance of leisure, and small variety of reading; but their wits being shut up in the cells of a few authors (chiefly Aristotle their dictator) as their persons were shut up in the cells of monasteries and colleges; and knowing little history, either of nature or time; did out of no great quantity of matter, and infinite agitation of wit, spin out unto us those laborious webs of learning which are extant in their books. For the wit and mind of man, if it work upon matter, which is the contemplation of the creatures of God, worketh according to the stuff, and is limited thereby; but if it work upon itself, as the spider worketh his web, then it is endless, and brings forth indeed cobwebs of learning, admirable for the fineness of thread and work, but of no substance or profit.

This same unprofitable subtility or curiosity is of two sorts; either in the subject itself that they handle, when it is a fruitless speculation or controversy, (whereof there are no small number both in divinity and philosophy,) or in the manner or method of handling of a knowledge; which amongst them was this; upon every particular position or assertion to frame objections, and to those objections, solutions; which solutions were for the most part not confutations, but distinctions: whereas indeed the strength of all sciences is, as the strength of the old man's faggot, in the bond. For the harmony of a science, supporting each part the other, is and ought to be the true and brief confutation and suppression of all the smaller sort of objections; but on the other side, if you take

18 In the translation he mentions another vanity of style, though not of so bad a kind, as commonly succeeding the last in point of time,—a style in which all the study is to have the words pointed, the sentences concise, and the whole composition rather twisted into shape than allowed to flow (*oratio denique potius versa quam fusa*): a trick which has the effect of making everything seem more ingenious than it really is. Such a style (he says) is found largely in Seneca, less in Tacitus and the second Pliny, and has found favour of late with the ears of our own time; but though it is agreeable to ordinary understandings and so procures some respect for literature, yet to more exact judgments it is deservedly distasteful, and may be set down among the distempers of learning, being, as well as the other, a kind of hunting after words and verbal prettiness.

out every axiom, as the sticks of the faggot, one by one, you may quarrel with them and bend them and break them at your pleasure: so that as was said of Seneca, *Verborum minutiis rerum frangit pondera* [that he broke up the weight and mass of the matter by verbal points and niceties]; so a man may truly say of the schoolmen, *Quæstionum minutiis scientiarum frangunt soliditatem* [they broke up the solidity and coherency of the sciences by the minuteness and nicety of their questions]. For were it not better for a man in a fair room to set up one great light, or branching candlestick of lights, than to go about with a small watch candle into every corner? And such is their method, that rests not so much upon evidence of truth proved by arguments, authorities, similitudes, examples, as upon particular confutations and solutions of every scruple, cavillation, and objection; breeding for the most part one question as fast it solveth another; even as in the former resemblance, when you carry the light into one corner, you darken the rest: so that the fable and fiction of Scylla seemeth to be a lively image of this kind of philosophy or knowledge; which was transformed into a comely virgin for the upper parts; but then *Candida succinctam latrantibus inguina monstris* [there were barking monsters all about her loins]: so the generalities of the schoolmen are for a while good and proportionable; but then when you descend into their distinctions and decisions, instead of a fruitful womb for the use and benefit of man's life, they end in monstrous altercations and barking questions. So as it is not possible but this quality of knowledge must fall under popular contempt, the people being apt to contemn truth upon occasion of controversies and altercations, and to think they are all out of their way which never meet: and when they see such digladiation about subtilities and matter of no use nor moment, they easily fall upon that judgment of Dionysius of Syracusa, *Verba ista sunt senum otiosorum* [it is the talk of old men that have nothing to do].

Notwithstanding certain it is, that if those schoolmen to their great thirst of truth and unwearied travail of wit had joined variety and universality of reading and contemplation, they had proved excellent lights, to the great advancement of all learning and knowledge. But as they are, they are great undertakers indeed, and fierce with dark keeping; [19] but as in the inquiry of the divine truth their pride inclined to leave the oracle of God's word and to vanish in the mixture of their own inventions, so in the inquisition of nature they ever left the oracle of God's works and adored the deceiving and deformed images which the unequal mirror of their own minds or a few received authors or principles did represent unto them. And thus much for the second disease of learning.

For the third vice or disease of learning, which concerneth deceit or untruth, it is of all the rest the foulest; as that which doth destroy the essential form of knowledge, which is nothing but a representation of truth: for the truth of being and the truth of knowing are one, differing no more than the direct beam and the beam reflected. This vice therefore brancheth itself into two sorts; delight in deceiving, and aptness to be deceived; imposture and credulity; which, although they appear to be of a diverse nature, the one seeming to proceed of cunning, and the other of simplicity, yet certainly they do for the most part concur: for as the verse noteth,

"Percontatorem fugito, nam garrulus idem est *,"

an inquisitive man is a prattler, so upon the like reason a credulous man is a deceiver: as we see it in fame, that he that will easily believe rumours will as easily augment rumours and add somewhat to them of his own; which Tacitus wisely noteth, when he saith, *Fingunt simul creduntque* [as fast as they believe one tale they make another [20]]: so great an affinity hath fiction and belief.

This facility of credit, and accepting or admitting things weakly authorized

[19] That is, fierce from being kept in the dark; the allusion being, as we see more clearly from a corresponding passage in the early Latin fragment [*ferocitatem autem et confidentiam quæ illos qui pauca norunt sequi solet* (*ut animalia in tenebris educata*) etc.—*Cog. de Sci. Hum.* 1st fragm. § 10], to the effect of darkness on the temper of animals.—*R. L. E.* The rest of this sentence, from "but as they are" is omitted in the translation. See note p. 50. [* Hor. *Ep.* i. 18, 69.]

[20] I think this is the sense in which Bacon must have understood these words; but it

or warranted, is of two kinds, according to the subject: for it is either a belief of history (as[21] the lawyers speak, matter of fact), or else of matter of art and opinion. As to the former, we see the experience and inconvenience of this error in ecclesiastical history; which hath too easily received and registered reports and narrations of miracles wrought by martyrs, hermits, or monks of the deserts, and other holy men, and their relics, shrines, chapels, and images[22]: which though they had a passage for a time, by the ignorance of the people, the superstitious simplicity of some, and the politic toleration of others, holding them but as divine poesies; yet after a period of time, when the mist began to clear up, they grew to be esteemed but as old wives' fables, impostures of the clergy, illusions of spirits, and badges of antichrist, to the great scandal and detriment of religion.

So in natural history, we see there hath not been that choice and judgment used as ought to have been; as may appear in the writings of Plinius, Cardanus, Albertus, and divers of the Arabians; being fraught with much fabulous matter, a great part not only untried but notoriously untrue, to the great derogation of the credit of natural philosophy with the grave and sober kind of wits. Wherein the wisdom and integrity of Aristotle is worthy to be observed; that having made so diligent and exquisite a history of living creatures, hath mingled it sparingly with any vain or feigned matter; and yet on the other side[23] hath cast all prodigious narrations which he thought worthy the recording into one book; excellently discerning that matter of manifest truth, such whereupon observation and rule was to be built, was not to be mingled or weakened with matter of doubtful credit; and yet again that rarities and reports that seem uncredible are not to be suppressed or denied to the memory of men.

And as for the facility of credit which is yielded to arts and opinions, it is likewise of two kinds; either when too much belief is attributed to the arts themselves, or to certain authors in any art. The sciences themselves which have had better intelligence and confederacy with the imagination of man than with his reason, are three in number; Astrology, Natural Magic, and Alchemy; of which sciences nevertheless the ends or pretences are noble. For astrology pretendeth to discover that correspondence or concatenation which is between the superior globe and the inferior: natural magic pretendeth to call and reduce natural philosophy from a variety of speculations to the magnitude of works: and alchemy pretendeth to make separation of all the unlike parts of bodies which in mixtures of nature are incorporate. But the derivations and prosecutions to these ends, both in the theories and in the practices, are full of error and vanity; which the great professors themselves have sought to veil over and conceal by enigmatical writings, and referring themselves to auricular traditions, and such other devices to save the credit of impostures. And yet surely to alchemy this right is due, that it may be compared to the husbandman whereof Æsop makes the fable, that when he died told his sons that he had left unto them gold buried under ground in his vineyard; and they digged all over the ground, and gold they found none, but by reason of their stirring and digging the mould about the roots of their vines, they had a great vintage the year following: so assuredly the search and stir to make gold hath brought to light a great number of good and fruitful inventions and experiments, as well for the disclosing of nature as for the use of man's life.

And as for the overmuch credit that hath been given unto authors in sciences, in making them dictators, that their words should stand, and not counsels[24] to give advice; the damage is infinite that sciences have received thereby, as the principal cause that hath kept them low, at a stay without growth or advance-

is not the sense in which Tacitus employs them (An. v. 10.). He meant that they at once invented the tale and believed it: they "credited their own lie".—*J. S.*

[21] So the original. Edd. 1629 and 1633 have *or as*.

[22] The rest of the paragraph is omitted in the translation. See note p. 50.

[23] *Sake* in the original, and also in edd. 1629 and 1633.

[24] So the original. Edd. 1629 and 1633 have *consuls*. The translation has *dictatoria quadam potestate munivit ut edicant, non senatoria ut consulant*. Bacon probably wrote *counsell^rs*.

ment. For hence it hath come that in arts mechanical the first deviser comes shortest, and time addeth and perfecteth; but in sciences the first author goeth furthest, and time leeseth and corrupteth. So we see, artillery, sailing, printing, and the like, were grossly managed at the first, and by time accommodated and refined; but contrariwise the philosophies and sciences of Aristotle, Plato, Democritus, Hippocrates, Euclides, Archimedes, of most vigour at the first, and by time degenerate and imbased; whereof the reason is no other, but that in the former many wits and industries have contributed in one; and in the later many wits and industries have been spent about the wit of some one, whom many times they have rather depraved than illustrated. For as water will not ascend higher than the level of the first spring-head from whence it descendeth, so knowledge derived from Aristotle, and exempted from liberty of examination, will not rise again higher than the knowledge of Aristotle. And therefore, although the position be good, *Oportet discentum credere* [a man who is learning must be content to believe what he is told], yet it must be coupled with this, *Oportet edoctum judicare* [when he has learned it he must exercise his judgment and see whether it be worthy of belief]; for disciples do owe unto masters only a temporary belief and a suspension of their own judgment until they be fully instructed, and not an absolute resignation or perpetual captivity: and therefore to conclude this point, I will say no more but, so let great authors have their due, as time which is the author of authors be not deprived of his due, which is further and further to discover truth. Thus have I gone over these three diseases of learning; besides the which, there are some other rather peccant humours than formed diseases, which nevertheless are not so secret and intrinsic but that they fall under a popular observation and traducement, and therefore are not to be passed over.

The first of these is the extreme affecting of two extremities; the one Antiquity, the other Novelty: wherein it seemeth the children of time do take after the nature and malice of the father. For as he devoureth his children, so one of them seeketh to devour and suppress the other; while antiquity envieth there should be new additions, and novelty cannot be content to add but it must deface. Surely the advice of the prophet is the true direction in this matter, *State super vias antiquas, et videte quænam sit via recta et bona, et ambulate in ea* [stand ye in the old ways, and see which is the good way, and walk therein]. Antiquity deserveth that reverence, that men should make a stand thereupon, and discover what is the best way; but when the discovery is well taken, then to make progression. And to speak truly, *Antiquitas sæculi juventus mundi*. These times are the ancient times, when the world is ancient, and not those which we account ancient *ordine retrograde*, by a computation backward from ourselves.

Another error, induced by the former, is a distrust that any thing should be now to be found out, which the world should have missed and passed over so long time; as if the same objection were to be made to time that Lucian maketh to Jupiter and other heathen gods, of which he wondereth that they begot so many children in old time and begot none in his time, and asketh whether they were become septuagenary, or whether the law *Pappia*, made against old men's marriages, had restrained them*. So it seemeth men doubt lest time is become past children and generation; wherein contrariwise we see commonly the levity and unconstancy of men's judgments, which, till a matter be done, wonder that it can be done; and as soon as it is done, wonder again that it was no sooner done; as we see in the expedition of Alexander into Asia, which at first was prejudged as a vast and impossible enterprise; and yet afterwards it pleaseth Livy to make no more of it than this, *Nil aliud quàm bene ausus vana contemnere* [it was but taking courage to despise vain apprehensions]. And the same happened to Columbus in the western navigation. But in intellectual matters it is much more common; as may be seen in most of the propositions of Euclid, which till they be demonstrate, they seem strange to our assent; but being demonstrate, our mind accepteth of them by a kind of relation (as the lawyers speak) as if we had known them before.

Another error, that hath also some affinity with the former, is a conceit that of former opinions or sects, after variety and examination, the best hath still pre-

[* Said not by Lucian but by Seneca. Lactant. *De falsa relig.* i. 16.]

vailed and suppressed the rest; so as if a man should begin the labour of a new search, he were but like to light upon somewhat formerly rejected, and by rejection brought into oblivion: as if the multitude, or the wisest for the multitude's sake, were not ready to give passage rather to that which is popular and superficial than to that which is substantial and profound; for the truth is, that time seemeth to be of the nature of a river or stream, which carrieth down to us that which is light and blown up, and sinketh and drowneth that which is weighty and solid.

Another error, of a diverse nature from all the former, is the over-early and peremptory reduction of knowledge into arts and methods; from which time commonly sciences receive small or no augmentation. But as young men when they knit and shape perfectly, do seldom grow to a further stature; so knowledge, while it is in aphorisms and observations, it is in growth; but when it once is comprehended in exact methods, it may perchance be further polished and illustrate [25], and accommodated for use and practice; but it increaseth no more in bulk and substance.

Another error, which doth succeed that which we last mentioned, is that after the distribution of particular arts and sciences, men have abandoned universality, or *philosophia prima;* which cannot but cease and stop all progression. For no perfect discovery can be made upon a flat or a level: neither is it possible to discover the more remote and deeper parts of any science, if you stand but upon the level of the same science, and ascend not to a higher science.

Another error hath proceeded from too great a reverence, and a kind of adora tion of the mind and understanding of man; by means whereof men have withdrawn themselves too much from the contemplation of nature and the observations of experience, and have tumbled up and down in their own reason and conceits. Upon these intellectualists, which are notwithstanding commonly taken for the most sublime and divine philosophers, Heraclitus gave a just censure, saying, *Men sought truth in their own little worlds, and not in the great and common world;* for they disdain to spell and so by degrees to read in the volume of God's works; and contrariwise by continual meditation and agitation of wit do urge and as it were invocate their own spirits to divine and give oracles unto them, whereby they are deservedly deluded.

Another error that hath some connexion with this latter is, that men have used to infect their meditations, opinions, and doctrines, with some conceits which they have most admired, or some sciences which they have most applied; and given all things else a tincture according to them, utterly untrue and unproper. So hath Plato intermingled his philosophy with theology, and Aristotle with logic, and the second school of Plato, Proclus and the rest, with the mathematics. For these were the arts which had a kind of primogeniture with them severally. So have the alchemists made a philosophy out of a few experiments of the furnace; and Gilbertus, our countryman, hath made a philosophy out of the observations of a loadstone. So Cicero, when, reciting the several opinions of the nature of the soil, he found a musician that held the soil was but a harmony, saith pleasantly, *Hic ab arte sua non recessit*, etc. [he was constant to his own art]. But of these conceits Aristotle speaketh seriously and wisely, when he saith, *Qui respiciunt ad pauca de facili pronunciant* [they who take only few points into account find it easy to pronounce judgment].

Another error is an impatience of doubt, and haste to assertion without due and mature suspension of judgment. For the two ways of contemplation are not unlike the two ways of action commonly spoken of by the ancients; the one plain and smooth in the beginning, and in the end impassable; the other rough and troublesome in the entrance, but after a while fair and even. So it is in contemplation; if a man will begin with certainties, he shall end in doubts; but if he will be content to begin with doubts, he shall end in certainties.

Another error is in the manner of the tradition and delivery of knowledge, which is for the most part magistral and peremptory, and not ingenuous and faithful; in a sort as may be soonest believed, and not easiliest examined. It is true

[25] So the original. Ed. 1633 has *illustrated.*

that in compendious treatises for practice that form is not to be disallowed. But in the true handling of knowledge, men ought not to fall either on the one side into the vein of Velleius the Epicurean, *Nil tam metuens, quàm ne dubitare aliqua de re videretur* [who feared nothing so much as the seeming to be in doubt about anything], nor on the other side into Socrates his ironical doubting of all things, but to propound things sincerely, with more or less asseveration, as they stand in a man's own judgment proved more or less.

Other errors there are in the scope that men propound to themselves, whereunto they bend their endeavours; for whereas the more constant and devote [26] kind of professors of any science ought to propound to themselves to make some additions to their science, they convert their labours to aspire to certain second prizes; as to be a profound interpreter or commenter, to be a sharp champion or defender, to be a methodical compounder or abridger; and so the patrimony of knowledge cometh to be sometimes improved, but seldom augmented.

But the greatest error of all the rest is the mistaking or misplacing of the last or furthest end of knowledge. For men have entered into a desire of learning and knowledge, sometimes upon a natural curiosity and inquisitive appetite; sometimes to entertain their minds with variety and delight; sometimes for ornament and reputation; and sometimes to enable them to victory of wit and contradiction; and most times for lucre and profession; and seldom sincerely to give a true account of their gift of reason, to the benefit and use of men: as if there were sought in knowledge a couch, whereupon to rest a searching and restless spirit; or a terrace, for a wandering and variable mind to walk up and down with a fair prospect; or a tower of state, for a proud mind to raise itself upon; or a fort or commanding ground, for strife and contention; or a shop, for profit or sale; and not a rich storehouse, for the glory of the Creator and the relief of man's estate. But this is that which will indeed dignify and exalt knowledge, if contemplation and action may be more nearly and straitly conjoined and united together than they have been; a conjunction like unto that of the two highest planets, Saturn the planet of rest and contemplation, and Jupiter the planet of civil society and action. Howbeit, I do not mean, when I speak of use and action, that end before-mentioned of the applying of knowledge to lucre and profession: for I am not ignorant how much that diverteth and interrupteth the prosecution and advancement of knowledge; like unto the golden ball thrown before Atalanta, which while she goeth aside and stoopeth to take up, the race is hindered,

"Declinat cursus, aurumque volubile tollit."

Neither is my meaning, as was spoken of Socrates, to call philosophy down from heaven to converse upon the earth; that is, to leave natural philosophy aside, and to apply knowledge only to manners and policy. But as both heaven and earth do conspire and contribute to the use and benefit of man, so the end ought to be, from both philosophies to separate and reject vain speculations and whatsoever is empty and void, and to preserve and augment whatsoever is solid and fruitful; that knowledge may not be as a curtesan, for pleasure and vanity only, or as a bond-woman, to acquire and gain to her master's use; but as a spouse, for generation, fruit, and comfort.

Thus have I described and opened, as by a kind of dissection, those peccant humours (the principal of them) which have [27] not only given impediment to the proficience of learning but have given also occasion to the traducement thereof: wherein if I have been too plain, it must be remembered *Fidelia vulnera amantis, sed dolosa oscula malignantis* [faithful are the wounds of a friend, but the kisses of an enemy are deceitful]. This I think I have gained, that I ought to be the better believed in that which I shall say pertaining to commendation, because I have proceeded so freely in that which concerneth censure. And yet I have no purpose to enter into a laudative of learning, or to make a hymn to the muses (though I am of opinion that it is long since their rites were duly celebrated): but my intent is, without varnish or amplification, justly to weigh the dignity of knowledge in the balance with other things, and

[26] So the original. Ed. 1633 has *devoute*. [27] *hath* in all the old editions.

to take the true value thereof by testimonies and arguments divine and human.

First, therefore, let us seek the dignity of knowledge in the arch-type or first platform, which is in the attributes and acts of God, as far as they are revealed to man and may be observed with sobriety; wherein we may not seek it by the name of learning; for all learning is knowledge acquired, and all knowledge in God is original: and therefore we must look for it by another name, that of wisdom or sapience, as the Scriptures call it.

It is so then, that in the work of the creation we see a double emanation of virtue from God; the one referring more properly to power, the other to wisdom; the one expressed in making the subsistence of the matter, and the other in disposing the beauty of the form. This being supposed, it is to be observed, that for anything which appeareth in the history of the creation, the confused mass and matter of heaven and earth was made in a moment, and the order and disposition of that chaos or mass was the work of six days; such a note of difference it pleased God to put upon the works of power and the works of wisdom; wherewith concurreth, that in the former it is not set down that God said, *Let there be heaven and earth*, as it is set down of the works following; but actually, that God made heaven and earth: the one carrying the style of a manufacture, and the other of a law, decree, or counsel.

To proceed to that which is next in order, from God to spirits; we find as far as credit is to be given to the celestial hierarchy of that supposed Dionysius the senator of Athens [28], the first place or degree is given to the angels of love which are termed Seraphim; the second to the angels of light, which are termed Cherubim; and the third and so following places to thrones, principalities and the rest, which are all angels of power and ministry; so as the angels of knowledge and illumination are placed before the angels of office and domination.

To descend from spirits and intellectual forms to sensible and material forms; we read the first form that was created was light, which hath a relation and correspondence in nature and corporal things, to knowledge in spirits and incorporal things.

So in the distribution of days, we see the day wherein God did rest and contemplate his own works, was blessed above all the days wherein he did effect and accomplish them.

After the creation was finished, it is set down unto us that man was placed in the garden to work therein; which work so appointed to him could be no other than work of contemplation; that is, when the end of work is but for exercise and experiment, not for necessity; for there being then no reluctation of the creature, nor sweat of the brow, man's employment must of consequence have been matter of delight in the experiment, and not matter of labour for the use. Again, the first acts which man performed in Paradise consisted of the two summary parts of knowledge; the view of creatures, and the imposition of names. As for the knowledge which induced the fall, it was, as was touched before, not the natural knowledge of creatures, but the moral knowledge of good and evil; wherein the supposition was, that God's commandments or prohibitions were not the originals of good and evil, but that they had other beginnings, which man aspired to know, to the end to make a total defection from God, and to depend wholly upon himself.

To pass on: in the first event or occurrence after the fall of man, we see (as the Scriptures have infinite mysteries, not violating at all the truth of the story or letter), an image of the two estates, the contemplative state and the active state, figured in the two persons of Abel and Cain, and in the two simplest and most primitives trades of life; that of the shepherd, (who, by reason of his leisure, rest in a place, and living in view of heaven, is a lively image of a contemplative life), and that of the husbandman: where we see again the favour and election of God went to the shepherd, and not to the tiller of the ground.

[28] *quæ Dionysii Areopagitæ nomine evulgatur*, are the words of the translation: the insinuation implied in the word *supposed*, being withdrawn, or at least not so strongly expressed. See note p. 50.

So in the age before the flood, the holy records within those few memorials which are there entered and registered have vouchsafed to mention and honour the name of the inventors and authors of music and works in metal. In the age after the flood, the first great judgment of God upon the ambition of man was the confusion of tongues; whereby the open trade and intercourse of learning and knowledge was chiefly imbarred.

To descend to Moses the lawgiver, and God's first pen: he is adorned by the Scriptures with this addition and commendation, that he was *seen in all the learning of the Egyptians;* which nation we know was one of the most ancient schools of the world: for so Plato brings in the Egyptian priest saying unto Solon: *You Grecians are ever children; you have no knowledge of antiquity, nor antiquity of knowledge.* Take a view of the ceremonial law of Moses; you shall find, besides the prefiguration of Christ, the badge or difference of the people of God, the exercise and impression of obedience, and other divine uses and fruits thereof, that some of the most learned Rabbins have travelled profitably and profoundly to observe, some of them a natural, some of them a moral, sense or reduction of many of the ceremonies and ordinances. As in the law of the leprosy, where it is said, *If the whiteness have overspread the flesh, the patient may pass abroad for clean; but if there be any whole flesh remaining, he is to be shut up for unclean;* one of them noteth a principle of nature, that putrefaction is more contagious before maturity than after: and another noteth a position of moral philosophy, that men abandoned to vice do not so much corrupt manners, as those that are half good and half evil. So in this and very many other places in that law, there is to be found, besides the theological sense, much aspersion of philosophy.

So likewise in that excellent book of Job, if it be revolved with diligence, it will be found pregnant and swelling with natural philosophy; as for example, cosmography and the roundness of the world; *Qui extendit aquilonem super vacuum, et appendit terram super nihilum* [who stretches out the north upon the empty space, and hangeth the earth upon nothing]; wherein the pensileness of the earth, the pole of the north, and the finiteness or convexity of heaven are manifestly touched. So again the matter of astronomy; *Spiritus ejus ornavit cœlos, et obstetricante manu ejus eductus est Coluber tortuosus* [by his spirit he hath garnished the heavens; his hand hath formed the crooked Serpent]. And in another place; *Nunquid conjungere valebis micantes stellas Pleiadas, aut gyrum Arcturi poteris dissipare?* [canst thou bring together the glittering stars of the Pleiades, or scatter the array of Arcturus?] where the fixing of the stars, ever standing at equal distance, is with great elegancy noted. And in another place, *Qui facit Arcturum, et Oriona, et Hyadas, et interiora Austri* [which maketh Arcturus, Orion, and Hyades, and the secrets of the South]; where again he takes knowledge of the depression of the southern pole, calling it the secrets of the south, because the southern stars were in that climate unseen. Matter of generation; *Annon sicut lac mulsisti me, et sicut caseum coagulasti me?* etc. [hast thou not drawn me forth like milk, and curdled me like cheese?]. Matter of minerals; *Habet argentum venarum suarum principia: et auro locus est in quo conflatur, ferrum de terra tollitur, et lapis solutus calore in æs vertitur* [surely there is a vein for the silver, and a place for gold where they fine it. Iron is taken out of the earth, and brass is molten out of the stone]: and so forwards in that chapter.

So likewise in the person of Salomon the king, we see the gift or endowment of wisdom and learning, both in Salomon's petition and in God's assent thereunto, preferred before all other terrene and temporal felicity. By virtue of which grant or donative of God, Salomon became enabled not only to write those excellent parables or aphorisms concerning divine and moral philosophy, but also to compile a natural history of all verdure [29], from the cedar upon the mountain to the moss upon the wall, (which is but a rudiment between putrefaction and an herb), and also of all things that breathe or move. Nay, the same Salomon the king, although he excelled in the glory of treasure and magnificent

[29] *Verdor* in edd. 1605, 1629, 1633; which perhaps ought to be retained, as another form of the word rather than another way of spelling it.

buildings, of shipping and navigation, of service and attendance, of fame and renown and the like, yet he maketh no claim to any of those glories, but only to the glory of inquisition of truth; for so he saith expressly, *The glory of God is to conceal a thing, but the glory of the king is to find it out;* as if, according to the innocent play of children, the Divine Majesty took delight to hide his works to the end to have them found out; and as if kings could not obtain a greater honour than to be God's playfellows in that game, considering the great commandment of wits and means, whereby nothing needeth to be hidden from them.

Neither did the dispensation of God vary in the times after our Saviour came into the world; for our Saviour himself did first show his power to subdue ignorance, by his conference with the priests and doctors of the law, before he shewed his power to subdue nature by his miracles. And the coming of the Holy Spirit was chiefly figured and expressed in the similitude and gift of tongues, which are but *vehicula scientiæ* [carriers of knowledge].

So in the election of those instruments which it pleased God to use for the plantation of the faith, notwithstanding that at the first he did employ persons altogether unlearned otherwise than by inspiration, more evidently to declare his immediate working, and to abase all human wisdom or knowledge; yet nevertheless that counsel of his was no sooner performed, but in the next vicissitude and succession he did send his divine truth into the world waited on with other learnings as with servants or handmaids: for so we see St. Paul, who was only learned amongst the apostles, had his pen most used in the Scriptures of the New Testament.

So again we find that many of the ancient bishops and fathers of the Church were excellently read and studied in all the learning of the heathen; insomuch that the edict of the emperor Julianus, (whereby it was interdicted unto Christians to be admitted into schools, lectures, or exercises of learning*) was esteemed and accounted a more pernicious engine and machination against the Christian faith, than were all the sanguinary prosecutions of his predecessors; neither could the emulation and jealousy of Gregory the first of that name, Bishop of Rome, ever obtain the opinion of piety or devotion; but contrariwise received the censure of humour, malignity, and pusillanimity [30], even amongst holy men; in that he designed to obliterate and extinguish the memory of heathen antiquity and authors. But contrariwise it was the Christian Church, which amidst the inundations of the Scythians on the one side from the north-west, and the Saracens from the east, did preserve in the sacred lap and bosom thereof the precious relics even of heathen learning, which otherwise had been extinguished as if no such thing had ever been.

And we see before our eyes, that in the age of ourselves and our fathers, when it pleased God to call the church of Rome to account for their degenerate manners and ceremonies, and sundry doctrines obnoxious and framed to uphold the same abuses; at one and the same time it was ordained by the Divine Providence that there should attend withal a renovation and new spring of all other knowledges [31], and on the other side we see the Jesuits, who partly in themselves and partly by the emulation and provocation of their example, have much quickened and strengthened the state of learning,—we see (I say) what notable service and reparation they have done to the Roman see.

Wherefore to conclude this part, let it be observed that there be two principal duties and services, besides ornament and illustration, which philosophy and human learning do perform to faith and religion. The one, because they are an effectual inducement to the exaltation of the glory of God: For as the Psalms and other Scriptures do often invite us to consider and magnify the great and wonderful works of God, so if we should rest only in the contemplation of the exterior of them as they first offer themselves to our senses, we should do a like injury unto the majesty of God as if we should judge or construe of the store

[30] This clause is omitted in the translation; and the words *cætera viri egregii* are introduced after the name of Gregory. See note p. 50.

[31] All this, from the beginning of the paragraph, is omitted in the translation.

[* As Gibbon points out, the edict only forbids Christian professors to teach.]

of some excellent jeweller by that only which is set out toward the street in his shop. The other, because they minister a singular help and preservative against unbelief and error: For our Saviour saith, *You err, not knowing the Scriptures, nor the power of God;* laying before us two books or volumes to study, if we will be secured from error; first the Scriptures, revealing the will of God, and then the creatures expressing his power; whereof the later is a key unto the former; not only opening our understanding to conceive the true sense of the Scriptures, by the general notions of reason and rules of speech; but chiefly opening our belief, in drawing us into a due meditation of the omnipotency of God, which is chiefly signed and engraven upon his works. Thus much therefore for divine testimony and evidence concerning the true dignity and value of learning.

As for human proofs, it is so large a field, as in a discourse of this nature and brevity it is fit rather to use choice of those things which we shall produce, than to embrace the variety of them. First therefore, in the degrees of human honour amongst the heathen it was the highest, to obtain to a veneration and adoration as a God. This unto the Christians is as the forbidden fruit. But we speak now separately of human testimony: according to which that which the Grecians call *apotheosis*, and the Latins *relatio inter divos*, was the supreme honour which man could attribute unto man; specially when it was given, not by a formal decree or act of state, as it was used among the Roman emperors, but by an inward assent and belief; which honour being so high, had also a degree or middle term; for there were reckoned above human honours, honours [32] heroical and divine; in the attribution and distribution of which honours we see antiquity made this difference: that whereas founders and uniters of states and cities, lawgivers, extirpers of tyrants, fathers of the people, and other eminent persons in civil merit, were honoured but with the titles of worthies or demi-gods; such as were Hercules, Theseus, Minos, Romulus, and the like; on the other side, such as were inventors and authors of new arts, endowments, and commodities towards man's life, were ever consecrated amongst the gods themselves; as was Ceres, Bacchus, Mercurius, Apollo, and others; and justly; for the merit of the former is confined within the circle of an age or a nation; and is like fruitful showers, which though they be profitable and good, yet serve but for that season, and for a latitude of ground where they fall; but the other is indeed like the benefits of heaven, which are permanent and universal. The former again is mixed with strife and perturbation; but the later hath the true character of divine presence, coming [33] in *aura leni*, without noise or agitation.

Neither is certainly that other merit of learning, in repressing the inconveniences which grow from man to man, much inferior to the former, of relieving the necessities which arise from nature; which merit was lively set forth by the ancients in that feigned relation of Orpheus theatre; where all beasts and birds assembled, and forgetting their several appetites, some of prey, some of game, some of quarrel, stood all sociably together listening unto the airs and accords of the harp; the sound whereof no sooner ceased, or was drowned by some louder noise, but every beast returned to his own nature: wherein is aptly described the nature and condition of men; who are full of savage and unreclaimed desires, of profit, of lust, of revenge, which as long as they give ear to precepts, to laws, to religion, sweetly touched with eloquence and persuasion of books, of sermons, of harangues, so long is society and peace maintained; but if these instruments be silent, or that sedition and tumult make them not audible, all things dissolve into anarchy and confusion.

But this appeareth more manifestly, when kings themselves, or persons of authority under them, or other governors in commonwealths and popular estates, are endued with learning. For although he might be thought partial to his own profession, that said *Then should people and estates be happy, when either kings were philosophers, or philosophers kings;* yet so much is verified by experience,

[32] *honour* in edd. 1605, 1629, 1633.

[33] *commonly* in edd. 1629 and 1633. In the original, *com-* ends a line and the rest of the word has accidentally dropped out. [Quotation from 1 Kings, xix. 12. *Vulgate.*]

that under learned princes and governors there have been ever the best times * ; for howsoever kings may have their imperfections in their passions and customs, yet if they be illuminate by learning, they have those notions of religion, policy, and morality, which do preserve them and refrain them from all ruinous and peremptory errors and excesses ; whispering evermore in their ears, when counsellors and servants stand mute and silent. And senators or counsellors likewise which be learned, do proceed upon more safe and substantial principles than counsellors which are only men of experience ; the one sort keeping dangers afar off, whereas the other discover them not till they come near hand, and then trust to the agility of their wit to ward or avoid them.

Which felicity of times under learned princes (to keep still the law of brevity, by using the most eminent and selected examples) doth best appear in the age which passed from the death of Domitianus the emperor until the reign of Commodus ; comprehending a succession of six princes [34], all learned or singular favourers and advancers of learning ; which age, for temporal respects, was the most happy and flourishing that ever the Roman empire (which then was a model of the world) enjoyed : a matter revealed and prefigured unto Domitian in a dream the night before he was slain ; for he thought there was grown behind upon his shoulders a neck and a head of gold, which came accordingly to pass in those golden times which succeeded : of which princes we will make some commemoration ; wherein although the matter will be vulgar, and may be thought fitter for a declamation than agreeable to a treatise infolded as this is, yet because it is pertinent to the point in hand, *neque semper arcum tendit Apollo* [and Apollo does not keep his bow always bent], and to name them only were too naked and cursory, I will not omit it altogether [35].

The first was Nerva ; the excellent temper of whose government is by a glance in Cornelius Tacitus touched to the life : *Postquam divus Nerva res olim insociabiles miscuisset, imperium et libertatem* [he united and reconciled two things which used not to go together—government and liberty] [36]. And in token of his learning, the last act of his short reign left to memory was a missive to his adopted son Trajan, proceeding upon some inward discontent at the ingratitude of the times, comprehended in a verse of Homer's ;

> Telis, Phœbe, tuis lacrymas ulciscere nostras†.
> [O Phœbus, with thy shafts avenge these tears].

Trajan, who succeeded, was for his person not learned : but if we will hearken to the speech of our Saviour, that saith, *He that receiveth a prophet in the name of a prophet, shall have a prophet's reward*, he deserveth to be placed amongst the most learned princes : for there was not a greater admirer of learning or benefactor of learning ; a founder of famous libraries, a perpetual advancer of learned men to office, and a familiar converser with learned professors and preceptors, who were noted to have then most credit in court. On the other side, how much Trajan's virtue and government was admired and renowned, surely no testimony of grave and faithful history doth more lively set forth, than that legend tale of Gregorius Magnus, bishop of Rome, who was noted for the extreme envy he bare towards all heathen excellency : and yet he is reported, out of the love and estimation of Trajan's moral virtues, to have made unto God passionate and fervent prayers for the delivery of his soul out of hell ; and to have obtained it,

[* Plato, *Republic*, B.v.] [34] So edd. 1629 and 1633. The original has *sciences*.

[35] In the *De Augmentis* he merely says " *de quibus*," *i.e.* the golden times, " *sigillatim sed brevissime verba faciam* ". And the next five paragraphs are condensed into one.

[36] Agric. 3. : *Quanquam . . . Nerva Cæsar res olim dissociabiles miscuerit, principatum ac libertatem.* This quotation is omitted in the translation, where nothing is said of the character of Nerva's government except that he was *clementissimus imperator, quique, si nihil aliud, orbi Trajanum dedit ;* from which it would almost seem that Bacon thought it hardly deserved the praise which Tacitus bestows upon it. In evidence of his *learning* he adds that he was the friend, and as it were the disciple, of Apollonius the Pythagorean. [† *Iliad*, i. 42 ; Xiphilinus *in Nerva*.]

with a caveat that he should make no more such petitions [37]. In this prince's time also the persecutions against the Christians received intermission, upon the certificate of Plinius Secundus, a man of excellent learning and by Trajan advanced.

Adrian, his successor, was the most curious man that lived, and the most universal inquirer; insomuch as it was noted for an error in his mind, that he desired to comprehend all things, and not to reserve himself for the worthiest things; falling into the like humour that was long before noted in Philip of Macedon, who when he would needs over-rule and put down an excellent musician in an argument touching music, was well answered by him again, *God forbid, Sir* (saith he), *that your fortune should be so bad, as to know these things better than I* [38]. It pleased God likewise to use the curiosity of this emperor as an inducement to the peace of his church in those days. For having Christ in veneration, not as a God or Saviour, but as a wonder or novelty, and having his picture in his gallery matched with Appollonius (with whom in his vain imagination he thought he had some conformity), yet it served the turn to allay the bitter hatred of those times against the Christian name; so as the church had peace during his time [39]. And for his government civil, although he did not attain to that of Trajan's [40] in glory of arms or perfection of justice, yet in deserving of the weal of the subject he did exceed him. For Trajan erected many famous monuments and buildings; insomuch as Constantine the Great in emulation was wont to call him *Parietaria*, wall flower*, because his name was upon so many walls: but his buildings and works were more of glory and triumph than use and necessity. But Adrian spent his whole reign, which was peaceable, in a perambulation or survey of the Roman empire; giving order and making assignation where he went for re-edifying of cities, towns, and forts decayed, and for cutting of rivers and streams, and for making bridges and passages, and for policing [41] of cities and commonalties with new ordinances and constitutions, and granting new franchises and incorporations; so that his whole time was a very restoration of all the lapses and decays of former times.

Antoninus [42] Pius, who succeeded him, was a prince excellently learned; and had the patient and subtile wit of a schoolman; insomuch as in common speech (which leaves no virtue untaxed) he was called *cymini sector*†, a carver or divider of cummin seed, which is one of the least seeds; such a patience he had and settled spirit to enter into the least and most exact differences of causes; a fruit no doubt of the exceeding tranquillity and serenity of his mind; which being

37 To this story Dante alludes in the tenth canto of *Purgatory*; taking it apparently from the life of Gregory by Paul the Deacon. It seems first to have been mentioned by John Damascene in his discourse "De iis qui in fide dormierunt"; from whom St. Thomas Aquinas quotes it in his *Supplementary Questions*, 71. 5. The hymn sung in the fourteenth century in the Cathedral of Mantua on St. Paul's day, is another curious instance of the appreciation of Heathen worth in the middle ages. It is there said of St. Paul,

> Ad Maronis mausoleum
> Ductus fudit super eum
> Piæ rorem lacrymæ;
> Quem te, inquit, reddidissem
> Si te vivum invenissem
> Poetarum maxime!

See Schœll's *Histoire de la Littérature Romaine*.—*R. L. E.* This whole passage is omitted in the translation.

38 Plutarch, Apoph.

39 There seems here a confusion of two stories. It was Alexander Severus who according to Lampridius had a picture of our Saviour "matched with Apollonius" and with some others. Hadrian however did not honour Apollonius, and is said to have thought of dedicating a temple to Christ, which, if I remember rightly, Alexander actually did. —*R. L. E.*

40 So in all three editions. Qy. Trajan? [* Aurel. Victor, Ejust. c. 41.]

41 *pollicing*, edd. 1605 and 1629. *pollishing*, ed. 1633.

42 *Antonius*, edd. 1605, 1629, 1633. [† Xiph. *in Anton. Pio.*]

no ways charged or incumbered either with fears, remorses, or scruples, but having been noted for a man of the purest goodness, without all fiction or affectation, that hath reigned or lived, made his mind continually present and entire. He likewise approached a degree nearer unto Christianity, and became, as Agrippa said unto St. Paul, *half a Christian ;* holding their religion and law in good opinion, and not only ceasing persecution, but giving way to the advancement of Christians.

There succeeded him the first *Divi fratres*, the two adoptive brethren, Lucius Commodus Verus, son to Ælius Verus, who delighted much in the softer kind of learning, and was wont to call the poet Martial his Virgil ; and Marcus Aurelius Antoninus ; whereof the later, who obscured his colleague [43] and survived him long, was named the Philosopher : who as he excelled all the rest in learning, so he excelled them likewise in perfection of all royal virtues ; insomuch as Julianus the emperor, in his book intitled *Cæsares*, being as a pasquil or satire to deride all his predecessors, feigned that they were all invited to a banquet of the gods, and Silenus the jester sat at the nether end of the table and bestowed a scoff on every one as they came in ; but when Marcus Philosophus came in, Silenus was gravelled and out of countenance, not knowing where to carp at him ; save at the last he gave a glance at his patience towards his wife. And the virtue of this prince, continued with that of his predecessor, made the name of Antoninus so sacred in the world, that though it were extremely dishonoured in Commodus, Caracalla, and Heliogabalus, who all bare the name, yet when Alexander Severus refused the name because he was a stranger to the family, the Senate with one acclamation said, *Quomodo Augustus, sic et Antoninus* [let the name of Antoninus be as the name of Augustus] : in such renown and veneration was the name of these two princes in those days, that they would have it as a perpetual addition in all the emperors' style. In this emperor's time also the church for the most part was in peace ; so as in this sequence of six princes we do see the blessed effects of learning in sovereignty, painted forth in the greatest table of the world..

But for a tablet or picture of smaller volume, (not presuming to speak of your Majesty that liveth,) in my judgment the most excellent is that of queen Elizabeth, your immediate predecessor in this part of Britain ; a prince that, if Plutarch were now alive to write lives [44] by parallels, would trouble him, I think, to find for her a parallel amongst women. This lady was endued with learning in her sex singular, and rare [45] even amongst masculine princes ; whether we speak of learning of [46] language or of science ; modern or ancient ; divinity or humanity. And unto the very last year of her life she accustomed to appoint set hours for reading, scarcely any young student in an university more daily or more duly. As for her [47] government, I assure myself I shall not exceed if I do affirm that this part of the island never had forty-five years of better times ; and yet not through the calmness of the season, but through the wisdom of her regiment. For if there be considered of the one side, the truth of religion established ; the constant peace and security ; the good administration of justice ; the temperate use of the prerogative, not slackened, nor much strained ; the flourishing state of learning, sortable to so excellent a patroness ; the convenient estate of wealth and means, both of crown and subject ; the habit of obedience, and the moderation of discontents ; and there be considered on the other side, the differences of religion, the troubles of neighbour countries, the ambition of Spain, and opposition of Rome ; and then that she was solitary and of herself : these things I say considered, as I could not have chosen an instance so recent and so proper, so I

[43] In the translation he says that Lucius though not so good as his brother was better than most of the other emperors. (*Fratri quidem bonitate cedens, reliquos imperatores plurimos superans*).

[44] *lynes*, ed. 1605 and 1629. *lines* ed. 1633.

[45] So edd. 1629 and 1633. Ed. 1605 has *grace.*

[46] Edd. 1629 and 1633 have *or ;* with a semicolon after *learning*, where the original has a comma ; the omission of which makes the meaning and construction clear.

[47] So edd. 1629 and 1633. The original has *the.*

suppose I could not have chosen one more remarkable or eminent, to the purpose now in hand; which is concerning the conjunction of learning in the prince with felicity in the people [48].

Neither hath learning an influence and operation only upon civil merit and moral virtue, and the arts or temperature of peace and peaceable government; but likewise it hath no less power and efficacy in enablement towards martial and military virtue and prowess; as may be notably represented in the examples of Alexander the Great and Cæsar the Dictator, mentioned before, but now in fit place to be resumed; of whose virtues and acts in war there needs no note or recital, having been the wonders of time in that kind; but of their affection towards learning, and perfections in learning, it is pertinent to say somewhat.

Alexander was bred and taught under Aristotle the great philosopher, who dedicated divers of his books of philosophy unto him. He was attended with Callisthenes and divers other learned persons, that followed him in camp, throughout his journeys and conquests. What price and estimation he had learning in doth notably appear in these three particulars: first, in the envy he used to express that he bare towards Achilles, in this that he had so good a trumpet of his praises as Homer's verses; secondly, in the judgment or solution he gave touching that precious cabinet of Darius, which was found among his jewels, whereof question was made what thing was worthy to be put into it, and he gave his opinion for Homer's works*; thirdly, in his letter to Aristotle, after he had set forth his books of nature, wherein he expostulateth with him for publishing the secrets or mysteries of philosophy, and gave him to understand that himself esteemed it more to excel other men in learning and knowledge than in power and empire†. And what use he had of learning doth appear, or rather shine, in all his speeches and answers, being full of science and use of science, and that in all variety.

And herein again it may seem a thing scholastical, and somewhat idle, to recite things that every man knoweth; but yet since the argument I handle leadeth me thereunto, I am glad that men shall perceive I am as willing to flatter (if they will so call it) an Alexander or a Cæsar or an Antoninus, that are dead many hundred years since, as any that now liveth: for it is the displaying of the glory of learning in sovereignty that I propound to myself, and not an humour of declaiming in any man's praises [49]. Observe then the speech he used of Diogenes, and see if it tend not to the true state of one of the greatest questions of moral philosophy; whether the enjoying of outward things or the contemning of them be the greatest happiness; for when he saw Diogenes so perfectly contented with so little, he said to those that mocked at his condition, *Were I not Alexander, I would wish to be Diogenes.* But Seneca inverteth it, and saith, *Plus erat quod hic nollet accipere, quam quod ille posset dare* ‡. There were more things which Diogenes would have refused, than those were which Alexander could have given or enjoyed.

Observe again that speech which was usual with him, *That he felt his mortality chiefly in two things, sleep and lust*; and see if it were not a speech extracted out of the depth of natural philosophy, and liker to have comen out of the mouth of Aristotle or Democritus than from Alexander [50].

See again that speech of humanity and poesy; when upon the bleeding of his wounds, he called unto him one of his flatterers that was wont to ascribe to him divine honour, and said, *Look, this is very blood; this is not such a liquor as Homer speaketh of, which ran from Venus' hand when it was pierced by Diomedes.*

See likewise his readiness in reprehension of logic, in the speech he used to Cassander upon a complaint that was made against his father Antipater: for

[48] This paragraph is entirely omitted in the *De Augmentis*; no doubt as one which would not be allowed at Rome and might lead to the proscription of the book. See note p. 50. [* Pliny, vii. 19.] [† Plutarch *in Alex.* c. 7.]

[49] This sentence is omitted in the translation. [‡ Seneca, *De Benef.* v. 4.]

[50] ***cum tam indigentia tam redundantia naturæ per illa duo designata, mortis sint tanquam arrhabones***; the two opposite imperfections of nature, deficiency and superfluity, exhaustion and incontinence, being as it were earnests of mortality.

when Alexander happed to say, *Do you think these men would have come from so far to complain, except they had just cause of grief?* and Cassander answered, *Yea, that was the matter, because they thought they should not be disproved;* said Alexander laughing, *See the subtilties of Aristotle, to take a matter both ways,* pro et contra, etc.

But note again how well he could use the same art which he reprehended, to serve his own humour, when bearing a secret grudge to Callisthenes because he was against the new ceremony of his adoration, feasting one night where the same Callisthenes was at the table, it was moved by some after supper, for entertainment sake, that Callisthenes who was an eloquent man might speak of some theme or purpose at his own choice; which Callisthenes did; choosing the praise of the Macedonian nation for his discourse, and performing the same with so good manner as the hearers were much ravished; whereupon Alexander, nothing pleased, said, *It was easy to be eloquent upon so good a subject*; but saith he, *Turn your style, and let us hear what you can say against us:* which Callisthenes presently undertook, and did with that sting and life, that Alexander interrupted him, and said, *The goodness of the cause made him eloquent before, and despite made him eloquent then again.*

Consider further, for tropes of rhetoric, that excellent use of a metaphor or translation, wherewith he taxed Antipater, who was an imperious and tyrannous governor: for when one of Antipater's friends commended him to Alexander for his moderation, that he did not degenerate, as his other lieutenants did, into the Persian pride, in use of purple, but kept the ancient habit of Macedon, of black; *True,* (saith Alexander,) *but Antipater is all purple within.* Or that other, when Parmenio came to him in the plain of Arbella, and shewed him the innumerable multitude of his enemies, specially as they appeared by the infinite number of lights, as it had been a new firmament of stars, and thereupon advised him to assail them by night: whereupon he answered, *That he would not steal the victory.*

For matter of policy, weigh that significant distinction, so much in all ages embraced, that he made between his two friends Hephæstion and Craterus, when he said, *That the one loved Alexander, and the other loved the king;* describing the principal difference of princes' best servants, that some in affection love their person, and others in duty love their crown.

Weigh also that excellent taxation of an error ordinary with counsellors of princes, that they counsel their masters according to the model of their own mind and fortune, and not of their masters; when upon Darius' great offers Parmenio had said, *Surely I would accept these offers, were I as Alexander;* saith Alexander, *So would I, were I as Parmenio.*

Lastly, weigh that quick and acute reply which he made when he gave so large gifts to his friends and servants, and was asked what he did reserve for himself, and he answered, *Hope;* weigh, I say, whether he had not cast up his account aright, because *hope* must be the portion of all that resolve upon great enterprises. For this was Cæsar's portion when he went first into Gaul, his estate being then utterly overthrown with largesses. And this was likewise the portion of that noble prince, howsoever transported with ambition, Henry Duke of Guise, of whom it was usually said, that he was the greatest usurer in France, because he had turned all his estate into obligations.

To conclude therefore: as certain critics are used to say hyperbolically, *That if all sciences were lost, they might be found in Virgil;* so certainly this may be said truly, there are the prints and footsteps of learning in those few speeches which are reported of this prince: the admiration of whom, when I consider him not as Alexander the Great, but as Aristotle's scholar, hath carried me too far.

As for Julius Cæsar, the excellency of his learning needeth not to be argued from his education, or his company, or his speeches; but in a further degree doth declare itself in his writings and works; whereof some are extant and permanent, and some unfortunately perished. For first, we see there is left unto us that excellent history of his own wars, which he intitled only a Commentary, wherein all succeeding times have admired the solid weight of matter, and the real passages and lively images of actions and persons, expressed in the greatest propriety of words and perspicuity of narration that ever was ; which that it was not the effect of a natural gift, but of learning and precept, is well witnessed by that work of his intitled *De Analogia*, being a grammatical philosophy, wherein he did labour to make this same *vox ad placitum* to become *vox ad licitum*, and to reduce custom of speech to congruity of speech ; and took as it were the picture of words from the life of reason [51].

So we receive from him, as a monument both of his power and learning, the then reformed computation of the year ; well expressing, that he took it to be as great a glory to himself to observe and know the law of the heavens as to give law to men upon the earth.

So likewise in that book of his *Anti-Cato*, it may easily appear that he did aspire as well to victory of wit as victory of war ; undertaking therein a conflict against the greatest champion with the pen that then lived, Cicero the orator.

So again in his book of *Apophthegms* which he collected, we see that he esteemed it more honour to make himself but a pair of tables to take the wise and pithy words of others, than to have every word of his own to be made an apophthegm or an oracle ; as vain princes, by custom of flattery, pretend to do. And yet if I should enumerate divers of his speeches, as I did those of Alexander, they are truly such as Salomon noteth, when he saith, *Verba sapientum tanquam aculei, et tanquam clavi in altum defixi* [the words of the wise are as goads, and as nails fixed deep in] : whereof I will only recite three, not so delectable for elegancy, but admirable for vigour and efficacy.

At first, it is reason he be thought a master of words, that could with one word appease a mutiny in his army ; which was thus. The Romans, when their generals did speak to their army, did use the word *Milites* ; but when the magistrates spake to the people, they did use the word *Quirites*. The soldiers were in tumult, and seditiously prayed to be cashiered ; not that they so meant, but by expostulation thereof to draw Cæsar to other conditions ; wherein he being resolute not to give way, after some silence, he began his speech, *Ego, Quirites* ; which did admit them already cashiered ; wherewith they were so surprised, crossed, and confused, as they would not suffer him to go on in his speech, but relinquished their demands, and made it their suit to be again called by the name of *Milites**.

The second speech was thus : Cæsar did extremely affect the name of king ; and some were set on, as he passed by, in popular acclamation to salute him king ; whereupon, finding the cry weak and poor, he put it off thus in a kind of jest, as if they had mistaken his surname ; *Non Rex sum, sed Cæsar* [I am not King, but Cæsar] : a speech, that if it be searched, the life and fulness of it can scarce be expressed : for first it was a refusal of the name, but yet not serious : again it did signify an infinite confidence and magnanimity, as if he presumed Cæsar was the greater title ; as by his worthiness it is come to pass till this day : but chiefly it was a speech of great allurement towards his own purpose ; as if the state did strive with him but for a name, whereof mean families were vested ; for Rex was a surname with the Romans, as well as King is with us.

The last speech which I will mention, was used to Metellus ; when Cæsar, after war declared, did possess himself of the city of Rome ; at which time entering into the inner treasury to take the money there accumulate, Metellus being tribune forbade him : whereto Cæsar said, *That if he did not desist, he would lay him dead in the place* ; and presently taking himself up, he added, *Young man, it is harder for me to speak it than to do it. Adolescens, durius est mihi hoc dicere quàm facere.* A speech compounded of the greatest terror and greatest clemency that could proceed out of the mouth of man.

But to return and conclude with him : it is evident himself knew well his own perfection in learning, and took it upon him ; as appeared when upon occasion

51 This passage is translated without addition or alteration. But Bacon seems to have changed his opinion afterwards upon the point in question. For in the sixth book of the *De Augmentis*, c. i., he intimates a suspicion that Cæsar's book was not a grammatical philosophy, but only a set of precepts for the formation of a pure, perfect, and unaffected style. See p. 122. [* Sueton, *in Julio*, c. 70. Cf. Appian, *Bel, Civ.* ii. 93.]

that some spake what a strange resolution it was in Lucius Sylla to resign his dictature, he scoffing at him, to his own advantage, answered, *That Sylla could not skill of letters, and therefore knew not how to dictate.*

And here it were fit to leave this point touching the concurrence of military virtue and learning; (for what example would come with any grace after those two of Alexander and Cæsar?) were it not in regard of the rareness of circumstance that I find in one other particular, as that which did so suddenly pass from extreme scorn to extreme wonder; and it is of Xenophon the philosopher, who went from Socrates' school into Asia, in the expedition of Cyrus the younger against king Artaxerxes. This Xenophon at that time was very young, and never had seen the wars before; neither had any command in the army, but only followed the war as a voluntary, for the love and conversation of Proxenus his friend. He was present when Falinus came in message from the great king to the Grecians, after that Cyrus was slain in the field, and they a handful of men left to themselves in the midst of the king's territories, cut off from their country by many navigable rivers, and many hundred miles. The message imported that they should deliver up their arms, and submit themselves to the king's mercy. To which message before answer was made, divers of the army conferred familiarly with Falinus; and among the rest Xenophon happened to say, *Why Falinus, we have now but these two things left, our arms and our virtue; and if we yield up our arms, how shall we make use of our virtue?* Whereto Falinus smiling on him, said, *If I be not deceived, young gentleman, you are an Athenian; and I believe you study philosophy, and it is pretty that you say; but you are much abused if you think your virtue can withstand the king's power**. Here was the scorn; the wonder followed: which was, that this young scholar or philosopher, after all the captains were murdered in parley by treason, conducted those ten thousand foot through the heart of all the king's high countries from Babylon to Græcia in safety, in despite of all the king's forces, to the astonishment of the world, and the encouragement of the Grecians in time succeeding to make invasion upon the kings of Persia; as was after purposed by Jason the Thessalian, attempted by Agesilaus the Spartan, and achieved by Alexander the Macedonian; all upon the ground of the act of that young scholar.

To proceed now from imperial and military virtue to moral and private virtue: first, it is an assured truth which is contained in the verses,

> Scilicet ingenuas didicisse fideliter artes
> Emollit mores, nec sinit esse feros†;

[a true proficiency in liberal learning softens and humanises the manners]. It taketh away the wildness and barbarism and fierceness of men's minds: but indeed the accent had need be upon *fideliter* [it must be a *true* proficiency]: for a little superficial learning[52] doth rather work a contrary effect. It taketh away all levity, temerity, and insolency, by copious suggestion of all doubts and difficulties, and acquainting the mind to balance reasons on both sides, and to turn back the first offers and conceits of the mind, and to accept of nothing but examined and tried. It taketh away vain admiration of any thing, which is the root of all weakness. For all things are admired, either because they are new, or because they are great. For novelty, no man that wadeth in learning or contemplation throughly, but will find that printed in his heart *Nil novi super terram* [there is nothing new under the sun]. Neither can any man marvel at the play of puppets, that goeth behind the curtain and adviseth well of the motion. And for magnitude, as Alexander the Great after that he was used to great armies and the great conquests of the spacious provinces in Asia, when he received letters out of Greece of some fights and services there, which were commonly for a passage or a fort or some walled town at the most, he said, *It seemed to him that he was advertised of the battles of the frogs and the mice, that the old tales went of*‡: so certainly if a man meditate much upon the universal frame of nature, the earth with men upon it (the divineness of souls except) will not seem much other than an ant-hill, whereas some ants carry corn, and some carry

[52] *tumultuaria cognitio.* [* Xen. *Anab.* ii. 1, 12.] Text inaccurate. [† Ovid. *Ex. Pont* ii. 9, 47. Inaccurate.] [‡ Plut. *in Agesil.* c. 15.]

their young, and some go empty, and all to and fro a little heap of dust. It taketh away or mitigateth fear of death or adverse fortune; which is one of the greatest impediments of virtue and imperfections of manners. For if a man's mind be deeply seasoned with the consideration of the mortality and corruptible nature of things, he will easily concur with Epictetus, who went forth one day and saw a woman weeping for her pitcher of earth that was broken, and went forth the next day and saw a woman weeping for her son that was dead; and thereupon said, *Heri vidi fragilem frangi, hodie vidi mortalem mori* [yesterday I saw a brittle thing broken, to-day a mortal dead]. And therefore Virgil did excellently and profoundly couple the knowledge of causes and the conquest of all fears together, as *concomitantia*.

> Felix qui potuit rerum cognoscere causas
> Quique metus omnes et inexorabile fatum
> Subjecit pedibus, strepitumque Acherontis avari.

> [Happy the man who doth the causes know
> Of all that is: serene he stands, above
> All fears; above the inexorable Fate,
> And that insatiate gulph that roars below.]

It were too long to go over the particular remedies which learning doth minister to all the diseases of the mind; sometimes purging the ill humours, sometimes opening the obstructions, sometimes helping digestion, sometimes increasing appetite, sometimes healing the wounds and exulcerations thereof, and the like; and therefore I will conclude with that which hath *rationem totius*; which is, that it disposeth the constitution of the mind not to be fixed or settled in the defects thereof, but still to be capable and susceptible of growth and reformation. For the unlearned man knows not what it is to descend into himself or to call himself to account, nor the pleasure of that *suavissima vita, indies sentire se fieri meliorem* [to feel himself each day a better man than he was the day before]. The good parts he hath he will learn to shew to the full and use them dexterously, but not much to increase them: the faults he hath he will learn how to hide and colour them, but not much to amend them; like an ill mower, that mows on still and never whets his scythe: whereas with the learned man it fares otherwise, that he doth ever intermix the correction and amendment of his mind with the use and employment thereof. Nay further, in general and in sum, certain it is that *veritas* and *bonitas* differ but as the seal and the print; for truth prints goodness, and they be the clouds of error which descend in the storms of passions and perturbations.

From moral virtue let us pass on to matter of power and commandment, and consider whether in right reason there be any comparable with that wherewith knowledge investeth and crowneth man's nature. We see the dignity of the commandment is according to the dignity of the commanded: to have commandment over beasts, as herdsmen have, is a thing contemptible; to have commandment over children, as schoolmasters have, is a matter of small honour; to have commandment over galley-slaves is a disparagement rather than an honour. Neither is the commandment of tyrants much better, over people which have put off the generosity of their minds: and therefore it was ever holden that honours in free monarchies and commonwealths had a sweetness more than in tyrannies; because the commandment extendeth more over the wills of men, and not only over their deeds and services. And therefore when Virgil putteth himself forth to attribute to Augustus Cæsar the best of human honours, he doth it in these words:

> victorque volentes
> Per populos dat jura, viamque affectat Olympo

> [Moving in conquest onward, at his will
> To willing peoples he gives laws, and shapes
> Through worthiest deeds on earth his course to Heaven].

But yet the commandment of knowledge is yet higher than the commandment over the will, for it is a commandment over the reason, belief, and understanding of man, which is the highest part of the mind, and giveth law to the will itself. For there is no power on earth which setteth up a throne or chair of estate in the spirits and souls of men, and in their cogitations, imaginations, opinions, and beliefs, but knowledge and learning. And therefore we see the detestable and extreme pleasure that arch-heretics and false prophets and impostors are transported with, when they once find in themselves that they have a superiority in the faith and conscience of men ; so great, that if they have once tasted of it, it is seldom seen that any torture or persecution can make them relinquish or abandon it. But as this is that which the author of the Revelation calleth the depth or profoundness of Satan ; so by argument of contraries, the just and lawful sovereignty over men's understanding, by force [53] of truth rightly interpreted, is that which approacheth nearest to the similitude of the divine rule.

As for fortune and advancement, the beneficence of learning is not so confined to give fortune only to states and commonwealths, as it doth not likewise give fortune to particular persons. For it was well noted long ago, that Homer hath given more men their livings than either Sylla or Cæsar or Augustus ever did, notwithstanding their great largesses and donatives and distributions of lands to so many legions. And no doubt it is hard to say whether arms or learning have advanced greater numbers. And in case of sovereignty, we see that if arms or descent have carried away the kingdom, yet learning hath carried the priesthood, which ever hath been in some competition with empire.

Again, for the pleasure and delight of knowledge and learning, it far surpasseth all other in nature : for shall the pleasures of the affections so exceed the senses, as much as the obtaining of desire or victory exceedeth a song or a dinner ; and must not of consequence the pleasures of the intellect or understanding exceed the pleasures of the affections ? We see in all other pleasures there is satiety, and after they be used, their verdure [54] departeth ; which sheweth well they be but deceits of pleasure, and not pleasures ; and that it was the novelty which pleased, and not the quality. And therefore we see that voluptuous men turn friars, and ambitious princes turn melancholy. But of knowledge there is no satiety, but satisfaction and appetite are perpetually interchangeable ; and therefore appeareth to be good in itself simply, without fallacy or accident. Neither is that pleasure of small efficacy and contentment to the mind of man, which the poet Lucretius describeth elegantly,

Suave mari magno, turbantibus æquora ventis, etc.

It is a view of delight (saith he) *to stand or walk upon the shore side, and to see a ship tossed with tempest upon the sea ; or to be in a fortified tower, and to see two battles join upon a plain. But it is a pleasure incomparable, for the mind of man to be settled, landed, and fortified in the certainty of truth ; and from thence to descry and behold the errors, perturbations, labours, and wanderings up and down of other men.*

Lastly, leaving the vulgar arguments, that by learning man excelleth man in that wherein man excelleth beasts ; that by learning man ascendeth to the heavens and their motions, where in body he cannot come ; and the like ; let us conclude with the dignity and excellency of knowledge and learning in that whereunto man's nature doth most aspire ; which is immortality or continuance ; for to this tendeth generation, and raising of houses and families ; to this buildings, foundations, and monuments ; to this tendeth the desire of memory, fame, and celebration ; and in effect, the strength of all other human desires. We see then how far the monuments of wit and learning are more durable than the monuments of power or of the hands. For have not the verses of Homer continued twenty-five hundred years or more, without the loss of a syllable or letter ; during which time infinite palaces, temples, castles, cities, have been decayed

[53] So edd. 1629 and 1633. The original has *face*.
[54] *verdour* in the original and also in edd. 1629 and 1633. See p. 62.

and demolished ? It is not possible to have the true pictures or statues of Cyrus, Alexander, Cæsar, no nor of the kings or great personages of much later years ; for the originals cannot last, and the copies cannot but leese of the life and truth. But the images of men's wits and knowledges remain in books, exempted from the wrong of time and capable of perpetual renovation. Neither are they fitly to be called images, because they generate still, and cast their seeds in the minds of others, provoking and causing infinite actions and opinions in succeeding ages. So that if the invention of the ship was thought so noble, which carrieth riches and commodities from place to place, and consociateth the most remote regions in participation of their fruits, how much more are letters to be magnified, which as ships pass through the vast seas of time, and make ages so distant to participate of the wisdom, illuminations, and inventions, the one of the other ? Nay further, we see some of the philosophers which were least divine and most immersed in the senses and denied generally the immortality of the soul, yet came to this point, that whatsoever motions the spirit of man could act and perform without the organs of the body they thought might remain after death ; which were only those of the understanding, and not of the affection : so immortal and incorruptible a thing did knowledge seem unto them to be*. But we, that know by divine revelation that not only the understanding but the affections purified, not only the spirit but the body changed, shall be advanced to immortality, do disclaim in [55] these rudiments of the senses. But it must be remembered both in this last point, and so it may likewise be needful in other places, that in probation of the dignity of knowledge or learning I did in the beginning separate divine testimony from human ; which method I have pursued, and so handled them both apart.

Nevertheless I do not pretend, and I know it will be impossible for me by any pleading of mine, to reverse the judgment, either of Æsop's cock, that preferred the barleycorn before the gem ; or of Midas, that being chosen judge between Apollo president of the Muses, and Pan god of the flocks, judged for plenty ; or of Paris, that judged for beauty and love against wisdom and power ; or of Agrippina, *occidat matrem, modo imperet* † [let him kill his mother so he be emperor], that preferred empire with condition never so detestable ; or of Ulysses, *qui vetulam prætulit immortalitati* [that preferred an old woman to an immortality], being a figure of those which prefer custom and habit before all excellency ; or of a number of the like popular judgments. For these things continue as they have been : but so will that also continue whereupon learning hath ever relied, and which faileth not : *Justificata est sapientia a filiis suis* [wisdom is justified of her children].

[* Mr. Ellis notes that "the doctrine of the soul's immortality here referred to is that which was attributed to Aristotle and his followers, who are here contrasted with the Platonists, as being more 'immersed in the senses' ".]

55 So all three editions. The translation has *nos autem . . . . conculcantes hæc rudimenta atque offucias sensuum, novimus* etc.

[† "Occidat dum imperet." Tac. Ann. xiv. 9.]

THE SECOND BOOK.

TO THE KING

It might seem to have more convenience, though it come often otherwise to pass, (excellent King,) that those which are fruitful in their generations, and have in themselves the foresight of immortality in their descendants, should likewise be more careful of the good estate of future times; unto which they know they must transmit and commend over their dearest pledges. Queen Elizabeth was a sojourner in the world in respect of her unmarried life; and was a blessing to her own times; and yet so as the impression of her good government, besides her happy memory, is not without some effect which doth survive her[1]. But to your Majesty, whom God hath already blessed with so much royal issue, worthy to continue and represent you for ever, and whose youthful and fruitful bed doth yet promise many the like renovations, it is proper and agreeable to be conversant not only in the transitory parts of good government, but in those acts also which are in their nature permanent and perpetual. Amongst the which (if affection do not transport me) there is not any more worthy than the further endowment of the world with sound and fruitful knowledge: for why should a few received authors stand up like Hercules' Columns, beyond which there should be no sailing or discovering, since we have so bright and benign a star as your Majesty to conduct and prosper us? To return therefore where we left, it remaineth to consider of what kind those acts are, which have been undertaken and performed by kings and others for the increase and advancement of learning: wherein I purpose to speak actively without digressing or dilating.

Let this ground therefore be laid, that all works are overcomen by amplitude of reward, by soundness of direction, and by the conjunction of labours. The first multiplieth endeavour, the second preventeth error, and the third supplieth the frailty of man. But the principal of these is direction: for *claudus in via antevertit cursorem extra viam* [the cripple that keeps the way gets to the end of the journey sooner than the runner who goes aside]; and Salomon excellently setteth it down, *If the iron be not sharp, it requireth more strength; but wisdom is that which prevaileth;* signifying that the invention or election of the mean is more effectual than any inforcement or accumulation of endeavours. This I am induced to speak, for that (not derogating from the noble intention of any that have been deservers towards the state of learning) I do observe nevertheless that their works and acts are rather matters of magnificence and memory than of progression and proficience, and tend rather to augment the mass of learning in the multitude of learned men than to rectify or raise the sciences themselves.

The works or acts of merit towards learning are conversant about three objects; the places of learning, the books of learning, and the persons of the learned. For as water, whether it be the dew of heaven or the springs of the earth, doth scatter and leese itself in the ground, except it be collected into some receptacle, where it may by union comfort and sustain itself; and for that cause the industry of man hath made and framed spring-heads, conduits, cisterns, and pools, which men have accustomed likewise to beautify and adorn with accomplishments of magnificence and state, as well as of use and necessity; so this excellent liquor of knowledge, whether it descend from divine inspiration or spring from human sense, would soon perish and vanish to oblivion, if it were not preserved

[1] This last clause is omitted in the translation. See note p. 50.

in books, traditions, conferences, and places appointed, as universities, colleges and schools, for the receipt and comforting of the same.

The works which concern the seats and places of learning are four; foundations and buildings, endowments with revenues, endowments with franchises and privileges, institutions and ordinances for government; all tending to quietness and privateness of life, and discharge of cares and troubles; much like the stations which Virgil prescribeth for the hiving of bees:

> Principio sedes apibus statioque petenda,
> Quo neque sit ventis aditus, etc.
> [First for thy bees a quiet station find,
> And lodge them under covert of the wind [2].]

The works touching books are two: first libraries, which are as the shrines where all the relics of the ancient saints, full of true virtue and that without delusion or imposture[3], are preserved and reposed; secondly, new editions of authors, with more correct impressions, more faithful translations, more profitable glosses, more diligent annotations, and the like.

The works pertaining to the persons of learned men (besides the advancement and countenancing of them in general) are two; the reward and designation of writers and inquirers concerning any parts of learning not sufficiently laboured and prosecuted.

These are summarily the works and acts, wherein the merits of many excellent princes and other worthy personages have been conversant. As for any particular commemorations, I call to mind what Cicero said, when he gave general thanks; *Difficile non aliquem, ingratum quenquam præterire* [it were hard to remember all, and yet ungracious to forget any]. Let us rather, according to the Scriptures, look unto that part of the race which is before us than look back to that which is already attained.

First therefore, amongst so many great foundations of colleges in Europe, I find it strange that they are all dedicated to professions, and none left free to arts and sciences at large. For if men judge that learning should be referred to action, they judge well; but in this they fall into the error described in the ancient fable; in which the other parts of the body did suppose the stomach had been idle, because it neither performed the office of motion, as the limbs do, nor of sense, as the head doth; but yet notwithstanding it is the stomach that digesteth and distributeth to all the rest. So if any man think philosophy and universality to be idle studies, he doth not consider that all professions are from thence served and supplied. And this I take to be a great cause that hath hindered the progression of learning, because these fundamental knowledges have been studied but in passage. For if you will have a tree bear more fruit than it hath used to do, it is not any thing you can do to the boughs, but it is the stirring of the earth and putting new mould about the roots that must work it. Neither is it to be forgotten that this dedicating of foundations and dotations to professory learning hath not only had a malign aspect and influence upon the growth of sciences, but hath also been prejudicial to states and governments. For hence it proceedeth that princes find a solitude in regard of able men to serve them in causes of estate, because there is no education collegiate which is free; where such as were so disposed might give themselves to histories, modern languages, books of policy and civil discourse, and other the like enablements unto service of estate.

And because founders of colleges do plant and founders of lectures do water, it followeth well in order to speak of the defect which is in public lectures; namely, in the smallness and meanness of the salary or reward which in most places is assigned unto them[4]; whether they be lectures of arts, or of professions. For it is necessary to the progression of sciences that readers[5] be of the most able and sufficient men; as those which are ordained for generating and propagating of sciences, and not for transitory use. This cannot be, except

[2] Dryden. [3] This clause is omitted in the *De Augmentis*. See note p. 50.
[4] In the *De Augmentis* he adds *præsertim apud nos.* [5] *i.e.* lecturers.

their condition and endowment be such as may content the ablest man to appropriate his whole labour and continue his whole age in that function and attendance; and therefore must have a proportion answerable to that mediocrity or competency of advancement which may be expected from a profession or the practice of a profession. So as, if you will have sciences flourish, you must observe David's military law, which was, *That those which staid with the carriage should have equal part with those which were in the action;* else will the carriages be ill attended: So readers in sciences are indeed the guardians of the stores and provisions of sciences whence men in active courses are furnished, and therefore ought to have equal entertainment with them; otherwise if the fathers in sciences be of the weakest sort or be ill-maintained,

Et patrum invalidi referent jejunia nati:

[the poor keeping of the parents will appear in the poor constitution of the offspring].

Another defect I note, wherein I shall need some alchemist to help me, who call upon men to sell their books and to build furnaces; quitting and forsaking Minerva and the Muses as barren virgins, and relying upon Vulcan. But certain it is that unto the deep, fruitful, and operative study of many sciences, specially natural philosophy and physic [6], books be not only the instrumentals; wherein also the beneficence of men hath not been altogether wanting; for we see spheres, globes, astrolabes, maps, and the like, have been provided as appurtenances to astronomy and cosmography, as well as books: we see likewise that some places instituted for physic have annexed the commodity of gardens for simples of all sorts, and do likewise command the use of dead bodies for anatomies. But these do respect but a few things. In general, there will hardly be any main proficience in the disclosing of nature, except there be some allowance for expenses about experiments; whether they be experiments appertaining to Vulcanus or Dædalus, furnace or engine, or any other kind; and therefore as secretaries and spials of princes and states bring in bills for intelligence, so you must allow the spials and intelligencers of nature to bring in their bills, or else you shall be ill advertised.

And if Alexander made such a liberal assignation to Aristotle of treasure for the allowance of hunters, fowlers, fishers, and the like, that he might compile an History of nature, much better do they deserve it that travail [7] in Arts of nature [8].

Another defect which I note, is an intermission or neglect in those which are governors in universities of consultation, and in princes or superior persons of visitation; to enter into account and consideration, whether the readings, exercises, and other customs appertaining unto learning, anciently begun and since continued, be well instituted or no; and thereupon to ground an amendment or reformation in that which shall be found inconvenient. For it is one of your Majesty's own most wise and princely maxims, *that in all usages and precedents, the times be considered wherein they first began; which if they were weak or ignorant, it derogateth from the authority of the usage, and leaveth it for suspect.* And therefore in as much as most of the usages and orders of the universities were derived from more obscure times, it is the more requisite they be re-examined. In this kind I will give an instance or two for example sake, of things that are the most obvious and familiar. The one is a matter which, though it be ancient and general, yet I hold to be an error; which is, that scholars in universities come too soon and too unripe to logic and rhetoric; arts fitter for graduates than children and novices: for these two, rightly taken, are the gravest of sciences; being the arts of arts, the one for judgment, the other for ornament; and they be the rules and directions how to set forth and

[6] *i.e.* medicine. [7] *travailes* in the original, and also in edd. 1629 and 1633.

[8] *i.e.* in working upon and altering nature by art. The meaning is expressed more clearly in the translation: *majus quiddum debetur iis qui non in saltibus naturæ pererrant, sed in labyrinthis artium viam aperiunt*: the compiler of a history of nature being likened to a wanderer through the woods, the "travailer in arts and nature" to one who makes his way through a labyrinth.

dispose matter; and therefore for minds empty and unfraught with matter, and which have not gathered that which Cicero calleth *sylva* and *supellex*, stuff and variety, to begin with those arts (as if one should learn to weigh or to measure or to paint the wind), doth work but this effect, that the wisdom of those arts, which is great and universal, is almost made contemptible, and is degenerate into childish sophistry and ridiculous affectation. And further, the untimely learning of them hath drawn on by consequence the superficial and unprofitable teaching and writing of them, as fitteth indeed to the capacity of children. Another is a lack I find in the exercises used in the universities, which do make too great a divorce between invention and memory; for their speeches are either premeditate *in verbis conceptis*, where nothing is left to invention, or merely *extemporal*, where little is left to memory: whereas in life and action there is least use of either of these, but rather of intermixtures of premeditation and invention, notes and memory; so as the exercise fitteth not the practice, nor the image the life; and it is ever a true rule in exercises, that they be framed as near as may be to the life and practice; for otherwise they do pervert the motions and faculties of the mind, and not prepare them. The truth whereof is not obscure, when scholars come to the practices of professions, or other actions of civil life; which when they set into, this want is soon found by themselves, and sooner by others. But this part, touching the amendment of the institutions and orders of universities, I will conclude with the clause of Cæsar's letter to Oppius and Balbus, *Hoc quemadmodum fieri possit, nonnulla mihi in mentem veniunt, et multa reperiri possunt; de iis rebus rogo vos ut cogitationem suscipiatis* [how this may be done, some things occur to me and more may be thought of. I would have you take these matters into consideration].

Another defect which I note, ascendeth a little higher than the precedent. For as the proficience of learning consisteth much in the orders and institutions of universities in the same states and kingdoms, so it would be yet more advanced, if there were more intelligence mutual between the universities of Europe than now there is. We see there be many orders and foundations, which though they be divided under several sovereignties and territories, yet they take themselves to have a kind of contract, fraternity, and correspondence one with the other, insomuch as they have Provincials and Generals [9]. And surely as nature createth brotherhood in families, and arts mechanical contract brotherhoods in communalties, and the anointment of God superinduceth a brotherhood in kings and bishops; so in like manner there cannot but be a fraternity in learning and illumination, relating to that paternity which is attributed to God, who is called the Father of illuminations or lights. The last defect which I will note is, that there hath not been, or very rarely been, any public designation of writers or inquirers concerning such parts of knowledge as may appear not to have been already sufficiently laboured or undertaken; unto which point it is an inducement, to enter into a view and examination what parts of learning have been prosecuted, and what omitted; for the opinion of plenty is amongst the causes of want, and the great quantity of books maketh a shew rather of superfluity than lack; which surcharge nevertheless is not to be remedied by making no more books, but by making more good books, which, as the serpent of Moses [10], might devour the serpents of the enchanters.

The removing of all the defects formerly enumerate, except the last, and of the active part also of the last, (which is the designation of writers), are *opera basilica* [works for a king]; towards which the endeavours of a private man may be but as an image in a crossway, that may point at the way but cannot go it. But the inducing part of the latter (which is the survey of learning) may be set forward by private travel. Wherefore I will now attempt to make a general and faithful perambulation of learning, with an inquiry what parts thereof lie fresh and waste, and not improved and converted by the industry of man; to the end that such a plot made and recorded to memory may both

9 *Præfectos (alios provinciales, alios generales) quibus omnes parent.*—De Aug.

10 Not Moses, but Aaron. Ex. i. 17.—*R. L. E.*

minister light to any public designation, and also serve to excite voluntary endeavours; wherein nevertheless my purpose is at this time to note only omissions and deficiencies, and not to make any redargution of errors or incomplete prosecutions[11]; for it is one thing to set forth what ground lieth unmanured, and another thing to correct ill husbandry in that which is manured[12].

In the handling and undertaking of which work I am not ignorant what it is that I do now move and attempt, nor insensible of mine own weakness to sustain my purpose; but my hope is that if my extreme love to learning carry me too far, I may obtain the excuse of affection; for that *it is not granted to man to love and to be wise*. But I know well I can use no other liberty of judgment than I must leave to others; and I for my part shall be indifferently glad either to perform myself or accept from another that duty of humanity, *Nam qui erranti comiter monstrat viam*, etc. [to put the wanderer in the right way]. I do foresee likewise that of those things which I shall enter and register as deficiencies and omissions, many will conceive and censure that some of them are already done and extant; others to be but curiosities, and things of no great use; and others to be of too great difficulty and almost impossibility to be compassed and effected. But for the two first, I refer myself to the particulars. For the last, touching impossibility, I take it those things are to be held possible which may be done by some person, though not by everyone; and which may be done by many, though not by any one; and which may be done in succession of ages, though not within the hour-glass of one man's life; and which may be done by public designation, though not by private endeavour. But notwithstanding if any man will take to himself rather that of Salomon, *Dicit piger, Leo est in via* [the slothful man saith there is a lion in the path], than that of Virgil, *Possunt quia posse videntur* [they find it possible because they think it possible], I shall be content that my labours be esteemed but as the better sort of wishes; for as it asketh some knowledge to demand a question not impertinent, so it requireth some sense to make a wish not absurd.

¶[13] The parts of human learning have reference to the three parts of Man's Understanding, which is the seat of learning: History to his Memory, Poesy to his Imagination, and Philosophy to his Reason. Divine learning receiveth the same distribution; for the spirit of man is the same, though the revelation of oracle and sense be diverse: so as theology consisteth also of History of the Church; of Parables, which is divine poesy; and of holy Doctrine or precept. For as for that part which seemeth supernumerary, which is Prophecy, it is but divine history; which hath that prerogative over human, as the narration may be before the fact as well as after.

¶[14] History is Natural, Civil, Ecclesiastical, and Literary; whereof the three first I allow as extant, the fourth I note as deficient. For no man hath propounded to himself the general state of learning to be described and represented from age to age, as many have done the works of nature and the state civil and ecclesiastical; without which the history

Historia Literarum.

[11] *infelicitates.*—De Aug.

[12] *i.e.* cultivated.

[13] De Aug. ii. 1. The substance of the following paragraph will be found considerably expanded in the first chapter of the *Descriptio Globi Intellectualis*, and set forth much more clearly and orderly in the first chapter of the second book of the *De Augmentis;* which begins here; the previous observations being introductory. As it may be convenient to the reader to have the means of referring at once to the corresponding passages of the more finished work, I shall mark with a ¶ the places where the several chapters begin; adding (where the case admits of it) some notice, more or less complete, of the differences between the two. See Preface.

[14] De Aug. ii. 4. In the translation the divisions are altered: History being divided into Natural and Civil,—History of Nature and History of Man; and Literary and Ecclesiastical History being considered as separate departments of the latter. See chap. 2. paragraph 1. This alteration induces an alteration in the order of treatment; the precedence being given to the History of Nature, which is the subject of the second chapter.

of the world seemeth to me to be as the statue of Polyphemus with his eye out; that part being wanting which doth most shew the spirit and life of the person. And yet I am not ignorant that in divers particular sciences, as of the jurisconsults, the mathematicians, the rhetoricians, the philosophers, there are set down some small memorials of the schools, authors, and books; and so likewise some barren relations touching the invention of arts or usages. But a just story of learning, containing the antiquities and originals of knowledges, and their sects; their inventions, their traditions; their diverse administrations and managings; their flourishings, their oppositions, decays, depressions, oblivions, removes; with the causes and occasions of them, and all other events concerning learning, throughout the ages of the world [15]; I may truly affirm to be wanting. The use and end of which work I do not so much design for curiosity, or satisfaction of those that are the lovers of learning; but chiefly for a more serious and grave purpose, which is this in few words, that it will make learned men wise in the use and administration of learning. For it is not St. Augustine's nor St. Ambrose works that will make so wise a divine, as ecclesiastical history throughly read and observed; and the same reason is of learning.

¶ [16] History of Nature is of three sorts; of nature in course, of nature erring or varying, and of nature altered or wrought; that is, history of Creatures, history of Marvels, and history of Arts.[17] The first of these no doubt is extant, and that in good perfection; the two later are handled so weakly and unprofitably, as I am moved to note them as deficient. For I find no sufficient or competent collection of the works of nature which have a digression and deflexion from the ordinary course of generations, productions, and motions; whether they be singularities of place and region, or the strange events of time and chance, or the effects of yet unknown proprieties, or the instances of exception to general kinds. It is true, I find a number of books of fabulous experiments and secrets, and frivolous impostures for pleasure and strangeness. But a substantial and severe collection of the Heteroclites or Irregulars of nature, well examined and described, I find not; specially not with due rejection of fables and popular errors; for as things now are, if an untruth in nature be once on foot, what by reason of the neglect of examination and countenance of antiquity, and what by reason of the use of the opinion in similitudes and ornaments of speech, it is never called down.

Historia Naturæ Errantis.

The use of this work, honoured with a precedent in Aristotle [18], is nothing less than to give contentment to the appetite of curious and vain wits, as the manner of Mirabilaries is to do; but for two reasons, both of great weight; the one to correct the partiality of axioms and opinions, which are commonly framed only upon common and familiar examples; the other because from the wonders of nature is the nearest intelligence and passage towards the wonders of art; for it is no more but by following and as it were hounding Nature in her wanderings, to be able to lead her afterwards to the same place again. Neither am I of opinion, in this History of Marvels, that superstitious narrations of sorceries, witchcrafts, dreams, divinations, and the like, where there is an assurance and clear evidence of the fact, be altogether excluded. For it is not yet known in what cases, and how far, effects attributed to superstition do participate of natural causes; and therefore howsoever the practice of such things is to be condemned, yet from the speculation and consideration of them light may be taken, not only for the discerning of the offences, but for the further disclosing of nature. Neither ought a man to make scruple of entering into these things for inquisition of truth, as your Majesty hath shewed in your own example; who with the two clear eyes of religion and natural philosophy have looked deeply

[15] The description of the required history is set forth much more particularly in the translation; and the whole paragraph rewritten and enlarged. [16] De Aug. ii. 2.

[17] This division is retained in the translation, but the exposition of it is extended into a long paragraph.

[18] De Miris Auscultationibus; which is now however generally admitted to be not Aristotle's.—*R. L. E.* See De Aug. ii. 2. Mr. Blakesley is of opinion that the nucleus of it was probably Aristotle's, but that it has been added to by subsequent writers.

and wisely into these shadows, and yet proved yourself to be of the nature of the sun, which passeth through pollutions and itself remains as pure as before. But this I hold fit, that these narrations which have mixture with superstition be sorted by themselves, and not to be mingled with the narrations which are merely and sincerely natural. But as for the narrations touching the prodigies and miracles of religions, they are either not true or not natural; and therefore impertinent for the story of nature.

Historia Mechanica.

For History of Nature Wrought or Mechanical, I find some collections made of agriculture, and likewise of manual arts; but commonly with rejection of experiments familiar and vulgar. For it is esteemed a kind of dishonour unto learning to descend to inquiry or meditation upon matters mechanical, except they be such as may be thought secrets, rarities and special subtilties; which humour of vain and supercilious arrogancy is justly derided in Plato; where he brings in Hippias, a vaunting sophist, disputing with Socrates, a true and unfeigned inquisitor of truth; where the subject being touching beauty, Socrates, after his wandering manner of inductions, put first an example of a fair virgin,and then of a fair horse, and then of a fair pot well glazed, whereat Hippias was offended, and said, *More than for courtesy's sake, he did think much to dispute with any that did allege such base and sordid instances*: whereunto Socrates answereth, *You have reason, and it becomes you well, being a man so trim in your vestiments*, etc. and so goeth on in an irony. But the truth is, they be not the highest instances that give the securest information; as may be well expressed in the tale so common of the philosopher, that while he gazed upwards to the stars fell into the water; for if he had looked down he might have seen the stars in the water, but looking aloft he could not see the water in the stars. So it cometh often to pass that mean and small things discover great better than great can discover the small; and therefore Aristotle noteth well, *that the nature of every thing is best seen in his smallest portions*, and for that cause he inquireth the nature of a commonwealth, first in a family, and the simple conjugations of man and wife, parent and child, master and servant, which are in every cottage: even so likewise the nature of this great city of the world and the policy thereof must be first sought in mean concordances and small portions. So we see how that secret of nature, of the turning of iron touched with the loadstone towards the north, was found out in needles of iron, not in bars of iron.

But if my judgment be of any weight, the use of History Mechanical is of all others the most radical and fundamental towards natural philosophy; such natural philosophy as shall not vanish in the fume of subtile, sublime, or delectable speculation, but such as shall be operative to the endowment and benefit of man's life; for it will not only minister and suggest for the present many ingenious practices in all trades, by a connexion and transferring of the observations of one art to the use of another, when the experiences of several mysteries shall fall under the consideration of one man's mind; but further it will give a more true and real illumination concerning causes and axioms than is hitherto attained. For like as a man's disposition is never well known till he be crossed, nor Proteus ever changed shapes till he was straitened and held fast; so the passages and variations of nature cannot appear so fully in the liberty of nature, as in the trials and vexations of art [19].

¶ [20] For Civil History, it is of three kinds [21]; not unfitly to be compared with the three kinds of pictures or images. For of pictures or images, we see some

[19] A paragraph is added in the translation, to say that not the mechanical arts only, but also the practical part of the liberal sciences, as well as many crafts which have not grown into formal arts (such, he means, as hunting, fishing, &c.), are to be included in the History Mechanical.

[20] De Aug. ii. 6. The 3rd chapter, concerning the two uses of natural history, and the 5th, concerning the dignity and difficulty of civil history, have nothing corresponding to them here.

[21] " I am not altogether ignorant in the laws of history and of the kinds. The same

are unfinished, some are perfect [22], and some are defaced. So of histories we may find three kinds, Memorials, Perfect Histories, and Antiquities; for Memorials are history unfinished, or the first or rough draughts of history, and Antiquities are history defaced, or some remnants of history which have casually escaped the shipwreck of time.

Memorials, or Preparatory History, are of two sorts; whereof the one may be termed Commentaries, and the other Registers. Commentaries are they which set down a continuance of the naked events and actions, without the motives or designs, the counsels, the speeches, the pretexts, the occasions, and other passages of action: for this is the true nature of a Commentary; though Cæsar, in modesty mixed with greatness, did for his pleasure apply the name of a Commentary to the best history of the world. Registers are collections of public acts, as decrees of council, judicial proceedings, declarations and letters of estate, orations, and the like, without a perfect continuance or contexture of the thread of the narration.

Antiquities or Remnants of History are, as was said, *tanquam tabula naufragii* [like the planks of a shipwreck]; when industrious persons by an exact and scrupulous diligence and observation, out of monuments, names, words, proverbs, traditions, private records and evidences, fragments of stories, passages of books that concern not story, and the like, do save and recover somewhat from the deluge of time.

In these kinds of unperfect histories I do assign no deficience, for they are *tanquam imperfecte mista* [things imperfectly compounded]; and therefore any deficience in them is but their nature. As for the corruptions and moths of history, which are Epitomes, the use of them deserveth to be banished, as all men of sound judgment have confessed; as those that have fretted and corroded the sound bodies of many excellent histories, and wrought them into base and unprofitable dregs.

¶ [23] History which may be called Just and Perfect History is of three kinds, according to the object which it propoundeth, or pretendeth to represent: for it either representeth a Time, or a Person, or an Action. The first we call Chronicles, the second Lives, and the third Narrations or Relations. Of these, although the first be the most complete and absolute kind of history and hath most estimation and glory, yet the second excelleth it in profit and use, and the third in verity and sincerity. For History of Times representeth the magnitude of actions and the public faces and deportments of persons, and passeth over in silence the smaller passages and motions of men and matters. But such being the workmanship of God as he doth hang the greatest weight upon the smallest wires, *maxima e minimis suspendens*, it comes therefore to pass, that such histories do rather set forth the pomp of business than the true and inward resorts thereof [24]. But Lives, if they be well written [25], propounding to themselves a person to represent in whom actions both greater and smaller, public and private, have a commixture, must of necessity contain a more true, native, and lively representation. So again Narrations and Relations of actions, as the War of Peloponnesus, the Expedition of Cyrus Minor, the Conspiracy of Catiline, cannot but be more purely and exactly true than Histories of Times, because they may choose an argument comprehensible within the notice and instruc-

hath been taught by many, but by no man better and with greater brevity than by that excellent learned gentleman Sir Francis Bacon."—*Raleigh*: Preface to the *History of the World*.—*R. L. E.*

[22] *parfite* in the original; the form in which the word was commonly written in Bacon's time.

[23] De Aug. ii. 7.

[24] And even (he adds in the translation) where they attempt to give the counsels and motives, yet still out of the same love of dignity and greatness they introduce into men's actions more gravity and wisdom than they really have; insomuch that you may find a truer picture of human life in some satires than in such histories.

[25] *i.e.* not mere eulogies. The tranlations adds: "neque enim de elogiis et hujusmodi commemorationibus jejunis loquimur."

tions of the writer: whereas he that undertaketh the story of a time, especially of any length, cannot but meet with many blanks and spaces which he must be forced to fill up out of his own wit and conjecture [26].

For the History of Times, (I mean of civil history) the providence of God hath made the distribution: for it hath pleased God to ordain and illustrate two exemplar states of the world, for arms, learning, moral virtue, policy, and laws; the state of Græcia, and the state of Rome; the histories whereof occupying the middle part of time, have more ancient to them, histories which may by one common name be termed the Antiquities of the World; and after them, histories which may be likewise called by the name of Modern History [27].

Now to speak of the deficiencies. As to the Heathen Antiquities of the world, it is in vain to note them for deficient. Deficient they are no doubt, consisting most of fables and fragments; but the deficience cannot be holpen; for antiquity is like fame, *caput inter nubila condit*, her head is muffled from our sight. For the History of the Exemplar States, it is extant in good perfection. Not but I could wish there were a perfect course of history for Græcia from Theseus to Philopœmen, (what time the affairs of Græcia drowned and extinguished in the affairs of Rome); and for Rome from Romulus to Justinianus, who may be truly said to be *ultimus Romanorum*. In which sequences of story the text of Thucydides and Xenophon in the one, and the texts of Livius, Polybius, Sallustius, Cæsar, Appianus, Tacitus, Herodianus in the other, to be kept entire without any diminution at all, and only to be supplied and continued. But this is matter of magnificence, rather to be commended than required: and we speak now of parts of learning supplemental, and not of supererogation.

But for Modern Histories, whereof there are some few very worthy, but the greater part beneath mediocrity, leaving the care of foreign stories to foreign states, because I will not be *curiosus in aliena republica* [a meddler in other nations' matters], I cannot fail to represent to your Majesty the unworthiness of the history of England in the main continuance thereof, and the partiality and obliquity of that of Scotland in the latest and largest author that I have seen; supposing that it would be honour for your Majesty and a work very memorable, if this island of Great Britain [28], as it is now joined in monarchy for the ages to come, so were joined in one history for the times passed; after the manner of the sacred history, which draweth down the story of the ten Tribes and of the Two tribes as twins together. And if it shall seem that the greatness of this work may make it less exactly performed, there is an excellent period of a much smaller compass of time as to the story of England; that is to say, from the Uniting of the Roses to the Uniting of the Kingdoms; a portion of time wherein, to my understanding, there hath been the rarest varieties that in like number of successions of any hereditary monarchy hath been known.

For it beginneth with the mixed adeption of a crown, by arms and title; an entry by battle, an establishment by marriage; and therefore times answerable, like waters after a tempest, full of working and swelling, though without extremity of storm; but well passed through by the wisdom of the pilot, being one of the most sufficient kings of all the number. Then followeth the reign of a king, whose actions, howsoever conducted [29], had much intermixture with

[26] On the other hand it must be confessed (he reminds us in the translation,—I give only the general import of the passage, which is of considerable length) that relations of this kind, especially if published near the time to which they refer, are in one respect of all narratives the most to be suspected; being commonly written either in favour or in spite. But then again it seldom happens that they are all on one side, so that the extreme views of each party being represented, an honest and judicious historian may, when the violence of faction has cooled down with time, find the truth among them.

[27] This paragraph and the next are omitted in the translation, and their place supplied by a general complaint that very many particular histories are still wanting; much to the injury in honour and reputation of the kingdoms and commonwealths which they concern.

[28] Spelt *Brittanie* in the original; *Brittany* in edd. 1629 and 1633.

[29] The distinction between the father and the son is more clearly marked in the trans-

the affairs of Europe, balancing and inclining them variably; in whose time also began that great alteration in the state ecclesiastical, an action which seldom cometh upon the stage: then the reign of a minor: then an offer of an usurpation, though it was but as *febris ephemera* [a diary ague]: then the reign of a queen matched with a foreigner: then of a queen that lived solitary and unmarried, and yet her government so masculine as it had greater impression and operation upon the states abroad than it any ways received from thence[30]: and now last, this most happy and glorious event, that this island of Britain, divided from all the world, should be united in itself; and that oracle of rest given to Æneas, *Antiquam exquirite matrem* [seek out your ancient mother], should now be performed and fulfilled upon the nations of England and Scotland, being now reunited in the ancient mother name of Britain, as a full period of all instability and peregrinations: so that as it cometh to pass in massive bodies, that they have certain trepidations and waverings before they fix and settle; so it seemeth that by the providence of God this monarchy, before it was to settle in your Majesty and your generations, (in which I hope it is now established for ever), it had these prelusive changes and varieties.

For Lives, I do find strange that these times have so little esteemed the virtues of the times, as that the writing of lives should be no more frequent. For although there be not many sovereign princes or absolute commanders, and that states are more collected into monarchies, yet are there many worthy personages that deserve better than dispersed report or barren eulogies. For herein the invention of one of the late poets[31] is proper, and doth well enrich the ancient fiction: for he feigneth that at the end of the thread or web of every man's life there was a little medal containing the person's name, and that Time waited upon the shears, and as soon as the thread was cut, caught the medals and carried them to the river of Lethe; and about the bank there were many birds flying up and down, that would get the medals and carry them in their beak a little while, and then let them fall into the river: only there were a few swans, which if they got a name, would carry it to a temple where it was consecrated. And although many men more mortal in their affections than in their bodies, do esteem desire of name and memory but as a vanity and ventosity,

Animi nil magnæ laudis egentes

[souls that have no care for praise]; which opinion cometh from that root, *non puris laudes contempsimus, quam laudanda facere desivimus* [men hardly despise praise till they have ceased to deserve it]; yet that will not alter Salomon's judgment, *Memoria justi cum laudibus, at impiorum nomen putrescet* [the memory of the just is blessed; but the name of the wicked shall rot]; the one flourisheth, the other either consumeth to present oblivion, or turneth to an ill odour. And therefore in that style or addition, which is and hath been long well received and brought in use, *felicis memoriæ, piæ memoriæ, bonæ memoriæ* [of happy, of pious, of good memory], we do acknowledge that which Cicero saith, borrowing it from Demosthenes, that *bona fama propria possessio defunctorum*[32] [good fame is all that a dead man can possess]; which possession I cannot but note that in our times it lieth much waste, and that therein there is a deficience.

For Narrations and Relations of particular actions, there were also to be wished a greater diligence therein; for there is no great action but hath some good pen which attends it. And because it is an ability not common to write a good history, as may well appear by the small number of them; yet if particularity of actions memorable were but tolerably reported as they pass, the com-

ation. Of Henry VII. he says *qui unus inter antecessores reges* consilio *enituit;* of Henry VIII.'s actions, *licet magis impetu quam consilio administratæ.* Had Bacon gone on with his history of Henry VIII. it would have been curious to contrast the portrait of the son governing more by passion than policy, with that of the father governing by policy without passion.

[30] This last clause is omitted in the *De Augmentis*. See note p. 50.

[31] Ariosto, *Orlando Furioso*, at the end of the 34th and the beginning of the 35th books.

[32] Compare Cicero, Philippic. 9. 5., with the opening of the λόγος ἐπιτάφιος, 1389–10.

piling of a complete History of Times might be the better expected, when a writer should arise that were fit for it; for the collection of such relations might be as a nursery garden, whereby to plant a fair and stately garden when time should serve.

¶ [33] There is yet another portion of history which Cornelius Tacitus maketh, which is not to be forgotten, specially with that application which he accoupleth it withal, Annals and Journals: appropriating to the former matters of estate, and to the later acts and accidents of a meaner nature. For giving but a touch of certain magnificent buildings, he addeth, *Cum ex dignitate populi Romani repertum sit, res illustres annalibus, talia diurnis urbis actis mandare* [that it had been thought suitable to the dignity of the Roman people to enter in their *annals* only matters of note and greatness; leaving such things as these to the *journal records* of the city]. So as there is a kind of contemplative heraldry, as well as civil. And as nothing doth derogate from the dignity of a state more than confusion of degrees; so it doth not a little embase the authority of an history, to intermingle matters of triumph or matters of ceremony or matters of novelty with matters of state. But the use of a Journal hath not only been in the history of times [34], but likewise in the history of persons, and chiefly of actions; for princes in ancient time had, upon point of honour and policy both, journals kept of what passed day by day; for we see the Chronicle which was read before Ahasuerus [35], when he could not take rest, contained matter of affairs indeed, but such as had passed in his own time, and very lately before: but the Journal of Alexander's house expressed a very small particularity, even concerning his person and court [36]; and it is yet an use well received in enterprises memorable, as expeditions of war, navigations, and the like, to keep diaries of that which passeth continually.

¶ [37] I cannot likewise be ignorant of a form of writing which some grave and wise men have used, containing a scattered history of those actions which they have thought worthy of memory, with politic discourse and observation thereupon; not incorporate into the history, but separately, and as the more principal in their intention; which kind of Ruminated History I think more fit to place amongst books of policy, whereof we shall hereafter speak, than amongst books of history [38]; for it is the true office of history to represent the events themselves together with the counsels, and to leave the observations and conclusions thereupon to the liberty and faculty of every man's judgment. But mixtures are things irregular, whereof no man can define.

So also is there another kind of history manifoldly mixed, and that is History of Cosmography: being compounded of natural history, in respect of the regions themselves; of history civil, in respect of the habitations, regiments, and manners of the people; and the mathematics, in respect of the climates and configurations towards the heavens: which part of learning of all others in this latter time hath obtained most proficience. For it may be truly affirmed to the honour of these times, and in a virtuous emulation with antiquity, that this great building of the world had never through-lights made in it, till the age of us and our fathers; for although they had knowledge of the antipodes,

[33] De Aug. ii. 9. Between this paragraph and the last there is introduced in the translation a chapter on the advantages and disadvantages of histories of the world, as distinguished from histories of particular countries.

[34] *time* in the original and also in edd. 1629 and 1633. The translation omits this clause. [35] Esther, vi. 1.

[36] Not that greater matters were excluded; but great and small were entered promiscuously as they occurred. (*Neque enim sicut annales tantum gravia. ita diaria tantum levia complexa sunt; sed omnia promiscue et cursim diariis excipiebantur, seu majoris seu minoris momenti.*) [37] De Aug. ii. 10.

[38] This remark is omitted in the translation, and another substituted, to the effect that this kind of ruminated history is an excellent thing, provided it be understood that the matter in hand is not history but observations upon history (*modo hujusmodi scriptor hoc agat et hoc se agere confiteatur*); for in a regular history the narrative ought not, he says, to be interrupted by comments of this kind. It should be pregnant with politic precepts, but the writer should not play the midwife.

> Nosque ubi primus equis oriens afflavit anhelis,
> Illic sera rubens accendit lumina Vesper
>
> [And while on us the fresh East breathes from far,
> For them the red West lights her evening star]:

yet that might be by demonstration, and not in fact; and if by travel, it requireth the voyage but of half the globe. But to circle the earth, as the heavenly bodies do, was not done nor enterprised till these later times: and therefore these times may justly bear in their word, not only *plus ultra*, in precedence of the ancient *non ultra*, and *imitabile fulmen* in precedence of the ancient *non imitabile fulmen*,

> Demens qui nimbos et non imitabile fulmen, etc.

but likewise *imitabile cœlum*; in respect of the many memorable voyages, after the manner of heaven, about the globe of the earth.

And this proficience in navigation and discoveries may plant also an expectation of the further proficience and augmentation of all sciences; because it may seem they are ordained by God to be coevals, that is, to meet in one age. For so the prophet Daniel speaking of the latter times foretelleth, *Plurimi pertransibunt, et multiplex erit scientia* [many shall pass to and fro, and knowledge shall be multiplied]: as if the openness and through passage of the world and the increase of knowledge were appointed to be in the same ages; as we see it is already performed in great part; the learning of these later times not much giving place to the former two periods or returns of learning, the one of the Grecians, the other of the Romans.

¶ [39] History Ecclesiastical receiveth the same divisions with History Civil: but further in the propriety thereof may be divided into History of the Church, by a general name; History of Prophecy; and History of Providence. The first describeth the times of the militant church; whether it be fluctuant, as the ark of Noah; or moveable, as the ark in the wilderness; or at rest, as the ark in the temple; that is, the state of the church in persecution, in remove, and in peace. This part I ought in no sort to note as deficient; only I would that the virtue and sincerity of it were according to the mass and quantity. But I am not now in hand with censures, but with omissions.

The second, which is History of Prophecy, consisteth of two relatives, the prophecy and the accomplishment; and therefore the nature of such a work ought to be, that every prophecy of the scripture be sorted with the event fulfilling the same, throughout the ages of the world; both for the better confirmation of faith, and for the better illumination of the Church touching those parts of prophecies which are yet unfulfilled; allowing nevertheless that latitude which is agreeable and familiar unto divine prophecies; being of the nature of their author, with whom a thousand years are but as one day; and therefore are not fulfilled punctually at once, but have springing and germinant accomplishment throughout many ages, though the height or fulness of them may refer to some one age. This is a work which I find deficient, but is to be done with wisdom, sobriety, and reverence, or not at all.

Historia Prophetica

The third, which is History of Providence, containeth that excellent correspondence which is between God's revealed will and his secret will; which though it be so obscure as for the most part it is not legible to the natural man; no, nor many times to those that behold it from the tabernacle; yet at some times it pleaseth God, for our better establishment and the confuting of those which are as without God in the world, to write it in such text and capital letters that, as the prophet saith, *he that runneth by may read it* [40]; that is, mere sensual per-

[39] De Aug. ii. 11.

[40] Habak. ii. 2. Mr. Ellis has remarked in his note on the corresponding passage in the *De Augmentis* that this expression, now so familiar and almost proverbial, is in fact a misquotation of the text and a misrepresentation of the meaning of the prophet.

sons, which hasten by God's judgments and never bend or fix their cogitations upon them, are nevertheless in their passage and race urged to discern it. Such are the notable events and examples of God's judgments, chastisements, deliverances, and blessings. And this is a work which hath passed through the labour of many [41], and therefore I cannot present as omitted.

¶ [42] There are also other parts of learning which are Appendices to history. For all the exterior proceedings of man consist of words and deeds; whereof history doth properly receive and retain in memory the deeds; and if words, yet but as inducements and passages to deeds; so are there other books and writings, which are appropriate to the custody and receit of words only; which likewise are of three sorts; Orations, Letters, and Brief Speeches or Sayings. Orations are pleadings, speeches of counsel; laudatives, invectives, apologies, reprehensions; orations of formality or ceremony, and the like. Letters are according to all the variety of occasions; advertisements, advices, directions, propositions, petitions, commendatory, expostulatory, satisfactory, of compliment, of pleasure, of discourse, and all other passages of action. And such as are written from wise men are, of all the words of man, in my judgment the best; for they are more natural than orations and public speeches, and more advised than conferences or present speeches. So again letters of affairs from such as manage them or are privy to them are of all others the best instructions for history, and to a diligent reader the best histories in themselves. For Apophthegms, it is a great loss of that book of Cæsar's; for as his history and those few letters of his which we have and those apophthegms which were of his own excel all men's else, so I suppose would his collection of Apophthegms have done; for as for those which are collected by others, either I have no taste in such matters, or else their choice hath not been happy [43]. But upon these three kinds of writings I do not insist, because I have no deficiences to propound concerning them.

Thus much therefore concerning History; which is that part of learning which answereth to one of the cells, domiciles, or offices of the mind of man; which is that of the Memory.

¶ [44] POESY is a part of learning in measure of words for the most part restrained, but in all other points extremely licensed, and doth truly refer to the Imagination; which, being not tied to the laws of matter, may at pleasure join that which nature hath severed, and sever that which nature hath joined, and so make unlawful matches and divorces of things: *Pictoribus atque poetis*, etc. [Painters and Poets have always been allowed to take what liberties they would]. It is taken in two senses, in respect of words or matter. In the first sense it is but a character of style, and belongeth to arts of speech, and is not pertinent for the present [45]. In the later, it is (as hath been said) one of the principal portions of learning, and is nothing else but Feigned History, which may be styled as well in prose as in verse.

"Write the vision and make it plain upon the tables that he may run that readeth it." It would be a curious inquiry, who first made this mistake.

[41] In the translation he says, "*sane in calamos nonnullorum piorum virorum incidit*, sed non sine partium studio." Indeed it is difficult to see how, without partiality, such a history of Providence could be written at all. For take any signal calamity and look at it in its historical character only,—who shall say whether it is a chastisement or a martyrdom? a judgment upon the sinner, or a trial of the saint? [42] De Aug. ii. 12.

[43] Some further remarks upon the value and use of Apophthegms are introduced in the *De Augmentis*; of these, a translation will be given in my preface to Bacon's own collection of Apophthegms.

[44] De Aug. ii. 13. The arrangement is partly altered in the translation, and much new matter introduced: among the rest, a whole paragraph concerning the true use and dignity of dramatic poetry, as a vehicle of moral instruction; which is connected in striking manner with the remark that men in bodies are more open to impressions than when alone.

[45] A sentence is added in the translation to explain that under this head satires, elegies, epigrams, and odes are included.

The use of this Feigned History hath been to give some shadow of satisfaction to the mind of man in those points wherein the nature of things doth deny it; the world being in proportion inferior to the soul; by reason whereof there is agreeable to the spirit of man a more ample greatness, a more exact goodness, and a more absolute variety, than can be found in the nature of things. Therefore, because the acts or events of true history have not that magnitude which satisfieth the mind of man, poesy feigneth acts and events greater and more heroical; because true history propoundeth the successes and issues of actions not so agreeable to the merits of virtue and vice, therefore poesy feigns them more just in retribution, and more according to revealed providence; because true history representeth actions and events more ordinary and less interchanged, therefore poesy endueth them with more rareness, and more unexpected and alternative variations. So as it appeareth that poesy serveth and conferreth to magnanimity, morality, and to delectation. And therefore it was ever thought to have some participation of divineness, because it doth raise and erect the mind, by submitting the shews of things to the desires of the mind; whereas reason doth buckle and bow the mind unto the nature of things. And we see that by these insinuations and congruities with man's nature and pleasure, joined also with the agreement and consort it hath with music, it hath had access and estimation in rude times and barbarous regions, where other learning stood excluded.

The division of poesy which is aptest in the propriety thereof, (besides those divisions which are common unto it with history, as feigned chronicles, feigned lives; and the appendices of history, as feigned orations, feigned epistles, and the rest), is into Poesy Narrative, Representative, and Allusive. The Narrative is a mere imitation of history, with the excesses before remembered; choosing for subject commonly wars and love, rarely state, and sometimes pleasure or mirth [46]. Representative is as a visible history, and is an image of actions as if they were present, as history is of actions in nature as they are, (that is) past. Allusive or Parabolical is a narration applied only to express some special purpose or conceit [47]. Which later kind of parabolical wisdom was much more in use in the ancient times, as by the fables of Æsop and the brief sentences of the Seven and the use of hieroglyphics may appear. And the cause was, for that it was then of necessity to express any point of reason which was more sharp or subtile than the vulgar in that manner; because men in those times wanted both variety of examples and subtilty of conceit: and as hieroglyphics were before letters, so parables were before arguments: and nevertheless now and at all times they do retain much life and vigour, because reason cannot be so sensible nor examples so fit.

But there remaineth yet another use of Poesy Parabolical, opposite to that which we last mentioned: for that tendeth to demonstrate and illustrate that which is taught or delivered, and this other to retire and obscure it: that is when the secrets and mysteries of religion, policy, or philosophy are involved in fables or parables. Of this in divine poesy we see the use is authorized. In heathen poesy we see the exposition of fables doth fall out sometimes with great felicity; as in the fable that the giants being overthrown in their war against the gods, the Earth their mother in revenge thereof brought forth Fame:

> Illam Terra parens, irâ irritata deorum,
> Extremam, ut perhibent, Cœo Enceladoque sororem
> Progenuit:

expounded that when princes and monarchs have suppressed actual and open rebels, then the malignity of people (which is the mother of rebellion) doth bring forth libels and slanders and taxations of the state, which is of the same kind with rebellion, but more feminine. So in the fable that the rest of the gods

[46] The last clause of this sentence is omitted in the translation.

[47] This obscure sentence is explained in the translation to mean that Parabolic Poesy is *historia cum typo, quæ intellectualia deducit ad sensum*,—typical history, by which ideas that are objects of the Intellect are represented in forms that are objects of the Sense.

having conspired to bind Jupiter, Pallas called Briareus with his hundred hands to his aid : expounded that monarchies need not fear any curbing of their absoluteness by mighty subjects, as long as by wisdom they keep the hearts of the people, who will be sure to come in on their side. So in the fable that Achilles was brought up under Chiron the Centaur, who was part a man and part a beast : expounded ingeniously but corruptly by Machiavel, that it belongeth to the education and discipline of princes to know as well how to play the part of the lion in violence and the fox in guile, as of the man in virtue and justice [48]. Nevertheless in many the like encounters, I do rather think that the fable was first, and the exposition devised, than that the moral was first, and thereupon the fable framed. For I find it was an ancient vanity in Chrysippus, that troubled himself with great contention to fasten the assertions of the Stoics upon the fictions of the ancient poets. But yet that all the fables and fictions of the poets were but pleasure and not figure, I interpose no opinion. Surely of those poets which are now extant, even Homer himself, (notwithstanding he was made a kind of Scripture by the later schools of the Grecians), yet I should without any difficulty pronounce that his fables had no such inwardness in his own meaning ; but what they might have upon a more original tradition, is not easy to affirm ; for he was not the inventor of many of them [49].

In this third part of learning, which is poesy, I can report no deficience. For being as a plant that cometh of the lust of the earth, without a formal seed, it hath sprung up and spread abroad more than any other kind. But to ascribe unto it that which is due ; for the expressing of affections, passions, corruptions, and customs, we are beholding to poets more than to the philosophers' works ; and for wit and eloquence not much less than to orators' harangues [50]. But it is not good to stay too long in the theatre. Let us now pass on to the judicial place or palace of the mind, which we are to approach and view with more reverence and attention.

¶ [51] The knowledge of man is as the waters, some descending from above, and some springing from beneath ; the one informed by the light of nature, the other inspired by divine revelation. The light of nature consisteth in the notions of the mind and the reports of the senses ; for as for knowledge which man receiveth by teaching, it is cumulative and not original ; as in a water that besides his own spring-head is fed with other springs and streams. So then according to these two different illuminations or originals, knowledge is first of all divided into Divinity and Philosophy.

In philosophy, the contemplations of man do either penetrate unto God, or are circumferred to Nature, or are reflected or reverted upon Himself. Out of which several inquiries there do arise three knowledges, Divine philosophy, Natural philosophy, and Human philosophy or Humanity. For all things are marked and stamped with this triple character, of the power of God, the difference of nature, and the use of man. But because the distributions and partitions of knowledge are not like several lines that meet in one angle, and so touch but in a point ; but are like branches of a tree that meet in a stem, which hath a dimension and quantity of entireness and continuance, before it come to discontinue and break itself into arms and boughs ;

[48] The Prince, c. 18. As two of the animals are the same, it is possible that Machiavelli was thinking of what was said of Boniface VIII. by the predecessor whom he forced to abdicate,—that he came in like a fox, would reign like a lion, and die like a dog.—*R. L. E.*

[49] For these examples there is substituted in the translation a full exposition of the three fables of Pan, Perseus, and Dionysus. And it is worth observing that, upon the question whether there was really a mystic sense at the bottom of the ancient fables, Bacon expresses in the translation a more decided inclination to the affirmative than he does here.

[50] This sentence is omitted in the translation.

[51] De Aug. iii. 1. The order of this chapter is changed in the translation, and a good deal added.

therefore it is good, before we enter into the former distribution, to erect and constitute one universal science, by the name of *Philosophia Prima*, Primitive or Summary Philosophy, as the main and common way, before we come where the ways part and divide themselves; which science whether I should report as deficient or no, I stand doubtful. For I find a certain rhapsody of Natural Theology, and of divers parts of Logic; and of that part of Natural Philosophy which concerneth the Principles, and of that other part of Natural Philosophy which concerneth the Soul or Spirit; all these strangely commixed and confused; but being examined, it seemeth to me rather a depredation of other sciences, advanced and exalted unto some height of terms[52], than any thing solid or substantive of itself. Nevertheless I cannot be ignorant of the distinction which is current, that the same things are handled but in several respects; as for example, that logic considereth of many things as they are in notion, and this philosophy as they are in nature; the one in appearance, the other in existence. But I find this difference better made than pursued. For if they had considered Quantity, Similitude, Diversity, and the rest of those Extern Characters of things, as philosophers, and in nature, their inquiries must of force have been of a far other kind than they are. For doth any of them, in handling Quantity, speak of the force of union, how and how far it multiplieth virtue? Doth any give the reason, why some things in nature are so common and in so great mass, and others so rare and in so small quantity? Doth any, in handling Similitude and Diversity, assign the cause why iron should not move to iron, which is more like, but more to the loadstone, which is less like? Why in all diversities of things there should be certain participles in nature, which are almost ambiguous to which kind they should be referred? But there is a mere and deep silence touching the nature and operation of those Common Adjuncts of things, as in nature; and only a resuming and repeating of the force and use of them in speech or argument. Therefore, because in a writing of this nature I avoid all subtility, my meaning touching this original or universal philosophy is thus, in a plain and gross description by negative: *That it be a receptacle for all such profitable observations and axioms as fall not within the compass of any of the special parts of philosophy or sciences, but are more common and of a higher stage.*

Now that there are many of that kind need not be doubted. For example is not the rule, *Si inæqualibus æqualia addas, omnia erunt inæqualia* [if equals be added to unequals, the wholes will be unequal], an axiom as well of justice as of the mathematics[53]? And is there not a true coincidence between commutative and distributive justice, and arithmetical and geometrical proportion? Is not that other rule, *Quæ in eodem tertio conveniunt, et inter se conveniunt*

[52] *Et sublimitate quadam sermonis hominum qui se ipsos admirari amant tanquam in vertice scientiarum collocatam.*—De Aug. The substance of the rest of this paragraph, till we come to the last sentence, is transferred to the end of the chapter in the *De Augmentis* and set forth more fully and clearly.

[53] This clause is printed out of its place both in the original and in the editions of 1629 and 1633; being inserted after the next sentence. It is obviously an error of the printer; but worth noticing as the evidence of imperfection of the arrangements then made for correcting the press. I am inclined to think that in Bacon's time the proof-sheets were never revised by the author.

In the translation we are told that the axiom holds with regard to *distributive* justice only. (*Eadem in Ethicis obtinet quatenus ad justitiam distributivam; siquidem in justitiâ commutativa, ut paria imparibus tribuantur ratio æquitatis postulat; at in distributiva, nisi imparia imparibus præstentur, iniquitas fuerit maxima.*) Equal measure distributed to unequal conditions produces an unequal result; a truth of which many striking illustrations are furnished by the operations of our own laws as between the rich and the poor, when the same penalty inflicted for the same offence falls heavily on the one and lightly on the other. In matter of *commutation*,—as in a question, for instance, of compensation for property destroyed,—this of course does not hold. The coincidence between commutative and distributive justice and arithmetical and geometrical proportion is not alluded to in the translation. But this may have been by accident; the translator perhaps not having observed where the misplaced sentence was meant to come in.

[things that are equal to the same are equal to each other], a rule taken from the mathematics, but so potent in logic as all syllogisms are built upon it? Is not the observation, *Omnia mutantur, nil interit* [all things change, but nothing is lost], a contemplation in philosophy thus, That the *quantum* of nature is eternal? in natural theology thus, That it requireth the same omnipotence to make somewhat nothing, which at the first made nothing somewhat? according to the scripture, *Didici quod omnia opera quæ fecit Deus perseverent in perpetuum; non possumus eis quicquam addere nec auferre* [I know that whatsoever God doeth, it shall be for ever; nothing can be put to it, nor anything taken from it]. Is not the ground, which Machiavel wisely and largely discourseth concerning governments, that the way to establish and preserve them is to reduce them *ad principia* [54], a rule in religion and nature [55] as well as in civil administration? Was not the Persian Magic a reduction or correspondence of the principles and architectures of nature to the rules and policy of governments? Is not the precept of a musician, to fall from a discord or harsh accord upon a concord or sweet accord, alike true in affection? Is not the trope of music, to avoid or slide from the close or cadence, common with the trope of rhetoric of deceiving expectation? Is not the delight of the quavering upon a stop in music the same with [56] the playing of light upon the water?

Splendet tremulo sub lumine pontus
[Beneath the trembling light glitters the sea].

Are not the organs of the senses of one kind with the organs of reflexion, the eye with a glass, the ear with a cave or strait determined and bounded [57]? Neither are these only similitudes, as men of narrow observation may conceive them to be, but the same footsteps of nature, treading or printing upon several subjects or matters. This science therefore (as I understand it) I may justly report as deficient; for I see sometimes the profounder sort of wits, in handling some particular argument, will now and then draw a bucket of water out of this well for their present use; but the springhead thereof seemeth to me not to have been visited, being of so excellent use both for the disclosing of nature and the abridgment of art.

Philosophia Prima, sive de Fontibus Scientiarum.

¶ [58] This science being therefore first placed as a common parent, like unto Berecynthia, which had so much heavenly issue,

Omnes cœlicolas, omnes supera alta tenentes
[All dwellers in the heaven and upper sky]:

we may return to the former distribution of the three philosophies; Divine, Natural, and Human. And as concerning Divine philosophy or Natural Theology, it is that knowledge or rudiment of knowledge concerning God which may be obtained by the contemplation of his creatures; which knowledge may be truly termed divine in respect of the object, and natural in respect of the light. The bounds of this knowledge are, that it sufficeth to convince atheism, but not to inform religion: and therefore there was never miracle wrought by God to convert an atheist, because the light of nature might have led him to confess a God: but miracles have been wrought to convert idolaters and the superstitious, because no light of nature extendeth to declare the will and true worship of God. For as all works do shew forth the power and skill of the workman, and not his image; so it is of the works of God; which do shew the omnipotency and wisdom of the maker, but not his image: and therefore therein the heathen opinion differeth from the sacred truth; for they supposed the world to be the image of God, and man to be an extract or compendious image of the world; but the Scriptures never vouchsafe to attribute to the world that honour, as to be the image of God, but only *the work of his hands*; neither do they speak of any other image of God, but man. Wherefore by the contemplation of nature to induce

54 Discorsi, iii. 1. 55 The translation says *in physics*, omitting the word *religion*.
56 So ed. 1633. The original and the ed. 1629 have *which*.
57 Some other instances are added in the translation. 58 De Aug. iii. 2.

and inforce the acknowledgement of God, and to demonstrate his power, providence, and goodness, is an excellent argument, and hath been excellently handled by divers. But on the other side, out of the contemplation of nature, or ground of human knowledges, to induce any verity or persuasion concerning the points of faith, is in my judgment not safe: *Da fidei quæ fidei sunt* [give unto Faith that which is Faith's]. For the Heathen themselves conclude as much in that excellent and divine fable of the golden chain: *That men and gods were not able to draw Jupiter down to the earth; but contrariwise, Jupiter was able to draw them up to heaven.* So as we ought not to attempt to draw down or submit the mysteries of God to our reason; but contrariwise to raise and advance our reason to the divine truth. So as in this part of knowledge touching divine philosophy, I am so far from noting any deficience, as I rather note an excess: whereunto I have digressed, because of the extreme prejudice which both religion and philosophy have received and may receive by being commixed together; as that which undoubtedly will make an heretical religion, and an imaginary and fabulous philosophy.

Otherwise it is of the nature of angels and spirits, which is an appendix of theology both divine and natural, and is neither inscrutable nor interdicted; for although the Scripture saith, *Let no man deceive you in sublime discourse touching the worship of angels, pressing into that he knoweth not,* etc. yet notwithstanding if you observe well that precept, it may appear thereby that there be two things only forbidden, adoration of them, and opinion fantastical of them; either to extol them further than appertaineth to the degree of a creature, or to extol a man's knowledge of them further than he hath ground. But the sober and grounded inquiry which may arise out of the passages of holy Scriptures, or out of the gradations of nature, is not restrained. So of degenerate and revolted spirits, the conversing with them or the employment of them is prohibited, much more any veneration towards them. But the contemplation or science of their nature, their power, their illusions, either by Scripture or reason, is a part of spiritual wisdom. For so the apostle said, *We are not ignorant of his stratagems*; and it is no more unlawful to inquire the nature of evil spirits than to enquire the force of poisons in nature, or the nature of sin and vice in morality. But this part touching angels and spirits, I cannot note as deficient, for many have occupied themselves in it; I may rather challenge it, in many of the writers thereof, as fabulous and fantastical.

¶ [59] Leaving therefore Divine Philosophy or Natural Theology (not Divinity or Inspired Theology, which we reserve for the last of all, as the haven and sabbath of all man's contemplations), we will now proceed to Natural Philosophy. If then it be true that Democritus said, *That the truth of nature lieth hid in certain deep mines and caves*; and if it be true likewise that the Alchemists do so much inculcate, that Vulcan is a second nature, and imitateth that dexterously and compendiously which nature worketh by ambages and length of time; it were good to divide natural philosophy into the mine and the furnace, and to make two professions or occupations of natural philosophers, some to be pioneers and some smiths; some to dig, and some to refine and hammer. And surely I do best allow of a division of that kind, though in more familiar and scholastical terms; namely, that these be the two parts of natural philosophy, — the Inquisition of Causes, and the Production of Effects; Speculative, and Operative; Natural Science, and Natural Prudence. For as in civil matters there is a wisdom of discourse and a wisdom of direction; so is it in natural. And here I will make a request, that for the latter (or at least for a part thereof) I may revive and reintegrate the misapplied and abused name of Natural Magic; which in the true sense is but Natural Wisdom, or Natural Prudence; taken according to the ancient acception, purged from vanity and superstition [60]. Now although it be true, and I know it well, that there is an intercourse between Causes and Effects, so as both these knowledges, Speculative and Operative, have a great connexion between themselves; yet because all true and fruitful Natural Philosophy hath a double scale or ladder, ascendent and descendent; ascending from

[59] De Aug. iii. 3.

[60] This request is omitted in the translation.

experiments to the invention of causes, and descending from causes to the invention of new experiments; therefore I judge it most requisite that these two parts be severally considered and handled.

¶ [61] Natural Science or Theory is divided into Physic and Metaphysic: wherein I desire it may be conceived that I use the word Metaphysic in a differing sense from that that is received: and in like manner I doubt not but it will easily appear to men of judgment that in this and other particulars, wheresoever my conception and notion may differ from the ancient, yet I am studious to keep the ancient terms. For hoping well to deliver myself from mistaking by the order and perspicuous expressing of that I do propound, I am otherwise zealous and affectionate to recede as little from antiquity, either in terms or opinions, as may stand with truth and the proficience of knowledge. And herein I cannot a little marvel at the philosopher Aristotle, that did proceed in such a spirit of difference and contradiction towards all antiquity; undertaking not only to frame new words of science at pleasure, but to confound and extinguish all ancient wisdom; insomuch as he never nameth or mentioneth an ancient author or opinion, but to confute and reprove; wherein for glory, and drawing followers and disciples, he took the right course. For certainly there cometh to pass and hath place in human truth, that which was noted and pronounced in the highest truth: *Veni in nomine Patris, nec recipitis me; si quis venerit in nomine suo, eum recipietis* [I have come in my Father's name, and ye receive me not; if one come in his own name, him ye will receive]. But in this divine aphorism (considering to whom it was applied, namely to Antichrist, the highest deceiver), we may discern well that *the coming in a man's own name*, without regard of antiquity or paternity, is no good sign of truth; although it be joined with the fortune and success of an *Eum recipietis*. But for this excellent person [62] Aristotle, I will think of him that he learned that humour of his scholar, with whom it seemeth he did emulate, the one to conquer opinions as the other to conquer all nations. Wherein nevertheless, it may be, he may at some men's hands that are of a bitter disposition get a like title as his scholar did;

> Felix terrarum prædo, non utile mundo
> Editus exemplum, etc.

[a fortunate robber, who made prize of nations]; so

> Felix doctrinæ prædo

[a fortunate robber, who made prize of learning]. But to me on the other side that do desire, as much as lieth in my pen, to ground a sociable intercourse [63] between antiquity and proficience, it seemeth best to keep way with antiquity *usque ad aras* [as far as may be without violating higher obligations]; and therefore to retain the ancient terms, though I sometimes alter the uses and definitions; according to the moderate proceeding in civil government, where although there be some alteration, yet that holdeth which Tacitus wisely noteth, *eadem magistratuum vocabula* [the name of the magistracies are not changed].

To return therefore to the use and acception of the term Metaphysic, as I do now understand the word: It appeareth by that which hath been already said, that I intend Philosophia Prima, Summary Philosophy, and Metaphysic, which heretofore have been confounded as one, to be two distinct things. For the one I have made as a parent or common ancestor to all knowledge, and the other I have now brought in as a branch or descendent of Natural Science. It appeareth likewise that I have assigned to Summary Philosophy the common principles and axioms which are promiscuous and indifferent to several sciences. I have assigned unto it likewise the inquiry *touching the operation of the relative and adventive characters of essences, as Quantity, Similitude, Diversity, Possibility*, and the rest; with this distinction and provision; that they be handled as they have efficacy in nature, and not logically. It appeareth likewise that Natural

[61] De Aug. iii. 4. [62] *viro tam eximio certe, et ob acumen ingenii mirabili.*—De Aug.

[63] *entercourse* in the original,—the form of the word commonly used by Bacon.

Theology, which heretofore hath been handled confusedly with Metaphysic, I have inclosed and bounded by itself. It is therefore now a question, what is left remaining for Metaphysic; wherein I may without prejudice preserve thus much of the conceit of antiquity, that Physic should contemplate that which is inherent in matter and therefore transitory, and Metaphysic that which is abstracted and fixed. And again that Physic should handle that which supposeth in nature only a being and moving [64], and Metaphysic should handle that which supposeth further in nature a reason, understanding, and platform [65]. But the difference, perspicuously expressed, is most familiar and sensible. For as we divided Natural Philosophy in general into the Inquiry of Causes and Productions of Effects; so that part which concerneth the Inquiry of Causes we do subdivide, according to the received and sound division of Causes; the one part, which is Physic, enquireth and handleth the Material and Efficient Causes; and the other, which is Metaphysic, handleth the Formal and Final Causes.

Physic (taking it according to the derivation, and not according to our idiom for Medicine,) is situate in a middle term or distance between Natural History and Metaphysic. For Natural History describeth the *variety of things*; Physic the causes, but *variable or respective causes*; and Metaphysic, the *fixed and constant causes.*

> Limus ut hic durescit, et hæc ut cera liquescit,
> Uno eodemque igni
> [As the same fire which makes the soft clay hard
> Makes hard wax soft]:

Fire is the cause of induration, but respective to clay; fire is the cause of colliquation, but respective to wax; but fire is no constant cause either of induration or colliquation. So then the physical causes are but the efficient and the matter. Physic hath three parts; whereof two respect nature *united* or *collected*, the third contemplateth nature *diffused* or *distributed*. Nature is collected either into one entire *total*, or else into the same *principles* or *seeds*. So as the first doctrine is touching the Contexture or Configuration of things, as *de mundo, de universitate rerum*. The second is the doctrine concerning the Principles or Originals of things. The third is the doctrine concerning all Variety and Particularity of things, whether it be of the differing substances, or their differing qualities and natures; whereof there needeth no enumeration, this part being but as a gloss or paraphrase, that attendeth upon the text of Natural History [66]. Of these three I cannot report any as deficient. In what truth or perfection they are handled, I make not now any judgment: but they are parts of knowledge not deserted by the labour of man.

For Metaphysic, we have assigned unto it the inquiry of Formal and Final Causes; which assignation, as to the former of them, may seem to be nugatory and void, because of the received and inveterate opinion that the inquisition of man is not competent to find out *essential forms* or *true differences*; of which opinion we will take this hold; that the invention of Forms is of all other parts of knowledge the worthiest to be sought, if it be possible to be found. As for the possibility, they are ill discoverers that think there is no land when they can see nothing but sea. But it is manifest that Plato in his opinion of Ideas, as one that had a wit of elevation situate as upon a cliff, did descry *that forms were the true object of knowledge*; but lost the real fruit of his opinion, by considering of forms as absolutely abstracted from matter, and not confined and determined by matter; and so turning his opinion upon Theology, wherewith all his natural philosophy is infected. But if any man shall keep a continual watchful and severe eye upon action, operation, and the use of knowledge, he may advise and take notice what are the Forms, the disclosures whereof are fruitful and important to the state of man. For as to the Forms of substances—Man only except, of whom it is said, *Formavit hominem de limo terræ, et spiravit in faciem ejus spiraculum*

[64] The translation adds "and natural necessity."

[65] *ideam.* [66] On this branch of the subject there is a large addition of ten or twelve pages in the *De Augmentis.*

vitæ [He formed man of the dust of the ground, and breathed into his nostrils the breath of life], and not as of all other creatures, *Producant aquæ, producat terra* [let the waters bring forth, let the earth bring forth],—the Forms of Substances I say (as they are now by compounding and transplanting multiplied) are so perplexed, as they are not to be enquired [67] ; no more than it were either possible or to purpose to seek in gross *the forms of those sounds which make words*, which by composition and transposition of letters are infinite. But on the other side, to enquire *the form of those sounds or voices which make simple letters* is easily comprehensible, and being known, induceth and manifesteth the forms of all words, which consist and are compounded of them. In the same manner to enquire the Form of a lion, of an oak, of gold, nay of water, of air, is a vain pursuit : but to enquire the Forms of sense, of voluntary motion, of vegetation, of colours, of gravity and levity, of density, of tenuity, of heat, of cold, and all other natures and qualities, which like an alphabet are not many, and of which the essences (upheld by matter) of all creatures do consist ; to enquire I say the *true forms* of these, is that part of Metaphysic which we now define of. Not but that Physic doth make inquiry and take consideration of the same natures : but how ? Only as to the Material and Efficient Causes of them, and not as to the Forms. For example ; if the cause of Whiteness in snow or froth be enquired, and it be rendered thus, *that the subtile intermixture of air and water is the cause*, it is well rendered ; but nevertheless, is this the Form of Whiteness ? No ; but it is the Efficient, which is ever but *vehiculum formæ* [the carrier of the Form] [68]. This part of Metaphysic I do not find laboured and performed ; whereat I marvel not, because I hold it not possible to be invented by that course of invention which hath been used ; in regard that men (which is the root of all error) have made too untimely a departure and too remote a recess from particulars.

Metaphysica, sive De Formis et Finibus Rerum.

But the use of this part of Metaphysic which I report as deficient, is of the rest the most excellent in two respects ; the one, because it is the duty and virtue of all knowledge to abridge the infinity of individual experience as much as the conception of truth will permit, and to remedy the complaint of *vita brevis, ars longa* [life is short and art is long] ; which is performed by uniting the notions and conceptions of sciences [69]. For knowledges are as pyramides, whereof history is the basis : so of Natural Philosophy the basis is Natural History ; the stage next the basis is Physic ; the stage next the vertical point is Metaphysic. As for the vertical point, *Opus quod operatur Deus à principio usque ad finem* [the work which God worketh from the beginning to the end], the Summary Law of Nature, we know not whether man's inquiry can attain unto it. But these three be the true *stages* of knowledge ; and are to them that are depraved no better than the giants' hills [Pelion, Ossa, and Olympus, piled upon each other],

> Ter sunt conati imponere Pelio Ossam,
> Scilicet atque Ossæ frondosum involvere Olympum :

but to those which refer all things to the glory of God, they are as the three acclamations, *Sancte, sancte, sancte ;* holy in the description or dilatation of his works, holy in the connection or concatenation of them, and holy in the union of them in a perpetual and uniform law. And therefore the speculation was excellent in Parmenides and Plato, although but a speculation in them, That all things by scale did ascend to unity. So then always that knowledge is worthiest, which is charged with least multiplicity ; which appeareth to be Metaphysic ; as that which considereth the Simple Forms of Differences of things, which are few in number, and the degrees and co-ordinations whereof make all this variety. The second respect which valueth and commendeth this part of Metaphysic, is that it

[67] Or at least (adds the translation) the enquiry must be put off till forms of simpler nature have been discovered.

[68] A sentence is added here in the translation ; see note on *Valerius Terminus*, c. 11.

[69] *i.e.* collecting them into axioms more general, applicable to all the individual varieties : (*axiomata scientiarum in magis generalia, et quæ omni materiæ rerum individuarum competant, colligendo et uniendo*).

doth enfranchise the power of man unto the greatest liberty and possibility of works and effects. For Physic carrieth men in narrow and restrained ways, subject to many accidents of impediments, imitating the ordinary flexuous courses of nature; but *latæ undique sunt sapientibus viæ*: to sapience (which was anciently defined to be *rerum divinarum et humanarum scientia* [the knowledge of things human and divine]), there is ever choice of means. For physical causes give light to new invention *in simili materia*; but whosoever knoweth any *form* knoweth the utmost possibility of *superinducing that nature upon any variety of matter*, and so is less restrained in operation, either to the basis of the Matter, or the condition of the Efficient: which kind of knowledge Salomon likewise, though in a more divine sense, elegantly described: *Non arctabuntur gressus tui, et currens non habebis offendiculum* [thy steps shall not be straitened; thou shalt run and not stumble]. The ways of sapience are not much liable either to particularity or chance [70].

The second part of Metaphysic is the inquiry of *final* causes, which I am moved to report not as omitted, but as misplaced [71]. And yet if it were but a fault in order, I would not speak of it; for order is matter of illustration, but pertaineth not to the substance of sciences: but this misplacing hath caused a deficience, or at least a great inproficience in the sciences themselves. For the handling of final causes mixed with the rest in physical inquiries, hath intercepted the severe and diligent inquiry of all real and physical causes, and given men the occasion to stay upon these satisfactory and specious causes, to the great arrest and prejudice of further discovery. For this I find done not only by Plato, who ever anchoreth upon that shore, but by Aristotle, Galen, and others, which do usually likewise fall upon these flats of *discoursing causes*. For to say that *the hairs of the eye-lids are for a quickset and fence about the sight*; or that *the firmness of the skins and hides of living creatures is to defend them from the extremities of heat or cold*; or that *the bones are for the columns or beams, whereupon the frames of the bodies of living creatures are built*; or that *the leaves of trees are for protecting of the fruit*; or that *the clouds are for watering of the earth*; or that *the solidness of the earth is for the station and mansion of living creatures*, and the like, is well enquired and collected in Metaphysic; but in Physic they are impertinent. Nay, they are indeed but remoras and hinderances to stay and slug the ship from further sailing, and have brought this to pass, that the search of the Physical Causes hath been neglected and passed in silence. And therefore the natural philosophy of Democritus and some others, who did not suppose a mind or reason in the frame of things, but attributed *the form thereof able to maintain itself to infinite essays or proofs of nature*, which they term *fortune*, seemeth to me (as far as I can judge by the recital and fragments which remain unto us) in particularities of physical causes more real and better enquired than that of Aristotle and Plato; whereof both intermingled final causes, the one as a part of theology, and the other as a part of logic, which were the favourite studies respectively of both those persons. Not because those final causes are not true, and worthy to be enquired, being kept within their own province; but because their excursions into the limits of physical causes hath bred a vastness and solitude in that track. For otherwise keeping their precincts and borders, men are extremely deceived if they think there is an enmity or repugnancy at all between them. For the cause rendered, that *the hairs about the eye-lids are for the safeguard of the sight*, doth not impugn the cause rendered, that *pilosity is incident to orifices of moisture*; *Muscosi fontes* [the mossy springs], etc. Nor the cause rendered, that *the firmness of hides is for the armour of the body again against extremities of heat or cold*, doth not impugn the cause rendered, that *contraction of pores is incident to the outwardest parts, in regard of their adjacence to foreign or unlike bodies*; and so of the rest: both causes being true and compatible, the one declaring an intention, the other a consequence only. Neither doth this call in question or derogate from divine providence, but highly confirm and

[70] *i.e.* neither confined to particular methods, nor liable to be defeated by accidental obstructions. (*Nec angustiis nec obicibus obnoxias esse.*)

[71] *i.e.* placed in the department of Physic instead of Metaphysic. (*Solent enim inquiri inter Physica, non inter Metaphysica*).

exalt it. For as in civil actions he is the greater and deeper politique, that can make other men the instruments of his will and ends and yet never acquaint them with his purpose, so as they shall do it and yet not know what they do, than he that imparteth his meaning to those he employeth ; so is the wisdom of God more admirable, when nature intendeth one thing and providence draweth forth another, than if he had communicated to particular creatures and motions the characters and impressions of his providence. And thus much for Metaphysic ; the later part whereof I allow as extant, but wish it confined to its proper place.

¶ [72] Nevertheless there remaineth yet another part of Natural Philosophy, which is commonly made a principal part, and holdeth rank with Physic special and Metaphysic ; which is Mathematic ; but I think it more agreeable to the nature of things and to the light of order to place it as a branch of Metaphysic ; for the subject of it being Quantity, not Quantity indefinite, which is but a relative and belongeth to *philosophia prima* (as hath been said), but Quantity determined or proportionable, it appeareth to be one of the Essential Forms of things ; as that that is causative in nature of a number of effects ; insomuch as we see in the schools both of Democritus and of Pythagoras, that the one *did ascribe figure to the first seeds of things*, and the other *did suppose numbers to be the principles and originals of things* ; and it is true also that of all other forms (as we understand forms) it is the most abstracted and separable from matter, and therefore most proper to Metaphysic ; which hath likewise been the cause why it hath been better laboured and enquired than any of the other forms, which are more immersed into matter. For it being the nature of the mind of man (to the extreme prejudice of knowledge) to delight in the spacious liberty of generalities, as in a champion * region, and not in the inclosures of particularity ; the Mathematics of all other knowledge were the goodliest fields to satisfy that appetite. But for the placing of this science, it is not much material [73] : only we have endeavoured in these our partitions to observe a kind of perspective, that one part may cast light upon another.

The Mathematics are either Pure or Mixed. To the Pure Mathematics are those sciences belonging which handle Quantity Determinate, merely severed from any axioms of natural philosophy ; and these are two, Geometry and Arithmetic ; the one handling Quantity continued, and the other dissevered. Mixed hath for subject some axioms or parts of natural philosophy, and considereth Quantity determined, as it is auxiliary and incident unto them. For many parts of nature can neither be invented with sufficient subtilty nor demonstrated with sufficient perspicuity nor accommodated unto use with sufficient dexterity, without the aid and intervening of the Mathematics : of which sort are Perspective, Music, Astronomy, Cosmography, Architecture, Enginery, and divers others. In the Mathematics I can report no deficience, except it be that men do not sufficiently understand the excellent use of the Pure Mathematics, in that they do remedy and cure many defects in the wit and faculties intellectual. For if the wit be too dull, they sharpen it ; if too wandering, they fix it ; if too inherent in the sense, they abstract it. So that as tennis is a game of no use in itself, but of great use in respect it maketh a quick eye and a body ready to put itself into all postures ; so in the Mathematics, that use which is collateral and intervenient is no less worthy than that which is principal and intended [74]. And as for the Mixed Ma-

[72] De Aug. iii. 6. Observe that in translating this part of the work Bacon has not only made great additions, but changed the order. * [*i.e.* "champain"].

[73] In the *De Augmentis* he concludes by placing it as an appendix and auxiliary to Natural Philosophy, in order to mark more distinctly its proper function ; which he complains that the mathematicians are apt to forget, and to exalt it, as the logicians exalt logic, above the sciences which it is its business to serve.

[74] The whole of this passage relating to the use of pure mathematics in the training of the intellect is omitted in the translation ; and the omission has been represented as indicating a change in Bacon's opinion either as to the value of this particular study in that respect, or as to the expediency of encouraging any study which is "useful only to the mind" of the student. This conjecture is hardly reconcilable however with the fact that the same recommendation of mathematics as a cure for certain defects of the

thematics, I may only make this prediction, that there cannot fail to be more kinds of them, as nature grows further disclosed. Thus much of Natural Science or the part of Nature Speculative.

¶ [75] For Natural Prudence, or the part Operative of Natural Philosophy, we will divide it into three parts, Experimental, Philosophical, and Magical,[76] which three parts active have a correspondence and analogy with the three parts Speculative Natural History, Physic, and Metaphysic. For many operations have been invented, sometimes by a casual incidence and occurrence, sometimes by a purposed experiment; and of those which have been found by an intentional experiment, some have been found out by varying or extending the same experiment, some by transferring and compounding divers experiments the one into the other, which kind of invention an empiric may manage [77]. Again, by the knowledge of physical causes there cannot fail to follow many indications and designations of new particulars, if men in their speculation will keep one eye upon use and practice. But these are but coastings along the shore, *premendo littus iniquum*; for it seemeth to me there can hardly be discovered any radical or fundamental alterations and innovations in nature, either by the fortune and essays of experiments, or by the light and direction of physical causes. If therefore we have reported Metaphysic deficient, it must follow that we do the like of Natural Magic, which hath relation thereunto. For as for the Natural Magic whereof now there is mention in books, containing certain credulous and superstitious conceits and observations of Sympathies and Antipathies and hidden proprieties, and some frivolous experiments, strange rather by disguisement than in themselves; it is as far differing in truth of nature from such a knowledge as we require, as the story of King Arthur of Britain, or Hugh of Bourdeaux, differs from Cæsar's commentaries in truth of story. For it is manifest that Cæsar did greater things *de vero* than those imaginary heroes were feigned to do. But he did them not in that fabulous manner. Of this kind of learning the fable of Ixion was a figure, who designed to enjoy Juno, the goddess of power; and instead of her had copulation with a cloud, of which mixture were begotten centaurs and chimeras. So whosoever shall entertain high and vaporous imaginations instead of a laborious and sober inquiry of truth, shall beget hopes and beliefs of strange and impossible shapes. And therefore we may note in these sciences which hold so much of imagination and belief, as this degenerate Natural Magic, Alchemy, Astrology, and the like, that in their propositions the description of the means is ever more monstrous than the pretence or end. For it is a thing more probable, that he that knoweth well the natures of Weight, of Colour, of Pliant and Fragile in respect of the hammer, of Volatile and Fixed in respect of the fire, and the rest, may superinduce upon some metal the nature and form of gold by such mechanique as belongeth to the production of the natures afore rehearsed, than that some

Naturalis Magia, sive Physica Operativa Major.

ntellect is repeated both in a later chapter of the *De Augmentis* (vi. 4.; to which place indeed the observation properly belongs), and in the *Essay on Studies* as published in 1625. Nor is there any difficulty in accounting for the omission of it here. When Bacon wrote the *Advancement* in 1605, he had no deficiency to report in the department of Mathematics; he could not name any branch of the study which had not been properly pursued, and merely took the opportunity of observing by the way that the study of the pure mathematics had a collateral and incidental value as an instrument of education: an observation very good and just in itself, but not at all to the purpose of the argument. When he revised the work in 1622 he knew more about mathematics, and was able to point out certain deficiencies which were very much to the purpose,—especially as to the doctrine of Solids in Geometry and of Series in Arithmetic; and in introducing a relevant observation he naturally struck out the irrelevant one.

[75] De Aug. iii. 5.

[76] In the translation the name *Natural Prudence* is omitted; the *part operative* is divided into two parts, instead of three; viz. Mechanic and Magic; and the whole exposition is much altered and enlarged.

[77] Being a matter of ingenuity and sagacity, rather than philosophy (*quæ magis ingeniosa res est et sagax, quam philosophica*). This is in fact the *Experientia Literata* of which we hear more further on.

grains of the medicine projected should in a few moments of time turn a sea of quicksilver or other material into gold. So it is more probable, that he that knoweth the nature of arefaction, the nature of assimilation of nourishment to the thing nourished, the manner of increase and clearing of spirits, the manner of the depredations which spirits make upon the humours and solid parts, shall by ambages of diets, bathings, anointings, medicines, motions, and the like, prolong life or restore some degree of youth or vivacity, than that it can be done with the use of a few drops or scruples of a liquor or receit. To conclude therefore, the true Natural Magic, which is that great liberty and latitude of operation which dependeth upon the knowledge of Forms, I may report deficient, as the relative thereof is. To which part, if we be serious and incline not to vanities and plausible discourse, besides the deriving and deducing the operations themselves from Metaphysic, there are pertinent two points of much purpose, the one by way of preparation, the other by way of caution. The first is, that there be made a *Calendar resembling an inventory* [78] of the estate of man, containing all the inventions (being the works or fruits of nature or art) which are now extant and whereof man is already possessed; out of which doth naturally result a note, what things are yet held impossible, or not invented; which calendar will be the more artificial and serviceable, if to every reputed impossibility you add what thing is extant which cometh the neareth in degree to that impossibility; to the end that by these optatives and potentials man's inquiry may be the more awake in deducing direction of works from the speculation of causes. And secondly, that those experiments be not only esteemed which have an immediate and present use, but those principally which are of most universal consequence for invention of other experiments, and those which give most light to the invention of causes; for the invention of the mariner's needle, which giveth the direction, is of no less benefit for navigation than the invention of the sails, which give the motion [79].

Inventarium Opum humanarum.

[80] Thus have I passed through Natural Philosophy, and the deficiences thereof; wherein if I have differed from the ancient and received doctrines, and thereby shall move contradiction; for my part, as I affect not to dissent, so I purpose not to contend. If it be truth,

> Non canimus surdis, respondent omnia sylvæ
> [All as we sing the listening woods reply]:

the voice of nature will consent, whether the voice of man do or no. And as Alexander Borgia was wont to say of the expedition of the French for Naples, that they came with chalk in their hands to mark up their lodgings, and not with weapons to fight; so I like better that entry of truth which cometh peaceably with chalk to mark up their minds which are capable to lodge and harbour it, than that which cometh with pugnacity and contention.

[81] But there remaineth a division of Natural Philosophy according to the *report of the inquiry*, and nothing concerning the matter or subject; and that is positive and Considerative; when the inquiry reporteth either an Assertion or a Doubt. These doubts or *non liquets* are of two sorts, Particular and Total. For the first, we see a good example thereof in Aristotle's Problems, which deserved to have had a better continuance, but so nevertheless as there is one point whereof warning is to be given and taken. The registering of doubts hath two excellent uses: the one, that it saveth philosophy from errors and falsehoods; when that which

[78] This is the Inventory which (as I think) was to be contained in the tenth chapter of the *Valerius Terminus*. See my note on Mr. Ellis's preface.

[79] This example is omitted in the translation, to make room for a better (with which Bacon was probably not acquainted in 1605)—the artificial congelation of water; an experiment which he especially valued as giving light to the secret process of condensation.

[80] The passage corresponding to this paragraph concludes the third book of the *De Augmentis*. That which follows is transferred to the middle of the fourth chapter.

[81] The substance of this paragraph will be found in the middle of the fourth chapter of the third book of the *De Augmentis*.

is not fully appearing is not collected into assertion, whereby error might draw error, but reserved in doubt: the other, that the entry of doubts are as so many suckers or spunges to draw use [82] of knowledge; insomuch as that which if doubts had not preceded a man should never have advised but passed it over without note, by the suggestion and solicitation of doubts is made to be attended and applied. But both these commodities do scarcely countervail an inconvenience which will intrude itself, if it be not debarred; which is, that when a doubt is once received men labour rather how to keep it a doubt still than how to solve it, and accordingly bend their wits [83]. Of this we see the familiar example in lawyers and scholars, both which if they have once admitted a doubt, it goeth ever after authorized for a doubt. But that use of wit and knowledge is to be allowed which laboureth to make doubtful things certain, and not those which labour to make certain things doubtful. Therefore these *calendars of doubts* I commend as excellent things, so that there be this caution used, that when they be throughly sifted and brought to resolution, they be from thenceforth omitted, decarded,* and not continued to cherish and encourage men in doubting. To which calendar of doubts or problems I advise be annexed another calendar, as much or more material, which is a calendar of popular error: I mean chiefly in natural history,[84] such as pass in speech and conceit, and are nevertheless apparently detected and convicted of untruth; that man's knowledge be not weakened nor imbased by such dross and vanity. As for the *doubts or non liquets general or total*, I understand those differences of opinions touching the principles of nature and the fundamental points of the same, which have caused the diversity of sects, schools, and philosophies; as that of Empedocles, Pythagoras, Democritus, Parmenides, and the rest [85]. For although Aristotle, as though he had been of the race of the Ottomans, thought he could not reign except the first thing he did he killed all his brethren; yet to those that seek truth and not magistrality, it cannot but seem a matter of great profit to see before them the several opinions touching the foundations of nature; not for any exact truth that can be expected in those theories; for as the same phænomena in astronomy are satisfied by the received astronomy of the diurnal motion and the proper motions of the planets with their eccentrics and epicycles, and likewise by the theory of Copernicus who supposed the earth to move; and the calculations are indifferently agreeable to both; so the ordinary face and view of experience is many times satisfied by several theories and philosophies; whereas to find the real truth requireth another manner of severity and attention. For as Aristotle saith that children at the first will call every woman mother, but afterwards they come to distinguish according to truth; so experience, if it be in childhood, will call every philosophy mother, but when it cometh to ripeness it will discern the true mother. So as in the mean time it is good to see the several glosses and opinions upon nature, whereof it may be every one in some one point hath seen clearer than his fellows. Therefore I wish some collection to be made painfully and understandingly *de antiquis philosophiis*, out of all the possible light which remaineth to us of them [86]. Which kind of work I find deficient. But here I must give warning, that it be done distinctly and severely [87]; the philosophies of every one throughout by themselves; and not by titles packed and faggoted up together, as hath been done by Plutarch. For it is the harmony

Continuatio Problematum in Natura.

Catalogus Falsitatum grassantium in historia Naturæ.

De Antiquis Philosophiis.

[82] *i.e.* increase. (*quæ incrementa scientiæ perpetuo ad se sugant et alliciant*).

[83] This is explained in the translation by adding that the recognition of the doubt has the effect of raising champions to maintain each side, and so keeping it up.

* [*i.e.* discarded]. [84] *vel in Historia Naturali*, vel in Dogmatibus.—*De Aug.*

[85] In the translation *Empedocles* is omitted; and *Philolaus, Xenophanes, Anaxagoras, Leucippus*, added.

[86] Such (according to the translation) as the Lives of the ancient Philosophers, Plutarch's collection of placita, Plato's quotations, Aristotle's confutations, and the scattered notices in Lactantius, Philo, Philostratus, &c.

[87] So both in the original and in ed. 1633; perhaps a misprint for "severally." Ed. 1629 has *severedly*. The translation has *distinctè* only.

of a philosophy in itself which giveth it light and credence ; whereas if it be singled and broken, it will seem more foreign and dissonant. For as when I read in Tacitus the actions of Nero or Claudius, with circumstances of times, inducements, and occasions, I find them not so strange ; but when I read them in Suetonius Tranquillus gathered into titles and bundles, and not in order of time, they seem more monstrous and incredible ; so is it of any philosophy reported entire, and dismembered by articles. Neither do I exclude opinions of latter times to be likewise represented in this calendar of sects of philosophy, as that of Theophrastus Paracelsus, eloquently reduced into an harmony by the pen of Severinus the Dane ; and that of Telesius, and his scholar Donius, being as a pastoral philosophy, full of sense but of no great depth ; and that of Fracastorius, who though he pretended not to make any new philosophy, yet did use the absoluteness of his own sense upon the old ; and that of Gilbertus our countryman, who revived, with some alterations and demonstrations, the opinions of Xenophanes [88] ; and any other worthy to be admitted.

Thus have we now dealt with two of the three beams of man's knowledge ; that is *Radius Directus*, which is referred to nature, *Radius Refractus*, which is referred to God, and cannot report truly because of the inequality of the medium. There resteth *Radius Reflexus* whereby Man beholdeth and contemplateth himself.

¶ [89] We come therefore now to that knowledge whereunto the ancient oracle directeth us, which is the *knowledge of ourselves ;* which deserveth the more accurate handling, by how much it toucheth us more nearly. This knowledge, as it is the end and term of natural philosophy in the intention of man, so notwithstanding it is but a portion of natural philosophy in the continent of nature. And generally let this be a rule, that all partitions of knowledges be accepted rather for lines and veins, than for sections and separations ; and that the continuance and entireness of knowledge be preserved. For the contrary hereof hath made particular sciences to become barren, shallow, and erroneous ; while they have not been nourished and maintained from the common fountain. So we see Cicero the orator complained of Socrates and his school, that he was the first that separated philosophy and rhetoric ; whereupon rhetoric became an empty and verbal art. So we may see that the opinion of Copernicus touching the rotation of the earth [90], which astronomy itself cannot correct because it is not repugnant to any of the phænomena, yet natural philosophy may correct. So we see also that the science of medicine, if it be destituted and forsaken by natural philosophy, it is not much better than an empirical practice. With this reservation therefore we proceed to Human Philosophy or Humanity, which hath two parts : the one considereth man segregate, or distributively ; the other congregate, or in society. So as Human Philosophy is either Simple and Particular, or Conjugate and Civil. Humanity Particular consisteth of the same parts whereof man consisteth ; that is, of knowledges which respect the Body, and of knowledges that respect the Mind. But before we distribute so far, it is good to constitute. For I do take the consideration in general and at large of Human Nature to be fit to be emancipate and made a knowledge by itself ; not so much in regard of those delightful and elegant discourses which have been made of the dignity of man [91], of

[88] This passage is considerably altered in the translation, and the differences are worth noticing as bearing upon the course of Bacon's reading and the development of his views in the interval. After the notice of Paracelsus the translation proceeds " or of Telesius of Consentium, *who revived the philosophy of Parmenides and so turned the arms of the Peripatetics against themselves* ; or of *Patricius the Venetian, who sublimated the fumes of the Platonists ;* or of our countryman Gilbert, who set up again the *doctrines of Philolaus* ". The names of *Donius, Fracastorius,* and *Xenophanes* are entirely omitted. I do not know whether Mr. Ellis's attention had been directed to these changes.

[89] De Aug. iv. 1. The whole of this chapter is much altered and enlarged ; re-written rather than translated.

[90] The translation adds, *quæ nunc quoque invaluit.*

[91] In the *De Augmentis* this part is numbered among the Desiderata. The *miseries* of man, he says, have been well set forth both by philosophers and theologians ; but

his miseries, of his state and life, and the like *adjuncts of his common and undivided nature ;* but chiefly in regard of the knowledge concerning the *sympathies and concordances between the mind and body*, which, being mixed, cannot be properly assigned to the sciences of either.

This knowledge hath two branches : for as all leagues and amities consist of mutual Intelligence and mutual Offices, so this league of mind and body hath these two parts ; *how the one discloseth the other*, and *how the one worketh upon the other ;* Discovery, and Impression. The former of these hath begotten two arts, both of Prediction or Prenotion ; whereof the one is honoured with the inquiry of Aristotle, and the other of Hippocrates. And although they have of later time been used to be coupled with superstitious and fantastical arts, yet being purged and restored to their true state, they have both of them a solid ground in nature, and a profitable use in life. The first is Physiognomy, which discovereth the disposition of the mind by the lineaments of the body. The second is the Exposition of Natural Dreams, which discovereth the state of the body by the imaginations of the mind. In the former of these I note a deficience.[92] For Aristotle hath very ingeniously and diligently handled the factures of the body, but not the gestures of the body, which are no less comprehensible by art, and of greater use and advantage. For the Lineaments of the body do disclose the disposition and inclination of the mind in general ; but the Motions of the countenance and parts do not only so, but do further disclose the present humour and state of the mind and will. For as your Majesty saith most aptly and elegantly, *As the tongue speaketh to the ear, so the gesture speaketh to the eye.* And therefore a number of subtile persons, whose eyes do dwell upon the faces and fashions of men, do well know the advantage of this observation, as being most part of their ability ; neither can it be denied but that it is a great discovery of dissimulations, and a great direction in business.

The latter branch, touching Impression, hath not been collected into art, but hath been handled dispersedly ; and it hath the same relation or antistrophe that the former hath. For the consideration is double : Either *how and how far the humours and affects* [93] *of the body do alter or work upon the mind ;* or again, *how and how far the passions or apprehensions of the mind do alter or work upon the body*. The former of these hath been inquired and considered as a part and appendix of Medicine, but much more as a part of Religion or Superstition. For the physician prescribeth cures of the mind in phrensies and melancholy passions ; and pretendeth also to exhibit medicines to exhilarate the mind, to confirm the courage, to clarify the wits, to corroborate the memory, and the like ; but the scruples and superstitions of diet and other regiment of the body in the sect of the Pythagoreans, in the heresy of the Manicheans, and in the law of Mahomet, do exceed. So likewise the ordinances in the Ceremonial Law, interdicting the eating of the blood and the fat, distinguishing between beasts clean and unclean for meat, are many and strict. Nay the faith itself, being clear and serene from all clouds of Ceremony, yet retaineth the use of fastings, abstinences, and other macerations and humiliations of the body, as things real, and not figurative.[94] The root and life of all which prescripts is, (besides the ceremony [95]), the consideration of the dependency which the affections of the mind are submitted unto upon the state and disposition of the body, And if any

of what he calls the *triumphs* of man, (that is, instances of the highest perfection which human faculties, mental or bodily, have exhibited), he wishes a collection to be made from history ; and gives a page or two of anecdotes by way of example.

[92] With regard to the latter, of which nothing more is said here, he observes in the *De Augmentis* that the treatment it has received is full of follies, and not grounded upon the most solid basis,—which is that when the same sensation is produced in the sleeper by an internal cause which is usually the effect of some external act, he will dream of that act ; as in the case of nightmare, where the sensation of oppression on the stomach created by the fumes of indigestion makes a man dream that his body is oppressed by a weight superimposed.

[93] *temperamentum.*—De Aug.

[94] *tanquam rerum non mere ritualium sed etiam fructuosarum.*—De Aug.

[95] The translation adds, " and the exercise of obedience ".

man of weak judgment do conceive that this suffering of the mind from the body doth either question the immortality or derogate from the sovereignty of the soul, he may be taught in easy instances, that the infant in the mother's womb is compatible with the mother[96] and yet separable; and the most absolute monarch is sometimes led by his servants and yet without subjection. As for the reciprocal knowledge, which is the operation of the conceits and passions of the mind upon the body, we see all wise physicians in the prescriptions of their regiments to their patients do ever consider *accidentia animi*, as of great force to further or hinder remedies or recoveries; and more specially it is an inquiry of great depth and worth concerning Imagination, how and how far it altereth the body proper of the imaginant. For although it hath a manifest power to hurt, it followeth not it hath the same degree of power to help; no more than a man can conclude, that because there be pestilent airs, able suddenly to kill a man in health, therefore there should be sovereign airs, able suddenly to cure a man in sickness. But the inquisition of this part is of great use, though it needeth, as Socrates said, *a Delian diver*, being difficult and profound. But unto all this knowledge *de communi vinculo*, of the concordances between the mind and the body, that part of inquiry is most necessary, which considereth of the *seats* and *domiciles* which the several faculties of the mind do take and occupate in the organs of the body; which knowledge hath been attempted, and is controverted, and deserveth to be much better enquired. For the opinion of Plato, who placed *the understanding in the brain, animosity* (which he did unfitly call *anger*, having a greater mixture with *pride*) *in the heart*, and *concupiscence or sensuality in the liver*, deserveth not to be despised; but much less to be allowed[97]. So then we have constituted (as in our own wish and advice) the inquiry *touching human nature entire*, as a just portion of knowledge to be handled apart.

¶ [98] The knowledge that concerneth man's body is divided as the good of man's body is divided, unto which it referreth. The good of man's body is of four kinds, Health, Beauty, Strength, and Pleasure: so the knowledges are Medicine, or art of Cure; art of Decoration, which is called Cosmetic; art of Activity, which is called Athletic; and art Voluptuary, which Tacitus truly calleth *eruditus luxus*, [educated luxury]. This subject of man's body is of all other things in nature most susceptible of remedy, but then that remedy is most susceptible of error. For the same subtility of the subject doth cause large possibility and easy failing; and therefore the inquiry ought to be more exact.

To speak therefore of Medicine, and to resume that we have said, ascending a little higher: The ancient opinion that man was Microcosmus, an abstract or model of the world, hath been fantastically strained by Paracelsus and the alchemists, as if there were to be found in man's body certain correspondences and parallels, which should have respect to all varieties of things, as stars, planets, minerals, which are extant in the great world. But thus much is evidently true, that of all substances which nature hath produced, man's body is the most extremely compounded. For we see herbs and plants are nourished by earth and water; beasts for the most part by herbs and fruits; man by the flesh of beasts, birds, fishes, herbs, grains, fruits, water, and the manifold alterations, dressings, and preparations of these several bodies, before they come to be his food and aliment. Add hereunto that beasts have a more simple order of life, and less change of affections to work upon their bodies; whereas man in his mansion, sleep, exercise, passions, hath infinite variations; and it cannot be denied but that the Body of man of all other things is of the most compounded mass. The Soul on the other side is the simplest of substances, as is well expressed,

[96] *i.e.* suffers together with the mother: *simul cum matribus affectibus compatitur*.

[97] Neither (he adds in the translation) is that other arrangement free from error, which places the several intellectual faculties, Imagination, Reason, and Memory, in the several ventricles of the brain.

[98] De Aug. iv. 2.

Purumque reliquit
Æthereum sensum atque auraï simplicis ignem:
[Pure and unmixed
The ethereal sense is left—mere air and fire].

So that it is no marvel that the soul so placed enjoys no rest, if that principle be true that *Motus rerum est rapidus extra locum, placidus in loco* [things move rapidly to their place and calmly in their place]. But to the purpose. This variable composition of man's body hath made it as an instrument easy to distemper; and therefore the poets did well to conjoin Music and Medicine in Apollo: because the office of medicine is but to tune this curious harp of man's body and to reduce it to harmony. So then the subject being so variable hath made the art by consequent more conjectural; and the art being conjectural hath made so much the more place to be left for imposture. For almost all other arts and sciences are[99] judged by acts or masterpieces[100] as I may term them, and not by the successes and events. The lawyer is judged by the virtue of his pleading, and not by the issue of the cause. The master in the ship is judged by the directing his course aright, and not by the fortune of the voyage. But the physician, and perhaps the politique, hath no particular acts demonstrative of his ability, but is judged most by the event; which is ever but as it is taken: for who can tell, if a patient die or recover, or if a state be preserved or ruined, whether it be art or accident? And therefore many times the impostor is prized, and the man of virtue taxed. Nay, we see [the[101]] weakness and credulity of men is such, as they will often prefer a montabank[102] or witch before a learned physician. And therefore the poets were clear-sighted in discerning this extreme folly, when they made Æsculapius and Circe brother and sister, both children of the sun, as in the verses,

Ipse repertorem medicinæ talis et artis
Fulmine *Phœbigenam* Stygias detrusit ad undas
[*Apollo's son* from whom that art did grow
Jove struck with thunder to the shades below].

And again,

Dives inaccessos ubi *Solis filia* lucos etc.
[Now by the shelves of Circe's coast they run,—
Circe the rich, the *daughter of the sun*[103]].

For in all times, in the opinion of the multitude, witches and old women and impostors have had a competition with physicians. And what followeth? Even this, that physicians say to themselves, as Salomon expresseth it upon an higher occasion; *If it befal to me as befalleth to the fools, why should I labour to be more wise?* And therefore I cannot much blame physicians, that they use commonly to intend some other art or practice, which they fancy, more than their profession. For you shall have of them antiquaries, poets, humanists, statesmen, merchants, divines, and in every of these better seen than in their profession; and no doubt upon this ground, that they find that mediocrity and excellency in their art maketh no difference in profit or reputation towards their fortune; for the weakness of patients and sweetness of life and nature of hope[104] maketh men depend upon physicians with all their defects. But nevertheless these things which we have spoken of are courses begotten between a little occasion and a great deal of sloth and default; for if we will excite and awake our observation, we shall see in familiar instances what a predominant faculty the *subtilty of spirit*[105] hath over the *variety of matter or form*. Nothing more variable than faces

[99] So edd. 1629 and 1633. The original omits *are*.
[100] *virtute sua et functione.*—De Aug.
[101] *the* omitted both in the original and in edd. 1629 and 1633.
[102] This is the spelling of the old editions; and ought apparently to be revived by those who believe that our orthography is the guardian of our etymologies.
[103] Dryden.
[104] The translation adds *et amicorum commendatio.*
[105] *i.e.* of the understanding: *intellectus subtilitas et acumen.*

and countenances; yet men can bear in memory the infinite distinctions of them; nay, a painter with a few shells of colours, and the benefit of his eye and habit of his imagination, can imitate them all that ever have been, are, or may be, if they were brought before him. Nothing more variable than voices; yet men can likewise discern them personally; nay, you shall have a *buffon* or *pantomimus* will express as many as he pleaseth. Nothing more variable than the differing sounds of words; yet men have found the way to reduce them to a few simple letters. So that it is not *the insufficiency or incapacity of man's mind*, but it is *the remote standing or placing thereof*, that breedeth these mazes and incomprehensions: for as the sense afar off is full of mistaking but is exact at hand, so is it of the understanding; the remedy whereof is not to quicken or strengthen the organ, but to go nearer to the object; and therefore there is no doubt but if the physicians will learn and use the true approaches and avenues of nature, they may assume as much as the poet saith:

> Et quoniam variant morbi, variabimus artes;
> Mille mali species, mille salutis erunt

[varying their arts according to the variety of diseases,—for a thousand forms of sickness a thousand methods of cure]. Which that they should do, the nobleness of their art doth deserve; well shadowed by the poets, in that they made Æsculapius to be the son of the Sun, the one being the fountain of life, the other as the second stream; but infinitely more honoured by the example of our Saviour, who made the body of man the object of his miracles, as the soul was the object of his doctrine. For we read not that ever he vouchsafed to do any miracle about honour, or money (except that one for giving tribute to Cæsar), but only about the preserving, sustaining, and healing the body of man.

Medicine is a science which hath been (as we have said) more professed than laboured, and yet more laboured than advanced: the labour having been, in my judgment, rather in circle than in progression. For I find much iteration, but small addition. It considereth *causes of diseases*, with the *occasions or impulsions*; the *diseases themselves*, with the *accidents*; and the *cures*, with the *preservations*[106]. The deficiencies which I think good to note being a few of many, and those such as are of a more open and manifest nature, I will enumerate, and not place.

Narrationes medicinales.

The first is the discontinuance of the ancient and serious diligence of Hippocrates, which used to set down a narrative of the special cases of his patients, and how they proceeded, and how they were judged by recovery or death. Therefore having an example proper in the father of the art, I shall not need to allege an example foreign, of the wisdom of the lawyers, who are careful to report new cases and decisions for the direction of future judgments. This continuance of Medicinal History I find deficient; which I understand neither to be so infinite as to extend to every common case, nor so reserved as to admit none but wonders: for many things are new in the manner, which are not new in the kind; and if men will intend to observe, they shall find much worthy to observe.

Anatomia comparata.

In the inquiry which is made by Anatomy I find much deficienee: for they inquire of the *parts*, and their *substances*, *figures*, and *collocations*; but they inquire not of the *diversities of the parts*[107], the *secrecies of the passages*, and the *seats or nestling of the humours*, nor much of

[106] Here the translation departs widely from the original. The parts, or offices, into which Medicine is divided in the *De Augmentis* are: 1. the preservation of health; 2. the cure of diseases; 3. the prolongation of life; with regard to the first of which Bacon complains that physicians have treated it in several respects unskilfully or imperfectly; and with regard to the last that they have not recognized the prolongation of natural life as a principal part of their science, being satisfied if they can prevent it from being shortened by diseases. Under the second he includes the whole doctrine of diseases,—the causes, the symptoms, and the remedies, all in fact that is here included under the general head of Medicine,—and so strikes again into the text.

[107] *i.e.* they inquire of the parts, etc., of the human body in general, but not of the

the *footsteps and impressions of diseases*: the reason of which omission I suppose to be, because the first inquiry may be satisfied in the view of one or a few anatomies; but the latter, being comparative and casual, must arise from the view of many. And as to the diversity of parts, there is no doubt but the facture or framing of the inward parts is as full of difference as the outward, and in that is the *cause continent* of many diseases; which not being observed, they quarrel many times with the humours, which are not in fault; the fault being in the very frame and mechanic of the part, which cannot be removed by medicine alterative, but must be accommodate and palliate by diets and medicines familiar. And for the passages and pores, it is true which was anciently noted that the more subtile of them appear not in anatomies, because they are shut and latent in dead bodies, though they be open and manifest in live: which being supposed, though the inhumanity of *anatomia vivorum* [anatomy of the living subject] was by Celsus justly reproved; yet in regard of the [108] great use of this observation, the inquiry needed not by him so slightly to have been relinquished altogether, or referred to the casual practices of surgery; but might have been well diverted upon the dissection of beasts alive, which notwithstanding the dissimilitude of their parts, may sufficiently satisfy this inquiry. And for the humours, they are commonly passed over in anatomies as purgaments; whereas it is most necessary to observe what cavities, nests, and receptacles the humours do find in the parts, with the differing kind of the humour so lodged and received. And as for the footsteps of diseases, and their devastations of the inward parts, imposthumations, exulcerations, discontinuations, putrefactions, comsumptions, contractions, extensions, convulsions, dislocations, obstructions, repletions, together with all preternatural substances, as stones, carnosities, excrescences, worms, and the like; they ought to have been exactly observed by multitude of anatomies and the contribution of men's several experiences, and carefully set down both historically according to the appearances, and artificially with a reference to the diseases and symptoms which resulted from them, in case where the anatomy is of a defunct patient; whereas now upon opening of bodies they are passed over slightly and in silence.

Inquisitio ulterior de Morbis insanabilibus.

In the inquiry of diseases, they do abandon the cures of many, some as in their nature incurable, and others as past the period of cure; so that Sylla and the triumvirs never proscribed so many men to die, as they do by their ignorant edicts; whereof [109] numbers do escape with less difficulty than they did in the Roman proscriptions. Therefore I will not doubt to note as a deficience, that they inquire not the perfect cures of many diseases, or extremities of diseases, but pronouncing them incurable do enact a law of neglect, and exempt ignorance from discredit.

De Euthanasia exteriore.

Nay further, I esteem it the office of a physician not only to restore health, but to mitigate pain and dolors; and not only when such mitigation may conduce to recovery, but when it may serve to make a fair and easy passage: for it is no small felicity which Augustus Cæsar was wont to wish to himself, that same *Euthanasia*; and which was specially noted in the death of Antoninus Pius, whose death was after the fashion and semblance of a kindly and pleasant sleep. So it is written of Epicurus, that after his disease was judged desperate, he drowned his stomach and senses with a large draught and ingurgitation of wine; whereupon the epigram was made, *Hinc stygias ebrius hausit aquas*; he was not sober enough to taste any bitterness of the Stygian water. But the physicians contrariwise do make a kind of scruple and religion to stay with the patient after the disease is deplored; whereas, in my judgment, they ought both to enquire the skill and to give the attendances for the facilitating and assuaging of the pains and agonies of death.

In the consideration of the Cures of diseases, I find a deficience in the receipts of

diversities of the parts in different bodies,—of simple, but not of comparative, anatomy. This whole paragraph is much enlarged in the translation, and the order changed.

108 So edd. 1629 and 1633. The original omits *the*.

109 *i.e.* of whom nevertheless: *quorum tamen plurimi etc.*—De Aug.

Medicinæ Experimentales. propriety respecting the particular cures of diseases [110]: for the physicians have frustrated the fruit of tradition and experience by their magistralities in adding and taking out and changing *quid pro quo* in their receipts, at their pleasures; commanding so over the medicine as the medicine cannot command over the disease. For except it be treacle and mithridatum, and of late *diascordium* [111], and a few more, they tie themselves to no receipts severely and religiously: for as to the confections of sale which are in the shops they are for readiness and not for propriety; for they are upon general intentions of purgings, opening, comforting, altering, and not much appropriate to particular diseases: and this is the cause why empirics and old women are more happy many times in their cures than learned physicians, because they are more religious in holding their medicines. Therefore here is the deficience which I find, that physicians have not, partly out of their own practice, partly out of the constant probations reported in books, and partly out of the traditions of empirics, set down and delivered over certain *experimental medicines* for the cure of particular diseases, besides their own *conjectural* and *magistral descriptions*. For as they were the men of the best composition in the state of Rome, which either being consuls inclined to the people, or being tribunes inclined to the senate; so in the matter we now handle, they be the best physicians, which being learned incline to the traditions of experience, or being empirics incline to the methods of learning.

Imitatio Naturæ in Balneis et Aquis Medicinalibus. In preparation of Medicines, I do find strange, specially considering how mineral medicines have been extolled, and that they are safer for the outward than inward parts, that no man hath sought to make an imitation by art of Natural Baths and Medicinal Fountains, which nevertheless are confessed to receive their virtues from minerals: and not so only, but discerned and distinguished from what particular mineral they receive tincture, as sulphur, vitriol, steel, or the like; which nature if it may be reduced to compositions of art, both the variety of them will be increased, and the temper of them will be more commanded [112].

Filium Medicinale, sive de vicibus Medicinarum. But lest I grow to be more particular than is agreeable either to my intention or to proportion, I will conclude this part with the note of one deficience more, which seemeth to me of greatest consequence; which is, that the prescripts in use are too compendious to attain their end: for to my understanding, it is a vain and flattering opinion to think any medicine can be so sovereign or so happy, as that the receipt or use of it can work any great effect upon the body of man. It were a strange speech which spoken, or spoken oft, should reclaim a man from a vice to which he were by nature subject. It is order, pursuit, sequence, and interchange of application, which is mighty in nature; which although it require more exact knowledge in prescribing and more precise obedience in observing, yet is recompensed with the magnitude of effects. And although a man would think, by the daily visitations of the physicians, that there were a pursuance in the cure; yet let a man look into their prescripts and ministrations, and he shall find them but inconstancies and every day's devices, without any settled providence or project. Not that every scrupulous or superstitious prescript is effectual, no more than every straight way is the way to heaven; but the *truth of the direction* must precede *severity of observance* [113].

110 *i.e.* the particular medicines proper for particular diseases, as distinguished from "general intentions".

111 In the translation he adds "the confection of Alkermes".

112 So edd. 1629 and 1633. The original has *commended.*

113 The latter part of this paragraph is considerably enlarged in the translation, rather however by way of explanation than addition, till he comes to the end; when in closing his account of the *Desiderata* in the science of curing diseases, he adds that there is however one other remaining which is of more consequence than all the rest—namely that of a true and active Natural Philosophy for the Science of Medicine to be built upon.

Between this paragraph and the next is interposed a long passage upon the prolongation of life of which there are no traces at all here.

For Cosmetic, it hath parts civil, and parts effeminate: for cleanness of body was ever esteemed to proceed from a due reverence to God, to society, and to ourselves [114]. As for artificial decoration, it is well worthy of the deficiencies which it hath; being neither fine enough to deceive, nor handsome to use, nor wholesome to please [115].

For Athletic, I take the subject of it largely; that is to say, for any point of ability whereunto the body of man may be brought, whether it be of *activity* or of *patience;* whereof activity hath two parts, *strength* and *swiftness;* and patience likewise hath two parts, *hardness against wants and extremities,* and *indurance of pain or torment.* Whereof we see the practices in tumblers, in savages [116], and in those that suffer punishment: nay, if there be any other faculty which falls not within any of the former divisions, as in those that dive, that obtain a strange power of containing respiration, and the like, I refer it to this part. Of these things the practices are known, but the philosophy that concerneth them is not much enquired; the rather, I think, because they are supposed to be obtained either by an aptness of nature, which cannot be taught, or only by continual custom, which is soon prescribed; which though it be not true, yet I forbear to note any deficiencies; for the Olympian Games are down long since, and the mediocrity of these things is for use; as for the excellency of them, it serveth for the most part but for mercenary ostentation.

For Arts of Pleasure Sensual, the chief deficience in them is of laws to repress them [117]. For as it hath been well observed that the arts which flourish in times while virtue is in growth, are military; and while virtue is in state, are liberal; and while virtue is in declination, are voluptuary; so I doubt that this age of the world is somewhat upon the descent of the wheel. With arts *voluptuary* I couple practices *joculary*; for the deceiving of the sense is one of the pleasures of the senses. As for games of recreation, I hold them to belong to civil life and education [118]. And thus much of that particular Human Philosophy which concerns the Body, which is but the tabernacle of the mind.

114 To whom (he adds in the translation) we owe no less reverence—nay even more—than to others. So in the *New Atlantis,* "and they say (*i.e.* the people of Bensalem) that the reverence of a man's self is, next to Religion, the chiefest bridle of all vices".

115 So all the editions. He must have meant to write, "handsome to please, nor wholesome to use".

By artificial decoration he means painting the face, as we learn from the translation; where he expresses wonder that this *prava consuetudo fucandi* is not prohibited by the laws, along with sumptuous apparel and lovelocks.

116 The translation adds "in the stupendous strength shown by maniacs."

117 Here we have an important addition in the translation. Whether when he wrote the *Advancement of Learning* Bacon had forgotten Painting and Music or meant to find another place for them, I cannot say; but in the *De Augmentis* he includes them among the *Artes Voluptariæ;* which he cannot have intended to do when he wrote this sentence. The passage in which they are introduced is to this effect:—The arts of pleasure, he says, are as many as the senses themselves are. To the eye belongs Painting, with innumerable other arts of magnificence in matter of Buildings, Gardens, Dresses, Vases; Gems, etc.; to the ear Music, with its various apparatus of voices, wind, and strings; and of all the sensual arts those which relate to Sight and Hearing are accounted the most liberal; for as these two senses are the purest and most chaste, so the sciences which belong to them are the most learned; both being waited upon by the Mathematics, and one having some relation to memory and demonstrations, the other to manners and affections of the mind, The rest of the sensual pleasures, with the arts appertaining to them, are held in less honour, as being nearer akin to luxury and magnificence. Unguents, perfumes, delicacies of the table, and especially stimulants of lust, stand more in need of a censor to repress than a master to teach them; and as it has been well observed, etc.

118 This observation is omitted in the translation; and a new paragraph is introduced stating that everything that relates to the body of man (though there be some which do not properly belong to either of the three offices above mentioned, viz., the preservation of health, the cure of diseases, and the prolongation of life) is to be considered as included in Medicine.

¶ [119] For Human Knowledge which concerns the Mind, it hath two parts; the one that inquireth of *the substance or nature of the soul or mind*, the other that inquireth of *the faculties or functions thereof*. Unto the first of these, the considerations of *the original of the soul*, whether it be *native or adventive*, and *how far it is exempted from laws of matter*, and of the *immortality thereof*, and many other points, do appertain: which have been not more laboriously inquired than variously reported; so as the travail therein taken seemeth to have been rather in a maze than in a way. But although I am of opinion that this knowledge may be more really and soundly inquired, even in nature, than it hath been; yet I hold that in the end it must be bounded by religion, or else it will be subject to deceit and delusion; for as the substance of the soul in the creation was not extracted out of the mass of heaven and earth by the benediction of a *producat*, but was immediately inspired from God; so it is not possible that it should be (otherwise than by accident) subject to *the laws of heaven and earth*, which are *the subject of philosophy*; and therefore the true knowledge of the nature and state of the soul must come by the same inspiration that gave the substance [120]. Unto this part of knowledge touching the soul there be two appendices; which, as they have been handled, have rather vapoured forth fables than kindled truth; Divination and Fascination.

Divination hath been anciently and fitly divided into *artificial* and *natural*; whereof *artificial* is when the mind maketh a prediction by argument, concluding upon signs and tokens; *natural* is when the mind hath a presentation by an internal power, without the inducement of a sign. Artificial is of two sorts; either when the argument is coupled with a derivation of causes, which is *rational*; or when it is only grounded upon a coincidence of the effect, which is *experimental*: whereof the later for the most part is superstitious; such as were the heathen observations upon the inspection of sacrifices, the flights of birds, the swarming of bees; and such as was the Chaldean Astrology, and the like. For *artificial divination*, the several kinds thereof are distributed amongst particular knowledges. The Astronomer hath his predictions, as of conjunctions, aspects, eclipses, and the like. The Physician hath his predictions, of death, of recovery, of the accidents and issues of diseases. The Politique hath his predictions; *O urbem venalem, et cito perituram, si emptorem invenerit!* [a city in which all things are for sale and which will fall to the first purchaser], which stayed not long to be performed, in Sylla first, and after in Cæsar. So as these predictions are now impertinent, and to be referred over. But the divination which springeth from the internal nature of the soul, is that which we now speak of; which hath been made to be of two sorts, *primitive* and by *influxion*. Primitive is grounded upon the supposition that the mind, when it is withdrawn and collected into itself and not diffused into the organs of the body, hath some extent and latitude of prenotion; which therefore appeareth most in sleep, in extasies, and near death; and more rarely in waking apprehensions; and is induced and furthered by those abstinences and observances which make the mind most to consist in itself. By influxion is grounded upon the conceit that the mind, as a mirror or glass, should take illumination from the foreknowledge of God and spirits; unto which the same regiment doth likewise conduce. For the retiring of the mind within itself is the state which is most susceptible of divine influxions; save that it is accompanied in this case with a fervency and

[119] De Aug. iv. 3.

[120] In the translation a new division is introduced which does not appear to be distinctly recognized here—the human soul being divided into Rational and Irrational; the one divine and peculiar to humanity, the other (which is merely its instrument) being of the earth and common to man and brute; and the remark in the text is confined to the first of these only. The other soul, which he calls the *anima sensibilis sive producta*, is represented as a fit subject of physical enquiry, in its nature and substance as well as in its faculties; though the enquiry has not been well pursued with regard to either. Concerning the doctrine of the Duality of the Soul see Mr. Ellis's General Introduction, § 14.

elevation (which the ancients noted by *fury*), and not with a repose and quiet as it is in the other.

Fascination is the power and act of imagination, intensive upon other bodies than the body of the imaginant: for of that we spake in the proper place: wherein the school of Paracelsus and the disciples of pretended Natural Magic have been so intemperate, as they have exalted the power of the imagination to be much one with the power of miracle-working faith; others that draw nearer to probability, calling to their view the secret passages of things, and especially of the contagion that passeth from the body [121], do conceive it should likewise be agreeable to nature that there should be some transmissions and operations from spirit to spirit, without the mediation of the senses; whence the conceits have grown (now almost made civil) of the Mastering Spirit, and the force of confidence, and the like. Incident unto this is the inquiry how to raise and fortify the imagination; for if the imagination fortified have power, then it is material to know how to fortify and exalt it. And herein comes in crookedly and dangerously a palliation of a great part of Ceremonial Magic. For it may be pretended that Ceremonies, Characters, and Charms, do work not by a tacit or sacramental contract with evil spirits, but serve only to strengthen the imagination of him that useth it; as images are said by the Roman church [122] to fix the cogitations and raise the devotions of them that pray before them. But for mine own judgment, if it be admitted that imagination hath power, and that Ceremonies fortify imagination, and that they be used sincerely and intentionally for that purpose [123]; yet I should hold them unlawful, as opposing to that first edict which God gave unto man, *In sudore vultus comedes panem tuum* [in the sweat of thy brow shalt thou eat thy bread]. For they propound those noble effects which God hath set forth unto man to be bought at the price of labour, to be attained by a few easy and slothful observances. Deficiences in these knowledges I will report none, other than the general deficience, that it is not known how much of them is verity and how much vanity [124].

¶ [125] The knowledge which respecteth the Faculties of the Mind of man is of two kinds; the one respecting his Understanding and Reason, and the other his Will, Appetite, and Affection; whereof the former produceth Position or Decree, the latter Action or Execution. It is true that the Imagination is an agent or *nuncius* in both provinces, both the judicial and the ministerial. For Sense sendeth over to Imagination before Reason have judged: and Reason sendeth over to Imagination before the Decree can be acted; for Imagination ever precedeth Voluntary Motion: saving that this Janus of Imagination hath differing faces; for the face towards Reason hath the print of Truth, but the face towards Action hath the print of Good; which nevertheless are faces

Quales decet esse sororum,

[sister-faces]. Neither is the Imagination simply and only a messenger; but is invested with or at leastwise usurpeth no small authority in itself, besides the duty of the message. For it was well said by Aristotle, *That the mind hath over the body that commandment, which the lord hath over a bondman; but that reason hath over the imagination that commandment which a magistrate hath over a free citizen*; who may come also to rule in his turn. For we see that in matters of Faith and Religion we raise our Imagination above our Reason [126]; which is

[121] In the translation he adds " the irradiations of the senses, and the conveyance of magnetic virtues ".

[122] In the translation, the words " said by the Roman church " are omitted, and *in Religione usus imaginum . . . . invaluit* are substituted. See note p. 50.

[123] *i.e.* as a physical remedy, without any thought of inviting thereby the assistance of spirits,—as explained in the translation.

[124] This sentence is omitted in the translation altogether; and the chapter concludes with a notice at considerable length of two *Desiderata* not mentioned here; the doctrine of Voluntary Motion, and the doctrine of Sense and the Sensible.

[125] De Aug. v. 1.

[126] Not (he adds in the translation), that the divine illumination resides in the Ima-

the cause why Religion sought ever access to the mind by similitudes, types, parables, visions, dreams. And again in all persuasions that are wrought by eloquence and other impressions of like nature, which do paint and disguise the true appearance of things, the chief recommendation unto Reason is from the Imagination [127]. Nevertheless, because I find not any science that doth properly or fitly pertain to the Imagination, I see no cause to alter the former division. For as for Poesy, it is rather a pleasure or play of imagination, than a work or duty thereof. And if it be a work, we speak not now of such parts of learning as the Imagination produceth, but of such sciences as handle and consider of the Imagination; no more than we shall speak now of such knowledges as Reason produceth, (for that extendeth to all philosophy,) but of such knowledges as do handle and inquire of the faculty of Reason: so as Poesy had his true place [128]. As for the power of the Imagination in nature, and the manner of fortifying the same, we have mentioned it in the doctrine *De Anima*, whereunto most fitly it belongeth. And lastly, for Imaginative or Insinuative Reason, which is the subject of Rhetoric, we think it best to refer it to the Arts of Reason. So therefore we content ourselves with the former division, that Human Philosophy which respecteth the faculties of the mind of man hath two parts, Rational and Moral.

The part of Human Philosophy which is rational, is of all knowledges, to the most wits, the least delightful; and seemeth but a net of subtility and spinosity. For as it was truly said, that knowledge is *pabulum animi* [the food of the mind]; so in the nature of men's appetite to this food, most men are of the taste and stomach of the Israelites in the desert, that would fain have returned *ad ollas carnium* [to the flesh-pots], and were weary of manna; which, though it were celestial, yet seemed less nutritive and comfortable. So generally men taste well knowledges that are drenched in flesh and blood, Civil History, Morality, Policy, about the which men's affections, praises, fortunes, do turn and are conversant; but this same *lumen siccum* [this dry light], doth parch and offend most men's watery and soft natures. But to speak truly of things as they are in worth, Rational Knowledges are the keys of all other arts; for as Aristotle saith aptly and elegantly, *That the hand is the Instrument of Instruments, and the mind is the Form of Forms*: so these be truly said to be the Art of Arts: neither do they only direct, but likewise confirm and strengthen; even as the habit of shooting doth not only enable to shoot a nearer shoot, but also to draw a stronger bow.

The Arts Intellectual are four in number; divided according to the ends whereunto they are referred: for man's labour is to *invent* [129] that which is *sought* or *propounded*; or to *judge* that which is *invented*; or to *retain* that which is *judged*; or to *deliver over* that which is *retained*. So as the arts must be four; Art of Inquiry or Invention: Art of Examination or Judgment; Art of Custody or Memory; and Art of Elocution or Tradition.

130 ¶ Invention is of two kinds, much differing; the one, of Arts and Sciences;

gination,—its seat being rather in the very citadel of the mind and understanding;—but that the divine grace uses the motions of the Imagination as an instrument of illumination, just as it uses the motions of the will as an instrument of virtue.

127 This is better explained in the translation; where it is observed that the arts of speech by which men's minds are soothed, inflamed, or carried away, consist in exciting the Imagination till it gets the better of the Reason.

128 This whole sentence is omitted in the translation; the reason for not altering the former division being stated simply thus: *Nam Phantasia scientias fere non parit; siquidem Poesis (quæ a principio Phantasiæ attributa est) pro lusu potius ingenii quam pro scientia habenda.* Poesy, which belongs properly to Imagination, is not to be considered as a part of *knowledge*; and the two other offices of the Imagination belong, one to the doctrine *de anima*, the other to Rhetoric. There is no occasion therefore to make a place for Imagination among the parts of knowledge which concern the faculties of the human mind.

129 It may perhaps be worth while to observe that Bacon uses the word *invent* simply as equivalent to *invenire*—to find out.

130 De Aug. v. 2.

and the other, of Speech and Arguments. The former of these I do report deficient; which seemeth to me to be such a deficience as if in the making of an inventory touching the estate of a defunct it should be set down *that there is no ready money*. For as money will fetch all other commodities, so this knowledge is that which should purchase all the rest. And like as the West-Indies had never been discovered if the use of the mariner's needle had not been first discovered, though the one be vast regions and the other a small motion; so it cannot be found strange if sciences be no further discovered, if the art itself of invention and discovery hath been passed over.

That this part of knowledge is wanting, to my judgment standeth plainly confessed: for first, Logic doth not pretend to invent Sciences or the Axioms of Sciences, but passeth it over with a *cuique in sua arte credendum* [the knowledge that pertains to each art must be taken on trust from those that profess it]. And Celsus acknowledgeth it[131] gravely, speaking of the empirical and dogmatical sects of physicians, *That medicines and cures were first found out, and then after the reasons and causes were discoursed; and not the causes first found out, and by light from them the medicines and cures discovered.* And Plato in his Theætetus[132] noteth well, *That particulars are infinite, and the higher generalities give no sufficient direction; and that the pith of all sciences, which maketh the arts-man differ from the inexpert, is in the middle propositions, which in every particular knowledge are taken from tradition and experience.* And therefore we see that they which discourse of the inventions and originals of things, refer them rather to chance than to art, and rather to beasts, birds, fishes, serpents, than to men.

Dictamnum genetrix Cretæa carpit ab Ida,
Puberibus caulem foliis et flore comantem
Purpureo: non illa feris incognita capris
Gramina, cum tergo volucres hæsere sagittæ
[A sprig of dittany his mother brought,
Gathered by Cretan Ide; a stalk it is
Of woolly leaf, crested with purple flower;
Which well the wild-goat knows when in his side
Sticks the winged shaft].

So that it was no marvel (the manner of antiquity being to consecrate inventors) that the Ægyptians had so few human idols in their temples, but almost all brute:

Omnigenumque Deum monstra, et latrator Anubis,
Contra Neptunum et Venerem, contraque Minervam, etc.
[All kinds and shapes of Gods, a monstrous host,
The dog Anubis foremost, stood arrayed
'Gainst Neptune, Venus, Pallas, etc.].

And if you like better the tradition of the Grecians, and ascribe the first inventions to men, yet you will rather believe that Prometheus first struck the flints, and marvelled at the spark, than that when he first struck the flints he expected the spark; and therefore we see the West-Indian Prometheus had no intelligence with the European, because of the rareness with them of flint, that gave the first occasion[133]. So as it should seem that hitherto men are

131 See note on Nov. Org. i. 73.

132 Instead of "Plato *in his Theætetus* noteth" the translation has *Plato non semel innuit.*

133 This curious passage, which is omitted in the *De Augmentis*, must refer to what Bacon had read in Ramusio of the way in which the natives of the West Indian islands kindled their fires, by rubbing pieces of wood together. Several passages in Bacon's writings show that he was a reader of Ramusio. See Ramusio, vol. iii. p. 103. a. for Oviedo's description of the method.

In reality the coincidence between the customary mode of kindling fire in the West Indies and the superstitious usages of Europe is remarkable. The latter seem to point

rather beholden to a wild goat for surgery, or to a nightingale for music, or to the Ibis for some part of physic,[134] or to the pot lid that flew open for artillery, or generally to chance or anything else, than to Logic, for the invention of arts and sciences. Neither is the form of invention which Virgil describeth much other:

> Ut varias usus meditando extunderet artes
> Paulatim

[that practice with meditation might by degrees hammer out the arts]. For if you observe the words well, it is no other method than that which brute beasts are capable of, and do put in ure; which is *a perpetual intending or practising some one thing, urged and imposed by an absolute necessity of conservation of being*: for so Cicero saith very truly, *Usus uni rei deditus et naturam et artem sæpe vincit* [practice applied constantly to one thing will often do more than either nature or art can]. And therefore if it be said of men,

> Labor omnia vincit
> Improbus, et duris urgens in rebus egestas
> [Stern labour masters all,
> And want in poverty importunate],

it is likewise said of beasts, *Quis psittaco docuit suum* χαῖρε? [who taught the parrot to say how d'ye do?] Who taught the raven in a drowth to throw pebbles into an hollow tree where she spied water, that the water might rise so as she might come to it? Who taught the bee to sail through such a vast sea of air, and to find the way from a field in flower a great way off to her hive? Who taught the ant to bite every grain of corn that she burieth in her hill, lest it should take root and grow? And then the word *extundere*, which importeth the extreme difficulty, and the word *paulatim*, which importeth the extreme slowness, and we are where we were, even amongst the Ægyptians' gods; there being little left to the faculty of Reason, and nothing to the duty of Art, for matter of invention.

Secondly, the induction which the logicians speak of, and which seemeth familiar with Plato[135], whereby the Principles of sciences may be pretended

back to a time when the use of steel and flint was unknown. The Noth-feuer of the Germans was kindled by rubbing pieces of wood together. This fire, originally connected with the worship of Fro, was lighted when cattle were threatened with murrain, and they were made to pass through it. Dr. Jamieson in his Scottish Dictionary mentions precisely the same practice at a comparatively recent period in Scotland in a case in which the murrain had done great mischief. The long continuance of this practice is a sort of illustration of Spinoza's bitter remark that Superstition is the child of Adversity, there being no man, he observes, who in prosperity does not think himself wise enough to take care of himself. See Spinoza, *Trac. Theol. Politicus*, chap. i.: and for the German superstition Wolf's *Die deutsche Götterlehre*, pp. 27. 83.

The holy fire of Vesta, according to Festus (in voce *Ignis*), was rekindled when it had been allowed to go out, by friction of two pieces of wood. Plutarch's statement that the rays of the sun concentrated by reflection were employed for the purpose seems improbable, and is apparently founded on a misconception or mistranslation of some earlier account of the matter. Pliny mentions, but without reference to Vesta, this mode of kindling fire, and states that the best combination is laurel wood with ivy.—*R. L. E.*

It is worth observing that though the passage in the text is omitted in the *De Augmentis*, the substance of it is retained in the *Cogitata et Visa*. *Nam ideo in ignis invento Prometheum Novæ ab Europæo dissensisse, quod apud eos silicis non est copia.—J. S.*

[134] *pro lavationibus intestinorum.*—De Aug.

[135] This reference to Plato is omitted in the translation, as well as the allusion to the derivation of the middle propositions. The induction in question is merely described as "the form of induction which Logic proposes, whereby to discover and prove the principles of sciences".

to be invented, and so the middle propositions by derivation from the principles, —their form of induction, I say, is utterly vicious and incompetent: wherein their error is the fouler, because it is the duty of Art to perfect and exalt Nature; but they contrariwise have wronged, abused, and traduced nature. For he that shall attentively observe how the mind doth gather this excellent dew of knowledge, like unto that which the poet speaketh of, *Aërei mellis cœlestia dona* [the gift of heaven, aërial honey], distilling and contriving it out of particulars natural and artificial, as the flowers of the field and garden, shall find that the mind of herself by nature doth manage and act an induction much better than they describe it. For to conclude *upon an enumeration of particulars without instance contradictory* is no conclusion, but a conjecture; for who can assure (in many subjects) upon those particulars which appear of a side, that there are not other on the contrary side which appear not? As if Samuel should have rested upon those sons of Issay [136] which were brought before him, and failed of David, which was in the field. And this form (to say truth) is so gross, as it had not been possible for wits so subtile as have managed these things to have offered it to the world, but that they hasted to their *theories* and *dogmaticals*, and were imperious and scornful toward particulars; which their manner was to use but as *lictores* and *viatores*, for sergeants and whifflers, *ad summovendam turbam*, to make way and make room for their opinions, rather than in their true use and service. Certainly it is a thing may touch a man with a religious wonder, to see how the footsteps of seducement are the very same in divine and human truth: for as in divine truth man cannot endure to become as a child; so in human, they reputed the attending the Inductions (whereof we speak) as if it were a second infancy or childhood.

Thirdly, allow some Principles or Axioms were rightly induced, yet nevertheless certain it is that Middle Propositions cannot be deduced from them in subject of nature [137] by Syllogism, that is *by touch and reduction of them to principles in a middle term.* It is true that in sciences popular, as moralities, laws, and the like, yea and divinity (because it pleaseth God to apply himself to the capacity of the simplest), that form may have use; and in natural philosophy likewise, by way of argument or satisfactory reason, *quæ assensum parit, operis effœta est* [which procures assent but can do no work]: but the subtilty of nature and operations will not be enchained in those bonds: for Arguments consist of Propositions, and Propositions of Words; and Words are but the current tokens or marks of Popular Notions of things; which notions, if they be grossly and variably collected out of particulars, it is not the laborious examination either of consequences of arguments, or of the truth of propositions, that can ever correct that error; being (as the physicians speak) in the first digestion: and therefore it was not without cause, that so many excellent philosophers became Sceptics and Academics, and denied any certainty of knowledge or comprehension, and held opinion that the knowledge of man extended only to appearances and probabilities. It is true that in Socrates it was supposed to be but a form of irony, *Scientiam dissimulando simulavit* [an affection of knowledge under pretence of ignorance]: for he used to disable his knowledge, to the end to enhance his knowledge [138]; like the humour of Tiberius in his beginnings, that would reign, but would not acknowledge so much [139]; and in the later Academy, which Cicero embraced, this opinion also of *acatalepsia* (I doubt) was not held sincerely: for that all those which excelled in copie of speech seem to have chosen that sect, as that which was fittest to give glory to their eloquence and variable discourses; being rather like progresses of pleasure than journeys to an end. But assuredly many scattered in both Academies did hold it in subtilty and integrity. But here was their chief

[136] So in all three editions. The *De Augmentis* has *Isai.*

[137] *in rebus naturalibus, quæ participant ex materiâ.*—De Aug.

[138] *i.e.* pretended not to know what it was plain he knew, that he might be thought to know likewise what he knew not—*renunciando scilicet iis quæ manifesto sciebat ut eo modo ea etiam quæ nesciebat scire putaretur.*

[139] This allusion to Tiberius is omitted in the translation.

error; they charged the deceit upon the Senses; which in my judgment (notwithstanding all their cavillations) are very sufficient to certify and report truth, though not always immediately, yet by comparison [140], by help of instrument, and by producing and urging such things as are too subtile for the sense to some effect comprehensible by the sense, and other like assistance. But they ought to have charged the deceit upon *the weakness* [141] *of the intellectual powers and upon the manner of collecting and concluding upon the reports of the senses.* This I speak not to disable the mind of man, but to stir it up to seek help: for no man, be he never so cunning or practised, can make a straight line or perfect circle by steadiness of hand, which may be easily done by help of a ruler or compass [142].

Experientia literata, et interpretatio Naturæ.

This part of invention, concerning the invention of sciences, I purpose (if God give me leave) hereafter to propound; having digested it into two parts; whereof the one I term *Experientia literata*, and the other *Interpretatio Naturæ* [143]: the former being but a degree and rudiment of the latter. But I will not dwell too long, nor speak too great upon a promise.

¶ [144] The invention of speech or argument is not properly an invention: for to invent is to discover that we know not, and not to recover or resummon that which we already know; and the use of this invention is no other but *out of the knowledge whereof our mind is already possessed, to draw forth or call before us that which may be pertinent to the purpose which we take into our consideration.* So as, to speak truly, it is no *Invention*, but a *Remembrance* or *Suggestion*, with an application; which is the cause why the schools do place it after judgment, as subsequent and not precedent. Nevertheless, because we do account it a Chase as well of deer in an inclosed park as in a forest at large, and that it hath already obtained the name, let it be called invention: so as it be perceived and discerned, that the scope and end of this invention is readiness and present use of our knowledge, and not addition or amplification thereof.

To procure this ready use of knowledge there are two courses, Preparation and Suggestion [145]. The former of these seemeth scarcely a part of Knowledge, consisting rather of diligence than of any artificial erudition. And herein Aristotle wittily, but hurtfully, doth deride the sophists near his time, saying, *they did as if one that professed the art of shoe-making should not teach how to make up a shoe, but only exhibit in a readiness a number of shoes of all fashions and sizes.* But yet a man might reply, that if a shoemaker should have no shoes in

[140] There is nothing about comparison in the translation.

[141] In the translation he adds *contumacy—tum erroribus tum contumaciæ* (*quæ rebus ipsis morigera esse recusat*)—and also *pravis demonstrationibus;* an insertion which (though the observation is implied perhaps in the English) I have thought worth noticing; because these *pravæ demonstrationes* were Idols of the *Theatre*, of which in the *Advancement of Learning* there is no mention.

[142] This it is then (he adds, writing eighteen years later) which I have in hand, and am labouring with mighty effort to accomplish—namely to make the mind of man by help of art a match for the nature of things,—to discover an art of Indication and Direction whereby all other arts with their axioms and works may be detected and brought to light.

[143] The one being the method of inquiry which proceeds from one experiment to another by a kind of natural sagacity; the other that which proceeds from experiments to axioms, and thence by the light of the axioms to new experiments. *Aut enim defertur indicium ab experimentis ad experimenta, aut ab experimentis ad axiomata quæ et ipsa nova experimenta designent.* Of this *Experientia literata* there follows in the *De Augmentis* an exposition at considerable length; in which the several methods of experimenting are described, with illustrations. And this concludes the chapter, the exposition of the other part, the Interpretatio Naturæ, being reserved for the *Novum Organum.*

[144] De Aug. v. 3.

[145] In the translation he calls these respectively *Promptuaria* and *Topica*: the one being a collection of arguments such as you are likely to want, laid up ready for use; the other a system of directions to help you in looking for the thing you want to find.

his shop, but only work as he is bespoken, he should be weakly customed. But our Saviour, speaking of Divine Knowledge, saith, *that the kingdom of heaven is like a good householder, that bringeth forth both new and old store ;* and we see the ancient writers of rhetoric do give it in precept, that pleaders should have the Places whereof they have most continual use ready handled in all the variety that may be ; as that, to speak for the literal interpretation of the law against equity, and contrary ; and to speak for presumptions and inferences against testimony, and contrary. And Cicero himself, being broken unto it by great experience, delivereth it plainly, that whatsoever a man shall have occasion to speak of (if he will take the pains), he may have it in effect premeditate, and handled *in thesi ;* so that when he cometh to a particular, he shall have nothing to do but to put to names and times and places, and such other circumstances of individuals. We see likewise the exact diligence of Demosthenes ; who in regard of the great force that the entrance and access into causes hath to make a good impression, had ready framed a number of prefaces for orations and speeches. All which authorities and precedents may overweigh Aristotle's opinion, that would have us change a rich wardrobe for a pair of shears.

But the nature of the collection of this provision or preparatory store, though it be common both to logic and rhetoric, yet having made an entry of it here, where it came first to be spoken of, I think fit to refer over the further handling of it to rhetoric.

The other part of Invention, which I term Suggestion, doth assign and direct us to certain *marks* or *places*, which may excite our mind to return and produce such knowledge as it hath formerly collected, to the end we may make use thereof. Neither is this use (truly taken) only to furnish argument to dispute probably with others, but likewise to minister unto our judgment to conclude aright within ourselves. Neither may these Places serve only to apprompt our invention, but also to direct our inquiry. For a faculty of wise interrogating is half a knowledge. For as Plato saith, *Whosoever seeketh, knoweth that which he seeketh for in a general notion ; else how shall he know it when he hath found it ?* And therefore the larger [146] your Anticipation is, the more direct and compendious is your search. But the same Places which will help us what to produce of that which we know already, will also help us, if a man of experience were before us, what questions to ask ; or if we have books and authors to instruct us, what points to search and revolve : so as I cannot report [147] that this part of invention, which is that which the schools call Topics, is deficient.

Nevertheless Topics are of two sorts, *general* and *special.* The general we have spoken to ; but the particular hath been touched by some, but rejected generally as inartificial and variable. But leaving the humour which hath reigned too much in the schools, (which is to be vainly subtile in a few things which are within their command, and to reject the rest), I do receive particular

[146] *amplior et* certior.—De Aug.

[147] Thus the sentence stands both in the original and in the editions of 1629 and 1633 ; though I do not understand the connexion between it and the sentence preceding. Possibly an intermediate sentence has dropped out, or some alteration has been inadvertently made which disturbs the construction. In the translation the arrangement of the whole passage is changed, and all is made clear. He begins by dividing Topics into two kinds, General and Particular. The General (he says) has been sufficiently handled in Logic, and therefore he leaves it with a passing remark (*illud tamen obiter monendum videtur*) to the effect of that in the text ; " neither is this use," etc. down to " search and revolve ". But Particular Topics, he proceeds, are more to the purpose and of great value, and have not received the attention they deserve. He then goes on to explain at length what he means ; repeating the observations in the next paragraph with some amplification and greater clearness, and then giving a specimen of the thing, in a series of Particular Topics or articles of inquiry concerning Heavy and Light ; with which the chapter concludes. With regard to the importance of these *Topicæ* as a part of Bacon's method of inquiry—an importance so considerable that he meant to devote a special work to the subject,—see my prefaces to the *Parasceve* and to the *Topica Inquisitionis de Luce et Lumine.*

Topics, that is places or directions of invention and inquiry in every particular knowledge, as things of great use; being mixtures of Logic with the matter of sciences; for in these it holdeth, *Ars inveniendi adolescit cum inventis* [every act of discovery advances the art of discovery]; for as in going of a way we do not only gain that part of the way which is passed, but we gain the better sight of that part of the way which remaineth; so every degree of proceeding in a science giveth a light to that which followeth; which light if we strengthen, by drawing it forth into questions or places of inquiry, we do greatly advance our pursuit.

¶ [148] Now we pass unto the arts of Judgment, which handle the natures of Proofs and Demonstrations; which as to Induction hath a coincidence with Invention: *for in all inductions, whether in good or vicious form, the same action of the mind which inventeth, judgeth; all one as in the sense;* but otherwise it is in proof by syllogism; for the proof being not immediate but by mean, *the invention of the mean* is one thing, and the *judgment of the consequence* is another; the one exciting only, the other examining. Therefore for the real and exact form of judgment we refer ourselves to that which we have spoken of *Interpretation of Nature.*

For the other judgment by Syllogism, as it is a thing most agreeable to the mind of man, so it hath been vehemently and excellently laboured. For the nature of man doth extremely covet to have somewhat in his understanding fixed and immoveable, and as a rest and support of the mind. And therefore as Aristotle endeavoureth to prove that in all motion there is some point quiescent; and as he elegantly expoundeth the ancient fable of Atlas (that stood fixed and bare up the heaven from falling) to be meant of the poles or axle-tree of heaven, whereupon the conversion is accomplished; so assuredly men have a desire to have an Atlas or axle-tree within to keep them from fluctuation, which is like to a perpetual peril of falling; therefore men did hasten to set down some Principles about which the variety of their disputations might turn.

So then this art of Judgment is but *the reduction of propositions to principles in a middle term;* the Principles to be agreed by all and exempted from argument; the Middle Term to be elected at the liberty of every man's invention; the Reduction to be of two kinds, direct and inverted; the one when the proposition is reduced to the principle, which they term a *Probation ostensive;* the other when the contradictory of the proposition is reduced to the contradictory of the principle, which is that which they call *per incommodum*, or *pressing an absurdity;* the number of middle terms to be [149] as the proposition standeth degrees more or less removed from the principle.

But this art hath two several methods of doctrine; the one by way of direction, the other by way of caution: the former frameth and setteth down a true form of consequence, by the variations and deflexions from which errors and inconsequences may be exactly judged; toward the composition and structure of which form, it is incident to handle the parts thereof, which are propositions, and the parts of propositions, which are simple words [150]; and this is that part of logic which is comprehended in the Analytics.

The second method of doctrine was introduced for expedite use and assurance sake; discovering the more subtile forms of sophisms and illaqueations with their redargutions, which is that which is termed *Elenches.* For although in the more gross sorts of fallacies it happeneth (as Seneca maketh the comparison well) as in juggling feats, which though we know not how they are done, yet we know well it is not as it seemeth to be; yet the more subtile sort of them doth not only put a man besides his answer, but doth many times abuse his judgment.

This part concerning *Elenches* [151] is excellently handled by Aristotle in pre-

[148] De Aug. v. 4. [149] *i.e.* to be more or fewer.

[150] This clause is omitted in the translation; and a new observation is introduced in its place; viz. that though this direction contains in itself a kind of Elenche or confutation (for the straight indicates the crooked), yet it is safest to employ Elenches (that is, Elenches properly so called) as monitors, for the better detection of fallacies by which the judgment would otherwise be ensnared.

[151] In the translation the Doctrine of Elenches is divided into three kinds—*Elenchos*

cept, but more excellently by Plato in example, not only in the persons of the Sophists, but even in Socrates himself; who professing to affirm nothing, but to infirm that which was affirmed by another, hath exactly expressed all the forms of objection, fallace [152], and redargution. And although we have said that the use of this doctrine is for redargution, yet it is manifest the degenerate and corrupt use is for caption and contradiction [153]; which passeth for a great faculty, and no doubt is of very great advantage: though the difference be good which was made between orators and sophisters, that the one is as the greyhound, which hath his advantage in the race, and the other as the hare, which hath her advantage in the turn, so as it is the advantage of the weaker creature.

But yet further, this doctrine of *Elenches* hath a more ample latitude and extent than is perceived; namely, unto divers parts of knowledge; whereof some are laboured and other omitted. For first, I conceive (though it may seem at first somewhat strange) that that part which is variably referred sometimes to Logic sometimes to Metaphysic, touching the *common adjuncts* of *essences*, is but an elenche [154]; for the great sophism of all sophisms being equivocation or ambiguity of words and phrase, specially of such words as are most general and intervene in every inquiry, it seemeth to me that the true and fruitful use (leaving vain subtilties and speculations) of the inquiry of majority, minority, priority, posteriority, identity, diversity, possibility, act, totality, parts, existence, privation, and the like, are but wise cautions against ambiguities of speech. So again the distribution of things into certain tribes, which we call categories or predicaments, are but cautions against the confusion of definitions and divisions.

Secondly, there is a seducement that worketh by the strength of the impression and not by the subtilty of the illaqueation; not so much perplexing the reason as overruling it by power of the imagination. But this part I think more proper to handle when I shall speak of Rhetoric [155].

But lastly, there is yet a much more important and profound kind of fallacies in the mind of man, which I find not observed or enquired at all, and think good to place here, as that which of all others appertaineth most to rectify judgment: the force whereof is such, as it doth not dazzle or snare the understanding in some particulars, but doth more generally and inwardly infect and corrupt the state thereof [156]. For the mind of man is far from the nature of a clear and equal glass, wherein the beams of things should reflect according to their true incidence; nay, it is rather like an enchanted glass, full of superstition and imposture, if it be not delivered and reduced. For this purpose, let us consider the false appearances that are imposed upon us by the general nature of the mind [157], beholding them in an example or two; as first, in that instance which is the root of all superstition, namely, *That to the nature of the mind of all men it is consonant for the affirmative or active to affect more than the negative or privative:*

Sophismatum, Elenchos Hermeniæ, Elenchos imaginum sive Idolorum: *i.e.* Cautions against Sophisms, against ambiguity of words, against Idols or false appearances; and it is to the first only that the observation which follows is applied.

[152] So in all the editions; and not (I think) a misprint for *fallacie*, but another word, formed not from *fallacia* but from *fallax*. Compare "Colours of Good and Evil", § 1. "The *fallax* of this colour," etc.

[153] *i.e.* the true sense is to answer sophistical arguments, the corrupt use to invent sophistical objections.

[154] This is the part which in the translation he calls *Elenchos Hermeniæ*; and explains much more clearly and fully.

[155] This paragraph was omitted altogether in the translation.

[156] Here we have the doctrine of Idols, in its earliest form; the names not being yet given, and the Idols of the Theatre not yet introduced into the company. For the history of this doctrine see preface to the *Novum Organum*, note C. In the *De Augmentis* the names are given, and the fourth kind mentioned, though only to be set aside as not belonging to the present argument. The exposition of the three first is also considerably fuller than here, though not nearly so full as in the *Novum Organum*, to which we are referred.

[157] These are the Idols of the Tribe.

so that a few times hitting or presence, countervails oft-times failing or absence[158]; as was well answered by Diagoras to him that shewed him in Neptune's temple the great number of pictures of such as had scaped shipwrack and had paid their vows to Neptune, saying, *Advise now, you that think it folly to invocate Neptune in tempest: Yea but* (saith Diagoras) *where are they painted that are drowned?* Let us behold it in another instance, namely, *That the spirit of man, being of an equal and uniform substance, doth usually suppose and feign in nature a greater equality and uniformity than is in truth.* Hence it cometh that the mathematicians cannot satisfy themselves, except they reduce the motions of the celestial bodies to perfect circles, rejecting spiral lines, and labouring to be discharged of eccentrics. Hence it cometh, that whereas there are many things in nature as it were *monodica*[159], *sui juris* [singular, and like nothing but themselves]; yet the cogitations of man do feign unto them relatives, parallels, and conjugates, whereas no such thing is; as they have feigned an element of Fire, to keep square with Earth, Water, and Air, and the like: nay, it is not credible, till it be opened, what a number of fictions and fancies the similitude of human actions and arts[160], together with the making of man *communis mensura*, have brought into Natural Philosophy; not much better than the heresy of the Anthropomorphites, bred in the cells of gross and solitary monks, and the opinion of Epicurus, answerable to the same in heathenism, who supposed the gods to be of human shape. And therefore Velleius the Epicurian[161] needed not to have asked, why God should have adorned the heavens with stars, as if he had been an Ædilis, one that should have set forth some magnificent shews or plays. For if that great work-master had been of an human disposition, he would have cast the stars into some pleasant and beautiful works and orders, like the frets in the roofs of houses; whereas one can scarce find a posture in square or triangle or straight line amongst such an infinite number; so differing an harmony there is between the spirit of Man and the spirit of Nature.

Let us consider again the false appearances imposed upon us by every man's own individual nature and custom[162], in that feigned supposition that Plato maketh of the cave: for certainly if a child were continued in a grot or cave under the earth until maturity of age, and came suddenly abroad, he would have strange and absurd imaginations; so in like manner, although our persons live in the view of heaven, yet our spirits are included in the caves of our own complexions and customs; which minister unto us infinite errors and vain opinions, if they be not recalled to examination[163]. But hereof we have given many examples in one of the errors, or peccant humours, which we ran briefly over in our first book.

And lastly, let us consider the false appearances that are imposed upon us by words[164], which are framed and applied according to the conceit and capacities of the vulgar sort: and although we think we govern our words, and prescribe it

[158] which (he adds in the translation) is the root of all superstition and vain credulity, in matters of astrology, dreams, omens, &c.

[159] So the word is spelt throughout Bacon's writings, as observed by Mr. Ellis, *infra*. The introduction here of *sui juris* as the Latin equivalent seems to show that the error arose from a mistake as to the etymology of the Greek word.

[160] *i.e.* the supposed resemblance between the arts and actions of M. n and the operations of Nature: *naturalium operationum ad similitudinem actionum humanarum reductio: hoc ipsum inquam, quod putetur talia naturam facere qualia Homo facit.*

[161] So in the original: the word being pronounced in Bacon's time *Epicurian*. See *Walker on Shakespeare's versification*, p. 211.

[162] These are the Idols of the Cave.

[163] *i.e.* if they be not corrected by the continual contemplation of nature at large: *sive specu sua raro tantum et ad breve aliquod tempus prodeant, et non in contemplatione naturæ perpetuo, tanquam sub dio, morentur.*

It may be worth observing that Bacon guards himself against being supposed to represent the full intention of Plato's parable, by adding in a parenthesis *missa illa exquisita parabolæ subtilitate.*

[164] These are the Idols of the Market-place.

well, *Loquendum ut vulgus, sentiendum ut sapientes* [a man should speak like the vulgar, and think like the wise], yet certain it is that words, as a Tartar's bow, do shoot back upon the understanding of the wisest, and mightily entangle and pervert the judgment; so as it is almost necessary in all controversies and disputations to imitate the wisdom of the Mathematicians, in setting down in the very beginning the definitions of our words and terms, that others may know how we accept and understand them, and whether they concur with us or no[165]. For it cometh to pass for want of this, that we are sure to end there where we ought to have begun, which is in questions and differences about words. To conclude therefore, it must be confessed that it is not possible to divorce ourselves from these fallacies and false appearances, because they are inseparable from our nature and condition of life; so yet nevertheless the caution of them (for all elenches, as was said, are but cautions) doth extremely import the true conduct of human judgment. The particular elenches or cautions against these three false appearances I find altogether deficient.

Elenchi magni, sive de Idolis animi humani, nativis et adventitiis.

There remaineth one part of judgment of great excellency, which to mine understanding is so slightly touched, as I may report that also deficient; which is the application of the differing kinds of proofs to the differing kinds of subjects; for there being but four kinds of demonstrations, that is, by the immediate consent of the mind or sense; by induction; by sophism; and by congruity, which is that which Aristotle calleth *demonstration in orb or circle*, and not *a notioribus*[166]; every of these hath certain subjects in the matter of sciences, in which respectively they have chiefest use; and certain other, from which respectively they ought to be excluded: and the rigour and curiosity in requiring the more severe proofs in some things, and chiefly the facility in contenting ourselves with the more remiss proofs in others, hath been amongst the greatest causes of detriment and hindrance to knowledge. The distributions and assignations of demonstrations, according to the analogy of sciences, I note as deficient.

De Analogia Demonstrationum.

¶[167] The custody or retaining of knowledge is either in Writing or Memory; whereof Writing hath two parts, the nature of the *character*, and the order of the *entry*. For the art of *characters*, or other visible notes of words or things, it hath nearest conjugation with grammar, and therefore I refer it to the due place[168]. For the *disposition* and *collocation* of that knowledge which we preserve in writing, it consisteth in a good digest of commonplaces; wherein I am not ignorant of the prejudice imputed to the use of common-place books, as causing a retardation of reading, and some sloth or relaxation of memory. But because it is but a counterfeit thing in knowledges to be forward and pregnant, except a man be deep and full, I hold the entry of common-places to be a matter of great use and essence in studying; as that which assureth copie of invention, and contracteth judgment to a strength. But this is true, that of the *methods* of common-places that I have seen, there is none of any sufficient worth; all of them carrying merely the face of a *school*, and not of a *world*; and referring to vulgar matters and pedantical divisions without all life or respect to action.

For the other principal part of the custody of knowledge, which is Memory,

[165] It might seem from this that Bacon thought the premising of definitions would be a sufficient remedy for the evil. But in the translation he changes the sentence and expressly warns us that it is not; for the definitions themselves, he says, are made of *words*; and though we think to remove ambiguities by the use of technical terms, etc., yet all is not enough, and we must look for a remedy which goes deeper.

[166] *non a notioribus scilicet, sed tanquam de plano.*—De Aug.

[167] De Aug. v. 5.

[168] All this is omitted in the translation. The art of retaining knowledge is divided into two doctrines: viz. concerning the helps (*adminicula*) of memory, and concerning Memory itself. The only help of memory which is mentioned is writing; concerning which, after remarking that without this help the memory cannot be trusted to deal with matters of length and requiring exactness, especially such as the interpretation of nature, he insists upon the value of a good digest of common-places even in the old and popular sciences, and so proceeds as in the text.

I find that faculty in my judgment weakly enquired of. An art there is extant of it; but it seemeth to me that there are better precepts than that art, and better practices of that art than those received. It is certain the art (as it is) may be raised to points of ostentation prodigious: but in use (as it is now managed) it is barren; not burdensome nor dangerous to natural memory, as is imagined, but barren; that is, not dexterous to be applied to the serious use of business and occasions. And therefore I make no more estimation of repeating a great number of names or words upon once hearing, or the pouring forth of a number of verses or rhymes *ex tempore*, or the making of a satirical simile of every thing, or the turning of every thing to a jest, or the falsifying or contradicting of every thing by cavil, or the like, (whereof in the faculties of the mind there is great copie, and such as by device and practice may be exalted to an extreme degree of wonder,) than I do of the tricks of tumblers, funambuloes, baladines; the one being the same in the mind that the other is in the body; matters of strangeness without worthiness.

This art of Memory is but built upon two intentions; the one Prenotion, the other Emblem. Prenotion dischargeth the indefinite seeking of that we would remember, and directeth us to seek in a narrow compass; that is, somewhat that hath congruity with our *place of memory*. Emblem reduceth conceits intellectual to images sensible, which strike the memory more; out of which axioms may be drawn much better practique than that in use; and besides which axioms, there are divers more touching help of memory, not inferior to them [169]. But I did in the beginning distinguish, not to report those things deficient, which are but only ill managed.

¶ [170] There remaineth the fourth kind of Rational Knowledge, which is transitive, concerning the expressing or transferring our knowledge to others; which I will term by the general name of Tradition or Delivery. Tradition hath three parts; the first concerning the *organ* of tradition; the second concerning the *method* of tradition; and the third concerning the *illustration* of tradition [171].

For the organ of tradition, it is either Speech or Writing: for Aristotle saith well, *Words are the images of cogitations, and letters are the images of words*; but yet it is not of necessity that cogitations be expressed by the medium of words. For *whatsoever is capable of sufficient differences* [172], *and those perceptible by the sense, is in nature competent to express* [173] *cogitations.* And therefore we see in the commerce of barbarous [174] people that understand not one another's language, and in the practice of divers that are dumb and deaf, that men's minds are expressed in gestures, though not exactly, yet to serve the turn. And we understand further [175] that it is the use of China and the kingdoms of the high Levant to write in Characters Real, which express neither letters nor words in gross, but Things or Notions; insomuch as countries and provinces, which understand not one another's language, can nevertheless read one another's writings, because the characters are accepted more generally than the languages do extend; and therefore they have a vast multitude of characters; as many, I suppose, as radical words [176].

These Notes of Cogitations are of two sorts; the one when the note hath some similitude or congruity with the notion; the other *ad placitum*, having force

[169] The nature and use of these prænotions and emblems is explained and illustrated in the translation by several examples; but the substance of the observation is not altered. [170] De Aug. vi. 1.

[171] In the *De Augmentis*, *tradition* (in these last three cases) is translated *sermo*; which appears to be used in the general sense of *communication*.

[172] *i.e.* sufficient to explain the variety of notions.

[173] *i.e.* to convey the cogitations of one man to another (*fieri posse vehiculum cogitationum de homine in hominem*), and so to be an organ of *tradition* (*traditivæ*).

[174] *Barbarous* is omitted in the translation: the thing being equally seen in civilized people who know no common language.

[175] *notissimum fieri jam cœpit.*

[176] This observation is transferred in the *De Augmentis* to the next paragraph, and applied generally to all systems of writing in Characters Real.

only by contract or acceptation. Of the former sort are Hieroglyphics and Gestures. For as to Hieroglyphics, (things of ancient use, and embraced chiefly by the Ægyptians, one of the most ancient nations), they are but as continued impresses and emblems. And as for Gestures, they are as transitory Hieroglyphics, and are to Hieroglyphics as words spoken are to words written, in that they abide not; but they have evermore, as well as the other, an affinity with the things signified: as Periander, being consulted with how to preserve a tyranny newly usurped, bid the messenger attend and report what he saw him do; and went into his garden and topped all the highest flowers; signifying, that it consisted in the cutting off and keeping low of the nobility and *grandes*[177]. *Ad placitum* are the Characters Real before mentioned, and Words: although some have been willing by curious inquiry, or rather by apt feigning, to have derived imposition of names from reason and intendment; a speculation elegant, and, by reason it searcheth into antiquity, reverent; but sparingly mixed with truth, and of small fruit[178]. This portion of knowledge, touching the Notes of Things and cogitations in general, I find not enquired, but deficient. And although it may seem of no great use, considering that words and writings by letters do far excel all the other ways; yet because this part concerneth as it were the mint of knowledge, (for words are the tokens current and accepted for conceits, as moneys are for values, and that it is fit men be not ignorant that moneys may be of another kind than gold and silver), I thought good to propound it to better enquiry.

De Notis Rerum.

Concerning Speech and Words, the consideration of them hath produced the science of Grammar: for man still striveth to reintegrate himself in those benedictions, from which by his fault he hath been deprived; and as he hath striven against the first general curse by the invention of all other arts, so hath he sought to come forth of the second general curse (which was the confusion of tongues) by the art of Grammar: whereof the use in a mother[179] tongue is small; in a foreign tongue more; but most in such foreign tongues as have ceased to be vulgar tongues, and are turned only to learned tongues. The duty of it is of two natures; the one popular[180], which is for the speedy and perfect attaining languages, as well for intercourse of speech as for understanding of authors; the other philosophical, examining the power and nature of words as they are the footsteps and prints of reason: which kind of analogy between words and reasons is handled *sparsim*, brokenly, though not entirely[181]; and therefore I cannot report it deficient, though I think it very worthy to be reduced into a science by itself.

Unto Grammar also belongeth, as an appendix, the consideration of the Accidents of Words; which are measure, sound, and elevation or accent, and the sweetness and harshness of them; whence hath issued some curious observations

177 So in the original; and I believe always in Bacon; the Spanish word being still treated as a foreigner, and the accent falling no doubt upon the first syllable.

178 The substance of this remark is introduced in the translation in another place. Here it is merely said that Characters Real have nothing emblematic in them; but are merely *surds*, framed *ad placitum* and silently agreed upon by custom.

179 *in another tongue* ed. 1605: *in mother tongue* edd. 1629 and 1633. The translation has *in linguis quibusque vernaculis*.

180 In the translation he substitutes *literary* for *popular*.

181 Here are introduced in the translation some interesting remarks on the subject of the analogy between words and reason; in which it is worth observing, among other things, that Bacon appears to have changed his opinion as to the nature of Cæsar's book *De Analogia*, since he wrote the first book of the *Advancement*. See above, p. 70. There he describes it as " a grammatical philosophy, wherein he did labour to make this same *vox ad placitum* to become *vox ad licitum*, and to reduce custom of speech to congruity of speech, and took as it were the picture of words from the life of reason ". Here he says he has doubted whether that book of Cæsar's treated of such a grammatical philosophy as he is speaking of; but that he rather suspects it contained nothing very high or subtile, but only precepts for the formation of a chaste and perfect style, free from vulgarity and affectation.

in Rhetoric, but chiefly Poesy, as we consider it in respect of the verse and not of the argument: wherein though men in learned tongues do tie themselves to the ancient measures, yet in modern languages it seemeth to me as free to make new measures of verses as of dances; for a dance is a measured pace, as a verse is a measured speech [182]. In these things the sense is better judge than the art

Cœnæ fercula nostræ
Mallem convivis quam placuisse cocis

[the dinner is to please the guests that eat it, not the cook that dresses it]. And of the servile expressing antiquity in an unlike and an unfit subject, it is well said, *Quod tempore antiquum videtur, id incongruitate est maxime novum* [there is nothing more new than an old thing that has ceased to fit].

For Ciphers, they are commonly in letters or alphabets, but may be in words. The kinds of Ciphers (besides the simple ciphers with changes and intermixtures of nulls and non-significants) are many, according to the nature or rule of the infolding; Wheel-ciphers, Key-ciphers, Doubles, etc. But the virtues of them, whereby they are to be preferred, are three; that they be not laborious to write and read; that they be impossible to decipher; and, in some cases, that they be without suspicion. The highest degree whereof is to write *omnia per omnia;* which is undoubtedly possible, with a proportion quintuple at most of the writing infolding to the writing infolded, and no other restraint whatsoever [183]. This art of Ciphering, hath for relative an art of Discipering; by supposition [184] unprofitable; but, as things are, of great use. For suppose that ciphers were well managed, there be multitudes of them which exclude the discipherer. But in regard of the rawness and unskilfulness of the hands through which they pass, the greatest matters are many times carried in the weakest ciphers.

In the enumeration of these private and retired arts, it may be thought I seek to make a great muster-roll of sciences; naming them for shew and ostentation, and to little other purpose. But let those which are skilful in them judge whether I bring them in only for appearance, or whether in that which I speak of them (though in few marks) there be not some seed of proficience. And this must be remembered, that as there be many of great account in their countries and provinces, which when they come up to the Seat of the Estate are but of mean rank and scarcely regarded; so these arts being here placed with the principal and supreme sciences, seem petty things; yet to such as have chosen them to spend their studies in them [185], they seem great matters.

¶ [186] For the Method of Tradition, I see it hath moved a controversy in our

[182] This observation is omitted in the translation, and instead we have a censure of the attempts (made not long before Bacon's time) to force the modern languages into the ancient measures; measures (he says) which are incompatible with the frame of the languages themselves, and not less offensive to the ear. But this censure may perhaps be considered as a development of the remark which concludes this paragraph, and which is also omitted. Certainly there is no English metre which represents the metrical effect of the Virgilian hexameter worse than the English hexameter as people write it now; and if any one would try to write it so as to represent the metrical effect truly, by attending to the distinction between accent and quantity, and distributing them according to the same laws, he would find the truth of Bacon's remark that *ipsæ linguæ fabrica respuit;* the English language does not supply the materials.

[183] In the *De Augmentis* he gives a specimen of a cipher by which this feat of writing *omnia per omnia* (that is of conveying any words you please under cover of any other words you please, provided only that they contain not less than five times as many letters) may be accomplished; a cipher invented, he says, by himself when he was at Paris.

[184] *i.e.* if things were as they might be: *attamen præcautione solerti fieri possit inutilis.*

[185] *qui operam illis præcipue impenderint.*—De Aug. The original edition and that of 1629 have "to spend their labours studies in them",—which is also the reading of the edition 1633, except that it has a comma after "labours". "Labours and studies" is the reading of modern editions; but I think it is more likely that one of the words was meant to be substituted for the other.

[186] De Aug. vi. 2.

time[187]. But as in civil business, if there be a meeting and men fall at words there is commonly an end of the matter for that time and no proceeding at all; so in learning, where there is much controversy there is many times little inquiry. For this part of knowledge of Method seemeth to me so weakly enquired as I shall report it deficient.

Method hath been placed, and that not amiss, in Logic, as a part of Judgment: for as the doctrine of Syllogisms comprehendeth the rules of judgment upon that which is invented, so the doctrine of Method containeth the rules of judgment upon that which is to be delivered; for judgment precedeth Delivery, as it followeth Invention[188]. Neither is the method or the nature of the tradition material only to the *use* of knowledge, but likewise to the *progression* of knowledge: for since the labour and life of one man cannot attain to perfection of knowledge, the wisdom of the Tradition is that which inspireth the felicity of continuance and proceeding. And therefore the most real diversity of method is of method referred to Use, and method referred to Progression; whereof the one may be termed Magistral, and the other of Probation[189].

The later whereof seemeth to be *via deserta et interclusa* [a way that is abandoned and stopped up]. For as knowledges are now delivered, there is a kind of contract of error between the deliverer and the receiver: for he that delivereth knowledge desireth to deliver it in such form as may be best believed, and not as may be best examined; and he that receiveth knowledge desireth rather present satisfaction than expectant inquiry; and so rather not to doubt than not to err; glory making the author not to lay open his weakness, and sloth making the disciple not to know his strength.

But knowledge that is delivered as a thread to be spun on, ought to be delivered and intimated[190], if it were possible, *in the same method wherein it was invented;* and so is it possible of knowledge induced. But in this same anticipated and prevented knowledge, no man knoweth how he came to the knowledge which he hath obtained. But yet nevertheless, *secundum majus et minus*, a man may revisit and descend unto the foundations of his knowledge and consent; and so transplant it into another as it grew in his own mind. For it is in knowledges as it is in plants: if you mean to use the plant, it is no matter for the roots; but if you mean to remove it to grow, then it is more assured to rest upon roots than slips. So the delivery of knowledges (as it is now used) is as of fair bodies of trees without the roots; good for the carpenter, but not for the planter; but if you will have sciences grow, it is less matter for the shaft or body of the tree, so you look well to the taking up of the roots. Of which kind of delivery the method of the mathematiques, in that subject, hath some shadow; but generally I see it neither put in ure nor put in inquisition, and therefore note it for deficient.

De Methodo sincera, sive ad filios Scientiarum[191].

Another diversity of Method there is, which hath some affinity with the former, used in some cases by the discretion of the ancients, but disgraced since by the impostures of many vain persons, who have made it as a false light for their counterfeit merchandises; and that is, Enigmatical and Disclosed[192]. The pretence

[187] Besides Ramus himself and Carpentier, one of the principal persons in this controversy was the Cardinal D'Ossat, of whom some account will be found in De Thou's memoirs.—*R. L. E.*

[188] So edd. 1629 and 1633. The original has *Inventions.*

[189] Called *Initiativa* in the translation; and explained to mean the method which discloses the inner mysteries of science; and distinguished from the other not as more secret but as more profound; the one announcing the results of enquiry, the other exhibiting the method and process which led to them.

[190] So in all the editions; but probably a misprint for *insinuated.* The translation has *insinuanda.*

[191] In the translation he gives it the additional name of *Traditio Lampadis;* alluding to the transmission of the lighted torch from one to another in the Greek torch-race. See Preface to *Nov. Org.* p. 222, note.

[192] In the translation he calls the latter *exoterica*, the former *acroamatica;* and explains that the affinity between the *acroamatica* and the *initiativa* lies in this only—that

whereof [193] is to remove the vulgar capacities from being admitted to the secrets of knowledges, and to reserve them to selected auditors, or wits of such sharpness as can pierce the veil.

Another diversity of Method, whereof the consequence is great, is the delivery of knowledge in Aphorisms, or in Methods ; wherein we may observe that it hath been too much taken into custom, out of a few Axioms or observations upon any subject to make a solemn and formal art ; filling it with some discourses, and illustrating it with examples, and digesting it into a sensible Method ; but the writing in Aphorisms hath many excellent virtues, whereto the writing in Method doth not approach.

For first, it trieth the writer, whether he be superficial or solid : for Aphorisms, except they should be ridiculous, cannot be made but of the pith and heart of sciences ; for discourse of illustration is cut off ; recitals of examples are cut off discourse of connexion and order is cut off ; descriptions of practice are cut off ; so there remaineth nothing to fill the Aphorisms but some good quantity of observation : and therefore no man can suffice, nor in reason will attempt, to write aphorisms, but he that is sound and grounded. But in Methods,

> Tantum series juncturaque pollet,
> Tantum de medio sumptis accedit honoris

[the arrangement and connexion and joining of the parts has so much effect], as a man shall make a great shew of an art, which if it were disjointed would come to little. Secondly, Methods are more fit to win consent or belief, but less fit to point to action ; for they carry a kind of demonstration in orb or circle, one part illuminating another, and therefore satisfy ; but particulars, being dispersed, do best agree with dispersed directions. And lastly, Aphorisms, representing a knowledge broken, do invite men to enquire farther ; whereas Methods, carrying the shew of a total, do secure men, as if they were at furthest.

Another diversity of Method, which is likewise of great weight, is the handling of knowledge by Assertions and their Proofs, or by Questions and their Determinations ; the latter kind whereof, if it be immoderately followed, is as prejudicial to the proceeding of learning, as it is to the proceeding of an army to go about to besiege every little fort or hold. For if the field be kept and the sum of the enterprise pursued, those smaller things will come in of themselves : indeed [194] a man would not leave some important piece enemy at his back. In like manner, the use of confutation in the delivery of sciences ought to be very sparing ; and to serve to remove strong preoccupations and prejudgments, and not to minister and excite disputations and doubts.

Another diversity of Methods is *according to the subject or matter which is handled ;* for there is a great difference in delivery of the Mathematics, which are the most abstracted of knowledges, and Policy, which is the most immersed : and howsoever contention hath been moved touching an uniformity of method in multiformity of matter, yet we see how that opinion, besides the weakness of it, hath been of ill desert towards learning, as that which taketh the way to reduce learning to certain empty and barren generalities ; being but the very husks and shells of sciences, all the kernel being forced out and expulsed with the torture and press of the method ; [195] and therefore as I did allow well of particular Topics for invention, so I do allow likewise of particular Methods of tradition.

each addresses itself to a select audience ; for in themselves (*re ipsa*) they are opposite ; the *initiativa* adopting a method of delivery more open than ordinary ; the *acroamatica*, one more obscure ; the " vulgar capacities " being excluded in the one case by the necessary subtilty of the argument, in the other by an affected obscurity in the exposition. Concerning the latter method, see Preface to the *Novum Organum*, note B.

193 *i.e.* of the enigmatical method.

194 *i.e.* " although indeed " : (*illud tamen inficias non iverim*, etc.).

195 This observation is introduced in the translation at the beginning of the chapter, and applied particularly to the method of *dichotomies ;* which are not mentioned, I think, by name in the *Advancement.*

Another diversity of judgment [196] in the delivery and teaching of knowledge is *according unto the light and presuppositions of that which is delivered ;* for that knowledge which is new and foreign from opinions received, is to be delivered in another form than that that is agreeable [197] and familiar ; and therefore Aristotle, when he thinks to tax Democritus, doth in truth commend him, where he saith, *If we shall indeed dispute, and not follow after similitudes,* etc. For those whose conceits are seated in popular opinions, need only but to prove or dispute ; but those whose conceits are beyond popular opinions, have a double labour ; the one to make themselves conceived, and the other to prove and demonstrate ; so that it is of necessity with them to have recourse to similitudes and translations to express themselves. And therefore in the infancy of learning, and in rude times, when those conceits which are now trivial were then new, the world was full of Parables and Similitudes ; for else would men either have passed over without mark or else rejected for paradoxes that which was offered, before they had understood or judged. So in divine learning we see how frequent Parables and Tropes are [198] : for it is a rule, *That whatsoever science is not consonant to presuppositions, must pray in aid of similitudes.*

There be also other diversities of Methods, vulgar and received ; as that of Resolution or Analysis, of Constitution or Systasis, of Concealment or Cryptic [199], &c. which I do allow well of ; though I have stood upon those which are least handled and observed. All which I have remembered to this purpose, because I would erect and constitute one general inquiry, which seems to me deficient, touching the Wisdom of Tradition. *De prudentia Traditionis.*

But unto this part of knowledge concerning Method doth further belong not only the Architecture of the whole frame of a work, but also the several beams and columns thereof ; not as to their stuff, but as to their quantity and figure ; and therefore Method considereth not only the disposition of the Argument or Subject, but likewise the Propositions ; not as to their truth or matter, but as to their limitation and manner. For herein Ramus merited better a great deal in reviving the good rules of Propositions, Καθόλου πρῶτον, κατὰ παντός, etc.[200] than he did in introducing the canker of Epitomes [201] ; and yet (as it is the condition of human things that, according to the ancient fables, *The most precious things have the most pernicious keepers*) ; it was so, that the attempt of the one made him fall upon the other [202]. For he had need be well conducted that should design to make Axioms *convertible*, if he make them not withal *circular*, and *non-promovent*, or *incurring into themselves* : but yet the intention was excellent.

The other considerations of Method concerning Propositions are chiefly touching the utmost propositions, which limit the dimensions of sciences ; for every knowledge may be fitly said, besides the profundity, (which is the truth and substance of it, that makes it solid), to have a longitude and a latitude ; accounting the latitude towards other sciences, and the longitude towards action ; that is, from the greatest generality to the most particular precept : the one giveth rule how far one knowledge ought to intermeddle within the province of another,

[196] *i.e.* a diversity of method to be used with judgment. (*Sequitur aliud methodi discrimen in tradendis scientiis cum judicio adhibendum*). This may perhaps be an error of the press or of the transcriber, some words having accidentally dropped out. It may however be merely an effect of hasty composition, of which there are many evidences in this part of the work.

[197] *i.e.* in accordance with received opinions. (*Opinionibus jam pridem imbibitis et receptis affinis*).

[198] This allusion to divine learning is omitted in the translation.

[199] In the translation he adds *Diæretica* and *Homerica*, and observes that he does not dwell upon these because they have been rightly invented and distributed.

[200] That they should be true generally, primarily, and essentially.—*R. L. E.*

[201] Instead of " the canker of Epitomes ", the translation substitutes " his peculiar method and dichotomies ".

[202] The attempt to amend propositions cast him upon those epitomes and shallows of knowledge, as they are called in the translation—*epitomas illas et scientiarum vada.*

which is the rule they call Καθαυτὸ [203]; the other giveth rule unto what degree of particularity a knowledge should descend: which latter I find passed over in silence, being in my judgment the more material; for certainly there must be somewhat left to practice [204]; but how much is worthy the inquiry. We see remote and superficial generalities do but offer knowledge to scorn of practical men; and are no more aiding to practice, than an Ortelius' universal map is to direct the way between London and York. The better sort of rules have been not unfitly compared to glasses of steel unpolished, where you may see the images of things, but first they must be filed: so the rules will help, if they be laboured and polished by practice. But how chrystalline they may be made at the first, and how far forth they may be polished aforehand, is the question; the inquiry whereof seemeth to me deficient.

De productione Axiomatum.

There hath been also laboured and put in practice a method, which is not a lawful method, but a method of imposture; which is to deliver knowledges in such manner, as men may speedily come to make a shew of learning who have it not: such was the travail of Raymundus Lullius, in making that art which bears his name, not unlike to some books of Typocosmy which have been made since; being nothing but a mass of words of all arts, to give men countenance that those which use the terms might be thought to understand the art; which collections are much like a fripper's or broker's shop, that hath ends of every thing, but nothing of worth.

¶ [205] Now we descend to that part which concerneth the Illustration of Tradition, comprehended in that science which we call Rhetoric, or Art of Eloquence; a science excellent, and excellently well laboured. For although in true value it is inferior to wisdom, as it is said by God to Moses, when he disabled himself for want of this faculty, *Aaron shall be thy speaker, and thou shalt be to him as God;* yet with people it is the more mighty: for so Salomon saith, *Sapiens corde appellabitur prudens, sed dulcis eloquio majora reperiet,* [the wise in heart shall be called prudent, but he that is sweet of speech shall compass greater things]; signifying that profoundness of wisdom will help a man to a name or admiration, but that it is eloquence that prevaileth in an active life. And as to the labouring of it, the emulation of Aristotle with the rhetoricians of his time, and the experience of Cicero, hath made them in their works of Rhetorics exceed themselves. Again, the excellency of examples of eloquence in the orations of Demosthenes and Cicero, added to the perfection of the precepts of eloquence, hath doubled the progression in this art; and therefore the deficiencies which I shall note will rather be in some collections which may as handmaids attend the art, than in the rules or use of the art itself.

Notwithstanding, to stir the earth a little about the roots of this science, as we have done of the rest: The duty and office of Rhetoric is *to apply Reason to Imagination* [206] for the better moving of the will. For we see Reason is disturbed in the administration thereof by three means; by Illaqueation or Sophism, which pertains to Logic; by Imagination or Impression [207], which pertains to Rhetoric; and by Passion or Affection, which pertains to Morality [208]. And as in negotiation with others men are wrought by cunning, by importunity, and by vehemency; so in this negotiation within ourselves men are undermined by Inconsequences, solicited and importuned by Impressions or Observations, and transported by Passions. Neither is the nature of man so unfortunately built, as that those powers and arts should have force to disturb reason, and not to

[203] This is omitted in the translation. "The rule they call καθαυτὸ," is the rule that propositions should be true *essentially.*

[204] For we must not fall into the error of Antoninus Pius (he adds in the translation)—to become *Cymini Sectores,* multiplying divisions to the last degree of minuteness.

[205] De Aug. vi. 3.

[206] Rhetoric being to the Imagination what Logic is to the Understanding.—*De Aug.*

[207] In the translation he substitutes *per præstigias verborum;* false impressions produced by words on the imagination.

[208] *i.e.* moral philosophy. (*Ethica.*)

establish and advance it: for the end of Logic is to teach a form of argument to secure reason, and not to entrap it; the end of Morality is to procure the affections to obey[209] reason, and not to invade it; the end of Rhetoric is to fill the imagination to second reason, and not to oppress it: for these abuses of arts come in but *ex obliquo*, for caution.

And therefore it was great injustice in Plato, though springing out of a just hatred of the rhetoricians of his time, to esteem of Rhetoric but as a voluptuary art, resembling it to cookery, that did mar wholesome meats, and help unwholesome by variety of sauces to the pleasure of the taste. For we see that speech is much more conversant in adorning that which is good than in colouring that which is evil; for there is no man but speaketh more honestly than he can do or think: and it was excellently noted by Thucydides in Cleon[210], that because he used to hold on the bad side in causes of estate, therefore he was ever inveighing against eloquence and good speech; knowing that no man can speak fair of courses sordid and base. And therefore as Plato said elegantly, *That virtue, if she could be seen, would move great love and affection*; so seeing that she cannot be shewed to the Sense by corporal shape, the next degree is to shew her to the Imagination in lively representation: for to shew her to Reason only in subtilty of argument, was a thing ever derided in Chrysippus and many of the Stoics; who thought to thrust virtue upon men by sharp disputations and conclusions, which have no sympathy with the will of man.

Again, if the affections in themselves were pliant and obedient to reason, it were true there should be no great use of persuasions and insinuations to the will, more than of naked proposition and proofs; but in regard of the continual mutinies and seditions of the affections,

Video meliora, proboque;
Deteriora sequor

[whereby they who not only see the better course, but approve it also, nevertheless follow the worse], reason would become captive and servile, if Eloquence of Persuasions did not practise and win the Imagination from the Affection's part, and contract a confederacy between the Reason and Imagination against the Affections. For the affections themselves carry ever an appetite to good, as reason doth; the difference is, that *the affection beholdeth merely the present; reason beholdeth the future and sum of time*; and therefore the present filling the imagination more, reason is commonly vanquished; but after that force of eloquence and persuasion hath made things future and remote appear as present, then upon the revolt of the imagination reason prevaileth.

We conclude therefore, that Rhetoric can be no more charged with the colouring of the worse part, than Logic with Sophistry, or Morality with Vice[211]. For we know the doctrines of contraries are the same, though the use be opposite. It appeareth also that Logic differeth from Rhetoric, not only as the fist from the palm, the one close, the other at large; but much more in this, that Logic handleth reason exact and in truth, and Rhetoric handleth it as it is planted in popular opinions and manners. And therefore Aristotle doth wisely place Rhetoric as between Logic on the one side and moral or civil knowledge on the

209 In the translation he says *ut rationi militent*; to fight on the side of reason.

210 In the translation he says, more correctly, "it was noted by Thucydides *as a censure passed upon Cleon*" (*tale quidpiam solitum fuisse objici Cleoni*); for the observation is made by Diodotus in his answer to Cleon's speech, iii. 42.

211 The last clause is omitted in the translation. I do not know why. For according to Bacon's doctrine, expounded originally in the *Meditationes Sacræ* upon the text *non accipit stultus verba prudentiæ nisi ea dixeris quæ versantur in corde ejus*, and repeated here a little further on,—namely, that a man can neither protect his own virtue against evil arts, nor reclaim others from vice, without the help of the knowledge of evil,—Morality has a relation to Vice exactly corresponding with that of Logic to Sophistry; unless it be maintained that the Logician ought to be prepared to practise Sophistry as well as to detect and defeat it.

other, as participating of both : for the proofs and demonstrations of Logic are toward all men indifferent and the same ; but the proofs and persuasions of Rhetoric ought to differ according to the auditors :

Orpheus in sylvis, inter delphinas Arion

[to be in the woods an Orpheus, among the dolphins an Arion] : which application, in perfection of idea, ought to extend so far, that if a man should speak of the same thing to several persons, he should speak to them all respectively and several ways : though this *politic part of eloquence in private speech* it is easy for the greatest orators to want, whilst by the observing their well-graced forms of speech they leese the volubility of application : and therefore it shall not be amiss to recommend this to better inquiry[212] ; not being curious whether we place it here, or in that part which concerneth policy.

De prudentia Sermonis privati.

Now therefore will I descend to the deficiences, which (as I said) are but attendances[213] : and first, I do not find the wisdom and diligence of Aristotle well pursued, who began to make a collection of *the popular signs and colours of good and evil, both simple and comparative*, which are as the Sophisms of Rhetoric (as I touched before). For example :

Colores boni et mali, simplicis et comparati.

SOPHISMA.

Quod laudatur, bonum : quod vituperatur, malum.

REDARGUTIO.

Laudat venales qui vult extrudere merces.
Malum est, malum est, inquit emptor : sed cum recesserit, tum gloriabitur[214].

The defects in the labour of Aristotle are three : one, that there be but a few of many ; another, that their Elenches are not annexed[215] : and the third, that he conceived but a part of the use of them : for their use is not only in probation, but much more in impression. For many forms are equal in signification which are differing in impression ; as the difference is great in the piercing of that which is sharp and that which is flat, though the strength of the percussion be the same ; for there is no man but will be a little more raised by hearing it said, *your enemies will be glad of this :*

Hoc Ithacus velit, et magno mercentur Atridæ :

than by hearing it said only, *This is evil for you.*

Secondly, I do resume also that which I mentioned before touching Provision or Preparatory store for the furniture of speech and readiness of invention ; which appeareth to be of two sorts ; the one in resemblance to a shop of pieces unmade up, the other to a shop of things ready made up ; both to be applied to that which is frequent and most in request : the former of these I will call *Antitheta*, and the latter *Formulæ*.

Antitheta are Theses argued *pro et contra ;* wherein men may be more large

212 Being a thing which the more it is considered the more it will be valued (*rem certe quam quo attentius quis recogitet, eo pluris faciet*).

213 and which are all of the nature of collections for store (*pertinent omnia ad promptuariam*).

214 SOPHISM.—That which people praise is good, that which they blame is bad.
ELENCHE.—He praises his wares who wants to get them off his hands.
It is naught, it is naught, sayth the buyer ; but when he is gone he will vaunt.

215 In the translation, instead of the single example given above, he inserts a collection of twelve, by way of specimen ; each having the elenche annexed and completely explained. This collection is a translation, with corrections and additions, of the English tract entitled " Colours of Good and Evil," which was printed along with the Essays in 1597.

Antitheta Rerum. and laborious: but (in such as are able to do it) to avoid prolixity of entry, I wish the seeds of the several arguments to be cast up into some brief and acute sentences; not to be cited, but to be as skeins or bottoms of thread, to be unwinded at large when they come to be used; supplying authorities and examples by reference.

PRO VERBIS LEGIS.

Non est interpretatio, sed divinatio, quæ recedit a literâ.
Cum receditur a literâ, judex transit in legislatorem.

PRO SENTENTIA LEGIS.

Ex omnibus verbis est eliciendus sensus qui interpretatur singula[216].

[217] *Formulæ* are but decent and apt passages or conveyances of speech which may serve indifferently for differing subjects; as of preface, conclusion, digression, transition, excusation, &c. For as in buildings there is great pleasure and use in the well-casting of the stair-cases, entries, doors, windows, and the like; so in speech the conveyances and passages are of special ornament and effect.

A CONCLUSION IN A DELIBERATIVE.

So may we redeem the faults passed, and prevent the inconveniences future.

¶ [218] There remain two appendices touching the tradition of knowledge, the one Critical, the other Pedantical.[219] For all knowledge is either delivered by teachers, or attained by men's proper endeavours: and therefore as the principal part of tradition of knowledge concerneth chiefly writing [220] of books, so the relative part thereof concerneth reading of books. Whereunto appertain incidently these considerations. The first is concerning the true correction and edition of authors; wherein nevertheless rash diligence hath done great prejudice. For these critics have often presumed that that which they understand not is false set down: as the Priest that where he found it written of St. Paul, *Demissus est per sportam* [he was let down in a basket], mended his book, and made it *Demissus est per portam* [he was let out by the gate]; because *sporta* was an hard word, and out of his reading[221]; and surely their errors, though

[216] For the Words of the Law.—Interpretation which departs from the letter, is not interpretation but divination.

When the letter is departed from the Judge becomes the Lawgiver.

For the Intention of the Law.—The sense according to which each word is to be interpreted must be collected from all the words together.

Of these *antitheta* a large collection will be found in the *De Augmentis*, set forth by way of specimen in the manner here recommended.

[217] Of these *formulæ*—or *formulæ minores* as he afterwards called them—three other examples are given in the *De Augmentis*, all from Cicero. Bacon's own speeches and narrative writings would supply many very good ones.

[218] De Aug. vi. 4. [219] *Pædagogica*, in the translation.

[220] *in writing*, in the original; and also in the editions 1629 and 1633. The translation has *in lectione librorum consistit*.

[221] For this illustration, which as reflecting upon a Priest might have been offensive at Rome, another is substituted in the *De Augmentis*, which is "not so palpable and ridiculous." A striking instance of the same kind occurs in two recent editions of this very work. In an edition of the *Advancement of Learning* published by J. W. Parker in 1852, *Orosius* is substituted for *Osor us* in the passage (p. 283), "Then grew the flowing and watery vein of Osorius, the Portugal Bishop, to be in price"; with the following note: "All the editions have *Osorius*, which however must be a mere misprint. He was not a Portuguese, but a Spaniard, born at Tarragona, nor indeed ever a bishop. He was sent by St. Augustine on a mission to Jerusalem, and is supposed to have died in Africa in the earlier part of the fifth century." In the following year Mr. H. Bohn published a translation of the *De Augmentis*, which is little more than a reprint of Shaw's transla-

they be not so palpable and ridiculous, are yet of the same kind. And therefore as it hath been wisely noted, the most corrected copies are commonly the least correct.

The second is concerning the exposition and explication of authors, which resteth in annotations and commentaries; wherein it is over usual to blanch the obscure places, and discourse upon the plain.

The third is concerning the times, which in many cases give great light to true interpretations [222].

The fourth is concerning some brief censure and judgment of the authors; that men thereby may make some election unto themselves what books to read.

And the fifth is concerning the syntax and disposition of studies; that men may know in what order or pursuit to read [223].

For Pedantical knowledge, it containeth that difference of Tradition which is proper for youth; whereunto appertain divers considerations of great fruit.

As first, the timing and seasoning of knowledges; as with what to initiate them, and from what for a time to refrain them.

Secondly, the consideration where to begin with the easiest and so proceed to the more difficult; and in what courses [224] to press the more difficult and then to turn them to the more easy: for it is one method to practise swimming with bladders, and another to practise dancing with heavy shoes.

A third is the application of learning according unto the propriety of the wits; for there is no defect in the faculties intellectual but seemeth to have a proper cure contained in some studies: as for example, if a child be bird-witted, that is, hath not the faculty of attention, the Mathematics giveth a remedy thereunto; for in them, if the wit be caught away but a moment, one is new to begin. And as sciences have a propriety towards faculties for cure and help, so faculties or powers have a sympathy towards sciences for excellency or speedy profiting; and therefore it is an inquiry of great wisdom, what kinds of wits and natures are most apt and proper for what sciences.

Fourthly, the ordering of exercises is matter of great consequence to hurt or help; for as is well observed by Cicero, men in exercising their faculties, if they be not well advised, do exercise their faults and get ill habits as well as good; so as there is a great judgment to be had in the continuance and intermission of exercises. It were too long to particularise a number of other considerations of this nature, things but of mean appearance, but of singular efficacy. For as the wronging or cherishing of seeds or young plants is that that is most important to their thriving; and as it was noted that the first six kings being in truth as tutors of the state of Rome in the infancy thereof, was the principal cause of the immense greatness of that state which followed: so the culture and manurance of minds in youth hath such a forcible (though unseen) operation, as hardly any length of time or contention of labour can countervail it afterwards. And it is not amiss to observe also how small and mean faculties gotten

tion, revised and edited by Mr. Joseph Devey. In this edition *Orosius* is silently substituted for *Osorius* in the same passage, with this note: "Neither a Portuguese, nor a bishop, but a Spanish monk born at Tarragona, and sent by St. Augustine on a mission to Jerusalem in the commencement of the fifth century". The mistake is the more remarkable because the passage in Bacon refers obviously and unmistakably to the period of the Reformation.

222 This point is omitted in the translation, except in so far as it is involved in an observation which is added under the next head—viz. that editors besides giving "some brief censure and judgment of their authors" should compare them with other writers on the same subjects. But I am inclined to suspect that the omission was accidental for the truth is, that without constant reference to the times and circumstances in which he wrote hardly any author can be properly understood.

223 This point is also omitted in the translation; perhaps as included in the "censure and judgment"; which (he adds) is as it were the Critic's chair; an office ennobled in his time by some great men, *majores certe nostro judicio quam pro modulo criticorum*,—men above the stature of critics.

224 So all the editions: probably a misprint for *cases*.

by education, yet when they fall unto great men or great matters, do work great and important effects; whereof we see a notable example in Tacitus of two stage-players, Percennius and Vibulenus, who by their faculty of playing put the Pannonian armies into an extreme tumult and combustion. For there arising a mutiny amongst them upon the death of Augustus Cæsar, Blæsus the lieutenant had committed some of the mutiners; which were suddenly rescued; whereupon Vibulenus got to be heard speak, which he did in this manner:—*These poor innocent wretches, appointed to cruel death, you have restored to behold the light. But who shall restore my brother to me, or life unto my brother? that was sent hither in message from the legions of Germany to treat of the common cause, and he hath murdered him this last night by some of his fencers and ruffians, that he hath about him for his executioners upon soldiers. Answer, Blæsus, what is done with his body? The mortalest enemies do not deny burial. When I have performed my last duties to the corpse with kisses, with tears, command me to be slain besides him; so that these my fellows, for our good meaning and our true hearts to the legions, may have leave to bury us.*[225] With which speech he put the army into an infinite fury and uproar; whereas truth was he had no brother, neither was there any such matter, but he played it merely as if he had been upon the stage.

[225] The last clause does not give the exact meaning of the original, from which it may seem that Bacon was reporting the speech from memory; unless it be that a line has accidentally dropped out. By inserting after "fellows" the words "seeing us put to death for no crime, but only for", &c. the sense would be represented with sufficient accuracy.

In the translation, this passage relating to "Pedantical knowledge",—that is the knowledge which concerns the instruction of youth—is considerably enlarged, and a distinct opinion is expressed upon many of the points which are here only noticed as worthy of inquiry. He begins by recommending the schools of the Jesuits as the best model,—an opinion which he had already intimated in the first book of the *Advancement*. He approves of a collegiate education both for boys and young men, as distinguished from a private education under masters. He wishes compendiums to be avoided, and the system which, aiming at precocity, produces overconfidence and a mere show of proficiency. He would encourage independence of mind, and if any one shows a taste for studies which lie out of the regular course, and can find time to pursue them, he would by no means have him restrained. Of the two methods mentioned in the text, one beginning with the easiest tasks, the other with the most difficult, he recommends a judicious intermixture, as best for the advancement of the powers both of mind and body. With regard to the "application of learning according unto the propriety of the wits," he observes (besides its use as a corrective of mental defects) that masters ought to attend to it for the guidance of the parents in choosing their sons' course of life; and also because a man will advance so much faster in studies for which he has a natural aptitude than in any others. With regard to the "ordering of exercises" he recommends the system of intermissions. (*Itaque tutius est intermittere exercitia et subinde repetere, quam assidue continuare et urgere*). Lastly he would decidedly have the art of *acting* (*actio theatralis*) made a part of the education of youth. The Jesuits, he says, do not despise it; and he thinks they are right; for though it be of ill repute as a profession (*si sit professoria, infamis est*) yet as a part of *discipline* it is of excellent use. It strengthens the memory, it regulates the tone and effect of the voice and pronunciation, it teaches a decent carriage of the countenance and gesture, it begets no small degree of confidence, and accustoms young men to bear being looked at. In Bacon's time, when masques acted by young gentlemen of the Universities or Inns of Court were the favourite entertainment of princes, these things were probably better attended to than they are now—and he could have pointed no doubt to many living examples in illustration of his remark. The examples which modern experience supplies are all of the negative kind, but not therefore the less significant. The art of speaking, of recitation, even of reading aloud, is not now taught at all; and the consequence is that even among men otherwise accomplished not many will be found who can either speak a speech of their own, or recite the speech of another, or read a book aloud, so as to be listened to with pleasure in a mixed company for a quarter of an hour together.

But to return : we are now come to a period of Rational Knowledges ; wherein if I have made the divisions other than those that are received, yet would I not be thought to disallow all those divisions which I do not use. For there is a double necessity imposed upon me of altering the divisions. The one, because it differeth in end and purpose, to sort together those things which are next in nature, and those things which are next in use. For if a secretary of state should sort his papers, it is like in his study or general cabinet he would sort together things of a nature, as treaties, instructions, etc. but in his boxes or particular cabinet he would sort together those that he were like to use together, though of several natures ; so in this general cabinet of knowledge it was necessary for me to follow the divisions of the nature of things ; whereas if myself had been to handle any particular knowledge, I would have respected the divisions fittest for use. The other, because the bringing in of the deficiences did by consequence alter the partitions of the rest : for let the knowledge extant (for demonstration sake) be fifteen ; let the knowledge with the deficiences be twenty ; the parts of fifteen are not the parts of twenty ; for the parts of fifteen are three and five ; the parts of twenty are two, four, five, and ten. So as these things are without contradiction, and could not otherwise be.

¶ [226] We proceed now to that knowledge which considereth of the Appetite [227] and Will of Man ; whereof Salomon saith, *Ante omnia, fili, custodi cor tuum ; nam inde procedunt actiones vitæ* [keep thy heart with all diligence, for thereout come the actions of thy life]. In the handling of this science, those which have written seem to me to have done as if a man that professeth to teach to write did only exhibit fair copies of alphabets and letters joined, without giving any precepts or directions for the carriage of the hand and framing of the letters. So have they made good and fair exemplars and copies, carrying the draughts and portraitures of Good, Virtue, Duty, Felicity ; propounding them well described as the true objects and scopes of man's will and desires ; but how to attain these excellent marks, and how to frame and subdue the will of man to become true and conformable to these pursuits, they pass it over altogether, or slightly and unprofitably. For it is not the disputing *that moral virtues are in the mind of man by habit and not by nature*, or the distinguishing *that generous spirits are won by doctrines and persuasions, and the vulgar sort by reward and punishment* [228], and the like scattered glances and touches, that can excuse the absence of this part.

The reason of this omission I suppose to be that hidden rock whereupon both this and many other barks of knowledge have been cast away ; which is, that men have despised to be conversant in ordinary and common matters : the judicious direction whereof nevertheless is the wisest doctrine (for life consisteth not in novelties or subtilities) ; but contrariwise they have compounded sciences chiefly of a certain resplendent or lustrous mass of matter, chosen to give glory either to the subtility of disputations or to the eloquence of discourses.¶ But Seneca giveth an excellent check to eloquence ; *Nocet illis eloquentia, quibus non rerum cupiditatem facit, sed sui* [eloquence does mischief when it draws men's attention away from the matter to fix it on itself]. Doctrines should be such as should make men in love with the lesson, and not with the teacher ; being directed to the auditor's benefit, and not to the author's commendation : and therefore those are of the right kind which may be concluded as Demosthenes concludes his counsel, *Quæ si feceritis, non oratorem duntaxat in præsentia laudabitis, sed vosmetipsos etiam non ita multo post statu rerum vestrarum meliore* [if you follow this advice you will do a grace to yourselves no less than to the speaker,—to him by your vote to-day, to yourselves by the improvement which you will presently find in your affairs].

Neither needed men of so excellent parts to have despaired of a fortune which

[226] De Aug. vii. 1.

[227] In the translation the word *Appetite* is omitted ; and the Will is described as governed by right reason, seduced by apparent good, having the passions for spurs, the organs and voluntary motions for ministers.

[228] Or the giving it in precept (he adds in the translation) that if you would rectify the mind you must bend it like a wand in the direction contrary to its inclination.

the poet Virgil promised himself, (and indeed obtained,) who got as much glory of eloquence, wit, and learning in the expressing of the observations of husbandry, as of the heroical acts of Æneas :—

> Nec sum animi dubius, verbis ea vincere magnum
> Quam sit, et angustis his addere rebus honorem.
> [How hard the task, alas! full well I know,
> With charm of words to grace a theme so low.]

And surely if the purpose be in good earnest not to write at leisure that which men may read at leisure, but really to instruct and suborn action and active life, these Georgics of the mind, concerning the husbandry and tillage thereof, are no less worthy than the heroical descriptions of Virtue, Duty, and Felicity. Wherefore the main and primitive division of moral knowledge seemeth to be into the Exemplar or Platform of Good, and the Regiment or Culture of the Mind ; the one describing the nature of good, the other prescribing rules how to subdue, apply, and accommodate the will of man thereunto.

The doctrine touching the Platform or Nature of Good considereth it either Simple or Compared : either the kinds of good, or the degrees of good : in the later whereof those infinite disputations which were touching the supreme degree thereof, which they term felicity, beatitude, or the highest good, the doctrines concerning which were as the heathen divinity, are by the Christian faith discharged. And as Aristotle saith, *That young men may be happy, but not otherwise but by hope* ; so we must all acknowledge our minority, and embrace the felicity which is by hope of the future world.

Freed therefore and delivered from this doctrine of the philosophers' heaven, whereby they feigned an higher elevation of man's nature than was, (for we see in what an height of style Seneca writeth, *Vere magnum, habere fragilitatem hominis, securitatem Dei* [it is true greatness to have in one the frailty of a man and the security of a God], we may with more sobriety and truth receive the rest of their inquiries and labours. Wherein for the Nature of Good Positive or Simple, they have set it down excellently, in describing the forms of Virtue and Duty, with their situations and postures, in distributing them into their kinds, parts, provinces, actions, and administrations, and the like : nay farther, they have commended them to man's nature and spirit with great quickness of argument and beauty of persuasions ; yea, and fortified and intrenched them (as much as discourse can do) against corrupt and popular opinions. Again, for the Degrees and Comparative Nature of Good, they have also excellently handled it in their triplicity of Good, in the comparisons between a contemplative and an active life, in the distinction between virtue with reluctation and virtue secured, in their encounters between honesty and profit, in their balancing of virtue with virtue, and the like ; so as this part deserveth to be reported for excellently laboured [229].

Notwithstanding, if before they had comen to the popular and received notions of virtue and vice, pleasure and pain, and the rest, they had stayed a little longer upon the inquiry concerning the roots of good and evil, and the strings of these roots, they had given, in my opinion, a great light to that which followed ; and specially if they had consulted with nature, they had made their doctrines less prolix and more profound ; which being by them in part omitted and in part handled with much confusion, we will endeavour to resume and open in a more clear manner.

There is formed in every thing a double nature of good : the one, as every thing is a total or substantive in itself ; the other, as it is a part or member of a greater body ; whereof the later is in degree the greater and the worthier, because it tendeth to the conservation of a more general form. Therefore we see the iron in particular sympathy moveth to the loadstone ; but yet if it exceed a certain quantity, it forsaketh the affection to the loadstone, and like a good patriot moveth to the earth, which is the region and country of massy bodies ; so may we go forward, and see that water and massy bodies move to the centre of the earth;

[229] Well by the ancient philosophers, but still better (according to the translation) by the divines in their discussions of moral duties and virtues, cases of conscience, sins, etc.

but rather than to suffer a divulsion in the continuance of nature, they will move upwards from the centre of the earth, forsaking their duty to the earth in regard of their duty to the world. This double nature of good, and the comparative thereof, is much more engraven upon man, if he degenerate not; unto whom the conservation of duty to the public ought to be much more precious than the conservation of life and being: according to that memorable speech of Pompeius Magnus, when being in commission of purveyance for a famine at Rome, and being dissuaded with great vehemency and instance by his friends about him that he should not hazard himself to sea in an extremity of weather, he said only to them, *Necesse est ut eam, non ut vivam* [it is needful that I go, not that I live]. But it may be truly affirmed that there was never any philosophy, religion, or other discipline, which did so plainly and highly exalt the good which is communicative, and depress the good which is private and particular, as the Holy Faith; well declaring, that it was the same God that gave the Christian law to men, who gave those laws of nature to inanimate creatures that we spake of before; for we read that the elected saints of God have wished themselves anathematized and razed out of the book of life, in an ecstasy of charity and infinite feeling of communion.

This being set down and strongly planted, doth judge and determine most of the controversies wherein Moral Philosophy is conversant. For first it decideth the question touching the preferment of the contemplative or active life, and decideth it against Aristotle. For all the reasons which he bringeth for the contemplative are private, and respecting the pleasure and dignity of a man's self, (in which respects no question the contemplative life hath the pre-eminence): not much unlike to that comparison which Pythagoras made for the gracing and magnifying of philosophy and contemplation; who being asked what he was, answered, *That if Hiero were ever at the Olympian games, he knew the manner, that some came to try their fortune for the prizes, and some came as merchants to utter their commodities, and some came to make good cheer and meet their friends, and some to look on; and that he was one of them that came to look on.* But men must know, that in this theatre of man's life it is reserved only for God and Angels to be lookers on. Neither could the like question ever have been received in the church, notwithstanding their *Pretiosa in oculis Domini mors sanctorum ejus* [precious in the sight of the Lord is the death of his saints], by which place they would exalt their civil death and regular professions, but upon this defence, that the monastical life is not simple [230] contemplative, but performeth the duty either of incessant prayers and supplications, which hath been truly esteemed as an office in the church, or else of writing or taking [231] instructions for writing concerning the law of God, as Moses did when he abode so long in the mount. And so we see Henoch the seventh from Adam, who was the first Contemplative and walked with God, yet did also endow the church with prophecy, which St. Jude citeth. But for contemplation which should be finished in itself without casting beams upon society, assuredly divinity knoweth it not.

It decideth also the controversies between Zeno and Socrates and their schools and successions on the one side, who placed felicity in virtue simply or attended; the actions and exercises whereof do chiefly embrace and concern society; and on the other side [232], the Cyrenaics and Epicureans, who placed it in pleasure, and made virtue (as it is used in some comedies of errors, wherein the mistress and the maid change habits), to be but as a servant, without which pleasure cannot be served and attended; and the reformed school of the Epicureans, which placed it in serenity of mind and freedom from perturbation; as if they would have deposed Jupiter again, and restored Saturn and the first age, when there was no

230 Edd. 1629 and 1633 have *simply*.

231 So edd. 1629 and 1633. The original has *in taking*. In the translation the words "taking instructions for writing" are omitted; as applicable, I suppose, to the case of Moses only, not of the Church; and *multo in otio* substituted.

232 *Et reliquas complures sectas et scholas, ex altera parte: veluti,* etc. All the opinions which are about to be cited belong to "the other side"—*i.e.* the side opposed to that of Zeno and Socrates; a point which from the careless composition of the English is not immediately clear.

summer nor winter, spring nor autumn, but all after one air and season; and Herillus [233], which placed felicity in extinguishment of the disputes of the mind, making no fixed nature of good and evil, esteeming things according to the clearness of the desires, or the reluctation [234]; which opinion was revived in the heresy of the Anabaptists, measuring things according to the motions of the spirit, and the constancy or wavering of belief: all which are manifest to tend to private repose and contentment, and not to point of society.

It censureth also the philosophy of Epictetus, which presupposeth that felicity must be placed in those things which are in our power, lest we be liable to fortune and disturbance: as if it were not a thing much more happy to fail in good and virtuous ends for the public, than to obtain all that we can wish to ourselves in our proper fortune; as Consalvo said to his soldiers, shewing them Naples, and protesting he had rather die one foot forwards than to have his life secured for long by one foot of retreat; whereunto the wisdom of that heavenly leader hath signed, who hath affirmed that *a good conscience is a continual feast*: shewing plainly that the conscience of good intentions, howsoever succeeding, is a more continual joy to nature than all the provision which can be made for security and repose.

It censureth likewise that abuse of philosophy which grew general about the time of Epictetus, in converting it into an occupation or profession; as if the purpose had been, not to resist and extinguish perturbations, but to fly and avoid the causes of them, and to shape a particular kind and course of life to that end; introducing such an health of mind, as was that health of body of which Aristotle speaketh of Herodicus, who did nothing all his life long but intend his health: whereas if men refer themselves to duties of society, as that health of body is best which is ablest to endure all alterations and extremites, so likewise that health of mind is most proper [235] which can go through the greatest temptations and perturbations. So as Diogenes' opinion is to be accepted, who commended not them which abstained, but them which sustained, and could refrain their mind *in præcipitio*, and could give unto the mind (as is used in horsemanship) the shortest stop or turn.

Lastly, it censureth the tenderness and want of application [236] in some of the most ancient and reverend philosophers and philosophical men, that did retire too easily from civil business, for avoiding of indignities and perturbations; whereas the resolution of men truly moral ought to be such as the same Consalvo said the honour of a soldier should be, *e telâ crassiore* [of a stouter web], and not so fine that every thing should catch in it and endanger it.

¶ [237] To resume Private or Particular Good, it falleth into the division of Good Active and Passive: for this difference of Good (not unlike to that which amongst the Romans was expressed in the familiar or household terms of Promus and Condus) is formed also in all things; and is best disclosed in the two several appetites in creatures, the one to preserve or continue themselves, and the other to dilate or multiply themselves; whereof the later seemeth to be the worthier. For in nature, the heavens, which are the more worthy, are the agent; and the earth, which is the less worthy, is the patient. In the pleasures of living creatures, that of generation is greater than that of food. In divine doctrine, *Beatius est dare quam accipere* [it is more blessed to give than to receive]. And in life, there is no man's spirit so soft, but esteemeth the effecting of somewhat that he hath fixed in his desire more than sensuality. Which priority of the Active Good is much upheld by the consideration of our estate to be mortal and exposed to fortune;

233 The translation has "and lastly that exploded school of Pyrrho and Herillus".

234 That is, esteeming those actions good which are attended with clearness and composure of mind, those bad which proceed with dislike and reluctation—(*actiones pro bonis aut malis habentes, prout ex animo, motu puro et irrefracto, aut contra cum aversatione et reluctatione, prodirent*).

235 *i.e.* that mind is to be considered truly and properly healthy—(*animus ille demum vere et propria sanus et validus censendus est*).

236 meaning what we should now rather call want of compliance or accommodation—(*ineptitudinem ad morigerandum*).

237 De Aug. vii. 2.

for if we might have a perpetuity and certainty in our pleasures, the *state* [238] of them would advance their price; but when we see it is but *Magni æstimamus mori tardius* [we think it a great matter to be a little longer in dying], and *Ne glorieris de crastino, nescis partum diei* [boast not thyself of to-morrow, thou knowest not what the day may bring forth], it maketh us to desire to have somewhat secured and exempted from time; which are only our deeds and works; as it is said *Opera eorum sequuntur eos* [their works follow them]. The pre-eminence likewise of this Active Good is upheld by the affection which is natural in man towards variety and proceeding; which in the pleasures of the sense (which is the principal part of Passive Good) can have no great latitude: *Cogita quamdiu eadem feceris; cibus, somnus, ludus; per hunc circulum curritur; mori velle non tantum fortis, aut miser, aut prudens, sed etiam fastidiosus potest* [if you consider, says Seneca, how often you do the same thing over and over; food, sleep, exercise, and then food, sleep, exercise again, and so round and round; you will think that there needs neither fortitude nor misery nor wisdom to reconcile a man to death; one might wish to die for mere weariness of being alive]. But in enterprises, pursuits, and purposes of life, there is much variety; whereof men are sensible with pleasure in their inceptions, progressions, recoils, reintegrations, approaches, and attainings to their ends: so as it was well said, *Vita sine proposito languida et vaga est* [life without an object to pursue is a languid and tiresome thing]. Neither hath this Active Good any [239] identity with the good of society, though in some case it hath an incidence into it: for although it do many times bring forth acts of beneficence, yet it is with a respect private to a man's own power, glory, amplification, continuance; as appeareth plainly when it findeth a contrary subject. For that gigantine state of mind which possesseth the troublers of the world, such as was Lucius Sylla, and infinite other in smaller model, who would have all men happy or unhappy as they were their friends or enemies, and would give form to the world according to their own humours, (which is the true Theomachy), pretendeth and aspireth to active good [240], though it recedeth furthest from good of society, which we have determined to be the greater.

To resume Passive Good, it receiveth a subdivision of Conservative and Perfective. For let us take a brief review of that which we have said: we have spoken first of the Good of Society, the intention thereof embraceth the form of Human Nature, whereof we are members and portions, and not our own proper and individual form; we have spoken of Active Good, and supposed it as a part of Private and Particular Good; and rightly [241]; for there is impressed upon all things a triple desire or appetite proceeding from love to themselves; one of preserving and continuing their form; another of advancing and perfecting their form; and a third of multiplying and extending their form upon other things; whereof the multiplying or signature of it upon other things is that which we handled by the name of Active Good. So as there remaineth the conserving of it, and perfecting or raising of it; which later is the highest degree of Passive Good. For to preserve in state is the less, to preserve with advancement is the greater. So in man,

> Igneus est ollis vigor, et cœlestis origo [242]
> [The living fire that glows those seeds within
> Remembers its celestial origin].

His approach or assumption to divine or angelical nature is the perfection of his form; the error or false imitation of which good is that which is the tempest of human life; while man, upon the instinct of an advancement formal and essen-

238 *i.e.* the stability—(*securitas et mora*).

239 So edd. 1629 and 1633. The original has *and*.

240 *i.e.* apparent good of the individual—(*bonum activum individuale saltem apparens*).

241 This passage, from *for let us take* etc. to *rightly*, is omitted in the translation; and the argument proceeds more clearly without it.

242 The connexion of this with the preceding sentence is made clearer in the translation by the remark that there are found throughout the universe certain nobler natures which inferior natures recognise as their origin and towards which they aspire.

tial, is carried to seek an advancement local. For as those which are sick, and find no remedy, do tumble up and down and change place, as if by a remove local they could obtain a remove internal; so is it with men in ambition, when failing of the mean to exalt their nature, they are in a perpetual estuation to exalt their place. So then Passive Good is, as was said, either Conservative or Perfective.

To resume the good of Conservation or Comfort, which consisteth *in the fruition of that which is agreeable to our natures;* it seemeth to be the most pure and natural of pleasures, but yet the softest and the lowest. And this also receiveth a difference, which hath neither been well judged of nor well enquired. For the good of fruition or contentment is placed either in the sincereness of the fruition, or in the quickness and vigour of it; the one superinduced by the equality, the other by vicissitude; the one having less mixture of evil, the other more impression of good. Whether of these is the greater good, is a question controverted; but whether man's nature may not be capable of both, is a question not enquired.

The former question being debated between Socrates and a Sophist, Socrates placing felicity in an equal and constant peace of mind, and the Sophist in much desiring and much enjoying, they fell from argument to ill words: the Sophist saying that Socrates' felicity was the felicity of a block or stone; and Socrates saying that the Sophist's felicity was the felicity of one that had the itch, who did nothing but itch and scratch. And both these opinions do not want their supports. For the opinion of Socrates is much upheld by the general consent even of the Epicures themselves, that virtue beareth a great part in felicity; and if so, certain it is that virtue hath more use in clearing perturbations than in compassing desires. The Sophist's opinion is much favoured by the assertion we last spake of, that good of advancement is greater than good of simple preservation; because every obtaining a desire hath a shew of advancement [243], as motion though in a circle hath a shew of progression.

But the second question, decided the true way, maketh the former superfluous. For can it be doubted but that there are some who take more pleasure in enjoying pleasures than some other, and yet nevertheless are less troubled with the loss or leaving of them? so as this same *Non uti ut non appetas, non appetere ut non metuas, sunt animi pusilli et diffidentis* [to abstain from the use of a thing that you may not feel a want of it; to shun the want that you may not fear the loss of it; are the precautions of pusillanimity and cowardice]. [244] And it seemeth to me, that most of the doctrines of the philosophers are more fearful and cautionary than the nature of things requireth. So have they increased the fear of death in offering to cure it. For when they would have a man's whole life to be but a discipline or preparation to die, they must needs make men think that it is a terrible enemy against whom there is no end of preparing. Better saith the poet:

> Qui finem vitæ extremum inter munera ponat
> Naturæ

[the end of life is to be counted among the boons of nature]. So have they sought to make men's minds too uniform and harmonical, by not breaking them sufficiently to contrary motions: the reason whereof I suppose to be, because they themselves were men dedicated to a private, free, and unapplied course of life. For as we see, upon the lute or like instrument, a *ground,* though it be sweet and have shew of many changes, yet breaketh not the hand to such strange and hard stops and passages as a *set song* or *voluntary;* much after the same manner was the diversity between a philosophical and a civil life [245]. And therefore men are to imitate the wisdom of jewellers; who, if there be a grain or a cloud or an ice

[243] *i.e.* towards the perfection of nature; only a *show* of advancement, however, not necessarily a real one—(*quia rerum cupitarum adeptiones naturam videantur sensim perficere; quod licet vere non faciant, tamen,* etc.).

[244] Compare Shakespeare's sonnet—

> I cannot chuse
> But weep to have that which I fear to lose.

[245] This illustration is omitted in the translation.

which may be ground forth without taking too much of the stone, they help it; but if it should lessen or abate the stone too much, they will not meddle with it: so ought men so to procure serenity as they destroy not magnanimity.

Having therefore deduced the Good of Man which is Private and Particular as far as seemeth fit, we will now return to that good of man which respecteth and beholdeth society, which we may term Duty; because the term of Duty is more proper to a mind well framed and disposed towards others, as the term of Virtue is applied to a mind well formed and composed in itself; though neither can a man understand Virtue without some relation to society, nor Duty without an inward disposition. This part may seem at first to pertain to science civil and politic; but not if it be well observed. For it concerneth the regiment and government of every man over himself, and not over others. And as in architecture the direction of framing the posts, beams, and other parts of building, is not the same with the manner of joining them and erecting the building; and in mechanicals, the direction how to frame an instrument or engine, is not the same with the manner of setting it on work and employing it; and yet nevertheless in expressing of the one you incidently express the aptness towards the other; so the doctrine of conjugation of men in society differeth from that of their conformity thereunto [246].

This part of Duty is subdivided into two parts; the common duty of every man as a man or member of a state, the other, the respective or special duty of every man, in his profession, vocation, and place. The first of these is extant and well laboured, as hath been said. The second likewise I may report rather dispersed than deficient; which manner of dispersed writing in this kind of argument I acknowledge to be best. For who can take upon him to write of the proper duty, virtue, challenge, and right of every several vocation, profession and place? For although sometimes a looker on may see more than a gamester, and there be a proverb more arrogant than sound, *That the vale best discovereth the hill;* yet there is small doubt but that men can write best and most really and materially in their own professions; and that the writing of speculative men of active matter for the most part doth seem to men of experience, as Phormio's argument of the wars seemed to Hannibal, to be but dreams and dotage. Only there is one vice which accompanieth them that write in their own professions, that they magnify them in excess. But generally it were to be wished (as that which would make learning indeed solid and fruitful) that active men would or could become writers.

In which kind I cannot but mention, *honoris causa,* your Majesty's excellent book touching the duty of a king: a work richly compounded of divinity, morality, and policy, with great aspersion of all other arts; and being in mine opinion one of the most sound and healthful writings that I have read; not distempered in the heat of invention, nor in the coldness of negligence; not sick of dizziness [247], as those are who leese themselves in their order; nor of convulsions [248], as those which cramp in matters impertinent; not savouring of perfumes and paintings, as those who seek to please the reader more than nature [249] beareth; and chiefly well disposed in the spirits thereof, being agreeable to truth and apt for action; and far removed from that natural infirmity, whereunto I noted those that write in their own professions to be subject, which is, that they exalt it above measure. For your Majesty hath truly described, not a king of Assyria or Persia

246 *i.e.* of the conformation of men to the business of society—(*quæ eos reddit ad hujusmodi societatis commoda conformes et bene affectos*).

247 *Dusinesse* in the original. *Businesse* in edd. 1629 and 1623. *Vertigine* in De Aug.

248 The words "convulsion" and "cramp" seem to describe a forced and abrupt style; an idea not implied in the words of the translation, which may be retranslated thus: "not distracted in digressions, as those which wind about to take in matters impertinent"—(*ut illa quæ nihil ad rhombum sunt expatiatione aliqua flexuosa complectatur*).

249 *i.e.* the nature of the argument.—(*qui lectorum potius delectationi quam argumenti naturæ inserviunt*).

in their extern glory, but a Moses or a David, pastors of their people. Neither can I ever leese out of my remembrance what I heard your Majesty in the same sacred spirit of government deliver in a great cause of judicature, which was, *That Kings ruled by their laws as God did by the laws of nature, and ought as rarely to put in use their supreme prerogative as God doth his power of working miracles.* And yet notwithstanding, in your book of a free monarchy, you do well give men to understand, that you know the plenitude of the power and right of a King, as well as the circle of his office and duty. Thus have I presumed to allege this excellent writing of your Majesty, as a prime or eminent example of tractates concerning special and respected duties; wherein I should have said as much, if it had been written a thousand years since. Neither am I moved with certain courtly decencies, which esteem it flattery to praise in presence. No, it is flattery to praise in absence; that is, when either the virtue is absent, or the occasion is absent; and so the praise is not natural, but forced, either in truth or in time. But let Cicero be read in his oration *pro Marcello*, which is nothing but an excellent table of Cæsar's virtue, and made to his face; besides the example of many other excellent persons, wiser a great deal than such observers [250]; and we will never doubt, upon a full occasion, to give just praises to present or absent.

But to return: there belongeth further to the handling of this part [251] touching the duties of professions and vocations, a Relative or opposite, touching the frauds, cautels, impostures, and vices of every profession; which hath been likewise handled: but how? rather in a satire and cynically, than seriously and wisely: for men have rather sought by wit to deride and traduce much of that which is good in profession, than with judgment to discover and sever that which is corrupt. For, as Salomon saith, He that cometh to seek after knowledge with a mind to scorn and censure, shall be sure to find matter for his humour, but no matter for his instruction: *Quærenti derisori scientiam ipsa se abscondit; sed studioso fit obviam.* But the managing of this argument with integrity and truth, which I note as deficient, seemeth to me to be one of the best fortifications for honesty and virtue that can be planted. For as the fable goeth of the Basilisk, that if he see you first you die for it, but if you see him first he dieth; so is it with deceits and evil arts; which if they be first espied they leese their life, but if they prevent they endanger. So that we are much beholden to Machiavel and others, that write what men do and not what they ought to do. For it is not possible to join serpentine wisdom with the columbine innocency, except men know exactly all the conditions of the serpent; his baseness and going upon his belly, his volubility and lubricity, his envy and sting, and the rest; that is, all forms and natures of evil. For without this, virtue lieth open and unfenced. Nay an honest man can do no good upon those that are wicked to reclaim them, without the help of the knowledge of evil. For men of corrupted minds presuppose that honesty groweth out of simplicity of manners, and believing of preachers, schoolmasters, and men's exterior language: so as, except you can make them perceive that you know the utmost reaches of their own corrupt opinions, they despise all morality. *Non recipit stultus verba prudentiæ, nisi ea dixeris quæ versantur in corde ejus* [the fool will not listen to the words of the wise, unless you first tell him what is in his own heart].[252]

Unto this part touching Respective Duty doth also appertain the duties between husband and wife, parent and child, master and servant: so likewise the laws of friendship and gratitude, the civil bond of companies, colleges, and politic bodies, of neighbourhood, and all other proportionate duties; not as they are parts of government and society, but as to the framing of the mind of particular persons.

[250] In the translation he merely adds the single example of Pliny the younger in his Panegyric on Trajan. When he wrote the *Advancement of Learning*, he appears to have been under the impression that Pliny's Panegyric was spoken after Trajan's death. See below, p. 147.

[251] So edd. 1629 and 1633. The original has *partie*.

[252] In the translation this is set down as a *desideratum* under the title of *Satira Seria sive tractatus de interioribus rerum.*

The knowledge concerning good respecting Society doth handle it also not simply alone, but comparatively; whereunto belongeth the weighing of duties between person and person, case and case, particular and public: as we see in the proceeding [253] of Lucius Brutus against his own sons, which was so much extolled; yet what was said?

Infelix, utcunque ferent ea facta [254] minores

[unhappy man! whatever judgment posterity shall pass upon that deed, etc.]. So the case was doubtful, and had opinion on both sides. Again, we see when M. Brutus and Cassius invited to a supper certain whose opinions they meant to feel, whether they were fit to be made their associates, and cast forth the question touching the killing of a tyrant being an usurper, they were divided in opinion; some holding that servitude was the extreme of evils, and others that tyranny was better than a civil war: and a number of the like cases there are of comparative duty. Amongst which that of all others is the most frequent, where the question is of a great deal of good to ensue of a small injustice. Which Jason of Thessalia determined against the truth: *Aliqua sunt injuste facienda, ut multa justa fieri possint* [that there may be justice in many things there must be injustice in some]. But the reply is good, *Authorem præsentis justitiæ habes, sponsorem futuræ non habes* [the justice that is to be done now is in your power, but where is your security for that which is to be done hereafter?] Men must pursue things which are just in present, and leave the future to the divine Providence. So then we pass on from this general part touching the exemplar and description of good.

De Cultura Animi.

¶ [255] Now therefore that we have spoken of this fruit of life, it remaineth to speak of the husbandry that belongeth thereunto; without which part the former seemeth to be no better than a fair image or statue which is beautiful to contemplate, but is without life and motion: whereunto Aristotle himself subscribeth in these words: *Necesse est scilicet de virtute dicere, et quid sit, et ex quibus gignatur. Inutile enim fere fuerit virtutem quidem nosse, acquirendæ autem ejus modos et vias ignorare. Non enim de virtute tantum, qua specie sit, quærendum est, sed et quomodo sui copiam faciat; utrumque enim volumus, et rem ipsam nosse, et ejus compotes fieri: hoc autem ex voto non succedet, nisi sciamus et ex quibus et quomodo.* [It is necessary to determine concerning virtue not only what it is but whence it proceeds. For there would be no use in knowing virtue without knowing the ways and means of acquiring it. For we have to consider not only what it is, but how it is to be had. For we want both to know virtue and to be virtuous; which we cannot be without knowing both the whence and the how.] In such full words and with such iteration doth he inculcate this part. So saith Cicero in great commendation of Cato the second that he had applied himself to philosophy *non ita disputandi causa, sed ita vivendi'* [not that he might talk like a philosopher, but that he might live like one]. And although the neglect of our times, wherein few men do hold any consultations touching the reformation of their life, (as Seneca excellently saith, *De partibus vitæ quisque deliberat, de summâ nemo,*) [every man takes thought about the parts of his life, no man about the whole], may make this part seem superfluous; yet I must conclude with that aphorism of Hippocrates, *Qui gravi morbo correpti dolores non sentiunt, iis mens ægrotat* [they that are sick and yet feel no pain are sick in their minds]; they need medicine not only to assuage the disease but to awake the sense. And if it be said that the cure of men's minds belongeth to sacred Divinity, it is most true: but yet Moral Philosophy may be preferred unto her as a wise servant and humble handmaid. For as the Psalm saith, *that the eyes of the handmaid look perpetually towards the mistress,* and yet no doubt many things are left to the discretion of the handmaid to discern of the mistress' will; so ought Moral Philosophy to give a constant attention to the doctrines of

[253] *in animadversione illa severa et atroci.*—De Aug.

[254] *Fata* both in the *Advancement* and in the *De Augmentis*.

[255] De Aug. vii. 3.

Divinity, and yet so as it may yield of herself (within due limits) many sound and profitable directions.

This part therefore, because of the excellency thereof, I cannot but find exceeding strange that it is not reduced to written inquiry; the rather because it consisteth of much matter wherein both speech and action is often conversant, and such wherein the common talk of men (which is rare, but yet cometh sometimes to pass) is wiser than their books. It is reasonable therefore that we propound it in the more particularity, both for the worthiness, and because we may acquit ourselves for reporting it deficient; which seemeth almost incredible, and is otherwise conceived and presupposed by those themselves that have written. We will therefore enumerate some heads or points thereof, that it may appear the better what it is, and whether it be extant.

First therefore, in this, as in all things which are practical, we ought to cast up our account, what is in our power and what not; for the one may be dealt with by way of alteration, but the other by way of application only. The husbandman cannot command neither the nature of the earth nor the seasons of the weather; no more can the physician the constitution of the patient nor the variety of accidents. So in the culture and cure of the mind of man, two things are without our command; points of nature, and points of fortune; for to the basis of the one, and the conditions of the other, our work is limited and tied. In these things therefore it is left unto us to proceed by application:

Vincenda est omnis fortuna ferendo

[all fortune may be overcome by endurance or suffering]; and so likewise,

Vincenda est omnis natura ferendo

[all nature may be overcome by suffering]. But when that we speak of suffering, we do not speak of a dull and neglected suffering, but of a wise and industrious suffering, which draweth and contriveth use and advantage out of that which seemeth adverse and contrary; which is that property which we call Accommodating or Applying[256]. Now the wisdom of application resteth principally in the exact and distinct knowledge of the precedent state or disposition unto which we do apply: for we cannot fit a garment, except we first take measure of the body.

So then the first article of this knowledge is to set down sound and true distributions and descriptions of the several characters and tempers of men's natures and dispositions, specially having regard to those differences which are most radical in being the fountains and causes of the rest, or most frequent in concurrence or commixture [257]; wherein it is not the handling of a few of them in passage, the better to describe the mediocrities of virtues, that can satisfy this intention; for if it deserve to be considered, *that there are minds which are proportioned to great matters, and others to small* (which Aristotle handleth or ought to have handled by the name of Magnanimity), doth it not deserve as well to be considered, *that there are minds proportioned to intend many matters, and others to few?* [258] so that some can divide themselves, others can perchance do exactly well, but it must be but in few things at once; and so there cometh to be a *narrowness of mind*, as well as a *pusillanimity*. And again, *that some minds are proportioned to that which may be dispatched at once, or within a short return of time; others to that which begins afar off, and is to be won with length of pursuit;*

Jam tum tenditque fovetque

[he begins to attend and nurse his project while it is yet in the cradle]; so that

[256] These observations are omitted in the translation, and the whole passage is rewritten, though rather with a view of expressing the meaning more clearly than of altering it.

[257] It is remarkable that the observations which follow, down to "benignity or malignity", are entirely omitted in the translation.

[258] So all the editions: a second *intend* having probably dropped out accidentally.

there may be fitly said to be a *longanimity* ; which is commonly also ascribed to God as a *magnanimity*. So further deserved it to be considered by Aristotle *that there is a disposition in conversation (supposing it in things which do in no sort touch or concern a man's self) to soothe and please, and a disposition contrary to contradict and cross* ; and deserveth it not much better to be considered, *that there is a disposition, not in conversation or talk but in matter of more serious nature, (and supposing it still in things merely indifferent), to take pleasure in the good of another, and a disposition contrariwise to take distaste at the good of another* ; which is that property [259] which we call good-nature or ill-nature, benignity or malignity? And therefore I cannot sufficiently marvel that this part of knowledge touching the several characters of natures and dispositions should be omitted both in morality and policy, considering it is of so great ministery and suppeditation to them both. A man shall find in the traditions of astrology some pretty and apt divisions of men's natures, according to the predominances of the planets; *lovers of quiet, lovers of action, lovers of victory, lovers of honour, lovers of pleasure, lovers of arts, lovers of change*, and so forth. A man shall find in the wisest sort of these Relations which the Italians make touching Conclaves, the natures of the several Cardinals handsomely and lively painted forth. A man shall meet with in every day's conference the denominations of *sensitive, dry, formal, real, humorous, certain, huomo di prima impressione, huomo di ultima impressione* and the like [260] : and yet nevertheless this kind of observations wandereth in words, but is [261] not fixed in inquiry. For the distinctions are found (many of them), but we conclude no precepts upon them ; wherein our fault is the greater, because both history, poesy, and daily experience are as goodly fields where these observations grow ; whereof we make a few posies to hold in our hands, but no man bringeth them to the confectionary, that receits might be made of them for use of life[262].

Of much like kind are those impressions of nature, which are imposed upon the mind *by the sex, by the age, by the region, by health and sickness, by beauty and deformity*, and the like, which are inherent and not extern ; and again those which are caused by extern fortune ; as *sovereignity, nobility, obscure birth, riches, want, magistracy, privateness, prosperity, adversity, constant fortune, variable fortune, rising per saltum, per gradus*, and the like. And therefore we see that Plautus maketh it a wonder to see an old man beneficent : *benignitas hujus ut adolescentuli est* [he is as generous as if he were a young man] : St. Paul concludeth that severity of discipline was to be used to the Cretans, *Increpa eos durè* [rebuke them sharply], upon the disposition of their country ; *Cretenses*

[259] *properly* both in the original, and in edd. 1629 and 1633.

[260] This sentence is omitted in the translation ; perhaps from the difficulty of finding equivalent terms in Latin ; but the substance of the observation is contained in the remark (transplanted from a former paragraph) that in this matter the common talk of men is wiser than their books.

[261] *as* both in the original and in edd. 1629 and 1633.

[262] In place of this we have in the translation a passage of considerable length recommending the wiser sort of historians as supplying the best material for this kind of treatise ; not only in the formal character which they commonly give of any principal personage on recording his death, but still more in the occasional observations interwoven into the body of the narrative, when in relating any of his actions they introduce some remark upon his nature and disposition. Bacon instances the character of Africanus and the elder Cato as drawn by Livy ; of Tiberius, Claudius, and Nero, in Tacitus ; of Septimius Severus, in Herodian ; of Louis XI. in Philip de Comines ; of Ferdinand, Maximilian, Leo, and Clement, in Guicciardini. (His own Henry VII. would have furnished another instance, as good as any). Of these he would have a full and careful analysis made, exhibiting not the entire character, but the several features and individual peculiarities of mind and disposition which make it up (*imaginum ipsarum lineæ et ductus magis simplices*), with their connexion and bearing one upon another :—a kind of moral and mental anatomy, as a basis for a system of moral and mental medicine. He prefers the historians to the poets for this purpose, because in the poets the characters are commonly drawn with exaggeration.

semper mendaces, malæ bestiæ, ventres pigri [the Cretans are always liars, evil beasts, slow bellies]: Sallust noteth that it is usual with Kings to desire contradictories; *Sed plerumque regiæ voluntates, ut vehementes sunt, sic mobiles, sæpeque ipsæ sibi adversæ* [royal desires, as they are violent, so are they changeable, and often incompatible with each other]: Tacitus observeth how rarely raising of the fortune mendeth the disposition; *Solus Vespasianus mutatus in melius* [Vespasian the only one of the emperors that changed for the better]: Pindarus maketh an observation that great and sudden fortune for the most part defeateth men [263]; *Qui magnam felicitatem concoquere non possunt* [that cannot digest great felicity]: so the Psalm sheweth it is more easy to keep a measure in the enjoying [264] of fortune than in the increase of fortune; *Divitiæ si affluant, nolite cor apponere* [if riches increase, set not your heart upon them]. These observations and the like I deny not but are touched a little by Aristotle as in passage in his Rhetorics, and are handled in some scattered discourses; but they were never incorporate into Moral Philosophy, to which they do essentially appertain; as the knowledge of the diversity of grounds and moulds doth to agriculture, and the knowledge of the diversity of complexions and constitutions doth to the physician; except we mean to follow the indiscretion of empirics, which minister the same medicines to all patients.

Another article of this knowledge is the inquiry touching the affections; for as in medicining of the body it is in order first to know the divers complexions and constitutions, secondly the diseases, and lastly the cures; so in medicining of the mind, after knowledge of the divers characters of men's natures, it followeth in order to know the diseases and infirmities of the mind, which are no other than the perturbations and distempers of the affections. For as the ancient politiques [265] in popular estates were wont to compare the people to the sea and the orators to the winds, because as the sea would of itself be calm and quiet if the winds did not move and trouble it, so the people would be peaceable and tractable if the seditious orators did not set them in working and agitation; so it may be fitly said, that the mind in the nature thereof would be temperate and stayed, if the affections, as winds, did not put it into tumult and perturbation. And here again I find strange, as before, that Aristotle should have written divers volumes of Ethics, and never handled the affections, which is the principal subject thereof; and yet in his Rhetorics, where they are considered but collaterally and in second degree (*as they may be moved by speech*), he findeth place for them, and handleth them well for the quantity; but where their true place is, he pretermitteth them. For it is not his disputations about pleasure and pain that can satisfy this inquiry, no more than he that should generally handle the nature of light can be said to handle the nature of colours; for pleasure and pain are to the particular affections as light is to particular colours. Better travails I suppose had the Stoics taken in this argument, as far as I can gather by that which we have at second hand: but yet it is like it was after their manner, rather in subtilty of definitions (which in a subject of this nature are but curiosities) than in active and ample descriptions and observations. So likewise I find some particular writings of an elegant nature touching some of the affections, as of *anger*, of *comfort upon adverse accidents* [266], of *tenderness of countenance* [267], and other. But the poets and writers of histories are the best doctors of this knowledge; where we may find painted forth with great life, how affections are kindled and incited; and how pacified and refrained; and how again contained from act and further degree; how they disclose themselves, how they work, how they vary, how they gather and fortify [268], how they are inwrapped one within another, and how they do fight and encounter one with another, and other the like particularities: amongst the which this last is of special use in moral and civil matters;

[263] *animos plerumque enervare et solvere.*—De Aug. [264] *statu.*—De Aug.

[265] So edd. 1629 and 1633. The original has *in politiques.*

[266] This is omitted in the translation.

[267] This I suppose is what the French call *mauvaise honte.* The translation is *De inutili verecundia*, which is the Latin rendering of περὶ δυσωπίας, the title of a tract by Plutarch.

[268] This is omitted in the translation.

how (I say) to set affection against affection, and to master one by another; even as we use to hunt beast with beast and fly bird with bird, which otherwise percase we could not so easily recover: upon which foundation is erected that excellent use of *præmium* and *pæna*, whereby civil states consist; employing the predominant affections of *fear* and *hope*, for the suppressing and bridling the rest. For as in the government of states it is sometimes necessary to bridle one faction with another, so it is in the government within.

Now come we to those points which are within our own command, and have force and operation upon the mind to affect the will and appetite and to alter manners: wherein they ought to have handled *custom, exercise, habit, education, example, imitation, emulation, company, friends, praise, reproof, exhortation, fame, laws, books, studies*: these as [269] they have determinate use in moralities, from these the mind suffereth, and of these are such receipts and regiments compounded and described, as may seem to recover or preserve the health and good estate of the mind, as far as pertaineth to human medicine: of which number we will visit [270] upon some one or two as an example of the rest, because it were too long to prosecute all; and therefore we do resume Custom and Habit to speak of.

The opinion of Aristotle seemeth to me a negligent opinion, that of those things which consist by nature nothing can be changed by custom; using for example, that if a stone be thrown ten thousand times up, if will not learn to ascend; and that by often seeing or hearing, we do not learn to see or hear the better. For though this principle be true in things wherein nature is *peremptory*, (the reason whereof we cannot now stand to discuss), yet it is otherwise in things wherein nature admitteth *a latitude*. For he might see that a strait glove will come more easily on with use, and that a wand will by use bend otherwise than it grew, and that by use of the voice we speak louder and stronger, and that by use of enduring heat or cold we endure it the better, and the like: which later sort have a nearer resemblance unto that subject of manners he handleth than those instances which he allegeth. But allowing his conclusion, *that virtues and vices consist in habit*, he ought so much the more to have taught the manner of superinducing that habit: for there be many precepts of the wise ordering the exercises of the mind, as there is of ordering the exercises of the body; whereof we will recite a few.

The first shall be, that we beware we take not at the first either too *high* a strain or too *weak*: for if too high, in a diffident [271] nature you discourage; in a confident nature you breed an opinion of facility, and so a sloth; and in all natures you breed a further expectation than can hold out, and so an insatisfaction [272] on the end: if too weak of the other side, you may not look to perform and overcome any great task.

Another precept is, to practise all things chiefly at two several times, the one when the mind is best disposed, the other when it is worst disposed; that by the one you may gain a great step, by the other you may work out the knots and stonds of the mind, and make the middle times the more easy [273] and pleasant.

Another precept is, that which Aristotle mentioneth by the way, which is to bear ever towards the contrary extreme of that whereunto we are by nature inclined: like unto the rowing against the stream, or making a wand straight by bending [274] him contrary to his natural crookedness.

Another precept is, that the mind is brought to any thing better, and with more sweetness and happiness, if that whereunto you pretend be not first in the

[269] So in all the editions. Perhaps it should be *are*. (*Hæc enim sunt illa quæ regnant in moralibus*). If *as* be right, we should properly read, *for from these* etc.

[270] So the original. Edd. 1629 and 1633 have *insist*: perhaps rightly. The translation has *unum aut alterum deligemus in quibus paullulum immorabimur*.

[271] So edd. 1629 and 1633. The original has *different*.

[272] And thence a discouragement—(*id quod animum semper dejicit et confundit*).

[273] So edd. 1629 and 1633. The original has *easily*. Possibly Bacon wrote *run more easily*. The translation has *facile et placide delabentur*. This part of the original edition is carelessly printed.

[274] So ed. 1633. The original has *bynding*, and ed. 1629 *binding*.

intention, but *tanquam aliud agendo*, because of the natural hatred of the mind against necessity and constraint. Many other axioms there are touching the managing of *Exercise* and *Custom*; which being so conducted, doth prove indeed another nature: but being governed by chance, doth commonly prove but an ape of nature, and bringeth forth that which is lame and counterfeit.

So if we should handle *books* and *studies*, and what influence and operation they have upon manners, are there not divers precepts of great caution and direction appertaining thereunto? Did not one of the fathers in great indignation call Poesy *vinum dæmonum*, because it increaseth temptations, perturbations, and vain opinions? Is not the opinion of Aristotle worthy to be regarded, wherein he saith that young men are no fit auditors of moral philosophy [275], because they are not settled from the boiling heat of their affections, nor attempered with time and experience? And doth it not hereof come, that those excellent books and discourses of the ancient writers (whereby they have persuaded unto virtue most effectually, by representing her in state and majesty, and popular opinions against virtue in their parasites' coats, fit to be scorned and derided,) are of so little effect towards honesty of life, because they are not read and revolved by men in their mature and settled years, but confined almost to boys and beginners? But is it not true also, that much less young men are fit auditors of matters of policy, till they have been thoroughly seasoned in religion and morality; lest their judgments be corrupted, and made apt to think that there are no true differences of things, but according to utility and fortune; as the verse describes it, *Prosperum et felix scelus virtus vocatur* [a crime that is successful is called a virtue]; and again, *Ille crucem pretium sceleris tulit, hic diadema* [the same crime is rewarded in one man with a gibbet and in another with a crown]; which the poets do speak satirically, and in indignation on virtue's behalf; but books of policy do speak it seriously and positively; for so it pleaseth Machiavel to say, *that if Cæsar had been overthrown he would have been more odious than ever was Catiline*; as if there had been no difference but in fortune, between a very fury of lust and blood, and the most excellent spirit (his ambition reserved) of the world? Again, is there not a caution likewise to be given of the doctrines of moralities themselves (some kinds of them), lest they make men too precise, arrogant, incompatible; as Cicero saith of Cato, *In Marco Catone hæc bona quæ videmus divina et egregia, ipsius scitote esse propria; quæ nonnunquam requirimus, ea sunt omnia non a naturâ, sed a magistro* [his excellences were his own, his defects came from the schoolmaster]? Many other axioms and advices there are touching those proprieties and effects which studies do infuse and instil into manners. And so likewise is there touching the use of all those other points, of company, fame, laws, and the rest, which we recited in the beginning in the doctrine of morality.

But there is a kind of Culture of the Mind that seemeth yet more accurate and elaborate than the rest, and is built upon this ground; that the minds of all men are at some times in a state more perfect, and at other times in a state more depraved. The purpose therefore of this practice [276] is to fix and cherish the good hours of the mind, and to obliterate and take forth the evil. The fixing of the good hath been practised by two means; vows or constant resolutions; and observances and exercises; which are not to be regarded so much in themselves, as because they keep the mind in continual obedience. The obliteration of the evil hath been practised by two means; some kind of redemption or expiation of that which is past; and an inception or account *de novo* for the time to come. But this part seemeth sacred and religious, and justly; for all good Moral Philosophy (as was said) is but an handmaid to religion.

[275] Not of *moral* but of *political* philosophy. See Mr. Ellis's note to *De Aug.* B. vii., near end. That in the passage there quoted from *Troilus and Cressida* the observation and the error were both derived directly from the *Advancement of Learning* admits of little doubt. But how came Virgilio Malvezzi, in his *Discorsi sopro Cornelio Tacito* published in 1622, to make the same mistake? "E non è discordante da questa mia opinione Aristotele, il qual dice, che i giovani non sono buoni ascultatori delle *morali*." I quote from ed. 1635. The passage occurs in the address to the reader, p. 3.

[276] *i.e.* method of culture (*hujus culturæ intentio et institutum*).

Wherefore we will conclude with that last point which is of all other means the most compendious and summary, and again the most noble and effectual, to the reducing of the mind unto virtue and good estate; which is the electing and propounding unto a man's self good and virtuous ends of his life, such as may be in a reasonable sort within his compass to attain. For if these two things be supposed, that a man set before him honest and good ends, and again that he be resolute, constant, and true unto them, it will follow that he shall mould himself into all virtue at once. And this is indeed like the work of nature; whereas the other course is like the work of the hand. For as when a carver makes an image, he shapes only that part whereupon he worketh; as if he be upon the face, that part which shall be the body is but a rude stone still, till such times as he comes to it; but contrariwise when nature makes a flower or living creature, she formeth rudiments of all the parts at one time; so in obtaining virtue by *habit*, while a man practiseth temperance, he doth not profit much to fortitude, nor the like; but when he dedicateth and applieth himself to *good ends*, look what virtue soever the pursuit and passage towards those ends doth commend unto him, he is invested of a precedent disposition to conform himself thereunto; which state of mind Aristotle doth excellently express himself, that it ought not be to called *virtuous*, but *divine*; his words are these: *Immanitati autem consentaneum est opponere eam, quæ supra humanitatem est, heroicam sive divinam virtutem*: and a little after, *Nam ut feræ neque vitium neque virtus est, sic neque Dei: sed hic quidem status altius quiddam virtute est, ille aliud quiddam a vitio* [that which answers to the brutal degree of vice is the heroical or divine degree of virtue. . . . For as neither virtue nor vice can be predicated of a brute, so neither can it of a God: the divine condition being something higher than virtue, the brutal something different from vice]. And therefore we may see what celsitude of honour Plinius Secundus attributeth to Trajan in his funeral oration [277], where he said, *that men needed to make no other prayers to the gods, but that they would continue as good lords to them as Trajan had been*; as if he had not been only an imitation of divine nature but a pattern of it. But these be heathen and profane passages, having but a shadow of that divine state of mind which religion and the holy faith doth conduct men unto, by imprinting upon their souls Charity, which is excellently called the bond of Perfection, because it comprehendeth and fasteneth all virtues together. And as [278] it is elegantly said by Menander of vain love, which is but a false imitation of divine love, *Amor melior sophista lævo ad humanam vitam*, that love teacheth a man to carry himself better than the sophist or preceptor, which he calleth *left-handed*, because with all his rules and preceptions he cannot form a man so *dexterously*, nor with that facility to prize himself and govern himself, as love can do; so certainly if a man's mind be truly inflamed with charity, it doth work him suddenly into greater perfection than all the doctrine of morality can do, which is but a sophist in comparison of the other. Nay further, as Xenophon observed truly that all other affections, though they raise the mind, yet they do it by distorting and uncomeliness of ecstasies or excesses; but only love doth exalt the mind, and nevertheless at the same instant doth settle and compose it; so in all other excellencies, though they advance nature, yet they are subject to excess; only charity admitteth no excess: for so we see, aspiring to be like God in power, the angels transgressed and fell; *Ascendam, et ero similis Altissimo* [I will ascend and be like unto the Highest]: by aspiring to be like God in knowledge, man transgressed and fell; *Eritis sicut Dii, scientes bonum et malum* [ye shall be as Gods, knowing good and evil]; but by aspiring to a similitude of God in goodness or love, neither man nor angel ever transgressed or shall transgress. For unto that imitation we are called: *Diligite inimicos vestros, benefacite eis qui oderunt vos, et orate pro persequentibus et calumniantibus vos, ut sitis filii Patris vestri qui in cœlis est, qui solem suum oriri facit super bonos et malos, et pluit super justos et injustos*; [love your enemies, do good to them that hate you, and pray for them which despitefully use you and persecute you, that ye may

277 The words "funeral oration" are omitted in the translation. It was not a funeral oration, but a Panegyric spoken in Trajan's presence. See above, p. 140.

278 So edd. 1629 and 1633. The original omits *as*.

be the children of your Father which is in heaven, who maketh his sun to rise on the evil and on the good, and sendeth rain on the just and on the unjust]. So in the first platform of the divine nature itself, the heathen religion speaketh thus, *Optimus Maximus* [Best and Greatest]: and the sacred Scriptures thus, *Misericordia ejus super omnia opera ejus* [his mercy is over all his works].

Wherefore I do conclude this part of moral knowledge, concerning the Culture and Regiment of the Mind; wherein if any man, considering the parts thereof which I have enumerated, do judge that my labour is but to collect into an Art or Science that which hath been pretermitted by others as matter of common sense and experience, he judgeth well. But as Philocrates sported with Demosthenes, *You may not marvel (Athenians), that Demosthenes and I do differ, for he drinketh water, and I drink wine*; and like as we read of an ancient parable of *the two gates of sleep*,

> Sunt geminæ somni portæ: quarum altera fertur
> Cornea, qua veris facilis datur exitus umbris:
> Altera candenti perfecta nitens elephanto,
> Sed falsa ad cœlum mittunt insomnia manes:
> [Two gates there are of sleep; of horn the one,
> By which the true shades pass; of ivory
> Burnished and white the other, but through it
> Into the upper world false dreams are sent]:

so if we put on sobriety and attention, we shall find it a sure maxim in knowledge that the more pleasant liquor (*of wine*) is the more vaporous, and the braver gate (*of ivory*) sendeth forth the falser dreams [279].

But we have now concluded *that general part of Human Philosophy, which contemplateth man segregate, and as he consisteth of body and spirit.* Wherein we may further note, that there seemeth to be a relation or conformity between the good of the mind and the good of the body. For as we divided the good of the body into *health, beauty, strength,* and *pleasure*; so the good of the mind, inquired in rational and moral knowledge[280], tendeth to this, to make the mind *sound*, and without perturbation; *beautiful*, and graced with decency; and *strong* and *agile* for all duties of life. These three, as in the body so in the mind, seldom meet, and commonly sever. For it is easy to observe that many have strength of wit and courage, but have neither health from perturbations, nor any beauty or decency in their doings: some again have an elegancy and fineness of carriage, which have neither soundness of honesty, nor substance of sufficiency: and some again have honest and reformed minds, that can neither become themselves nor manage business: and sometimes two of them meet, and rarely all three. As for pleasure, we have likewise determined that the mind ought not to be reduced to stupidity, but to retain pleasure; confined rather in the subject of it, than in the strength and vigour of it [281].

279 The allusion to Philocrates and Demosthenes and to the difference between wine and water is omitted in the translation; probably because Bacon had since used the same illustration in an opposite sense (see *Nov. Org.* i. 123.), taking the wine to represent his own philosophy, with its variety of material and elaborate process of manufacture, and the water to represent the popular philosophy of his time which was content with what came; and the present passage reads the clearer and better for the omission. After "he judgeth well", yet let him remember (he says) that the object I am in pursuit of is not beauty and fair appearance, but utility and truth; and let him a little call to mind the meaning of that ancient parable, *Sunt geminæ somni portæ*, etc. Great no doubt is the magnificence of the ivory gate, but the true dreams pass by the gate of horn.

280 *i.e.* considered with reference to reason and morals—(*si juxta moralis doctrinæ scita illud contemplemur*).

281 For in a mind properly disposed, the act and exercise of virtue ought to be accompanied with a sense of pleasure; as is more clearly expressed in the translation. There are some, he says, who have both health, beauty, and strength of mind; and so perform their duties well; but, from a kind of Stoical severity and insensibility, take no pleasure

¶ [282] Civil Knowledge is conversant about a subject which of all others is most immersed in matter, and hardliest reduced to axiom. Nevertheless, as Cato the censor said, *That the Romans were like sheep, for that a man might better drive a flock of them, than one of them ; for in a flock, if you could get but some few go right, the rest would follow* : so in that respect moral philosophy is more difficile than policy. Again, moral philosophy propoundeth to itself the framing of internal goodness ; but civil knowledge requireth only an external goodness ; for that as to society sufficeth ; and therefore it cometh oft to pass that there be evil times in good governments : for so we find in the holy story, when the kings were good, yet it is added, *Sed adhuc populus non direxerat* [283] *cor suum ad Dominum Deum patrum suorum* [but as yet the people had not turned their hearts towards the Lord God of their fathers]. Again, States, as great engines, move slowly, and are not so soon put out of frame : for as in Egypt the seven good years sustained the seven bad, so governments for a time well grounded do bear out errors following : but the resolution of particular persons is more suddenly subverted. These respects do somewhat qualify the extreme difficulty of civil knowledge.

This knowledge hath three parts, according to the three summary actions of society ; which are Conversation, Negotiation, and Government. For man seeketh in society comfort, use and protection: and they be three wisdoms of divers natures, which do often sever; wisdom of the behaviour, wisdom of business, and wisdom of state.

The wisdom of Conversation ought not to be over much affected, but much less despised ; for it hath not only an honour in itself, but an influence also into business and government [284]. The poet saith,

Nec vultu destrue verba tuo :

a man may destroy the force of his words with his countenance : so may he of his deeds, saith Cicero ; recommending to his brother affability and easy access ; *Nil interest habere ostium apertum, vultum clausum ;* it is nothing won to admit men with an open door, and to receive them with a shut and reserved countenance. So we see Atticus, before the first interview between Cæsar and Cicero, the war depending, did seriously advise Cicero touching the composing and ordering of his countenance and gesture. And if the government of the countenance be of such effect, much more is that of the speech, and other carriage appertaining to conversation ; the true model whereof seemeth to me well expressed by Livy, though not meant for this purpose ; *Ne aut arrogans videar, aut obnoxius ; quorum alterum est alienæ libertatis obliti, alterum suæ :* the sum of behaviour is to retain a man's own dignity, without intruding upon the liberty of others. On the other side, if behaviour and outward carriage be intended too much, first it may pass into affection [285], and then *quid deformius quam scenam in vitam transferre* [what more unseemly than to be always playing a part] ; to act a man's life ? But although it proceed not to that extreme, yet it consumeth time, and employeth the mind too much. And therefore as we use to advise young students from company keeping, by saying, *Amici fures temporis* [friends are thieves of time] ; so certainly the intending of the discretion of behaviour is a great thief of meditation. Again, such as are accomplished in that honor [286] of urbanity

in them (*sed tamen Stoica quadam tristitia et stupiditate præditi, virtutis quidem actiones exercent, gaudiis non perfruuntur*).

[282] De Aug. viii. 1.

[283] *dixerat* in the original and also in edd. 1629 and 1633. *direxerat.*—De Aug.

[284] In the translation he compares the value of Conversation in business to that of action in oratory.

[285] So the original. Edd. 1629 and 1633 have *affectation ;* which is the more modern form of the word. But the other was I think the more common when the *Advancement* was written.

[286] *howr* in original : *hour* in ed. 1633. Ed. 1629 has *forme ;* which is the reading of all the modern editions. But *fourme* could not easily be mistaken for *howr*, whereas *honor* carelessly written would be hardly distinguishable from it. The translation also,

please themselves in name [287], and seldom aspire to higher virtue ; whereas those that have defect in it do seek comeliness by reputation : for where reputation is, almost every thing becometh ; but where that is not, it must be supplied by *puntos* and compliments. Again, there is no greater impediment of action than an over-curious observance of decency, and the guide of decency, which is time and season. For as Salomon sayeth, *Qui respicit ad ventos, non seminat ; et qui respicit ad nubes non metet;* [he that looketh to the winds doth not sow, and he that regardeth the clouds shall not reap] : a man must make his opportunity, as oft as find it. To conclude ; Behaviour seemeth to me as a garment of the mind, and to have the conditions of a garment. For it ought to be made in fashion; it ought not to be too curious ; it ought to be shaped so as to set forth any good making of the mind, and hide any deformity ; and above all, it ought not to be too strait or restrained for exercise or motion. But this part of civil knowledge hath been elegantly handled, and therefore I cannot report it for deficient.

¶ [288] The wisdom touching Negotiation or Business hath not been hitherto collected into writing, to the great derogation of learning and the professors of learning. For from this root springeth chiefly that note or opinion, which by us is expressed in adage to this effect, that there is no great concurrence between learning and wisdom. For of the three wisdoms which we have set down to pertain to civil life, for wisdom of Behaviour, it is by learned men for the most part despised, as an inferior to virtue and an enemy to meditation ; for wisdom of Government, they acquit themselves well when they are called to it, but that happeneth to few ; but for the wisdom of Business, wherein man's life is most conversant, there be no books of it, except some few scattered advertisements that have no proportion to the magnitude of this subject. For if books were written of this as the other, I doubt not but learned men with mean experience would far excel men of long experience without learning, and outshoot them in their own bow.

Neither needeth it at all to be doubted that this knowledge should be so variable as it falleth not under precept ; for it is much less infinite than science of Government, which we see is laboured and in some part reduced. Of this wisdom it seemeth some of the ancient Romans in the saddest and wisest times were professors ; for Cicero reporteth that it was then [289] in use for senators that had name and opinion for general wise men, as Coruncanius, Curius, Lælius, and many others, to walk at certain hours in the Place, and to give audience to those that would use their advice ; and that the particular citizens would resort unto them, and consult with them of the marriage of a daughter, or of the employing of a son, or of a purchase or bargain, or of an accusation, and every other occasion incident to man's life ; so as there is a wisdom of counsel and advice even in private causes, arising out of an universal insight into the affairs of the world ; which is used indeed upon particular cases [290] propounded, but is gathered by general observation of causes of like nature. For so we see in the book which Q. Cicero [291] writeth to his brother *De petitione consulatus* (being the only book of business that I know written by the ancients), although it concerned a particular action then on foot, yet the substance thereof consisteth of many wise and politic axioms, which contain not a temporary but a perpetual direction in the case of popular elections. But chiefly we may see in those aphorisms which have place amongst divine writings, composed by Salomon the king, of whom

though the expression is altered, preserves the idea of honour. *Qui primas adeo in urbanitate obtinent et ad hanc rem unam quasi nati videntur.*

[287] So both the original and ed. 1633. Ed. 1629 has " in *it* " ; which has been followed by modern editors. The translation has *ut sibi ipsis in illa sola complaceant.* If *name* be the right word (which I doubt) the meaning must be that they are satisfied with the good report which it procures them. Perhaps it should be " please themselves in the same ".

[288] De Aug. viii. 2.

[289] *i.e.* in the times of which he writes,—a little before his own. (*paulo ante sua secula*).

[290] So the original. Edd. 1629 and 1633 have *causes*.

[291] So edd. 1629 and 1633 and *De Aug.* The original omits *Q.*

the Scriptures testify that his heart was as the sands of the sea, encompassing the world and all worldly matters; we see, I say, not a few profound and excellent cautions, precepts, positions, extending to much variety of occasions; whereupon we will stay awhile, offering to consideration some number of examples [292].

Sed et cunctis sermonibus qui dicuntur ne accommodes aurem tuam, ne forte audias servum tuum maledicentem tibi [Hearken not unto all words that are spoken, lest thou hear thy servant curse thee]. Here is concluded the provident stay of inquiry of that which we would be loth to find [293]: as it was judged great wisdom in Pompeius Magnus that he burned Sertorius' papers unperused.

Vir sapiens si cum stulto contenderit, sive irascatur sive rideat, non inveniet requiem [A wise man if he contend with a fool, whether he be angry or whether he laugh, shall find no rest]. Here is described the great disadvantage which a wise man hath in undertaking a lighter person than himself; which is such an engagement as whether a man turn the matter to jest, or turn it to heat, or howsoever he change copy, he can no ways quit himself well of it.

Qui delicatè a pueritia nutrit servum suum, postea sentiet eum contumacem [He that delicately bringeth up his servant from a child shall have him become froward at the length]. Here is signified, that if a man begin too high a pitch in his favours, it doth commonly end in unkindness and unthankfulness.

Vidisti virum velocem in opere suo? Coram regibus stabit, nec erit inter ignobiles [Seest thou a man that is quick in his business? He shall stand before kings; his place shall not be among mean men]. Here is observed that, of all virtues for rising to honour, quickness of dispatch is the best; for superiors many times love not to have those they employ too deep or too sufficient, but ready and diligent.

Vidi cunctos viventes qui ambulant sub sole, cum adolescente secundo qui consurgit pro eo [I beheld all the living which walk under the sun, with the second youth that shall stand in his place]. Here is expressed that which was noted by Sylla first, and after him by Tiberius: *Plures adorant solem orientem quam occidentem vel meridianum* [294] [there be more that worship the rising sun than the sun setting or at mid-day].

Si spiritus potestatem habentis ascenderit super te, locum tuum ne dimiseris; quia curatio faciet cessare peccata maxima [If the spirit of the ruler rise up against thee, leave not thy place; for observance will remove great offences]. Here caution is given that upon displeasure, retiring is of all courses the unfittest; for a man leaveth things at worst, and depriveth himself of means to make them better.

Erat civitas parva, et pauci in ea viri: venit contra eam rex magnus, et vadavit eam, intruxitque munitiones per gyrum, et perfecta est obsidio: inventusque est in ea vir pauper et sapiens, et liberavit eam per sapientiam suam; et nullus deinceps recordatus est hominis illius pauperis [There was a little city and few men within it; and there came a great king against it and besieged it and raised great bulwarks round about it: and there was found in it a poor wise man, and he by his wisdom delivered the city; yet no man remembered that same poor man]. Here the corruption [295] of states is set forth, that esteem not virtue or merit longer than they have use of it.

Mollis responsio frangit iram [A soft answer defeateth wrath]. Here is noted that silence or rough answer exasperateth; but an answer present and temperate pacifieth.

Iter pigrorum quasi sepes spinarum [The way of the slothful is as an hedge

[292] This is what he calls in the translation *Doctrina de Negotiis Sparsis*. The example which follows is greatly enlarged: the number of proverbs commented upon being increased by a third, and the comments being much fuller.

[293] Compare L'Estrange's *Fables and storyes moralized*, vol. ii. p. 6, ed. 1708.

[294] The words *vel meridianum* are omitted in the translation; and it is difficult to understand how they got in; for they are not to be found in either of the passages alluded to, and they seem to carry the observation beyond the truth.

[295] So edd. 1629 and 1633. The original has *corruptions*.

of thorns]. Here is lively represented how laborious sloth proveth in the end; for when things are deferred till the last instant and nothing prepared beforehand, every step findeth a brier or an impediment, which catcheth or stoppeth.

Melior est finis orationis quam principium [Better is the end of a speech than the beginning thereof]. Here is taxed the vanity of formal speakers, that study more about prefaces and inducements than upon the conclusions and issues of speech.

Qui cognoscit in judicio faciem, non bene facit; iste et pro buccella panis deseret veritatem [He that respecteth persons in judgment doth not well; even for a piece of bread will that man depart from the truth]. Here is noted, that a judge were better be a briber than a respecter of persons; for a corrupt judge offendeth not so lightly [296] as a facile.

Vir pauper calumnians pauperes similis est imbri vehementi, in quo paratur fames [A poor man that beareth witness against the poor is like a sweeping rain which leaveth no food]. Here is expressed the extremity of necessitous extortions, figured in the ancient fable of the full and hungry horse-leech.

Fons turbatus pede, et vena corrupta, est justus cadens coram impio [A righteous man falling down before the wicked is as a troubled fountain and a corrupt spring]. Here is noted, that one judicial and exemplar iniquity in the face of the world, doth trouble the fountains of justice more than many particular injuries passed over by connivance.

[297] *Qui subtrahit aliquid a patre et a matre, et dicit hoc non esse peccatum, particeps est homicidii* [Whoso robbeth his father and his mother, and saith it is no transgression, is the companion of a destroyer]. Here is noted, that whereas men in wronging their best friends use to extenuate their fault, as if they might presume or be bold upon them, it doth contrariwise indeed aggravate their fault, and turneth it from injury to impiety.

Noli esse amicus homini iracundo, nec ambulato cum homine furioso [Make no friendship with an angry man, neither go with a furious man]. Here caution is given, that in the election of our friends we do principally avoid those which are impatient, as those that will espouse us to many factions and quarrels.

Qui conturbat domum suam, possidebit ventum [He that troubleth his own house shall inherit the wind]. Here is noted, that in domestical separations and breaches men do promise to themselves quieting of their mind and contentment; but still they are deceived of their expectation, and it turneth to wind.

Filius sapiens lætificat patrem: filius vero stultus mæstitia est matri suæ [A wise son maketh a glad father, but a foolish son is the heaviness of his mother]. Here is distinguished, that fathers have most comfort of the good proof of their sons; but mothers have most discomfort of their ill proof, because women have little discerning of virtue, but of fortune [298].

Qui celat delictum, quærit amicitiam; sed qui altero sermone repetit, separat fœderatos [He that covereth a transgression seeketh love, but he that repeateth a matter separateth very friends]. Here caution is given, that reconcilement is better managed by an *amnesty*, and passing over that which is past, than by apologies and excusations.

In omni opere bono erit abundantia; ubi autem verba sunt plurima, ibi frequenter

[296] So the original. Edd. 1629 and 1633 have *highly*: a conjectural emendation probably, by some critic who did not know that *lightly* meant *easily, readily, upon slight occasion*; or did not observe that the point of the observation rests entirely upon this word. The corrupt judge does not offend less highly than the facile; but less frequently.

[297] This proverb is omitted in the translation.

[298] In the translation he adds two other causes—the greater tenderness of the mother's affection, and (perhaps) a consciousness that her own indulgence has spoiled her son; and instead of saying that the mother has "little discerning of virtue", he only says that the father understands its value better. The allusion to fortune is omitted altogether; and indeed it is not easy to see how it bears upon the case in point; the son in question being by the supposition not unfortunate but foolish. I thought it right to mention this alteration, because it is more than a development of the remark in the text; it is a correction of the opinion implied in it.

egestas [In every good work there shall be abundance, but where there are many words there is penury]. Here is noted that words and discourse abound most where there is idleness and want.

Primus in sua causa justus ; sed venit altera pars, et inquirit in eum [He that is first in his own cause seemeth just ; but the other party cometh and searcheth him]. Here is observed, that in all causes the first tale possesseth much ; in sort [299] that the prejudice thereby wrought will be hardly removed, except some abuse or falsity in the information be detected.

[300] *Verba bilinguis quasi simplicia, et ipsa perveniunt ad interiora ventris* [The words of the double-tongued man which seem artless are they that go down to the innermost parts of the belly]. Here is distinguished, that flattery and insinuation which seemeth set and artificial sinketh not far ; but that entereth deep which hath shew of nature, liberty, and simplicity.

Qui erudit derisorem, ipse sibi injuriam facit ; et qui arguit impium, sibi maculam generat [He that reproveth a scorner doth himself wrong, and he that rebuketh a wicked man getteth himself a blot]. Here caution is given how we tender reprehension to arrogant and scornful natures, whose manner is to esteem it for contumely, and accordingly to return it.

Da sapienti occasionem, et addetur ei sapientia [Give opportunity to a wise man, and he will be yet wiser]. Here is distinguished the wisdom brought into habit, and that which is but verbal and swimming only in conceit ; for the one upon the occasion presented is quickened and redoubled, the other is amazed and confused.

Quomodo in aquis resplendent vultus prospicientium, sic corda hominum manifesta sunt prudentibus [As the face of one that looketh upon the water is reflected therein, so the hearts of men are manifest unto the wise]. Here the mind of a wise man is compared to a glass, wherein the images of all diversity of natures and customs are represented ; from which representation proceedeth that application,

Qui sapit, innumeris moribus aptus erit

[a wise man will know how to apply himself to all sorts of characters].

Thus have I staid somewhat longer upon these sentences politic of Salomon than is agreeable to the proportion of an example ; led with a desire to give authority to this part of knowledge, which I noted as deficient, by so excellent a precedent ; and have also attended them with brief observations, such as to my understanding offer no violence to the sense, though I know they may be applied to a more divine use : but it is allowed even in divinity, that some interpretations, yea and some writings, have more of the Eagle than others. But taking them as instructions for life, they might have received large discourse, if I would have broken them and illustrated them by deducements and examples.

Neither was this in use only with the Hebrews ; but it is generally to be found in the wisdom of the more ancient times, that as men found out any observation that they thought was good for life, they would gather it and express it in parable or aphorism or fable. But for fables, they were vicegerents and supplies where examples failed : now that the times abound with history, the aim is better when the mark is alive. And therefore the form of writing which of all others is fittest for this variable argument of negotiation and occasions is that which Machiavel chose wisely and aptly for government ; namely, *discourse upon histories or examples*. For knowledge drawn freshly and in our view out of particulars, knoweth the best way to particulars again. And it hath much greater life for practice when the discourse attendeth upon the example, than when the example attendeth upon the discourse. For this is no point of order, as it seemeth at first, but of substance. For when the example is the ground, being set down in an history at large, it is set down with all circumstances. which may sometimes control the discourse thereupon made and sometimes supply it, as a very pattern

[299] So the original. Edd. 1629 and 1633 have *in such sort :* an attempt at correction where none was wanted. [300] This proverb is omitted in the translation.

for action [301]; whereas the examples alleged for the discourse's sake are cited succinctly and without particularity, and carry a servile aspect toward the discourse which they are brought in to make good.

But this difference is not amiss to be remembered, that as history of Times is the best ground for discourse of government, such as Machiavel handleth, so histories of Lives is the most proper for discourse of business, as [302] more conversant in private actions. Nay there is a ground of discourse for this purpose fitter than them both, which is *discourse upon letters*, such as are wise and weighty, as many are of Cicero *ad Atticum* and others. For letters have a great [303] and more particular representation of business than either Chronicles or Lives. Thus have we spoken both of the matter and form of this part of civil knowledge touching Negotiation [304], which we note to be deficient.

But yet there is another part of this part, which differeth as much from that whereof we have spoken as *sapere* and *sibi sapere* [*to be wise* and *to be wise for oneself*], the one moving as it were to the circumference, the other to the centre. For there is a wisdom of counsel, and again there is a wisdom of pressing a man's own fortune; and they do sometimes meet, and often sever. For many are wise in their own ways that are weak for government or counsel; like ants, which is a wise creature for itself, but very hurtful for the garden. This wisdom the Romans did take much knowledge of [305]: *Nam pol sapiens* (saith the comical poet) *fingit fortunam sibi* [the wise man fashions his fortune for himself]; and it grew to an adage, *Faber quisque fortunæ propriæ* [every man has tools to make his own fortune with], and Livy attributeth it to Cato the first, *In hoc viro tanta vis animi et ingenii inerat, ut quocunque loco natus esset, sibi ipse fortunam facturus videretur*, [such was his force of mind and genius that in whatever state he had been born he would have made himself a fortune].

This conceit or position [306] if it be too much declared and professed, hath been thought a thing impolitic and unlucky; as was observed in Timotheus the Athenian; who having done many great services to the estate in his government, and giving an account thereof to the people as the manner was, did conclude every particular with this clause, and in this fortune had no part. And it came so to pass that he never prospered in any thing he took in hand afterward: for this is too high and too arrogant, savouring of that which Ezekiel saith of Pharaoh, *Dicis, Fluvius est meus, et ego feci memet ipsum* [thou sayest the river is mine, and I made myself]; or of that which another prophet speaketh, that men offer sacrifices to their nets and snares; and that which the poet expresseth,

> Dextra mihi Deus, et telum quod missile [307] libro,
> Nunc adsint!

[301] So edd. 1629 and 1633. The original has *gaine*. I doubt whether *action* be the right word, and should rather suspect *aime*, which might look very like *gaine* if the tail of a letter from the line above happened to strike through the *a*. The translation has *unde fit loco exemplaris ad imitationem et practicam*.

[302] *is* both in orig. and in edd. 1629 and 1633. Blackbourne substituted *because it is*. Instead of "private actions", the translation substitutes "actions of all kinds great and small"—(*quoniam omnem occasionum et negotiorum, tam grandium quam leviorum, varietatem complectuntur*).

[303] So all three editions, though *great* can hardly be the right word. I should suspect *nearer*. The translation has *magis in proximo et ad vivum negotia solent repræsentare*.

[304] i.e. *de negotiis sparsis*.

[305] And yet (he adds in the translation) there were no better patriots,—*licet patriæ optimis curatoribus*.

[306] The translation has *hoc genus prudentiæ*.

[307] *inutile* in the original, and also in ed. 1633: obviously a misprint. Ed. 1629 and the *De Augmentis* have it right.

In addition to these instances he cites in the translation another from Julius Cæsar himself. When the soothsayer reported the auspices unfavourable, he was heard to mutter "they will be more favourable when I will". The anecdote comes from Suetonius. It was the only occasion (Bacon adds) on which Cæsar so far forgot himself as to betray

[my right hand and my spear are the God I trust in]. For these confidences were ever unhallowed, and unblessed. And therefore those that were great politiques indeed ever ascribed their successes to their felicity, and not to their skill or virtue. For so Sylla surnamed himself *Felix*, not *Magnus* [the Fortunate, not the Great]. So Cæsar said [308] to the master of the ship, *Cæsarem portas et fortunam ejus* [you carry Cæsar and his fortune].

But yet nevertheless these positions, *Faber quisque fortunæ suæ ; Sapiens dominabitur astris ; Invia virtuti nulla est via* [every man should be the maker of his own fortune; the wise man will command his stars; nothing impossible to virtue]: and the like, being taken and used as spurs to industry, and not as stirrups to insolency, rather for resolution than for presumption or outward declaration, have been ever thought sound and good, and are no question imprinted in the greatest minds; who are so sensible of this opinion as they can scarce contain it within. As we see in Augustus Cæsar, (who was rather diverse from his uncle than inferior in virtue [309]), how when he died, he desired his friends about him to give him a *Plaudite ;* as if he were conscient to himself that he had played his part well upon the stage. This part of knowledge we do report also as deficient: not but that it is practised too much, but it hath not been reduced to writing. And therefore lest it should seem to any that it is not comprehensible by axiom, it is requisite, as we did in the former, that we set down some heads or passages of it.

Faber Fortunæ, sive de Ambitu vitæ

Wherein it may appear at the first a new and unwonted argument to teach men how to raise and make their fortune; a doctrine wherein every man perchance will be ready to yield himself a disciple, till he see the difficulty: for Fortune layeth as heavy impositions as Virtue; and it is as hard and severe a thing to be a true politique, as to be truly moral. But the handling hereof concerneth learning greatly, both in honour and in substance: in honour, because pragmatical men may not go away with an opinion that learning is like a lark that can mount and sing and please herself, and nothing else; but may know that she holdeth as well of the hawk, that can soar aloft, and can also descend and strike upon the prey: in substance, because it is the perfect law of inquiry of truth, *that nothing be in the globe of matter, which should not be likewise in the globe of crystal, or form ;* that is that there be not anything in being and action, which should not be drawn and collected into contemplation and doctrine. Neither doth learning admire or esteem of this architecture of fortune otherwise than as of an inferior work: for no man's fortune can be an end worthy of his being, and many times the worthiest men do abandon their fortune willingly for better respects: but nevertheless fortune as an organ of virtue and merit deserveth the consideration.

First therefore the precept which I conceive to be most summary towards the prevailing in fortune, is to obtain that window which Momus did require, who seeing in the frame of man's heart such angles and recesses, found fault there was not a window to look into them; that is, to procure good informations of particulars touching persons, their natures, their desires and ends, their customs and fashions, their helps and advantages, and whereby they chiefly stand; so again their weaknesses and disadvantages, and where they lie most open and obnoxious; their friends, factions, dependances; and again their opposites, enviers, competitors, their moods and times, *Sola viri molies aditus et tempora noras ;* their principles, rules, and observations, and the like: and this not only of persons, but of actions; what are on foot from time to time, and how they are conducted, favoured, opposed; and how they import, and the like. For the

his secret thoughts—(*nunquam, quod memini, impotentiam cogitationum suarum arcanarum prodidit nisi simili dicto*); and his death followed soon after.

[308] better (adds the translation) than in the instance above mentioned.

[309] *sed vir certe paulo moderatior.* In Bacon's character of Augustus—the fragment entitled *Imago Civilis Augusti Cæsaris*—he acknowledges that he was inferior to Julius in strength of mind, but asserts that he was superior in beauty and health of mind; Julius's aspirations being restless, boundless, and inordinate; those of Augustus sober, well ordered, and within compass.

knowledge of present actions is not only material in itself, but without it also the knowledge of persons is very erroneous: for men change with the actions; and whiles they are in pursuit they are one, and when they return to their nature they are another. These informations of particulars touching persons and actions are as the minor propositions in every active syllogism; for no excellency of observations (which are as the major propositions) can suffice to ground a conclusion, if there be error and mistaking in the minors.

That this knowledge is possible, Salomon is our surety; who saith, *Consilium in corde viri tanquam aqua profunda; sed vir prudens exhauriet illud* [counsel in the heart of man is like deep water; but a man of understanding will draw it out]. And although the knowledge itself falleth not under precept, because it is of individuals, yet the instructions for the obtaining of it may.

We will begin therefore with this precept, according to the ancient opinion, that the sinews of wisdom are slowness of belief and distrust; that more trust be given to countenances and deeds than to words; and in words, rather to sudden passages and surprised words, than to set and purposed words. Neither let that be feared which is said, *fronti nulla fides* [no trusting to the face]: which is meant of a general outward behaviour, and not of the private and subtile motions and labours of the countenance and gesture; which as Q. Cicero elegantly saith, is *animi janua*, the gate of the mind. None more close than Tiberius, and yet Tacitus saith of Gallus, *Etenim vultu offensionem conjectaverat* [he had seen displeasure in his countenance]. So again, noting the differing character and manner of his commending Germanicus and Drusus in the senate, he saith touching his fashion wherein he carried his speech of Germanicus, thus; *Magis in speciem adornatis verbis, quam ut penitus sentire videretur* [it was in words too laboured and specious to be taken for what he really felt]; but of Drusus thus; *Paucioribus, sed intentior, et fida oratione* [he said less, but more earnestly, and in a style of sincerity]; and in another place, speaking of his character of speech when he did anything that was gracious and popular, he saith that in other things he was *velut eluctantium verborum* [of a kind of struggling speech]; but then again, *solutius loquebatur quando subveniret* [he spoke with more freedom when he was speaking in a man's favour]. So that there is no such artificer of dissimulation, nor no such commanded countenance (*vultus jussus*) that can sever from a feigned tale some of these fashions, either a more slight and careless fashion, or more set and formal, or more tedious and wandering, or coming from a man more drily and hardly.

Neither are *deeds* such assured pledges, as that they may be trusted without a judicious consideration of their magnitude and nature: *Fraus sibi in parvis fidem præstruit, ut majore emolumento fallat* [it is a trick of treachery to win itself credit at the first by fidelity in small things, that being thereupon trusted in greater it may deceive with more advantage]; and the Italian thinketh himself upon the point to be bought and sold, when he is better used than he was wont to be without manifest cause. For small favours, they do but lull men asleep, both as to caution and as to industry, and are as Demosthenes calleth them, *Alimenta socordiæ* [sops to feed sloth]. So again we see how false the nature of some deeds are, in that particular which Mutianus practised upon Antonius Primus, upon that hollow and unfaithful reconcilement which was made between them; whereupon Mutianus advanced many of the friends of Antonius: *simul amicis ejus præfecturas et tribunatus largitur* [making them prefects and tribunes]: wherein under pretence to strengthen him, he did desolate him, and won from him his dependances.

As for *words*, (though they be like waters to physicians, full of flattery and uncertainty), yet they are not to be despised, specially with the advantage of passion and affection. For so we see Tiberius upon a stinging and incensing speech of Agrippina came a step forth of his dissimulation, when he said, *You are hurt because you do not reign;* of which Tacitus saith, *Audita hæc raram occulti pectoris vocem elicuere; correptamque Græco versu admonuit, ideo lædi quia non regnaret,* [these words drew from Tiberius the voice, so rarely heard, of his secret heart: he retorted upon her with a Greek verse, that she was hurt, etc.]. And therefore the poet doth elegantly call passions tortures, that urge men to confess their secrets :

Vino tortus et ira.

And experience sheweth, there are few men so true to themselves and so settled, but that, sometimes upon heat, sometimes upon bravery, sometimes upon kindness, sometimes upon trouble of mind and weakness, they open themselves ; specially if they be put to it with a counter-dissimulation, according to the proverb of Spain, *Di mentira, y sacaras verdad, Tell a lie and find a truth.*

As for the knowing of men which is at second hand from reports ; men's weaknesses [310] and faults are best known from their enemies, their virtues and abilities from their friends, their customs and times from their servants, their conceits and opinions from their familiar friends with whom they discourse most. General fame is light, and the opinions conceived by superiors or equals [311] are deceitful ; for to such men are more masked : *Verior fama e domesticis emanat* [the truer kind of report comes from those who see them at home].

But the soundest disclosing and expounding of men is by their natures and ends ; wherein the weakest sort of men are best interpreted by their natures and the wisest by their ends [312]. For it was both pleasantly and wisely said (though I think very untruly) by a nuncio of the pope, returning from a certain nation where he served as lieger ; whose opinion being asked touching the appointment of one to go in his place, he wished that in any case they did not send one that was too wise ; because no very wise man would ever imagine what they in that country were like to do. And certainly it is an error frequent for men to shoot over, and to suppose deeper ends and more compass reaches than are : the Italian proverb being elegant, and for the most part true :

Di danari, di senno, e di fede,
Cè nè manco che non credi :

There is commonly less money, less wisdom, and less good faith, than men do account upon.

But Princes upon a far other reason are best interpreted by their natures, and private persons by their ends ; for princes being at the top of human desires, they have for the most part no particular ends whereto they aspire [313], by distance from which a man might take measure and scale of the rest of their actions and desires ; which is one of the causes that maketh their hearts more inscrutable [314]. Neither is it sufficient to inform ourselves in men's ends and natures of the variety of them only, but also of the predominancy, what humour reigneth most, and what end is principally sought. For so we see, when Tigellinus saw himself outstripped by Petronius Turpilianus in Nero's humours of pleasures, *metus ejus rimatur* [315], he wrought upon Nero's fears, whereby he brake the other's neck.

But to all this part of inquiry the most compendious way resteth in three things. The first, to have general acquaintance and inwardness with those which have general acquaintance and look most into the world ; and specially according to the diversity of business and the diversity of persons, to have privacy and conversation with some one friend at least which is perfect and well intelligenced in every several kind. The second is to keep a good mediocrity in liberty of speech and secrecy ; in most things liberty ; secrecy where it im-

310 So ed. 1633. The original and ed. 1629 have *weaknesse.*

311 The translation omits *equals :* a correction no doubt of Bacon's own.

312 According to the translation, the weaker *and the more simple* by their natures ; the wiser *and the more close* by their ends.

313 *i.e.* not earnestly and constantly—(*ad quos, præsertim vehementer et constanter, aspirent*).

314 Whereas private persons are almost all like travellers making for their journey's end ; and if you know what they are aiming at, you may guess by that what they are likely to do and what not to do.

315 So edd. 1629 and 1633. The original has *rinaeur.*

porteth; for liberty of speech inviteth and provoketh liberty to be used again, and so bringeth much to a man's knowledge; and secrecy, on the other side, induceth trust and inwardness. The last is the reducing of a man's self to this watchful and serene habit, as to make account and purpose, in every conference and action, as well to observe as to act. For as Epictetus would have a philosopher in every particular action to say to himself, *Et hoc volo, et etiam institutum servare* [I would do this and keep my course too]; so a politic man in every thing should say to himself, *Et hoc volo, ac etiam aliquid addiscere* [I would do it and also learn something from it[316]]. I have stayed the longer upon this precept of obtaining good information, because it is a main part by itself, which answereth to all the rest. But, above all things, caution must be taken that men have a good stay and hold of themselves, and that this much knowledge do not draw on much meddling; for nothing is more unfortunate than light and rash intermeddling in many matters; so that this variety of knowledge tendeth in conclusion but only to this, to make a better and freer choice of those actions which may concern us, and to conduct them with the less error and the more dexterity.

The second precept concerning this knowledge is, for men to take good information touching their own person, and well to understand themselves; knowing that, as St. James saith, though men look oft in a glass, yet they do suddenly forget themselves; wherein as the divine glass is the word of God, so the politic glass is the state of the world or times wherein we live, in the which we are to behold ourselves.

For men ought to take an unpartial view of their own abilities and virtues; and again of their wants and impediments; accounting these with the most, and those other with the least; and from this view and examination to frame the considerations following.

First, to consider how the constitution of their nature sorteth with the general state of the times; which if they find agreeable and fit, then in all things to give themselves more scope and liberty; but if differing and dissonant, then in the whole course of their life to be more close, retired, and reserved: as we see in Tiberius, who was never seen at a play and came not into the senate in twelve of his last years; whereas Augustus Cæsar lived ever in men's eyes, which Tacitus observeth: *Alia Tiberio morum via* [Tiberius's ways were different [317]].

Secondly, to consider how their nature sorteth with professions and courses of life, and accordingly to make election, if they be free; and, if engaged, to make the departure at the first opportunity: as we see was done by duke Valentine, that was designed by his father to a sacerdotal profession, but quitted it soon after in regard of his parts and inclination; being such nevertheless, as a man cannot tell well whether they were worse for a prince or for a priest.

Thirdly to consider how they sort with those whom they are like to have competitors and concurrents, and to take that course wherein there is most solitude, and themselves like to be most eminent: as Cæsar Julius did, who at first was an orator or pleader; but when he saw the excellency of Cicero, Hortensius, Catulus, and others, for eloquence, and saw there was no man of reputation for the wars but Pompeius, upon whom the state was forced to rely, he forsook his course begun toward a civil and popular greatness, and transferred his designs to a martial greatness.

Fourthly, in the choice of their friends and dependances, to proceed according to the composition of their own nature; as we may see in Cæsar, all whose friends and followers were men active and effectual, but not solemn or of reputation [318].

316 *i.e.* something which may be of use hereafter. And therefore (adds the translation) those who are so intent on the business in hand that, like Montaigne, they pay no attention to anything that turns up by the way, make excellent ministers for Kings and Commonwealths, but bad managers of their own fortune.

317 In the translation Pericles is mentioned as another instance—(*eadem et Periclis ratio fuit*).

318 And men (the translation adds) who were infinitely loyal to Cæsar himself, but arrogant and contemptuous towards all men else; such as Antonius, Hirtius, Pansa, Oppius, Balbus, Dolabella, Pollio, and the rest.

Fifthly, to take special heed how they guide themselves by examples, in thinking they can do as they see others do ; whereas perhaps their natures and carriages are far differing ; in which error it seemeth Pompey was, of whom Cicero saith, that he was wont often to say, *Sylla potuit, ego non potero ?* [Sylla could do it, why not I ?] wherein he was much abused, the natures and proceedings of himself and his example being the unlikest in the world ; the one being fierce, violent, and pressing the fact ; the other solemn, and full of majesty and circumstance, and therefore the less effectual.

But this precept touching the politic knowledge of ourselves hath many other branches whereupon we cannot insist.

Next to the well understanding and discerning of a man's self, there followeth the well opening and revealing [319] a man's self ; wherein we see nothing more usual than for the more able man to make the less shew. For there is a great advantage in the well setting forth of a man's virtues, fortunes, merits ; and again in the artificial covering of a man's weaknesses, defects, disgraces ; staying upon the one, sliding from the other ; cherishing the one by circumstances, gracing the other by exposition, and the like : wherein we see that Tacitus saith of Mutianus, who was the greatest politique of his time, *Omnium quæ dixerat feceratque arte quâdam ostentator* [having a certain art of displaying to advantage all he said and did] ; which requireth indeed some art, lest it turn tedious and arrogant ; but yet so as ostentation (though it be to the first degree of vanity) seemeth to me rather a vice in manners than in policy : for as it is said, *Audacter calumniare* [320], *semper aliquid hæret* [slander boldly, there is ever some that sticks] ; so, except it be in a ridiculous degree of deformity, *Audacter te vendita, semper aliquid hæret* [put forward your own pretensions boldly—something always sticks]. For it will stick with the more ignorant and inferior sort of men, though men of wisdom and rank do smile at it and despise it ; and yet the authority won with many doth countervail the disdain of a few. But if it be carried with decency and government, as with a natural, pleasant, and ingenious [321] fashion ; or at times when it is mixed with some peril and unsafety, (as in military persons) ; or at times when others are most envied ; or with easy and careless passage to it and from it, without dwelling too long or being too serious ; or with an equal freedom of taxing a man's self as well as gracing himself ; or by occasion of repelling or putting down others' injury or insolency ; it doth greatly add to reputation ; and surely not a few solid natures, that want this ventosity and cannot sail in the height of the winds, are not without some prejudice and disadvantage by their moderation.

But for these flourishes and enhancements of virtue, as they are not perchance unnecessary, so it is at least necessary that virtue be not disvalued and imbased under the just price ; which is done in three manners : by offering and obtruding a man's self ; wherein men think he is rewarded, when he is accepted : by doing too much [322] ; which will not give that which is well done leave to settle, and in the end induceth satiety : and by finding too soon the fruit of a man's virtue, in commendation, applause, honour, favour ; wherein if a man be pleased with a little, let him hear what is truly said, *Cave ne insuetus rebus majoribus videaris, si hæc te res parva sicuti magna delectat* [if he take so much delight in a little thing, he will be thought unused to greater things].

But the covering of defects is of no less importance than the valuing of good parts ; which may be done likewise in three manners ; by Caution, by Colour, and by Confidence. Caution is when men do ingeniously and discreetly avoid to be put into those things for which they are not proper : whereas contrariwise bold

319 In the translation this part of the subject is distributed into three separate heads ; —the art of setting a man's self forth to advantage (*se ostentare*)—of making himself understood (*se declarare*)—of turning and shaping himself according to occasion (*flectere se et effingere*) ; and the order of the precepts which follow is changed to suit this arrangement. The three next paragraphs belong to the first head,—the art of ostentation.

320 *calumniari* in the original.

321 *i.e.* ingenuous.

322 Especially in the beginning, and at once—(*quando quis in principio rei gerendæ viribus suis nimium abutitur, et quod sensim erat præstandum uno impetu effundit*).

and unquiet spirits will thrust themselves into matters without difference, and so publish and proclaim all their wants. Colour is when men make a way for themselves to have a construction made of their faults or wants as proceeding from a better cause, or intended for some other purpose : for of the one it is well said, *Sæpe latet vitium proximitate boni* [a vice will often hide itself under the shadow of a neighbouring virtue] ; and therefore whatsoever want a man hath, he must see that he pretend the virtue that shadoweth it ; as if he be dull, he must affect gravity ; if a coward, mildness ; and so the rest : for the second, a man must frame some probable cause why he should not do his best, and why he should dissemble his abilities ; and for that purpose must use to dissemble those abilities which are notorious in him [323], to give colour that his true wants are but industries and dissimulations. For Confidence, it is the last [324] but the surest remedy ; namely, to depress and seem to despise whatsoever a man cannot attain ; observing the good [325] principle of the merchants, who endeavour to raise the price of their own commodities, and to beat down the price of others. But there is a confidence that passeth [326] this other ; which is, to face out a man's own defects, in seeming to conceive that he is best in those things wherein he is failing ; and, to help that again, to seem on the other side that he hath least opinion of himself in those things wherein he is best : like as we shall see it commonly in poets, that if they shew their verses, and you except to any, they will say *that that line cost them more labour than any of the rest ;* and presently will seem to disable and suspect rather some other line, which they know well enough to be the best in the number. But above all, in this righting and helping of a man's self in his own carriage, he must take heed he shew not himself dismantled and exposed to scorn and injury, by too much dulceness, goodness, and facility of nature, but shew some sparkles of liberty, spirit, and edge : which kind of fortified carriage, with a ready rescuing [327] of a man's self from scorns, is sometimes of necessity imposed upon men by somewhat in their person or fortune [328] ; but it ever succeedeth with good felicity [329].

Another precept of this knowledge is, by all possible endeavour to frame the mind to be pliant and obedient to occasion ; for nothing hindereth men's fortunes so much as this *Idem manebat neque idem decebat* [continuing the same when the same is no longer fit] ; men are where they were, when occasions turn : and therefore to Cato, whom Livy maketh such an architect of fortune, he addeth that he had *versatile ingenium* [a wit that could turn well]. And thereof it cometh that these grave solemn wits, which must be like themselves and cannot make departures, have more dignity than felicity. But in some it is nature to be somewhat viscous and inwrapped, and not easy to turn. In some it is a conceit that is almost a nature, which is, that men can hardly make themselves believe that they ought to change their course, when they have found good by it in former experience. For Machiavel noteth wisely, how Fabius Maximus

323 This clause is omitted in the translation ; which says only *ut quod non possimus nolle videamur*).

324 Meaning, I think, the least worthy—the last to be resorted to. The translation has *impudens certe est remedium, sed tamen* etc.

325 *i.e.* prudent—*mercatorum prudentium more, quibus solenne est et proprium, ut* etc.

326 *i.e.* in impudence—(*hoc ipso impudentius*).

327 So ed. 1633. The original and ed. 1629 have *resussing.*

328 As in the case of deformed persons, and bastards, and persons disgraced—(*veluti fit in deformibus, et spuriis, et ignominia aliqua mulctatis*).

329 According to the arrangement adopted in the translation, the observations on the first head—the art of ostentation—end here ; and the art of *declaration*, that is of making oneself understood, is next handled. The substance of the remarks on this head will be found in the next page, in the paragraph beginning " Another part of this knowledge is the observing a good mediocrity ", etc. Then follows the paragraph which stands next in the text ; which refers to the third head,—*quod ad animum flectendum et effingendum attinet.* And with this he concludes what he has to say of " the two *summary* precepts concerning the architecture of Fortune ". The rest he gives as a sample of particular precepts (*præcepta sparsa*) on the same subject.

would have been temporizing still, according to his old bias, when the nature of the war was altered and required hot pursuit. In some other it is want of point and penetration in their judgment, that they do not discern when things have a period, but come in too late after the occasion ; as Demosthenes compareth the people of Athens to country fellows when they play in a fence school, that if they have a blow, then they remove their weapon to that ward, and not before. In some other it is a lothness to leese labours passed, and a conceit that they can bring about occasions to their ply[330] ; and yet in the end, when they see no other remedy, then they come to it with disadvantage; as Tarquinius, that gave for the third part of Sibylla's books the treble price, when he might at first have had all three for the simple. But from whatsoever root or cause this restiveness of mind proceedeth, it is a thing most prejudicial ; and nothing is more politic than to make the wheels of our mind concentric and voluble with the wheels of fortune.

[331] Another precept of this knowledge, which hath some affinity with that we ast spake of, but with difference, is that which is well expressed, *Fatis accede Deisque* [take the way which the Fates and the Gods offer] ; that men do not only turn with the occasions but also run with the occasions, and not strain their credit or strength to over hard or extreme points, but choose in their actions that which is most passable : for this will preserve men from foil, not occupy them too much about one matter, win opinion of moderation, please the most [332], and make a shew of a perpetual felicity in all they undertake ; which cannot but mightily increase reputation.

Another part of this knowledge seemeth to have some repugnancy with the former two, but not as I understand it ; and it is that which Demosthenes uttereth in high terms ; *Et quemadmodum receptum est, ut exercitum ducat imperator, sic et a cordatis viris res ipsæ ducendæ ; ut quæ ipsis videntur, ea gerantur, et non ipsi eventus persequi cogantur* [as the captain leads the army, so should wise men lead affairs ; they should get that done which they think good to be done, and not be forced to follow at the heels of events]. For if we observe, we shall find two differing kinds of sufficiency in managing of business : some can make use of occasions aptly and dexterously, but plot little ; some can urge and pursue their own plots well, but cannot accommodate nor take in [333] ; either of which is very unperfect without the other.

Another part of this knowledge is the observing a good mediocrity in the declaring or not declaring a man's self : for although depth of secrecy, and making way *qualis est via navis in mari* [like the way of a ship through the water], (which the French calleth *sourdes menées*, when men set things in work without opening themselves at all), be sometimes both prosperous and admirable ; yet many times *Dissimulatio errores parit qui dissimulatorem ipsum illaqueant* [dissimulation breeds mistakes in which the dissembler himself is caught]. And therefore we see the greatest politiques have in a natural and free manner professed their desires, rather than been reserved and disguised in them. For so we see that Lucius Sylla made a kind of profession, *that he wished all men happy or unhappy as they stood his friends or enemies*. So Cæsar, when he went first into Gaul, made no scruple to profess *that he had rather be first in a village than second at Rome*. So again as soon as he had begun the war, we see what Cicero saith of

330 The rest of this sentence is omitted in the translation.

331 This, in the translation, stands as the second of the *præcepta sparsa ;* that of accustoming the mind to value things according as they conduce to our particular ends being placed first. Throughout this part of the work the meaning is expressed more fully and clearly in the Latin, but where no material alteration or addition is introduced, and where the meaning of the English is plain enough, I do not stay to point out the differences.

332 That is, I suppose, by bringing us less into collision with them—(*pauciores offendemus*).

333 So in all three editions, though the sentence seems to be imperfect. The meaning must be that they cannot seize and turn to advantage accidents which fall out unexpectedly in their favour. The translation has *alii toti sunt in machinando, qui occasiones quæ opportune incidunt non arripiunt.*

him; *Alter* (meaning of Cæsar) *non recusat, sed quodammodo postulat, ut (ut est) sic appelletur tyrannus* [he does not refuse, but in a manner demands, to be called what he is—tyrant]. So we may see in a letter of Cicero to Atticus, that Augustus Cæsar in his very entrance into affairs, when he was a dearling of the senate, yet in his harangues to the people would swear *Ita parentis honores consequi liceat,* [as I hope to attain my father's honours]; which was no less than the tyranny, save that, to help it, he would stretch forth his hand towards a statua of Cæsar's that was erected in the place: and [334] men laughed and wondered and said Is it possible? or Did you ever hear the like [335]? and yet thought [336] he meant no hurt, he did it [337] so handsomely and ingenuously. And all these were prosperous: whereas Pompey, who tended to the same end but in a more dark and dissembling manner, as Tacitus saith of him, *Occultior non melior* [having his intentions better concealed but not better], wherein Sallust concurreth, *ore probo, animo inverecundo* [an honest tongue but a shameless mind], made it his design by infinite secret engines to cast the state into an absolute anarchy and confusion, that the state might cast itself into his arms for necessity and protection, and so the sovereign power be put upon him, and he never seen in it: and when he had brought it (as he thought) to that point, when he was chosen consul alone, as never any was, yet he could make no great matter of it, because men understood him not; but was fain in the end to go the beaten track of getting arms into his hands, by colour of the doubt of Cæsar's designs: so tedious, casual, and unfortunate are these deep dissimulations; whereof it seemeth Tacitus made this judgment, that they were a cunning of an inferior form in regard of true policy; attributing the one to Augustus, the other to Tiberius, where speaking of Livia he saith, *Et cum artibus mariti simulatione filii bene composita* [that she was of a happy composition, uniting the arts of her husband with the dissimulation of her son]; for surely the continual habit of dissimulation is but a weak and sluggish cunning, and not greatly politic.

Another precept of this Architecture of Fortune is to accustom our minds to judge of the proportion or value of things as they conduce and are material to our particular ends; and that to do substantially, and not superficially. For we shall find the logical part (as I may term it) of some men's minds good, but the mathematical part erroneous; that is, they can well judge of consequences, but not of proportions and comparison [338]; preferring things of shew and sense before things of substance and effect. So some fall in love with access to princes, others with popular fame and applause, supposing they are things of great purchase; when in many cases they are but matters of envy, peril, and impediment. So some measure things according to the labour and difficulty or assiduity which are spent about them; and think if they be ever moving, that they must needs advance and proceed; as Cæsar saith in a despising manner of Cato the second, when he describeth how laborious and indefatigable he was to no great purpose; *Hæc omnia magno studio agebat.* So in most things men are ready to abuse themselves in thinking the greatest means [339] to be best, when it should be the fittest.

As for the true marshalling of men's pursuits towards their fortune as they are more or less material, I hold them to stand thus. First the amendment of their own minds; for the remove of the impediments of the mind will sooner clear the passages of fortune, than the obtaining fortune will remove the impediments of the mind. In the second place I set down wealth and means; which I know most men would have placed first, because of the general use which it beareth towards all variety of occasions. But that opinion I may condemn with like reason as Machiavel doth that other, that moneys were the sinews of the wars; whereas (saith he) the true sinews of the wars are the sinews of men's

[334] So the original; edd. 1629 and 1633 have *whereat many men.*

[335] So the original; edd. 1629 and 1633 have *like to this.* [336] *though* in orig.

[337] *i.e.* he seemed to say what he felt—(*nihil malitiæ in eo suspicabantur qui tam candide et ingenue quid sentiret loqueretur*).

[338] *De* pretiis *vero imperitissime.*—De Aug.

[339] *i.e.* the greatest persons used as means—(*si magni alicujus aut honorati viri opera utantur*).

arms, that is, a valiant, populous, and military nation; and he voucheth aptly the authority of Solon, who when Crœsus shewed him his treasury of gold said to him, that if another came that had better iron he would be master of his gold. In like manner it may be truly affirmed that it is not moneys that are the sinews of fortune, but it is the sinews and steel of men's minds, wit, courage, audacity, resolution, temper, industry, and the like. In third [340] place I set down reputation, because of the peremptory tides and currents it hath; which if they be not taken in their due time are seldom recovered, it being extremely hard to play an after-game of reputation. And lastly I place honour, which is more easily won by any of the other three, much more by all, than any of them can be purchased by honour. To conclude this precept, as there is order and priority in matter, so is there in time, the preposterous placing whereof is one of the commonest errors; while men fly to their ends when they should intend their beginnings, and do not take things in order of time as they come on, but marshal them according to greatness and not according to instance; not observing the good precept, *Quod nunc instat agamus*

[Despatch we now what stands us now upon].

Another precept of this knowledge is, not to embrace any matters which do occupy too great a quantity of time, but to have that sounding in a man's ears, *Sed fugit interea, fugit irreparabile tempus* [while he is making ready to do it the time for doing it is gone]; and that is the cause why those which take their course of rising by professions of burden, as lawyers, orators, painful divines, and the like, are not commonly so politic for their own fortune [341], otherwise than in their ordinary way, because they want time to learn particulars, to wait occasions, and to devise plots [342].

Another precept of this knowledge is to imitate nature which doth nothing in vain; which surely a man may do, if he do well interlace his business, and bend not his mind too much upon that which he principally intendeth [343]. For a man ought in every particular action so to carry the motions of his mind, and so to have one thing under another, as if he cannot have that he seeketh in the best degree, yet to have it in a second, or so in a third; and if he can have no part of that which he purposed, yet to turn the use of it to somewhat else [344]; and if he cannot make anything of it for the present, yet to make it as a seed of somewhat in time to come; and if he can contrive no effect or substance from it, yet to win some good opinion by it, or the like; so that he should exact an account [345] of himself, of every action to reap somewhat, and not to stand amazed and confused if he fail of that he chiefly meant: for nothing is more impolitic than to mind actions wholly one by one; for he that doth so leeseth infinite occasions which intervene, and are many times more proper and propitious for somewhat that he shall need afterwards, than for that which he urgeth for the present; and therefore men must be perfect in that rule, *Hæc oportet facere, et illa non omittere* [these things ought ye to do, and not to leave the other undone].

Another precept of this knowledge is, not to engage a man's self peremptorily in any thing, though it seem not liable to accident; but ever to have a window to fly out at, or a way to retire [346]; following the wisdom in the ancient fable of the two frogs, which consulted when their plash was dry whither they should go; and the one moved to go down into a pit, because it was not likely the water

[340] So the original and ed. 1629. Ed. 1633 has *the third.*

[341] So the original. Edd. 1629 and 1633 have *fortunes.*

[342] Whereas (he adds in the translation) you will find in courts and commonwealths that the best promoters of their own fortune are those who have no public duty to discharge, and make their own rising their only business.

[343] This last clause is omitted in the translation.

[344] *i.e.* to turn his labour taken therein to some other use—(*ad alium quempiam præter destinatum finem operam impensam flectamus*).

[345] So the original. Edd. 1629 and 1633 omit *an.*

[346] The rest of this paragraph is omitted in the translation.

would dry there; but the other answered, *True, but if it do, how shall we get out again?*

Another precept of this knowledge is that ancient precept of Bias, construed not to any point of perfidiousness but only to caution and moderation, *Et ama tanquam inimicus futurus, et odi tanquam amaturus* [love your friend as you would love one who may hereafter be your enemy; hate your enemy as one who may hereafter be your friend]; for it utterly betrayeth all utility for men to embark themselves too far in unfortunate friendships, troublesome spleens, and childish and humorous envies or emulations.

But I continue this beyond the measure of an example; led, because I would not have such knowledges which I note as deficient to be thought things imaginative or in the air, or an observation or two much made of; but things of bulk and mass, whereof an end is hardlier made than a beginning. It must be likewise conceived, that in these points which I mention and set down, they are far from complete tractates of them, but only as small pieces for patterns. And lastly, no man I suppose will think that I mean fortunes are not obtained without all this ado; for I know they come tumbling into some men's laps; and a number obtain good fortunes by diligence in a plain way, little intermeddling, and keeping themselves from gross errors.

But as Cicero, when he setteth down an Idea of a perfect Orator, doth not mean that every pleader should be such; and so likewise, when a Prince or a Courtier hath been described by such as have handled those subjects, the mould hath used to be made according to the perfection of the art, and not according to common practice: so I understand it that it ought to be done in the description of a Politic man; I mean politic for his own fortune.

But it must be remembered all this while, that the precepts which we have set down are of that kind which may be counted and called *bonæ artes* [honest arts]. As for evil arts, if a man would set down for himself that principle of Machiavel, *that a man seek not to attain virtue itself, but the appearance only thereof; because the credit of virtue is a help, but the use of it is cumber*; or that other of his principles, *that he presuppose that men are not fitly to be wrought otherwise but by fear, and therefore that he seek to have every man obnoxious, low, and in strait*, which the Italians call *seminar spine*, to sow thorns; or that other principle contained in the verse which Cicero citeth, *Cadant amici, dummodo inimici intercidant* [down with friends so enemies go down with them], as the Triumvirs, which sold every-one to other the lives of their friends for the deaths of their enemies; or that other protestation of L. Catilina, to set on fire and trouble states, to the end to fish in droumy waters, and to unwrap their fortunes; *Ego si quid in fortunis meis excitatum sit incendium, id non aqua sed ruina restinguam*, [if my fortunes be set on fire I will put it out not with water but with demolition]: or that other principle of Lysander *that children are to be deceived with comfits, and men with oaths*: and the like evil and corrupt positions, whereof (as in all things) there are more in number than of the good: certainly with these dispensations from the laws of charity and integrity the pressing of a man's fortune may be more hasty and compendious. But it is in life as it is in ways; the shortest way is commonly the foulest, and surely the fairer way is not much about.

But men if they be in their own power and do bear and sustain themselves and be not carried away with a whirlwind or tempest of ambition, ought in the pursuit of their own fortune to set before their eyes not only that general map of the world, that *all things are vanity and vexation of spirit*, but many other more particular cards and directions: chiefly that, that Being without well-being is a curse and the greater being the greater curse, and that all virtue is most rewarded and all wickedness most punished in itself: according as the poet saith excellently:

> Quæ vobis, quæ digna, viri, pro laudibus istis
> Præmia posse rear solvi? pulcherrima primum
> Dii *moresque* dabunt vestri
> [What recompence, O friends, can I hold out
> Worthy such deeds? The best is that ye have.—
> God's blessing and your proper nobleness]:

and so of the contrary. And secondly they ought to look up to the eternal providence and divine judgment, which often subverteth the wisdom of evil plots and imaginations, according to that Scripture, *He hath conceived mischief, and shall bring forth a vain thing.* And although men should refrain themselves from injury and evil arts, yet this incessant and sabbathless pursuit of a man's fortune leaveth not tribute which we owe to God of our time; who (we see) demandeth a tenth of our substance, and a seventh, which is more strict, of our time: and it is to small purpose to have an erected face towards heaven, and a perpetual grovelling spirit upon earth, eating dust as doth the serpent; *Atque affigit humo divinæ particulam auræ* [fixing to earth the ethereal spark divine]. And if any man flatter himself that he will employ his fortune well though he should obtain it ill, as was said concerning Augustus Cæsar, and after of Septimius Severus, *that either they should never have been born or else they should never have died,* they did so much mischief in the pursuit and ascent of their greatness, and so much good when they were established; yet these compensations and satisfactions are good to be used, but never good to be purposed. And lastly, it is not amiss for men in their race toward their fortune to cool themselves a little with that conceit which is elegantly expressed by the emperor Charles the Fifth in his instructions to the king his son, *that fortune hath somewhat of the nature of a woman, that if she be too much wooed she is the farther off.* But this last is but a remedy for those whose tastes are corrupted: let men rather build upon that foundation which is as a corner-stone of divinity and philosophy, wherein they join close, namely that same *Primum quærite.* For divinity saith, *Primum quærite regnum Dei, et ista omnia adjicientur vobis* [seek ye first the kingdom of God, and all these things shall be added unto you]: and philosophy saith, *Primum quærite bona animi, cætera aut aderunt aut non oberunt* [seek ye first the good things of the mind, all other good things will either come or not be wanted]. And although the human foundation hath somewhat of the sand [347], as we see in M. Brutus when he brake forth into that speech,

Te colui, Virtus, ut rem; at tu nomen inane es

[I took thee, Virtue, for a reality, but I find thee an empty name]; yet the divine foundation is upon the rock. But this may serve for a taste of that knowledge which I noted as deficient.

¶ [348] Concerning Government, it is a part of knowledge secret and retired, in both these respects in which things are deemed secret; for some things are secrets because they are hard to know, and some because they are not fit to utter. We see all governments are obscure and invisible.

Totamque infusa per artus
Mens agitat molem, et magno se corpore miscet
[In every pore diffused the great mind works,
Stirs all the mass, and thro' the huge frame lives].

Such is the description of governments. We see the government of God over the world is hidden, insomuch as it seemeth to participate of much irregularity and confusion. The government of the Soul in moving the Body is inward and profound, and the passages thereof hardly to be reduced to demonstration.

347 *same* in the original: *sands* in edd. 1629 and 1633.

348 De Aug. viii. 3. The first part of this chapter is entirely altered in the translation; the remarks on the secret nature of Government, as a subject not proper for scrutiny, being omitted altogether; and the complimentary excuse for not entering upon it himself being transferred to the opening of the book. In this place indeed he speaks of it as a subject which his own long experience as an officer of state qualified him to handle, and on which he had some work in contemplation, though he thought it would be either abortive or posthumous; alluding probably to the *New Atlantis,* in which we know from Dr. Rawley that he did intend to exhibit a model of a perfect government. For the present however he confines himself to two treatises, given by way of example; one on the art of extending the bounds of Empire (which is a translation of the twenty-ninth Essay); the other on Universal Justice.

Again, the wisdom of antiquity (the shadows whereof are in the poets) in the description of torments and pains, next unto the crime of rebellion which was the Giants' offence, doth detest the offence of futility [349]; as in Sisyphus and Tantalus. But this was meant of particulars: nevertheless even unto the general rules and discourses of policy and government there is due a reverent and reserved handling.

But contrariwise in the governors toward the governed all things ought, as far as the frailty of man permitteth, to be manifest and revealed. For so it is expressed in the Scriptures touching the government of God, that this globe which seemeth to us a dark and shady body, is in the view of God as crystal: *Et in conspectu sedis tanquam mare vitreum simile crystallo* [and before the Throne there was a sea of glass, like unto crystal]. So unto princes and states, and specially towards wise senates and councils, the natures and dispositions of the people, their conditions and necessities, their factions and combinations, their animosities and discontents, ought to be, in regard of the variety of their intelligences, the wisdom of their observations, and the height of their station where they keep sentinel, in great part clear and transparent. Wherefore, considering that I write to a king that is a master of this science, and is so well assisted, I think it decent to pass over this part in silence, as willing to obtain the certificate which one of the ancient philosophers aspired unto; who being silent, when others contended to make demonstration of their abilities by speech, desired it might be certified for his part, *that there was one that knew how to hold his peace*.

Notwithstanding, for the more public part of government, which is Laws, I think good to note only one deficience; which is, that all those which have written of laws, have written either as philosophers or as lawyers, and none as statesmen. As for the philosophers, they make imaginary laws for imaginary commonwealths; and their discourses are as the stars, which give little light because they are so high. For the lawyers, they write according to the states where they live, what is received law, and not what ought to be law: for the wisdom of a lawmaker is one, and of a lawyer is another. For there are in nature certain fountains of justice, whence all civil laws are derived but as streams; and like as waters do take tinctures and tastes from the soils through which they run, so do civil laws vary according to the regions and governments where they are planted, though they proceed from the same fountains. Again, the wisdom of a lawmaker consisteth not only in a platform of justice, but in the application thereof; taking into consideration by what means laws may be made certain, and what are the causes and remedies of the doubtfulness and incertainty of law; by what means laws may be made apt and easy to be executed, and what are the impediments and remedies in the execution of laws; what influence laws touching private right of *meum* and *tuum* have into the public state, and how they may be made apt and agreeable; how laws are to be penned and delivered, whether in Texts or in Acts; brief or large; with preambles or without; how they are to be pruned and reformed from time to time; and what is the best means to keep them from being too vast in volumes or too full of multiplicity and crossness; how they are to be expounded, when upon causes emergent and judicially discussed, and when upon responses and conferences touching general points or questions; how they are to be pressed, rigorously or tenderly; how they are to be mitigated by equity and good conscience; and whether discretion and strict law are to be mingled in the same courts or kept apart in several courts; again, how the practice, profession, and erudition of law is to be censured and governed; and many other points touching the administration and (as I may term it) animation of laws. Upon which I insist the less, because I purpose (if God give me leave), having begun a work of this nature in aphorisms [350], to propound it hereafter, noting it in the mean time for deficient.

De prudentia legislatoria, sive de fontibus Juris.

And for your Majesty's laws of England, I could say much of their

[349] So edd. 1629 and 1633. The original has *facilitie*. By *futility* I understand *idle curiosity*.

[350] This was no doubt the treatise which is given by way of specimen in the *De Aug-*

dignity, and somewhat of their defect; but they cannot but excel the civil laws in fitness for the government; for the civil law was *non hos quæsitum munus in usus*; it was not made for the countries which it governeth. Hereof I cease to speak, because I will not intermingle matter of action with matter of general learning [351].

THUS have I concluded this portion of learning touching Civil Knowledge; and with Civil knowledge have concluded Human Philosophy; and with human philosophy, Philosophy in General. And being now at some pause, looking back into that I have passed through, this writing seemeth to me (*si nunquam fallit imago*) as far as a man can judge of his own work, not much better than that noise or sound which musicians make while they are tuning their instruments; which is nothing pleasant to hear, but yet is a cause why the music is sweeter afterwards. So have I been content to tune the instruments of the muses that they may play that have better hands. And surely, when I set before me the condition of these times, in which learning hath made her third visitation or circuit, in all the qualities thereof; as the excellency and vivacity of the wits of this age; the noble helps and lights which we have by the travails of ancient writers; the art of printing, which communicateth books to men of all fortunes; the openness of the world by navigation, which hath disclosed multitudes of experiments, and a mass of natural history; the leisure wherewith these times abound, not employing men so generally in civil business, as the states of Græcia did in respect of their popularity, and the state of Rome in respect of the greatness of their monarchy; the present disposition of these times at this instant to peace [352]; the consumption of all that ever can be said in controversies of religion, which have so much diverted men from other sciences; the perfection of your Majesty's learning, which as a phœnix may call whole vollies of wits to follow you; and the inseparable propriety of time, which is ever more and more to disclose truth; I cannot but be raised to this persuasion, that this third period of time will far surpass that of the Græcian and Roman learning: only if men will know their own strength and their own weakness both; and take one from the other light of invention, and not fire of contradiction; and esteem of the inquisition of truth as of an enterprise, and not as of a quality or ornament; and employ wit and magnificence to things of worth and excellency, and not to things vulgar and of popular estimation. As for my labours, if any man shall please himself or others in the reprehension of them, they shall make that ancient and patient request, *Verbera sed audi* [strike me if you will, only hear me]; let men reprehend them, so they observe and weigh them. For the appeal is (lawful though it may be it shall not be needful) from the first cogitations of men to their second, and from the nearer times to the times further off. Now let us come to that learning, which both the former times were not so blessed as to know, sacred and inspired Divinity, the Sabaoth and port of all men's labours and peregrinations.

¶ [353] THE prerogative of God extendeth as well to the reason as to the will of

mentis. The perfection of a law is there described as consisting in five things: it must be certain in its meaning; just in its rules; convenient in execution; agreeable to the form of government; and productive of virtue in the governed. Of these heads the first only is discussed; but under it almost all the points enumerated in the text come under consideration, more or less completely.

351 This paragraph is omitted in the translation.

352 This was written just after the conclusion of peace between England and Spain; when the translation was published the *disposition* of the times was less peaceable, but a greater part of Europe was actually at peace; and accordingly instead of the expression in the text he substitutes, "the peace which is at this time enjoyed by Britain, Spain, Italy, France too at last, and other regions not a few".

353 De Aug. ix. 1. This chapter is greatly altered in the translation; much of it being entirely omitted, much condensed, and a little added. In the exordium he announces the subject of the book as one which does not belong to human reason and natural philo-

man; so that as we are to obey his law though we find a reluctation in our will, so we are to believe his word though we find a reluctation in our reason. For if we believe only that which is agreeable to our sense, we give consent to the matter and not to the author; which is no more than we would do towards a suspected and discredited witness; but that faith which was accounted to Abraham for righteousness was of such a point as whereat Sarah laughed, who therein was an image of natural reason.

Howbeit (if we will truly consider it) more worthy it is to believe than to know as we now know. For in knowledge man's mind suffereth from sense, but in belief it suffereth from spirit, such one as it holdeth for more authorised than itself [354], and so suffereth from the worthier agent. Otherwise it is of the state of man glorified; for then faith shall cease, and we shall know as we are known.

Wherefore we conclude that sacred Theology (which in our idiom we call Divinity) is grounded only upon the word and oracle of God, and not upon the light of nature: for it is written, *Cœli enarrant gloriam Dei* [the Heavens declare the glory of God], but it is not written, *Cœli enarrant voluntatem Dei* [the Heavens declare the will of God], but of that it is said, *Ad legem et testimonium: si non fecerint secundum verbum istud,* etc. [to the law and to the testimony: if they do not according to this word, etc.]. This holdeth not only in those points of faith which concern the great mysteries of the Deity, of the Creation, of the Redemption, but likewise those which concern the law moral truly interpreted: *Love your enemies: do good to them that hate you; be like to your heavenly Father, that suffereth his rain to fall upon the just and unjust.* To this it ought to be applauded, *Nec vox hominem sonat:* it is a voice beyond the light of nature. So we see the heathen poets, when they fall upon a libertine passion, do still expostulate with laws and moralities, as if they were opposite and malignant to nature: *Et quod natura remittit, invida jura negant* [what Nature suffers envious laws forbid]. So said Dendamis the Indian unto Alexander's messengers, That he had heard somewhat of Pythagoras and some other of the wise men of Græcia, and that he held them for excellent men: but that they had a fault, which was that they had in too great reverence and veneration a thing they called law and manners. So it must be confessed that a great part of the law moral is of that perfection, whereunto the light of nature cannot aspire. How then is it that man is said to have by the light and law of nature some notions and conceits of virtue and vice, justice and wrong, good and evil? Thus; because the light of nature is used in two several senses; the one, that which springeth from reason, sense, induction, argument, according to the laws of heaven and earth; the other, that which is imprinted upon the spirit of man by an inward instinct, according to the law of conscience, which is a sparkle of the purity of his first estate: in which later sense only he is participant of some light and discerning touching the perfection of the moral law: but how? sufficient to check the vice, but not to inform the duty. So then the doctrine of religion, as well moral as mystical, is not to be attained but by inspiration and revelation from God.

sophy. He will not therefore attempt to lay out the "partitions" of it, but merely offer a few suggestions, concerning not the matter revealed by Theology, but the manner of the revelation. These suggestions, which are but three in number, together with the remarks by which they are introduced, agree substantially with those in the text: all that does not bear immediately upon them being omitted. And I think all the differences may be sufficiently accounted for by the change of design; while the change of design itself may probably have been suggested by the difficulty of expounding the subject of theology on a scale similar to that adopted with regard to other subjects, without introducing matter which might have caused the work to be proscribed in Italy. See note p. 50.

354 In the translation this is expressed rather differently. *In scientia enim* mens humana *patitur a sensu, qui a rebus materialis resilit; in fide autem* anima *patitur ab anima quæ est agens dignius:* Knowledge being (if I understand the meaning rightly) a function of the *anima sensibilis*, faith of the *anima rationalis;* the one receiving its impressions from things material, the other from things spiritual.

The use notwithstanding of reason in spiritual things, and the latitude thereof, is very great and general: for it is not for nothing that the apostle calleth religion *our reasonable service of God*; insomuch as the very ceremonies and figures of the old law were full of reason and signification, much more than the ceremonies of idolatry and magic, that are full of nonsignificants and surd characters. But most specially the Christian Faith, as in all things so in this, deserveth to be highly magnified; holding and preserving the golden mediocrity in this point between the law of the Heathen and the law of Mahomet, which have embraced the two extremes. For the religion of the Heathen had no constant belief or confession, but left all to the liberty of argument; and the religion of Mahomet on the other side interdicteth argument altogether: the one having the very face of error, and the other of imposture: whereas the Faith doth both admit and reject disputation with difference.

The use of human reason in religion is of two sorts: the former, in the conception and apprehension of the mysteries of God to us revealed; the other in the inferring and deriving of doctrine and direction thereupon. The former extendeth to the mysteries themselves; but how? by way of illustration, and not by way of argument. The later consisteth indeed of probation and argument. In the former we see God vouchsafeth to descend to our capacity, in the expressing of his mysteries in sort as may be sensible unto us; and doth grift[355] his revelations and holy doctrine upon the notions of our reason, and applieth his inspirations to open our understanding, as the form of the key to the ward of the lock[356]: for the later, there is allowed us an use of reason and argument secondary and respective, although not original and absolute. For after the articles and principles of religion are placed, and exempted from examination of reason, it is then permitted unto us to make derivations and inferences from and according to the analogy of them for our better direction. In nature this holdeth not; for both the principles are examinable by induction, though not by a medium or syllogism; and besides, those principles or first positions have no discordance with that reason which draweth down and deduceth the inferior positions. But yet it holdeth not in religion alone, but in many knowledges both of greater and smaller nature, namely wherein there are not only *posita* but *placita*; for in such there can be no use of absolute reason. We see it familiarly in games of wit, as chess, or the like; the draughts and first laws of the game are positive, but how? merely *ad placitum*, and not examinable by reason; but then how to direct our play thereupon with best advantage to win the game, is artificial and rational. So in human laws there be many grounds and maxims which are *placita juris*, positive upon authority and not upon reason, and therefore not to be disputed; but what is most just, not absolutely, but relatively and according to those maxims, that affordeth a long field of disputation. Such therefore is that secondary reason which hath place in divinity, which is grounded upon the *placets* of God.

De usu legitimo rationis humanæ in divinis.

Here therefore I note this deficience, that there hath not been to my understanding sufficiently inquired and handled *the true limits and use of reason in spiritual things*, as a kind of divine dialectic: which for that it is not done, it seemeth to me a thing usual, by pretext of true conceiving that which is revealed, to search and mine into that which is not revealed; and by pretext of enucleating inferences and contradictories, to examine that which is positive; the one sort falling into the error of Nicodemus, demanding to have things made more sensible than it pleaseth God to reveal them: *Quomodo possit homo nasci cum sit senex* [how can a man be born when he is old?]; the other sort into the error of the disciples

355 So the original and ed. 1629. Ed. 1633 has *graft*.

356 It being our own duty at the same time to open and enlarge our understanding that it may be capable of receiving them. *Qua tamen in parte nobis ipsis deesse minime debemus; cum enim Deus ipse opera rationis nostræ in illuminationibus suis utatur, etiam nos eandem in omnes partes versare debemus quo magis capaces simus ad mysteria recipienda et imbibenda: modo animus ad amplitudinem mysteriorum pro modulo suo dilatetur, non mysteria ad angustias animi constringantur.*

which were scandalized at a show of contradiction; *Quid est hoc quod dicit nobis? Modicum, et non videbitis me; et iterum, modicum, et videbitis me*, etc. [what is this that he saith unto us? a little while and ye shall not see me, and again a little while and ye shall see me, etc.].

Upon this I have insisted the more in regard of the great and blessed use thereof; for this point well laboured and defined of would in my judgment be an opiate to stay and bridle not only the vanity of curious speculations, wherewith the schools labour, but the fury of controversies, wherewith the church laboureth. For it cannot but open men's eyes, to see that many controversies do merely pertain to that which is either not revealed or positive; and that many others do grow upon weak and obscure inferences or derivations: which latter sort, if[357] men would revive the blessed style of that great doctor of the Gentiles, would be carried thus, *Ego, non Dominus* [I, not the Lord], and again, *Secundum consilium meum* [according to my counsel]; in opinions and counsels, and not in positions and oppositions. But men are now over-ready to usurp the style *Non ego, sed Dominus* [not I, but the Lord]; and not so only, but to bind it with the thunder and denunciation of curses and anathemas, to the terror of those which have not sufficiently learned out of Salomon that *the causeless curse shall not come*[358].

Divinity hath two principal parts; the matter informed or revealed, and the nature of the information or revelation: and with the later we will begin[359], because it hath most coherence with that which we have now last handled. The nature of the information consisteth of three branches; the limits of the information, the sufficiency of the information, and the acquiring or obtaining the information. Unto the limits of the information belong these considerations; how far forth particular persons continue to be inspired; how far forth the church is inspired; and how far forth reason may be used: the last point whereof I have noted as deficient. Unto the sufficiency of the information belong two considerations; what points of religion are fundamental, and what perfective, being matter of further building and perfection upon one and the same foundation; and again, how the gradations of light according to the dispensation of times are material to the sufficiency of belief.

De gradibus unitatis in Civitate Dei.

Here again I may rather give it in advice than note it as deficient, that the points fundamental, and the points of further perfection only, ought to be with piety and wisdom distinguished: a subject tending to much like end as that I noted before; for as that other were likely to abate the number of controversies, so this is like to abate the heat of many of them. We see Moses when he saw the Israelite and the Ægyptian fight, he did not say, *Why strive you?* but drew his sword and slew the Ægyptian: but when he saw the two Israelites fight, he said, *You are brethren, why strive you?* If the point of doctrine be an Ægyptian, it must be slain by the sword of the Spirit, and not reconciled; but if it be an Israelite, though in the wrong, then, *Why strive you?* We see of the fundamental points, our Saviour penneth the league thus, *He that is not with us, is against us;* but of points not fundamental, thus, *He that is not against us, is with us.* So we see the coat of our Saviour was entire without seam, and so is the doctrine of the Scriptures in itself; but the garment of the Church was of divers colours, and yet not divided. We see the chaff may and ought to be severed from the corn in the ear, but the tares may not be pulled up from the corn in the field: so as it is a thing of great use well to define what and of what latitude those points are, which do make men merely aliens and disincorporate from the Church of God[360].

357 The original and also edd. 1629 and 1633 have *of*.

358 In the translation this last sentence is omitted, and the substance both of this and of the preceding paragraph is set forth in a better order and more concisely, though to the same general effect.

359 In the translation he expressly confines himself to the latter only, and the rest of the paragraph is omitted.

360 Of this paragraph again the substance is given in the translation, though in a some-

For the obtaining of the information, it resteth upon the true and sound interpretation of the Scriptures, which are the fountains of the water of life. The interpretations of the Scriptures [361] are of two sorts; methodical, and solute or at large. For this divine water, which excelleth so much that of Jacob's well, is drawn forth much in the same kind as natural water useth to be out of wells and fountains; either it is first forced up into a cistern, and from thence fetched and derived for use; or else it is drawn and received in buckets and vessels immediately where it springeth. The former sort whereof, though it seem to be the more ready, yet in my judgment is more subject to corrupt [362]. This is that method which hath exhibited unto us the scholastical divinity; whereby divinity hath been reduced into an art, as into a cistern, and the streams of doctrine or positions fetched and derived from thence.

In this men have sought three things, a summary brevity, a compacted strength, and a complete perfection; whereof the two first they fail to find, and the last they ought not to seek. For as to brevity, we see in all summary methods, while men purpose to abridge they give cause to dilate. For the sum or abridgment by contraction becometh obscure, the obscurity requireth exposition, and the exposition is deduced into large commentaries, or into common places and titles, which grow to be more vast than the original writings whence the sum was at first extracted. So we see the volumes of the schoolmen are greater much than the first writings of the fathers, whence the Master of the Sentences [363] made his sum or collection. So in like manner the volumes of the modern doctors of the civil law exceed those of the ancient jurisconsults, of which Tribonian compiled the digest [364]. So as this course of sums and commentaries is that which doth infallibly make the body of sciences more immense in quantity, and more base in substance.

And for strength, it is true that knowledges reduced into exact methods have a shew of strength, in that each part seemeth to support and sustain the other; but this is more satisfactory than substantial; like unto buildings which stand by architecture and compaction, which are more subject to ruin than those which are built more strong in their several parts, though less compacted. But it is plain that the more you recede from your grounds the weaker do you conclude; and as in nature the more you remove yourself from particulars the

what different order; and a sentence is added to the following effect: If any one thinks (he says) that this has been done already, let him consider again and again how far it has been done with sincerity and moderation. In the mean time he who speaks of peace is like enough to receive the answer which Jehu gave to the messenger—*Is it peace, Jehu? What hast thou to do with peace? Get thee behind me.* For it is not peace between the contending opinions that most men have at heart, but the establishment of their own opinions (*cum non pax, sed partes, plerisque cordi sint*).

361 A sentence is introduced here in the translation, to say that he speaks only of the *method* of interpretation, not of the authority: the ground of the authority being the consent of the Church.

362 This censure, as well as the remarks upon the methodical system which are contained in the three following paragraphs, are omitted in the translation; probably as involving matter which would not have been allowed at Rome.

363 Peter the Lombard, Bishop of Paris, wrote a Sum of Theology in four books entitled "The Sentences"; and according to the taste of the middle ages acquired the title of "Master of the Sentences". Many of these scholastic titles are curious. Thus Thomas Aquinas is Doctor Angelicus; Buonaventura, Doctor Seraphicus; Alexander Hales, Doctor Irrefragabilis; Duns Scotus, Doctor Subtilis; Raymund Lully, Doctor Illuminatus; Roger Bacon, Doctor Mirabilis; Occam, Doctor Singularis.—*R. L. E.*

364 Compare with this remark that of Maphæus Vegius—"Existimabas, ut opinor", —he is apostrophising Tribonian—"plurimum conducere utilitati studentium, si quod antea in multitudine tractatuum tardius effecerunt coangustatis postea libris citius adsequi possunt. . . . Sed longe secus ac persuadebas tibi cessit. Quis namque nesciat infinitas et nonnunquam ineptas vanasque interpretationes quibus nulla fere lex exempta est?" See Maphæus Vegius de Verborum significatione, xiv. 77., apud Savigny History of Roman Law in the Middle Ages, ch. 59.—*R.L.E.*

greater peril of error you do incur, so much more in divinity the more you recede from the Scriptures by inferences and consequences, the more weak and dilute are your positions.

And as for perfection or completeness in divinity, it is not to be sought; which makes this course of artificial divinity the more suspect. For he that will reduce a knowledge into an art will make it round and uniform; but in divinity many things must be left abrupt and concluded with this: *O altitudo sapientiæ et scientiæ Dei! quam incomprehensibilia sunt judicia ejus, et non investigabiles viæ ejus!* [O the depth of the wisdom and knowledge of God! How incomprehensible are his judgments, and his ways past finding out!]. So again the apostle saith, *Ex parte scimus* [we know in part], and to have the form of a total where there is but matter for a part cannot be without supplies by supposition and presumption. And therefore I conclude, that the true use of these Sums and Methods hath place in institutions or introductions preparatory unto knowledge; but in them, or by deducement from them, to handle the main body and substance of a knowledge, is in all sciences prejudicial, and in divinity dangerous.

As to the interpretation of the Scriptures solute and at large, there have been divers kinds introduced and devised; some of them rather curious and unsafe, than sober and warranted. Notwithstanding thus much must be confessed, that the Scriptures, being given by inspiration and not by human reason, do differ from all other books in the author; which by consequence doth draw on some difference to be used by the expositor. For the inditer of them did know four things which no man attains to know; which are, the mysteries of the kingdom of glory; the perfection of the laws of nature; the secrets of the heart of man; and the future succession of all ages [365]. For as to the first, it is said, *He that presseth into the light shall be oppressed of the glory:* and again, *No man shall see my face and live.* To the second, *When he prepared the heavens I was present, when by law and compass he inclosed the deep.* To the third, *Neither was it needful that any should bear witness to him of Man, for he knew well wha was in Man.* And to the last, *From the beginning are known to the Lord all his works.*

From the former two [366] of these have been drawn certain senses and expositions of Scriptures, which had need be contained within the bounds of sobriety; the one anagogical, and the other philosophical. But as to the former, man is not to prevent his time: *Videmus nunc per speculum in ænigmate, tunc autem facie ad faciem* [now we see through a glass darkly, but then face to face]: wherein nevertheless there seemeth to be a liberty granted, as far forth as the polishing of this glass, or some moderate explication of this ænigma. But to press too far into it, cannot but cause a dissolution and overthrow of the spirit of man. For in the body there are three degrees of that we receive into it; Aliment, Medicine, and Poison; whereof aliment is that which the nature of man can perfectly alter and overcome: medicine is that which is partly converted by nature, and partly converteth nature; and poison is that which worketh wholly upon nature, without that that nature can in any part work upon it. So in the mind whatsoever knowledge reason cannot at all work upon and convert, is a mere intoxication, and endangereth a dissolution of the mind and understanding.

But for the latter [367], it hath been extremely set on foot of late time by the

[365] Of these four things he mentions in the translation only the two last; introducing the mention of them in the next paragraph but three, and in the mean time omitting altogether both this and the following paragraph.

[366] *i.e.* from the intimations in the Scriptures concerning the Kingdom of Glory and the Laws of Nature. Edd. 1629 and 1633 have "from the former of these two"; obviously a misprint, though adopted in all modern editions.

[367] *i.e.* the *philosophical* exposition. The "*former*", *i.e.* the *anagogical* exposition, is not mentioned in the translation; which only says that the method of interpretation solute and at large has been carried to excess in two ways; first in supposing such perfection in the Scriptures that all philosophy is to be sought there, secondly in interpreting

school of Paracelsus, and some others, that have pretended to find the truth of all natural philosophy in the Scriptures; scandalizing and traducing all other philosophy as heathenish and profane. But there is no such enmity between God's word and his works. Neither do they give honour to the Scriptures, as they suppose, but much imbase them. For to seek heaven and earth in the word of God, whereof it is said, *Heaven and earth shall pass, but my word shall not pass,* is to seek temporary things amongst eternal: and as to seek divinity in philosophy is to seek the living amongst the dead, so to seek philosophy in divinity is to seek the dead amongst the living[368]: neither are the pots or lavers whose place was in the outward part of the temple to be sought in the holiest place of all, where the ark of the testimony was seated. And again, the scope or purpose of the Spirit of God is not to express matters of nature in the Scriptures, otherwise than in passage, and for application to man's capacity and to matters moral or divine. And it is a true rule *Authoris aliud agentis parva authoritas* [what a man says incidentally about matters which are not in question has little authority]; for it were a strange conclusion, if a man should use a similitude for ornament or illustration sake, borrowed from nature or history according to vulgar conceit, as of a Basilisk, an Unicorn, a Centaur, a Briareus, an Hydra, or the like, that therefore he must needs be thought to affirm the matter thereof positively to be true. To conclude therefore, these two interpretations, the one by reduction or ænigmatical, the other philosophical or physical, which have been received and pursued in imitation of the rabbins and cabalists, are to be confined with a *Noli altum sapere, sed time* [be not overwise, but fear].

But the two later points, known to God and unknown to man, touching *the secrets of the heart, and the successions of time,* doth make a just and sound difference between the manner of the exposition of the Scriptures, and all other books. For it is an excellent observation which hath been made upon the answers of our Saviour Christ to many of the questions which were propounded to him, how that they are impertinent to the state of the question demanded; the reason whereof is, because not being like man, which knows man's thoughts by his words, but knowing man's thoughts immediately, he never answered their words, but their thoughts[369]: much in the like manner it is with the Scriptures, which being written to the thoughts of men, and to the succession of all ages, with a foresight of all heresies, contradictions, differing estates of the church, yea and particularly of the elect, are not to be interpreted only according to the latitude of the proper sense of the place, and respectively towards that present occasion whereupon the words were uttered; or in precise congruity or contexture with the words before or after; or in contemplation of the principal scope of the place; but have in themselves, not only totally or collectively, but distributively in clauses and words, infinite springs and streams of doctrine to water the church in every part[370]; and therefore as the literal sense is as it were the main stream or river; so the moral sense chiefly, and sometimes the allegorical or typical, are they whereof the church has most use: not that I wish men to be bold in allegories, or indulgent or light in allusions; but that I do much condemn that interpretation of the Scripture which is only after the manner as men use to interpret a profane book.

In this part touching the exposition of the Scriptures, I can report no deficience; but by way of remembrance this I will add: In perusing books of divinity, I find many[371] books of controversies; and many of common places and treatises[372]; a mass of positive divinity, as it is made an art; a number

them in the same manner as one would interpret an uninspired book. The remarks on the first of these excesses coincide with the first half of this paragraph (the rest being omitted), those on the second with the next paragraph.

368 The rest of this paragraph is omitted in the translation.

369 And also (the translation adds) because he addressed himself not solely to those present, but to men of all times and places to whom the gospel was to be preached.

370 The rest of the paragraph is omitted in the translation.

371 In the translation he says *too many*.

372 also "cases of conscience"—which he especially commends further on, in a passage not translated.

of sermons and lectures, and many prolix commentaries upon the Scriptures, with harmonies and concordances: but that form of writing in divinity, which in my judgment is of all others most rich and precious, is positive divinity collected upon particular texts of Scriptures in brief observations; not dilated into common places, not chasing after controversies, not reduced into method of art; a thing abounding in sermons, which will vanish, but defective in books which will remain; and a thing wherein this age excelleth. For I am persuaded, and I may speak it with an *Absit invidia verbo* [meaning no offence], and no ways in derogation of antiquity, but as in a good emulation between the vine and the olive, that if the choice and best of those observations upon texts of Scriptures which have been made dispersedly in sermons within this your Majesty's island[373] of Britain by the space of these forty years and more (leaving out the largeness of exhortations and applications thereupon) had been set down in a continuance, it had been the best work in divinity which had been written since the apostles' times[374].

Emanationes Scripturarum in doctrinas positivas.

The matter informed by divinity is of two kinds; matter of belief and truth of opinion, and matter of service and adoration; which is also judged and directed by the former; the one being as the internal soul of religion, and the other as the external body thereof. And therefore the heathen religion was not only a worship of idols, but the whole religion was an idol in itself; for it had no soul, that is, no certainty of belief or confession; as a man may well think, considering the chief doctors of their church were the poets; and the reason was, because the heathen gods were no jealous gods, but were glad to be admitted into part, as they had reason. Neither did they respect the pureness of heart, so they might have external honour and rites.

But out of these two do result and issue four main branches of divinity; Faith, Manners, Liturgy, and Government. Faith containeth the doctrine of the nature of God, of the attributes of God, and of the works of God. The nature of God consisteth of three persons in unity of Godhead. The attributes of God are either common to the Deity, or respective to the persons. The works of God summary are two, that of the Creation, and that of the Redemption; and both these works, as in total they appertain to the unity of the Godhead, so in their parts they refer to the three persons: that of the Creation, in the mass of the matter to the Father; in the disposition of the form to the Son; and in the continuance and conservation of the being to the Holy Spirit: so that of the Redemption, in the election and counsel to the Father; in the whole act and consummation to the Son; and in the application to the Holy Spirit; for by the Holy Ghost was Christ conceived in flesh, and by the Holy Ghost are the elect regenerate in spirit. This work likewise we consider either effectually in the elect; or privatively[375] in the reprobate; or according to appearance in the visible church.

For Manners, the doctrine thereof is contained in the law, which discloseth sin. The law itself is divided, according to the edition thereof, into the law of Nature, the law Moral, and the law Positive; and according to the style, into Negative and Affirmative, Prohibitions and Commandments. Sin, in the matter

[373] So edd. 1629 and 1633. The original has *ilands*.

[374] This last sentence is omitted in the translation,—no doubt as being inadmissible at Rome. But in its place is introduced one of Bacon's happiest illustrations, and one which is not, I think, to be found anywhere in his own English. "Certainly (he says) as we find it in wines, that those which flow freely from the first treading of the grape are sweeter than those which are squeezed out by the wine-press, because the latter taste somewhat of the stone and the rind; so are those doctrines most wholesome and sweet which ooze out of the Scriptures when gently crushed, and are not forced into controversies and common places."

The next six paragraphs are entirely omitted,—as belonging to that part of the subject with which he has professed in the beginning that he will not meddle.

[375] The original, and also edd. 1629 and 1633, have *privately*.

and subject thereof, is divided according to the commandments; in the form thereof, it referreth to the three persons, in Deity: sins of Infirmity against the Father, whose more special attribute is Power; sins of Ignorance against the Son, whose attribute is Wisdom; and sins of Malice against the Holy Ghost, whose attribute is Grace or Love. In the motions of it, it either moveth to the right hand or to the left; either to blind devotion, or to profane and libertine transgression; either in imposing restraint where God granteth liberty, or in taking liberty where God imposeth restraint. In the degrees and progress of it, it divideth itself into thought, word, or act. And in this part I commend much the deducing of the law of God to cases of conscience; for that I take indeed to be a breaking, and not exhibiting whole, of the bread of life. But that which quickeneth both these doctrines of faith and manners, is the elevation and consent of the heart; whereunto appertain books of exhortation, holy meditation, Christian resolution, and the like.

For the Liturgy or service, it consisteth of the reciprocal acts between God and man; which, on the part of God, are the preaching of the word and the sacraments, which are seals to the covenant, or as the visible word; and on the part of man [376], invocation of the name of God, and under the law, sacrifices, which were as visible prayers or confessions: but now the adoration being *in spiritu et veritate* [in spirit and in truth], there remaineth only *vituli labiorum* [offerings of the lips]; although the use of holy vows of thankfulness and retribution may be accounted also as sealed petitions.

And for the Government of the church, it consisteth of the patrimony of the church, the franchises of the church, and the offices and jurisdictions of the church, and the laws of the church directing the whole; all which have two considerations, the one in themselves, the other how they stand compatible and agreeable to the civil estate.

This matter of divinity is handled either in form of instruction of truth, or in form of confutation of falsehood. The declinations from religion, besides the privative [377], which is atheism and the branches thereof, are three; Heresies, Idolatry, and Witchcraft; Heresies, when we serve the true God with a false worship; Idolatry, when we worship false gods, supposing them to be true; and Witchcraft, when we adore false gods, knowing them to be wicked and false. For so your Majesty doth excellently well observe, that Witchcraft is the height of Idolatry. And yet we see though these be true degrees, Samuel teacheth us that they are all of a nature, when there is once a receding from the word of God; for so he saith, *Quasi peccatum ariolandi est repugnare, et quasi scelus idololatriæ nolle acquiescere* [rebellion is as the sin of Witchcraft, and Stubbornness as the crime of Idolatry].

These things I have passed over so briefly because I can report no deficience concerning them: for I can find no space or ground that lieth vacant and unsown in the matter of divinity; so diligent have men been, either in sowing of good seed or in sowing of tares.

THUS have I made as it were a small Globe of the Intellectual World, as truly and faithfully as I could discover; with a note and description of those parts which seem to me not constantly occupate, or not well converted by the labour of man. In which, if I have in any point receded from that which is commonly received, it hath been with a purpose of proceeding *in melius*, and not *in aliud*; a mind of amendment and proficience, and not of change and difference. For I could not be true and constant to the argument I handle, if I were not willing to go beyond others; but yet not more willing than to have others go beyond me again: which may the better appear by this, that I have propounded my opinions naked and unarmed, not seeking to preoccupate the liberty of men's judgments by confutation. For in any thing which is well set down, I am in good hope that if the first reading move an objection, the second reading will make an answer. And in those things wherein I have erred, I am sure I have not prejudiced the right by litigious arguments; which certainly have

[376] So edd. 1629 and 1633. The original has *mans*.
[377] So edd. 1629 and 1633. The original has *primitive*.

this contrary effect and operation, that they add authority to error, and destroy the authority of that which is well invented: for question is an honour and preferment to falsehood, as on the other side it is a repulse to truth. But the errors I claim and challenge to myself as mine own. The good, if any be, is due *tanquam adeps sacrificii* [as the fat of the sacrifice], to be incensed to the honour, first of the Divine Majesty, and next of your Majesty, to whom on earth I am most bounden.

VALERIUS TERMINUS

Preface by ROBERT LESLIE ELLIS.

THE following fragments of a great work on the Interpretation of Nature were first published in Stephens's Letters and Remains [1734]. They consist partly of detached passages, and partly of an epitome of twelve chapters of the first book of the proposed work. The detached passages contain the first, sixth, and eighth chapters, and portions of the fourth, fifth, seventh, ninth, tenth, eleventh, and sixteenth. The epitome contains an account of the contents of all the chapters from the twelfth to the twenty-sixth inclusive, omitting the twentieth, twenty-third, and twenty-fourth. Thus the sixteenth chapter is mentioned both in the epitome and among the detached passages, and we are thus enabled to see that the two portions of the following tract belong to the same work, as it appears from both that the sixteenth chapter was to treat of the doctrine of idola.

It is impossible to ascertain the motive which determined Bacon to give to the supposed author the name of Valerius Terminus, or to his commentator, of whose annotations we have no remains, that of Hermes Stella. It may be conjectured that by the name Terminus he intended to intimate that the new philosophy would put an end to the wandering of mankind in search of truth, that it would be the *terminus ad quem* in which when it was once attained the mind would finally acquiesce.

Again, the obscurity of the text was to be in some measure removed by the annotations of Stella; not however wholly, for Bacon in the epitome of the eighteenth chapter commends the manner of publishing knowledge "whereby it shall not be to the capacity nor taste of all, but shall as it were single and adopt his reader." Stella was therefore to throw a kind of starlight on the subject, enough to prevent the student's losing his way, but not much more.

However this may be, the tract is undoubtedly obscure, partly from the style in which it is written, and partly from its being only a fragment. It is at the same time full of interest, inasmuch as it is the earliest type of the *Instauratio*. The first book of the work ascribed to Valerius Terminus would have corresponded to the *De Augmentis* and to the first book of the *Novum Organum*, the plan being that it should contain whatever was necessary to be known before the new method could be stated. In the second book, as in the second book of the *Novum Organum*, we should have found the method itself.

The *Advancement of Learning*, which was developed into the *De Augmentis*, corresponds to the first ten chapters of *Valerius Terminus*, and especially to the first and tenth. To the remainder of the book (a few chapters are clearly wanted after the last mentioned in the epitome) corresponds the first book of the *Novum Organum*. The tenth chapter, of which we have only a small fragment, is entitled "The Inventory, or an Enumeration and View of Inventions already discovered and in use; together with a note of the wants, and the nature of the supplies." It therefore corresponds to the second book of the *Advancement*, and to the last eight books of the *De Augmentis*, but would doubtless have been a mere summary.[1] When Bacon subsequently determined to give more development to this part of the subject, he was naturally led to make a break after the inventory, and thus we get the origin of the separation between the *De Augmentis* and the *Novum Organum*.

[1] See my note at the end of this Preface. *J.S.*

The most important portion of *Valerius Terminus* is the eleventh chapter, which contains a general statement of the problem to be solved. It corresponds to the opening axioms of the second book of the *Novum Organum*, but differs from them in containing very little on the subject of forms. What Bacon afterwards called the investigation of the form he here calls the freeing of a direction. The object to be sought for is, he says, " the revealing and discovering of new inventions and operations."—" This to be done without the errors and conjectures of art, or the length or difficulties of experience." In order to guide men's travels, a full direction must be given to them, and the fullness of a direction consists in two conditions, certainty and liberty. Certainty is when the direction is infallible; liberty when it comprehends all possible ways and means. Both conditions are fulfilled by the knowledge of the form, to which the doctrine of direction entirely corresponds. This correspondency Bacon recognises towards the end of the chapter, but in illustrating the two conditions of which we have been speaking he does not use the word form. The notion of the form or formal cause comes into his system only on historical grounds. In truth, in *Valerius Terminus* he is disposed to illustrate the doctrine of direction not so much by that of the formal cause as by two rules which are of great importance in the logical system of Ramus. " The two commended rules by him set down," that is by Aristotle, " whereby the axioms of science are precepted to be made convertible, and which the latter men have not without elegancy surnamed, the one the rule of truth because it preventeth deceipt; the other the rule of prudence because it freeth election; are the same thing in speculation and affirmation, which we now affirm." And then follows an example, of which Bacon says that it " will make my meaning attained, and yet percase make it thought that they attained it not." In this example the effect to be produced is whiteness, and the first direction given is to intermingle air and water; of this direction it is said that it is certain, but very particular and restrained, and he then goes on to free it by leaving out the unessential conditions. Of this however it is not now necessary to speak at length; but the " two commended rules " may require some illustration.

In many passages of his works Peter Ramus condemns Aristotle for having violated three rules which he had himself propounded. To these rules Ramus gives somewhat fanciful names. The first is the rule of truth, the second the rule of justice, and the third the rule of wisdom. These three rules are all to be fulfilled by the principles of every science (axiomata artium). The first requires the proposition to be in all cases true, the second requires its subject and predicate to be essentially connected together, and the third requires the converse of the proposition to be true as well as the proposition itself. The whole of this theory, to which Ramus and the Ramistæ seem to have ascribed much importance, is founded on the fourth chapter of the first book of the *Posterior Analytics*. Aristotle in speaking of the principles of demonstration explains the meaning of three phrases, *κατὰ παντός*, *de omni*; *καθ' αὐτό*, *per se*; and *καθόλου*, *universaliter*. When the predicate can be affirmed in all cases and at all times of the subject of a proposition, the predication is said to be *de omni* or *κατὰ παντός*. Again, whatever is so connected with the essence of a thing as to be involved in its definition is said to belong to it *per se*, *καθ' αὐτό* and the same phrase is applicable when the thing itself is involved in the definition of that which we refer to it. Thus a line belongs *per se* to the notion of a triangle, because the definition of a triangle involves the conception of a line, and odd and even belong *per se* to the notion of number, because the definition of odd or even introduces the notion of a number divisible or not divisible into equal parts.[2] Lastly, that which always belongs to any given subject, and belongs to it inasmuch as it is that which it is, is said to belong to it *καθόλου*, *universaliter*. Thus to have angles equal to two right angles does not belong to any figure taken at random, it is not true of figure *κατὰ παντός*, and though it is true of any isosceles triangle yet it is not true of it in the first instance [3] nor inasmuch as it is isosceles. But it is true of a triangle in all cases and

[2] Aristotle mentions a third sense of *κατὰ παντός*, which it is not here necessary to mention.

[3] *ἀλλ' οὐ πρῶτον, ἀλλὰ τὸ τρίγωνον πρότερον.*

because it is a triangle, and therefore belongs to it καθόλου, *universaliter*. It is manifest that whenever this is the case the proposition is convertible. Thus a figure having angles equal to two right angles is a triangle.

Aristotle is not laying down three general rules, but he was understood to do so by Ramus—whose rules of truth, justice, and wisdom respectively correspond to the three phrases of which we have been speaking.

Bacon, adopting two of these rules (he makes no allusion to that of justice), compares them with the two conditions which a direction ought to fulfil. If it be certain, the effect will follow from it at all times and in all cases. And this corresponds to the rule of truth. If it be free, then whenever the effect is present the direction must have been complied with. The presence of either implies that of the other. And this is the practical application of the rule of wisdom.

I have thought it well to enter into this explanation, because it shows in the first place that the system of Peter Ramus had considerable influence on Bacon's notions of logic, and in the second that he had formed a complete and definite conception of his own method before he had been led to connect it with the doctrine of forms.

At the end of the eleventh chapter Bacon proposes to give three cautions whereby we may ascertain whether what seems to be a direction really is one. The general principle is that the direction must carry you a degree or remove nearer to action, operation, or light ; else it is but an abstract or varied notion. The first of the three particular cautions is " that the nature discovered be more original than the nature supposed, and not more secondary or of the like degree " : a remark which, taken in conjunction with the illustrations by which it is followed, serves to confirm what I have elsewhere endeavoured to show, that Bacon's idea of natural philosophy was the explanation of the secondary qualities of bodies by means of the primary. The second caution is so obscurely expressed that I can only conjecture that it refers to the necessity of studying abstract qualities before commencing the study of concrete bodies. Composition subaltern and composition absolute are placed in antithesis to each other. The latter phrase apparently describes the synthesis of abstract natures by which an actual ultimate species is formed, and the former [refers] to the formation of a class of objects which all agree in possessing the nature which is the subject of inquiry. The fragment breaks off before the delivery of this second caution is completed, and we therefore know nothing of the third and last.

NOTE

The manuscript from which Robert Stephens printed these fragments was found among some loose papers placed in his hands by the Earl of Oxford, and is now in the British Museum; Harl. MSS. 6462. It is a thin paper volume of the quarto size, written in the hand of one of Bacon's servants, with corrections, erasures, and interlineations in his own.

The chapters of which it consists are both imperfect in themselves (all but three),—some breaking off abruptly, others being little more than tables of contents,—and imperfect in their connexion with each other; so much so as to suggest the idea of a number, of separate papers loosely put together. But it was not so (and the fact is important) that the volume itself was actually made up. However they came together, they are here fairly and consecutively copied out. Though it be a collection of fragments therefore it is such a collection as Bacon thought worthy not only of being preserved, but of being transcribed into a volume; and a particular account of it will not be out of place.

The contents of the manuscript before Bacon touched it may be thus described.

1. A titlepage, on which is written "VALERIUS TERMINUS of the Interpretation of Nature, with the annotations of HERMES STELLA."
2. "Chapter I. Of the limits and end of knowledge;" with a running title, "Of the Interpretation of Nature."
3. "The chapter immediately following the Inventory; being the 11th in order."
4. "A part of the 9th chapter, immediately precedent to the Inventory, and inducing the same."
5. "The Inventory, or an enumeration and view of inventions already discovered and in use, together with a note of the wants and the nature of the supplies; being the 10th chapter, and this a fragment only of the same."
6. Part of a chapter, not numbered, "Of the internal and profound errors and superstitions in the nature of the mind, and of the four sorts of Idols or fictions which offer themselves to the understanding in the Inquisition of knowledge."
7. "Of the impediments of knowledge; being the third chapter, the preface only of it."
8. "Of the impediments which have been in the times and in diversion of wits; being the fourth chapter."
9. "Of the impediments of knowledge for want of a true succession of wits, and that hitherto the length of one man's life hath been the greatest measure of knowledge; being the fifth chapter."
10. "That the pretended succession of wits hath been evil placed, forasmuch as after variety of sects and opinions the most popular and not the truest prevaileth and weareth out the rest; being the sixth chapter."
11. "Of the impediments of knowledge in handling it by parts, and in slipping off particular sciences from the root and stock of universal knowledge; being the seventh chapter."
12. "That the end and scope of knowledge hath been generally mistaken, and that men were never well advised what it was they sought" (part of a chapter not numbered).
13. "An abridgment of divers chapters of the first book;" namely, the 12th, 13th and 14th (over which is a running title "Of active knowledge"); and (without any running title) the 15th, 16th, 17th, 18th, 19th, 21st, 22nd, 25th, and 26th. These abridgments have no headings; and at the end is written, "The end of the Abridgment of the first book of the Interpretation of Nature".

Such was the arrangement of the manuscript as the transcriber left it; which I have thought worth preserving, because I seem to see traces in it of two separate stages in the development of the work; the order of the chapters as they are transcribed being probably the same in which Bacon wrote them; and the numbers inserted at the end of the headings indicating the order in which, when he placed them in the transcriber's hands, it was his intention to arrange them; and because it proves at any rate that at that time the design of the whole book was clearly laid out in his mind.

There is nothing, unfortunately, to fix the *date* of the transcript, unless it be implied in certain astronomical or astrological symbols written on the blank outside of the volume; in which the figures 1603 occur.[1] This may possibly be the transcriber's note of the time when he finished his work; for which (but for one circumstance which I shall mention presently) I should think the year 1603 as likely a date as any; for we know from a letter of Bacon's, dated 3rd July 1603, that he had at that time resolved

[1] See the second page of the facsimile in vol. iii. of ed. Ellis and Spedding. The writing in the original is on the outside of the last leaf, which is in fact the cover. The front cover, if there ever was one, is lost. The ink with which the line containing the symbols is written corresponds with that in the body of the MS.; and the line itself is placed symmetrically in the middle of the page, near the top. The two lower lines are apparently by another hand, probably of later date, certainly in ink of a different colour, and paler. The word "Philosophy" is in Bacon's own hand, written lightly in the upper corner at the left, and is no doubt merely a docket inserted afterwards when he was sorting his papers. What connexion there was between the note and the MS. it is impossible to say. But it is evidently a careful memorandum of something, set down by somebody when the MS. was at hand; and so many of the characters resemble those adopted to represent the planets and the signs of the zodiac, that one is led to suspect in it a note of the positions of the heavenly bodies at the time of some remarkable accident;—perhaps the plague, of which 30,578 persons died in London, during the year ending 22nd December, 1603. The period of the commencement, the duration, or the cessation of such an epidemic might naturally be so noted. Now three of the characters clearly represent respectively Mercury, Aquarius, and Sagittarius. The sign for Jupiter as we find it in old books, is so like a 4, that the first figure of 45 may very well have been meant for it. The monogram at the beginning of the line bears a near resemblance to the sign of Capricorn in its most characteristic feature. And the mark over the sign of Aquarius appears to be an abbreviation of that which usually represents the Sun. (The blot between 1603 and B is nothing; being only meant to represent a figure 6 blotted out with the finger before the ink was dry). Suspecting therefore that the writing contained a note of the positions of Mercury and Jupiter in the year 1603, I sent a copy to a scientific friend and asked him if from such data he could determine the month indicated. He found upon a rough calculation (taking account of mean motions only) that Jupiter did enter the sign of Sagittarius about the 10th of August, 1603, and continued there for about a twelvemonth; that the Sun entered Aquarius about the 12th or 13th of January, 1603–4; and that Mercury was about the 16th or 17th of the same month in the 26th or 27th degree of Capricorn:—coincidences which would have been almost conclusive as to the date indicated, if Capricorn had only stood where Aquarius does, and vice versâ. But their position as they actually stood in the MS. is a formidable, if not fatal, objection to the interpretation.

According to another opinion with which I have been favoured, the first monogram is a *nota bene;* the next group may mean *Dies Mercurii* (Wednesday) *26th January,* 1603; and the rest refers to something not connected with astronomy. But to this also there is a serious objection. The 26th of January, 1603–4, was a Friday; and it seems to me very improbable that any Englishman would have described the preceding January as belonging to the year 1603. Bacon himself invariably dated according to the civil year, and the occasional use of the historical year in loose memoranda would have involved all his dates in confusion. I should think it more probable that the writer (who may have been copying a kind of notation with which he was not familiar) miscopied the sign of Venus into that of Mercury; in which case it would mean *Friday,* 26th January, 1603–4. But even then the explanation would be unsatisfactory, as leaving so much unexplained. Those however who are familiar with old MSS. relating to such subjects may probably be able to interpret the whole.

" to meddle as little as possible in the King's causes," and to " put his ambition wholly upon his pen" ; and we know from the *Advancement of Learning* that in 1605 he was engaged upon a work entitled " The Interpretation of Nature " : to which I may add that there is in the Lambeth Library a copy of a letter from Bacon to Lord Kinlosse, dated 25th March, 1603, and written in the same hand as this manuscript.

Bacon's corrections, if I may judge from the character of the handwriting, were inserted a little later ; for it is a fact that about the beginning of James' reign his writing underwent a remarkable change, from the hurried Saxon hand full of large sweeping curves, and with letters imperfectly formed and connected, which he wrote in Elizabeth's time, to a small, neat, light, and compact one, formed more upon the Italian model which was then coming into fashion ; and when these corrections were made it is evident that this new character had become natural to him and easy. It is of course impossible to fix the precise date of such a change,—the more so because his autographs of this period are very scarce,—but whenever it was that he corrected this manuscript, it is evident that he then considered it worthy of careful revision. He has not merely inserted a sentence here and there, altered the numbers of the chapters, and added words to the headings in order to make the description more exact ; but he has taken the trouble to add the running title wherever it was wanting, thus writing the words " of the Interpretation of Nature " at full length not less than eighteen times over ; and upon the blank space of the titlepage he has written out a complete table of contents [2]. In short, if he had been preparing the manuscript for the press or for a fresh transcript, he could not have done it more completely or carefully,—only that he has given no directions for altering the order of the chapters so as to make it correspond with the numbers. And hence I infer that up to the time when he made these corrections, this was the form of the great work on which he was engaged : it was a work concerning the Interpretation of Nature ; which was to begin where the *Novum Organum* begins ; and of which the first book was to include all the preliminary considerations preparatory to the exposition of the formula.

I place this fragment here in deference to Mr. Ellis's decided opinion that it was written before the *Advancement of Learning*. The positive ground indeed which he alleges in support of that conclusion I am obliged to set aside, as founded, I think, upon a misapprehension ; and the supposition that no part of it was written later involves a difficulty which I cannot yet get over to my own satisfaction. But that the body of it was written earlier I see no reason to doubt ; and if so, this is its proper place. [3]

The particular point on which I venture to disagree with Mr. Ellis I have stated in a note upon his preface to the *Novum Organum*, promising at the same time a fuller explanation of the grounds of my own conclusion, which I will now give.

The question is, whether the " Inventory " in the 10th chapter of *Valerius Terminus* was to have exhibited a general survey of the state of knowledge corresponding with that which fills the second book of the *Advancement of Learning*. I think not.

It is true indeed that the title of that 10th chapter,—namely, " The Inventory, or an enumeration and view of inventions already discovered and in use, with a note of the wants and the nature of the supplies "—has at first sight a considerable resemblance to the description of the contents of the second book of the *Advancement of Learning*,—namely, " A general and faithful perambulation of learning, with an inquiry what parts thereof lie fresh and waste, and not improved and converted by the industry of Man ; . . . wherein nevertheless my purpose is at this time to note only omissions and deficiencies, and not to make any redargutions of errors," and so on. But an " enumeration of *Inventions* " is not the same thing as " a perambulation of *Learning* " ; and it will be found upon closer examination that the " Inventory " spoken of in *Valerius Terminus* does really correspond to one, and one only, of the fifty-one *Desiderata* set down at the end of the *De Augmentis ;* viz. that *Inventarium opum humanarum*, which was to be an appendix to the *Magia naturalis*. See *De Aug.* iii. 5. This will appear clearly by comparing the descriptions of the two.

In the *Advancement of Learning* Bacon tells us that there are two points of much purpose pertaining to the department of Natural Magic : the first of which is, " That there

[2] I am inclined to think that there was an interval between the writing of the first eleven titles and the last two; during which the Italian character had become more familiar to him.

[3] The present reprint follows a different order.—Ed.

be made a calendar resembling an Inventory of the *estate of man*, containing *all the Inventions, being the works or fruits of nature or art*, which are now extant *and of which man is already possessed ;* out of which doth naturally result a note what things are yet held impossible or not invented ; which calendar will be the more artificial and serviceable if to every reputed impossibility you add what thing is extant which cometh the nearest in degree to that impossibility : to the end that by these optatives and essentials man's inquiry may be the more awake in deducing direction of works from the speculation of causes ".

The inventory which was to have been inserted in the 10th chapter of *Valerius Terminus* is thus introduced :—" The plainest method and most directly pertinent to this intention will be to make distribution of *sciences, arts, inventions, works,* and their portions, *according to the use and tribute which they yield and render to the condition of man's life ;* and under those several uses, being as several offices of provisions, to charge and tax what may be reasonably exacted or demanded, ... and then upon those charges and taxations to distinguish and present as it were in several columns *what is extant and already found*, and what is *defective and further to be provided.* Of which provisions because in many of them, after the manner of slothful and faulty accomptants, it will be returned by way of excuse that no such are to be had, it will be fit to give some light *of the nature of the supplies ;* whereby it will evidently appear that they are to be compassed and procured ". And that the calendar was to deal, not with knowledge in general, but only with arts and sciences of *invention* in its more restricted sense—the *pars operativa de natura* (*De Aug.* iii. 5.)—appears no less clearly from the opening of the 11th chapter, which was designed immediately to follow the "Inventory". "It appeareth then what is now in proposition, not by general circumlocution but by particular note. No former philosophy," etc., etc., "but the revealing and discovering of *new inventions and operations*, ... the nature and kinds of which inventions *have been described* as they could be discovered ", etc. If further evidence were required of the exact resemblance between the Inventory of *Valerius Terminus* and the Inventarium of the *Advancement* and the *De Augmentis*, I might quote the end of the 9th chapter, where the particular expressions correspond, if possible, more closely still. But I presume that the passages which I have given are enough ; and that the opinion which I have elsewhere expressed as to the origin of the *Advancement of Learning*,—namely, that the writing of it was a by-thought and no part of the work on the Interpretation of Nature as originally designed,—will not be considered inconsistent with the evidence afforded by these fragments.

That the *Valerius Terminus* was composed before the *Advancement*, though a conclusion not deducible from the Inventory, is nevertheless probable : but to suppose that it was so composed *exactly in its present form*, involves, as I said, a difficulty ; which I will now state. The point is interesting, as bearing directly upon the development in Bacon's mind of the doctrine of Idols ; concerning which see preface to *Novum Organum*, note C. But I have to deal with it here merely as bearing upon the probable date of this fragment.

In treating of the department of Logic in the *Advancement*, Bacon notices as altogether wanting " the particular elenches or cautions against *three* false appearances " or fallacies by which the mind of man is beset : the " caution " of which, he says, " doth extremely import the true conduct of human judgment ". These false appearances he describes, though he does not give their names ; and they correspond respectively to what he afterwards called the Idols of the Tribe, the Cave, and the Forum. But he makes no mention of the fourth ; namely, the Idols of the Theatre. Now in *Valerius Terminus* we find two separate passages in which the Idols are mentioned ; and in both all four are enumerated, and all by name ; though what he afterwards called Idols of the Forum, he there calls Idols of the Palace ; and it seems to me very unlikely that, if when he wrote the *Advancement* he had already formed that classification, he should have omitted all mention of the Idols of the Theatre ; for though it is true that that was not the place to discuss them, and therefore in the corresponding passage of the *De Augmentis* they are noticed as to be passed by " for the present," yet they *are* noticed by name, and in all Bacon's later writings the confutation of them holds a very prominent place.

To me the most probable explanation of the fact is this. I have already shown that between the composition and the transcription of these fragments the design of the work appears to have undergone a considerable change ; the order of the chapters being entirely altered. We have only to suppose therefore that they were composed before the *Advancement* and transcribed after, and that in preparing them for the transcriber Bacon made the same kind of alterations in the originals which he afterwards made upon the transcript, and the difficulty disappears. Nothing would be easier than to correct " three " into " four ", and insert " the Idols of the Theatre " at the end of the sentence.

And this reminds me (since I shall have so much to do with these questions of date) to suggest a general caution with regard to them all ; namely, that in the case of fragments like these, the comparison of isolated passages can hardly ever be relied upon for evidence of the date or order of composition, or of the progressive development of the writer's views : and for this simple reason,—we can never be sure that the passages as they now stand formed part of the original writing. The copy of the fragment which we have may be (as there is reason to believe this was) a transcript from several loose papers, written at different periods and containing alterations or additions made from time to time. We may know perhaps that when Bacon published the *Advancement of Learning* he was ignorant of some fact with which he afterwards became acquainted ; we may find in one of these fragments,—say the *Temporis Partus Masculus,*—a passage implying acquaintance with that fact. Does it follow that the *Temporis Partus Masculus* was written after the *Advancement of Learning ?* No ; for in looking over the manuscript long after it was written, he may have observed and corrected the error. And we cannot conclude that he at the same time altered the whole composition so as to bring it into accordance with the views he then held ; for that might be too long a work. He may have inserted a particular correction, but meant to rewrite the whole ; and if so, in spite of the later date indicated by that particular passage, the body of the work would still represent a stage in his opinions anterior to the *Advancement of Learning.*

I have felt some doubt whether, in printing this fragment, I should follow the example of Stephens, who gave it exactly as he found it ; or that of later editors, who have altered the order of the chapters so as to make it agree with the numbers. The latter plan will perhaps, upon the whole, be the more convenient. There can be little doubt that the numbers of the chapters indicate the order in which Bacon meant them to be read ; and if any one wishes to compare it with the order in which they seem to have been written, he has only to look at Bacon's table of contents, which was made with reference to the transcript, and which I give unaltered, except as to the spelling.

The notes to this piece are mine.—*J. S.*

VALERIUS TERMINUS

OF THE INTERPRETATION OF NATURE: WITH THE ANNOTATIONS OF HERMES STELLA.[1]

A few fragments of the first book, viz.

1. The first chapter entire. [Of the ends and limits of knowledge].
2. A portion of the 11th chapter. [Of the scale].
3. A small portion of the 9th chapter [being an Inducement to the Inventory].
4. A small portion of the 10th chapter, being the preface to the Inventory.
5. A small portion of the 16th chapter [being a preface to the inward elenches of the mind].
6. A portion of the 4th chapter. [Of the impediments of knowledge in general].
7. A small portion of the 5th chapter. [Of the diversion of wits].
8. The 6th chapter entire. [Of]
9. A portion of the 7th chapter.
10. The 8th chapter entire.
11. Another portion of the 9th chapter.
12. The Abridgment of the 12. 13. 14. 15. 16. 17. 18. 19. 21. 22. 25. 26th chapters of the first book.
13. The first chapter of [the] a book of the same argument written in Latin and destined [for] to be [traditionary] separate and not public[2].

    None of the Annotations of Stella are set down in these fragments.

[1] This is written in the transcriber's hand: all that follows in Bacon's. The words between brackets have a line drawn through them.

[2] This refers to the first chapter of the *Temporis Partus Masculus ;* which follows in the MS. volume, but not here. It is important as bearing upon the date of that fragment.

Cap. 1.

Of the limits and end of knowledge.

In the divine nature both religion and philosophy hath acknowledged goodness in perfection, science or providence comprehending all things, and absolute sovereignty or kingdom. In aspiring to the throne of power the angels transgressed and fell; in presuming to come within the oracle of knowledge man transgressed and fell[1]; but in pursuit towards the similitude of God's goodness or love (which is one thing, for love is nothing else but goodness put in motion or applied) neither man or spirit ever hath transgressed, or shall transgress.

The angel of light that was, when he presumed before his fall, said within himself, *I will ascend and be like unto the Highest;* not God, but the highest. To be like to God in goodness, was no part of his emulation; knowledge, being in creation an angel of light, was not the want which did most solicit him; only because he was a minister he aimed at a supremacy; therefore his climbing or ascension was turned into a throwing down or precipitation.

Man on the other side, when he was tempted before he fell, had offered unto him this suggestion, *that he should be like unto God.* But how? Not simply, but in this part, *knowing good and evil.* For being in his creation invested with sovereignty of all inferior creatures, he was not needy of power or dominion; but again, being a spirit newly inclosed in a body of earth, he was fittest to be allured with appetite of light and liberty of knowledge; therefore this approaching and intruding into God's secrets and mysteries was rewarded with a further removing and estranging from God's presence. But as to the goodness of God, there is no danger in contending or advancing towards a similitude thereof, as that which is open and propounded to our imitation. For that voice (whereof the heathen and all other errors of religion have ever confessed that it sounds not like man), *Love your enemies; be you like unto your heavenly Father, that suffereth his rain to fall both upon the just and the unjust,* doth well declare, that we can in that point commit no excess; so again we find it often repeated in the old law, *Be you holy as I am holy;* and what is holiness else but goodness, as we consider it separate and guarded from all mixture and all access of evil?

Wherefore seeing that knowledge is of the number of those things which are to be accepted of with caution and distinction; being now to open a fountain, such as it is not easy to discern where the issues and streams thereof will take and fall; I thought it good and necessary in the first place to make a strong and sound head or bank to rule and guide the course of the waters; by setting down this position or firmament, namely, *That all knowledge is to be limited by religion, and to be referred to use and action.*

For if any man shall think by view and inquiry into these sensible and material things, to attain to any light for the revealing of the nature or will of God, he shall dangerously abuse himself. It is true that the contemplation of the creatures of God hath for end (as to the natures of the creatures themselves) knowledge, but as to the nature of God, no knowledge, but wonder; which is nothing else but contemplation broken off, or losing itself. Nay further, as it was aptly said by one of Plato's school *the sense of man resembles the sun, which openeth and revealeth the terrestrial globe, but obscureth and concealeth the celestial;* so doth the sense dis-

[1] This clause is repeated in the margin, in the transcriber's hand.

cover natural things, but darken and shut up divine. And this appeareth sufficiently in that there is no proceeding in invention of knowledge but by similitude; and God is only self-like, having nothing in common with any creature, otherwise than as in shadow and trope. Therefore attend his will as himself openeth it, and give unto faith that which unto faith belongeth; for more worthy it is to believe than to think or know, considering that in knowledge (as we now are capable of it) the mind suffereth from inferior natures; but in all belief it suffereth from a spirit which it holdeth superior and more authorized than itself.

To conclude, the prejudice hath been infinite that both divine and human knowledge hath received by the intermingling and tempering of the one with the other; as that which hath filled the one full of heresies, and the other full of speculative fictions and vanities.

But now there are again which, in a contrary extremity to those which give to contemplation an over-large scope, do offer too great a restraint to natural and lawful knowledge, being unjustly jealous that every reach and depth of knowledge wherewith their conceits have not been acquainted, should be too high an elevation of man's wit, and a searching and ravelling too far into God's secrets; an opinion that ariseth either of envy (which is proud weakness and to be censured and not confuted), or else of a deceitful simplicity. For if they mean that the ignorance of a second cause doth make men more devoutly to depend upon the providence of God, as supposing the effects to come immediately from his hand, I demand of them, as Job demanded of his friends, *Will you lie for God as man will for man to gratify him?* But if any man without any sinister humour doth indeed make doubt that this digging further and further into the mine of natural knowledge is a thing without example and uncommended in the Scriptures, or fruitless; let him remember and be instructed; for behold it was not that pure light of natural knowledge, whereby man in Paradise was able to give unto every living creature a name according to his propriety, which gave occasion to the fall; but it was an aspiring desire to attain to that part of moral knowledge which defineth of good and evil, whereby to dispute God's commandments and not to depend upon the revelation of his will, which was the original temptation. And the first holy records, which within those brief memorials of things which passed before the flood entered few things as worthy to be registered but only lineages [2] and propagations, yet nevertheless honour the remembrance of the inventor both of music and works in metal. Moses again (who was the reporter) is said to have been seen in all the Egyptian learning, which nation was early and leading in matter of knowledge. And Salomon the king, as out of a branch of his wisdom extraordinarily petitioned and granted from God, is said to have written a natural history of all that is green, from the cedar to the moss (which is but a rudiment between putrefaction and an herb), and also of all that liveth and moveth. And if the book of Job be turned over, it will be found to have much aspersion of natural philosophy. Nay, the same Salomon the king affirmeth directly that the glory of God *is to conceal a thing, but the glory of the king is to find it out,* as if according to the innocent play of children the divine Majesty took delight to hide his works, to the end to have them found out; for in naming the king he intendeth man, taking such a condition of man as hath most excellency and greatest commandment of wits and means, alluding also to his own person, being truly one of those clearest burning lamps, whereof himself speaketh in another place, when he saith *The spirit of man is as the lamp of God, wherewith he searcheth all inwardness;* which nature of the soul the same Salomon holding precious and inestimable, and therein conspiring with the affection of Socrates, who scorned the pretended learned men of his time for raising great benefit of their learning (whereas Anaxagoras contrariwise and divers others being born to ample patrimonies decayed them in contemplation), delivereth it in precept yet remaining, *Buy the truth, and sell it not; and so of wisdom and knowledge.*

And lest any man should retain a scruple as if this thirst of knowledge were rather an humour of the mind than an emptiness or want in nature and an instinct from God, the same author defineth of it fully, saying, *God hath made every thing*

[2] *linages* in original. See a note to *The New Atlantis*, near middle.

in beauty according to season ; also he hath set the world in man's heart, yet can he not find out the work which God worketh from the beginning to the end ; declaring not obscurely that God hath framed the mind of man as a glass capable of the image of the universal world, joying to receive the signature thereof as the eye is of light, yea not only satisfied in beholding the variety of things and vicissitude of times, but raised also to find out and discern those ordinances and decrees which throughout all these changes are infallibly observed. And although the highest generality of motion or summary law of nature God should still reserve within his own curtain, yet many and noble are the inferior and secondary operations which are within man's sounding. This is a thing which I cannot tell whether I may so plainly speak as truly conceive, that as all knowledge appeareth to be a plant of God's own planting, so it may seem the spreading and flourishing, or at least the bearing and fructifying of this plant, by a providence of God, nay not only by a general providence but by a special prophecy, was appointed to this autumn of the world : for to my understanding it is not violent to the letter, and safe now after the event, so to interpret that place in the prophecy of Daniel where speaking of the latter times it is said, *Many shall pass to and fro, and science shall be increased ;* as if the opening of the world by navigation and commerce and the further discovery of knowledge should meet in one time or age.

But howsoever that be, there are besides the authorities of Scriptures before recited, two reasons of exceeding great weight and force why religion should dearly protect all increase of natural knowledge : the one, because it leadeth to the greater exaltation of the glory of God ; for as the Psalms and other Scriptures do often invite us to consider and to magnify the great and wonderful works of God, so if we should rest only in the contemplation of those shews which first offer themselves to our senses, we should do a like injury to the majesty of God, as if we should judge of the store of some excellent jeweller by that only which is set out to the street in his shop. The other reason is, because it is a singular help and a preservative against unbelief and error ; for, saith our Saviour, *You err, not knowing the Scriptures nor the power of God ;* laying before us two books or volumes to study if we will be secured from error ; first the Scriptures revealing the will of God, and then the creatures expressing his power ; for that latter book will certify us that nothing which the first teacheth shall be thought impossible. And most sure it is, and a true conclusion of experience, that a little natural philosophy inclineth the mind to atheism, but a further proceeding bringeth the mind back to religion.

To conclude then, let no man presume to check the liberality of God's gifts, who, as was said, *hath set the world in man's heart.* So as whatsoever is not God but parcel of the world, he hath fitted it to the comprehension of man's mind, if man will open and dilate the powers of his understanding as he may.

But yet evermore it must be remembered that the least part of knowledge passed to man by this so large a charter from God must be subject to that use for which God hath granted it ; which is the benefit and relief of the state and society of man ; for otherwise all manner of knowledge becometh malign and serpentine, and therefore as carrying the quality of the serpent's sting and malice it maketh the mind of man to swell ; as the Scripture saith excellently, *knowledge bloweth up, but charity buildeth up.* And again the same author doth notably disavow both power and knowledge such as is not dedicated to goodness or love, for saith he, *If I have all faith so as I could remove mountains* (there is power active), *if I render my body to the fire* (there is power passive), *if I speak with the tongues of men and angels* (there is knowledge, for language is but the conveyance of knowledge), *all were nothing.*

And therefore it is not the pleasure of curiosity, nor the quiet of resolution, nor the raising of the spirit, nor victory of wit, nor faculty of speech, nor lucre of profession, nor ambition of honour or fame, nor inablement for business, that are the true ends of knowledge ; some of these being more worthy than other, though all inferior and degenerate : but it is a restitution and reinvesting (in great part) of man to the sovereignty and power (for whensoever he shall be able to call the creatures by their true names he shall again command them) which he had in his first state of creation. And to speak plainly and clearly, it is a discovery of al operations and possibilities of operations from immortality (if it were possible) to the meanest mechanical practice. And therefore knowledge that tendeth but to satisfaction is but as a courtesan, which is for pleasure and not for fruit or generation. And knowledge that tendeth to profit or profession or glory is but as the golden ball thrown before Atalanta, which while she goeth aside and stoopeth to take up she hindereth the race. And knowledge referred to some particular point of use is but as Harmodius which putteth down one tyrant, and not like Hercules who did perambulate the world to suppress tyrants and giants and monsters in every part [3]. It is true, that in two points the curse is peremptory and not to be removed; the one that vanity must be the end in all human effects, eternity being resumed, though the revolutions and periods may be delayed. The other that the consent of the creature being now turned into reluctation, this power cannot otherwise be exercised and administered but with labour, as well in inventing as in executing; yet nevertheless chiefly that labour and travel which is described by the sweat of the brows more than of the body; that is such travel as is joined with the working and discursion of the spirits in the brain: for as Salomon saith excellently, *The fool putteth to more strength, but the wise man considereth which way*, signifying the election of the mean to be more material than the multiplication of endeavour. It is true also that there is a limitation rather potential than actual, which is when the effect is possible, but the time or place yieldeth not the matter or basis whereupon man should work. But notwithstanding these precincts and bounds, let it be believed, and appeal thereof made to *Time* (with renunciation nevertheless to all the vain and abusing promises of Alchemists and Magicians, and such like light, idle, ignorant, credulous, and fantastical wits and sects), that the new-found world of land was not greater addition to the ancient continent than there remaineth at this day a world of inventions and sciences unknown, having respect to those that are known, with this difference, that the ancient regions of knowledge will seem as barbarous compared with the new, as the new regions of people seem barbarous compared to many of the old.

The dignity of this end (of endowment of man's life with new commodities) appeareth by the estimation that antiquity made of such as guided thereunto. For whereas founders of states, lawgivers, extirpers of tyrants, fathers of the people, were honoured but with the titles of Worthies or Demigods, inventors were ever consecrated amongst the Gods themselves. And if the ordinary ambitions of men lead them to seek the amplification of their own power in their countries, and a better ambition than that hath moved men to seek the amplification of the power of their own countries amongst other nations, better again and more worthy must that aspiring be which seeketh the amplification of the power and kingdom of mankind over the world; the rather because the other two prosecutions are ever culpable of much perturbation and injustice; but this is a work truly divine, which cometh *in aura leni*, without noise or observation.

The access also to this work hath been by that port or passage, which the divine Majesty (who is unchangeable in his ways) doth infallibly continue and observe; that is the felicity wherewith he hath blessed an humility of mind, such as rather laboureth to spell and so by degrees to read in the volumes of his creatures, than to solicit and urge and as it were to invocate a man's own spirit to divine and give oracles unto him. For as in the inquiry of divine truth, the pride of men hath ever inclined to leave the oracles of God's word and to vanish in the mixture of their own inventions; so in the self-same manner, in inquisition of nature they have ever left the oracles of God's works, and adored the deceiving and deformed imagery which the unequal mirrors of their own minds have represented unto them. Nay it is a point fit and necessary in the front and beginning of this work without hesitation or reservation to be professed, that it is no less true in this kingdom of knowledge than in God's kingdom of heaven, that no man shall enter into it *except he become first as a little child* [4].

[3] The words "that is, man's miseries and necessities," which followed in the transcript, have a line drawn through them.

[4] This chapter ends at the top of a new page. The rest is left blank.

Of the impediments of knowledge, being the 4th[5] *chapter, the preface only of it.*

In some things it is more hard to attempt than to achieve, which falleth out when the difficulty is not so much in the matter or subject, as it is in the crossness and indisposition of the mind of man to think of any such thing, to will or to resolve it. And therefore Titus Livius in his declamatory digression wherein he doth depress and extenuate the honour of Alexander's conquests saith, *Nihil aliud quam bene ausus vana contemnere:* in which sort of things it is the manner of men first to wonder that any such thing should be possible, and after it is found out to wonder again how the world should miss it so long. Of this nature I take to be the invention and discovery of knowledge, etc.

The impediments which have been in the times, and in diversion of wits, being the 5th chapter[6], a small fragment in the beginning of that chapter.

The encounters of the times have been nothing favourable and prosperous for the invention of knowledge; so as it is not only the daintiness of the seed to take, and the ill mixture and unliking of the ground to nourish or raise this plant, but the ill season also of the weather by which it hath been checked and blasted. Especially in that the seasons have been proper to bring up and set forward other more hasty and indifferent plants, whereby this of knowledge hath been starved and overgrown; for in the descent of times always there hath been somewhat else in reign and reputation, which hath generally aliened and diverted wits and labours from that employment.

For as for the uttermost antiquity, which is like fame that muffles her head and tells tales, I cannot presume much of it; for I would not willingly imitate the manner of those that describe maps, which when they come to some far countries whereof they have no knowledge, set down how there be great wastes and deserts there: so I am not apt to affirm that they knew little, because what they knew is little known to us. But if you will judge of tnem by the last traces that remain to us, you will conclude, though not so scornfully as Aristotle doth, that saith our ancestors were extreme gross, as those that came newly from being moulded out of the clay or some earthly substance; yet reasonably and probably thus, that it was with them in matter of knowledge but as the dawning or break of day. For at that time the world was altogether home-bred, every nation looked little beyond their own confines or territories, and the world had no through lights then, as it hath had since by commerce and navigation, whereby there could neither be that contribution of wits one to help another, nor that variety of particulars for the correcting of customary conceits.

And as there could be no great collection of wits of several parts or nations, so neither could there be any succession of wits of several times, whereby one might refine the other, in regard they had not history to any purpose. And the manner of their traditions was utterly unfit and unproper for amplification of knowledge. And again the studies of those times, you shall find, besides wars, incursions, and rapines, which were then almost every where betwixt states adjoining (the use of leagues and confederacies being not then known), were to populate by multitude of wives and generation, a thing at this day in the waster part of the West-Indies principally affected; and to build sometimes for habitation towns and cities, sometimes for fame and memory monuments, pyramids, colosses, and the like. And if there happened to rise up any more civil wits[7]; then would he found and erect some new laws, customs, and usages, such as now of late years, when the world was revolute almost to the like rudeness and obscurity, we see both in our own nation and abroad many examples of, as well in a number of tenures reserved upon men's lands, as in divers customs of towns and manors, being the devices that such wits wrought upon in such times of deep ignorance, etc.[8].

[5] The word "third" has a line drawn through it, and 4th is written over it in Bacon's hand.

[6] Originally "being the fourth chapter the beginning": the correction all in Bacon's hand.

[7] *witts* in MS. Probably a mistake for *witte*.

[8] The "etc." in Bacon's hand.

The impediments of knowledge for want of a true succession of wits, and that hitherto the length of one man's life hath been the greatest measure of knowledge, being the 6th chapter, the whole chapter [9].

In arts mechanical the first device comes shortest and time addeth and perfecteth. But in sciences of conceit the first author goeth furthest and time leeseth and corrupteth. Painting, artillery, sailing, and the like, grossly managed at first, by time accommodate and refined. The philosophies and sciences of Aristotle, Plato, Democritus, Hippocrates, of most vigour at first, by time degenerated and imbased. In the former many wits and industries contributed in one: In the latter many men's wits spent to deprave the wit of one.

The error is both in the deliverer and in the receiver. He that delivereth knowledge desireth to deliver it in such form as may be soonest believed, and not as may be easiliest examined. He that receiveth knowledge desireth rather present satisfaction than expectant search, and so rather not to doubt than not to err. Glory maketh the author not to lay open his weakness, and sloth maketh the disciple not to know his strength.

Then begin men to aspire to the second prizes; to be a profound interpreter and commenter, to be a sharp champion and defender, to be a methodical compounder and abridger. And this is the unfortunate succession of wits which the world hath yet had, whereby the patrimony of all knowledge goeth not on husbanded or improved, but wasted and decayed. For knowledge is like a water that will never arise again higher than the level from which it fell; and therefore to go beyond Aristotle by the light of Aristotle is to think that a borrowed light can increase the original light from whom it is taken. So then no true succession of wits having been in the world, either we must conclude that knowledge is but a task for one man's life, and then vain was the complaint that *life is short, and art is long*: or else, that the knowledge that now is, is but a shrub, and not that tree which is never dangerous, but where it is to the purpose of knowing Good and Evil; which desire ever riseth upon an appetite to elect and not to obey, and so containeth in it a manifest defection.

That the pretended succession of wits hath been evil placed, forasmuch as after variety of sects and opinions, the most popular and not the truest prevaileth and weareth out the rest; being the 7th chapter; a fragment [10].

It is sensible to think that when men enter first into search and inquiry, according to the several frames and compositions of their understanding they light upon different conceits, and so all opinions and doubts are beaten over, and then men having made a taste of all wax weary of variety, and so reject the worst and hold themselves to the best, either some one if it be eminent, or some two or three if they be in some equality, which afterwards are received and carried on, and the rest extinct.

But truth is contrary, and that time is like a river which carrieth down things which are light and blown up, and sinketh and drowneth that which is sad and weighty. For howsoever governments have several forms, sometimes one governing, sometimes few, sometimes the multitude; yet the state of knowledge is ever a *Democratie*, and that prevaileth which is most agreeable to the senses and conceits of people. As for example there is no great doubt but he that did put the beginning of things to be *solid, void, and motion to the centre*, was in better earnest than he that put *matter, form, and shift*; or he that put *the mind, motion and matter*. For no man shall enter into inquisition of nature, but shall pass by that opinion of Democritus, whereas he shall never come near the other two opinions, but leave them aloof for the schools and table-talk. Yet those of Aristotle and Plato, because they be both agreeable to popular sense, and the one was

9 Originally "the fifth chapter": "6th" substituted, and "the whole chapter" added, in Bacon's hand.

10 Originally "the sixth chapter": "7th" substituted, and "a fragment" added in Bacon's hand.

uttered with subtilty and the spirit of contradiction, and the other with a style of ornament and majesty, did hold out, and the other gave place, etc. [11].

Of the impediments of knowledge in handling it by parts, and in slipping off particular sciences from the root and stock of universal knowledge, being the 8th[12] *chapter*, the whole chapter.

Cicero, the orator, willing to magnify his own profession, and thereupon spending many words to maintain that eloquence was not a shop of good words and elegancies but a treasury and receipt of all knowledges, so far forth as may appertain to the handling and moving of the minds and affections of men by speech, maketh great complaint of the school of Socrates; that whereas before his time the same professors of wisdom in Greece did pretend to teach an universal *Sapience* and knowledge both of matter and words, Socrates divorced them and withdrew philosophy and left rhetoric to itself, which by that destitution became but a barren and unnoble science. And in particular sciences we see that if men fall to subdivide their labours, as to be an oculist in physic, or to be perfect in some one title of the law, or the like, they may prove ready and subtile, but not deep or sufficient, no not in that subject which they do particularly attend, because of that consent which it hath with the rest. And it is a matter of common discourse of the chain of sciences how they are linked together, insomuch as the Grecians, who had terms at will, have fitted it of a name of *Circle Learning*. Nevertheless I that hold it for a great impediment towards the advancement and further invention of knowledge, that particular arts and sciences have been disincorporated from general knowledge, do not understand one and the same thing which Cicero's discourse and the note and conceit of the Grecians in their word *Circle Learning* do intend. For I mean not that use which one science hath of another for ornament or help in practice, as the orator hath of knowledge of affections for moving, or as military science may have use of geometry for fortifications; but I mean it directly of that use by way of supply of light and information which the particulars and instances of one science do yield and present for the framing or correcting of the axioms of another science in their very truth and notion. And therefore that example of *oculists* and *title lawyers* doth come nearer my conceit than the other two; for sciences distinguished have a dependence upon universal knowledge to be augmented and rectified by the superior light thereof, as well as the parts and members of a science have upon the *Maxims* of the same science, and the mutual light and consent which one part receiveth of another. And therefore the opinion of Copernicus in astronomy, which astronomy itself cannot correct because it is not repugnant to any of the appearances, yet natural philosophy doth correct. On the other side if some of the ancient philosophers had been perfect in the observations of astronomy, and had called them to counsel when they made their principles and first axioms, they would never have divided their philosophy as the Cosmographers do their descriptions by globes, making one philosophy for heaven and another for under heaven, as in effect they do.

So if the moral philosophers that have spent such an infinite quantity of debate touching Good and the highest good, had cast their eye abroad upon nature and beheld the appetite that is in all things to receive and to give; the one motion affecting preservation and the other multiplication; which appetites are most evidently seen in living creatures in the pleasure of nourishment and generation; and in man do make the aptest and most natural division of all his desires, being either of sense of pleasure or sense of power; and in the universal frame of the world are figured, the one in the beams of heaven which issue forth, and the other in the lap of the earth which takes in: and again if they had observed the motion of congruity or situation of the parts in respect of the whole, evident in so many particulars; and lastly if they had considered the motion (familiar in attraction

11 The "etc." in Bacon's hand.

12 Originally "seventh": "8th" substituted, and "the whole chapter" added in Bacon's hand.

of things) to approach to that which is higher in the same kind; when by these observations so easy and concurring in natural philosophy, they should have found out this quaternion of good, in enjoying or fruition, effecting or operation, consenting or proportion, and approach or assumption; they would have saved and abridged much of their long and wandering discourses of pleasure, virtue, duty, and religion. So likewise in this same logic and rhetoric, or arts [13] of argument and grace of speech, if the great masters of them would but have gone a form lower, and looked but into the observations of Grammar concerning the kinds of words, their derivations, deflexions, and syntax; especially enriching the same with the helps of several languages, with their differing proprieties of words, phrases, and tropes; they might have found out more and better footsteps of common reason, help of disputation, and advantages of cavillation, than many of these which they have propounded. So again a man should be thought to dally, if he did note how the figures of rhetoric and music are many of them the same. The repetitions and traductions in speech and the reports and hauntings of sounds in music are the very same things. Plutarch hath almost made a book of the Lacedæmonian kind of jesting, which joined ever pleasure with distaste. *Sir* (saith a man of art to Philip king of Macedon when he controlled him in his faculty), *God forbid your fortune should be such as to know these things better than I.* In taxing his ignorance in his art he represented to him the perpetual greatness of his fortune, leaving him no vacant time for so mean a skill. Now in music it is one of the ordinariest flowers to fall from a discord or hard tune upon a sweet accord. The figure that Cicero and the rest commend as one of the best points of elegancy, which is the fine checking of expectation, is no less well known to the musicians when they have a special grace in flying the close or cadence. And these are no allusions but direct communities, the same delights of the mind being to be found not only in music, rhetoric, but in moral philosophy, policy, and other knowledges, and that obscure in the one, which is more apparent in the other, yea and that discovered in the one which is not found at all in the other, and so one science greatly aiding to the invention and augmentation of another. And therefore without this intercourse the axioms of sciences will fall out to be neither full nor true; but will be such opinions as Aristotle in some places doth wisely censure, when he saith *These are the opinions of persons that have respect but to a few things.* So then we see that this note leadeth us to an administration of knowledge in some such order and policy as the king of Spain in regard of his great dominions useth in state; who though he hath particular councils for several countries and affairs, yet hath one council of State or last resort, that receiveth the advertisements and certificates from all the rest. Hitherto of the diversion, succession, and conference of wits.

That the end and scope of knowledge hath been generally mistaken, and that men were never well advised what it was they sought; being the 9th chapter, whereof a fragment (which is the end of the same chapter is before [14].

It appeareth then how rarely the wits and labours of men have been converted to the severe and original inquisition of knowledge; and in those who have pretended, what hurt hath been done by the affectation of professors and the distraction of such as were no professors [15]; and how there was never in effect any conjunction or combination of wits in the first and inducing search, but that every man wrought apart, and would either have his own way or else would go no further than his guide, having in the one case the honour of a first, and in the other the ease of a second; and lastly how in the descent and continuance of wits and labours the succession hath been in the most popular and weak opinions, like unto

[13] *acts* in MS., I think.

[14] See p. 194. note 17.; and compare Table of Contents (p. 185.) No. 3.
The number of this chapter was not stated in the transcript as it originally stood: the words in Roman characters are all added in Bacon's hand, at the end of the title: nothing is struck out.

[15] This clause is repeated in the margin and marked for insertion in its proper place

the weakest natures which many times have most children, and in them also the condition of succession hath been rather to defend and to adorn than to add; and if to add, yet that addition to be rather a refining of a part than an increase of the whole. But the impediments of time and accidents, though they have wrought a general indisposition, yet are they not so peremptory and binding as the internal impediments and clouds in the mind and spirit of man, whereof it now followeth to speak.

The Scripture speaking of the worst sort of error saith, *Errare fecit eos in invio et non in via.* For a man may wander in the way, by rounding up and down. But if men have failed in their very direction and address that error will never by good fortune correct itself. Now it hath fared with men in their contemplations as Seneca saith it fareth with them in their actions, *De partibus vitæ quisque deliberat, de summa nemo.* A course very ordinary with men who receive for the most part their final ends from the inclination of their nature, or from common example and opinion, never questioning or examining them, nor reducing them to any clear certainty; and use only to call themselves to account and deliberation touching the means and second ends, and thereby set themselves in the right way to the wrong place. So likewise upon the natural curiosity and desire to know, they have put themselves in way without foresight or consideration of their journey's end.

For I find that even those that have sought knowledge for itself, and not for benefit or ostentation or any practical enablement in the course of their life, have nevertheless propounded to themselves a wrong mark, namely satisfaction (which men call truth) and not operation. For as in the courts and services of princes and states it is a much easier matter to give satisfaction than to do the business; so in the inquiring of causes and reasons it is much easier to find out such causes as will satisfy the mind of man and quiet objections, than such causes as will direct him and give him light to new experiences and inventions. And this did Celsus note wisely and truly, how that the causes which are in use and whereof the knowledges now received do consist, were in time minors and subsequents to the knowledge of the particulars out of which they were induced and collected; and that it was not the light of those causes which discovered particulars, but only the particulars being first found, men did fall on glossing and discoursing of the causes; which is the reason why the learning that now is hath the curse of barrenness, and is courtesan-like, for pleasure, and not for fruit [16]. Nay to compare it rightly, the strange fiction of the poets of the transformation of Scylla seemeth to be a lively emblem of this philosophy and knowledge; a fair woman upwards in the parts of show, but when you come to the parts of use and generation, Barking Monsters; for no better are the endless distorted questions, which ever have been, and of necessity must be, the end and womb of such knowledge.

But yet nevertheless [17] here I may be mistaken, by reason of some which have much in their pen the referring sciences to action and the use of man, which mean quite another matter than I do. For they mean a contriving of directions and precepts for readiness of practice, which I discommend not, so it be not occasion that some quantity of the science be lost; for else it will be such a piece of husbandry as to put away a manor lying somewhat scattered, to buy in a close that lieth handsomely about a dwelling. But my intention contrariwise is to increase and multiply the revenues and possessions of man, and not to trim up only or order with conveniency the grounds whereof he is already stated. Wherefore the better to make myself understood that I mean nothing less than words, and directly to demonstrate the point which we are now upon, that is, what is the true end, scope, or office of knowledge, which I have set down to consist not in any plausible, delectable, reverend, or admired discourse, or any satisfactory arguments, but in effecting and working, and in discovery of particulars not revealed

[16] Here in the transcript the chapter ended. The next sentence is written in the margin in Bacon's own hand.

[17] This paragraph, which stands as the third fragment in the order of the transcript, is headed in the transcriber's hand, "*A part of the 9th chapter immediately precedent to the Inventory and inducing the same*".

before for the better endowment and help of man's life ; I have thought good to make as it were a Kalendar or Inventory of the wealth, furniture, or means of man according to his present estate, as far as it is known ; which I do not to shew any universality of sense or knowledge, and much less to make a satire of reprehension in respect of wants and errors, but partly because cogitations new had need of some grossness and inculcation to make them perceived ; and chiefly to the end that for the time to come (upon the account and state now made and cast up) it may appear what increase this new manner of use and administration of the stock (if it be once planted) shall bring with it hereafter ; and for the time present (in case I should be prevented by death to propound and reveal this new light [18] as I purpose) yet I may at the least give some awaking note both of the wants in man's present condition and the nature of the supplies to be wished ; though for mine own part neither do I much build upon my present *anticipations*, neither do I think ourselves yet learned or wise enough to wish reasonably : for as it asks some knowledge to demand a question not impertinent, so it asketh some sense to make a wish not absurd [19].

The Inventory, or an enumeration and view of inventions already discovered and in use, together with a note of the wants and the nature of the supplies, being the 10*th chapter; and this* a small fragment thereof, being the preface to the Inventory[20].

The plainest method and most directly pertinent to this intention, will be to make distribution of sciences, arts, inventions, works, and their portions, according to the use and tribute which they yield and render to the conditions of man's life, and under those several uses, being as several offices of provisions, to charge and tax what may be reasonably exacted or demanded ; not guiding ourselves neither by the poverty of experiences and probations, nor according to the vanity of credulous imaginations ; and then upon those charges and taxations to distinguish and present, as it were in several columns, what is extant and already found, and what is defective and further to be provided. Of which provisions, because in many of them after the manner of slothful and faulty officers and accomptants it will be returned (by way of excuse) that no such are to be had, it will be fit to give some light of the nature of the supplies, whereby it will evidently appear that they are to be compassed and procured [21]. And yet nevertheless on the other side again it will be as fit to check and control the vain and void assignations and gifts whereby certain ignorant, extravagant, and abusing wits have pretended to indue the state of man with wonders, differing as much from truth in nature as Cæsar's Commentaries differeth from the acts of King Arthur or Huon of Bourdeaux in story. For it is true that Cæsar did greater things than those idle wits had the audacity to feign their supposed worthies to have done ; but he did them not in that monstrous and fabulous manner.

The chapter immediately following the Inventory ; being the 11*th in order ;* a part thereof [22].

It appeareth then what is now in proposition not by general circumlocution but by particular note. No former philosophy varied in terms or method ; no new *placet* or speculation upon particulars already known ; no referring to action by any manual of practice ; but the revealing and discovering of new inventions and operations. This to be done without the errors and conjectures of art, or the length or difficulties of experience ; the nature and kinds of which inventions have been described as they could be discovered ; for your eye cannot pass one kenning without further sailing ; only we have stood upon the best advantages

[18] *direction* had been written first.

[19] The chapter ends before the bottom of the page ; leaving about a fifth of it blank.

[20] The words *fragment only of the same,* with which the original heading ended, have a line drawn through them, and the words in Roman character are added in Bacon's hand.

[21] The concluding sentence, which is crowded into the page and overflows into the margin, has evidently been inserted subsequently to the original transcript. After "procured" there seems to be an "etc.".

[22] The words in Roman letters are inserted in Bacon's hand.

of the notions received, as upon a mount, to shew the knowledges adjacent and confining. If therefore the true end of knowledge not propounded hath bred large error, the best and perfectest condition of the same end not perceived will cause some declination. For when the butt is set up men need not rove, but except the white be placed men cannot level. This perfection we mean not in the worth of the effect, but in the nature of the direction; for our purpose is not to stir up men's hopes, but to guide their travels. The fulness of direction to work and produce any effect consisteth in two conditions, certainty and liberty. Certainty is when the direction is not only true for the most part, but infallible. Liberty is when the direction is not restrained to some definite means, but comprehendeth all the means and ways possible; for the poet saith well *Sapientibus undique latæ sunt viæ*, and where there is the greatest plurality of change, there is the greatest singularity of choice. Besides as a conjectural direction maketh a casual effect, so a particular and restrained direction is no less casual than an uncertain. For those particular means whereunto it is tied may be out of your power or may be accompanied with an overvalue of prejudice; and so if for want of certainty in direction you are frustrated in success, for want of variety in direction you are stopped in attempt. If therefore your direction be certain, it must refer you and point you to somewhat which, if it be present, the effect you seek will of necessity follow, else may you perform and not obtain. If it be free, then must it refer you to somewhat which if it be absent the effect you seek will of necessity withdraw, else may you have power and not attempt. This notion Aristotle had in light, though not in use. For the two commended rules by him set down, whereby the *axioms* of sciences are precepted to be made convertible, and which the latter men have not without elegancy surnamed the one the rule of truth because it preventeth deceit, the other the rule of prudence because it freeth election, are the same thing in speculation and affirmation which we now observe. An example will make my meaning attained, and yet percase make it thought that they attained it not. Let the effect to be produced be *Whiteness;* let the first direction be that if air and water be intermingled or broken in small portions together, whiteness will ensue, as in snow, in the breaking of the waves of the sea and rivers, and the like. This direction is certain, but very particular and restrained, being tied but to air and water. Let the second direction be, that if air be mingled as before with any transparent body, such nevertheless as is uncoloured and more grossly transparent than air itself, that then etc. as glass or crystal, being beaten to fine powder, by the interposition of the air becometh white; the white of an egg being clear of itself, receiving air by agitation becometh white, receiving air by concoction becometh white; here you are freed from water, and advanced to a clear body, and still tied to air. Let the third direction exclude or remove the restraint of an uncoloured body, as in amber, sapphires, etc. which beaten to fine powder become white; in wine and beer, which brought to froth become white. Let the fourth direction exclude the restraint of a body more grossly transparent than air, as in flame, being a body compounded between air and a finer substance than air; which flame if it were not for the smoke, which is the third substance that incorporateth itself and dyeth the flame, would be more perfect white. In all these four directions air still beareth a part. Let the fifth direction then be, that if any bodies, both transparent but in an unequal degree, be mingled as before, whiteness will follow; as oil and water beaten to an ointment, though by settling the air which gathereth in the agitation be evaporate, yet remaineth white; and the powder of glass or crystal put into water, whereby the air giveth place, yet remaineth white, though not so perfect. Now are you freed from air, but still you are tied to transparent bodies. To ascend further by scale I do forbear, partly because it would draw on the example to an over-great length, but chiefly because it would open that which in this work I determine to reserve; for to pass through the whole history and observation of colours and objects visible were too long a digression; and our purpose is now to give an example of a free direction, thereby to distinguish and describe it; and not to set down a form of interpretation how to recover and attain it. But as we intend not now to reveal, so we are circumspect not to mislead and therefore (this warning being given) returning to our purpose in hand, we admit the sixth direction to be, that all bodies or parts of bodies which are unequal equally, that is in a simple proportion, do represent whiteness [23]; we will explain this, though we induce it not. It is then to be understood, that absolute equality produceth transparence, inequality in simple order or proportion produceth whiteness, inequality in compound or respective order or proportion produceth all other colours, and absolute or orderless inequality produceth blackness; which diversity, if so gross a demonstration be needful, may be signified by four tables; a blank, a chequer, a fret, and a medley; whereof the fret is evident to admit great variety. Out of this assertion are satisfied a multitude of effects and observations, as that whiteness and blackness are most incompatible with transparence; that whiteness keepeth light, and blackness stoppeth light, but neither passeth it; that whiteness or blackness are never produced in rainbows, diamonds, crystals, and the like; that white giveth no dye, and black hardly taketh dye; that whiteness seemeth to have an affinity with dryness, and blackness with moisture; that adustion causeth blackness, and calcination whiteness; that flowers are generally of fresh colours, and rarely black, etc. All which I do now mention confusedly by way of derivation and not by way of induction. This sixth direction, which I have thus explained, is of good and competent liberty for whiteness fixed and inherent, but not for whiteness fantastical or appearing, as shall be afterwards touched. But first do you need a reduction back to certainty or verity; for it is not all position or contexture of unequal bodies that will produce colour; for *aqua fortis*, oil of *vitriol*, etc. more manifestly, and many other substances more obscurely, do consist of very unequal parts, which yet are transparent and clear. Therefore the reduction must be, that the bodies or parts of bodies so intermingled as before be of a certain grossness or magnitude; for the unequalities which move the sight must have a further dimension and quantity than those which operate many other effects. Some few grains of saffron will give a tincture to a ton of water; but so many grains of civet will give a perfume to a whole chamber of air. And therefore when Democritus (from whom Epicurus did borrow it) held that the position of the solid portions was the cause of colours, yet in the very truth of his assertion he should have added, that the portions are required to be of some magnitude. And this is one cause why colours have little inwardness and necessitude with the nature and proprieties of things, those things resembling in colour which otherwise differ most, as salt and sugar, and contrariwise differing in colour which otherwise resemble most, as the white and blue violets, and the several veins of one agate or marble, by reason that other virtues consist in more subtile proportions than colours do; and yet are there virtues and natures which require a grosser magnitude than colours, as well as scents and divers other require a more subtile; for as the portion of a body will give forth scent which is too small to be seen, so the portion of a body will shew colours which is too small to be endued with weight; and therefore one of the prophets with great elegancy describing how all creatures carry no proportion towards God the creator, saith, *That all the nations in respect of him are like the dust upon the balance*, which is a thing appeareth but weigheth not. But to return, there resteth a further freeing of this sixth direction; for the clearness of a river or stream sheweth white at a distance, and crystalline glasses deliver the face or any other object falsified in whiteness, and long beholding the snow to a weak eye giveth an impression of azure rather than of whiteness. So as for whiteness in apparition only and representation by the qualifying of the light, altering the *intermedium*, or affecting the eye itself, it reacheth not. But you must free your direction to the producing of such an incidence, impression, or operation, as may cause a precise and determinate passion of the eye; a matter which is much more easy to induce than that which we have passed through; but yet because it hath a full coherence both with that act of radiation (which hath hitherto been con-

23 Compare *De Aug.* iii. 4. "At in Metaphysicâ, si fiat inquisitio, hujusmodi quidpiam reperies; Corpora duo Diaphana intermixta, Portionibus eorum Opticis simplici ordine sive æqualiter collocatis, constituere Albedinem." And observe that this sentence is not to be found in the corresponding passage of the *Advancement of Learning*, but is interpolated in the translation.

ceived and termed so unproperly and untruly by some an effluxion of spiritual species and by others an investing of the *intermedium* with a motion which successively is conveyed to the eye) and with the act of sense, wherein I should likewise open that which I think good to withdraw, I will omit. Neither do I contend but that this motion which I call the freeing of a direction, in the received philosophies (as far as a swimming anticipation could take hold) might be perceived and discerned; being not much other matter than that which they did not only aim at in the two rules of *Axioms* before remembered, but more nearly also in [24] that which they term the form or formal cause, or that which they call the true difference; both which nevertheless it seemeth they propound rather as impossibilities and wishes than as things within the compass of human comprehension. For Plato casteth his burden and saith *that he will revere him as a God, that can truly divide and define* [25]; which cannot be but by true forms and differences. Wherein I join hands with him, confessing as much as yet assuming to myself little; for if any man can by the strength of his *anticipations* find our forms, I will magnify him with the foremost. But as any of them would say that if divers things which many men know by instruction and observation another knew by revelation and without those means, they would take him for somewhat supernatural and divine; so I do acknowledge that if any man can by anticipations reach to that which a weak and inferior wit may attain to by interpretation, he cannot receive too high a title. Nay I for my part do indeed admire to see how far some of them have proceeded by their *anticipations;* but how? it is as I wonder at some blind men, to see what shift they make without their eye-sight; thinking with myself that if I were blind I could hardly do it. Again, Aristotle's school confesseth that there is no true knowledge but by causes, no true cause but the form, no true form known except one, which they are pleased to allow; and therefore thus far their evidence standeth with us, that both hitherto there hath been nothing but a shadow of knowledge, and that we propound now that which is agreed to be worthiest to be sought, and hardest to be found. There wanteth now a part very necessary, not by way of supply but by way of caution; for as it is seen for the most part that the outward tokens and badges of excellency and perfection are more incident to things merely counterfeit than to that which is true, but for [26] a meaner and baser sort; as a dubline is more like a perfect ruby than a spinel, and a counterfeit angel is made more like a true angel than if it were an angel coined of China gold; in like manner the direction carrieth a resemblance of a true direction in verity and liberty which indeed is no direction at all. For though your direction seem to be certain and free by pointing you to a nature that is unseparable from the nature you inquire upon, yet if it do not carry you on a degree or remove nearer to action, operation, or light to make or produce, it is but superficial and counterfeit. Wherefore to secure and warrant what is a true direction, though that general note I have given be perspicuous in itself (for a man shall soon cast with himself whether he be ever the nearer [27] to effect and operate or no, or whether he have won but an abstract or varied notion) yet for better instruction I will deliver three particular notes of caution. The first is that the nature discovered be more original than the nature supposed, and not more secondary or of the like degree; as to make a stone bright or to make it smooth it is a good direction to say, make it even; but to make a stone even it is no good direction to say, make it bright or make it smooth; for the rule is that the disposition of any thing referring to the state of it in itself or the parts, is more original than that which is relative or transitive towards another thing. So evenness is the disposition of the stone in itself, but smooth is to the hand and bright to the eye, and yet nevertheless they all cluster and concur; and yet the direction is more unperfect, if it do appoint you to such a relative as is in the same kind and not in a diverse. For in the direction to produce brightness by smoothness, although properly it win no degree, and will never teach you any new particulars before unknown; yet by way of suggestion or bringing to mind it may draw your consideration to some particulars known but not remembered; as you shall sooner

[24] *than* in MS.
[25] See *Nov. Org.* ii. 26.
[26] So MS. qu. *of*?
[27] *neare* MS.

remember some practical means of making smoothness, than if you had fixed your consideration only upon brightness ; but if the direction had been to make brightness by making reflexion, as thus, make it such as you may see your face in it, this is merely secondary, and helpeth neither by way of informing nor by way of suggestion. So if in the inquiry of whiteness you were directed to make such a colour as should be seen furthest in a dark light ; here you are advanced nothing at all. For these kinds of natures are but proprieties, effects, circumstances, concurrences, or what else you shall like to call them, and not radical and formative natures towards the nature supposed. The second caution is that the nature inquired be collected by division before composition, or to speak more properly, by composition subaltern before you ascend to composition absolute, etc.[28].

Of the internal and profound errors and superstitions in the nature of the mind and of the four sorts of idols or fictions which offer themselves to the understanding in the inquisition of knowledge ; being the 16th chapter, and this a small fragment thereof, being a preface to the inward elenches of the mind[29].

The opinion of Epicurus that the gods were of human shape, was rather justly derided than seriously confuted by the other sects, demanding whether every kind of sensible creatures did not think their own figure fairest, as the horse, the bull, and the like, which found no beauty but in their own forms, as in appetite of lust appeared. And the heresy of the Anthropomorphites was ever censured for a gross conceit bred in the obscure cells of solitary monks that never looked abroad. Again the fable so well known of *Quis pinxit leonem*, doth set forth well that there is an error of pride and partiality, as well as of custom and familiarity. The reflexion also from glasses so usually resembled to the imagery of the mind, every man knoweth to receive error and variety both in colour, magnitude and shape, according to the quality of the glass. But yet no use hath been made of these and many the like observations, to move men to search out and upon search to give true cautions of the native and inherent errors in the mind of man which have coloured and corrupted all his notions and impressions.

I do find therefore in this enchanted glass four Idols or false appearances of several and distinct sorts, every sort comprehending many subdivisions : the first sort, I call Idols of the *Nation* or *Tribe ;* the second, idols of the *Palace ;* the third, idols of the *Cave ;* and the fourth, idols of the *Theatre*, etc.[30].

Here followeth an abridgment of divers chapters of the first book of Interpretation of Nature [31].

CAP. 12.

That in deciding and determining of the truth of knowledge, men have put themselves upon trials not competent. That antiquity and authority ; common and confessed notions ; natural and yielding consent of the mind ; the harmony and coherence of a knowledge in itself ; the establishing of principles with the touch and reduction of other propositions unto them ; inductions without instances contradictory ; and the report of the senses ; are none of them absolute and infallible evidence of truth, and bring no security sufficient for effects and operations. That the discovery of new works and active directions not known before, is the only trial to be accepted of ; and yet not that neither, in case where one particular giveth light to another ; but where particulars induce an axiom of observation, which axiom found out discovereth and designeth new particulars. That the nature of this trial is not only upon the point, whether the knowledge be

28 The word " subaltern " (for which a blank was left by the transcriber) and the " etc." have been inserted by Bacon. The chapter ends nearly at the bottom of the page.

29 The words in Roman character have been added by Bacon.

30 The " etc." in Bacon's hand. The chapter ends in the middle of the second page, and the heading of the next (which is the 4th), follows immediately ; whence I infer that the whole formed part of the original transcript.

31 The words " Interpretation of Nature " added in Bacon's hand.

profitable or no, but even upon the point whether the knowledge be true or no; not because you may always conclude that the Axiom which discovereth new instances is true, but contrariwise you may safely conclude that if it discover not any new instance it is in vain and untrue. That by new instances are not always to be understood new recipes but new assignations, and of the diversity between these two. That the subtilty of words, arguments, notions, yea of the senses themselves, is but rude and gross in comparison of the subtilty of things; and of the slothful and flattering opinions of those which pretend to honour the mind of man in withdrawing and abstracting it from particulars, and of the inducements and motives whereupon such opinions have been conceived and received.

Cap. 13.

Of the error in propounding chiefly the search of causes and productions of things concrete, which are infinite and transitory, and not of abstract natures, which are few and permanent. That these natures are as the alphabet or simple letters, whereof the variety of things consisteth; or as the colours mingled in the painter's shell, wherewith he is able to make infinite variety of faces or shapes [32]. An enumeration of them according to popular note. That at the first one would conceive that in the schools by natural philosophy were meant the knowledge of the efficients of things concrete; and by metaphysic the knowledge of the forms of natures simple; which is a good and fit division of knowledge: but upon examination there is no such matter by them intended. That the little inquiry into the production of simple natures sheweth well that works were not sought; because by the former knowledge some small and superficial deflexions from the ordinary generations and productions may be found out, but the discovery of all profound and radical alteration must arise out of the latter knowledge.

Cap. 14.

Of the error in propounding the search of the materials or dead beginnings of principles of things, and not the nature of motions, inclinations, and applications That the whole scope of the former search is mpertinent and vain; both because there are no such beginnings, and if there were they could not be known. That the latter manner of search (which is all) they pass over compendiously and slightly as a by-matter. That the several conceits in that kind, as that the lively and moving beginnings of things should be shift or appetite of matter to privation; the spirit of the world working in matter according to platform: the proceeding or fructifying of distinct kinds according to their proprieties; the intercourse of the elements by mediation of their common qualities; the appetite of like portions to unite themselves; amity and discord, or sympathy and antipathy; motion to the centre, with motion of stripe or press; the casual agitation, aggregation, and essays of the solid portions in the void space; motion of shuttings and openings; are all mere nugations; and that the calculating and ordination of the true degrees, moments, limits, and laws of motions and alterations (by means whereof all works and effects are produced), is a matter of a far other nature than to consist in such easy and wild generalities.

Cap. 15.

Of the great error of inquiring knowledge in Anticipations. That I call Anticipations the voluntary collections that the mind maketh of knowledge, which is every man's reason. That though this be a solemn thing, and serves the turn to negotiate between man and man (because of the conformity and participation of men's minds in the like errors), yet towards inquiry of the truth of things and works it is of no value. That civil respects are a lett that this pretended reason should not be so contemptibly spoken of as were fit and medicinable, in regard that [33] hath been too much exalted and glorified, to the

[32] This last illustration is added in the margin in Bacon's hand.

[33] So MS. by mistake probably for *it*; the transcriber taking *yt* for *y^t*.

infinite detriment of man's estate. Of the nature of words and their facility and aptness to cover and grace the defects of Anticipations. That it is no marvel if these Anticipations have brought forth such diversity and repugnance in opinions, theories, or philosophies, as so many fables [34] of several arguments. That had not the nature of civil customs and government been in most times somewhat adverse to such innovations, though contemplative, there might have been and would have been many more. That the second school of the Academics and the sect of Pyrrho, or the considerers that denied comprehension, as to the disabling of man's knowledge (entertained in Anticipations) is well to be allowed, but that they ought when they had overthrown and purged the floor of the ruins to have sought to build better in place. And more especially that they did unjustly and prejudicially to charge the deceit upon the report of the senses, which admitteth very sparing remedy; being indeed to have been charged upon the Anticipations of the mind, which admitteth a perfect remedy. That the information of the senses is sufficient, not because they err not, but because the use of the sense in discovering of knowledge is for the most part not immediate. So that it is the work, effect, or instance, that trieth the Axiom, and the sense doth but try the work done or not done, being or not being. That the mind of man in collecting knowledge needeth great variety of helps, as well as the hand of man in manual and mechanical practices needeth great variety of instruments. And that it were a poor work that, if instruments were removed, men would overcome with their naked hands. And of the distinct points of want and insufficiency in the mind of man.

Cap. 16.

That the mind of a man, as it is not a vessel of that content or receipt to comprehend knowledge without helps and supplies, so again it is not sincere, but of an ill and corrupt tincture. Of the inherent and profound errors and superstitions in the nature of the mind, and of the four sorts of Idols or false appearances that offer themselves to the understanding in the inquisition of knowledge; that is to say, the Idols of the Tribe, the Idols of the Palace, the Idols of the Cave, and the Idols of the Theatre. That these four, added to the incapacity of the mind and the vanity and malignity of the affections, leave nothing but impotency and confusion. A recital of the particular kinds of these four Idols, with some chosen examples of the opinions they have begot, such of them as have supplanted the state of knowledge most.

Cap. 17.

Of the errors of such as have descended and applied themselves to experience, and attempted to induce knowledge upon particulars. That they have not had the resolution and strength of mind to free themselves wholly from Anticipations, but have made a confusion and intermixture of Anticipations and observations, and so vanished. That if any have had the strength of mind generally to purge away and discharge all Anticipations, they have not had the greater and double strength and patience of mind, as well to repel new Anticipations after the view and search of particulars, as to reject old which were in their mind before; but have from particulars and history flown up to principles without the mean degrees, and so framed all the middle generalities or axioms, not by way of scale or ascension from particulars, but by way of derivation from principles; whence hath issued the infinite chaos of shadows and notions [35], wherewith both books and minds have [36] been hitherto, and may be yet hereafter much more pestered. That in the course of those derivations, to make them yet the more unprofitable, they have used when any light of new instance opposite to any assertion appeared, rather to reconcile the instance than to amend the rule. That if any have had

[34] *fable* in MS.

[35] This word is written between the lines in Bacon's hand, and I am not sure that I read it right. Stephens read it *moths*, which is certainly wrong. It is more like *nocons* than any word I can think of.

[36] *hath* in MS.

or shall have the power and resolution to fortify and inclose his mind against all Anticipations, yet if he have not been or shall not be cautioned by the full understanding of the nature of the mind and spirit of man, and therein of the seats pores and passages both of knowledge and error, he hath not been nor shall not be possibly able to guide or keep on his course aright. That those that have been conversant in experience and observation have used, when they have intended to discover the cause of any effect, to fix their consideration narrowly and exactly upon that effect itself with all the circumstances thereof, and to vary the trial thereof as many ways as can be devised; which course amounteth but to a tedious curiosity, and ever breaketh off in wondering and not in knowing ; and that they have not used to enlarge their observation to match and sort that effect with instances of a diverse subject, which [37] must of necessity be before any cause be found out. That they have passed over the observation of instances vulgar and ignoble, and stayed their attention chiefly upon instances of mark ; whereas the other sort are for the most part more significant and of better light and information. That every particular that worketh any effect is a thing compounded (more or less) of diverse single natures (more manifest and more obscure), and that it appeareth not to whether of the natures the effect is to be ascribed, and yet notwithstanding they have taken a course without breaking particulars and reducing them by exclusions and inclusions to a definite point, to conclude upon inductions in gross, which empirical course is no less vain than the scholastical. That all such as have sought action and work out of their inquiry have been hasty and pressing to discover some practices for present use, and not to discover Axioms, joining with them the new assignations as their sureties. That the forerunning of the mind to frame recipes upon Axioms at the entrance is like Atalanta's golden ball that hindereth and interrupteth the course, and is to be inhibited till you have ascended to a certain stage and degree of generalities ; which forbearance will be liberally recompensed in the end ; and that chance discovereth new inventions by one and one, but science by knots and clusters. That they have not collected sufficient quantity of particulars, nor them in sufficient certainty and subtilty, nor of all several kinds, nor with those advantages and discretions in the entry and sorting which are requisite ; and of the weak manner of collecting natural history which hath been used. Lastly that they had no knowledge of the formulary of interpretation, the work whereof is to abridge experience and to make things as certainly found out by Axiom in short time, as by infinite experience in ages.

Cap. 18.

That the cautels and devices put in practice in the delivery of knowledge for the covering and palliating of ignorance, and the gracing and overvaluing of that they utter, are without number ; but none more bold and more hurtful than two ; the one that men have used of a few observations upon any subject to make a solemn and formal art, by filling it up with discourse, accommodating it with some circumstances and directions to practice, and digesting it into method, whereby men grow satisfied and secure, as if no more inquiry were to be made of that matter ; the other, that men have used to discharge ignorance with credit, in defining all those effects which they cannot attain unto to be out of the compass of art and human endeavour. That the very styles and forms of utterance are so many characters of imposture, some choosing a style of pugnacity and contention, some of satire and reprehension, some of plausible and tempting similitudes and examples, some of great words and high discourse, some of short and dark sentences, some of exactness of method, all of positive affirmation, without disclosing the true motives and proofs of their opinions, or free confessing their ignorance or doubts, except it be now and then for a grace, and in cunning to win the more credit in the rest, and not in good faith. That although men be free from these errors and incumbrances in the will and affection, yet it is

[37] The words " according to their own rules " follow in the MS., but a line is drawn through them.

not a thing so easy as is conceived to convey the conceit of one man's mind into the mind of another without loss or mistaking, specially in notions new and differing from those that are received. That never any knowledge was delivered in the same order it was invented, no not in the mathematic, though it should seem otherwise in regard that the propositions placed last do use the propositions or grants placed first for their proof and demonstration. That there are forms and methods of tradition wholly distinct and differing, according to their ends whereto they are directed. That there are two ends of tradition of knowledge, the one to teach and instruct for use and practice, the other to impart or intimate for re-examination and progression. That the former of these ends requireth a method not the same whereby it was invented and induced, but such as is most compendious and ready whereby it may be used and applied. That the latter of the ends, which is where a knowledge is delivered to be continued and spun on by a succession of labours, requireth a method whereby it may be transposed to another in the same manner as it was collected, to the end it may be discerned both where the work is weak, and where it breaketh off. That this latter method is not only unfit for the former end, but also impossible for all knowledge gathered and insinuated by Anticipations, because the mind working inwardly of itself, no man can give a just account how he came to that knowledge which he hath received, and that therefore this method is peculiar for knowledge gathered by interpretation. That the discretion anciently observed, though by the precedent of many vain persons and deceivers disgraced, of publishing part, and reserving part to a private succession, and of publishing in a manner whereby it shall not be to the capacity nor taste of all, but shall as it were single and adopt his reader, is not to be laid aside, both for the avoiding of abuse in the excluded, and the strengthening of affection in the admitted. That there are other virtues of tradition, as that there be no occasion given to error, and that it carry a vigour to root and spread against the vanity and injuries of time; all which if they were ever due to any knowledge delivered, or if they were never due to any human knowledge heretofore delivered, yet are now due to the knowledge propounded.

Cap. 19.

Of the impediments which have been in the affections, the principle whereof hath been despair or diffidence, and the strong apprehension of the difficulty, obscurity, and infiniteness which belongeth to the invention of knowledge, and that men have not known their own strength, and that the supposed difficulties and vastness of the work is rather in shew and muster than in state or substance where the true way is taken. That this diffidence hath moved and caused some never to enter into search, and others when they have been entered either to give over or to seek a more compendious course than can stand with the nature of true search. That of those that have refused and prejudged inquiry, the more sober and grave sort of wits have depended upon authors and traditions, and the more vain and credulous resorted to revelation and intelligence with spirits and higher natures. That of those that have entered into search, some having fallen upon some conceits which they after consider to be the same which they have found in former authors, have suddenly taken a persuasion that a man shall but with much labour incur and light upon the same inventions which he might with ease receive from others; and that it is but a vanity and self-pleasing of the wit to go about again, as one that would rather have a flower of his own gathering, than much better gathered to his hand. That the same humour of sloth and diffidence suggesteth that a man shall but revive some ancient opinion which was long ago propounded, examined, and rejected. And that it is easy to err in conceit that a man's observation or notion is the same with a former opinion, both because new conceits must of necessity be uttered in old words, and because [38] upon true and erroneous grounds men may meet in consequence or

[38] A parenthesis " (as the Schools well know) " which follows here, has a line drawn through it.

conclusion, as several lines or circles that cut in some one point. That the greatest part of those that have descended into search have chosen for the most artificial and compendious course to induce principles out of particulars, and to reduce all other propositions unto principles ; and so instead of the nearest way, have been led to no way or a mere labyrinth. That the two contemplative ways have some resemblance with the old parable of the two moral ways, the one beginning with incertainty and difficulty and ending in plainness and certainty, and the other beginning with shew of plainness and certainty, and ending in difficulty and incertainty. Of the great and manifest error and untrue conceit or estimation of the infiniteness of particulars, whereas indeed all prolixity is in discourse and derivations ; and of the infinite and most laborious expence of wit that hath been employed upon toys and matters of no fruit or value. That although the period of one age cannot advance men to the furthest point of interpretation of nature (except the work should be undertaken with greater helps than can be expected,) yet it cannot fail in much less space of time to make return of many singular commodities towards the state and occasions of man's life. That there is less reason of distrust in the course of interpretation now propounded than in any knowledge formerly delivered, because this course doth in sort equal men's wits, and leaveth no great advantage or pre-eminence to the perfect and excellent motions of the spirit. That to draw a straight line or to make a circle perfect round by aim of hand only, there must be a great difference between an unsteady and unpractised hand and a steady and practised, but to do it by rule or compass it is much alike.

Cap. 21.

Of the impediments which have been in the two extreme humours of admiration of antiquity and love of novelty, and again of over-servile reverence or over light scorn of the opinions of others.

Cap. 22.

Of the impediments which have been in the affection of pride, specially of one kind, which is the disdain of dwelling and being conversant much in experiences and particulars, specially such as are vulgar in occurrency, and base and ignoble in use. That besides certain higher mysteries of pride, generalities seem to have a dignity and solemnity, in that they do not put men in mind of their familiar actions, in that they have less affinity with arts mechanical and illiberal, in that they are not so subject to be controuled by persons of mean observation, in that they seem to teach men that they know not, and not to refer them to that they know. All which conditions directly feeding the humour of pride, particulars do want. That the majesty of generalities and the divine nature of the mind in taking them (if they be truly collected, and be indeed the direct reflexions of things), cannot be too much magnified. And that it is true that interpretation is the very natural and direct intention, action, and progression of the understanding delivered from impediments. And that all Anticipation is but a deflexion or declination by accident.

Cap. 25.

Of the impediments which have been in the state of heathen religion and other superstitions and errors of religion. And that in the true religion there hath not [39] nor is any impediment, except it be by accident or intermixture of humour. That a religion which consisteth in rites and forms of adoration, and not in confessions and beliefs, is adverse to knowledge ; because men having liberty to inquire and discourse of Theology at pleasure, it cometh to pass that all inquisition of nature endeth and limiteth itself in such metaphysical or theological discourse ; whereas if men's wits be shut out of that part, it turneth them again to discover and so to seek reason of reason more deeply. And that such was the religion of the Heathen. That a religion that is jealous of the variety of learning,

[39] So MS.

discourse, opinions, and sects, (as misdoubting it may shake the foundations), or that cherisheth devotion upon simplicity and ignorance, as ascribing ordinary effects to the immediate working of God, is adverse to knowledge. That such is the religion of the Turk, and such hath been the abuse of Christian religion at some several times, and in some several factions. And of the singular advantage which the Christian religion hath towards the furtherance of true knowledge, in that it excludeth and interdicteth human reason, whether by interpretation or anticipation, from examining or discussing of the mysteries and principles of faith.

CAP. 26.

Of the impediments which have been in the nature of society and the policies of state. That there is no composition of estate or society, nor order or quality of persons, which have not some point of contrariety towards true knowledge. That monarchies incline wits to profit and pleasure, and commonwealths to glory and vanity. That universities incline wits to sophistry and affectation, cloisters to fables and unprofitable subtilty, study at large to variety ; and that it is hard to say, whether mixture of contemplations with an active life, or retiring wholly to contemplations, do disable and hinder the mind more.

FILUM LABYRINTHI, SIVE FORMULA INQUISITIONIS

PREFACE.

THE following fragment was first printed in Stephens's second collection (1734), from a manuscript belonging to Lord Oxford, which is now in the British Museum (Harl. MSS. 6797. fo. 139). As far as it goes, it agrees so nearly with the *Cogitata et Visa* that either might be taken for a free translation of the other, with a few additions and omissions. But I think the English was written first; probably at the time when the idea first occurred to Bacon of drawing attention to his doctrine by exhibiting a specimen of the process and the result in one or two particular cases. The *Cogitata et Visa* professes to be merely a preface framed to prepare the way for an example of a legitimate philosophical investigation proceeding regularly by Tables. Such an example, or at least the plan and skeleton of it, will be found further on,[1] with the title *Filum Labyrinthi, sive Inquisitio legitima de Motu;* and the title prefixed to this fragment is most easily explained by supposing that a specimen of an *Inquisitio legitima* was meant to be included in it.

It is here printed from the original MS. which is a fair copy in the hand of one of Bacon's servants, carefully corrected in his own.

J. S.

[[1] A Latin treatise, not given in this reprint.—ED.]

AD FILIOS [1].

PARS PRIMA.

1. FRANCIS BACON thought in this manner. The knowledge whereof the world is now possessed, especially that of nature, extendeth not to magnitude and certainty of works. The Physician pronounceth many diseases incurable, and faileth oft in the rest. The Alchemists wax old and die in hopes. The Magicians perform nothing that is permanent and profitable. The Mechanics take small light from natural philosophy, and do but spin on their own little threads. Chance sometimes discovereth inventions; but that worketh not in years, but ages. So he saw well, that the inventions known are very unperfect; and that new are not like to be brought to light but in great length of time; and that those which are, came not to light by philosophy.

2. He thought also this state of knowledge was the worse, because men strive (against themselves) to save the credit of ignorance, and to satisfy themselves in this poverty. For the Physician, besides his cauteles of practice, hath this general cautele of art, that he dischargeth the weakness of his art upon supposed impossibilities: neither can his art be condemned, when itself judgeth. That philosophy also, out of which the knowledge of physic, which now is in use, is hewed, receiveth certain positions and opinions, which (if they be well weighed) induce this persuasion, that no great works are to be expected from art, and the hand of man; as in particular that opinion, *that the heat of the sun and fire differ in kind;* and that other, *that Composition is the work of man, and Mixture is the work of nature*, and the like; all tending to the circumscription of man's power, and to artificial despair; killing in men, not only the comfort of [2] imagination but the industry of trial; only upon vain glory to have their art thought perfect, and that all is impossible that is not already found. The Alchemist dischargeth his art upon his own errors, either supposing a misunderstanding of the words of his authors, which maketh him listen after auricular traditions; or else a failing in the true proportions and scruples of practice, which maketh him renew infinitely his trials; and finding also that he lighteth upon some mean experiments and conclusions by the way, feedeth upon them, and magnifieth them to the most, and supplieth the rest in hopes. The Magician, when he findeth something (as he conceiveth) above nature effected, thinketh, when a breach is once made in nature, that it is all one to perform great things and small; not seeing that they are but subjects of a certain kind, wherein magic and superstition hath played in all times. The Mechanical person, if he can refine an invention, or put two or three observations or practices together in one, or couple things better with their use, or make the work in less or greater volume, taketh himself for an inventor. So he saw well, that men either persuade themselves of new inventions as of impossibilities; or else think they are already extant, but in secret and in few hands; or that they account of those little industries and additions, as of inventions: all which turneth to the averting of their minds from any just and constant labour to invent further in any quantity.

3. He thought also, when men did set before themselves the variety and perfection of works produced by mechanical arts, they are apt rather to admire the provisions of man, than to apprehend his wants; not considering, that the original inventions and conclusions of nature which are the life of all that variety, are not many nor deeply fetched; and that the rest is but the subtile and ruled

[1] This is written at the top of the page, in the left-hand corner, in Bacon's hand.

[2] *of* is omitted in the MS.

motion of the instrument and hand; and that the shop therein is not unlike the library, which in such number of books containeth (for the far greater part) nothing but iterations, varied sometimes in form, but not new in substance. So he saw plainly, that opinion of store was a cause of want; and that both works and doctrines appear many and are few.

4. He thought also, that knowledge is uttered to men, in a form as if every thing were finished; for it is reduced into arts and methods, which in their divisions do seem to include all that may be. And how weakly soever the parts are filled, yet they carry the show and reason of a total; and thereby the writings of some received authors go for the very art: whereas antiquity used to deliver the knowledge which the mind of man had gathered, in observations, aphorisms, or short and dispersed sentences, or small tractates of some parts that they had diligently meditated and laboured; which did invite men, both to ponder that which was invented, and to add and supply further. But now sciences are delivered to be believed and accepted, and not to be examined and further discovered; and the succession is between master and disciple, and not between inventor and continuer or advancer: and therefore sciences stand at a stay, and have done for many ages, and that which is positive is fixed, and that which is question is kept question, so as the columns of no further proceeding are pitched. And therefore he saw plainly, men had cut themselves off from further invention; and that it is no marvel that that is not obtained, which hath not been attempted, but rather shut out and debarred.

5. He thought also, that knowledge is almost generally sought either for delight and satisfaction, or for gain and profession, or for credit and ornament, and that every one of these are as Atalanta's balls, which hinder the race of invention. For men are so far in these courses from seeking to increase the mass of knowledge, as of that mass which is they will take no more than will serve their turn: and if any one amongst so many seeketh knowledge for itself, yet he rather seeketh to know the variety of things, than to discern of the truth and causes of them; and if his inquisition be yet more severe, yet it tendeth rather to judgment than to invention; and rather to discover truth in controversy than new matter; and if his heart be so large as he propoundeth to himself further discovery or invention, yet it is rather of new discourse and speculation of causes, than of effects and operations: and as for those that have so much in their mouths, action and use and practice and the referring of sciences thereunto, they mean it of application of that which is known, and not of a discovery of that which is unknown. So he saw plainly, that this mark, namely invention of further means to endow the condition and life of man with new powers or works, was almost never yet set up and resolved in man's intention and inquiry.

6. He thought also, that, amongst other knowledges, natural philosophy hath been the least followed and laboured. For since the Christian faith, the greatest number of wits have been employed, and the greatest helps and rewards have been converted upon divinity. And beforetime likewise, the greatest part of the studies of philosophers was consumed in moral philosophy, which was as the heathen divinity. And in both times a great part of the best wits betook themselves to law, pleadings, and causes of estate; specially in the time of the greatness of the Romans, who by reason of their large empire needed the service of all their able men for civil business. And the time amongst the Grecians in which natural philosophy seemed most to flourish, was but a short space; and that also rather abused in differing sects and conflicts of opinions, than profitably spent: since which time, natural philosophy was never any profession, nor never possessed any whole man, except perchance some monk in a cloister, or some gentleman in the country, and that very rarely; but became a science of passage, to season a little young and unripe wits, and to serve for an introduction to other arts, specially physic and the practical mathematics. So as he saw plainly, that natural philosophy hath been intended by few persons, and in them hath occupied the least part of their time, and that in the weakest of their age and judgment.

7. He thought also, how great opposition and prejudice natural philosophy had received by superstition, and the immoderate and blind zeal of religion; for

he found that some of the Grecians which first gave the reason of thunder, had been condemned of impiety; and that the cosmographers which first discovered and described the roundness of the earth, and the consequence thereof touching the *Antipodes*, were not much otherwise censured by the ancient fathers of the Christian Church; and that the case is now much worse, in regard of the boldness of the schoolmen and their dependances in the monasteries, who having made divinity into an art, have almost incorporated the contentious philosophy of Aristotle into the body of Christian religion. And generally he perceived in men of devout simplicity, this opinion, that the secrets of nature were the secrets of God and part of that glory whereinto the mind of man if it seek to press shall be oppressed; and that the desire in men to attain to so great and hidden knowledge, hath a resemblance with that temptation which caused the original fall: and on the other side in men of a devout policy, he noted an inclination to have the people depend upon God the more, when they are less acquainted with second causes; and to have no stirring in philosophy, lest it may lead to an innovation in divinity, or else should discover matter of further contradiction to divinity. But in this part resorting to the authority of the Scriptures, and holy examples, and to reason, he rested not satisfied alone, but much confirmed. For first he considered that the knowledge of nature, by the light whereof man discerned of every living creature, and imposed names according to their propriety, was not the occasion of the fall; but the moral knowledge of good and evil, affected to the end to depend no more upon God's commandments, but for man to direct himself; neither could he find in any Scripture, that the inquiry and science of man in any thing, under the mysteries of the Deity, is determined and restrained, but contrariwise allowed and provoked; for concerning all other knowledge teh Scripture pronounceth, *That it is the glory of God to conceal, but it is the glory of man* (*or of the king*, for the king is but the excellency of man) *to invent;* and again, *The spirit of man is as the lamp of God, wherewith he searcheth every secret;* and again most effectually, *That God hath made all things beautiful and decent, according to the return of their seasons; also that he hath set the world in man's heart, and yet man cannot find out the work which God worketh from the beginning to the end;* shewing that the heart of man is a continent of that concave or capacity, wherein the content of the world (that is, all forms of the creatures and whatsoever is not God) may be placed or received; and complaining that through the variety of things and vicissitudes of times (which are but impediments and not impuissances) man cannot accomplish his invention. In precedent also he set before his eyes, that in those few memorials before the flood, the Scripture honoureth the name of the inventors of music and works in metal; that Moses had this addition of praise, that he was seen in all the learning of the Egyptians; that Solomon [3], in his grant of wisdom from God, had contained as a branch thereof, that knowledge whereby he wrote a natural history of all verdor, from the cedar to the moss, and of all that breatheth; that the book of Job, and many places of the prophets, have great aspersion of natural philosophy; that the Church in the bosom and lap thereof, in the greatest injuries of times, ever preserved (as holy relics) the books of philosophy and all heathen learning; and that when Gregory the bishop of Rome became adverse and unjust to the memory of heathen antiquity, it was censured for pusillanimity in him, and the honour thereof soon after restored, and his own memory almost persecuted by his successor Sabinian; and lastly in our times and the ages of our fathers, when Luther and the divines of the Protestant Church on the one side, and the Jesuits on the other, have enterprised to reform, the one the doctrine, the other the discipline and manners of the Church of Rome, he saw well how both of them have awaked to their great honour and succour all human learning. And for reason, there cannot be a greater and more evident than this; that all knowledge and specially that of natural philosophy tendeth highly to the magnifying of the glory of God in his power, providence, and benefits; appearing and engraven in his works, which without this knowledge are beheld but as through a veil; for if the heavens in the body of them do declare the glory of God to the eye, much more do they

[3] So spelt in MS

in the rule and decrees of them declare it to the understanding. And another reason not inferior to this is, that the same natural philosophy principally amongst all other human knowledge doth give an excellent defence against both extremes of religion, superstition and infidelity ; for both it freeth the mind from a number of weak fancies and imaginations, and it raiseth the mind to acknowledge that to God all things are possible : for to that purpose speaketh our Saviour in that first canon against heresies delivered upon the case of the resurrection, *You err, not knowing the Scriptures, nor the power of God ;* teaching that there are but two fountains of heresy, not knowing the will of God revealed in the Scriptures, and not knowing the power of God revealed or at least made most sensible in his creatures. So as he saw well, that natural philosophy was of excellent use to the exaltation of the Divine Majesty ; and that which is admirable, that being a remedy of superstition, it is nevertheless an help to faith. He saw likewise, that the former opinions to the prejudice thereof had no true ground ; but must spring either out of mere ignorance, or out of an excess of devotion, to have divinity all in all, whereas it should be only above all (both which states of mind may be best pardoned) ; or else out of worse causes, namely out of envy, which is proud weakness and deserveth to be despised ; or out of some mixture of imposture, to tell a lie for God's cause ; or out of an impious diffidence, as if men should fear to discover some things in nature which might subvert faith. But still he saw well, howsoever these opinions are in right reason reproved, yet they leave not to be most effectual hindrances to natural philosophy and invention.

8. He thought also, that there wanted not great contrariety to the further discovery of sciences, in regard of the orders and customs of universities, and also in regard of common opinion. For in universities and colleges men's studies are almost confined to certain authors, from which if any dissenteth or propoundeth matter of redargution, it is enough to make him thought a person turbulent; whereas if it be well advised, there is a great difference to be made between matter contemplative and active. For in government change is suspected, though to the better ; but it is natural to arts to be in perpetual agitation and growth ; neither is the danger alike of new light, and of new motion or remove. And for vulgar and received opinions, nothing is more usual nor more usually complained of, than that it is imposed [4] for arrogancy and presumption for men to authorise themselves against antiquity and authors, towards whom envy is ceased, and reverence by time amortised ; it not being considered what Aristotle himself did (upon whom the philosophy that now is chiefly dependeth) ; who came with a professed contradiction to all the world, and did put all his opinions upon his own authority and argument, and never so much as nameth an author but to confute and reprove him ; and yet his success well fulfilled the observation of Him that said, *If a man come in his own name, him will you receive.* Men think likewise that if they should give themselves to the liberty of invention and travail of inquiry, that they shall light again upon some conceits and contemplations which have been formerly offered to the world, and have been put down by better, which have prevailed and brought them to oblivion ; not seeing that howsoever the property and breeding of knowledges is in great and excellent wits, yet the estimation and price of them is in the multitude, or in the inclinations of princes and great persons meanly learned. So as those knowledges are like to be received and honoured, which have their foundation in the subtility or finest trial of common sense, or such as fill the imagination ; and not such knowledge as is digged out of the hard mine of history and experience, and falleth out to be in some points as adverse to common sense or popular reason, as religion, or more. Which kind of knowledge, except it be delivered with strange advantages of eloquence and power, may be likely to appear and disclose a little to the world and straight to vanish and shut again. So that time seemeth to be of the nature of a river or flood, that bringeth down to us that which is light and blown up, and sinketh and drowneth that which is solid and grave. So he saw well, that both in the state of religion, and in the administration of learning, and in common opinion, there were many and continual stops and traverses to the course of invention.

[4] So MS. : a miscopy, I suspect, for *imputed.*

9. He thought also, that the invention of works and further possibility was prejudiced in a more special manner than that of speculative truth ; for besides the impediments common to both, it hath by itself been notably hurt and discredited by the vain promises and pretences of Alchemy, Magic, Astrology, and such other arts, which (as they now pass) hold much more of imagination and belief than of sense and demonstration. But to use the poet's language, men ought to have remembered that although Ixion of a cloud in the likeness of Juno begat Centaurs and Chimæras, yet Jupiter also of the true Juno begat Vulcan and Hebe. Neither is it just to deny credit to the greatness of the acts of Alexander, because the like or more strange have been feigned of an Amadis or an Arthur, or other fabulous worthies. But though this in true reason should be, and that men ought not to make a confusion of unbelief ; yet he saw well it could not otherwise be in event, but that experience of untruth had made access to truth more difficult, and that the ignominy of vanity had abated all greatness of mind.

10. He thought also, there was found in the mind of man an affection naturally bred, and fortified and furthered by discourse and doctrine, which did pervert the true proceeding towards active and operative knowledge. This was a false estimation, that it should be as a diminution to the mind of man to be much conversant in experiences and particulars subject to sense and bound in matter, and which are laborious to search, ignoble to meditate, harsh to deliver, illiberal to practise, infinite as is supposed in number, and no ways accommodate to the glory of arts. This opinion or state of mind received much credit and strength by the school of Plato, who thinking that particulars rather revived the notions or excited the faculties of the mind, than merely informed ; and having mingled his philosophy with superstition, which never favoureth the sense ; extolleth too much the understanding of man in the inward light thereof. And again Aristotle's school, which giveth the due to the sense in assertion, denieth it in practice much more than that of Plato. For we see the schoolmen, Aristotle's succession, which were utterly ignorant of history, rested only upon agitation of wit ; whereas Plato giveth good example of inquiry by induction and view of particulars ; though in such a wandering manner as is of no force or fruit. So that he saw well, that the supposition of the sufficiency of man's mind hath lost the means thereof [5].

[5] Here the MS. ends abruptly in the middle of the page. At the top is written in Bacon's hand " The English as much as was parfited ". The blank part of the last page seems to have formed the outside of a miscellaneous bundle, and bears the following docket, also in Bacon's hand, " Severall fragments of discourses ".

PREFACE TO THE NOVUM ORGANUM

BY ROBERT LESLIE ELLIS.[1]

THE *Novum Organum* was published in 1620. Certain prolegomena to the whole of the Instauratio were prefixed to it, namely a Proœmium beginning " Franciscus de Verulamio sic cogitavit," a dedication to King James, a general preface, and an account, entitled Distributio Operis, of the parts of which the Instauratio was to consist. Of these the *Novum Organum* is the second ; the *De Augmentis*, which was not then published, occupying the place of the first. Accordingly in most editions of Bacon's works the prolegomena are prefixed, not to the *Novum Organum*, but to the *De Augmentis ;* and this is doubtless their natural place. Nevertheless as Bacon's general design was not completed, it seems better to allow them to remain in their original position, especially as in the Proœmium Bacon explains why he publishes one portion of the Instauratio apart from the rest. " Decrevit," he there says, speaking of himself, " prima quæque quæ perficere licuit in publicum edere. Neque hæc festinatio ambitiosa fuit, sed sollicita, ut si quid illi humanitus accideret, exstaret tamen designatio quædam ac destinatio rei quam animo complexus est," etc.

After the Proœmium and the dedication we come to the Præfatio Generalis, in which Bacon speaks of the unprosperous state of knowledge and of the necessity of a new method ; and then follows the Distributio Operis. The Instauratio is to be divided into six portions, of which the first is to contain a general survey of the present state of knowledge. In the second men are to be taught how to use their understanding aright in the investigation of Nature. In the third all the phenomena of the universe are to be stored up as in a treasure-house, as the materials on which the new method is to be employed. In the fourth examples are to be given of its operation and of the results to which it leads. The fifth is to contain what Bacon had accomplished in natural philosophy without the aid of his own method, but merely " ex eodem intellectûs usu quem alii in inquirendo et inveniendo adhibere consueverunt ". It is therefore less

[1] Mr. Ellis's preface to the *Novum Organum* was written when he was travelling abroad and had not his books of reference about him. He was at work upon it the night he was taken ill at Mentone, and was not afterwards able either to finish or to revise it. I have added a page or two at the end, by which the analysis of the first book is completed. Of the second book it was not necessary to say anything ; the subject of it being Bacon's *method*, which has been fully discussed in the General Preface. A few bibliographical inaccuracies of little consequence in themselves I have corrected, either in notes or by the insertion of words within brackets. These were merely oversights, hardly avoidable in the first draft of a work written in such circumstances. But there are also a few opinions expressed incidentally in which I cannot altogether concur, though they have evidently been adopted deliberately. With regard to these (Mr. Ellis not being in a condition to enter into a discussion of them) I had no course but to explain the grounds of my dissent, and leave every man to decide for himself upon the questions at issue. To avoid inconvenient interruptions however, I have thrown my arguments into an appendix, and contented myself in the foot notes with marking the particular expressions which I hold to be questionable.—*J. S.*

important than the rest, and Bacon declares that he will not bind himself to the conclusions it contains. Moreover its value will altogether cease when the sixth part can be completed, wherein will be set forth the new philosophy—the result of the application of the new method to all the phenomena of the universe. But to complete this, the last part of the Instauratio, Bacon does not hope : he speaks of it as a thing " et supra vires et ultra spes nostras collocata."

The greater part of the plan traced in the Distributio remained unfulfilled. Not to speak of the last division of the Instauratio, no part of Bacon's writings can properly be referred either to the fourth or fifth, except two prefaces which are found among the fragments published by Gruter.[2] To the fifth division however M. Bouillet [3] is disposed to refer several of Bacon's philosophical writings ; as, for instance, the tracts entitled *De Fluxu et Refluxu Maris*, and *Thema Cœli*. But though they correspond with the description which Bacon gives of the contents of the fifth part of the Instauratio, there is no reason to suppose that they would have been comprised in it. They were written a considerable time before the publication of the *Novum Organum ;* the *Thema Cœli* being clearly of the same date as the *Descriptio Globi intellectualis*, written in 1612 [4], and the *De Fluxu et Refluxu Maris* being probably written before Bacon had become acquainted with Galileo's theory of the tides. This theory was published in 1616 ; and it is reasonable to suppose that Bacon, who speaks of it in the *Novum Organum*, would have mentioned it in the *De Fluxu*, if the latter had not been written either before it was published, or but a short time afterwards [5]. These tracts, and the others which M. Bouillet mentions, are clearly occasional writings not belonging to the circuit of the Instauratio.

To the fourth part have been referred the *Historia Ventorum*, the *Historia Vitæ et Mortis*, etc. This however is contrary to Bacon's description of them in the dedication to Prince Charles prefixed to the *Historia Ventorum*. They are there spoken of as the " primitiæ Historiæ nostræ naturalis." Even the general title with which the *Historia Ventorum* and the titles of five other Historiæ were published, shows that they belong not to the fourth but to the third part of the Instauratio. It is as follows :—*Historia Naturalis ad condendam Philosophiam, sive Phænomena Universi, quæ est Instaurationis Magnæ pars tertia.* It is moreover manifest that as the fourth part was to contain applications to certain subjects of Bacon's method of induction, these treatises, in which the method is nowhere employed, cannot belong to it. M. Bouillet, though he justly dissents from Shaw's [6] arrangement, by whom they are referred to the fourth part, nevertheless commits an error of the same kind by introducing into this division of the Instauratio a fragment on Motion, published by Gruter with the title *Filum Labyrinthi, sive Inquisitio legitima de Motu.* This fragment, which is doubtless anterior to the *Novum Organum*, contains many thoughts and expressions which are found more perfectly developed either in the *Novum Organum* itself, or in the Distributio Operis. It is not to be supposed that Bacon, after thus expressing

[2] Francisci Baconi de Verulamio Scripta in naturali et universali Philosophia. Amst. 1653.—*J. S.*

[3] Œuvres Philosophiques de Bacon, publiées d'après les textes originaux, avec notice, sommaires et éclaircissemens, par M. N. Bouillet. Paris, 1834.—*J. S.*

[4] See the Preface to the *Descriptio Globi intellectualis.*—*J.S.*

[5] That the *De Fluxu* was written before the *Thema Cœli* is almost proved by the allusion to it in the following passage : " Verum hujusce rei demonstrationes et evidentias in anticipatione nostrâ de fluxu et refluxu maris plene tractavimus ". I say *almost* proved, because Bacon in writing a piece which was designed to come after another which was not yet written, would sometimes refer to that other as if it were already done. But it is not likely that he should have done so here ; for in any general scheme the *Thema Cœli* would have come before the *De Fluxu*. In a letter to Bacon, dated 14th April 1619, Tobie Matthew speaks of Galileo's having *answered* Bacon's discourse touching the flux and reflux of the sea : but he alludes apparently to a discourse of Galileo's on that subject which had never been printed.—*J. S.*

[6] The Philosophical Works of Francis Bacon, Baron of Verulam, etc. ; methodised and made English from the Originals, by Peter Shaw, M.D., London, 1733.—*J. S.*

himself in the Distributio—" Neque enim hoc siverit Deus ut phantasiæ nostræ somnium pro exemplari mundi edamus ; sed potius benigne faveat ut apocalypsim ac veram visionem vestigiorum et sigillorum Creatoris super creaturas scribamus "—would have repeated this remarkable sentence with scarcely any alteration in another part of the Instauratio [7] ; nor that he would have repeated in a somewhat less finished form the whole substance of the hundred and twenty-fifth aphorism of the first book of the *Novum Organum*. Yet we must admit this improbable supposition, if we decide on giving to the *Inquisitio legitima* the place which M. Bouillet has assigned to it. The truth is, that many of Bacon's shorter tracts preserved by Gruter and others are merely, so to speak, experimental fragments, of which the substance is embodied in his more finished writings.

Of the fourth and fifth parts of the Instauratio nothing, as I have already remarked, has been preserved except the prefaces, if indeed any other portion of them ever existed. But of the third, though it is altogether incomplete, we have nevertheless large fragments. Two years after the publication of the *Novum Organum* Bacon published the *Historia Naturalis ad condendam Philosophiam*, which has been already mentioned. In this however only the *Historia Ventorum* is contained in extenso ; and of the five other Historiæ of which Bacon speaks in the dedication, and of which he proposed to publish one every month, only two are now in existence, namely the *Historia Vitæ et Mortis*, published in 1623, and the *Historia Densi et Rari* which is contained in Rawley's *Opuscula varia posthuma*, published in 1658. Of the other three, namely the Historiæ *Gravis et Levis*, *Sympathiæ et Antipathiæ Rerum*, and *Sulphuris Mercurii et Salis*, we have only the prefaces, which were published in the same volume as the *Historia Ventorum*.

These Historiæ, and the *Sylva Sylvarum*, published soon after Bacon's death by Rawley, are the only works which we are entitled to refer to the third part of the Instauratio. With respect to the former we have the authority of Bacon's own title page and dedication ; and Rawley's dedication of the latter to King Charles shows that it is included under the general designation of *Historia Naturalis ad condendam Philosophiam* [8].

Other tracts however, of more or less importance, have been placed in the third part of the Instauratio, as for instance a fragment, published by Rawley in 1658, entitled *Historia et Inquisitio prima de Sono et Auditu et de Formâ Soni et latente processu Soni, sive Sylva Soni Auditus*. But the substance of this fragment occurs also in the *Sylva Sylvarum*, and therefore it cannot have been Bacon's intention to publish both as portions of his *Historia Naturalis*. It is probable that the *Historia de Sono et Auditu* was originally written as a portion of the general scheme of natural history [9] which was to form the third part of the In-

[7] I doubt whether this argument can be safely relied upon. Among the works which were certainly meant to stand as part of the Instauratio several remarkable passages occur twice and more than twice. But there are other grounds for concluding that the *Inquisitio de Motu* was written soon after the *Cogitata et Visa* (1607). In the *Commentarius solutus*, a kind of diary which will be printed among the Occasional Works, I find the following entry under the date July 26. 1608 :—" The finishing the 3 tables *De Motu, De Calore et Frigore, De Sono*." After which follow (July 27.) several pages of notes for an *Inquisitio legitima de Motu*. It would seen that this *Inquisitio* was designed originally to be the example in which the new method was to be set forth (see last section of *Cogitata et Visa*), but that the *Inquisitio de Calore et Frigore* was afterwards preferred ; probably as more manageable.—*J. S.*

[8] " The whole body of the *Natural History*, either designed or written by the late Lord Viscount St. Albans, was dedicated to Your Majesty in the book *De Ventis*, about four years past, when Your Majesty was prince, so as there needed no new dedication of this work, but only in all humbleness to let Your Majesty know that it is yours ".—*Dedication to the King of the Sylva Sylvarum.*

[9] It was probably the table *De Sono* referred to in the *Commentarius solutus*, July 26. 1608 (see note 7, above), and designed, like the tables *De Motu* and *De Calore et Frigore*, for an example of the new method.—*J. S.*

stauratio ; but it is certainly superseded by the *Sylva Sylvarum*, and is therefore not entitled to the position which has generally been assigned to it. So, too, the *Historiæ Naturalis ad condendam Philosophiam Præfatio destinata* [10], published by Gruter, is clearly irreconcilable with the plan laid down in the dedication to Prince Charles of the *Historia Naturalis*. For Bacon's intention when he wrote the preface which Gruter has published was plainly to commence his *Natural History* by treating of density and rarity, and not of the natural history of the winds. Subsequently he changed his plan ; and the first published portion of the third part of the Instauratio is, as we have seen, the *Historia Ventorum*. But this change of plan plainly shows that he had determined to cancel the fragment preserved by Gruter. Whenever what an author publishes or prepares for publication supersedes or contradicts unpublished and unfinished papers, these ought beyond all question to be set aside, and if published at all to be published apart from his other writings. Against some of the other fragments included in the third part of the Instauratio there is no such direct evidence as there is against those of which we have been speaking ; but it only gives rise to needless confusion to mix up with what we know it was Bacon's intention to publish as portions of his *Historia Naturalis*, loose fragments touching which we have no information whatever.

From what has been said it is manifest that what we possess of the third part of the Instauratio is merely a fragment—for the *Sylva Sylvarum*, a miscellaneous collection of observations gathered for the most part out of books, nowise completes Bacon's general design. In truth it is a design which cannot be completed, there being no limit to the number of the " Phænomena universi " which are potentially if not actually cognisable ; and it is to be observed that even if all the facts actually known at any instant could be collected and systematised (and even this is plainly impossible), yet still Bacon's aim would not be attained. For these facts alone would be insufficient as materials for the sixth part of the Instauratio, in which was to be contained all the knowledge of Nature man is capable of. Every day brings new facts to light not less entitled than those previously known to find a place in a complete description of the phenomena of the universe [11]. From many places in Bacon's writings it appears, as I have elsewhere remarked, that he had formed no adequate conception of the extent and variety of Nature. In a letter to R. P. Baranzan, who had apparently remarked by way of objection to Bacon's scheme of philosophy that a complete natural history would be a work of great extent and labour, Bacon observes that it would perhaps be sixfold as voluminous as that of Pliny. We have here therefore a sort of estimate of the limits which, in his judgment, the third part of the Instauratio would not exceed. What now exists of it is perhaps one twentieth in magnitude of this estimate.

Even the second part of the Instauratio, the *Novum Organum* itself, is incomplete. The second book concludes with the doctrine of prerogative instances,

10 See Bouillet, vol. ii. p. 264. The preface in question is the introduction to the *Tabula Exporrectionis et Expansionis Materiæ*, a rudiment of the *Historia Densi et Rari*. It was published by Gruter, before the *Historia Densi et Rari* appeared, among the *Impetus Philosophici :* with the title, *Phænomena Universi ; sive Historia Naturalis ad condendam Philosophiam. Præfatio*. M. Bouillet gives the preface only.

11 This would be true, I think, of all new facts which were not obviously reconcilable with laws previously known. But is it not conceivable that so complete a knowledge might be attained of the *laws* of Nature, that it could not be increased or affected by the discovery of any new *fact* in Nature ? If we had as complete a knowledge of other laws of Nature as we have of gravitation, for instance, new facts would still come to light, but with respect to the laws themselves they would all say the same thing, and therefore bring no new knowledge. Every new application of mechanical power contains some new *fact* more or less connected with gravitation ; yet unless a machine can be made which shall produce results not only new (i.e. such as had never been produced before) but inexplicable by the received theory of gravitation, are we not entitled to say that we know all that can be known about gravitation ?—*J. S.*

But in its twenty-first aphorism a number of subjects are mentioned of which this doctrine is the first, the last being the " Scala ascensoria et descensoria axiomatum." Neither this, nor any of these subjects after the first, except the last but one, is anywhere discussed in Bacon's writings ; and our knowledge of his method is therefore incomplete. Even the penultimate division of the *Novum Organum* which was published along with the first two books, and which treats " de parascevis ad inquisitionem," has all the appearance of being a fragment, or at least of being less developed than Bacon had intended it to be.

The first part of the Instauratio is represented, not inadequately, by the *De Augmentis*, published about three years after the Distributio Operis and the *Novum Organum*. It is a translation with large additions of the *Advancement of Learning*, published in 1605 ; and if we regard the latter as a development of the ninth chapter of *Valerius Terminus*, which is an early fragment containing the germ of the whole of the Instauratio [12], the *De Augmentis* will appear to belong naturally to the great work of which it now forms the first and only complete portion. In the preface prefixed to it by Rawley it is said that Bacon, finding " the part relating to the Partitions of the Sciences already executed, though less solidly than the dignity of the argument demanded, . . . thought the best thing he could do would be to go over again what he had written, and to bring it to the state of a satisfactory and completed work. And in this way he considers that he fulfils the promise which he has given respecting the first part of the Instauration." [13]

From this general view of the different parts of the Instauratio, as described in the Distributio Operis, we proceed to consider more particularly the *Novum Organum*. Although it was left incomplete, it is nevertheless of all Bacon's works that upon which he bestowed the most pains. In the first book especially every word seems to have been carefully weighed ; and it would be hard to omit or to change anything without injuring the meaning which Bacon intended to convey. His meaning is not always obvious, but it is always expressed with singular precision and felicity. His chaplain, Rawley, says that he had seen among his papers at least twelve yearly revisions of the *Novum Organum* [14]. Assuming, which there is no reason to doubt, that this statement may be relied upon, it would seem to follow that the composition of the *Novum Organum* commenced in 1608. And this agrees tolerably well with the circumstance that the *Cogitata et Visa* was sent to Bodley in 1607, as we learn from the date of Bodley's reply to it. If we suppose that the tract published with this title by Gruter is the same as that which was sent to Bodley, a passage near the end acquires a significance which has not I think been remarked. In the *Cogitata et Visa* Bacon speaks of the considerations whereby he had been led to perceive the necessity of a reform in philosophy, and goes on to say that the question as to how his new method might be most fitly given to the world had been much in his thoughts.

[12] I should rather say, the germ of all that part of the Instauratio which treated of the Interpretation of Nature. For I cannot find in the *Valerius Terminus* any traces of the *first* part, of which the *Advancement of Learning* was the germ. See Note A. at the end.—*J. S.*

[13] My own reasons for thinking that the *De Augmentis* did *not* form part of the original design, together with the circumstances which, as I suppose, determined Bacon to enlarge that design so as to take it in, will be explained in the preface to the *De Augmentis*.—*J. S.*

[14] " Ipse reperi in archivis Dominationis suæ autographa plus minus duodecim *Organi novi*, de anno in annum elaborati et ad incudem revocati ; et singulis annis ulteriore limâ subinde politi et castigati." In the preceding sentence, he calls it " multorum annorum et laboris improbi proles ".—*Auctoris Vita*, prefixed to the *Opuscula varia posthuma*, 1658. In the English life prefixed to the *Resuscitatio*, which was published the year before, he says, " I myself have seen at the least twelve copies of the Instauration ; revised year by year, one after another ; and every year altered and amended in the frame thereof ". I doubt whether we can fairly infer from these expressions that these twelve several copies were made in twelve several years ; but substantially they bear out the inference drawn from them.—*J. S.*

"Atque diu," he proceeds, "et acriter rem cogitanti et perpendenti ante omnia visum est ei tabulas inveniendi, sive legitimæ inquisitionis formulas . . . in aliquibus subjectis proponi tanquam ad exemplum et operis descriptionem fere visibilem [15]. . . . Visum est autem, nimis abruptum esse ut à tabulis ipsis docendi initium sumatur. Itaque idonea quædam præfari oportuisse, quod et jam se fecisse arbitratur." It was Bacon's intention therefore when he wrote the *Cogitata et Visa*, and when apparently some years later [16] he communicated it to Bodley, to publish an example of the application of his method to some particular subject—an intention which remained unfulfilled until the publication of the *Novum Organum*. We may therefore conjecture that it was about this time that Bacon addressed himself to the great work of composing the *Novum Organum* [17]; and this agrees with what Rawley says of its having been twelve years in hand. This view also explains why the whole substance of the *Cogitata et Visa* is reproduced in the first book of the *Novum Organum*; for this tract was designed to be an introduction to a particular example of the new method of induction, such as that which we find near the beginning of the second book. Bacon's purpose in writing it was therefore the same as that which he had in view in the first book of the *Novum Organum*,—namely to procure a favourable reception for an example and illustration of his method. What has been said may be in some measure confirmed by comparing the *Cogitata et Visa* with an earlier tract,—namely the *Partis secundæ Delineatio et Argumentum*. When he wrote this tract Bacon did not propose to set forth his method merely by means of an example; on the contrary, the three ministrations to the sense, to the memory, and to the reason, of which the last is the new method of induction, were to be set forth in order and didactically. Whereas in the *Novum Organum* Bacon remarks, "incipiendum est à fine" (that is, the method of induction must be set forth before the method of collecting facts and that of arranging them so as best to assist the memory); and having said this, he goes on at once to his example,—namely, the investigation of the Form of heat. Thus it appears that after Bacon had not only decided on writing a great work on the reform of philosophy, but had also determined on dividing it into parts of which the second was to contain the exposition of his new method, he in some measure changed his plan, and resolved to set forth the essential and operative part of his system chiefly by means of an example. This change of plan appears to be marked by the *Cogitata et Visa*,—a circumstance which makes this tract one of the most interesting of the precursors of the *Novum Organum*.

15 In the *Commentarius solutus*, under date July 26. 1608, I find the following memorandum :—"Seeing and trying whether the B. of Canterb. may not be affected in it, being single and glorious, and believing the sense".

"Not desiring to draw in the Bp. Awnd. [Bishop Andrews, probably] being single, rich, sickly, and professor to some experiments: *this after the table of motion or some other in part set in forwardness*."

Some other memoranda in the same place relate to the gaining of *physicians*, and learning from them experiments of surgery and physic; which explains the epithet "sickly" in the above extract.—*J. S.*

16 Bodley's answer is dated Feb. 19. 1607; i.e. 1607–8; in which he says, "I must tell you, to be plain, that you have very much wronged yourself and the world, to smother such a treasure so long in your coffer". But I do not think we can infer from this that the *Cogitata et Visa* had been written "some years" before. Bodley may only allude to his having kept such thoughts so long to himself.—*J. S.*

17 In the *Commentarius solutus*, under date July 26. 1608, I find the following memorandum :—"The finishing the Aphorisms, *Clavis interpretationis*, and then setting forth the book," and in the same page, a little after, "Imparting my *Cogitata et Visa*, with choice, ut videbitur". The aphorisms here spoken of may have been the "Aphorismi et Consilia de auxiliis mentis et accensione luminis naturalis"; a fragment containing the substance of the first, second, and third aphorisms of the first book of the *Novum Organum*, and the first, third, and sixteenth of the second. *Clavis interpretationis* was probably the name which was afterwards exchanged for *Novum Organum*. —*J. S.*

That the *Partis secundæ Delineatio* is earlier than the *Cogitata et Visa* appears plainly from several considerations which M. Bouillet, who expresses a contrary opinion, seems to have overlooked. In the first place, whole sentences and even paragraphs of the *Cogitata et Visa* are reproduced with scarcely any alteration in the *Novum Organum* ; whereas this is by no means the case with any passage of the *Partis secundæ Delineatio*. But as it may be said that this difference arises from the different character of the two tracts, of which the one is simply a summary of a larger work, whereas the more developed style of the other resembles that of the *Novum Organum*, it may be well to compare them somewhat in detail.

In speaking of the prospects which the reform of philosophy was to open to mankind, Bacon thus expresses himself in the *Novum Organum* :—" Quinetiam prudentia civilis ad consilium vocanda est et adhibenda, quæ ex præscripto diffidit, et de rebus humanis in deterius conjicit ". The corresponding sentence in the *Cogitata et Visa* is, " Consentaneum enim esse, prudentiam civilem in hâc parte adhibere, quæ ex præscripto diffidit et de humanis in deterius conjicit ". Again, in the *Partis secundæ Delineatio* the same idea is thus expressed, " Si quis sobrius (ut sibi videri possit,) et civilis prudentiæ diffidentiam ad hæc transferens, existimet hæc quæ dicimus votis similia videri," etc. Here the somewhat obscure phrase " civilis prudentiæ diffidentiam " is clearly the germ of that by which it is replaced in the other two passages, namely, " prudentia civilis quæ ex præscripto diffidit." Again, in the *Partis secundæ Delineatio* Bacon affirms that ordinary induction " puerile quiddam est et precario concludit, periculo ab instantiâ contradictoriâ exposita " : in the *Cogitata et Visa*, that the logicians have devised a form of induction " admodum simplicem et plane puerilem, quæ per enumerationem tantum procedat, atque propterea precario non necessario concludat ". The clause " quæ per enumerationem tantum procedat ", which adds greatly to the distinctness of the whole sentence, is retained in the Distributio Operis, in which it is said that the induction of the logicians, " quæ procedit per enumerationem simplicem, puerile quiddam est, precario concludit, et periculo ab instantiâ contradictoriâ exponitur ". To take another case : in the *Partis secundæ Delineatio*, Bacon, speaking of those who might object to his frequent mention of practical results as a thing unworthy of the dignity of philosophy, affirms that they hinder the accomplishment of their own wishes. " Quin etiam illis, quibus in contemplationis amorem effusis frequens apud nos operum mentio asperum quiddam atque ingratum et mechanicum sonat, monstrabimus quantum illi desideriis suis propriis adversentur, quum puritas contemplationum atque substructio et inventio operum prorsus eisdem rebus nitantur, ac simul perficiantur." In the *Cogitata et Visa*, this sentence recurs in a modified and much neater form :—" Si quis autem sit cui in contemplationis amorem et venerationem effuso ista operum frequens et cum tanto honore mentio quiddam asperum et ingratum sonet, is pro certo sciat se propriis desideriis adversari ; etenim in naturâ, opera non tantum vitæ beneficia, sed et veritatis pignora esse ". On comparing these two sentences, it is difficult to believe that Bacon would have omitted the antithesis with which the latter ends in order to introduce the somewhat cumbrous expressions which correspond to it in the former, especially as we find this antithesis reproduced, though with another context, in the *Novum Organum*. " Opera ipsa," it is there said, " pluris facienda sunt quatenus sunt veritatis pignora quam propter vitæ commoda" [18].

These instances will probably be thought sufficient to justify us in concluding that the *Partis secundæ Delineatio*, in which no mention is made of the plan of setting forth the new method of induction by means of an example, is of earlier date than the *Cogitata et Visa*, in which this plan, actually employed in the *Novum Organum*, is spoken of as that which Bacon had decided on adopting. This

[18] Nov. Org. i. 124. It is well to mention that some of the expressions in this aphorism which do not occur in the *Cogitata et Visa* will be found in the *Partis secundæ Delineatio*. But it will be observed that I am only comparing passages which occur in all three works. Of the greater general resemblance of the *Cogitata et Visa* to the *Novum Organum* there can be no question.

question of priority is not without interest; for if the *Partis secundæ Delineatio* is anterior to the *Cogitata et Visa*, the general plan of the Instauratio must have been formed a considerable time before 1607, about which time Bacon probably commenced the composition of the *Novum Organum*. If we could determine the date of *Valerius Terminus*, we should be able to assign limits within which the formation of this plan, so far as relates to the division of the work into six portions, may be supposed to lie. For the first book of *Valerius Terminus* was to include all that was to precede the exposition of the new method of induction, which was to be the subject of the second; that is, it was to comprehend, along with the first part of the Instauratio [19], the general reflexions and precepts which form the subject of the first book of the *Novum Organum*. Nor does it appear that *Valerius Terminus* was to contain anything corresponding to the last four parts of the Instauratio [20]; it was a work, as its title [21] shows, on the Interpretation of Nature; that is, it was to be a statement of Bacon's method, without professing either to give the collection of facts to which the method was to be applied, or the results thereby obtained. Unfortunately, there appears to be no evidence tending to enable us to assign the time at which (or not long after it) *Valerius Terminus* was written. That it is earlier than the *Advancement of Learning* seems to follow from the circumstance that Bacon, when he wrote it, designed to include in a single chapter the general survey of human knowledge which in the *Advancement* is developed into two books [22]. Bacon has on all occasions condemned epitomes, and it is therefore altogether improbable that after writing the *Advancement of Learning* he would have endeavoured to compress its contents, or even those of the second book, within the limits proposed in *Valerius Terminus*. On the other hand, we may suppose that before writing the *Advancement* he had not seen how much he had to say on the subject to which it relates. We may conclude therefore, on these and other grounds, that *Valerius Terminus* was written some time before 1605; how much before cannot be known; but as by comparing the *Partis secundæ Delineatio* and the *Cogitata et Visa* with the *Novum Organum* we have seen reason to conclude that the general plan of the Instauratio was formed before Bacon had decided on propounding his method by means of an example, so by comparing the first-named of these three works with *Valerius Terminus*, we perceive that the idea of the work on the Interpretation of Nature, that is, on the new method of induction, was anterior in Bacon's mind to that of the Instauratio.

And this conclusion is confirmed by all we know of Bacon's early writings. In the earliest of all (if we assume that the *Temporis Partus Masculus*, published by Gruter [23], is the same as the *Temporis Partus Maximus* mentioned by Bacon in his letter to Fulgenzio), the most prominent notion is that true science consists in the interpretation of Nature—a phrase by which Bacon always designates a just method of induction. But nothing is said either there or in any early fragment whereby we are led to suppose that Bacon then thought of producing a great work like the Instauratio. On the contrary, in the *De Interpretatione Naturæ Procemium* he proposes to communicate his peculiar method and the results to which it was to lead, only to chosen followers; giving to the world merely an exoteric doctrine, namely the general views of science which afterwards formed the substance of the *Cogitata et Visa* and ultimately of the first book of the *Novum Organum* [24].

From what has been said it follows that we should form an inadequate conception of the *Novum Organum* if we were to regard it merely as a portion of the Instauratio. For it contains the central ideas of Bacon's system, of which the

[19] Query. See Note A. at the end, § 1.—*J. S.*

[20] Query. See Note A. at the end, § 2.—*J. S.*

[21] "Valerius Terminus of the Interpretation of Nature; with the Annotations of Hermes Stella. A few fragments of the first book, viz.," etc.

[22] Query. See Note A. at the end, § 1.—*J. S.*

[23] Say rather, "the several tracts collected by M. Bouillet under the title *Temporis Partus Masculus*". See Note A. at the end, § 3.—*J. S.*

[24] See Note A. at the end, § 4.—*J. S.*

whole of the Instauratio is only the developement. In his early youth Bacon formed the notion of a new method of induction, and from that time forth this notion determined the character of all his speculations. Later in life he laid the plan of a great work, within the limits of which the materials to which his method was to be applied and the results thereby to be obtained might be stored up, together with a statement of the method itself. But of this great plan the interpretation of Nature was, so to speak, the soul,—the formative and vivifying principle; not only because Bacon conceived that the new method only could lead to the attainment of the great ends which he had in view, but also because it was the possession of this method which had suggested to him the hopes which he entertained [25]. There seems some reason to believe that his confidence in his peculiar method of induction did not increase as he grew older; that is to say, he admits in the *Novum Organum* that the interpretation of Nature is not so much an artificial process as the way in which the mind would naturally work if the obstacles whereby it is hindered in the pursuit of truth were once set aside [26]. So that his precepts are, he says, not of absolute necessity: "necessitatem ei (arti interpretationis scilicet) ac si absque eâ nil agi possit, aut etiam perfectionem non attribuimus,"—an admission not altogether in the spirit of the earlier writings in which the art of interpretation is spoken of as a secret of too much value to be lightly revealed [27].

If it be asked why Bacon determined on propounding his method by means of an example, the answer is to be sought for in the last paragraphs of the *Cogitata et Visa*. He seems to have thought that it would thus obtain a favourable reception, because its value would be to a certain extent made manifest by the example itself. Likewise he hoped in this way to avoid all occasion of dispute and controversy, and thought that an example would be enough to make his meaning understood by all who were capable of understanding it. "Fere enim se in eâ esse opinione, nempe (quod quispiam dixit) prudentibus hæc satis fore, imprudentibus autem ne plura quidem."

His expectations have not been fulfilled, for very few of those who have spoken of Bacon have understood his method, or have even attempted to explain its distinguishing characteristics, namely the certainty of its results, and its power of reducing all men to one common level.

Another reason for the course which he followed may not improbably have been that he was more or less conscious that he could not demonstrate the validity, or at least the practicability, of that which he proposed. The fundamental principle in virtue of which alone a method of exclusions can necessarily lead to a positive result, namely that the subject matter to which it is applied consists of a finite number of elements, each of which the mind can recognise and distinguish from the rest, cannot, it is manifest, be for any particular case demonstrated à priori. Bacon's method in effect assumes that substances can always

[25] I quite agree in this, but not quite on the same grounds. In Note A. at the end of this preface, the reader will find a statement, too long for a foot-note, of such points in the foregoing argument as I consider disputable. It was the more necessary to point them out, because the arrangement of the pieces in this edition, for which I am responsible, will otherwise create a difficulty; being in some respects inconsistent with the opinions here expressed.—*J. S.*

[26] Nov. Org. i. 130. "Est enim Interpretatio verum et naturale opus mentis, demptis iis quæ obstant". But compare the following passage in *Valerius Terminus*, c. 22: "that it is true that interpretation is the very natural and direct intention, action, and progression of the understanding, delivered from impediments. And that all anticipation is but a reflexion or declination by accident". So that if we may infer from the passage in the *Novum Organum* that his confidence had abated, we must suppose that when he wrote the *Valerius Terminus* it had not risen to its height. But for my own part I doubt whether his opinion on this point ever changed.—*J. S.*

[27] Not, I think, as a secret of too much value to be revealed, but as an argument too abstruse to be made popular. See Note B. at the end, where I have endeavoured to bring together all the evidence upon which the presumption in the text is founded, and to show that it proves either too much or too little.—*J. S.*

be resolved into an aggregation of a certain number of abstract qualities, and that their essence is adequately represented by the result of this analysis. Now this assumption or postulate cannot be made the subject of a direct demonstration, and probably Bacon came gradually to perceive more or less the difficulties which it involves. But these difficulties are less obvious in special cases than when the question is considered generally, and on this account Bacon may have decided to give instead of a demonstration of his method an example of its use. He admits at the close of the example that the operation of the method is imperfect, saying that at first it could not but be so, and implying that its defects would be removed when the process of induction had been applied to rectify our notions of simple natures. He thus seems to be aware of the inherent defect of his method, namely that it gives no assistance in the formation of conceptions, and at the same time to hope that this would be corrected by some modification of the inductive process. But of what nature this modification is to be he has nowhere stated ; and it is to be remarked that in his earliest writings the difficulty here recognised is not even mentioned. In *Valerius Terminus* nothing is said of the necessity of forming correct notions of simple natures,—the method of exclusions then doubtless appearing to contain all that is necessary for the investigation of Nature.

Bacon may also have been influenced by other considerations. We have seen that he was at first unwilling that his peculiar method should become generally known. In the *De Interpretatione Naturæ Procæmium* he speaks of its being a thing not to be published, but to be communicated orally to certain persons [28]. In *Valerius Terminus* his doctrine was to be veiled in an abrupt and obscure style[29], such as, to use his own expression, would choose its reader,—that is, would remain unread except by worthy recipients of its hidden meaning. This affected obscurity appears also in the *Temporis Partus Masculus*. In this unwillingness openly to reveal his method Bacon coincided with the common feeling of his own and earlier times. In the middle ages no new discovery was freely published. All the secrets, real or pretended, of the alchemists were concealed in obscure and enigmatic language ; and to mention a well-known instance, the anagram in which Roger Bacon is supposed to have recorded his knowledge of the art of making gunpowder is so obscure, that its meaning is even now more or less doubtful. In Bacon's own time one of the most remarkable discoveries of Galileo—that of the phases of Venus—was similarly hidden in an anagram, though the veil in this case was more easily seen through. This disposition to conceal scientific discoveries and methods is connected with the views which in the middle ages were formed of the nature of science. To know that which had previously been unknown was then regarded as the result not so much of greater industry or acuteness as of some fortunate accident, or of access to some hidden source of information : it was like finding a concealed treasure, of which the value would be decreased if others were allowed to share in it. Moreover the love of the marvellous inclined men to believe in the existence of wonderful secrets handed down by tradition from former ages, and any new discovery acquired something of the same mysterious interest by being kept back from the knowledge of the vulgar. Other causes, which need not here be detailed, increased this kind of reserve ; such as the dread of the imputation of unlawful knowledge, the facility which it gave to deception and imposture, and the like.

The manner in which Bacon proposed at one time to perpetuate the knowledge of his method is also in accordance with the spirit of the middle ages. In the writings of the alchemists we meet continually with stories of secrets transmitted by their possessor to one or more disciples. Thus Artefius records the conversation wherein his master, Boemund, transmitted to him the first prin-

28 See Note B. at the end, extract 4th, and the concluding remarks in which I have explained my own view of the kind of reserve which Bacon at this time meditated. —*J. S.*

29 See the same note, extract 1st. I cannot think it was by " abruptness and obscurity " that he proposed to effect the desired separation of readers either in *Valerius Terminus* or in the *Temporis Partus Masculus*.—*J. S.*

ciples of all knowledge; and it is remarkable that in this and similar cases the disciple is called "mi fili" by his instructor—a circumstance which shows from what source Bacon derived the phrase "ad filios," which appears in the titles of several of his early pieces. Even in the *De Augmentis* the highest and most effectual form of scientific teaching is called the "methodus ad filios"[30].

When he wrote the *Cogitata et Visa*, Bacon seems to have perceived[31] how much of vanity and imposture had always been mixed up with this affectation of concealment and reserve. "Reperit autem", he there says, "homines in rerum scientiâ quam sibi videntur adepti, interdum proferendâ interdum occultandâ, famæ et ostentationi servire; quin et eos potissimum qui minus solida proponunt, solere ea quæ afferunt obscurâ et ambiguâ luce venditare, ut facilius vanitati suæ velificare possint". The matter which he has in hand, he goes on to say, is one which it were nowise fitting to defile by affectation or vain glory; but yet it cannot be forgotten that inveterate errors, like the delusions of madmen, are to be overcome by art and subtlety, and are always exasperated by violence and opposition. The result of this kind of dilemma is that the method is to be propounded in an example,—a decision in which it is probable that he was still more or less influenced by the example of those whom he here condemns.

Thus much of the connection between the plan of the *Novum Organum* and that which Bacon laid down in the *Cogitata et Visa*. That there is no didactic exposition of his method in the whole of his writings has not been sufficiently remarked by those who have spoken of his philosophy; probably because what he himself regarded as a sort of exoteric doctrine, namely the views of science contained in the first book of the *Novum Organum*, have received much more attention than the method itself, which is nevertheless the cardinal point of his whole system. Bacon is to be regarded, not as the founder of a new philosophy, but as the discoverer of a new method; at least we must remember that this was his own view of himself and of his writings.

I proceed to give some account of the structure of the *Novum Organum* and of the parts into which it may be most conveniently divided.

[30] Lib. vi. c. 2. I cannot think however that the merit of this method had anything to do with *secresy*. For the distinctive object of it is stated to be the "continuatio et ulterior progressus" of knowledge; and its distinctive characteristic, the being "solito *apertior*". Its aim was to transfer knowledge into the mind of the disciple in the same form in which it grew in the teacher's mind, like a plant with its roots on, that it might continue to grow. Its other name is "traditio lampadis", alluding to the Greek torch-race; which was run, as I understand it, not between individuals, but between what we call *sides*. Each side had a lighted torch they were so arranged that each bearer, as he began to slacken, handed it to another who was fresh; and the side whose torch first reached the goal, still a-light, was the winner. The term "filii", therefore, alludes, I think, to the successive generations, not who should inherit the secret, but who should carry on the work. Compare the remarks in the *Sapientia Veterum* (Fab. xxvi. near the end), upon the torch-races in honour of Prometheus. "Atque continet in se monitum, idque prudentissimum, ut perfectio scientiarum a successione, non ab unius alicujus pernicitate aut facultate, expectetur. . . . . Atque optandum esset ut isti ludi in honorem Promethei, sive humanæ naturæ, instaurarentur, atque res *certamen, et æmulationem, et bonam fortunam* reciperet; neque ex unius cujuspiam face tremulâ atque agitatâ penderet." To me, I must confess, the explanation above given of Bacon's motives for desiring a select audience seems irreconcilable both with the objects which he certainly had in view and with the spirit in which he appears to have pursued them. "Fit audience, though few," he no doubt desired; and I can easily believe that he wished not only to find the fit, but also to exclude the unfit. But the question is, whether his motive in so selecting and so limiting his audience was unwillingness to part with his treasure, or solicitude for the furtherance of his work. To decide this question I have brought together all the passages in which he speaks of the "singling and adopting" of the "fit and legitimate reader". But the collection, with the remarks which it suggests, being too long for a foot-note, I have placed them at the end of this preface. See Note B.—*J. S.*

[31] See Note B., extract 7th. But observe that in the 1st, 3rd, and 4th, he shows himself quite as sensible of the vanity and imposture which such secrecy had been made to subserve.—*J. S.*

After the preface, in which Bacon professes that it is not his intention to destroy the received philosophy, but rather that from henceforth there should be two co-existing and allied systems,—the one sufficient for the ordinary purposes of life, and such as would satisfy those who are content with probable opinions and commonly received notions ; the other for the sons of science, who desire to attain to certainty and to an insight into the hidden things of Nature,—we come to the *Novum Organum* itself ; which commences with some weighty sentences concerning the relation of Man to Nature. The first aphorism, perhaps the most often quoted sentence in the *Novum Organum*, occurs twice in the fragments published by Gruter ; namely in the *Aphorismi et Consilia de Auxiliis Mentis*, and again in a less perfect form in the *De Interpretatione Naturæ Sententiæ* XII., both which fragments are included [by M. Bouillet] [32] under the title *Temporis Partus Masculus*, though they are clearly of different dates. The wording of the aphorism in the former is almost precisely the same as in the *Novum Organum*. In all three places man is styled "naturæ minister et interpres". He is naturæ interpres, because in every object which is presented to him there are two things to be considered, or rather two aspects of the same thing,—one the phenomenon which Nature presents to the senses—the other the inward mechanism and action, of which the phenomenon in question is not only the result but also the outward sign. To pass therefore from the phenomenon to its hidden cause is to interpret the signs which enable us to become acquainted with the operations of Nature. Again, he is the minister naturæ, because in all his works he can only arrange the things with which he deals in the order and form which Nature requires. All the rest comes from her only ; the conditions she requires having been fulfilled, she produces new phenomenon according to the laws of her own action. Thus the two words *minister* and *interpres* refer respectively to works and contemplation—to power and knowledge—the substance of Bacon's theory of both being compressed into a single phrase. The third and fourth aphorisms are developments of the first ; the second relating not to the theory of knowledge, but to the necessity of providing helps for the understanding.

Then follow (5–10) reflections on the sterility of the existing sciences, and (11–17) remarks on the inutility of logic. In (14) Bacon asserts that everything must depend on a just method of induction. From (18) to (37) he contrasts the only two ways in which knowledge can be sought for ; namely anticipations of Nature and the interpretation of Nature. In the former method men pass at once from particulars to the highest generalities, and thence deduce all intermediate propositions ; in the latter they rise by gradual induction and successively, from particulars to axioms of the lowest generality, then to intermediate axioms, and so ultimately to the highest. And this is the true way, but as yet untried.

Then from (38) to (68) Bacon developes the doctrine of idols. It is to be remarked that he uses the word idolon in antithesis to idea, the first place where it occurs being the twenty-third aphorism. "Non leve quiddam interest", it is there said, "inter humanæ mentis idola et divinæ mentis ideas". He nowhere refers to the common meaning of the word, namely the image of a false god. Idols are with him "placita quædam inania", or more generally, the false notions which have taken possession of men's minds. The doctrine of idols stands [he says] in the same relation to the interpretation of Nature, as the doctrine of fallacies to ordinary logic.

Of idols Bacon enumerates four kinds,—the idols of the tribe, of the cave, of the market-place, and of the theatre ; and it has been supposed that this classification is borrowed from Roger Bacon, who in the beginning of the *Opus Majus* speaks of four hindrances whereby men are kept back from the attainment of true knowledge. But this supposition is for several reasons improbable. The *Opus Majus* was not printed until the eighteenth century, and it is unlikely that Francis Bacon would have taken the trouble of reading it, or any part of it, in manuscript [33]. In the first place there is no evidence in any part of his works of this

[32] Not so included by *Gruter*. See note A. at the end, § 3.—*J. S.*

[33] I can hardly think that he would have omitted to look into a work like the *Opus Majus*, if he had had the opportunity. But it is very probable that no copy of it was

kind of research, and in the second he had no high opinion of his namesake, of whom he has spoken with far less respect than he deserves. The only work of Roger Bacon's which there is any good reason for believing that he was acquainted with is a tract on the art of prolonging life, which was published at Paris in 1542, and of which an English translation appeared in 1617. The general resemblance between the spirit in which the two Bacons speak of science and of its improvement is, notwithstanding what has sometimes been said, but slight. Both no doubt complain that sufficient attention has not been paid to observation and experiment, but that is all; and these complaints may be found in the writings of many other men, especially in the time of Francis Bacon. Nothing is more clear than that the essential doctrines of his philosophy—among which that of idols is to be reckoned—are, so far as he was aware, altogether his own. There is moreover but little analogy between his idols and his namesake's hindrances to knowledge. The principle of classification is altogether different, and the notion of a real connection between the two was probably suggested simply by there being the same number of idols as of hindrances[34]. It is therefore well to remark that in the early form of the doctrine of idols there were only three. In the *Partis secundæ Delineatio* the idols wherewith the mind is beset are said to be of three kinds: they either are inherent and innate or adscititious; and if the latter, arise either from received opinions in philosophy or from wrong principles of demonstration. This classification occurs also in *Valerius Terminus*[35].

The first of these three classes corresponds to the first and second of those spoken of in the *Novum Organum*. The idols of the tribe are those which belong, as Aristotle might have said, to the human mind as it is human,—the erroneous tendencies common more or less to all mankind. The idols of the cave arise from each man's mental constitution: the metaphor being suggested by a passage in the [opening of the seventh book of Plato's *Republic*[36].] Both classes of extraneous idols mentioned in the *Partis secundæ Delineatio* are included in the idola theatri, and the idola fori correspond to nothing in the earlier classification[37]. They also are extraneous idols, but result neither from received opinions nor erroneous forms of demonstration, but from the influence which words of necessity exert. They are called idols of the market-place because they are caused by the daily intercourse of common life. "Verba," remarks Bacon, "ex captu vulgi imponuntur."

It is only when we compare the later with the earlier form of the doctrine of idols that we perceive the principle of classification which Bacon was guided by, namely the division of idols according as they come from the mind itself or from

procurable; possible that he did not even know of its existence. The manner in which he speaks of Roger Bacon in the *Temporis Partus Masculus*, as belonging to the "utile genus" of experimentalists, "qui de theoriis non admodum soliciti *mechanicâ quâdam subtilitate rerum inventarum extensiones prehendunt*", seems rather to imply that he knew of him at that time chiefly by his reputation for mechanical inventions.—*J. S.*

[34] That the two may be the more conveniently compared, I have quoted Roger Bacon's exposition of his "offendicula", in a note upon the 39th aphorism, in which the names of the four "Idols" first occur. How slight the resemblance is between the two may be ascertained by a very simple test. If you are already acquainted with Francis Bacon's classification, try to assign each of the "offendicula" to its proper class. If not, try by the help of Roger's classification to find out Francis's.—*J. S.*

[35] Not in *Valerius Terminus*. It occurs in the *Distributio Operis*, and may be traced though less distinctly in the *Advancement* and the *De Augmentis*. See Note C. at the end.—*J. S.*

[36] Mr. Ellis had written "in the   of Aristotle". But the words of the *De Augmentis* (v. 4)., "de specu *Platonis*", prove that it was the passage in Plato which suggested the metaphor.—*J. S.*

[37] i.e. in the classification adopted in the *Partis secundæ Delineatio*; for they correspond exactly with the third kind of fallacies or false appearances mentioned in the *Advancement*, and with the idols of the palace in *Valerius Terminus*. And I think they were meant to be included among the "Inhærentia et Innata" of the *Delineatio*. See Note C.—*J. S.*

without [38]. In the *Novum Organum* two belong to the former class and two to the latter, so that the members of the classification are better balanced [39] than in the previous arrangement: in both perhaps we perceive a trace of the dichotomizing principle of Ramus, one of the seeming novelties which he succeeded in making popular [40].

After enumerating the four kinds of idols, Bacon gives instances of each (45–67); and speaking in (62) of idols of the theatre, introduces a triple classification of false philosophies, to which he seems to have attached much importance, as we find it referred to in many parts of his writings. False philosophy is sophistical, empirical, or superstitious; sophistical, when it consists of dialectic subtleties built upon no better foundation than common notions and every-day observation; empirical, when it is educed out of a few experiments, however accurately examined; and superstitious, when theological traditions are made its basis. In the *Cogitata et Visa* he compares the rational philosophers (that is, those whose system is sophistical,—the name implying that they trust too much to reason and despise observation) to spiders whose webs are spun out of their own bodies, and the empirics to the ant which simply lays up its store and uses it. Whereas the true way is that of the bee, which gathers its materials from the flowers of the field and of the garden, and then, ex propriâ facultate, elaborates and transforms them [41]. The third kind of false philosophy is not here mentioned. In the *Novum Organum* Bacon perhaps intended particularly to refer to the Mosaical philosophy of Fludd, who is one of the most learned of the Cabalistic writers [42].

In (69) Bacon speaks of faulty demonstrations as the defences and bulwarks of idols, and divides the common process for the establishment of axioms and conclusions into four parts, each of which is defective. He here describes in general terms the new method of induction. In the next aphorism, which concludes this part of his object, he condemns the way in which experimental researches have commonly been carried on.

The doctrine of idols seems, when the *Novum Organum* was published, to have been esteemed one of its most important portions. Mersenne at least, the earliest critic on Bacon's writings, his *Certitude des Sciences* having been published in 1625 [43], speaks of the four idols, or rather of Bacon's remarks upon them, as the four buttresses of his philosophy. In Bacon's own opinion this doctrine was of much importance. Thus in the *De Interpretatione Naturæ Sententiæ Duodecim* he says, in the abrupt style of his earlier philosophical writings, "Qui primum et ante alia omnia animi motus humani penitus non explorarit, ibique scientiæ meatus et errorum sedes accuratissime descriptas non habuerit, is omnia

[38] Rather, I think, as they are separable or inseparable from our nature and condition in life. See Note C.—*J. S.*

[39] Compare the Distributio Operis, where the classification is retained, with the *Novum Organum*, where it is not alluded to, and I think it will be seen that Bacon did not intend to balance the members in this way. See Note C. at the end.—*J. S.*

[40] Bacon alludes to Ramus in the *De Augmentis* vi. 2., "De unicâ methodo et dichotomiis perpetuis nil attinet dicere. Fuit enim nubecula quædam doctrinæ quæ cito transiit: res certe simul et scientiis damnosissima," etc.

[41] In the *Advancement of Learning* and the *De Augmentis*, the schoolmen in particular are compared to the spider; a passage which has been misunderstood by a distinguished writer, whose judgments seem not unfrequently to be as hastily formed as they are fluently expressed, and who conceives that Bacon intended to condemn the study of psychology.

In speaking of the field and the garden, Bacon refers repectively to observations of Nature and artificial experiment; an instance of the "curiosa felicitas" of his metaphors.

[42] Fludd's work, entitled *Philosophia Moysaica*, was published in 1638.

[43] In the *Biographie Universelle* (Mersenne) it is incorrectly said that this work was published in 1636, and an idle story is mentioned that it was in reality written, not by Mersenne, but by Lord Herbert of Cherbury,—a story sufficiently refuted by its scrupulous and submissive orthodoxy.

larvata et veluti incantata reperiet; fascinum ni solverit interpretari non poterit"[44].

From (71) to (78) he speaks of the signs and tokens whereby the defects and worthlessness of the received sciences are made manifest. The origin of these sciences, the scanty fruits they have borne, the little progress they have made, all testify against them; as likewise the confessions of the authors who have treated of them, and even the general consent with which they have been received. "Pessimum," says Bacon, "omnium est augurium, quod ex consensu capitur in rebus intellectualibus"[45].

From (78) to (92) Bacon speaks of the causes of the errors which have hindered the progress of science; intending thereby to show that there is no reason to doubt the value of the reform which he is about to propose, because though in itself seemingly plain and obvious it has nevertheless remained so long unthought of. On the contrary, there is, he affirms, good reason for being surprised that even now any one should have thought of it.

The first of these causes is the comparative shortness of the periods which, out of the twenty-five centuries which intervene between Thales and Bacon's own time, have been really favourable to the progress of science. The second, that even during the more favourable times natural philosophy, the great mother of the sciences, has been for the most part neglected; men having of late chiefly busied themselves with theology, and among the Greeks and Romans with moral philosophy, "quæ ethnicis vice theologiæ erat". Moreover, even when men occupied themselves the most with natural philosophy (Bacon refers to the age of the early Greek physicists), much time was wasted through controversies and vain glory. Again, even those who have bestowed pains upon natural philosophy have seldom, especially in these latter times, given themselves wholly up to it. Thus, natural philosophy having been neglected and the sciences thereby severed from their root, it is no wonder that their growth has been stopped.

Another cause of their scanty progress, is that their true end, the benefit and relief of man's estate, has not been had in remembrance. This error Bacon speaks of in the *Advancement* as the greatest of all, coupling however therewith the relief of man's estate the glory of the Creator. Again, the right path for the advancement of knowledge has not only been neglected but blocked up, men having come not only to neglect experience but also to despise it. Also the reverence for antiquity has hindered progress; and here Bacon repeats the remark he had made in the *Advancement*, that antiquity was the world's youth, and the latter times its age[46].

Again, the progress of science has been hindered by too much respect for what

[44] So also in the *Valerius Terminus*, c. 17.: "That if any have had or shall have the power and resolution to fortify and inclose his mind against all anticipations, yet if he have not been or shall not be cautioned by the full understanding of the nature of the mind and spirit of man, and therein of the seats, pores, and passages both of knowledge and error, he hath not been, nor shall not be, possibly able to guide or keep on his course aright".—*J. S.*

[45] He however excepts matters political and religious.

[46] This remark is in itself not new; we read, for instance, in the book of Esdras, that the world has lost its youth, and that the times begin to wax old. Nor is it new in the application here made of it. Probably several writers in the age which preceded Bacon's had already made it, for in that age men were no longer willing to submit to the authority of antiquity, and still felt bound to justify their dissent. Two writers may at any rate be mentioned by whom the thought is as distinctly expressed as by Bacon, namely Giordano Bruno and Otto Casmann; the former in the *Cena di Cenere*, the latter in the preface to his *Problemata Marina*, which was published in 1596, and therefore a few years later than the *Cena*, with which however it is not likely that Casmann was acquainted. Few writers of celebrity comparable to Bruno's appear to have been so little read.

I have quoted both passages in a note on the corresponding passage in [the first book of] the *De Augmentis*: that in the *Cena di Cenere* was first noticed by Dr. Whewell. See his *Philosophy of the Inductive Sciences*, ii. 198.

has been already accomplished. And this has been increased by the appearance of completeness which systematic writers on science have given to their works, and also by the vain and boastful promises of some who have pretended to reform philosophy. Another reason why more has not been accomplished, is that so little has been attempted.

To these hindrances Bacon adds three others,—superstitious bigotry, the constitution of schools, universities, and colleges, and the lack of encouragement; and then concludes this part of the subject with that which he affirms to have been the greatest obstacle of all, namely despair of the possibility of progress. To remove this, he goes on to state the grounds of hope for the future,—a discussion which extends from (93) to (115).

" Principium autem," he begins, " sumendum a Deo " ; that is to say, the excellence of the end proposed is in itself an indication that the matter in hand is from God, nor is the prophecy of Daniel concerning the latter times to be omitted, namely that many shall go to and fro and knowledge shall be increased. Again, the errors committed in time past are a reason for hoping better things in the time to come. He therefore sets forth these errors at some length (95–107). This enumeration begins with the passage already mentioned [as occurring in the *Cogitata et Visa*], in which the true method is spoken of as intermediate to those of the dogmatici or rationales, and of the empirici. There will be, he concludes, good ground for hope when the experimental and reasoning faculties are more intimately united than they have ever yet been. So likewise when natural philosophy ceases to be alloyed with matter extraneous to it, and when any one can be found content to begin at the beginning and, putting aside all popularly received notions and opinions, to apply himself afresh to experiences and particulars. And here Bacon introduces an illustration which he has also employed elsewhere, comparing the regeneration of the sciences to the exploits of Alexander, which were at first esteemed portentous and more than human, and yet afterwards it was Livy's judgment that he had done no more than despise a vain show of difficulty. Bacon then resumes his enumeration of the improvements which are to be made, each of which will be a ground of hope. The first is a better natural history than has yet been composed; and it is to be observed that a natural history which is designed to contain the materials for the instauration of philosophy differs essentially from a natural history which has no such ulterior end: the chief difference is, that an ordinary natural history does not contain the experimental results furnished by the arts. In the second place, among these results themselves there is a great lack of experimenta lucifera, that is of experiments which, though not practically useful, yet serve to give light for the discovery of causes and axioms: hitherto men have busied themselves for the most part with experimenta fructifera, that is experiments of use and profit. Thirdly, experimental researches must be conducted orderly and according to rule and law, and not as hitherto in a desultory and irregular manner. Again, when the materials required have been collected, the mind will not be able to deal with them without assistance and memoriter: all discoveries ought to be based upon written records—" nulla nisi de scripto inventio probanda est". This is what Bacon calls experientia literata [47], his meaning apparently being that out of the storehouse of natural history all the facts connected with any proposed subject of investigation should be extracted and reduced to writing before anything else is done. Furthermore, all these facts must not only be reduced to writing, but arranged tabularly. In dealing with facts thus collected and arranged, we are to regard them chiefly as the materials for the construction of axioms, our path leading us upwards from particulars to axioms, and then downwards from axioms to works;

[47] " Illâ vero in usum veniente, ab experientiâ *factâ demum literatâ*, melius sperandum ". In Montagu's edition *literatâ* is printed incorrectly with a capital letter; which makes it seem as if the *experientia facta literata* here spoken of were the same as the *experientia quam vocamus literatam* in Aph. 103. But they are, in fact, two different things; the one being opposed to experience which proceeds without any written record of its results; the other to *vaga experientia et se tantum sequens*—experience which proceeds without any method in its inquiries. See my note on Aph. 101.—*J. S.*

and the ascent from particulars to axioms must be gradual, that is axioms of a less degree of generality must always be established before axioms of a higher. Again a new form of induction is to be introduced; for induction by simple enumeration is childish and precarious. But true induction analyses nature by rejections and exclusions, and concludes affirmatively after a sufficient number of negatives. And our greatest hope rests upon this way of induction. Also the axioms thus established are to be examined whether they are of wider generality than the particulars employed in their construction, and if so, to be verified by comparing them with other facts, " per novorum particularium designationem [48], quasi fidejussione quâdam". Lastly, the sciences must be kept in connection with natural philosophy.

Bacon then goes on (108–114) to state divers grounds of hope derived from other sources than those of which he has been speaking, namely, the errors hitherto committed. The first is that without any method of invention men have made certain notable discoveries; how many more, then, and greater, by the method now to be proposed. Again, of discoveries already made, there are many which before they were made would never have been conceived of as possible, which is a reason for thinking that many other things still remain to be found out of a nature wholly unlike any hitherto known. In the course of ages these too would doubtless some time or other come to light; but by a regular method of discovery they will be made known far more certainly and in far less time,—*propere et subito et simul.* Bacon mentions particularly, as discoveries not likely to have been thought of beforehand, gun-powder, silk, and the mariner's compass; remarking that if the conditions to be fulfilled had been stated, men would have sought for something far more akin than the reality to things previously known in the case of gun-powder, if its effects only had been described, they would have thought of some modification of the battering-ram or the catapult, and not of an expansive vapour; and so in the other cases. He also mentions the art of printing as an invention perfectly simple when once made, and which nevertheless was only made after a long course of ages. Again, we may gain hope from seeing what an infinity of pains and labour men have bestowed on far less matters than that now in hand, of which if only a portion were given to the advancement of sound and real knowledge, all difficulties might be overcome. This remark Bacon makes with reference to his natural and experimental history, which he admits will be a great and royal work, and of much labour and cost. But the number of particulars to be observed ought not to deter us; on the contrary, if we consider how much smaller it is than that of the figments of the understanding, we shall find even in this grounds for hope. To these figments, commenta ingenii, the phænomena of Nature and the arts are but a mere handful. Some hope too, Bacon thinks, may be derived from his own example; for if, though of weak health, and greatly hindered by other occupations, and moreover in this matter altogether " protopirus " and following no man's track nor even communicating these things with any, he has been able somewhat to advance therein, how much may not be hoped for from the conjoined and successive labours of men at leisure from all other business? Lastly, though the breeze of hope from that new world were fainter than it is, still it were worth while to follow the adventure, seeing how great a reward success would bring.

And here (115), Bacon says, concludes the pulling-down part, pars destruens, of the Instauration. It consists of three confutations; namely, of the natural working of the mind, of received methods of demonstration, and of received theories or philosophies. In this division we perceive the influence of the first form of the doctrine of Idols. As the *Novum Organum* now stands, the pars destruens cannot be divided into three portions, each containing one of the confutations just mentioned. Thus, for instance, the doctrine of Idols, which undoubtedly forms a distinct section of the whole work, relates to all three. Errors natural to the mind, errors of demonstration, errors of theory, are all therein treated of;

[48] I understand *designatio* here to mean *discovery*. The test of the truth of the axiom was to be the discovery by its light of *new* particulars. See *Valerius Terminus*, ch. xii., quoted in note on Aph. 106.—*J. S.*

and Bacon then goes on to another part of the subject, in which, though from a different point of view, they are all again considered. The sort of cross division here introduced is explained by a passage in the *Partis secundæ Delineatio*, in which the doctrine of Idols is introduced by the remark, " Pars destruens triplex est secundum triplicem naturam idolorum quæ mentem obsident". And then, after dividing idols into the three classes already mentioned, he proceeds thus:— " Itaque pars ista quam destruentem appellamus tribus redargutionibus absolvitur, redargutione philosophiarum, redargutione demonstrationum, et redargutione rationis humanæ nativæ". When the doctrine of Idols was thrown into its present form it ceased to afford a convenient basis for the pars destruens ; and accordingly the substance of the three redargutiones is in the *Novum Organum* less systematically set forth than Bacon purposed that it should be when he wrote the *Partis secundæ Delineatio* [49]. It is to be remarked that *Redargutio Philosophiarum* is the title of one of the chapters in the third and last of the tracts published by Gruter with the title *Temporis Partus Masculus* [50], and that it is also the title of a tract published [by Stephens in 1734, and reprinted] by Mallet [in 1760 [51]], and evidently of a later date than the other of the same name.

From (116) to (128) Bacon endeavours to obviate objections and unfavourable opinions of his design. In the first place he plainly declares that he is no founder of a sect or school,—therein differing from the ancient Greeks, and from certain new men, namely Telesius, Patricius, and Severinus. Abstract opinions on nature and first principles are in his judgment of no great moment. Nor again does he promise to mankind the power of accomplishing any particular or special works —for with him works are not derived from works nor experiments from experiments, but causes and axioms are derived from both, and from these new works and experiments are ultimately deduced ; and at present the natural history of which he is in possession is not sufficient for the purposes of legitimate interpretation, that is, for the establishment of axioms. Again, that his *Natural History* and *Tables of Invention* are not free from errors, which at first they cannot be, is not a matter of much importance. These errors, if not too numerous, will readily be corrected when causes and axioms have been discovered, just as errors in a manuscript or printed book are easily corrected by the meaning of the passage in which they occur. Again, it may be said that the *Natural History* contains many commonplace things ; also many things mean and sordid ; and lastly many things too subtle to be of any use. To this a threefold answer is to be given. In the first place, rare and notable things cannot be understood, much less new things brought to light, unless the causes of common things and their causes' causes be duly examined and searched out. Secondly, whatever is worthy of existence is also worthy to be known ; for knowledge represents and is the image of existence.

[49] I think this apparent discrepancy may be better explained. It appears to me that the number of idols was originally three,—the Tribe, the Cave, and the Market-place ; all belonging to the *ratio humana nativa ;* fallacies innate or inherent in the human understanding,—to be guarded against, but not to be got rid of ; and that a fourth was added afterwards, but of quite a different kind ; consisting of fallacies which have no natural affinity to the understanding, but come from without and may be turned out again ; impressions derived from the systems which men have been taught to accept as true, or from the methods of demonstration which they have been taught to rely upon as conclusive. These are the Idols of the Theatre, and the sole objects of the two Redargutiones which stand first in the *Delineatio*, and last in the *Novum Organum*. If this be true, the *Redargutio rationis humanæ nativæ* (or I should rather say, the part of the *Novum Organum* which belongs to it) extends from the 40th to the 60th aphorism ; and the *Redargutio Philosophiarum* and *Demonstrationum* from the 61st to the 115th. For a fuller explanation and justification of this view, see Note C.—*J. S.*

[50] Say rather, " is the title prefixed by M. Bouillet to the second chapter of the fragment printed by Gruter with the heading *Tradendi modus legitimus* ". I cannot find that M. Bouillet had any authority for giving it this title, more than the tenor of the chapter itself, which shows that it fits.—*J. S.*

[51] A small portion of it was printed by Gruter at the end of the *Partis secundæ Delineatio* [and it seems to have been the beginning of the *Pars secunda* itself].

Lastly, things apparently useless are in truth of the greatest use. No one will deny that light is useful, though it is not tangible or material. And the accurate knowledge of simple natures is as light, and gives access to all the secrets on which works depend, though in itself it is of no great use.

Again it may be thought a hard saying that all sciences and authors are at once to be set aside together. But in reality this is both a more modest censure and one that carries with it a greater show of reason than any partial condemnation. It implies only that the errors hitherto committed are fundamental, and that they have not been corrected because as yet they have not been sufficiently examined. It is no presumption if any man asserts that he can draw a circle more truly with a pair of compasses than another can without; and the new method puts men's understandings nearly on the same level, because everything is to be done by definite rules and demonstrations. Bacon anticipates also another objection, that he has not assigned to the sciences their true and highest aim; which is the contemplation of truth,—not works, however great or useful. He affirms that he values works more inasmuch as they are signs and evidences of truth than for their practical utility It may also, he continues, be alleged that the method of the ancients was in reality the same as ours, only that after they had constructed the edifice of the sciences they took away the scaffolding. But this is refuted both by what they themselves say of their method [52], and by what is seen of it in their writings. Again he affirms that he does not inculcate, as some might suppose, a [53] [final suspension of judgment, as if the mind were incapable of knowing anything; that if he enjoins caution and suspense it is not as doubting the competency of the senses and understanding, but for their better information and guidance; that the method of induction which he proposes is applicable not only to what is called natural philosophy, as distinguished from logic, ethics, and politics, but to every department of knowledge; the aim being to obtain an insight into the nature of things by processes varied according to the conditions of the subject; and that in declaring that no great progress can be expected either in knowledge of truth or in power of operation by the methods of inquiry hitherto employed, he means no disrespect to the received arts and sciences, but fully recognises them as excellent in their proper place and use, and would have them honoured and cultivated accordingly.

These explanations,—together with some remarks (129), by way of encouragement to followers and fellow-labourers, on the dignity, importance, and grandeur of the end in view,—bring the preliminary considerations to a close, and clear the way for the exposition of the art of interpretation itself; which is commenced, but not completed, in the second book. What this art was, has been fully discussed in the general preface, and it is not necessary therefore to follow the subject further here. Only it is important to remark that whatever value Bacon may have attached to it, he certainly did not at this time profess to consider it either as a thing absolutely necessary, or even as the thing most necessary, for any real progress in science. In the concluding aphorism of the first book he distinctly warns the reader that the precepts which he is about to give, though he believes them to be very useful and sound, and likely to prove a great help, are not offered either as perfect in themselves or as so indispensable that nothing can be done without them. Three things only he represents as indispensable: 1st, ut "justam naturæ et experientiæ historiam præsto haberent homines atque in eâ sedulo versarentur;" 2nd, "ut receptas opiniones et notiones deponerent;" 3rd, "ut mentem a generalissimis et proximis ab illis ad tempus cohiberent". These three conditions being secured, the art of interpretation (being indeed the true and natural operation of the mind when freed from impediments) might, he thinks, suggest itself without a teacher: "fore ut etiam vi propriâ et genuinâ mentis, absque aliâ arte, in formam nostram interpretandi incidere possent; est enim interpretatio verum et naturale opus mentis, demptis iis quæ obstant": an admission which helps to account for the fact that during the five years which he

[52] I have adopted here the correction introduced into the text of the present edition.

[53] Mr. Ellis had written thus far when the fever seized him. The remaining pages which complete the analysis of the first book, are mine.—*J. S.*

afterwards devoted to the development of his philosophy, he applied himself almost exclusively to the natural history; leaving the exposition of his method of interpretation still incomplete. For it cannot be denied that, among the many things which remained to be done, the setting forward of the *Natural History* was, according to this view, the one which stood next in order of importance. In furtherance of the two other principal requisites, he had already done what he could. Every motive by which men could be encouraged to lay prejudices aside, and refrain from premature generalisations, and apply themselves to the sincere study of Nature, had already been laid before them. It remained to be seen whether his exhortations would bring other labourers into the field; but in the mean time the question lay between the completion of the *Novum Organum*, which was not indispensable, and the commencement of the collection of a *Natural History*, which was; and when he found that other labourers did *not* come forward to help, he naturally applied himself to the latter.]

NOTES

BY JAMES SPEDDING.

Note A.

I THOUGHT it better not to interrupt the reader with notes during the progress of the foregoing argument, but as some points are assumed in it upon which I shall have to express a different opinion hereafter, it may be well to notice them here; the rather because I fully concur in the conclusion notwithstanding.

1. It is assumed that the first book of *Valerius Terminus* was designed to comprehend a general survey of knowledge, such as forms the subject of the second book of the *Advancement of Learning* and of the last eight books of the *De Augmentis Scientiarum*, as well as the general reflexions and precepts, which form the subject of the first book of the *Novum Organum*;—to comprehend in short the whole first part of the Instauratio, together with the introductory portion of the second.

This is inferred from the description of the "Inventary" which was to be contained in the tenth chapter of *Valerius Terminus*, as compared with the contents of the second book of the *Advancement of Learning*.

Now my impression is that this Inventary would have corresponded, not to the second book of the *Advancement*, but only to a certain *Inventarium opum humanarum* which is there, and also in the *De Augmentis* (iii. 5), set down as a *desideratum*; and which was to be, not a general survey of all the departments of knowledge, but merely an appendix to one particular department; that, namely, which is called in the *Advancement Naturalis Magia, sive Physica operativa major*[1]; and in the *Catalogus Desideratorum* at the end of the *De Augmentis, Magia Naturalis, sive Deductio formarum ad opera*.

The grounds of this conclusion will be explained fully in their proper place.[2] It is enough at present to mark the point as disputable; and to observe that if this argument fails, there seems to be no reason for thinking that anything corresponding to the *first* part of the Instauratio entered into the design of *Valerius Terminus*; also that the principal ground here alleged for concluding that *Valerius Terminus* was written some time before the *Advancement*—a conclusion which involves one considerable difficulty —is taken away.

2. It is assumed also that *Valerius Terminus* was *not* to contain anything corresponding to the last four parts of the Instauratio, but was to be merely "a statement of Bacon's method, without professing to give either the collection of facts to which the method was to be applied, or the results thereby obtained".

This appears to be inferred chiefly from the title—viz. "Of the Interpretation of Nature".

Now it seems to me that this argument proves too much. For I find the same title given to another unfinished work—the *Temporis Partus Masculus*—of which we happen to know that it was meant to be in three books; the first to be entitled *Perpolitio et applicatio mentis*; the second, *Lumen Naturæ, seu formula Interpretationis*; the third, *Natura illuminata, sive Veritas Rerum.* The first would have corresponded therefore to the first book of the *Novum Organum*; the second, being a statement of the new method, to the second and remaining books; the third, being a statement of the application of the new method, to the sixth and last part of the Instauratio. It would seem from this that when Bacon designed the *Temporis Partus Masculus*, he had conceived

[1] See margin. It is to be observed that in Montagu's edition of the *Advancement* the titles in the margin are by some strange negligence omitted; so that the correspondence between the two Inventaries was the more easily overlooked.

[2] See my note at the end of Mr. Ellis's preface to *Valerius Terminus*.

the idea of a work embracing the entire field of the Instauratio (the first part only excepted), though less fully developed and differently distributed. And I see no sufficient reason for supposing that the design of the *Valerius Terminus* was less extensive.

3. " The *Temporis Partus Masculus* published by Gruter " is spoken of as probably or possibly " the same as the *Temporis Partus Maximus* mentioned by Bacon in his letter to Fulgenzio ", and if so, the earliest of all his writings.

Now the writing or rather collection of writings here alluded to is that published not by Gruter but by M. Bouillet; in whose edition of the " Œuvres Philosophiques " the title *Temporis Partus Masculus* is prefixed to four distinct pieces. 1. A short prayer. 2. A fragment headed *Aphorismi et Consilia de auxiliis mentis et accensione luminis naturalis*. 3. A short piece entitled *De Interpretatione Naturæ sententiæ duodecim*. 4. A fragment in two chapters headed *Tradendi modus legitimus*. It is true that from the manner in which M. Bouillet has printed them, any one would suppose that he had Gruter's authority for collecting them all under the same general title. But it is not so. In Gruter's *Scripta philosophica* the title *Temporis Partus Masculus* appears in connexion with the first, and the first only. The last has indeed an undoubted claim to it upon other and better authority. But I can find no authority whatever for giving it to the other two. If therefore the resemblance of the names be thought a sufficient reason for identifying the *Partus Masculus* with the *Partus Maximus*, that identity must be understood as belonging to the first and fourth only. The grounds of that opinion and of my own dissent from it will be discussed in the proper place. With regard to the argument now in hand,—(viz. whether Bacon, when he wrote the *Temporis Partus Masculus*, had yet thought of producing a great work like the Instauratio)—it is enough perhaps to observe that at whatever period or periods of his life these four pieces were composed, they all belong to the second part of the Instauratio; not as prefaces or prospectuses, but as portions of the work itself; and that if none of them contains any allusion to the other parts, the same may be said of the first book of the *Novum Organum* itself; and therefore that we cannot be warranted in concluding from that fact that the plan of the Instauratio had not yet been conceived.

4. It is assumed that the work which Bacon contemplated when he wrote the *De Interpretatione Naturæ Procemium* would not have contained the new method and its results (these being, according to his then intention, to be communicated only to chosen followers), but merely the general views of science which form the subject of the first book of the *Novum Organum*.

This seems to be gathered from what he says in the Procemium concerning the manner in which the several parts of the work were to be published: " Publicandi autem ista ratio ea est, ut quæ ad ingeniorum correspondentias captandas et mentium areas purgandas pertinent, edantur in vulgus et per ora volitent: *reliqua per manus tradantur cum electione et judicio* ": the " reliqua " being, as appears a little farther on, " ipsa Interpretationis formula et inventa per eandem ": from which it seems to be inferred that the exposition of the new method was not only not to be *published* along with the rest of the work, but to be excluded from it altogether;—to be kept as a secret, and transmitted *orally*. The grounds of this opinion I shall examine more particularly in a subsequent note with reference to another question. The question with which we are now dealing is only whether at that time Bacon can be supposed to have " thought of producing a great work like the Instauratio " : upon which I will only say that as an intention not to *publish* does not imply an intention not to *write*, so neither does an intention to write imply an intention to publish. And since there is nothing in the *Partis secundæ Delineatio* from which we can infer that even then he intended to *publish* the whole, I do not see how we can infer that the design of *composing* a great work like the Instauratio had been conceived in the interval between the writing of these two pieces. For as in the one case he may not have intended to publish what we know he did intend to write, so in the other he may have intended to write what we know he did not intend to publish. And indeed though the Procemium stands in Gruter's volume by itself and we cannot know to which of Bacon's projected works on the Interpretation of Nature it was meant to be prefixed, there is none which it seems to fit so well as the *Temporis Partus Masculus*. Now the *Temporis Partus Masculus*, as we know from the titles of the three books above quoted, *was* to contain both the *formula Interpretationis* and the *inventa per eandem*.

All these points will be considered more at large when I come to state the grounds upon which I have assigned to each tract its place in this edition. In the meantime I am unwilling to let any conclusion of importance appear to rest upon them; and in the present case all inferences which are in any way dependent upon the assumptions which I have noticed as questionable may I think be freely dispensed with. That to bring in a new method of Induction was Bacon's central idea and original design, and that the idea of an *Instauratio Magna* came after, may in the absence of all evidence to the contrary be safely enough inferred from his own words in the *Advancement of Learning;* where after reporting a deficiency of the first magnitude in that department of knowledge which concerns the invention of sciences,—a deficiency proved by the barrenness and accounted for by the viciousness and incompetency of the *method of induction then in use*,—he adds, "This part of Invention, concerning the Invention of Sciences, I purpose, if God give me leave, hereafter to propound; having digested into two parts; whereof the one I term *Experientia Literata*, and the other *Interpretatio Naturæ* [3]; the former being but a degree and rudiment of the latter. But I will not dwell too long nor speak too great upon a promise." This "Interpretatio Naturæ" can have been nothing else therefore than a new method of induction to supply the place of the vicious and incompetent method then in use; and since among all the reported "deficiencies" this is the only one which he himself proposes to supply,—for of the others he merely gives specimens to make his meaning clear,—we may, I think, safely conclude that this and no other was the great work which he was meditating when he wrote the *Advancement of Learning.* His expressions moreover seem to imply that this work was already begun and in progress; and seeing that the *Valerius Terminus* answers the description both in title and (so far as the first book goes, which is all we know of) in contents also, why may we not suppose that it was a commencement or a sketch of the very work he speaks of, and that of the fragment which has been preserved part was written before and part after? a supposition probable enough in itself, and by which at least one difficulty, which I shall mention hereafter [4], is effectually removed.

As an additional reason for thinking that the Idea of the Instauratio Magna was of later date than that of a work on the Interpretation of Nature, I may observe that the *name* Instauratio does not occur in any of Bacon's letters earlier than 1609. The earliest of his compositions in which it appears was probably the *Partis Instaurationis secundæ Delineatio et Argumentum;* but of this the date cannot be fixed with any certainty; and as Gruter is our only authority for it, and the word Instauratio appears in the title only, not in the body of the work, we cannot even be sure that it was originally there. If Gruter found a manuscript headed "Partis secundæ Delineatio, etc.", and evidently referring to the parts of the Instauratio Magna, he was likely enough to insert the word silently by way of explanation.

Note B.

The question is, how far, by what means, and with what motive, Bacon at one time wished to keep his system secret.

Let us first compare all the passages in which such an intention appears to be intimated, or such a practice alluded to; taking them in chronological order, as far as our knowledge of the dates of his various writings enables us to do so. These which follow are all that I have been able to find.

1. *Valerius Terminus.* Ch. 18.

"That the discretion anciently observed, though by the precedent of many vain persons and deceivers abused, of publishing part and reserving part to a private succession, and of publishing in such a manner whereby it may not be to the taste or capacity of all, *but shall as it were single and adopt his reader*, is not to be laid aside; both for the avoiding of abuse in the excluded, and the strengthening of affection in the admitted".

And again (Ch. 11.), "To ascend further by scale I do forbear, partly because it would

[3] The corresponding passage in the *De Augmentis* calls it "Interpretatio Naturae *sive Novum Organum*".

[4] See my note at the end of Mr. Ellis's Preface to the *Valerius Terminus.*

draw on the example to an over-great length, but chiefly because it would open that *which in this work I determine to reserve.*"

2. *Advancement of Learning.*

"And as Alexander Borgia was wont to say of the expedition of the French for Naples, that they came with chalk in their hands to mark up their lodgings, and not with weapons to fight; so I like better that entry of truth which cometh peaceably with chalk to *mark up those minds which are capable to lodge and harbour it,* than that which cometh with pugnacity and contention."

3. *Advancement of Learning.*

"Another diversity of method there is," [he is speaking of the different methods of "tradition", i.e. of communicating and transmitting knowledge] which hath some affinity with the former, used in some cases by the discretion of the ancients, but disgraced since by the impostures of many vain persons, who have made it as a false light for their counterfeit merchandises; and that is, *enigmatical* and *disclosed.* The pretence whereof [that is, of the enigmatical method] is to *remove the vulgar capacities* from being admitted to the secrets of knowledges, and to *reserve them to selected auditors,* or wits of such sharpness as can pierce the veil."

4. *Prooemium de Interpretatione Naturæ.*

"Publicandi autem ista ratio ea est, ut quæ ad ingeniorum correspondentias captandas et mentium areas purgandas pertinent, edantur in vulgus et per ora volitent; reliqua *per manus tradantur cum electione et judicio.* Nec me latet usitatum et tritum esse impostorum artificium, ut quædam a vulgo secernant nihilo iis ineptiis quas vulgo propinant meliora. Sed ego sine omni impostura, ex providentiâ sanâ prospicio, ipsam interpretationis formulam et inventa per eandem, *intra legitima et optata ingenia clausa,* vegetiora et munitiora futura."

5. *De Interpretatione Naturæ Sententiæ* XII.
De moribus Interpretis.

"Sit etiam in scientiâ quam adeptus est nec occultandâ nec proferendâ vanus, sed ingenuus et prudens: tradatque inventa non ambitiosè aut malignè, sed modo primum maxime vivaci et vegeto, id est ad injurias temporis munitissimo, et ad scientiam propagandam fortissimo, deinde ad errores pariendos innocentissimo, et ante omnia *qui sibi legitimum lectorem seponat.*"

6. *Temporis Partus Masculus.* C. I.

An tu censes cum omnes omnium mentium aditus ac meatus obscurissimis idolis, iisdemque alte hærentibus et inustis, obsessi et obstructi sint, veris Rerum et nativis radiis sinceras et politas areas adesse? Nova invenienda est ratio quâ mentibus obductissimis illabi possimus. Ut enim phreneticorum deliramenta arte et ingenio subvertuntur, vi et contentione efferantur, omnino ita in hâc universali insaniâ mos gerendus est. Quid? leviores illæ conditiones, quæ ad legitimum scientiæ tradendæ modum pertinent, an tibi tam expeditæ et faciles videntur? ut modus innocens sit; d est nulli prorsus errori ansam et occasionem præbeat? ut vim quandam insitam et innatam habeat tum ad fidem conciliandam, tum ad pellendas injurias temporis, adeo ut scientia ita tradita, veluti planta vivax et vegeta, quotidie serpat et adolescat? ut *idoneum et legitimum sibi lectorem seponat et quasi adoptet?*"

7. *Cogitata et visa.*

"Itaque de re non modo perficiendâ sed et communicandâ et tradendâ (quâ par est curâ) cogitationem suscipiendam esse. Reperit autem homines in rerum scientiâ quam sibi videntur adepti, interdum proferendâ interdum occultandâ, famæ et ostentationi servire: quin et eos potissimum qui minus solida proponunt solere ea quæ adferunt obscurâ et ambiguâ luce venditare, ut facilius vanitati suæ velificare possint. Putare autem se id tractare quod ambitione aliquâ aut affectatione polluere minime dignum sit; sed tamen necessario eo decurrendum esse (nisi forte rerum et animorum valde imperitus esset, et prorsus inexplorato viam inire vellet) ut satis meminerit, inveteratos semper errores, tanquam phreneticorum deliramenta, arte et ingenio subverti, vi et contentione efferari. Itaque prudentiâ et morigeratione quâdam utendum (quanta cum simplicitate et candore conjungi potest) ut contradictiones ante extinguentur quam excitentur. . . . . Venit ei itaque in mentem posse aliquid simplicius proponi, quod *in vulgus non editum*, saltem tamen ad rei tam salutaris abortum arcendum satis fortasse esse possit. Ad hunc finem parare se de naturâ opus quod errores minimâ asperitate destruere, et ad hominum mentes non turbide accedere possit; quod et facilius fore, quod non se pro duce gesturus, sed ex naturâ lucem præbiturus et sparsurus sit, ut duce postea non sit opus."

8. *Redargutio Philosophiarum* (the beginning of the *Pars secunda*, following the *Delineatio*).

" Omnem violentiam (ut jam ab initio professi sumus) abesse volumus: atque quod Borgia facetè de Caroli octavi expeditione in Italiam dixit; Gallos venisse in manibus cretam tenentes quâ diversoria notarent, non arma quibus perrumperent; similem quoque inventorum nostrorum et rationem et successum animo præcipimus; nimirum ut potius *animos hominum capaces et idoneos seponere et subire possint*, quam contra sentientibus molesta sint."

9. *Novum Organum*. I. 35.

" Dixit Borgia de expeditione Gallorum in Italiam, eos venisse cum cretâ in manibus, ut diversoria notarent, non cum armis, ut perrumperent: Itidem et nostra ratio est; ut doctrina nostra *animos idoneos et capaces subintret*; confutationum enim nullus est usus, ubi de principiis et ipsis notionibus atque etiam de formis demonstrationum dissentimus."

10. *De Augmentis Scientiarum*. VI. 2.

" Sequitur aliud methodi discrimen, priori [methodo ad filios, etc.], intentione affine, reipsâ fere contrarium. Hoc enim habet utraque methodus commune, ut *vulgus auditorum a selectis separet*; illud oppositum, quod prior introducit modum tradendi solito apertiorem; altera, de quâ jam dicemus, occultiorem. Sit igitur discrimen tale, ut altera methodus sit exoterica, altera acroamatica. Etenim quam antiqui adhibuerunt præcipue in edendis libris differentiam, eam nos transferimus ad ipsum modum tradendi. Quin etiam acroamatica ipsa apud veteres in usu fuit, atque prudentur et cum judicio adhibita. At acroamaticum sive ænigmaticum istud dicendi genus posterioribus temporibus dehonestatum est a plurimis, qui eo tanquam lumine ambiguo et fallaci abusi sunt ad merces suas adulterinas extrudendas. Intentio autem ejus ea esse videtur, *ut traditionis involucris vulgus (profanum scilicet) a secretis scientiarum summoveatur; atque illi tantum admittantur qui aut per manus magistrorum parabolarum interpretationem nacti sunt, aut proprio ingenii acumine et subtilitate intra velum penetrare possint*."

These are all the passages I have been able to find, in which the advantage of keeping certain parts of knowledge reserved to a select audience is alluded to. And the question is whether the reserve which Bacon contemplated can be justly compared with that practised by the alchemists and others, who concealed their discoveries as " treasures of which the value would be decreased if others were allowed to share in it ".

Now I would observe in the first place that though the expression " single and adopt his reader ", or its equivalent, occurs in all these passages, and that too in immediate reference to the method of delivery or transmission, yet in many of them the object of so singling and adopting the reader was certainly *not* to keep the knowledge secret; for many, indeed most, of them relate to that part of the subject which Bacon never proposed to reserve, but which was designed " edi in vulgus et per ora volitare ". The part which he proposed to reserve is distinctly defined in the fourth extract as " ipsa interpretationis formula et inventa per eandem " ; the part to be published is " ea quæ ad ingeniorum correspondentias captandas et mentium areas purgandas pertinent ". Now it is unquestionably to this latter part that the second, the eighth, and the ninth extracts refer. " Primo enim," he says in the *Partis secundæ Delineatio*, " mentis area æquanda et liberanda ab eis quæ hactenus recepta sunt ". This he calls *Pars destruens*; and he proposes to begin with the *Redargutio Philosophiarum*, from the introduction to which the eighth extract is taken. And the other two must of course be classed with it. Thus the " animi capaces et idonei " which he wishes " seponere et subire ", are clearly identified with the minds marked up with chalk as capable of lodging and harbouring the truth, which are spoken of in the *Advancement*.

Next to the *Pars destruens* came the *Pars præparans*, the object of which was to prepare men's expectations for what was coming, and by dislodging erroneous preconceptions to make their minds ready for the reception of the truth. To this part belongs the seventh extract; and if the seventh, then the sixth, which evidently corresponds to it; and if the sixth, then the fifth, which is but the sixth condensed. Or if there be any doubt about the correspondence between the seventh and sixth, it will I think be removed by comparing them both with the following passage which winds up the description of the *Pars præparans* in the *Partis secundæ Delineatio.*

"Quod si cui supervacua videatur accurata ista nostra quam adhibemus ad mentes præparandas diligentia, atque cogitet hoc quiddam esse ex pompâ et in ostentationem compositum; itaque cupiat rem ipsam missis ambagibus et præstructionibus simpliciter exhiberi; certe optabilis nobis foret (si vera esset) hujusmodi insimulatio. Utinam enim tam proclive nobis esset difficultates et impedimenta vincere quam fastum inanem et falsum apparatum deponere. Verum hoc velimus homines existiment, nos haud inexplorato viam in tantâ solitudine inire, præsertim cum argumentum hujusmodi præ manibus habeamus quod tractandi imperitiâ perdere et veluti exponere nefas sit. Itaque ex perpenso et perspecto tam rerum quam animorum statu, duriores fere aditus ad hominum mentes quam ad res ipsas invenimus, ac tradendi labores inveniendi laboribus haud multo leviores experimur, atque, quod in intellectualibus res nova fere est, morem gerimus, et tam nostras cogitationes quam aliorum simul bajulamus. Omne enim idolum vanum arte atque obsequio ac debito accessu subvertitur, vi et contentione atque incursione subitâ et abruptâ efferatur. . . . . Quâ in re accedit et alia quædam difficultas ex moribus nostris non parva, quod constantissimo decreto nobis ipsi sancivimus, ut candorem nostrum et simplicitatem perpetuo retineamus, nec per vana ad vera aditum quæramus; sed ita obsequio nostro moderemur ut tamen non per artificium aliquod vafrum aut imposturam aut aliquid simile imposturæ, sed tantummodo per ordinis lumen et novorum super saniorem partem veterum sollertem insitionem, nos nostrorum votorum compotes fore speremus".

Now all this was to precede and prepare for the exposition of the method of induction itself—the "formula ipsa interpretationis"—which alone it was proposed to reserve; and therefore we must understand the *legitimus lector* of the fifth and sixth extracts, as corresponding with the "*animus capax et idoneus*" of the eighth and ninth; and with the mind "chalked and marked up" by truth as "capable to lodge and harbour it", of the second; and we must not suppose that the process of singling and adopting the fit reader was to be effected by any restraint in communication, or any obscurity in style, which should exclude others; but by presenting the truth in such a shape as should be least likely to shock prejudice or awaken contradiction, and most likely to win its way into those minds which were best disposed to receive it. The object was to propagate knowledge so that it should grow and spread: the difficulty anticipated was not in excluding auditors, but in finding them [5].

Thus I conceive that six out of the ten passages under consideration must be set aside as not bearing at all upon the question at issue. Of the four that remain, two must be set aside in like manner, because, though they directly allude to the practice of transmitting knowledge as a secret from hand to hand, they contain no evidence that Bacon approved of it. These are the third and the last, and come respectively from the *Advancement of Learning*, one of his earliest works, and from the *De Augmentis Scientiarum*, one of his latest. In both these works the object being to show in what departments the stock of knowledge then existing was defective, the various methods which have been or may be adopted for the transmission of knowledge are pointed out as a fit subject of inquiry, and the secret or enigmatical or acroamatic method is described among the rest; but it is described only, not recommended.

There remain therefore only the first and the fourth extracts to be considered; and it is true that in both of these Bacon intimates an intention to reserve the communica-

[5] It may be worth while perhaps to compare with these passages an expression which Bacon uses in his letter to Dr. Playfere,—proposing to him to translate the *Advancement of Learning* into Latin; where a similar meaning is conveyed under another image. "Wherefore since I have only taken upon me to ring a bell to call other wits together, which is the meanest office, it cannot but be consonant to my desire to have that bell heard as far as can be. And since they are but sparks which can work but upon matter prepared, I have the more reason to wish that those sparks may fly abroad, *that they may the better find and light upon those minds and spirits that are apt to be kindled.*"

tion of one part of his philosophy—the "formula ipsa interpretationis et inventa per eandem"—to certain fit and chosen persons. May we infer from the expressions which he there uses, that his object was to prevent it from becoming generally known, as being a treasure which would lose its value by being divulged? Such a supposition seems to me inconsistent not only with all we know of his proceedings, purposes, and aspirations, but with the very explanation with which he himself accompanies the suggestion. The fruits which he anticipated from his philosophy were not only intended for the benefit of all mankind, but were to be gathered in another generation. Is it conceivable that at any time of his life he would have willingly foregone the aid of any single fellow-labourer, or that anything could have been more welcome than the prospect of a rapid and indefinite increase of those "legitima et optata ingenia" in whose hands it might be expected to thrive and spread? But setting general probabilities aside, let us look at the reasons which he himself assigns for the precaution which he meditates. Ask why in *Valerius Terminus* he proposes to reserve part of his discovery to "a private succession?" His answer is, first, "for the *prevention of abuse* in the excluded"; that is, because if it should fall into incapable and unfit hands it will be misused and mismanaged: secondly, "for the *strengthening of affection* in the admitted"; that is, because the fit and capable will take more interest in the work when they feel that it is committed to their charge. Ask again why in the *Procœmium* he proposes to keep the Formula of interpretation private,—"intra legitima et optata ingenia clausa?" The answer is to the same effect—it will be "vegetior et munitior"; it will flourish better and be kept safer. And certainly if we refer to any of the many passages in which he has either enumerated the obstructions which had hitherto hindered the progress of knowledge, or described the qualifications, moral and intellectual, and the order of proceeding, which he considered necessary for the successful prosecution of the new philosophy, we may easily understand why he anticipated more hindrance than help from a popular audience.

Upon a review of the evidence therefore I see no reason to suspect that he had any other motive for his proposed reserve than that which he himself assigns; and I think we may conclude that he meant to withhold the publication of his Formula, not "as a secret of too much value to be lightly revealed", but as a subject too abstruse to be handled successfully except by the fit and few.

Note C.

On some changes in Bacon's treatment of his doctrine of Idols.

"When the doctrine of Idols" (says Mr. Ellis) "was thrown into its present form" [i.e. the form in which it appears in the *Novum Organum*, as contrasted with that in which it appears in the *Partis secundæ Delineatio*], "it ceased to afford a convenient basis for the *pars destruens*, and accordingly the substance of the three *Redargutiones* is in the *Novum Organum* less systematically set forth than Bacon purposed that it should be when he wrote the *Partis secundæ Delineatio*".

That the argument is set forth in the *Novum Organum* less systematically than Bacon originally intended, is no doubt true; for when he wrote the "Partis secundæ *Delineatio* et *Argumentum*", he meant to handle the subject regularly and completely, or (as he would himself have expressed it) "in Corpore tractatus justi"; and this in the entrance of the *Novum Organum*, which is the "Pars secunda" itself, we are expressly warned not to expect. "Sequitur secunda pars Instaurationis, quæ artem ipsam interpretandi Naturam et verioris adoperationis Intellectûs exhibet: *neque eam ipsam tamen in Corpore tractatus justi;* sed tantûm *digestam per summas, in Aphorismos*." A succession of aphorisms, not formally connected with each other, was probably the most convenient form for setting forth all that was most important in those parts of his work which he had ready; for without binding him to exhibit them in regular and apparent connexion, it left him at liberty to make the connexion as perfect and apparent as he pleased. But it has one disadvantage: the divisions between aphorism and aphorism tend to conceal from the eye the larger divisions between subject and subject. And hence arises the appearance (for I think it is only an appearance) of a deviation from the plan originally marked out for the treatment of the *pars destruens*. Between the publication of the *Advancement of Learning* and the composition of the *Novum Organum*, the doctrine of

Idols underwent one considerable modification; but not, I think, the one here supposed. That modification was introduced *before* the *Partis secundæ Delineatio* was drawn up; and after that I cannot find evidence of any substantial change.

I will first exhibit the successive aspects which the doctrine assumes, and then give what I suppose to be the true history of them.

In the *Advancement of Learning*, the Idols, native and adventitious, of the human mind are distributed into three kinds; not distinguished as yet by names, but corresponding respectively to those of the Tribe, the Cave, and the Market-place. In *Valerius Terminus*, they are distributed into four kinds: the Tribe, the Palace (corresponding with the Market-place), the Cave, and the Theatre. In the *Partis secundæ Delineatio* they are distributed again into three, but classified quite differently. The two great divisions of Adventitious and Native are retained: "aut adscititia sunt...nimirum quæ immigrârunt in mentem, etc., aut ea qua menti ipsi et substantiæ ejus inhærentia sunt et innata"; but the subdivisions are entirely changed;—the Adventitious being here divided into two kinds, neither of which is recognized at all in the *Advancement*; the Native, which are divided into two kinds in the *Advancement*, not being divided at all here, but classed together as one. In the *Advancement* we find nothing corresponding to the Idols of the Theatre, to which belong *both* the kinds of adventitious Idols mentioned in the *Delineatio*—those derived *ex philosophorum placitis*, and those derived *ex perversis legibus demonstrationum*;—in the *Delineatio* we find nothing corresponding to the Idols of the Market-place, which among those mentioned in the *Advancement* are alone entitled to be classed as adventitious. Thus the difference between the two appears at first to be total and radical, amounting to an entire rearrangement of all the classes. Instead of Idols of the Tribe, the Cave, and the Market-place, we find Idols of the Philosophies, the Demonstrations, and the Human Mind.

But the truth is that Bacon, being now engaged in laying out the large outlines of his subject, omits the minor distinctions which belong to the development of it in detail, and leaves the particular distribution and description of those "fallacies and false appearances" which are "inseparable from our nature and condition in life"—those namely which he had spoken of in the *Advancement*—to be handled in the work itself. Having, however, as he came into closer contact with his subject, foreseen the opposition which he must expect from prejudices and false appearances of another kind—prejudices which had no root in the mind itself, which were *not* "inseparable from our nature and condition of life",—mere immigrants and strangers that had come in and might be turned out,—namely, the belief in received systems and attachment to received methods of demonstration,—he had resolved to deal with these first; and therefore introduces them as a separate class, dividing them into two parts and assigning to each what we may call a separate chapter. These he afterwards called Idols of the Theatre, and treated them in the manner proposed; with this difference only—that he placed them last instead of first, and ran the two chapters into one.

This being allowed, it will be found that the one substantial change which the doctrine of Idols underwent was the admission of these Idola Theatri into the company, and that there is no real difference between the form of that doctrine as indicated in the *Delineatio* and as developed in the *Novum Organum*.

The only difficulty which this view of the subject presents is one which may be probably enough accounted for as an oversight of Bacon's own. I mean the classification of the Idola Fori, the source of which is no doubt extraneous, among the natives. Bacon was never very careful about subtle logical distinctions, and in this case his attention had not as yet been specially called to the point. For in the *Advancement of Learning*, though the great division between Native and Adventitious appears to be recognized in the margin, there is no hint of it in the text,—the particular Idols not being arranged with any reference to those two general heads; while in *Valerius Terminus* the larger division is not alluded to at all, and the order in which the four Idols are there enumerated,—the first and third being of one class, the second and fourth of the other,—seems to prove that no such classification was then in his mind. Besides, it is to be remembered that the Idola Fori, however distinct in their origin, are in their nature and qualities much nearer akin to the other two than to the Idola Theatri. For though they come from without, yet when they are once in they naturalise themselves and take up their abode along with the natives, produce as much confusion, and can as hardly be expelled. Philosophical systems may be exploded, false methods of demonstration may be discarded, but intercourse of words is "inseparable from our condition in life."

At any rate, let the logical error implied be as large as it may, it is certain that Bacon did in fact always class these three together. Wherever he mentions the Idols of the Market-place with any reference to classification, they are grouped with those of the Tribe and the Cave, and distinguished from those of the Theatre. In the *Temporis Partus Masculus*, c. 2. (which is, I think, the earliest form of the *Redargutio Philosophiarum*, though probably of later date than the *Delineatio*) we find "Nam Idola quisque sua (*non* jam *scenæ* dico, sed præcipue *fori et specus*"), etc. In the *De Augmentis Scientiarum* where the four kinds of Idols are enumerated by name and in order, the line of separation is drawn not between the two first and the two last (as it would have been if Bacon had meant to balance the members of his classification on the "dichotomising principle", as suggested by Mr. Ellis, p. 225), but between the three first and the fourth; the Idola Fori being classed along with the Idola Tribûs and Specûs, as "quæ plane obsident mentem, neque evelli possunt", the Idola Theatri being broadly distinguished from them, as "quæ abnegari possunt et deponi", and which may therefore for the present be set aside. In the *Novum Organum* itself, though the divisions between aphorism and aphorism tend, as I have said, to obscure the divisions of subject, yet if we look carefully we shall see that the line of demarcation is drawn exactly in the same place, and almost as distinctly. For after speaking of the *three* first kinds of Idol, Bacon proceeds (Aph. 61.), "At Idola *Theatri* innata non sunt [like those of the Tribe and Cave] *nec occulto insinuata in Intellectum* [like those of the Market-place], sed ex fabulis theoriarum et perversis legibus demonstrationum plane indita et recepta". Lastly, in the *Distributio Operis*, where the particular Idols are not mentioned by name, but the more general classification of the *Delineatio* is retained, it is plain that under the class *Adscititia* he meant to include the Idols of the Theatre only—("adscititia vero immigrârunt in mentes hominum, vel ex philosophorum placitis et sectis, vel ex perversis legibus demonstrationum")—and therefore he must still have meant to include the Idols of the Market-place, along with the two first, under the class *Innata*.

It is worthy of remark however that, in the *Novum Organum* itself, the distinction between Adscititia and Innata disappears. And the fact probably is that when he came to describe the several Idols one by one, he became aware both of the logical inconsistency of classing the Idola Fori among the Innata, and of the practical inconvenience of classing them among the Adscititia, and therefore resolved to drop the dichotomy altogether and range them in four co-ordinate classes. And it is the removal of this boundary line which makes it *seem* at first sight as if the arrangement were quite changed, whereas it is in fact only inverted. According to the plan of the *Partis secundæ Delineatio* and also of the *Distributio Operis*, the confutation of the Immigrants,—that is, the *Redargutio Philosophiarum* and *Redargutio Demonstrationum*,—was to have the precedence, and the confutation of the Natives,—that is, the *Redargutio Rationis humanæ nativæ*,—was to follow. As it is, he begins with the last and ends with the first. And the reason of this change of plan is not difficult to divine. The *Redargutio Philosophiarum*, as he handles it, traverses a wider and more various field, and rises gradually into a strain of prophetic anticipation, after which the *Redargutio Rationis* would have sounded flat.

THE GREAT INSTAURATION
PROŒMIUM

FRANCIS OF VERULAM REASONED THUS WITH HIMSELF, AND JUDGED IT TO BE FOR THE INTEREST OF THE PRESENT AND FUTURE GENERATIONS THAT THEY SHOULD BE MADE ACQUAINTED WITH HIS THOUGHTS.

BEING convinced that the human intellect makes its own difficulties, not using the true helps which are at man's disposal soberly and judiciously; whence follows manifold ignorance of things, and by reason of that ignorance mischiefs innumerable; he thought all trial should be made, whether that commerce between the mind of man and the nature of things, which is more precious than anything on earth, or at least than anything that is of the earth, might by any means be restored to its perfect and original condition, or if that may not be, yet reduced to a better condition than that in which it now is. Now that the errors which have hitherto prevailed, and which will prevail for ever, should (if the mind be left to go its own way), either by the natural force of the understanding or by help of the aids and instruments of Logic, one by one correct themselves, was a thing not to be hoped for: because the primary notions of things which the mind readily and passively imbibes, stores up, and accumulates (and it is from them that all the rest flow) are false, confused, and over-hastily abstracted from the facts; nor are the secondary and subsequent notions less arbitrary and inconstant; whence it follows that the entire fabric of human reason which we employ in the inquisition of nature is badly put together and built up, and like some magnificent structure without any foundation. For while men are occupied in admiring and applauding the false powers of the mind, they pass by and throw away those true powers which, if it be supplied with the proper aids and can itself be content to wait upon nature instead of vainly affecting to overrule her, are within its reach. There was but one course left, therefore,—to try the whole thing anew upon a better plan, and to commence a total reconstruction of sciences, arts, and all human knowledge, raised upon the proper foundations. And this, though in the project and undertaking it may seem a thing infinite and beyond the powers of man, yet when it comes to be dealt with it will be found sound and sober, more so than what has been done hitherto. For of this there is some issue; whereas in what is now done in the matter of science there is only a whirling round about, and perpetual agitation, ending where it began. And although he was well aware how solitary an enterprise it is, and how hard a thing to win faith and credit for, nevertheless he was resolved not to abandon either it or himself; nor to be deterred from trying and entering upon that one path which is alone open to the human mind. For better it is to make a beginning of that which may lead to something, than to engage in a perpetual struggle and pursuit in courses which have no exit. And certainly the two ways of contemplation are much like those two ways of action, so much celebrated, in this—that the one, arduous and difficult in the beginning, leads out at last into the open country; while the other, seeming at first sight easy and free from obstruction, leads to pathless and precipitous places.

Moreover, because he knew not how long it might be before these things would occur to any one else, judging especially from this, that he has found no man hitherto who has applied his mind to the like, he resolved to publish at once so much as he has been able to complete. The cause of which haste was not ambition for himself, but solicitude for the work; that in case of his death there might remain some outline and project of that which he had conceived, and some evidence likewise of his honest mind and inclination towards the benefit of the human race. Certain it is that all other ambition whatsoever seemed poor in his eyes compared with the work which he had in hand; seeing that the matter at issue is either nothing, or a thing so great that it may well be content with its own merit, without seeking other recompence.

EPISTLE DEDICATORY

TO OUR MOST GRACIOUS AND MIGHTY PRINCE AND LORD JAMES, BY THE GRACE OF GOD OF GREAT BRITAIN, FRANCE, AND IRELAND KING, DEFENDER OF THE FAITH, ETC.

Most Gracious and Mighty King,

YOUR Majesty may perhaps accuse me of larceny, having stolen from your affairs so much time as was required for this work. I know not what to say for myself. For of time there can be no restitution, unless it be that what has been abstracted from your business may perhaps go to the memory of your name and the honour of your age; if these things are indeed worth anything. Certainly they are quite new; totally new in their very kind: and yet they are copied from a very ancient model; even the world itself and the nature of things and of the mind. And to say truth, I am wont for my own part to regard this work as a child of time rather than of wit; the only wonder being that the first notion of the thing, and such great suspicions concerning matters long established, should have come into any man's mind. All the rest follows readily enough. And no doubt there is something of accident (as we call it) and luck as well in what men think as in what they do or say. But for this accident which I speak of, I wish that if there be any good in what I have to offer, it may be ascribed to the infinite mercy and goodness of God, and to the felicity of your Majesty's times; to which as I have been an honest and affectionate servant in my life, so after my death I may yet perhaps, through the kindling of this new light in the darkness of philosophy, be the means of making this age famous to posterity; and surely to the times of the wisest and most learned of kings belongs of right the regeneration and restoration of the sciences. Lastly, I have a request to make—a request no way unworthy of your Majesty, and which especially concerns the work in hand; namely, that you who resemble Solomon in so many things—in the gravity of your judgments, in the peacefulness of your reign, in the largeness of your heart, in the noble variety of the books which you have composed—would further follow his example in taking order for the collecting and perfecting of a Natural and Experimental History true and severe (unincumbered with literature and book-learning), such as philosophy may be built upon,—such, in fact, as I shall in its proper place describe: that so at length, after the lapse of so many ages, philosophy and the sciences may no longer float in air, but rest on the solid foundation of experience of every kind, and the same well examined and weighed. I have provided the machine, but the stuff must be gathered from the facts of nature. May God Almighty long preserve your Majesty!

Your Majesty's
Most bounden and devoted
Servant,
FRANCIS VERULAM,
CHANCELLOR.

THE GREAT INSTAURATION

[AUTHOR'S] PREFACE.

That the state of knowledge is not prosperous nor greatly advancing ; and that a way must be opened for the human understanding entirely different from any hitherto known, and other helps provided, in order that the mind may exercise over the nature of things the authority which properly belongs to it.

It seems to me that men do not rightly understand either their store or their strength, but overrate the one and underrate the other. Hence it follows, that either from an extravagant estimate of the value of the arts which they possess, they seek no further ; or else from too mean an estimate of their own powers, they spend their strength in small matters and never put it fairly to the trial in those which go to the main. These are as the pillars of fate set in the path of knowledge ; for men have neither desire nor hope to encourage them to penetrate further. And since opinion of store is one of the chief causes of want, and satisfaction with the present induces neglect of provision for the future, it becomes a thing not only useful, but absolutely necessary, that the excess of honour and admiration with which our existing stock of inventions is regarded be in the very entrance and threshold of the work, and that frankly and without circumlocution, stripped off, and men be duly warned not to exaggerate or make too much of them. For let a man look carefully into all that variety of books with which the arts and sciences abound, he will find everywhere endless repetitions of the same thing, varying in the method of treatment, but not new in substance, insomuch that the whole stock, numerous as it appears at first view, proves on examination to be but scanty. And for its value and utility it must be plainly avowed that that wisdom which we have derived principally from the Greeks is but like the boyhood of knowledge, and has the characteristic property of boys : it can talk, but it cannot generate ; for it is fruitful of controversies but barren of works. So that the state of learning as it now is appears to be represented to the life in the old fable of Scylla, who had the head and face of a virgin, but her womb was hung round with barking monsters, from which she could not be delivered. For in like manner the sciences to which we are accustomed have certain general positions which are specious and flattering ; but as soon as they come to particulars, which are as the parts of generation, when they should produce fruit and works, then arise contentions and barking disputations, which are the end of the matter and all the issue they can yield. Observe also, that if sciences of this kind had any life in them, that could never have come to pass which has been the case now for many ages—that they stand almost at a stay, without receiving any augmentations worthy of the human race ; insomuch that many times not only what was asserted once is asserted still, but what was a question once is a question still, and instead of being resolved by discussion is only fixed and fed ; and all the tradition and succession of schools is still a succession of masters and scholars, not of inventors and those who bring to further perfection the things invented. In the mechanical arts we do not find it so ; they, on the contrary, as having in them some breath of life, are continually growing and becoming more perfect. As originally invented they are commonly rude, clumsy, and shapeless ; afterwards they acquire new powers and more commodious arrangements and constructions ; in so far that men shall sooner leave the study and pursuit of them and turn to something else, than they arrive at the ultimate perfection of which they are capable. Philosophy and the intellectual sciences, on the contrary, stand like statues, worshipped and celebrated, but not moved or advanced. Nay, they sometimes flourish most in the hands of the first author, and afterwards degenerate. For when men have once made over their judgments to others' keeping, and (like those senators whom they called *Pedarii*) have agreed to support some one person's opinion, from that time they make no enlargement of the sciences themselves, but fall to the servile office of embellishing certain individual authors and increasing their retinue. And let it not be said that the sciences have been growing gradually till they have at last reached their full stature, and so (their course being completed) have settled in the works of a few writers; and that there being now no room for the invention of better, all that remains is to embellish and cultivate those things which have been invented already. Would it were so! But the truth is that this appropriating of the sciences has its origin in nothing better than the confidence of a few persons and the sloth and indolence of the rest. For after the sciences had been in several parts perhaps cultivated and handled diligently, there has risen up some man of bold disposition, and famous for methods and short ways which people like, who has in appearance reduced them to an art, while he has in fact only spoiled all that the others had done. And yet this is what posterity like, because it makes the work short and easy, and saves further inquiry, of which they are weary and impatient. And if any one take this general acquiescence and consent for an argument of weight, as being the judgment of Time, let me tell him that the reasoning on which he relies is most fallacious and weak. For, first, we are far from knowing all that in the matter of sciences and arts has in various ages and places been brought to light and published; much less, all that has been by private persons secretly attempted and stirred; so neither the births nor the miscarriages of Time are entered in our records. Nor, secondly, is the consent itself and the time it has continued a consideration of much worth. For however various are the forms of civil polities, there is but one form of polity in the sciences; and that always has been and always will be popular. Now the doctrines which find most favour with the populace are those which are either contentious and pugnacious, or specious and empty; such, I say, as either entangle assent or tickle it. And therefore no doubt the greatest wits in each successive age have been forced out of their own course; men of capacity and intellect above the vulgar having been fain, for reputation's sake, to bow to the judgment of the time and the multitude; and thus if any contemplations of a higher order took light anywhere, they were presently blown out by the winds of vulgar opinions. So that Time is like a river, which has brought down to us things light and puffed up, while those which are weighty and solid have sunk. Nay, those very authors who have usurped a kind of dictatorship in the sciences and taken upon them to lay down the law with such confidence, yet when from time to time they come to themselves again, they fall to complaints of the subtlety of nature, the hiding-places of truth, the obscurity of things, the entanglement of causes, the weakness of the human mind; wherein nevertheless they show themselves never the more modest, seeing that they will rather lay the blame upon the common condition of men and nature than upon themselves. And then whatever any art fails to attain, they ever set it down upon the authority of that art itself as impossible of attainment; and how can art be found guilty when it is judge in its own cause? So it is but a device for exempting ignorance from ignominy. Now for those things which are delivered and received, this is their condition: barren of works, full of questions; in point of enlargement slow and languid; carrying a show of perfection in the whole, but in the parts ill filled up; in selection popular, and unsatisfactory even to those who propound them; and therefore fenced round and set forth with sundry artifices. And if there be any who have determined to make trial for themselves, and put their own strength to the work of advancing the boundaries of the sciences, yet have they not ventured to cast themselves completely loose from received opinions or to seek their knowledge at the fountain; but they think they have done some great thing if they do but add and introduce into the existing sum of science something of their own ; prudently considering with themselves that by making the addition they can assert their liberty, while they retain the credit of modesty by assenting to the rest. But these mediocrities and middle ways so much praised, in deferring to opinions and customs, turn to the great detriment of the sciences. For it is hardly possible at once to admire an author and to go beyond him ; knowledge being as water, which will not rise above the level from which it fell. Men of this kind, therefore, amend some things, but advance little ; and improve the condition of knowledge, but do not extend its range. Some, indeed, there have been who have gone more boldly to work, and taking it all for an open matter and giving their genius full play, have made a passage for themselves and their own opinions by pulling down and demolishing former ones ; and yet all their stir has but little advanced the matter ; since their aim has been not to extend philosophy and the arts in substance and value, but only to change doctrines and transfer the kingdom of opinions to themselves ; whereby little has indeed been gained, for though the error be the opposite of the other, the causes of erring are the same in both. And if there have been any who, not binding themselves either to other men's opinions or to their own, but loving liberty, have desired to engage others along with themselves in search, these, though honest in intention, have been weak in endeavour. For they have been content to follow probable reasons, and are carried round in a whirl of arguments, and in the promiscuous liberty of search have relaxed the severity of inquiry. There is none who has dwelt upon experience and the facts of nature as long as is necessary. Some there are indeed who have committed themselves to the waves of experience, and almost turned mechanics ; yet these again have in their very experiments pursued a kind of wandering inquiry, without any regular system of operations. And besides they have mostly proposed to themselves certain petty tasks, taking it for a great matter to work out some single discovery ;—a course of proceeding at once poor in aim and unskilful in design. For no man can rightly and successfully investigate the nature of anything in the thing itself ; let him vary his experiments as laboriously as he will, he never comes to a resting-place, but still finds something to seek beyond. And there is another thing to be remembered ; namely, that all industry in experimenting has begun with proposing to itself certain definite works to be accomplished, and has pursued them with premature and unseasonable eagerness ; it has sought, I say, experiments of Fruit, not experiments of Light ; not imitating the divine procedure, which in its first day's work created light only and assigned to it one entire day ; on which day it produced no material work, but proceeded to that on the days following. As for those who have given the first place to Logic, supposing that the surest helps to the sciences were to be found in that, they have indeed most truly and excellently perceived that the human intellect left to its own course is not to be trusted ; but then the remedy is altogether too weak for the disease ; nor is it without evil in itself. For the Logic which is received, though it be very properly applied to civil business and to those arts which rest in discourse and opinion, is not nearly subtle enough to deal with nature ; and in offering at what it cannot master, has done more to establish and perpetuate error than to open the way to truth.

Upon the whole therefore, it seems that men have not been happy hitherto either in the trust which they have placed in others or in their own industry with regard to the sciences ; especially as neither the demonstrations nor the experiments as yet known are much to be relied upon. But the universe to the eye of the human understanding is framed like a labyrinth ; presenting as it does on every side so many ambiguities of way, such deceitful resemblances of objects and signs, natures so irregular in their lines, and so knotted and entangled. And then the way is still to be made by the uncertain light of the sense, sometimes shining out, sometimes clouded over, through the woods of experience and particulars ; while those who offer themselves for guides are (as was said) themselves also puzzled, and increase the number of errors and wanderers. In circumstances so difficult neither the natural force of man's judgment nor even any accidental felicity offers any chance of success. No excellence of wit, no repetition of chance experiments, can overcome such difficulties as these. Our steps must be guided by a clue, and the whole way from the very first perception of the senses must be laid out upon a sure plan. Not that I would be understood to mean that nothing whatever has been done in so many ages by so great labours. We have no reason to be ashamed of the discoveries which have been made, and no doubt the ancients proved themselves in everything that turns on wit and abstract meditation, wonderful men. But as in former ages, when men sailed only by observation of the stars, they could indeed coast along the shores of the old continent or cross a few small and mediterranean seas; but before the ocean could be traversed and the new world discovered, the use of the mariner's needle, as a more faithful and certain guide, had to be found out; in like manner the discoveries which have been hitherto made in the arts and sciences are such as might be made by practice, meditation, observation, argumentation, —for they lay near to the senses, and immediately beneath common notions; but before we can reach the remoter and more hidden parts of nature, it is necessary that a more perfect use and application of the human mind and intellect be introduced.

For my own part at least, in obedience to the everlasting love of truth, I have committed myself to the uncertainties and difficulties and solitudes of the ways, and relying on the divine assistance have upheld my mind both against the shocks and embattled ranks of opinion, and against my own private and inward hesitations and scruples, and against the fogs and clouds of nature, and the phantoms flitting about on every side; in the hope of providing at last for the present and future generations guidance more faithful and secure. Wherein if I have made any progress, the way has been opened to me by no other means than the true and legitimate humiliation of the human spirit. For all those who before me have applied themselves to the invention of arts have but cast a glance or two upon facts and examples and experience, and straightway proceeded, as if invention were nothing more than an exercise of thought, to invoke their own spirits to give them oracles. I, on the contrary, dwelling purely and constantly among the facts of nature, withdraw my intellect from them no further than may suffice to let the images and rays of natural objects meet in a point, as they do in the sense of vision; whence it follows that the strength and excellency of the wit has but little to do in the matter. And the same humility which I use in inventing I employ likewise in teaching. For I do not endeavour either by triumphs of confutation, or pleadings of antiquity, or assumption of authority, or even by the veil of obscurity, to invest these inventions of mine with any majesty; which might easily be done by one who sought to give lustre to his own name rather than light to other men's minds. I have not sought (I say) nor do I seek either to force or ensnare men's judgments, but I lead them to things themselves and the concordances of things, that they may see for themselves what they have, what they can dispute, what they can add and contribute to the common stock. And for myself, if in anything I have been either too credulous or too little awake and attentive, or if I have fallen off by the way and left the inquiry incomplete, nevertheless I so present these things naked and open, that my errors can be marked and set aside before the mass of knowledge be further infected by them; and it will be easy also for others to continue and carry on my labours. And by these means I suppose that I have established for ever a true and lawful marriage between the empirical and the rational faculty, the unkind and ill-starred divorce and separation of which has thrown into confusion all the affairs of the human family.

Wherefore, seeing that these things do not depend upon myself, at the outset of the work I most humbly and fervently pray to God the Father, God the Son, and God the Holy Ghost, that remembering the sorrows of mankind and the pilgrimage of this our life wherein we wear out days few and evil, they will vouchsafe through my hands to endow the human family with new mercies. This likewise I humbly pray, that things human may not interfere with things divine, and that from the opening of the ways of sense and the increase of natural light there may arise in our minds no incredulity or darkness with regard to the divine mysteries; but rather that the understanding being thereby purified and purged of fancies and vanity, and yet not the less subject and entirely submissive to the divine oracles, may give to faith that which is faith's. Lastly, that knowledge being now discharged of that venom which the serpent infused into it, and which makes the mind of man to swell, we may not be wise above measure and sobriety, but cultivate truth in charity.

And now having said my prayers I turn to men; to whom I have certain salutary admonitions to offer and certain fair requests to make. My first admonition (which was also my prayer) is that men confine the sense within the limits of duty in respect of things divine: for the sense is like the sun, which reveals the face of earth, but seals and shuts up the face of heaven. My next, that in flying from this evil they fall not into the opposite error, which they will surely do if they think that the inquisition of nature is in any part interdicted or forbidden. For it was not that pure and uncorrupted natural knowledge whereby Adam gave names to the creatures according to their propriety, which gave occasion to the fall. It was the ambitious and proud desire of moral knowledge to judge of good and evil, to the end that man may revolt from God and give laws to himself, which was the form and manner of the temptation. Whereas of the sciences which regard nature, the divine philosopher declares that "it is the glory of God to conceal a thing, but it is the glory of the King to find a thing out". Even as though the divine nature took pleasure in the innocent and kindly sport of children playing at hide and seek, and vouchsafed of his kindness and goodness to admit the human spirit for his playfellow at that game. Lastly, I would address one general admonition to all; that they consider what are the true ends of knowledge, and that they seek it not either for pleasure of the mind, or for contention or for superiority to others, or for profit, or fame, or power, or any of these inferior things; but for the benefit and use of life; and that they perfect and govern it in charity. For it was from lust of power that the angels fell, from lust of knowledge that man fell; but of charity there can be no excess, neither did angel or man ever come in danger by it.

The requests I have to make are these. Of myself I say nothing; but in behalf of the business which is in hand I entreat men to believe that it is not an opinion to be held, but a work to be done; and to be well assured that I am labouring to lay the foundation, not of any sect of doctrine, but of human utility and power. Next, I ask them to deal fairly by their own interests, and laying aside all emulations and prejudices in favour of this or that opinion, to join in consultation for the common good; and being now freed and guarded by the securities and helps which I offer from the errors and impediments of the way, to come forward themselves and take part in that which remains to be done. Moreover, to be of good hope, nor to imagine that this Instauration of mine is a thing infinite and beyond the power of man, when it is in fact the true end and termination of infinite error; and seeing also that it is by no means forgetful of the conditions of mortality and humanity, (for it does not suppose that the work can be altogether completed within one generation, but provides for its being taken up by another); and finally that it seeks for the sciences not arrogantly in the little cells of human wit, but with reverence in the greater world. But it is the empty things that are vast: things solid are most contracted and lie in little room. And now I have only one favour more to ask (else injustice to me may perhaps imperil the business itself)—that men will consider well how far, upon that which I must needs assert (if I am to be consistent with myself), they are entitled to judge and decide upon these doctrines of mine; inasmuch as all that premature human reasoning which anticipates inquiry, and is abstracted from the facts rashly and sooner than is fit, is by me rejected (so far as the inquisition of nature is concerned), as a thing uncertain, confused, and ill built up; and I cannot be fairly asked to abide by the decision of a tribunal which is itself on its trial.

THE PLAN OF THE WORK.

The work is in six Parts :—

1. *The Divisions of the Sciences.*
2. *The New Organon ; or Directions concerning the Interpretation of Nature.*
3. *The Phenomena of the Universe ; or a Natural and Experimental History for the foundation of Philosophy.*
4. *The Ladder of the Intellect.*
5. *The Forerunners ; or Anticipations of the New Philosophy.*
6. *The New Philosophy ; or Active Science.*

The Arguments of the several Parts.

It being part of my design to set everything forth, as far as may be, plainly and perspicuously (for nakedness of the mind is still,as nakedness of the body once was, the companion of innocence and simplicity), let me first explain the order and plan of the work. I distribute it into six parts.

The first part exhibits a summary or general description of the knowledge which the human race at present possesses. For I thought it good to make some pause upon that which is received ; that thereby the old may be more easily made perfect and the new more easily approached. And I hold the improvement of that which we have to be as much an object as the acquisition of more. Besides which it will make me the better listened to ; for " He that is ignorant (says the proverb) receives not the words of knowledge, unless thou first tell him that which is in his own heart". We will therefore make a coasting voyage along the shores of the arts and sciences received ; not without importing into them some useful things by the way.

In laying out the division of the sciences, however, I take into account not only things already invented and known, but likewise things omitted which ought to be there. For there are found in the intellectual as in the terrestial globe waste regions as well as cultivated ones. It is no wonder therefore if I am sometimes obliged to depart from the ordinary divisions. For in adding to the total you necessarily alter the parts and sections ; and the received divisions of the sciences are fitted only to the received sum of them as it stands now.

With regard to those things which I shall mark as omitted, I intend not merely to set down a simple title or a concise argument of that which is wanted. For as often as I have occasion to report anything as deficient, the nature of which is at all obscure, so that men may not perhaps easily understand what I mean or what the work is which I have in my head, I shall always (provided it be a matter of any worth) take care to subjoin either directions for the execution of such work, or else a portion of the work itself executed by myself as a sample of the whole ; thus giving assistance in every case either by work or by counsel. For if it were for the sake of my own reputation only and other men's interests were not concerned in it, I would not have any man think that in such cases merely some light and vague notion has crossed my mind, and that the things which I desire and offer at are no better than wishes ; when they are in fact things which men may certainly command if they will, and of which I have formed in my own mind a clear and detailed conception. For I do not propose merely to survey these regions in my mind, like an augur taking auspices, but to enter them like a general who means to take possession.—So much for the first part of the work.

Having thus coasted past the ancient arts, the next point is to equip the intellect for passing beyond. To the second part therefore belongs the doctrine concerning the better and more perfect use of human reason in the inquisition of things, and the true helps of the understanding: that thereby (as far as the condition of mortality and humanity allows) the intellect may be raised and exalted, and made capable of overcoming the difficulties and obscurities of nature. The art which I introduce with this view (which I call *Interpretation of Nature*) is a kind of logic; though the difference between it and the ordinary logic is great; indeed immense. For the ordinary logic professes to contrive and prepare helps and guards for the understanding, as mine does; and in this one point they agree. But mine differs from it in three points especially; viz. in the end aimed at; in the order of demonstration; and in the starting point of the inquiry.

For the end which this science of mine proposes is the invention not of arguments but of arts; not of things in accordance with principles, but of principles themselves; not of probable reasons, but of designations and directions for works. And as the intention is different, so accordingly is the effect; the effect of the one being to overcome an opponent in argument, of the other to command nature in action.

In accordance with this end is also the nature and order of the demonstrations. For in the ordinary logic almost all the work is spent about the syllogism. Of induction the logicians seem hardly to have taken any serious thought, but they pass it by with a slight notice, and hasten on to the formulæ of disputation. I on the contrary reject demonstration by syllogism, as acting too confusedly, and letting nature slip out of its hands. For although no one can doubt that things which agree in a middle term agree with one another (which is a proposition of mathematical certainty), yet it leaves an opening for deception; which is this. The syllogism consists of propositions; propositions of words; and words are the tokens and signs of notions. Now if the very notions of the mind (which are as the soul of words and the basis of the whole structure) be improperly and over-hastily abstracted from facts, vague, not sufficiently definite, faulty in short in many ways, the whole edifice tumbles. I therefore reject the syllogism; and that not only as regards principles (for to principles the logicians themselves do not apply it) but also as regards middle propositions; which, though obtainable no doubt by the syllogism, are, when so obtained, barren of works, remote from practice, and altogether unavailable for the active department of the sciences. Although therefore I leave to the syllogism and these famous and boasted modes of demonstration their jurisdiction over popular arts and such as are matter of opinion (in which department I leave all as it is), yet in dealing with the nature of things I use induction throughout, and that in the minor propositions as well as the major. For I consider induction to be that form of demonstration which upholds the sense and closes with nature, and comes to the very brink of operation, if it does not actually deal with it.

Hence it follows that the order of demonstration is likewise inverted. For hitherto the proceeding has been to fly at once from the sense and particulars up to the most general propositions, as certain fixed poles for the argument to turn upon, and from these to derive the rest by middle terms: a short way, no doubt, but precipitate; and one which will never lead to nature, though it offers an easy and ready way to disputation. Now my plan is to proceed regularly and gradually from one axiom to another, so that the most general are not reached till the last; but then when you do come to them you find them to be not empty notions, but well defined, and such as nature would really recognize as her first principles, and such as lie at the heart and marrow of things.

But the greatest change I introduce is in the form itself of induction and the judgment made thereby. For the induction of which the logicians speak, which proceeds by simple enumeration, is a puerile thing; concludes at hazard; is always liable to be upset by a contradictory instance; takes into account only what is known and ordinary; and leads to no result.

Now what the sciences stand in need of is a form of induction which shall analyse experience and take it to pieces, and by a due process of exclusion and rejection lead to an inevitable conclusion. And if that ordinary mode of judgment practised by the logicians was so laborious, and found exercise for such great wits, how much more labour must we be prepared to bestow upon this other, which is extracted not merely out of the depths of the mind, but out of the very bowels of nature.

Nor is this all. For I also sink the foundations of the sciences deeper and firmer; and I begin the inquiry nearer the source than men have done heretofore; submitting to examination those things which the common logic takes on trust. For, first, the logicians borrow the principles of each science from the science itself; secondly, they hold in reverence the first notions of the mind; and lastly, they receive as conclusive the immediate informations of the sense, when well disposed. Now upon the first point, I hold that true logic ought to enter the several provinces of science armed with a higher authority than belongs to the principles of those sciences themselves, and ought to call those putative principles to account until they are fully established. Then with regard to the first notions of the intellect; there is not one of the impressions taken by the intellect when left to go its own way, but I hold it for suspected, and no way established, until it has submitted to a new trial and a fresh judgment has been thereupon pronounced. And lastly, the information of the sense itself I sift and examine in many ways. For certain it is that the senses deceive; but then at the same time they supply the means of discovering their own errors; only the errors are here, the means of discovery are to seek.

The sense fails in two ways. Sometimes it gives no information, sometimes it gives false information. For first, there are very many things which escape the sense, even when best disposed and no way obstructed; by reason either of the subtlety of the whole body, or the minuteness of the parts, or distance of place; or slowness or else swiftness of motion, or familiarity of object, or other causes. And again when the sense does apprehend a thing its apprehension is not much to be relied upon. For the testimony and information of the sense has reference always to man, not to the universe; and it is a great error to assert that the sense is the measure of things.

To meet these difficulties, I have sought on all sides diligently and faithfully to provide helps for the sense—substitutes to supply its failures, rectifications to correct its errors; and this I endeavour to accomplish not so much by instruments as by experiments. For the subtlety of experiments is far greater than that of the sense itself, even when assisted by exquisite instruments; such experiments, I mean, as are skilfully and artificially devised for the express purpose of determining the point in question. To the immediate and proper perception of the sense therefore I do not give much weight; but I contrive that the office of the sense shall be only to judge of the experiment, and that the experiment itself shall judge of the thing. And thus I conceive that I perform the office of a true priest of the sense (from which all knowledge in nature must be sought, unless men mean to go mad) and a not unskilful interpreter of its oracles; and that while others only profess to uphold and cultivate the sense, I do so in fact. Such then are the provisions I make for finding the genuine light of nature and kindling and bringing it to bear. And they would be sufficient of themselves, if the human intellect were even, and like a fair sheet of paper with no writing upon it. But since the minds of men are strangely possessed and beset, so that there is no true and even surface left to reflect the genuine rays of things, it is necessary to seek a remedy for this also.

Now the idols, or phantoms, by which the mind is occupied are either adventitious or innate. The adventitious come into the mind from without; namely, either from the doctrines and sects of philosophers, or from perverse rules of demonstration. But the innate are inherent in the very nature of the intellect, which is far more prone to error than the sense is. For let men please themselves as they will in admiring and almost adoring the human mind, this is certain: that as an uneven mirror distorts the rays of objects according to its own figure and section, so the mind, when it receives impressions of objects through the sense, cannot be trusted to report them truly, but in forming its notions mixes up its own nature with the nature of things.

And as the first two kinds of idols are hard to eradicate, so idols of this last kind cannot be eradicated at all. All that can be done is to point them out so that this insidious action of the mind may be marked and reproved (else as fast as old errors are destroyed new ones will spring up out of the ill complexion of the mind itself, and so we shall have but a change of errors, and not a clearance); and to lay it down once for all as a fixed and established maxim, that the intellect is not qualified to judge except by means of induction, and induction in its legitimate form. This doctrine then of the expurgation of the intellect to qualify it for dealing with truth, is comprised in three refutations: the refutation of the Philosophies; the refutation of the Demonstrations; and the refutation of the Natural Human Reason. The explanation of which things, and of the true relation between the nature of things and the nature of the mind, is as the strewing and decoration of the bridal chamber of the Mind and the Universe, the Divine Goodness assisting; out of which marriage let us hope (and be this the prayer of the bridal song) there may spring helps to man, and a line and race of inventions that may in some degree subdue and overcome the necessities and miseries of humanity. This is the second part of the work.

But I design not only to indicate and mark out the ways, but also to enter them. And therefore the third part of the work embraces the Phenomena of the universe; that is to say, experience of every kind, and such a natural history as may serve for a foundation to build philosophy upon. For a good method of demonstration or form of interpreting nature may keep the mind from going astray or stumbling, but it is not any excellence of method that can supply it with the material of knowledge. Those however who aspire not to guess and divine but to discover and know; who propose not to devise mimic and fabulous worlds of their own, but to examine and dissect the nature of this very world itself; must go to facts themselves for everything. Nor can the place of this labour and search and worldwide perambulation be supplied by any genius or meditation or argumentation; no, not if all men's wits could meet in one. This therefore we must have, or the business must be for ever abandoned. But up to this day such has been the condition of men in this matter, that it is no wonder if nature will not give herself into their hands.

For first, the information of the sense itself, sometimes failing, sometimes false; observation, careless, irregular, and led by chance; tradition, vain and fed on humour; practice, slavishly bent upon its work; experiment, blind, stupid, vague, and prematurely broken off; lastly, natural history trivial and poor;—all these have contributed to supply the understanding with very bad materials for philosophy and the sciences.

Then an attempt is made to mend the matter by a preposterous subtlety and winnowing of argument. But this comes too late, the case being already past remedy; and is far from setting the business right or sifting away the errors. The only hope therefore of any greater increase or progress lies in a reconstruction of the sciences.

Of this reconstruction the foundation must be laid in natural history, and that of a new kind and gathered on a new principle. For it is in vain that you polish the mirror if there are no images to be reflected; and it is as necessary that the intellect should be supplied with fit matter to work upon, as with safeguards to guide its working. But my history differs from that in use (as my logic does) in many things,—in end and office, in mass and composition, in subtlety, in selection also and setting forth, with a view to the operations which are to follow.

For first, the object of the natural history which I propose is not so much to delight with variety of matter or to help with present use of experiments, as to give light to the discovery of causes and supply a suckling philosophy with its first food. For though it be true that I am principally in pursuit of works and the active department of the sciences, yet I wait for harvest-time, and do not attempt to mow the moss or to reap the green corn. For I well know that axioms once rightly discovered will carry whole troops of works along with them, and produce them, not here and there one, but in clusters. And that unseasonable and puerile hurry to snatch by way of earnest at the first works which come within reach, I utterly condemn and reject, as an Atalanta's apple that hinders the race. Such then is the office of this natural history of mine.

Next, with regard to the mass and composition of it : I mean it to be a history not only of nature free and at large (when she is left to her own course and does her work her own way),—such as that of the heavenly bodies, meteors, earth and sea, minerals, plants, animals,—but much more of nature under constraint and vexed ; that is to say, when by art and the hand of man she is forced out of her natural state, and squeezed and moulded. Therefore I set down at length all experiments of the mechanical arts, of the operative part of the liberal arts, of the many crafts which have not yet grown into arts properly so called, so far as I have been able to examine them and as they conduce to the end in view. Nay (to say the plain truth) I do in fact (low and vulgar as men may think it) count more upon this part both for helps and safeguards than upon the other ; seeing that the nature of things betrays itself more readily under the vexations of art than in its natural freedom.

Nor do I confine the history to Bodies ; but I have thought it my duty besides to make a separate history of such Virtues as may be considered cardinal in nature. I mean those original passions or desires of matter which constitute the primary elements of nature ; such as Dense and Rare, Hot and Cold, Solid and Fluid, Heavy and Light, and several others.

Then again, to speak of subtlety : I seek out and get together a kind of experiments much subtler and simpler than those which occur accidentally. For I drag into light many things which no one who was not proceeding by a regular and certain way to the discovery of causes would have thought of inquiring after ; being indeed in themselves of no great use ; which shows that they were not sought for on their own account ; but having just the same relation to things and works which the letters of the alphabet have to speech and words—which, though in themselves useless, are the elements of which all discourse is made up.

Further, in the selection of the relations and experiments I conceive I have been a more cautious purveyor than those who have hitherto dealt with natural history. For I admit nothing but on the faith of eyes, or at least of careful and severe examination ; so that nothing is exaggerated for wonder's sake, but what I state is sound and without mixture of fables or vanity. All received or current falsehoods also (which by strange negligence have been allowed for many ages to prevail and become established) I proscribe and brand by name ; that the sciences may be no more troubled with them. For it has been well observed that the fables and superstitions and follies which nurses instil into children do serious injury to their minds ; and the same consideration makes me anxious, having the management of the childhood as it were of philosophy in its course of natural history, not to let it accustom itself in the beginning to any vanity. Moreover, whenever I come to a new experiment of any subtlety (though it be in my own opinion certain and approved), I nevertheless subjoin a clear account of the manner in which I made it ; that men knowing exactly how each point was made out, may see whether there be any error connected with it, and may arouse themselves to devise proofs more trustworthy and exquisite, if such can be found ; and finally, I interpose everywhere admonitions and scruples, and cautions, with a religious care to eject, repress, and as it were exorcise every kind of phantasm.

Lastly, knowing how much the sight of man's mind is distracted by experience and history, and how hard it is at the first (especially for minds either tender or preoccupied) to become familiar with nature, I not unfrequently subjoin observations of my own, being as the first offers, inclinations, and as it were glances of history towards philosophy ; both by way of an assurance to men that they will not be kept for ever tossing on the waves of experience, and also that when the time comes for the intellect to begin its work, it may find everything the more ready. By such a natural history then as I have described, I conceive that a safe and convenient approach may be made to nature, and matter supplied of good quality and well prepared for the understanding to work upon.

And now that we have surrounded the intellect with faithful helps and guards, and got together with most careful selection a regular army of divine works, it may seem that we have no more to do but to proceed to philosophy itself.

And yet in a matter so difficult and doubtful there are still some things which it seems necessary to premise, partly for convenience of explanation, partly for present use.

Of these the first is to set forth examples of inquiry and invention according to my method, exhibited by anticipation in some particular subjects; choosing such subjects as are at once the most noble in themselves among those under inquiry, and most different one from another; that there may be an example in every kind. I do not speak of those examples which are joined to the several precepts and rules by way of illustration (for of these I have given plenty in the second part of the work); but I mean actual types and models, by which the entire process of the mind and the whole fabric and order of invention from the beginning to the end, in certain subjects, and those various and remarkable, should be set as it were before the eyes. For I remember that in the mathematics it is easy to follow the demonstration when you have a machine besides you; whereas without that help all appears involved and more subtle than it really is. To examples of this kind,—being in fact nothing more than an application of the second part in detail and at large,—the fourth part of the work is devoted.

The fifth part is for temporary use only, pending the completion of the rest; like interest payable from time to time until the principal be forthcoming. For I do not make so blindly for the end of my journey, as to neglect anything useful that may turn up by the way. And therefore I include in this fifth part such things as I have myself discovered, proved, or added,—not however according to the true rules and methods of interpretation, but by the ordinary use of the understanding in inquiring and discovering. For besides that I hope my speculations may in virtue of my continual conversancy with nature have a value beyond the pretensions of my wit, they will serve in the meantime for wayside inns, in which the mind may rest and refresh itself on its journey to more certain conclusions. Nevertheless I wish it to be understood in the meantime that they are conclusions by which (as not being discovered and proved by the true form of interpretation) I do not at all mean to bind myself. Nor need any one be alarmed at such suspension of judgment, in one who maintains not simply that nothing can be known, but only that nothing can be known except in a certain course and way; and yet establishes provisionally certain degrees of assurance, for use and relief until the mind shall arrive at a knowledge of causes in which it can rest. For even those schools of philosophy which held the absolute impossibility of knowing anything were not inferior to those which took upon them to pronounce. But then they did not provide helps for the sense and understanding, as I have done, but simply took away all their authority: which is quite a different thing—almost the reverse.

The sixth part of my work (to which the rest is subservient and ministrant) discloses and sets forth that philosophy which by the legitimate, chaste, and severe course of inquiry which I have explained and provided is at length developed and established. The completion however of this last part is a thing both above my strength and beyond my hopes. I have made a beginning of the work —a beginning, as I hope, not unimportant:—the fortune of the human race will give the issue;—such an issue, it may be, as in the present condition of things and men's minds cannot easily be conceived or imagined. For the matter in hand is no mere felicity of speculation, but the real business and fortunes of the human race, and all power of operation. For man is but the servant and interpreter of nature: what he does and what he knows is only what he has observed of nature's order in fact or in thought; beyond this he knows nothing and can do nothing. For the chain of causes cannot by any force be loosed or broken, nor can nature be commanded except by being obeyed. And so those twin objects, human Knowledge and human Power, do really meet in one; and it is from ignorance of causes that operation fails.

And all depends on keeping the eye steadily fixed upon the facts of nature and so receiving their images simply as they are. For God forbid that we should give out a dream of our own imagination for a pattern of the world; rather may he graciously grant to us to write an apocalypse or true vision of the footsteps of the Creator imprinted on his creatures.

Therefore do thou, O Father, who gavest the visible light as the first fruits of creation, and didst breathe into the face of man the intellectual light as the crown and consummation thereof, guard and protect this work, which coming from thy goodness returneth to thy glory. Thou when thou turnedst to look upon the works which thy hands had made, sawest that all was very good, and didst rest from thy labours. But man, when he turned to look upon the work which his hands had made, saw that all was vanity and vexation of spirit, and could find no rest therein. Wherefore if we labour in thy works with the sweat of our brows thou wilt make us partakers of thy vision and thy sabbath. Humbly we pray that this mind may be steadfast in us, and that through these our hands, and the hands of others to whom thou shalt give the same spirit, thou wilt vouchsafe to endow the human family with new mercies.

THE FIRST PART OF THE INSTAURATION,

WHICH COMPRISES THE DIVISIONS OF THE SCIENCES, IS WANTING.

But some account of them will be found in the Second Book of the "Proficience and Advancement of Learning, Divine and Human." [1]

Next comes

THE SECOND PART OF THE INSTAURATION,

WHICH EXHIBITS THE ART ITSELF OF INTERPRETING NATURE, AND OF THE TRUER EXERCISE OF THE INTELLECT;

Not however in the form of a regular Treatise, but only a Summary digested into Aphorisms.[2]

[*Notes by Mr. Spedding to the titles of the Latin originals.*]

1 This is omitted in the common editions of Bacon's collected works (in all, I believe, except Montagu's); the *De Augmentis Scientiarum*, with the title "*Instaurationis Magnæ pars prima*" prefixed on a separate leaf, being substituted for it. And it is true that Bacon did afterwards decide upon supplying this deficiency by a translation of the *Advancement of Learning* enlarged; that he produced the *De Augmentis Scientiarum* with that intention and understanding; and that though the original edition does not bear "*Instaurationis Magnæ pars prima*" on the titlepage, yet in Dr. Rawley's reprint of it in 1638 those words were inserted. Nevertheless this notice is of importance, as showing that when Bacon published the *Novum Organum* he did not look to a mere *enlargement* of the *Advancement of Learning* as satisfying the intention of the *pars prima;* for if he had, he would have referred to the work itself, not to the second book only. He meant, no doubt, to reproduce the substance of it in a different form. And my own impression is that the *Descriptio Globi Intellectualis* was originally designed for this place, and that he had not yet abandoned the hope of completing it; but that soon after, —fortune gone, health shaken, assistance not to be commanded, and things of more importance remaining to be done,—he found he had not time to finish it on so large a scale, and therefore resolved to enlarge the old house instead of building a new one.—*J. S.*

2 This explains a certain discrepancy between the design of the second part, as set forth in the *Distributio Operis*, and the execution of it in the *Novum Organum*. The *Distributio*, like the *Delineatio*, was probably written when Bacon intended to work it out in a regular and consecutive treatise, and represents the *idea* of the work more perfectly than the work itself.—*J. S.*

THE SECOND PART OF THE WORK, WHICH IS CALLED

THE NEW ORGANON;

OR, TRUE DIRECTIONS CONCERNING THE INTERPRETATION OF NATURE.

[AUTHOR'S] PREFACE.

THOSE who have taken upon them to lay down the law of nature as a thing already searched out and understood, whether they have spoken in simple assurance or professional affectation, have therein done philosophy and the sciences great injury. For as they have been successful in inducing belief, so they have been effective in quenching and stopping inquiry; and have done more harm by spoiling and putting an end to other men's efforts than good by their own. Those on the other hand who have taken a contrary course, and asserted that absolutely nothing can be known—whether it were from hatred of the ancient sophists, or from uncertainty and fluctuation of mind, or even from a kind of fulness of learning, that they fell upon this opinion,—have certainly advanced reasons for it that are not to be despised; but yet they have neither started from true principles nor rested in the just conclusion, zeal and affectation having carried them much too far. The more ancient of the Greeks (whose writings are lost) took up with better judgment a position between these two extremes,—between the presumption of pronouncing on everything, and the despair of comprehending anything; and though frequently and bitterly complaining of the difficulty of inquiry and the obscurity of things, and like impatient horses champing the bit, they did not the less follow up their object and engage with Nature; thinking (it seems) that this very question,—viz. whether or no anything can be known,—was to be settled not by arguing, but by trying. And yet they too, trusting entirely to the force of their understanding, applied no rule, but made everything turn upon hard thinking and perpetual working and exercise of the mind.

Now my method, though hard to practise, is easy to explain; and it is this. I propose to establish progressive stages of certainty. The evidence of the sense, helped and guarded by a certain process of correction, I retain. But the mental operation which follows the act of sense I for the most part reject; and instead of it I open and lay out a new and certain path for the mind to proceed in, starting directly from the simple sensuous perception. The necessity of this was felt no doubt by those who attributed so much importance to Logic; showing thereby that they were in search of helps for the understanding, and had no confidence in the native and spontaneous process of the mind. But this remedy comes too late to do any good, when the mind is already, through the daily intercourse and conversation of life, occupied with unsound doctrines and beset on all sides by vain imaginations. And therefore that art of Logic, coming (as I said) too late to the rescue, and no way able to set matters right again, has had the effect of fixing errors rather than disclosing truth. There remains but one course for the recovery of a sound and healthy condition,—namely, that the entire work of the understanding be commenced afresh, and the mind itself be from the very outset not left to take its own course, but guided at every step; and the business be done as if by machinery. Certainly if in things mechanical men had set to work with their naked hands, without help or force of instruments, just as in things intellectual they have set to work with little else than the naked forces of the understanding, very small would the matters have been which, even with their best efforts applied in conjunction, they could have attempted or accomplished. Now (to pause awhile upon this example and look in it as in a glass) let us suppose that some vast obelisk were (for the decoration of a triumph or some such magnificence) to be removed from its place, and that men should set to work upon it with their naked hands ; would not any sober spectator think them mad ? And if they should send then for more people, thinking that in that way they might manage it, would he not think them all the madder ? And if they then proceeded to make a selection, putting away the weaker hands, and using only the strong and vigorous, would he not think them madder than ever ? And if lastly, not content with this, they resolved to call in aid the art of athletics, and required all their men to come with hands, arms, and sinews well anointed and medicated according to the rules of art, would he not cry out that they were only taking pains to show a kind of method and discretion in their madness ? Yet just so it is that men proceed in matters intellectual,— with just the same kind of mad effort and useless combination of forces,—when they hope great things either from the number and co-operation or from the excellency and acuteness of individual wits ; yea, and when they endeavour by Logic (which may be considered as a kind of athletic art) to strengthen the sinews of the understanding ; and yet with all this study and endeavour it is apparent to any true judgment that they are but applying the naked intellect all the time ; whereas in every great work to be done by the hand of man it is manifestly impossible, without instruments and machinery, either for the strength of each to be exerted or the strength of all to be united.

Upon these premises two things occur to me of which, that they may not be overlooked, I would have men reminded. First, it falls out fortunately as I think for the allaying of contradictions and heart-burnings, that the honour and reverence due to the ancients remains untouched and undiminished ; while I may carry out my designs and at the same time reap the fruit of my modesty. For if I should profess that I, going the same road as the ancients, have something better to produce, there must needs have been some comparison or rivalry between us (not to be avoided by any art of words) in respect of excellency or ability of wit ; and though in this there would be nothing unlawful or new (for if there be anything misapprehended by them, or falsely laid down, why may not I, using a liberty common to all, take exception to it ?) yet the contest, however just and allowable, would have been an unequal one perhaps, in respect of the measure of my own powers. As it is however,—my object being to open a new way for the understanding, a way by them untried and unknown,—the case is altered ; party zeal and emulation are at an end ; and I appear merely as a guide to point out the road ; an office of small authority, and depending more upon a kind of luck than upon any ability or excellency. And thus much relates to the persons only. The other point of which I would have men reminded relates to the matter itself.

Be it remembered then that I am far from wishing to interfere with the philosophy which now flourishes, or with any other philosophy more correct and complete than this which has been or may hereafter be propounded. For I do not object to the use of this received philosophy, or others like it, for supplying matter for disputations or ornaments for discourse,—for the professor's lecture and for the business of life. Nay more, I declare openly that for these uses the philosophy which I bring forward will not be much available. It does not lie in the way. It cannot be caught up in passage. It does not flatter the understanding by conformity with preconceived notions. Nor will it come down to the apprehension of the vulgar except by its utility and effects.

Let there be therefore (and may it be for the benefit of both) two streams and two dispensations of knowledge ; and in like manner two tribes or kindreds of students in philosophy—tribes not hostile or alien to each other, but bound together by mutual services ;—let there in short be one method for the cultivation, another for the invention, of knowledge.

And for those who prefer the former, either from hurry or from considerations of business or for want of mental power to take in and embrace the other (which must needs be most men's case), I wish that they may succeed to their desire in what they are about, and obtain what they are pursuing. But if any man there be who, not content to rest in and use the knowledge which has already been discovered, aspires to penetrate further ; to overcome, not an adversary in argument, but nature in action ; to seek, not pretty and probable conjectures, but certain and demonstrable knowledge ;—I invite all such to join themselves, as true sons of knowledge, with me, that passing by the outer courts of nature, which numbers have trodden, we may find a way at length into her inner chambers. And to make my meaning clearer and to familiarise the thing by giving it a name, I have chosen to call one of these methods or ways *Anticipation of the Mind*, the other *Interpretation of Nature*.

Moreover I have one request to make. I have on my own part made it my care and study that the things which I shall propound should not only be true, but should also be presented to men's minds, how strangely soever preoccupied and obstructed, in a manner not harsh or unpleasant. It is but reasonable however (especially in so great a restoration of learning and knowledge) that I should claim of men one favour in return ; which is this ; If any one would form an opinion or judgment either out of his own observation, or out of the crowd of authorities, or out of the forms of demonstration (which have now acquired a sanction like that of judicial laws,) concerning these speculations of mine, let him not hope that he can do it in passage or by the by ; but let him examine the thing thoroughly ; let him make some little trial for himself of the way which I describe and lay out ; let him familiarize his thoughts with that subtlety of nature to which experience bears witness ; let him correct by seasonable patience and due delay the depraved and deep-rooted habits of his mind ; and when all this is done and he has begun to be his own master, let him (if he will) use his own judgment.

APHORISMS CONCERNING THE INTERPRETATION OF NATURE AND THE KINGDOM OF MAN.

APHORISM

I.

MAN, being the servant and interpreter of Nature,[1] can do and understand so much and so much only as he has observed in fact or in thought of the course of nature: beyond this he neither knows anything nor can do anything.

II.

Neither the naked hand nor the understanding left to itself can effect much. It is by instruments and helps that the work is done, which are as much wanted for the understanding as for the hand. And as the instruments of the hand either give motion or guide it, so the instruments of the mind supply either suggestions for the understanding or cautions.

III.

Human knowledge and human power meet in one; for where the cause is not known the effect cannot be produced. Nature to be commanded must be obeyed[2]; and that which in contemplation is as the cause is in operation as the rule.

IV.

Towards the effecting of works, all that man can do is to put together or put asunder natural bodies. The rest is done by nature working within[3].

V.

The study of nature with a view to works is engaged in by the mechanic, the mathematician, the physician, the alchemist, and the magician; but by all (as things now are) with slight endeavour and scanty success.[4]

VI.

It would be an unsound fancy and self-contradictory to expect that things which have never yet been done can be done except by means which have never yet been tried.

[1] That the physician is "naturæ minister", φυεσως ὑπηρέτης, is quoted more than once from Hippocrates by Galen, xv. 369, xvi. 35 (Kuhn): the first passage in his commentary on Hippoc. *De Aliment.* iii., the second in his do. *De Humor.* i.

[2] This antithesis was probably suggested by Publius Syrus's gnome:—"Casta ad virum matrona parendo imperat".

[The phrase occurs above, it will be noted, towards the end of the "Plan of the Work" and again below, Aph. 129. Dean Kitchin cites the parallel of Livy's account of Hannibal (xxi. 4) as "parendum atque imperandum" in his difficulties.—ED.]

[3] For some remarks upon the first four Aphorisms, see the Preface, p. 223.—*J. S.* [It will be observed that below, Aph. 75, Bacon appears to reject the formula here set down. In the parallel passage in the *De Augmentis* (B. II. c. ii.) he attempts to combine the two positions.—ED.]

[4] [Dean Kitchin has pointed out that "Bacon's remarks were being falsified at the very time he wrote. Mechanics had produced fly-clocks, telescopes, and other useful contrivances. Mathematics boasted of Kepler and Galileo; and the discoveries of Harvey and Gilbert were opening out a new world for medical research; but see p. 243".—ED.]

VII.

The productions of the mind and hand seem very numerous in books and manufactures. But all this variety lies in an exquisite subtlety and derivations from a few things already known ; not in the number of axioms.

VIII.

Moreover the works already known are due to chance and experiment rather than to sciences ; for the sciences we now possess are merely systems for the nice ordering and setting forth of things already invented ; not methods of invention or directions for new work.

IX.

The cause and root of nearly all evils in the sciences is this — that while we falsely admire and extol the powers of the human mind we neglect to seek for its true helps.

X.

The subtlety of nature is greater many times over than the subtlety of the senses and understanding ; so that all those specious meditations, speculations, and glosses in which men indulge are quite from the purpose [5], only there is no one by to observe it.

XI.

As the sciences which we now have do not help us in finding out new works, so neither does the logic which we now have help us in finding out new sciences.

XII.

The logic now in use serves rather to fix and give stability to the errors which have their foundation in commonly received notions than to help the search after truth. So it does more harm than good.

XIII.

The syllogism is not applied to the first principles of sciences, and is applied in vain to intermediate axioms ; being no match for the subtlety of nature. It commands assent therefore to the proposition, but does not take hold of the thing.

XIV.

The syllogism consists of propositions, propositions consist of words, words are symbols of notions. Therefore if the notions themselves (which is the root of the matter) are confused and over-hastily abstracted from the facts, there can be no firmness in the superstructure. Our only hope therefore lies in a true induction.

XV.

There is no soundness in our notions whether logical or physical. Substance, Quality, Action, Passion, Essence itself, are not sound notions : much less are Heavy, Light, Dense, Rare, Moist, Dry, Generation, Corruption, Attraction, Repulsion, Element, Matter, Form, and the like ; but all are fantastical and ill defined.

XVI.

Our notions of less general species, as Man, Dog, Dove, and of the immediate perceptions of the sense, as Hot, Cold, Black, White, do not materially mislead us ; yet even these are sometimes confused by the flux and alteration of matter and the mixing of one thing with another. All the others which men have hitherto adopted are but wanderings, not being abstracted and formed from things by proper methods.

[5] Literally, " are a thing insane ". The meaning appears to be, that these speculations being founded upon such an inadequate conception of the case, must necessarily be so wide of the truth that they would seem like mere madness if we could only compare them with it : like the aim of a man blindfolded to bystanders looking on.—*J. S.*

XVII.

Nor is there less of wilfulness and wandering in the construction of axioms than in the formation of notions; not excepting even those very principles which are obtained by common induction; but much more in the axioms and lower propositions educed by the syllogism.

XVIII

The discoveries which have hitherto been made in the sciences are such as lie close to vulgar notions, scarcely beneath the surface. In order to penetrate into the inner and further recesses of nature, it is necessary that both notions and axioms be derived from things by a more sure and guarded way; and that a method of intellectual operation be introduced altogether better and more certain.

XIX.

There are and can be only two ways of searching into and discovering truth. The one flies from the senses and particulars to the most general axioms, and from these principles, the truth of which it takes for settled and immoveable, proceeds to judgment and to the discovery of middle axioms. And this way is now in fashion. The other derives axioms from the senses and particulars, rising by a gradual and unbroken ascent, so that it arrives at the most general axioms last of all. This is the true way, but as yet untried [6].

XX.

The understanding left to itself takes the same course (namely, the former) which it takes in accordance with logical order. For the mind longs to spring up to positions of higher generality, that it may find rest there; and so after a little while wearies of experiment. But this evil is increased by logic, because of the order and solemnity of its disputations.

XXI.

The understanding left to itself, in a sober, patient, and grave mind, especially if it be not hindered by received doctrines, tries a little that other way, which is the right one, but with little progress; since the understanding, unless directed and assisted, is a thing unequal, and quite unfit to contend with the obscurity of things.

XXII.

Both ways set out from the senses and particulars, and rest in the highest generalities; but the difference between them is infinite. For the one just glances at experiment and particulars in passing, the other dwells duly and orderly among them. The one, again, begins at once by establishing certain abstract and useless generalities, the other rises by gradual steps to that which is prior and better known in the order of nature.

XXIII.

There is a great difference between the idols [7] of the human mind and the ideas of the divine. That is to say, between certain empty dogmas, and the true signatures and marks set upon the works of creation as they are found in nature.

XXIV.

It cannot be that axioms established by argumentation should avail for the discovery of new works; since the subtlety of nature is greater many times over than

[6] [Compare the criticism of J. S. Mill on this Aphorism, in his *Logic*, B. VI. ch. v. § 5. —ED.]

[7] [See above pp. 118, 223, as to the signification of "Idols." The word *idola*, as used by Bacon, means (as the context here shows) not objects of worship, but illusions or false appearances—the original sense of the Greek word. As Professor Fowler notes, Bacon in his *Cogitata et Visa* (14th par.) uses the word *spectra* with the same force as elsewhere *idola*. Compare Hallam, *Introduction to the Literature of Europe*, pt. III. ch. iii. § 60, where it is pointed out that the error of reading "idol" as "worshipped object" has been fallen into by Playfair, Brown, Dugald Stewart, and others.—ED.]

the subtlety of argument. But axioms duly and orderly formed from particulars easily discover the way to new particulars, and thus render sciences active.

XXV.

The axioms now in use, having been suggested by a scanty and manipular experience and a few particulars of most general occurrence, are made for the most part just large enough to fit and take these in: and therefore it is no wonder if they do not lead to new particulars. And if some opposite instance, not observed or not known before, chance to come in the way, the axiom is rescued and preserved by some frivolous distinction; whereas the truer course would be to correct the axiom itself.

XXVI.

The conclusions of human reason as ordinarily applied in matter of nature, I call for the sake of distinction *Anticipations of Nature* (as a thing rash or premature.) That reason which is elicited from facts by a just and methodical process, I call *Interpretation of Nature.*

XXVII.

Anticipations are a ground sufficiently firm for consent; for even if men went mad all after the same fashion, they might agree one with another well enough.

XXVIII.

For the winning of assent, indeed, anticipations are far more powerful than interpretations; because being collected from a few instances, and those for the most part of familiar occurrence, they straightway touch the understanding and fill the imagination; whereas interpretations on the other hand, being gathered here and there from very various and widely dispersed facts, cannot suddenly strike the understanding; and therefore they must needs, in respect of the opinions of the time, seem harsh and out of tune; much as the mysteries of faith do.

XXIX.

In sciences founded on opinions and dogmas, the use of anticipations and logic is good; for in them the object is to command assent to the proposition, not to master the thing.

XXX.

Though all the wits of all the ages should meet together and combine and transmit their labours, yet will no great progress ever be made in science by means of anticipations; because radical errors in the first concoction of the mind are not to be cured by the excellence of functions and remedies subsequent.

XXXI.

It is idle to expect any great advancement in science from the superinducing and engrafting of new things upon old. We must begin anew from the very foundations, unless we would revolve for ever in a circle with mean and contemptible progress.

XXXII.

The honour of the ancient authors, and indeed of all, remains untouched; since the comparison I challenge is not of wits or faculties, but of ways and methods, and the part I take upon myself is not that of a judge, but of a guide.

XXXIII.

This must be plainly avowed: no judgment can be rightly formed either of my method or of the discoveries to which it leads, by means of anticipations (that is to say, of the reasoning which is now in use); since I cannot be called on to abide by the sentence of a tribunal which is itself on its trial.

XXXIV.

Even to deliver and explain what I bring forward is no easy matter; for things in themselves new will yet be apprehended with reference to what is old.

XXXV.

It was said by Borgia of the expedition of the French into Italy, that they came with chalk in their hands to mark out their lodgings, not with arms to force their way in [8]. I in like manner would have my doctrine enter quietly into the minds that are fit and capable of receiving it ; for confutations cannot be employed, when the difference is upon first principles and very notions and even upon forms of demonstration.

XXXVI.

One method of delivery alone remains to us ; which is simply this : we must lead men to the particulars themselves, and their series and order ; while men on their side must force themselves for awhile to lay their notions by and begin to familiarise themselves with facts.

XXXVII.

The doctrine of those who have denied that certainty could be attained at all, has some agreement with my way of proceeding at the first setting out ; but they end in being infinitely separated and opposed. For the holders of that doctrine assert simply that nothing can be known ; I also assert that not much can be known in nature by the way which is now in use. But then they go on to destroy the authority of the senses and understanding ; whereas I proceed to devise and supply helps for the same.

XXXVIII.

The Idols and false notions which are now in possession of the human understanding, and have taken deep root therein, not only so beset men's minds that truth can hardly find entrance, but even after entrance obtained, they will again in the very instauration of the sciences meet and trouble us, unless men being forewarned of the danger fortify themselves as far as may be against their assaults.

XXXIX.

There are four classes of Idols which beset men's minds. To these for distinction's sake I have assigned names, — calling the first class *Idols of the Tribe* ; the second, *Idols of the Cave* ; the third, *Idols of the Market-place* ; the fourth, *Idols of the Theatre* [9].

[8] " Diceva in quei tempi Papa Alessandro sesto che i Francesi havevano corso l'Italia con gli speroni di legno et presola col gesso : dicendo cosi perchè pigliando essi gli alloggiamenti nelle città loro furieri segnavano le porte delle case col gesso ; et cavalcando per loro diporto i gentil' huomini per le terre à sollazzo usavano di portare nelle scarpette à calcagni certi stecchi di legno appuntati, delli quali in vece di speroni si servivano per andare le cavalcature."—*Nardi, Vita di Malespini*, [1597,] p. 18.

In an epitome of the history of Charles the Eighth, which will be found in the " Archives curieuses " of Cember, vol. i. p. 197, and which was apparently written about the beginning of the seventeenth century, the remark ascribed to Alexander the Sixth by Nardi and Bacon is mentioned as a popular saying.

[9] These four idols have been compared to the four hindrances to truth enumerated by Roger Bacon. These are, the use of insufficient authority, custom, popular opinions, and the concealment of ignorance and display of apparent knowledge. The last two may be likened to the idols of the market-place and the theatre. But the principle of the classification is different. [See on this subject the Preface, p. 223. Roger Bacon's words are as follows :—

" Quatuor vero maxima sunt comprehendendæ veritatis offendicula, quæ omnem quemcunque sapientem impediunt, et vix aliquem permittunt ad verum titulum sapientiæ pervenire : viz. fragilis et indignæ auctoritatis exemplum, consuetudinis diuturnitas vulgi sensus imperiti, et propriæ ignorantiæ occultatio cum ostentatione sapientiæ apparentis. His omnis homo involvitur, omnis status occupatur. Nam quilibet singulis artibus vitæ et studii et omnis negotii tribus pessimis ad eandem conclusionem utitur argumentis : scil. hoc exemplificatum est per majores, hoc consuetum est, hoc vulgatum est, ergo tenendum. . . . Si vero hæc tria refellantur aliquando magnificâ rationis potentia, quartum semper in promptu est et in ore cujuslibet, ut quilibet ignorantiam suam

XL.

The formation of ideas and axioms by true induction is no doubt the proper remedy to be applied for the keeping off and clearing away of idols. To point them out, however, is of great use ; for the doctrine of Idols is to the Interpretation of Nature what the doctrine of the refutation of Sophisms is to common Logic.

XLI.

The Idols of the Tribe have their foundation in human nature itself, and in the tribe or race of men. For it is a false assertion that the sense of man is the measure of things [10]. On the contrary, all perceptions as well of the sense as of the mind are according to the measure of the individual and not according to the measure of the universe. And the human understanding is like a false mirror, which, receiving rays irregularly, distorts and discolours the nature of things by mingling its own nature with it.

XLII.

The Idols of the Cave[11] are the Idols of the individual man. For every one (besides the errors common to human nature in general) has a cave or den of his own, which refracts and discolours the light of nature ; owing either to his own proper and peculiar nature ; or to his education and conversation with others ; or to the reading of books, and the authority of those whom he esteems and admires ; or to the differences of impressions, accordingly as they take place in a mind preoccupied and predisposed or in a mind indifferent and settled ; or the like. So that the spirit of man (according as it is meted out to different individuals [12]) is in fact a thing variable and full of perturbation, and governed as it were by chance. Whence it was well observed by Heraclitus that men look for sciences in their own lesser worlds, and not in the greater or common world [13].

XLIII.

There are also Idols formed by the intercourse and association of men with each other, which I call Idols of the Market-place, on account of the commerce and consort of men there. For it is by discourse that men associate ; and words are imposed according to the apprehension of the vulgar. And therefore the ill and unfit choice of words wonderfully obstructs the understanding. Nor do the definitions or explanations wherewith in some things learned men are wont to guard and defend themselves, by any means set the matter right. But words plainly force and overrule the understanding, and throw all into confusion, and lead men away into numberless empty controversies and idle fancies.

XLIV.

Lastly, there are Idols which have immigrated into men's minds from the various dogmas of philosophies, and also from wrong laws of demonstration. These I call Idols of the Theatre ; because in my judgment all the received systems are but so many stage-plays, representing worlds of their own creation after an unreal and scenic fashion. Nor is it only of the systems now in vogue, or only of the ancient sects and philosophies, that I speak ; for many more plays of the same kind

excuset, et licet nihil dignum sciat illud tamen magnificet imprudenter [impudenter ?] et sic saltem suæ stultitiæ infelici solatio veritatem opprimat et elidat."—*Opus Majus*, l. i.—*J. S.*]

10 Protagoras. See Hippias major. [Professor Fowler justly notes that " if this is meant to represent the dictum of Protagoras it does so most inadequately ". Protagoras was simply asserting the relativity of knowledge.—ED.]

11 [Compare Plato, *Republic*, B. vii.—ED.]

12 This was Mr. Ellis's translation of *prout disponitur in hominibus singulis ;* supposing Bacon to allude to Averroës' doctrine of one intellect, whereof each man had an undivided share. I should myself have understood *disponitur* as referring to the disposition of the parts of the spirit in itself, not to the distribution of it in different persons ; as in the expression *well disposed, ill disposed, etc.*—*J. S.*

13 See Sextus Empiricus, *Adversus Logicos*, i. § 133 ; and compare ii. 286, of the same treatise.

may yet be composed and in like artificial manner set forth; seeing that errors the most widely different have nevertheless causes for the most part alike. Neither again do I mean this only of entire systems, but also of many principles and axioms in science, which by tradition, credulity, and negligence have come to be received.

But of these several kinds of Idols I must speak more largely and exactly, that the understanding may be duly cautioned.

XLV.

The human understanding is of its own nature prone to suppose the existence of more order and regularity in the world than it finds. And though there be many things in nature which are singular and unmatched, yet it devises for them parallels and conjugates and relatives which do not exist. Hence the fiction that all celestial bodies move in perfect circles [14]; spirals and dragons [15] being (except in name) utterly rejected. Hence too the element of Fire with its orb is brought in, to make up the square with the other three which the sense perceives [16]. Hence also the ratio of density of the so-called elements is arbitrarily fixed at ten to one [17]. And so on of other dreams. And these fancies affect not dogmas only, but simple notions also.

XLVI.

The human understanding when it has once adopted an opinion (either as being the received opinion or as being agreeable to itself) draws all things else to support and agree with it. And though there be a greater number and weight of instances to be found on the other side, yet these it either neglects and despises, or else by some distinction sets aside and rejects; in order that by this great and pernicious pre-determination [18] the authority of its former conclusions may remain inviolate. And therefore it was a good answer that was made by one who when they showed him hanging in a temple a picture of those who had paid their vows as having escaped shipwreck, and would have him say whether he did not now acknowledge the power of the gods,—" Aye," asked he again, " but where are they painted that were drowned after their vows?" [19] And such is the way of all superstition,

[14] [The Bohn editor remarks that " though Kepler had, when Bacon wrote this, already demonstrated his three great laws concerning the elliptical path of the planets, neither Bacon nor Descartes seems to have known or assented to his discoveries".—Ed.]

[15] It does not appear in what sense Bacon uses the word " draco ". In its ordinary acceptation in old astronomy, it denoted the great circle which is approximately the projection on the sphere of the moon's orbit. The ascending node was called the caput draconis, and the descending the cauda draconis. The same terms were occasionally applied to the nodes of the planetary orbits. It is not improbable that Bacon intended to complain of the rejection of spirals of double curvature, or helices, which traced on the surface of the sphere might represent inequalities in latitude. Compare (Nov. Org. II. 48) what is said of the variations of which the " motus rotationis spontaneus " admits.

[16] The orb of the element of fire was supposed to lie above that of the element of air, and therefore might be said " non subjici sensui." The quaternion of elements follows directly from the quaternion of elementary qualities: namely, hot, cold, moist, dry. For these may be combined two and two in six different ways; two of these combinations are rejected as simply contradictory (viz. hot and cold, moist and dry); and to each of the other combinations corresponds one of the four elements.

[17] This doctrine of the decupla ratio of density of the elements was suggested by a passage in Aristotle [De Gen. et Cor. ii. 6.]. It is found in all books of mediæval physics. Cf. the *Margarita Philosophiæ*, ix. c. 4.. or Alsted's *Encyclopædia*, where it is thus expressed: " Proportio elementorum ad se invicem ratione transmutationis est decupla, ratione magnitudinis non satis explorata." The transmutability of one element into another is an essential part of the Peripatetic doctrine of elements. It is found also in the *Timæus*.

[18] Rather perhaps " prejudging the matter to a great and pernicious extent, in order that," etc. (*non sine magno et pernicioso præjudicio, quo, etc.*)—*J. S.*

[19] This story is told of Diagoras by Cicero, *De Nat. Deor.* iii., and of Diogenes the Cynic by Diogenes Laërtius.

whether in astrology, dreams, omens, divine judgments, or the like ; wherein men, having a delight in such vanities, mark the events where they are fulfilled, but where they fail, though this happen much oftener, neglect and pass them by. But with far more subtlety does this mischief insinuate itself into philosophy and the sciences ; in which the first conclusion colours and brings into conformity with itself all that come after, though far sounder and better. Besides, independently of that delight and vanity which I have described, it is the peculiar and perpetual error of the human intellect to be more moved and excited by affirmatives than by negatives ; whereas it ought properly to hold itself indifferently disposed towards both alike. Indeed in the establishment of any true axiom, the negative instance is the more forcible of the two.

XLVII.

The human understanding is moved by those things most which strike and enter the mind simultaneously and suddenly, and so fill the imagination ; and then it feigns and supposes all other things to be somehow, though it cannot see how, similar to those few things by which it is surrounded. But for that going to and fro to remote and heterogeneous instances, by which axioms are tried as in the fire, the intellect is altogether slow and unfit, unless it be forced thereto by severe laws and overruling authority.

XLVIII.

The human understanding is unquiet ; it cannot stop or rest, and still presses onward, but in vain. Therefore it is that we cannot conceive of any end or limit to the world ; but always as of necessity it occurs to us that there is something beyond [20]. Neither again can it be conceived how eternity has flowed down to the present day ; for that distinction which is commonly received of infinity in time past and in time to come can by no means hold ; for it would thence follow that one infinity is greater than another, and that infinity is wasting away and tending to become finite. The like subtlety arises touching the infinite divisibility of lines, from the same inability of thought to stop [21]. But this inability interferes more mischievously in the discovery of causes : for although the most general principles in nature ought to be held merely positive, as they are discovered, and cannot with truth be referred to a cause ; nevertheless the human understanding being unable to rest still seeks something prior in the order of nature. And then it is that in struggling towards that which is further off it falls back upon that which is more nigh at hand ; namely, on final causes : which have relation clearly to the nature of man rather than to the nature of the universe [22], and from this source have strangely defiled philosophy. But he is no less an unskilled and shallow philosopher who seeks causes of that which is most general, than he who in things subordinate and subaltern omits to do so.

20 Thus Leibnitz derived from the principle of sufficient reason a proof of the infinite extent of the universe, alleging that if it were of finite dimensions no reason could be given for its occupying any one region of space rather than any other.

21 In the phrase " subtilitas de lineis semper divisibilibus," reference is made to Aristotle, who in several places in his writings (particularly in the tract περὶ ἀτόμων γραμμάτων maintains that in theory every magnitude is divisible without limit.

22 This censure appears to be expressed without sufficient limitation ; for it is difficult to assent to the assertion that the notion of the final cause, considered generally, is more ex naturâ hominis than that of the efficient. The subject is one of which it is difficult to speak accurately ; but it may be said that wherever we think that we recognise a tendency towards a fulfilment or realisation of an idea, there the notion of the final cause comes in. It can only be from inadvertence that Professor Owen has set the doctrine of the final cause as it were in antithesis to that of the unity of type ; by the former he means the doctrine that the suitability of an animal to its mode of life is the one thing aimed at or intended in its structure. It cannot be doubted that Aristotle would have recognised the preservation of the type as not less truly a final cause than the preservation of the species or than the well-being of the individual. The final cause connects tself with what in the language of modern German philosophy is expressed by the phrase ' the Idea in Nature ".

XLIX.

The human understanding is no dry light [23], but receives an infusion from the will and affections; whence proceed sciences which may be called "sciences as one would." For what a man had rather were true he more readily believes. Therefore he rejects difficult things from impatience of research; sober things, because they narrow hope; the deeper things of nature, from superstition; the light of experience, from arrogance and pride, lest his mind should seem to be occupied with things mean and transitory; things not commonly believed, out of deference to the opinion of the vulgar. Numberless in short are the ways, and sometimes imperceptible, in which the affections colour and infect the understanding.

L.

But by far the greatest hindrance and aberration of the human understanding proceeds from the dulness, incompetency, and deceptions of the senses; in that things which strike the sense outweigh things which do not immediately strike it, though they be more important. Hence it is that speculation commonly ceases where sight ceases; insomuch that of things invisible there is little or no observation. Hence all the working of the spirits inclosed in tangible bodies lies hid and unobserved of men. So also all the more subtle changes of form in the parts of coarser substances (which they commonly call alteration, though it is in truth local motion through exceedingly small spaces) is in like manner unobserved. And yet unless these two things just mentioned be searched out and brought to light, nothing great can be achieved in nature, as far as the production of works is concerned. So again the essential nature of our common air, and of all bodies less dense than air (which are very many), is almost unknown. For the sense by itself is a thing infirm and erring; neither can instruments for enlarging or sharpening the senses do much; but all the truer kind of interpretation of nature is effected by instances and experiments fit and apposite; wherein the sense decides touching the experiment only, and the experiment touching the point in nature and the thing itself.

LI.

The human understanding is of its own nature prone to abstractions, and gives a substance and reality to things which are fleeting. But to resolve nature into abstractions is less to our purpose than to dissect her into parts; as did the school of Democritus, which went further into nature than the rest. Matter rather than forms should be the object of our attention, its configurations and changes of configuration, and simple action, and law of action or motion; for forms are figments of the human mind, unless you will call those laws of action forms.

LII.

Such then are the idols which I call *Idols of the Tribe*; and which take their rise either from the homogeneity of the substance of the human spirit [24], or from its preoccupation, or from its narrowness, or from its restless motion, or from an infusion of the affections, or from the incompetency of the senses, or from the mode of impression.

[23] Heraclitus apud Plut., *De Esu Carnium*. This doctrine of Idols is spoken of with great disrespect by Spinoza. He asserts that neither Des Cartes nor Bacon ever perceived the true source of error, and adds: "De Bacone parum dicam, qui de hâc re admodum confuse loquitur, et fere nihil probat, sed tantum narrat:" and concludes by saying, "quas adhuc alias causas adsignat (he has just enumerated three of the Idols of the Tribe) facile omnes ad unicam Cartesii reduci possunt; scilicet quia voluntas humana est libera et latior intellectu; sive, ut ipse Verulamius magis confuse loquitur, quia intellectus luminis sicci non est, sed recipit infusionem a voluntate". See *Spinoza to Oldenburg*, ep. 2. vol. ii. p. 146, of Bruder's edition.

[24] Compare *Advanc. of Learning*: "That the spirit of man *being of an equal and uniform substance* doth usually suppose and feign in nature a greater equality and uniformity than is in truth".—*J. S.*

LIII.

The *Idols of the Cave* take their rise in the peculiar constitution, mental or bodily, of each individual; and also in education, habit, and accident. Of this kind there is a great number and variety; but I will instance those the pointing out of which contains the most important caution, and which have most effect in disturbing the clearness of the understanding.

LIV.

Men become attached to certain particular sciences and speculations, either because they fancy themselves the authors and inventors thereof, or because they have bestowed the greatest pains upon them and become most habituated to them. But men of this kind, if they betake themselves to philosophy and contemplations of a general character, distort and colour them in obedience to their former fancies; a thing especially to be noticed in Aristotle, who made his natural philosophy a mere bond-servant to his logic, thereby rendering it contentious and well nigh useless. The race of chemists again out of a few experiments of the furnace have built up a fantastic philosophy, framed with reference to a few things; and Gilbert also, after he had employed himself most laboriously in the study and observation of the loadstone, proceeded at once to construct an entire system in accordance with his favourite subject [25].

LV.

There is one principal and as it were radical distinction between different minds, in respect of philosophy and the sciences; which is this: that some minds are stronger and apter to mark the differences of things, others to mark their resemblances. The steady and acute mind can fix its contemplations and dwell and fasten on the subtlest distinctions: the lofty and discursive mind recognises and puts together the finest and most general resemblances. Both kinds however easily err in excess, by catching the one at gradations the other at shadows.

LVI.

There are found some minds given to an extreme admiration of antiquity, others to an extreme love and appetite for novelty; but few so duly tempered that they can hold the mean, neither carping at what has been well laid down by the ancients, nor despising what is well introduced by the moderns. This however turns to the great injury of the sciences and philosophy; since these affectations of antiquity and novelty are the humours of partisans rather than judgments; and truth is to be sought for not in the felicity of any age, which is an unstable thing, but in the light of nature and experience, which is eternal. These factions therefore must be abjured, and care must be taken that the intellect be not hurried by them into assent.

LVII.

Contemplations of nature and of bodies in their simple form break up and distract the understanding, while contemplations of nature and bodies in their composition and configuration overpower and dissolve the understanding: a distinc-

[25] [This is one of several instances (Aph. 64, 70) in which Bacon speaks slightingly of Gilbert, whose work on the Magnet (1600) has stood scientific test so much better than most of Bacon's own scientific speculation. Whewell, in a letter to Spedding of Nov. 17, 848, says of Bacon: "Almost the only matter for which I find reason to blame him is his injustice to Gilbert, whom he scarcely ever mentions, except to blame him for the narrowness of his method, but whose philosophy was really almost as wide as Bacon's own, and solid precisely on account of his starting from such a reality as magnetic forces" (*Life of Whewell*, 2nd ed. 1882, p. 355). Spedding in reply offers an interesting defence of Bacon (*Id.* pp. 358, 361–2). In the third edition of his *History of the Inductive Sciences* (1857) Whewell does not take up the same position; he even speaks of "Bacon's contemporary, Gilbert, whom he frequently praises as a philosopher" (i. 297), though in the later edition of the *Philosophy of Discovery* (1860, pp. 114–115) he leaves standing his blame of Bacon's unfairness.—ED.]

tion well seen in the school of Leucippus and Democritus as compared with the other philosophies. For that school is so busied with the particles that it hardly attends to the structure; while the others are so lost in admiration of the structure that they do not penetrate to the simplicity of nature. These kinds of contemplation should therefore be alternated and taken by turns; that so the understanding may be rendered at once penetrating and comprehensive, and the inconveniences above mentioned, with the idols which proceed from them, may be avoided.

LVIII.

Let such then be our provision and contemplative prudence for keeping off and dislodging the *Idols of the Cave*, which grow for the most part either out of the predominance of a favourite subject, or out of an excessive tendency to compare or to distinguish, or out of partiality for particular ages, or out of the largeness or minuteness of the objects contemplated. And generally let every student of nature take this as a rule,—that whatever his mind seizes and dwells upon with peculiar satisfaction is to be held in suspicion, and that so much the more care is to be taken in dealing with such questions to keep the understanding even and clear.

LIX.

But the *Idols of the Market-place* are the most troublesome of all: idols which have crept into the understanding through the alliances of words and names. For men believe that their reason governs words; but it is also true that words react on the understanding; and this it is that has rendered philosophy and the sciences sophistical and inactive. Now words, being commonly framed and applied according to the capacity of the vulgar, follow those lines of division which are most obvious to the vulgar understanding. And whenever an understanding of greater acuteness or a more diligent observation would alter those lines to suit the true divisions of nature, words stand in the way and resist the change. Whence it comes to pass that the high and formal discussions of learned men end oftentimes in disputes about words and names; with which (according to the use and wisdom of the mathematicians) it would be more prudent to begin, and so by means of definitions reduce them to order. Yet even definitions cannot cure this evil in dealing with natural and material things; since the definitions themselves consist of words, and those words beget others: so that it is necessary to recur to individual instances, and those in due series and order; as I shall say presently when I come to the method and scheme for the formation of notions and axioms.

LX.

The idols imposed by words on the understanding are of two kinds. They are either names of things which do not exist (for as there are things left unnamed through lack of observation, so likewise are there names which result from fantastic suppositions and to which nothing in reality corresponds), or they are names of things which exist, but yet confused and ill-defined, and hastily and irregularly derived from realities. Of the former kind are Fortune, the Prime Mover, Planetary Orbits, Element of Fire, and like fictions which owe their origin to false and idle theories. And this class of idols is more easily expelled, because to get rid of them it is only necessary that all theories should be steadily rejected and dismissed as obsolete.

But the other class, which springs out of a faulty and unskilful abstraction, is intricate and deeply rooted. Let us take for example such a word as *humid*; and see how far the several things which the word is used to signify agree with each other; and we shall find the word *humid* to be nothing else than a mark loosely and confusedly applied to denote a variety of actions which will not bear to be reduced to any constant meaning. For it both signifies that which easily spreads itself round any other body; and that which in itself is indeterminate and cannot solidise; and that which readily yields in every direction; and that which easily divides and scatters itself; and that which easily unites and collects itself; and that which readily flows and is put in motion; and that which readily clings to another body and wets it; and that which is easily reduced to a liquid, or being solid easily melts. Accordingly when you come to apply the word,—if you take it in one sense, flame is humid ; if in another, air is not humid ; if in another, fine dust is humid ; if in another, glass is humid. So that it is easy to see that the notion is taken by abstraction only from water and common and ordinary liquids, without any due verification.

There are however in words certain degrees of distortion and error. One of the least faulty kinds is that of names of substances, especially of lowest species and well-deduced (for the notion of *chalk* and of *mud* is good, of *earth* bad) ; a more faulty kind is that of actions, as to *generate, to corrupt, to alter ;* the most faulty is of qualities (except such as are the immediate objects of the sense) as *heavy, light, rare, dense,* and the like. Yet in all these cases some notions are of necessity a little better than others, in proportion to the greater variety of subjects that fall within the range of the human sense.

LXI.

But the *Idols of the Theatre* are not innate, nor do they steal into the understanding secretly, but are plainly impressed and received into the mind from the play-books of philosophical systems and the perverted rules of demonstration. To attempt refutations in this case would be merely inconsistent with what I have already said : for since we agree neither upon principles nor upon demonstrations there is no place for argument. And this is so far well, inasmuch as it leaves the honour of the ancients untouched. For they are no wise disparaged—the question between them and me being only as to the way. For as the saying is, the lame man who keeps the right road outstrips the runner who takes a wrong one. Nay it is obvious that when a man runs the wrong way, the more active and swift he is the further he will go astray.

But the course I propose for the discovery of sciences is such as leaves but little to the acuteness and strength of wits, but places all wits and understandings nearly on a level [26]. For as in the drawing of a straight line or a perfect circle, much depends on the steadiness and practice of the hand, if it be done by aim of hand only, but if with the aid of rule or compass, little or nothing ; so is it exactly with my plan. But though particular confutations would be of no avail, yet touching the sects and general divisions of such systems I must say something ; something also touching the external signs which show that they are unsound ; and finally something touching the causes of such great infelicity and of such lasting and general agreement in error ; that so the access to truth may be made less difficult, and the human understanding may the more willingly submit to its purgation and dismiss its idols.

LXII.

Idols of the Theatre, or of Systems, are many, and there can be and perhaps will be yet many more. For were it not that now for many ages men's minds have been busied with religion and theology ; and were it not that civil governments, especially monarchies, have been averse to such novelties, even in matters speculative ; so that men labour therein to the peril and harming of their fortunes, not only unrewarded, but exposed also to contempt and envy ; doubtless there would have arisen many other philosophical sects like to those which in great variety flourished once among the Greeks. For as on the phenomena of the heavens many hypotheses may be constructed, so likewise (and more also) many various dogmas may be set up and established on the phenomena of philosophy. And in the plays of this philosophical theatre you may observe the same thing which is found in the theatre of the poets, that stories invented for the stage are more compact and elegant, and more as one would wish them to be, than true stories out of history.

In general however there is taken for the material of philosophy either a great deal out of a few things, or a very little out of many things ; so that on both sides philosophy is based on too narrow a foundation of experiment and natural history, and decides on the authority of too few cases. For the Rational School of

[26] [This is one of the passages of Bacon on which Macaulay is justly severe. " Bacon's promise," adds Professor Fowler, " never has been and never can be fulfilled."—ED.]

philosophers snatches from experience a variety of common instances, neither duly ascertained nor diligently examined and weighed, and leaves all the rest to meditation and agitation of wit.

There is also another class of philosophers, who having bestowed much diligent and careful labour on a few experiments, have thence made bold to educe and construct systems; wresting all other facts in a strange fashion to conformity therewith.

And there is yet a third class, consisting of those who out of faith and veneration mix their philosophy with theology and traditions; among whom the vanity of some has gone so far aside as to seek the origin of sciences among spirits and genii. So that this parent stock of errors—this false philosophy—is of three kinds; the Sophistical, the Empirical, and the Superstitious.

LXIII.

The most conspicuous example of the first class was Aristotle, who corrupted natural philosophy by his logic: fashioning the world out of categories; assigning to the human soul, the noblest of substances, a genus from words of the second intention [27]; doing the business of density and rarity (which is to make bodies of greater or less dimensions, that is, occupy greater or less spaces), by the frigid distinction of act and power; asserting that single bodies have each a single and proper motion [28], and that if they participate in any other, then this results from an external cause; and imposing countless other arbitrary restrictions on the nature of things; being always more solicitous to provide an answer to the question and affirm something positive in words, than about the inner truth of things; a failing best shown when his philosophy is compared with other systems of note among the Greeks. For the Homœomera of Anaxagoras; the Atoms of Leucippus and Democritus; the Heaven and Earth of Parmenides; the Strife and Friendship of Empedocles; Heraclitus's doctrine how bodies are resolved into the indifferent nature of fire, and remoulded into solids; have all of them some taste of the natural philosopher,—some savour of the nature of things, and experience, and bodies; whereas in the physics of Aristotle you hear hardly anything but the words of logic; which in his metaphysics also, under a more imposing name, and more forsooth as a realist than a nominalist, he has handled over again. Nor let any weight be given to the fact, that in his books on animals and his problems, and other of his treatises, there is frequent dealing with experiments. For he had come to his conclusion before; he did not consult experience, as he should have done, in order to the framing of his decisions and axioms; but having first determined the question according to his will, he then resorts to experience, and bending her into conformity with his placets leads her about like a captive in a procession; so that even on this count he is more guilty than his modern followers, the schoolmen, who have abandoned experience altogether.

LXIV.

But the Empirical school of philosophy gives birth to dogmas more deformed and monstrous than the Sophistical or Rational school. For it has its foundations not in the light of common notions, (which though it be a faint and superficial light, is yet in a manner universal, and has reference to many things [29],) but in the narrowness and darkness of a few experiments. To those therefore who are daily busied with these experiments, and have infected their imagination with them, such a philosophy seems probable and all but certain; to all men else incredible and vain. Of this there is a notable instance in the alchemists and their dogmas; though it is hardly to be found elsewhere in these times, except perhaps in the philosophy of Gilbert. Nevertheless with regard to philosophies

[27] This censure refers to Aristotle's definition of the soul (*De Anima*, ii. 1).

[28] "Simplicis corporis simplicem esse motum" is an important principle in Aristotelian physics, as one of the bases on which the system of the universe was made to depend. See, for instance, Melanchthon's *Initia Doctr. Physicæ*, p. 41.

[29] *Ad multa pertinens.* In the formation of such notions many things have been taken into account.—*J. S.*

of this kind there is one caution not to be omitted; for I foresee that if ever men are roused by my admonitions to betake themselves seriously to experiment and bid farewell to sophistical doctrines, then indeed through the premature hurry of the understanding to leap or fly to universals and principles of things, great danger may be apprehended from philosophies of this kind; against which evil we ought even now to prepare.

LXV.

But the corruption of philosophy by superstition and an admixture of theology is far more widely spread, and does the greatest harm, whether to entire systems or to their parts. For the human understanding is obnoxious to the influence of the imagination no less than to the influence of common notions. For the contentious and sophistical kind of philosophy ensnares the understanding; but this kind, being fanciful and tumid and half poetical, misleads it more by flattery. For there is in man an ambition of the understanding, no less than of the will, especially in high and lofty spirits.

Of this kind we have among the Greeks a striking example in Pythagoras, though he united with it a coarser and more cumbrous superstition; another in Plato and his school, more dangerous and subtle. It shows itself likewise in parts of other philosophies, in the introduction of abstract forms and final causes and first causes, with the omission in most cases of causes intermediate, and the like. Upon this point the greatest caution should be used. For nothing is so mischievous as the apotheosis of error; and it is a very plague of the understanding for vanity to become the object of veneration. Yet in this vanity some of the moderns [30] have with extreme levity indulged so far as to attempt to found a system of natural philosophy on the first chapter of Genesis, on the book of Job, and other parts of the sacred writings; seeking for the dead among the living [31]: which also makes the inhibition and repression of it the more important, because from this unwholesome mixture of things human and divine there arises not only a fantastic philosophy but also an heretical religion. Very meet it is therefore that we be sober-minded, and give to faith that only which is faith's [32].

LXVI.

So much then for the mischievous authorities of systems, which are founded either on common notions, or on a few experiments, or on superstition. It remains to speak of the faulty subject-matter of contemplations, especially in natural philosophy. Now the human understanding is infected by the sight of what takes place in the mechanical arts, in which the alteration of bodies proceeds chiefly by composition or separation, and so imagines that something similar goes on in the universal nature of things. From this source has flowed the fiction of elements, and of their concourse for the formation of natural bodies. Again, when man contemplates nature working freely, he meets with different species of things, of animals, of plants, of minerals; whence he readily passes into the opinion that there are in nature certain primary forms which nature intends to educe, and that the remaining variety proceeds from hindrances and aberrations of nature in the fulfilment of her work, or from the collision of different species and the transplanting of one into another. To the first of these speculations we owe our primary qualities of the elements [33]; to the other our occult properties [34] and

[30] [Professor Fowler notes that the allusion here is probably to Dr. Robert Fludd. See notes to B. II. Aph. xiii. § 38, and to *De Augmentis*, B. II. c. xiii.—Ed.]

[31] [See Luke xxiv. 5. The phrase occurs several times in Bacon.—Ed.]

[32] [Professor Fowler remarks that "We must recollect that sentiments of this kind, which with us have become commonplaces, were in Bacon's time novel and almost paradoxical".—Ed.]

[33] The elementary qualities are four in number,—hot, cold, dry, moist; and it is by combining them two and two that the Peripatetic conception of the nature of each element is formed. Thus fire is hot and dry, water cold and moist, etc. All the other qualities of bodies, which result from the combination and mutual modification of the elementary and primary qualities, were called secondary qualities.

[34] [Compare Newton's *Optics*, B. iii.—Ed.]

specific virtues ; and both of them belong to those empty *compendia* of thought wherein the mind rests, and whereby it is diverted from more solid pursuits. It is to better purpose that the physicians bestow their labour on the secondary qualities of matter, and the operations of attraction, repulsion, attenuation, conspissation, dilatation, astriction, dissipation, maturation, and the like ; and were it not that by those two compendia which I have mentioned (elementary qualities, to wit, and specific virtues) they corrupted their correct observations in these other matters,—either reducing them to first qualities and their subtle and incommensurable mixtures, or not following them out with greater and more diligent observation to third and fourth qualities, but breaking off the scrutiny prematurely,—they had made much greater progress. Nor are powers of this kind (I do not say the same, but similar) to be sought for only in the medicines of the human body, but also in the changes of all other bodies.

But it is a far greater evil that they make the quiescent principles, *wherefrom*, and not the moving principles, *whereby*, things are produced, the object of their contemplation and inquiry. For the former tend to discourse, the latter to works. Nor is there any value in those vulgar distinctions of motion which are observed in the received system of natural philosophy, as generation, corruption, augmentation, diminution, alteration, and local motion. What they mean no doubt is this : —If a body, in other respects not changed, be moved from its place, this is *local motion ;* if without change of place or essence, it be changed in quality, this is *alteration ;* if by reason of the change the mass and quantity of the body do not remain the same, this is *augmentation* or *diminution ;* if they be changed to such a degree that they change their very essence and substance and turn to something else, this is *generation* and *corruption*. But all this is merely popular, and does not at all go deep into nature ; for these are only measures and limits, not kinds of motion. What they intimate is *how far*, not *by what means*, or *from what source*. For they do not suggest anything with regard either to the desires of bodies or to the development of their parts : it is only when that motion presents the thing grossly and palpably to the sense as different from what it was, that they begin to mark the division. Even when they wish to suggest something with regard to the causes of motion, and to establish a division with reference to them, they introduce with the greatest negligence a distinction between motion natural and violent ; a distinction which is itself drawn entirely from a vulgar notion, since all violent motion is also in fact natural ; the external efficient simply setting nature working otherwise than it was before. But if, leaving all this, any one shall observe (for instance) that there is in bodies a desire of mutual contact, so as not to suffer the unity of nature to be quite separated or broken and a vacuum thus made ; or if any one say that there is in bodies a desire of resuming their natural dimensions or tension, so that if compressed within or extended beyond them, they immediately strive to recover themselves, and fall back to their old volume and extent ; or if any one say that there is in bodies a desire of congregating towards masses of kindred nature,—of dense bodies, for instance, towards the globe of the earth, of thin and rare bodies towards the compass of the sky ; all these and the like are truly physical kinds of motion ;—but those others are entirely logical and scholastic, as is abundantly manifest from this comparison.

Nor again is it a less evil, that in their philosophies and contemplations their labour is spent in investigating and handling the first principles of things and the highest generalities of nature ; whereas utility and the means of working result entirely from things intermediate. Hence it is that men cease not from abstracting nature till they come to potential and uninformed matter, nor on the other hand from dissecting nature till they reach the atom ; things which, even if true, can do but little for the welfare of mankind.

LXVII.

A caution must also be given to the understanding against the intemperance which systems of philosophy manifest in giving or withholding assent ; because intemperance of this kind seems to establish Idols and in some sort to perpetuate them, leaving no way open to reach and dislodge them.

This excess is of two kinds: the first being manifest in those who are ready in deciding, and render sciences dogmatic and magisterial; the other in those who deny that we can know anything, and so introduce a wandering kind of inquiry that leads to nothing; of which kinds the former subdues, the latter weakens the understanding. For the philosophy of Aristotle, after having by hostile confutations destroyed all the rest (as the Ottomans serve their brothers), has laid down the law on all points; which done, he proceeds himself to raise new questions of his own suggestion, and dispose of them likewise; so that nothing may remain that is not certain and decided: a practice which holds and is in use among his successors.

The school of Plato, on the other hand, introduced *Acatalepsia*, at first in jest and irony, and in disdain of the older sophists, Protagoras, Hippias, and the rest, who were of nothing else so much ashamed as of seeming to doubt about anything [35]. But the new Academy made a dogma of it, and held it as a tenet. And though theirs is a fairer seeming way than arbitrary decisions; since they say that they by no means destroy all investigation, like Pyrrho and his Refrainers, but allow of some things to be followed as probable, though of none to be maintained as true; yet still when the human mind has once despaired of finding truth, its interest in all things grows fainter; and the result is that men turn aside to pleasant disputations and discourses and roam as it were from object to object, rather than keep on a course of severe inquisition. But, as I said at the beginning, and am ever urging, the human senses and understanding, weak as they are, are not to be deprived of their authority, but to be supplied with helps.

LXVIII.

So much concerning the several classes of Idols, and their equipage: all of which must be renounced and put away with a fixed and solemn determination, and the understanding thoroughly freed and cleansed; the entrance into the kingdom of man, founded on the sciences, being not much other than the entrance into the kingdom of heaven, whereinto none may enter except as a little child.

LXIX.

But vicious demonstrations are as the strongholds and defences of Idols; and those we have in logic do little else than make the world the bond-slave of human thought, and human thought the bond-slave of words. Demonstrations truly are in effect the philosophies themselves and the sciences. For such as *they* are, well or ill established, such are the systems of philosophy and the contemplations which follow. Now in the whole of the process which leads from the sense and objects to axioms and conclusions, the demonstrations which we use are deceptive and incompetent. This process consists of four parts, and has as many faults. In the first place, the impressions of the sense itself are faulty; for the sense both fails us and deceives us. But its shortcomings are to be supplied, and its deceptions to be corrected. Secondly, notions are ill drawn from the impressions of the senses, and are indefinite and confused, whereas they should be definite and distinctly bounded. Thirdly, the induction is amiss which infers the principles of sciences by simple enumeration, and does not, as it ought, employ exclusions and solutions (or separations) of nature. Lastly, that method of discovery and proof according to which the most general principles are first established, and then intermediate axioms are tried and proved by them, is the parent of error and the curse of all science. Of these things however, which now I do but touch upon, I will speak more largely when, having performed these expiations and purgings of the mind, I come to set forth the true way for the interpretation of nature.

LXX.

But the best demonstration by far is experience, if it go not beyond the actual experiment. For if it be transferred to other cases which are deemed similar, unless such transfer be made by a just and orderly process, it is a fallacious thing. But the manner of making experiments which men now use is blind and stupid.

[35] See Cicero. *De Nat. Deor.* i. c. 8.

And therefore, wandering and straying as they do with no settled course, and taking counsel only from things as they fall out, they fetch a wide circuit and meet with many matters, but make little progress; and sometimes are full of hope, sometimes are distracted; and always find that there is something beyond to be sought. For it generally happens that men make their trials carelessly, and as it were in play; slightly varying experiments already known, and, if the thing does not answer, growing weary and abandoning the attempt. And even if they apply themselves to experiments more seriously and earnestly and laboriously, still they spend their labour in working out some one experiment, as Gilbert with the magnet, and the chemist with gold; a course of proceeding not less unskilful in the design than small in the attempt. For no one successfully investigates the nature of a thing in the thing itself; the inquiry must be enlarged, so as to become more general.

And even when they seek to educe some science or theory from their experiments, they nevertheless almost always turn aside with overhasty and unseasonable eagerness to practice; not only for the sake of the uses and fruits of the practice, but from impatience to obtain in the shape of some new work an assurance for themselves that it is worth their while to go on; and also to show themselves off to the world, and so raise the credit of the business in which they are engaged. Thus, like Atalanta, they go aside to pick up the golden apple, but meanwhile they interrupt their course, and let the victory escape them. But in the true course of experience, and in carrying it on to the effecting of new works, the divine wisdom and order must be our pattern. Now God on the first day of creation created light only, giving to that work an entire day, in which no material substance was created. So must we likewise from experience of every kind first endeavour to discover true causes and axioms; and seek for experiments of Light, not for experiments of Fruit. For axioms rightly discovered and established supply practice with its instruments, not one by one, but in clusters, and draw after them trains and troops of works. Of the paths however of experience, which no less than the paths of judgment are impeded and beset, I will speak hereafter here I have only mentioned ordinary experimental research as a bad kind of demonstration. But now the order of the matter in hand leads me to add something both as to those *signs* which I lately mentioned,—(signs that the systems of philosophy and contemplation in use are in a bad condition)—and also as to the *causes* of what seems at first so strange and incredible. For a knowledge of the signs prepares assent; an explanation of the causes removes the marvel: which two things will do much to render the extirpation of Idols from the understanding more easy and gentle.

LXXI.

The sciences which we possess come for the most part from the Greeks. For what has been added by Roman, Arabic, or later writers is not much nor of much importance [36]; and whatever it is, it is built on the foundation of Greek discoveries [37]. Now the wisdom of the Greeks was professorial and much given to disputations; a kind of wisdom most adverse to the inquisition of truth. Thus that name of Sophists which by those who would be thought philosophers was in contempt cast back upon and so transferred to the ancient rhetoricians, Gorgias, Protagoras, Hippias, Polus, does indeed suit the entire class, Plato, Aristotle, Zeno, Epicurus, Theophrastus, and their successors Chrysippus, Carneades, and the rest. There was this difference only, that the former class was wandering and mercenary, going about from town to town, putting up their wisdom to sale, and taking a price for it; while the latter was more pompous and dignified, as composed of men who had fixed abodes, and who opened schools and taught their

[36] [As Professor Fowler notes, this is unjust as regards the Arabs, whose services to chemistry and to mathematics (as regards algebra) are of real importance.—ED.]

[37] M. Chasles appears to have shown this with respect to the principle of position in arithmetic. We derive it, according to him, not from the Hindoos or Arabs, but from the Greeks. It is remarkable that the Chinese have from the earliest times known how to express any number by means of a few characters.

philosophy without reward. Still both sorts, though in other respects unequal, were professorial; both turned the matter into disputations, and set up and battled for philosophical sects and heresies; so that their doctrines were for the most part (as Dionysius not unaptly rallied Plato) " the talk of idle old men to ignorant youths" [38]. But the elder of the Greek philosophers, Empedocles, Anaxagoras, Leucippus, Democritus, Parmenides, Heraclitus, Xenophanes, Philolaus, and the rest (I omit Pythagoras as a mystic), did not, so far as we know, open schools; but more silently and severely and simply—that is, with less affectation and parade—betook themselves to the inquisition of truth. And therefore they were in my judgment more successful; only that their works were in the course of time obscured by those slighter persons who had more which suits and pleases the capacity and tastes of the vulgar: time, like a river, bringing down to us things which are light and puffed up, but letting weighty matters sink. Still even they were not altogether free from the failing of their nation; but leaned too much to the ambition and vanity of founding a sect and catching popular applause. But the inquisition of truth must be despaired of when it turns aside to trifles of this kind. Nor should we omit that judgment, or rather divination, which was given concerning the Greeks by the Ægyptian priest,—that " they were always boys, without antiquity of knowledge or knowledge of antiquity" [39]. Assuredly they have that which is characteristic of boys; they are prompt to prattle but cannot generate; for their wisdom abounds in words but is barren of works. And therefore the signs which are taken from the origin and birth-place of the received philosophy are not good.

LXXII.

Nor does the character of the time and age yield much better signs than the character of the country and nation. For at that period there was but a narrow and meagre knowledge either of time or place; which is the worst thing that can be, especially for those who rest all on experience. For they had no history, worthy to be called history, that went back a thousand years; but only fables and rumours of antiquity. And of the regions and districts of the world they knew but a small portion; giving indiscriminately the name of Scythians to all in the North, of Celts to all in the West; knowing nothing of Africa beyond the hither side of Æthiopia, of Asia beyond the Ganges; much less were they acquainted with the provinces of the New World, even by hearsay or any well-founded rumour; nay, a multitude of climates and zones, wherein innumerable nations breathe and live, were pronounced by them to be uninhabitable; and the travels of Democritus, Plato, and Pythagoras, which were rather suburban excursions than distant journeys, were talked of as something great. In our times on the other hand both many parts of the New World and the limits on every side of the Old World are known, and our stock of experience has increased to an infinite amount. Wherefore if (like astrologers) we draw signs from the season of their nativity or birth, nothing great can be predicted of those systems of philosophy.

LXXIII.

Of all signs there is none more certain or more noble than that taken from fruits. For fruits and works are as it were sponsors and sureties for the truth of philosophies. Now, from all these systems of the Greeks, and their ramifications through particular sciences, there can hardly after the lapse of so many years be adduced a single experiment which tends to relieve and benefit the condition of man, and which can with truth be referred to the speculations and theories of philosophy. And Celsus ingenuously and wisely owns as much, when he tells us that the experimental part of medicine was first discovered, and that afterward men philosophised about it, and hunted for and assigned causes; and not by an inverse process that philosophy and the knowledge of causes led to the discovery and development of the experimental part [40]. And therefore it was not strange

[38] Diog. Laert. in Platon. c. 18.

[39] Plato, *Timaeus*.

[40] " Repertis deinde medicinæ remediis homines de rationibus eorum disserere cœpisse :

that among the Ægyptians, who rewarded inventors with divine honours and sacred rites, there were more images of brutes than of men ; inasmuch as brutes by their natural instinct have produced many discoveries, whereas men by discussion and the conclusions of reason have given birth to few or none.

Some little has indeed been produced by the industry of chemists ; but it has been produced accidentally and in passing, or else by a kind of variation of experiments, such as mechanics use ; and not by any art or theory, for the theory which they have devised rather confuses the experiments than aids them. They too who have busied themselves with Natural Magic, as they call it, have but few discoveries to show, and those trifling and imposture-like. Wherefore, as in religion we are warned to show our faith by works, so in philosophy by the same rule the system should be judged of by its fruits, and pronounced frivolous if it be barren ; more especially if, in place of fruits of grape and olive, it bear thorns and briars of dispute and contention.

LXXIV.

Signs also are to be drawn from the increase and progress of systems and sciences. For what is founded on nature grows and increases ; while what is founded on opinion varies but increases not. If therefore those doctrines had not plainly been like a plant torn up from its roots, but had remained attached to the womb of nature and continued to draw nourishment from her, that could never have come to pass which we have seen now for twice a thousand years ; namely, that the sciences stand where they did and remain almost in the same condition ; receiving no noticeable increase, but on the contrary, thriving most under their first founder, and then declining. Whereas in the mechanical arts, which are founded on nature and the light of experience, we see the contrary happen, for these (as long as they are popular) are continually thriving and growing, as having in them a breath of life ; at first rude, then convenient, afterwards adorned, and at all times advancing.

LXXV.

There is still another sign remaining (if sign it can be called, when it is rather testimony, nay, of all testimony the most valid) ; I mean the confession of the very authorities whom men now follow. For even they who lay down the law on all things so confidently, do still in their more sober moods fall to complaints of the subtlety of nature, the obscurity of things, and the weakness of the human mind. Now if this were all they did, some perhaps of a timid disposition might be deterred from further search, while others of a more ardent and hopeful spirit might be whetted and incited to go on farther. But not content to speak for themselves, whatever is beyond their own or their master's knowledge or reach they set down as beyond the bounds of possibility, and pronounce, as if on the authority of their art, that it cannot be known or done ; thus most presumptuously and invidiously turning the weakness of their own discoveries into a calumny on nature herself, and the despair of the rest of the world. Hence the school of the New Academy, which held *Acatalepsia* as a tenet and doomed men to perpetual darkness. Hence the opinion that Forms or true differences of things (which are in fact laws of pure act) are past finding out and beyond the reach of man. Hence too those opinions in the department of action and operation ; as that the heat of the sun and of fire are quite different in kind,—lest men should imagine that by the operations of fire anything like the works of nature can be educed and formed. Hence the notion that composition only is the work of man,

nec post rationem medicinam esse inventam, sed post inventam medicinam rationem esse quæsitam."—*Celsus, Præfatio.*

But this remark is not made by Celsus as the expression of his own opinion ; on the contrary it occurs in his statement of the views entertained by the empirical school of medicine, to which he is decidedly opposed. The error of citing Celsus as an authority for it is repeated in several parts of Bacon's works. [See among others *De Augmentis,* v. 2.—*J. S.*]

and mixture of none but nature [41],—lest men should expect from art some power of generating or transforming natural bodies. By this sign, therefore, men will easily take warning not to mix up their fortunes and labours with dogmas not only despaired of but dedicated to despair.

LXXVI.

Neither is this other sign to be omitted ;—that formerly there existed among philosophers such great disagreement, and such diversities in the schools themselves ; a fact which sufficiently shows that the road from the senses to the understanding was not skilfully laid out, when the same groundwork of philosophy (the nature of things to wit) was torn and split up into such vague and multifarious errors. And although in these times disagreements and diversities of opinion on first principles and entire systems are for the most part extinguished, still on parts of philosophy there remain innumerable questions and disputes, so that it plainly appears that neither in the systems themselves nor in the modes of demonstration is there anything certain or sound.

LXXVII.

And as for the general opinion that in the philosophy of Aristotle at any rate there is great agreement ; since after its publication the systems of older philosophers died away, while in the times which followed nothing better was found ; so that it seems to have been so well laid and established as to have drawn both ages in its train ; I answer in the first place, that the common notion of the falling off of the old systems upon the publication of Aristotle's works is a false one ; for long afterwards, down even to the times of Cicero and subsequent ages, the works of the old philosophers still remained. But in the times which followed, when on the inundation of barbarians into the Roman empire human learning had suffered shipwreck, then the systems of Aristotle and Plato, like planks of lighter and less solid material, floated on the waves of time, and were preserved. Upon the point of consent also men are deceived, if the matter be looked into more keenly. For true consent is that which consists in the coincidence of free judgments, after due examination. But far the greater number of those who have assented to the philosophy of Aristotle have addicted themselves thereto from prejudgment and upon the authority of others ; so that it is a following and going along together, rather than consent. But even if it had been a real and widespread consent, still so little ought consent to be deemed a sure and solid confirmation, that it is in fact a strong presumption the other way. For the worst of all auguries is from consent in matters intellectual (divinity excepted, and politics where there is right of vote) [42]. For nothing pleases the many unless it strikes the imagination, or binds the understanding with the bonds of common notions, as I have already said. We may very well transfer therefore from moral to intellectual matters, the saying of Phocion, that if the multitude assent and applaud, men ought immediately to examine themselves as to what blunder or fault they may have committed [43]. This sign therefore is one of the most unfavourable. And so much for this point ; namely, that the signs of truth and soundness in the received systems and sciences are not good ; whether they be drawn from their origin, or from their fruits, or from their progress, or from the confessions of their founders, or from general consent.

LXXVIII.

I now come to the *causes* of these errors, and of so long a continuance in them through so many ages ; which are very many and very potent ;—that all wonder how these considerations which I bring forward should have escaped men's notice

[41] The reference is to Galen, who in his treatise *De Natural. Facultatibus* contrasts the inwardly formative powers of nature with the external operations of art. [Compare Aph. iv., where Bacon advances the very proposition he here disparages.—Ed.]

[42] Bacon does not mean that the votes of a majority are necessarily valid in matters of divinity or politics, but merely that, from the nature of the case, the argument from consent has more weight in these than in purely scientific questions.

[43] [See Plutarch's *Life of Phocion*, c. 8.—Ed.]

till now, may cease; and the only wonder be, how now at last they should have entered into any man's head and become the subject of his thoughts; which truly I myself esteem as the result of some happy accident, rather than of any excellence of faculty in me; a birth of Time rather than a birth of Wit. Now, in the first place, those so many ages, if you weigh the case truly, shrink into a very small compass. For out of the five and twenty centuries over which the memory and learning of men extends, you can hardly pick out six that were fertile in sciences or favourable to their development. In times no less than in regions there are wastes and deserts. For only three revolutions and periods of learning can properly be reckoned; one among the Greeks, the second among the Romans, and the last among us, that is to say, the nations of Western Europe, and to each of these hardly two centuries can justly be assigned. The intervening ages of the world, in respect of any rich or flourishing growth of the sciences, were unprosperous. For neither the Arabians nor the Schoolmen need be mentioned; who in the intermediate times rather crushed[44] the sciences with a multitude of treatises, than increased their weight. And therefore the first cause of so meagre a progress in the sciences is duly and orderly referred to the narrow limits of the time that has been favourable to them.

LXXIX.

In the second place there presents itself a cause of great weight in all ways; namely, that during those very ages in which the wits and learning of men have flourished most, or indeed flourished at all, the least part of their diligence was given to natural philosophy. Yet this very philosophy it is that ought to be esteemed the great mother of the sciences. For all arts and all sciences, if torn from this root, though they may be polished and shaped and made fit for use, yet they will hardly grow. Now it is well known that after the Christian religion was received and grew strong, by far the greater number of the best wits applied themselves to theology: that to this both the highest rewards were offered, and helps of all kinds most abundantly supplied; and that this devotion to theology chiefly occupied that third portion or epoch of time among us Europeans of the West; and the more so because about the same time both literature began to flourish and religious controversies to spring up. In the age before, on the other hand, during the continuance of the second period among the Romans, the meditations and labours of philosophers were principally employed and consumed on moral philosophy, which to the Heathen was as theology to us. Moreover in those times the greatest wits applied themselves very generally to public affairs; the magnitude of the Roman empire requiring the services of a great number of persons. Again, the age in which natural philosophy was seen to flourish most among the Greeks, was but a brief particle of time; for in early ages the Seven Wise Men, as they were called, (all except Thales) applied themselves to morals and politics; and in later times, when Socrates had drawn down philosophy from heaven to earth, moral philosophy became more fashionable than ever, and diverted the minds of men from the philosophy of nature.

Nay, the very period itself in which inquiries concerning nature flourished, was by controversies and the ambitious display of new opinions corrupted and made useless. Seeing therefore that during those three periods natural philosophy was in a great degree either neglected or hindered, it is no wonder if men made but small advance in that to which they were not attending.

LXXX.

To this it may be added that natural philosophy, even among those who have attended to it, has scarcely ever possessed, especially in these later times, a disengaged and whole man (unless it were some monk studying in his cell, or some gentleman in his country-house), but that it has been made merely a passage and bridge to something else. And so this great mother of the sciences has with strange indignity been degraded to the offices of a servant; having to attend on the business of medicine or mathematics, and likewise to wash and imbue youthful

[44] *Contriverunt*: wore them hard, I suppose; like a path much trodden.—*J. S.*

and unripe wits with a sort of first dye, in order that they may be the fitter to receive another afterwards. Meanwhile let no man look for much progress in the sciences—especially in the practical part of them—unless natural philosophy be carried on and applied to particular sciences, and particular sciences be carried back again to natural philosophy. For want of this, astronomy, optics, music, a number of mechanical arts, medicine, itself—nay, what one might more wonder at, moral and political philosophy, and the logical sciences,—altogether lack profoundness, and merely glide along the surface and variety of things ; because after these particular sciences have been once distributed and established, they are no more nourished by natural philosophy ; which might have drawn out of the true contemplation of motions, rays, sounds, texture and configuration of bodies, affections, and intellectual perceptions, the means of imparting to them fresh strength and growth. And therefore it is nothing strange if the sciences grow not, seeing they are parted from their roots.

LXXXI.

Again there is another great and powerful cause why the sciences have made but little progress ; which is this. It is not possible to run a course aright when the goal itself has not been rightly placed. Now the true and lawful goal of the sciences is none other than this : that human life be endowed with new discoveries and powers. But of this the great majority have no feeling, but are merely hireling and professorial ; except when it occasionally happens that some workman of acuter wit and covetous of honour applies himself to a new invention ; which he mostly does at the expense of his fortunes. But in general, so far are men from proposing to themselves to augment the mass of arts and sciences, that from the mass already at hand they neither take nor look for anything more than what they may turn to use in their lectures, or to gain, or to reputation, or to some similar advantage. And if any one out of all the multitude court science with honest affection and for her own sake, yet even with him the object will be found to be rather the variety of contemplations and doctrines than the severe and rigid search after truth. And if by chance there be one who seeks after truth in earnest, yet even he will propose to himself such a kind of truth as shall yield satisfaction to the mind and understanding in rendering causes for things long since discovered, and not the truth which shall lead to new assurance of works and new light of axioms. If then the end of the sciences has not as yet been well placed, it is not strange that men have erred as to the means.

LXXXII.

And as men have misplaced the end and goal of the sciences ; so again, even if they had placed it right, yet they have chosen a way to it which is altogether erroneous and impassable. And an astonishing thing it is to one who rightly considers the matter, that no mortal should have seriously applied himself to the opening and laying out of a road for the human understanding direct from the sense, by a course of experiment orderly conducted and well built up ; but that all has been left either to the mist of tradition, or the whirl and eddy of argument, or the fluctuations and mazes of chance and of vague and ill-digested experience. Now let any man soberly and diligently consider what the way is by which men have been accustomed to proceed in the investigation and discovery of things ; and in the first place he will no doubt remark a method of discovery very simple and inartificial ; which is the most ordinary method, and is no more than this. When a man addresses himself to discover something, he first seeks out and sees before him all that has been said about it by others ; then he begins to meditate for himself ; and so by much agitation and working of the wit solicits and as it were evokes his own spirit to give him oracles : which method has no foundation at all, but rests only upon opinions and is carried about with them.

Another may perhaps call in logic to discover it for him ; but that has no relation to the matter except in name. For logical invention does not discover principles and chief axioms, of which arts are composed, but only such things as appear to be consistent with them. For if you grow more curious and importunate and busy, and question her of probations and invention of principles or primary axioms, her answer is well known ; she refers you to the faith you are bound to give to the principles of each separate art.

There remains simple experience ; which, if taken as it comes, is called accident ; if sought for, experiment. But this kind of experience is no better than a broom without its band, as the saying is ;—a mere groping, as of men in the dark, that feel all round them for the chance of finding their way ; when they had much better wait for daylight, or light a candle, and then go. But the true method of experience on the contrary first lights the candle, and then by means of the candle shows the way ; commencing as it does with experience duly ordered and digested, not bungling or erratic, and from it educing axioms, and from established axioms again new experiments ; even as it was not without order and method that the divine word operated on the created mass. Let men therefore cease to wonder that the course of science is not yet wholly run, seeing that they have gone altogether astray ; either leaving and abandoning experience entirely, or losing their way in it and wandering round and round as in a labyrinth ; whereas a method rightly ordered leads by an unbroken route through the woods of experience to the open ground of axioms.

LXXXIII.

This evil however has been strangely increased by an opinion or conceit, which though of long standing is vain and hurtful ; namely, that the dignity of the human mind is impaired by long and close intercourse with experiments and particulars, subject to sense and bound in matter ; especially as they are laborious to search, ignoble to meditate, harsh to deliver, illiberal to practise, infinite in number, and minute in subtlety. So that it has come at length to this, that the true way is not merely deserted, but shut out and stopped up ; experience being, I do not say abandoned or badly managed, but rejected with disdain.

LXXXIV.

Again, men have been kept back as by a kind of enchantment from progress in the sciences by reverence for antiquity, by the authority of men accounted great in philosophy, and then by general consent. Of the last I have spoken above.

As for antiquity, the opinion touching it which men entertain is quite a negligent one, and scarcely consonant with the word itself. For the old age of the world is to be accounted the true antiquity ; and this is the attribute of our own times, not of that earlier age of the world in which the ancients lived ; and which, though in respect of us it was the elder, yet in respect of the world it was the younger[45]. And truly as we look for greater knowledge of human things

[45] This remark is not, I think, given by Bacon as a quotation, and it is probable that he did not derive it from any earlier writer. But in the works of several of the scientific reformers we find similar reflexions. Of writers earlier than Bacon or contemporary with him, we may refer to Gilbert, to Galileo, to the *Apologia pro Galileo* of Campanella, and particularly to the *Cena di Cenere* of Giordano Bruno. The following passage from the last-named writer, in which he appears to have anticipated Bacon, has been referred to by Dr. Whewell in the *Philosophy of the Inductive Sciences.* " Sia come la si vuole," says one of the interlocutors in Bruno's dialogue, " io non voglio discostar mi dal parer degli antichi, perche dice il saggio, Ne l'antiquità è la sapienza." To which another replies : " E soggiunge ' In molti anni la prudenza '. Se voi intendeste bene qualche dite, vedreste che dal vostro fondamento s'inferisce il contrario di quel che pensate. Voglio dire che noi siamo più vecchi ed abbiamo più lunga età, che i nostri predecessori."—*Cena di Cenere*, i. p. 132 of Wagner's edition of G. Bruno.

The idea that the early ages were the world's youth is to be found in the second book of Esdras, or is at any rate directly suggested by an expression which occurs there : " Seculum perdidit juventutem suam, et tempora appropinquant senescere."—2 *Esdras*, xiv. 10. The same idea occurs in Casmann's *Problemata Marina*, which was published in 1546. " Si . . . antiquiorum dignitas ex tempore major videtur, id nostros qui hodie docent posteriores unice commendabit, nam tempus . . . doctius et prudentius evadit ex continuo progressu, ut senescens judicio sit acriore, solidiore, et maturiore." [Note to parallel passage in *De Augmentis*, B. i.]

and a riper judgment in the old man than in the young, because of his experience and of the number and variety of the things which he has seen and heard and thought of; so in like manner from our age, if it but knew its own strength and chose to essay and exert it, much more might fairly be expected than from the ancient times, inasmuch as it is a more advanced age of the world, and stored and stocked with experiments and observations.

Nor must it go for nothing that by the distant voyages and travels which have become frequent in our times, many things in nature have been laid open and discovered which may let in new light upon philosophy. And surely it would be disgraceful if, while the regions of the material globe,—that is, of the earth, of the sea, and of the stars,— have been in our times laid widely open and revealed, the intellectual globe should remain shut up within the narrow limits of old discoveries.

And with regard to authority, it shows a feeble mind to grant so much to authors and yet deny time his rights, who is the author of authors, nay rather of all authority. For rightly is truth called the daughter of time [46], not of authority. It is no wonder therefore if those enchantments of antiquity and authority and consent have so bound up men's powers that they have been made impotent (like persons bewitched) to accompany with the nature of things [47].

LXXXV.

Nor is it only the admiration of antiquity, authority, and consent, that has forced the industry of man to rest satisfied with the discoveries already made; but also an admiration for the works themselves of which the human race has long been in possession. For when a man looks at the variety and the beauty of the provision which the mechanical arts have brought together for men's use, he will certainly be more inclined to admire the wealth of man than to feel his wants: not considering that the original observations and operations of nature (which are the life and moving principle of all that variety) are not many nor deeply fetched, and that the rest is but patience, and the subtle and ruled motion of the hand and instruments;—as the making of clocks (for instance) is certainly a subtle and exact work: their wheels seem to imitate the celestial orbs, and their alternating and orderly motion, the pulse of animals: and yet all this depends on one or two axioms of nature.

Again, if you observe the refinement of the liberal arts, or even that which relates to the mechanical preparation of natural substances; and take notice of such things as the discovery in astronomy of the motions of the heavens, of harmony in music, of the letters of the alphabet (to this day not in use among the Chinese) in grammar: or again in things mechanical, the discovery of the works of Bacchus and Ceres—that is, of the arts of preparing wine and beer, and of making bread; the discovery once more of the delicacies of the table, of distillations and the like; and if you likewise bear in mind the long periods which it has taken to bring these things to their present degree of perfection (for they are all ancient except distillation [48]), and again (as has been said of clocks) how little they owe to observations and axioms of nature, and how easily and obviously and as it were by casual suggestion they may have been discovered; you will easily cease from wondering, and on the contrary will pity the condition of mankind, seeing that in a course of so many ages there has been so great a dearth and barrenness of arts and inventions. And yet these very discoveries which we

46 [See Aulus Gellius, *Noctes Atticæ*, lib. xii. c. 11.—Ed.]

47 Compare *Sylv. Sylvar.* § 888.—*J. S.*

48 It has been said that Porson affirmed that distillation was known to the ancients. Dutens of course maintains that it was: but the passage he quotes from Dioscorides merely refers to sublimation. The word alembic is, as he remarks, a compound of the Arabic article with the Greek word ἄμβιξ, operculum; thus resembling in formation the word "almagest" and some others. But no valid conclusion can be drawn from hence. See Dutens, *Origine des Découvertes*, etc., p. 187 of the London edition. See a very interesting account of the history of distillation in Humboldt's *Examen critique de l'Histoire de la Géographie*, etc., vol. ii. p. 306.

have just mentioned, are older than philosophy and intellectual arts. So that, if the truth must be spoken, when the rational and dogmatical sciences began the discovery of useful works came to an end [49].

And again, if a man turn from the workshop to the library, and wonder at the immense variety of books he sees there, let him but examine and diligently inspect their matter and contents, and his wonder will assuredly be turned the other way; for after observing their endless repetitions, and how men are ever saying and doing what has been said and done before, he will pass from admiration of the variety to astonishment at the poverty and scantiness of the subjects which till now have occupied and possessed the minds of men.

And if again he descend to the consideration of those arts which are deemed curious rather than safe, and look more closely into the works of the Alchemists or the Magicians, he will be in doubt perhaps whether he ought rather to laugh over them or to weep. For the Alchemist nurses eternal hope, and when the thing fails, lays the blame upon some error of his own; fearing either that he has not sufficiently understood the words of his art or of his authors (whereupon he turns to tradition and auricular whispers), or else that in his manipulations he has made some slip of a scruple in weight or a moment in time (whereupon he repeats his trials to infinity); and when meanwhile among the chances of experiment he lights upon some conclusions either in aspect new or for utility not contemptible, he takes these for earnest of what is to come, and feeds his mind upon them, and magnifies them to the most, and supplies the rest in hope. Not but that Alchemists have made a good many discoveries, and presented men with useful inventions. But their case may be well compared to the fable of the old man, who bequeathed to his sons gold buried in a vineyard, pretending not to know the exact spot; whereupon the sons applied themselves diligently to the digging of the vineyard, and though no gold was found there, yet the vintage by that digging was made more plentiful.

Again, the students of natural magic, who explain everything by Sympathies and Antipathies, have in their idle and most slothful conjectures ascribed to substances wonderful virtues and operations; and if ever they have produced works, they have been such as aim rather at admiration and novelty than at utility and fruit.

In superstitious magic on the other hand (if of this also we must speak), it is especially to be observed that they are but subjects of a certain and definite kind wherein the curious and superstitious arts, in all nations and ages, and religions also, have worked or played. These therefore we may pass. Meanwhile it is nowise strange if opinion of plenty has been the cause of want.

LXXXVI.

Further, this admiration of men for knowledges and arts,—an admiration in itself weak enough, and well-nigh childish—has been increased by the craft and artifices of those who have handled and transmitted sciences. For they set them forth with such ambition and parade, and bring them into the view of the world so fashioned and masked, as if they were complete in all parts and finished. For if you look at the method of them and the divisions, they seem to embrace and comprise everything which can belong to the subject. And although these divisions are ill filled out and are but as empty cases, still to the common mind they present the form and plan of a perfect science. But the first and most ancient seekers after truth were wont, with better faith and better fortune too, to throw the knowledge which they gathered from the contemplation of things, and which they meant to store up for use, into aphorisms; that is, into short and scattered sentences, not linked together by an artificial method; and did not pretend or profess to embrace the entire art. But as the matter now is, it is nothing strange if men do not seek to advance in things delivered to them as long since perfect and complete.

[49] Thus we find Aristotle speaks of philosophy as having sprung up after all the wants of life were satisfied. See the beginning of the *Metaphysics*.

LXXXVII.

Moreover the ancient systems have received no slight accession of reputation and credit from the vanity and levity of those who have propounded new ones; especially in the active and practical department of natural philosophy. For there have not been wanting talkers and dreamers who, partly from credulity, partly in imposture, have loaded mankind with promises, offering and announcing the prolongation of life, the retardation of age, the alleviation of pain, the repairing of natural defects, the deceiving of the senses; arts of binding and inciting the affections, of illuminating and exalting the intellectual faculties, of transmuting substances, of strengthening and multiplying motions at will, of making impressions and alterations in the air, of bringing down and procuring celestial influences; arts of divining things future, and bringing things distant near, and revealing things secret; and many more. But with regard to these lavish promisers, this judgment would not be far amiss; that there is as much difference in philosophy between their vanities and true arts, as there is in history between the exploits of Julius Cæsar or Alexander the Great, and the exploits of Amadis of Gaul or Arthur of Britain. For it is true that those illustrious generals really did greater things than these shadowy heroes are even feigned to have done; but they did them by means and ways of action not fabulous or monstrous. Yet surely it is not fair that the credit of true history should be lessened because it has sometimes been injured and wronged by fables. Meanwhile it is not to be wondered at, if a great prejudice is raised against new propositions, especially when works are also mentioned, because of those impostors who have attempted the like; since their excess of vanity, and the disgust it has bred, have their effect still in the destruction of all greatness of mind in enterprises of this kind.

LXXXVIII.

Far more, however, has knowledge suffered from littleness of spirit and the smallness and slightness of the tasks which human industry has proposed to itself. And what is worst of all, this very littleness of spirit comes with a certain air of arrogance and superiority.

For in the first place there is found in all arts one general device, which has now become familiar,—that the author lays the weakness of his art to the charge of nature: whatever his art cannot attain he sets down on the authority of the same art to be in nature impossible. And truly no art can be condemned if it be judge itself. Moreover the philosophy which is now in vogue embraces and cherishes certain tenets, the purpose of which (if it be diligently examined) is to persuade men that nothing difficult, nothing by which nature may be commanded and subdued, can be expected from art or human labour; as with respect to the doctrine that the heat of the sun and of fire differ in kind, and to that other concerning mixture, has been already observed. Which things, if they be noted accurately, tend wholly to the unfair circumscription of human power, and to a deliberate and factitious despair; which not only disturbs the auguries of hope, but also cuts the sinews and spurs of industry, and throws away the chances of experience itself; and all for the sake of having their art thought perfect, and for the miserable vainglory of making it believed that whatever has not yet been discovered and comprehended can never be discovered or comprehended hereafter.

And even if a man apply himself fairly to facts, and endeavour to find out something new, yet he will confine his aim and intention to the investigation and working out of some one discovery and no more; such as the nature of the magnet, the ebb and flow of the sea, the system of the heavens, and things of this kind, which seem to be in some measure secret, and have hitherto been handled without much success. Whereas it is most unskilful to investigate the nature of anything in the thing itself; seeing that the same nature which appears in some things to be latent and hidden is in others manifest and palpable; wherefore in the former it produces wonder, in the latter excites no attention; as we find it in the nature of consistency, which in wood or stone is not observed, but is passed over under the appellation of solidity, without further inquiry as to why separation or solution of continuity is avoided ; while in the case of bubbles, which form themselves into certain pellicles, curiously shaped into hemispheres, so that the solution of continuity is avoided for a moment, it is thought a subtle matter. In fact what in some things is accounted a secret has in others a manifest and well known nature, which will never be recognized as long as the experiments and thoughts of men are engaged on the former only.

But generally speaking, in mechanics old discoveries pass for new, if a man does but refine or embellish them, or unite several in one, or couple them better with their use, or make the work in greater or less volume than it was before, or the like.

Thus then it is no wonder if noble inventions and worthy of mankind have not been brought to light, when men have been contented and delighted with such trifling and puerile tasks, and have even fancied that in them they have been endeavouring after, if not accomplishing, some great matter.

LXXXIX.

Neither is it to be forgotten that in every age Natural Philosophy has had a troublesome adversary and hard to deal with ; namely, superstition, and the blind and immoderate zeal of religion. For we see among the Greeks that those who first proposed to men's then uninitiated ears the natural causes for thunder and for storms, were thereupon found guilty of impiety [50]. Nor was much more forbearance shown by some of the ancient fathers of the Christian church to those who on most convincing grounds (such as no one in his senses would now think of contradicting) maintained that the earth was round and of consequence asserted the existence of the antipodes [51].

Moreover, as things now are, to discourse of nature is made harder and more perilous by the summaries and systems of the schoolmen ; who having reduced theology into regular order as well as they were able, and fashioned it into the shape of an art, ended in incorporating the contentious and thorny philosophy of Aristotle, more than was fit, with the body of religion [52].

To the same result, though in a different way, tend the speculations of those who have taken upon them to deduce the truth of the Christian religion from the principles of philosophers, and to confirm it by their authority ; pompously solemnising this union of the sense and faith as a lawful marriage, and entertaining men's minds with a pleasing variety of matter, but all the while disparaging things divine by mingling them with things human. Now in such mixtures of theology with philosophy only the received doctrines of philosophy are included ; while new ones, albeit changes for the better, are all but expelled and exterminated.

Lastly, you will find that by the simpleness of certain divines, access to any philosophy, however pure, is well nigh closed. Some are weakly afraid lest a deeper search into nature should transgress the permitted limits of sobermindedness ; wrongfully wresting and transferring what is said in holy writ against those who pry into sacred mysteries, to the hidden things of nature, which are barred by no prohibition. Others with more subtlety surmise and reflect that if second causes are unknown everything can more readily be referred to the divine hand and rod ; a point in which they think religion greatly concerned ; which is in fact nothing else but to seek to gratify God with a lie. Others fear

50 [See Aristophanes, *Clouds*, 372–383.—Ed.]

51 [Cf. Lactantius, *Div. Instit.* iii. 24 ; Augustine, *De civ. Dei*, xvi. 9.—Ed.]

52 Compare Kepler in the introduction to his great work *De Stellâ Martis* :—" In theologiâ quidem authoritatum, in Philosophiâ vero rationum esse momenta ponderanda. Sanctus igitur Lactantius qui terram negavit esse rotundam : Sanctus Augustinus qui rotunditate concessâ negavit tamen Antipodas : Sanctum Officium hodiernorum qui exilitate terræ concessâ negant tamen ejus motum : at magis mihi sancta Veritas qui terram et rotundam et Antipodibus circumhabitam et contemptissimæ parvitatis esse et denique per sidera ferri, salvo Doctorum ecclesiæ respectu, ex philosophiâ demonstro." See for a defence of St. Boniface, touching the story of the Antipodes and Virgilius Bishop of Saltzburg, Fromondus *De Orbe Terræ Immobili*, c. 4.

from past example that movements and changes in philosophy will end in assaults on religion. And others again appear apprehensive that in the investigation of nature something may be found to subvert or at least shake the authority of religion, especially with the unlearned. But these two last fears seem to me to savour utterly of carnal wisdom ; as if men in the recesses and secret thoughts of their hearts doubted and distrusted the strength of religion and the empire of faith over the sense, and therefore feared that the investigation of truth in nature might be dangerous to them. But if the matter be truly considered, natural philosophy is after the word of God at once the surest medicine against superstition, and the most approved nourishment for faith, and therefore she is rightly given to religion as her most faithful handmaid, since the one displays the will of God, the other his power. For he did not err who said " Ye do err, not knowing the Scriptures, nor the power of God," thus coupling and blending in an indissoluble bond information concerning his will and meditation concerning his power. Meanwhile it is not surprising if the growth of Natural Philosophy is checked, when religion, the thing which has most power over men's minds, has by the simpleness and incautious zeal of certain persons been drawn to take part against her.

XC.

Again, in the customs and institutions of schools, academies, colleges, and similiar bodies destined for the abode of learned men and the cultivation of learning, everything is found adverse to the progress of science. For the lectures and exercises there are so ordered, that to think or speculate on anything out of the common way can hardly occur to any man. And if one or two have the boldness to use any liberty of judgment, they must undertake the task all by themselves ; they can have no advantage from the company of others. And if they can endure this also, they will find their industry and largeness of mind no slight hindrance to their fortune. For the studies of men in these places are confined and as it were imprisoned in the writings of certain authors, from whom if any man dissent he is straightway arraigned as a turbulent person and an innovator. But surely there is a great distinction between matters of state and the arts ; for the danger from new motion and from new light is not the same. In matters of state a change even for the better is distrusted, because it unsettles what is established ; these things resting on authority, consent, fame and opinion, not on demonstration. But arts and sciences should be like mines, where the noise of new works and further advances is heard on every side. But though the matter be so according to right reason, it is not so acted on in practice ; and the points above mentioned in the administration and government of learning put a severe restraint upon the advancement of the sciences.

XCI.

Nay, even if that jealousy were to cease, still it is enough to check the growth of science, that efforts and labours in this field go unrewarded. For it does not rest with the same persons to cultivate sciences and to reward them. The growth of them comes from great wits; the prizes and rewards of them are in the hands of the people, or of great persons, who are but in very few cases even moderately learned. Moreover this kind of progress is not only unrewarded with prizes and substantial benefits ; it has not even the advantage of popular applause. For it is a greater matter than the generality of men can take in, and is apt to be overwhelmed and extinguished by the gales of popular opinions. And it is nothing strange if a thing not held in honour does not prosper.

XCII.

But by far the greatest obstacle to the progress of science and to the undertaking of new tasks and provinces therein, is found in this—that men despair and think things impossible. For wise and serious men are wont in these matters to be altogether distrustful ; considering with themselves the obscurity of nature, the shortness of life, the deceitfulness of the senses, the weakness of the judgment, the difficulty of experiment and the like ; and so supposing that in the revolution of time and of the ages of the world the sciences have their ebbs and flows; that at one season they grow and flourish, at another wither and decay, yet in such sort that when they have reached a certain point and condition they can advance no further. If therefore any one believes or promises more, they think this comes of an ungoverned and unripened mind, and that such attempts have prosperous beginnings, become difficult as they go on, and end in confusion. Now since these are thoughts which naturally present themselves to grave men and of great judgment, we must take good heed that we be not led away by our love for a most fair and excellent object to relax or diminish the severity of our judgment; we must observe diligently what encouragement dawns upon us and from what quarter; and, putting aside the lighter breezes of hope, we must thoroughly sift and examine those which promise greater steadiness and constancy. Nay, and we must take state-prudence too into our counsels, whose rule is to distrust, and to take the less favourable view of human affairs. I am now therefore to speak touching Hope; especially as I am not a dealer in promises, and wish neither to force nor to ensnare men's judgments, but to lead them by the hand with their good will. And though the strongest means of inspiring hope will be to bring men to particulars; especially to particulars digested and arranged in my Tables of Discovery (the subject partly of the second but much more of the fourth part of my Instauration), since this is not merely the promise of the thing but the thing itself; nevertheless that everything may be done with gentleness, I will proceed with my plan of preparing men's minds; of which preparation to give hope is no unimportant part. For without it the rest tends rather to make men sad (by giving them a worse and a meaner opinion of things as they are than they now have, and making them more fully to feel and know the unhappiness of their own condition) than to induce any alacrity or to whet their industry in making trial. And therefore it is fit that I publish and set forth those conjectures of mine which make hope in this matter reasonable; just as Columbus did, before that wonderful voyage of his across the Atlantic, when he gave the reasons for his conviction that new lands and continents might be discovered besides those which were known before; which reasons, though rejected at first, were afterwards made good by experience, and were the causes and beginnings of great events.

XCIII.

The beginning is from God [53]: for the business which is in hand, having the character of good so strongly impressed upon it, appears manifestly to proceed from God, who is the author of good, and the Father of Lights. Now in divine operations even the smallest beginnings lead of a certainty to their end. And as it was said of spiritual things, "The kingdom of God cometh not with observation," so is it in all the greater works of Divine Providence; everything glides on smoothly and noiselessly, and the work is fairly going on before men are aware that it has begun. Nor should the prophecy of Daniel be forgotten, touching the last stages of the world:—"Many shall go to and fro, and knowledge shall be increased"; clearly intimating that the thorough passage of the world (which now by so many distant voyages seems to be accomplished, or in course of accomplishment), and the advancement of the sciences, are destined by fate, that is, by Divine Providence, to meet in the same age.

XCIV.

Next comes a consideration of the greatest importance as an argument of hope; I mean that drawn from the errors of past time, and of the ways hitherto trodden. For most excellent was the censure once passed upon a government that had been unwisely administered. "That which is the worst thing in reference to the past, ought to be regarded as best for the future. For if you had done all that your duty demanded, and yet your affairs were no better, you would not have even a hope left you that further improvement is possible. But now, when your misfortunes are owing, not to the force of circumstances, but to your own errors, you may hope that by dismissing or correcting these errors, a great change may

53 'Εκ Διός αρχωμεσθα—Aratus, *Phenom.* i. 1.

be made for the better "[54]. In like manner, if during so long a course of years men had kept the true road for discovering and cultivating sciences, and had yet been unable to make further progress therein, bold doubtless and rash would be the opinion that farther progress is possible. But if the road itself has been mistaken, and men's labour spent on unfit objects, it follows that the difficulty has its rise not in things themselves, which are not in our power, but in the human understanding, and the use and application thereof, which admits of remedy and medicine. It will be of great use therefore to set forth what these errors are; for as many impediments as there have been in times past from this cause, so many arguments are there of hope for the time to come. And although they have been partly touched before, I think fit here also, in plain and simple words, to represent them.

XCV.

Those who have handled sciences have been either men of experiment or men of dogmas. The men of experiment are like the ant; they only collect and use; the reasoners resemble spiders, who make cobwebs out of their own substance[55]. But the bee takes a middle course; it gathers its material from the flowers of the garden and of the field, but transforms and digests it by a power of its own. Not unlike this is the true business of philosophy; for it neither relies solely or chiefly on the powers of the mind, nor does it take the matter which it gathers from natural history and mechanical experiments and lay it up in the memory whole, as it finds it; but lays it up in the understanding altered and digested. Therefore from a closer and purer league between these two faculties, the experimental and the rational, (such as has never yet been made) much may be hoped.

XCVI.

We have as yet no natural philosophy that is pure; all is tainted and corrupted; in Aristotle's school by logic; in Plato's by natural theology; in the second school of Platonists, such as Proclus and others, by mathematics, which sought only to give definiteness to natural philosophy, not to generate or give it birth. From a natural philosophy pure and unmixed, better things are to be expected.

XCVII.

No one has yet been found so firm of mind and purpose as resolutely to compel himself to sweep away all theories and common notions, and to apply the understanding, thus made fair and even, to a fresh examination of particulars. Thus it happens that human knowledge, as we have it, is a mere medley and ill-digested mass, made up of much credulity and much accident, and also of the childish notions which we at first imbibed.

Now if any one of ripe age, unimpaired senses, and well-purged mind, apply himself anew to experience and particulars, better hopes may be entertained of that man. In which point I promise to myself a like fortune to that of Alexander the great; and let no man tax me with vanity till he have heard the end; for the thing which I mean tends to the putting off of all vanity. For of Alexander and his deeds Æschines spake thus: " Assuredly we do not live the life of mortal men; but to this end were we born, that in after ages wonders might be told of us"; as if what Alexander had done seemed to him miraculous[56]. But in the next age Titus Livius took a better and a deeper view of the matter, saying in effect, that Alexander " had done no more than take courage to despise vain apprehensions "[57]. And a like judgment I suppose may be passed on myself in future ages: that I did no great things, but simply made less account of things that were accounted great. In the meanwhile, as I have already said, there is no hope except in a new birth of science; that is, in raising it regularly up from experience and building it afresh; which no one (I think) will say has yet been done or thought of.

[54] Demosthenes: see the first Philippic, p. 40, and the third, p. 112. Ed. Reisk.

[55] Stobæus, *Florileg.* § 82. Compare *De Augmentis*, v. 2.

[56] Æschines, *De Corona*, p. 72. Ed. H. Stephen. [57] Lib. ix. c. 17.

XCVIII.

Now for grounds of experience—since to experience we must come—we have as yet had either none or very weak ones; no search has been made to collect a store of particular observations sufficient either in number, or in kind, or in certainty, to inform the understanding, or in any way adequate. On the contrary, men of learning, but easy withal and idle, have taken for the construction or for the confirmation of their philosophy certain rumours and vague fames or airs of experience, and allowed to these the weight of lawful evidence. And just as if some kingdom or state were to direct its counsels and affairs, not by letters and reports from ambassadors and trustworthy messengers, but by the gossip of the streets; such exactly is the system of management introduced into philosophy with relation to experience. Nothing duly investigated, nothing verified, nothing counted, weighed, or measured, is to be found in natural history: and what in observation is loose and vague, is in information deceptive and treacherous. And if any one thinks that this is a strange thing to say, and something like an unjust complaint, seeing that Aristotle, himself so great a man, and supported by the wealth of so great a king, has composed so accurate a history of animals; and that others with greater diligence, though less pretence, have made many additions; while others, again, have compiled copious histories and descriptions of metals, plants, and fossils; it seems that he does not rightly apprehend what it is that we are now about. For a natural history which is composed for its own sake is not like one that is collected to supply the understanding with information for the building up of philosophy. They differ in many ways, but especially in this; that the former contains the variety of natural species only, and not experiments of the mechanical arts. For even as in the business of life a man's disposition and the secret workings of his mind and affections are better discovered when he is in trouble than at other times; so likewise the secrets of nature reveal themselves more readily under the vexations of art than when they go their own way. Good hopes may therefore be conceived of natural philosophy, when natural history, which is the basis and foundation of it, has been drawn up on a better plan; but not till then.

XCIX.

Again, even in the great plenty of mechanical experiments, there is yet a greater scarcity of those which are of most use for the information of the understanding. For the mechanic, not troubling himself with the investigation of truth, confines his attention to those things which bear upon his particular work, and will not either raise his mind or stretch out his hand for anything else. But then only will there be good ground of hope for the further advance of knowledge, when there shall be received and gathered together into natural history a variety of experiments, which are of no use in themselves, but simply serve to discover causes and axioms; which I call "*Experimenta lucifera*", experiments of *light*, to distinguish them from those which I call "*fructifera*", experiments of *fruit*.

Now experiments of this kind have one admirable property and condition; they never miss or fail. For since they are applied, not for the purpose of producing any particular effect, but only of discovering the natural cause of some effect, they answer the end equally well whichever way they turn out; for they settle the question.

C.

But not only is a greater abundance of experiments to be sought for and procured, and that too of a different kind from those hitherto tried; an entirely different method, order, and process for carrying on and advancing experience must also be introduced. For experience, when it wanders in its own track, is, as I have already remarked, mere groping in the dark, and confounds men rather than instructs them. But when it shall proceed in accordance with a fixed law, in regular order, and without interruption, then may better things be hoped of knowledge.

CI.

But even after such a store of natural history and experience as is required for the work of the understanding, or of philosophy, shall be ready at hand, still the understanding is by no means competent to deal with it offhand and by memory alone; no more than if a man should hope by force of memory to retain and make himself master of the computation of an ephemeris. And yet hitherto more has been done in matter of invention by thinking than by writing; and experience has not yet learned her letters. Now no course of invention can be satisfactory unless it be carried on in writing. But when this is brought into use, and experience has been taught to read and write, better things may be hoped.

CII.

Moreover, since there is so great a number and army of particulars, and that army so scattered and dispersed as to distract and confound the understanding, little is to be hoped for from the skirmishings and slight attacks and desultory movements of the intellect, unless all the particulars which pertain to the subject of inquiry shall, by means of Tables of Discovery, apt, well arranged, and as it were animate, be drawn up and marshalled; and the mind be set to work upon the helps duly prepared and digested which these tables supply.

CIII.

But after this store of particulars has been set out duly and in order before our eyes, we are not to pass at once to the investigation and discovery of new particulars or works; or at any rate if we do so we must not stop there. For although I do not deny that when all the experiments of all the arts shall have been collected and digested, and brought within one man's knowledge and judgment, the mere transferring of the experiments of one art to others may lead, by means of that experience which I term *literate*, to the discovery of many new things of service to the life and state of man, yet it is no great matter that can be hoped from that; but from the new light of axioms, which having been educed from those particulars by a certain method and rule, shall in their turn point out the way again to new particulars, greater things may be looked for. For our road does not lie on a level, but ascends and descends; first ascending to axioms, then descending to works.

CIV.

The understanding must not however be allowed to jump and fly from particulars to remote axioms and of almost the highest generality (such as the first principles, as they are called, of arts and things), and taking stand upon them as truths that cannot be shaken, proceed to prove and frame the middle axioms by reference to them; which has been the practice hitherto; the understanding being not only carried that way by a natural impulse, but also by the use of syllogistic demonstration trained and inured to it. But then, and then only, may we hope well of the sciences, when in a just scale of ascent, and by successive steps not interrupted or broken, we rise from particulars to lesser axioms; and then to middle axioms, one above the other; and last of all to the most general. For the lowest axioms differ but slightly from bare experience, while the highest and most general (which we now have) are notional and abstract and without solidity. But the middle are the true and solid and living axioms, on which depend the affairs and fortunes of men; and above them again, last of all those which are indeed the most general; such I mean as are not abstract, but of which those intermediate axioms are really limitations[58].

The understanding must not therefore be supplied with wings, but rather hung with weights, to keep it from leaping and flying. Now this has never yet been done; when it is done, we may entertain better hopes of the sciences.

CV.

In establishing axioms, another form of induction must be devised than has hitherto been employed; and it must be used for proving and discovering not

[58] *i.e.* particular cases.

first principles (as they are called) only, but also the lesser axioms, and the middle, and indeed all. For the induction which proceeds by simple enumeration is childish; its conclusions are precarious, and exposed to peril from a contradictory instance; and it generally decides on too small a number of facts, and on those only which are at hand. But the induction which is to be available for the discovery and demonstration of sciences and arts, must analyse nature by proper rejections and exclusions; and then, after a sufficient number of negatives, come to a conclusion on the affirmative instances: which has not yet been done or even attempted, save only by Plato, who does indeed employ this form of induction to a certain extent for the purpose of discussing definitions and ideas [59]. But in order to furnish this induction or demonstration well and duly for its work, very many things are to be provided which no mortal has yet thought of; insomuch that greater labour will have to be spent in it than has hitherto been spent on the syllogism. And this induction must be used not only to discover axioms, but also in the formation of notions. And it is in this induction that our chief hope lies.

CVI.

But in establishing axioms by this kind of induction, we must also examine and try whether the axiom so established be framed to the measure of those particulars only from which it is derived, or whether it be larger and wider. And if it be larger and wider, we must observe whether by indicating to us new particulars it confirm that wideness and largeness as by a collateral security; that we may not either stick fast in things already known, or loosely grasp at shadows and abstract forms; not at things solid and realised in matter. And when this process shall have come into use, then at last shall we see the dawn of a solid hope.

CVII.

And here also should be remembered what was said above concerning the extending of the range of natural philosophy to take in the particular sciences, and the referring or bringing back of the particular sciences to natural philosophy; that the branches of knowledge may not be severed and cut off from the stem. For without this the hope of progress will not be so good.

CVIII.

So much then for the removing of despair and the raising of hope through the dismissal or rectification of the errors of past time. We must now see what else there is to ground hope upon. And this consideration occurs at once—that if many useful discoveries have been made by accident or upon occasion, when men were not seeking for them but were busy about other things; no one can doubt but that when they apply themselves to seek and make this their business, and that too by method and in order and not by desultory impulses, they will discover far more. For although it may happen once or twice that a man shall stumble on a thing by accident which, when taking great pains to search for it, he could not find; yet upon the whole it unquestionably falls out the other way. And therefore far better things, and more of them, and at shorter intervals, are to be expected from man's reason and industry and direction and fixed application than from accident and animal instinct and the like, in which inventions have hitherto had their origin.

CIX.

Another argument of hope may be drawn from this,—that some of the inventions already known are such as before they were discovered it could hardly have entered any man's head to think of; they would have been simply set aside as impossible. For in conjecturing what may be men set before them the example

59 This is one of many passages which show that Bacon was very far from asserting that he was the first to propose an inductive method. It is remarkable that M. de St. Hilaire in his translation of the treatise *De Animâ* of Aristotle has repeated the popular assertion that Bacon claimed to be the first discoverer of induction.

of what has been, and divine of the new with an imagination preoccupied and coloured by the old; which way of forming opinions is very fallacious; for streams that are drawn from the springheads of nature do not always run in the old channels.

If, for instance, before the invention of ordnance, a man had described the thing by its effects, and said that there was a new invention, by means of which the strongest towers and walls could be shaken and thrown down at a great distance; men would doubtless have begun to think over all the ways of multiplying the force of catapults and mechanical engines by weights and wheels and such machinery for ramming and projecting; but the notion of a fiery blast suddenly and violently expanding and exploding would hardly have entered into any man's imagination or fancy; being a thing to which nothing immediately analogous had been seen, except perhaps in an earthquake or in lightning, which as *magnalia* or marvels of nature, and by man not imitable, would have been immediately rejected.

In the same way, if before the discovery of silk, any one had said there was a kind of thread discovered for the purposes of dress and furniture, which far surpassed the thread of linen or of wool in fineness and at the same time in strength, and also in beauty and softness; men would have begun immediately to think of some silky kind of vegetable, or of the finer hair of some animal, or of the feathers and down of birds; but of a web woven by a tiny worm, and that in such abundance, and renewing itself yearly, they would assuredly never have thought. Nay, if any one had said anything about a worm, he would no doubt have been laughed at as dreaming of a new kind of cobwebs.

So again, if before the discovery of the magnet, any one had said that a certain instrument had been invented by means of which the quarters and points of the heavens could be taken and distinguished with exactness; men would have been carried by their imagination to a variety of conjectures concerning the more exquisite construction of astronomical instruments; but that anything could be discovered agreeing so well in its movements with the heavenly bodies, and yet not a heavenly body itself, but simply a substance of metal or stone, would have been judged altogether incredible. Yet these things and others like them lay for so many ages of the world concealed from men, nor was it by philosophy or the rational arts that they were found out at last, but by accident and occasion; being indeed, as I said, altogether different in kind and as remote as possible from anything that was known before; so that no preconceived notion could possibly have led to the discovery of them.

There is therefore much ground for hoping that there are still laid up in the womb of nature many secrets of excellent use, having no affinity or parallelism with any thing that is now known, but lying entirely out of the beat of the imagination, which have not yet been found out. They too no doubt will some time or other, in the course and revolution of many ages, come to light of themselves, just as the others did; only by the method of which we are now treating they can be speedily and suddenly and simultaneously presented and anticipated.

CX.

But we have also discoveries to show of another kind, which prove that noble inventions may be lying at our very feet, and yet mankind may step over without seeing them. For however the discovery of gunpowder, of silk, of the magnet, of sugar, of paper, or the like, may seem to depend on certain properties of things themselves and nature, there is at any rate nothing in the art of printing which is not plain and obvious. Nevertheless for want of observing that although it is more difficult to arrange types of letters than to write letters by the motion of the hand, there is yet this difference between the two, that types once arranged serve for innumerable impressions, but letters written with the hand for a single copy only; or perhaps again for want of observing that ink can be so thickened as to colour without running (particularly when the letters face upwards and the impression is made from above)—for want, I say, of observing these things, men went for so many ages without this most beautiful discovery, which is of so much service in the propagation of knowledge.

But such is the infelicity and unhappy disposition of the human mind in this course of invention, that it first distrusts and then despises itself: first will not believe that any such thing can be found out; and when it is found out, cannot understand how the world should have missed it so long. And this very thing may be justly taken as an argument of hope; namely, that there is a great mass of inventions still remaining, which not only by means of operations that are yet to be discovered, but also through the transferring, comparing, and applying of those already known, by the help of that Learned Experience of which I spoke, may be deduced and brought to light.

CXI.

There is another ground of hope that must not be omitted. Let men but think over their infinite expenditure of understanding, time, and means on matters and pursuits of far less use and value; whereof if but a small part were directed to sound and solid studies, there is no difficulty that might not be overcome. This I thought good to add, because I plainly confess that a collection of history natural and experimental, such as I conceive it and as it ought to be, is a great, I may say a royal work, and of much labour and expense.

CXII.

Meantime, let no man be alarmed at the multitude of particulars, but let this rather encourage him to hope. For the particular phenomena of art and nature are but a handful to the inventions of the wit, when disjoined and separated from the evidence of things. Moreover this road has an issue in the open ground and not far off; the other has no issue at all, but endless entanglement. For men hitherto have made but short stay with experience, but passing her lightly by, have wasted an infinity of time on meditations and glosses of the wit. But if some one were by that could answer our questions and tell us in each case what the fact in nature is, the discovery of all causes and sciences would be but the work of a few years.

CXIII.

Moreover I think that men may take some hope from my own example. And this I say not by way of boasting, but because it is useful to say it. If there be any that despond, let them look at me, that being of all men of my time the most busied in affairs of state, and a man of health not very strong (whereby much time is lost), and in this course altogether a pioneer, following in no man's track, nor sharing these counsels with any one, have nevertheless by resolutely entering on the true road, and submitting my mind to Things, advanced these matters, as I suppose, some little way. And then let them consider what may be expected (after the way has been thus indicated) from men abounding in leisure, and from association of labours, and from successions of ages: the rather because it is not a way over which only one man can pass at a time (as is the case with that of reasoning), but one in which the labours and industries of men (especially as regards the collecting of experience), may with the best effect be first distributed and then combined. For then only will men begin to know their strength when instead of great numbers doing all the same things, one shall take charge of one thing and another of another.

CXIV.

Lastly, even if the breath of hope which blows on us from that New Continent were fainter than it is and harder to perceive [60]; yet the trial (if we would not bear a spirit altogether abject) must by all means be made. For there is no comparison between that which we may lose by not trying and by not succeeding; since by not trying we throw away the chance of an immense good; by not succeeding we only incur the loss of a little human labour. But as it is, it appears to me from what has been said, and also from what has been left unsaid, that

[60] Bacon refers to what Peter Martyr Anghiera has related, that Columbus, observing the west-winds which blow at certain times of the year on the coast of Portugal, came to the conclusion that there must be land to generate them.

there is hope enough and to spare, not only to make a bold man try, but also to make a sober-minded and wise man believe.

CXV.

Concerning the grounds then for putting away despair, which has been one of the most powerful causes of delay and hindrance to the progress of knowledge, I have now spoken. And this also concludes what I had to say touching the *signs* and *causes* of the errors, sluggishness, and ignorance which have prevailed; especially since the more subtle causes, which do not fall under popular judgment and observation, must be referred to what has been said on the Idols of the human mind.

And here likewise should close that part of my Instauration, which is devoted to pulling down: which part is performed by three refutations; first, by the refutation of the *natural human reason*, left to itself; secondly, by the refutation of the *demonstrations;* and thirdly, by the refutation of the *theories*, or the received systems of philosophy and doctrine. And the refutation of these has been such, as alone it could be: that is to say, by signs and the evidence of causes since no other kind of confutation was open to me, differing as I do from others; both on first principles and on rules of demonstration.

It is time therefore to proceed to the art itself and rule of interpreting nature; still however there remains something to be premised. For whereas in this first book of aphorisms I proposed to prepare men's minds as well for understanding as for receiving what is to follow; now that I have purged and swept and levelled the floor of the mind, it remains that I place the mind in a good position and as it were in a favourable aspect towards what I have to lay before it. For in a new matter, it is not only the strong preoccupation of some old opinion that tends to create a prejudice, but also a false preconception or prefiguration of the new thing which is presented. I will endeavour therefore to impart sound and true opinions as to the things I propose, although they are to serve only for the time, and by way of interest (so to speak), till the thing itself, which is the principal, be fully known.

CXVI.

First, then, I must request men not to suppose that after the fashion of ancient Greeks, and of certain moderns, as Telesius, Patricius, Severinus [61], I wish to found a new sect in philosophy. For this is not what I am about; nor do I think that it matters much to the fortunes of men what abstract notions one may entertain concerning nature and the principles of things; and no doubt many old theories of this kind can be revived and many new ones introduced; just as many theories of the heavens may be supposed, which agree well enough with the phenomena and yet differ with each other.

But for my part I do not trouble myself with any such speculative and withal unprofitable matters. My purpose, on the contrary, is to try whether I cannot in very fact lay more firmly the foundations, and extend more widely the limits, of the power and greatness of man. And although on some special subjects and in an incomplete form I am in possession of results which I take to be far more true and more certain and withal more fruitful than those now received, (and these I have collected into the fifth part of my Instauration), yet I have no entire or universal theory to propound. For it does not seem that the time is come for such an attempt. Neither can I hope to live to complete the sixth part of the Instauration (which is destined for the philosophy discovered by the legitimate interpretation of nature), but hold it enough if in the intermediate business I bear myself soberly and profitably, sowing in the meantime for future ages the seeds of a purer truth, and performing my part towards the commencement of the great undertaking.

[61] See *De Aug.* iv. 3. for a rather fuller mention of these philosophers, and the note upon the passage. See also, for Telesius, the preface to *De Principiis atque Originibus;* for Patricius, the *Descriptio Globi intellectualis;* for Severinus, the *Temporis Partus Masculus.—J.S.*

CXVII.

And as I do not seek to found a school, so neither do I hold out offers or promises of particular works. It may be thought indeed, that I who make such frequent mention of works and refer everything to that end, should produce some myself by way of earnest. But my course and method, as I have often clearly stated and would wish to state again, is this,—not to extract works from works or experiments from experiments (as an empiric), but from works and experiments to extract causes and axioms, and again from those causes and axioms new works and experiments, as a legitimate interpreter of nature. And although in my tables of discovery (which compose the fourth part of the Instauration), and also in the examples of particulars (which I have adduced in the second part), and moreover in my observations on the history (which I have drawn out in the third part), any reader of even moderate sagacity and intelligence will everywhere observe indications and outlines of many noble works; still I candidly confess that the natural history which I now have, whether collected from books or from my own investigations, is neither sufficiently copious nor verified with sufficient accuracy to serve the purposes of legitimate interpretation.

Accordingly, if there be any one more apt and better prepared for mechanical pursuits, and sagacious in hunting out works by the mere dealing with experiment, let him by all means use his industry to gather from my history and tables many things by the way, and apply them to the production of works, which may serve as interest until the principal be forthcoming. But for myself, aiming as I do at greater things, I condemn all unseasonable and premature tarrying over such things as these; being (as I often say) like Atalanta's balls. For I do not run off like a child after golden apples, but stake all on the victory of art over nature in the race; nor do I make haste to mow down the moss or the corn in blade, but wait for the harvest in its due season.

CXVIII.

There will be found, no doubt, when my history and tables of discovery are read, some things in the experiments themselves that are not quite certain, or perhaps that are quite false; which may make a man think that the foundations and principles upon which my discoveries rest are false and doubtful. But this is of no consequence; for such things must needs happen at first. It is only like the occurrence in a written or printed page of a letter or two mistaken or misplaced; which does not much hinder the reader, because such errors are easily corrected by the sense. So likewise may there occur in my natural history many experiments which are mistaken and falsely set down, and yet they will presently by the discovery of causes and axioms be easily expunged and rejected. It is nevertheless true that if the mistakes in natural history and experiments are important, frequent, and continual, they cannot possibly be corrected or amended by any felicity of wit or art. And therefore, if in my natural history, which has been collected and tested with so much diligence, severity, and I may say religious care, there still lurk at intervals certain falsities or errors in the particulars,—what is to be said of common natural history, which in comparison with mine is so negligent and inexact? and what of the philosophy and sciences built on such a sand (or rather quicksand)? Let no man therefore trouble himself for this.

CXIX.

There will be met with also in my history and experiments many things which are trivial and commonly known; many which are mean and low; many, lastly, which are too subtle and merely speculative, and that seem to be of no use; which kind of things may possibly avert and alienate men's interests.

And first for those things which seem common; let men bear in mind that hitherto they have been accustomed to do no more than refer and adapt the causes of things which rarely happen to such as happen frequently; while of those which happen frequently they never ask the cause, but take them as they are for granted. And therefore they do not investigate the causes of weight, of the rotation of heavenly bodies, of heat, cold, light, hardness, softness, rarity, density, liquidity, solidity, animation, inanimation, similarity, dissimilarity, organization, and the like; but admitting these as self-evident and obvious, they dispute and decide on other things of less frequent and familiar occurrence.

But I, who am well aware that no judgment can be passed on uncommon or remarkable things, much less anything new brought to light, unless the causes of common things, and the causes of those causes, be first duly examined and found out, am of necessity compelled to admit the commonest things into my history. Nay, in my judgment philosophy has been hindered by nothing more than this,—that things of familiar and frequent occurrence do not arrest and detain the thoughts of men, but are received in passing without any inquiry into their causes; insomuch that information concerning things which are not known is not oftener wanted than attention concerning things which are.

CXX.

And for things that are mean or even filthy,—things which (as Pliny says) must be introduced with an apology [62],—such things, no less than the most splendid and costly, must be admitted into natural history. Nor is natural history polluted thereby; for the sun enters the sewer no less than the palace, yet takes no pollution. And for myself, I am not raising a capitol or pyramid to the pride of man, but laying a foundation in the human understanding for a holy temple after the model of the world. That model therefore I follow. For whatever deserves to exist deserves also to be known, for knowledge is the image of existence; and things mean and splendid exist alike. Moreover as from certain putrid substances—musk, for instance, and civet—the sweetest odours are sometimes generated, so too from mean and sordid instances there sometimes emanates excellent light and information. But enough and more than enough of this; such fastidiousness being merely childish and effeminate.

CXXI.

But there is another objection which must be more carefully looked to: namely, that there are many things in this History which to common apprehension, or indeed to any understanding accustomed to the present system, will seem to be curiously and unprofitably subtle. Upon this point therefore above all I must say again what I have said already,—that at first and for a time I am seeking for experiments of light, not for experiments of fruit; following therein, as I have often said, the example of the divine creation; which on the first day produced light only, and assigned to it alone one entire day, nor mixed up with it on that day any material work.

To suppose therefore that things like these are of no use is the same as to suppose that light is of no use, because it is not a thing solid or material. And the truth is that the knowledge of simple natures well examined and defined is as light; it gives entrance to all the secrets of nature's workshop, and virtually includes and draws after it whole bands and troops of works, and opens to us the sources of the noblest axioms; and yet in itself it is of no great use. So also the letters of the alphabet in themselves and apart have no use or meaning, yet they are the subject-matter for the composition and apparatus of all discourse. So again the seeds of things are of much latent virtue, and yet of no use except in their development. And the scattered rays of light itself, until they are made to converge, can impart none of their benefit.

But if objection be taken to speculative subtleties, what is to be said of the schoolmen, who have indulged in subtleties to such excess? in subtleties too that were spent on words, or at any rate on popular notions (which is much the same thing), not on facts of nature; and such as were useless not only in their origin but also in their consequences; and not like those I speak of, useless indeed for the present, but promising infinite utility hereafter. But let men be assured of this, that all subtlety of disputation and discourse, if not applied till after axioms are discovered, is out of season and preposterous; and that the true and proper or at any rate the chief time for subtlety is in weighing experience and in found-

[62] Plin. *Hist. Nat.* i. ad init. Compare also Aristotle, *De Part. Animal.* i. 5.

ing axioms thereon ; for that other subtlety, though it grasps and snatches at nature, yet can never take hold of her. Certainly what is said of opportunity or fortune is most true of nature ; she has a lock in front, but is bald behind.

Lastly, concerning the disdain to receive into natural history things either common, or mean, or over-subtle and in their original condition useless, the answer of the poor woman to the haughty prince [63], who had rejected her petition as an unworthy thing and beneath his dignity, may be taken for an oracle,— " Then leave off being king ". For most certain it is that he who will not attend to things like these, as being too paltry and minute, can neither win the kingdom of nature nor govern it.

CXXII.

It may be thought also a strange and harsh thing that we should at once and with one blow set aside all sciences and all authors ; and that too without calling in any of the ancients to our aid and support, but relying on our own strength.

And I know that if I had chosen to deal less sincerely, I might easily have found authority for my suggestions by referring them either to the old times before the Greeks (when natural science was perhaps more flourishing, though it made less noise, not yet having passed into the pipes and trumpets of the Greeks), or even, in part at least, to some of the Greeks themselves ; and so gained for them both support and honour ; as men of no family devise for themselves by the good help of genealogies the nobility of a descent from some ancient stock. But for my part, relying on the evidence and truth of things, I reject all forms of fiction and imposture ; nor do I think that it matters any more to the business in hand, whether the discoveries that shall now be made were long ago known to the ancients, and have their settings and their risings according to the vicissitude of things and course of ages, than it matters to mankind whether the new world be that island of Atlantis with which the ancients were acquainted, or now discovered for the first time. For new discoveries must be sought from the light of nature, not fetched back out of the darkness of antiquity.

And as for the universality of the censure, certainly if the matter be truly considered, such a censure is not only more probable but more modest too, than a partial one would be. For if the errors had not been rooted in primary notions, there must have been some true discoveries to correct the false. But the errors being fundamental, and not so much of false judgment as of inattention and oversight, it is no wonder that men have not obtained what they have not tried for, nor reached a mark which they never set up, nor finished a course which they never entered on or kept.

And as for the presumption implied in it ; certainly if a man undertakes by steadiness of hand and power of eye to describe a straighter line or more perfect circle than any one else, he challenges a comparison of abilities ; but if he only says that he with the help of a rule or a pair of compasses can draw a straighter line or a more perfect circle than any one else can by eye and hand alone, he makes no great boast. And this remark, be it observed, applies not merely to this first and inceptive attempt of mine, but to all that shall take the work in hand hereafter. For my way of discovering sciences goes far to level men's wits, and leaves but little to individual excellence ; because it performs everything by the surest rules and demonstrations. And therefore attribute my part in all this, as I have often said, rather to good luck than to ability, and account it a birth of time rather than of wit. For certainly chance has something to do with men's thoughts, as well as with their works and deeds.

CXXIII.

I may say then of myself that which one said in jest (since it marks the distinction so truly), " It cannot be that we should think alike, when one drinks water and the other drinks wine " [64]. Now other men, as well in ancient as in modern times, have in the matter of sciences drunk a crude liquor like water, either flowing spontaneously from the understanding, or drawn up by logic, as

[63] [Philip of Macedon. See Plutarch, *Apophthegmata.*—Ed.]

[64] [Said by Philocrates of Demosthenes. Demos. *De Falsa Legatione.*—Ed.]

by wheels from a well. Whereas I pledge mankind in a liquor strained from countless grapes, from grapes ripe and fully seasoned, collected in clusters, and gathered, and then squeezed in the press, and finally purified and clarified in the vat. And therefore it is no wonder if they and I do not think alike.

CXXIV.

Again, it will be thought, no doubt, that the goal and mark of knowledge which I myself set up (the very point which I object to in others) is not the true or the best; for that the contemplation of truth is a thing worthier and loftier than all utility and magnitude of works; and that this long and anxious dwelling with experience and matter and the fluctuations of individual things, drags down the mind to earth, or rather sinks it to a very Tartarus of turmoil and confusion; removing and withdrawing it from the serene tranquillity of abstract wisdom, a condition far more heavenly. Now to this I readily assent; and indeed this which they point at as so much to be preferred, is the very thing of all others which I am about. For I am building in the human understanding a true model of the world, such as it is in fact, not such as a man's own reason would have it to be; a thing which cannot be done without a very diligent dissection and anatomy of the world. But I say that those foolish and apish images of worlds which the fancies of men have created in philosophical systems, must be utterly scattered to the winds. Be it known then how vast a difference there is (as I said above) between the Idols of the human mind and the Ideas of the divine. The former are nothing more than arbitrary abstractions; the latter are the creator's own stamp upon creation, impressed and defined in matter by true and exquisite lines. Truth therefore and utility are here the very same things [65]: and works themselves are of greater value as pledges of truth than as contributing to the comforts of life.

CXXV.

It may be thought again that I am but doing what has been done before; that the ancients themselves took the same course which I am now taking; and that it is likely therefore that I too, after all this stir and striving, shall come at last to some one of those systems which prevailed in ancient times. For the ancients too, it will be said, provided at the outset of their speculations a great store and abundance of examples and particulars, digested the same into note books under heads and titles, from them completed their systems and arts, and afterwards, when they understood the matter, published them to the world,—adding a few examples here and there for proof and illustration; but thought it superfluous and inconvenient to publish their notes and minutes and digests or particulars; and therefore did as builders do,—after the house was built they removed the scaffolding and ladders out of sight. And so no doubt they did. But this objection (or scruple rather) will be easily answered by any one who has not quite forgotten what I have said above. For the form of inquiry and discovery that was in use among the ancients is by themselves professed, and appears on the very face of their writings. And that form was simply this. From a few examples and particulars (with the addition of common notions and perhaps of some portion of the received opinions which have been most popular) they flew at once to the most general conclusions, or first principles of science: taking the truth of these as fixed and immovable, they proceeded by means of intermediate propositions to educe and prove from them the inferior conclusions; and out of these they framed the art. After that, if any new particulars and examples repugnant to their dogmas were mooted and adduced, either they subtly moulded them into their system by distinctions or explanations of their rules, or else coarsely got rid of them by exceptions; while to such particulars as were not repugnant they laboured to assign causes in conformity with those of their principles. But this was not the natural history and experience that was wanted; far from it; and besides, that flying off to the highest generalities ruined all.

[65] *Ipsissimæ res.* I think this must have been Bacon's meaning, though not a meaning which the word can properly bear.—*J. S.*

CXXVI.

It will also be thought that by forbidding men to pronounce and to set down principles as established until they have duly arrived through the intermediate steps at the highest generalities, I maintain a sort of suspension of the judgment and bring it to what the Greeks call *Acatalepsia*,—a denial of the capacity of the mind to comprehend truth. But in reality that which I meditate and propound is not *Acatalepsia*, but *Eucatalepsia* ; not denial of the capacity to understand, but provision for the understanding truly ; for I do not take away authority from the senses, but supply them with helps ; I do not slight the understanding, but govern it. And better surely it is that we should know all we need to know, and yet think our knowledge imperfect, than that we should think our knowledge perfect, and yet not know anything we need to know.

CXXVII.

It may also be asked (in the way of doubt rather than objection) whether I speak of natural philosophy only, or whether I mean that the other sciences, logic, ethics, and politics, should be carried on by this method. Now I certainly mean what I have said to be understood of them all ; and as the common logic, which governs by the syllogism, extends not only to natural but to all sciences ; so does mine also, which proceeds by induction, embrace everything. For I form a history and tables of discovery for anger, fear, shame, and the like ; for matters political ; and again for the mental operations of memory, composition and division [66], judgment and the rest ; not less than for heat and cold, or light, or vegetation, or the like [67]. But nevertheless since my method of interpretation, after the history has been prepared and duly arranged, regards not the working and discourse of the mind only (as the common logic does) but the nature of things also, I supply the mind with such rules and guidance that it may in every case apply itself aptly to the nature of things. And therefore I deliver many and diverse precepts in the doctrine of Interpretation, which in some measure modify the method of invention according to the quality and condition of the subject of the inquiry.

CXXVIII.

On one point not even a doubt ought to be entertained ; namely, whether I desire to pull down and destroy the philosophy and arts and sciences which are at present in use. So far from that, I am most glad to see them used, cultivated, and honoured. There is no reason why the arts which are now in fashion should not continue to supply matter for disputation and ornaments for discourse, to be employed for the convenience of professors and men of business ; to be in short like current coin, which passes among men by consent. Nay I frankly declare that what I am introducing will be but little fitted for such purposes as these, since it cannot be brought down to common apprehension, save by effects and works only. But how sincere I am in my professions of affection and good will towards the received sciences, my published writings, especially the books on the Advancement of Learning, sufficiently show ; and therefore I will not attempt to prove it further by words. Meanwhile I give constant and distinct warning that by the methods now in use neither can any great progress be made in the doctrines and contemplative part of sciences, nor can they be carried out to any magnitude of works.

[66] [Mr. Ellis suggested that the meaning here may be "synthesis and analysis". Professor Fowler, however, has pointed out that what is meant is " affirmation and negation", following the terminology of Aristotle.—ED.]

[67] This passage is important because it shows that Bacon proposed to apply his method to mental phenomena ; which is in itself a sufficient refutation of M. Cousin's interpretation of the passage in which, when censuring the writings of the schoolmen, he compares them to the self-evolved web of the spider. I have elsewhere spoken more at length of this passage. [See preface, p. 230.]

CXXIX.

It remains for me to say a few words touching the excellency of the end in view. Had they been uttered earlier, they might have seemed like idle wishes; but now that hopes have been raised and unfair prejudices removed, they may perhaps have greater weight. Also if I had finished all myself, and had no occasion to call in others to help and take part in the work, I should even now have abstained from such language, lest it might be taken as a proclamation of my own deserts. But since I want to quicken the industry and rouse and kindle the zeal of others, it is fitting that I put men in mind of some things.

In the first place then, the introduction of famous discoveries appears to hold by far the first place among human actions; and this was the judgment of the former ages. For to the authors of inventions they awarded divine honours; while to those who did good service in the state (such as founders of cities and empires, legislators, saviours of their country from long endured evils, quellers of tyrannies, and the like) they decreed no higher honours than heroic. And certainly if a man rightly compare the two, he will find that this judgment of antiquity was just. For the benefits of discoveries may extend to the whole race of man, civil benefits only to particular places; the latter last not beyond a few ages, the former through all time. Moreover the reformation of a state in civil matters is seldom brought in without violence and confusion; but discoveries carry blessings with them, and confer benefits without causing harm or sorrow to any.

Again, discoveries are as it were new creations, and imitations of God's works; as well sang the poet:—

> " To man's frail race great Athens long ago
> First gave the seed whence waving harvests grow,
> And *re-created* all our life below " [68].

And it appears worthy of remark in Solomon, that though mighty in empire and in gold; in the magnificence of his works, his court, his household, and his fleet; in the lustre of his name and the worship of mankind; yet he took none of these to glory in, but pronounced that " The glory of God is to conceal a thing; the glory of the king to search it out " [69].

Again, let a man only consider what a difference there is between the life of men in the most civilized province of Europe, and in the wildest and most barbarous districts of New India; he will feel it be great enough to justify the saying that " man is a god to man ", not only in regard of aid and benefit, but also by a comparison of condition. And this difference comes not from soil, not from climate, not from race, but from the arts.

Again, it is well to observe the force and virtue and consequences of discoveries; and these are to be seen nowhere more conspicuously than in those three which were unknown to the ancients, and of which the origin, though recent, is obscure and inglorious; namely, printing, gunpowder, and the magnet. For these three have changed the whole face and state of things throughout the world; the first in literature, the second in warfare, the third in navigation; whence have followed innumerable changes; insomuch that no empire, no sect, no star seems to have exerted greater power and influence in human affairs than these mechanical discoveries.

Further, it will not be amiss to distinguish the three kinds and as it were grades of ambition in mankind. The first is of those who desire to extend their own power in their native country; which kind is vulgar and degenerate. The second is of those who labour to extend the power of their country and its dominion among men. This certainly has more dignity, though not less covetousness. But if a man endeavour to establish and extend the power and dominion of the human race itself over the universe, his ambition (if ambition it can be called) is without doubt both a more wholesome thing and a more noble than the other two. Now the empire of man over things depends wholly on the arts and sciences. For we cannot command nature except by obeying her.

[68] Lucretius, vi. 1–3.

[69] Prov. xxv. 2.

Again, if men have thought so much of some one particular discovery as to regard him as more than man who has been able by some benefit to make the whole human race his debtor, how much higher a thing to discover that by means of which all things else shall be discovered with ease! And yet (to speak the whole truth), as the uses of light are infinite, in enabling us to walk, to ply our arts, to read, to recognise one another; and nevertheless the very beholding of the light is itself a more excellent and a fairer thing than all the uses of it;— so assuredly the very contemplation of things, as they are, without superstition or imposture, error or confusion, is in itself more worthy than all the fruit of inventions.

Lastly, if the debasement of arts and sciences to purposes of wickedness, luxury, and the like, be made a ground of objection, let no one be moved thereby. For the same may be said of all earthly goods; of wit, courage, strength, beauty, wealth, light itself, and the rest. Only let the human race recover that right over nature which belongs to it by divine bequest, and let power be given it; the exercise thereof will be governed by sound reason and true religion.

CXXX.

And now it is time for me to propound the art itself of interpreting nature; in which, although I conceive that I have given true and most useful precepts, yet I do not say either that it is absolutely necessary (as if nothing could be done without it) or that it is perfect. For I am of opinion that if men had ready at hand a just history of nature and experience, and laboured diligently thereon; and if they could bind themselves to two rules,—the first, to lay aside received opinions and notions; and the second, to refrain the mind for a time from the highest generalisations, and those next to them,—they would be able by the native and genuine force of the mind, without any other art, to fall into my form of interpretation. For interpretation is the true and natural work of the mind when freed from impediments [70]. It is true however that by my precepts everything will be in more readiness, and much more sure.

Nor again do I mean to say that no improvement can be made upon these. On the contrary, I that regard the mind not only in its own faculties, but in its connection with things, must needs hold that the art of discovery may advance as discoveries advance.

70 Compare *Valerius Terminus*, ch. 22:—"That it is true that interpretation is the very natural and direct intention, action, and progression of the understanding, delivered from impediments; and that all anticipation is but a deflexion or declination by accident". Also *Adv. of Learn.* (2d book):—"For he that shall attentively observe how the mind doth gather this excellent dew of knowledge, like unto that which the poet speaketh of, *Aërii mellis cœlestia dona*, distilling and contriving it out of particulars natural and artificial, as the flowers of the field and garden, shall find that the mind of herself by nature doth manage and act an induction much better than they describe it".—*J. S.*

THE SECOND BOOK OF APHORISMS CONCERNING THE INTERPRETATION OF NATURE AND THE KINGDOM OF MAN.

APHORISM

I.

ON a given body to generate and superinduce a new nature or new natures, is the work and aim of Human Power. Of a given nature to discover the form, or true specific difference, or nature-engendering nature [1], or source of emanation (for these are the terms which come nearest to a description of the thing), is the work and aim of Human Knowledge [2]. Subordinate to these primary works are two others that are secondary and of inferior mark ; to the former, the transformation of concrete bodies, so far as this is possible [3] ; to the latter, the discovery, in every case of generation and motion, of the *latent process* carried on from the manifest efficient and the manifest material to the form which is engendered ; and in like manner the discovery of the *latent configuration* of bodies at rest and not in motion.

II.

In what an ill condition human knowledge is at the present time, is apparent even from the commonly received maxims. It is a correct position that " true knowledge is knowledge by causes." And causes again are not improperly distributed into four kinds ; the material, the formal, the efficient, and the final. But of these the final cause rather corrupts than advances the sciences, except such as have to do with human action. The discovery of the formal is despaired of. The efficient and the material (as they are investigated and received, that is, as remote causes, without reference to the latent process leading to the form)

[1] This is the only passage in which I have met with the phrase *natura naturans* used as it is here. With the later schoolmen, as with Spinoza, it denotes God considered as the *causa immanens* of the universe, and therefore, according to the latter at least, not hypostatically distinct from it. (On the Pantheistic tendency occasionally perceptible among the schoolmen, see Neander's Essay on Scotus Erigena in the *Berlin Memoirs.*) Bacon applies it to the Form, considered as the *causa immanens* of the properties of the body. I regret not having been able to trace the history of this remarkable phrase. It does not occur, I think, in St. Thomas Aquinas, though I have met with it in an index to his *Summa ;* the passage referred to containing a quotation from St. Augustine, in which the latter speaks of "ea natura quæ creavit omnes cæteras instituitque naturas". (*V.* St. Aug., *De Trin.* xiv. 9.) Neither does it occur, so far as I am aware, where we might have expected it, in the *De Divisione Naturæ* of Scotus Erigena. Vossius, *De Vitiis Latini Sermonis*, notices its use among the schoolmen, but gives no particular reference. [The phrase in question *does* occur in Thomas Aquinas ;—" Etiam Deus a quibusdam dicitur. *natura naturans*" (*Summa : Prima Secundae*, Quæst. 85, art. 6). And see Hauréau, *Histoire de la philosophie scolastique*, 1872–80, i. 189, for a trace of the idea in John Scotus and others in the ninth century.—ED.]

[2] See General Preface, § 7, p. 15.

[3] The *possibility* of transmutation, long and strenuously denied, though certainly on no sufficient grounds, is now generally admitted. " There was a time when this fundamental doctrine of the alchemists was opposed to known analogies. It is now no longer so opposed to them, only some stages beyond their present development."—Faraday, *Lectures on Non-Metallic Elements*, p. 106.

are but slight and superficial, and contribute little, if anything, to true and active science. Nor have I forgotten that in a former passage I noted and corrected as an error of the human mind the opinion that Forms give existence. For though in nature nothing really exists beside individual bodies, performing pure individual acts according to a fixed law, yet in philosophy this very law, and the investigation, discovery, and explanation of it, is the foundation as well of knowledge as of operation. And it is this law, with its clauses, that I mean when I speak of *Forms* ; a name which I the rather adopt because it has grown into use and become familiar.

III.

If a man be acquainted with the cause of any nature (as whiteness or heat) in certain subjects only, his knowledge is imperfect ; and if he be able to superinduce an effect on certain substances only (of those susceptible of such effect), his power is in like manner imperfect. Now if a man's knowledge be confined to the efficient and material causes (which are unstable causes, and merely vehicles, or causes which convey the form in certain cases) he may arrive at new discoveries in reference to substances in some degree similar to one another, and selected beforehand ; but he does not touch the deeper boundaries of things. But whosoever is acquainted with Forms, embraces the unity of nature in substances the most unlike ; and is able therefore to detect and bring to light things never yet done, and such as neither the vicissitudes of nature, nor industry in experimenting, nor accident itself, would ever have brought into act, and which would never have occurred to the thought of man. From the discovery of Forms therefore results truth in speculation and freedom in operation.

IV.

Although the roads to human power and to human knowledge lie close together, and are nearly the same, nevertheless on account of the pernicious and inveterate habit of dwelling on abstractions, it is safer to begin and raise the sciences from those foundations which have relation to practice, and to let the active part itself be as the seal which prints and determines the contemplative counterpart. We must therefore consider, if a man wanted to generate and superinduce any nature upon a given body, what kind of rule or direction or guidance he would most wish for, and express the same in the simplest and least abstruse language. For instance, if a man wishes to superinduce upon silver the yellow colour of gold or an increase of weight (observing the laws of matter), or transparency on an opaque stone, or tenacity on glass, or vegetation on some substance that is not vegetable,—we must consider, I say, what kind of rule or guidance he would most desire. And in the first place, he will undoubtedly wish to be directed to something which will not deceive him in the result, nor fail him in the trial. Secondly, he will wish for such a rule as shall not tie him down to certain means and particular modes of operation. For perhaps he may not have those means, nor be able conveniently to procure them. And if there be other means and other methods for producing the required nature (beside the one prescribed) these may perhaps be within his reach ; and yet he shall be excluded by the narrowness of the rule, and get no good from them. Thirdly, he will desire something to be shown him, which is not as difficult as the thing proposed to be done, but comes nearer to practice.

For a true and perfect rule of operation then the direction will be *that it be certain, free, and disposing or leading to action.* And this is the same thing with the discovery of the true Form. For the Form of a nature is such, that given the Form the nature infallibly follows. Therefore it is always present when the nature is present, and universally implies it, and is constantly inherent in it. Again, the Form is such, that if it be taken away the nature infallibly vanishes. Therefore it is always absent when the nature is absent, and implies its absence, and inheres in nothing else. Lastly, the true Form is such that it deduces the given nature from some source of being which is inherent in more natures, and which is better known in the natural order of things than the Form itself. For a true and perfect axiom of knowledge then the direction and precept will be, *that another nature be discovered which is convertible with the given nature, and yet is a limitation of a more general nature, as of a true and real genus* [4]. Now these two directions, the one active the other contemplative, are one and the same thing; and what in operation is most useful, that in knowledge is most true.

V.

The rule or axiom for the transformation of bodies is of two kinds. The first regards a body as a troop or collection of simple natures. In gold, for example, the following properties meet. It is yellow in colour; heavy up to a certain weight; malleable or ductile to a certain degree of extension; it is not volatile, and loses none of its substance by the action of fire; it turns into a liquid with a certain degree of fluidity; it is separated and dissolved by particular means; and so on for the other natures which meet in gold. This kind of axiom, therefore, deduces the thing from the forms of simple natures. For he who knows the forms of yellow, weight, ductility, fixity, fluidity, solution, and so on, and the methods for superinducing them, and their gradations and modes, will make it his care to have them joined together in some body, whence may follow the transformation of that body into gold [5]. And this kind of operation pertains to the first kind of action. For the principle of generating some one simple nature is the same as that of generating many; only that a man is more fettered and tied down in operation, if more are required, by reason of the difficulty of combining into one so many natures; which do not readily meet, except in the beaten and ordinary paths of nature. It must be said however that this mode of operation (which looks to simple natures though in a compound body) proceeds from what in nature is constant and eternal and universal, and opens broad roads to human power, such as (in the present state of things) human thought can scarcely comprehend or anticipate.

The second kind of axiom, which is concerned with the discovery of the *latent process*, proceeds not by simple natures, but by compound bodies, as they are found in nature in its ordinary course. As, for instance, when inquiry is made, from what beginnings, and by what method and by what process, gold or any other metal or stone is generated, from its first menstrua and rudiments up to the perfect mineral; or in like manner by what process herbs are generated, from

[4] Let us adopt, for distinctness of expression, the theory commonly known as Boscovich's,—a theory which forms the basis of the ordinary mathematical theories of light, of heat, and of electricity. This theory supposes all bodies to be constituted of inextended atoms or centres of force, each of which attracts or repels and is attracted or repelled by all the rest. All the phenomena of nature are thus ascribed to mechanical forces, and all the differences which can be conceived to exist between two bodies,—gold say, and silver,—can only arise either from the different configuration of the centres of force, or from the different law by which they act on one another.

Assuming the truth of this theory, the question, why are some bodies transparent and others not so—in other words, what is the essential cause of transparency, which is precisely what Bacon would call the form of transparency,—is to be answered by saying that a certain configuration of the centres of force, combined with the existence of a certain law of force, constitutes such a system that the vibrations of the luminiferous ether pass through it. What this configuration or this law may be, is a question which the present state of mathematical physics does not enable us to answer; but there is no reason *a priori* why in time to come it may not receive a complete solution. If it does we shall then have arrived at a knowledge, on Boscovich's theory, of the form of transparency. Those who are acquainted with the recent progress of physical science know that questions of this kind, so far from being rejected as the questions of a mere dreamer, are thought to be of the highest interest and importance, and that no inconsiderable advance has already been made towards the solution of some at least among them.

[5] "On pourroit trouver le moyen de contrefaire l'or en sorte qu'il satisferoit à toutes les épreuves qu'on en a jusqu'ici; mais on pourroit aussi découvrir alors une nouvelle manière d'essai, qui donneroit le moyen de distinguer l'or naturel de cet or fait par artifice . . . nous pourrions avoir une définition plus parfaite de l'or que nous n'en avons présentement."—Leibnitz, *Nouv. Ess. sur l'Entendement*, c. 2.

the first concretion of juices in the ground or from seeds up to the formed plant, with all the successive motions and diverse and continued efforts of nature. So also in the inquiry concerning the process of development in the generation of animals, from coition to birth ; and in like manner of other bodies.

It is not however only to the generations of bodies that this investigation extends, but also to other motions and operations of nature. As, for instance, when inquiry is made concerning the whole course and continued action of nutrition, from the first reception of the food to its complete assimilation ; or again, concerning the voluntary motion of animals, from the first impression on the imagination and the continued efforts of the spirit up to the bendings and movements of the limbs ; or concerning the motion of the tongue and lips and other instruments, and the changes through which it passes till it comes to the utterance of articulate sounds. For these inquiries also relate to natures concrete or combined into one structure, and have regard to what may be called particular and special habits of nature, not to her fundamental and universal laws which constitute Forms. And yet it must be confessed that this plan appears to be readier and to lie nearer at hand and to give more ground for hope than the primary one.

In like manner the operative which answers to this speculative part, starting from the ordinary incidents of nature, extends its operation to things immediately adjoining, or at least not far removed. But as for any profound and radical operations on nature, they depend entirely on the primary axioms. And in those things too where man has no means of operating, but only of knowing, as in the heavenly bodies (for these he cannot operate upon or change or transform), the investigation of the fact itself or truth of the thing, no less than the knowledge of the causes and consents, must come from those primary and catholic axioms concerning simple natures ; such as the nature of spontaneous rotation, of attraction or magnetism, and of many others which are of a more general form than the heavenly bodies themselves. For let no one hope to decide the question whether it is the earth or heaven that really revolves in the diurnal motion, until he has first comprehended the nature of spontaneous rotation.

VI.

But this Latent Process, of which I speak, is quite another thing than men, pre-occupied as their minds now are, will easily conceive. For what I understand by it is not certain measures or signs or successive steps of process in bodies which can be seen ; but a process perfectly continuous, which for the most part escapes the sense.

For instance ; in all generation and transformation of bodies, we must inquire what is lost and escapes ; what remains, what is added ; what is expanded, what contracted ; what is united, what separated ; what is continued, what cut off ; what propels, what hinders ; what predominates, what yields ; and a variety of other particulars.

Again, not only in the generation or transformation of bodies are these points to be ascertained, but also in all other alterations and motions it should in like manner be inquired what goes before, what comes after ; what is quicker, what more tardy ; what produces, what governs motion ; and like points ; all which nevertheless in the present state of the sciences (the texture of which is as rude as possible and good for nothing) are unknown and unhandled. For seeing that every natural action depends on things infinitely small, or at least too small to strike the sense, no one can hope to govern or change nature until he has duly comprehended and observed them.

VII.

In like manner the investigation and discovery of the *latent configuration* in bodies is a new thing, no less than the discovery of the Latent Process and of the Form [6]. For as yet we are but lingering in the outer courts of nature, nor

[6] The distinction between the Latent Process and Latent Schematism in the absolute way in which it is here stated, involves an assumption which the progress of science will

are we preparing ourselves a way into her inner chambers. Yet no one can endow a given body with a new nature, or successfully and aptly transmute it into a new body, unless he has attained a competent knowledge of the body so to be altered or transformed. Otherwise he will run into methods which, if not useless, are at any rate difficult and perverse and unsuitable to the nature of the body on which he is operating. It is clear therefore that to this also a way must be opened and laid out.

And it is true that upon the anatomy of organized bodies (as of man and animals) some pains have been well bestowed and with good effect; and a subtle thing it seems to be, and a good scrutiny of nature. Yet this kind of anatomy is subject to sight and sense, and has place only in organized bodies. And besides it is a thing obvious and easy, when compared with the true anatomy of the Latent Configuration in bodies which are thought to be of uniform structure; especially in things that have a specific character[7] and their parts, as iron, stone; and again in parts of uniform structure in plants and animals, as the root the leaf, the flower, flesh, blood, and bones. But even in this kind, human industry has not been altogether wanting; for this is the very thing aimed at in the separation of bodies of uniform structure by means of distillations and other modes of analysis; that the complex structure of the compound may be made apparent by bringing together its several homogeneous parts. And this is of use too, and conduces to the object we are seeking; although too often fallacious in its results, because many natures which are in fact newly brought out and superinduced by fire and heat and other modes of solution, are taken to be the effect of separation merely, and to have subsisted in the compound before. And after all, this is but a small part of the work of discovering the true Configuration in the compound body; which Configuration is a thing far more subtle and exact, and such as the operation of fire rather confounds than brings out and makes distinct.

Therefore a separation and solution of bodies must be effected, not by fire indeed, but by reasoning and true induction, with experiments to aid; and by a comparison with other bodies, and a reduction to simple natures and their Forms, which meet and mix in the compound. In a word we must pass from Vulcan to Minerva, if we intend to bring to light the true textures and configurations of bodies; on which all the occult and, as they are called, specific properties and virtues in things depend; and from which too the rule of every powerful alteration and transformation is derived.

For example, we must inquire what amount of spirit there is in every body, what of tangible essence; and of the spirit, whether it be copious and turgid, or meagre and scarce; whether it be fine or coarse, akin to air or to fire, brisk or sluggish, weak or strong, progressive or retrograde, interrupted or continuous, agreeing with external and surrounding objects or disagreeing, etc. In like manner we must inquire into the tangible essence (which admits of no fewer differences than the spirit), into its coats, its fibres, its kinds of texture. Moreover the disposition of the spirit throughout the corporeal frame, with its pores, passages,

probably show to be unfounded; namely, that bodies apparently at rest are so molecularly. Whereas all analogy and the fact that they act on the senses by acting mechanically on certain deferent media combine to show that we ought to consider bodies even at rest as dynamical and not as statical entities. On this view there is no difficulty in understanding the nature of what appear to be spontaneous changes, because every dynamical system carries within itself the seeds of its own decay, except in particular cases; that is, the type of motion so alters, with greater or less rapidity, that the sensible qualities associated with it pass away. The introduction of the idea of unstable equilibrium in connexion with organic chemistry, was a step in the direction which molecular Physics will probably soon take.

[7] In Bacon's time only certain things were supposed to belong to natural *species*, all others being merely *elementary*. A ruby has a specific character, is *specificatum;* common stone or rock *non ita;*—they are mere modifications of the element earth, etc. A "specific virtue" is a virtue given by a thing's specific character, transcending the qualities of the elements it consists of. [See note on *De Augm.* ii. 3.]

veins and cells, and the rudiments or first essays of the organized body, fall under the same investigation. But on these inquiries also, and I may say on all the discovery of the Latent Configuration, a true and clear light is shed by the primary axioms, which entirely dispels all darkness and subtlety.

VIII.

Nor shall we thus be led to the doctrine of atoms, which implies the hypothesis of a vacuum and that of the unchangeableness of matter (both false assumptions); we shall be led only to real particles, such as really exist. Nor again is there any reason to be alarmed at the subtlety of the investigation, as if it could not be disentangled; on the contrary, the nearer it approaches to simple natures, the easier and plainer will everything become; the business being transferred from the complicated to the simple; from the incommensurable to the commensurable; from surds to rational quantities; from the infinite and vague to the finite and certain; as in the case of the letters of the alphabet and the notes of music. And inquiries into nature have the best result, when they begin with physics and end in mathematics. Again, let no one be afraid of high numbers or minute fractions. For in dealing with numbers it is as easy to set down or conceive a thousand as one, or the thousandth part of an integer as an integer itself.

IX.

From the two kinds of axioms which have been spoken of, arises a just division of philosophy and the sciences; taking the received terms (which come nearest to express the thing) in a sense agreeable to my own views. Thus, let the investigation of Forms, which are (in the eye of reason at least, and in their essential law) eternal and immutable, constitute *Metaphysics;* and let the investigation of the Efficient Cause, and of Matter, and of the Latent Process, and the Latent Configuration (all of which have reference to the common and ordinary course of nature, not to her eternal and fundamental laws) constitute *Physics.* And to these let there be subordinate two practical divisions: to Physics, *Mechanics;* to Metaphysics, what (in a purer sense of the word) I call *Magic,* on account of the broadness of the ways it moves in, and its greater command over nature.

X.

Having thus set up the mark of knowledge, we must go on to precepts, and that in the most direct and obvious order. Now my directions for the interpretation of nature embrace two generic divisions; the one how to educe and form axioms from experience; the other how to deduce and derive new experiments from axioms. The former again is divided into three ministrations; a ministration to the sense, a ministration to the memory, and a ministration to the mind or reason.

For first of all we must prepare a *Natural and Experimental History,* sufficient and good; and this is the foundation of all; for we are not to imagine or suppose, but to discover, what nature does or may be made to do.

But natural and experimental history is so various and diffuse, that it confounds and distracts the understanding, unless it be ranged and presented to view in a suitable order. We must therefore form *Tables and Arrangements of Instances,* in such a method and order that the understanding may be able to deal with them.

And even when this is done, still the understanding, if left to itself and its own spontaneous movements, is incompetent and unfit to form axioms, unless it be directed and guarded. Therefore in the third place we must use *Induction,* true and legitimate induction, which is the very key of interpretation. But of this, which is the last, I must speak first, and then go back to the other ministrations.

XI.

The investigation of Forms proceeds thus: a nature being given, we must first of all have a muster or presentation before the understanding of all known instances which agree in the same nature, though in substances the most unlike.

And such collection must be made in the manner of a history, without premature speculation, or any great amount of subtlety. For example, let the investigation be into the Form of Heat.

Instances Agreeing in the Nature of Heat.

1. The rays of the sun, especially in summer and at noon.
2. The rays of the sun reflected and condensed, as between mountains, or on walls, and most of all in burning-glasses and mirrors.
3. Fiery meteors.
4. Burning thunderbolts.
5. Eruptions of flame from the cavities of mountains.
6. All flame.
7. Ignited solids.
8. Natural warm-baths.
9. Liquids boiling or heated.
10. Hot vapours and fumes, and the air itself, which conceives the most powerful and glowing heat, if confined; as in reverberatory furnaces.
11. Certain seasons that are fine and cloudless by the constitution of the air itself, without regard to the time of year.
12. Air confined and underground in some caverns, especially in winter.
13. All villous substances, as wool, skins of animals, and down of birds, have heat.
14. All bodies, whether solid or liquid, whether dense or rare (as the air itself is), held for a time near the fire.
15. Sparks struck from flint and steel by strong percussion.
16. All bodies rubbed violently, as stone, wood, cloth, etc., insomuch that poles and axles of wheels sometimes catch fire; and the way they kindled fire in the West Indies was by attrition.
17. Green and moist vegetables confined and bruised together, as roses packed in baskets; insomuch that hay, if damp when stacked, often catches fire [8].
18. Quick lime sprinkled with water.
19. Iron, when first dissolved by strong waters in glass, and that without being put near the fire. And in like manner tin, etc., but not with equal intensity.
20. Animals, especially and at all times internally; though in insects the heat is not perceptible to the touch by reason of the smallness of their size.
21. Horse-dung and like excrements of animals when fresh.
22. Strong oil of sulphur and of vitriol has the effect of heat in burning linen.

[8] "That seeds when germinating, as they lie heaped in large masses, evolve a considerable degree of heat, is a fact long known from the malting of grain; but the cause of it was incorrectly sought for in a process of fermentation. To Göppert (*Ueber Wärmeentwickelung in der lebenden Pflanze*) is due the merit of having demonstrated that such is not the case, but that the evolution of heat is connected with the process of germination. Seeds of very different chemical composition (of different grains, of Hemp, Clover, *Spergula, Brassica*, etc.), made to germinate in quantities of about a pound, became heated, at a temperature of the air of 48°—66°, to 59°—120° Fahr.

"It was likewise shown by Göppert that full-grown plants also, such as Oats, Maize, *Cyperus esculentus, Hyoscyamus, Sedum acre*, etc., laid together in heaps and covered with bad conductors of heat, cause a thermometer placed among them to rise about 2°—7° (*Spergula* as much as 22°) above the temperature of the air . . .

"A very great evolution of heat occurs in the blossom of the *Aroideæ*. This is considerable even in our *Arum maculatum*, and according to Dutrochet's researches (*Comptes Rendus*, 1839, 695.) rises to 25°—27° above the temperature of the air. But this phenomenon is seen in a far higher degree in *Colocasia odora*, in which plant it has been investigated by Brougniart (*Nouv. Ann. d. Muséum*, iii.). Vrolik and Vriese (*Ann. des Sc. Nat.*, sec. ser. v. 134.) and Van Beek and Bersgma (*Obs. thermo-élect. s. l'élév. de températ. des Fleurs d. Colocas. odor.* 1838). These last observers found the maximum of heat 129°, when the temperature of the air was 79°."—Mohl *On the Vegetable Cell*, translated by Arthur Henfrey, Lond. 1852, pp. 101 and 102.

23. Oil of marjoram and similar oils have the effect of heat in burning the bones of the teeth.

24. Strong and well rectified spirit of wine has the effect of heat; insomuch that the white of an egg being put into it hardens and whitens almost as if it were boiled; and bread thrown in becomes dry and crusted like toast [9].

25. Aromatic and hot herbs, as *dracunculus, nasturtium vetus*, etc., although not warm to the hand (either whole or in powder), yet to the tongue and palate, being a little masticated, they feel hot and burning.

26. Strong vinegar, and all acids, on all parts of the body where there is no epidermis, as the eye, tongue; or on any part when wounded and laid bare of the skin; produce a pain but little differing from that which is created by heat.

27. Even keen and intense cold produces a kind of sensation of burning. Nec Boreæ penetrabile frigus adurit [10].

28. Other instances.

This table I call the *Table of Essence and Presence.*

XII.

Secondly, we must make a presentation to the understanding of instances in which the given nature is wanting; because the Form, as stated above, ought no less to be absent when the given nature is absent, than present when it is present. But to note all these would be endless.

The negatives should therefore be subjoined to the affirmatives, and the absence of the given nature inquired of in those subjects only that are most akin to the others in which it is present and forthcoming. This I call the *Table of Deviation, or of Absence in Proximity.*

Instances in Proximity where the Nature of Heat is Absent.

Answering to the first affirmative instance. 1. The rays of the moon and of stars and comets are not found to be hot to the touch [11]; indeed the severest colds are observed to be at the full moons.

The larger fixed stars, however, when passed or approached by the sun, are supposed to increase and give intensity to the heat of the sun; as is the case when the sun is in the sign Leo, and in the Dog-days.

To the 2nd. 2. The rays of the sun in what is called the middle region of the air do not give heat; for which there is commonly assigned not a bad reason, viz. that that region is neither near enough to the body of the sun from which the rays emanate, nor to the earth from which they are reflected. And this appears from the fact that on the tops of mountains, unless they are very high, there is perpetual snow. On the other hand it has been observed that on the peak of Teneriffe, and among the Andes of Peru, the very tops of

[9] The analogy which Bacon here remarks, arises probably, in the second instance from the desiccative power due to the strong affinity of alcohol for water. The French chemist Lassaigne found, I believe, that alcohol extracted a red colouring matter from unboiled lobster shells; but I am not aware that the modus operandi has in this case been explained. But by far the most remarkable case of what may be called simulated heat, is furnished by the action of carbonic acid gas on the skin. Of late years baths of this gas have been used medicinally; but M. Boussingault long since remarked the sensation of heat which it produces. He states that at Quindiu in New Granada there are sulphur works, and that at various points nearly pure carbonic gas escapes from shallow excavations in the surface, containing, however, a trace of hydro-sulphuric acid; that the temperature of this issuing stream of gas is lower than the external air, but that the sensation is the same as that produced by a hot-air bath of perhaps from 40° to 45° or 48° Centigrade (104° to 118° Fahr.). As this effect has not been noticed in carbonic acid gas prepared artificially, it is probable that it requires for its production the gas to be in motion; so that the necessary conditions are not present when the hand is inserted into a jar of the gas.

[10] Nor burns the sharp cold of the northern blast. Virgil, *Georg.* i. 93.

[11] M. Melloni has recently succeeded in making sensible the moon's calorific rays.

the mountains are free from snow ; which lies only somewhat lower down. Moreover the air itself at the very top is found to be by no means cold, but only rare and keen ; insomuch that on the Andes it pricks and hurts the eyes by its excessive keenness, and also irritates the mouth of the stomach, producing vomiting. And it was observed by the ancients that on the top of Olympus the rarity of the air was such that those who ascended it had to carry sponges with them dipped in vinegar and water, and to apply them from time to time to their mouth and nose, the air being from its rarity not sufficient to support respiration [12] ; and it was further stated that on this summit the air was so serene, and so free from rain and snow and wind, that letters traced by the finger in the ashes of the sacrifices on the altar of Jupiter remained there until the next year without being at all disturbed. And at this day travellers ascending to the top of the Peak of Teneriffe make the ascent by night and not by day ; and soon after the rising of the sun are warned and urged by their guides to come down without delay, on account of the danger they run lest the animal spirits should swoon and be suffocated by the tenuity of the air.

To the 2nd. 3. The reflexion of the rays of the sun in regions near the polar circles is found to be very weak and ineffective in producing heat ; insomuch that the Dutch who wintered in Nova Zembla [13], and expected their ship to be freed from the obstructions of the mass of ice which hemmed her in by the beginning of July, were disappointed of their expectation, and obliged to take to their boat. Thus the direct rays of the sun seem to have but little power, even on the level ground ; nor have the reflex much, unless they are multiplied and combined ; which is the case when the sun tends more to the perpendicular ; for then the incident rays make acuter angles, so that the lines of the rays are nearer each other ; whereas on the contrary, when the sun shines very obliquely, the angles are very obtuse, and thus the lines of rays are at a greater distance from each other. Meanwhile it should be observed that there may be

[12] Aristotle seems to be the first person who mentions this notion. See the ***Problems***, xxvi. 36 ; where, however, he speaks of Athos and οἱ τοιοῦτοι, and not of Olympus. The passages on the subject are to be found in Ideler's *Meteorologia veterum Græcorum et Romanorum* (Berlin, 1832), at p. 81. Compare his edition of the *Meteorologies* of Aristotle, where he has given *in extenso* the passage in which Geminus speaks in the same manner of Mount Cyllene in Arcadia, and also a similar statement made by Philoponus with respect to Olympus. The whole class of stories seem (as Ideler following Lobeck remarks) to have somewhat of a mythical character. G. Bruno apparently confounded Philoponus with Alexander Aphrodisiensis, when in the *Cena di Cenere* he asserted that the latter mentions the sacrifices on the top of Olympus. In the passage on the subject in which we might expect to find him doing so, namely in his Commentary on the *Meteorologies*, i. c. 3, he does not specify any particular mountain.

That there is no wind nor rain on Olympus is mentioned as a common opinion by St. Augustine, *De Civ. Dei*, xvi. 27. Compare Dante, *Purg.* xxviii. 112.

[13] This of course refers to Barentz's expedition in search of a North-East passage. He passed the winter 1596–7 at Nova Zembla. [In Barentz's first voyage, 1594, he was stopped by the ice on the 13th of July, and obliged to return. In his third voyage, 1596, his first considerable check was on the 19th of July ; after which he only succeeded in coasting round the northern point of Nova Zembla till the 26th of August, where the ship stuck fast and they were forced to leave her and winter on the island, and return in their boats in the beginning of June, 1597. See the letter signed by the company: "Three Voyages by the North East," etc., Hakluyt Society, 1853, p. 191. This letter was begun on the 1st of June : " Having till this day stayed for the time and opportunity in hope to get our ship loose, and now are clean out of hope thereof, for that it lieth shut up and enclosed in the ice," etc. : and ended on the 13th, " notwithstanding that while we were making ready to be gone, we had great wind out of the west and north-west, and yet find no alteration nor bettering in the weather, and therefore in the last extremity we left it ". This narrative, written by Gerrit de Veer, one of the party, was first published in Dutch in 1598 ; translated into Latin and French the same year ; into Italian in 1599 ; into English in 1609. See Introduction, p. cxviii. " Per initia mensis *Junii* " would have been more accurate.—*J. S.*]

many operations of the sun, and those too depending on the nature of heat, which are not proportioned to our touch ; so that in respect of us their action does not go so far as to produce sensible warmth, but in respect of some other bodies they have the effect of heat.

To the 2nd. 4. Try the following experiment. Take a glass fashioned in a contrary manner to a common burning-glass, and placing it between your hand and the rays of the sun, observe whether it diminishes the heat of the sun, as a burning-glass increases and strengthens it. For it is evident in the case of optical rays that according as the glass is made thicker or thinner in the middle as compared with the sides, so do the objects seen through it appear more spread or more contracted. Observe therefore whether the same is the case with heat.

To the 2nd. 5. Let the experiment be carefully tried, whether by means of the most powerful and best constructed burning glasses, the rays of the moon can be so caught and collected as to produce even the least degree of warmth. But should this degree of warmth prove too subtle and weak to be perceived and apprehended by the touch, recourse must be had to those glasses which indicate the state of the atmosphere in respect of heat and cold. Thus let the rays of the moon fall through a burning-glass on the top of a glass of this kind, and then observe whether there ensues a sinking of the water through warmth.

To the 2nd. 6. Let a burning-glass also be tried with a heat that does not emit rays or light [14], as that of iron or stone heated but not ignited, boiling water, and the like ; and observe whether there ensue an increase of the heat, as in the case of the sun's rays.

To the 2nd. 7. Let a burning-glass also be tried with common flame.

To the 3rd. 8. Comets (if we are to reckon these too among meteors) [15] are not found to exert a constant or manifest effect in increasing the heat of the season, though it is observed that they are often followed by droughts. Moreover bright beams and pillars and openings in the heavens appear more frequently in winter than in summer time, and chiefly during the intensest cold, but always accompanied by dry weather. Lightning, however, and coruscations and thunder, seldom occur in the winter, but about the time of great heat. Falling stars, as they are called, are commonly supposed to consist rather of some bright and lighted viscous substance, than to be of any strong fiery nature. But on this point let further inquiry be made.

To the 4th. 9. There are certain coruscations which give light but do not burn. And these always come without thunder.

To the 5th. 10. Eructations and eruptions of flame are found no less in cold than in warm countries, as in Iceland and Greenland. In cold countries too the trees are in many cases more inflammable and more pitchy and resinous than in warm ; as the fir, pine, and others. The situations however and the nature of the soil in which eruptions of this kind usually occur have not been carefully enough ascertained to enable us to subjoin a Negative to this Affirmative Instance.

To the 6th. 11. All flame is in all cases more or less warm ; nor is there any Negative to be subjoined. And yet they say that the *ignis fatuus* (as it is called), which sometimes even settles on a wall, has not much heat ; perhaps as much as the flame of spirit of wine, which is mild and soft. But still milder must that flame be, which according to certain grave and trustworthy histories has been seen shining about the head and locks of boys and girls, without at all burning

[14] Mersenne says the greater number of the experiments mentioned in the second book of the *Novum Organum* had already been made, and mentions particularly, as if he had himself tried it, the reflexion of all kinds of heat by a burning mirror. He also asserts that light is always accompanied by heat. *De la Vérité des Sciences* (1625), p. 210.

[15] That there was no reason for supposing comets to be more than merely meteoric exhalations is the thesis maintained, and doubtless with great ability, by Galileo in his *Saggiatore*,—the true view, or at least a nearer approach to it, having been propounded by the Jesuit Grassi. Bacon perhaps alludes to this controversy.

the hair, but softly playing round it. It is also most certain that about a horse, when sweating on the road, there is sometimes seen at night, and in clear weather, a sort of luminous appearance without any manifest heat. And it is a well known fact, and looked upon as a sort of miracle, that a few years ago a girl's stomacher, on being slightly shaken or rubbed, emitted sparks; which was caused perhaps by some alum or salts used in the dye, that stood somewhat thick and formed a crust, and were broken by the friction. It is also most certain that all sugar, whether refined or raw, provided only it be somewhat hard, sparkles when broken or scraped with a knife in the dark. In like manner sea and salt water is sometimes found to sparkle by night when struck violently by oars [16]. And in storms too at night time, the foam of the sea when violently agitated emits sparks, and this sparkling the Spaniards call *Sea Lung* [17]. With regard to the heat of the flame which was called by ancient sailors Castor and Pollux, and by moderns St. Elmo's Fire, no sufficient investigation thereof has been made.

To the 7th. 12. Every body ignited so as to turn to a fiery red, even if unaccompanied by flame, is always hot; neither is there any Negative to be subjoined to this Affirmative. But that which comes nearest seems to be rotten wood, which shines by night, and yet is not found to be hot; and the putrefying scales of fish, which also shine in the dark, and yet are not warm to the touch; nor again is the body of the glow-worm, or of the fly called *Luciola*, found to be warm to the touch.

To the 8th. 13. In what situation and kind of soil warm baths usually spring has not been sufficiently examined; and therefore no Negative is subjoined.

To the 9th. 14. To warm liquids I subjoin the Negative Instance of liquid itself in its natural state. For we find no tangible liquid which is warm in its own nature and remains so constantly; but the warmth is an adventitious nature, superinduced only for the time being; so that the liquids which in power and operation are hottest, as spirit of wine, chemical oil of spices, oil of vitriol and sulphur, and the like, which burn after a while, are at first cold to the touch. The water of natural warm baths on the other hand, if received into a vessel and separated from its springs, cools just like water that has been heated on a fire. But it is true that oily substances are less cold to the touch than watery, oil being less cold than water, and silk than linen. But this belongs to the Table of Degrees of Cold.

To the 10th. 15. In like manner to hot vapour I subjoin as a Negative the nature of vapour itself, such as we find it with us. For exhalations from oily substances, though easily inflammable, are yet not found to be warm, unless newly exhaled from the warm body.

To the 10th. 16. In like manner I subjoin as a Negative to hot air the nature of air itself. For we do not find here any air that is warm, unless it has either been confined, or compressed, or manifestly warmed by the sun, fire, or some other warm substance.

To the 11th. 17. I here subjoin the Negative of colder weather than is suitable to the season of the year, which we find occurs during east and north winds; just as we have weather of the opposite kind with the south and west winds. So a tendency to rain, especially in winter time, accompanies warm weather; while frost accompanies cold.

To the 12th. 18. Here I subjoin the Negative of air confined in caverns during the summer. But the subject of air in confinement should by all

[16] [This false explanation is one of Bacon's most gratuitous miscarriages. As the phenomenon admittedly does not always result on concussion, the inference breaks down. —Ed.]

[17] The phrase "pulmo marino" is as much Italian as Spanish,—except of course, that in Italian "pulmo" is replaced by "polmo,"—and is merely a translation of πνεύμων θαλάσσιος, which is used by Dioscorides, *De Materiâ Medicâ*, ii. 39. The luminous appearance arises apparently from serpent medusæ, which in texture are like the substance of the lungs, from which circumstance they derive the name which Dioscorides gives them. Cf. *De Aug.* iv. 3.

means be more diligently examined. For in the first place it may well be matter of doubt what is the nature of air in itself with regard to heat and cold. For air manifestly receives warmth from the influence of the heavenly bodies, and cold perhaps from the exhalations of the earth ; and again in the middle region of air, as it is called, from cold vapours and snow ; so that no opinion can be formed as to the nature of air from the examination of air that is at large and exposed ; but a truer judgment might be made by examining it when confined. It is however necessary for the air to be confined in a vessel of such material as will not itself communicate warmth or cold to the air by its own nature, nor readily admit the influence of the outer atmosphere. Let the experiment therefore be made in an earthen jar wrapped round with many folds of leather to protect it from the outward air, and let the vessel remain tightly closed for three or four days ; then open the vessel and test the degree of heat or cold by applying either the hand or a graduated glass.

To the 13th. 19. In like manner a doubt suggests itself, whether the warmth in wool, skins, feathers, and the like, proceeds from a faint degree of heat inherent in them, as being excretions from animals ; or from a certain fat and oiliness, which is of a nature akin to warmth ; or simply, as surmised in the preceding article, from the confinement and separation of the air. For all air that is cut off from connexion with the outer air seems to have some warmth. Try the experiment therefore with fibrous substances made of linen ; not of wool, feathers, or silk, which are excretions from animals. It should also be observed that all powders (in which there is manifestly air enclosed) are less cold than the whole substances they are made from ; as likewise I suppose that all froth (as that which contains air) is less cold than the liquor it comes from.

To the 14th. 20. To this no Negative is subjoined. For there is nothing found among us either tangible or spirituous which does not contract warmth when put near fire. There is this difference however, that some substances contract warmth more quickly, as air, oil, and water ; others more slowly, as stone and metal [18]. But this belongs to the Table of Degrees.

To the 15th. 21. To this Instance I subjoin no Negative, except that I would have it well observed that sparks are produced from flint and steel, or any other hard substance, only when certain minute particles are struck off from the substance of the stone or metal ; and that the attrition of the air does not of itself ever produce sparks, as is commonly supposed. And the sparks themselves too, owing to the weight of the ignited body, tend rather downwards than upwards ; and on going out become a tangible sooty substance.

To the 16th. 22. There is no Negative, I think, to be subjoined to this Instance. For we find among us no tangible body which does not manifestly gain warmth by attrition ; insomuch that the ancients fancied that the heavenly bodies had no other means or power of producing warmth than by the attrition of the air in their rapid and hurried revolution. But on this subject we must further inquire whether bodies discharged from engines, as balls from cannon, do not acquire some degree of heat from the very percussion, so as to be found somewhat warm when they fall. Air in motion, however, rather chills than warms, as appears from wind, bellows, and blowing with the mouth contracted. But motion of this kind is not so rapid as to excite heat, and is the motion of a mass, and not of particles ; so that it is no wonder if it does not generate heat.

To the 17th. 23. On this Instance should be made more diligent inquiry. For herbs and vegetables when green and moist seem to contain some latent heat, though so slight that it is not perceptible to the touch when they are single ; but only when they are collected and shut up together, so that their spirits may not breathe out into the air, but may mutually cherish each other ; whereupon there arises a palpable heat, and sometimes flame in suitable matter.

To the 18th. 24. On this Instance too should be made more diligent inquiry. For quick lime sprinkled with water seems to contract heat, either by the concentration of heat before dispersed, as in the above-mentioned case

[18] [This is of course an error.—Ed.]

of confined herbs, or because the igneous spirit is irritated and exasperated by the water, so as to cause a conflict and reaction. Which of these two is the real cause will more readily appear if oil be poured on instead of water; for oil will serve equally well with water to concentrate the enclosed spirit, but not to irritate it. We should also extend the experiment both by employing the ashes and rusts of different bodies, and by pouring in different liquids.

To the 19th. 25. To this Instance is subjoined the Negative of other metals which are softer and more fusible. For gold-leaf dissolved by *aqua regia* gives no heat to the touch; no more does lead dissolved in *aqua fortis;* neither again does quicksilver (as I remember); but silver itself does, and copper too (as I remember); tin still more manifestly; and most of all iron and steel, which not only excite a strong heat in dissolution, but also a violent ebullition[19]. It appears therefore that the heat is produced by conflict; the strong waters penetrating, digging into, and tearing asunder the parts of the substance, while the substance itself resists. But where the substances yield more easily, there is hardly any heat excited.

To the 20th. 26. To the heat of animals no Negative is subjoined, except that of insects (as above-mentioned), on account of their small size. For in fishes, as compared with land animals, it is rather a low degree than an absence of heat that is noted. But in vegetables and plants there is no degree of heat perceptible to the touch, either in their exudations or in their pith when freshly exposed. In animals however is found a great diversity of heat, both in their parts (there being different degrees of heat about the heart, in the brain, and on the skin) and in their accidents, as violent exercise and fevers.

To the 22nd 27. To this Instance it is hard to subjoin a Negative. Indeed the excrements of animals when no longer fresh have manifestly a potential heat, as is seen in the enriching of soil.

To the 22nd and 23rd. 28. Liquids, whether waters or oils, which possess a great and intense acridity, act like heat in tearing asunder bodies, and burning them after some time; yet to the touch they are not hot at first. But their operation is relative and according to the porosity of the body to which they are applied. For *aqua regia* dissolves gold but not silver [20]; *aqua fortis*, on the contrary, dissolves silver, but not gold; neither dissolves glass, and so on with others.

To the 24th. 29. Let trial be made of spirit of wine on wood; and also on butter, wax, or pitch; and observe whether by its heat it in any degree melts them. For the twenty-fourth instance exhibits a power in it that resembles heat in producing incrustation. In like manner therefore try its power in producing liquefaction. Let trial also be made with a graduated or calendar glass, hollow at the top; pour into the hollow spirit of wine well rectified, cover it up that the spirit may better retain its heat, and observe whether by its heat it makes the water sink.

To the 25th. 30. Spices and acrid herbs strike hot on the palate, and much hotter on the stomach. Observe therefore on what other substances they produce the effects of heat. Sailors tell us that when large parcels and masses of spices are, after being long kept close, suddenly opened, those who first stir and take them out run the risk of fever and inflammation [21]. It can also be tried whether such spices and herbs when pounded would not dry bacon and meat hung over them, as smoke does.

[19] This ebullition is of course not the *result* of the heat, but arises from the disengagement of gas during the action of the acid on the metal.

[20] *Aqua regia* is a mixture of nitric and hydrochloric acids. Its power of dissolving gold is ascribed by Davy to the liberation of chlorine by the neutral action of the two acids. The different result in the case of silver arises from the insolubility of chloride of silver.

[21] In the *Annals of Philosophy* a case is mentioned in which the effluvia arising on the opening of a large bark store at Guayra were sufficiently powerful to cure a bad fever. [Mr. Ellis puts this note to the original without remarking that it is of contrary effect to Bacon's statement.—Ed.]

To the 26th. 31. There is an acridity or pungency both in cold things, as vinegar and oil of vitriol, and in hot, as oil of marjoram and the like. Both alike therefore cause pain in animate substances, and tear asunder and consume the parts in such as are inanimate. To this Instance again there is no Negative subjoined. Moreover we find no pain in animals, save with a certain sensation of heat.

To the 27th. 32. There are many actions common both to heat and cold, though in a very different manner. For boys find that snow after a while seems to burn their hands; and cold preserves meat from putrefaction, no less than fire; and heat contracts bodies, which cold does also. But these and similar instances may more conveniently be referred to the inquiry concerning Cold.

XIII.

Thirdly, we must make a presentation to the understanding of instances in which the nature under inquiry is found in different degrees, more or less; which must be done by making a comparison either of its increase and decrease in the same subject, or of its amount in different subjects, as compared one with another. For since the Form of a thing is the very thing itself, and the thing differs from the form no otherwise than as the apparent differs from the real, or the external from the internal, or the thing in reference to man from the thing in reference to the universe; it necessarily follows that no nature can be taken as the true form, unless it always decrease when the nature in question decreases, and in like manner always increase when the nature in question increases. This Table therefore I call the *Table of Degrees* or the *Table of Comparison.*

Table of Degrees or Comparison in Heat.

I will therefore first speak of those substances which contain no degree at all of heat perceptible to the touch, but seem to have a certain potential heat only or disposition and preparation for hotness. After that I shall proceed to substances which are hot actually, and to the touch, and to their intensities and degrees.

1. In solid and tangible bodies we find nothing which is in its nature originally hot. For no stone, metal, sulphur, fossil, wood, water, or carcass of animal is found to be hot. And the hot water in baths seems to be heated by external causes; whether it be by flame or subterraneous fire, such as is thrown up from Ætna and many other mountains, or by the conflict of bodies, as heat is caused in the dissolutions of iron and tin. There is therefore no degree of heat palpable to the touch in animate substances; but they differ in degree of cold, wood not being equally cold with metal[22]. But this belongs to the Table of Degrees in Cold.

2. As far however as potential heat and aptitude for flame is concerned, there are many inanimate substances found strongly disposed thereto, as sulphur, naphtha, rock oil.

3. Substances once hot, as horse-dung from animal heat, and lime or perhaps ashes and soot from fire, retain some latent remains of their former heat. Hence certain distillations and resolutions of bodies are made by burying them in horse-dung, and heat is excited in lime by sprinkling it with water, as already mentioned.

4. In the vegetable creation we find no plant or part of plant (as gum or pitch) which is warm to the human touch. But yet, as stated above, green herbs gain warmth by being shut up; and to the internal touch, as the palate or stomach, and even to external parts, after a little time, as in plasters and ointments, some vegetables are perceptibly warm and others cold.

5. In the parts of animals after death or separation from the body, we find nothing warm to the human touch. Not even horse-dung, unless enclosed and buried, retains its heat. But yet all dung seems to have a potential heat, as is seen in the fattening of the land. In like manner carcasses of animals have some such latent and potential heat; insomuch that in burying grounds, where

22 [Another erroneous explanation. The difference is now known to be one of conductivity.—ED.]

burials take place daily, the earth collects a certain hidden heat, which consumes a body newly laid in it much more speedily than pure earth. We are told too that in the East there is discovered a fine soft texture, made of the down of birds, which by an innate force dissolves and melts butter when lightly wrapped in it.

6. Substances which fatten the soil, as dung of all kinds, chalk, sea-sand, salt, and the like, have some disposition to heat.

7. All putrefaction contains in itself certain elements of a slight heat [23], though not so much as to be perceived by the touch. For not even those substances which on putrefaction turn to animalculæ, as flesh, cheese, etc., feel warm to the touch; no more does rotten wood, which shines in the dark. Heat however in putrid substances sometimes betrays itself by foul and powerful odours.

8. The first degree of heat therefore among those substances which feel hot to the touch, seems to be the heat of animals, which has a pretty great extent in its degrees. For the lowest, as in insects, is hardly perceptible to the touch; but the highest scarce equals the sun's heat in the hottest countries and seasons, nor is it too great to be borne by the hand. It is said however of Constantius [24], and some others of a very dry constitution and habit of body, that in violent fevers they became so hot as somewhat to burn the hand that touched them.

9. Animals increase in heat by motion and exercise, wine, feasting, venus, burning fevers, and pain.

10. When attacked by intermittent fevers, animals are at first seized with cold and shivering, but soon after they become exceedingly hot, which is their condition from the first in burning and pestilential fevers.

11. Let further inquiry be made into the different degrees of heat in different animals, as in fishes, quadrupeds, serpents, birds; and also according to their species, as in the lion, the kite, the man; for in common opinion fish are the least hot internally, and birds the hottest; especially doves, hawks, and sparrows [25].

12. Let further inquiry be made into the different degrees of heat in the different parts and limbs of the same animal. For milk, blood, seed, eggs, are found to be hot only in a moderate degree, and less hot than the outer flesh of the animal when in motion or agitated. But what the degree of heat is in the brain, stomach, heart, etc., has not yet been in like manner inquired.

13. All animals in winter and cold weather are cold externally, but internally they are thought to be even hotter.

14. The heat of the heavenly bodies, even in the hottest countries, and at the hottest times of the year and day, is never sufficiently strong to set on fire or burn the driest wood or straw, or even tinder, unless strengthened by burning-glasses or mirrors. It is however able to extract vapour from moist substances.

15. By the tradition of astronomers some stars are hotter than others. Of planets, Mars is accounted the hottest after the sun; then comes Jupiter, and then Venus [26]. Others, again, are set down as cold; the moon, for instance, and above all Saturn. Of fixed stars, Sirius is said to be the hottest, then Cor Leonis or Regulus, then Canicula, and so on.

16. The sun gives greater heat the nearer he approaches to the perpendicular or zenith; and this is probably true of the other planets also, according to the proportion of their heat. Jupiter, for instance, is hotter, probably, to us when under Cancer or Leo than under Capricorn or Aquarius.

17. We must also believe that the sun and other planets give more heat in

[23] This is true of eremacausis rather than of real putrefaction. But the distinction belongs to the recent history of chemistry.

[24] The person here referred to is Constantius II., the son of Constantine the Great. The burning heat of the fever of which he died is mentioned by Ammianus Marcellinus, l. xxi. c. 15.

[25] [Orig. *struthiones*. *Struthio* commonly means an ostrich, but Mr. Spedding surmised it to stand here for *strutheus*, the sparrow.—Ed.]

[26] By some Venus was accounted cold and moist. Vide *Margarita Phil.* p. 627. Ptolemy, however, confirms what Bacon says of her.

perigee, from their proximity to the earth, than they do in apogee. But if it happens that in some region the sun is at the same time in perigee and near the perpendicular, his heat must of necessity be greater than in a region where he is also in perigee, but shining more obliquely. And therefore the altitude of the planets in their exaltation in different regions ought to be noted, with respect to perpendicularity or obliquity.

18. The sun and other planets are supposed to give greater heat when nearer to the larger fixed stars. Thus when the sun is in Leo he is nearer Cor Leonis, Cauda Leonis, Spica Virginis, Sirius and Canicula, than when he is in Cancer, in which sign however he is nearer to the perpendicular [27]. And it must be supposed that those parts of the heavens shed the greatest heat (though it be not at all perceptible to the touch) which are the most adorned with stars, especially of a larger size.

19. Altogether, the heat of the heavenly bodies is increased in three ways; first, by perpendicularity; secondly, by proximity or perigee; thirdly, by the conjunction or combination of stars.

20. The heat of animals, and of the rays of the heavenly bodies also (as they reach us), is found to differ by a wide interval from flame, though of the mildest kind, and from all ignited bodies; and from liquids also, and air itself when highly heated by fire. For the flame of spirit of wine, though scattered and not condensed, is yet sufficient to set paper, straw, or linen on fire; which the heat of animals will never do, or of the sun without a burning-glass or mirror.

21. There are however many degrees of strength and weakness in the heat of flame and ignited bodies. But as they have never been diligently inquired into, we must pass them lightly over. It appears however that of all flame that of spirit of wine is the softest, unless perhaps *ignis fatuus* be softer, and the flames or sparklings arising from the sweat of animals. Next to this, as I suppose, comes flame from light and porous vegetable matter, as straw, reeds, and dried leaves; from which the flame from hairs or feathers does not much differ. Next perhaps comes flame from wood, especially such as contains but little rosin or pitch; with this distinction however, that the flame from small pieces of wood (such as are commonly tied up in fagots) is milder than the flame from trunks and roots of trees. And this you may try any day in furnaces for smelting iron, in which a fire made with fagots and boughs of trees is of no great use. After this I think comes flame from oil, tallow, wax, and such like fat and oily substances, which have no great acrimony. But the most violent heat is found in pitch and rosin; and yet more in sulphur, camphor, naphtha, rock-oil, and salts (after the crude matter is discharged), and in their compounds, as gunpowder, Greek fire (commonly called wild fire), and its different kinds, which have so stubborn a heat that they are not easily extinguished by water.

22. I think also that the flame which results from some imperfect metals is very strong and eager. But on these points let further inquiry be made.

23. The flame of powerful lightning seems to exceed in strength all the former; for it has even been known to melt wrought iron into drops; which those other flames cannot do.

24. In ignited bodies too there are different degrees of heat, though these again have not yet been diligently examined. The weakest heat of all, I think, is that from tinder, such as we use to kindle flame with; and in like manner that of touchwood or tow, which is used in firing cannon. After this comes ignited wood or coal, and also bricks, and the like heated to ignition. But of all ignited substances, the hottest, as I take it, are ignited metals; as iron, copper, etc. But these require further investigation.

25. Some ignited bodies are found to be much hotter than some flames. Ignited iron, for instance, is much hotter and more consuming than flame of spirit of wine.

27 This astrological fancy was probably suggested by a wish to explain why July is hotter than June. In the division of the Zodiac into trigons each of which corresponds to one of the elements, Leo forms one of the corners of the fiery trigon; and it is moreover the sun's proper sign.

26. Of substances also which are not ignited but only heated by fire, as boiling water and air confined in furnaces, some are found to exceed in heat many flames and ignited substances.

27. Motion increases heat, as you may see in bellows, and by blowing; insomuch that the harder metals are not dissolved or melted by a dead or quiet fire, till it be made intense by blowing.

28. Let trial be made with burning-glasses, which (as I remember) act thus. If you place a burning-glass at the distance of (say) a span from a combustible body, it will not burn or consume it so easily as if it were first placed at the distance of (say) half a span, and then moved gradually and slowly to the distance of the whole span. And yet the cone and union of rays are the same; but the motion itself increases the operation of the heat [28].

29. Fires which break out during a strong wind are thought to make greater progress against than with it; because the flame recoils more violently when the wind gives way than it advances while the wind is driving it on.

30. Flame does not burst out, nor is it generated, unless some hollow space be allowed it to move and play in; except the explosive flame of gunpowder, and the like, where compression and imprisonment increase its fury.

31. An anvil grows very hot under the hammer, insomuch that if it were made of a thin plate it might, I suppose, with strong and continuous blows of the hammer, grow red like ignited iron. But let this be tried by experiment.

32. But in ignited substances which are porous, so as to give the fire room to move, if this motion be checked by strong compression, the fire is immediately extinguished. For instance, when tinder, or the burning wick of a candle or lamp, or even live charcoal or coal, is pressed down with an extinguisher, or with the foot, or any similar instrument, the operation of the fire instantly ceases.

33. Approximation to a hot body increases heat in proportion to the degree of approximation. And this is the case also with light; for the nearer an object is brought to the light, the more visible it becomes.

34. The union of different heats increases heat, unless the hot substances be mixed together. For a large fire and a small fire in the same room increase one another's heat; but warm water plunged into boiling water cools it.

35. The continued application of a hot body increases heat, because heat perpetually passing and emanating from it mingles with the previously existing heat, and so multiplies the heat. For a fire does not warm a room as well in half an hour as it does if continued through the whole hour. But this is not the case with light; for a lamp or candle gives no more light after it has been long lighted, than it did at first.

36. Irritation by surrounding cold increases heat, as you may see in fires during a sharp frost. And this I think is owing not merely to the confinement and contraction of the heat, which is a kind of union, but also to irritation. Thus when air or a stick is violently compressed or bent, it recoils not merely to the point it was forced from, but beyond it on the other side. Let trial therefore be carefully made by putting a stick or some such thing into flame, and observing whether it is not burnt more quickly at the sides than in the middle of the flame.

37. There are many degrees in susceptibility of heat. And first of all it is to be observed how slight and faint a heat changes and somewhat warms even those bodies which are least of all susceptible of heat. Even the heat of the hand communicates some heat to a ball of lead or any metal, if held in it a little while. So readily and so universally is heat transmitted and excited, the body remaining to all appearance unchanged.

38. Of all substances that we are acquainted with, the one which most readily receives and loses heat is air; as is best seen in calendar glasses [air thermoscopes], which are made thus [29]. Take a glass with a hollow belly, a thin and oblong neck; turn it upside down and lower it, with the mouth downwards and

[28] The only explanation of this is, that the focal length of the lens lay between a span and half a span.

[29] I am very much inclined to think that Bacon heard of the vitrum calendare from Fludd, or à Fluctibus, as he is called in Latin, who returned from Italy in [1605], and

the belly upwards, into another glass vessel containing water ; and let the mouth of the inserted vessel touch the bottom of the receiving vessel, and its neck lean slightly against the mouth of the other, so that it can stand. And that this may be done more conveniently, apply a little wax to the mouth of the receiving glass, but not so as to seal its mouth quite up ; in order that the motion, of which we are going to speak, and which is very facile and delicate, may not be impeded by want of a supply of air.

The lowered glass, before being inserted into the other, must be heated before a fire in its upper part, that is its belly. Now when it is placed in the position I have described, the air which was dilated by the heat will, after a lapse of time sufficient to allow for the extinction of that adventitious heat, withdraw and contract itself to the same extension or dimension as that of the surrounding air at the time of the immersion of the glass ; and will draw the water upwards to a corresponding height. To the side of the glass there should be affixed a strip of paper, narrow and oblong, and marked with as many degrees as you choose. You will then see, according as the day is warm or cold, that the air contracts under the action of cold, and expands under the action of heat ; as will be seen by the water rising when the air contracts, and sinking when it dilates. But the air's sense of heat and cold is so subtle and exquisite as far to exceed the perception of the human touch, insomuch that a ray of sunshine, or the heat of the breath, much more the heat of one's hand placed on the top of the glass, will cause the water immediately to sink in a perceptible degree [30]. And yet I think that animal spirits have a sense of heat and cold more exquisite still, were it not that it is impeded and deadened by the grossness of the body.

39. Next to air, I take those bodies to be most sensitive of heat which have been recently changed and compressed by cold, as snow and ice ; for they begin to dissolve and melt with any gentle heat. Next to them, perhaps, comes quicksilver. After that follow greasy substances, as oil, butter, and the like ; then comes wood ; then water ; and lastly stones and metals, which are slow to heat, especially in the inside. These, however, when once they have acquired heat retain it very long ; in so much that an ignited brick, stone, or piece of iron, when plunged into a basin of water, will remain for a quarter of an hour, or thereabouts, so hot that you cannot touch it.

40. The less the mass of a body, the sooner is it heated by the approach of a

in whose philosophy, built upon certain abstract notions of rarefaction and condensation, perpetual reference is made to the air-thermometer, to which he gives the same name.

[30] In consequence of this description of the Vitrum Calendare, the invention of the Thermometer has been ascribed to Bacon ; but without good reason. Fludd was the first to publish an account of the Thermometer ; but Nelli says, and (admitting his authorities) truly, that Galileo's invention was anterior to any publication of Fludd's. Nelli speaks of a letter preserved in the library of his family "in copiâ", which Castelli addressed to Cesarina in 1638. Castelli says that, more than thirty-five years before, Galileo had shown him an experiment which he describes ; namely, the rise of the water into an inverted tube with a bulb at one extremity, when the open end of the tube is put into a vessel of water, and goes on, " del quale effetto il medesimo Signor Galileo si era servito per fabbricare un Istromento da esaminare i gradi del caldo e del freddo". Thus far Castelli ; but how long after the original experiment the instrument was made, does not appear from his statement. Nelli also refers to Viviani's Life of Galileo, wherein it is said that Galileo invented the Thermometer between 1593 and 1597. It has not, I think, been remarked that the rise of water under the circumstances of Galileo's original experiment had already been described in Porta's *Natural Magic ;* though, as is usually the case with Porta, one cannot be sure whether he had ever actually seen it. " Possumus etiam solo calore aquam ascendere facere. Sit dolium supra turrim, vel ligneum, vel argillaceum aut æreum, quod melius erit, et canalem habeat in medio, qui descendat inferius usque ad aquam, et in eâ submersus sit, sed adglutinatus, ne respiret. Calefiat vas superius vel sole vel igne, nam aër, qui in alvo continetur, rarefit et foras prolabitur, unde aquam in bullas tumere videbimus, mox absentiâ solis ubi vas refrigescit, aër condensatur, et quum non sufficiat inclusus aër vacuum replere, accersitur aqua et ascendit supra."—*Porta's Magic*, book xix. chap. 4.

hot body; which shows that all heat of which we have experience is in some sort opposed to tangible matter.

41. Heat, as far as regards the sense and touch of man, is a thing various and relative; insomuch that tepid water feels hot if the hand be cold, but cold if the hand be hot.

XIV.

How poor we are in history any one may see from the foregoing tables; where I not only insert sometimes mere traditions and reports (though never without a note of doubtful credit and authority) in place of history proved and instances certain, but am also frequently forced to use the words "Let trial be made", or "Let it be further inquired".

XV.

The work and office of these three tables I call the Presentation of Instances to the Understanding. Which presentation having been made, Induction itself must be set at work; for the problem is, upon a review of the instances, all and each, to find such a nature as is always present or absent with the given nature, and always increases and decreases with it; and which is, as I have said, a particular case of a more general nature. Now if the mind attempt this affirmatively from the first, as when left to itself it is always wont to do, the result will be fancies and guesses and notions ill defined, and axioms that must be mended every day; unless like the schoolmen we have a mind to fight for what is false; though doubtless these will be better or worse according to the faculties and strength of the understanding which is at work. To God, truly, the Giver and Architect of Forms, and it may be to the angels and higher intelligences, it belongs to have an affirmative knowledge of forms immediately, and from the first contemplation[31]. But this assuredly is more than man can do, to whom it is granted only to proceed at first by negatives, and at last to end in affirmatives, after exclusion has been exhausted.

XVI.

We must make therefore a complete solution and separation of nature, not indeed by fire, but by the mind, which is a kind of divine fire. The first work therefore of true induction (as far as regards the discovery of Forms) is the rejection or exclusion of the several natures which are not found in some instance where the given nature is present, or are found in some instance where the given nature is absent, or are found to increase in some instance when the given nature decreases, or to decrease when the given nature increases. Then indeed after the rejection and exclusion has been duly made, there will remain at the bottom, all light opinions vanishing into smoke, a Form affirmative, solid and true and well defined. This is quickly said; but the way to come at it is winding and intricate. I will endeavour however not to overlook any of the points which may help us towards it.

XVII.

But when I assign so prominent a part to Forms, I cannot too often warn and admonish men against applying what I say to those forms to which their thoughts and contemplations have hitherto been accustomed.

For in the first place I do not at present speak of Compound Forms, which are, as I have remarked, combinations of simple natures according to the common course of the universe; as of the lion, eagle, rose, gold, and the like. It will be time to treat of these when we come to the Latent Processes and Latent Configurations, and the discovery of them, as they are found in what are called substances or natures concrete.

And even in the case of simple natures I would not be understood to speak of abstract forms and ideas, either not defined in matter at all, or ill defined.

[31] It was, I apprehend, the received doctrine that whatever knowledge the angelic nature is capable of it attains at once. See St. Thomas Aquinas, *Summa Theol.* Ima., q. 45, a. 2.

For when I speak of Forms, I mean nothing more than those laws and determinations of absolute actuality, which govern and constitute any simple nature, as heat, light, weight, in every kind of matter and subject that is susceptible of them. Thus the Form of Heat or the Form of Light is the same thing as the Law of Heat or the Law of Light. Nor indeed do I ever allow myself to be drawn away from things themselves and the operative part. And therefore when I say (for instance) in the investigation of the form of heat, "reject rarity," or "rarity does not belong to the form of heat", it is the same as if I said, "It is possible to superinduce heat on a dense body"; or, "It is possible to take away or keep out heat from a rare body".

But if any one conceive that my Forms too are of a somewhat abstract nature, because they mix and combine things heterogeneous (for the heat of heavenly bodies and the heat of fire seem to be very heterogeneous; so do the fixed red of the rose or the like, and the apparent red in the rainbow, the opal, or the diamond; so again do the different kinds of death; death by drowning, by hanging, by stabbing, by apoplexy, by atrophy; and yet they agree severally in the nature of heat, redness, death); if any one, I say, be of this opinion, he may be assured that his mind is held in captivity by custom, by the gross appearance of things, and by men's opinions [32]. For it is most certain that these things, however heterogeneous and alien from each other, agree in the Form or Law which governs heat, redness and death; and that the power of man cannot possibly be emancipated and freed from the common course of nature, and expanded and exalted to new efficients and new modes of operation, except by the revelation and discovery of Forms of this kind. And yet, when I have spoken of this union of nature, which is the point of most importance, I shall proceed to the divisions and veins of nature, as well the ordinary as those that are more inward and exact, and speak of them in their place.

XVIII.

I must now give an example of the Exclusion or Rejection of natures which by the Tables of Presentation are found not to belong to the Form of Heat; observing in the meantime that not only each table suffices for the rejection of any nature, but even any one of the particular instances contained in any of the tables. For it is manifest from what has been said that any one contradictory instance overthrows a conjecture as to the Form. But nevertheless for clearness' sake and that the use of the tables may be more plainly shown, I sometimes double or multiply an exclusion.

An Example of Exclusion, or Rejection of Natures from the Form of Heat.

1. On account of the rays of the sun, reject the nature of the elements [33].
2. On account of common fire, and chiefly subterraneous fires (which are the most remote and most completely separate from the rays of heavenly bodies), reject the nature of heavenly bodies.
3. On account of the warmth acquired by all kinds of bodies (minerals, vegetables, skin of animals, water, oil, air, and the rest) by mere approach to a fire, or other hot body, reject the distinctive or more subtle texture of bodies.

[32] The objection here anticipated has actually been made. It has been said that we cannot be sure that any quality always proceeds from the same cause. And in truth, though the axiom "like causes produce like effects", and vice versâ, seems to be inseparable from the idea of causation, yet the force of the objection remains. For the reference of sensible qualities to outward objects involves a subjective element. The same colour, as referred to a substance as the object in which it resides, is a different thing as it is a fixed colour, or prismatic, or epipolar, etc. They agree, it may be said, in the type of undulation; but viewed as properties of bodies, or with reference to operations on them, they are distinct. And if we could go further into the mechanism of sensation, we should probably recede further both from concrete bodies and from practice.

[33] This refers to the antithesis, almost fundamental in Peripatetic physics, of the celestial and the elementary. Heat, since the sun's rays are hot, cannot depend on elemental as contradistinguished from the celestial nature.

4. On account of ignited iron and other metals, which communicate heat to other bodies and yet lose none of their weight or substance, reject the communication or admixture of the substance of another hot body.

5. On account of boiling water and air, and also on account of metals and other solids that receive heat but not to ignition or red heat, reject light or brightness.

6. On account of the rays of the moon and other heavenly bodies, with the exception of the sun, also reject light and brightness.

7. By a comparison of ignited iron and the flame of spirit of wine (of which ignited iron has more heat and less brightness, while the flame of spirit of wine has more brightness and less heat), also reject light and brightness.

8. On account of ignited gold and other metals, which are of the greatest density as a whole, reject rarity.

9. On account of air, which is found for the most part cold and yet remains rare, also reject rarity.

10. On account of ignited iron, which does not swell in bulk, but keeps within the same visible dimensions [34], reject local or expansive motion of the body as a whole.

11. On account of the dilation of air in calendar glasses and the like, wherein the air evidently moves locally and expansively and yet acquires no manifest increase of heat, also reject local or expansive motion of the body as a whole.

12. On account of the ease with which all bodies are heated, without any destruction or observable alteration, reject a destructive nature, or the violent communication of any new nature.

13. On account of the agreement and conformity of the similar effects which are wrought by heat and cold, reject motion of the body as a whole, whether expansive or contractive.

14. On account of heat being kindled by the attrition of bodies, reject a principial nature. By principial nature I mean that which exists in the nature of things positively, and not as the effect of any antecedent nature [35].

There are other natures beside these; for these tables are not perfect, but meant only for examples.

All and each of the above mentioned natures do not belong to the Form of Heat. And from all of them man is freed in his operations on Heat.

XIX.

In the process of exclusion are laid the foundations of true Induction, which however is not completed till it arrives at an Affirmative. Nor is the exclusive part itself at all complete, nor indeed can it possibly be so at first. For exclusion is evidently the rejection of simple natures; as if we do not yet possess sound and true notions of simple natures, how can the process of exclusion be made accurate? Now some of the above-mentioned notions (as that of the nature of the elements, of the nature of heavenly bodies, or rarity) are vague and ill-defined. I therefore, well knowing and nowise forgetting how great a work I am about (viz. that of rendering the human understanding a match for things and nature),

[34] [This is of course a blunder.—Ed.]

[35] Bacon here anticipates not merely the essential character of the most recent theory of heat, but also the kind of evidence by which it has been established. The proof that caloric does not exist,—in other words that heat is not the manifestation of a peculiar substance diffused through nature,—rests mainly on experiments of friction.

Mr. Joule and Professor Thomson ascribe the discovery of this proof chiefly to Sir Humphrey Davy (see Beddoes's *Contributions to Physical and Medical Knowledge*, p. 14): but though Davy's experiments guard against sources of error of which Bacon takes no notice, the merit of having perceived the true significance of the production of heat by friction belongs of right to Bacon.

It is curious that in the essay in which he opposes the doctrine of caloric, Davy endeavours to introduce a new error of the same kind, and to show that light really is a natura principialis,' a peculiar substance which in combination with oxygen properly so called constitutes oxygen gas, which he accordingly calls phosoxygen.

do not rest satisfied with the precepts I have laid down; but proceed further to devise and supply more powerful aids for the use of the understanding; which I shall now subjoin. And assuredly in the Interpretation of Nature the mind should by all means be so prepared and disposed, that while it rests and finds footing in due stages and degrees of certainty, it may remember withal (especially at the beginning) that what it has before it depends in great measure upon what remains behind.

XX.

And yet since truth will sooner come out from error than from confusion, I think it expedient that the understanding should have permission, after the three Tables of First Presentation (such as I have exhibited) have been made and weighed, to make an essay of the Interpretation of Nature in the affirmative way; on the strength both of the instances given in the tables, and of any others it may meet with elsewhere. Which kind of essay I call the *Indulgence of the Understanding*, or the *Commencement of Interpretation*, or the *First Vintage*.

First Vintage concerning the Form of Heat.

It is to be observed that the Form of a thing is to be found (as plainly appears from what has been said) in each and all the instances in which the thing itself is to be found; otherwise it would not be the Form. It follows therefore that there can be no contradictory instance. At the same time the Form is found much more conspicuous and evident in some instances than in others; namely in those wherein the nature of the Form is less restrained and obstructed and kept within bounds by other natures. Instances of this kind I call Shining or Striking Instances. Let us now therefore proceed to the First Vintage concerning the Form of Heat.

> From a survey of the instances, all and each, the nature of which Heat is a particular case appears to be Motion. This is displayed most conspicuously in flame, which is always in motion, and in boiling or simmering liquids, which also are in perpetual motion. It is also shown in the excitement or increase of heat caused by motion, as in bellows and blasts; on which see Tab. 3, Inst. 29; and again in other kinds of motion, on which see Tab. 3, Inst. 28 and 31. Again it is shown in the extinction of fire and heat by any strong compression, which checks and stops the motion; on which see Tab. 3, Inst. 30 and 32. It is shown also by this, that all bodies are destroyed, or at any rate notably altered by all strong and vehement fire and heat; whence it is quite clear that heat causes a tumult and confusion and violent motion in the internal parts of a body which perceptibly tends to its dissolution.

When I say of Motion that it is as the genus of which heat is a species, I would be understood to mean, not that heat generates motion or that motion generates heat (though both are true in certain cases), but that Heat itself, its essence and quiddity, is Motion and nothing else; limited however by the specific difference which I will presently subjoin, as soon as I have added a few cautions for the sake of avoiding ambiguity.

Sensible heat is a relative notion, and has relation to man, not to the universe; and is correctly defined as merely the effect of heat on the animal spirits. Moreover, in itself it is variable, since the same body, according as the senses are predisposed, induces a perception of cold as well as of heat. This is clear from Inst. 41, Tab. 3.

Nor again must the communication of Heat, or its transitive nature, by means of which a body becomes hot when a hot body is applied to it, be confounded with the Form of Heat. For heat is one thing, heating another. Heat is produced by the motion of attrition without any preceding heat, an instance which excludes heating from the Form of Heat. And even when heat is produced by the approach of a hot body, this does not proceed from the Form of Heat, but depends entirely on a higher and more general nature, viz. on the nature of assimilation or self-multiplication, a subject which requires a separate inquiry.

Again, our notion of fire is popular, and of no use ; being made up of the combination in any body of heat and brightness, as in common flame and bodies heated to redness.

Having thus removed all ambiguity, I come at length to the true specific differences which limit Motion and constitute it the Form of Heat.

The first difference then is this. Heat is an expansive motion, whereby a body strives to dilate and stretch itself to a larger sphere or dimension than it had previously occupied. This difference is most observable in flame, where the smoke or thick vapour manifestly dilates and expands itself into flame.

It is shown also in all boiling liquid, which manifestly swells, rises, and bubbles; and carries on the process of self-expansion, till it turns into a body far more extended and dilated than the liquid itself, namely, into vapour. smoke, or air.

It appears likewise in all wood and combustibles, from which there generally arises exudation and always evaporation.

It is shown also in the melting of metals, which, being of the compactest texture, do not readily swell and dilate ; but yet their spirit being dilated in itself, and thereupon conceiving an appetite for further dilation, forces and agitates the grosser parts into a liquid state. And if the heat be greatly increased it dissolves and turns much of their substance to a volatile state.

It is shown also in iron or stones, which, though not melted or dissolved. are yet softened. This is the case also with sticks, which when slightly heated in hot ashes become flexible.

But this kind of motion is best seen in air, which continuously and manifestly dilates with a slight heat, as appears in Inst. 38, Tab. 3.

It is shown also in the opposite nature of cold. For cold contracts all bodies[36] and makes them shrink ; insomuch that in intense frosts nails fall out from walls, brazen vessels crack, and heated glass on being suddenly placed in the cold cracks and breaks. In like manner air is contracted by a slight chill, as in Inst. 38, Tab. 3. But on these points I shall speak more at length in the inquiry concerning Cold.

Nor is it surprising that heat and cold should exhibit many actions in common (for which see Inst. 32, Tab. 2), when we find two of the following specific differences (of which I shall speak presently) suiting either nature ; though in this specific difference (of which I am now speaking) their actions are diametrically opposite. For heat gives an expansive and dilating, cold a contractive and condensing motion.

The second difference is a modification of the former ; namely, that heat is a motion expansive or towards the circumference, but with this condition, that the body has at the same time a motion upwards. For there is no doubt that there are many mixed motions. For instance, an arrow or dart turns as it goes forward, and goes forward as it turns. And in like manner the motion of heat is at once a motion of expansion and a motion upwards. This difference is shown by putting a pair of tongs or a poker in the fire. If you put it in perpendicularly and hold it by the top, it soon burns your hand ; if at the side or from below, not nearly so soon.

It is also observable in distillations *per descensorium* ; which men use for delicate flowers, that soon lose their scent. For human industry has discovered the plan of placing the fire not below but above, that it may burn the less. For not only flame tends upwards, but also all heat.

But let trial be made of this in the opposite nature of cold ; viz. whether cold does not contract a body downwards, as heat dilates a body upwards. Take therefore two iron rods, or two glass tubes, exactly alike ; warm them a little, and place a sponge steeped in cold water or snow at the bottom of the one, and the same at the top of the other. For I think that the extremities

[36] [This overlooks the familiar contrary case of water.—Ed.]

of the rod which has the snow at the top will cool sooner than the extremities of the other which has the snow at the bottom ; just as the opposite is the case with heat.

The third specific difference is this ; that heat is a motion of expansion not uniformly of the whole body together, but in the smaller parts of it ; and at the same time checked, repelled, and beaten back, so that the body acquires a motion alternative, perpetually quivering, striving and struggling, and irritated by repercussion, whence springs the fury of fire and heat.

This specific difference is most displayed in flame and boiling liquids, which are perpetually quivering and swelling in small portions, and again subsiding.

It is also shown in those bodies which are so compact that when heated or ignited they do not swell or expand in bulk ; as ignited iron, in which the heat is very sharp.

It is shown also in this, that a fire burns most briskly in the coldest weather.

Again, it is shown in this, that when the air is extended in a calendar glass without impediment or repulsion,—that is to say, uniformly and equably—there is no perceptible heat. Also when wind escapes from confinement, although it burst forth with the greatest violence, there is no very great heat perceptible ; because the motion is of the whole, without a motion alternating in the particles. And with a view to this, let trial be made whether flame does not burn more sharply towards the sides than in the middle of the flame.

It is also shown in this, that all burning acts on minute pores of the body burnt ; so that burning undermines, penetrates, pricks, and stings the body like the points of an infinite number of needles. It is also an effect of this, that all strong waters (if suited to the body on which they are acting) act as fire does in consequence of their corroding and pungent nature.

And this specific difference (of which I am now speaking) is common also to the nature of cold ; for in cold the contractive motion is checked by a resisting tendency to expand, just as in heat the expansive motion is checked by a resisting tendency to contract. Thus, whether the particles of a body work inward or outward, the mode of action is the same, though the degree of strength be very different ; because we have not here on the surface of the earth anything that is intensely cold. See Inst. 27, Tab. 9.

The fourth specific difference is a modification of the last ; it is, that the preceding motion of stimulation or penetration must be somewhat rapid and not sluggish, and must proceed by particles, minute indeed, yet not the finest of all, but a degree larger.

This difference is shown by a comparison of the effects of fire with the effects of time or age. Age or time dries, consumes, undermines and reduces to ashes, no less than fire ; indeed with an action far more subtle ; but because such motion is very sluggish, and acts on particles very small, the heat is not perceived.

It is also shown by comparing the dissolution of iron and gold. Gold is dissolved without any heat being excited, while the dissolution of iron is accompanied by a violent heat, though it takes place in about the same time. The reason is that in gold the separating acid enters gently and works with subtlety, and the parts of the gold yield easily ; whereas in iron the entrance is rough and with conflict, and the parts of the iron have greater obstinacy.

It is shown also to some degree in some gangrenes and mortifications, which do not excite great heat or pain on account of the subtle nature of putrefaction.

Let this then be the First Vintage or Commencement of Interpretation concerning the Form of Heat, made by way of indulgence to the understanding.

Now from this our First Vintage it follows that the Form or true definition of heat (heat, that is, in relation to the universe, not simply in relation to man) is in few words as follows : *Heat is a motion, expansive, restrained, and acting in its strife upon the smaller particles of bodies.* But the expansion is thus modified ; *while it expands all ways, it has at the same time an inclination upwards.* And the struggle in the particles is modified also ; *it is not sluggish, but hurried and with violence* [37].

[37] The Inquisition into the form of heat suggests these remarks :—

1st. A great part of it conduces in no way to the result. This may be said to be the natural consequence of the method of inquiry.

2nd. Heat (caloric) is confounded with the effects of chemical agencies, which are said " exequi opera caloris ".

3rd. A greater source of confusion is the complete absence of any recognition of the principle that all bodies tend to acquire the temperature of those about them, and that the difference ad tactum which makes one body feel hotter or colder than another depends not on its being hotter or colder, but on the different degree of facility which they have in communicating their own respective temperature. In consequence of this, it had always been taught that one class of bodies were in their own nature cold, another hot, and so on. All liquids were cold. Experiments with a thermometer would have shown that they were not ; but these Bacon did not try,—an instance among others how far he was from rejecting all he had been taught.

Of which remarks we may observe that, of the " Instantiæ convenientes ", 13 is an instance of the third, while from 22 to the end exemplify the second ;—of the " Instantiæ in proximo ", 14–19 are to be referred to the third ; from 27 to the end, to the second.

4th. Calidum and Frigidum seem to be considered distinct and not correlative qualities.

5th. The adoption of astrological fables about the hot and cold influence of the stars and planets [is to be remarked in the *Tabula Graduum*, 15 et seqq.]

Then comes the result, that the natura calidi is a motus expansivus. This is seen [in air], " Optime cernitur in aëre qui per exiguum calorem se dilatat continuo et manifesto, ut per Inst. 38 Tab. 3 " : that is, by the instance of a vitrum calendare, or air-thermometer. And this is beyond question a good instance. But then in the " exemplum exclusivæ ", § 11, we read " Per dilatationem aëris in vitris calendariis et similibus, qui movetur localiter et expansive manifesto, neque tamen colligit manifestum augmentum caloris, rejice etiam motum localem aut expansivum secundum totum ". How is this passage to be reconciled with the preceding ? For if the example of the vitrum calendare proves anything, it proves a motus expansivus secundum totum ; and if on account of our having no manifest evidence that the air waxes hot when it expands, the example does not prove this, why is it adduced ? The source of this confusion I believe to be that, though Bacon saw reason to affirm expansion to be the essence of the hot, yet he was perplexed by examples of two kinds ; (*a*) bodies which do not visibly expand when they are heated, e.g. red hot iron ; (*b*) bodies which expand without becoming heated, e.g. compressed air when relieved from pressure. For the first difficulty, it might have occurred to him that the hot iron does expand, though not enough to be perceived (except by accurate measurement) to do so ; and if he had followed the indication thus given, he might have been the discoverer of a general and most important law. The difficulty which the second class of phenomena creates ought to have prevented Bacon from assigning expansion as the forma calidi,—as being that which must always make a body hot, and without which it could not become so. For it would be too liberal an interpretation to say that the expressions " motus cohibitus et refrænatus ", whereby the idea of expansion is qualified, refer to a condition essential in the case of elastic fluids,—namely that the expansion in becoming heated is due to an increased elasticity, and not to any decrease of external pressure. Even had the modification required by this class of cases been introduced, there still remains that of liquids whose temperature is below that of maximum density, which is altogether intractable. Of this phenomenon, however, it would be unreasonable to expect Bacon to have known anything. But setting it aside, if it were affirmed that Bacon, after having had a glimpse of the truth suggested by some obvious phenomena, had then recourse, as he himself expresses it, to certain " differentiæ inanes " in order to save the phenomena, I think it would be hard to dispute the truth of this censure.

Nevertheless, of the matters contained in the investigation, there are several of considerable interest, though, as has been said, they are not connected with the final result,

Viewed with reference to operation it is the same thing. For the direction is this : *If in any natural body you can excite a dilating or expanding motion, and can so repress this motion and turn it back upon itself, that the dilation shall not proceed equably, but have its way in one part and be counteracted in another, you will undoubtedly generate heat ;* without taking into account whether the body be elementary (as it is called) or subject to celestial influence ; whether it be luminous or opaque ; rare or dense ; locally expanded or confined within the bounds of its first dimension ; verging to dissolution or remaining in its original state ; animal, vegetable, or mineral, water, oil or air, or any other substance whatever susceptible of the above-mentioned motion. Sensible heat is the same thing ; only it must be considered with reference to the sense. Let us now proceed to further aids.

XXI.

The Tables of First Presentation and the Rejection or process of Exclusion being completed, and also the First Vintage being made thereupon, we are to proceed to the other helps of the understanding in the Interpretation of Nature and true and perfect Induction. In propounding which, I mean, when Tables are necessary, to proceed upon the Instances of Heat and Cold ; but when a smaller number of examples will suffice, I shall proceed at large ; so that the inquiry may be kept clear, and yet more room be left for the exposition of the system.

I propose to treat then in the first place of *Prerogative instances* [38] ; secondly, of the *Supports of Induction ;* thirdly, of the *Rectification of Induction ;* fourthly, of *Varying the Investigation according to the nature of the Subject ;* fifthly, of *Prerogative Natures* with respect to Investigation, or of what should be inquired first and what last ; sixthly, of the *Limits of Investigation*, or a Synopsis of all Natures in the Universe ; seventhly, of the *Application to Practice*, or of things in their relation to Man ; eighthly, of *Preparations for Investigation ;* and lastly, of the *Ascending and Descending Scale of Axioms*.

XXII.

Among Prerogative Instances I will place first *Solitary Instances*. Those are Solitary Instances which exhibit the nature under investigation in subjects which have nothing in common with other subjects except that nature ; or, again, which do not exhibit the nature under investigation in subjects which resemble other subjects in every respect except in not having that nature. For it is clear that such instances make the way short, and accelerate and strengthen the process of exclusion ; so that a few of them are as good as many.

For instance, if we are inquiring into the nature of Colour, prisms, crystals, which show colours not only in themselves but externally on a wall, dews, etc., are Solitary Instances. For they have nothing in common with the colours fixed in flowers, coloured stones, metals, woods, etc., except the colour. From which we easily gather that colour is nothing more than a modification of the image of light received upon the object [39], resulting in the former case from the different

The relation between heat and mechanical action has recently become the subject of some very remarkable speculations, derived from the views suggested by S. Carnot in his *Reflections sur la Puissance Motrice du Feu*. Two views have been propounded. In one (that of S. Carnot himself), mechanical action is regarded as convertible with the transference from body to body of caloric. The other rejects the notion of caloric (the substance of heat) altogether. On this view mechanical action is convertible with the generation of heat, i.e. the raising of a given quantity of a given body from one given temperature to another. Both make use of the axiom " ex nihilo nihil " : and the conclusions thus obtained, especially in the second way of considering the subject, which I cannot doubt is the true one, are most remarkable, and the more interesting because they are, so to speak, the interpretation of a maxim whose truth is admitted à priori.

[38] [As to the meaning of this term see above, p. 25-26.—ED.]

[39] Reference is made to Telesius's system of vision. " Lux donata est facultate sese effundendi multiplicandique et aërem propriâ specie afficiendi, itaque et oculos subę-

degrees of incidence, in the latter from the various textures and configurations of the body. These instances are Solitary in respect of resemblance.

Again, in the same investigation, the distinct veins of white and black in marble, and the variegation of colour in flowers of the same species, are Solitary Instances. For the black and white streaks in marble, or the spots of pink and white in a pink[40], agree in everything almost except the colour. From which we easily gather that colour has little to do with the intrinsic nature of a body, but simply depends on the coarser and as it were mechanical arrangement of the parts. These instances are Solitary in respect of difference. Both kinds I call *Solitary Instances*, or *Ferine*,[41] to borrow a term from astronomers.

XXIII.

Among Prerogative Instances I will next place *Migratory Instances.* They are those in which the nature in question is in the process of being produced when it did not previously exist, or on the other hand of disappearing when it existed before. And therefore, in either transition, such instances are always twofold, or rather it is one instance in motion or passage, continued till it reaches the opposite state. Such instances not only accelerate and strengthen the exclusive process, but also drive the affirmative or Form itself into a narrow compass. For the Form of a thing must necessarily be something which in the course of this migration is communicated, or on the other hand which in the course of this migration is removed and destroyed. And though every exclusive promotes the affirmative, yet this is done more decidedly when it occurs in the same than in different subjects. And the betrayal of the form in a single instance leads the way (as is evident from all that has been said) to the discovery of it in all. And the simpler the Migration, the more must the instance be valued. Besides Migratory Instances are of great use with a view to operation; because in exhibiting the form in connexion with that which causes it to be or not to be, they supply a clear direction for practice in some cases; whence the passage is easy to the cases that lie next. There is however in these instances a danger which requires caution; viz. lest they lead us to connect the Form too much with the efficient, and so possess the understanding, or at least touch it, with a false opinion concerning the Form, drawn from a view of the efficient. But the efficient is always understood to be merely the vehicle that carries the Form. This is a danger however easily remedied by the process of exclusion legitimately conducted.

I must now give an example of a Migratory Instance. Let the nature to be investigated be Whiteness; an instance migrating to production or existence is glass whole and pounded. Again, simple water and water agitated into froth. For glass and water in their simple state are transparent, not white; whereas pounded glass and water in froth are white, not transparent. We must therefore inquire what has happened to the glass or water from this Migration. For it is obvious that the Form of Whiteness is communicated and conveyed by that pounding of the glass and that agitation of the water. We find, however, that nothing has been added except the breaking up of the glass and water into small parts, and the introduction of air. But we have made no slight advance to the

undi." . . Again, "lux quæ res quibus insunt [colores] permeat . . . ab ipsarum intingitur coloribus, et eas transvecta oculos subit."—*De Rerum Nat.* vii. 31. See also other passages of the same book. Bacon uses "imago" as equivalent to "species", the word used in the preceding quotation.

40 [Orig. *Garophylli.*] Caryophyllea was a flower much cultivated in Holland in the sixteenth century; see Lemmius, *De Miraculis* (1581), p. 107. (The description seems more applicable to the tulip.) The flowers meant are pinks and carnations.

41 I believe the word which Bacon here employs is at least very much less used than another of perhaps the same origin for which he has perhaps accidentally substituted it. "Feralis," we read in the *Lexicon Mathematicum* of Vitalis (1668), which appears to give a tolerably complete vocabulary of astrological words, "apud astronomos dicitur planeta, quando fuerit in loco ubi nullam cum reliquis familiaritatem habet: quod quidem maximum est detrimentum," etc.

discovery of the Form of Whiteness when we know that two bodies, both transparent but in a greater or less degree (viz. air and water, or air and glass), do when mingled in small portions together exhibit whiteness, through the unequal refraction of the rays of light [42].

But an example must at the same time be given of the danger and caution to which I alluded. For at this point it might readily suggest itself to an understanding led astray by efficient causes of this kind, that air is always required for the Form of Whiteness, or that Whitness is generated by transparent bodies only ; notions entirely false, and refuted by numerous exclusions. Whereas it will be found that (setting air and the like aside) bodies entirely even in the particles which affect vision are transparent, bodies simply uneven are white ; bodies uneven and in a compound yet regular texture are all colours except black ; while bodies uneven and in a compound, irregular, and confused texture are black [43]. Here then I have given an example of an Instance Migrating to production or existence in the proposed nature of Whiteness. An Instance Migrating to destruction in the same nature of Whiteness, is froth or snow in dissolution. For the water puts off Whiteness and puts on transparency, on returning to its integral state without air.

Nor must I by any means omit to mention that under Migratory Instances are to be included not only those which are passing towards production and destruction, but also those which are passing towards increase and decrease ; since these also help to discover the Form, as is clear from the above definition of Form and the Table of Degrees. Thus paper, which is white when dry, but when wetted (that is, when air is excluded and water introduced) is less white and approaches nearer to the transparent, is analogous to the above given Instances.

XXIV.

Among Prerogative Instances I will put in the third place *Striking Instances*, of which I have made mention in the First Vintage concerning Heat, and which I also call *Shining Instances*, or *Instances Freed and Predominant*. They are those which exhibit the nature in question naked and standing by itself, and also in its exaltation or highest degree of power ; as being disenthralled and freed from all impediments, or at any rate by virtue of its strength dominant over, suppressing and coercing them. For since every body contains in itself many forms of natures united together in a concrete state, the result is that they severally crush, depress, break, and enthrall one another, and thus the individual forms are obscured. But certain subjects are found wherein the required nature appears more in its vigour than in others, either through the absence of impediments or the predominance of its own virtue. And instances of this kind strikingly display the Form. At the same time in these instances also we must use caution, and check the hurry of the understanding. For whatever displays the Form too conspicuously, and seems to force it on the notice of the understanding, should be held suspect, and recourse be had to a rigid and careful exclusion.

To take an example ; let the nature inquired into be Heat. A Striking Instance of the motion of expansion, which (as stated above) is the main elèment in the Form of Heat, is a calendar glass of air. For flame, though it manifestly exhibits expansion, still, as susceptible of momentary extinction, does not display the progress of expansion. Boiling water too, on account of the easy transition of water to vapour or air, does not so well exhibit the expansion of water in its own body. Again, ignited iron and like bodies are so far from displaying the progress of expansion, that in consequence of their spirit being crushed and broken by the coarse and compact particles which curb and subdue

[42] Bacon would perhaps have given as another illustration of what he has here said the beautiful whiteness of frosted silver, if he had been aware that it is in reality silver foam. It appears that when silver is in a state of fusion a very large quantity of oxygen is condensed on and within its surface, the whole of which escapes at the moment of solidification. This explanation of the appearance of granulated silver is due, I believe, to Gay Lussac.

[43] Compare *Valerius Terminus*, ch. xi.

it, the expansion itself is not at all conspicuous to the senses. But a calendar glass strikingly displays expansion in air, at once conspicuous, progressive, permanent, and without transition.

To take another example; let the nature inquired into be Weight. A Striking Instance of weight is quicksilver. For it far surpasses in weight all substances but gold, and gold itself is not much heavier [44]. But quicksilver is a better instance for indicating the Form of Weight than gold; because gold is solid and consistent, characteristics which seem related to density; whereas quicksilver is liquid and teeming with spirit, and yet is heavier by many degrees than the diamond and other bodies that are esteemed the most solid. From which it is obvious that the Form of Heaviness or Weight depends simply on quantity of matter and not on compactness of frame.

XXV.

Among Prerogative Instances I will put in the fourth place *Clandestine Instances*, which I also call *Instances of the Twilight*, and which are pretty nearly the opposites of Striking Instances. For they exhibit the nature under investigation in its lowest degree of power, and as it were in its cradle and rudiments; striving indeed and making a sort of first attempts, but buried under and subdued by a contrary nature. Such instances however are of very great service for the discovery of Forms; because as Striking Instances lead easily to specific differences, so are Clandestine Instances the best guides to *genera*, that is, to those common natures whereof the natures proposed are nothing more than particular cases.

For example, let the nature proposed be Consistency, or the nature of that which determines its own figure; opposed to which is Fluidity. Those are Clandestine Instances which exhibit some feeble and low degree of consistency in a fluid; as a bubble of water, which is a sort of consistent pellicle of determined figure, made of the body of the water. Of a similar kind are the droppings from a house, which if there be water to follow, lengthen themselves out into a very thin thread, to preserve the continuity of the water; but if there be not water enough to follow, then they fall in round drops, which is the figure that best preserves the water from a solution of continuity. But at the very moment of time when the thread of water ceases and the descent in drops begins, the water itself recoils upwards to avoid discontinuation. Again in metals, which in fusion are liquid but more tenacious, the molten drops often fly to the top and stick there. A somewhat similar instance is that of children's looking-glasses, which little boys make on rushes with spittle; where also there is seen a consistent pellicle of water. This however is much better shown in that other childish sport, when they take water, made a little more tenacious by soap, and blow it through a hollow reed, and so shape the water into a sort of castle of bubbles; which by the interposition of the air become so consistent as to admit of being thrown some distance without discontinuation [45]. But best of all is it seen in frost and snow, which assume such a consistency that they can be almost cut with a knife, although they are formed out of water, both fluids. All which facts not obscurely intimate that Consistent and Fluid are only vulgar notions, and relative to the sense; and that in fact there is inherent in all bodies a disposition to shun and escape discontinuation; but that it is faint and feeble in homogeneous bodies (as fluids), more lively and strong in bodies compounded of heterogeneous matter; the reason being that the approach of heterogeneous matter binds bodies together, while the insinuation of homogeneous matter dissolves and relaxes them.

To take another instance, let the proposed nature be the attraction or coming

[44] This mistake occurs also in the *Historia Densi et Rari*. According to Bacon, the density of mercury is to that of gold as thirty-nine is to forty, nearly; the real ratio being little more than as seven to ten. The way in which his experiments were made account for a large part of this error.

[45] Far tougher bubbles than the ordinary kind may be blown in water in which silk cocoons have been steeped. Some curious experiments on this subject are mentioned in Porter on *Silk Manufactures* (Lardner's Cyclop.).

together of bodies. In the investigation of its Form the most remarkable Striking Instance is the magnet. But there is a contrary nature to the attractive; namely, the non-attractive, which exists in a similar substance. Thus there is iron, which does not attract iron, just as lead does not attract lead, nor wood wood, nor water water. Now a Clandestine Instance is a magnet armed with iron, or rather the iron in an armed magnet. For it is a fact in nature that an armed magnet at some distance off does not attract iron more powerfully than an unarmed magnet. But if the iron be brought so near as to touch the iron in the armed magnet, then the armed magnet supports a far greater weight of iron than a simple and unarmed magnet; on account of the similarity of substance between the pieces of iron; an operation altogether clandestine and latent in the iron before the magnet was applied [46]. Hence it is manifest that the Form of Coition is something which is lively and strong in the magnet, feeble and latent in iron. Again, it has been observed that small wooden arrows without an iron point, discharged from large engines, pierce deeper into wooden materials (say the sides of ships, or the like) than the same arrows tipped with iron, an account of the similarity of substance between the two pieces of wood; although this property had previously been latent in the wood. In like manner, although air does not manifestly attract air, or water water in entire bodies, yet a bubble is more easily dissolved on the approach of another bubble then if that other bubble were away, by reason of the appetite of coition between water and water, and between air and air. Such Clandestine Instances (which, as I have said, are of the most signal use) exhibit themselves most conspicuously in small and subtle portions of bodies; the reason being that larger masses follow more general forms; as shall be shown in the proper place.

XXVI.

Among Prerogative Instances I will put in the fifth place *Constitutive Instances*, which I also call *Manipular*. They are those which constitute a single species of the proposed nature a sort of Lesser Form. For since the genuine Forms (which are always convertible with the proposed natures) lie deep and are hard to find, it is required by the circumstances of the case and the infirmity of the human understanding that particular Forms, which collect together certain groups of instances (though not all) into some common notion, be not neglected, but rather be diligently observed. For whatever unites nature, though imperfectly, paves the way to the discovery of Forms. Instances therefore which are useful in this regard are of no despicable power, but have a certain prerogative.

But great caution must here be employed, lest the human understanding, after having discovered many of those particular Forms and thereupon established partitions or divisions of the nature in question, be content to rest therein, and instead of proceeding to the legitimate discovery of the great Form, take it for granted that the nature from its very roots is manifold and divided, and so reject and put aside any further union of the nature, as a thing of superfluous subtlety and verging on mere abstraction.

For example, let the proposed nature be Memory, or that which excites and aids the memory. Constitutive Instances are, order or distribution, which clearly aids the memory; also topics or "places" in artificial memory; which may either be *places* in the proper sense of the word, as a door, angle, window, and the like; or familiar and known persons; or any other things at pleasure (provided they be placed in a certain order), as animals, vegetables; words too, letters, characters, historical persons, and the like; although some of these are more suitable and convenient than others. Such artificial places help the memory wonderfully, and exalt it far above its natural powers. Again, verse is learnt and remembered more easily than prose. From this group of three instances, viz. order, artificial places, and verse, one species of aid to the memory

[46] This explanation of the effect of arming a magnet is wholly unsatisfactory. Before the *Novum Organum* was published, Galileo had shown that the armature acts by producing a more perfect contact. See the *Dialogi dei Sistemi massimi*, Giorn. 3^a, p. 140. I quote from the new edition, Firenze, 1842.

is constituted. And this species may with propriety be called the cutting off of infinity. For when we try to recollect or call a thing to mind, if we have no prenotion or perception of what we are seeking, we seek and toil and wander here and there, as if in infinite space. Whereas if we have any sure prenotion, infinity is at once cut off, and the memory has not so far to range. Now in the three foregoing instances the prenotion is clear and certain. In the first it must be something which suits the order ; in the second it must be an image which bears some relation or conformity to the places fixed ; in the third, it must be words that fall into the verse ; and thus infinity is cut off. Other instances, again, will give us this second species ; that whatever brings the intellectual conception into contact with the sense (which is indeed the method most used in mnemonics) assists the memory. Other instances will give us this third species ; that things which make their impression by way of a strong affection, as by inspiring fear, admiration, shame, delight, assist the memory. Other instances will give us this fourth species ; that things which are chiefly imprinted when the mind is clear and not occupied with anything else either before or after, as what is learnt in childhood, or what we think of before going to sleep, also things that happen for the first time, dwell longest in the memory. Other instances will give us this fifth species ; that a multitude of circumstances or points to take hold of aids the memory ; as writing with breaks and divisions, reading or reciting aloud. Lastly other instances will give us this sixth species ; that things which are waited for and raise the attention dwell longer in the memory than what flies quickly by. Thus, if you read anything over twenty times, you will not learn it by heart so easily as if you were to read it only ten, trying to repeat it between whiles, and when memory failed looking at the book. It appears then that there are six Lesser Forms of aids to the memory ; viz. the cutting off of infinity ; the reduction of the intellectual to the sensible ; impression made on the mind in a state of strong emotion ; impression made on the mind disengaged ; multitude of points to take hold of ; expectation beforehand.

To take another example, let the proposed nature be Taste or Tasting. The following instances are Constitutive. Persons who are by nature without the sense of smell cannot perceive or distinguish by taste food that is rancid or putrid, nor food that is seasoned with garlic, or with roses, or the like. Again, persons whose nostrils are accidentally obstructed by a catarrh cannot distinguish or perceive anything putrid or rancid or sprinkled with rosewater. Again, persons thus affected with catarrh if while they have something fetid or perfumed in their mouth or palate they blow their nose violently, immediately perceive the rancidity or the perfume. These instances then will give and constitute this species, or rather division, of taste ; that the sense of taste is in part nothing else than an internal smell, passing and descending from the upper passages of the nose to the mouth and palate. On the other hand the tastes of salt, sweet, sour, acid, rough, bitter, and the like, are as perceptible to those in whom the sense of smell is wanting or stopped as to any one else ; so that it is clear that the sense of taste is a sort of compound of an internal smell and a delicate power of touch ; of which this is not the place to speak.

To take another example, let the proposed nature be the communication of quality without admixture of substance. The instance of light will give or constitute one species of communication ; heat and the magnet another. For the communication of light is momentaneous, and ceases at once on the removal of the original light. But heat and the virtue of the magnet, after they have been transmitted to or rather excited in a body, lodge and remain there for a considerable time after the removal of the source of motion.

Very great in short is the prerogative of Constitutive Instances ; for they are of much use in the forming of definitions (especially particular definitions) and in the division and partition of natures ; with regard to which it was not ill said by Plato, " That he is to be held as a god who knows well how to define and to divide " [47].

[47] In the *Phaedrus*, 266 B. [Noted by Dean Kitchin in his ed. of the *Novum Organum*.]

XXVII.

Among Prerogative Instances I will put in the sixth place *Instances Conformable*, or *of Analogy* ; which I also call *Parallels*, or *Physical Resemblances*. They are those which represent the resemblances and conjugations of things, not in Lesser Forms (as Constitutive Instances do) but merely in the concrete. Hence they may be called the first and lowest steps toward the union of nature. Nor do they constitute any axiom immediately from the beginning, but simply point out and mark a certain agreement in bodies. But although they are of little use for the discovery of forms, they nevertheless are very serviceable in revealing the fabric of the parts of the universe, and anatomising its members ; from which they often lead us along to sublime and noble axioms, especially those which relate to the configuration of the world rather than to simple forms and natures.

For example, these following are instances of Conformity ; a looking-glass and the eye ; and again, the construction of the ear and places returning an echo. From which conformity, to say nothing of the mere observation of the resemblance which is in many respects useful, it is easy to gather and form this axiom, —that the organs of the senses, and bodies which produce reflexions to the senses, are of a like nature. Again, upon this hint the understanding easily rises to a higher and nobler axiom, which is this : that there is no difference between the consents or sympathies of bodies endowed with sensation and those of inanimate bodies without sensation, except that in the former an animal spirit is added to the body so disposed, but is wanting in the latter. Whence it follows that there might be as many senses in animals as there are sympathies between inanimate bodies, if there were perforations in the animate body, allowing the animal spirit to pass freely into a member rightly disposed, as into a fit organ. Again, as many as are the senses in animals, so many without doubt are the motions in an inanimate body where animal spirit is wanting ; though necessarily there are many more motions in inanimate bodies than there are senses in animate, on account of the paucity of organs of sense. And of this a manifest example is exhibited in pain. For though there are many kinds and varieties of pain in animals (as the pain of burning, for one, of intense cold for another ; again, of pricking, squeezing, stretching, and the like), it is yet most certain that all of them, as far as the motion is concerned, exist in inanimate substances ; for example, in wood or stone, when it is burnt or frozen or pricked or cut or bent or stretched, and so on ; though they do not enter the senses for want of the animal spirit.

Again, the roots and branches of plants (which may seem strange) are Conformable Instances. For all vegetable matter swells and pushes out its parts to the surface, as well upwards as downwards. Nor is there any other difference between roots and branches than that the root is buried in the ground, while the branches are exposed to the air and sun [48]. For if you take a tender and flourishing branch of a tree, and bend it down into a clod of earth, although it does not cohere with the ground itself, it presently produces not a branch but a root. And *vice versâ*, if earth be placed at the top, and so kept down with a stone or any hard substance as to check the plant and prevent it from shooting upwards, it will put forth branches into the air downwards.

Again, the gums of trees, and most rock gems, are Conformable Instances. For both of these are nothing else than exudations and filterings of juices ; the former from trees, the latter from rocks : whence is produced the splendour and clearness in each ; that is, by the fine and delicate filtering. Hence too it is that the hairs of animals are not generally so beautiful and of so vivid a colour as the feathers of birds ; viz. because the juices do not filter so finely through skin as through quills.

Again, the scrotum in males and the matrix in females are Conformable Instances. So that the great organic difference between the sexes (in land animals

[48] In many plants part of the stem grows underground, while in others part at least of the root is above the surface. The true distinction has relation to the functions of the two organs. There is nothing in the root analogous (except under special circumstances) to buds or nodes, and consequently no true ramification.

at least) appears to be nothing more than that the one organization is external, and the other internal [49]. That is to say, the greater force of heat in the male thrusts the genitals outwards; whereas in the female the heat is too feeble to effect this, and thus they are contained within.

The fins of fish again and the feet of quadrupeds, or the feet and wings of birds, are Conformable Instances; to which Aristotle has added the four folds in the motions of serpents [50]. Whence it appears that in the structure of the universe the motions of living creatures are generally effected by a quaternion of limbs or of bendings.

Again, the teeth of land animals and the beaks of birds are Conformable Instances; from which it is manifest that in all perfect animals there is a determination of some hard substance to the mouth.

Nor is that an absurd similitude or conformity which has been remarked between man and a plant inverted. For the root of the nerves and faculties in animals is the head, while the seminal parts are the lowest,—the extremities of the legs and arms not reckoned. In a plant on the other hand, the root (which answers to the head) is regularly placed in the lowest part, and the seeds in the highest [51].

To conclude, it cannot too often be recommended and enjoined, that men's diligence in investigating and amassing natural history be henceforward entirely changed, and turned into the direction opposite to that now in use. For hitherto men have used great and indeed over-curious diligence in observing the variety of things, and explaining the exact specific differences of animals, herbs, and fossils; most of which are rather sports of nature than of any serious use towards science. Such things indeed serve to delight, and sometimes even give help in practice; but for getting insight into nature they are of little service or none. Men's labour therefore should be turned to the investigation and observation of the resemblances and analogies of things, as well in wholes as in parts. For these it is that detect the unity of nature, and lay a foundation for the constitution of sciences [52].

But here must be added a strict and earnest caution, that those only are to be taken for Conformable and Analogous Instances which indicate (as I said at the beginning) Physical Resemblances; that is, real and substantial resemblances; resemblances grounded in nature, not accidental or merely apparent; much less superstitious or curious resemblances, such as the writers on natural magic (very frivolous persons, hardly to be named in connexion with such serious matters as we are now about) are everywhere parading; similitudes and sympathies of

[49] This remark seems to have been suggested by a similar passage in Telesius, *De Rerum Naturâ*, vi. 18.:—"Masculo . . . magnus datus est calor, qui et membrum genitale foras propellat et sanguinem multum beneque omnem compactum conficiat, etc. Fœminæ autem . . . languens inditus est calor, qui neque genitale vas foras propellere nec è semine spiritum educere queat." The doctrine however of this passage was first taught by Galen, from whom Telesius derived it. See Galen, *De Usu Partium*, xiv. 6.

[50] De Anim. Incessu, i. 7.

[51] On the other hand, one is tempted to trace an analogy between the flower in plants and the skull in man and vertebrate animals in general: each occurring at the end of the axis of development, and each consisting of four segments—whorls or vertebræ. But by far the most remarkable analogy between plants and animals relates to the mode of development of their tissues, which, there is reason to believe, were all primarily formed from cells. The evidence in favour of this proposition is perhaps not yet quite complete.

It is curious that, after it had been established in the case of plants, Schleiden conceived that in this unity of original structure he had found a character peculiar to vegetable life, so that the analogy between plants and animals seemed to be impaired by the discovery.

[52] "Natura infinita est, sed qui symbola animadverterit omnia intelliget, licet non omnino," are the words of a great poet, who perhaps also is entitled to be called a great philosopher. They form the motto of one of the happiest illustrations of what Bacon meant by *instantia conformis*,—the Parthenogenesis of Professor Owen.

things that have no reality, which they describe and sometimes invent with great vanity and folly.

But to leave these ; the very configuration of the world itself in its greater parts presents Conformable Instances which are not to be neglected. Take for example Africa and the region of Peru with the continent stretching to the Straits of Magellan, in each of which tracts there are similar isthmuses and similar promontories ; which can hardly be by accident [53].

Again, there is the Old and New World ; both of which are broad and extended towards the north, narrow and pointed towards the south.

We have also most remarkable Instances of Conformity in the intense cold existing in what is called the middle region of the air and the violent fires which are often found bursting forth from beneath the ground ; which two things are *ultimities* and extremes ; that is to say, the extreme of the nature of cold towards the circumferance of the sky, of heat towards the bowels of the earth ; by *antiperistasis* or the rejection of the contrary nature.

Lastly, the Conformity of Instances in the axioms of science is deserving of notice. Thus the rhetorical trope of deceiving expectation is conformable with the musical trope of avoiding or sliding from the close or cadence : the mathematical postulate that if two things are equal to the same thing they are equal to one another, is conformable with the rule of the syllogism in logic which unites propositions agreeing in a middle term [54]. In fine, a certain sagacity in investigating and hunting out Physical Conformities and Similitudes is of very great use in very many cases.

XXVIII.

Among Prerogative Instances I will put in the seventh place *Singular Instances ;* which I also call *Irregular* or *Heteroclite ;* to borrow a term from grammarians. They are such as exhibit bodies in the concrete, which seem to be out of the course and broken off from the order of nature, and not agreeing with other bodies of the same kind. For Conformable Instances are like each other ; Singular Instances are like themselves alone. The use of Singular Instances is the same as that of Clandestine ; namely to raise and unite nature for the purpose of discovering kinds or common natures, to be afterwards limited by true specific differences. For we are not to give up the investigation, until the properties and qualities found in such things as may be taken for miracles of nature be reduced and comprehended under some Form or fixed Law ; so that all the irregularity or singularity shall be found to depend on some common Form, and the miracle shall turn out to be only in the exact specific differences, and the degree, and the rare concurrence ; not in the species itself ; whereas now the thoughts of men go no further than to pronounce such things the secrets and mighty works of nature, things as it were causeless, and exceptions to general rules.

Examples of Singular Instances are the sun and moon among stars ; the magnet among stones ; quicksilver among metals ; the elephant among quadrupeds ; the venereal sense among kinds of touch ; the scent of hounds among kinds of smell.

[53] A. von Humboldt has pointed out the conformity of the opposite shores of the Atlantic—the approximate correspondence between the projections on each side and the recesses on the other. But Bacon apparently compares not the opposite but the corresponding coasts of Africa and America. C. Concepcion would correspond to C. Negro ; but the parallelism is not very close.

[54] The importance of the parallel here suggested was never understood until the present time, because the language of mathematics and of logic has hitherto not been such as to permit the relation between them to be recognised. Mr. Boole's *Laws of Thought* contains the first development of ideas of which the germ is to be found in Bacon and Leibnitz ; to the latter of whom the fundamental principle that in logic $a^2=a$ was known (v. Leibnitz, *Philos. Works*, by Erdmann, 1840, p. 130). It is not too much to say that Mr. Boole's treatment of the subject is worthy of these great names.

Other calculuses of inference (using the word in its widest sense), besides the mathematical and the logical, yet perhaps remain to be developed ; but this is a subject on which it is impossible here to enter.

So among grammarians the letter S is held singular, on account of its easy combination with consonants, sometimes with two, sometimes even with three; which property no other letter has. Such instances must be regarded as most valuable, because they sharpen and quicken investigation, and help to cure the understanding depraved by custom and the common course of things.

XXIX.

Among Prerogative Instances I will put in the eighth place *Deviating Instances ;* that is, errors, vagaries and prodigies of nature, wherein nature deviates and turns aside from her ordinary course. Errors of nature differ from Singular Instances in this, that the latter are prodigies of species, the former of individuals. Their use is pretty nearly the same; for they correct the erroneous impressions suggested to the understanding by ordinary phenomena, and reveal Common Forms. For in these also we are not to desist from inquiry, until the cause of the deviation is discovered. This cause however does not rise properly to any Form, but simply to the latent process that leads to the Form. For he that knows the ways of nature will more easily observe her deviations ; and on the other hand he that knows her deviations will more accurately describe her ways [55].

They differ in this also from Singular Instances, that they give much more help to practice and the operative part. For to produce new species would be very difficult ; but to vary known species, and thereby produce many rare and unusual results is less difficult. Now it is an easy passage from miracles of nature to miracles of art. For if nature be once detected in her deviation, and the reason thereof made evident, there will be little difficulty in leading her back by art to the point whither she strayed by accident ; and that not only in one case, but also in others ; for errors on one side point out and open the way to errors and deflexions on all sides. Under this head there is no need of examples ; they are so plentiful. For we have to make a collection or particular natural history of all prodigies and monstrous births of nature ; of everything in short that is in nature new, rare, and unusual. This must be done however with the strictest scrutiny, that fidelity may be ensured. Now those things are to be chiefly suspected which depend in any way on religion ; as the prodigies of Livy ; and those not less which are found in writers on natural magic or alchemy, and men of that sort ; who are a kind of suitors and lovers of fables. But whatever is admitted must be drawn from grave and credible history and trustworthy reports.

XXX.

Among Prerogative Instances I will put in the ninth place *Bordering Instances ;* which I also call *Participles.* They are those which exhibit species of bodies that seem to be composed of two species, or to be rudiments between one species and another. These instances might with propriety be reckoned among Singular or Heteroclite Instances ; for in the whole extent of nature they are of rare and extraordinary occurrence. But nevertheless for their worth's sake they should be ranked and treated separately ; for they are of excellent use in indicating the composition and structure of things, and suggesting the causes of the number and quality of the ordinary species in the universe, and carrying on the understanding from that which is to that which may be.

Examples of these are: moss, which holds a place between putrescence and a plant ; some comets, between stars and fiery meteors ; flying fish, between birds and fish ; bats, between birds and quadrupeds : also the ape, between man and beast,—

"Simia quam similis turpissima bestia nobis" [56] ;

likewise the biformed births of animals, mixed of different species, and the like.

XXXI.

Among Prerogative Instances I will put in the tenth place *Instances of Power*, or of the *Fasces* (to borrow a term from the badges of empire) ; which I also call

[55] See Owen, *On the Nature of Limbs*, p. 54.

[56] Ennius, quoted by Cicero.

Instances of the Wit, or *Hands of Man.* These are the noblest and most consummate works in each art, exhibiting the ultimate perfection of it. For since our main object is to make nature serve the business and conveniences of man, it is altogether agreeable to that object that the works which are already in man's power should (like so many provinces formerly occupied and subdued) be noted and enumerated, especially such as are the most complete and perfect ; because starting from them we shall find an easier and nearer passage to new works hitherto unattempted. For if from an attentive contemplation of these a man pushes on his work with zeal and activity, he will infallibly either advance them a little further, or turn them aside to something in their neighbourhood, or even apply and transfer them to some more noble use.

Nor is this all. But as by rare and extraordinary works of nature the understanding is excited and raised to the investigation and discovery of Forms capable of including them ; so also is this done by excellent and wonderful works of art ; and that in a much greater degree, because the method of creating and constructing such miracles of art is in most cases plain, whereas in the miracles of nature it is generally obscure. But with these also we must use the utmost caution, lest they depress the understanding and fasten it as it were to the ground.

For there is danger lest the contemplation of such works of art, which appear to be the very summits and crowning points of human industry, may so astonish and bind and bewitch the understanding with regard to them, that it shall be incapable of dealing with any other, but shall think that nothing can be done in that kind except by the same way in which these were done ; only with the use of greater diligence and more accurate preparation.

Whereas on the contrary this is certain ; that the ways and means of achieving the effects and works hitherto discovered and observed are for the most part very poor things ; and that all power of a high order depends on Forms, and is derived in order from the sources thereof ; not one of which has yet been discovered.

And therefore (as I have said elsewhere [57]) if a man had been thinking of the war engines and battering-rams of the ancients, though he had done it with all his might and spent his whole life in it, yet he would never have lighted on the discovery of cannon acting by means of gunpowder. Nor again, if he had fixed his observation and thought on the manufacture of wool and cotton, would he ever by such means have discovered the nature of the silkworm or of silk.

Hence it is that all the discoveries which can take rank among the nobler of their kind, have (if you observe) been brought to light, not by small elaborations and extensions of arts, but entirely by accident. Now there is nothing which can forestall or anticipate accident (which commonly acts only at long intervals) except the discovery of Forms.

Particular examples of such instances it is unnecessary to adduce, there is such plenty of them. For what we have to do is simply this :—to seek out and thoroughly inspect all mechanical arts, and all liberal too (as far as they deal with works), and make therefrom a collection or particular history of the great and masterly and most perfect works in every one of them, together with the mode of their production or operation.

And yet I do not tie down the diligence that should be used in such a collection to those works only which are esteemed the masterpieces and mysteries of any art, and which excite wonder. For wonder is the child of rarity ; and if a thing be rare, though in kind it be no way extraordinary, yet it is wondered at. While on the other hand things which really call for wonder on account of the difference in species which they exhibit as compared with other species, yet if we have them by us in common use, are but slightly noticed.

Now the singularities of art deserve to be noticed no less than those of nature, of which I have already spoken [58]. And as among the singularities of nature I placed the sun, the moon, the magnet, and the like,—things in fact most familiar, but in nature almost unique ; so also must we do with the singularities of art.

For example, a Singular Instance of art is paper, a thing exceedingly common. Now if you observe them with attention, you will find that artificial materials

[57] I. § 109. [58] II. § 28.

are either woven in upright and transverse threads, as silk, woollen or linen cloth, and the like ; or cemented of concreted juices, as brick, earthenware, glass, enamel, porcelain, etc., which are bright if well united, but if not, are hard indeed but not bright. But all things that are made of concrete juices are brittle, and no way cohesive or tenacious. On the contrary paper is a tenacious substance, that may be cut or torn ; so that it imitates and almost rivals the skin or membrane of an animal, the leaf of a vegetable, and the like pieces of Nature's workmanship. For it is neither brittle like glass, nor woven as cloth ; but is in fibres, not distinct threads, just like natural materials ; so that among artificial materials you will hardly find anything similar : but it is altogether singular [59]. And certainly among things artificial those are to be preferred which either come nearest to an imitation of nature, or on the contrary overrule and turn her back.

Again, as instances of the Wit and Hand of Man, we must not altogether contemn juggling and conjuring tricks. For some of them, though in use trivial and ludicrous, yet in regard to the information they give may be of much value.

Lastly, matters of superstition and magic (in the common acceptation of the word) must not be entirely omitted. For although such things lie buried deep beneath a mass of falsehood and fable, yet they should be looked into a little ; for it may be that in some of them some natural operation lies at the bottom ; as in fascination, strengthening of the imagination, sympathy of things at a distance, transmission of impressions from spirit to spirit no less than from body to body, and the like.

XXXII.

From what has been said it is clear that the five classes of instances last mentioned (namely, Instances Conformable, Singular, Deviating, Bordering, and of Power) ought not to be reserved until some certain nature be in question (as the other instances which I have placed first, and most of those that are to follow should), but a collection of them must be begun at once, as a sort of particular history ; because they serve to digest the matters that enter the understanding, and to correct the ill complexion of the understanding itself, which cannot but be tinged and infected, and at length perverted and distorted, by daily and habitual impressions.

These instances therefore should be employed as a sort of preparative for setting right and purging the understanding. For whatever withdraws the understanding from the things to which it is accustomed, smooths and levels its surface for the reception of the dry and pure light of true ideas.

Moreover such instances pave and prepare the way for the operative part ; as will be shown in the proper place, when I come to speak of deductions leading to Practice.

XXXIII.

Among Prerogative Instances I will put in the eleventh place *Instances of Companionship and of Enmity ;* which I also call Instances of *Fixed Propositions*. They are those which exhibit a body or concrete substance, in which the nature inquired into constantly attends, as an inseparable companion ; or in which on the contrary it constantly retreats, and is excluded from companionship, as an enemy and foe. For from such instances are formed certain and universal propositions, either affirmative or negative ; in which the subject will be a body in concrete, and the predicate the nature itself that is in question. For particular propositions are in no case fixed ; I mean propositions in which the nature in question is found in any concrete body to be fleeting and moveable, that is to say accruing or acquired, or on the other hand departing or put away. Wherefore particular propositions have no prerogative above others, save only in the case of Migration, of which I have already spoken. Nevertheless even these particular propositions being prepared and collated with universal propositions are of great use, as shall be shown in the proper place. Nor even in the universal propositions do we require exact or absolute affirmation or negation. For it is

[59] It is curious that Bacon should not have remarked that all the qualities here mentioned belong to felt as well as to paper.

sufficient for the purpose in hand, even if they admit of some rare and singular exception.

The use of Instances of Companionship is to bring the Affirmative of the Form within narrow limits. For as by Migratory Instances the Affirmative of the Form is narrowed to this,—that the Form of the thing must needs be something which by the act of Migration is communicated or destroyed ; so in Instances of Companionship, the Affirmative of the Form is narrowed to this,—that the Form of the thing must needs be something which enters as an element into such a concretion of body, or contrariwise which refuses to enter ; so that he who well knows the constitution or configuration of such a body will not be far from bringing to light the Form of the nature under inquiry.

For example, let the nature in question be Heat. An Instance of Companionship is Flame. For in water, air, stone, metal, and most other substances, heat is variable, and may come and go ; but all flame is hot, so that heat is always in attendance on the concretion of flame. But no Hostile Instance of Heat is to be found here. For the senses know nothing of the bowels of the earth, and of all the bodies which we do know, there is not a single concretion that is not susceptible of heat.

But to take another instance ; let the nature in question be Consistency. A Hostile Instance is Air. For metal can be fluid and can also be consistent ; and so can glass ; water also can be consistent, when it is frozen ; but it is impossible that air should ever be consistent, or put off its fluidity.

But with regard to such Instances of Fixed Propositions I have two admonitions to give, which may help the business in hand. The first is, that if a universal affirmative or negative be wanting, that very thing be carefully noted as a thing that is not ; as we have done in the case of Heat, where a universal negative (as far as the essences that have come under our knowledge are concerned) is not to be found in the nature of things. In like manner, if the nature in question be Eternity or Incorruptibility, no universal affirmative is to be found here. For Eternity or Incorruptibility cannot be predicated of any of the bodies lying below the heavens and above the bowels of the earth. The other admonition is, that to universal propositions, affirmative or negative, concerning any concrete body, there be subjoined those concretes which seem to approach most nearly to that which is not ; as in heat, the gentlest and least burning flames ; in incorruptibility, gold, which comes nearest to it. For all such indicate the limits of nature between that which is and that which is not, and help to circumscribe Forms, and prevent them from escaping and straying beyond the conditions of matter.

XXXIV.

Among Prerogative Instances I will put in the twelfth place those *Subjunctive Instances* mentioned in the last aphorism, which I otherwise call *Instances of Ultimity or Limit.* For such instances are not only useful when subjoined to fixed propositions, but also by themselves and in their own properties. For they point out not obscurely the real divisions of nature and measures of things, and how far in any case nature may act or be acted upon, and then the passages of nature into something else. Of this kind are gold in weight ; iron in hardness ; the whale in animal bulk ; the dog in scent ; the combustion of gunpowder in rapid expansion ; and the like. Nor should extremes in the lowest degree be less noticed than extremes in the highest ; such as spirit of wine in weight [60] ; silk in softness ; the worms of the skin in animal bulk ; and the like.

XXXV.

Among Prerogative Instances I will put in the thirteenth place *Instances of Alliance or Union.* They are those which mingle and unite natures supposed to be heterogeneous, and marked and set down as such in the received divisions.

[60] Although precise directions for making ether were given by Valerius Cordus in 1544, yet it is said to have remained unnoticed until it was rediscovered in the eighteenth century. Bacon's want of acquaintance with it, implied in this and other passages, is therefore not surprising.

Instances of Alliance show that operations and effects attributed to some one heterogeneous nature as peculiar to it, may belong also to other heterogeneous natures; so that this supposed heterogeneity is proved to be not real or essential, but only a modification of a common nature. They are therefore of most excellent use in raising and elevating the understanding from specific differences to *genera*, and in dispelling phantoms and false images of things, which in concrete substances come before us in disguise. For example, let the nature in question be heat. We are told (and it seems to be a division quite received and authorized) that there are three kinds of heat; the heat of heavenly bodies, the heat of animals and the heat of fire; and that these heats (especially one of them as compared with the other two) are in their very essence and species—that is to say, in their specific nature—distinct and heterogeneous; since the heat of heavenly bodies and of animals generates and cherishes, while the heat of fire wastes and destroys. We have, therefore, an Instance of Alliance in that common case, when the branch of a vine is brought within a house where a fire is constantly kept up, and the grapes ripen on it a whole month sooner than they do out of doors; so that the ripening of fruit, even while it hangs on the tree, may be brought about by fire, though such ripening would seem to be the proper work of the sun [61]. From this beginning therefore, the understanding, rejecting the notion of essential heterogeneity, easily rises to inquire what are in reality those points of difference between the heat of the sun and of fire which cause their operations to be so dissimilar, however they may themselves partake of a common nature.

These differences will be found to be four. The first is, that the heat of the sun compared with the heat of fire is far milder and softer in degree; the second is, that in quality (at least as it reaches us through the air) it is far moister; the third (and this is the main point) is, that it is exceedingly unequal, now approaching and increased, now receding and diminished; which thing chiefly contributes to the generation of bodies. For Aristotle was right in asserting [62] that the principal cause of the generations and corruptions which are going on here on the surface of the earth is the oblique course of the sun through the zodiac; whence the heat of the sun, partly by the alternation of day and night, partly by the succession of summer and winter, becomes strangely unequal. And yet this great man must go on at once to corrupt and deprave what he has rightly discovered. For laying down the law to nature (as his way is), he very dictatorially assigns as the cause of generation the approach of the sun, and as the cause of corruption his retreat; whereas both together (the approach of the sun and his retreat), not respectively, but as it were indifferently, afford a cause both for generation and production; since inequality of heat ministers to generation and corruption, equality to conservation only. There is also a fourth specific differ-

[61] The regular use of artificial heat in greenhouses and conservatories was not known in Bacon's time. In the *Maison Champêtre*, an encyclopædia of gardening and agriculture published in 1607, nothing is said of it; nor is there anything on the subject in the writings of Porta, though in his *Nat. Mag.* he has spoken of various modes of accelerating the growth of fruits and flowers. In the *Sylva Sylvarum* (412), however, Bacon speaks of housing hot-country plants to save them, and, in the *Essay on Gardens*, of stoving myrtles. The idea of what are now called greenhouses was introduced into England from Holland about the time of the Revolution. The orangery at Heidelberg, formed, I believe, about the middle of the seventeenth century, is said to be the earliest conservatory on record.

It is related that Albertus Magnus, entertaining the emperor at Cologne during the winter, selected for the place of entertainment the garden of his monastery. Everything was covered with snow, and the guests were much inclined to be discontented; but when the feast began, the snow cleared away; the trees put forth, first leaves, then blossoms, then fruit; and the climate became that of summer, This glorious summer, which had thus abruptly succeeded to the winter of their discontent, lasted only till the conclusion of the feast, when everything resumed its former aspect. It would be a fanciful explanation, and I know not whether it has ever been suggested, to say that Albertus Magnus really entertained the emperor in a conservatory, and only led his guests through the garden. See, for the story, Grimm's *Deutsche Sagen*.

[62] *Meteorologia*, i. 14.

ence between the heat of the sun and of fire, and one of very great moment ; viz. that the sun operates by gentle action through long spaces of time, whereas the operations of fire, urged on by the impatience of man, are made to finish their work in shorter periods. But if any one were to set to work diligently to temper the heat of fire and reduce it to a milder and more moderate degree, as is easily done in many ways ; and were then to sprinkle and intermix a little moisture ; and if above all he were to imitate the heat of the sun in its inequality ; and lastly if he could submit to a slow procedure, not indeed corresponding to the operations of the sun, but yet slower than men generally adopt in working with fire ; he would speedily get rid of the notion of different kinds of heat, and would attempt to imitate, if not equal or in some cases even surpass, the works of the sun by the heat of fire. We have a similar Instance of Alliance in the revival of butterflies stupefied and half dead with cold, by slightly warming them at a fire ; so that you may easily see that fire is no more without the power of giving life to animals than of ripening vegetables. Thus also Fracastorius's celebrated invention of the heated pan with which doctors cover the heads of apoplectic patients who are given over [63], manifestly expands the animal spirits, compressed and all but extinguished by the humours and obstructions of the brain, and exciting them to motion, just as fire acts on air or water, by consequence quickens and gives them life. Eggs also are sometimes hatched by the heat of fire, which thus exactly imitates animal heat ; and there are many instances of the same kind ; so that no one can doubt that the heat of fire may in many subjects be modified so as to resemble the heat of heavenly bodies and of animals [64].

Again, let the natures in question be Motion and Rest. It appears to be a received division and drawn from the depths of philosophy, that natural bodies either move in circle, or move straight forward, or remain at rest. For there is either motion without limit, or rest at a limit, or progress towards a limit. Now that perpetual motion of rotation seems to be proper to the heavenly bodies ; station or rest seems to belong to the globe of the earth ; while other bodies (which they call heavy or light, being indeed placed out of the region to which they naturally belong) are carried towards the masses or congregations of their likes ; light bodies upwards towards the circumference of the heaven ; heavy bodies downwards towards the earth. And this is pretty talk.

But we have an Instance of Alliance in one of the lower comets, which though far below the heaven, nevertheless revolve. And Aristotle's fiction of a comet being tied to or following some particular star has long been exploded, not only because the reason for it is not probable, but because we have manifest experience of the discursive and irregular motion of comets through various parts of the sky.

Again, another Instance of Alliance on this subject is the motion of air, which within the tropics, where the circles of rotation are larger, seems itself also to revolve from east to west.

Again, another Instance would be the ebb and flow of the sea, if it be found that the waters themselves are carried in a motion of rotation (however slow and evanescent) from east to west ; though subject to the condition of being driven back twice in the day. For if things be so, it is manifest that that motion of rotation is not limited to heavenly bodies, but is shared also by air and water.

Even that property of light substances, viz. that they tend upwards, is somewhat at fault. And on this point a bubble of water may be taken as an Instance

[63] It is mentioned in the life of Fracastorius, that when dying of apoplexy, and speechless, he made signs for the application of a cucurbita (or cupping-vessel) to his head, remembering the remarkable cure which he had effected in the case of a nun at Verona. It is scarcely necessary to remark that " dry cupping," as it is called, acts simply by partially removing the pressure of the atmosphere : the heat applied to the vessel has no other effect than that of rarefying the air it contains.

[64] Bacon's rejection of the essential heterogeneity of the three species of heat is apparently taken from Telesius, *De Rerum Nat.* vi. 20. Telesius remarks, as Bacon does, that eggs may be hatched, and insects apparently dead restored to life, by means of artificial heat.

of Alliance. For if there be air under the water, it rapidly ascends to the surface, by that motion of percussion (as Democritus calls it) by which the descending water strikes and raises the air upwards; not by any effort or struggle of the air itself. And when it is come to the surface of the water, then the air is stopped from further ascent by a slight resistance it meets with in the water, which does not immediately allow itself to be separated; so that the desire of air to ascend must be very slight.

Again, let the nature in question be Weight. It is quite a received division, that dense and solid bodies move toward the centre of the earth, rare and light toward the circumference of the heaven, as to their proper places. Now as for this notion of places, though such things prevail in the schools, it is very silly and childish to suppose that place has any power. Therefore philosophers do but trifle when they say that if the earth were bored through, heavy bodies would stop on reaching the centre. Certainly it would be a wonderful and efficacious sort of nothing, or mathematical point, which could act on bodies, or for which bodies could have desire; for bodies are not acted on except by bodies. But this desire of ascending and descending depends either on the configuration of the body moved or on its sympathy or consent with some other body. Now if there be found any body which, being dense and solid, does not move to the earth, there is an end of this division. But if Gilbert's opinion be received, that the earth's magnetic power of attracting heavy bodies does not extend beyond the orb of its virtue (which acts always to a certain distance and no more)[65], and if this opinion be verified by a single instance, in that we shall have got at last an Instance of Alliance on the subject of weight. But at present there does not occur any instance on this subject certain and manifest. What seems to come nearest to one is that of the waterspouts, often seen in the voyage over the Atlantic Ocean towards either of the Indies. For so great is the quantity and mass of water suddenly discharged by these waterspouts, that they seem to have been collections of water made before, and to have remained hanging in these places; and afterwards to have been rather thrown down by some violent cause, than to have fallen by the natural motion of gravity; so that it may be conjectured that a dense and compact mass, at a great distance from the earth would hang like the earth itself, and not fall unless thrust down. But on this point I affirm nothing certain. Meanwhile in this and many other cases it will easily be seen how poor we are in natural history, when in place of certain instances I am sometimes compelled to adduce as examples bare suppositions.

Again, let the nature in question be Discourse of Reason. The distinction between human reason and the sagacity of brutes appears to be a perfectly correct one. Yet there are certain instances of actions performed by animals, by which it seems that brutes too have some power of syllogising; as in the old story of the crow which, in a time of great drought being half dead with thirst, saw some water in the hollow trunk of a tree; and finding it too narrow to get in, proceeded to drop in a number of pebbles, till the water rose high enough for it to drink; which thing afterwards passed into a proverb.

Again, let the nature in question be Visibility. It appears to be a very correct and safe division which regards light as primarily visible, and affording the power of seeing; while colour is secondarily visible, and cannot be seen without light, so that it appears to be nothing more than an image or modification of

[65] In Gilbert's philosophy, the earth's magnetic action is not distinguished from gravity. See *De Mundo*, ii. c. 3. Again, that the magnetic action of the earth or of a magnet is confined to a definite orb appears from a variety of passages. See *De Magnete*, ii. c. 7., and the definitions prefixed to this work. Gilbert distinguished between the "orb of virtue", which includes the whole space through which any magnetic action extends, and the "orb of coition", which is "totum illud spatium per quod minimum magneticum per magnetem movetur". He asserts that the orb of the magnetic virtue extends to the moon, and ascribes the moon's inequalities to the effects it produces (*De Mundo*, ii. c. 19). In the preceding chapter he remarks, "Luna magnetice alligatur terræ, quia facies ejus semper versus terram".

light.[66] And yet there appear to be instances of alliance on either side, viz., snow in great quantities, and the flame of sulphur; in one of which there appears to be a colour primarily giving light, in the other a light verging on colour.

XXXVI.

Among Prerogative Instances I will put in the fourteenth place *Instances of the Fingerpost*[67]; borrowing the term from the fingerposts[68] which are set up where roads part, to indicate the several directions. These I also call *Decisive and Judicial*, and in some cases, *Oracular* and *Commanding Instances*. I explain them thus. When in the investigation of any nature the understanding is so balanced as to be uncertain to which of two or more natures the cause of the nature in question should be assigned, on account of the frequent and ordinary concurrence of many natures, Instances of the Fingerpost show the union of one of the natures with the nature in question to be sure and indissoluble, of the other to be varied and separable; and thus the question is decided, and the former nature is admitted as the cause, while the latter is dismissed and rejected. Such instances afford very great light, and are of high authority, the course of interpretation sometimes ending in them and being completed. Sometimes these Instances of the Fingerpost meet us accidentally among those already noticed; but for the most part they are new, and are expressly and designedly sought for and applied, and discovered only by earnest and active diligence.

For example, let the nature in question be the Ebb and Flow of the Sea; each of which is repeated twice a day, and takes six hours each time; subject to some slight difference which coincides wth the motion of the moon. The following will be a case of the parting of the roads.

This motion must necessarily be caused either by the advance and retreat of the waters; as water shaken in a basin leaves one side when it washes the other; or else by a lifting up of the waters from the bottom and falling again: as water in boiling rises and falls. The question is to which of these two causes the ebb and flow should be assigned. Now, if we take the first, it follows that when there is a flood on one side of the sea, there must be at the same time an ebb somewhere on the other. To this point therefore the inquiry is brought. Now it has been observed by Acosta and others, after careful research, that on the shores of Florida and the opposite shores of Spain and Africa the floods take place at the same times, and the ebbs take place at the same times also; and not that there is an ebb from the shores of Spain and Africa when there is a flood on the shores of Florida[69]. And yet if you look at it more closely, this does not prove the case in favour of the rising and against the progressive motion. For waters may move in progression, and yet rise upon the opposite shores of the same channel at the same time; as when they are thrust together and driven on from some other quarter. For so it is with rivers, which rise and fall on both banks at the same hours; and yet that motion is clearly one of progression; namely, of the waters entering the mouth of the rivers from the sea. It may therefore happen in a like manner that waters coming in a vast mass from the East Indian Ocean are driven together and pushed into the channel of the Atlantic, and on that account flood both sides at once. We must inquire therefore whether there be any other channel, in which the water can be retreating and ebbing at that same time; and we have the South Sea, a sea at least as wide, indeed wider and larger than the Atlantic, which is sufficient for the purpose.

At length then we have come to an Instance of the Fingerpost in this case: and it is this. If we find for certain that when there is a flood on the opposite coasts of Florida and Spain in the Atlantic, there is also a flood on the coasts of Peru

[66] The doctrine of this passage seems to be taken from Telesius, *De Rerum Naturâ*. vii. c. 31.:—"Sensus ipse primo illam [lucem] et per se visilem colores siquidem visiles, at secundo a luce loco et lucis omnino opera visiles declarat."

[67] [Orig. *Instantias Crucis*.]

[68] [*I.e.* "crosses."]

[69] Compare the *De Fluxu et Refluxu Maris*. I have not been able to find this statement in Acosta, who speaks of the synchronism of the tides on the opposite sides of South America, as shown by the meeting of the tidal waves in the Straits of Magellan. (iii. 14.)

and the back of China in the South Sea, then indeed on the authority of this Decisive Instance we must reject the assertion that the ebb and flow of the sea, which is the thing inquired into, takes place by a progressive motion ; for there is no sea or place left in which the retreat or ebbing can be going on at the same time. And this may be most conveniently ascertained by asking the inhabitants of Panama and Lima (where the two Oceans, the Atlantic and Pacific, are separated by a small isthmus) whether the ebb and flow of the sea takes place on the opposite sides of the isthmus at the same time ; or contrariwise, when it is ebbing on one side it is flowing on the other. Now this decision or rejection appears to be certain, if we take it for granted that the earth is immoveable. But if the earth revolves, it is perhaps possible that in consequence of the unequal rotation (in point of speed) of the earth and waters of the sea, the waters are violently driven upwards into a heap, which is the flood, and (when they can bear no more piling) released and let down again, which is the ebb. But on this inquiry should be made separately. Still even on this hypothesis, our position remains equally fixed, that there must of necessity be an ebb of the sea going on in some parts at the same time that a flood is going on in others.

Again, let the nature in question be the latter of the two motions we have supposed ; namely, the rising and sinking motion ; if on careful examination we reject the former motion of which I spoke,—the progressive. With regard to this nature the road branches into three. For the motion by which the waters rise in the flood and sink in the ebb without any accession of other waters rolling in, must necessarily be brought about in one of these three ways. Either there is an accession of water poured out from the interior of the earth, and again retreating into it ; or there is no accession to the mass of water, but the same waters (without increase of quantity) are extended or rarefied, so as to occupy a greater space and dimension, and again contract themselves ; or there is no increase either of supply or of extension, but the same waters (the same in quantity as in density) are raised by some magnetic force attracting them from above, and by consent therewith, and then fall back again. Let us now dismiss the two former causes of motion and reduce our inquiry to the last ; that is to say, let us inquire whether any such raising by consent or magnetic force may happen. Now in the first place it is evident that the waters, as they lie in the trench or hollow of the sea, cannot all be raised at once, for want of something to take their place at the bottom ; so that even if there were in water any such desire to rise, it would be barred and checked by the cohesion of things, or (as it is commonly called) the abhorrence of a vacuum. It remains that the waters must be raised in one part, and thereby be diminished and retreat in another. Again, it will follow of necessity that the magnetic force, since it cannot act upon the whole, will act with the greatest intensity on the middle, so as to raise up the water in the middle ; upon which the rest must follow and fall away from the sides.

Thus at length we come to an Instance of the Fingerpost on this subject. For if we find that in the ebb of the sea the surface of the water is more arched and round, the waters rising in the middle of the sea and falling away from the sides, that is, the shores ; and that in the flood the same surface is more even and level, the waters returning to their former position ; then indeed on the authority of this Decisive Instance the raising by magnetic force may be admitted ; otherwise it must be utterly rejected. And this would not be difficult to ascertain by trial in straits with sounding lines [70] ; viz. whether during ebbs the sea be not higher or deeper towards the middle than during floods. It is to be observed however that, if this be the case, the waters must (contrary to the common opinion) rise in ebbs and sink in floods, so as to clothe and wash the shores.

Again, let the nature investigated be the Spontaneous Motion of Rotation ; and in particular, whether the Diurnal Motion, whereby to our eyes the sun and stars rise and set, be a real motion of rotation in the heavenly bodies, or a motion apparent in the heavenly bodies, and real in the earth. We may here take for

[70] It is scarcely necessary to remark that wherever soundings are possible, tidal phenomena are derivative, and give no direct information as to the form the ocean would assume if the hypothesis of the equilibrium theory represented the reality.

an Instance of the Fingerpost the following. If there be found in the ocean any motion from east to west, however weak and languid ; if the same motion be found a little quicker in the air, especially within the tropics, where because of the larger circles it is more perceptible ; if the same motion be found in the lower comets, but not lively and vigorous ; if the same motion be found in planets, but so distributed and graduated, that the nearer a planet is to the earth its motion is slower, the further a planet is distant from the earth its motion is quicker, and quickest of all in the starry sphere ; then indeed we should receive the diurnal motion as real in the heavens, and deny such motion to the earth ; because it will be manifest that motion from east to west is perfectly cosmical, and by consent of the universe ; being most rapid in the highest parts of the heavens, and gradually falling off, and finally stopping and becoming extinct in the immoveable,—that is, the earth [71].

Again, let the nature in question be that other Motion of Rotation so much talked of by philosophers, the Resistant and Contrary Motion to the Diurnal, viz. from west to east ; which old philosophers attribute to the planets ; also to the starry sphere ; but Copernicus and his followers to the earth as well ; and let us inquire whether any such motion be found in nature, or whether it be not rather a thing invented and supposed for the abbreviation and convenience of calculation, and for the sake of that pretty notion of explaining celestial motions by perfect circles. For this motion in the heavens is by no means proved to be true and real, either by the failing of a planet to return in its diurnal motion to the same point of the starry sphere, or by this, that the poles of the zodiac differ from the poles of the world ; to which two things we owe this idea of motion. For the first phenomenon is well accounted for by supposing that the fixed stars outrun the planets, and leave them behind ; the second, by supposing a motion in spiral lines ; so that the inequality of return and the declination to the tropics may rather be modifications of the one diurnal motion, than motions contrary or round different poles. And most certain it is, if one may but play the plain man for a moment (dismissing the fancies of astronomers and schoolmen, whose way it is to overrule the senses, often without reason, and to prefer what is obscure), that this motion does actually appear to the sense such as I have described ; for I once had a machine made with iron wires to represent it [72].

The following would be an Instance of the Fingerpost on this subject. If it be found in any history worthy of credit, that there has been any comet, whether high or low, which has not revolved in manifest agreement (however irregular) with the diurnal motion, but has revolved in the opposite direction, then certainly we may set down thus much as established, that there *may be* in nature some such motion. But if nothing of the kind can be found, it must be regarded as questionable, and recourse be had to other Instances of the Fingerpost about it.

Again, let the nature in question be Weight or Heaviness. Here the road will branch into two, thus. It must needs be that heavy and weighty bodies

[71] Nothing shows better than an instance of this kind, the impossibility of reducing philosophical reasoning to a uniform method of exclusion. How could the analogical argument in the text be stated in accordance with what Bacon seems to recognise as the only true form of induction,—that, namely, which proceeds by exclusion ? The argument depends on a wholly non-logical element, the conviction of the unity and harmony of nature.

[72] This passage does the author little credit. He does not seem to have perceived that the resolution of the apparent motion into other simpler motions was an essentially necessary step before the phenomena could be grouped together in any general law. The transition from the apparent motion to the real motions could never have been made unless the former had been resolved in the manner which Bacon here condemns. From the concluding remark no astronomer would have dissented, " talem esse motum ad sensum, qualem diximus ". About this there can be no question ; but the whole passage shows how little Bacon understood the scope and the value of the astronomy of his own time.

either tend of their own nature to the centre of the earth, by reason of their proper configuration ; or else that they are attracted by the mass and body of earth itself as by the congregation of kindred substances, and move to it by sympathy. If the latter of these be the cause, it follows that the nearer heavy bodies approach to the earth, the more rapid and violent is their motion to it ; and that the further they are from the earth, the feebler and more tardy is their motion (as is the case with magnetic attraction) ; and that this action is confined to certain limits ; so that if they were removed to such a distance from the earth that the earth's virtue could not act upon them, they would remain suspended like the earth itself, and not fall at all. With regard to this then, the following would be an Instance of the Fingerpost. Take a clock moved by leaden weights, and another moved by the compression of an iron spring ; let them be exactly adjusted, that one go not faster or slower than the other : then place the clock moving by weights on the top of a very high steeple, keeping the other down below ; and observe carefully whether the clock on the steeple goes more slowly than it did, on account of the diminished virtue of its weights. Repeat the experiment in the bottom of a mine, sunk to a great depth below the ground ; that is, observe whether the clock so placed does not go faster than it did, on account of the increased virtue of its weights. If the virtue of the weights is found to be diminished on the steeple, and increased in the mine, we may take the attraction of the mass of the earth as the cause of weight [73].

Again, let the nature investigated be the Polarity of the Iron Needle when touched with the magnet. With regard to this nature the road will branch into two, thus. Either the touch of the magnet of itself invests the iron with polarity to the north and south ; or it simply excites and prepares the iron, while the actual motion is communicated by the presence of the earth ; as Gilbert thinks, and labours so strenuously to prove. To this point therefore tend the observations which he has collected with great sagacity and industry. One is, that an iron nail, which has lain for a long time in a direction between north and south, gathers polarity without the touch of the magnet by its long continuance in this position ; as if the earth itself, which on account of the distance acts but feebly (the surface or outer crust of the earth being destitute, as he insists, of magnetic power), were yet able by this long continuance to supply the touch of the magnet, and excite the iron, and then shape and turn it when excited. Another is, that if iron that has been heated white-hot, be while cooling laid length-wise between north and south, it also acquires polarity without the touch of the magnet ; as if the parts of the iron, set in motion by ignition, and afterwards recovering themselves, were at the very moment of cooling more susceptible and sensitive of the virtue emanating from the earth than at other times, and thus became excited by it. But these things, though well observed, do not quite prove what he asserts [74].

[73] Nothing can be more ingenious than the *instantia crucis* here proposed. A series of observations were made by Dr. Whewell and Mr. Airy to determine the effect on the time of vibration of a pendulum, produced by carrying it to the bottom of a mine ; but, probably from the effect of local attractions, the results were scarcely as satisfactory as might have been expected. In the autumn of 1854, Mr. Airy instituted similar experiments in the Harton Colliery. They appear likely to afford more satisfactory results than the older series made at Dolcoath.

Voltaire cites the passage in the text in support of his remark that " le plus grand service, peut-être, que F. Bacon ait rendu à la philosophie a été de deviner l'attraction ". But in reality the notion of attraction in one form or other (e.g the attraction of the sea by the moon) sprang up in the infancy of physical speculation ; and it cannot be affirmed that Bacon's ideas on the subject were as clear as those of his predecessor William Gilbert. (See note on *De Aug.* ii. 13) By an error similar to Voltaire's, some of Dante's commentators have claimed for him the credit of being the first to indicate the true cause of the tides. The passage on which this claim is founded is in the *Paradiso*, xvi. 82.

[74] See, for these two remarks, the twelfth chapter of the third book of Gilbert's treatise *De Magnete*. It is illustrated by a curious woodcut representing the smith forging a bar of iron, and holding it, as he does so, in the plane of the meridian.

Now with regard to this question an Instance of the Fingerpost would be the following. Take a magnetic globe [75] and mark its poles ; and set the poles of the globe towards the east and west, not towards the north and south, and let them remain so ; then place at the top an untouched iron needle, and allow it to remain in this position for six or seven days.

The needle while over the magnet (for on this point there is no dispute) will leave the poles of the earth and turn towards the poles of the magnet ; and therefore, as long as it remains thus, it points east and west. Now if it be found that the needle, on being removed from the magnet and placed on a pivot, either starts off at once to the north and south, or gradually turns in that direction, then the presence of the earth must be admitted as the cause ; but if it either points as before east and west, or loses its polarity, this cause must be regarded as questionable, and further inquiry must be made.

Again, let the nature in question be the Corporeal Substance of the Moon ; that is, let us inquire whether it be rare, consisting of flame or air, as most of the old philosophers opined ; or dense and solid, as Gilbert and many moderns, with some ancients, maintain [76]. The reasons for the latter opinion rest chiefly on this, that the moon reflects the rays of the sun ; nor does light seem to be reflected except by solid bodies. Therefore Instances of the Fingerpost on this question will (if any) be those which prove that reflexion may take place from a rare body, as flame, provided it be of sufficient denseness. Certainly one cause of twilight, among others, is the reflexion of the rays of the sun from the upper part of the air. Likewise we occasionally see rays of the sun in fine evenings reflected from the fringes of dewy clouds with a splendour not inferior to that reflected from the body of the moon, but brighter and more gorgeous [77], and yet there is no proof that these clouds have coalesced into a dense body of water. Also we observe that the dark air behind a window at night reflects the light of a candle, just as a dense body would [78]. We should also try the experiment of allowing the sun's rays to shine through a hole on some dusky blueish flame. For indeed the open rays of the sun, falling on the duller kinds of flame, appear to deaden them, so that they seem more like white smoke than flame. These are what occur to me at present as Instances of the Fingerpost, with reference to this question ; and better may perhaps be found. But it should always be observed that reflexion from flame is not to be expected, except from a flame of some depth, for otherwise it borders on transparency. This however may be set down as certain,—that light on an even body is always either received and transmitted or reflected.

Again let the nature in question be the Motion of Projectiles (darts, arrows, balls, etc.) through the air. This motion the schoolmen, as their way is, explain in a very careless manner ; thinking it enough to call it a violent motion as distinguished from what they call a natural motion ; and to account for the first percussion or impulse by the axiom that two bodies cannot occupy the same place on account of the impenetrability of matter ; and not troubling themselves at all how the motion proceeds afterwards. But with reference to this inquiry the road branches into two in this way. Either this motion is caused by the air carrying the projected body and collecting behind it, as the stream in the case of a boat, or the wind in that of straws ; or it is caused by the parts of the body itself not enduring the impression, but pushing forward in succession to relieve

[75] Orig. *Terrella*. This is a word used by Gilbert to denote a spherical magnet. One of the fundamental ideas of his philosophy was that the earth was a great magnet ; and a magnet of the same form was therefore called a " little earth," or *terrella*. See, for instance, his treatise *De Magnete*, ii. cc. 7, 8.

[76] See Gilbert's *De Mundo*, ii. c. 13 *et seqq*. [Bacon here appears to lean to the view that the moon is a vapour, see also p. 704.—Ed.]

[77] The comparison of the brightness of the moon in the daytime with that of a cloud was ingeniously applied by Bouguer to determine the ratio of the moon's light to the sun's.

[78] [It is hardly necessary to remark that it is not the air but the glass that causes the reflection.—Ed.]

themselves from it. The former of these explanations is adopted by Fracastorius and almost all who have entered into the investigation with any subtlety [79], and there is no doubt that the air *has* something to do with it : but the other motion is undoubtedly the true one, as is shown by countless experiments. Among others the following would be an Instance of the Fingerpost on this subject : that a thin iron plate or stiffish iron wire, or even a reed or pen split in half, when pressed into a curve between the finger and thumb, leaps away. For it is obvious that this motion cannot be imputed to the air gathering behind the body, because the source of motion is in the middle of the plate or reed, not in the extremities.

Again, let the nature in question be the rapid and powerful motion of the Expansion of Gunpowder into flame ; by which such vast masses are upheaved, such great weights discharged, as we see in mines and mortars. With respect to this nature the road branches into two in this way. The motion is excited either by the mere desire of the body to expand when set on fire, or partly by that and partly by the desire of the crude spirit in the body, which flies rapidly away from the fire, and bursts violently from its embrace as from a prison-house. The schoolmen and common opinion deal only with the former desire. For men fancy themselves very fine philosophers, when they assert that the flame is endowed by its elementary form with a necessity of occupying a larger space than the body had filled when in the form of powder, and that hence the motion ensues. Meanwhile they forget to notice, that although this be true on the supposition that flame is generated, it is yet possible for the generation of flame to be hindered by a mass of matter sufficient to suppress and choke it ; so that the case is not reduced to the necessity they insist on. For that expansion must necessarily take place, and that there must needs follow thereon a discharge or removal of the opposing body, if flame be generated, they rightly judge. But this necessity is altogether avoided if the solid mass suppress the flame before it be generated. And we see that flame, especially in its first generation, is soft and gentle, and requires a hollow space wherein to play and try its strength. Such violence therefore cannot be attributed to flame by itself. But the fact is, that the generation of these windy flames, or fiery winds as they may be called, arises from a conflict of two bodies of exactly opposite natures ; the one being highly inflammable, which is the nature of sulphur, the other abhorring flame, as the crude spirit in nitre ; so that there ensues a strange conflict, the sulphur kindling into flame with all its might (for the third body, the willow-charcoal, does no more than incorporate and combine the other two), while the spirit of the nitre bursts forth with all its might, and at the same time dilates itself (as air, water, and all crude bodies do when affected by heat), and by thus flying and bursting out fans meanwhile the flame of the sulphur on all sides as with hidden bellows.

On this subject we may have Instances of the Fingerpost of two kinds. The first, of those bodies which are most highly inflammable, as sulphur, camphor, naphtha and the like, with their compounds ; which catch fire more quickly and easily than gunpowder if not impeded (from which it appears that the desire of bursting into flame does not produce by itself that stupendous effect) ; the other, of those bodies which shun and abhor flame, as all salts. For we find that if salts are thrown into the fire, their aqueous spirit bursts out with a crackling noise,

[79] See Fracastorius, *De Sympathiâ et Antipathiâ*, c. 4. The notion that the air concurred in producing the continued motion of projectiles is found in the *Timæus*, p. 80. Plato has been speaking of respiration, of which his theory is, that the expiration of air through the nostrils and mouth pushes the contiguous external air from its place, which disturbs that near it, and so on until a circle is formed, whereby, by antiperistasis, air is forced in through the flesh, to fill up the cavity of the chest—a circulation of air through the body, in short. On the same principle he would have explained a variety of other phenomena—the action of cupping instruments, swallowing, the motion of projectiles, etc., etc. All these, however, after suggesting the explanation, he leaves unexplained. But Plutarch, *Quæst. Platon.* x. (p. 177. of Reiske's *Plutarch*) developes a similar explanation in each case. This explanation, however, is not Plato's but Plutarch's, though it is probably what Plato would himself have said.

before flame is caught; which is the case also, though in a milder degree, with the stiffer kinds of leaves: the aqueous part escaping before the oily catches fire. But this is best seen in quicksilver, which is not inaptly called mineral water [80]. For quicksilver, without bursting into flame, by mere eruption and expansion almost equals the force of gunpowder, and is also said when mixed with gunpowder to increase its strength.

Again, let the nature in question be the Transitory Nature of Flame and its momentaneous extinction. For the nature of flame appears to have no fixed consistency here with us, but to be every moment generated and every moment extinguished; for it is clear that in flames which continue and last, the continuance we see is not of the same individual flame, but is caused by a succession of new flame regularly generated; nor does the flame remain numerically identical; as is easily seen from this, that if the food or fuel of flame be taken away, the flame instantly goes out. With reference to this nature the roads branch into two, thus. The momentaneous nature proceeds either from a cessation of the cause which at first produced the flame, as in light, sound, and the motion called "violent"; or from this, that the flame, though able by its own nature to remain with us, suffers violence and is destroyed by the contrary natures that surround it.

On this subject therefore we may take the following as an Instance of the Fingerpost. We see in large fires how high the flames ascend; for the broader the base of the flame, the higher is its vertex. Thus extinction appears to commence at the sides, where the flame is compressed and troubled by the air. But the heart of the flame, which is not touched by the air but surrounded by other flame on all sides, remains numerically identical; nor is it extinguished until gradually compressed by the surrounding air. Thus all flame is in the form of a pyramid, being broader at the base where the fuel is, but sharp at the vertex, where the air is antagonistic and fuel is wanting. But smoke is narrow at the base and grows broader as it ascends, like an inverted pyramid; the reason being that the air admits smoke and compresses flame. For let no one dream that lighted flame is air, when in fact they are substances quite heterogeneous.

But we may have an Instance of the Fingerpost more nicely adapted to this purpose, if the thing can be made manifest with bicoloured lights. Fix a lighted wax taper in a small metal stand; place the stand in the middle of a bowl, and pour round it spirit of wine, but not enough to reach the top of the stand. Then set fire to the spirit of wine. The spirit of wine will yield a blueish, the taper a yellow flame. Observe therefore whether the flame of the taper (which is easily distinguished by its colour from the flame of the spirit of wine, since flames do not mix at once, as liquids do) remains in a conical or rather tends to a globular form, now that there is nothing to destroy or compress it [81]. If the latter is found to be the case, it may be set down as certain that flame remains numerically identical as long as it is enclosed within other flame and feels not the antagonistic action of the air.

Let this suffice for Instances of the Fingerpost. I have dwelt on them at some length, to the end that men may gradually learn and accustom themselves to judge of nature by Instances of the Fingerpost and Experiments of Light, and not by probable reasonings.

XXXVII.

Among Prerogative Instances I will put in the fifteenth place *Instances of Divorce;* which indicate the separation of natures of most familiar occurrence. They differ from the instances subjoined to the Instances of Companionship, in that the latter indicate the separation of a nature from some concrete substance with which it is ordinarily in conjunction, while these instances indicate the separation of one nature from another. They differ from Instances of the Fingerpost, in that they determine nothing, but simply notify the separability of one

[80] It is well known that the expansive force of the vapour of mercury at high temperatures is enormous.

[81] This experiment is mentioned as actually *tried* in *Sylva Sylvarum*, 31.

nature from another. Their use is to detect false forms, and to dissipate slight theories suggested by what lies on the surface, and so serve as ballast to the understanding.

For example, let the natures investigated be those four natures which Telesius accounts as messmates and chamber-fellows [82]; namely, Heat, Brightness, Rarity, Mobility or promptness to motion. We find however many Instances of Divorce between them. For air is rare and mobile, not hot or bright; the moon is bright without heat; boiling water is hot without light; the motion of an iron needle on a pivot is quick and nimble, and yet the body is cold, dense and opaque; and there are many more of the kind.

Again, let the natures investigated be Corporeal Nature and Natural Action. For it seems that natural action is not found except as subsisting in some body. Yet in this case also we shall perhaps be able to find some Instance of Divorce; such, for example, as magnetic action, by which iron is drawn to the magnet, heavy bodies to the globe of the earth. There may also be added some other operations performed at a distance. For such action takes place both in time, occupying moments not a mere instant of time; and in space, passing through degrees and distances. There is therefore some moment of time, and some distance of space, in which the virtue or action remains suspended between the two bodies which produce the motion. The question therefore is brought to this; whether the bodies which are the limits of the motion dispose or alter the intermediate bodies, so that by a succession of actual contacts the virtue passes from limit to limit, meanwhile subsisting in the intermediate body; or whether there is no such thing, but only the bodies, the virtue, and the distances. In rays of light indeed, and sounds, and heat, and certain other things acting at a distance it is probable that the intermediate bodies are disposed and altered; the more so, because they require a medium qualified for carrying on the operation. But that magnetic or attractive virtue admits of media without distinction, nor is the virtue impeded in any kind of medium. And if the virtue or action has nothing to do with the intermediate body, it follows that there is a natural virtue or action subsisting for a certain time and in a certain space without a body; since it neither subsists in the limiting nor in the intermediate bodies. And therefore magnetic action may be an Instance of Divorce between corporeal nature and natural action. To which may be appended as a corollary or advantage not to be omitted, that here is a proof furnished by merely human philosophy of the existence of essences and substances separate from matter and incorporeal. For allow that a natural virtue and action, emanating from a body, can exist for a certain time and in a certain space altogether without a body, and you are not far from allowing that it can also emanate originally from an incorporeal substance. For corporeal nature appears to be no less requisite for sustaining and conveying natural action, than for exciting or generating it.

XXXVIII.

Now follow five classes of instances which under one general name I call *Instances of the Lamp*, or *of First Information*. They are those which aid the senses. For since all Interpretation of Nature commences with the senses, and leads from the perceptions of the senses by a straight, regular, and guarded path to

[82] The fundamental idea of Telesius's philosophy is, that heat and cold are the great constituent principles of the universe, and that the antithesis between them corresponds to that which he recognises between the sun and the earth:—" Omnino calidus, tenuis, candidus, mobilisque est Sol; Terra contra frigida, crassa, immobilis, tenebricosaque . . . unum Sol in terram emittens calorem ejus naturam facultatesque et conditiones ex eâ deturbat omnes, suasque ei indit; et eodem ferme modo quo Sol terram, etiam calor quivis, vel qui e commotis contritisque enascitur rebus, quæ corripit exuperatque immutare videtur; frigus scilicet ex iis, ejusque facultates conditionesque omnes, crassitiem, obscuritatem, immobilitatem, deturbare, et se ipsum iis, propriasque facultates conditionesque omnes, tenuitatem, albedinem et mobilitatem, indere . . . . videtur."—*De Rerum Naturâ*, i. c. I.

the perceptions of the understanding, which are true notions and axioms, it follows of necessity that the more copious and exact the representations of the senses, the more easily and prosperously will everything proceed.

Of these five Instances of the Lamp, the first strengthen, enlarge, and rectify the immediate actions of the senses ; the second make manifest things which are not directly perceptible by means of others which are ; the third indicate the continued processes or series of those things and motions which are for the most part unobserved except in their end or periods ; the fourth provide the sense with some substitute when it utterly fails ; the fifth excite the attention and notice of the sense, and at the same time set bounds to the subtlety of things. Of these I shall now speak in their order.

XXXIX.

Among Prerogative Instances I will put in the sixteenth place *Instances of the Door or Gate*, this being the name I give to instances which aid the immediate actions of the senses. Now of all the senses it is manifest that sight has the chief office in giving information. This is the sense therefore for which we must chiefly endeavour to procure aid. Now the aids to sight are of three kinds ; it may be enabled to perceive objects that are not visible ; to perceive them further off ; and to perceive them more exactly and distinctly.

Of the first kind (not to speak of spectacles and the like, which serve only to correct or relieve the infirmity of a defective vision, and therefore give no more information) are those recently invented glasses which disclose the latent and invisible minutiæ of bodies, and their hidden configurations and motions, by greatly increasing their apparent size ; instruments by the aid of which the exact shape and outline of body in a flea, a fly, a worm, and also colours and motions before unseen, are not without astonishment discerned. It is also said [83] that a straight line drawn with a pen or pencil is seen through such glasses to be very uneven and crooked ; the fact being that neither the motion of the hand, though aided by a ruler, nor the impression of the ink or colour, is really even ; although the unevenness is so minute that it cannot be detected without such glasses. And here (as is usual in things new and wonderful) a kind of superstitious observation has been added ; viz. that glasses of this sort do honour to the works of nature, but dishonour to the works of art [84]. The truth however is only this, that natural textures are far more subtle than artificial. For the microscope, the instrument I am speaking of, is only available for minute objects ; so that if Democritus had seen one, he would perhaps have leaped for joy, thinking a way was now discovered of discerning the atom, which he had declared to be altogether invisible [85]. The incompetency however of such glasses, except for minutiæ alone, and even for them when existing in a body of considerable size, destroys the use of the invention. For if it could be extended to larger bodies, or to the minutiæ of larger bodies, so that the texture of a linen cloth could be seen like network, and thus the latent minutiæ and inequalities of gems, liquors, urine, blood, wounds, &c. could be distinguished, great advantages might doubtless be derived from the discovery.

Of the second kind are those other glasses discovered by the memorable efforts of Galileo, by the aid of which, as by boats or vessels, a nearer intercourse with the heavenly bodies can be opened and carried on. For these show us that the milky way is a group or cluster of small stars entirely separate and distinct ; of which

83 Compare *Aph.* xiii. § 28. It would appear from the passage in the text that Bacon had not even seen one of the newly invented microscopes.—*J. S.*

84 Leibnitz goes as far as to say, " La matière arrangée par une sagesse divine doti être essentiellement organisée partout ; . . . il y a machine dans les parties de la machine naturelle à l'infini ".—*Sur le Principe de Vie*, p. 431 of Erdmann's edition.

85 Democritus maintained that the atom was wholly incognisable by the senses. Thus Sextus Empiricus mentions him along with Plato as having held the doctrine. *μόνα τὰ νοητὰ ἀληθῆ εἶναι* ; the reason in the case of Democritus being that his atoms, which alone he recognised as realities, possessed *πάσης αἰσθητῆς ποιότητος ἔρημον φύσιν*. —Sext. Emp. *Advers. Logicos*, ii. § 6.

fact there was but a bare suspicion among the ancients. They seem also to point out that the spaces of the planetary orbits, as they are called, are not altogether destitute of other stars, but that the heaven begins to be marked with stars before we come to the starry sphere itself; although with stars too small to be seen without these glasses. With this instrument we can descry those small stars wheeling as in a dance round the planet Jupiter; whence it may be conjectured that there are several centres of motion among the stars. With this the inequalities of light and shade in the moon are more distinctly seen and placed; so that a sort of selenography can be made. With this we descry spots in the sun, and similar phenomena; all indeed noble discoveries, so far as we may safely trust to demonstrations of this kind; which I regard with suspicion chiefly because the experiment stops with these few discoveries, and many other things equally worthy of investigation are not discovered by the same means [86].

Of the third kind are measuring rods, astrolabes, and the like; which do not enlarge the sense of sight, but rectify and direct it. And if there are other instances which aid the remaining senses in their immediate and individual actions, and yet are of a kind which add nothing to the information already possessed, they are not to the present purpose; and therefore I have omitted to mention them.[87]

XL

Among Prerogative Instances I will put in the seventeenth place *Summoning Instances*, borrowing the name from the courts of law; because they summon objects to appear which have not appeared before; I also call them *Evoking Instances*. They are those which reduce the non-sensible to the sensible; that is, make manifest things not directly perceptible by means of others which are.

An object escapes the senses, either on account of its distance; or on account of the interposition of intermediate bodies; or because it is not fitted for making an impression on the sense; or because it is not sufficient in quantity to strike the sense; or because there is not time enough for it to act on the sense; or because the impression of the object is such as the sense cannot bear; or because the sense has been previously filled and occupied by another object, so that there is not room for a new motion. These cases have reference principally to the sight and secondarily to the touch. For these two senses give information at large and concerning objects in general; whereas the other three give no information hardly, but what is immediate and relates to their proper objects.

In the first kind, where an object is imperceptible by reason of its distance, there is no way of manifesting it to the sense but by joining to it or substituting for it some other object which may challenge and strike the sense from a greater distance; as in communication by beacons, bells, and the like.

[86] Galileo often mentions the attempt which many of the Peripaticians made to set aside all arguments founded on his discoveries with the telescope, by saying that they were mere optical delusions. J. C. La Galla, in his dissertation *De Phænominis in Orbe Lunæ*, has a section entitled "De Telescopii Veritate", in which, though an Aristotelian, he has nevertheless admitted that this objection is untenable.

[87] Compare this with the passage in the *Descriptio Globi Intellectualis* (c. v.) where Bacon speaks of Galileo's invention and discoveries (the firstfruits of which had just been announced) in a strain of more sanguine expectation. From that passage, written eight years before, we may learn (I think) why it was that Bacon had now begun to doubt how far these observations could be trusted. Believing as he did that all the received theories of the heavens were full of error, as soon as he heard that by means of the telescope men could really see so much further into the heavens than before, he was prepared to hear of a great number of new and unexpected phenomena; and his only fear was that the observers, instead of following out their observations patiently and carefully, would begin to form new theories. But now that nine years had passed since the discovery of Jupiter's satellites, the spots in the sun, etc., and no new discovery of importance had been announced, he wondered how it could be that men seeing so much further should be able to see so little more than they did, and began to suspect that it was owing to some defect either in the instrument or in the methods of observation.—*J. S.*

In the second kind, this reduction or secondary manifestation is effected when objects that are concealed by the interposition of bodies within which they are enclosed, and cannot conveniently be opened out, are made manifest to the sense by means of those parts of them which lie on the surface, or make their way from the interior. Thus the condition of the human body is known by the state of the pulse, urine, and the like.

In the third and fourth kind, reductions are applicable to a great many things, and in the investigations of nature should be sought for on all sides. For example, it is obvious that air and spirit, and like bodies, which in their entire substance are rare and subtle, can neither be seen nor touched. Therefore in the investigation of bodies of this kind, it is altogether necessary to resort to reductions.

Thus let the nature in question be the Action and Motion of the Spirit enclosed in tangible bodies. For everything tangible that we are acquainted with contains an invisible and intangible spirit, which it wraps and clothes as with a garment. Hence that three-fold source, so potent and wonderful, of the process of the spirit in a tangible body. For the spirit in a tangible substance, if discharged, contracts bodies and dries them up; if detained, softens and melts them; if neither wholly discharged nor wholly detained, gives them shape, produces limbs, assimilates, digests, ejects, organises, and the like. And all these processes are made manifest to the sense by conspicuous effects.

For in every tangible inanimate body the enclosed spirit first multiplies itself, and as it were feeds upon those tangible parts which are best disposed and prepared for that purpose; and so digests and elaborates and turns them into spirit; and then they escape together. Now this elaboration and multiplication of the spirit is made manifest to the sense by diminution of weight. For in all desiccation there is some decrease of quantity; not only of the quantity of spirit previously existing in the body, but also of the body itself, which was before tangible and is newly changed; for spirit is without weight. Now the discharge or emission of the spirit is made manifest to the sense in the rust of metals and other similar putrefactions, which stop short before they come to the rudiments of life; for these belong to the third kind of process. For in compact bodies the spirit finds no pores or passages through which to escape, and is therefore compelled to push and drive before it the tangible parts themselves, so that they go out along with it; whence proceed rust and the like. On the other hand the contraction of the tangible parts, after some of the spirit is discharged (upon which desiccation ensues), is made manifest to the sense not only by the increased hardness of the body, but much more by the rents, contractions, wrinklings, and shrivellings, in the body which thereupon take place. For the parts of wood split asunder and are contracted; skins shrivel; and not only that, but if the spirit is suddenly discharged by the heat of fire, they hasten so fast to contraction as to curl and roll themselves up.

On the contrary, where the spirit is detained, and yet expanded and excited by heat or something analogous thereto (as happens in the more solid or tenacious bodies), then are bodies softened, as iron white hot; or they become fluid, as metals; or liquid, as gums, wax, and the like. Thus the contrary operations of heat, which hardens some substances and melts others, are easily reconciled; since in the former the spirit is discharged, in the latter it is excited and detained; whereof the melting is the proper action of the heat and spirit; the hardening is the action of the tangible parts only on occasion of the discharge of the spirit.

But when the spirit is neither wholly detained nor wholly discharged, but only makes trials and experiments within its prison-house, and meets with tangible parts that are obedient and ready to follow, so that whithersoever the spirit leads they go along with it, then ensues the forming of an organic body, and the development of organic parts, and all the other vital actions as well in vegetable as in animal substances. And these operations are made manifest to the sense chiefly by careful observation of the first beginnings and rudiments or essays of life in animalculæ generated from putrefaction; as in ants' eggs, worms, flies, frogs after rain, etc. There is required however for the production of life both mildness in the heat and pliancy in the substance, that the spirit may neither be so hurried as to break out, nor be confined by the obstinacy of the parts ; but may rather be able to mould and model them like wax.

Again, that most noble distinction of spirit which has so many applications (viz. spirit cut off ; spirit simply branching ; spirit at once branching and cellulate ; of which the first is the spirit of all inanimate substances, the second of vegetables, the third of animals), is brought as it were before the eyes by several instances of this kind of reduction.

In like manner it appears that the more subtle textures and configurations of things (though the entire body be visible or tangible) are perceptible neither to the sight nor touch. And therefore in these also our information comes by way of reduction. Now the most radical and primary difference between configurations is drawn from the abundance or scantiness of the matter occupying the same space or dimensions. For all other configurations (which have reference to the dissimilarity of the parts contained in the same body, and to their collocation and position) are but secondary in comparison with the former.

Thus let the nature in question be the Expansion or Coition of Matter in bodies compared one with another ; viz. how much matter occupies how much space in each. For there is nothing more true in nature than the twin propositions, that "nothing is produced from nothing," and "nothing is reduced to nothing," but that the absolute quantum or sum total of matter remains unchanged, without increase or diminution.[88] Nor is it less true that of that quantum of matter more or less is contained under the same space or dimensions according to the diversity of bodies ; as in water more, in air less ; so that to assert that a given volume of water can be changed into an equal volume of air is as much as to say that something can be reduced to nothing ; as on the other hand to maintain that a given volume of air can be turned into an equal volume of water, is the same as to say that something can be produced out of nothing. And it is from this abundance and scantiness of matter, that the abstract notions of Dense and Rare, though variously and promiscuously used, are properly speaking derived. We must also take for granted a third proposition which is also sufficiently certain ; viz. that this greater or less quantity of matter in this or that body is capable of being reduced by comparison to calculation and to exact or nearly exact proportions. Thus one would be justified in asserting that in any given volume of gold there is such an accumulation of matter, that spirit of wine, to make up an equal quantity of matter, would require twenty-one times the space occupied by the gold.

Now the accumulation of matter and its proportions are made manifest to the sense by means of weight. For the weight answers to the quantity of matter, in the parts of a tangible body ; whereas spirit and the quantum of matter which it contains cannot be computed by weight ; for it rather diminishes the weight than increases it. But I have drawn up a very accurate Table on this subject ; in which I have noted down the weights and volumes of all the metals, the principal stones, woods, liquors, oils, and many other bodies as well natural as artificial ; a thing of great use in many ways, as well for light of information as for direction in practice ; and one that discloses many things quite beyond expectation. Not the least important of which is this—it shows that all the variety in tangible bodies known to us (such bodies I mean as are tolerably compact and not quite spongy and hollow, and chiefly filled with air) does not exceed the limit of the ratio of 1 to 21 :—so limited is nature, or at any rate that part of it with which we have principally to do.

I have also thought it worth while to try whether the proportions can be calculated which intangible or pneumatic bodies bear to bodies tangible. This I attempted by the following contrivance. I took a glass phial, capable of holding about an ounce ; using a small vessel, that less heat might be required to produce evaporation. This phial I filled with spirit of wine almost to the neck ; selecting spirit of wine, because I found by the former Table that of all tangible bodies (which are well united and not hollow) this is the rarest, and contains the least

88 It is worth remarking that Bacon here asserts as absolutely certain a maxim which is assuredly no result of experience. The same doctrine is as distinctly though not so emphatically asserted by Telesius, I. c. 5.

quantity of matter in a given space. After that, I noted exactly the weight of the spirit and phial together. I then took a bladder, capable of holding about a quart; from which I squeezed out, as well as I could, all the air, until the two sides of the bladder met. The bladder I had previously rubbed over gently with oil, to make it closer; and having thus stopped up the pores, if there were any, I inserted the mouth of the phial within the mouth of the bladder, and tied the latter tightly round the former with a thread, smeared with wax in order that it might stick more closely and tie more firmly. After this I set the phial on a chafing dish of hot coals. Presently the steam or breath of the spirit of wine, which was dilated and rendered pneumatic by the heat, began gradually to expand the bladder, and swelled it out on all sides like a sail. When this took place, I immediately took the glass off the fire placing it on a carpet, that it might not crack with the cold; at the same time making a hole in the bladder, lest the steam should turn liquid again on the cessation of the heat, and so disturb the calculations. I then removed the bladder, and weighing the spirit of wine which remained, computed how much had been converted into steam or air. Then comparing the space which the body had occupied while it was spirit of wine in the phial, with the space which it afterwards occupied when it had become pneumatic in the bladder, I computed the results; which showed clearly, that the body had acquired by the change a degree of expansion a hundred times greater that it had had before.

Again, let the nature in question be Heat or Cold, in a degree too weak to be perceptible to the sense. These are made manifest to the sense by a calendar glass such as I have described above. For the heat and cold are not themselves perceptible to the touch, but the heat expands the air, and the cold contracts it. Nor again is this expansion and contraction of the air perceptible to the sight; but the expansion of the air depresses the water, the contraction raises it; and so at last is made manifest to the sight; not before, nor otherwise.

Again, let the nature in question be the Mixture of Bodies; viz. what they contain of water, oil, spirit, ash, salt, and the like; or (to take a particular instance) what quantity of butter, curd, whey, etc., is contained in milk. These Mixtures, so far as relates to tangible elements, are made manifest to the sense by artificial and skilful separations. But the nature of the spirit in them, though not immediately perceived, is yet discovered by the different motions and efforts of the tangible bodies in the very act and process of their separation; and also by the acridities and corrosions, and by the different colours, smells, and tastes of the same bodies after separation. And in this department men have laboured hard, it is true, with distillations and artificial separations, but not with much better success than in the other experiments which have been hitherto in use; for they have but groped in the dark, and gone by blind ways, and with efforts painstaking rather than intelligent; and (what is worst of all) without attempting to imitate or emulate nature, but rather destroying by the use of violent heats and over-strong powers all that more subtle configuration in which the occult virtues and sympathies of things chiefly reside. Nor do they remember or observe, while making such separations, the circumstance which I have elsewhere pointed out,—namely, that when bodies are tormented by fire or other means, many qualities are communicated by the fire itself, and by the bodies employed to effect the separation, which did not exist previously in the compound; whence strange fallacies have arisen. For it must not be supposed that all the vapour which is discharged from water by the action of fire, was formerly vapour or air in the body of the water; the fact being that the greatest part of it was created by the expansion of the water from the heat of the fire.

So in general, all the nice tests of bodies whether natural or artificial, by which the genuine are distinguished from the adulterated, the better from the viler sort, should be referred to this division; for they make manifest to the sense things not directly perceptible, by means of those which are. They should therefore be sought and collected from all quarters with diligent care.

With regard to the fifth way in which objects escape the sense, it is obvious that the action of sense takes place in motion, and that motion takes place in time. If therefore the motion of any body be either so slow or so quick that it bears no proportion to the moments which the sense takes to act in, the object is not perceived at all; as in the motion of the hand of a clock, and again in the motion of a musket ball. Now motion which is too slow to be perceived is easily and usually made manifest to the sense by means of aggregates of motion. Motion which is too quick has not hitherto been competently measured; and yet the investigation of nature requires that this be done in some cases.

In the sixth kind, where the sense is hindered by the too great power of the object, the reduction may be effected either by removing the object to a greater distance from the sense; or by deadening its effects by the interposition of a medium which will weaken without annihilating the object; or by admitting and receiving the reflexion of the object, where the direct impression is too powerful; as that of the sun for instance in a basin of water.

The seventh cause, where the sense is so charged with one object that it has no room for the admission of another, is almost wholly confined to the sense of smell, and has little to do with the matter in hand. So much then for the reduction of the non-sensible to the sensible,—or the modes of making manifest to the sense things not directly perceptible by means of others which are [89].

Sometimes however the reduction is made not to the sense of a man, but of some other animal, whose sense in some cases is keener than man's; as of certain scents to the sense of a dog; of the light which is latent in air when not illumined from without to the sense of a cat, owl, and similar animals, which see in the dark. For Telesius has justly observed, that there is in the air itself a certain original light, though faint and weak, and hardly of any use to the eyes of men and most animals; inasmuch as animals to whose sense this light is adapted see in the dark, which it is hardly to be believed they do either without light, or by a light within.

Observe also, that at present I am dealing with the deficiencies of the senses and their remedies. The deceptions of the senses must be referred to the particular inquiries concerning sense and the objects of sense; excepting only that grand deception of the senses, in that they draw the lines of nature with reference to man and not with reference to the universe; and this is not to be corrected except by reason and universal philosophy.

XLI.

Among Prerogative Instances I will put in the eighteenth place *Instances of the Road*; which I also call *Travelling Instances* and *Articulate Instances*. They are those which point out the motions of nature in their gradual progress. This class of instances escapes the observation rather than the sense. For it is strange how careless men are in this matter; for they study nature only by fits and at intervals, and when bodies are finished and completed, not while she is at work upon them. Yet if any one were desirous of examining and studying the contrivances and industry of an artificer, he would not be content with beholding merely the rude materials of the art, and then the completed works; but would rather wish to be present when the artificer was at his labours and carrying his work on. And a like course should be taken with the investigation of nature. For instance, if we are inquiring into the vegetation of plants, we must begin from the very sowing of the seed, and observe (as we may easily do, by taking out day after day the seeds that have lain in the ground two days, three days, four days, and so on, and carefully examining them) how and when the seed begins to puff and swell, and to be as it were filled with spirit; secondly, how it begins to burst the skin and put forth fibres, at the same time raising itself slightly upwards, unless the ground

89 An excellent instance of the "deductio nonsensibilis ad sensibile" [in the second kind] occurs in the experiments recently made by Messrs. Hopkins and Joule for determining the melting-point of substances subjected to great pressure. The substance acted on is enclosed in a tube out of reach and sight. But a bit of magnetized steel has previously been introduced into it, and is supported by it as long as it remains solid. A magnetic needle is placed beside the apparatus, a certain amount of deviation being, of course, produced by the steel within the tube. The moment the temperature reaches the melting-point, the steel sinks; and its doing so is indicated by the motion of the needle.

be very stiff ; also, how it puts forth its fibres, some for the root downwards, and some for the stem upwards, and sometimes also creeping sideways, if it there finds the ground more open and yielding ; with many other things of the kind. In the same way we should examine the hatching of eggs, in which we might easily observe the whole process of vivification and organisation, and see what parts proceed from the yolk, and what from the white of the egg, and so forth. A similar course should be taken with animals generated from putrefaction[90]. For to prosecute such inquiries concerning perfect animals by cutting out the fœtus from the womb would be too inhuman, except when opportunities are afforded by abortions, the chase, and the like. There should therefore be set a sort of night watch over nature, as showing herself better by night than by day. For these may be regarded as night studies by reason of the smallness of our candle and its continual burning.

The same too should be attempted with inanimate substances ; as I have done myself in investigating the expansion of liquids by fire. For there is one mode of expansion in water, another in wine, another in vinegar, another in verjuice[91], and quite another in milk and oil ; as was easily to be seen by boiling them over a slow fire, and in a glass vessel, in which everything may be clearly distinguished. These matters however I touch but briefly ; meaning to treat of them more fully and exactly when I come to the discovery of the *Latent Process* of things. For it should all along be borne in mind that in this place I am not handling the things themselves, but only giving examples.

XLII.

Among Prerogative Instances I will put in the nineteenth place *Supplementary* or *Substitutive Instances ;* which I also call *Instances of Refuge*. They are those which supply information when the senses entirely fail us ; and therefore we fly to them when appropriate instances are not to be had. Now substitution is made in two ways ; either by gradual approximation or by analogy. To take an example. There is no medium known by the interposition of which the operation of the magnet in drawing iron is entirely prevented. Gold placed between does not stop it, nor silver, nor stone, nor glass, wood, water, oil, cloth or fibrous substances, air, flame, etc. But yet by nice tests some medium may possibly be found to deaden its virtue more than any other ; comparatively, that is, and in some degree. Thus it may be that the magnet would not attract iron as well through a mass of gold, as through an equal space of air, or through ignited silver as well as through cold ; and so in other cases. For I have not made the trial myself in these cases. It is enough to propose such experiments by way of example. Again there is no body we are acquainted with which does not contract heat on being brought near the fire. And yet air contracts heat much more quickly than stone. Such is the Substitution which is made by gradual approximation.

Substitution by analogy is doubtless useful, but is less certain, and should therefore be applied with some judgment. It is employed, when things not directly perceptible are brought within reach of the sense, not by perceptible

[90] The epithet perfecta is generally given to those animals which cannot result from putrefaction. Cæsalpinus, in the *Quæstiones Peripat.* v. 1., maintains that all animals may result from putrefaction, and that this was the doctrine of Aristotle. The same opinion had, I believe, been advanced by Averroës. That mice may be produced by equivocal generation is asserted, as a matter not admitting of dispute, by Cardan, *De Rerum Varietate.* Cæsalpinus refers to the same instance, but less confidently than Cardan.. It is worth remarking that Aristotle, though he speaks of the great fecundity of mice, and even of their being impregnated by licking salt, does not mention the possibility of their being produced by putrefaction. (*De Hist. Animal.* vi. 37 ; *Problem.* x 64.) Paracelsus, *De Rerum Generatione*, affirms that all animals produced from putrefaction are more or less venomous. Telesius's opinion is that the more perfect animals cannot result from putrefaction, because the condition of temperature necessary to their production cannot be fulfilled except by means of animal heat.

[91] Wine made of sour grapes (Pliny, xiv. 18, and elsewhere).

operations of the imperceptible body itself, but by observation of some cognate body which is perceptible [92]. For example, suppose we are inquiring into the Mixture of Spirits, which are invisible bodies; there seems to be a certain affinity between bodies and the matter that feeds or nourishes them. Now the food of flame seems to be oil and fat substances; of air, water and watery substances; for flame multiplies itself over exhalations of oil, air over the vapour of water. We should therefore look to the mixture of water and oil, which manifests itself to the sense; since the mixture of air and flame escapes the sense. Now oil and water, which are mingled together very imperfectly by composition or agitation, are in herbs and blood and the parts of animals, very subtly and finely mingled. It is possible therefore that something similar may be the case with the mixture of flame and air in pneumatic bodies; which, though not readily mingling by simple commixture, yet seem to be mingled together in the spirits of plants and animals; especially as all animate spirit feeds on moist substances of both kinds, watery and fat, as its proper food.

Again, if the inquiry be not into the more perfect mixtures of pneumatic bodies, but simply into their composition; that is, whether they be readily incorporated together; or whether there be not rather, for example, certain winds and exhalations or other pneumatic bodies which do not mix with common air but remain suspended and floating therein in globules and drops; and are rather broken and crushed by the air than admitted into or incorporated with it; this is a thing which cannot be made manifest to the senses in common air and other pneumatic bodies, by reason of their subtlety; yet how far the thing may take place we may conceive, by way of image or representation, from what takes place in such liquids as quicksilver, oil, or water; and likewise from the breaking up of air when it is dispersed in water and rises in little bubbles; and again in the thicker kinds of smoke; and lastly in dust raised and floating in the air; in all of which cases no incorporation takes place. Now the representation I have described is not a bad one for the matter in question, provided that diligent inquiry has been first made whether there can be such a heterogeneity in pneumatic bodies as we find there is in liquids; for if there can, then these images by analogy may not inconveniently be substituted.

But with regard to these Supplementary Instances, although I stated that information was to be derived from them in the absence of instances proper, as a last resource; yet I wish it to be understood that they are also of great use, even when proper instances are at hand; for the purpose, I mean, of corroborating the information which the others supply. But I shall treat of them more fully, when I come in due course to speak of the *Supports of Induction.*

XLIII.

Among Prerogative Instances I will put in the twentieth place *Dissecting Instances;* which I also call *Awakening Instances*, but for a different reason. I call them Awakening, because they awaken the understanding; Dissecting, because they dissect nature; for which reason also I sometimes call them *Democritean.* They are those which remind the understanding of the wonderful and exquisite subtlety of nature, so as to stir it up and awaken it to attention and observation and due investigation. Such, for example, as these following: that a little drop of ink spreads to so many letters or lines; that silver gilt stretches to such a length of gilt wire [93]; that a tiny worm, such as we find in the skin, possesses in itself both spirit and a varied organization; that a little saffron tinges

[92] Du Bois Reymond's *Researches in Animal Electricity* give a good example of this. He constructed what may be called an electrical model of a muscle, and succeeded in obtaining an illustration not only of his fundamental result, namely, that any transverse section is negative with respect to any longitudinal one, but also of the more complicated relations between two different portions of the same section.

[93] Dr. Woolaston's method for obtaining wires of extreme fineness was perhaps suggested by the circumstance mentioned in the text. He enclosed a gold wire in a cylinder of silver, drew them out together, and then dissolved away the silver by means of warm nitrous acid.

a whole hogshead of water; that a little civet or musk scents a much larger volume of air; that a little incense raises such a cloud of smoke; that such exquisite differences of sounds, as articulate words, are carried in every direction through the air, and pierce even, though considerably weakened, through the holes and pores of wood and water; and are moreover echoed back, and that too with such distinctness and velocity; that light and colour pass through the solid substances of glass and water so speedily, and in so wide an extent, and with such copious and exquisite variety of images, and are also refracted and reflected; that the magnet acts through bodies of all sorts, even the most compact; and yet (which is more strange) that in all these, passing as they do through an indifferent medium (such as the air is), the action of one does not much interfere with the action of another; that is to say, that at the same time there are carried through spaces of air so many images of visible objects, so many impressions of articulate sound, so many distinct odours, as of a violet, rose, etc.; moreover heat and cold, and magnetic influences; all (I say) at once without impeding one another, just as if they had their own roads and passages set apart, and none ever struck or ran against other.

To these Dissecting Instances it is useful however to subjoin instances which I call Limits of Dissection; as that in the cases above mentioned, though one action does not disturb or impede another action of a different kind; yet one action does overpower and extinguish another action of the same kind; as the light of the sun extinguishes that of a glowworm; the report of a cannon drowns the voice; a strong scent overpowers a more delicate one; an intense heat a milder one; a plate of iron interposed between a magnet and another piece of iron destroys the action of the magnet. But this subject also will find its proper place among the Supports of Induction.

XLIV.

So much for instances which aid the senses; instances which are chiefly useful for the Informative Part of our subject. For information commences with the senses. But the whole business terminates in Works; and as the former is the beginning, so the latter is the end of the matter. I will proceed therefore with the instances which are pre-eminently useful for the Operative Part. They are of two kinds, and seven in number, though I call them all by the general name of *Practical Instances*. In the Operative Part there are two defects, and two corresponding prerogatives of instances. For operation either fails us or it overtasks us. The chief cause of failure in operation (especially after natures have been diligently investigated) is the ill determination and measurement of the forces and actions of bodies. Now the forces and actions of bodies are circumscribed and measured, either by distances of space, or by moments of time, or by concentration of quantity, or by predominance of virtue; and unless these four things have been well and carefully weighed, we shall have sciences, fair perhaps in theory, but in practice inefficient. The four instances which are useful in this point of view I class under one head as *Mathematical Instances* and *Instances of Measurement*.

Operation comes to overtask us, either through the admixture of useless matters, or through the multiplicity of instruments, or through the bulk of the material and of the bodies that may happen to be required for any particular work. Those instances therefore ought to be valued, which either direct practice to the objects most useful to mankind; or which save instruments; or which spare material and provision. The three instances which serve us here, I class together as *Propitious* or *Benevolent Instances*. These seven Instances I will now discuss separately, and with them conclude that division of my subject which relates to the Prerogatives or Rank of Instances.

XLV.

Among Prerogative Instances I will put in the twenty-first place *Instances of the Rod or Rule*; which I also call *Instances of Range* or of *Limitation*. For the powers and motions of things act and take effect at distances, not indefinite or accidental, but finite and fixed; so that to ascertain and observe these distances in the investigation of the several natures is of the greatest advantage to practice, not only to prevent its failure but also to extend and increase its power. For we are sometimes enabled to extend the range of powers, and as it were to diminish distances ; as for instance by the use of telescopes.

Most of these powers act and take effect only by manifest contact ; as in the impact of two bodies, where the one does not move the other from its place unless they touch each other. Also medicines that are applied externally, as ointments or plasters, do not exert their virtues without touching the body. Finally the objects of the taste and touch do not strike those senses unless they be contiguous to the organs.

There are also powers which act at a distance, though a very small one ; and of these only a few have been hitherto observed, albeit there are many more than men suspect ; as (to take common examples) when amber or jet attracts straws ; bubbles dissolve bubbles on being brought together ; certain purgative medicines draw humours downwards [94], and the like. So too the magnetic power by which iron and a magnet, or two magnets, are made to meet, operates within a fixed but narrow sphere of action ; but if there be any magnetic virtue flowing from the earth (a little below the surface), and acting on a steel needle in respect of its polarity, the action operates at a great distance.

Again, if there be any magnetic power which operates by consent between the globe of the earth and heavy bodies, or between the globe of the moon and the waters of the sea (as seems highly probable in the semimenstrual ebbs and floods [95]), or between the starry sphere and the planets, whereby the latter are attracted to their apogees ; all these must operate at very great distances. There are found also certain materials which catch fire a long way off ; as we are told the naphtha of Babylon does [96]. Heat also insinuates itself at great distances ; as also does cold ; insomuch that by the inhabitants of Canada the masses of ice that break loose and float about the northern ocean, and are borne through the Atlantic towards that coast, are perceived at a great distance by the cold they give out. Perfumes also (though in these there appears to be always a certain corporeal discharge) act at remarkable distances ; as those find who sail along the coasts of Florida, or some parts of Spain, where there are whole woods of lemon and orange and like odoriferous trees, or thickets of rosemary, marjoram, and the like [97]. Lastly the radiations of light and impressions of sound operate at vast distances.

But whether the distances at which these powers act be great or small, it is certain that they are all finite and fixed in the nature of things, so that there is a certain limit never exceeded ; and a limit which depends either on the mass or quantity of matter in the bodies acted on ; or on the strength or weakness of the powers acting ; or on the helps or hindrances presented by the media in which they act ; all which things should be observed and brought to computation. Moreover

[94] Bacon here speaks in accordance with the medical theory in which the brain is the origin and seat of the rheum, which descends from thence and produces disease in other organs—a theory preserved in the word catarrh. Certain purgatives were supposed to draw the rheum down.

[95] It is worth remarking that Galileo speaks contemptuously of the notion that the moon exerts any influence on the tides. His strong wish to explain everything mechanically led him in this instance wrong, as a similar wish has led many others. It arose, not unnaturally, from a reaction against the unsatisfactory explanations which the schoolmen were in the habit of deducing from the specific or occult properties of bodies. Even Leibnitz, in his controversy with Clarke, shows a tendency towards an exclusive preference of a mechanical system of physics, though in other parts of his writings he had spoken favourably of the doctrine of attraction, and though his whole philosophy ought, one would think, to have made him indifferent to the point in dispute. In a system of pre-established harmony, action by contact is as merely apparent as action at a distance.

[96] Strabo, xvi. p. 472. Pliny, ii. § 109.

[97] To the same purpose Milton, *Paradise Lost*, iv. 99 :—

"As when to them who sail
Beyond the Cape of Hope," etc.

the measurements of violent motions (as they are called), as of projectiles, guns, wheels, and the like, since these also have manifestly their fixed limits, should be observed and computed.

There are found also certain motions and virtues of a contrary nature to those which operate by contact and not at a distance; namely, those which operate at a distance and not by contact; and again those which operate more feebly at a less distance, and more powerfully at a greater. The act of sight for instance is not well performed in contact, but requires a medium and a distance. Yet I remember being assured by a person of veracity that he himself under an operation for the cataract, when a small silver needle was inserted within the first coat of the eye in order to remove the pellicle of the cataract and push it into a corner, saw most distinctly the needle passing over the very pupil. But though this may be true, it is manifest that large bodies are not well or distinctly seen except at the vertex of a cone [98], the rays from the object converging at a certain distance from it. Moreover, old people see objects better at a little distance than if quite close. In projectiles too it is certain that the impact is not so violent at too small a distance as it is a little further off. These therefore and like things should be observed in the measurements of motions with regard to distances.

There is also another kind of local measurement of motions which must not be omitted. This has to do with motions not progressive, but spherical; that is, with the expansion of bodies into a greater sphere or their contraction into a less. For among our measurements of motions we must inquire what degree of compression or extension bodies (according to their nature) easily and freely endure, and at what point they begin to resist, till at last they will bear no more. Thus when a blown bladder is compressed, it allows a certain compression of the air, but if the compression be increased, the air does not endure it, and the bladder bursts.

But this same thing I have tested more accurately by a subtle experiment. I took a small bell of metal, light and thin, such as is used for holding salt, and plunged it into a basin of water, so that it carried down with it the air contained in its cavity to the bottom of the basin; where I had previously placed a small globe, on which the bell was to light. I found then that if the globe was small enough in proportion to the cavity, the air contracted itself into a less space, and was simply squeezed together, not squeezed out. But if it was too large for the air to yield freely, then the air impatient of greater pressure raised the bell on one side, and rose to the surface in bubbles.

Again to test the extension, as well as compression, of which air was susceptible, I had recourse to the following device. I took a glass egg, with a small hole at one end of it, and having drawn out the air through the hole by violent suction, I immediately stopt up the hole with my finger, and plunged the egg into water, and then took away my finger. The air having been extended by the suction and dilated beyond its natural dimensions, and therefore struggling to contract itself again (so that if the egg had not been plunged into the water, it would have drawn in air with a hissing sound), now drew in water in sufficient quantities to allow the air to recover its old sphere or dimension [99].

Now it is certain that the rarer bodies (such as air) allow a considerable degree of contraction, as has been stated; but that tangible bodies (such as water) suffer compression with much greater difficulty, and to a less extent. How far they do suffer it, I have investigated in the following experiment. I had a hollow globe of lead made, capable of holding about two pints, and sufficiently thick to bear considerable force. Having made a hole in it, I filled it with water, and

[98] That is, the eye being at the apex of the visual cone.

[99] This explanation is wholly unsatisfactory. The principle upon which the true explanation depends, namely the pressure of the atmosphere, was, it seems tolerably certain, first suggested by Torricelli. If the experiment were performed in vacuo, no water would enter the egg, unless the egg were plunged to a considerable depth into the water, or unless the vacuum within it were more perfect than could be produced in the manner described.

then stopt up the hole with melted lead, so that the globe became quite solid. I then flattened two opposite sides of the globe with a heavy hammer, by which the water was necessarily contracted into less space; a sphere being the figure of largest capacity. And when the hammering had no more effect in making the water shrink, I made use of a mill or press; till the water impatient of further pressure exuded through the solid lead like a fine dew. I then computed the space lost by the compression, and concluded that this was the extent of compression which the water had suffered; but only when constrained by great violence[100].

But the compression or extension endured by more solid, dry, or more compact bodies, such as wood, stones and metals, is still less than this, and scarcely perceptible. For they free themselves either by breaking, or by moving forward, or by other efforts; as is apparent in the bending of wood or metal, in clocks moving by springs, in projectiles, hammerings, and numberless other motions. And all these things with their measures should in the investigation of nature be explored and set down, either in their certitude, or by estimate, or by comparison, as the case will admit.

XLVI.

Among Prerogative Instances I will put in the twenty-second place *Instances of the Course*; which I also call *Instances of the Water*; borrowing the term from the hourglasses of the ancients, which contained water instead of sand. These measure nature by periods of time, as the Instances of the Rod by degress of space. For all motion or natural action is performed in time; some more quickly, some more slowly, but all in periods determined and fixed in the nature of things. Even those actions which seem to be performed suddenly and (as we say) in the twinkling of an eye, are found to admit of degree in respect of duration.

First then we see that the revolutions of heavenly bodies are accomplished in calculated times; as also the flux and reflux of the sea. The motion of heavy bodies to the earth, and of light bodies towards the heavens, is accomplished in definite periods, varying with the bodies moved and the medium through which they move[101]. The sailing of ships, the movements of animals, the transmission of missiles, are all performed likewise in times which admit (in the aggregate) of measurement. As for heat, we see boys in winter time bathe their hands in flame without being burned, and jugglers by nimble and equable movements turn vessels full of wine or water upside down and then up again, without spilling

[100] This is perhaps the most remarkable of Bacon's experiments; and it is singular that it was so little spoken of by subsequent writers. Nearly fifty years after the publication of the *Novum Organum*, an account of a similar experiment was published by Megalotti, who was secretary of the Accademia del Cimento at Florence; and it has since been familiarly known as the Florentine experiment. I quote his account of it. "Facemmo lavorar di getto una grande ma sottil palla d'argento, e quella ripiena d'acqua raffreddata col ghiaccio serramo con saldissime vite. Di poi cominciammo a martellarla leggiermente per ogni verso, onde ammaccato l'argento (il quale per la sua crudezza non comporta d'assottigliarsi e distendersi come farebbe l'oro raffinato, o il piombo, o altro metallo più dolce) veniva a ristrignersi, e scemare la sua interna capacità, senza che l'acqua patisse una minima compressione, poichè ad ogni colpo si videa trasudare per tutti i pori del metallo a guisa d'argento vivo il quale da alcuna pelle premuto minutamente sprizzasse."—*Saggi di naturali Esperienze fatte nell' Accademia del Cimento*, p. 204. Firenze, 1667. The writer goes on to remark that the absolute incompressibility of water is not proved by this experiment, but merely that it is not to be compressed in the manner described. But the experiment is on other grounds inconclusive.

It is to be remarked that Leibnitz, *Nouveaux Essais*, in mentioning the Florentine experiment, says that the globe was of gold (p. 229, Erdmann), whereas the Florentine academicians expressly say why they preferred silver to either gold or lead.

[101] Galileo had shown, before the year 1592, that the resistance of the air being set aside, all bodies fall with equal velocity. He left Pisa in that year in consequence of the disputes which were occasioned by this refutation of the Aristotelian doctrine, that the velocity is as the weight.

the liquid ; and many other things of a similar kind. The compressions also and expansions and eruptions of bodies are performed, some more quickly, some more slowly, according to the nature of the body and motion, but in certain periods. Moreover in the explosion of several guns at once, which are heard sometimes to the distance of thirty miles, the sound is caught by those who are near the spot where the discharge is made, sooner than by those who are at a greater distance. Even in sight, whereof the action is most rapid, it appears that there are required certain moments of time for its accomplishment ; as is shown by those things which by reason of the velocity of their motion cannot be seen—as when a ball is discharged from a musket. For the ball flies past in less time than the image conveyed to the sight requires to produce an impression.

This fact, with others like it, has at times suggested to me a strange doubt, viz. whether the face of a clear and starlight sky be seen at the instant at which it really exists, and not a little later ; and whether there be not, as regards our sight of heavenly bodies, a real time and an apparent time, just like the real place and apparent place which is taken account of by astronomers in the correction for parallaxes. So incredible did it appear to me that the images or rays of heavenly bodies could be conveyed at once to the sight through such an immense space, and did not rather take a perceptible time in travelling to us. But this suspicion as to any considerable interval between the real time and the apparent afterwards vanished entirely [102], when I came to think of the infinite loss and diminution of quantity which distance causes in appearance between the real body of the star and its seen image ; and at the same time when I observed the great distance (sixty miles at the least) at which bodies merely white are instantly seen here on earth ; while there is no doubt that the light of heavenly bodies exceeds many times over in force of radiation, not merely the vivid colour of whiteness, but also the light of every flame that is known to us. Again the immense velocity in the body itself as discerned in its daily motion (which has so astonished certain grave men that they preferred believing that the earth moved) renders this motion of ejaculation of rays therefrom (although wonderful, as I have said, in speed) more easy of belief. But what had most weight of all with me was, that if any perceptible interval of time were interposed between the reality and the sight, it would follow that the images would oftentimes be intercepted and confused by clouds rising in the meanwhile, and similar disturbances in the medium [103]. And thus much for the simple measures of time.

But not only must we seek the measure of motions and actions by themselves, but much more in comparison ; for this is of excellent use and very general application. Now we find that the flash of a gun is seen sooner than its report is heard ; although the ball must necessarily strike the air before the flame behind it can get out ; and this is owing it seems to the motion of light being more

[102] [It will be observed that, with unlucky ingenuity, Bacon here lets the truth slip from his hands after he has glimpsed it. The Bohn editor notes that Dominic Cassini let it slip in the same way.—Ed.]

[103] I do not know how to understand this passage without attributing to Bacon a confusion of ideas which seems hardly credible. For surely the very thing which he supposes *would* happen if there were a perceptible interval between the *veritas* and the *visus*, that is to say, between the time when a star (for instance) *is* at a given point and the time when we see it there,—in other words, if the image took any time in coming to the eye,—this very thing does actually happen as often as the star is hidden by a cloud or dimmed by a vapour : the species, to use his own word, are intercepted or confused. If indeed, the *force* of the rays were diminished,—and this I suppose would be one consequence of diminished velocity,—the thing would happen more frequently, because there would be more obstructions which they could not overcome : they would be intercepted or confused by media which they now pass through. But the force being the same, and the stream continuous, the *time* of passage could make no difference in this respect. In another respect, namely, the facility of observation, it would make a very great difference ; and it is remarked by Brinkley that, if the velocity of light had been much less than it is, astronomy would have been all but an impossible science. But that is another matter.—*J. S.*

rapid than that of sound. We find too that visible images are received by the sight faster than they are dismissed ; thus the strings of a violin, when struck by the finger, are to appearance doubled or tripled, because a new image is received before the old one is gone ; which is also the reason why rings being spun round look like globes, and a lighted torch, carried hastily at night, seems to have a tail[104]. And it was upon this inequality of motions in point of velocity that Galileo built his theory of the flux and reflux of the sea ; supposing that the earth revolved faster than the water could follow ; and that the water therefore first gathered in a heap and then fell down, as we see it do in a basin of water moved quickly[105]. But this he devised upon an assumption which cannot be allowed, viz. that the earth moves ; and also without being well informed as to the sexhorary motion of the tide.

But an example of the thing I am treating of, to wit, the comparative measures of motions—and not only of the thing itself, but also of its eminent use (of which I spoke just now)—is conspicuous in mining with gunpowder ; where vast masses of earth, buildings, and the like are upset and thrown into the air, by a very small quantity of powder. The cause of which is doubtless this : that the motion of expansion in the impelling powder is quicker many times over than the motion of the resisting gravity ; so that the first motion is over before the counter motion is begun, and thus at first the resistance amounts to nothing. Hence too it happens that in projectiles it is not the strong blow, but the sharp and quick, that carries the body furthest. Nor would it be possible for the small quantity of animal spirit in animals, especially in such huge creatures as the whale or elephant, to bend and guide such a vast mass of body, were it not for the velocity of the spirit's motion, and the slowness of the bodily mass in exerting its resistance.

This one thing indeed is a principal foundation of the experiments in natural magic, of which I shall speak presently ; wherein a small mass of matter overcomes and regulates a far larger mass ; I mean the contriving that of two motions one shall by its superior velocity get the start and take effect before the other has time to act.

Lastly, this distinction of foremost and hindmost ought to be observed in every natural action. Thus in an infusion of rhubarb the purgative virtue is extracted first, the astringent afterwards. And something of the kind I have found on steeping violets in vinegar, where the sweet and delicate scent of the flower is extracted first, and then the more earthy part of the flower, which mars the scent. Therefore, if violets be steeped in vinegar for a whole day, the scent is extracted much more feebly ; but if you keep them in for a quarter of an hour only and then take them out, and (since the scented spirit in violets is small) put in fresh violets every quarter of an hour as many as six times, the infusion is at last so enriched that although there have not been violets in the vinegar, however renewed, for more than an hour and a half altogether, there nevertheless remains in it a most grateful odour, as strong as the violet itself, for an entire

[104] Of the phenomena which he here enumerates Bacon undoubtedly gives the right explanation, though in the case of vibrating strings his explanation is not altogether complete. The distinct or quasi-distinct images to which he refers correspond to limiting positions of the vibrating string.

[105] This account of Galileo's theory of the tides is inaccurate. In this theory the tides are caused by the varying velocity of different points of the earth's surface, arising from the composition of the earth's two motions, namely that about its axis, and that in its orbit. Bacon does not seem to have perceived that both these motions are essential to the explanation. That the earth's being in motion might be the cause of the tides, had been suggested before the time of Galileo by Cæsalpinus in the *Quæstiones Peripateticæ*, iii. 5. It is odd that Patricius, in giving an account of all the theories which had in his time been devised to explain the cause of the tides (see his *Pancosmia*, I. 28.), does not mention Cæsalpinus's, though it was published some years before his own work. Galileo perhaps alludes to Cæsalpinus in his letter to Cardinal Orsino, dated 8th January, 1616. See, for remarks on Cæsalpinus's doctrine, the *Problemata Marina* of Casmann, published in 1596. Casmann's own theory is that of expansion.

year. It should be observed however that the odour does not gather its full strength till after a month from the time of infusion. In the distillation too of aromatic herbs crushed in spirit of wine, it appears that there first rises an aqueous and useless phlegm ; then a water containing more of the spirit of wine ; and lastly, a water containing more of the aroma. And of this kind there are to be found in distillations a great many facts worthy of notice. But let these suffice for examples.

XLVII.

Among Prerogative Instances I will put in the twenty-third place *Instances of Quantity ;* which (borrowing a term from medicine) I also call *Doses of Nature.* These are they which measure virtues according to the *quantity* of the bodies in which they subsist, and show how far the *mode* of the virtue depends upon the *quantity* of the body. And first there are certain virtues, which subsist only in a cosmical quantity ; that is, such a quantity as has consent with the configuration and fabric of the universe. The earth for instance stands fast ; its parts fall. The waters in seas ebb and flow ; but not in rivers, except through the sea coming up. Secondly, almost all particular virtues act according to the greater or less quantity of the body. Large quantities of water corrupt slowly, small ones quickly. Wine and beer ripen and become fit to drink much more quickly in bottles than in casks. If an herb be steeped in a large quantity of liquid, infusion takes place rather than impregnation ; if in a small, impregnation rather than infusion. Thus in its effect on the human body a bath is one thing, a slight sprinkling another. Light dews again never fall in the air, but are dispersed and incorporated with it. And in breathing on precious stones you may see the slight moisture instantly dissolved, like a cloud scattered by the wind. Once more, a piece of a magnet does not draw so much iron as the whole magnet. On the other hand there are virtues in which smallness of quantity has more effect ; as in piercing, a sharp point pierces more quickly than a blunt one ; a pointed diamond cuts glass, and the like.

But we must not stay here among indefinites, but proceed to inquire what *proportion* the Quantity of a Body bears to the Mode of its Virtue. For it would be natural to believe that the one was equal to the other ; so that if a bullet of an ounce weight falls to the ground in a given time, a bullet of two ounces ought to fall twice as quickly ; which is not the fact. Nor do the same proportions hold in all kinds of virtues, but widely different. These measures therefore must be sought from experiment, and not from likelihood or conjecture.

Lastly, in all investigation of nature the quantity of body —the Dose, as it were—required to produce any effect must be set down ; and cautions as to the Too Little and Too Much be interspersed.

XLVIII.

Among Prerogative Instances I will put in the twenty-fourth place *Instances of Strife ;* which I also call *Instances of Predominance.* These indicate the mutual predominance and subjection of virtues ; which of them is stronger and prevails, which of them is weaker and gives way. For the motions and efforts of bodies are compounded, decomposed, and complicated, no less than the bodies themselves. I will therefore first propound the principal kinds of motions or active virtues ; in order that we may be able more clearly to compare them together in point of strength, and thereby to point out and designate more clearly the Instances of Strife and Predominance.

Let the First Motion be that motion of *Resistance* [106] in matter which is inherent in each several portion of it, and in virtue of which it absolutely refuses to be annihilated ; so that no fire, no weight or pressure, no violence, no length of time can reduce any portion of matter, be it ever so small, to nothing ; but it will ever be something, and occupy some space ; and, to whatever straits it may be brought, will free itself by changing either its form or its place ; or if this may not be, will subsist as it is ; and will never come to such a pass as to be either

[106] Orig. *Motus Antitypiæ.* This term was first used by Aristotle.

nothing or nowhere. This motion the Schoolmen (who almost always name and define things rather by effects and incapacities than by inner causes) either denote by the axiom "Two bodies cannot be in one place", or call "The motion to prevent penetration of dimensions". Of this motion it is unnecessary to give examples, as it is inherent in every body.

Let the Second Motion be what I call Motion of *Connexion ;* by which bodies do not suffer themselves to be separated at any point from contact with another body ; as delighting in mutual connexion and contact. This motion the Schoolmen call "Motion to prevent a vacuum," as when water is drawn up by suction or in a pump ; the flesh by cupping-glasses ; or when water stops without running out in perforated jars, unless the mouth of the jar be opened to let in the air ; and in numberless instances of a similar kind.

Let the Third Motion be what I call Motion of *Liberty ;* by which bodies strive to escape from preternatural pressure or tension, and to restore themselves to the dimensions suitable to their nature. Of this motion also we have innumerable examples ; such as (to speak first of escape from pressure) the motion of water in swimming, of air in flying, of water in rowing, of air in the undulations of winds, of a spring in clocks ; of which we have also a pretty instance in the motion of the air compressed in children's popguns, when they hollow out an alder twig or some such thing, and stuff it up at both ends with a piece of pulpy root or the like, and then with a ramrod thrust one of the roots or whatever the stuffing be towards the other hole, from which the root at the further end is discharged with a report ; and that before it is touched by the nearer root or the ramrod. As for bodies escaping from tension, this motion displays itself in air remaining in glass eggs after suction ; in strings, in leather and in cloth, which recoil after tension, unless it has gained too great strength by continuance ; and in similar phenomena. This motion the Schoolmen refer to under the name of "Motion in accordance with the form of the Element" ; an injudicious name enough, since it is a motion which belongs not only to fire, air and water, but to every variety of solid substance, as wood, iron, lead, cloth, parchment, etc. ; each of which bodies has its own proper limit of dimension, out of which it cannot easily be drawn to any considerable extent. But since this Motion of Liberty is of all the most obvious, and is of infinite application, it would be a wise thing to distinguish it well and clearly. For some very carelessly confuse this motion with the two former motions of Resistance and Connexion ; the motion, that is, of escape from pressure with the Motion of Resistance ; of escape from tension with the Motion of Connexion ; just as if bodies when compressed yield or expand, that there may not ensue penetration of dimensions; and, when stretched, recoil and contract, that there may not ensue a vacuum. Whereas if air when compressed had a mind to contract itself to the density of water, or wood to the density of stone, there would be no necessity for penetration of dimensions ; yet there might be a far greater compression of these bodies, than they ever do actually sustain. In the same way, if water had a mind to expand to the rarity of air, or stone to the rarity of wood, there would be no need for a vacuum to ensue ; and yet there might be effected a far greater extension of these bodies than they ever do actually sustain. Thus the matter is never brought to a penetration of dimensions or to a vacuum, except in the extreme limits of condensation and rarefaction ; whereas the motions of which I speak stop far short of these limits, and are nothing more than desires which bodies have for preserving themselves in their consistencies (or, if the Schoolmen like, in their forms), and not suddenly departing therefrom, unless they be altered by gentle means, and with consent. But it is far more necessary (because much depends upon it) that men should know, that violent motion (which we call Mechanical, but which Democritus, who in expounding his primary motions is to be ranked even below second rate philosophers, called Motion of Stripe) is nothing more than this Motion of Liberty, that is, of escape from compression to relaxation. For either in a mere thrust, or in flight through the air, there occurs no movement or change of place, until the parts of the body moved are acted upon and compressed by the impelling body more than their nature will bear. Then indeed, when each part pushes against the next, one after the other, the whole is moved; and it not only moves forward, but revolves at the same time; the parts seeking in that way also to free themselves or to distribute the pressure more equally. And so much for this Motion.

Let the Fourth Motion be that to which I have given the name of the Motion of *Matter;* which is in some sort the converse of the last-named motion. For in the Motion of Liberty bodies dread, loathe, and shun a new dimension, or a new sphere, or new expansion or contraction (which are all names for the same thing), and strive with all their might to recoil, and recover their old consistency. On the contrary in this Motion of Matter, bodies desire a new sphere or dimension, and aspire thereto readily and quickly, and sometimes, as in the case of gunpowder, with most violent effort. Now the instruments of this motion, not indeed the sole, but the most potent, or at any rate the most common, are heat and cold. For instance; air, if expanded by tension, as by suction in glass eggs, labours under a strong desire to recover itself. But if heat be applied, it longs on the contrary to expand, and desires a new sphere, and passes into it readily, as into a new form (so they phrase it); and after a certain degree of expansion cares not to return, unless invited thereto by the application of cold; which is not a return, but a renewed transmutation. In the same way water, if made to contract by pressure, resists and wishes to become such as it was, that is, larger. But if there intervene intense and continued cold, it changes itself spontaneously and gladly to the density of ice; and if the cold be continued long, without interruption from heat, as in grottoes and caverns of some depth, it turns to crystal [107] or some similar material, and never recovers its form.

Let the Fifth Motion be the Motion of *Continuity;* by which I do not mean simple and primary continuity with some other body (for that is the Motion of Connexion), but self-continuity in a given body. For it is most certain that all bodies dread a solution of continuity; some more, some less, but all to a certain extent. For while in hard bodies, as steel or glass, the resistance to discontinuity is exceedingly strong; even in liquids, where it seems to disappear, or at all events to be very feeble, it is not altogether absent, but is certainly there, though in its lowest degree of power, and betrays itself in very many experiments; as in bubbles, in the roundness of drops, in the thin threads of droppings from roofs, in the tenacity of glutinous bodies, and the like. But most of all does this appetite display itself, if an attempt be made to extend the discontinuity to minute fragments. For in a mortar, after a certain amount of pulverisation the pestle produces no further effect; water does not penetrate into minute chinks; even air itself, notwithstanding its subtlety, does not suddenly pass through the pores of solid vessels, but only after long insinuation.

Let the Sixth Motion be that which I call Motion *for Gain,* or Motion *of Want.* It is that by which bodies, when placed among quite heterogeneous and hostile bodies, if they find an opportunity of escaping from these and uniting themselves to others more cognate, (though these others be such as have no close union with them), do nevertheless embrace the latter and choose them as preferable; and seem to view this connexion in the light of a *gain* (whence the term), as though they stood in need of such bodies. For instance, gold or any other metal in the leaf does not like the surrounding air. If therefore it meet with any thick tangible body, (as a finger, paper, what you will,) it instantly sticks to it and is not easily torn away. So too paper, cloth, and the like do not agree well with the air which is lodged in their pores. They are therefore glad to imbibe water or other moisture and eject the air. A piece of sugar too, or a sponge, if dipped at one end in water or wine, while the other stands out far above the surface, draws the water or the wine gradually upwards.

Hence we derive an excellent rule for opening and dissolving bodies. For (to

107 Pliny, xxxvii. 9. Also Seneca, *Natural Questions.* Though this account of the origin of crystals is of course erroneous, yet there is a class of crystals which have been shown to occupy the volume which their water of crystallisation would in the state of ice; so that their other constituents may in some sort be said to take up no space. This curious analogy with ice was proved by Playfair and Joule in a very considerable number of cases. See *Phil. Mag.*, Dec. 1845.

say nothing of corrosives and strong waters which open for themselves a way) if there can be found a body proportioned to and more in harmony and affinity with a given solid body, than that with which it is, as of necessity, mixed, the solid body immediately opens and relaxes itself, and shutting out or ejecting the latter, receives the former into itself. Nor does this Motion for Gain act or exist only in immediate contact. For electricity (of which Gilbert and others after him have devised such stories) is nothing else than the appetite of a body when excited by gentle friction—an appetite which does not well endure the air, but prefers some other tangible body, if it be found near at hand.

Let the Seventh Motion be what I call the Motion *of the Greater Congregation ;* by which bodies are carried towards masses of a like nature with themselves ; heavy bodies to the globe of the earth, light to the compass of the heaven. This the Schoolmen have denoted by the name of *Natural Motion ;* from superficial considerations ; either because there was nothing conspicuous externally which could produce such motion (and therefore they supposed it to be innate and inherent in things themselves) ; or perhaps because it never ceases. And no wonder ; for the earth and heaven are ever there, whereas the causes and origins of most other motions are sometimes absent, sometimes present. Accordingly this motion, because it ceases not, but when others cease is felt instantly, they deem perpetual and proper ; all others adscititious. This motion however in point of fact is sufficiently weak and dull, being one which, except in bodies of considerable bulk, yields and succumbs to all other motions, as long as they are in operation. And though this motion has so filled men's thoughts as to have put all others almost out of sight, yet it is but little that they know about it, being involved in many errors with regard to it.

Let the Eighth Motion be the Motion of *the Lesser Congregation ;* by which the homogeneous parts in a body separate themselves from the heterogeneous, and combine together ; by which also entire bodies from similarity of substance embrace and cherish each other, and sometimes are attracted and collected together from a considerable distance ; as when in milk, after it has stood awhile, the cream rises to the top ; while in wine the dregs sink to the bottom. For this is not caused by the motion of heaviness and lightness only, whereby some parts rise up and some sink down ; but much more by a desire of the homogeneous parts to come together and unite in one. Now this motion differs from the Motion of Want in two points. One is that in the latter there is the stronger stimulus of a malignant and contrary nature ; whereas in this motion (provided there be nothing to hinder or fetter it) the parts unite from friendship, even in the absence of a foreign nature to stir up strife. The other point is, that the union is here closer, and as it were with greater choice. In the former, if only the hostile body be avoided, bodies not closely related come together ; whereas in the latter, substances are drawn together by the tie of close relationship, and as it were combine into one. And this motion resides in all composite bodies, and would readily show itself were it not bound and restrained by other appetites and necessities in the bodies, which interfere with the union in question.

Now the binding of this motion takes place generally in three ways ; by the torpor of bodies ; by the check of a dominant body ; and by external motions. Now, for the torpor of bodies, it is certain that there resides in tangible substances a certain sluggishness, more or less, and an aversion from change of place ; insomuch that, unless they be excited, they had rather remain as they are than change for the better. Now this torpor is shaken off by the help of three things ; either by heat, or by the eminent virtue of some cognate body, or by lively and powerful motion. And as for the help of heat, it is for this reason that heat has been defined to be " that which separates Heterogeneous and congregates Homogeneous parts " ; a definition of the Peripatetics justly derided by Gilbert, who says it is much the same as if a man were to be defined as that which sows wheat and plants vines—for that it is a definition simply by effects, and those particular [108]. But the definition has a worse fault ; inasmuch as these effects, such

108 For the definition we may refer to the *Margarita Philosophiæ*, xi. 3. It is founded on a passage in the *De Gen. et Corr.* ii. 2. Gilbert's censure on it is to be found in his

as they are, arise not from a peculiar property of heat, but only indirectly (for cold does the same, as I shall afterwards show); being caused by the desire of homogeneous parts to unite; heat simply aiding to shake off the torpor which had previously bound the desire. As for the help derived from the virtue of a cognate body, it is well seen in an armed magnet, which excites in iron the virtue of detaining iron by similarity of substance; the torpor of the iron being cast off by the virtue of the magnet. And as for help derived from motion, it is shown in wooden arrows, having their points also of wood, which penetrate more deeply into wood than if they were tipped with steel, owing to the similarity of substance; the torpor of the wood being shaken off by the rapid motion. Of these two experiments I have spoken also in the Aphorism on Clandestine Instances.

That binding of the motion of the Lesser Congregation which is caused by the restraint of a dominant body, is seen in the resolution of blood and urine by cold. For as long as those bodies are filled with the active spirit, which, as lord of the whole, orders and restrains the several parts of whatsoever sort, so long the homogeneous parts do not meet together on account of the restraint; but as soon as the spirit has evaporated, or been choked by cold, then the parts being freed from restraint meet together in accordance with their natural desire. And thus it happens that all bodies which contain an eager spirit (as salts and the like) remain as they are, and are not resolved; owing to the permanent and durable restraint of a dominant and commanding spirit.

That binding of the motion of Lesser Congregation which is caused by external motion, is most conspicuous in the shaking of bodies to prevent putrefaction. For all putrefaction depends on the assembling together of homogeneous parts; whence there gradually ensues the corruption of the old form, as they call it, and the generation of a new. For putrefaction, which paves the way for the generation of a new form, is preceded by a dissolution of the old; which is itself a meeting together of homogeneous parts. That indeed, if not impeded, is simple resolution; but if it be met by various obstacles, there follow putrefactions, which are the rudiments of a new generation. But if (which is the present question) a frequent agitation be kept up by external motion, then indeed this motion of uniting (which is a delicate and tender one, and requires rest from things without) is disturbed and ceases; as we see happen in numberless instances. For example, the daily stirring or flowing of water prevents it from putrefying; winds keep off pestilence in the air; corn turned and shaken in the granary remains pure; all things in short that are shaken outwardly are the slower to putrefy inwardly.

Lastly, I must not omit that meeting of the parts of bodies, which is the chief cause of induration and desiccation. For when the spirit, or moisture turned to spirit, has escaped from some porous body (as wood, bone, parchment, and the like), then the grosser parts are with stronger effort drawn and collected together; whence ensues induration or desiccation; which I take to be owing not so much to the Motion of Connexion, to prevent a vacuum, as to this motion of friendship and union.

As for the meeting of bodies from a distance, that is a rare occurrence, and yet it exists in more cases than are generally observed. We have illustrations of it when bubble dissolves bubble; when medicine draws humours by similarity of substance; when the chord of one violin makes the chord of another sound an

posthumous work *De Mundo nostro sublunari Philosophia nova*, which was published by Gruter in 1651, long after the death of Bacon. It seems, however, as Gruter remarks, that the work, which he suggests may have been written before the treatise *De Magnete*, published in 1600, had been read in manuscript by "viri magni et famæ celeberrimæ". "Illi perspicace in Physicis præsertim ingenio haud pœnitendæ in evolvendo operæ testimonium dederunt, quod integrum excussisse censeantur, et aliqua a vulgaribus opinionibus abhorrentia calculo suo comprobata hinc sparsim citent"; in which I do not doubt that Gruter refers to Bacon. Bacon's quotation seems to have been made from imperfect memory, as the words of the original are:—"quid illud ostendit aut quæ illa differentia ab effectu tantum in quibusdam corporibus, *congregans homogenea et disgregans heterogenea?* ac si diceres hominem animal esse carduos et sentes evellens, et fruges serens, cum istud sit agricolæ studium".—*De Mundo*, etc., i. c. 26.

unison, and the like. I suspect also that this motion prevails in the spirits of animals, though it be altogether unknown. At any rate it exists conspicuously in the magnet and magnetised iron. And now that we are speaking of the motions of the magnet, they ought to be carefully distinguished. For there are four virtues or operations in the magnet, which should not be confounded but kept apart; although the wonder and admiration of men have mixed them up together. The first is, the attraction of magnet to magnet, or of iron to magnet, or of magnetised iron to iron. The second is its polarity, and at the same time its declination. The third, its power of penetrating through gold, glass, stone, everything. The fourth, its power of communicating its virtue from stone to iron, and from iron to iron, without communication of substance. In this place however I am speaking only of the first of these virtues —that is, its attractive power. Remarkable also is the motion of attraction between quicksilver and gold; insomuch that gold attracts quicksilver, though made up into ointments; and men who work amid the vapours of quicksilver usually hold a piece of gold in their mouths, to collect the exhalations which would otherwise penetrate into their skulls and bones; by which also the piece of gold is presently turned white. And so much for the motion of the Lesser Congregation.

Let the Ninth Motion be the *Magnetic;* which, though it be of the same genus with the Motion of the Lesser Congregation, yet if it operates at great distances and on large masses, deserves a separate investigation; especially if it begin not with contact, as most, nor lead to contact, as all motions of congregation do; but simply raise bodies or make them swell, and nothing more. For if the moon raises the waters, or makes moist things swell; if the starry heaven attracts planets to their apogees; if the sun holds Venus and Mercury so that their elongations never exceed a certain distance; these motions seem to fall properly neither under the Greater nor the Lesser Congregation, but to be of a sort of intermediate and imperfect Congregation, and therefore ought to constitute a species by themselves.

Let the Tenth Motion be that of *Flight;* a motion the exact opposite of that of the Lesser Congregation; by which bodies from antipathy flee from and put to flight hostile bodies, and separate themselves from them, or refuse to mingle with them. For although in some cases this motion may seem to be an accident or a consequence of the motion of the lesser congregation, because the homogeneous parts cannot meet without dislodging and ejecting the heterogeneous, still it is a motion that should be classed by itself, and formed into a distinct species, because in many cases the appetite of Flight is seen to be more dominant than the appetite of Union.

This motion is eminently conspicuous in the excretions of animals; and not less in objects odious to some of the senses, especially the smell and the taste. For a fetid odour is so rejected by the sense of smell as to induce by consent in the mouth of the stomach a motion of expulsion; a rough and bitter taste is so rejected by the palate or throat as to induce by consent a shaking of the head and a shudder. But this motion has place in other things also. It is observed in certain forms of reaction; as in the middle region of the air, where the cold seems to be the effect of the rejection of the nature of cold from the confines of the heavenly bodies; as also the great heats and burnings which are found in subterranean places, appear to be rejections of the nature of heat from the inner parts of the earth. For heat and cold, in small quantities, kill one another; but if they be in large masses, and as it were in regular armies, the result of the conflict is that they displace and eject each other in turn. It is also said that cinnamon and other perfumes retain their scent longer when placed near sinks and foul smelling places, because they refuse to come out and mingle with stenches. It is certain that quicksilver, which of itself would reunite into an entire mass, is kept from doing so by spittle, hog's lard, turpentine, and the like; owing to the ill consent which its parts have with such bodies, from which when spread around them they draw back; so that their desire to fly from these intervening bodies is more powerful than their desire of uniting with parts like themselves. And this is called the *mortification* of quicksilver. The fact also that oil does not mix with water is not simply owing to the difference of weight, but to the ill consent of these fluids; as may be seen from the fact that spirit of wine, though lighter than oil, yet mixes well enough with water. But most of all is the Motion of Flight conspicuous in nitre and such like crude bodies, which abhor flame; as in gunpowder, quicksilver, and gold. But the flight of iron from one pole of the magnet is well observed by Gilbert to be not a Flight strictly speaking, but a conformity and meeting in a more convenient situation [109].

Let the Eleventh Motion be that of *Assimilation*, or of *Self-Multiplication*, or again of simple *Generation*. By which I mean not the generation of integral bodies, as plants or animals, but of bodies of uniform texture. That is to say, by this motion such bodies convert others which are related, or at any rate well disposed to them, into their own substance and nature. Thus flame over vapours and oily substances multiplies itself and generates new flame; air over water and watery substances multiplies itself and generates new air; spirit, vegetable and animal, over the finer parts as well of watery as of oily substance in its food, multiplies itself and generates new spirit; the solid parts of plants and animals, as the leaf, flower, flesh, bone, and the like, severally assimilate new substance to follow and supply what is lost out of the juices of their food. For let no one adopt the wild fancy of Paracelsus, who (blinded I suppose by his distillations) will have it that nutrition is caused only by separation; and that in bread and meat lie eye, nose, brain, liver [110]; in the moisture of the ground, root, leaf, and flower. For as the artist out of the rude mass of stone or wood educes, by separation and rejection of what is superfluous, leaf, flower, eye, nose, hand, foot, and the like; so, he maintains, does Archæus, the internal artist, educe out of food by separation and rejection the several members and parts of our body. But to leave such trifles, it is most certain that the several parts, as well similar as organic, in vegetables and animals do first attract with some degree of selection the juices of their food, which are alike or nearly so for all, and then assimilate them and turn them into their own nature. Nor does this Assimilation or simple Generation take place only in animate bodies, but inanimate also participate therein, as has been stated of flame and air. Moreover, the non-vital spirit, which is contained in every tangible animated substance, is constantly at work to digest the coarser parts and turn them into spirit, to be afterwards discharged; whence ensues diminution of weight and desiccation, as I have stated elsewhere. Nor must we set apart from Assimilation that accretion which is commonly distinguished from alimentation; as when clay between stones concretes and turns into a stony substance, or the scaly substance on the teeth turns into a

[109] Gilbert, *De Magnete*, ii. c. 4.

[110] I have not been able to find any passage in Paracelsus which altogether corresponds to this remark; and in his *Modus Pharmacandi* the process of digestion is described without reference to the Archeus; nor is it said that each member "latet in pane vel cibo." "Hoc scimus, quod cujusque membri nutrimentum latitet in pane, carne, et in aliis similiter." "Quot vero modis et quibus, necnon quâ ratione membris corporis nutrimentum dividatur, nos ignoramus; hoc tantùm scimus, rem ita se habere ut diximus."—*De Mod. Pharm.* v. p. 233. (I use the edition of 1603.)

Bacon has, however, correctly stated the general doctrine that alimentation is by separation; and again Paracelsus affirms that "officium vero Archei est in microcosmo purum ab impuro separare".—*De Morbis Tartareis*, iii. 195. The truth is that Paracelsus's views are so often repeated and varied in the course of his writings, that it is difficult to know how far his opinions are represented by any particular passage.

It is well to remark that, to a certain extent, the theory here so decidedly condemned has, by the recent progress of organic chemistry, been shown to be true. Nothing seems better established than that the nitrogenised components of animal bodies are derived from the corresponding elements of their food. With respect to fat, it is, I believe, a prevailing opinion at present, that animals have the power of converting into it the starch or sugar of their food; and the production of butyric acid by fermentation has been regarded as at least an illustration of the transformation. One of the highest authorities on such a subject, however, I mean M. Boussingault, was, at least a few years ago, of a different opinion. He regarded animal fat as the representative of the fatty matters contained in the food.

substance as hard as the teeth themselves, and so on. For I am of opinion that there resides in all bodies a desire for Assimilation, as well as for uniting with homogeneous substances; but this virtue is bound, as is the other, though not by the same means. But these means, as well as the way of escape from them, ought to be investigated with all diligence, because they pertain to the rekindling of the vital power in old age. Lastly, it seems worthy of observation that in the nine motions of which I have spoken bodies seem to desire only the preservation of their nature, but in this tenth the propagation of it.

Let the Twelfth Motion be that of *Excitation*; a motion which seems to belong to the genus of Assimilation, and which I sometimes call by that name. For it is a motion diffusive, communicative, transitive, and multiplicative, as is the other; and agreeing with it generally in effect, though differing in the mode of effecting and in the subject matter. For the Motion of Assimilation proceeds as it were with authority and command; it orders and forces the assimilated body to turn into the assimilating. But the Motion of Excitation proceeds, so to speak, with art and by insinuation, and stealthily; simply inviting and disposing the excited body to the nature of the exciting. Again, the Motion of Assimilation multiplies and transforms bodies and substances; thus, more flame is produced, more air, more spirit, more flesh. But in the Motion of Excitation, virtues only are multiplied and transferred; more heat being engendered, more magnetic power, more putrefying. This motion is particularly conspicuous in heat and cold. For heat does not diffuse itself, in heating a body, by communication of the original heat, but simply by exciting the parts of the body to that motion which is the Form of Heat; of which I have spoken in the First Vintage concerning the Nature of Heat. Consequently heat is excited far more slowly and with far greater difficulty in stone or metal than in air, owing to the unfitness and unreadiness of those bodies to receive the motion; so that it is probable that there may exist materials in the bowels of the earth which altogether refuse to be heated, because through their greater condensation they are destitute of that spirit with which this Motion of Excitation generally begins. In like manner the magnet endues iron with a new disposition of its parts and a conformable motion, but loses nothing of its own virtue. Similarly leaven, yeast, curd, and certain poisons excite and invite a successive and continued motion in dough, beer, cheese, or the human body, not so much by the force of the exciting as by the predisposition and easy yielding of the excited body[111].

Let the Thirteenth Motion be the Motion of *Impression*; which also is of the same genus with the Motion of Assimilation, and is of diffusive motions the most subtle. I have thought fit however to make a distinct species of it, on account of a remarkable difference between it and the two former. For the simple Motion of Assimilation actually transforms the bodies themselves; so that you may take away the first mover, and there will be no difference in what follows. For the first kindling into flame, or the first turning into air, has no effect on the

[111] The theory here proposed is nearly equivalent to the most recent views on the same subject, as the following passage will sufficiently show.—It is obvious that both statements, however much of truth they may involve, are indefinite and unsatisfactory. It is not said whether the new properties engendered depend upon new types of motion or new arrangements, though the latter is probably Liebig's opinion.

"All the phenomena of fermentation, when taken together, establish the correctness of the principle long since recognised by Laplace and Berthollet, namely, *that an atom or molecule, put in motion by any power whatever, may communicate its own motion to another atom in contact with it.*

"This is a dynamical law of the most general application, manifested everywhere when the resistance or force opposing the motion, such as the vital principle, the force of affinity, electricity, cohesion, etc., is not sufficiently powerful to arrest the motion imparted.

"This law has only recently been recognised as a cause of the alterations in forms and properties which occur in our chemical combinations; and its establishment is the greatest and most enduring acquisition which chemical science has derived from the study of fermentation."—*Liebig's Letters on Chemistry*, p. 209.

flame or air next generated. In like manner, the Motion of Excitation continues, after the first mover is withdrawn, for a very considerable time; as in a heated body, when the primary heat has been removed; in magnetised iron, when the magnet has been put away; in dough, when the leaven has been taken out. But the Motion of Impression, though diffusive and transitive, seems to depend for ever on the prime mover; so that if that be taken away or cease to act, it immediately fails and comes to an end; and therefore the effect must be produced in a moment, or at any rate in a very brief space of time. The Motions therefore of Assimilation and Excitation I call Motions of the *Generation of Jupiter*, because the generation continues; but this, the Motion of the *Generation of Saturn*, because the birth is immediately devoured and absorbed. It manifests itself in three things; in rays of light, in the percussions of sounds, and in magnetism, as regards the communication of the influence. For if you take away light, colours and its other images instantly disappear; if you take away the original percussion and the vibration of the body thence produced, the sound soon after dies away. For though sounds are troubled as they pass through their medium by winds, as if by waves, yet it must be carefully noted that the original sound does not last all the time the resonance goes on. For if you strike a bell, the sound seems to be continued for a good long time; whereby we might easily be led into the error of supposing that during the whole of the time the sound is as it were floating and hanging in the air; which is quite untrue. For the resonance is not the same identical sound, but a renewal of it; as is shown by quieting or stopping the body struck. For if the bell be held tight so that it cannot move, the sound at once comes to an end, and resounds no more; as in stringed instruments, if after the first percussion the string be touched, either with the finger, as in the harp, or with the quill, as in the spinnet, the resonance immediately ceases. Again, when the magnet is removed, the iron immediately drops. The moon indeed cannot be removed from the sea, nor the earth from the falling body, and therefore we can try no experiment in these cases, but the principle is the same.

Let the Fourteenth Motion be the Motion of *Configuration* or *Position;* by which bodies seem to desire not union or separation, but position, collocation and configuration with respect to others. This motion is a very abtruse one, and has not been well investigated. In some cases indeed it seems to be without a cause, though not, I believe, really so. For if it be asked, why the heavens revolve rather from east to west than from west to east; or why they turn on poles placed near the Bears, rather than about Orion, or in any other part of heaven; such questions seem to border on insanity, since these phenomena ought rather to be received as results of observation, and merely positive facts. But though there are no doubt in nature certain things ultimate and without cause, this does not appear to me to be one of them, being caused in my opinion by a certain harmony and consent of the universe, which has not yet fallen under observation[112]. And if we admit the motion of the earth from west to east, the same questions remain. For it also moves on certain poles. And why, it might be asked, should these poles be placed where they are, rather than anywhere else?[113] Again the polarity, direction and declination of the magnet are referable to this motion. There are also found in bodies as well natural as artificial, especially in solids, a certain collocation and position of parts, and a kind of threads and fibres, which ought to be carefully investigated; since, until they are under-

[112] The most striking instance of this kind of harmony is the circumstance that all the movements of the solar system are in the same general direction, viz., from west to east. Laplace has attempted to calculate the probability that this uniformity is the result of a common cause determining the direction of their movements; but these numerical estimations of the probability of the truth of any induction are, on several accounts, altogether unsatisfactory.

[113] This passage shows that Bacon was not aware that the poles are not fixed (collocati) anywhere; in other words, that he was not acquainted with the precession of the equinoxes;—an additional proof how little of his attention had been given to mathematical physics.

stood, these bodies cannot be conveniently managed or controlled. But those eddyings in fluids, by which when pressed, before they can free themselves, they relieve each other, that they may all have a fair share of the pressure, belong more properly to the Motion of Liberty.

Let the Fifteenth Motion be the Motion of *Transition*, or Motion *according to the Passages* ; by which the virtues of bodies are more or less impeded or promoted by their media, according to the nature of the body and of the acting virtues, and also of the medium. For one medium suits light, another sound, another heat and cold, another magnetic virtues, and so on.

Let the Sixteenth Motion be the *Royal* (as I call it) or *Political* Motion ; by which the predominant and commanding parts in any body curb, tame, subdue and regulate the other parts, and compel them to unite, separate, stand still, move, and range themselves, not in accordance with their own desires, but as may conduce to the well being of the commanding part ; so that there is a sort of Government and Polity exerted by the ruling over the subject parts. This motion is eminently conspicuous in the spirits of animals, where, as long as it is in vigour, it tempers all the motions of the other parts. It is found however in other bodies in a lower degree ; as I said of blood and urine, which are not decomposed till the spirit, which mixes and keeps together their parts, be discharged or quenched. Nor is this motion confined to spirits, though in most bodies the spirits are masters owing to their rapid and penetrating motion. But in bodies of greater density, and not filled with a lively and quickening spirit (such as there is in quicksilver and vitriol), the thicker parts are the masters ; so that unless this yoke and restraint be by some expedient shaken off, there is very little hope of any new transformation of such bodies. But let no one suppose that I am forgetful of the point at issue, because while this series and distribution of motions tends to nothing else but the better investigation of their Predominancy by Instances of Strife, I now make mention of Predominancy among the motions themselves. For in describing this Royal Motion, I am not treating of the Predominancy of motions or virtues, but of the Predominancy of parts in bodies ; such being the Predominancy which constitutes the peculiar species of motion in question.

Let the Seventeenth Motion be the *Spontaneous Motion of Rotation*, by which bodies delighting in motion and favourably placed for it enjoy their own nature and follow themselves, not another body ; and court (so to speak) their own embraces. For bodies seem either to move without limit, or to remain altogether at rest, or to tend to a limit, at which according to their nature they either revolve or rest. Those which are favourably placed, if they delight in motion, move in a circle ; with a motion, that is, eternal and infinite. Those which are favourably placed, and abhor motion, remain at rest. Those which are not favourably placed move in a right line (as the shortest path) to consort with bodies of their own nature[114]. But this Motion of Rotation admits of nine differences ; regarding, 1. the centre round which the bodies move ; 2. the poles on which they move ; 3. their circumference or orbit, according to their distance

[114] This passage is wholly in accordance with the Peripatetic system of physics. But the modifications which Bacon goes on to enumerate, to which, as he conceives, the eternal circular motions of the heavenly bodies may be subject, are sufficient to destroy the whole *à priori* argument in favour of such a system of astronomy as that which we find in the twelfth book of the *Metaphysics*. It has not been sufficiently observed that the Ptolemaic system is no less at variance with the Peripatetic philosophy than the heliocentrical. The attempts of Turrianus and Fracastorius to construct what may be called an orthodox system of astronomy—that is one in which all the motions should take place in circles of which the earth is the centre—was suggested chiefly, as we learn from the *Homocentrica* of the latter, by the wish to reconcile astronomy and philosophy. It had no scientific value, since it left all the phenomena of variations of parallax and apparent diameter unexplained, or, at any rate, gave an explanation of them which no astronomer would accept. It was nevertheless favourably received by the systematic Peripaticians. See, for instance, Flaminius, *De prima Philosoph. Paraph.* p. 119. (I quote the Basle edition of 1557.)

from the centre; 4. their velocity, according to the greater or less rapidity of their rotation; 5. the course of their motion, as from east to west, or from west to east; 6. their declination from a perfect circle by spiral lines more or less distant from their centre; 7. their declination from a perfect circle by spiral lines more or less distant from their poles; 8. the greater or less distance of these spirals from each other; 9. and lastly, the variation of the poles themselves, if they be movable; which however has nothing to do with rotation, unless it be circular [115]. This motion in common and long received opinion is looked upon as the proper motion of heavenly bodies; though there is a grave dispute with regard to it among some both of the ancients and of the moderns, who have attributed rotation to the earth. But a juster question perhaps arises upon this (if it be not past question), namely, whether this motion (admitting that the earth stands still) is confined to the heavens, and does not rather descend and communicate itself to the air and waters. The Motion of Rotation in missiles, as in darts, arrows, musket-balls, and the like, I refer to the Motion of Liberty.

Let the Eighteenth Motion be the Motion of *Trepidation*, to which, as understood by astronomers, I do not attach much credit [116]. But in searching carefully everywhere for the appetites of natural bodies, this motion comes before us, and ought, it seems, to constitute a species by itself. It is a motion of what may be called perpetual captivity, and occurs when bodies that have not quite found their right place, and yet are not altogether uneasy, keep for ever trembling and stirring themselves restlessly, neither content as they are nor daring to advance further. Such a motion is found in the heart and pulses of animals, and must of necessity occur in all bodies which so exist in a mean state between conveniences and inconveniences, that when disturbed they strive to free themselves, and being again repulsed, are yet for ever trying again.

Let the Nineteenth and last Motion be one which, though it hardly answers to the name, is yet indisputably a motion; and let us call it the Motion of *Repose*, or of *Aversion to Move*. It is by this motion that the earth stands still in its mass, while its extremities are moving towards the middle; not to an imaginary centre, but to union. By this appetite also all bodies of considerable density abhor motion; indeed the desire of not moving is the only appetite they have; and though in countless ways they be enticed and challenged to motion, they yet, as far as they can, maintain their proper nature. And if compelled to move they nevertheless seem always intent on recovering their state of rest, and moving no more. While thus engaged indeed they show themselves active, and struggle for it with agility and swiftness enough, as weary and impatient of all delay. Of this appetite, but a partial representation can be seen; since here with us, from the subduing and concocting power of the heavenly bodies, all tangible substances are not only not condensed to their utmost, but are even mixed with some portion of spirit.

Thus then have I set forth the species or simple elements of motions, appetites, and active virtues, which are in nature most general. And under these heads,

[115] I believe the sense is that unless we restrict ourselves to circular motion, that is unless we reject the sixth and seventh species of variation, it will not be necessary for us to suppose the poles themselves to be movable: in other words, that the phenomena of which we could by this hypothesis give an account may be adequately represented without it by means of spirals.

[116] The name of trepidation was given by the Alphonsine astronomers to a motion by which they imagined the starry heaven to be affected, and in virtue of which its equinoxes described small circles of nine degrees radius about those of the ninth or next superior orb. To account for this motion they introduced a tenth orb. The phenomenon, however, thus accounted for was altogether imaginary, although it is true that the length of the tropical year, by supposed variations of which the idea of trepidation was suggested, is not rigorously constant. It may be questioned whether Bacon's hesitation to accept the astronomical motion of trepidation had any better foundation than his doubts whether the proper motions of the planetary orbs were anything more than "res confictæ et suppositæ". The question of the existence or non-existence of trepidation could only be decided by a person conversant with the details of the received system of astronomy.

no small portion of natural science is sketched out. I do not however mean to say that other species may not be added, or that the divisions I have made may not be drawn more accurately according to the true veins of nature, or reduced to a smaller number. Observe nevertheless that I am not here speaking of any abstract divisions; as if one were to say that bodies desire either the exaltation or the propagation or the fruition of their nature; or again, that the motions of things tend to the preservation and good either of the universe, as Resistance and Connexion; or of great wholes, as the Motions of the Greater Congregation, Rotation, and Aversion to Move; or of special forms, as the rest. For though these assertions be true, yet unless they be defined by true lines in matter and the fabric of nature, they are speculative and of little use. Meanwhile these will suffice, and be of good service in weighing the Predominances of Virtues and finding out Instances of Strife; which is our present object.

For of the motions I have set forth some are quite invincible; some are stronger than others, fettering, curbing, arranging them; some carry farther than others; some outstrip others in speed; some cherish, strengthen, enlarge, and accelerate others.

The Motion of Resistance is altogether adamantine and invincible. Whether the Motion of Connexion be so, I am still undecided. For I am not prepared to say for certain whether or no there be a vacuum, either collected in one place or interspersed in the pores of bodies [117]. But of one thing I am satisfied, that the reason for which a vacuum was introduced by Leucippus and Democritus (namely, that without it the same bodies could not embrace and fill sometimes larger and sometimes smaller spaces) is a false one. For matter is clearly capable of folding and unfolding itself in space, within certain limits, without the interposition of a vacuum; nor is there in air two thousand times as much of vacuity as there is in gold; which on their hypothesis there should be [118]. Of this I am sufficiently convinced by the potency of the virtues of pneumatical bodies (which otherwise would be floating in empty space like fine dust), and by many other proofs. As for the other motions, they rule and are ruled in turn, in proportion to their vigour, quantity, velocity, force of projection, and also to the helps and hindrances they meet with.

117 "Vacuum permistum," κενὸν ἀχώριστον, is vacuum diffused through the interstices of any portion of matter. By "vacuum coacervatum," κενὸν κεχωρισμένον, is meant clear empty space. See, for this distinction, Aristotle, *Phys.* iv. 7. Hero of Alexandria, whom Bacon mentions more than once, approves of those who admit the former kind of vacuum and reject the latter. See the Introduction to his *Spiritalia*. [It is perhaps worth observing that in the fable entitled: "Cupido sive Atomus (*De Sap. Vet.* xvii.), where the theory of a vacuum is mentioned, this distinction was not introduced till Bacon revised the work in his later years. The passage which stands thus in the original edition (1609)—"Quisquis autem atomum ponit et vacuum, necessario virtutem atomi ad distans introducit"—is altered, in the edition published by Rawley after Bacon's death, to "Quisquis autem atomum asserit atque vacuum (licet istud vacuum intermistum ponat, non segregatum) necessario," etc.—*J. S.*]

118 "Ex vacuo bis millies" is to be rendered "two thousand times as much of vacuity". Bacon thought spirit of wine a hundred times denser than its own vapour, and gold twenty-one times denser than spirit of wine. In the *Historia Densi et Rari*, he remarks that air is at least a hundred-fold rarer than water; and from the table there given it appears that the specific density of gold is to that of water as 1,000 to 56, nearly. Hence he must have estimated the density of gold at 1900-fold that of air. Now, if we take the same weight of air and of gold, it is clear that, neglecting the space occupied by the solid matter, supposed equally dense, of each, the ratio of their densities is the same as that of the "vacua permista" which they respectively contain, and that if we take the solid matter into account the "ex vacuo" in the case of air must bear a larger ratio than that of the densities to the "ex vacuo" of gold; so that we may take it in round numbers to be as two thousand to one, as in the text.

The passage is important as showing that Bacon, notwithstanding his frequent mention of Democritus, did not adopt the atomic philosophy, though he did not absolutely reject the physical part of it.

For instance, there are some armed magnets that hold and suspend iron of sixty times their own weight; so far does the motion of the Lesser prevail over the motion of the Greater Congregation; but if the weight be increased, it is overcome. A lever of given strength will raise a given weight; so far does the Motion of Liberty prevail over that of the Greater Congregation; but if the weight be increased, it is overcome. Leather stretches to a certain extent without breaking; so far does the Motion of Continuity prevail over the Motion of Tension; but if the tension be increased, the leather breaks, and the Motion of Continuity is overcome. Water runs out at a crack of a certain size; so far does the motion of the Greater Congregation prevail over the Motion of Continuity; but if the crack be smaller, it gives way, and the Motion of Continuity prevails. If you charge a gun with ball and sulphur only, and apply the match, the ball is not discharged; the Motion of the Greater Congregation overcoming in this case the Motion of Matter. But if you charge with gunpowder, the Motion of Matter in the sulphur prevails, being aided by the Motions of Matter and of Flight in the nitre. And so of other cases. Instances of Strife, therefore, which point out the Predominances of Virtues, together with the manner and proportion in which they predominate or give place, should be sought and collected from all quarters with keen and careful diligence.

Nor should we examine less carefully the modes in which these motions give way. That is to say, whether they stop altogether, or whether they continue to resist, but are overpowered. For in bodies here with us there is no real rest, either in wholes or in parts; but only in appearance. And this apparent rest is caused either by equilibrium, or by absolute predominancy of motions; by equilibrium, as in scales, which stand still if the weights be equal; by predominancy, as in watering-pots with holes in them, where the water rests and is kept from falling out by the predominancy of the Motion of Connexion. But it should be observed, as I have said, how far these yielding motions carry their resistance. For if a man be pinned to the ground, tied hand and foot, or otherwise held fast, and yet struggle to rise with all his might, the resistance is not the less, though it be unsuccessful. But the real state of the case (I mean whether by predominancy the yielding motion is, so to speak, annihilated, or whether rather a resistance is continued, though we cannot see it) will perhaps, though latent in the conflicts of motions, be apparent in their concurrence. For example, let trial be made in shooting. See how far a gun will carry a ball straight, or as they say point blank; and then try whether, if it be fired upwards, the stroke will be feebler than when it is fired downwards, where the Motion of Gravity concurs with the blow.

Lastly, such Canons of Predominance as we meet with should be collected; for instance, that the more common the good sought, the stronger the motion. Thus the Motion of Connexion, which regards communion with the universe, is stronger than the Motion of Gravity, which regards only communion with dense bodies. Again, that appetites which aim at a private good seldom prevail against appetites which aim at a more public good, except in small quantities; rules which I wish held good in politics.

XLIX.

Among Prerogative Instances I will put in the twenty-fifth place *Intimating Instances;* those I mean, which intimate or point out what is useful to man. For mere Power and mere Knowledge exalt human nature, but do not bless it. We must therefore gather from the whole store of things such as make most for the uses of life. But a more proper place for speaking of these will be when I come to treat of Applications to Practice. Besides in the work itself of Interpretation in each particular subject, I always assign a place to the *Human Chart*, or *Chart of things to be wished for*. For to form judicious wishes is as much a part of knowledge as to ask judicious questions.

L.

Among Prerogative Instances I will put in the twenty-sixth place *Polychrest Instances* or *Instances of General Use.* They are those which relate to a variety of cases and occur frequently; and therefore save no small amount of labour and fresh demonstration. Of the instruments and contrivances themselves the proper place for speaking will be when I come to speak of Applications to Prac tice and Modes of Experimenting. Moreover those which have been already discovered and come into use will be described in the particular histories of the several arts. At present I will subjoin a few general remarks on them as examples merely of this General Use.

Besides the simple bringing together and putting asunder of them, man operates upon natural bodies chiefly in seven ways: viz. either by exclusion of whatever impedes and disturbs; or by compressions, extensions, agitations, and the like; or by heat and cold; or by continuance in a suitable place; or by the checking and regulation of motion; or by special sympathies; or by the seasonable and proper alternation, series, and succession of all these ways, or at any rate of some of them.

With regard to the first; the common air, which is everywhere about us and pressing in, and the rays of the heavenly bodies, cause much disturbance. Whatever therefore serves to exclude them, may justly be reckoned among things of General Use. To this head belong the material and thickness of the vessels in which the bodies are placed on which we are going to operate; also the perfect stopping up of vessels by consolidation and *lutum sapientiæ*, as the chemists call it. Also the closing in of substances by liquids poured on the outside is a thing of very great use; as when they pour oil on wine or juices of herbs, which spreading over the surface like a lid preserves them excellently from the injury of the air. Nor are powders bad things; for though they contain air mixed up with them, they yet repel the force of the body of air round about; as we see in the preservation of grapes and other fruits in sand and flour. It is good too to spread bodies over with wax, honey, pitch, and like tenacious substances, for the more perfect enclosure of them, and to keep off the air and heavenly bodies. I have sometimes tried the effect of laying up a vessel or some other body in quicksilver, which of all substances that can be poured round another is far the densest. Caverns again and subterraneous pits are of great use in keeping off the heat of the sun and that open air which preys upon bodies; and such are used in the North of Germany as granaries. The sinking of bodies in water has likewise the same effect; as I remember to have heard of bottles of wine being let down into a deep well to cool; but through accident or neglect being left there for many years, and then taken out; and that the wine was not only free from sourness or flatness, but much finer tasted; owing, it would seem, to a more exquisite commixture of its parts. And if the case require that bodies be let down to the bottom of the water, as in a river or the sea, without either touching the water or being enclosed in stopped vessels, but surrounded by air alone; there is good use in the vessel which has been sometimes employed for working under water on sunk ships, whereby divers are enabled to remain a long while below, and take breath from time to time. This machine was a hollow bell made of metal, which being let down parallel to the surface of the water, carried with it to the bottom all the air it contained. It stood on three feet (like a tripod) the height of which was somewhat less than that of a man, so that the diver, when his breath failed, could put his head into the hollow of the bell, take breath, and then go on with his work. I have heard also of a sort of machine or boat capable of carrying men under water for some distance [119]. Be that as it may, under such a vessel as I have described bodies of any sort can easily be suspended; and it is on that account that I have mentioned this experiment.

There is also another advantage in the careful and complete closing of bodies; for not only does it keep the outer air from getting in (of which I have already

[119] According to Beckmann, the first distinct mention of the diving-bell, at least in modern times, is to be found in Fainsius, as quoted by Schott. Fainsius gives an account of some Greeks who exhibited a diving-bell at Toledo, before Charles the Fifth and his court, in 1538. [Bacon's words in the text specify in addition to the diving-bell a submarine boat, such as that exhibited by Drebbel in 1620. See below an allusion to the same subject in the *New Atlantis*, and note to the *De Augmentis*, v. 2.—ED.]

spoken), but also it keeps the spirits of the body, on which the operation is going on inside, from getting out. For it is necessary for one who operates on natural bodies to be certain of his total quantities; that is, that nothing evaporates or flows away. For then, and then only, are profound alterations made in bodies, when, while nature prevents annihilation, art prevents also the loss or escape of any part. On this subject there has prevailed a false opinion, which, if true, would make us well nigh despair of preserving the perfect quantity without diminution; namely, that the spirits of bodies, and air when rarefied by a high degree of heat, cannot be contained in closed vessels, but escape through the more delicate pores. To this opinion men have been led by the common experiment of an inverted cup placed on water with a candle in it or a piece of paper lighted; the consequence of which is that the water is drawn up; and also by the similar experiment of cupping-glasses, which when heated over flame draw up the flesh. For in each of these experiments they imagine that the rarefied air escapes, and that its quantity being thereby diminished, the water or flesh comes up into its place by the Motion of Connexion. But this is altogether a mistake. For the air is not diminished in quantity, but contracted in space; nor does the motion of the rising of the water commence till the flame is extinguished or the air cooled; and therefore physicians, to make their cupping glasses draw better, lay on them cold sponges dipped in water. And therefore there is no reason why men should be much afraid of the easy escape of air or spirits. For though it be true that the most solid bodies have pores, still air or spirit does not eaily submit to such extremely fine comminution; just as water refuses to run out at very small chinks.

With regard to the second of the seven modes of operating above mentioned, it is particularly to be observed, that compression and such violent means have indeed with respect to local motion and the like a most powerful effect; as in machines and projectiles; an effect which even causes the destruction of organic bodies, and of such virtues as consist altogether in motion. For all life, nay all flame and ignition, is destroyed by compression; just as every machine is spoilt or deranged by the same. It causes the destruction likewise of virtues which consist in the position and coarser dissimilarity of parts. This is the case with colours; for the whole flower has not the same colour as when it is bruised, nor the whole piece of amber as the same piece pulverised. So also it is with tastes; for there is not the same taste in an unripe pear as there is in a squeezed and softened one; for it manifestly contracts sweetness by the process. But for the more remarkable transformations and alterations of bodies of uniform structure such violent means are of little avail; since bodies do not acquire thereby a new consistency that is constant and quiescent, but one that is transitory, and ever striving to recover and liberate itself. It would not be amiss however to make some careful experiments for the purpose of ascertaining whether the condensation or the rarefaction of a body of nearly uniform structure (as air, water, oil, and the like), being induced by violence, can be made to be constant and fixed, and to become a kind of nature. This should first be tried by simple continuance, and then by means of helps and consents. And the trial might easily have been made (if it had but occurred to me) when I was condensing water, as mentioned above, by hammer and press, till it burst forth from its enclosure. For I should have left the flattened sphere to itself for a few days, and after that drawn off the water; that so I might have seen whether it would immediately occupy the same dimensions, which it had before condensation. If it had not done so, either immediately or at any rate soon after, we might have pronounced the condensation a constant one; if it had, it would have appeared that a restoration had taken place, and that the compression was transitory. Something of a similar kind I might have tried also with the expansion of air in the glass eggs. For after powerful suction I might have stopped them suddenly and tightly; I might have left the eggs so stopped for some days; and then tried whether on opening the hole the air would be drawn up with a hissing noise; whether on plunging them into water, as much water would be drawn up as there would have been at first without the delay. For it is probable—at least it is worth trying—that this might have been, and may be, the case; since in bodies of structure not quite so uniform, the lapse of time does produce such effects. For a stick bent for some time by compression does not recoil; and this must not be imputed to any loss of quantity in the wood through the lapse of time; since the same will be the case with a plate of steel, if the time be increased, and steel does not evaporate. But if the experiment succeed not with mere continuance, the business must not be abandoned, but other aids must be employed. For it is no small gain if by the application of violence we can communicate to bodies fixed and permanent natures. For thus air can be turned into water by condensation, and many other effects of the kind can be produced; man being more the master of violent motions than of the rest.

The third of the seven modes above mentioned relates to that which whether in Nature or in Art is the great instrument of operation, viz. heat and cold. And herein man's power is clearly lame on one side. For we have the heat of fire, which is infinitely more potent and intense than the heat of the sun as it reaches us, or the warmth of animals. But we have no cold save such as is to be got in winter time, or in caverns, or by application of snow and ice; which is about as much perhaps in comparison as the heat of the sun at noon in the torrid zone, increased by the reflexions of mountains and walls; for such heat as well as such cold can be endured by animals for a short time. But they are nothing to be compared to the heat of a burning furnace, or with any cold corresponding to it in intensity. Thus all things with us tend to rarefaction, and desiccation, and consumption; nothing hardly to condensation and inteneration, except by mixtures and methods that may be called spurious. Instances of cold therefore should be collected with all diligence; and such it seems may be found by exposing bodies on steeples in sharp frosts; by laying them in subterranean caverns; by surrounding them with snow and ice in deep pits dug for the purpose; by letting them down into wells; by burying them in quicksilver and metals; by plunging them into waters which petrify wood; by burying them in the earth, as the Chinese are said to do in the making of porcelain, where masses made for the purpose are left, we are told, under ground for forty or fifty years, and transmitted to heirs, as a kind of artificial minerals; and by similar processes. And so too all natural condensations caused by cold should be investigated, in order that, their causes being known, they may be imitated by art. Such we see in the sweating of marble and stones; in the dews condensed on the inside of window panes, towards morning, after a night's frost; in the formation and gathering of vapours into water under the earth, from which springs often bubble up. Everything of this kind should be collected.

Besides things which are cold to the touch, there are found others having the power of cold, which also condense; but which seem to act on the bodies of animals only, and hardly on others. Of this sort we have many instances in medicines and plasters; some of which condense the flesh and tangible parts, as astringent and inspissatory medicaments; while others condense the spirits, as is most observable in soporifics. There are two ways in which spirits are condensed by medicaments soporific, or provocative of sleep; one by quieting their motion, the other by putting them to flight.. Thus violets, dried rose leaves, lettuce, and like benedict or benignant medicaments, by their kindly and gently cooling fumes invite the spirits to unite, and quiet their eager and restless motion. Rose water too, applied to the nose in a fainting fit, causes the resolved and too relaxed spirits to recover themselves, and as it were cherishes them. But opiates and kindred medicaments put the spirits utterly to flight, by their malignant and hostile nature. And therefore if they be applied to an external part, the spirits immediately flee away from that part, and do not readily flow into it again; if taken internally, their fumes, ascending to the head, disperse in all directions the spirits contained in the ventricles of the brain; and these spirits thus withdrawing themselves, and unable to escape into any other part, are by consequence brought together and condensed, and sometimes are utterly choked and extinguished; though on the other hand these same opiates taken in moderation do by a secondary accident (namely, the condensation which succeeds the coming together) comfort the spirits, and render them more robust, and check their useless and inflammatory motions; whereby they contribute no little to the cure of diseases, and prolongation of life.

Nor should we omit the means of preparing bodies to receive cold. Among others I may mention that water slightly warm is more easily frozen than quite cold.

Besides, since nature supplies cold so sparingly, we must do as the apothecaries do, who when they cannot get a simple, take its succedaneum or *quid pro quo*, as they call it ; such as aloes for balsam, cassia for cinnamon. In like manner we should look round carefully to see if there be anything that will do instead of cold ; that is to say, any means by which condensations can be effected in bodies otherwise than by cold, the proper office of which is to effect them. Such condensations, as far as yet appears, would seem to be limited to four. The first of these is caused by simple compression, which can do but little for permanent density, since bodies recoil, but which perhaps may be of use as an auxiliary. The second is caused by the contraction of the coarser parts in a body, after the escape of the finer ; such as takes place in indurations by fire, in the repeated quenchings of metals, and like processes. The third is caused by the coming together of those homogeneous parts in a body which are the most solid, and which previously had been dispersed, and mixed with the less solid ; as in the restoration of sublimated mercury, which occupies a far greater space in powder than as simple mercury, and similarly in all purging of metals from their dross. The fourth is brought about through sympathy, by applying substances which from some occult power condense. These sympathies or consents at present manifest themselves but rarely ; which is no wonder, since before we succeed in discovering Forms and Configurations, we cannot hope for much from an inquiry into sympathies. With regard to the bodies of animals indeed, there is no doubt that there are many medicines, whether taken internally or externally, which condense as it were by consent, as I have stated a little above. But in the case of inanimate substances such operation is rare. There has indeed been spread abroad, as well in books as in common rumour, the story of a tree in one of the Tercera or Canary Isles (I do not well remember which) which is constantly dripping ; so as to some extent to supply the inhabitants with water [120]. And Paracelsus says that the herb called *Ros Solis* is at noon and under a burning sun filled with dew, while all the other herbs round it are dry [121]. But both of these stories I look upon as fabulous. If they were true, such instances would be of most signal use, and most worthy of examination. Nor do I conceive that those honey-dews, like manna, which are found on the leaves of the oak in the month of May, are formed and condensed by any peculiar property in the leaf of the oak, but that while they fall equally on all leaves, they are retained on those of the oak, as being well united, and not spongy as most of the others are.

As regards heat, man indeed has abundant store and command thereof ; but observation and investigation are wanting in some particulars, and those the most necessary, let the alchemists say what they will. For the effects of intense heat are sought for and brought into view ; but those of a gentler heat, which fall in most with the ways of nature, are not explored, and therefore are unknown. And therefore we see that by the heats generally used the spirits of bodies are greatly exalted, as in strong waters, and other chemical oils ; that the tangible parts are hardened, and, the volatile being discharged, sometimes fixed ; that the homogeneous parts are separated, while the heterogeneous are in a coarse way incorporated and mixed up together ; above all that the junctures of composite bodies, and their more subtle configurations, are broken up and confounded.

[120] This wonderful tree is described in Johnston's *Dendrographia*, published at Frankfort in 1669. See book the tenth, c. 4. One of the authorities he refers to is Cardan (*De variet. rerum*), from whom not improbably Bacon derived the story. The tree is said to be found in the island of Ferro. Cardan, with more than usual caution, remarks, at the close of the account he gives of it : " Sed postquam hoc tot scriptores affirmant, fieri potest ut tale aliquid contingat, sed modus nondum perspectus est".—*De rerum variet.* vi. c. 22. Compare *Oviedo in Ramusio*, iii. 71. a.

[121] I have not been able to find this in Paracelsus. It seems, however, to accord with his theory of dew,—namely that it is an exudation from the sun and stars ; the suppression of which would lead to the formation of additional suns.

Whereas the operations of a gentler heat ought to have been tried and explored, whereby more subtle mixtures and regular configurations might be generated and educed, after the model of nature, and in imitation of the works of the sun; as I have shadowed forth in the Aphorism on Instances of Alliance. For the operations of nature are performed by far smaller portions at a time, and by arrangements far more exquisite and varied than the operation of fire, as we use it now. And it is then that we shall see a real increase in the power of man, when by artificial heats and other agencies the works of nature can be represented in form, perfected in virtue, varied in quantity, and, I may add, accelerated in time. For the rust of iron is slow in forming, but the turning into *Crocus Martis* is immediate; and it is the same with verdigris and ceruse; crystal is produced by a long process, while glass is blown at once; stones take a long time to grow, while bricks are quickly baked. Meanwhile (to come to our present business), heats of every kind, with their effects, should be diligently collected from all quarters and investigated,—the heat of heavenly bodies by their rays direct, reflected, refracted, and united in burning-glasses and mirrors; the heat of lightning, of flame, of coal fire; of fire from different materials; of fire close and open, straightened and in full flow, modified in fine by the different structures of furnaces; of fire excited by blowing; of fire quiescent and not excited; of fire removed to a greater or less distance; of fire passing through various media; moist heats, as of a vessel floating in hot water [122], of dung, of external and internal animal warmth, of confined hay; dry heats, as of ashes, lime, warm sand; in short heats of all kinds with their degrees.

But above all, we must try to investigate and discover the effects and operations of heat when applied and withdrawn gradually, orderly, and periodically, at due distances and for due time. For such orderly inequality is in truth the daughter of the heavens, and mother of generation; nor is anything great to be expected from a heat either vehement or precipitate or that comes by fits and starts. In vegetables this is most manifest; and also in the wombs of animals there is a great inequality of heat, from the motion, sleep, food and passions of the female in gestation; lastly in the wombs of the earth itself, those I mean in which metals and fossils are formed, the same inequality has place and force. Which makes the unskilfulness of some alchemists of the reformed school all the more remarkable,—who have conceived that by the equable warmth of lamps and the like, burning uniformly, they can attain their end. And so much for the operations and effects of heat. To examine them thoroughly would be premature, till the Forms of things and the Configurations of bodies have been further investigated and brought to light. For it will then be time to seek, apply, and adapt our instruments, when we are clear as to the pattern.

The fourth mode of operating is by continuance, which is as it were the steward and almoner of nature. Continuance I call it, when a body is left to itself for a considerable time, being meanwhile defended from all external force. For then only do the internal motions exhibit and perfect themselves, when the extraneous and adventitious are stopped. Now the works of time are far subtler than those of fire. For wine cannot be so clarified by fire, as it is by time; nor are the ashes produced by fire so fine as the dust into which substances are resolved and wasted by ages. So too the sudden incorporations and mixtures precipitated by fire are far inferior to those which are brought about by time. And the dissimilar and varied configurations which bodies by continuance put on, such as putrefactions, are destroyed by fire or any violent heat. Meanwhile it would not be out of place to observe that the motions of bodies when quite shut up have in them

[122] Orig. *Balnei Mariæ.* This is properly "balneum maris;" that is, a mode of communicating heat to any substance by putting it into a vessel which is placed in another containing water. The latter being put on the fire, the former and its contents become gradually and moderately heated. The reason of the name is obvious. From "balneum maris" the French made by a kind of translation (the final *s* not being sounded) "bain marie;" and the form in the text is, I think, merely a retranslation of the French phrase, the meaning of the second word being mistaken. Balneum Mariæ is however, I believe, a common phrase with old writers on chemistry.

something of violence. For such imprisonment impedes the spontaneous motions of the body. And therefore continuance in an open vessel is best for separations ; in a vessel quite closed for commixtures ; in a vessel partly closed, but with the air entering, for putrefactions. But indeed instances showing the effects and operations of continuance should be carefully collected from all quarters.

The regulation of motion (which is the fifth mode of operating) is of no little service. I call it regulation of motion, when one body meeting another impedes, repels, admits or directs its spontaneous motion. It consists for the most part in the shape and position of vessels. Thus the upright cone in alembics helps the condensation of vapours ; the inverted cone in receivers helps the draining off of the dregs of sugar. Sometimes a winding form is required, and one that narrows and widens in turn, and the like. For all percolation depends on this, that the meeting body opens the way to one portion of the body met and shuts it to another. Nor is the business of percolation or other regulation of motion always performed from without ; it may also be done by a body within a body ; as when stones are dropped into water to collect its earthy parts ; or when syrups are clarified with the whites of eggs, that the coarser parts may adhere thereto, after which they may be removed. It is also to this regulation of motion that Telesius has rashly and ignorantly enough attributed the shapes of animals, which he says are owing to the channels and folds in the womb[123]. But he should have been able to show the like formation in the shells of eggs, in which there are no wrinkles or inequalities. It is true however that the regulation of motion gives the shapes in moulding and casting.

Operations by consents or aversions (which is the sixth mode) often lie deeply hid. For what are called occult and specific properties, or sympathies and antipathies, are in great part corruptions of philosophy. Nor can we have much hope of discovering the consents of things before the discovery of Forms and Simple Configurations. For consent is nothing else than the adaptation of Forms and Configurations to each other.

The broader and more general consents of things are not however quite so obscure. I will therefore begin with them. Their first and chief diversity is this, that some bodies differ widely as to density and rarity, but agree in configurations, while others agree as to density and rarity but differ in configurations. For it has not been ill observed by the chemists in their triad of first principles, that sulphur and mercury[124] run through the whole universe. (For what they add about salt is absurd, and introduced merely to take in bodies earthy, dry, and fixed.) But certainly in these two one of the most general consents in nature does seem to be observable. For there is consent between sulphur, oil and greasy exhalation, flame, and perhaps the body of a star. So is there between mercury, water and watery vapours, air, and perhaps the pure and intersidereal ether. Yet these two quaternions or great tribes of things (each within its own limits) differ immensely in quantity of matter and density, but agree very well in configuration ; as appears in numerous cases. On the other hand metals agree well together in quantity and density, especially as compared with vegetables, etc., but differ very widely in configuration ; while in like manner vegetables and animals vary almost infinitely in their configurations, but in quantity of matter or density their variation is confined to narrow limits.

The next most general consent is that between primary bodies and their

123 Telesius's doctrine of the formation of the embryo is essentially the same as Galen's, namely that a system of arteries, etc., must be first of all formed in the germ, and that these, by applying themselves to corresponding parts on the surface of the matrix, determine the channels through which nourishment is supplied, and therefore (mediately) the development of the different members of the fœtus. But it does not seem that he would have admitted that the smoothness of the shells of eggs was an objection to his theory. At any rate, he illustrates it by reference to the appearances presented by an egg opened during incubation. *De rerum naturâ*, vi. c. 4 and 40.

124 This triad is the fundamental point of Paracelsus's chemical and medical philosophy. See his works throughout, and particularly the tract *De tribus primis essentiis*, contained in the third book of his philosophical works.

supports; that is, their menstrua[125] and foods. We must therefore inquire, under what climates, in what earth, and at what depth, the several metals are generated; and so of gems, whether produced on rocks or in mines; also in what soil the several trees and shrubs and herbs thrive best, and take, so to speak, most delight; moreover what manurings, whether by dung of any sort, or by chalk, sea-sand, ashes, etc., do the most good; and which of them are most suitable and effective according to the varieties of soil. Again, the grafting and inoculating of trees and plants, and the principle of it, that is to say, what plants prosper best on what stocks, depends much on sympathy. Under this head it would be an agreeable experiment, which I have heard has been lately tried, of engrafting forest-trees (a practice hitherto confined to fruit-trees); whereby the leaves and fruit are greatly enlarged, and the trees made more shady. In like manner the different foods of animals should be noted under general heads, and with their negatives. For carnivorous animals cannot live on herbs, whence the order of Feuillans (though the will in man has more power over the body than in other animals) has after trial (they say) well nigh disappeared; the thing not being endurable by human nature[126]. Also the different materials of putrefaction, whence animalculæ are generated, should be observed.

The consents of primary bodies with their subordinates (for such those may be considered which I have noted) are sufficiently obvious. To these may be added the consents of the senses with their objects. For these consents, since they are most manifest, and have been well observed and keenly sifted, may possibly shed great light on other consents also which are latent.

But the inner consents and aversions or friendships and enmities of bodies (for I am almost weary of the words sympathy and antipathy an account of the superstitions and vanities associated with them) are either falsely ascribed,

125 By "menstrua" are meant the substances out of which any species of mineral is generated, or, in other words, the causa materialis of its existence. See, on the generation of metals and other minerals, the fourth and fifth books of Agricola's work *De ortu et causis fossilium*. He gives an account of the opinions of Aristotle, Theophrastus, etc. In modern chemistry the word menstruum is nearly equivalent to solvent. By the school of Paracelsus the word is used so vaguely that it is difficult to determine what idea they attached to it, or how they derived their sense of the word from its original signification. When the word is used as in the text, the metaphor seems to be taken from the Aristotelian theory of generation, in which *κατὰ τὴν πρώτην ὑλήν ἐστιν ἡ τῶν καταμηνίων φύσις*.

126 Orig. *Ordo Folitanorum*. Bacon doubtless refers to the order of Feüillans. Jean de la Barrière, after holding the Cistercian abbey of Feüillans in commendam for eleven years, renounced the world in 1573, and in the course of a few years introduced a most austere rule of life into the abbey of which he was the head. His monks knelt on the floor during their refections, and some of them were in the habit of drinking out of skulls. They abstained from eggs, fish, butter, oil, and even salt, and confined themselves to pottage made of herbs boiled in water, and bread so coarse and black that beasts refused to eat of it. After a while they gave up wine also. Clement VIII. permitted the society to draw up constitutions for the establishment of their rule. By these the excessive rigour of their way of life was checked, which was done in obedience to the Pope, and in consequence of the deaths of fourteen monks in a single week at Feüillans. These constitutions were ratified in 1595. Assuming, of which there seems no doubt, that the Folitani of Bacon are the Feüillans, I may remark that the latinised form of Feüillans used is Fuliensis, as an adjective; the proper style of the society being "Congregatio Cistertiomonastica B. Mariæ Fuliensis". I have not seen the work of Morotius to which Helyot, from whom the preceding account is taken, refers, but in that of C. Henrique, also mentioned by Helyot, I do not find any authority for *Folitani*. It is probable that Bacon's chief information on the subject was gathered orally during his residence in France, before the Feüillans had ceased from their first love. The expression "ordo . . . fere evanuit" must be taken to mean that the severe rule that they had at first was given up. See Helyot, *Hist. des Ordres Monastiques*, ivme partie, c. 38. Spondanus, *An.* 1586, iv. For some particulars of the early history of the Abbey of Feüillans, and especially for the will of Jean de la Barrière, see *Voyage Littéraire de deux Bénédictins*, ii. p. 16.

mixed with tables, or from want of observation very rarely met with. For if it be said that there is enmity between the vine and colewort, because when planted near each other they do not thrive; the reason is obvious—that both of these plants are succulent and exhaust the ground, and thus one robs the other. If it be said that there is consent and friendship between corn and the corn-cockle or the wild poppy, because these herbs hardly come up except in ploughed fields; it should rather be said that there is enmity between them, because the poppy and corn-cockle are emitted and generated from a juice in the earth which the corn has left and rejected; so that sowing the ground with corn prepares it for their growth. And of such false ascriptions there is a great number. As for fables, they should be utterly exterminated. There remains indeed a scanty store of consents which have been approved by sure experiment; such as those of the magnet and iron, of gold and quicksilver, and the like. And in chemical experiments on metals there are found also some others worthy of observation, but they are found in greatest abundance (if one may speak of abundance in such a scarcity) in certain medicines, which by their occult (as they are called) and specific properties have relation either to limbs, or humours, or diseases, or sometimes to individual natures. Nor should we omit the consents between the motions and changes of the moon and the affections of bodies below; such as may be gathered and admitted, after strict and honest scrutiny, from experiments in agriculture, navigation, medicine, and other sciences. But the rarer all the instances of more secret consents are, the greater the diligence with which they should be sought after, by means of faithful and honest traditions and narrations; provided this be done without any levity or credulity, but with an anxious and (so to speak) a doubting faith. There remains a consent of bodies, inartificial perhaps in mode of operation, but in use a Polychrest, which should in no wise be omitted, but examined into with careful attention. I mean the proneness or reluctance of bodies to draw together or unite by composition or simple apposition. For some bodies are mixed together and incorporated easily, but others with difficulty and reluctance. Thus powders mix best with water; ashes and lime with oils, and so on. Nor should we merely collect instances of the propensity or aversion of bodies for mixture, but also of the collocation of their parts, of their distribution and digestion when they are mixed, and finally of their predominancy after the mixture is completed.

There remains the seventh and last of the seven modes of operation, namely the means of operating by the alternation of the former sex; but it would not be seasonable to bring forward examples of this, till our search has been carried somewhat more deeply into the others singly. Now a series or chain of such alternations, adapted to particular effects, is a thing at once more difficult to discover, and most effective to work with. But men are utterly impatient both of the inquiry and the practice; though it is the very thread of a labyrinth, as regards works of any magnitude. Let this suffice to exemplify the Polychrest Instances.

LI.

Among Prerogative Instances I will put in the twenty-seventh and last place *Instances of Magic;* by which I mean those wherein the material or efficient cause is scanty or small, as compared with the work and effect produced; so that, even where they are common, they seem like miracles; some at first sight, others even after attentive consideration. These indeed nature of herself supplies sparingly; but what she may do when her folds have been shaken out, and after the discovery of Forms and Processes and Configurations, time will show. But these magical effects (according to my present conjecture) are brought about in three ways; either by self-multiplication, as in fire, and in poisons called specific, and also in motions which are increased in power by passing from wheel to wheel; or by excitation or invitation in another body, as in the magnet, which excites numberless needles without losing any of its virtue, or in yeast and the like; or by anticipation of motion, as in the case already mentioned of gunpowder and cannons and mines. Of which ways the two former require a knowledge of consents; the third, a knowledge of the measurement of motions. Whether there be any mode of changing bodies *per minima* (as they call it) and of transposing the subtler configurations of matter (a thing required in every sort of transformation of bodies) so that art may be enabled to do in a short time that which nature accomplishes by many windings, is a point on which I have at present no sure indications. And as in matters solid and true I aspire to the ultimate and supreme, so do I for ever hate all things vain and tumid, and do my best to discard them.

LII.

So much then for the Dignities or Prerogatives of Instances. It must be remembered however that in this Organum of mine I am handling logic, not philosophy. But since my logic aims to teach and instruct the understanding, not that it may with the slender tendrils of the mind snatch at and lay hold of abstract notions (as the common logic does), but that it may in very truth dissect nature, and discover the virtues and actions of bodies, with their laws as determined in matter; so that this science flows not merely from the nature of the mind, but also from the nature of things; no wonder that it is everywhere sprinkled and illustrated with speculations and experiments in nature, as examples of the art I teach. It appears then from what has been said that there are twenty-seven Prerogative Instances; namely, Solitary Instances; Migratory Instances; Striking Instances; Clandestine Instances; Constitutive Instances; Conformable Instances; Singular Instances; Deviating Instances; Bordering Instances; Instances of Power; Instances of Companionship and of Enmity; Subjunctive Instances; Instances of Alliance; Instances of the Fingerpost; Instances of Divorce; Instances of the Door; Summoning Instances; Instances of the Road; Instances Supplementary; Dissecting Instances; Instances of the Rod; Instances of the Course; Doses of Nature; Instances of Strife; Intimating Instances; Polychrest Instances; Magical Instances. Now the use of these instances, wherein they excel common instances, is found either in the Informative part or in the Operative, or in both. As regards the Informative, they assist either the senses or the understanding; the senses, as the five Instances of the Lamp; the understanding, either by hastening the Exclusion of the Form, as Solitary Instances;—or by narrowing and indicating more nearly the Affirmative of the Form, as Instances Migratory, Striking, of Companionship, and Subjunctive;—or by exalting the understanding and leading it to genera and common natures; either immediately, as Instances Clandestine, Singular, and of Alliance; or in the next degree, as Constitutive; or in the lowest, as Conformable;—or by setting the understanding right when led astray by habit, as Deviating Instances; —or by leading it to the Great Form or Fabric of the Universe, as Bordering Instances;—or by guarding it against false forms and causes, as Instances of the Fingerpost and of Divorce. In the Operative Part, they either point out, or measure, or facilitate practice. They point it out, by showing with what we should begin, that we may not go again over old ground, as Instances of Power; or to what we should aspire if means be given, as Intimating Instances. The four Mathematical Instances measure practice: Polychrest and Magical Instances facilitate it.

Again out of these twenty-seven instances there are some of which we must make a collection at once, as I said above, without waiting for the particular investigation of natures. Of this sort are Instances Conformable, Singular, Deviating, Bordering, of Power, of the Dose, Intimating, Polychrest, and Magical. For these either help and set right the understanding and senses, or furnish practice with her tools in a general way. The rest need not be inquired into till we come to make Tables of Presentation for the work of the Interpreter concerning some particular nature. For the instances marked and endowed with these Prerogatives are as a soul amid the common instances of Presentation, and as I said at first, a few of them do instead of many; and therefore in the formation of the Tables they must be investigated with all zeal, and set down therein. It was necessary to handle them beforehand because I shall have to speak of them in what follows. But now I must proceed to the supports and rectifications of Induction, and then to concretes, and Latent Processes, and Latent

Configurations, and the rest, as set forth in order in the twenty-first Aphorism; that at length (like an honest and faithful guardian) I may hand over to men their fortunes, now their understanding is emancipated and come as it were of age; whence there cannot but follow an improvement in man's estate, and an enlargement of his power over nature. For man by the fall fell at the same time from his state of innocency and from his dominion over creation. Both of these losses however can even in this life be in some part repaired; the former by religion and faith, the latter by arts and sciences. For creation was not by the curse made altogether and for ever a rebel, but in virtue of that charter, "In the sweat of thy face shalt thou eat bread," it is now by various labours (not certainly by disputations or idle magical ceremonies, but by various labours) at length and in some measure subdued to the supplying of man with bread; that is, to the uses of human life.

END OF THE SECOND BOOK OF THE NEW ORGANUM.

PREPARATIVE TOWARDS A NATURAL AND EXPERIMENTAL HISTORY[1]

PREFACE.

BY JAMES SPEDDING.

AMONG the eight subjects which were to have been handled in the remaining books of the *Novum Organum* (see ii. 21), the last but one is entitled *De parascevis ad inquisitionem*, under which head Bacon intended (as appears by the introduction to the following treatise) to set forth the character of the *Natural and Experimental History*, which was to form the third part of the Instauratio.

What may have been the logical connexion between these eight subjects which determined him to reserve this for the penultimate place, it seems impossible, by the help of the titles alone, to divine. But whatever the order in which he thought advisable to approach it, there can be no doubt that this Natural and Experimental History was always regarded by him as a part of his system both fundamental and indispensable. So earnestly indeed and so frequently does he insist on the importance of it, that I once believed it to be the one real novelty which distinguished his philosophy from those of his contemporaries and immediate predecessors. And even now, though Mr. Ellis's analysis of the Baconian Induction has given me much new light and considerably modified my opinion in that matter, I am still inclined to think that Bacon himself regarded it not only as a novelty, but as *the* novelty from which the most important results were to be expected ; and however experience may have proved that his expectations were in great part vain and his scheme impracticable, I cannot help suspecting that more of it is practicable than has yet been attempted, and that the greatest results of science are still to be looked for from a further proceeding in this direction.

The grounds of this opinion will be explained most conveniently in connexion with the following treatise ; a treatise published by Bacon (on account of the exceeding importance of the subject) out of its proper place and incomplete ; and to which I find nothing among Mr. Ellis's papers that can serve as preface.

In what the distinctive peculiarity of the Baconian philosophy really consisted, is a question to which every fresh inquirer gives a fresh answer. Before I was acquainted with Mr. Ellis's, which is the latest, and formed upon the largest survey and subtlest scrutiny of the evidence, I had endeavoured to find one for myself, and had come to a conclusion which, though quite different from his, is not I think irreconcilable with it, but contains (as I still venture to believe) a part, though a part only, of the truth. And the question which I wish now to raise is whether, as my solution was imperfect from not taking any account of the novelty contained in the *method of Induction* as Bacon understood it, Mr. Ellis's be not likewise imperfect from not taking sufficient account of the novelty contained in the *Natural History* as Bacon intended it to be employed ; and whether there be not room for a third solution more complete than either, as including both.

That the philosophy which Bacon meant to announce was in some way essentially different not only from any that had been before but from any that has been

1 [Translation of *Parasceve ad historiam naturalem et experimentalem*. Published in 1620 in the same volume with the *Novum Organum*.]

since, is a position from which in both cases the inquiry sets out; and since it is one which will not perhaps be readily granted by everybody, it may be worth while to explain the considerations which led me to it; the rather because Mr. Ellis and myself, though proceeding not only independently but by entirely different roads and in pursuit of different objects—he endeavouring to penetrate the secret of Bacon's philosophy, I endeavouring to understand the objects and purposes of his life—meet nevertheless at this point in the same conclusion.

The process by which I arrived at it myself, I cannot explain better than by transcribing a paper which I wrote on the subject in 1847; at which time I had not seen any part of Mr. Ellis's argument, or heard his opinion upon the question at issue. What my own opinion is now, I will state afterwards; but first I give the paper exactly as I then wrote it; the length of the extract being justified—at least if there be any truth in the conclusion—by the importance of the questions at issue; for it bears upon the business of the present and future quite as much as on the knowledge of the past. The form in which it is written,—that of a familiar conversation between two friends,—happened to be the most convenient for the business I was then about; and as I could not present the argument more clearly in any other, I leave it as it is.

A.

Before you go on I wish you would satisfy me on one point, upon which I have hitherto sought satisfaction in vain. What after all was it that Bacon did for philosophy? In what did the wonder and in what did the benefit consist? I know that people have all agreed to call him the Father of the Inductive Philosophy; and I know that the sciences made a great start about his time and have in some departments made great progress since. But I could never yet hear what one thing he discovered that would not have been discovered just as soon without his help. It is admitted that he was not fortunate in any of his attempts to apply his principles to practice. It is admitted that no actual scientific discovery of importance was made by him. Well, he might be the father of discovery for all that. But among all the important scientific discoveries which have been made by others since his time, is there any one that can be traced to his teaching? traced to any principles of scientific investigation originally laid down by him, and by no other man before him or contemporary with him? I know very well that he did lay down a great many just principles;—principles which must have been acted upon by every man that ever pursued the study of Nature with success. But what of that? It does not follow that we *owe* these principles to him. For I have no doubt that I myself,—I that cannot tell how we know that the earth goes round, or why an apple falls or why the antipodes do not fall,—I have no doubt (I say) that if I sat down to devise a course of investigation for the determination of these questions, I should discover a great many just principles which Herschel and Faraday must hereafter act upon, as they have done heretofore. Nay if I should succeed in setting them forth more exactly, concisely, impressively, and memorably, than any one has yet done, they might soon come to be called *my* principles. But if that were all, I should have done little or nothing for the advancement of science. I should only have been finding for some of its processes a better name. I want to know whether Bacon did anything more than this; and if so, what. In what did the principles laid down by him essentially differ from those on which (while he was thus labouring to expound them) Galileo was already acting? From all that I can hear, it seems evident that the Inductive Philosophy received its great impulse, not from the great prophet of new principles, but from the great discoverers of new facts; not from Bacon, but from Galileo and Kepler. And I suppose that, with regard to those very principles even, if you wanted illustrations of what is commonly called the Baconian method, you would find some of the very best among the works of Gilbert and Galileo. What was it then that Bacon did which entitles him to be called the Regenerator of Philosophy? or what was it that he dreamt he was doing which made him think the work so entirely his own, so immeasurably important, and likely to be received with such incredulity by at least one generation of mankind?

B.

A pertinent question; for there is no doubt that he was under that impression. "*Cum argumentum hujusmodi præ manibus habeam* (says he) *quod tractandi imperitiâ perdere et veluti exponere* NEFAS *sit.*" He was persuaded that the argument he had in charge was of such value that to risk the loss of it by unskilful handling would be not only a pity but an impiety. You wish to know, and the wish is reasonable, what it was. For answer I would refer you to the philosophers; only I cannot say that their answers are satisfactory to myself. The old answer was that Bacon was the first to break down the dominion of Aristotle. This is now, I think, generally given up. His opposition to Aristotle was indeed conceived in early youth, and (though he was not the first to give utterance to it) I dare say it was not the less his own, and in the proper sense of the word, original. But the real overthrower of Aristotle was the great stir throughout the intellectual world which followed the Reformation and the revival of learning. It is certain that his authority had been openly defied some years before the publication of Bacon's principal writings; and it could not in the nature of things have survived much longer. Sir John Herschel however, while he freely admits that the Aristotelian philosophy had been effectually overturned without Bacon's aid, still maintains Bacon's title to be looked upon in all future ages as the great Reformer of Philosophy; not indeed that he *introduced* inductive reasoning as a new and untried process, but on account of his "keen perception and his broad and spirit-stirring, almost enthusiastic announcement of its paramount importance, as the alpha and omega of science, as the grand and only chain for linking together of physical truths, and the eventual key to every discovery and every application".

A.

That is all very fine; but it seems to me rather to account for his having the title than to justify his claim to it;—rather to explain how he comes by his reputation than to prove that he deserves it. Try the question upon a modern case. We are now standing upon the threshold of a new era in the science of History. It is easy to see that the universal study of History must be begun afresh upon a new method. Tales, traditions, and all that has hitherto been accounted most authentic in our knowledge of past times, must be set aside as doubtful; and the whole story must be spelt out anew from charters, names, inscriptions, monuments, and such like contemporary records. Now an eloquent man might easily make a broad and spirit-stirring announcement of the paramount importance of this process, as the only key by which the past can be laid open to us as it really was,—the grand and only chain for linking historical truths, and so forth. But would he thereby entitle himself to be called the great reformer of History? Surely not. Such a man might perhaps get the credit, but it is Niebuhr that has done the thing: for Niebuhr was the first both to see the truth and to set the example.

B.

So, I confess, it seems to me. And if I thought that Bacon had aimed at no more than that, I should not think that his time had been altogether well-employed, or his sense of the importance of his own mission to mankind altogether justified. For surely a single great discovery made by means of the inductive process would have done more to persuade mankind of the paramount importance of it, than the most eloquent and philosophical exposition. Therefore in forsaking his experiments about gravitation, light, heat, etc., in order to set forth his classification of the "Prerogatives of Instances", and to lay down general principles of philosophy, he would have been leaving the effectual promotion of his work to secure the exaltation of his name, than which nothing could be more opposite both to his principles and his practice. If his ambition had been only to have his picture stand as the frontispiece of the new philosophy, he could not have done better indeed than come forward as the most eloquent expounder of its principles. But if he wanted (as undoubtedly he did above all other things) to set it on work and bring it into fashion, his business was to produce the most striking illustration of its powers,—the most striking practical proof of what it could do.

Therefore if I thought, as Herschel seems to think, that there was no essential or considerable difference between the doctrines which Bacon preached and those which Galileo practised;—that Galileo was as the Niebuhr of the new philosophy (according to your own illustration), and Bacon only as your supposed eloquent man;—I should agree with you that Bacon's right to be called the Reformer of Philosophy is not made out. But when I come to look at Bacon's own exposition of his views and compare them with the latest and most approved account I have met with of Galileo's works, I cannot but think that the difference between what Galileo was doing and what Bacon wanted to be done is not only essential but immense.

A.

Nay, if the difference be immense, how comes it to be overlooked? It is from no want of the wish to claim for Bacon all the credit he deserves in that line.

B.

No. Rather perhaps from the wish to claim too much. We are so anxious to give him his due that we must needs ascribe to him all that has been done since his time; from which it seems to follow that we are practising his precepts, and that the Baconian philosophy has in fact been flourishing among us for the last 200 years. You believe this, don't you?

A.

People tell me so; and I suppose the only doubt is whether it be exclusively and originally his;—there is no doubt, I fancy, that it *is* his.

B.

Certainly that appears to be the general opinion; and it may seem an audacious thing in me to say that it is a mistake. But I cannot help it. It is true that a new philosophy is flourishing among us which was born about Bacon's time; and Bacon's name (as the brightest which presided at the time of its birth) has been inscribed upon it.

"Hesperus, that led
The starry host, rode brightest:"

not that Hesperus did actually *lead* the other stars; he and they were moving under a common force, and they would have moved just as fast if he had been away; but because he shone brightest, he *looked* as if he led them. But if I may trust Herschel, I must think that it is the *Galilean* philosophy that has been flourishing all these years; and if I may trust my own eyes and power of construing Latin, I must think that the *Baconian* philosophy has yet to come.

If Bacon were to reappear among us at the next meeting of the Great British Association,—or say rather if he had appeared there two or three years ago (for there seems to be something great and new going on now), I think he would have shaken his head. I think he would have said, "Here has been a great deal of very good diligence used by several persons; but it has not been used upon a well-laid plan. These solar systems, and steam-engines, and Daguerreotypes, and electric telegraphs, are so many more pledges of what might be expected from an instauration of philosophy such as I recommended more than 200 years ago; why have you not tried that? You have been acting all the time like a king who should attempt to conquer a country by encouraging private adventurers to make incursions each on his own account, without any system of combined movements to subdue and take possession. I see that wherever you have the proper materials and plenty of them your work is excellent; so was Gilbert's in my time; so was Galileo's; nay even Kepler—though his method was as unskilful as that of the boy who in doing a long-division sum would first guess at the quotient and then multiply it into the divisor to see whether it were true, and if it came out wrong would make another guess and multiply again, and so on till he guessed right at last,—yet because he had a copious collection of materials ready to his hand, and enormous perseverance however perversely applied, and a religious veracity, did at last hit upon one of the greatest discoveries ever made by one man. But what could Kepler have done without Tycho Brahe's tables of observation ? And what might Galileo not have done if he had had a large enough collection of facts ? This therefore it is that disappoints me. I do not see any sufficient collection made of materials,—that is, of facts in nature—or any effectual plan on foot for making one. You are scarcely better off in that respect than I was ; you have each to gather the materials upon which you are to work. You cannot build houses, or weave shirts, or learn languages so. If the builder had to make his own bricks, the weaver to grow his own flax, the student of a dead language to make his own concordance, where would be your houses, your shirts, or your scholars ? And by the same rule if the interpreter of Nature is to forage for his facts, what progress can you expect in the art of interpretation ? Your scholar has his dictionary provided to his hand ; but your natural philosopher has still to make his dictionary for himself.

" And I wonder the more at this, because this is the very thing of all others which I myself pointed out as absolutely necessary to be supplied,—as the thing which was to be set about in the first place,—the thing *without* which no great things could possibly be done in philosophy. And since you have done me the honour to think so very highly of my precepts, I am a little surprised that you have not thought it worth while in so very essential a point to follow them. And to say the truth, I could wish for my own reputation (if that were of any consequence) that you had either honoured me a little more in that way, or not honoured me quite so much in other ways. You call me the Father of your Philosophy, meaning it for the greatest compliment you can pay. I thank you for the compliment, but I must decline the implied responsibility. I assure you this is none of mine.—May I ask whether any attempt has been made to collect that '*Historiam naturalem et experimentalem quæ sit in ordine ad condendam philosophiam*', concerning which I did certainly give some very particular directions ;—which I placed as conspicuously as I could in the very front and entrance of my design ;—of which I said that all the genius and meditation and argumentation in the world could not do instead of it ; no, not if all men's wits could meet in one man's head ; therefore that this we must have, or else the business must be given up ? [2] —If this has been fairly tried and found impracticable or ineffectual, blot me out of your books as a dreamer that thought he had found out a great thing but it turned out nothing. If not, I still think it would be worth your while to try it."

A.

I partly comprehend your meaning ; but I should prefer it in a less dramatic form. You think that the difference between what Galileo did and what Bacon wanted to be done, lay in this—that Bacon's plan presupposed a history (or dictionary as you call it) of Universal Nature, as a storehouse of facts to work upon ; whereas Galileo was content to work upon such facts and observations as he collected for himself. But surely this is only a difference in degree. Both used the facts in the same way ; only Bacon wanted a larger collection of them.

B.

Say rather, Bacon wanted a collection large *enough* to give him the command of all the avenues to the secrets of Nature. You might as well say that there is only a difference of degree between the method of the man who runs his single head against a fortress, and the man who raises a force strong enough to storm it,—because each uses the force he has in the same way, only one wants more of it than the other :—or between stopping *all* the leaks in a vessel and stopping as

[2] Neque huic labori et inquisitioni et mundanæ perambulationi, ulla ingenii aut meditationis aut argumentationis substitutio aut compensatio sufficere potest, non si omnia omnium ingenia coierint. Itaque aut hoc prorsus habendum aut negotium in perpetuum deserendum.

many as you conveniently can. The truth is, that though the difference between a *few* and *a few more* is only a difference of degree, the difference between *enough* and *not enough* is a difference in kind. According to Galileo's method, the work at best could be done but partially. According to Bacon's (so at least he believed) it would be done effectually and altogether.

I will put you a case by way of illustration. Two men (call them James and John) find a manuscript in a character unknown to either of them. James, being skilled in languages and expert at making out riddles, observes some characters similar to those of one of the languages which he understands ; immediately sets himself to guess what they are ; and succeeds in puzzling out here a name and there a date, with plausibility. Each succeeding guess, if it be right, makes the next easier ; and there is no knowing precisely how much may be made out in this manner, or with what degree of certainty. The process is inductive, and the results, so far as they go, are discoveries. John seeing him thus employed comes up and says : " This is all very ingenious and clever, and far more than I could do by the same process. But you are not going the right way to work. You will never be able to decipher the manuscript in this way. I will tell you what we must do. Here (you see) are certain forms of character which continually recur. Here is one that comes more than once in every line ; here another that comes once in every two or three lines ; a third that comes only twice or thrice in a page ; and so on. Let us have a list made of these several forms, with an index showing where and how often they occur. In the meantime I will undertake, upon a consideration of the general laws of language, to tell you, by the comparative frequency of their recurrence, what parts of speech most of these are. So we shall know which of them are articles, which conjunctions, which relatives, which auxiliaries, and so on. Setting these apart we shall be better able to deal with the nouns and verbs ; and then by comparing the passages in which each occurs we shall be able, with the help of your language learning, to make out the meaning first of one, then of another. As each is determined, the rest will be easier to determine ; and by degrees we shall come to know them all. It is a slow process compared with yours, and will take time and labour and many hands. But when it is done we shall be able to read the whole book."

Here I think you have a picture in little of the difference between Bacon's project for the advancement of philosophy and that which was carried into effect (certainly with remarkable success) by the new school of inductive science which flourished in his time. If we want to pursue the parallel further, we have only to suppose that John, after completing in a masterly manner a great portion of his work on the universal laws of language ; after giving particular directions for the collection, arrangement, and classification of the index, and even doing several pages of it himself by way of example ; is called away, and obliged to leave the completion of the work to his successors ; and that his successors (wanting diligence to finish, patience to wait, or ability to execute) immediately fall back to the former method ;—in which they make such progress and take such pride, that they never think of following out John's plan, but leave it exactly where he left it. And here I think you have a true picture of the state in which the matter now rests.

A.

I see. The manuscript is the volume of Nature. The learned linguist and expert maker-out of puzzles is Galileo or one of his school. The work on the laws of language is the *Novum Organum.* The index is the Natural and experimental History *quæ sit in ordine ad condendam Philosophiam.* The making-out of the words one by one is the Interpretation of Nature—

B.

And the ultimate reading of the whole book is the " *Historia Illuminata sive Veritas Rerum* " ; the " *Philosophia Secunda* " ; the sixth and last part of the Instauration ; the consummation which Bacon knew he was not to be permitted himself to see, but trusted that (if men were true to themselves) the Fortune of the Human Race would one day achieve.

A.

And you think that they have not been true to themselves ?

B.

Why, what have they done with this work since he left it ? There it lies to speak for itself, sticking in the middle of the *Novum Organum*. No attempt has been made, that I can hear of, to carry it out further. People seem hardly to know that it is not complete. John Mill observes that Bacon's method of inductive logic is defective, but does not advert to the fact that of *ten* separate processes which it was designed to include, the first only has been explained. The other nine he had in his head, but did not live to set down more of them than the names. And the particular example which he has left of an inductive inquiry does not profess to be carried beyond the first stage of generalization,—the *vindemiatio prima* as he calls it.

A.

It may be so ; but *why* have they not attempted to carry his process out further ? Is it not because they have found that they can get on faster with their old tools ?

B.

Because they *think* they can get on faster ; you cannot say they have *found* it until they have tried.

A.

Have they not tried Bacon's way partially, and found it not so handy ? Has not Sir John Herschel, for instance, tried the use of his famous classification of Instances, and pronounced it " more apparent than real " ? And is it not a fact that no single discovery of importance has been actually made by proceeding according to the method recommended by Bacon ? I am sure I have heard as much reported upon the authority of a very eminent modern writer upon these subjects.

B.

So have I. And I can well believe that the use of Bacon's " Prerogatives of Instances ", *in the way they have been used*, is not much ; and for the reason given by Herschel, viz., because the same judgment which enables you to assign the Instance its proper *class*, enables you, without that assignation, to recognise its proper value. Therefore so long as the task of gathering his Instances as they grow wild in the woods is left to the Interpreter of Nature himself, there is little use in a formal classification ; he knows exactly what he wants ; what is not to his purpose he need not trouble himself with ; what is to his purpose he can apply to that purpose at once. And each several man of genius will no doubt acquire a knack of his own by which he will arrive at his results faster than by any formal method. But suppose the Interpreter wants to use the help of other people, to whom he cannot impart his own genius or his peculiar gift of knowing at first sight what is to the purpose and what not. He wants them to assist him in gathering materials. How shall he direct them in their task so that their labours may be available for himself ? I take it he must distribute the work among several and make it pass through several processes. One man may be used to make a rough and general collection,—what we call an *omnium gatherum*. Another must be employed to reduce the confused mass into some order fit for reference. A third to clear it of superfluities and rubbish. A fourth must be taught to classify and arrange what remains. And here I cannot but think that Bacon's arrangement of Instances according to what he calls their Prerogatives, or some better arrangement of the same kind which experience ought to suggest, would be found to be of great value ; especially when it is proposed to make through all the regions of Nature separate collections of this kind such as may combine into one general collection. For though it be true that as long as each man works only for himself he may trust to the *usus uni rei deditus* for finding out the method of proceeding which best suits the trick of his own mind,—and each will probably pursue a different method,—yet when many men's labours are to be gathered into one table, any collector of statistics will tell you that they must all work according to a common pattern. And in the subject we are speaking of, which is coextensive with the mind of man on one side and the nature of things on the other, that will undoubtedly be the best pattern which is framed upon the justest theory of the human understanding;—for which distinction Bacon's would seem to be no unlikely candidate.

However I am here again getting out of my province. It may be that Bacon's project was visionary; or it may be that it is only *thought* visionary, because since his death no heart has been created large enough to believe it practicable. The philosophers must settle that among themselves. But be the cause what it will, it is clear to me on the one hand that the thing has not been seriously attempted; and on the other, that Bacon was fully satisfied that nothing of worth could be hoped for without it; therefore that we have no right to impute to him either the credit of all that has been done by the new philosophy, or the discredit of all that has been left undone.

A.

Certainly not; if you are right as to the fact. But I still think there must be some mistake. How is it possible that among so many distinguished men as have studied Bacon's philosophy with so much reverence, such a large feature can have been overlooked?

B.

I cannot pretend to explain that. But an appeal to one's own eyes is always lawful. Here is one passage which is enough by itself to settle the question. If you are not satisfied with it, I can quote half a dozen more to the same effect; "*Illud interim quod sæpe diximus etiam hoc loco præcipue repetendum est*—"

A.

Translate; if you would have me follow.

B.

"I must repeat here again what I have so often said;—that though all the wits of all the ages should meet in one,—though the whole human race should make Philosophy their sole business,—though the whole earth were nothing but colleges and academies and schools of learned men,—yet without such a natural and experimental history as I am going to describe, no progress worthy of the human race in Philosophy and the Sciences could possibly be made: whereas if such a history were once provided and well ordered with the addition of such auxiliary and light-giving experiments as the course of Interpretation would itself suggest, the investigation of Nature and of all sciences would be the work only of a few years. Either this must be done, therefore, or the business must be abandoned. For in this way and in this way only can the foundation be laid of all true and active Philosophy."

A.

Where does he say that?

B.

In the Preface to what he calls the "*Parasceve ad Historiam naturalem et experimentalem*", which is in fact nothing more than a description of the sort of history which he wanted,—such a history as a true Philosophy might be built upon,—with directions to be observed in collecting it. He published it (somewhat out of its proper place) in the same volume with the *Novum Organum*, in order that, if possible, men might be set about the work at once; of such primary importance did he hold it to be. If you distrust my translation, take it in his own English. In presenting the *Novum Organum* to the King, after explaining the nature and objects of the work and his reason for publishing it in an imperfect shape, he adds, "There is another reason for my so doing; which is to try whether I can get help in one intended part of this work, namely the compiling of a natural and experimental history, *which must be the main foundation of a true and active philosophy*". And again about a week after, in reply to the King's gracious acknowledgement of the book,—"This comfortable beginning makes me hope further that your

Majesty will be aiding to me in setting men on work for the collecting of a natural and experimental history, which is *basis totius negotii*". And this was no afterthought, but an essential feature of his design as he had conceived it at least sixteen years before. There is extant a description of this proposed history, which appears to have been written as early as 1604; and though the only copy that I know of is in an imperfect and mutilated manuscript, enough remains to show that in all its material features it agreed exactly with the description set forth in the *Parasceve*.

Now you know I am not going to discuss the merit of his plan. It may (as I said) have been all a delusion. But grant it a delusion—still it was a delusion under which he was actually labouring. If every man of science that ever lived had considered it and pronounced it puerile and ridiculous, still their unanimous verdict could not, in the face of his own repeated and earnest declarations, persuade me that it was not an essential part of Bacon's scheme; that it was not (in his perfect and rooted judgment) the one key to the cipher in which the fortunes of the human race are locked up,—the one thing *with* which all might be done; *without* which nothing. And this is all that is necessary for our present business. For we are not discussing his philosophical capacity, but his personal character and purposes as illustrated by the tenour of his life.

Such in 1847 were my reasons for rejecting as unsatisfactory all the explanations I had then met with of the distinctive peculiarity of the Baconian philosophy, and such the result of my attempts to find a more satisfactory one for myself.

In rejecting former explanations as unsatisfactory, Mr. Ellis, it will be seen, concurs with me, and for much the same reason. According to them "it becomes", he says, "impossible to justify or to understand Bacon's assertion that his system was essentially new". He then proceeds to point out one great peculiarity by which it aspired to differ from all former systems—a peculiarity residing in the supposed perfection of the logical machinery; which, since it would of itself account for Bacon's belief of its importance no less than for his assertion of its novelty, does certainly supply a new explanation unencumbered with the difficulties pointed out in the foregoing extract. But there is another difficulty which it leaves behind. It is impossible, I think, to reconcile with this supposition the course which Bacon afterwards took in expounding and developing his system. For if the great secret which he had, or thought he had, in his keeping, lay only, or even chiefly, in the perfection of the logical machinery—in the *method* of induction; if this method was a kind of mechanical process—an *organum* or engine—at once "wholly new", "universally applicable," "in all cases infallible," and such as anybody might manage; if his explanation of this method in the second book of the *Novum Organum* is so incomplete that it leaves all the principal practical difficulties unexplained; and if it was a thing which nobody but himself had any notion of, or any belief in; how is it that, during the remaining five years of his life—years of eager and unremitting labour, devoted almost exclusively to the exposition of his philosophy—he made no attempt to complete the explanation of it? Why did he leave the *Novum Organum* as it was, being a work which he could have completed alone, and which indeed he only could have completed, and apply himself with advised and deliberate industry to the collection of Natural History; a work which he knew he could not carry to perfection himself, even in any of its parts; which he had once thought it a waste of time to employ himself upon, as being within every man's capacity; concerning the execution of which he had already given sufficient general directions; and of which, even when accomplished, the right use could not be made except in virtue of that very *method* or logical machinery, the constitution and management of which still remained to be explained? It was not that he had changed his opinion as to the value of it; his sense of the difficulties may have increased, his views as to details may have altered; but there is no reason to think that he ever lost any part of his faith either in the importance or in the practicability of it. It was not that when he came to closer quarters with the subject, he felt that he was himself unable to deal with it: Two years after the publication of the first part of the *Novum Organum*, and three years before his death, he speaks of the second part as a thing yet to be done, but adds, " quam tamen animo jam complexus et metitus sum " [3]. It was not that he thought the description he had already given sufficient. In the winter of 1622, he tells us that there are " haud pauca, *eaque ex præcipuis*," still wanting. It was not that he had found any disciple or fellow-labourer to whom he might entrust the completion of his unfinished task: to the very last he felt himself alone in his work. It was not from inadvertence: he left the *Novum Organum* for the *Natural History* deliberately, because it seemed upon consideration the better and more advisable course; " quare omnino et ante omnia in hoc incumbere *satius et consultius* visum est ". It was not that he wanted either time or industry; for during the five succeeding years he completed the *De Augmentis*, and composed his histories of the *Winds*, of *Life and Death*, of *Dense and Rare*; his lost treatise on *Heavy and Light*, his lost *Abecedarium Naturæ*, his *New Atlantis*, his *Sylva Sylvarum*. Why did he employ no part of that time in completing the description of the new machine? in explaining how he proposed to supply the defects [4] and rectify the errors [5] of the imperfect logical process which he had already exhibited; how to adapt the mode of inquiry to the nature of the subject [6]; how to determine what questions ought to be dealt with first,—what " natures " to have precedence in the order of inquiry [7]; above all, how to ascertain where the inquiry might safely terminate as having left no " nature " in the universe unchallenged [8],—a security without which the whole process must always have been in danger of vitiation from an " instance contradictory " remaining behind? To me the question appears to admit of but one answer. He considered the collection of natural history upon the plan he meditated, to be, in practice at least, a more important part of his philosophy than the *Organum* itself,—a work of which the nature and importance more needed to be pressed upon the attention of mankind,—of which the neglect would be more fatal to the progress of science. That this was in fact his opinion at the very time he was composing the *Novum Organum* may be inferred from the last aphorism of the first book, as I have pointed out at the end of the preface. That he was still of the same opinion two years after, we have his own express declaration in the *Auctoris monitum* prefixed to the *History of the Winds*, where he explains his motives for going on with the third part of the Instauratio, instead of finishing the second. It had occurred to him, he there tells us, that if the *Organum* should fall into the hands of some man of genius capable of understanding and willing to use it, still without a natural history of the proper kind provided to his hand, he would not know how to proceed; whereas if a full and faithful history of nature and the arts were set before him, he might succeed *even by the old method*—" licet via veteri pergere malint, nec via nostri organi (quæ ut nobis videtur aut unica est aut optima) uti "—in building upon it something of solid worth. " Itaque huc res redit ' he concludes; " ut *organum nostrum*, etiamsi fuerit absolutum, *absque historiâ naturali non multum, historia naturalis absque organo non parum*, instaurationem scientiarum sit provectura." I know not how therefore to escape the conclusion that, in Bacon's own estimate of his own system, the Natural History held the place of *first* importance. He regarded it as not less new [9] than the new *method*, and as more indispensable. Though the " *via nostri organi* " still appeared to him to be " *aut unica aut optima* ", something of substantial worth might, he thought, be accomplished without it. Without a natural history " tali qualem nunc præcipiemus ", he thought no advance of any value could possibly be made.

[3] Letter to Fulgenzio.

[4] De Adminiculis Inductionis.

[5] De Rectificatione Inductionis.

[6] De Variatione Inquisitionis pro naturâ subjecti.

[7] De Prærogativis Naturarum quatenus ad inquisitionem, sive de eo quod inquirendum est prius et posterius.

[8] De Terminis Inquisitionis, sive de Synopsi omnium naturarum in universo.

[9] His assertion of the *novelty* is as strong in the one case as in the other. *Atque hoc posterius* [viz. the use of natural history, " tanquam materia prima philosophiæ atque veræ inductionis supellex sive sylva "] *nunc agitur; nunc inquam*, NEQUE UNQUAM ANTEHAC."

What may be the real value of this part of Bacon's system is, of course, quite another question. The evidence just adduced goes only to show what was the value which he himself set upon it, and affects the question no otherwise than by giving it a new interest, and suggesting the expediency of considering more carefully than has yet, I think, been done, whether his advice on this head might not be followed—I do not say as far as he intended—but much further than has yet been tried; with effects—I do not say such as he anticipated—but larger than we are likely to get any other way.

That he himself indeed, even if all mankind had united to carry his plan into effect, would have been disappointed with the result, I have little doubt. For I suppose the collected observations of all the world,—reduced to writing, digested, and brought into his study,—would not have sufficed to give him that knowledge of the *forms* of nature which was to carry with it the command over her powers. He would have found no doubt, upon trial, that his scheme involved difficulties of which he had formed no conception. He would have found that the facts which must be known in order to complete the three tables of comparence, and to "perfect the *exclusiva*", were so infinite in number that to gather them by simple observation without some theoretic principle of selection would be an endless task, and to deal with them when gathered a hopeless one. He might still indeed have hoped to arrive ultimately at an *alphabet* of nature (her principles being probably few and simple, though her phenomena so enormously complex); but he would have found that a *dictionary* or *index* of nature (and such was to be the office of the *Natural History*), to be complete enough for the purposes of the *Novum Organum*, must be nearly as voluminous as Nature herself. He would have found it necessary, therefore (as I suppose all inventors have done both before and since his time), to make material changes in his original plan of operation, and to reduce his hopes far below their original dimensions. But a man may be in the right way to his end, though the end itself be further off than he imagines; and before we cast Bacon's plan finally aside, we may be fairly called upon to show either that the way he wanted us to go is in its nature impracticable or that there is better hope of arriving at the desired end by some other.

Mr. Ellis's judgment upon the first point may be partly gathered from his general remarks upon the third part of the Instauratio; but I am fortunately in possession of his opinion (called forth by the exposition of my own views in the dialogue above quoted) upon the specific practical question now under discussion. It was communicated to me in a letter dated 13th September, 1847, and appears to contain his deliberate judgment as to the *practicability* of making a collection of natural history, such as would be available for scientific purposes, in the manner in which Bacon proposed to have it made.

"That it is impossible (he says) to sever the business of experiment and observation from that of theorising, it would perhaps be rash to affirm. But it seems to me that such a severance could hardly be effected. A transcript of nature, if I may so express myself,—that is, such a collection of observed phenomena as would serve as the basis and materials of a system of natural philosophy,—would be like nature itself infinite in extent and variety. No such collection could be formed; and, were it formed, general laws and principles would be as much hidden in a mass of details as they are in the world of phenomena.

"The marshalling idea, teaching the philosopher what observations he is to make, what experiments to try, seems necessary in order to deliver him from this difficulty. Can we conceive that such experiments as those of Faraday could have preceded the formation of any hypothesis? You allude, I think, to what has been done in the way of systematic observation with reference to terrestrial magnetism. And beyond all doubt the division of labour is possible and necessary in many scientific inquiries. But then this separating of the observer from the theoriser is only possible (at least, in such a case as that of magnetism) when the latter can tell his "bajulus" what experiments he is to make, and how they are to be made. As a matter of fact, the memoirs of Gauss, which have done so much to encourage systematic observation of terrestrial magnetism, contain many results of theory directly bearing on observation; *e.g.*, the method of determining the absolute measure of magnetism.

* * * * *

"Of course I remember that Bacon speaks of experiments to be suggested by theory: as for instance in Solomon's house; all I mean is, that it seems doubtful whether a large collection of facts can in most sciences be made useful, unless some theory has guided its formation."

Now I am quite willing to accept this judgment as perfectly sound and just; as pointing truly at the practical difficulties involved in Bacon's scheme, and proving that it could not be carried out *completely* on the plan he proposed, or attain *completely* the end at which he aimed; and certainly, if I thought that such completeness was a condition absolutely essential,—that, unless observation could be carried on without any help whatever from theory, the work could not proceed at all; or that the results of observation so conducted could be of no scientific value unless they amounted to a perfect "transcript of nature;"—if I thought, in short, it was a scheme which, unless it led to everything, would lead to nothing,—I should accept these remarks as disposing finally of the whole question. But why should I think so? That the severance of theory and observation should be *absolute* does not appeal to me to be at all necessary for the practical prosecution of the enterprise; I can hardly think that it even formed part of the original design; and though it is true that the collection of natural history could not have been used *in the way Bacon proposed*, unless it were more complete than it ever could have been made, yet for use in the ordinary way (and this was certainly *one* of the uses he contemplated for it) its value would be increased by every new observation; and who can say at what point observations so conducted must necessarily stop?

That Bacon intended one set of men to be employed in collecting facts, and another in deriving consequences from them, is no doubt true. Unless theory and observation could be so far separated as to admit practically of such a distribution of parts, his plan must no doubt have been given up, and it is objected that his distribution is practically impossible, because the observers, unless they had some precedent theory to guide them, could never know what observations to make in order to bring out the facts which the theorist requires to know. I cannot but think, however, that this objection supposes a separation of the two functions far more complete than Bacon ever contemplated. He may have used words which in strict logical construction imply such a kind of separation; but if so, his words meant more than he himself meant. His intellect was remarkable for breadth rather than subtlety,—quicker, to use his own division, in perceiving resemblances than distinctions,—and in writing he always aimed at conciseness, force, point, picturesqueness, and at making himself plain to common understandings, far more than at metaphysical exactness of expression. Now, however true it may be, as a metaphysical proposition, that some amount of theory is involved in every observation, and still more in every series of observations, it is no less true, as a familiar fact, that observations made by one man, without conscious reference to any theory whatever, may be perfectly available to another with reference to theories of which the first never heard or dreamed. Colonel Reid's theory of storms, for instance, was worked out, I am told, not in the West Indies among the hurricanes, but at the Admiralty among the ship's logs. And though Bacon would never have denied that many results of theory go to the correct keeping of a ship's log, who can doubt that a collection of logs kept during hurricanes would have been accepted by him as a most valuable contribution to a history of the winds, and a good specimen of the very thing he wanted? It would be easy to add more instances; but I suppose nobody will deny that, *in this sense*, observation and theory *can* be carried on apart and by different persons. And if it be objected that the observers will never hit upon all the facts which are necessary to suggest or establish the theory, unless their observations be renewed again and again under directions devised by the theorist with special reference to what he wants to know, I reply by asking what is to prevent the renewal of them, under directions so devised, as often as necessary? a thing (I may observe) which Bacon himself distinctly intended. "Illud interim," he says, after giving an example of a "topica particularis" in the *De Augmentis*, "quod monere occœpimus iterum monemus, nempe ut homines debeant topicas particulares suas alternare, ita ut post majores progressus aliquos in inquisitione factos, *aliam et subinde aliam in-*

stituant topicam, si modo scientiarum fastigia conscendere cupiant." Now if the directions, judicious to begin with, be judiciously varied and repeated as the inquiry proceeds, an immense mass of observations of the greatest importance to science might surely be collected in this very way. Nay, in subjects which have their phenomena spread far and wide over the world (like winds, seasons, and oceanic or atmospheric currents), it is in the gradual accumulation of observations so made that our only hope lies of ever coming to understand their laws at all; and if we cannot cause them to be collected under direction and design, we must wait till they accumulate by accident. For it is manifestly impossible that in such subjects as these, philosophers should provide themselves with all the facts which they want unless they can use the help of those who are not philosophers. What science deals with phenomena more subtle and delicate than meteorology? Yet hear Sir John Herschel. "It happens fortunately that almost every *datum* which the scientific meteorologist can require is furnished in its best and most available state by that definite systematic process known as the '*keeping a meteorological register*', which consists in noting at stated hours of every day the readings of all the meteorological instruments at command, as well as all such facts or indications of wind and weather as are susceptible of being definitely described and estimated without instrumental aid. Occasional observations apply to occasional and remarkable phenomena, and are by no means to be neglected; *but it is to the regular meteorological register, steadily and perseveringly kept throughout the whole of every voyage, that we must look for the development of the great laws of this science* [10]."

Between the officers of Her Majesty's navy registering the readings of their instruments in all latitudes and longitudes, and the man of science in his study deducing the laws of meteorology from a comparison of the results, the division of labour is surely as complete as Bacon would have desired. Nor would the scientific directions previously furnished to the officers for their guidance, directions when, where, what, and how to observe and record,—though containing "many results of theory bearing upon observation,"—have seemed to him either objectionable or superfluous: on the contrary, such directions form part of his own design as explained by himself. In the concluding paragraph of the tract which has suggested these remarks he distinctly announces his intention to draw up certain heads of inquiry showing what points with reference to each subject were more particularly to be observed. And though he did not live to execute this part of his design, a few fragments remaining among his papers show in what manner he proposed to proceed. And (if an idle looker-on who can offer no help in the work may presume to offer an opinion) I could wish that men of science would apply themselves earnestly to the solution of this practical problem: What measures are to be taken in order that the greatest variety of judicious observations of nature all over the world may be carried on in concert upon a scientific plan, and brought to a common centre? With reference to some particular subjects, such measures have been of late years taken on a scale of Baconian magnitude. The system of observations instituted by the Great British Association with respect to Terrestrial Magnetism, if I am rightly informed as to the nature and scale of it, is one which Bacon would have welcomed as he welcomed the first tidings from Galileo's telescope; he would have accepted it as an enterprise "dignum humano genere". A similar system of concerted observations is now in contemplation with regard to oceanic currents. As a specimen of the same thing in a more general character, take the "Admiralty Manual of Scientific Inquiry", to which I have already referred; a book of practical directions drawn up by some of the most eminent scientific men of our day with special reference to the progress of science in several of its most important departments; directions addressed not to men who are themselves engaged in the theoretical investigation of the subjects, or guided by any "marshalling idea", but to "officers of the navy and travellers in general", telling them what things to observe, in order that their observations may be available for the pur-

10 "Manual of Scientific Inquiry, prepared for the use of officers in Her Majesty's navy and travellers in general." Edited by Sir John F. W. Herschel, Bt., p. 281.

poses of scientific inquiry. These are exactly what Bacon would have called "Topicæ Inquisitionis"—instructions for the examination of Nature "super articulos"; and the whole scheme is in perfect accordance, so far as it goes, with Bacon's notion of the way in which men might be set on work for the completing of a natural and experimental history. Why should it not go further? Who can believe that the subjects contained in this little volume are the only subjects to which this method of collecting observations can be applied? who venture to fix the limit beyond which, under such a system sagaciously devised, wisely administered, energetically carried out, and extended to all the departments of nature which admit of it, human discovery may not go?—*J. S.*

TRANSLATION OF THE PARASCEVE.

DESCRIPTION OF A NATURAL AND EXPERIMENTAL HISTORY, SUCH AS MAY SERVE FOR THE FOUNDATION OF A TRUE PHILOSOPHY.

My object in publishing my Instauration by parts is that some portion of it may be put out of peril. A similar reason induces me to subjoin here another small portion of the work, and to publish it along with that which has just been set forth. This is the description and delineation of a Natural and Experimental History such as may serve to build philosophy upon, and containing material true and copious and aptly digested for the work of the Interpreter which follows. The proper place for it would be when I come in due course to the *Preparatives* of Inquiry. I have thought it better however to introduce it at once without waiting for that. For a history of this kind, such as I conceive and shall presently describe, is a thing of very great size, and cannot be executed without great labour and expense; requiring as it does many people to help, and being (as I have said elsewhere) a kind of royal work. It occurs to me therefore that it may not be amiss to try if there be any others who will take these matters in hand; so that while I go on with the completion of my original design, this part which is so manifold and laborious may even during my life (if it so please the Divine Majesty) be prepared and set forth, others applying themselves diligently to it along with me; the rather because my own strength (if I should have no one to help me) is hardly equal to such a province. For as much as relates to the work itself of the intellect, I shall perhaps be able to master that by myself; but the materials on which the intellect has to work are so widely spread, that one must employ factors and merchants to go everywhere in search of them and bring them in. Besides I hold it to be somewhat beneath the dignity of an undertaking like mine that I should spend my own time in a matter which is open to almost every man's industry. That however which is the main part of the matter I will myself now supply, by diligently and exactly setting forth the method and description of a history of this kind, such as shall satisfy my intention; lest men for want of warning set to work the wrong way, and guide themselves by the example of the natural histories now in use, and so go far astray from my design. Meanwhile what I have often said I must here emphatically repeat; that if all the wits of all the ages had met, or shall hereafter meet together; if the whole human race had applied or shall hereafter apply themselves to philosophy, and the whole earth had been or shall be nothing but academies and colleges and schools of learned men; still without a natural and experimental history such as I am going to prescribe, no progress worthy of the human race could have been made or can be made in philosophy and the sciences. Whereas on the other hand, let such a history be once provided and well set forth, and let there be added to it such auxiliary and light-giving experiments as in the very course of interpretation will present themselves or will have to be found out; and the investigation of nature and of all sciences will be the work of a few years. This therefore must be done, or the business must be given up. For in this way, and in this way only, can the foundations of a true and active philosophy be established; and then will men wake as from deep sleep, and at once perceive what a difference there is between the dogmas and figments of the wit and a true and active philosophy, and what it is in questions of nature to consult nature herself.

First then I will give general precepts for the composition of this history; then I will set out the particular figure of it, inserting sometimes as well the purpose to which the inquiry is to be adapted and referred as the particular point to be inquired; in order that a good understanding and forecast of the mark aimed at may suggest to men's minds other things also which I may perhaps have overlooked. This history I call *Primary History*, or the *Mother History*.

APHORISMS ON THE COMPOSITION OF THE PRIMARY HISTORY.

Aphorism

I.

Nature exists in three states, and is subject as it were to three kinds of regimen. Either she is free, and develops herself in her own ordinary course; or she is forced out of her proper state by the perverseness and insubordination of matter and the violence of impediments; or she is constrained and moulded by art and human ministry. The first state refers to the *species* of things; the second to *monsters;* the third to *things artificial.* For in things artificial nature takes orders from man, and works under his authority: without man, such things would never have been made. But by the help and ministry of man a new face of bodies, another universe or theatre of things, comes into view. Natural History therefore is threefold. It treats of the *liberty* of nature, or the *errors* of nature, or the *bonds* of nature: so that we may fairly distribute it into history of *Generations*, of *Pretergenerations*, and of *Arts;* which last I also call *Mechanical* or *Experimental* history. And yet I do not make it a rule that these three should be kept apart and separately treated. For why should not the history of the monsters in the several species be joined with the history of the species themselves? And things artificial again may sometimes be rightly joined with the species, though sometimes they will be better kept separate. It will be best therefore to consider these things as the case arises. For too much method produces iterations and prolixity as well as none at all.

II.

Natural History, which in its subject (as I said) is threefold, is in its use twofold. For it is used either for the sake of the knowledge of the particular things which it contains, or as the primary material of philosophy and the stuff and subject-matter of true induction. And it is this latter which is now in hand: now, I say, for the first time: nor has it ever been taken in hand till now. For neither Aristotle, nor Theophrastus, nor Dioscorides, nor Caius Plinius, ever set this before them as the end of natural history. And the chief part of the matter rests in this: that they who shall hereafter take it upon them to write natural history should bear this continually in mind—that they ought not to consult the pleasure of the reader, no nor even that utility which may be derived immediately from their narrations; but to seek out and gather together such store and variety of things as may suffice for the formation of true axioms. Let them but remember this, and they will find out for themselves the method in which the history should be composed. For the end rules the method.

III.

But the more difficult and laborious the work is, the more ought it to be discharged of matters superfluous. And therefore there are three things upon which men should be warned to be sparing of their labour,—as those which will immensely increase the mass of the work, and add little or nothing to its worth.

First then, away with antiquities, and citations or testimonies of authors; also with disputes and controversies and differing opinions; everything in short which is philological. Never cite an author except in a matter of doubtful credit: never introduce a controversy unless in a matter of great moment. And for all that concerns ornaments of speech, similitudes, treasury of eloquence, and such like emptinesses, let it be utterly dismissed. Also let all those things which are admitted be themselves set down briefly and concisely, so that they may be nothing less than words. For no man who is collecting and storing up materials for ship-building or the like, thinks of arranging them elegantly, as in a shop, and displaying them so as to please the eye; all his care is that they be sound and good, and that they be so arranged as to take up as little room as possible in the warehouse. And this is exactly what should be done here.

Secondly, that superfluity of natural histories in descriptions and pictures of species, and the curious variety of the same, is not much to the purpose. For small varieties of this kind are only a kind of sports and wanton freaks of nature; and come near to the nature of individuals. They afford a pleasant recreation in wandering among them and looking at them as objects in themselves; but the information they yield to the sciences is slight and almost superfluous.

Thirdly, all superstitious stories (I do not say stories of prodigies, when the report appears to be faithful and probable; but superstitious stories) and experiments of ceremonial magic should be altogether rejected. For I would not have the infancy of philosophy, to which natural history is as a nursing-mother, accustomed to old wives' fables. The time will perhaps come (after we have gone somewhat deeper into the investigation of nature) for a light review of things of this kind; that if there remain any grains of natural virtue in these dregs, they may be extracted and laid up for use. In the meantime they should be set aside. Even the experiments of natural magic should be sifted diligently and severely before they are received; especially those which are commonly derived from vulgar sympathies and antipathies, with great sloth and facility both of believing and inventing.

And it is no small thing to relieve natural history from the three superfluities above mentioned, which would otherwise fill volumes. Nor is this all. For in a great work it is no less necessary that what is admitted should be written succinctly than that what is superfluous should be rejected; though no doubt this kind of chastity and brevity will give less pleasure both to the reader and the writer. But it is always to be remembered that this which we are now about is only a granary and storehouse of matters, not meant to be pleasant to stay or live in, but only to be entered as occasion requires, when anything is wanted for the work of the *Interpreter*, which follows.

IV.

In the history which I require and design, special care is to be taken that it be of wide range and made to the measure of the universe. For the world is not to be narrowed till it will go into the understanding (which has been done hitherto), but the understanding to be expanded and opened till it can take in the image of the world, as it is in fact. For that fashion of taking few things into account, and pronouncing with reference to a few things, has been the ruin of everything. To resume then the divisions of natural history which I made just now,—viz. that it is a history of Generations, Pretergenerations, and Arts,—I divide the History of Generations into five parts. The first, of Ether and things Celestial. The second, of Meteors and the regions (as they call them) of Air; viz. of the tracts which lie between the moon and the surface of the earth; to which part also (for order's sake, however the truth of the thing may be) I assign Comets of whatever kind, both higher and lower. The third, of Earth and Sea. The fourth, of the Elements (as they call them), flame or fire, air, water, earth; understanding however by Elements, not the first principles of things, but the greater masses of natural bodies. For the nature of things is so distributed that the quantity or mass of some bodies in the universe is very great, because their configurations require a texture of matter easy and obvious; such as are those four bodies which I have mentioned; while of certain other bodies the quantity is small and weakly supplied, because the texture of matter which they require is very complex and subtle, and for the most part determinate and organic; such as are the species of natural things,—metals, plants, animals. Hence I call the former kind of bodies the *Greater Colleges*, the latter the *Lesser Colleges*. Now the fourth part of the history is of those Greater Colleges—under the name of Elements, as I said. And let it not be thought that I confound this fourth part with the second and third, because in each of them I have mentioned air, water, and earth. For the history of these enters into the second and third, as they are integral parts of the world, and as they relate to the fabric and configuration of the universe. But in the fourth is contained the history of their own substance and nature, as it exists in their several parts of uniform structure, and without reference to the whole. Lastly, the fifth part of the history contains the Lesser Colleges, or Species; upon which natural history has hitherto been principally employed.

As for the history of Pretergenerations, I have already said that it may be most conveniently joined with the history of Generations; I mean the history of prodigies which are natural. For the superstitious history of marvels (of whatever kind) I remit to a quite separate treatise of its own; which treatise I do not wish to be undertaken now at first, but a little after, when the investigation of nature has been carried deeper.

History of Arts, and of Nature as changed and altered by Man, or Experimental History, I divide into three. For it is drawn either from mechanical arts, or from the operative part of the liberal arts; or from a number of crafts and experiments which have not yet grown into an art properly so called, and which sometimes indeed turn up in the course of most ordinary experience, and do not stand at all in need of art.

As soon therefore as a history has been completed of all these things which I have mentioned, namely, Generations, Pretergenerations, Arts and Experiments, it seems that nothing will remain unprovided whereby the sense can be equipped for the information of the understanding. And then shall we be no longer kept dancing within little rings, like persons bewitched, but our range and circuit will be as wide as the compass of the world.

V.

Among the parts of history which I have mentioned, the history of Arts is of most use, because it exhibits things in motion, and leads more directly to practice. Moreover it takes off the mask and veil from natural objects, which are commonly concealed and obscured under the variety of shapes and external appearance. Finally, the vexations of art are certainly as the bonds and handcuffs of Proteus, which betray the ultimate struggles and efforts of matter. For bodies will not be destroyed or annihilated; rather than that they will turn themselves into various forms. Upon this history therefore, mechanical and illiberal as it may seem (all fineness and daintiness set aside), the greatest diligence must be bestowed.

Again, among the particular arts those are to be preferred which exhibit, alter, and prepare natural bodies and materials of things; such as agriculture, cookery, chemistry, dyeing; the manufacture of glass, enamel, sugar, gunpowder, artificial fires, paper, and the like. Those which consist principally in the subtle motion of the hands or instruments are of less use; such as weaving, carpentry, architecture, manufacture of mills, clocks, and the like; although these too are by no means to be neglected, both because many things occur in them which relate to the alterations of natural bodies, and because they give accurate information concerning local motion, which is a thing of great importance in very many respects.

But in the whole collection of this history of Arts, it is especially to be observed and constantly borne in mind, that not only those experiments in each art which serve the purpose of the art itself are to be received, but likewise those which turn up anyhow by the way. For example, that locusts or crabs, which were before of the colour of mud, turn red when baked, is nothing to the table; but this very instance is not a bad one for investigating the nature of redness, seeing that the same thing happens in baked bricks. In like manner the fact that meat is sooner salted in winter than in summer, is not only important for the cook that he may know how to regulate the pickling, but is likewise a good instance for showing the nature and impression of cold. Therefore it would be an utter mistake to suppose that my intention would be satisfied by a collection of experiments of arts made only with the view of thereby bringing the several arts to greater perfection. For though this be an object which in many cases I do not despise, yet my meaning plainly is that all mechanical experiments should be as streams flowing from all sides into the sea of philosophy. But how to select the more important instances in every kind (which are principally and with the greatest diligence to be sought and as it were hunted out) is a point to be learned from the prerogatives of instances.

VI.

In this place also is to be resumed that which in the 99th, 119th, and 120th Aphorisms of the first book I treated more at large, but which it may be enough here to enjoin shortly by way of precept; namely, that there are to be received into this history, first, things the most ordinary, such as it might be thought superfluous to record in writing, because they are so familiarly known; secondly, things mean, illiberal, filthy (for "to the pure all things are pure," and if money obtained from Vespasian's tax smelt well, much more does light and information from whatever source derived); thirdly, things trifling and childish (and no wonder, for we are to become again as little children); and lastly, things which seem over subtle, because they are in themselves of no use. For the things which will be set forth in this history are not collected (as I have already said) on their own account; and therefore neither is their importance to be measured by what they are worth in themselves, but according to their indirect bearing upon other things, and the influence they may have upon philosophy.

VII.

Another precept is, that everything relating both to bodies and virtues in nature be set forth (as far as may be) numbered, weighed, measured, defined. For it is works we are in pursuit of, not speculations; and practical working comes of the due combination of physics and mathematics. And therefore the exact revolutions and distances of the planets—in the history of the heavenly bodies; the compass of the land and the superficial space it occupies in comparison of the waters—in the history of earth and sea; how much compression air will bear without strong resistance—in the history of air; how much one metal outweighs another—in the history of metals; and numberless other particulars of that kind are to be ascertained and set down. And when exact proportions cannot be obtained, then we must have recourse to indefinite estimates and comparatives. As for instanse (if we happen to distrust the calculations of astronomers as to the distances of the planets), that the moon is within the shadow of the earth; that Mercury is beyond the moon; and the like. Also when mean proportions cannot be had, let extremes be proposed; as that a weak magnet will raise so many times its own weight of iron, while the most powerful will raise sixty times its own weight (as I have myself seen in the case of a very small armed magnet). I know well enough that these definite instances do not occur readily or often, but that they must be sought for as auxiliaries in the course of interpretation itself when they are most wanted. But nevertheless if they present themselves accidentally, provided they do not too much interrupt the progress of the natural history, they should also be entered therein.

VIII.

With regard to the credit of the things which are to be admitted into the history; they must needs be either certainly true, doubtful whether true or not, or certainly not true. Things of the first kind should be set down simply; things of the second kind with a qualifying note, such as "it is reported", "they relate," "I have heard from a person of credit," and the like. For to add the arguments on either side would be too laborious and would certainly interrupt the writer too much. Nor is it of much consequence to the business in hand; because (as I have said in the 118th Aphorism of the first book) mistakes in experimenting, unless they abound everywhere, will be presently detected and corrected by the truth of axioms. And yet if the instance be of importance, either from its own use or because many other things may depend upon it, then certainly the name of the author should be given; and not the name merely, but it should be mentioned withal whether he took it from report, oral or written (as most of Pliny's statements are), or rather affirmed it of his own knowledge; also whether it was a thing which happened in his own time or earlier; and again whether it was a thing of which, if it really happened, there must needs have been many witnesses; and finally whether the author was a vainspeaking and light person, or sober and severe; and the like points, which bear upon the weight of the evidence. Lastly things which though certainly not true are yet current and much in men's mouths, having either through neglect or from the use of them in similitudes prevailed now for many ages, (as that the diamond binds the magnet, garlic weakens it; that amber attracts everything except basil; and other things of that kind) these it will not be enough to reject silently; they must be in express words proscribed, that the sciences may be no more troubled with them.

Besides, it will not be amiss, when the source of any vanity or credulity happens to present itself, to make a note of it; as for example, that the power of exciting Venus is ascribed to the herb *Satyrion*, because its root takes the shape of testicles; when the real cause of this is that a fresh bulbous root grows upon it every year, last year's root still remaining; whence those twin bulbs. And it is manifest that this is so; because the new root is always found to be solid and succulent, the old withered and spongy. And therefore it is no marvel fi one sinks in water and the other swims; which nevertheless goes for a wonder, and has added credit to the other virtues ascribed to this herb.

IX.

There are also some things which may be usefully added to the natural history, and which will make it fitter and more convenient for the work of the interpreter which follows. They are five.

First, questions (I do not mean as to causes but as to the fact) should be added, in order to provoke and stimulate further inquiry; as in the history of Earth and Sea, whether the Caspian ebbs and flows, and at how many hours' interval; whether there is any Southern Continent, or only islands; and the like.

Secondly, in any new and more subtle experiment the manner in which the experiment was conducted should be added, that men may be free to judge for themselves whether the information obtained from that experiment be trustworthy or fallacious; and also that men's industry may be roused to discover if possible methods more exact.

Thirdly, if in any statement there be anything doubtful or questionable, I would by no means have it suppressed or passed in silence, but plainly and perspicuously set down by way of note or admonition. For I want this primary history to be compiled with a most religious care, as if every particular were stated upon oath; seeing that it is the book of God's works, and (so far as the majesty of heavenly may be compared with the humbleness of earthly things) a kind of second Scripture.

Fourthly, it would not be amiss to intersperse observations occasionally, as Pliny has done; as in the history of Earth and Sea, that the figure of the earth (as far as it is yet known) compared with the seas, is narrow and pointed towards the south, wide and broad towards the north; the figure of the sea contrary:—that the great oceans intersect the earth in channels running north and south, not east and west; except perhaps in the extreme polar regions. It is also very good to add canons (which are nothing more than certain general and catholic observations); as in the history of the Heavenly Bodies, that Venus is never distant more than 46 parts from the sun; Mercury never more than 23; and that the planets which are placed above the sun move slowest when they are furthest from the earth, those under the sun fastest. Moreover there is another kind of observation to be employed, which has not yet come into use, though it be of small importance. This is, that to the enumeration of things which are should be subjoined an enumeration of things which are not. As in the history of the Heavenly Bodies, that there is not found any star oblong or triangular, but that every star is globular; either globular simply, as the moon; or apparently angular but globular in the middle, as the other stars; or apparently radiant but globular in the middle, as the sun;—or that the stars are scattered about the sky in no order at all; so that there is not found among them either quincunx or square, or any other regular figure (howsoever the names be given of Delta, Crown,

Cross, Chariot, etc.),—scarcely so much as a straight line; except perhaps in the belt and dagger of Orion.

Fifthly, that may perhaps be of some assistance to an inquirer which is the ruin and destruction of a believer; viz. a brief review, as in passage, of the opinions now received, with their varieties and sects; that they may touch and rouse the intellect, and no more.

X.

And this will be enough in the way of general precepts; which if they be diligently observed, the work of the history will at once go straight towards its object and be prevented from increasing beyond bounds. But if even as here circumscribed and limited it should appear to some poor-spirited person a vast work, let him turn to the libraries; and there among other things let him look at the bodies of civil and canonical law on one side, and at the commentaries of doctors and lawyers on the other; and see what a difference there is between the two in point of mass and volume. For we (who as faithful secretaries do but enter and set down the laws themselves of nature and nothing else) are content with brevity, and almost compelled to it by the condition of things; whereas opinions, doctrines, and speculations are without number and without end.

And whereas in the Plan of the Work I have spoken of the *Cardinal Virtues* in nature, and said that a history of these must also be collected and written before we come to the work of Interpretation; I have not forgotten this, but I reserve this part for myself; since until men have begun to be somewhat more closely intimate with nature, I cannot venture to rely very much on other people's industry in that matter.

And now should come the delineation of the particular histories. But I have at present so many other things to do that I can only find time to subjoin a Catalogue of their titles. As soon however as I have leisure for it, I mean to draw up a set of questions on the several subjects, and to explain what points with regard to each of the histories are especially to be inquired and collected, as conducing to the end I have in view,—like a kind of particular Topics. In other words, I mean (according to the practice in civil causes) in this great Plea or Suit granted by the divine favour and providence (whereby the human race seeks to recover its right over nature), to examine nature herself and the arts upon interrogatories.

CATALOGUE OF PARTICULAR HISTORIES BY TITLES.

1. History of the Heavenly Bodies; or Astronomical History.
2. History of the Configuration of the Heaven and the parts thereof towards the Earth and the parts thereof; or Cosmographical History.
3. History of Comets.
4. History of Fiery Meteors.
5. History of Lightnings, Thunderbolts, Thunders, and Coruscations.
6. History of Winds and Sudden Blasts and Undulations of the Air.
7. History of Rainbows.
8. History of Clouds, as they are seen above.
9. History of the Blue Expanse, of Twilight, of Mock-Suns, Mock-Moons, Haloes, various colours of the Sun; and of every variety in the aspect of the heavens caused by the medium.
10. History of Showers, Ordinary, Stormy, and Prodigious; also of Water-Spouts (as they are called); and the like.
11. History of Hail, Snow, Frost, Hoar-frost, Fog, Dew, and the like.
12. History of all other things that fall or descend from above, and that are generated in the upper region.
13. History of Sounds in the upper region (if there be any), besides Thunder.
14. History of Air as a whole, or in the Configuration of the World.
15. History of the Seasons or Temperatures of the Year, as well according to the variations of Regions as according to accidents of Times and periods of Years; of Floods, Heats, Droughts, and the like.
16. History of Earth and Sea; of the Shape and Compass of them, and their Configurations compared with each other; and of their broadening or narrowing; of Islands in the Sea; of Gulfs of the Sea, and Salt Lakes within the Land; Isthmuses and Promontories.
17. History of the Motions (if any be) of the Globe of Earth and Sea; and of the Experiments from which such motions may be collected.
18. History of the greater Motions and Perturbations in Earth and Sea; Earthquakes, Tremblings and Yawnings of the Earth, Islands newly appearing; Floating Islands; Breaking off of Land by entrance of the Sea, Encroachments and Inundations and contrariwise Recessions of the Sea; Eruptions of Fire from the Earth; Sudden Eruptions of Waters from the Earth; and the like.
19. Natural History of Geography; of Mountains, Vallies, Woods, Plains, Sands, Marches, Lakes, Rivers, Torrents, Springs, and every variety of their course, and the like; leaving apart Nations, Provinces, Cities, and such like matters pertaining to Civil life.
20. History of Ebbs and Flows of the Sea; Currents, Undulations, and other Motions of the Sea.
21. History of the other Accidents of the Sea; its Saltness, its various Colours, its Depth; also of Rocks, Mountains and Vallies under the Sea, and the like.

Next come Histories of the Greater Masses.

22. History of Flame and of things Ignited.
23. History of Air, in Substance, not in the Configuration of the World.
24. History of Water, in Substance, not in the Configuration of the World.
25. History of Earth and the diversity thereof, in Substance, not in the Configuration of the World.

Next come Histories of Species.

26. History of perfect Metals, Gold, Silver ; and of the Mines, Veins, Marcasites of the same ; also of the Working in the Mines.
27. History of Quicksilver.
28. History of Fossils ; as Vitriol, Sulphur, etc.
29. History of Gems ; as the Diamond, the Ruby, etc.
30. History of Stones ; as Marble, Touchstone, Flint, etc.
31. History of the Magnet.
32. History of Miscellaneous Bodies, which are neither entirely Fossil nor Vegetable ; as Salts, Amber, Ambergris, etc.
33. Chemical History of Metals and Minerals.
34. History of Planets, Trees, Shrubs, Herbs ; and of their parts, Roots, Stalks, Wood, Leaves, Flowers, Fruits, Seeds, Gums, etc.
35. Chemical History of Vegetables.
36. History of Fishes, and the Parts and Generation of them.
37. History of Birds, and the Parts and Generation of them.
38. History of Quadrupeds, and the Parts and Generation of them.
39. History of Serpents, Worms, Flies, and other insects ; and of the Parts and Generation of them.
40. Chemical History of the things which are taken by Animals.

Next come Histories of Man.

41. History of the Figure and External Limbs of Man, his Stature, Frame, Countenance and Features ; and of the variety of the same according to Races and Climates, or other smaller differences.
42. Physiognomical History of the same.
43. Anatomical History, or of the Internal Members of Man ; and of the variety of them, as it is found in the Natural Frame and Structure, and not merely as regards Diseases and Accidents out of the course of Nature.
44. History of the Parts of Uniform Structure in Man ; as Flesh, Bones, Membranes, etc.
45. History of Humours in Man ; Blood, Bile, Seed, etc.
46. History of Excrements ; Spittle, Urine, Sweats, Stools, Hair of the Head, Hairs of the Body, Whitlows, Nails, and the like.
47. History of Faculties ; Attraction, Digestion, Retention, Expulsion, Sanguification, Assimilation of Aliment into the members, conversion of Blood and Flower of Blood into Spirit, etc.
48. History of Natural and Involuntary Motions ; as Motion of the Heart, the Pulses, Sneezing, Lungs, Erection, etc.
49. History of Motions partly Natural and partly Violent ; as of Respiration, Cough, Urine, Stool, etc.
50. History of Voluntary Motions ; as of the Instruments of Articulation of Words ; Motions of the Eyes, Tongue, Jaws, Hands, Fingers ; of Swallowing, etc.
51. History of Sleep and Dreams.
52. History of different habits of Body—Fat, Lean ; of the Complexions (as they call them), etc.
53. History of the Generation of Man.
54. History of Conception, Vivification, Gestation in the Womb, Birth, etc.
55. History of the Food of Man ; and of all things Eatable and Drinkable ; and of all Diet ; and of the variety of the same according to nations and smaller differences.
56. History of the Growth and Increase of the Body, in the whole and in its parts.
57. History of the Course of Age ; Infancy, Boyhood, Youth, Old Age ; of Length and Shortness of Life, and the like, according to nations and lesser differences.
58. History of Life and Death.
59. History Medicinal of Diseases, and the Symptoms and Signs of them.

60. History Medicinal of the Treatment and Remedies and Cures of Diseases.
61. History Medicinal of those things which preserve the Body and the Health.
62. History Medicinal of those things which relate to the Form and Comeliness of the Body.
63. History Medicinal of those things which alter the Body, and pertain to Alterative Regimen.
64. History of Drugs.
65. History of Surgery.
66. Chemical History of Medicines.
67. History of Vision, and of things Visible.
68. History of Painting Sculpture, Modelling, etc.
69. History of Hearing and Sound.
70. History of Music.
71. History of Smell and Smells.
72. History of Taste and Tastes.
73. History of Touch, and the objects of Touch.
74. History of Venus, as a species of Touch.
75. History of Bodily Pains, as species of Touch.
76. History of Pleasure and Pain in general.
77. History of the Affections ; as Anger, Love, Shame, etc.
78. History of the Intellectual Faculties ; Reflexion, Imagination, Discourse, Memory, etc.
79. History of Natural Divinations.
80. History of Diagnostics, or Secret Natural Judgments.

81. History of Cookery, and the arts thereto belonging, as of the Butcher, Poulterer, etc.
82. History of Baking, and the Making of Bread, and the arts thereto belonging, as of the Miller, etc.
83. History of Wine.
84. History of the Cellar and of different kinds of Drink.
85. History of Sweetmeats and Confections.
86. History of Honey.
87. History of Sugar.
88. History of the Dairy.
89. History of Baths and Ointments.
90. Miscellaneous History concerning the care of the body,—as of Barbers, Perfumers, etc.
91. History of the working of Gold, and the arts thereto belonging.
92. History of the manufactures of Wool, and the arts thereto belonging.
93. History of the manufactures of Silk, and the arts thereto belonging.
94. History of manufactures of Flax, Hemp, Cotton, Hair, and other kinds of Thread, and the arts thereto belonging.
95. History of manufactures of Feathers.
96. History of Weaving, and the arts thereto belonging.
97. History of Dyeing.
98. History of Leather-making, Tanning, and the arts thereto belonging.
99. History of Ticking and Feathers.
100. History of working in Iron.
101. History of Stone-cutting.
102. History of the making of Bricks and Tiles.
103. History of Pottery.
104. History of Cements, etc.
105. History of working in Wood.
106. History of working in Lead.
107. History of Glass and all vitreous substances, and of Glass-making.
108. History of Architecture generally.
109. History of Waggons, Chariots, Litters, etc.
110. History of Printing, of Books, of Writing, of Sealing ; of Ink, Pen, Paper, Parchments, etc.

111. History of Wax.
112. History of Basket-making.
113. History of Mat-making, and of manufactures of Straw, Rushes, and the like.
114. History of Washing, Scouring, etc.

115. History of Agriculture, Pasturage, Culture of Woods, etc.
116. History of Gardening.
117. History of Fishing.
118. History of Hunting and Fowling.
119. History of the Art of War, and of the arts thereto belonging, as Armoury, Bow-making, Arrow-making, Musketry, Ordinance, Cross-bows, Machines, etc.
120. History of the Art of Navigation, and of the crafts and arts thereto belonging.
121. History of Athletics and Human Exercises of all kinds.
122. History of Horsemanship.
123. History of Games of all kinds.
124. History of Jugglers and Mountebanks.
125. Miscellaneous History of various Artificial Materials,—as Enamel, Porcelain, various Cements, etc.
126. History of Salts.
127. Miscellaneous History of various Machines and Motions.
128. Miscellaneous History of Common Experiments which have not grown into an Art.

Histories must also be written of Pure Mathematics ; though they are rather observations than experiments.

129. History of the Natures and Powers of Numbers.
130. History of the Natures and Powers of Figures.

It may not be amiss to observe that, whereas many of the experiments must come under more titles than one (as the History of Plants and the History of the Art of Gardening have many things in common), it will be more convenient to investigate them with reference to Arts, and to arrange them with reference to Bodies. For I care little about the mechanical arts themselves : only about those things which they contribute to the equipment of philosophy. But these things will be better regulated as the case arises.

OF THE DIGNITY AND ADVANCEMENT OF LEARNING[1]

Books II.—IX.

PREFACE.

BY JAMES SPEDDING.

In a letter dated June 30, 1622, Bacon speaks of the *De Augmentis Scientiarum* as a work already in the hands of translators, and likely to be finished by the end of the summer. "Librum meum de progressu Scientiarum traducendum commisi. Illa translatio, volente Deo, sub finem æstatis perficietur.[2]" Therefore though it was not published till the autumn of 1623, it may be considered as coming, in order of composition, next among the Philosophical works to the *Novum Organum* and *Parasceve.*

It was intended to serve for the first part of the Instauratio Magna, according to the plan laid out in the *Distributio Operis*—the part which is there entitled *Partitiones Scientiarum*, and described as exhibiting a complete survey of the world of human knowledge as it then was—"Scientiæ ejus sive doctrinæ in cujus possessione humanum genus hactenus versatur summam sive descriptionem universalem". The relation which it bears to the rest of the work is best explained in the dedicatory letter prefixed to the *Dialogue of a Holy War.* "And again, for that my book of Advancement of Learning may be some preparative or key for the better opening of the Instauration, because it exhibits a mixture of new conceits and old, whereas the Instauration gives the new unmixed, otherwise than with some aspersion of the old for taste's sake, I have thought good to procure a translation of that book into the general language, not without great and ample additions and enrichment thereof, especially in the second book, which handleth the partition of sciences; in such sort as I hold it [3] may serve in lieu of the first part of the Instauration, and acquit my promise in that part."

But why, when Bacon determined to fit this work for that part, did he not give it the proper title? Curious as he always was in the choice of names, why not call it "*Partitiones* Scientiarum", which describes the proper business of the first part of the Instauratio, instead of "De *dignitate et augmentis* Scientiarum", which passes it by?

The answer, I think, is that he felt it would be inappropriate. The form in which the *De Augmentis* was cast retained so strong an impress of the original design out of which it grew,—a design truly and exactly described in the title, and having no immediate reference to the ultimate plan of the Instauratio—that another title referring to another design would have been manifestly unfit. When he wrote the Advancement of Learning, he was already engaged upon a work concerning the Interpretation of Nature, which (to judge from the fragments and sketches that remain) was meant to begin at once where the *Novum Organum* begins, without any preliminary review of the existing condition of knowledge.

1 Translation of the *De Dignitate et Augmentis Scientiarum.*

2 Letter to Father Redempt. Baranzan.

3 That is, the second book; as appears more clearly from the Latin version of this letter, which was written later. "Idque ita cumulate præstiti ut judicem *librum illum jam in plures divisum,* pro primâ Instaurationis parte haberi posse, quam *Partitionum Scientiarum* nomine antea insignivi."

a work corresponding to that which in the foregoing extract he calls "the Instauration", as distinguished from the Advancement of Learning, which was to serve as "a preparative or key" to it; and the writing of a book which should exhibit a complete and particular survey of the state of knowledge then extant in the world was, I suspect, a by-thought suggested by a particular accident.

However Bacon may have underrated the difficulties of the reform which he proposed, he was well aware that it could not be carried into effect by a private man. A private man might suggest the course, and produce a specimen; but the execution of the work on a scale of adequate magnitude required the means and influence of a King or Pope. Now it happened, by a very singular accident, that while he was engaged in considering and maturing his plan, there succeeded to the throne of England a man whose tastes and previous training qualified him more than most other men to take an earnest, active, and intelligent interest in it. James the First was a man of peace by principle and inclination, of solid, various, and extensive learning, and of great intellectual activity. It is difficult even now to say why he might not have proved, in the province of letters, a great governor. At that time, when his faults were not yet known, he must have appeared like the very man for such an office. To Bacon it would naturally seem an object of the first importance to engage him, if possible, as a patron of the new philosophy; and, as men's minds are most impressible in times of transition, he would wish to lose no time in attempting to give his ambition a turn in that direction, while his fortune was fresh, his course unsettled, his imagination excited and open to great ideas. For this purpose, however, the work on the Interpretation of Nature was not forward enough to be available, nor very fit perhaps in itself, had it been more forward than it was. The idea was too new, the scheme too vast, the end too remote, to engage the serious attention of a king nearly forty years old, who had been bred in the ancient learning and attained a proficiency in it of which he was proud. "Restat unica salus ac sanitas ut opus mentis universum de integro resumatur" was an avowal which might well startle him. Not so a work representing the state of human science as it was, and the means of perfecting and extending it in many new directions. This lay in James's own province; of the review of what had been already done few men of his time were better qualified to judge; few perhaps were more likely to be attracted and excited by the prospect of doing more. Now Bacon's own travels in search of the light he had been looking for had carried him over the whole surface of the intellectual globe; and he was therefore well qualified to report upon the condition of it,—to declare how far and in what directions the dominion of knowledge had been already advanced, what regions were still unexplored and unsubdued, and what measures might best be taken to bring them into subjection. Such a representation was likely enough to make an impression on a mind constituted and trained like that of James the First. Possibly it might even rouse him to take up the extension of knowledge as a royal business; in which case the new philosophy would have started with advantages not otherwise to be hoped for.

This work therefore Bacon seems to have set about at once. There is reason to believe that the first book of the Advancement of Learning, which treats of the excellence and dignity of knowledge as a pursuit for kings and statesmen, was written in 1603, immediately after James's accession; and the second, which treats of the deficiencies remaining and the supplies required, in 1605; the intervening year of 1604 having been too much occupied with civil business to allow much leisure for the prosecution of a work of that kind. It was important to push it forward as fast as possible, even at the expense of completeness: for the very object for which I suppose it to have been undertaken,—that of making an impression on the king's mind while it was in the best state to receive impressions,—would have been lost by delay; and accordingly in the autumn of 1605 appeared "the Twoo Bookes of Francis Bacon, of the proficience and advancement of Learning, divine and humane"; with many marks of haste in form and composition, and even in substance not altogether adequate to the argument in hand, but nevertheless well enough adapted for its immediate purpose, if I have rightly conjectured what that purpose was.

If this be the true history of the Advancement of Learning, the rest follows naturally. The stroke, though well aimed, was not successful. The book may have raised James's opinion of Bacon, but it did not inspire him with any zeal for the Great Instauration. There it was, however; and it contained such a quantity of the best fruits of Bacon's mind, and so many new views bearing on the great reform which he meditated, that it seemed a pity not to find a place for it in the great work. This was easily done by enlarging the original design so as to include a preliminary survey of the existing state of knowledge; in which case the substance of the second book of the Advancement might do duty as the first part of the Instauratio Magna. If we knew when the fragment entitled *Partis Instaurationis Secundæ Delineatio* was written, we might almost fix the time at which this enlargement of the original design was resolved upon. For in that fragment Bacon proposes to distribute the whole subject of the Interpretation of Nature through the second, third, fourth, fifth, and sixth parts of the work, exactly as in the *Distributio Operis;* a place being reserved for a first part, though the nature of its contents is not specified. And from the *Descriptio Globi Intellectualis*, which was written in 1612, and appears, as I have elsewhere remarked, to be a commencement of the *Partitiones Scientiarum* itself, we may partly infer the form in which he then intended to cast that part.

Why he afterwards altered his intention and resolved to content himself with a mere translation of the two books of the Advancement with additions, it is not difficult to conjecture, if we take into account the circumstances of his life. When the *Novum Organum* was published in October 1620, the king had just resolved to call a new Parliament after six years' intermission, and questions of vital interest both at home and abroad hung upon the issue of it. The necessary preparations for the session, Bacon's own impeachment, which almost immediately followed, a severe illness consequent upon that, his condemnation and imprisonment, negotiations with importunate creditors, and the composition of the *History of Henry the Seventh*, which was finished in October 1621, must have given him occupation enough during the next twelve months. Then came the question, how he was to proceed with the Instauratio, so as to máke the most of such time and means as remained. Sixty-two years old, with health greatly impaired, an income scarcely sufficient to live upon, and an establishment of servants much reduced, he could not afford to waste labour upon things not essential. The Novum Organum was not half finished. The Natural History was not even begun, and no fellow-labourer had yet come forward to help in it [4]. It was only in the completion of the first of the six parts that he could hope for material assistance from others. Even this, if he had attempted to recast it in the form which I suppose him to have designed,—the form indicated in the *Descriptio Globi Intellectualis*,—he could hardly have executed by deputy; whereas a translation of the Advancement of Learning might be so executed, and would need only corrections and additions to make it a complete survey of the intellectual globe, adequate in substance to its place, though not symmetrical in form. Accordingly, "by help of some good pens which did not forsake him," he proceeded at once to put this in train, and then turned his own attention to the Natural History, which he considered as "*basis totius negotii*".

Concerning the causes which delayed the publication of the *De Augmentis* a twelvemonth beyond the expected time, I have no information. But it is probable that the additions which suggested themselves as he proceeded were far larger than he had anticipated; being indeed in the second book as much again as the original, and more. The measures which he took however were in this instance quite successful; and by sacrificing a little symmetry of form, he succeeded in effectually preserving the substance of this first part of his great work [5].

4 "Neque huic rei deero quantum in me est. Utinam habeam et adjutores idoneos." —*Letter to Father Redempt. Baranzan*, 30 June, 1622.

5 The volume in which it originally appeared bore the following general title-page: *Opera Francisci Baronis de Verulamio, vice-comitis Sancti Albani, Tomus primus. Qui continet* De Augmentis Scientiarum *libros IX. Ad regem suum. Londini, in officina*

Tenison mentions " Mr. Herbert "—that is, George Herbert, the poet—as one of the translators employed. But we have it upon Rawley's authority that Bacon took a great deal of pains with it himself (proprio marte plurimum desudavit)—so that we must consider the whole translation as stamped with his authority. Many years before he had asked Dr. Playfer to do it ; who (according to Tenison) sent him a specimen, but " of such superfine Latinity, that the Lord Bacon did not encourage him to labour further in that work, in the penning of which he desired not so much neat and polite, as clear masculine and apt expression " [6]. And it is not improbable that some such difficulty may have occurred. But Playfer's failure may be sufficiently accounted for by the state of his health. A memorandum in the *Commentarius Solutus* dated 26 July, 1608—" Proceeding with the translation of my book of Advancement of Learning—hearkening to some other if Playfer should fail,"—shows that at that time it was still in his hands ; and he died at the beginning of the next year.

I have only to add that all the notes to this work which bear no signature are Mr. Ellis's, except such parts of them as are inserted within brackets. These, as well as all notes signed J. S., are mine.

J. S.

Joannis Haviland, MDCXXIII. But this had reference to a collection (which he then meditated) of all his works, in Latin ; not to the order of the *Instauratio,* which was not in a condition to be published consecutively. See *Epistola ad Fulgentium* : Opuscula, p. 172.

[6] Baconiana, p. 26.

THE DIVISIONS OF THE SCIENCES, AND ARGUMENTS OF THE SEVERAL CHAPTERS.

BOOK THE SECOND.

CHAP. I.

DIVISION of all Human Learning into *History, Poesy, Philosophy*; with reference to the three Intellectual Faculties, *Memory, Imagination, Reason*; and that the same division holds good likewise in Theology.

CHAP. II.

Division of History into *Natural* and *Civil*; *Ecclesiastical* and *Literary* History being included in Civil. Division of Natural History, according to its subject, into History of *Generations, Preter-generations*, and *Arts*.

CHAP. III.

Second division of Natural History, according to its use and end, into *Narrative* and *Inductive*; and that the noblest end of Natural History is to minister and be in order to *the building up of Philosophy*; which end is aimed at by the *Inductive*. Division of the History of Generations into the History of the *Heavenly Bodies*, of *Meteors*, of *the Globe of Earth and Sea*, of *the Masses* or *Greater Colleges*, and of *the Species* or *Lesser Colleges*.

CHAP. IV.

Division of Civil History into *Ecclesiastical, Literary*, and *Civil* (properly so called): and that *Literary History* is wanting. Precepts for the construction thereof.

CHAP. V.

Of the dignity and difficulty of Civil History.

CHAP. VI.

First division of Civil History (properly so called) into *Memorials, Antiquities*, and *Perfect History*.

CHAP. VII.

Division of Perfect History into *Chronicles of Times, Lives of Persons, and Relations of Actions*. The explanation of these.

CHAP. VIII.

Division of History of Times into *Universal* and *Particular*. The advantages and disadvantages of each.

CHAP. IX.

Second division of History of Times into *Annals* and *Journals*.

CHAP. X.

Second division of Civil History (properly so called) into *Pure* and *Mixed*.

CHAP. XI.

Division of Ecclesiastical History, into History of the *Church*, History according to the *Prophecies*, and History of *Providence*.

CHAP. XII.

Of certain *Appendices to History*, which deal with the *words* of man, as History deals with their actions. Division of the same into *Orations, Letters*, and *Apophthegms*.

CHAP. XIII

Of the second principal branch of Learning, namely, *Poesy.* Division of Poesy into *Narrative, Dramatic,* and *Parabolical.* Three examples of Parabolical Poesy are propounded.

BOOK THE THIRD.

CHAP. I.

Division of Science into *Theology* and *Philosophy.* Division of Philosophy into three doctrines; concerning the *Deity,* concerning *Nature,* and concerning *Man.* Constitution of *Primary Philosophy* as the common mother of all.

CHAP. II.

Of *Natural Theology;* and the doctrine concerning *Angels* and *Spirits,* which is an appendix of the same.

CHAP. III.

Division of Natural Philosophy into *Speculative* and *Operative.* And that these two ought to be kept separate, both in the intention of the writer and in the body of the treatise.

CHAP. IV.

Division of Speculative doctrine concerning nature, into *Physic* (special) and *Metaphysic.* Whereof Physic inquires of the *Efficient Cause* and the *Material;* Metaphysic of the *Final Cause* and the *Form.* Division of Physic (special) into doctrine concerning *the Principles of Things,* concerning the *Fabric of Things,* or the world; and concerning the *Variety of Things.* Division of the doctrine concerning the Variety of Things into doctrine concerning things *concrete,* and doctrine concerning things *abstract.* The division of the doctrine concerning things concrete is referred to the same divisions which Natural History receives. Division of the doctrine concerning things abstract, into doctrine concerning the *Configuration of Matter,* and the doctrine concerning *motions.* Two appendices of Speculative Physic: *natural problems,* and *dogmas of the ancient philosophers.* Division of Metaphysic into doctrine concerning *Form,* and doctrine concerning *Final Causes.*

CHAP. V.

Division of the operative doctrine concerning Nature into *Mechanic* and *Magic,* which correspond to the divisions of the speculative doctrine: *Mechanic* answering to *Physic, Magic* to *Metaphysic.* Purification of the word Magic. Two appendices of the operative doctrine. *Inventory of the possessions of man;* and *Catalogue of Polychrests,* or things of general use.

CHAP. VI.

Of the great Appendix of Natural Philosophy, both speculative and operative, namely *Mathematics*: and that it ought rather to be placed among appendices than among substantive science. Division of Mathematics into *Pure* and *Mixed.*

BOOK THE FOURTH.

CHAP. I.

Division of the doctrine concerning Man into *Philosophy of Humanity* and *Philosophy Civil.* Division of the Philosophy of Humanity into doctrine concerning the *Body of Man,* and doctrine concerning the *Soul of Man.* Constitution of one general doctrine concerning the *Nature* or the *State of Man.* Division of the doctrine concerning the State of Man into doctrine concerning the *Person of Man,* and concerning the *League of Mind and Body.* Division of the doctrine concerning the Person of Man into doctrine concerning the *Miseries of Man,* and concerning his *Prerogatives.* Division of the doctrine concerning the League, into doctrine concerning *Indications* and concerning *Impressions.* Assignation of *Physiognomy* and *Interpretation of Natural Dreams* to the doctrine concerning Indications.

CHAP. II.

Division of the doctrine concerning the Body of Man into *Medicine, Cosmetic, Athletic,* and *Voluptuary.* Division of Medicine into three offices ; viz. the *Preservation of Health,* the *Cure of Diseases,* and the *Prolongation of Life.* And that the last division concerning the prolongation of Life ought to be kept separate from the other two.

CHAP. III.

Division of Human Philosophy relating to the Soul into doctrine concerning the *Breath of Life* and doctrine concerning the *Sensible or Produced Spirit.* Second division of the same Philosophy into doctrine concerning the *Substance and Faculties of the Soul,* and doctrine concerning the *Use and Objects of the Faculties.* Two appendices to the doctrine concerning the Faculties of the Soul : doctrine concerning *Natural Divination,* and doctrine concerning *Fascination.* Distribution of the faculties of the Sensible Soul into *Motion* and *Sense.*

BOOK THE FIFTH.

CHAP. I.

Division of the doctrine concerning the use and objects of the Faculties of the Human Soul into *Logic* and *Ethics.* Division of Logic into the arts of *Discovering,* of *Judging,* of *Retaining,* and of *Transmitting.*

CHAP. II.

Division of the Art of Discovering into discovery of *Arts* and discovery of *Arguments* and that the former of these (which is the most important) is wanting. Division of the art of discovery of Arts into *Learned Experience* and the *New Organon.* Description of *Learned Experience.*

CHAP. III.

Division of the art of discovery of Arguments into the *Promptuary,* and *Topics.* Division of Topics into *General* and *Particular.* Example of a Particular Topic, in an inquiry concerning Heavy and Light.

CHAP. IV.

Division of the art of Judging into judgment by *Induction* and judgment by *Syllogism.* The first whereof is referred to the New Organon. First division of Judgment by Syllogism into *Reduction Direct* and *Reduction Inverse.* Second division of the same into *Analytic,* and doctrine concerning *Detection of Fallacies.* Division of the doctrine concerning the detection of fallacies into detection of *sophistical* fallacies, fallacies *of interpretation,* and fallacies of *false appearances or Idols.* Division of Idols into Idols of the *Tribe,* Idols of the *Cave,* and Idols of the *Market-place.* Appendix to the Art of Judging ; viz. concerning *the Analogy of Demonstrations according to the nature of the subject.*

CHAP. V.

Division of the Art of Retaining into the doctrine concerning the *Helps of Memory* and the doctrine concerning *Memory itself.* Division of the doctrine concerning Memory itself into *Prenotion* and *Emblem.*

BOOK THE SIXTH.

CHAP. I.

Division of the art of Transmitting into the doctrine concerning the *Organ of Discourse,* the doctrine concerning the *Method of Discourse,* and the doctrine concerning the *Illustration of Discourse.* Division of the doctrine concerning the organ of discourse into the doctrine concerning the *Notations of Things,* concerning *Speech,* and concerning *Writing.* Whereof the two first constitute *Grammar* and are divisions of it. Division of the doctrine concerning the notations of things into *Hieroglyphics,* and *Real Characters.* Second division of Grammar into *Literary* and *Philosophic.* Reference of *Poesy in respect of metre* to the doctrine concerning *Speech.* Reference of the doctrine concerning *ciphers* to the doctrine concerning *Writing.*

CHAP. II.

The doctrine concerning the *Method of Discourse* is made a substantive and principal part of the art of *transmitting ;* and is named *Wisdom of Transmission.* Different kinds of method are enumerated, with a note of their advantages and disadvantages.

CHAP. III.

Of the foundations and office of the doctrine concerning Illustration of Discourse, or Rhetoric. Three appendices of Rhetoric, which relate only to the *Promptuary ; Colours of Good and Evil,* both simple and comparative ; *Antitheses of things ; Lesser Forms of Speeches.*

CHAP. IV.

Two general appendices of the Art of Transmission ; *Critical* and *Pedagogical.*

BOOK THE SEVENTH.

CHAP. I.

Division of Moral Knowledge into the *Exemplar* or *Platform of Good,* and the *Georgics* or *Culture of the Mind.* Division of the Platform of Good into *Simple* and *Comparative Good.* Division of Simple Good into *Individual* Good, and Good of *Communion.*

CHAP. II.

Division of Individual or Self-Good into *Active* and *Passive Good.* Division of *Passive Good* into *Conservative* and *Perfective Good.* Division of the *Good of Communion* into *General* and *Respective Duties.*

CHAP. III.

Division of the doctrine concerning the *Culture of the Mind* into the doctrine concerning the *Characters of Minds,* the *Affections,* and the *Remedies* or *Cures.* Appendix of this same doctrine, touching the *Congruity between the Good of the Mind and the Good of the Body.*

BOOK THE EIGHTH.

CHAP. I.

Division of Civil Knowledge into the doctrine concerning *Conversation, Negotiation,* and *Empire* or *State Government.*

CHAP. II.

Divison of the doctrine concerning Negotiation into the doctrine concerning *Scattered Occasions* and the doctrine concerning *Advancement in Life.* Example of the doctrine concerning Scattered Occasions from some of the Proverbs of Solomon. Precepts concerning Advancement in Life.

CHAP. III.

The divisions of the doctrine concerning Empire or Government are omitted ;—An Introduction only is made to two Deficients ; namely, the doctrine concerning the *Extension of the Bounds of Empire,* and the doctrine concerning *Universal Justice,* or the *Fountains of Law.*

BOOK THE NINTH.

CHAP. I.

The divisions of Inspired Divinity are omitted ;—Introduction only is made to three Deficients ; namely, the doctrine concerning the *Legitimate Use of the Human Reason in Divine Subjects ;* the doctrine concerning the *Degrees of Unity in the Kingdom of God ;* and the *Emanations of the Scriptures.*

pressions. Assignation of *Physiognomy* and *Interpretation of Natural Dreams* to the doctrine concerning Indications.

TRANSLATION OF THE

DE AUGMENTIS SCIENTIARUM

Book II.[1]

TO THE KING.[2]

It might seem to have more convenience, excellent King, though it come often otherwise to pass, that those who are fruitful in their generations, and have as it were the foresight of immortality in their descendants, should likewise be more careful than other men of the good estate of future times, to which they know they must transmit these their dearest pledges. Queen Elizabeth, rather a sojourner in the world than an inhabitant, in respect of her unmarried life, was an ornament to her own times and prospered them in many ways. But to your Majesty (whom God in His goodness has already blessed with so much royal issue, worthy to continue and represent you for ever, and whose youthful and fruitful bed still promises more[3]) it is proper and convenient not only to shed a lustre (as you do) on your own age, but also to extend your care to those things which all memory may preserve and which are in their nature eternal. Amongst which (if affection for learning transport me not) there is not any more noble or more worthy than the further endowment of the world with sound and fruitful knowledge. For how long shall we let a few received authors stand up like Hercules' columns, beyond which there shall be no sailing or discovery in science, when we have so bright and benign a star as your Majesty to conduct and prosper us?

To return then to the matter in hand: let us now review and consider with ourselves what has hitherto been done by kings and others for the increase and advancement of learning, and what has been left undone; and let us discuss the question solidly and distinctly, in a style active and masculine, without digressing or dilating. We may begin then by assuming (which will not be disputed) that all the greatest and most difficult works are overcome either by amplitude of reward, or by prudence and soundness of direction, or by conjunction of labours; whereof the first stimulates endeavour, the second removes uncertainty and error, and the third supplies the frailty of man. But of these three, prudence and soundness of direction,—that is, the pointing out and setting forth of the straight and ready way to the thing which is to be done,—must be placed first. For the cripple in the right way (as the saying is) outstrips the runner in the wrong. And Solomon observes, most aptly to the point in question, that "if the iron be blunt it requireth more strength, but wisdom is that which prevaileth"[4]; signifying that the prudent choice of the mean is more effectual for the purpose than either the enforcement or the accumulation of endeavours. This I am induced to say, for that (not derogating from the honour of those who have been in any way deservers towards the state of learning) I observe nevertheless that most of their works and acts have had in view rather their own magnificence and memory than the progress and advancement of learning, and have rather augmented the number of learned men than raised and rectified the sciences themselves.

[1] For the first book (which relates to the Dignity of Learning), see the "Advancement of Learning" above, pp. 43–74. The Latin differs so little from the English in that book, that a translation would be little else than a reprint. And the eight remaining books of the *De Augmentis Scientiarum*, considered as a treatise on the Divisions of the Sciences, are complete in themselves.—*J. S.*

[2] Here the first part of the *Instauratio Magna*, the *Partitiones Scientiarum*, properly begins; the nine following pages being the preface.—*J. S.*

[3] This passage, being transferred from the *Advancement of Learning*, must be considered of course as written in 1605.—*J. S.*

[4] Eccles. x. 10.

The works or acts which pertain to the advancement of learning are conversant about three objects; the places of learning, the books of learning, and the persons of the learned. For as water, whether it be the dew of Heaven or the springs of the earth, easily scatters and loses itself in the ground, except it be collected into some receptacle where it may by union and consort comfort and sustain itself (and for that cause the industry of man has devised aqueducts, cisterns, and pools, and likewise beautified them with various ornaments, for magnificence and state as well as for use and necessity); so this excellent liquor of knowledge, whether it descend from divine inspiration or spring from human sense, would soon perish and vanish into oblivion, if it were not preserved in books, traditions, and conferences; and especially in places appointed for such matters, as universities, colleges, and schools, where it may have both a fixed habitation and means and opportunity of increasing and collecting itself.

And first, the works which concern the *places of learning* are four; buildings, endowments with revenues, grants of franchises and privileges, and institutions and ordinances of government; all tending (for the most part) to retirement and quietness of life, and a release from cares and trouble; like the stations which Virgil prescribes for the hiving of honey bees.

Principio sedes apibus statioque petenda,
Quo neque sit ventis aditus, etc.[5]

The principal works touching *books* are two; first, libraries, which are as the shrines wherein all the relics of the ancient saints full of true virtue are preserved. Secondly, new editions of authors, with more correct impressions, more faithful translations, more profitable commentaries, more diligent annotations, and the like.

The works pertaining to the *persons of the learned* (besides the advancement and countenancing of them in general) are likewise two. The remuneration and designation of lecturers in arts already extant and invented; and the remuneration and appointment of writers and inquirers concerning those parts of learning not yet sufficiently laboured or prosecuted.

These are summarily the works and acts wherein the merits of many excellent princes and other illustrious personages towards learning have been manifested. As for the particular commemoration of any one who has deserved well of literature, I call to mind what Cicero said when, on his return from exile, he gave general thanks; "It is hard to remember all, ungrateful to pass by any." [6]. Let us rather (after the advice of Scripture) look forward to that part of the race which is still to be run, than look back to that which has been passed.

First therefore, among so many noble foundations of colleges in Europe, I find it strange that they are all dedicated to professions, and none left free to the study of arts and sciences at large. For if men judge that learning should be referred to use and action, they judge well; but it is easy in this to fall into the error pointed at in the ancient fable; in which the other parts of the body found fault with the stomach, because it neither performed the office of motion as the limbs do, nor of sense, as the head does; but yet notwithstanding it is the stomach which digests and distributes the aliment to all the rest. So if any man think that Philosophy and Universality are idle and unprofitable studies, he does not consider that all arts and professions are from thence supplied with sap and strength. And this I take to be a great cause, which has so long hindered the more flourishing progress of learning; because these fundamental knowledges have been studied but in passage, and not drunk deeper of. For if you will have a tree bear more fruit than it has used to do, it is not anything you can do to the boughs, but it is the stirring of the earth, and putting richer mould about the roots, that must work it. Neither is it to be forgotten that this dedication of colleges and societies to the use only of professory learning has not only

[5] Virg. *Georg.* iv. 8.:—First for thy bees a quiet station find,
And lodge them under covert of the wind.

[6] Cicero, *Post Red.* c. 12.

been inimical to the growth of the sciences, but has also been prejudicial to states and governments. For hence it proceeds that princes when they have to choose men for business of state find a wonderful dearth of able men around them; because there is no collegiate education designed for these purposes, where men naturally so disposed and affected might (besides other arts) give themselves especially to histories, modern languages, books of policy and civil discourse; whereby they might come better prepared and instructed to offices of state.

And because founders of Colleges do plant, and founders of Lectures do water, I must next speak of the deficiencies which I find in public lectures; wherein I especially disapprove of the smallness of the salary assigned to lecturers in arts and professions, particularly amongst ourselves. For it is very necessary to the progression of sciences that lecturers in every sort be of the most able and sufficient men; as those who are ordained not for transitory use, but for keeping up the race and succession of knowledge from age to age. This cannot be, except their condition and endowment be such that the most eminent professors may be well contented and willing to spend their whole life in that function and attendance, without caring for practice. And therefore if you will have sciences flourish, you must observe David's military law; which was, "That those who stayed with the baggage should have equal part with those who were in the action"[7]; else will the baggage be ill attended. So lecturers in sciences are as it were the keepers and guardians of the whole store and provision of learning, whence the active and militant part of the sciences is furnished; and therefore they ought to have equal entertainment and profit with the men of active life. Otherwise if the fathers in sciences be not amply and handsomely maintained, it will come to pass, as Virgil says of horses,—

Et patrum invalidi referent jejunia nati[8];

the poor keeping of the parents will be seen in the weakliness of the children.

I will now notice another defect, wherein I should call in some alchemist to help me; one of those who advise the studious to sell their books and build furnaces, and forsaking Minerva and the Muses as barren virgins, to rely upon Vulcan. But certain it is that for depth of speculation no less than for fruit of operation in some sciences (especially natural philosophy and physic) other helps are required besides books. Wherein also the beneficence of men has not been altogether wanting; for we see spheres, globes, astrolabes, maps, and the like have been provided and prepared as assistants to astronomy and cosmography, as well as books. We see likewise that some places instituted for physic have gardens for the examination and knowledge of simples of all sorts, and are not without the use of dead bodies for anatomical observations. But these respect but a few things. In general, it may be held for certain that there will hardly be any great progress in the unravelling and unlocking of the secrets of nature, except there be a full allowance for expenses about experiments; whether they be experiments appertaining to Vulcan or Dædalus (that is, the furnace or engine), or any other kind. And therefore as secretaries and emissaries of princes are allowed to bring in bills of expenses for their diligence in exploring and unravelling plots and civil secrets, so the searchers and spies of nature must have their expenses paid, or else you will never be well informed of a great number of things most worthy to be known. For if Alexander made such a liberal assignation of money to Aristotle, to support hunters, fowlers, fishers and the like, that he might be better furnished for compiling a History of Animals; certainly much more do they deserve it, who instead of wandering in the forests of nature, make their way through the labyrinths of arts.

Another defect to be noticed (and one of great importance) is a neglect of consultation in governors of universities, and of visitation in princes or superior

[7] 1 Sam. xxx. 24. Similarly it was provided by the laws of Alfonso the Wise, in accordance with earlier usage, that no division of spoil should be made until those in pursuit of the enemy had returned to the camp. See the *Siete Partidas*, ii. 26. 1.

[8] *Georg.* iii. 128.

persons, to enter into careful account and consideration whether the readings, disputations, and other scholastic exercises anciently begun, and since continued up to our time, may be profitably kept up, or whether we should rather abolish them and substitute better. For I find it is one of your Majesty's most wise maxims; "That in all usages or precedents the times be considered wherein they first begun; which, if they were disordered or ignorant, it derogates greatly from the authority of the precedents, and leaves all things for suspect". And therefore inasmuch as most of the institutions of the universities are derived from times a good deal more obscure and ignorant than our own, it is the more convenient that they be re-examined. In this kind I will give an instance or two, of things which appear the most obvious and familiar. It is a general custom (and yet I hold it to be an error) that scholars come too soon and too unripe to the study of logic and rhetoric, arts fitter for graduates than children and novices; for these two rightly taken are the gravest of sciences, being the arts of arts, the one for judgment, the other for ornament; besides they give the rule and direction how both to set forth and illustrate the subject matter. And therefore for minds empty and ignorant (and which have not yet gathered what Cicero calls "stuff [9]" or "furniture [10]", that is matter and variety) to begin with those arts (as if one should learn to weigh or to measure or to paint the wind), works but this effect, that the virtue and faculty of those arts (which are great and universal) are almost made contemptible, and either degenerate into childish sophistry and ridiculous affectation, or at least lose not a little of their reputation. And further, the premature and untimely learning of these arts has drawn on, by consequence, the superficial and unprofitable teaching and handling of them,—a manner of teaching suited to the capacity of children. Another instance of an error which has long prevailed in universities is this; that they make too great and mischievous a divorce between invention and memory. For most of the speeches there are either entirely premeditate, and delivered in preconceived words, where nothing is left to invention; or merely extempore, where little is left to memory; whereas in common life and action there is little use of either of these separately, but rather of intermixtures of them; that is of notes or commentaries and extempore speech; and thus the exercise fits not the practice, nor the image the life. But it must ever be observed as a rule in exercises, that they be made to represent in everything (as near as may be) the real actions of life; for otherwise they will pervert the motions and faculties of the mind, and not prepare them. The truth whereof appears clearly enough when scholars come to the practice of their professions, or other offices of civil life; which when they set into, this want I speak of is soon found out by themselves, but still sooner by others. But this part, touching the amendment of the Institutions and Orders of Universities, I will conclude with a sentence taken from one of Cæsar's letters to Oppius and Balbus; "How this may be done, some means occur to me, and many may be found; I beg you therefore to take these matters into consideration" [11].

Another defect which I note ascends a little higher than the preceding. For as the progress of learning consists not a little in the wise ordering and institutions of each several university; so it would be yet much more advanced if there were a closer connexion and relationship between all the different universities of Europe than now there is. For we see there are many orders and societies which, though they be divided under distant sovereignties and territories, yet enter into and maintain among themselves a kind of contact and fraternity, insomuch that they have governors (both provincial and general) whom they all obey. And surely as nature creates brotherhood in families, and arts mechanical contract brotherhoods in societies, and the anointment of God superinduces a brotherhood in kings and bishops, and vows and regulations make a brotherhood in religious orders; so in like manner there cannot but be a noble and generous brotherhood contracted among men by learning and illumination, seeing that God himself is called "the Father of Lights [12]".

9 *Sylva*, De Orator. iii. 26.

10 *Supellex*, Orator, c. 24.

11 Cic. Ep. ad Att. ix. 8.

12 St. James, i. 17.

The last defect I complain of (to which I have already alluded) is that there has not been, or very rarely been, any public designation of fit men either to write or to make inquiry concerning such parts of knowledge as have not been already sufficiently laboured. To which point it will greatly conduce, if a review and *census* be made of the sciences, and account be taken what parts of them are rich and well advanced, and what poor and destitute. For the opinion of plenty is amongst the causes of want; and the great quantity of books makes a show rather of superfluity than lack; of which surcharge nevertheless the true remedy is not to destroy the old books, but to make more good ones; of such a kind that like the serpent of Moses, they may devour the serpents of the enchanters [13].

The removal of all the defects formerly enumerated, except the last, and of the active part also of the last, which relates to the designation of writers, are truly works for a king; towards which the endeavours and industry of a private man can be but as an image in a crossway, that may point at the way but cannot go it. But the speculative part of it, which relates to the survey of knowledges to see what in each is deficient, is open likewise to private industry. Wherefore I now intend to make a general and faithful perambulation and survey of learning, with a very careful and accurate inquiry what parts thereof lie fresh and waste, and not yet improved and converted to use by the industry of man; to the end that such a plot marked out, and recorded to memory, may minister light both to public designations and voluntary endeavours. Wherein nevertheless my purpose is at this time to note only omissions and deficiencies, and not to make any redargution of errors and failures; for it is one thing to point out what parts lie untilled, and another thing to mend the manner of tillage.

In addressing myself to which task I am not ignorant how great a work I attempt, and how difficult a province I take upon me; nor again how far unequal my strength is to my will. Nevertheless I have great hope that if my extreme love to learning carry me too far, I may obtain the excuse of affection; for that "it is not granted to any man at the same time to love and to be wise" [14]. But I know well I can use no other liberty of judgment, than I must leave to others; and I for my part shall be equally glad either to perform myself or to accept from others that duty of humanity, to put the wanderer on the right way: *nam qui erranti comiter monstrat viam* [15], etc. I foresee likewise that many of those things which I shall think fit to enter in this registry of mine as omitted and deficient will incur censure on different accounts; some as being already done and extant; others as savouring of curiosity, and promising very scanty fruit; others as being too difficult and almost impossible to be compassed and effected by man. For the two first I refer myself to the particulars themselves. For the last, touching impossibility, I take it that all those things are to be held possible and performable, which may be done by some persons, though not by every one; and which may be done by many together, though not by one alone; and which may be done in the succession of ages, though not in one man's life; and lastly, which may be done by public designation and expense, though not by private means and endeavour. But notwithstanding if any man will take to himself rather the saying of Solomon, "The slothful man says there is a lion in the path", [16] than that of Virgil, *Possunt, quia posse videntur,* [17] "they find it possible because they think it possible," I shall be content that my labours be esteemed but as the better sort of wishes. For as it asks some knowledge of a thing to demand a question not impertinent, so it requires some sense to make a wish not absurd.

[13] Cic. *Ep. ad Att.* ix. 8. One of the earliest tracts on the subject of university reform is doubtless that which Peter Ramus (see his *Scholæ.* Basil. 1569, p. 1063) addressed to Charles the Ninth. It relates chiefly to the expenses arising from fees, etc., to the neglect of the civil law; which had always been coldly regarded at Paris, and to the trifling manner in which the scholastic disputations were conducted.

[14] Seneca, *Proverbiæ.* [Ascribed to Laberius.]

[15] Ennius, ap. Aul. Gell. xii. 4, and ap. Cic. *De Officiis*, i. 17.

[16] Prov. xxvi. 13.

[17] Virg. *Æn.* v. 231.

CHAPTER I.

The Division of all Human Learning into History, Poesy, Philosophy ; *with reference to the three Intellectual Faculties,* —Memory, Imagination, *and* Reason ; *and that the same division holds good likewise in* Theology.

THE best division of human learning is that derived from the three faculties of the rational soul, which is the seat of learning. History has reference to the Memory, poesy to the Imagination, and philosophy to the Reason. And by poesy here I mean nothing else than feigned history or fables ; for verse is but a character of style, and belongs to the arts of speech, whereof I will treat in its proper place.

History is properly concerned with individuals, which are circumscribed by place and time. For though Natural History may seem to deal with species, yet this is only because of the general resemblance which in most cases natural objects of the same species bear to one another ; so that when you know one, you know all. And if individuals are found, which are either unique in their species, like the sun and moon ; or notable deviations from their species, like monsters ; the description of these has as fit a place in Natural History as that of remarkable men has in Civil History. All this relates to the Memory.

Poesy, in the sense in which I have defined the word, is also concerned with individuals ; that is, with individuals invented in imitation of those which are the subject of true history ; yet with this difference, that it commonly exceeds the measure of nature, joining at pleasure things which in nature would never have come together, and introducing things which in nature would never have come to pass ; just as Painting likewise does. This is the work of Imagination.

Philosophy discards individuals ; neither does it deal with the impressions immediately received from them, but with abstract notions derived from these impressions ; in the composition and division whereof according to the law of nature and fact its business lies. And this is the office and work of Reason.

That these things are so, may be easily seen by observing the commencements of the intellectual process. The sense, which is the door of the intellect, is affected by individuals only. The images of those individuals—that is, the impressions which they make on the sense—fix themselves in the memory, and pass into it in the first instance entire as it were, just as they come. These the human mind proceeds to review and ruminate ; and thereupon either simply rehearses them, or makes fanciful imitations of them, or analyses and classifies them. Wherefore from these three fountains, Memory, Imagination, and Reason, flow these three emanations, History, Poesy, and Philosophy ; and there can be no others. For I consider history and experience to be the same thing, as also philosophy and the sciences.

Nor do I think that any other division is wanted for Theology. The information derived from revelation and the information derived from the sense differ no doubt both in the matter and in the manner of conveyance ; but the human mind is the same, and its repositories and cells the same. It is only like different liquids poured through different funnels into one and the same vessel. Theology therefore in like manner consists either of Sacred History, or of Parables, which are a divine poesy, or of Doctrines and Precepts, which are a perennial philosophy. For as for that part which seems supernumerary, which is Prophecy, it is but a kind of history : for divine history has this prerogative over human, that the narration may be before the event, as well as after.

CHAPTER II.

The Division of History into Natural *and* Civil; Ecclesiastical *and* Literary *History being included in Civil. Division of Natural History into* History of Generations, Pretergenerations, *and* Arts.

HISTORY is either Natural or Civil[1]. Natural History treats of the deeds and works of nature; Civil History of those of men. Matter of Divinity shows itself no doubt in both, but principally in the latter; so much so as to form a species of history proper to itself, which I call Sacred or Ecclesiastical. And a similar distinction is in my opinion also due to Learning and the Arts—their importance being such as to entitle them to a separate history of their own. And this (as well as the Ecclesiastical) I mean to be included in Civil History.

The division which I will make of Natural History is founded upon the state and condition of nature herself. For I find nature in three different states, and subject to three different conditions of existence. She is either free, and follows her ordinary course of development; as in the heavens, in the animal and vegetable creation, and in the general array of the universe; or she is driven out of her ordinary course by the perverseness, insolence, and frowardness of matter, and violence of impediments; as in the case of monsters; or lastly, she is put in constraint, moulded, and made as it were new by art and the hand of man; as in things artificial, Let Natural History therefore be divided into the History of Generations, of Pretergenerations, and of Arts; which last I also call Mechanical and Experimental History. Of these the first treats of the Freedom of Nature, the second of her Errors, the third of her Bonds. And I am the more induced to set down the History of the Arts as a species of Natural History, because an opinion has long been prevalent, that art is something different from nature, and things artificial different from things natural; whence this evil has arisen,—that most writers of Natural History think they have done enough when they have given an account of animals or plants or minerals, omitting all mention of the experiments of mechanical arts[2]. But there is likewise another and more subtle error which has crept into the human mind; namely, that of considering art as merely an assistant to nature, having the power indeed to finish what nature has begun, to correct her when lapsing into error, or to set her free when in bondage, but by no means to change, transmute, or fundamentally alter nature. And this has bred a premature despair in human enterprises. Whereas men ought on the contrary to be surely persuaded of this; that the artificial does not differ from the natural in form or essence, but only in the efficient, in that man has no power over nature except that of motion; he can put natural bodies together, and he can separate them; and therefore that wherever the case admits of the uniting or disuniting of natural bodies, by joining (as they say) actives with passives, man can do everything; where the case does not admit this, he can do nothing. Nor matters it, provided things are put in the way to produce an effect, whether it be done by human means or otherwise. Gold is sometimes refined in the fire and sometimes found pure in the sands, nature having done the work for herself. So also the rainbow is made in the sky out of a dripping cloud; it is also made here below with a jet of water. Still therefore it is nature which governs everything; but under nature are included these three; the *course* of nature, the *wanderings* of nature, and *art*, or nature with man to help; which three must therefore all be included in Natural History; as indeed they are in great measure by Pliny, the only person who ever under-

[1] In the *Advancement of Learning*, Bacon had given a quadripartite division of history—natural, civil, ecclesiastical, and literary. The third and fourth he now includes in the second.

[2] The antithesis of nature and art is a celebrated doctrine in the peripatetic philosophy. Natural things are distinguished from artificial, inasmuch as they have, what the latter are without, an intrinsic principle of formation. Aristotle, *De Gen. Anim.* ii. c. 1. The views which Bacon here expresses as to nature and art recur repeatedly in his writings. [See note on p. 259.]

took a Natural History according to the dignity of it [3]; though he was far from carrying out his undertaking in a manner worthy of the conception.

The first of these, the history of nature in course, is extant, and that in moderate perfection; but the two latter are so weakly and unprofitably handled that they may be set down as deficient. For you will find no sufficient and competent collection of those works of nature which have a digression and deflection from the ordinary course of generations, productions, and motions; whether they be singularities of place and region, or the strange events of time, or *casuum ingenia* (as they have been called)—devices of chance, or the effects of hidden properties, or productions of nature singular in their kind. It is true, I find books more than enough filled with fabulous experiments, idle secrets, and frivolous impostures, for pleasure and novelty; but a substantial and methodical collection of the Heteroclites or Irregulars of nature well examined and described I find not; especially not with due rejection and as it were public proscription of fables and popular errors. For as things now are, if an untruth in nature once get a footing and be made common, what by reason of men's reverence for antiquity, what by reason of the troublesomeness of putting it to the test anew, and what by reason of the use of the opinion in similitudes and ornaments of speech [4], it is never overthrown or retracted.

The end of this work, honoured with a precedent in Aristotle [5], is nothing less than to gratify the appetite of curious and vain wits, as the manner of mirabilaries is to do; but for two reasons, both of great weight; the one to correct the partiality of axioms and opinions, which are framed for the most part upon common and familiar examples; the other, because from the wonders of nature is the most clear and open passage to the wonders of art. For you have but to follow and as it were hound nature in her wanderings, and you will be able, when you like, to lead and drive her afterwards to the same place again. Neither am I of opinion, in this history of marvels, that superstitious narratives of sorceries, witchcrafts, charms, dreams, divinations, and the like, where there is an assurance and clear evidence of the fact, should be altogether excluded. For it is not yet known in what cases, and how far, effects attributed to superstition participate of natural causes; and therefore howsoever the use and practice of such arts is to be condemned, yet from the speculation and consideration of them (if they be diligently unravelled) a useful light may be gained, not only for the true judgment of the offences of persons charged with such practices, but likewise for the further disclosing of the secrets of nature. Neither ought a man to make scruple of entering and penetrating into these holes and corners, when the inquisition of truth is his sole object,—as your Majesty has shown in your own example; who, with the two clear and acute eyes of religion and natural philosophy, have looked deeply and wisely into those shadows, and yet proved yourself to be truly of the nature of the sun, which passes through pollutions and is not defiled [6]. I would recommend however that those narrations which are tinctured with superstition be sorted by themselves, and not mingled with those which are purely and sincerely natural. But as for narrations touching the prodigies and miracles of religions, they are either not true or not natural; and therefore impertinent for the story of nature.

For History of Nature Wrought, or Mechanical, as I also call it, I find some collections made of agriculture and likewise of many manual arts; but always

[3] Of Pliny's *Natural History* Humboldt has remarked that it is a book "which in richness of content no other work of antiquity comes near" —*Kosmos*, ii. 23. Sir T. Browne observes that there is scarcely any vulgar error which is not to be found in it.

[4] In Gilbert's work *De Magnete*, ii. 2, we find an amusing complaint of the same kind. It is worthy of remark that in the account Gilbert has given of the magnetical speculations of earlier writers, almost the only person of whose opinion he speaks with respect is S. Thomas Aquinas, among whose opuscula will be found one on the magnet.

[5] It is generally admitted that the *De Miris Auscultationibus* is not Aristotle's.

[6] The allusion is to King James's *Dæmonologie*, a work in three books, consisting of dialogues between Philomathes and Epistemon; the latter of whom represents the king's opinions on witchcraft.

(which is a great detriment in this kind of learning) with a neglect and rejection of experiments familiar and vulgar; which yet in the interpretation of nature are of equal, if not of more value than those which are less common. For it is esteemed a kind of dishonour upon learning for learned men to descend to inquiry or meditation upon matters mechanical; except they be such as may be thought secrets of art, or rarities and special subtleties. Which humour of vain and supercilious arrogancy is justly derided in Plato, where he brings in Hippias, a vaunting Sophist, disputing with Socrates, a true and unfeigned inquisitor of truth; where, the discourse being touching beauty, Socrates, after his loose and wandering manner of inductions, put first an example of a fair virgin, then of a fair horse, then of a fair pot well glazed. Whereat Hippias was offended, and said, "Were it not for courtesy's sake, I should be loth to dispute with one that did allege such base and sordid instances." Whereunto Socrates answered, "You have reason, and it becomes you well, being a man so trim in your vestments, and so fairly shod"; and so goes on in irony[7]. But the truth is, that they are not the highest instances which give the best or securest information; as is expressed not inelegantly in the common story of the philosopher[8] who, while he gazed upwards to the stars, fell into the water; for if he had looked down he might have seen the stars in the water, but looking aloft he could not see the water in the stars. So it often comes to pass that mean and small things discover great better than great can discover small, and therefore it was well observed by Aristotle, "that the nature of everything is best seen in its smallest portions". For which cause he inquires the nature of a commonwealth first in a family and the simplest conjugations of society—(man and wife, parent and child, master and servant)—which are present in every cottage[9]. Even so likewise the nature of this great city of the world, and the policy thereof, must be first sought in its primary concordances, and smallest portions; as we see that that secret of nature (esteemed one of the great mysteries) of the turning of iron touched with the loadstone towards the north, was found out not in bars of iron but in needles.

But if my judgment be of any weight, the use of History Mechanical is, of all others, the most radical and fundamental towards natural philosophy[10]; such natural philosophy I mean as shall not vanish in the fumes of subtle or sublime speculations, but such as shall be operative to relieve the inconveniences of man's estate. For it will not only be of immediate benefit, by connecting and transferring the observations of one art to the use of others, and thereby discovering new commodities; a result which must needs follow when the experience of different arts shall fall under the observation and consideration of one man's mind; but further, it will give a more true and real illumination concerning the investigation of causes of things and axioms of arts, than has hitherto shone upon mankind. For like as a man's disposition is never well known or proved till he be crossed, nor Proteus ever changed shapes till he was straitened and held fast; so nature exhibits herself more clearly under the trials and vexations of art than when left to herself.

Before I dismiss this part of Natural History (which I call mechanical and experimental) I must add that the body of this kind of history should not be made up from the mechanical arts alone, but also from the operative part of the

[7] See the *Hippias major*. The remark, however, which Hippias makes does not refer to what Socrates has said in his own character, but to what he supposes an imaginary interlocutor to say.

[8] Thales.

[9] *Politica*, i. 1. sub finem.

[10] Accordingly this was one of the first things which the Philosophical College which afterwards became the Royal Society attempted to accomplish. Oldenburg writes to Spinoza in September 1661:—"In our philosophical society we proceed diligently as far as opportunity offers with our experiments and observations, lingering over the compilation of the history of mechanic arts, with the idea that the forms and qualities of things can best be explained from mechanical principles, and that all natural effects can be produced through motion, shape, and consistency, without reference to inexplicable forms or occult qualities, which are but the refuge of ignorance."

liberal sciences, as well as from many other practices which have not as yet grown up into arts; so as to omit nothing which may tend to inform the intellect. And this is the first division of Natural History.

CHAPTER III.

The Second Division of Natural History, according to its Use and End, into Narrative *and* Inductive; *and that the noblest end of Natural History is to minister and be in order for the Foundation of Philosophy; which is the end aimed at in* Induction. *The Division of the History of Generations into the History of the* Heavenly Bodies, *the History of* Meteors, *the History of the* Globe of Earth and Sea, *the History of the* Masses or Greater Colleges, *and the History of the* Species or Lesser Colleges[1].

NATURAL HISTORY, which is threefold (as I said) in subject, is in use twofold. For it is used either for the sake of the knowledge of the things themselves that are committed to the history, or as the primary matter of philosophy. Now the first kind, which aims either to please by the agreeableness of the narrative or to help by the use of experiments, and is pursued for the sake of such pleasure or such profit, I account as far inferior in importance to that which is the stuff and material of a solid and lawful Induction, and may be called the nursing-mother of philosophy. Accordingly I shall now make a second division of Natural History into Narrative and Inductive; the latter whereof I sèt down as wanting. But let not any one be dazzled either by the great names of ancient philosophers or the great volumes of modern. For I well know that a natural history is extant, large in its bulk, pleasing in its variety, curious often in its diligence; but yet weed it of fables, antiquities, quotations, idle controversies, philology and ornaments (which are more fitted for table talk and the *noctes* of learned men than for the instauration of philosophy), and it will shrink into a small compass. Certainly it is very different from that kind of history which I have in view. For in the first place there are wanting those two parts of natural history which I have just mentioned, Pretergenerations and Arts, of which I make great account; and next, in the third and remaining part, that of Generations, only one out of five parts is sufficiently handled. For the history of Generations is composed of five subordinate parts. First, a history of the *Celestial Bodies*, exhibiting the actual phenomena simply and apart from theories. Second, a history of *Meteors* (including comets), and what they call the *Regions of the Air*; for there is no history of comets, fiery meteors, winds, rains, storms, and the like, which is of any value. Third, a history of the *Earth and Sea* (considered as integral parts of the universe), mountains, rivers, tides, sands, woods, islands, and the shapes of continents as they lie; in all these, inquiring and observing rather the laws of nature than cosmography. Fourth, a history of the *Common Masses of Matter*, which I call the *Greater Colleges* (commonly called the *Elements*); for I find there are no accounts of fire, air, earth, and water, with their natures, motions, operations, and impressions, such as to form a just body of history. Fifth and last, a history of the *Exquisite Collections of Matter*, which I call the *Lesser Colleges*, but which are generally called *Species*. Now it is only in this last that writers have shown any conspicuous industry; and yet in such sort that they have rather filled it with things superfluous (as figures of animals, plants, and the like), than enriched it with sound and careful observations, which should ever be annexed to natural history. And in a word all the natural history we have, whether in the mode of inquiry or in the matter collected, is quite unfit for the end which I have mentioned, namely, the Foundation of Philosophy. Wherefore I set down Inductive History as wanting. And so much for Natural History.

[1] This chapter is an addition to the *Advancement of Learning*.

CHAPTER IV.

The Division of Civil History into Ecclesiastical, Literary, *and* Civil (*which retains the name of the Genus*) *and that the History of* Literature *is wanting. Precepts for the Construction of it.*

CIVIL HISTORY may rightly be divided into three species. First, *Sacred* or *Ecclesiastical ;* next, that which we call *Civil History* (using the generic name specially) ; lastly, the History of *Learning* and *the Arts.* I will begin with the kind last-mentioned ; for the two former are extant, while the latter—the History of Learning—(without which the history of the world seems to me as the statue of Polyphemus without the eye ; that very feature being left out which most marks the spirit and life of the person), I set down as wanting. Not but I know that in the particular sciences of the jurisconsults, mathematicians, rhetoricians, philosophers, we have some slight mention or some barren narrations about the sects, schools, books, authors, and successions belonging to them ; also that there exist some meagre and unprofitable memoirs of the inventors of arts and usages ; but I say that a complete and universal History of Learning is yet wanting. Of this therefore I will now proceed to set forth the argument, the method of construction, and the use.

The *argument* is no other than to inquire and collect out of the records of all time what particular kinds of learning and arts have flourished in what ages and regions of the world ; their antiquities, their progresses, their migrations (for sciences migrate like nations) over the different parts of the globe ; and again their decays, disappearances, and revivals. The occasion and origin of the invention of each art should likewise be observed ; the manner and system of transmission, and the plan and order of study and practice. To these should be added a history of the sects, and the principal controversies in which learned men have been engaged, the calumnies to which they have been exposed, the praises and honours by which they have been rewarded ; an account of the principal authors, books, schools, successions, academies, societies, colleges, orders,—in a word, everything which relates to the state of learning. Above all things (for this is the ornament and life of Civil History), I wish events to be coupled with their causes. I mean, that an account should be given of the characters of the several regions and peoples ; their natural disposition, whether apt and suited for the study of learning, or unfitted and indifferent to it ; the accidents of the times, whether adverse or propitious to science ; the emulations and infusions of different religions ; the enmity or partiality of laws ; the eminent virtues and services of individual persons in the promotion of learning, and the like. Now all this I would have handled in a historical way, not wasting time, after the manner of critics, in praise and blame, but simply narrating the fact historically, with but slight intermixture of private judgment.

For the *manner* of compiling such a history I particularly advise that the matter and provision of it be not drawn from histories and commentaries alone ; but that the principal books written in each century,or perhaps in shorter periods, proceeding in regular order from the earliest ages, be themselves taken into consultation ; that so (I do not say by a complete perusal, for that would be an endless labour, but) by tasting them here and there, and observing their argument, style, and method, the Literary Spirit of each age may be charmed as it were from the dead.

With regard to the *use* of the work, it is not so much to swell the honour and pomp of learning with a profusion of images ; nor because out of my exceeding love for learning I wish the inquiry, knowledge, and preservation of everything that relates thereto to be pursued even to curiosity ; but chiefly for a purpose more serious and important ; which, in a word, is this : I consider that such a history as I have described would very greatly assist the wisdom and skill of learned men in the use and administration of learning ; that it would exhibit the movements and perturbations, the virtues and vices, which take place no less in intellectual than in civil matters ; and that from the observation of these the best system of government might be derived and established. For the works of St. Ambrose or St. Augustine will not make so wise a bishop or divine as a diligent examination and study of Ecclesiastical History; and the History of Learning would be of like service to learned men. For everything is subject to chance and error which is not supported by examples and experience. And so much for the History of Learning.

CHAPTER V.

On the Dignity and Difficulty of Civil History.[1]

I COME next to *Civil History*, properly so called, whereof the dignity and authority are pre-eminent among human writings. For to its fidelity are entrusted the examples of our ancestors, the vicissitudes of things, the foundations of civil policy, and the name and reputation of men. But the difficulty is no less than the dignity. For to carry the mind in writing back into the past, and bring it into sympathy with antiquity; diligently to examine, freely and faithfully to report, and by the light of words to place as it were before the eyes, the revolutions of times, the characters of persons, the fluctuations of counsels, the courses and currents of actions, the bottoms of pretences, and the secrets of governments; is a task of great labour and judgment—the rather because in ancient transactions the truth is difficult to ascertain, and in modern it is dangerous to tell. Hence Civil History is beset on all sides with faults; some (and these are the greater part) write only barren and commonplace narratives, a very reproach to history; others hastily and disorderly string together a few particular relations and trifling memoirs; others merely run over the heads of events: others, on the contrary, go into all the minutest particularities, and such as have no relation to the main action; some indulge their imaginations in bold inventions; while others impress on their works the image not so much of their minds as of their passions, ever thinking of their party, but no good witness as to facts; some are always inculcating their favourite political doctrines, and idly interrupting the narrative by going out of the way to display them; others are injudiciously prolix in reporting orations and harangues, and even in relating the actions themselves; so that, among all the writings of men, there is nothing rarer than a true and perfect Civil History. But my present purpose in this division of learning is to mark omissions, and not to censure faults. I will now pursue the divisions of Civil History, and those of the different kinds; for the species will be exhibited more clearly under several heads than under one head curiously traced through all its members.

CHAPTER VI.

The First Division of Civil History into Memorials, Antiquities, *and* Perfect History.

CIVIL HISTORY is of three kinds, not unfitly to be compared with the three kinds of pictures or images. For of pictures and images we see some are unfinished, and wanting the last touch; some are perfect; and some are mutilated and defaced by age. So Civil History (which is a kind of image of events and times) may be divided into three kinds, corresponding to these,—*Memorials, Perfect History*, and *Antiquities*. For Memorials are history unfinished, or the first rough drafts of history; and Antiquities are history defaced, or remnants of history which have casually escaped the shipwreck of time.

Memorials, or Preparatory History, are of two sorts, whereof the one may be termed *Commentaries*, the other *Registers*. Commentaries set down a bare continuance and tissue of actions and events without the causes and pretexts, the commencements and occasions, the counsels and orations, and other passages of action. For this is the true nature of a commentary, though Cæsar, in modesty mixed with greatness, chose to apply the name of a commentary to the best history extant. But Registers have a twofold character; for they either contain titles of things and persons in order of time, such as are called Annals and

[1] There is nothing corresponding to this chapter in the *Advancement of Learning*.—*J. S.*

Chronologies; or collections of public acts, such as edicts of princes, decrees of councils, judicial proceedings, public speeches, letters of state, and the like, without a perfect continuance or contexture of the thread of the narration.

Antiquities, or remnants of histories, are (as was said) like the spars of a shipwreck; when, though the memory of things be decayed and almost lost, yet acute and industrious persons, by a certain perseverance and scrupulous diligence, contrive out of genealogies, annals, titles, monuments, coins, proper names and styles, etymologies of words, proverbs, traditions, archives and instruments as well public as private, fragments of histories scattered about in books not historical,—contrive, I say, from all these things or some of them, to recover somewhat from the deluge of time; a work laborious indeed, but agreeable to men, and joined with a kind of reverence; and well worthy to supersede the fabulous accounts of the origins of nations, and to be substituted for fictions of that kind; entitled however to the less authority, because in things which few people concern themselves about, the few have it their own way.

In these kinds of Imperfect History I think no deficiency is to be assigned; for they are things, as it were, imperfectly compounded, and therefore any deficiency in them is but their nature. As for epitomes (which are certainly the corruptions and moths of histories) I would have them banished, whereto likewise most men of sound judgment agree, as being things that have fretted and corroded the bodies of many most excellent histories, and wrought them into base and unprofitable dregs.[1]

CHAPTER VII.

The Division of Perfect History into Chronicles, Lives, *and* Relations; *and the Explanation thereof.*

But *Perfect History* is of three kinds, according to the object which it propounds for representation. For it either represents a portion of time, or a person worthy of mention, or an action or exploit of the nobler sort. The first we call Chronicles or Annals; the second, Lives; the third, Narrations or Relations. Of these the first excels in estimation and glory; the second, in profit and examples; and the third in verity and sincerity. For History of Times represents the magnitude of public actions, and the public faces and deportments of persons, but omits and covers up in silence the smaller passages and motions of men and matters. But such being the workmanship of God, that he hangs the greatest weights upon the smallest wires, it comes commonly to pass that such a history, pursuing the greater things alone, rather sets forth the pomp and solemnity of business than the true and inward springs and resorts thereof. Moreover, when it does add and insert the counsels and motives, yet from its love of grandeur it introduces into human actions more gravity and prudence than they really have; so that a truer picture of human life may be found in a satire than in some histories of this kind. Whereas Lives, if they be well and carefully written (for I do not speak of elegies and barren commemorations of that sort), propounding to themselves a single person as their subject, in whom actions both trifling and important, great and small, public and private, must needs be united and mingled, certainly contain a more lively and faithful representation of things, and one which you may more safely and happily take for example in another case. But special Narrations and Relations of actions (as the Peloponnesian War, the Expedition of Cyrus, the Conspiracy of Catiline, and the like) cannot but be more purely and exactly true than the Perfect Histories of Times; because they may choose a manageable and definite argument, whereof a perfect knowledge and certainty and full information may be had; whereas the story of a time (especially if it be of a period much before the age of the writer) is sure to meet with many gaps in the records, and to contain empty spaces which must be filled up and supplied at pleasure by wit and conjecture. But this which I say touching the sincerity

[1] Bacon often condemns, and not altogether unjustly, the use of epitomes. The development of a liking for abridgments is certainly a remarkable feature in the decline of Roman literature.

of Relations, must be taken with reservation ; for (seeing that everything human is subject to imperfection, and good is almost always associated with evil) it must certainly be confessed that relations of this kind, especially if published near the time of the actions themselves (being commonly written either in favour or in spite), are of all other histories the most to be suspected. But then again the evil carries this remedy along with it ; that as these very relations are commonly put forth not by one side only, but by both, according to their several factions and parties, a way may be found to truth between the extremes on either hand ; and after party heat has cooled down, a good and prudent historian will obtain from them no bad materials and provision for a more perfect history.

With regard to the deficiencies of these three kinds of history, it is certain that there are many particular histories (I speak of such as may be of some moderate worth and dignity) which have been hitherto neglected, with the greatest detriment to the honour and name of the kings and states to which they belong ; though to mention them would take too much time. But leaving the care of foreign stories to foreign states (for I will not be a meddler in other nations' matters), I cannot fail to represent to your Majesty the unworthiness of the history of England as we now have it, in the main continuance thereof, and the partiality and obliquity of that of Scotland, in the latest and largest author that I have seen [1] ; supposing that it would be honour for your Majesty, and a work very acceptable to future ages, if this island of Great Britain, as it is now joined in one monarchy for the ages to come, so were joined in one history for the ages past ; after the manner of the Sacred History, which draws down the story of the ten tribes and of the two tribes as twins together. And if it shall seem that the greatness of this work (and great and difficult it is) may prevent it from being exactly and worthily performed, there is a memorable period of a much smaller compass of time, as to the history of England ; that is to say, from the Union of the Roses to the Union of the Kingdoms ; a portion of time wherein to my understanding there has been a greater variety of strange events than in like number of successions of any hereditary monarchy has ever been known. For it begins with the mixed obtaining of a crown, partly by arms, partly by title ; an entry by battle, an establishment by marriage ; and therefore times corresponding to these beginnings, like waters after a tempest, full of working and swelling, though without extremity of storm ; but well passed through by the wisdom of the pilot [2], who was the most conspicuous for policy of all the kings who preceded him. Then follows the reign of a king whose actions, though conducted more by impulse than policy, exercised no slight influence over the affairs of Europe ; balancing and inclining them variably. In whose reign also begun that great alteration in the State Ecclesiastical, an action which seldom comes upon the stage. Then the reign of a minor. Then an attempt at a usurpation, though it was but as a diary ague. Then the reign of a queen matched with a foreigner ; then of a queen that lived solitary and unmarried. And now, last, this most happy and glorious event, that this island of Britain, divided from all the world, should be united in itself, and that old oracle given to Æneas (Antiquam exquirite matrem [3]), which foreshowed the rest in store for him, should now be performed and fulfilled upon the most renowned nations of England and Scotland ; being now reunited in the ancient mother name of Britain, as a pledge and token of the end and period of all instability and peregrinations ; so that as it comes to pass in massive bodies, that they have certain trepidations and waverings before they fix and settle ; so it seems to have been ordained by the providence of God that this monarchy, before it settled and was confirmed in your Majesty and your royal generations (in which I hope it is now established for ever), should undergo these prelusive changes and varieties.

For Lives, I find it strange, when I think of it, that these our times have so

[1] Bacon alludes to Buchanan, of whom James speaks with much bitterness in the *Basilicon Doron*. It has been said that Buchanan's mind was failing when he wrote the concluding books of his history, in which Mary Queen of Scots is so much vilified.

[2] [Henry VII.] [3] Seek out your ancient mother. Virg. *Æn.* iii. 96.

little esteemed their own virtues, as that the commemoration and writings of the lives of those who have adorned our age should be no more frequent. For although there be but few sovereign kings or absolute commanders, and not many princes in free states (so many free states being now turned into monarchies), yet are there many worthy personages (even living under kings) that deserve better than dispersed report or dry and barren eulogy. For herein the invention of one of the later poets, by which he has enriched the ancient fiction, is not inelegant. He feigns that at the end of the thread or web of every man's life there hangs a little medal or collar, on which his name is stamped; and that Time waits upon the shears of Atropos, and as soon as the thread is cut, snatches the medals, carries them off, and presently throws them into the river Lethe; and about the river there are many birds flying up and down, who catch the medals, and after carrying them round and round in their beak a little while, let them fall into the river; only there are a few swans, which if they get a medal with a name immediately carry it off to a temple consecrated to immortality[4]. Now this kind of swan is for the most part wanting in our age. And although there are many men, more mortal in their cares and desires than in their bodies, who regard the desire of name and memory but as a vanity and ventosity,

Animi nil magnæ laudis egentes; [5]

whose philosophy and severity springs no doubt from that root "Non prius laudes contempsimus, quam laudanda facere desivimus[6]"—yet that will not alter Solomon's judgment, "The memory of the just is praised, but the name of the wicked shall rot[7]". The one flourishes for ever; the other either consumes to present oblivion, or turns to an ill odour. And therefore in that style or form of words which is well appropriated to the dead—(of happy memory, of pious memory, of blessed memory),—we seem to acknowledge that which Cicero says (having borrowed it from Demosthenes), "That good fame is the only possession a dead man has[8]"; which possession I cannot but note that in our times it lies in most part waste and neglected.

For *Narrations* and *Relations*, a greater diligence therein is also much to be wished; for there is hardly any great action which is not attended by some good pen that can describe it. And because it is an ability not common to write a perfect history as it ought to be written (as may well appear from the small number even of moderate historians), yet if particular actions were but tolerably reported as they pass, it might be expected that a writer would some time or other arise who by such help and assistance might compile a complete History of Times. For the collection of such Relations would be as a nursery, whereby to plant a fair and stately garden when time should serve.

CHAPTER VIII[1].

The Division of the History of Times into History Universal and Particular—*their Advantages and Disadvantages.*

History of Times is either Universal or Particular; whereof the latter contains the deeds of some kingdom, commonwealth, or people; the former those of the

[4] The poet referred to is Ariosto, *Orlando Furioso*, at the end of the 34th and beginning of the 35th Books. For this reference I am indebted to Mr. Singer, *Notes and Queries*, vol. v. p. 232. He remarks that the *Orlando Furioso* was then popular in the recent translation of Sir John Harrington. It would seem as if Bacon refers to the translation, which ascribes the power of giving immortality to "*Historians* learned and Poets rare," whereas the original speaks only of poets.

[5] "Souls that care not for praise."—Virg. *Æn.* v. 751.

[6] "When we have ceased to do things deserving of praise we find that praise is an idle thing."—*Plin. Ep.* iii. 91.

[7] Prov. x. 7.

[8] Cf. Cicero, *Philipp.* ix., and Demos. λογ. επιταφ. 1389, 10.

[1] There is nothing corresponding to this chapter in the *Advancement of Learning.*—*J. S.*

whole world. For there have been those who have affected to write the history of the world from its very beginning; exhibiting by way of history a medley of things and abridgments of narratives. Others have attempted to comprise, as in a perfect history, the memorable events of their own age all over the world; with noble enterprise, and no small result. For the affairs of men are not so far separated by the divisions of empires or countries, but they have a connexion in many things; and therefore it is certainly of use to have the fates, acts, and destinies of one age described and contained as it were on one tablet. It is true also that many writings of no contemptible character (such as are those Relations of which I previously spoke), which would otherwise perish and not be reprinted,—that these, or at all events the principal matters in them, find a place in a general history of this kind, and in this way are fixed and preserved. But if due attention be paid to the subject, it will be found that the laws of regular history are so strict, that they can scarce be observed in such a wide field of matter; so that the dignity of history is rather diminished than increased by the greatness of the mass of it. For the writer who has such a variety of things on all sides to attend to, will become gradually less scrupulous on the point of information; his diligence, grasping at so many subjects, will slacken in each; he will take up with rumours and popular reports, and thus construct his history from relations which are not authentic, or other frivolous materials of the kind. He will be obliged moreover (lest the work increase beyond measure) purposely to omit a number of things worthy of record, and often to sink into abridgments. He is liable likewise to another danger, not small, and diametrically opposed to the very utility which belongs to Universal History; for as Universal History preserves some narrations which would perhaps otherwise perish, so on the other hand it destroys many that are profitable enough in themselves and would otherwise live, for the sake of that compendious brevity of which men are so fond.

CHAPTER IX.

Another Division of the History of Times into Annals *and* Journals.

THE History of Times is also rightly divided into *Annals* and *Journals;* which division, though it take its name from periods of time, yet has also reference to the choice of subjects. For it is well observed by Cornelius Tacitus, after touching upon the magnificence of certain buildings, " That it was found suitable to the dignity of the Roman people to commit to Annals only matters of note, but such things as these to the Journals of the City " [1]; thus referring matters concerning the state to Annals, but the less important kind of actions or accidents to Journals. Certainly, in my judgment, there ought to be a kind of heraldry in arranging the precedence of books, no less than of persons. For as nothing derogates from the dignity of a state more than confusion of ranks and degrees, so it not a little embases the authority of a history to intermingle matters of lighter moment, such as triumphs, ceremonies, spectacles, and the like, with matters of state. And surely it were to be wished that this distinction came into fashion. But in our times journals are only used in sea-voyages and expeditions of war; whereas in ancient times it was a matter of honour with princes to keep journals of what passed day by day in their courts; as we see in the case of Ahasuerus, King of Persia, who, when he could not take rest, called for the Chronicles, where he read over again the account of the conspiracy of the Eunuchs [2]. But the journals of Alexander's house expressed every small particularity, so that even if he happened to sleep at table it was registered [3]. Not that, as none but grave matters were included in the Annals, so none but trifling ones were admitted into Journals; but everything, whether of greater or less concern, was promiscuously entered in the Journals as it passed.

[1] Tac. Ann. xiii. 31. [2] Esther, vi. 1. [3] Plut. Symp. i. 6.

CHAPTER X.

The Second Division of Civil History into Pure *and* Mixed.

THE last division of Civil History is into *Pure* and *Mixed.* Of the Mixed there are two principal kinds ; the one taken from Civil Science, the other principally from Natural. For some men have introduced a form of writing consisting of certain narratives not woven into a continuous history, but separate and selected according to the pleasure of the author ; which he afterwards reviews, and as it were ruminates over, and takes occasion from them to make politic discourse and observation [1]. Now this kind of Ruminated History I greatly approve, provided that the writer keep to it and profess it. But for a man who is professedly writing a Perfect History to be everywhere introducing political reflexions, and thereby interrupting the narrative, is unseasonable and wearisome. For though every wise history is pregnant (as it were) with political precepts and warnings, yet the writer himself should not play the midwife.

Another kind of Mixed History is the History of Cosmography ; which is indeed mixed of many things ; of Natural History, in respect of the regions themselves, their sites and products ; of History Civil, in respect of the habitations, governments, and manners of the people ; and of Mathematics, in respect of the climates and configurations of the heavens, beneath which the regions of the world lie. In which kind of history or science we may congratulate our own age. For this great building of the world has in our age been wonderfully opened and thorough-lighted ; and though the ancients had knowledge of the zones and the antipodes,

> Nosque ubi primus equis oriens afflavit anhelis,
> Illic sera rubens accendit lumina Vesper [2],

yet that might be by demonstration rather than by travel. But for a little vessel to emulate the heaven itself, and to circle the whole earth with a course even more oblique and winding than that of the heavenly bodies, is the privilege of our age ; so that these times may justly bear in their motto not only *plus ultra* [3]—further yet—in precedence of the ancient *non ultra*—no further ; and " Imitable Thunder " in precedence of the ancient " Inimitable Thunder ",

> (Demens qui nimbos, et non imitabile fulmen, etc.) [4],

but likewise, that which exceeds all admiration, " Imitable Heaven," in respect of our sea-voyages, by which the whole globe of earth has, after the manner of the heavenly bodies, been many times compassed and circumnavigated.

And this proficience in navigation and discovery may plant also great expectation of the further proficience and augmentation of the sciences ; especially as it may seem that these two are ordained by God to be coevals, that is, to meet in one age. For so the Prophet Daniel, in speaking of the latter times, foretells " That many shall go to and fro on the earth, and knowledge shall be increased " [5], as if the opening and thorough passage of the world, and the increase of knowledge, were appointed to be in the same age ; as we see it is already performed in great part ; the learning of these our times, not much giving place to the two former periods or returns of learning (the one of the Grecians, the other of the Romans), but in some respects far exceeding them.

1 The most celebrated work of this kind is one with which Bacon was familiar,—the *Discorsi* of Machiavelli, of which the narrative part is derived from Livy. Ammirati, who died in 1600, took Tacitus as his author. His *Discorsi* never attained the celebrity of those of Machiavelli.

2 And while on us the early morning breathes
With panting horses, there the blushing eve
Lights up her tardy signals.—Virgil, *Georg.* i. 250.

3 " *Plus ultra,*" which Bacon often quotes, was the motto adopted by the emperor Charles V.

4 Virg. *Æn.* vi. 590.

5 Daniel, xii. 4.

CHAPTER XI.

The Division of Ecclesiastical History into Ecclesiastical History Special, History of Prophecy, *and* History of Providence.

History Ecclesiastical receives nearly the same divisions as History Civil, for there are Ecclesiastical Chronicles, there are Lives of the Fathers, there are Relations of Synods and other things pertaining to the Church. But in itself it is properly divided into *History Ecclesiastical* (using the general name in a special sense), *History of Prophecy*, and *History of Divine Judgments or Providence.* The first describes the times of the Church Militant, and its different states; whether fluctuant, as the ark of Noah; or moveable, as the ark in the wilderness; or at rest, as the ark in the Temple; that is, the state of the Church in persecution, in remove, and in peace. In this part I find no deficiency, but rather superfluities; only I would that the virtue and sincerity of the relations were in accordance with the mass and quantity of the matter.

The second, which is History of Prophecy, consists of two relatives, the Prophecy and the Accomplishment; and therefore the plan of such a work ought to be, that every prophecy of Scripture be sorted with the event fulfilling the same, throughout all ages of the world; both for the better confirmation of faith, and for better instruction and skill in the interpretation of those parts of prophecies which are yet unfulfilled; allowing nevertheless that latitude which is agreeable and familiar to divine prophecies, that the fulfilments of them are taking place continually and not at the particular time only. For they are of the nature of their Author, "to whom a thousand years are but as one day, and one day as a thousand years"[1]; and though the height or fulness of them is commonly referred to some one age or particular period, yet they have at the same time certain gradations and processes of accomplishment through divers ages of the world. This is a work which I find deficient, but it is one that is to be done with great wisdom, sobriety, and reverence, or not at all.

The third part, which is History of Providence, has indeed been handled by the pens of some pious writers, but not without partiality. Its business is to observe that divine correspondence which sometimes exists between God's revealed and secret will. For though the judgments and counsels of God are so obscure that to the natural man they are altogether inscrutable, yea, and many times hidden from the eyes of those that behold them from the tabernacle, yet at some time it pleases the Divine Wisdom, for the better establishment of His people, and the confusion of those who are as without God in the world, to write it and report it to view in such capital letters that (as the Prophet saith) "He that runneth by may read it"[2]; that is, that mere sensual persons and voluptuaries, who hasten by God's judgments, and never bend or fix their thoughts upon them, are nevertheless, though running fast and busy about other things, forced to discern them. Such are late and unlooked for judgments; deliverances suddenly and unexpectedly vouchsafed; divine counsels, through tortuous labyrinths and by vast circuits, at length manifestly accomplishing themselves; and the like; all which things serve not only to console the minds of the faithful, but to strike and convince the consciences of the wicked.

[1] Psalm xc. 4., and 2 Pet. iii. 8.

[2] Habakkuk, ii. 2. Bacon seems to have misunderstood the meaning of the passage, the English translation of which is quite in accordance both with the Vulgate and with the Septuagint version. The meaning may be thus paraphrased: "Write so as that the message may be quickly read, in order that the reader may run at once and without loss of time". The idea of quick reading seems to have suggested that of a hasty and careless reader.

In my copy of Acosta's sermons for Advent, which has Bacon's autograph on the fly-leaf, and for which I am indebted to the kindness of the Rev. P. La Trobe, the following words are underlined: "Sed explanari in tabulis visio prophetica jubetur, ut possit, celeriter a legente percipi."—*Acostæ Conciones de Adventu,* (Col. Agrip. 1609) p. 178. Bacon perhaps connected *celeriter* with *legente* instead of with *percipi,* and was thus led to suppose that the passage was to be understood in the way in which he has taken it.

CHAPTER XII.

Of the Appendices to History ; which deal with the Words of Men (as History itself deals with their Actions). The Division thereof into Orations, Letters, *and* Apophthegms.

BUT not only man's *actions*, but his *words* also should be recorded. And these are no doubt sometimes inserted in history itself, so far as they contribute to the perspicuity and weight of the narrative. But the saying or words of men are properly preserved in books of *Speeches*, *Letters*, and *Apophthegms*. Certainly the Speeches of wise men on business and matters of grave and deep importance conduce greatly as well to the knowledge of the things themselves as to eloquence. But for instruction in civil prudence, still greater help is derived from Letters written by great men on weighty subjects. For of all the words of man nothing is more solid and excellent than letters of this kind ; for they are more natural than orations, and more advised than conferences on the sudden. And when there is a continued series of them in order of time (as we find in the letters of ambassadors, governors of provinces, and other ministers of state, to kings, senates, and other superior officers ; or, again, in the letters of rulers to their agents), they are of all others the most valuable materials for history. Neither are Apophthegms themselves only for pleasure and ornament, but also for use and action. For they are (as was said) " words which are as goads ",[1] words with an edge or point, that cut and penetrate the knots of business and affairs. Now occasions are continually returning, and what served once will serve again ; whether produced as a man's own or cited as an old saying. Nor can there be any question of the utility in civil matters of that which Cæsar himself thought worthy of his labour ; whose book of Apophthegms I wish were extant ; for all the collections which we have of this kind appear to me to have been compiled without much judgment.

And so much concerning History ; which is that part of learning which answers to one of the cells, domiciles, or offices of the mind of man, which is that of the Memory.

CHAPTER XIII.

On the second principal part of Learning, namely, Poesy. The Division of Poesy into Narrative, Dramatic, *and* Parabolical. *Three Examples of Parabolical Poesy are propounded.*

I NOW come to Poesy, which is a part of learning in measure of words for the most part restrained, but in all other points extremely free and licensed ; and therefore (as I said at first) it is referred to the imagination, which may at pleasure make unlawful matches and divorces of things. Now Poesy, as I have already observed, is taken in two senses ; in respect of words or matter. In the first sense it is but a character of speech ; for verse is only a kind of style and a certain form of elocution, and has nothing to do with the matter ; for both true history may be written in verse and feigned history in prose. But in the latter sense, I have set it down from the first as one of the principal branches of learning, and placed it by the side of history ; being indeed nothing else but an imitation of history at pleasure. And therefore, endeavouring as I do in these divisions to trace out and pursue the true veins of learning, without (in many points) following customs and the divisions which are received, I dismiss from the present discourse Satires, Elegies, Epigrams, Odes, and the like ; and refer them to philosophy and arts of speech. And under the name of Poesy, I treat only of feigned history.

The division of Poesy which is aptest and most according to the propriety thereof, besides those divisions which it has in common with History (for there are feigned Chronicles, feigned Lives, and feigned Relations), is into Poesy *Narrative*, *Dramatic*, and *Parabolical*. Narrative Poesy is a mere imitation of History, such as might pass for real, only that it commonly exaggerates things beyond probability. Dramatic Poesy is as History made visible ; for it represents ac-

[1] [Orig. *Secures aut mucrones verborum.* Cicero, *Epist. Fam.* ix.]

tions as if they were present, whereas History represents them as past. Parabolical Poesy is typical History, by which ideas that are objects of the intellect are represented in forms that are objects of the sense.

As for Narrative Poesy,—or Heroical, if you like so to call it (understanding it of the matter, not of the verse)—the foundation of it is truly noble, and has a special relation to the dignity of human nature. For as the sensible world is inferior in dignity to the rational soul, Poesy seems to bestow upon human nature those things which history denies to it ; and to satisfy the mind with the shadows of things when the substance cannot be obtained. For if the matter be attentively considered, a sound argument may be drawn from Poesy, to show that there is agreeable to the spirit of man a more ample greatness, a more perfect order, and a more beautiful variety than it can anywhere (since the Fall) find in nature. And therefore, since the acts and events which are the subjects of real history are not of sufficient grandeur to satisfy the human mind, Poesy is at hand to feign acts more heroical ; since the successes and issues of actions as related in true history are far from being agreeable to the merits of virtue and vice, Poesy corrects it, exhibiting events and fortunes as according to merit and the law of providence ; since true history wearies the mind with satiety of ordinary events, one like another, Poesy refreshes it, by reciting things unexpected and various and full of vicissitudes. So that this Poesy conduces not only to delight but also to magnanimity and morality. Whence it may be fairly thought to partake somewhat of a divine nature ; because it raises the mind and carries it aloft, accommodating the shows of things to the desires of the mind, not (like reason and history) buckling and bowing down the mind to the nature of things. And by these charms, and that agreeable congruity which it has with man's nature, accompanied also with music, to gain more sweet access, it has so won its way as to have been held in honour even in the rudest ages and among barbarous peoples, when other kinds of learning were utterly excluded.

Dramatic Poesy, which has the theatre for its world, would be of excellent use if well directed. For the stage is capable of no small influence both of discipline and of corruption. Now of corruptions in this kind we have enough ; but the discipline has in our times been plainly neglected. And though in modern states play-acting is esteemed but as a toy, except when it is too satirical and biting ; yet among the ancients it was used as a means of educating men's minds to virtue. Nay, it has been regarded by learned men and great philosophers as a kind of musician's bow by which men's minds may be played upon. And certainly it is most true, and one of the great secrets of nature, that the minds of men are more open to impressions and affections when many are gathered together than when they are alone [1].

[1] There is nothing in the *Advancement of Learning* corresponding to this paragraph. It is a curious fact that these remarks on the character of the modern drama were probably written, and were certainly first published, in the same year which saw the first collection of Shakespeare's plays ; of which, though they had been filling the theatre for the last thirty years, I very much doubt whether Bacon had ever heard. How little notice they attracted in those days as works of literary pretension, may be inferred from the extreme difficulty which modern editors have found in ascertaining the dates, or even the order, of their production. Though numbers of contemporary news-letters, filled with literary and fashionable intelligence, have been preserved, it is only in the Stationer's register and the accounts kept by the Master of the Revels that we find any notices of the publication or acting of Shakespeare's plays. In the long series of letters from John Chamberlain to Dudley Carleton, scattered over the whole period from 1598 to 1623,—letters full of the news of the month ; news of the court, the city, the pulpit, and the bookseller's shop ; in which court-masques are described in minute detail, author, actors, plot, performance, reception and all ;—we look in vain for the name of Shakespeare or of any one of his plays. And yet during that period Hamlet, Twelfth Night, Othello, Measure for Measure, the Merchant of Venice, Macbeth, Lear, The Tempest, the Winter's Tale, Coriolanus, and several more, must have appeared as novelties. And indeed that very letter without which we should hardly know that Shakespeare was personally known to any one in the great world as a distinguished dramatic writer,

But Parabolical Poesy is of a higher character than the others, and appears to be something sacred and venerable; especially as religion itself commonly uses its aid as a means of communication between divinity and humanity. But this too is corrupted by the levity and idleness of wits in dealing with allegory. It is of double use and serves for contrary purposes; for it serves for an infoldment; and it likewise serves for illustration. In the latter case the object is a certain method of teaching, in the former an artifice for concealment. Now this method of teaching, used for illustration, was very much in use in the ancient times. For the inventions and conclusions of human reason (even those that are now common and trite) being then new and strange, the minds of men were hardly subtle enough to conceive them, unless they were brought nearer to the sense by this kind of resemblances and examples. And hence the ancient times are full of all kinds of fables, parables, enigmas, and similitudes; as may appear by the numbers of Pythagoras, the enigmas of the Sphinx, the fables of Æsop, and the like. The Apophthegms too of the ancient sages commonly explained the matter by similitudes. Thus Menenius Agrippa among the Romans (a nation at that time by no means learned) quelled a sedition by a fable. In a word, as hieroglyphics were before letters, so parables were before arguments. And even now, and at all times, the force of parables is and has been excellent; because arguments cannot be made so perspicuous nor true examples so apt.

But there remains yet another use of Poesy Parabolical, opposite to the former; wherein it serves (as I said) for an infoldment; for such things, I mean, the dignity whereof requires that they should be seen as it were through a veil; that is when the secrets and mysteries of religion, policy, and philosophy are involved in fables or parables. Now whether any mystic meaning be concealed beneath the fables of the ancient poets is a matter of some doubt. For my own part I must confess that I am inclined to think that a mystery is involved in no small number of them [2]. Nor does the fact that they are left commonly to boys and grammarians, and held in slight repute, make me despise them; but rather, since it is evident that the writings in which these fables are related are, next to sacred story, the most ancient of human writings, and the fables themselves still more ancient (for they are related not as being invented by the writers, but as things believed and received from of old), I take them to be a kind of breath from the traditions of more ancient nations, which fell into the pipes of the Greeks. But since that which has hitherto been done in the interpretation of these parables, being the work of unskilful men, not learned beyond commonplaces, does not by any means satisfy me, I think fit to set down Philosophy according to the Ancient

—I mean Lord Southampton's letter in furtherance of a petition from him and Burbage to the Lord Chancellor Ellesmere—proves at the same time how little was known about him by people of that quality. "This other" (he writes, after describing him as his especial friend and the writer of some of our best English plays,) hath to name William Shakespeare. . . . Both are right famous in their qualities, though it longeth not of your lordship's gravity and wisdom to resort unto the places where they were wont to delight the public ear." This was in 1608; and yet only six years before, when Ellesmere received Elizabeth at Harewood, Othello had been acted there for her entertainment. Even now a writer otherwise unknown hardly becomes known as the author of a successful play. "At present," said Mr. Rogers, "new plays seem hardly to be regarded as literature; people may go to see them acted, but no one thinks of reading them. During the run of *Paul Pry*, I happened to be at a dinner-party, where everybody was talking about it,—that is, about Liston's performance of the hero. I asked first one person, then another, and then another, who was the author of it? Not a man or woman in the company knew that it was written by Poole!"—*Recollections of the Table-talk of Samuel Rogers*, p. 253.—*J. S.*

[2] The hesitating manner in which Bacon here expresses himself shows that he felt, what every one in modern times who has considered the subject must I think feel, how difficult it is to enter into the spirit of the ancient mythus. Its essence seems to consist in a half-conscious blending of an idea with something that was accepted as a fact. See particularly on this point Müller's *Introduction to Mythology*. The mythus degenerates into allegory when the idea and the fact are conceived of as antithetical.

Parables among the *desiderata*. Of which work I will subjoin one or two examples; not so much perhaps for the value of the thing as for the sake of carrying out my principle; which is this; whenever I set down a work among the desiderata (if there be anything obscure about it), I intend always to set forth either instructions for the execution of it, or an example of the thing; else it might be thought that it was merely some light notion that had glanced through my mind; or that I am like an augur measuring countries in thought, without knowing the way to enter them. I can report no other deficiency in Poesy; for being as a plant which comes from the lust of the earth without a formal seed, it has sprung up and spread abroad more than any other kind of learning. But I will now propound the examples, only three in number; one taken from things Natural, one from things Political, and one from things Moral.

The First Example of Philosophy according to the Fables of the Ancients, in Natural Philosophy. Of the Universe, *according to the Fable of* Pan.

THE ancients leave the parentage of Pan uncertain. Some call him the son of Mercury; others attribute to him a very different mode of generation, affirming that he sprang from the promiscuous intercourse of Penelope with all the suitors. There is also a third account, which must not be omitted; for some have declared that he was the son of Jupiter and Hybris (which signifies Contumely). Whatever his origin, the Fates are said to have been his sisters; who dwelt in a cave underground; while he himself lived in the open air. The person of Pan is described by ancient tradition as follows:—horns on his head, rising to a point and reaching up to heaven; his whole body rough and shaggy; his beard especially long; his figure biform, the upper part human, the lower part like a beast and ending in goat's feet. He carried as insignia of his office, in his left hand a pipe compact of seven reeds, in his right a shepherd's crook or staff, curved and bent at the upper end. His dress was a mantle of leopard's skin. The titles and offices attributed to him were these; he was the god of hunters; also of shepherds, and of all persons dwelling in the country; the president likewise of mountains. He was moreover next to Mercury the messenger of the Gods. He was regarded as the leader and commander of the Nymphs, who were always wont to dance and frisk around him. The Satyrs and their elders the Sileni were also of his company. He had besides the power to inspire sudden terrors, such especially as were vain and superstitious, which received the name of *Panics*. Not many actions are recorded of him. The chief one is that he challenged Cupid at wrestling and was overcome in the contest. He also caught the giant Typhon in nets and held him fast. They say likewise that when Ceres, in sorrow and passion for the rape of Proserpine, had hid herself, and all the gods were eagerly engaged in seeking her, and had dispersed themselves in different paths for the pursuit, it was reserved for Pan to meet with her by a happy accident as he was hunting, and inform the rest of her hiding-place. He presumed also to contend in music with Apollo, and in the judgment of Midas was pronounced victor; for which judgment Midas had to wear the ears of an ass, but not so as to be seen. No amours, or at least very few, are related of Pan; a strange thing for one of a crowd of Gods so profusely amorous. It is only said of him that he was the lover of Echo, who was also esteemed his wife; and of one other nymph besides, named Syringa; with desire for whom he was inflamed by the revengeful anger of Cupid, whom he had not scrupled to challenge to the wrestling. He is also said on one occasion to have drawn the Moon apart into deep woods. Moreover he had no issue (which likewise is a marvel, when the gods, especially those of the male kind, were so prolific), unless it were one daughter, a little handmaid named Iambe, who used to amuse strangers with ridiculous stories; and was supposed by some to be Pan's daughter by his wife Echo. The parable may be thus explained.

Pan (as the name itself imports) represents and denotes the Universe, or the All of Things. Concerning his origin there are only two opinions, nor can there indeed be more. For he either sprang from Mercury, that is, the Word of God (which the Holy Scripture places beyond question, and which was perceived also by those of the philosophers themselves who have been accounted most divine), or else from the seeds of things mixed and confused together. For some philosophers have set down the seeds of things as infinite in their substance; whence arose the doctrine of *Homœomeræ*, which Anaxagoras either invented or brought into repute. Some with greater penetration and judgment thought that the variety of things would be sufficiently explained, if the seeds were supposed to be in substance the same, but to take various, though certain and definite, figures; accounting for the rest by the position and connexion of the seeds one with the other[3]; from which opinion emanated the doctrine of Atoms invented by Leucippus, and sedulously followed out by Democritus. Others, though they asserted one principle of things (as, Thales, Water; Anaximenes, Air; Heraclitus Fire), yet maintained that principle itself to be actually one, but potentially[4] various and dispensable, as that which had latent within it the seeds of all things. But those who (like Plato and Aristotle) have represented Matter as entirely despoiled, shapeless, and indifferent to forms, have approached much nearer to the figure of the parable. For they have made Matter as a common harlot, and Forms as suitors[5]; so that all the opinions about the origins of things return to this point, and may be reduced to this distribution,—that the universe proceeds either from Mercury, or from Penelope and all her suitors. From the third story of Pan's origin, it would seem as if the Greeks, either by intercourse with the Egyptians or otherwise, had heard something of the Hebrew mysteries. For it relates to the state of the world, not at its very birth, but after the fall of Adam; exposed and made subject to death and corruption. For that state was and is the offspring of God and Sin (or Contumely). For the sin of Adam, when he wished to "become like God", was a kind of contumely. Therefore the threefold account of the birth of Pan may be allowed as true, if rightly distinguished with respect to facts and times. For this Pan (as we now view and understand him) is the offspring of the *Divine Word*, through the medium of *confused matter* (which itself however was the work of God), and with the help of *Sin*, and by Sin Corruption, entering in.

To the Nature of things, the *Fates* or Destinies of things are truly represented as sisters. For the beginnings, durations, and ends of things, as also their fallings, risings, labours, felicities, and in a word whatever may happen to an individual, are termed Fates; which, however, except it be in some noble individual (as a man, or a city, or a people), are commonly not observed and recognised. Now it is *Pan*, that is, the nature of things, that reduces these separate individuals to such various conditions; insomuch that the chain of nature and the thread of the Fates are (so far as individuals are concerned) the same thing. In addition to this the ancients feigned that Pan lived always in the open air, but the Fates in a huge subterranean cave, whence they suddenly flew to men with exceeding swiftness; because nature and the face of the universe is open and visible, whereas the fates of individuals are secret and rapid. But if Fate be taken in a wider acceptation, so as to signify every event of any kind, and not the more noble only, yet in this sense too it excellently answers to the universal frame of things; seeing that there is nothing in the order of nature so small as to be without a cause, nor again anything so great but it depends on something else; so that the fabric of nature contains in her own lap and bosom every event whatever, both small and great, and developes them in due season by a fixed law. Therefore no wonder that the Parcæ are represented as sisters of Pan, and certainly legitimate. For Fortune is the child of the vulgar, and has only found favour with the lighter kind of philosophers. Indeed Epicurus seems not only to be profane, but also foolish, when he says "That it is better to believe in the fable of the gods, than to

[3] To this opinion Bacon himself doubtless inclined, but he was not I think a believer in any atomic theory; that is to say, he seems to have rejected the idea of a vacuum. Of Democritus, however, so far as relates to his physical theories, he always speaks with respect. Leibnitz has remarked that the view which Bacon here mentions is common to all the scientific reformers of the early part of the seventeenth century.

[4] The antithesis of the actual and the potential is a fundamental doctrine in the peripatetic philosophy.

[5] See Arist. Physics, i. c. 9.

assert the power of fate "[6]; as if anything in the universe could be like an island, separated from connexion with the rest. But Epicurus, accommodating and subjecting his natural to his moral philosophy (as appears from his own words), would not willingly admit any opinion that depressed or hurt the mind, and troubled or disturbed that *Euthumia* of his, which he had adopted from Democritus. And so being more fond of enjoying the sweets of thought than patient of the truth, he fairly threw off the yoke, and rejected both the necessity of Fate and the fear of the gods. And so much for the relationship of Pan to the Fates.

Horns are attributed to the Universe, broad at the base and pointed at the top. For all nature rises to a point like a pyramid. Individuals, which lie at the base of nature, are infinite in number; these are collected into Species, which are themselves manifold; the Species rise again into Genera; which also by continual gradations are contracted into more universal generalities, so that at last nature seems to end as it were in unity; as is signified by the pyramidal form of the horns of Pan. Nor need we wonder if the horns of Pan reach even to the heaven, seeing that the transcendentals of nature, or universal ideas, do in a manner reach up to divinity. And hence the famous chain of Homer (that is, the chain of natural causes) was said to be fastened to the foot of Jupiter's throne; and we see that no one has handled metaphysics and the eternal and immovable in nature, and withdrawn his mind for awhile from the variable succession of things, without falling at once on Natural Theology; so easy and near a passage is it from the top of the pyramid to matters divine.

The body of nature is elegantly and truly represented as covered with hair; in allusion to the rays of things. For rays are as the hairs or bristles of nature, nor is there anything which is not more or less radiant. This is seen most evidently in the faculty of sight, and no less in all magnetic virtue, and every effect which takes place at a distance. For whatever produces an effect at a distance may be truly said to emit rays. But Pan's hair is especially long in the beard; because the rays of celestial bodies, especially of the sun, operate and pierce from a greater distance than any other; so that not only the surface, but even the interior of the Earth for some distance, is changed, wrought, and filled with spirit by them. And that figure of Pan's beard is the more elegant, because the sun himself, when the upper part is obscured by a cloud and his rays break out below, appears to the eye as bearded.

The body of nature is likewise most aptly described as biform, on account of the difference between bodies of the upper and lower world; whereof the former, from their beauty and regularity and constancy of motion, as well as their influence over the Earth and earthly things, are properly represented by the human figure, human nature participating of order and dominion. But the latter, by reason of their perturbation and irregular movements, and because they are for the most part ruled by the heavenly bodies, may be content with the figure of a brute beast. Moreover this same description of a biform body has reference to the participation of species; for there is no natural species which can be regarded as simple; every one seeming to participate and be made up of two. Thus man has somewhat of the brute, the brute somewhat of the plant, the plant somewhat of the body inanimate; so that all things are indeed biform, being compounded of a superior and inferior species. And it is a very acute allegory, that of the goat's feet; which refers to the upward motion of earthly bodies towards the regions of the air and heaven; where also they remain hanging, and whence they are rather forced down than descend of themselves. For the goat is a climbing animal, and loves to hang from rocks and cling to the sides of precipices; a tendency which is also exhibited in a wonderful manner by substances which belong properly to the lower world; as appears most plainly in clouds and meteors. Nay a question was raised not without reason by Gilbert, who has written upon the magnet most laboriously and after the experimental method, whether heavy bodies may not, when removed to a great distance from the earth, gradually lose their downward tendency[7]?

[6] Cf. Diog. Laërt. x. 134, the reference being, as Menage, following Gassendi, remarks, to the doctrines of the earlier physicists, Democritus, etc.

[7] Gilbert was of opinion that the earth is a great magnet which attracts all bodies

Of the two insignia which Pan bears in his hands, the one represents harmony, the other empire. For the pipe of seven reeds plainly denotes the consent and harmony of things, or concord mixed with discord (which is caused by the motion of the seven planets). For there are not found any other wanderings or manifest expatiations in the heavens, besides those of the planets, such as when combined and tempered with the regularity of the fixed stars and their eternal and invariable distance one from the other, may keep up and set in motion at once the constancy of species and the continual change of individuals. And if there be any lesser planets which are not visible, or any greater change in the heaven (as in some superlunary comets), it seems they are as pipes either entirely mute or vocal only for a season; inasmuch as their influences either do not approach so low as ourselves, or do not long interrupt the harmony of the seven pipes of Pan [8]. That sheephook also representing empire contains a noble metaphor, alluding to the mixture of straight and crooked in the ways of nature. And this rod or staff is crooked principally in the upper part; because all the works of Divine Providence in the world are mostly brought about in a mysterious and circuitous manner, so that while one thing appears to be doing another is doing really; as the selling of Joseph into Egypt, and the like. Moreover in all wise human governments, those who sit at the helm can introduce and insinuate what they desire for the good of the people more successfully by pretexts and indirect ways than directly. Nay (which perchance may seem strange), even in mere natural things you may deceive nature sooner than force her; so ineffectual and self-impeding are all things which are done directly; whereas on the other hand the indirect and insinuating way proceeds smoothly and gains its end. The cloak or mantle of Pan is ingeniously feigned to be the skin of a leopard; because it is full of spots. For the heavens are spotted with stars, the seas with islands, the earth with flowers; and even particular objects are commonly variegated on the surface, which may be regarded as their mantle.

The office of Pan could in no other way be so lively set forth and expressed, as by terming him the god of hunters. For every natural action, and indeed every motion and progression, is but a hunting. Arts and sciences hunt after their works; human counsels hunt after their ends; and all natural things hunt either after their food to preserve them, or after their pleasures and delights to perfect them (for all hunting is for the sake either of prey or pleasure); and this too by methods expert and sagacious:

> Torva leæna lupum sequitur, lupus ipse capellam:
> Florentem cytisum sequitur lasciva capella. [9]

Pan is also the god of all dwellers in the country, because such men live more according to nature than in cities and courts, where nature is corrupted by too

near its surface, although phenomena of polarity are only developed in a few cases. To every magnet he ascribed an "orb of virtue" beyond which it exerts no influence whatever, and also a smaller "orb of coition" such that the magnet cannot produce motion in any portion of matter which lies beyond it. As a heavy body therefore approaches the limit of the earth's orb of coition its downward tendency gradually diminishes. Imperfect as these views are, they show how clearly Gilbert had apprehended the general idea of attraction, and how little reason Voltaire had for his assertion that Bacon "a deviné l'attraction". [See note on *Nov. Org.* p. 346.]

[8] For dreams about the music of the spheres, see Robert Fludd's work *Utriusque Cosmi, majoris scilicet et minoris, metaphysica, physica, et technica Historia*, 1617. The third book of the first tractate is wholly *De Musicâ mundanâ*, and is illustrated by an engraving of a bass viol, of which the dimensions extend through the solar system. Bacon was, not improbably, acquainted with Fludd, who was one of the most learned of the cabalistic philosophers.

[9] Virg. *Eclog.*, ii. 63. :—The savage lioness the wolf pursues,
The wolf the kid, the kid the cytisus.

much cultivation; so that what the poet says of his mistress is by reason of such arts of luxury true likewise of nature,

Pars minima est ipsa puella sui.[10]

Pan is likewise termed the president of the mountains, because in mountains and high places nature is more open and exposed to sight and study. That Pan next to Mercury is the messenger of the gods is plainly a divine allegory; for next to the word of God, the image of the world is the herald of divine power and wisdom. "The heavens (says the Psalmist) declare the glory of God, and the firmament showeth his handiwork"[11].

Pan delights in the nymphs, that is in spirits; for the spirits of living creatures are the delight of the world. And with reason is he styled their leader, for each of them follows its own nature as a guide, round which after their own fashion they leap and frisk in endless variety and constant motion. And therefore one of the moderns has ingeniously referred all the powers of the soul to motion, and remarked on the conceit and precipitancy of some of the ancients, who in too eagerly fixing their eyes and thoughts on the memory, imagination, and reason, have neglected the Thinking Faculty, which holds the first place[12]. For he who remembers or recollects, thinks; he who imagines, thinks; he who reasons, thinks; and in a word the spirit of man, whether prompted by sense or left to itself, whether in the functions of the intellect, or of the will and affections, dances to the tune of the thoughts; and this is the frisking of the Nymphs. And in their company are ever found the Satyrs and Sileni, that is old age and youth. For all things have their merry and dancing time, and again their heavy and tippling time; and to one who truly considers them the pursuits of either age may appear perhaps ridiculous and deformed, like a Satyr or Silenus. As for the Panic terrors, a most wise doctrine is therein propounded. For nature has implanted in every living creature apprehension and fear, as the means of preserving its own life and essence, and avoiding and repelling the attacks of things hurtful. And yet this same nature knows not how to keep a mean, but is always intermixing vain and useless fears with such as are salutary; so that all things (if they might be seen within) are full of panic terrors; especially things human; and most of all among the common people, who are exceedingly troubled and agitated by superstition (which is nothing else but a panic terror), especially in hard and anxious and adverse times. Nor is this superstition confined to the vulgar, but it passes occasionally from them to the wiser sort; as Epicurus has said divinely (if only his other doctrines concerning the gods had breathed the same spirit). "It is not profane to deny the gods of the vulgar, but to apply the ideas of the vulgar to the gods"[13].

With regard to the presumption of Pan, and his challenging Cupid to wrestle, the meaning is that Matter is not devoid of an appetite and inclination to dissolve the world and fall back into the old Chaos, but that its force and malice is restrained and kept in order by the prevailing concord of things (which is signified by Love or Cupid). And therefore it falls out most luckily (or say rather by the infinite goodness of God) for man and the world, that Pan has the worst of that contest and goes away defeated. The same thing is alluded to in that other circumstance of catching Typhon in a net; because however it be that vast and strange swellings (for that is the meaning of Typhon) take place occasionally in nature,—whether of the sea or the clouds or the earth or any other body,—nevertheless all such exuberances and irregularities are by the nature of things

[10] Ov. *Rem. Amor.* 344:—So overlaid with ornament and art,
Herself is of herself the smallest part.

[11] Psalm xix. 1.

[12] The writer referred to is A. Donius. See his *De Naturâ Hominis*, 1581, the title of the twenty-first chapter of the second book of which is *Omnes Operationes Spiritus esse Motum et Sensum*. For an account of this "motus" see the sixteenth chapter of the second book. As might be supposed, Donius is altogether a materialist.

[13] Diogenes Laërt. x. 123.

caught and confined in an inextricable net, and bound down as with a chain of adamant.

As for the tale that the discovery of Ceres was reserved for this god, and that while he was hunting, and denied to the rest of the gods, though diligently and specially engaged in seeking her, it contains a very true and wise admonition, which is, not to look for the invention of things useful for life and civilisation from abstract philosophies, which are as it were the greater gods, even though they devote all their strength to the purpose ; but only from Pan, that is from sagacious experience and the universal knowledge of nature ; which oftentimes, by a kind of chance, and while engaged as it were in hunting, stumbles upon such discoveries. For the most useful inventions are due to experience, and have come to men like windfalls.

Again that contest in music and the issue of it exhibits a wholesome doctrine, and one which may well restrain and reduce to sobriety the pride and overweening confidence of human reason and judgment. For it seems that there are two kinds of harmony and music ; one of divine wisdom, the other of human reason. And to the human judgment, and the ears as it were of mortals, the government of the world and the more secret judgments of God sound somewhat harsh and untunable ; and though this be ignorance, such as deserves to be distinguished with the ears of an ass, yet those ears are worn secretly and not in the face of the world ; for it is not a thing observed or noticed as a deformity by the vulgar.

Lastly it is no marvel if no loves are attributed to Pan, besides his marriage with Echo. For the world enjoys itself, and in itself all things that are. Now he who is in love wants something ; and where there is plenty of everything there is no room for want. The world therefore can have no loves, nor any want (being content with itself), unless it be of *discourse*. Such is the nymph Echo, a thing not substantial but only a voice ; or if it be of the more exact and delicate kind, *Syringa*,—when the words and voices are regulated and modulated by numbers, whether poetical or oratorical. But it is well devised that of all words and voices Echo alone should be chosen for the world's wife ; for that is the true philosophy which echoes most faithfully the voices of the world itself, and is written as it were at the world's own dictation ; being nothing else than the image and reflexion thereof, to which it adds nothing of its own, but only iterates and gives it back.

The story that Pan once drew the Moon apart into deep woods, seems to have reference to the intercourse of sense with heavenly or divine things. For the case of Endymion is different from that of Pan. To Endymion the Moon descended of her own accord as he slept ; for divine influences sometimes steal spontaneously into the understanding when at rest, and withdrawn from the senses ; but if they are invoked and solicited by the sense, as by Pan, then they afford no other light but that,

> Quale per incertam Lunam sub luce maligna
> Est iter in silvis.[14]

That the world has no issue is another allusion to the sufficiency and perfection of it in itself. Generation goes on among the parts of the world ; but how can the whole generate, when no body exists out of itself ? As for that little woman, Iambe, Pan's putative daughter, it is an addition to the fable with a great deal of wisdom in it ; for by her are represented those vain babbling doctrines about the nature of things, which wander abroad in all times and fill the world ; doctrines barren in fact, counterfeit in breed, but by reason of their garrulity sometimes entertaining, and sometimes again troublesome and annoying.

[14] Virg. *Æn.* vi. 270 :—As by the wayward moon's inconstant light
A path through woods . . .

Another example of Philosophy according to the Ancient Parables, in Politics. *Of War according to the story of* Perseus.

PERSEUS, an Eastern man, was sent, it is said, by Pallas to destroy Medusa, who was a grievous plague to many nations of the West in the furthest parts of Spain. She was a monster, otherwise huge and savage, and of an aspect so foul and hideous that her look alone turned men into stones. Now Medusa was one of the Gorgons, and the only mortal amongst them, the others not being subject to death. Perseus then, equipping himself for so noble an enterprise, borrowed arms as presents from three of the gods; from Mercury wings,—fitted to the ankles, not the shoulders; from Pluto a helmet; from Pallas a shield and mirror. Nevertheless (though he was now so well furnished) he did not go direct to Medusa, but turned aside to the Græi. These were the half-sisters of the Gorgons; and were grey-headed from their birth, and like old women. They had but one eye and one tooth among them all; which, as they had occasion to go abroad, each wore by turns and put off again when she came back. This eye and this tooth they lent to Perseus. And now judging himself sufficiently armed to effect his purpose, he went against Medusa with all haste, flying. Her he found sleeping; but not daring to meet her gaze (in case she should wake), he turned his face away, and looking into the mirror of Pallas to direct his blow, cut off her head. From her blood spilt upon the ground immediately sprang forth Pegasus the winged horse. But the severed head Perseus transferred to the shield of Pallas, and fixed it there; where it still retained its former virtue, that whoever gazed upon it became as it were thunder or planet struck.

This fable seems to have been devised with reference to method and prudence in making war. And first, the undertaking of every war ought to be as a mission from Pallas; not from Venus (as the Trojan war was), or for any other slight motive; for resolutions respecting wars ought to be based on solid counsels. Secondly, with regard to the kind of war to be chosen, the fable propounds three very wholesome and important precepts. The first is, not to make too great a point of subjugating the neighbouring nations. For the method of enlarging a patrimony and an empire is not the same. In private estates contiguity of lands is taken into account, but in the extension of empire, occasion and facility for making war and fruit of conquest ought to be regarded in place of contiguity. And therefore Perseus, though in the East, did not shrink from an expedition even to the far West. Of this there is a notable instance in the different modes of war practised by Philip and Alexander, father and son. The former, engaging in wars with neighbouring countries, after much exertion and danger (for both at other times and especially at Chæronea he was reduced to extreme peril), added a few cities to his empire; whereas Alexander, with wise boldness undertaking a distant expedition into Persia, subjugated an infinite number of nations, and suffered more by his marches than his battles. But perhaps this difference is shown still more clearly in the increase of the empire of the Romans, who while they had scarce penetrated westward beyond Liguria, had already conquered and included within their empire eastern provinces as far off as Mount Taurus. So Charles the Eighth, King of France, having found the war with Bretagne (afterwards arranged by marriage [15]) no easy matter, undertook that distant enterprise against Naples, which he effected with wonderful ease and success. Certainly wars made upon distant nations have this advantage, that the invaders have to fight with those who have no experience of their mode of warfare and arms; whereas in a war with neighbours it is otherwise. Moreover the equipment of such expeditions is generally more perfect and better appointed, and the very boldness and confidence of the aggressor inspires greater terror into the enemy. Nor does it often happen in these distant expeditions that the enemy to whom the war is brought from such a distance can make diversions or counter-invasions, as is the case in wars between neighbours. But the chief point is

[15] In 1491 Charles the Eighth married Anne of Brittany, and thus put an end to the war which Bacon here speaks of.

that in subduing neighbouring states there is only a small choice of opportunities; whereas in distant enterprises the aggressor may carry the war at pleasure, either where military discipline is most relaxed, or the strength of a people is most weakened and impaired, or the rise of civil dissension and other like opportunities present themselves. The second precept is, that there must ever be a cause of war, just, pious, honourable, and popular. For this begets alacrity as well in the soldiers, as in those who provide the funds, opens the way to alliances, and conciliates friends, and has many other advantages. Now among the causes of war few are more popular than the putting down of tyrannies, beneath whose yoke the spirit and energy of the people are worn down and prostrated, as by the head of Medusa; a thing which gained Hercules divine honours. Certainly the Romans made it a great point of duty to hasten with all speed to succour their allies when in any way attacked. Wars also undertaken for a just revenge have almost always been successful; as the war against Brutus and Cassius to avenge the murder of Cæsar; of Severus to avenge the death of Pertinax; of Junius Brutus to avenge the death of Lucretia. In a word, whosoever either relieves or avenges by war the calamities and injuries of men, bears arms under Perseus. The third precept is, that in every war a true estimate of strength must be taken, and it must be duly considered whether the war be such as can be carried through and brought to an issue; so that one may not engage in pursuit of vast and boundless projects. For of the Gorgons (which are the representatives of war) Perseus wisely chose her alone who was of mortal nature, nor did he attempt impossibilities. Such then is the advice which the fable gives touching the things that require deliberation in undertaking war; the rest relates to the carrying it on.

In war those three gifts of the gods are of all things the most important; insomuch that they commonly command and carry with them fortune itself. For Perseus received speed from Mercury, secrecy of counsels from Orcus, and foresight from Pallas. And it is not without allegory, and that of the wisest sort, that those wings of speed (seeing speed is of much avail in war) were attached to the feet and not to the shoulders; because celerity is required not so much in the first onsets of war as in the pursuit and following up thereof. For no error in war is more common than this, that the prosecutions and subsidiary actions correspond not to the energy of the first commencements. And the helmet of Pluto (which used to render men invisible) is a manifest parable. For next to speed in war secrecy of counsels is of the greatest moment; of which indeed speed itself is a great part; for speed anticipates the disclosures of counsels. To the helmet of Pluto belongs also this: that there should be one commander in a war, with free instructions; for consultations held with many savour more of the crests of Mars than the helmet of Pluto. Variety of pretexts, ambiguous directions, rumours spread abroad, which either blind or avert men's eyes and involve the real design in obscurity, refer to the same. So also diligent and suspicious precautions respecting despatches, ambassadors, deserters, and many like matters, are wreathed round the helmet of Pluto. But it is of no less importance to discover the counsels of the enemy than to conceal our own. To the helmet of Pluto therefore must be added the mirror of Pallas, whereby to discern the strength or weakness of the enemy, their secret partisans, their discords and factions, their movements and designs. But since there is so much of chance in war, that no great confidence can be placed either in discovering the designs of the enemy, or in concealing our own, or even in speed itself, we must take special care to be armed with the shield of Pallas, that is, of foresight, so as to leave as little as possible to fortune. To this belong the exploring of roads before a march, the careful fortification of the camp (which in modern warfare has fallen almost into disuse, whereas the camps of the Romans were like a fortified town, to fall back upon in case of defeat), a firm and well drawn up line of battle, not trusting too much to light troops, or even to cavalry; in a word, everything which relates to a sound and careful system of defensive war; for the shield of Pallas is generally of more avail in war than the sword of Mars itself. But Perseus, however furnished with forces and courage, has still need of one thing more, of the greatest possible importance, before he commences the campaign; he must turn aside to the Græræ. Now the Græræ are Treasons, which are the Sisters of War, though not indeed own sisters, but as it were of less noble birth. For wars are noble and generous; treasons degenerate and base. They are portrayed appropriately as being grey-headed from their birth and like old women, by reason of the perpetual cares and anxieties attending traitors. Their power (before they openly desert) is in the eye or tooth; for all faction, when discontented and inclined to treason, is both watchful and biting. Moreover this eye and tooth are, as it were, common to them all; for whatever they learn and discover is handed from one to another, and circulates through the whole party. And with regard to the tooth, they all bite as it were with one mouth, and utter the same scandals; so that if you hear one, you hear all. Wherefore Perseus must conciliate these Græræ, and bring them into alliance with him, especially that they may lend him their eye and tooth; the eye to gain information; the tooth to spread rumours, raise envy, and gain over the minds of men. But when everything has been arranged in order for war, we must take special care, like Perseus, to find Medusa asleep; for he who undertakes a war wisely will almost always attack his enemy unprepared and in security. Lastly, in the very actions and onsets of war the mirror of Pallas must be resorted to; for there are many who before the time of danger can take a clear and accurate survey of the position of the enemy, but in the very moment of peril they are either stupefied with terror, or look their dangers too rashly in the face; and so rush madly into them, bent on overcoming, not on avoiding them. Neither of which things should be done; but we should turn aside the head and look into the mirror of Pallas, that the onset may be rightly directed without either terror or fury.

From the conclusion of the war and victory follow two effects, first, the birth and springing up of Pegasus, which evidently enough signifies Fame that flies abroad and proclaims the victory, and so makes what remains of the war easy and satisfactory; secondly, the carrying of Medusa's head on the shield; to which for excellence no other kind of defence can be compared. For one great and memorable enterprise successfully carried out paralyses every movement of the enemy, and stupefies disaffection itself.

The third Example of Philosophy according to the Ancient Fables, in Moral Philosophy. *Of Desire, according to the fable of* Dionysus.

They say that Semele, the mistress of Jupiter, having bound him by an inviolable oath to grant her a request whatever it might be, desired of him to come to her arms in the same form as he would to Juno; and so she was scorched to death in his embrace. The child which she bore in her womb was taken by his father and sewn up in his thigh, till the time of gestation was accomplished. And because the child, when in the thigh of Jupiter, pinched and galled him so as to make him limp, he received the name of Dionysus. After he was brought forth he was nursed for some years by Proserpine; and when he grew up his face was so like a woman's that it seemed doubtful of which sex he was. He was likewise once dead and buried for a time, but came to life again not long after. In his early youth he was the first to invent and explain the culture of the vine, and the making of wine, and its use; whereby becoming renowned and illustrious, he subdued the whole world and advanced to the furthest parts of India. He rode in a chariot drawn by tigers, round which danced certain deformed demons called Cobali; Acratus and others. The Muses also attended in his train. He took to wife Ariadne, whom Theseus had deserted and abandoned. His sacred tree was the ivy. He was regarded likewise as the inventor and institutor of sacred rites and orgies; but such as were fanatical and full of corruption and moreover cruel. He had also the power of exciting phrensy. At least it was by women excited to phrensy in his orgies that two renowned men, Pentheus and Orpheus, are said to have been torn to pieces; the one having climbed into a tree out of curiosity to see what they were doing; the other while playing sweetly and skilfully on the lyre. Moreover the actions of this god are often confounded with those of Jupiter.

The fable appears to relate to morals; and indeed there is scarcely anything better to be found in moral philosophy. Under the person of Bacchus is depicted the nature of Desire, or the passions and perturbations of the mind. First, therefore, with regard to the origin of Desire. The mother of all desire (though ever so hurtful) is nothing else than apparent good. For as the mother of virtue is real good, so the mother of desire is apparent good. One the lawful wife of Jupiter (in whose person the human soul is represented), the other his mistress; who nevertheless aspires, like Semele, to the honours of Juno. Now the conception of Desire is always in some unlawful wish, rashly granted before it has been understood and weighed; and as the passion warms, its mother (which is the nature and species of good), not able to endure the heat of it, is destroyed and perishes in the flame. Then the progress of Desire from its first conception is of this kind. It is both nursed and concealed in the human mind (which is its father); especially in the lower part of it, as in the thigh; where it causes such prickings, pains and depressions, that the actions and resolutions of the mind labour and limp with it. And even when it has grown strong with indulgence and custom, and breaks forth into acts (as if it had now accomplished its time and were fairly born and delivered), yet at first it is brought up for a time by Proserpine; that is, it seeks hiding-places and keeps itself secret, and as it were underground; until throwing off all restraints of shame and fear, and growing bolder and bolder, it either assumes the mask of some virtue, or sets infamy itself at defiance. And it is most true that every passion of the more violent kind is as it were of doubtful sex; for it has at once the force of a man and the weakness of a woman. It is well said likewise that Bacchus died and came to life again; for the passions seem sometimes lulled to sleep, and as it were dead; yet can they never be trusted, no not though they be buried. For give them matter and opportunity and they will rise again [15].

It is a wise allegory too, that of the invention of the vine. For every passion is very ingenious and sagacious in discovering the things which nourish and foster itself. Now of all things known to man wine is the most powerful and efficacious in stimulating and inflaming every kind of excitement; serving as a common fuel to desires in general. Very elegantly too is passion or desire described as the subduer of provinces and the undertaker of an endless course of conquests. For it is never content with what it has got, but with infinite and insatiable appetite tries for something more, and ever craves for new triumphs. Tigers likewise are kept in the stables of the passions, and at times yoked to their chariot; for when passion ceases to go on foot and comes to ride in its chariot, as in celebration of its victory and triumph over reason, then is it cruel, savage, and pitiless towards all that withstand or oppose it. Again there is humour in making those ridiculous demons dance about the chariot of Bacchus. For every passion of the more vehement kind produces motions in the eyes, and indeed in the whole countenance and gesture, which are uncomely, unsettled, skipping, and deformed; insomuch that when a man under the influence of any passion (as anger, scorn, love, or the like) seems most grand and imposing in his own eyes, to the lookers on he appears unseemly and ridiculous. It is true also that the Muses are seen in the train of passion; there being scarce any passion which has not some branch of learning to flatter it. For herein the majesty of the Muses suffers immensely from the license and wantonness of men's wits, turning those that should be the guides and standard-bearers of man's life into mere followers in the train and ministers to the pleasures of the passions.

Especially noble again is that part of the allegory which represents Bacchus as lavishing his love upon one whom another man had cast off. For most certain it is that passion ever seeks and aspires after that which experience has long since repudiated. And let all men who in pursuit and indulgence of their passions

[15] Yet Rochefoucauld has said " Il est impossible d'aimer une seconde fois ce qu'on a véritablement cessé d'aimer."—*Reflexions Morales*, 294. [The two observations are not, I think, incompatible with one another. Bacon speaks of the appetite rather than the sentiment; and Rochefoucauld does not say that a man cannot love again that which he *thinks* he has ceased to love.—*J. S.*]

care not what price they pay for the enjoyment of them, know this : that whatever be the object of their pursuit—be it honour or fortune or love or glory or knowledge, or what it may—they are paying court to things cast off,—things which many men in all times have tried and upon trial rejected with disgust.

Nor is the consecration of Ivy to Bacchus without its mystery. For this has a double propriety. First, because ivy flourishes in the winter ; next because it has the property of creeping and spreading about so many things, as trees, walls, buildings, etc. For as to the first, every passion flourishes and acquires vigour by being resisted and forbidden, as by reaction or *antiperistasis ;* like the ivy by the cold of winter. As to the second, any predominant passion in the human spirit spreads itself like ivy round all its actions and resolves, so that you cannot find anything free from the embrace of its tendrils. Neither is it to be wondered at if superstitious rites are attributed to Bacchus ; for almost every insane passion grows rank in depraved religions, insomuch that the pollutions of heretics are worse than the Bacchanalian orgies of the heathen ; whose superstitions likewise have been no less bloody than foul. Neither again is it wonderful that phrenises are thought to be inspired by Bacchus ; since every passion, in the excess thereof, is like a short madness, and if it continue vehement and obstinate, commonly ends in insanity. And that circumstance of the tearing to pieces of Pentheus and Orpheus amid the orgies of Bacchus, has an evident allegorical meaning ; for every ruling passion is extremely hostile and inveterate against two things ; whereof the one is curious inquisition ; the other, free and wholesome advice. Nor does it make any difference if that inquisition be merely for the sake of looking on, as from a tree, without any ill-feeling ; nor again if the advice be tendered ever so sweetly and skilfully ; for the orgies cannot upon any conditions endure either Pentheus or Orpheus. Lastly, the confusion of the persons of Jupiter and Bacchus may well be taken in an allegorical sense. For noble and illustrious actions and glorious and distinguished services proceed sometimes from virtue, right reason, and magnanimity ; and sometimes (however they are extolled and applauded without distinction) only from lurking passion and hidden desire ; and thus the deeds of Bacchus are not easily distinguished from the deeds of Jupiter [16].

But we stay too long in the theatre ; let us now pass to the palace of the mind, which we are to approach and enter with more reverence and attention.

[16] It seems not improbable that Bacon was led to consider the ancient mythology from the point of view which he has illustrated both here and in the *Wisdom of the Ancients*, by an author with many of whose writings he was familiar. Plutarch's treatise *De Iside et Osiride* is very much in the same manner.

Book III

CHAPTER I.

Division of Science into Theology *and* Philosophy. *Division of Philosophy into three doctrines; concerning the* Deity, *concerning* Nature, *and concerning* Man. *Constitution of* Primary Philosophy, *as the common mother of all.*

ALL History, excellent King, walks upon the earth, and performs the office rather of a guide than of a light; whereas Poesy is as a dream of learning; a thing sweet and varied, and that would be thought to have in it something divine; a character which dreams likewise affect. But now it is time for me to awake, and rising above the earth, to wing my way through the clear air of Philosophy and the Sciences.

The knowledge of man is as the waters. Some waters descend from above, and some spring from beneath; and in like manner the primary division of sciences is to be drawn from their sources; of which some are above in the heavens, and some here below. For all knowledge admits of two kinds of information; the one inspired by divine revelation, the other arising from the senses. For as to that knowledge which man receives by teaching, it is cumulative and not original; as it is likewise in waters, which beside their own springheads, are fed with other springs and streams. I will therefore divide knowledge into Divinity and Philosophy; meaning by Divinity Sacred or Inspired, not Natural Divinity; of which I will speak hereafter. But this (namely, Inspired Divinity) I will reserve to the end, that with it I may conclude my discourse; being as it is the haven and sabbath of all human contemplations.

The object of philosophy is threefold—God, Nature, and Man; as there are likewise three kinds of ray—direct, refracted, and reflected. For nature strikes the understanding with a ray direct; God, by reason of the unequal medium (viz. his creatures), with a ray refracted; man, as shown and exhibited to himself, with a ray reflected[1]. Philosophy may therefore be conveniently divided into three branches of knowledge: knowledge of God, knowledge of Nature, and knowledge of Man, or Humanity. But since the divisions of knowledge are not like several lines that meet in one angle; but are rather like branches of a tree that meet in one stem (which stem grows for some distance entire and continuous, before it divide itself into arms and boughs); therefore it is necessary, before we enter into the branches of the former division, to erect and constitute one universal science, to be as the mother of the rest, and to be regarded in the pro-

[1] The parallel which naturally suggests itself between light and knowledge has by several writers been traced in the modifications of which light is susceptible. Thus Roger Bacon, at the close of his *Perspectiva*, likens vision by direct light to divine knowledge, by refracted light to angelic knowledge, and by reflected light to human; and again to man's knowledge in the state of glory, "facie ad faciem," to his knowledge in the intermediate state, and to that which he has in this present life; "et hæc est recte per reflexionem, secundum quod dicit apostolus, Videmus nunc per speculum in ænigmate". And in this life also vision is triple: "scilicet recta in perfectis, fracta in imperfectis; et in malis et in negligentibus mandata Dei, est etiam per reflexionem," an assertion in support of which he quotes S. James, i. 23. and 24. But all these illustrations differ from that in the text, inasmuch as they relate to the different kinds of knowledge which appertain to different orders and states of being, and not to the differences which arise from the nature of the object. For a nearer parallel, at least with respect to the radius reflexus, see Plutarch *De Curiositate*, c. 3.

gress of knowledge as portion of the main and common way, before we come where the ways part and divide themselves. This science I distinguish by the name of *Philosophia Prima*, primitive or summary philosophy; or *Sapience*, which was formerly defined as the knowledge of things divine and human. To this no other is opposed; for it differs from the rest rather in the limits within which it ranges than in the subject matter; treating only of the highest stages of things. Which science whether I should report as deficient or not, I stand doubtful, though I rather incline to do so. For I find a certain rhapsody and incongruous mass of Natural Theology, of Logic, and of some parts of Natural Philosophy (as those concerning First Principles and the Soul), all mixed up and confused, and in the lofty language of men who take delight in admiring themselves advanced as it were to the pinnacle of the sciences. But setting all high conceits aside, my meaning is simply this: that a science be constituted, which may be a receptacle for all such axioms as are not peculiar to any of the particular sciences, but belong to several of them in common.

Now that there are very many axioms of that kind need not be doubted. For example, "if equals be added to unequals the wholes will be unequal," is a rule of mathematics. The same holds in ethics, as regards distributive justice; for in commutative justice the rule of equity requires that equals be given to unequals; whereas in distributive, if unequals be not given to unequals there is the greatest injustice [2]. Again "things that are equal to the same are equal to one another," is likewise a rule of mathematics; but it is at the same time so potent in logic as to be the basis of the syllogism. "The nature of everything is best seen in its smallest portions [3]," is a rule in Physics of such force that it produced the atoms of Democritus; and yet Aristotle made good use of it in his Politics, where he commences his inquiry of the nature of a commonwealth with a family. "All things are changed and nothing is lost [4]," is in like manner a rule in Physics, exhibited thus, "The Quantum of nature is neither diminished nor increased". The same holds in Natural Theology, with this variation, "It is the work of omnipotence to make somewhat nothing, and to make nothing somewhat;" which likewise the Scripture testifies; "I know that whatsoever God doeth, it shall be for ever; nothing can be put to it, nor anything taken from it [5]". "Things are preserved from destruction by bringing them back to their first principles," is a rule in Physics; the same holds good in Politics (as Machiavelli rightly observed), for there is scarcely anything which preserves states from destruction more than the reformation and reduction of them to their ancient manners [6]. "Putrefaction is more contagious before than after maturity," is a rule in Physics; the same is eminently true in Morals; for the men who are most wicked and profligate produce less corruption in the public manners than those who appear to have some soundness and virtue in them, and are only partly evil. "Whatever is preservative of a greater Form is more powerful in action," is a rule in Physics; for that the connexion of things should not be severed, nor a vacuum (as they call it) admitted, tends to preserve the fabric of the universe; whereas the collection of heavy bodies towards the mass of the earth tends to preserve only the region of dense bodies; and therefore the first motion overcomes the last. The same holds in Politics; for whatsoever contributes to preserve the whole state in its own nature, has greater power than that which only benefits the particular members of that state. It holds likewise in Theology; for of the theological virtues, charity, which is the virtue most communicative of good, excels all the rest. "The force of an agent is increased by the reaction of a contrary," is a rule in Physics [7]. The same has

[2] Cf. Arist. *Nic. Eth.* v. 3, 4, 5.

[3] This passage has been already quoted, B. II., c. 2.

[4] Ovid. *Metam.* xv. 165.

[5] Ecclesiastes, iii. 14.

[6] Machiavelli, *Discorsi*, § 1.

[7] The doctrine of Antiperistasis, that is of the increase of intensity of one of two contraries by the juxtaposition of the other, is applied by Aristotle, *Meteor.* i. c. 13, in the case of heat and cold, to explain the formation of hail. It is formally and generally stated in Averroës's commentary on this passage. See also Arist. *Probl.* ii. 16., and Plutarch's *Quæst. Naturales*.

wonderful efficacy in Politics, since every faction is violently irritated by the encroachment of a contrary faction. " A discord ending immediately in a concord sets off the harmony," is a rule in Music. The same holds in Ethics and in the affections. The trope of Music, to glide gently from the close of cadence (as they call it) when you seem to be on the point of it, resembles the trope of Rhetoric, of deceiving expectation. The quavering upon a stop in music gives the same pleasure to the ear as the playing of light on water or a diamond gives to the eye ;

—— splendet tremulo sub lumine pontus [8].

" The organs of the senses resemble the organs of reflexions," is a rule in Perspective ; for the eye is like to a glass [9], or to water ; and it is the same in Acoustics, for the instrument of hearing is like an obstruction in a cavern. These few cases are enough by way of examples. But indeed the chief business of the Persian magic (so much celebrated) was to note the correspondences between the architectures and fabrics of things natural and things civil [10]. Neither are all these which I have mentioned, and others of this kind, only similitudes (as men of narrow observation may perhaps conceive them to be), but plainly the same footsteps of nature treading or printing upon different subjects and matters. And it is a thing which has not as yet been carefully handled. You may perhaps find in the writings of the profounder sort of wits such axioms here and there sparingly inserted for the use of the argument they have in hand ; but for any body of such axioms, which should tend primitively and summarily to the advancement of the sciences, no one has as yet collected one ; though it is a thing of excellent use for displaying the unity of nature ; which is supposed to be the true office of Primitive Philosophy.

There is also another part of this philosophy, which, if you look to the terms, is ancient, if to the thing which I mean, is new. It is an inquiry with regard to the Adventitious Conditions of Essences (which we may call Transcendentals) as Much, Little ; Like, Unlike ; Possible, Impossible ; likewise Being and Not Being, and the like. For since these do not properly come under Physic, and the logical discussion concerning them belongs rather to the laws of reasoning than to the existence of things, it is very proper that the consideration of them (wherein there is no little dignity and profit) should not be altogether neglected, but should find at least some place in the divisions of the sciences. Nevertheless I mean that it should be handled in a way very different from the common. For example : no one who has treated of Much and Little has endeavoured to assign a reason why some things in nature are and can be so numerous and plentiful, others so few and scanty ; for it certainly cannot be that in the nature of things there should be as much gold as iron ; that roses should be as abundant as grass ;

[8] Virg. *Æn.* vii. 9. :—Beneath the trembling light glitters the sea.

[9] Orig. *Speculum.* That the word speculum is here used for " a glass " appears from the corresponding passage in the *Advancement of Learning.* This use of the word, though certainly uncommon, is sanctioned by the authority of C. Agrippa, who, distinguishing lenses from mirrors, calls the former " specula perspicua ". See his celebrated work, *De incertitudine et vanitate scientiarum,* with which Bacon seems, though he has spoken with undeserved contempt of its author, to have been familiar. The phrase used by S. Paul, " we see through a glass," is in the Vulgate " videmus per speculum," but it is at least doubtful whether in both versions it was not intended to suggest the idea of vision by reflected light ; so that the authority of the English translators cannot be cited in support of Bacon's use of the word " speculum ; " though on the other hand there are commentators who affirm that the word used in the original (ἐσόπτρον) means what in Latin is denoted by " speculare," in which case the vision δἰ ἐσόπτρου is of course by transmitted light.

[10] The system of Zoroaster, with which we are but imperfectly acquainted, was at one time the subject of almost as many idle fancies as the philosophy of Hermes Trismegistus. The first idea of the connexion between the Persian magic and the art of government was suggested by the circumstance mentioned in the *Alcibiades* of Plato—that the princes of Persia were by the same persons instructed in politics and in magic.

and that there should be as great variety of the specific as of the non-specific. In like manner no one in handling Similitude and Diversity has sufficiently explained why betwixt different species there almost always lie certain individuals which partake of the nature of both; as moss between corruption and a plant; fishes that stick to rocks and cannot move away, between a plant and an animal ; rats and mice, and some other things, between animals generated of putrefaction and of seed ; bats, between birds and beasts ; flying-fish (which are now well known), between birds and fishes ; seals, between fishes and quadrupeds ; and the like. Nor has any one inquired the reason why, seeing that likes delight in likes, iron does not attract iron, which the magnet does ; nor why gold does not attract gold, though it does attract quicksilver. With regard to these and similar things in the discussion of Transcendentals there is a deep silence ; for men have aimed rather at height of speech than at the subtleties of things. Wherefore I wish the real and solid inquiry, according to the laws of nature and not of language, concerning these Transcendentals or Adventitious Conditions of Essences, to have a place in Primitive or Summary Philosophy. And so much for *Philosophia Prima* (or Sapience), which I have with reason set down as deficient.

CHAPTER II.

Of Natural Theology ; *and the Doctrine concerning* Angels *and* Spirits, *which is an Appendix of the Same.*

THIS science being therefore first placed as a common parent, like unto Berecynthia, who had so much heavenly issue,

Omnes cœlicolas, omnes supera alta tenentes [1],

let us return to the former division of the three philosophies : Divine, Natural, and Human. For Natural Theology is also rightly called Divine Philosophy. It is defined as that knowledge, or rather rudiment of knowledge, concerning God, which may be obtained by the light of nature and the contemplation of his creatures ; and it may be truly termed divine in respect of the object, and natural in respect of the light. The bounds of this knowledge, truly drawn, are that it suffices to refute and convince Atheism, and to give information as to the law of nature ; but not to establish religion. And therefore there was never miracle wrought by God to convert an atheist ; because the light of nature might have led him to confess a God ; but miracles have been wrought to convert idolators and the superstitious, who acknowledged a deity but erred in his worship ; because no light of nature extends to declare the will and worship of God. For as all works show forth the power and skill of the workman, and not his image ; so it is of the works of God ; which show the omnipotency and wisdom, but do not portray the image of the Maker. And therefore therein the Heathen opinion differs from the sacred truth ; for they supposed the world to be the image of God, and man the image of the world ; whereas the Scriptures never vouchsafe to attribute to the world such honour as anywhere to call it the image of God, but only the work of his hands ; but man they directly term the image of God. Wherefore that God exists, that he governs the world, that he is supremely powerful, that he is wise and prescient, that he is good, that he is a rewarder, that he is an avenger, that he is an object of adoration—all this may be demonstrated from his works alone ; and there are many other wonderful mysteries concerning his attributes, and much more touching his regulations and dispensations over the universe, which may likewise be reasonably elicited and manifested from the same ; and this is an argument that has by some been excellently handled. But on the other side, out of the contemplation of nature and elements of human knowledge to induce any conclusion of reason or even any strong persuasion concerning the mysteries of faith, yea, or to inspect and sift them too curiously and search out the manner of the mystery, is in my opinion not safe. "Give unto faith the things which are faith's". For the Heathen themselves

[1] Virg. *Æn.* vii. 788 :—All gods, all dwelling in the heights of heaven.

concede as much, in that excellent and divine fable of the Golden Chain ; namely, that men and Gods were not able to draw Jupiter down to the earth ; but contrariwise, Jupiter was able to draw them up to heaven. And therefore it were a vain labour to attempt to adapt the heavenly mysteries of religion to our reason Fitter will it be that we raise our own minds to the adorable throne of heavenly truth. In this part therefore of Natural Theology I am so far from noting any deficience, that I rather find an excess; to note which I have a little digressed, because of the extreme prejudice and peril which is thereby threatened both to religion and philosophy ; as being that which will make at once an heretical religion and an imaginary and fabulous philosophy.

Otherwise it is of the nature of Angels and Spirits, which is neither inscrutable nor interdicted ; unto which likewise, from the affinity it bears to the human soul, the passage is in great part opened. Certainly the Scripture says, " Let no man deceive you in sublime discourse, touching the worship of angels, pressing into that he knoweth not " [2] ; yet notwithstanding if you observe well that precept, you will find that there are two things only forbidden therein : adoration of them, such as is only due to God, and opinion fantastical of them ; either to extol them further than appertains to the degree of a creature, or to extol a man's knowledge of them further than he has ground. But the sober inquiry about them, either ascending to the knowledge of their nature by the ladder of things corporeal, or beholding it in the soul of man as in a mirror, is nowise forbidden. So of unclean and fallen spirits [3], the conversing with them or the employment of them is prohibited ; much more any worship or veneration towards them. But the contemplation and knowledge of their nature, power and illusions, not only from passages of Scripture, but from reason or experience, is not the least part of spiritual wisdom. So certainly says the Apostle, " We are not ignorant of his stratagems [4] ". And it is no more unlawful to inquire the nature of evil spirits in Natural Theology, than to inquire the force of poisons in Physics, or the nature of vice in Ethics. But this part of knowledge touching angels and spirits I cannot note as deficient, seeing many have occupied themselves in it. I may rather challenge no small part of it, in many of the writers thereof, as superstitious, fabulous, and fantastical.

[2] Coloss. ii. 4. and 18.

[3] The theory of angels and that of fallen spirits form a large and not very profitable chapter in every scholastic Summa Theologiæ. The dogmatic basis of these speculations consists chiefly of spiritualising interpretations (sanctioned by the Fathers and especially by S. Augustine) of certain texts of Scripture and of the supposed visions of Dionysius the Areopagite. The theory of the angelic nature (both in its first and in its fallen state) which the ingenuity of the schoolmen elaborated from these data, is a most remarkable instance of metaphysical creation ; being no less than a determination of the conditions of thought and volition which exist among intelligences of a higher order than our own. That all such determinations are utterly unsatisfactory, both from the want of data and from the inherent and insurmountable difficulty of the problem to be solved, is not however to be denied.

I am not concerned to defend what the schoolmen have said upon the subject ; but I may be allowed to mention in connexion with it an instance of the flippant ignorance with which they are often spoken of. It is said in the history of Martinus Scriblerus that they discussed the question whether angels know things best in the morning. The assertion is of course founded on an absurd mistake of the meaning of the inquiry, " utrum matutina cognitio potior sit quam vespertina ". The doctrine of matutinal and vespertinal cognition the schoolmen derive from S. Augustine, and though neither its subtlety nor the eloquence with which it is expressed can prevent its being censured as an unauthorised speculation, yet no wise man will think it a matter to be jested with. I may refer with respect to it to Buonaventura's commentary on the second book of the *Sentences* of Peter Lombard : *Distinctio* 4 : *Quæstio* 2. The " conclusio " is, " Angelus bonus habet cum matutina vespertinam quoque cognitionem, quæ non temporis sed dignitatis inter se habent ordinem ".

[4] 2 Corinth. ii. 11.

CHAPTER III.

The division of Natural Philosophy into Speculative *and* Operative ; *and that these two should be kept separate, both in the intention of the writer and in the body of the treatise.*

LEAVING therefore Natural Theology (to which I refer the inquiry concerning Spirits as an appendix), let us now proceed to the second part ; namely, that concerning Nature and Natural Philosophy. It was well said by Democritus " That the truth of nature lies hid in certain deep mines and caves[1] ". It was not ill said by the alchemists, " That Vulcan is a second nature, and imitates that dexterously and compendiously which nature works circuitously and in length of time ". Why therefore should we not divide Natural Philosophy into two parts, the mine and the furnace ; and make two professions, or occupations of natural philosophers, some to be miners and some to be smiths ? And certainly though I may seem to say this in sport, yet I think a division of this kind most useful, when propounded in familiar and scholastical terms ; namely, that the doctrine of Natural Philosophy be divided into the Inquisition of Causes, and the Production of Effects ; Speculative and Operative. The one searching into the bowels of nature, the other shaping nature as on an anvil. And though I am well aware how close is the intercourse between causes and effects, so that the explanations of them must in a certain way be united and conjoined ; yet because all true and fruitful Natural Philosophy has a double scale or ladder, ascendent and descendent, ascending from experiments to axioms, and descending from axioms to the invention of new experiments ; therefore I judge it most requisite that these two parts, the Speculative and the Operative, be considered separately, both in the intention of the writer and in the body of the treatise.

CHAPTER IV.

The division of Speculative doctrine concerning nature into Physic (*special*) *and* Metaphysic. *Whereof Physic inquires of the* Efficient Cause *and the* Material ; *Metaphysic of the* Final Cause *and the* Form. *The division of Physic* (*special*) *into the doctrine concerning the* Principles of Things, *concerning the* Fabric of Things, *or the world, and concerning the* Variety of Things. *The division of the doctrine concerning the Variety of Things into doctrine concerning things* Concrete, *and doctrine concerning things* Abstract. *The division of the doctrine concerning things Concrete is referred to the same divisions which Natural History receives. The division of the doctrine concerning things Abstract into doctrine concerning the* Configurations of Matter *and doctrine concerning* Motions. *Two Appendices of Speculative Physic,* Natural Problems *and* Dogmas of the Ancient Philosophers. *The division of Metaphysic into doctrine concerning* Form *and the doctrine concerning* Final Causes.

THAT part of Natural Philosophy which is Speculative and Theoretical, we may divide into Physic special, and Metaphysic ; wherein I desire men to observe that I use the word *metaphysic* in a different sense from that which is commonly received. And here it may be convenient to explain my general purpose touching the use of terms ; which is, as well in this term of metaphysic, as in other cases where my conceptions and notions are novel and differ from the ancient, to retain with scrupulous care the ancient terms ; for hoping well that the very order of the matter and the clear explanation which I give of everything will prevent the words I use from being misunderstood, I am otherwise zealous (as far as may stand with truth and the proficience of knowledge) to recede as little as possible from antiquity, either in terms or opinions. And herein I cannot a little marvel at the boldness of Aristotle, who was stirred by such a spirit of difference and contradiction to wage war on all antiquity, undertaking not only to coin new words of science at pleasure, but to extinguish and obliterate all ancient wisdom ; insomuch that he never names or mentions an ancient author or opinion but to reprove the one and refute the other. For glory, indeed, and drawing followers and disciples, he took the right course therein. For certainly in the promulga-

[1] Diog. Laërt. in Pyrrho. c. 72.

tion and reception of philosophic truth the same thing comes to pass that was noted in the case of divine truth; "I came in my Father's name, and ye received me not; if one shall come in his own name, him ye will receive[1]". But in this divine aphorism, if we consider to whom it was applied (namely, to Anti-Christ, the highest deceiver of all ages), we may discern this well, that the coming in a man's own name, without regard of antiquity or (so to say), of paternity, is no good sign of truth, though it be oftentimes joined with the fortune and success of "Him ye will receive." But of Aristotle, so excellent a person as he was, and so wonderful for the acuteness of his mind, I can well believe that he learnt that humour from his scholar, whom perhaps he emulated; the one aspiring to conquer all nations, the other to conquer all opinions, and to establish for himself a kind of despotism in thought. Wherein nevertheless, it may be, he may at some men's hands, who are of a bitter temper and a sharp tongue, get a like title as his scholar did;

> Felix terrarum prædo, non utile mundo
> Editus exemplum[2]:

so

> Felix doctrinæ prædo, etc.

But to me on the other side (who desire, as much as lies in my pen, to ground a sociable intercourse between the old and the new in learning) it seems best to keep way with antiquity in all things lawful, and to retain the ancient terms, though I often alter their sense and definitions; according to the moderate and approved course of innovation in civil matters, by which, when the state of things is changed, yet the forms of words are kept; as Tacitus remarks, "The names of the magistrates are the same[3]".

To return therefore to the use and acceptation of the term *metaphysic*, as I understand the word. It appears by that which has been already said, that I intend Primitive or Summary Philosophy and Metaphysic, which heretofore have been confounded as one, to be two distinct things. For the one I have made a parent or common ancestor to all knowledge; the other, a branch or portion of Natural Philosophy. Now I have assigned to Primitive Philosophy the common principles and axioms which are promiscuous and indifferent to several sciences. I have assigned to it likewise the question of the Relative and Adventitious Conditions of Essences (which I have termed Transcendentals); as Much, Little Like, Unlike; Possible, Impossible, and the rest; with this provision alone, that they be handled as they have efficacy in nature, and not logically. But the inquiry concerning God, Unity, the nature of Good, Angels and Spirits, I have referred to Natural Theology. It may fairly therefore now be asked, what is left remaining for Metaphysic? Certainly nothing beyond nature; but of nature itself much the most excellent part. And herein without prejudice to truth I may preserve thus much of the conceit of antiquity, that Physic handles that which is most inherent in matter and therefore transitory, and Metaphysic that which is more abstracted and fixed. And again, that Physic supposes in nature only a being and moving and natural necessity; whereas Metaphysic supposes also a mind and idea. For that which I shall say comes perhaps to this. But avoiding all height of language, I will state the matter perspicuously and familiarly. I divided Natural Philosophy into the Inquiry of Causes and the Production of Effects. The Inquiry of Causes I referred to the Theoretical part of Philosophy. This I subdivide into Physic and Metaphysic. It follows that the true difference between them must be drawn from the nature of the causes that they inquire into. And therefore to speak plain and go no further about, Physic inquires and handles the Material and Efficient Causes, Metaphysic the Formal and Final.[4]

[1] St. John, v. 43.

[2] Cf. Lucan, x. 21.:—Great thief of nations, to the world sent forth
A dangerous precedent.
Great thief of learning, etc.
[Bacon has misquoted the passage.]

[3] Tac. Ann. i. 3.

[4] The classification of causes here referred to is Aristotle's. In the first book of the

Physic then comprehends causes vague, variable, respective; but does not aspire to the constant.

> Limus ut hic durescit, et hæc ut cera liquescit,
> Uno eodemque igne [5].

Fire is the cause of induration, but respective to clay; fire is the cause of colliquation, but respective to wax. Now I will divide Physic into three doctrines. For nature is either united and collected, or diffused and distributed. Nature is collected into one, either by reason of the community of the principles of all things, or by reason of the unity of the integral body of the universe. And thus this union of nature has begot two departments of Physic; the one concerning the first principles of things, the other concerning the structure of the universe or the world; which parts I have likewise usually termed the doctrines concerning the Sums of Things. The third doctrine (which handles nature diffused or distributed) exhibits all the varieties and lesser sums of things. Hence it appears that there are three physical doctrines in all: concerning the principles of things; concerning the world or structure of the universe; and concerning nature manifold or diffused. Which last, as I have said, includes all variety of things, and is but as a gloss or paraphrase attending upon the text of natural history. Of these three I cannot report any as totally deficient; but in what truth or perfection they are handled, I make not here any judgment.

But Physic diffused, which touches on the variety and particularity of things, I will again divide into two parts: Physic concerning things Concrete, and Physic concerning things Abstract; or Physic concerning Creatures and Physic concerning Natures. The one (to make use of logical terms) inquires concerning substances, with every variety of their accidents; and the other, concerning accidents, through every variety of substances. For example, if the inquiry be about a lion, or an oak, these support many different accidents; if contrariwise, it be about heat or gravity, these are found in many different substances. But as all Physic lies in a middle term between Natural History and Metaphysic, the former part (if you observe rightly) comes nearer to Natural History, the latter to Metaphysic. Concrete Physic is subject to the same division as Natural History; being conversant either with the heavens or meteors, or the globe of earth and sea, or the greater colleges, which they call the elements, or the lesser colleges or species, as also with pretergenerations and mechanics. For in all these Natural History investigates and relates the fact, whereas Physic likewise examines the causes; I mean the variable causes, that is, the Material and Efficient. Among these parts of Physic, that which inquires concerning the heavenly bodies is altogether imperfect and defective, though by reason of the dignity of

Metaphysics he has applied it, with singular felicity, to the history of philosophical speculation. In order to apprehend its nature, it is necessary to take the word cause in a wider signification than is ordinarily done.

The efficient cause is that which acts—the material cause that which is acted on, as when the fire melts wax, the former is the efficient, the latter the material cause of the effect produced. The formal cause is that which in the case of any object determines it to be that which it is, and is thus the cause of its various properties; it is thus the "ratio essentiæ", the "λόγος τῆς οὐσιας". The final cause is that for the sake of which any effect takes place, whether the agent is or is not intelligent; semper enim *intenditur* finis, non autem semper *cognoscitur*. These four kinds of causes may be divided into two classes, extrinsic and intrinsic; the efficient and final belonging to the first class, the material and formal to the second. It is obvious that these distinctions involve the postulate of what has been called the theory of physical influence, that is, that one substance really acts on another, and must at least be modified if we adopt any such theory on this subject as that of Leibnitz or of Herbart.

[5] Virg. *Ecl.* viii. 80:—

> As the same fire which makes the soft clay hard,
> Makes hard wax soft.

the subject it deserves special consideration. Astronomy has indeed a good foundation in phenomena, yet it is weak, and by no means sound; but astrology is in most parts without foundation even. Certainly astronomy offers to the human intellect a victim like that which Prometheus offered in deceit to Jupiter. Prometheus, in the place of a real ox, brought to the altar the hide of an ox of great size and beauty, stuffed with straw and leaves and twigs. In like manner astronomy presents only the exterior of the heavenly bodies (I mean the number of the stars, their positions, motions, and periods), as it were the hide of the heavens; beautiful indeed and skilfully arranged into systems; but the interior (namely the physical reasons) is wanting, out of which (with the help of astronomical hypotheses) a theory might be devised which would not merely satisfy the phenomena (of which kind many might with a little ingenuity be contrived), but which would set forth the substance, motion, and influence of the heavenly bodies as they really are. For long ago have those doctrines been exploded of the Force of the First Mover and the Solidity of the Heaven,—the stars being supposed to be fixed in their orbs like nails in a roof. And with no better reason is it affirmed, that there are different poles of the zodiac and of the world; that there is a Second Mover of counteraction to the force of the first; that all the heavenly bodies move in perfect circles; that there are eccentrics and epicycles whereby the constancy of motions in perfect circles is preserved; that the moon works no change or violence in the regions above it; and the like. And it is the absurdity of these opinions that has driven men to the diurnal motion of the earth; which I am convinced is most false. But there is scarce any one who has made inquiries into the physical causes, as well of the substance of the heavens both stellar and interstellar, as of the relative velocity and slowness of the heavenly bodies; of the different velocity of motion in the same planet; of the course of motions from east to west, and contrary; of their progressions, stationary positions, and retrogressions; of the elevation and fall of motions in apogee and perigee; of the obliquity of motions, either by spirals winding and unwinding towards the Tropics, or by those curves which they call *Dragons* [6]; of the poles of rotation, why they are fixed in such part of the heaven rather than in any other; and of some planets being fixed at a certain distance from the sun:—such an inquiry as this (I say) has hardly been attempted; but all the labour is spent in mathematical observations and demonstrations. Such demonstrations however only show how all these things may be ingeniously made out and disentangled, not how they may truly subsist in nature; and indicate the apparent motions only, and a system of machinery arbitrarily devised and arranged to produce them,—not the very causes and truth of things [7]. Wherefore astronomy,

[6] The word Draco is mostly used with reference to the Moon's orbit, and denotes the two zones included between it and the ecliptic; the nodes being respectively the Caput and Cauda Draconis. The symbols which are still used both for the nodes of the moon's orbit and for those of other orbits seem derived from this use of the word Draco.

[7] It is difficult to know what mode of investigation Bacon here intends to recommend. The problem of astronomy necessarily is, before any investigation as to the causes of the motions of the heavenly bodies can be undertaken, to determine what those motions really are. The distinction between real motions and apparent motions must be recognised before any progress can be made. And this distinction is not between a fact and a theory in the common acceptation of the words, but between a right theory and a wrong one. Bacon complains that the physical causes of the occasional immobility and regression of the planets have not been inquired into; but in this complaint is involved the theoretic assumption that the planets really are stationary and really do regress. This assumption is made in order to account for their appearing to us to change the direction of their motion. It is the obvious explanation, but nevertheless a wrong one; and if the phenomena in question are not physical phenomena but optical, to what purpose is it to attempt to assign physical causes for them? And so in the other cases which he mentions. The value of any hypothesis for the explanation of the phenomena of course depends on its simplicity and its completeness, and the attempt to reduce all the celestial motions to perfect circles was at the time at which it was made a great step in advance; though the idea of circular motion was unduly retained when it was found to

as it now is, is fairly enough ranked among the mathematical arts, not without disparagement to its dignity; seeing that, if it chose to maintain its proper office, it ought rather to be accounted as the noblest part of Physics. For whoever shall set aside the imaginary divorce between superlunary and sublunary things, and shall well observe the most universal appetites and passions of matter (which are powerful in both globes and make themselves felt through the universal frame of things), will obtain clear information of heavenly things from those which are seen amongst us; and on the other hand, from that which passes in the heavens he will gain no slight knowledge of some motions of the lower world as yet undiscovered; not only in as far as the latter are influenced by the former, but in as far as they have common passions[8]. Wherefore this, the physical part of astronomy, I pronounce deficient; giving it the name of *Living Astronomy*, in distinction from that stuffed ox of Prometheus, which was an ox in figure only.

As for Astrology, it is so full of superstition, that scarce anything sound can be discovered in it. Notwithstanding, I would rather have it purified than altogether rejected. If however anyone maintains that this science is not based on reason or physical speculations, but on blind experience and the observations of many ages, and on that ground refuses the test of physical reasons (as the Chaldeans professed to do): he may on the same grounds bring back auguries, and believe in divination, entrails, and all kinds of fables; for all these are set forth as the dictates of long experience and traditions passed from hand to hand. But for my part I admit astrology as a part of Physic, and yet attribute to it nothing more than is allowed by reason and the evidence of things, all fictions and superstitions being set aside. To consider the matter however a little more attentively. In the first place what an idle invention is that, that each of the planets reigns in turn for an hour, so that in the space of twenty-four hours each has three reigns, leaving three hours over! And yet this conceit was the origin of our division of the week (a thing so ancient and generally received); as is very evident from the alternation of days; for the ruling planet at the beginning of the succeeding day is always the fourth in order from the planet of the previous one, by reason of the three supernumerary hours of which I have spoken[9]. Secondly, I do not hesitate to reject as an idle superstition the doctrine of horoscopes and the distribution of houses; which is the very delight of astrology, and has held a sort of Bacchanalian revelry in the heavenly regions. Nor can I

be producing not simplicity but complication. But consciously or unconsciously the mind is always introducing principles of arrangement (ideas or hypotheses) among the objects of its attention, and the error of the passage in the text is in effect the common one of assuming that the form of hypothesis with which the mind happens to be familiar is on that account an absolute fact. It is well to remark, as the Newtonian philosophy is often spoken of as the great result of Bacon's methods, that none of Newton's astronomical discoveries could have been made, if astronomers had not continued to render themselves liable to Bacon's censure.

[8] This prediction has been fulfilled by the history of physical astronomy, and the information gained respecting the "motus inferiores" may be divided into two parts, "quatenus hi ab illis regantur" and "quatenus habeant passiones communes". To the first belong the theory of the tides and those of precession and nutation, to the second that of the earth's figure, which depends on the law of universal gravitation, and which therefore may be said to be a result of our knowledge of celestial phenomena. The way in which what takes place in one part of the solar system is, so to speak, reflected in others, is one of the most interesting subjects in physical astronomy.

[9] This explanation of the origin of the names of the days of the week is given by Dio Cassius, xxxvii. c. 21. He also gives another which is free from an objection which has been alleged against the first; namely that the names are older than the division of the day into twenty-four hours. It is that the successive days were assigned to the respective planets which are fourth in order from each other, from some notion of analogy in the divine harmony to a musical progression by fourths. Joseph Scaliger, as quoted by Selden, deduces the order of progression from the properties of a heptagon inscribed in a circle. See on this subject a very learned essay by Archdeacon Hare in the first volume of the *Philological Museum*.

sufficiently wonder how illustrious men and eminent in astrology have rested them on such slight foundations; for they say that as experience proves that the solstices, equinoxes, new moons, full moons, and the greater revolutions of the stars, exercise a great and manifest influence over natural bodies, it follows that the more exact and subtle positions of the stars must produce effects likewise more exquisite and secret. But they ought first to have excepted the operations of the sun by manifest heat, and likewise the magnetic influence of the moon on the half-monthly tides (for the daily ebb and flow of the sea is another thing), and then they will find the powers of the rest of the planets over natural things (as far as they are approved by experience) very weak and slight, and almost invisible, even in the greater revolutions. And therefore they should argue in a manner directly contrary; that as those greater revolutions have so little influence, these nice and minute differences of positions have no power at all. Thirdly, those fatalities, that the hour of nativity or conception influences the fortune of the birth, the hour of commencement the fortune of the enterprise, the hour of inquiry the fortune of the thing inquired into, and in short, the doctrines of *nativities*, *elections*, *inquiries*, and the like frivolities, have in my judgment for the most part nothing sure or solid, and are plainly refuted and convicted by physical reasons. It remains therefore to declare what I retain or approve of in astrology, and what is deficient in that which I approve. For this last it is (the pointing out of deficiencies) which is the object of this discourse; for otherwise (as I have often said) I cannot stay to censure. Among the received doctrines, then, I think that concerning *revolutions* has more soundness than the rest. But it will perhaps be better to lay down certain rules, as a standard by which we may weigh and examine astrological matters, so as to retain what is useful and to reject what is frivolous. First then, as I have before advised, let the greater revolutions be retained, but the smaller revolutions of horoscopes and houses be dismissed[10]. The former are like great guns, and can strike from afar; the latter are like little bows, and cannot transmit their force over much space. Secondly; the operation of the heavenly bodies does not affect all kinds of bodies, but only the more tender; such as humours, air and spirit; here however the operations of the heat of the sun and heavenly bodies must be excepted; which doubtless penetrates both to metals and to a great number of subterraneous bodies. Thirdly every operation of heavenly bodies extends rather to masses than to individuals; though it affects indirectly some individuals also; such, namely, as are more susceptible, and of softer wax as it were, than the rest of their species; as when a pestilent condition of air seizes on the less resisting bodies and passes by those which have more power of resistance. The fourth rule is not unlike the preceding; every operation of the heavenly bodies sheds its influence and power, not on small periods of time or within narrow limits, but upon the larger spaces. And therefore predictions of the temperature of the year may possibly be true; but those of particular days are rightly held of no account. The last rule (which has always been held by the wiser astrologers) is that there is no fatal necessity in the stars[11], but that they rather incline than compel. I will add one thing besides (wherein I shall certainly seem to take part with astrology, if it were reformed); which is, that I hold it for certain that the celestial bodies have in them certain other influences besides heat and light; which very influences however act by those rules laid down above, and not otherwise. But

[10] The heavens are in astrology divided into twelve compartments or houses, by means of six great circles which pass through the north and south points of the horizon, and divide the ecliptic into twelve equal portions. One of these circles coincides with the horizon, and the point of the ecliptic through which it passes at the moment of the nativity of the person whose destiny is to be ascertained, or of the commencement of the event whose fortunes are to be predicted, is called the horoscope. These divisions are spoken of by Sextus Empiricus, who with Julius Firmicus is our earliest authority on the subject of astrology. He seems rather to give the name of houses to definite signs of the Zodiac than to the divisions of which we have been speaking; a sense in which the term is also used by later writers.

[11] This saying is commonly ascribed to Ptolemy.

these lie concealed in the depths of Physic, and require a longer dissertation. I have thought fit therefore (on due consideration of what has been said) to set down as a desideratum an astrology framed in conformity with these principles; and as I have termed Astronomy based on Physical Reasons *Living Astronomy*, so Astrology similarly grounded I call *Sane Astrology*. And though what I have already said will in no slight degree contribute to the rectification and completion of this art, yet according to my custom I will add a few remarks which will clearly explain out of what materials it should be composed, and to what end it should be applied. In the first place, let there be received into Sane Astrology the doctrine concerning the commixture of rays; that is the conjunctions, oppositions, and other combinations or aspects of planets with regard to one another. And to this same part also I refer the passage of the planets through the signs of the zodiac, and their position under the same signs; for the position of a planet under a sign is a kind of conjunction of it with the stars of that sign. And in like manner also ought the oppositions and other combinations of the planets with regard to the stars of the signs to be observed; which has not hitherto been fully done. But these commixtures of the rays of fixed stars with one another, though useful in contemplating the structure of the universe and the nature of the regions lying below them, are of no avail for predictions, because they are always alike. Secondly, let there be received the approaches of each individual planet to the perpendicular, and its regressions from it, according to the climate of countries. For every planet, no less than the sun, has its summer and winter, in which as its rays fall more or less perpendicular, their force is stronger or weaker. For I have no doubt but that the moon in Leo has more power over natural bodies in our planet than when in Pisces; not because when in Leo the moon affects the heart, and when in Pisces the feet, as they talk; but by reason of her elevation towards the perpendicular and approximation to the larger stars, in the same manner as the sun [12]. Thirdly, let there be received the apogees and perigees of the planets, with a sufficient distinction as to what is due to the inherent vigour of the planet, and what to its proximity to us. For a planet is more active in its apogee or elevation, but more communicative in its perigee or descent. Fourthly, let there be received (to speak summarily) all the remaining accidents of the motions of planets; what are the accelerations and retardations of each in its course; what their progressions, actions, and regressions; what their distances from the sun, combustions, increases and diminutions of light, eclipses and the like; for all these things help to make the rays of the planets act more forcibly or more feebly, and in different modes and with different virtues. These four remarks relate to the radiations of the stars. Fifthly, let everything be received which may in any way disclose and explain the natures of the stars, whether erratic or fixed, in their proper essence and activity; as their size, their colour and aspect, their twinkling and vibration of light, their situation with reference to the poles or the exquinox, their asterisms; which of them are more mingled with other stars and which more solitary; which are higher, and which lower; which of the fixed stars are within the path of the sun and planets (that is within the zodiac), and which without; which of the planets is swifter, which slower; which of them move in the ecliptic, and which deviate to right or left of it; which of them may be retrograde, and which cannot; which of them may be at any distance from the sun, and which of them are confined to a certain limit; which of them move swifter in perigee, which in apogee; finally the anomalies of Mars, the wandering of Venus, the labours and wonderful passions which have been detected more than once both in the Sun and Venus [13]; and any other things of the like nature. Lastly let there be received also the particular natures and inclinations of the planets, and likewise of the fixed stars, as handed down by tradition; which as they are transmitted with very general consent, ought not (except when they are plainly at variance with physical reasons) to be lightly rejected. From such observations is Sane Astrology constructed, and by them alone should schemes of the heavens be formed and interpreted.

[12] The reason which Bacon rejects seems to be nearly as conclusive as that which he admits. [13] See the *Descriptio Globi Intellectualis* for some account of these passions.

Sane Astrology is applied more confidently to *predictions*, but more cautiously to *elections*; in both cases however within due limits. *Predictions* may be made of comets to come, which (I am inclined to think) may be foretold; of all kinds of meteors, of floods, droughts, heats, frosts, earthquakes, irruptions of water, eruptions of fire, great winds and rains, the various seasons of the year, plagues, epidemic diseases, plenty and dearth of grain, wars, seditions, schisms, transmigrations of peoples, and in short of all commotions or greater revolutions of things, natural as well as civil. But these predictions may also be made (though not so certainly) with reference to events more special and perhaps singular, if after the general inclinations of such times and seasons have been ascertained, they be applied with a clear judgment, either physical or political, to those species or individuals which are most liable to accidents of this nature; as for instance, if any one from a foreknowledge of the seasons of the year shall pronounce them more favourable or injurious to olives than to vines, to pulmonary than to liver complaints, to the inhabitants of hills than to those of valleys, to monks than to courtiers (by reason of their different manner of living); or if any one from knowledge of the influence which celestial bodies have upon human minds should discover it to be more favourable or more adverse to peoples than to kings, to learned and inquisitive men than to bold and warlike, to men of pleasure than to men of business or politicians. There are innumerable things of this kind; but they require (as I said before) not only that general knowledge derived from the stars (which are actives), but also a particular knowledge of the subjects (which are passives). Nor are *elections* to be altogether rejected; but less confidence is to be placed in them than in predictions. For we see that in planting and sowing and grafting, observation of the age of the moon is a thing not altogether frivolous. And there are many instances of the kind. But these elections also, even more than predictions, must be guided by our rules. And it must always be observed, that elections hold good in those cases only where both the virtue of the heavenly bodies is such as does not quickly pass, and the action of the inferior bodies is such as is not suddenly accomplished; which is the case in those examples cited above; for neither the changes of the moon nor the growth of plants are effected in an instant. As for those which depend upon exactness to a moment, they are to be rejected altogether. But many such cases are to be found likewise (though a man would not think it) in elections concerning civil matters. And if any one complains that while I have given some indication of the materials from which this improved astrology may be extracted, and likewise of the purpose for which it may be advantageously used, I have said nothing about the manner of extracting it, he does not deal fairly with me; for he requires of me the art itself, for which I am not accountable. Upon the question which he asks however I will say thus much. There are four ways only by which this science can be approached. First by future experiments; secondly by past experiments; thirdly by traditions; and lastly by physical reasons. With regard to future experiments, what need is there of saying anything? seeing it requires many ages to collect a sufficient number of them; so that it is useless to speculate about them. For past experiments, they are no doubt within man's reach; though to collect them is a work of great labour, and one requiring much leisure. For astrologers (if they would do themselves justice) may faithfully extract from history all the greater disasters (as inundations, plagues, battles, seditions, deaths of kings, and the like), and may examine (not according to the subtleties of horoscopes, but by those rules of revolutions which I have shadowed out) what the position of the heavenly bodies was at the times; so that where there is found a manifest agreement and coincidence of events, there a probable rule of predicted may be established. As to traditions, they must be carefully sifted, that what is plainly repugnant to physical reasons may be rejected, and what is in conformity with them may stand upon its own authority. Lastly of physical reasons, those are most adapted to this investigation which make inquiry into the universal appetites and passions of matter, and the simple and genuine motions of bodies. For upon these wings we ascend most safely to these celestial material substances. And so much for Sane Astrology.

Of astrological insanity (besides those fictions which I remarked above) there is another portion, which must not be omitted; though it ought properly to be excluded from astrology, and removed to what is called celestial magic. It rests upon a wonderful figment of the human mind,—namely, that any favourable position of the stars may be received on seals or signets (say of some metal or gem qualified for the purpose), by which the felicity of the hour, which would otherwise pass, may be arrested and as it were fixed as it flies. So the poet complains heavily that so noble an ancient art should have been lost.

> Annulus infuso non vivit mirus Olympo,
> Non magis ingentes humili sub lumine Phœbos
> Fert gemma, aut celso divulsas cardine Lunas [14].

It is true that the relics of saints and their virtues have been allowed by the Church of Rome (for in divine and immateriate things lapse of time does not matter); but to treasure up the relics of heaven, whereby the hour which is already past and as it were dead should revive and be continued, is mere superstition. Let these fancies then be dismissed, if the Muses be not turned to old women.

Abstract Physics may most rightly be divided into two parts—the doctrine concerning the Configurations of Matter, and the doctrine concerning Appetites and Motions. Both of these I will cursorily enumerate, and thence may be derived some shadow of the true Physic of Abstracts. The *Configurations of Matter* are Dense, Rare; Heavy, Light; Hot, Cold; Tangible, Pneumatic; Volatile, Fixed; Determinate, Fluid; Moist, Dry; Fat, Crude; Hard, Soft; Fragile, Tensile; Porous, Close; Spirituous, Jejune; Simple, Compound; Absolute, Imperfectly Mixed; Fibrous and Venous, Simple of Structure, or Equal; Similar, Dissimilar; Specific, Non-Specific; Organic, Inorganic; Animate, Inanimate. Further I do not go. For Sensible and Insensible, Rational and Irrational, refer to the doctrine concerning Man. Of *Appetites* and *Motions* there are two kinds. There are simple motions, in which lies the root of all natural actions, subject to the conditions of the configurations of matter. And there are Compound or Produced Motions; with which last the received philosophy (which takes but slight hold of the body of nature) commences. But compound motions of this kind (such as Generation, Corruption, and the rest) ought to be accounted as the sums or products of simple motions, rather than as primitive motions. The simple motions are, motions of *Resistance*—commonly called motion to prevent penetration of dimensions; motion of *Connexion*—which they call motion of abhorrence of a vacuum; motion of *Liberty*, to prevent preternatural compression or extension; motion into a *New Sphere*, or for the purpose of rarefaction or condensation; motion of the *Second Connexion*, or to prevent solution of continuity; motion of the *Greater Congregation*, or towards masses of connatural bodies, commonly called natural motion; motion of *Lesser Congregation*, commonly called motion of sympathy and antipathy; motion of *Disposition*, or for the ordering of the parts with reference to the whole; motion of *Assimilation*, or multiplication of its own nature upon another body; motion of *Excitation*, where the nobler agent excites a motion dormant and latent in another; motion of *Signature* or *Impression;* that is, operation without communication of substance;

[14] Not now the ring can in its circlet store
Heaven's living influence: the gem no more
Beneath its modest lustre bears the might
Of the great orbs that govern day and night.

I have not been able to discover whence these lines are taken. The notion they refer to gave rise to the word "Talisman," which seems to be a modification of the Greek word τέλεσμα, used like στοιχείωμα in the sense of a configuration of the heavenly bodies. See Salmasius *De Annis Climactericis*, and compare Von Hammer on Talismans, in the *Mines de l'Orient*. For this last reference I am indebted to the kindness of Mr. Scott, of Trinity College, Cambridge. See also Heyne, *Opuscula*, vol. 6., and the work to which he refers, namely the *Speculum Lapidum* of Camillus Leonardus, book 3rd. Some other references will be found in Le Roux de Lincy, *Livre des Légendes*.

motion of *Royalty*, or restraint of other motions by the motion predominant; motion *without limit*, or spontaneous rotation; motion of *Trepidation*, or Systole and Diastole, in bodies (that is) placed between things attractive and repugnant; lastly, motion of *Repose*, or *abhorrence of motion*, which is also causative of very many things. Such are Simple Motions; which truly proceed from the inward recesses of nature, and which by complication, continuation, alternation, restraint, repetition, and various modes of combination, form those compound motions or sums of motions which are generally received, or others like them. The sums of motion are those motions so much talked of,—generation, corruption; augmentation, diminution; alteration, and local motion; likewise mixture, separation; conversion. There remain as Appendices of Physic, the *measurements of motions*; namely, what is the effect of the *how much* or *dose* in nature; what of distance, which is not unfitly called the orb of virtue or activity[15]; what of rapidity or slowness; what of short or long delay; what of the force or dulness of the thing; what of the stimulus of surrounding things. And these are genuine parts of the true Physic of Abstracts; for in the configurations of matter in simple motions, in the sums or aggregates of motions, and in the measures of motions, the Physic of Abstracts is perfected. For voluntary motion in animals; the motion which takes place in the actions of the senses; motion of imagination, appetite, and will; motion of the mind, determination, and intellectual faculties; these I refer to their own proper doctrines. I repeat however that all these above mentioned are to be no further handled in Physic than the inquiry of their Material and Efficient causes; for as to their Formal and Final causes they are rehandled in Metaphysic.

I will subjoin two notable appendices of Physic, which regard not so much the matter as the manner of inquiry; namely *Problems of Nature* and *Dogmas of Ancient Philosophers*. The first is an appendix to nature manifold or scattered; the other, to nature united or summary. Both relate to the skilful proposing of *Doubts*; which is no despicable part of science. Problems deal with particular doubts; Dogmas with general ones, concerning first principles and the fabric of the universe. Of Problems there is a noble example in the books of Aristotle; a kind of work which certainly deserved not only to be honoured with the praises of posterity but to be continued by their labours; seeing that new doubts are daily arising. In this however there is a caution to be applied, which is of great importance. The registering and proposing of doubts has a double use; first it guards philosophy against errors, when upon a point not clearly proved no decision or assertion is made (for so error might beget error), but judgment is suspended and not made positive; secondly, doubts once registered are so many suckers or sponges which continually draw and attract increase of knowledge; whence it comes that things which, if doubts had not preceded, would have been passed by lightly without observation, are through the suggestion of doubt attentively and carefully observed. But these two advantages are scarcely sufficient to countervail one inconvenience which will intrude itself, if it be not carefully debarred; which is that a doubt if once allowed as just, and authorised as it were, immediately raises up champions on either side, by whom this same liberty of doubting is transmitted to posterity; so that men bend their wits rather to keep the doubt up than to determine and solve it. Of this examples everywhere occur both in lawyers and scholars, who when a doubt has been once admitted will have it remain for ever a doubt, and hold to authority in doubting as much as in asserting; whereas the legitimate use of reason is to make doubtful things certain and not certain things doubtful. Wherefore I say that a *calendar of doubts* or *problems in nature* is wanting, and I would wish it to be taken in hand; if only care be taken that as knowledge daily increases (which it certainly will, if men listen to me) those doubts which are clearly sifted and settled be blotted out from the list. And to this calendar I would annex another of no less utility; for seeing that in every inquiry there are found things plainly true, things doubtful, and things plainly false, it would be most advantageous to add to the calendar of doubts a calendar of *falsehoods and popular errors* prevalent

[15] The allusion is to Gilbert. See note pp. 444–5.

either in natural history or the dogmas of philosophers ; that the sciences may be no longer troubled with them.

With regard to the dogmas of the ancient philosophers, as those of Pythagoras, Philolaus, Xenophanes, Anaxagoras, Parmenides, Leucippus, Democritus, and the rest, (which men usually pass over with disdain), it will not be amiss to look upon them somewhat more modestly. For though Aristotle, after the Ottoman fashion, thought that he could not reign with safety unless he put all his brethren to death [16], yet for those who aim not at dominion or authority but at the inquiry and illustration of truth, it cannot but seem a useful thing to behold at one view the several opinions of different men touching the nature of things. Not however that there is any hope of gaining any truth of the purer kind from these or the like theories. For as the same phenomena, the same calculations, are compatible with the astronomical principles both of Ptolemy and Copernicus ; so this common experience of which we are now in possession, and the ordinary face of things, may adapt itself to many different theories ; whereas to find the real truth requires another manner of severity and attention. For as Aristotle says elegantly " that children when they begin to lisp call every woman mother, but afterwards come to distinguish their own," [17], so certainly experience when in childhood will call every philosophy mother, but when it comes to ripeness it will discern the true mother. In the meantime it will be good to peruse the several differing systems of philosophy, like different glosses upon nature ; whereof it may be that one is better in one place and another in another. Therefore I wish a work to be compiled with diligence and judgment out of the lives of the ancient philosophers, the collection of *placita* made by Plutarch, the citations of Plato, the confutations of Aristotle, and the scattered notices which we have in other books, both ecclesiastical and heathen (Lactantius, Philo, Philostratus, and the rest), concerning the *ancient philosophies*. For I do not find any such work extant. But here I must give warning that it be done distinctly, so that the several philosophies may be set forth each throughout by itself, and not by titles packed and faggoted up together, as has been done by Plutarch. For when a philosophy is entire, it supports itself, and its doctrines give light and strength the one to the other ; whereas if it be broken, it will seem more strange and dissonant. Certainly when I read in Tacitus of the actions of Nero or Claudius, invested with all the circumstances of times, persons, and occasions, I see nothing in them very improbable ; but when I read the same in Suetonius Tranquillus, gathered into titles and common places, and not presented in order of time, they seem something prodigious and quite incredible. And the case is the same in philosophy, when propounded entire and when dissected and dismembered. Neither do I exclude from this calendar of the dogmas of the old philosophers modern theories and doctrines ; such as that of Theophrastus Paracelsus, eloquently reduced into a body and harmony by Severinus the Dane ; or that of Telesius of Consentium, who revived the philosophy of Parmenides, and so turned the weapons of the Peripatetics against themselves ; or of Patricius the Venetian, who sublimated the

[16] Bacon, it is probable, alludes particularly to a memorable and then recent instance of this practice. Mahomet III., on becoming Sultan in 1595, put to death nineteen of his brothers and ten or twelve women supposed to be with child by his father. Pope, perhaps unconsciously, has imitated Bacon. In the character of Addison, he speaks of him as one who could

"Bear, like the Turk, no brother near the throne".

It is worthy of remark that the practice in question was established as a fundamental law of the state by Mahomet the Second. I quote his words from the French edition of Von Hammer's *History of the Ottoman Empire.* " La plupart des légistes ont déclaré que ces de mes illustres fils ou petits-fils qui monteront au trône pourront faire exécuter leurs frères afin d'assurer le repos du monde—ils devront agir en conséquence."—*L'Histoire de l'Empire Ottoman*, iii. p. 392.

A little further on Von Hammer remarks that " la légalité du meurtre est consacrée non seulement pour les frères du Sultan mais encore pour ses neveux et ses petits-fils ".

[17] *Physica.* i. c. 1.

fumes of the Platonists;[18] or of our countryman Gilbert, who revived the doctrines of Philolaus[19]; or of any other worthy to be admitted. Of these however (since their entire works are extant) I would only have summaries made therefrom and added to the rest. And so much for Physic and its Appendices.

For Metaphysic, I have already assigned to it the inquiry of Formal and Final Causes; which assignation, as far as it relates to Forms, may seem nugatory; because of a received and inveterate opinion that the Essential Forms or true differences of things cannot by any human diligence be found out; an opinion which in the meantime implies and admits that the invention of Forms is of all parts of knowledge the worthiest to be sought, if it be possible to be found. And as for the possibility of finding it, they are ill discoverers who think there is no land where they can see nothing but sea. But it is manifest that Plato, a man of sublime wit (and one that surveyed all things as from a lofty cliff)[20], did in his doctrine concerning Ideals descry that Forms were the true object of knowledge; howsoever he lost the fruit of this most true opinion by considering and trying to apprehend Forms as absolutely abstracted from matter; whence it came that he turned aside to theological speculations, wherewith all his natural philosophy is infected and polluted. But if we fix our eyes diligently, seriously and sincerely upon action and use, it will not be difficult to discern and understand what those Forms are the knowledge whereof may wonderfully enrich and benefit the condition of men. For as to the Forms of Substances (Man only excepted, of whom the Scripture saith, " That He made man of the dust of the earth and breathed into his nostrils the breath of life "[21], and not as of all other creatures, " Let the earth or the waters bring forth ",[22])—the Forms of Substances I say (as they are now by compounding and transplanting multiplied) are so perplexed and complicated, that it is either vain to inquire into them at all, or such inquiry as is possible should be put off for a time, and not entered upon until forms of a more simple nature have been rightly investigated and discovered. For as it would be neither easy nor of any use to inquire the form of the sound which makes any word, since words, by composition and transposition of letters, are infinite; whereas to inquire the form of the sound which makes any simple letter (that is, by what collision or application of the instruments of voice it is produced) is comprehensible, nay easy; and yet these forms of letters once known will lead us directly to the forms of words; so in like manner to inquire the form of a lion, of an oak, of gold, nay even of water or air, is a vain pursuit; but to inquire the form of dense, rare, hot, cold, heavy, light, tangible, pneumatic, volatile, fixed, and the like, as well configurations as motions, which in treating of Physic I have in great part enumerated (I call them *Forms of the First Class,*) and which (like the letters of the alphabet) are not many and yet make up and sustain the essences and forms of all substances[23];—this, I say, it is which I am attempting, and

18 Severinus was a Danish physician. He died in the year 1602, leaving several works on medical and philosophical subjects, in which he followed the opinions of Paracelsus. I am only acquainted with his *Idea Medicinæ Philosophicæ,* which there is reason to think Bacon had read. His writings are in point of style much superior to those of Paracelsus, who was however unquestionably a man of far more original genius.

Telesius's principal work is his *De Rerum Naturâ* [the first two books of which were published in 1565, and the whole in 1586]. Bacon derived more ideas from him than from any other of the " novelists," as he has somewhere called the philosophical innovators, and has written a separate treatise (the *Descriptio Globi Intellectualis*) on three systems of philosophy, of which his is one.

Patricius attempted to amalgamate the Platonic and Aristotelian philosophies. His principal work, entitled *Nova de Universis Philosophia,* was published in 1591.

19 As to the movement of the earth. See Diogenes Laërtius.

20 The reference is probably to the passage in the *Philebus* (p. 17 *et infra*) in which Plato speaks of the analysis of sounds into their constituent elements.

21 Gen. ii. 7. 22 Gen. i. 20. 24.

23 It clearly appears from this passage that Bacon's doctrine was that the forms of all substances might be determined by combining the results of a limited number of investigations of the forms of schematisms and motions, or as he elsewhere calls them, of

which constitutes and defines that part of Metaphysic of which we are now inquiring. Not but that Physic takes consideration of the same natures likewise (as has been said); but that is only as to their variable causes. For example; if the cause of whiteness in snow or froth be inquired, it is well rendered, that it is the subtle intermixture of air and water. But nevertheless this is far from being the form of whiteness, seeing that air intermixed with powdered glass or crystal would create a similar whiteness, no less than when mixed with water; it is only the efficient cause, which is nothing else than the vehicle of the form. But if the inquiry be made in Metaphysic you will find something of this sort; that two transparent bodies intermixed, with their optical portions arranged in a simple and regular order, constitute whiteness. This part of Metaphysic I find deficient; whereat I marvel not, because I hold it not possible that the Forms of things can be invented by that course of invention hitherto used; the root of the evil, of as all others, being this; that men have used to sever and withdraw their thoughts too soon and too far from experience and particulars, and have given themselves wholly up to their own meditations and arguments.

But the use of this part of Metaphysic, which I reckon amongst the deficient, is of the rest the most excellent in two respects; the one, because it is the duty and virtue of all knowledge to abridge the circuits and long ways of experience (as much as truth will permit), and to remedy the ancient complaint that "life is short and art is long"[24]. And this is best performed by collecting and uniting the axioms of sciences into more general ones, and such as may comprehend all individual cases. For knowledges are as pyramids, whereof history and experience are the basis. And so of Natural Philosophy the basis is Natural History; the stage next the basis is Physic; the stage next the vertical point is Metaphysic. As for the cone and vertical point ("the work which God worketh from the beginning to the end"[25], namely, the summary law of nature) it may fairly be doubted whether man's inquiry can attain to it. But these three are the true stages of knowledge; which to those that are puffed up with their own knowledge, and rebellious against God, are indeed no better than the giants' three hills;

> Ter sunt conati imponere Pelio Ossam,
> Scilicet atque Ossæ frondosum involvere Olympum[26]:

but to those who abasing themselves refer all things to the glory of God, they are as the three acclamations: Holy, Holy, Holy. For God is holy in the multitude of his works, holy in the order or connexion of them, and holy in the union of them. And therefore the speculation was excellent in Parmenides and Plato (although in them it was but a bare speculation), "that all things by a certain scale ascend to unity"[27]. So then always that knowledge is worthiest which

simple natures. (See *Novum Organum*, ii. 5.) For the phrase "Formæ primæ classis;" see infra p. 471. The difficulty of effecting this combination might be insuperable; he did not profess to be able to decide à priori that it was not so; but at any rate it would be only a synthetical difficulty and would not present itself until his analysis of nature was completed and the forms of her constituent elements determined. Of the possibility of attaining these two ends—namely (1) an analysis of nature resulting in the formation of a complete list of "naturæ simplices," and (2) the determination of their forms—he seems never to have doubted.

[24] Hippocrates, *Aph.* i. 1. [25] Eccles. iii. 11.

[26] Virg. *Georg.* i. 281.:—Mountain on mountain thrice they strove to heap.
Olympus, Ossa, piled on Pelion's steep.

[27] No such doctrine as this is to be found in the remains which have come down to us of the writings of Parmenides, and it is in effect inconsistent with what we know of his opinions. His fundamental dictum appears to have been that that which is, is one; incapable of change or motion. That visible things are in any sense parts or elements or attributes of the one immutable substance is, as far as we can judge, a later doctrine. To the question, what then are the phenomena of the visible universe, Parmenides gives no answer; unless we account as an answer what he says of their delusive and non-existent character. Even Plato was far from teaching the doctrine of an ascent to unity in the sense in which Bacon probably employed the terms. He no doubt adopted in

least burdens the intellect with multiplicity ; and this appears to be Metaphysic, as that which considers chiefly the simple forms of things (which I have above termed *forms of the first class*) ; since although few in number, yet in their commensurations and co-ordinations they make all this variety. The second respect which ennobles this part of Metaphysic, is that it enfranchises the power of men to the greatest liberty, and leads it to the widest and most extensive field of operation. For Physic carries men in narrow and restrained ways, imitating the ordinary flexuous courses of Nature ; but the ways of the wise are everywhere broad ; to wisdom (which was anciently defined to be the knowledge of things divine and human [28]) there is ever abundance and variety of means. For physical causes give light and direction to new inventions in similar matter. But whosoever knows any Form, knows also the utmost possibility of superinducing that nature upon every variety of matter, and so is less restrained and tied in operation, either to the basis of the matter or to the condition of the efficient ; which kind of knowledge Solomon likewise, though in a more divine sense, elegantly describes, " Thy steps shall not be straitened, and when thou runnest thou shalt not stumble " [29] ; meaning thereby that the ways of wisdom are not much liable either to straitness or obstructions.

The second part of Metaphysic is the inquiry of Final Causes, which I report not as omitted, but as misplaced. For they are generally sought for in Physic, and not in Metaphysic. And yet if it were but a fault in order I should not think so much of it ; for order is matter of illustration, but pertains not to the substance of sciences. But this misplacing has caused a notable deficience, and been a great misfortune to philosophy. For the handling of final causes in physics has driven away and overthrown the diligent inquiry of physical causes, and made men to stay upon these specious and shadowy causes, without actively pressing the inquiry of those which are really and truly physical ; to the great arrest and prejudice of science. For this I find done, not only by Plato, who ever anchors upon that shore, but also by Aristotle, Galen [30], and others, who also very frequently strike upon these shallows. For to introduce such causes as these, " that the hairs of the eyelids are for a quickset and fence about the sight " ; or " that the firmness of the skins and hides of living creatures is to defend them from the extremities of heat and cold " ; or " that the bones are for columns or beams, whereupon the frames of the bodies of living creatures are built " ; or " that the leaves of trees are for protecting the fruit from the sun and wind " ; or " that the clouds are formed above for watering the earth " ; or " that the thickness and solidity of the earth is for the station and mansion of living creatures ", and the like, is a proper inquiry in Metaphysic, but in Physic it is impertinent. Nay, as I was going to say, these discoursing causes (like those fishes they call *remoras*, which are said to stick to the sides of ships) have in fact hindered the voyage and progress of the sciences, and prevented them from holding on their course and advancing further ; and have brought it to pass that the inquiry of physical causes has been long neglected and passed in silence. And therefore the natural philosophy of Democritus and others, who removed God and Mind from the structure of things, and attributed the form thereof to infinite essays and proofs

his own sense the dictum of the Eleatæ, ἓν τὰ πάντα ; but with him as with them mere phenomena have no true existence. In later writers, however, Bacon may easily have found expressions derived from the authority of Plato and Parmenides, and more consonant with his own views of the nature of the universe. But so far as they themselves were concerned, it may I think be safely stated that though the latter affirmed the ἑνότης of that which exists, no doctrine of ἕνωσις entered into his teaching ; and that that which presents itself in the system of the former was essentially different from Bacon's ascent to unity.

28 See Cicero, *Tusc. Quæst.* iv. 26. 29 Prov. iv. 12.

30 See especially Galen's *De usu Partium*, which is in effect a treatise on the doctrine of final causes as exemplified in animal physiology. He calls the last book, which introduces the general considerations to which the subject leads, the Epode of the whole work ; explaining that he does so, because the Epode is sung while the chorus stands at the altar of the deity.

of nature [31] (which they termed by one name, Fate or Fortune), and assigned the causes of particular things to the necessity of matter, without any intermixture of final causes, seems to me (so far as I can judge from the fragments and relics of their philosophy) to have been, as regards physical causes, much more solid and to have penetrated further into nature than that of Aristotle and Plato; for this single reason, that the former never wasted time on final causes, while the latter were ever inculcating them. And in this Aristotle is more to be blamed than Plato, seeing that he left out the fountain of final causes, namely God, and substituted Nature for God; and took in final causes themselves rather as the lover of logic than of theology. And I say this, not because those final causes are not true and worthy to be inquired in metaphysical speculations; but because their excursions and irruptions into the limits of physical causes has bred a waste and solitude in that track. For otherwise, if they be but kept within their proper bounds, men are extremely deceived if they think there is any enmity or repugnancy at all between the two. For the cause rendered, "that the hairs about the eyelids are for the safeguard of the sight," does not impugn the cause rendered, "that pilosity is incident to orifices of moisture":

Muscosi fontes, etc.[32]

Nor the cause rendered, "that the firmness of hides in animals is for the armour of the body against extremities of heat or cold," does not impugn the cause rendered, "that this firmness is caused from the contraction of the pores in the outward parts by cold and depredation of the air;" and so of the rest; both causes being perfectly compatible, except that one declares an intention, the other a consequence only. Neither does this call in question or derogate from divine providence, but rather highly confirms and exalts it. For as in civil actions he is far the greater and deeper politician that can make other men the instruments of his ends and desires and yet never acquaint them with his purpose (so as they shall do what he wills and yet not know that they are doing it), than he that imparts his meaning to those he employs; so does the wisdom of God shine forth more admirably when nature intends one thing and Providence draws forth another, than if he had communicated to all natural figures and motions the characters and impressions of his providence. For instance, Aristotle, when he had made nature pregnant with final causes, laying it down that "Nature does nothing in vain [33], and always effects her will when free from impediments" and many other things of the same kind, had no further need of a God. But Democritus and Epicurus, when they proclaimed their doctrine of atoms, were tolerated so far by some of the more subtle wits; but when they proceeded to assert that the fabric of the universe itself had come together through the fortuitous concourse of the atoms, without a mind, they were met with universal ridicule. Thus so far are physical causes from withdrawing men from God and Providence, that contrariwise, those philosophers who have been occupied in searching them out can find no issue but by resorting to God and Providence at the last [34]. And so much for Metaphysic; the latter part whereof, concerning Final Causes, I allow to be extant in books both physical and metaphysical; in the latter rightly, in the former wrongly, by reason of the inconvenience that ensues thereon.

31 See in illustration of this phrase, Lucretius, v. 835. et seq.

32 Virg. *Eclog*. vii. 45. :—The mossy springs.

33 Aristotle, *Polit*. i. 2, and many other passages.

34 "C'est Dieu," affirms Leibnitz in a spirit not unlike that of the text, "qui est la dernière raison des choses; et la connoissance de Dieu n'est pas moins le principe des sciences que son essence et sa volonté sont les principes des êtres". And a little further on he remarks that "les principes généraux de la Physique et de la Mécanique même dépendent de la conduite d'une intelligence souveraine, et ne sauraient être expliqués sans le faire entrer en considération. C'est ainsi qu'il faut réconcilier la piété avec la raison, et qu'on pourra satisfaire aux gens de bien qui apprehendent les suites de la philosophie mécanique ou corpusculaire, comme si elle pouvait éloigner de Dieu, et des substances immatérielles, au lieu qu'avec les corrections requises, et tout bien entendu, elle doit nous y mener."—*Lettre à Bayle*, p. 106 of Erdmann's edition.

CHAPTER V.

Division of the operative doctrine concerning Nature into Mechanic *and* Magic, *which correspond with the divisions of the speculative doctrine;* Mechanic *answering to* Physic, Magic *to* Metaphysic. *Purification of the word Magic. Two appendices of the operative doctrine.* Inventory of the possessions of man, *and* Catalogue of Polychrests, *or things of general use.*

THE operative doctrine concerning nature I will likewise divide into two parts, and that by a kind of necessity, for this division is subject to the former division of the speculative doctrine; and as Physic and the inquisition of Efficient and Material causes produces Mechanic, so Metaphysic and the inquisition of Forms produces Magic. For the inquisition of Final Causes is barren, and like a virgin consecrated to God produces nothing[1]. I know that there is also a kind of Mechanic often merely empirical and operative, which does not depend on physic; but this I have remitted to Natural History, taking it away from Natural Philosophy. I speak only of that mechanic which is connected with physical causes. Nevertheless between these two kinds of mechanic there is also another which is not altogether operative, yet does not properly reach to philosophy. For all inventions of works which are known to men have either come by chance and so been handed down from one to another, or they have been purposely sought for. But those which have been found by intentional experiment have been either worked out by the light of causes and axioms, or detected by extending or transferring or putting together former inventions; which is a matter of ingenuity and sagacity rather than philosophy. And this kind, which I noways despise, I will presently touch on by the way, when I come to treat of *learned experience* among the parts of logic. But the mechanic of which I now treat is that which has been handled by Aristotle promiscuously, by Hero in spirituals, by Georgius Agricola, a modern writer, very diligently in minerals, and by many other writers in particular subjects[2]; so that I have no omissions to mark in this part, except that promiscuous mechanics, after the manner of Aristotle, ought to have been more diligently continued by the moderns, especially with selection of those

[1] No saying of Bacon's has been more often quoted and misunderstood than this. Carrying out his division of the *Doctrina de Naturâ*, which as we have seen depends upon Aristotle's quadripartite classification of causes, he remarks that to Physica corresponds Mechanica, and to Metaphysica, Magia. But Metaphysica contains two parts, the doctrine of forms and the doctrine of final causes. Bacon remarks that Magia corresponds to Metaphysica, inasmuch as the latter contains the doctrine of forms, that of final causes admitting from its nature of no practical applications. "Nihil parit," means simply "non parit opera", which though it would have been a more precise mode of expression would have destroyed the appositeness of the illustration. No one who fairly considers the context can, I think, have any doubts as to the limitation with which the sentence in question is to be taken. But it is often the misfortune of a pointed saying to be quoted apart from any context, and consequently to be misunderstood.

[2] The *Mechanical Problems* of Aristotle are here referred to. Of Hero, an Alexandrian physicist, who flourished about B.C. [100], Fludd makes frequent mention, and it is perhaps on this account that he is here introduced. It is remarkable that no notice is taken of Archimedes who, beyond all comparison, was the greatest mechanical philosopher of antiquity. With his writings however there is reason to think that Bacon had no acquaintance, and in the *Historia Densi et Rari* his most popularly known invention, that of the method of detecting the adulteration of Hiero's crown, is mentioned in a manner which seems to show that Bacon did not distinctly apprehend the principle on which it depends. With contemporary scientific writers, Bacon seldom appears to be acquainted, and it is therefore less remarkable that no mention is made of Stevinus, Galileo, Guldinus or Ghetaldus. Galileo's astronomical discoveries were of course more generally known than his mechanical researches.

The writings of Agricola, who has been called the German Pliny, are even now, it is said, of considerable value, and certainly entitle him to a high place among the scientific men of the sixteenth century. His greatest work is the *De re metallica*, in twelve books [published at Basle in 1555].

whereof either the causes are more obscure, or the effects more noble. But they who pursue these studies do but creep as it were along the shore,

premendo litus iniquum [3].

For it seems to me there can hardly be discovered any radical or fundamental alterations and innovations of nature, either by accidents or essays of experiments, or from the light and direction of physical causes; but only by the discovery of forms. If then I have set down that part of metaphysic which treats of forms as deficient, it must follow that I do the like of natural magic, which has relation thereunto. But I must here stipulate that magic, which has long been used in a bad sense, be again restored to its ancient and honourable meaning. For among the Persians magic was taken for a sublime wisdom, and the knowledge of the universal consents of things; and so the three kings who came from the east to worship Christ were called by the name of Magi. I however understand it as the science which applies the knowledge of hidden forms to the production of wonderful operations; and by uniting (as they say) actives with passives, displays the wonderful works of nature [4]. For as for that natural magic which flutters about so many books [5], embracing certain credulous and superstitious traditions and observations concerning sympathies and antipathies, and hidden and specific properties, with experiments for the most part frivolous, and wonderful rather for the skill with which the thing is concealed and masked than for the thing itself; it will not be wrong to say that it is as far differing in truth of nature from such a knowledge as we require, as the story of King Arthur of Britain, or Hugh of Bordeaux, and such like imaginary heroes, differs from Cæsar's Commentaries in truth of story. For it is manifest that Cæsar did greater things in reality than those imaginary heroes were feigned to do, but he did them not in that fabulous manner. Of this kind of learning the fable of Ixion was a figure; who designing to embrace Juno, the Goddess of Power, had intercourse with a fleeting cloud; out of which he begot Centaurs and Chimæras. So they who are carried away by insane and uncontrollable passion after things which they only fancy they see through the clouds and vapours of imagination, shall in place of works beget nothing else but empty hopes and hideous and monstrous spectres. But this popular and degenerate natural magic has the same kind of effect on men as some soporific drugs, which not only lull to sleep, but also during sleep instil gentle and pleasing dreams. For first it lays the understanding asleep by singing of specific properties and hidden virtues, sent as from heaven and only to be learned from the whispers of tradition; which makes men no longer alive and awake for the pursuit and inquiry of real causes, but to rest content with these slothful and credulous opinions; and then it insinuates innumerable fictions, pleasant to the mind, and such as one would most desire,—like so many dreams. And it is worth while to note that in these sciences which hold too much of imagination and belief (such as that light Magic of which I now speak, Alchemy, Astrology, and others the like) the means and theory are ever more monstrous than the end and action at which they aim. The conversion of silver, quicksilver, or any other metal into gold, is a thing difficult to believe; yet it is far more probable that a man who knows clearly the natures of weight, of the colour of yellow, of malleability and extension, of volatility and fixedness, and who has also made diligent search into the first seeds and menstruums of minerals, may at last by much and sagacious endeavour produce gold; than that a few grains of an elixir should in a few moments of time be able to turn other metals into gold by the agency of that elixir, as having power to perfect nature and free it from all impediments. So again the retarding of old age or the restoration of some degree of youth, are things hardly credible; yet it is far more

[3] Hor. Od. ii. 10:—"hugging the coast."

[4] Orig. *Magnalia naturæ*—a favourite phrase with Paracelsus. The word magnalia occurs in the Vulgate; see Ps. cvi. 22., where our version is "wondrous works".

[5] See for instance the *Natural Magic* of G. B. Porta, published in [1589]; which quite deserves the character here given of the class to which it belongs.

probable that a man who knows well the nature of arefaction and the depredations of the spirits upon the solid parts of the body, and clearly understands the nature of assimilation and of alimentation, whether more or less perfect, and has likewise observed the nature of the spirits, and the flame as it were of the body, whose office is sometimes to consume and sometimes to restore, shall by diets, bathings, anointings, proper medicines, suitable exercises, and the like, prolong life, or in some degree renew the vigour of youth; than that it can be done by a few drops or scruples of a precious liquor or essence. Again, that fates can be drawn from the stars is more than men will at once or lightly admit; but that the hour of nativity (which is very often either delayed or hastened by many natural accidents) should influence the fortune of a whole life; or that the hour of question has a fatal connexion with the subject of inquiry; these you may call mere follies. But such is the immoderation and intemperance of men that they not only promise to themselves things impossible, but expect to obtain the most difficult things without trouble or toil, as in a holiday recreation. And so much for Magic; whereof I have both vindicated the name itself from discredit, and separated the true kind from the false and ignoble.

But to this operative department of natural philosophy there belong two appendices, both of great value. The first is that there be made an Inventory of the Possessions of Man, wherein should be set down and briefly enumerated all the goods and possessions (whether derived from the fruits and proceeds of nature or of art) which men now hold and enjoy; with the addition of things once known but now lost; in order that those who address themselves to the discovery of new inventions may not waste their pains upon things already discovered and extant. Which calendar will be more workmanlike and more serviceable too, if you add to it a list of those things which are in common opinion reputed impossible in every kind, noting, in connexion with each, what thing is extant which comes nearest in degree to that impossibility; that by the one human invention may be stimulated, and by the other it may to a certain extent be directed; and that by these optatives and potentials active discoveries may the more readily be deduced. The second is, that there be also made a calendar of those experiments which are of most general use, and lead the way to the invention of others. For example, the experiment of the artificial freezing of water by the mixture of ice and bay salt[6] bears on an infinite number of things; for it reveals a secret method of condensation, than which nothing is more serviceable to man. For rarefactions we have fire at hand, but for the means of condensation we are in difficulty. Now it would greatly tend to abridge the work of invention if *Polychrests* of this kind were set down in a proper catalogue.

CHAPTER VI.

Of the great Appendix of Natural Philosophy, both Speculative and Operative, namely Mathematic; *and that it ought rather to be placed among Appendices than among Substantive Sciences. Division of Mathematic into* Pure *and* Mixed.

ARISTOTLE has well remarked that Physic and Mathematic produce Practice or Mechanic[1]; wherefore as we have already treated of the speculative and operative part of natural philosophy, it remains to speak of Mathematic, which is a science auxiliary to both. Now this in the common philosophy is annexed as a third part to Physic and Metaphysic; but for my part, being now engaged in

[6] The artificial congelation of water by snow and salt Bacon has elsewhere spoken of as a recent discovery. I have not been able to ascertain by whom it was made. In Boyle's *New Experiments of Cold*, it is said to be familiarly made use of in Italy, though scarcely known in England; and in the collection of experiments published by the Florentine Academicians in 1667 (in which collection the celebrated "Florentine experiment", which is in reality due to Bacon, is contained), artificial congelations are spoken of, but (probably because the subject was commonly known) without any reference to the history of the invention. "Sal nigrum," it may be well to mention, is Saltpetre.

[1] Arist. Praef. ad Quæst. Mechan.

reviewing and rehandling these things, if I meant to set it down as a substantive and principal science, I should think it more agreeable both to the nature of the thing and the clearness of order to place it as a branch of Metaphysic. For Quantity (which is the subject of Mathematic), when applied to matter, is as it were the dose of Nature, and is the cause of a number of effects in things natural; and therefore it must be reckoned as one of the Essential Forms of things. And so highly did the ancients esteem the power of figures and numbers, that Democritus ascribed to the figures of atoms the first principles of the variety of things; and Pythagoras asserted that the nature of things consisted of numbers. In the meantime it is true that of all natural forms (such as I understand them) Quantity is the most abstracted and separable from matter; which has likewise been the cause why it has been more carefully laboured and more acutely inquired into than any of the other forms, which are all more immersed in matter. For it being plainly the nature of the human mind, certainly to the extreme prejudice of knowledge, to delight in the open plains (as it were) of generalities rather than in the woods and inclosures of particulars, the mathematics of all other knowledge were the goodliest fields to satisfy that appetite for expatiation and meditation. But though this be true, regarding as I do not only truth and order but also the advantage and convenience of mankind, I have thought it better to designate Mathematics, seeing that they are of so much importance both in Physics and Metaphysics and Mechanics and Magic, as appendices and auxiliaries to them all. Which indeed I am in a manner compelled to do, by reason of the daintiness and pride of mathematicians, who will needs have this science almost domineer over Physic. For it has come to pass, I know not how, that Mathematic and Logic, which ought to be but the handmaids of Physic, nevertheless presume on the strength of the certainty which they possess to exercise dominion over it. But the place and dignity of this science is of less importance: let us now look to the thing itself.

Mathematic is either Pure or Mixed. To Pure Mathematic belong those sciences which handle Quantity entirely severed from matter and from axioms of natural philosophy. These are two, Geometry and Arithmetic; the one handling quantity continued, and the other dissevered. These two arts have been inquired into and handled with great wit and industry; and yet to the labours of Euclid in geometry no addition has been made by his successors worthy of so long an interval; nor has the doctrine of solids been sufficiently examined and advanced either by ancients or moderns, in proportion to the use and excellency of the subject [2]. And in arithmetic, neither have there been discovered formulas for the abridgment of computation sufficiently various and convenient, especially with regard to progressions, of which there is no slight use in Physics [3], nor has algebra been well perfected; and the mystic arithmetic of Pythagoras,

[2] We might here expect to find some mention of Archimedes and of Apollonius, whose labours contributed more to the progress of geometry than those of Euclid, who was rather a systematiser than an original discoverer, and whose Elements do not embrace the whole extent of the geometry of the Greeks. The doctrine of conic sections, which was commenced by Plato, and the method of limits of Archimedes, both most important portions of the Greek geometry, are of course not to be found in Euclid's Elements, not to mention a variety of isolated investigations. It is undoubtedly true that even long after Bacon's time geometry advanced more slowly beyond the limits it had attained in antiquity than other parts of mathematics, though in the present day it may be said to have become a new science. See on this head, the *Aperçu Historique des Méthodes de la Géométrie* of M. Chasles, himself one of those who have contributed the most to its recent progress.

[3] One would certainly not infer from this remark, to which there is nothing corresponding in the *Advancement of Learning* that Bacon was aware that in the interval which had elapsed since its publication, the greatest of all inventions for facilitating arithmetical computations had been made known. Napier's *Logarithms* were published in 1614, and reprinted on the continent in 1620; in which year Gunter's *Canon of Triangles* was also published. In 1618 Robert Napier's account of his father's method and Briggs's first table of Logarithms were both published. In the year succeeding that of

which has been revived of late from Proclus and fragments of Euclid, is a kind of wandering speculation : for it is incidental to the human mind, that when it cannot master the solid, it wastes itself on the superfluous. Mixed Mathematic has for its subject some axioms and parts of natural philosophy, and considers quantity in so far as it assists to explain, demonstrate, and actuate these. For many parts of nature can neither be invented with sufficient subtlety, nor demonstrated with sufficient perspicuity, nor accommodated to use with sufficient dexterity, without the aid and intervention of Mathematic : of which sort are Perspective, Music, Astronomy, Cosmography, Architecture, Machinery [4], and some others. In mixed Mathematics I do not find any entire parts now deficient, but I predict that hereafter there will be more kinds of them, if men be not idle. For as Physic advances farther and farther every day and develops new axioms, it will require fresh assistance from Mathematic in many things, and so the parts of Mixed Mathematics will be more numerous.

And now I have passed through the doctrine concerning Nature, and marked the deficiencies thereof. Wherein if I have differed from the ancient and received doctrines, and thereby given a handle to contradiction ; for my part, as I am far from wishing to dissent, so I purpose not to contend. If it be truth,

Non canimus surdis, respondent omnia silvae [5] :

the voice of nature will consent, whether the voice of man do or not. But as Alexander Borgia was wont to say of the expedition of the French to Naples, " that they came with chalk in their hands to mark up their lodgings, and not with weapons to break in " ; so I like better that entry of truth, which comes peaceably as with chalk to mark up those minds which are capable to lodge and harbour such a guest, than that which forces its way with pugnacity and contention. Having therefore gone through the two parts of philosophy respecting God and Nature, there remains the third, respecting Man.

the publication of the *De Augmentis* his larger tables, and probably those of Wingate made their appearance.

These dates are sufficient to show how much the attention of mathematicians was given to the subject. It would almost seem as if some one, possibly Savile, had told Bacon—what was no doubt true—that the application of the doctrine of series to arithmetical computation was not as yet brought to perfection, and that he had adopted the remark without understanding the importance of the discovery to which it referred, and perhaps without being aware that any such discovery had been made.

[4] Machinaria means the art of making machines, not mechanics in the common sense of the word. It therefore appears from this enumeration that Bacon was not acquainted with any application of mathematics to statics or dynamics, as he would certainly not have included these fundamental portions of mixed mathematics in the nonnullæ aliæ with which the list concludes. The omission of any reference to the mathematical doctrine of motion is not surprising, though Galileo's researches were known for many years before the publication of the *De Augmentis* , the theory of equilibrium, however, is as old as the time of Archimedes ; and we might therefore have expected that it would have been here mentioned.

[5] Virg. *Eclog*. x. 8. :—

To no deaf ears we sing, the echoing woods reply.

Book IV

CHAPTER I.

Division of the doctrine concerning Man into Philosophy of Humanity *and* Philosophy Civil. *Division of the Philosophy of Humanity into doctrine concerning the* Body of Man *and doctrine concerning the* Soul of Man. *Constitution of one general doctrine concerning the* Nature *or the* State of Man. *Division of the doctrine concerning the State of Man into doctrine concerning the* Person of Man, *and doctrine concerning the* League of Mind and Body. *Division of the doctrine concerning the Person of Man into doctrine concerning the* Miseries of Man, *and doctrine concerning his* Prerogatives. *Division of the doctrine concerning the League into doctrine concerning* Indications *and concerning* Impressions. *Assignation of* Physiognomy *and* Interpretation of Natural Dreams *to the doctrine concerning Indications.*

IF any one should aim a blow at me (excellent King) for anything I have said or shall hereafter say in this matter, (besides that I am within the protection of your Majesty,) let me tell him that he is acting contrary to the rules and practice of warfare. For I am but a trumpeter, not a combatant; one perhaps of those of whom Homer speaks,

Χαίρετε κήρυκες, Διὸς ἄγγελοι, ἠδὲ καὶ ἀνδρῶν [1] :

and such men might go to and fro everywhere unhurt, between the fiercest and bitterest enemies. Nor is mine a trumpet which summons and excites men to cut each other to pieces with mutual contradictions, or to quarrel and fight with one another; but rather to make peace between themselves, and turning with united forces against the Nature of Things, to storm and occupy her castles and strongholds, and extend the bounds of human empire, as far as God Almighty in his goodness may permit.

Let us now come to that knowledge whereunto the ancient oracle directs us, which is the knowledge of ourselves; which deserves the more accurate handling in proportion as it touches us more nearly. This knowledge is for man the end and term of knowledges; but of nature herself it is but a portion. And generally let this be a rule; that all divisions of knowledges be accepted and used rather for lines to mark or distinguish, than sections to divide and separate them; in order that solution of continuity in sciences may always be avoided. For the contrary hereof has made particular sciences to become barren, shallow, and erroneous; not being nourished and maintained and kept right by the common fountain and aliment. So we see Cicero the orator complaining of Socrates and his school, that he was the first who separated philosophy and rhetoric; whereupon rhetoric became an empty and verbal art [2]. So we may see that the opinion of Copernicus touching the rotation of the earth (which has now become prevalent) cannot be refuted by astronomical principles, because it is not repugnant to any of the phenomena; yet the principles of natural philosophy rightly laid down may correct it. Lastly we see that the science of medicine, if it be forsaken by natural philosophy, is not much better than an empirical practice. With this reservation therefore let us proceed to the doctrine concerning Man. It has two parts. For it considers man either segregate, or congregate and in society. The one I call the Philosophy of Humanity, the other Civil Philosophy. Philosophy of Humanity consists of parts similar to those of which man consists; that is, of know-

[1] Hom. *Il.* i. 334. :—Hail, heralds, messengers of Jove and men!

[2] Cicero *De Orat.* iii. c. 19.

ledges which respect the body, and of knowledges which respect the mind. But before we pursue the particular distributions let us constitute one general science concerning the Nature and State of Man; a subject which certainly deserves to be emancipated and made a knowledge of itself. It is composed of those things which are common as well to the body as the soul; and may be divided into two parts; the one regarding the nature of man undivided, and the other regarding the bond and connexion between the mind and body; the first whereof I will term the doctrine concerning the Person of Man, the second the doctrine concerning the League. But it is plain that these things, being common and mixed, could not all have been assigned to that first division, of sciences which regard the body and sciences which regard the mind.

The doctrine concerning the Person of Man takes into consideration two subjects principally; the Miseries of the human race, and the Prerogatives or Excellencies of the same. And for the miseries of humanity, the lamentation of them has been elegantly and copiously set forth by many, both in philosophical and theological writings. And it is an argument at once sweet and wholesome.

But that other subject of the Prerogatives of Man seems to me to deserve a place among the *desiderata*. Pindar in praising Hiero says most elegantly (as is his wont) that he "culled the tops of all virtues" [3]. And certainly I think it would contribute much to magnanimity and the honour of humanity, if a collection were made of what the schoolmen call the *ultimities*, and Pindar the *tops or summits*, of human nature, especially from true history; showing what is the ultimate and highest point which human nature has of itself attained in the several gifts of body and mind. What a wonderful thing, for example, is that which is told of Cæsar,—that he could dictate to five secretaries at once. And again those exercitations of the ancient rhetoricians, Protagoras and Gorgias, and of the philosophers, Callisthenes, Posidonius, Carneades,—who could speak elegantly and copiously, extempore, on either side of any subject,—is no small honour to the powers of the human wit. A thing inferior in use, but as a matter of display and ability perhaps still greater, is that which Cicero [4] relates of his master Archias—that he could speak extempore a great number of excellent verses about anything that happened to be going on at the time [5]. That Cyrus or Scipio could call so many thousands of men by name was a great feat of memory [6]. Nor are the triumphs of the moral virtues less famous than those of the intellectual. What a proof of patience is displayed in the story told of Anaxarchus, who, when questioned under torture, bit out his own tongue (the only hope of information), and spat it into the face of the tyrant [7]. Nor was it a less thing in point of endurance (however inferior in worthiness) which occurred in our own times in the case of the Burgundian who murdered the Prince of Orange [8]: being beaten with rods of iron and torn with red-hot pincers, he uttered not a single groan; nay, when something aloft broke and fell on the head of a bystander, the half-burnt wretch laughed in the midst of his torments, though but a little before he had wept at the cutting off of his curling locks. A wonderful composure and serenity of mind at the point of death has also been displayed by many; as in the case of the centurion related by Tacitus: when bidden by the soldier appointed to execute him to stretch out his neck boldly, "I wish," he replied, "that you may

[3] Pind. Olymp. i. 20. [4] Cicero, pro Archià, c. 8. [5] Cf. Laert. ix. 59.

[6] Xenophon says that Cyrus knew the names of all the officers (ἡγεμόνες) in his army; later writers go much farther, and affirm that he knew the names of all his soldiers. Compare Valerius Max. viii. 7. with Xenophon's statement, *Cyrop.* v. 3. The same exaggeration occurs in Solinus, c. 5.: "Cyrus memoriæ bono claruit, qui in exercitu cui numerosissimo præfuit nominatim singulos alloqueretur". The Scipio here mentioned is Lucius Scipio Asiaticus. Vide Solin. ubi sup. or Pliny, vii. 34.

[7] The story is somewhat differently told by Diogen. Laert. ix. 59; but in Pliny and Valerius Maximus we find it related as in the text. A similar story is told of Leæna in Ælian's *Hist. Var.*

[8] The person referred to is Balthazar Gerard of Franche Comté, who shot William IX., Prince of Orange, at Delft, in 1584. Vide *Histoire Générale des Pays Bas*, v. 384.

strike as boldly [9]." John Duke of Saxony, when the warrant was brought to him for his execution next day, was playing at chess. Calling a bystander to him, he said with a smile, "See whether I have not the best of the game; for when I am dead he (pointing to his adversary) will boast that he was winning" [10]. Our own More, too, Chancellor of England, when the day before he was to die a barber came to him (sent because his hair was long, which it was feared might make him more commiserated with the people) and asked him "whether he would be pleased to be trimmed," refused; and turning to the barber, "The King and I (said he) have a suit for my head, and till the title be cleared I will do no cost upon it". The same More, at the very instant of death, when he had already laid his head on the fatal block, lifted it up a little, and gently drew aside his beard, which was somewhat long, saying, "this at least hath not offended the King." But not to stay too long on the point, my meaning is sufficiently clear; namely, that the miracles of human nature, and its highest powers and virtues both in mind and body, should be collected into a volume, which should serve for a register of the Triumphs of Man. In which work I approve the design of Valerius Maximus and C. Pliny, and wish for their diligence and judgment.

With regard to the doctrine concerning the League or Common Bond between the soul and body, it is distributed into two parts. For as in all leagues and amities there is both mutual intelligence and mutual offices, so the description of this league of soul and body consists in like manner of two parts: namely, how these two (that is the Soul and the Body) disclose the one the other, and how they work the one upon the other; by knowledge or indication, and by impression. The former of these (that is, the description of what knowledge of the mind may be obtained from the habit of the body, or of the body from the accidents of the mind) has begotten two arts; both of prediction; whereof the one is honoured with the inquiry of Aristotle, and the other of Hippocrates. And although they have of later times been polluted with superstitious and fantastical arts, yet being purged and restored to their true state, they have both a solid ground in nature and a profitable use in life. The first is Physiognomy, which discovers the dispositions of the mind by the lineaments of the body; the second is the Interpretation of Natural Dreams, which discovers the state and disposition of the body by the agitations of the mind. In the former of these I note a deficience. For Aristotle has very ingeniously and diligently handled the structure of the body when at rest, but the structure of the body when in motion (that is, the gestures of the body) he has omitted; which nevertheless are equally within the observations of art, and of greater use and advantage [11]. For the lineaments of the body disclose the dispositions and inclinations of the mind in general; but the motions and gestures of the countenance and parts do not only so, but disclose likewise the seasons of access, and the present humour and state of the mind and will. For as your Majesty says most aptly and elegantly, "As the tongue speaketh to the ear so the gesture speaketh to the eye" [12]. And well is this known to a number of cunning and astute persons; whose eyes dwell upon the faces and gestures of men, and make their own advantage of it, as being most

[9] Tac. Ann. xv. 67. In the same spirit Giordano Bruno told his judges that it might well be that they had felt more fear in condemning him than he in hearing himself condemned.

[10] The Elector of Saxony, of whom this story is told, was, in 1547, irregularly condemned to death by Charles V. The sentence was not executed; and it seems doubtful whether the Emperor ever intended that it should be.

According to De Thou, the Elector, after making some remark on the Emperor's injustice, resumed and won the game.

[11] The physiognomical method of Aristotle consists chiefly in tracing the resemblances which exist between different kinds of animals and different individuals of the human species; a method followed by later writers, particularly G. B. Porta, and Lebrun, whose illustrations of his theory are well known, though the essay which they accompanied seems to have been lost.

[12] "For as the tongue speaketh to the eares, so doeth the gesture speake to the eyes of the auditour."—*Basilicon Doron*, book iii.—*J. S.*

part of their ability and wisdom. Neither indeed can it be denied, but that it is a wonderful index of simulation in another, and an excellent direction as to the choice of proper times and seasons to address persons ; which is no small part of civil wisdom. Nor let any one imagine that a sagacity of this kind may be of use with respect to particular persons, but cannot fall under a general rule ; for we all laugh and weep and frown and blush nearly in the same fashion ; and so it is (for the most part) in the more subtle motions. But if anyone be reminded here of chiromancy, let him know that it is a vain imposture, not worthy to be so much as mentioned in discourses of this nature. With regard to the Interpretation of Natural Dreams, it is a thing that has been laboriously handled by many writers, but it is full of follies. At present I will only observe that it is not grounded upon the most solid foundation of which it admits ; which is, that when the same sensation is produced in the sleeper by an internal cause which is usually the effect of some external act, that external act passes into the dream. A like oppression is produced in the stomach by the vapour of indigestion and by an external weight superimposed ; and therefore persons who suffer from the nightmare dream of a weight lying on them, with a great array of circumstances. A like pendulous condition of the bowels is produced by the agitation of the waves at sea, and by wind collected round the diaphragm ; therefore hypochondriacal persons often dream that they are sailing and tossing on the sea. There are likewise innumerable instances of this kind.

The latter branch of the doctrine of the League (which I have termed Impression) has not yet been collected into an art, but only comes in sometimes dispersedly in the course of other treatises. It has the same relation or antistrophe that the former has. For the consideration is twofold ; either how and how far the humours and temperament of the body alter and work upon the mind ; or again, how and how far the passions or apprehensions of the mind alter and work upon the body. For the physicians prescribe drugs to heal mental diseases, as in the treatment of phrensy and melancholy ; and pretend also to exhibit medicines to exhilarate the mind, to fortify the heart and thereby confirm the courage, to clarify the wits, to corroborate the memory, and the like. But the diets, and choice of meats and drinks, the ablutions and other observances of the body, in the sect of the Pythagoreans, in the heresy of the Manicheans, and in the law of Mahomet, exceed all measure [13]. So likewise the ordinances in the ceremonial law interdicting the eating of the blood and fat, and distinguishing between beasts clean and unclean for meat, are many and strict. Nay, the Christian faith itself (although clear and serene from all clouds of ceremony) yet retains the use of fastings, abstinences, and other macerations and humiliations of the body, as things not merely ritual, but also profitable. The root and life of all which prescripts (besides the ceremony and the exercise of obedience) consist in that of which we are speaking, namely the sympathy of the mind with the state and disposition of the body [14]. But if any man of weak judgment conceive that these impressions of the body on the mind either question the immortality of the soul, or derogate from its sovereignty over the body, a slight answer may serve for so slight a doubt. Let him take the case of an infant in the mother's womb, which is affected by that which affects the mother, and yet is in due time delivered and separated from her body ; or of monarchs who, though powerful, are sometimes controlled by their servants, and yet without abatement of their majesty royal.

As for the reciprocal part (which is the operation of the mind and its passions upon the body), it also has found a place in medicine. For there is no physician of any skill who does not attend to the accidents of the mind, as a thing most material towards recoveries, and of the greatest force to

[13] All these are probably surpassed by the Institutes of Menu, so far as they relate to the way of life of the Brahmins.

[14] The difficulty of conceiving the nature of the reciprocal influence of the mind and body led to its being altogether rejected by Malebranche and by Leibnitz. See the *Theodicæa* of the latter for a statement of the three theories, namely that of physical influence, that of occasional causes, and that of pre-established harmony.

further or hinder other remedies. But another question pertinent to this subject has been but sparingly inquired into, and nowise in proportion to its depth and worth; namely how far (setting the affections aside) the very imagination of the mind, or a thought strongly fixed and exalted into a kind of faith, is able to alter the body of the imaginant. For although it has a manifest power to hurt, yet it follows not that it has the same degree of power to help; no more indeed than a man can conclude, that because there are pestilent airs, able suddenly to kill a man in health, therefore there should be sovereign airs, able suddenly to cure a man in sickness. Such an inquiry would surely be of noble use; though it needs (as Socrates says[15]) a *Delian diver;* for it lies deep. Again, among those doctrines concerning the League, or the concordances between the mind and body, there is none more necessary than the inquiry concerning the proper seats and domiciles which the several faculties of the mind occupy in the body and its organs. Which kind of knowledge has not been without its followers; but what has been done in it is in most parts either disputed or slightly inquired; so that more diligence and acuteness is requisite. For the opinion of Plato[16], who placed the understanding in the brain, as in a castle; animosity (which he unfitly enough called anger, seeing it is more related to swelling and pride) in the heart; and concupiscence and sensuality in the liver; deserves neither to be altogether despised nor to be eagerly received. Neither again is that arrangement of the intelleetual faculties (imagination, reason, and memory) according to the respective ventricles of the brain, destitute of error. Thus then have I explained the doctrine concerning the nature of man undivided, and likewise the league between the mind and body.

CHAPTER II.

Division of the doctrine concerning the Body of Man into Medicine, Cosmetic, Athletic, and Voluptuary. *Division of Medicine into three offices;* viz. the Preservation of Health, the Cure of Diseases, and the Prolongation of Life; *and that the last division concerning the Prolongation of Life ought to be kept separate from the other two.*

THE doctrine that concerns man's body receives the same division as the good of man's body, to which it refers. The good of man's body is of four kinds; Health, Beauty, Strength, and Pleasure. The knowledges therefore are in number the same; Medicine, Cosmetic, Athletic, and Voluptuary, which Tacitus truly calls "educated luxury"[1].

Medicine is a most noble art, and according to the poets has a most illustrious pedigree. For they have represented Apollo as the primary god of medicine, and given him a son Æsculapius, likewise a god, professor of the same; seeing that in nature the sun is the author and source of life, the physician the preserver and as it were the second fountain thereof. But a far greater honour accrues to medicine from the works of our Saviour, who was the physician both of soul and body; and as he made the soul the peculiar object of his heavenly doctrine, so he made the body the peculiar object of his miracles. For we nowhere read of any miracle done by him with respect to honours or money (except that one, for giving tribute money to Cæsar), but only with respect to the body of man, for the preservation, support, or healing thereof.

This subject of medicine (namely man's body) is of all other things in nature most susceptible of remedy; but then that remedy is most susceptible of error. For the same subtlety and variety of the subject, as it supplies abundant means of healing, so it involves great facility of failing. And therefore as this art (especially as we now have it) must be reckoned as one of the most conjectural, so the inquiry of it must be accounted one of the most exact and difficult. Not that I share the idle notion of Paracelsus and the alchemists, that there are to be found in man's body certain correspondences and parallels which have respect to all the several species (as stars, planets, minerals) which are extant in the universe; foolishly and stupidly misapplying the ancient emblem (that man was a *microcosm*

[15] Diog. Laërt. ii. 22. and ix. 12.

[16] Plato, *Timæus*, p. 71.

[1] Tac. *Ann.* xvi. 18.

or epitome of the world) to the support of this fancy of theirs. But yet thus much is true, that (as I was going to say) of all substances which nature has produced man's body is the most multifariously compounded. For we see herbs and plants are nourished by earth and water; beasts for the most part by herbs and fruits; but man by the flesh of those beasts (quadrupeds, birds, and fishes), and also by herbs, grains, fruits, juices, and liquors of various kinds; not without manifold commixtures, dressings, and preparations of these several bodies, before they come to be his food and aliment. Add to this, that beasts have a more simple manner of life, and fewer affections to work upon their bodies, and those much alike in their operations; whereas man in his places of habitation, exercises, passions, sleep and watching, undergoes infinite variations; so that it is true that the body of man, of all other things, is of the most fermented and compounded mass. The soul on the other side is the simplest of substances; as is well expressed,

—— purumque reliquit
Æthereum sensum, atque auraï simplicis ignem [2].

Whence it is no marvel that the soul so placed enjoys no rest; according to the axiom that the motion of things out of their place is rapid, and in their place calm. But to return. This variable and subtle composition and structure of man's body has made it as a musical instrument of much and exquisite workmanship, which is easily put out of tune. And therefore the poets did well to conjoin music and medicine in Apollo; because the genius of both these arts is almost the same; for the office of the physician is but to know how to stretch and tune this harp of man's body that the harmony may be without all harshness or discord. So then the subject being so inconstant and variable has made the art by consequence more conjectural; and the art being so conjectural has made so much more room not only for error, but also for imposture. For almost all other arts and sciences are judged by their power and functions, and not by the successes and events. The lawyer is judged by the virtue of his pleading and speaking, not by the issue of the cause; the master of the ship is judged by his skill in steering, and not by the fortune of the voyage. But the physician, and perhaps the politician, have no particular acts whereby they may clearly exhibit their skill and ability; but are honoured or disgraced according to the event;—a most unfair way of judging. For who can know, if a patient die or recover, or if a state be preserved or ruined, whether it be art or accident? And therefore many times the impostor is prized, and the man of virtue censured. Nay, such is the weakness and credulity of men, that they will often prefer a witch or mountebank to a learned physician. And therefore the poets were clear-sighted when they made Circe sister of Æsculapius, and both children of the Sun; as is expressed in the verses,—respecting Æsculapius, that he was the son of Apollo.

Ipse repertorem medicinæ talis et artis
Fulmine *Phœbigenam* Stygias detrusit ad undas [3];

and again respecting Circe, that she was the daughter of the Sun,

Dives inaccessos ubi *Solis filia* lucos
Urit odoratam nocturna in lumina cedrum [4].

For in all times, in the opinion of the multitude, witches and old women and impostors have been the rivals in a manner of physicians, and almost contended

[2] Virg. *Æn*. vi. 747:—

—— pure and unmixed
The ethereal sense is left—mere air and fire.

[3] Virg. *Æn*. vii. 772:—

Apollo's son the healing art who gave
Jove hurled with thunder to the Stygian wave.

[4] Virg. *Æn*. vii. 11:—

Where *the Sun's daughter* in her deep retreat
Burns for her evening light the cedar sweet.

with them in celebrity for working cures. And what follows? Even this, that physicians say to themselves, as Solomon expresses it upon a higher occasion [5], "If it befall to me as befalleth to the fool, why should I labour to be more wise?" And therefore I can the less blame physicians that they commonly attend to some other art or practice, which they fancy more than their own. For you have among them poets, antiquaries, critics, rhetoricians, statesmen, divines; and in every one of these arts more learned than in their own profession. Nor does this happen, in my opinion, because (as a certain declaimer against the sciences objects to the physicians) they have so many sad and disgusting objects to deal with that they must needs withdraw their minds to other things for relief (for "he that is a man should not think anything that is human alien to him") [6]; but rather upon the ground we are now on, that they find that mediocrity and excellency in their art make no difference in profit or reputation towards their fortune. For the impatience of disease, the sweetness of life, the flattery of hope, the commendations of friends, make men depend upon physicians with all their defects. But yet if these things be more attentively considered, they tend rather to inculpate physicians than to excuse them. For instead of throwing away hope, they ought to put on more strength. For if any man will awake his observation and look a little about him, he will easily see from obvious and familiar examples what a mastery the subtlety and acuteness of the intellect has over the variety either of matter or of form. Nothing more variable than faces and countenances; yet men can bear in memory the infinite distinctions of them; nay, a painter with a few shells of colours, and the help of his eye, of the force of his imagination, and the steadiness of his hand, can imitate and draw the faces of all men that are, have been, or shall be, if they were only brought before him. Nothing more variable than the human voice, yet we easily distinguish the differences of it in different persons; nay and there are buffoons and pantomimes who will imitate and express to the life as many as they please. Nothing more variable than the differing sounds of words, yet men have found the way to reduce them to a few simple letters. And most true it is that perplexities and incomprehensions in science proceed commonly not from any want of subtlety or capacity in the mind, but from the object being placed too far off. For as the sense when at a distance from the object is full of mistaking, but when brought near enough does not much err, so is it with the understanding. But men are wont to look down upon nature as from a high tower and from a great distance, and to occupy themselves too much with generalities; whereas if they would come down and draw near to particulars and take a closer and more accurate view of things themselves, they would gain a more true and profitable knowledge of them. Wherefore the remedy of this evil is not merely to quicken or strengthen the organ, but also to go nearer to the object. And therefore there is no doubt but if the physicians would for a while set these generalities aside and go forth to meet Nature, they would obtain that of which the poet speaks,

> Et quoniam variant morbi, variabimus artes;
> Mille mali species, mille salutis erunt [7].

Which they should the rather do, because those very philosophies which physicians, whether regular practitioners or chemists, rely upon (and medicine not founded on philosophy is a weak thing) are themselves of little worth. Wherefore if generalities, though true, have the fault that they do not well lead the way to action; surely there is greater danger in those generalities which are in themselves false, and instead of leading mislead.

[5] Eccles. ii. 15.

[6] Ter. *Heauton.* i. 1. 25 :—Homo sum ; humani nihil a me alienum puto.

[7] Ovid. *Remed. Amor.* 525 :—

> Arts shall as various as diseases be;
> Though sickness take a thousand shapes, yet we
> Will find for each its several remedy.

Bacon deviates, probably intentionally, from the original, which begins : " Et quoniam variant *animi*, variamus et artes ".

Medicine therefore (as we have seen) is a science which has been hitherto more professed than laboured, and yet more laboured than advanced; the labours spent on it having been rather in a circle, than in progression. For I find in the writers thereon many iterations, but few additions. I will divide it into three parts, which I will term its three offices; the first whereof is the Preservation of Health, the second the Cure of Diseases, and the third the Prolongation of Life. But this last the physicians do not seem to have recognised as the primary part of their art, but to have confounded, ignorantly enough, with the other two. For they imagine that if diseases be repelled before they attack the body, and cured after they have attacked it, prolongation of life necessarily follows. But though there is no doubt of this, yet they have not penetration to see that these two offices pertain only to diseases, and such prolongation of life as is intercepted and cut short by them. But the lengthening of the thread of life itself, and the postponement for a time of that death which gradually steals on by natural dissolution and the decay of age, is a subject which no physician has handled in proportion to its dignity. And let not men make a scruple of it, as if this were a thing belonging to fate and Divine Providence which I am the first to bring within the office and function of art. For Providence no doubt directs all kinds of death alike, whether from violence or disease or the decay of age; yet it does not on that account exclude the use of preventions and remedies. But art and human industry do not command nature and destiny; they only serve and minister to them. Of this part however I will speak hereafter; having in the meantime premised thus much, lest any one should in ignorance confound this third office of medicine with the two former, as has been done hitherto.

With regard to the office of the preservation of health (the first of the three), many have written thereon, very unskilfully both in other respects and especially in attributing too much (as I think) to the choice of meats and too little to the quantity. Moreover with regard to quantity itself they have argued like moral philosophers, too much praising the mean; whereas both fasting, when made customary, and a generous diet, to which one is used, are better preservations of health than those mediocrities, which only make nature slothful and unable to bear either excess or want when it is necessary. Nor have the kinds of exercises which have most power to preserve health been by any physician well distinguished and pointed out; although there is scarcely any tendency to disease which may not be prevented by some proper exercise. Thus playing at bowls is good for diseases of the reins, archery for those of the lungs, walking and riding for weakness of the stomach, and the like. But as this part touching the preservation of health has been handled as a whole, it is not my plan to pursue the minor defects.

With regard to the cure of diseases, much labour has been bestowed on this part, but with slight profit. To it belongs the knowledge of the diseases to which the human body is subject; with their causes, symptoms, and remedies. In this second office of medicine there are many deficiencies; a few of which, but those the most glaring, I will propound; thinking it sufficient to enumerate them without any law of order or method.

The first is, the discontinuance of the very useful and accurate diligence of Hippocrates, who used to set down a narrative of the special cases of his patients; relating what was the nature of the disease, what the treatment, and what the issue [8]. Therefore having so notable and proper an example in a man who has been regarded as the father of his art, I shall not need to go abroad for an example from other arts; as from the wisdom of the lawyers, who have ever been careful to report the more important cases and new decisions, for instruction and direction in future cases. This continuance of medicinal history I find deficient; especially as carefully and judiciously digested into one body; which nevertheless I do not understand should be either so copious as to extend to every common case of daily occurrence (for that would be something infinite, and foreign to the purpose), or so reserved as to admit none but wonders and prodigies, as has been done by some. For many things are new in the manner and circumstances which are

[8] See Hippocrates' *De Epidemiis*.

not new in the kind; and if men will apply themselves to observe, they will find even in things which appear commonplace much that is worthy of observation.

Likewise in anatomical inquiries, those things which pertain to man's body in general are most diligently observed, even to curiosity and in the minutest particulars; but touching the varieties which are found in different bodies, the diligence of physicians falls short. And therefore I say that Simple Anatomy is handled most lucidly, but that Comparative Anatomy is wanting. For men inquire well of the several parts, and their substances, figures, and collocations; but the diversities of the figure and condition of those parts in different men they observe not. The reason of which omission I judge to be no other than that the former inquiry may be satisfied by the view of one or two anatomies, whereas the latter (being comparative and casual) requires the view and attentive study of many. The first likewise is a subject on which learned men may display their knowledge in lectures and before audiences; but the last is only to be gained by silent and long experience. Meanwhile there is no question but that the figure and structure of the inward parts is but little inferior in variety and lineaments to the outward; and that the hearts or livers or stomachs of men differ as much as their foreheads or noses or ears. And in these very differences of the internal parts are often found the "causes continent"[9] of many diseases; which not being observed by the physicians, they quarrel many times with the humours, which are not in fault, the fault being in the very mechanical frame of the part. In the cure of which diseases it is lost labour to employ medicines alterative (for the part admits not of alteration); but the thing must be corrected, and accommodated or palliated by diets and medicines familiar. To Comparative Anatomy belongs likewise the accurate observation as well of all kinds of humours, as of the footsteps and impressions of diseases in various dissected bodies. For the humours are commonly passed over in anatomies with disgust as purgaments; whereas it is of the first importance to observe of what sort and how manifold the different kinds of humours are (not relying too much on the common divisions of them) which are sometimes found in the human body; and in what cavities and receptacles each of them is most apt to lodge and nestle, and with what benefit or injury, and the like. So again the footsteps and impressions of diseases and the injuries and devastations they cause in the inward parts, ought in different anatomies to be diligently observed; namely imposthumations, exulcerations, discontinuations, putrefactions, corrosions, consumptions, contractions, extensions, convulsions, loosenings, dislocations, obstructions, repletions, tumours; together with all preternatural substances that are found in the human body (as stones, carnosities, excrescences, worms, and the like); all these, I say, and the like of them ought by that Comparative Anatomy which I have spoken of, and the collation of the several experiences of many physicians, to be carefully searched out and compared. But this variety of accidents is either slightly handled in anatomics or else passed over in silence.

Of that other defect in anatomy (that it has not been practised on live bodies) what need to speak? For it is a thing hateful and inhuman, and has been justly reproved by Celsus[10]. But yet it is no less true (as was anciently noted) that many of the more subtle passages, pores, and pertusions appear not in anatomical dissections, because they are shut and latent in dead bodies, though they be open and manifest in live[11]. Wherefore that utility may be considered as well as hu-

[9] The phrase is taken from Celsus, *præfatio*.

[10] "Insidere autem vivorum corpora et crudele et supervacuum est." *Id. ib.*

[11] This difficulty is almost entirely removed by the perfection to which the art of making anatomical preparations has been brought. Berengario of Carpi, who died at Ferrara in 1550, is said to have been the first person who made use of injections in order to render the vessels visible. He employed water (probably coloured) for this purpose. Swammerdam was the first to inject with wax. In one branch of anatomy, namely the doctrine of the development of the osseous parts, the use of madder in the food of the living animal has led to very curious results. It stains the portions of bone developed during its use of a bright red. Duhamel was the first to use this means of studying the growth of bone. Flourens has also employed it.

manity, the anatomy of the living subject is not to be relinquished altogether, nor referred (as it was by Celsus) to the casual practices of surgery; since it may be well discharged by the dissection of beasts alive, which, notwithstanding the dissimilitude of their parts to human, may, with the help of a little judgment, sufficiently satisfy this inquiry [12].

Again, in their inquiry concerning diseases, they find many which they pronounce incurable, some at their very commencement, and others after a certain period. So that the proscriptions of Sylla and the Triumvirs were as nothing to the proscriptions of physicians, wherein by most iniquitous edicts they give up so many to death; of whom nevertheless numbers escape with less difficulty than they did in the Roman proscriptions. Therefore I will not hesitate to set down among the *desiderata* a work on the cure of diseases which are held incurable; that so some physicians of eminence and magnanimity may be stirred up to take this work (as far as the nature of things permits) upon them; since the pronouncing these diseases incurable gives a legal sanction as it were to neglect and inattention, and exempts ignorance from discredit.

Again, to go a little further; I esteem it likewise to be clearly the office of a physician, not only to restore health, but also to mitigate the pains and torments of diseases; and not only when such mitigation of pain, as of a dangerous symptom, helps and conduces to recovery; but also when, all hope of recovery being gone, it serves only to make a fair and easy passage from life. For it is no small felicity which Augustus Cæsar was wont so earnestly to pray for, that same *Euthanasia* [13]; which likewise was observed in the death of Antoninus Pius, which was not so much like death as like falling into a deep and pleasant sleep. And it is written of Epicurus, that he procured the same for himself; for after his disease was judged desperate, he drowned his stomach and senses with a large draught and ingurgitation of wine; whereupon the epigram was made,

—— hinc Stygias ebrius hausit aquas [14].

He drowned in wine the bitterness of the Stygian water. But in our times, the physicians make a kind of scruple and religion to stay with the patient after he is given up, whereas in my judgment, if they would not be wanting to their office, and indeed to humanity, they ought both to acquire the skill and to bestow the attention whereby the dying may pass more easily and quietly out of life. This part I call the inquiry concerning *outward Euthanasia*, or the easy dying of the body (to distinguish it from that Euthanasia which regards the preparation of the soul); and set it down among the desiderata.

Again, in the cures of diseases I generally find this deficience; that the physicians of this age, though they pursue well enough the general intentions of cures, yet the particular receipts which are proper for the cure of particular diseases they either do not well understand or do not scrupulously observe. For physicians have frustrated and destroyed the fruit of tradition and experience by their magistralities, in adding and taking away and making changes in their receipts at their pleasure; and substituting *quid pro quo*, much like the chemist; usurping such command over the medicine, that the medicine loses all command over the disease. For except it be treacle and mithridate, and perhaps diascordium and the confection of alkermes [15], and a few other medicines, they tie themselves

12 Even this in the extent to which it has been carried appears to stand much in need of an apology; and it is satisfactory to find that one of our best anatomists seems to think so. I refer to Brodie's *Physiological Enquiries*.

13 Sueton. in August. c. 99.

14 Cf. Diog. Laërt. x. 16.

15 Theriaca, from which treacle is a corruption, is the name of a nostrum invented by Andromachus, who was physician to Nero. For an account of the history and composition of mithridaticum, see Celsus, v. 23. The invention of what was called diascordium is ascribed to Fracastorius, who speaks of it as "Diascordium nostrum" in his *De Cont. Morb. Cur.* iii. 7. The confection of Alkermes in its original form seems to have been invented by Mesné, an Arabian physician. About Bacon's time what was called mineral kermes, which was a preparation of antimony, was a popular medicine, but it is probable that he here refers either to the confection of Mesné or to some modification of it.

to scarce any certain receipts severely and religiously. For as to those confections which are for sale in the shops, they are rather in readiness for general intentions than accommodated and specially adapted to particular cures; for they do not specially regard any one disease, but relate generally to purging, opening, comforting, and altering. And this is principally the cause why empirics and old women are more happy many times in their cures than learned physicians, because they are more exact and religious in holding to the composition and confection of tried medicines. Indeed I remember a physician here in England, a famous practitioner, in religion almost a Jew, in reading a kind of Arab, who used to say, " Your physicians in Europe are indeed men of learning ; but they do not know the particular cures for disease ". He would also say in jest, not very reverently, " that our physicians are like bishops, who have the power of the keys, to bind and loose, and nothing more ". But to speak seriously ; I conceive that it would be of great use if some physicians, among the more distinguished both for learning and practice, would compose a work on medicines tried and approved by experiment for the cure of particular diseases. For if it be thought fitter for a learned physician (after taking account of the constitution and age of his patients, the season of the year, their customs, and the like) to apply his medicines according to the occasion, than to abide by any certain prescriptions, the opinion, though plausible, is fallacious, and allows too little weight to experience, and too much to judgment. For as they were the most useful citizens and of the best composition in the state of Rome, who either being consuls inclined to the people, or being tribunes inclined to the senate ; so in the matter we now handle, they are the best physicians, who being great in learning most incline to the traditions of experience, or being distinguished in practice do not reject the methods and generalities of art. As to the qualifying of medicines (if it be ever necessary), it ought rather to be done in the vehicles than in the body of the medicines, wherein nothing should be altered without evident necessity. This part therefore, which treats of authentic and positive medicines, I set down as wanting. But it is a thing that should not be undertaken without keen and severe judgment, and in synod, as it were, of select physicians.

Again, in preparation of medicines I find it strange (especially considering how mineral medicines have been so much lauded and extolled by the chemists [16], and that such medicines are safer applied outwardly than taken inwardly) that no man has endeavoured to make an imitation by art of natural baths and medicinal fountains ; although it is confessed that they receive their virtues from the mineral veins through which they flow ; and not only so, but as a manifest proof of the fact, human industry has found the way to discern and distinguish by analysis from what kind of mineral such waters receive tincture ; as sulphur, vitriol, steel, or the like. Which natural tincture if it might be reduced to compositions of art would put it in the power of man to make more kinds of them as occasion demands, and to regulate their temper at discretion. This part therefore, concerning the imitation of nature in artificial baths (an undertaking doubtless both easy and profitable), I judge to be deficient.

But lest I grow to be more particular than is agreeable either to my intention or to the nature of this treatise, I will conclude this part with the note of one deficience more, which seems to me of greatest consequence ; which is, that the method of treatment in use is too compendious to accomplish anything remarkable or difficult. For in my judgment it would be an opinion more flattering than true, to think any medicine can be so sovereign or so happy as that the simple use of it can work any great cure. It were a strange speech, which spoken once, or even spoken many times, should reclaim a man from a vice to which he is by nature subject. The thing is impossible. It is order, pursuit, sequence, and skilful interchange of application, which is mighty in nature. And these things, although they require greater judgment in prescribing and more constant obedience in observing, yet make up for it abundantly by the magnitude of the effects

[16] The school of medicine of which Paracelsus was the head distinguished itself from the Galenists, who had chiefly recourse to vegetable decoctions and infusions, by the use of mineral medicines. This school has been called that of the Iatro-chemists.

they produce. Now although a man would think, by the daily attentions which physicians pay to their patients,—their visitations, nursings, and prescriptions,—that they were pursuing the cure diligently and following it up by a certain path; yet let a man look more deeply into the prescripts and ministrations which physicians use and he shall find the most of them full of vacillation and inconstancy, devices of the moment, without any settled or foreseen course of cure; whereas they ought from the very first, as soon as ever the disease is fully discovered and known, to resolve upon some regular plan of treatment, and not to depart therefrom without grave reason. And let physicians be assured of this: that there may be (for example) three or four medicines rightly prescribed for the cure of some serious disease, which if taken in proper order and at proper intervals will work the cure; but if taken either singly, or in a different order, or without the interval, will prove most injurious. I do not mean that every scrupulous and superstitious prescript should be taken for the best (no more than every strait way is the way to heaven); the way must be the right way no less than the strait and difficult one. This part then, which I will call the *Physician's Clue*, I set down as deficient. And these are the things I find wanting in that part of medicine which relates to the cure of diseases: only there is one thing still remaining, which is of more consequence than all the rest;—namely, a true and active Natural Philosophy for the science of medicine to be built upon. But that does not belong to the present treatise.

The third part of medicine which I have set down is that which relates to the Prolongation of Life, which is new, and deficient; and the most noble of all. For if such a thing may be discovered, the business of medicine will no longer be confined to humble cures, nor will physicians be honoured only for necessity; but for a gift to men—of earthly gifts perhaps the greatest—of which, next to God, they may become the dispensers and administrators. For although to a Christian making for the Land of Promise the world is but a wilderness, yet even while we travel in the wilderness to have our shoes and garments (that is our bodies, which are as the clothing of the soul) not worn out by the way, must be accounted as a gift of divine grace. Upon this subject then, seeing it is of such excellence, and that I have set it down as wanting, I will after my manner give both admonitions, and directions, and precepts.

My first admonition is, that of the writers upon this argument there is none who has discovered anything great, not to say anything sound. Aristotle has indeed published a very short commentary upon it [17], in which there is some acuteness: which he, as usual, will have to be everything. But more modern writers have handled it so idly and superstitiously, that by reason of their vanity the argument itself has come to be reputed vain and senseless.

My second admonition is, that the very intentions of physicians in this matter are worth nothing, and rather serve to draw men's thoughts away from the point than to direct them to it. For they tell us that death consists in the destitution of warmth and moisture; and therefore that the natural warmth should be comforted, and the radical moisture cherished. Just as if this could be done by broths, or lettuces and mallows, or starch, or jujubes, or spices, or generous wines, or even spirits of wine and chemical oils; all of which are rather injurious than beneficial.

My third admonition is, that men should cease from trifling, nor be so credulous as to imagine that so great a work as this of delaying and turning back the course of nature can be effected by a morning draught or by the use of some precious

17 Aristotle's tract *De Long. et Brev. Vitæ*, which, as Bacon remarks, is very brief, relates to the length of life of all kinds of animals, and even of plants. Sanchez, a Spanish physician, who wrote a treatise on the same subject, thus remarks on Aristotle's: "Adeo longè breviterque disseruit Aristoteles, ut mirum sit tantum philosophum tam indignè rem hanc tractâsse". Not long before the publication of the *De Augmentis*, the *Methusala Vivax* of Dornavus was printed at Hanover; it contains an inquiry as to the causes of antediluvian longevity; Dornavus refutes the notion that the years in which the ages of the Patriarchs are stated are in reality only lunations, by referring to their ages when their first-born sons were begotten.

drug; by potable gold, or essence of pearls, or suchlike toys;—but be assured prolongation of life is a work of labour and difficulty and consisting of a great number of remedies, and those aptly connected one with another. For let no man be so dull as to believe that a thing which has never yet been done can be done now except by means yet unattempted.

My fourth admonition is, that men should rightly observe and distinguish between those things which conduce to a healthy life, and those which conduce to a long life. For there are some things which tend to exhilarate the spirits, strengthen the bodily functions, and keep off diseases, which yet shorten the sum of life, and without sickness hasten on the decay of old age. There are others also which are of service to prolong life and retard decay, which yet cannot be used without danger to health, so that they who use them for the prolongation of life should at the same time provide against such inconveniences as may arise from their use. And so much by way of admonition.

With regard to directions, the idea I have formed of the matter is this. Things are preserved and continued in two ways; either in their own identity, or by repair. In their own identity, as a fly or an ant in amber; a flower or an apple or wood in conservatories of snow; a corpse in balsam. By repair, as in flame, and in things mechanical. Now he that seeks to effect the prolongation of life must use both methods (for separate they have less power); and the human body must be preserved as bodies inanimate, and again as flame, and lastly to a certain degree as things mechanical are preserved. Therefore there are three intentions for the prolongation of life; prevention of waste, goodness of repair, and renewal of that which has begun to grow old. Waste is caused by two depredations; that of the native spirit, and that of the surrounding air. Both of these may be prevented in two ways; either by making those agents less predatory, or the patients (that is, the juices of the body) less susceptible of being preyed on. The spirit is made less predatory if it be either condensed in substance, as in the use of opiates and preparations of nitre, and in mortifications; or diminished in quantity, as in Pythagorean and monastic diets; or quieted in motion, as in leisure and tranquillity. The surrounding air becomes less predatory, when it is either less heated by the rays of the sun, as in cold climates, caves, mountains, and the columns of anchorites; or kept from the body, as by thick skins, the plumage of birds, and the use of oils and unguents without spices. The juices of the body are made less perceptible of depredation, by being rendered either hard, or roscid and oily: hard, as by rough diet, living in the open air, strong exercises, and some mineral baths; roscid, as by the use of sweet things, abstaining from salts and acids, and most of all by such a composition of drink as has very fine and subtle parts, yet free from all acrimony or acidity. Repair is produced by aliments. Now alimentation is promoted in four ways; by the digestion of the bowels to send out the nourishment, as is done by medicines comforting the principal bowels; by excitation of the external parts to attract the aliment, as by exercises, proper frictions, some proper unctions and baths; by preparation of the aliment itself, so that it may insinuate itself more easily and to a certain extent anticipate digestion, as in the various artificial modes of preparing food, mixing drink, fermenting bread, and combining together the virtues of these three; by comforting the last act of assimilation, as in seasonable sleep, and some external applications. The renovation of what has begun to grow old takes place in two ways; either by the inteneration of the habit of body itself, as in the use of baths, plasters, and unguents, which act so as to sink in without drawing anything out; or by draining out the old moisture and substituting new, as in seasonable and frequent purgings, lettings of blood, and attenuating diets, which restore the flower of the body. And so much for directions.

As for precepts, though many may be deduced from the directions themselves, I think fit to subjoin three as principal. The first is, that prolongation of life is to be expected rather from periodical diets, than from any familiar regimen of living, or even from the excellence of particular recipes. For things which have sufficient strength to turn back the course of nature are generally so strong, and produce such alterations, that they cannot be compounded with any medicine much less mixed with common food. It remains therefore, that they be used in series, and regularly, and at set times recurring at certain intervals.

The second is, that prolongation of life is to be expected rather from working on the spirits and from the softening of the parts, than from the modes of alimentation. For there being three things which act upon the human body and frame (not taking external accidents into account), namely the spirits, the parts, and the aliments; the way of prolonging life by the modes of alimentation is tedious and circuitous; whereas the ways by working on the spirits and on the parts are much shorter, and sooner attain the desired end; because the spirits are immediately affected both by vapours and passions, which have strange power upon them; and the parts by baths, unguents, and plasters, which also make sudden impressions.

The third is, that the softening of the parts from without should be effected by things of kindred substance, things that impress, and things that close up. For things of kindred substance are kindly and readily embraced and taken in by the parts, and perform the proper office of emollients: things that impress not only act as vehicles for the virtue of the emollients, making it sink more easily and deeper, but themselves also expand the parts a little; while things that close up retain and keep in and fix for awhile the virtue of both the others, and restrain perspiration, which is a thing opposed to the softening process, because it lets out the moisture. And so by these three (but rather disposed in order and succeeding each other, than mixed together) is the thing accomplished. At the same time I would have it understood that the intention of the softening is not to nourish the parts from without, but only to make them apter to receive nourishment. For whatever is more dry is less active in assimilating. And so much for the Prolongation of Life, now newly assigned to medicine, as the third part.

We come now to Cosmetic, which has parts civil and parts effeminate. For cleanness and decency of body is rightly esteemed to proceed from a modesty of manners, and from reverence, first of all towards God whose creatures we are; then towards society wherein we live; and then also towards ourselves, whom we ought to reverence not less, but rather more, than others. But that adulterate decoration, which makes use of dyes and pigments, is well worthy of the deficiencies which always attend it; being neither fine enough to deceive, nor convenient enough for use, nor safe and wholesome enough for health. And I wonder that this depraved custom of painting has been by the penal laws, both ecclesiastical and civil (which have been very severe against extravagance in apparel and effeminate dressing of the hair), so long overlooked. We read indeed of Jezebel, that she painted her face; but nothing of the kind is said of Esther or Judith.

Let us now proceed to Athletic. This I take in a sense somewhat larger than that in which it is usually understood. For to it I refer everything which conduces to the procuring of any kind of ability of which the human body is capable; whether of agility or of endurance. Agility has two parts, strength and swiftness; endurance has likewise two, patience of natural wants, and fortitude under torments. Of all which we often see remarkable examples, in the practice of tumblers, in the hard living of some savages, in the stupendous strength of maniacs, and in the constancy of some persons under exquisite tortures. And if there be found any other faculty not falling into the former divisions (such as the wonderful power of holding the breath, which is often seen in divers), I mean it to be referred to this art. Now that such things can sometimes be done, is very plain; but the philosophy and inquisition of causes relating to them is almost neglected; the rather, I think, because it is thought that such masteries of nature are only attained either by a peculiar aptness of nature in some men, which cannot be taught, or by continual custom from boyhood, a thing which depends upon authority rather than upon teaching. Which though it be not altogether true, yet of what avail is it to note defects in matters of this kind? For the Olympic Games are over long since; and besides in such things mediocrity is enough for use, excellency in them serving for the most part only for mercenary ostentation.

Lastly I come to Arts of Pleasure Sensual, which are divided according to the senses themselves. The pleasure of the eye is chiefly Painting, with a number of other arts (pertaining to magnificence) which respect houses, gardens, vestments, vases, cups, gems, and the like. The pleasure of the ears is Music, with its various apparatus of voices, wind, and strings; water instruments, once regarded as the leaders of this art, are now almost out of use. Of all these arts those which belong to the eye and ear are esteemed the most liberal; for these two senses are the purest; and the sciences thereof are the most learned, as having mathematics like a handmaid in their train. The one also has some reference to memory and demonstrations, the other to morality and the passions of the mind. The pleasures of the other senses, and the arts relating to them, are less esteemed; as being more allied to luxury than magnificence. For unguents, odours, the dainties and pleasures of the table, and most of all the stimulants of lust, need rather laws to repress than arts to teach them. It has been well observed by some that military arts flourish at the birth and rise of states; liberal arts when states are settled and at their height; and voluptuary arts when they are turning to decline and ruin. And I fear that this our age of the world, as being somewhat upon the descent of the wheel, inclines to arts voluptuary. Wherefore let these things pass. With arts voluptuary I couple arts jocular; for the deceiving of the senses is one of the pleasures of the senses.

And now having run over the doctrines concerning the body of man (Medicine, Cosmetic, Athletic, and the Art Voluptuary), I give this notice in passing; that whereas so many things come into consideration in the human body, parts, humours, functions, faculties, and accidents; and that (if it were a new matter) it would be fit that there should be a single body of learning touching the human body containing them all (like that doctrine concerning the soul, of which I shall soon come to speak); yet to avoid the too great multiplication of arts, or the transposition (more than need be) of their ancient limits, I receive the doctrine concerning the parts of the human body,—the functions, humours, respiration, sleep, generation, the fœtus and gestation in the womb, growth, puberty, old age, fatness, and the like,—into the body of medicine; not that they properly belong to those three offices, but because the human body is in everything the subject of medicine. But voluntary motion and sense I refer to the doctrine concerning the soul, because in these two the soul plays the principal part. And so much for the philosophy concerning the body of man; which is but the tabernacle of the mind.

CHAPTER III.

Division of Human Philosophy relating to the Soul into Doctrine concerning the Breath of Life *and Doctrine concerning the* Sensible or Produced Soul. *Second Division of the same Philosophy into Doctrine concerning the* Substance and Faculties of the Soul, *and Doctrine concerning the* Use and Objects of the Faculties. *Two Appendices of the Doctrine concerning the Faculties of the Soul; Doctrine concerning* Natural Divination *and Doctrine concerning* Fascination. *Distribution of the Faculties of the Sensible Soul into* Motion *and* Sense.

Let us now proceed to the doctrine which concerns the Human Soul, from the treasures whereof all other doctrines are derived. The parts thereof are two; the one treats of the rational soul, which is divine; the other of the irrational, which is common with brutes. I mentioned a little before (in speaking of Forms) the two different emanations of souls, which appear in the first creation thereof; the one springing from the breath of God, the other from the wombs of the elements. For touching the first generation of the rational soul, the Scripture says, " He hath made man of the dust of the earth, and breathed into his nostrils the breath of life "; whereas the generation of the irrational soul, or that of the brutes, was effected by the words, " Let the water bring forth; let the earth bring forth." [1] Now this soul (as it exists in man) is only the instrument of the

[1] To the same effect S. Thomas Aquinas says: " Anima brutorum producitur ex virtute aliquâ corporeâ, anima vero humana a Deo. Et ad hoc significandum dicitur Gen. i.

rational soul, and has its origin like that of the brutes in the dust of the earth. For it is not said that "He made the body of man of the dust of the earth," but that "He made man"; that is the entire man, excepting only the breath of life. Wherefore the first part of the general doctrine concerning the human soul I will term the doctrine concerning the Breath of Life; the other the doctrine concerning the Sensible or Produced Soul. But yet, as hitherto I handle philosophy only (for I have placed sacred divinity at the end of the work), I would not borrow this division from theology, if it were not consonant with the principle of philosophy also. For there are many and great excellencies of the human soul above the souls of brutes, manifest even to those who philosophise according to the sense. Now wherever the mark of so many and great excellencies is found, there also a specific difference ought to be constituted; and therefore I do not much like the confused and promiscuous manner in which philosophers have handled the functions of the soul; as if the human soul differed from the spirit of brutes in degree rather than in kind; as the sun differs from the stars, or gold from metals.

I must subjoin likewise another division of the general doctrine concerning the human soul before I speak more fully of the species. For that which I shall hereafter say of the species will concern both divisions alike; as well that which I have just set down, as that which I am now about to propose. Let this second division therefore be into the doctrine concerning the Substance and Faculties of the soul, and the doctrine concerning the Use and Objects of the Faculties.

Having therefore laid down these two divisions, let us now proceed to the species. The doctrine cocnerning the breath of life, as well as the doctrine concerning the substance of the rational soul, includes those inquiries touching its nature,—whether it be native or adventive, separable or inseparable, mortal or immortal, how far it is tied to the laws of matter, how far exempted from them; and the like. Which questions though even in philosophy they admit of an inquiry both more diligent and more profound than they have hitherto received, yet I hold that in the end all such must be handed over to religion to be determined and defined. Otherwise they will be subject to many errors and illusions of the sense. For since the substance of the soul in its creation was not extracted or produced out of the mass of heaven and earth, but was immediately inspired from God; and since the laws of heaven and earth are the proper subjects of philosophy; how can we expect to obtain from philosophy the knowledge of the substance of the rational soul? It must be drawn from the same divine inspiration, from which that substance first proceeded [2].

The doctrine concerning the sensible or produced soul, however, is a fit subject of inquiry even as regards its substance; but such inquiry appears to me to be deficient. For of what service are such terms as *ultimate act, form of the body*, and such toys of logic, to the doctrine concerning the substance of the soul [3]?

quantum ad alia animalia Producat terra animam viventem; Quantum vero ad hominem dicitur quod inspiravit in faciem ejus spiraculum vitæ."—*Sum. Theol.* i. 75, 6.

But the doctrine that in man there is an irrational soul, as in brutes, to which the rational soul is a distinct addition, is not only not countenanced as M. Bouillet supposes by S. Augustine and the schoolmen (see his edition of Bacon's philosophical works, ii. p. 531), but is distinctly condemned by them. Bacon derived it from Telesius. See General Preface, p. 30.

[2] The anima rationalis is immaterial,—the anima sensibilis is as much material as any other part of man's frame. To it however Telesius, whom Bacon here follows, ascribes sensation, imagination, etc., leaving the higher faculties, and especially the moral sense, as the portion of the anima rationalis. Donius, to whom Bacon refers a little further on, in effect rejects the anima rationalis altogether; admitting, in apparently insincere deference to received opinions, that it may exist; but holding that, if it does so, it is incognisable by human reason.

[3] Bacon refers to the Aristotelian definition of the soul, "Actus primus corporis physici organici vitam potentiâ habentis", and to the doctrine immediately connected with this definition that the soul is the form of man. It is obvious that the actus primus may also be spoken of as actus ultimus, according to the direction in which the arrangement

For the sensible soul—the soul of brutes—must clearly be regarded as a corporeal substance, attenuated and made invisible by heat; a breath (I say) compounded of the natures of flame and air, having the softness of air to receive impressions, and the vigour of fire to propagate its action; nourished partly by oily and partly by watery substances; clothed with the body, and in perfect animals residing chiefly in the head, running along the nerves, and refreshed and repaired by the spirituous blood of the arteries; as Bernardinus Telesius and his pupil Augustinus Donius have in part not altogether unprofitably maintained [4]. Let there be therefore a more diligent inquiry concerning this doctrine; the rather because the imperfect understanding of this has bred opinions superstitious and corrupt and most injurious to the dignity of the human mind, touching metempsychosis, and the purifications of souls in period of years, and indeed too near an affinity in all things between the human soul and the souls of brutes. For this soul is in brutes the principal soul, the body of the brute being its instrument whereas in man it is itself only the instrument of the rational soul, and may be more fitly termed not soul, but spirit. And so much for the substance of the soul.

The faculties of the soul are well known; understanding, reason, imagination, memory, appetite, will; in short all with which the logical and ethical sciences deal. But in the doctrine concerning the soul the origins of these faculties ought to be handled, and that physically, as they are innate and inherent in the soul; the uses only and objects of them being deputed to those other arts. In which part nothing of much value (in my opinion) has as yet been discovered; though I cannot indeed report it as deficient. This part touching the faculties of the mind has likewise two appendices, which themselves also, as they are handled, have rather produced smoke than any clear flame of truth. One of these is the doctrine of Natural Divination, the other of Fascination.

Divination has been anciently and not unfitly divided into two parts; Artificial and Natural. Artificial makes prediction by argument, concluding upon signs and tokens; Natural forms a presage from an inward presentiment of the mind, without the help of signs. Artificial is of two sorts; one argues from causes; the other only from experiments, by a kind of blind authority. Which latter is for the most part superstitious; such as were the heathen observations upon the inspection of entrails, the flights of birds, and the like. And the more solemn astrology of the Chaldeans was little better. But artificial divination of both kinds is dispersed among different knowledges. The astrologer has his predictions, from the position of the stars. The physician likewise has his predictions of approaching death, of recovery, of coming symptoms of diseases, from the urine, the pulse, the look of the patient, and the like. The politician also has his; "O venal city, that will quickly perish, if it finds a purchaser" [5]: which prediction was not long in being verified; being fulfilled in Sylla first, and afterwards in Cæsar. Predictions of this kind therefore are not to our present purpose, but are to be referred to their own arts. But Natural Divination, which springs from the inward power of the mind, is that which I now speak of. This is of two sorts; the one Primitive, the other by Influxion.

proceeds, but I do not know whether Bacon had any reason for deviating from the usual phraseology.

With respect to the phrase "forma corporis," it is to be remarked that the Scotists maintained the existence of a "forma corporis," that namely which gives the body corporeity distinct from the informing principle or soul of man;—a subtlety introduced to evade the difficulties which the gradual development of the body from its first rudiments to perfection,—that is, its gradual progress to corporeity,—appears to present when contrasted with the way in which the rational soul is infused. For it was a received opinion that the soul is not "ex traduce," that is not derived from that of the progenitor, but on the contrary is infused as it were ab extra into the body it informs.

[4] See the fifth book of Telesius *De Rerum Natura*, and the second book, particularly the fourth and fifth chapters, of Donius *De Naturâ Hominis*; and compare Campanella *De Sensu Rerum*, ii. 4. Campanella follows Telesius more closely than Donius does.

[5] Sallust, in Bell. Jugurth. 38.

Primitive is grounded upon the supposition that the mind, when it is withdrawn and collected into itself, and not diffused into the organs of the body, has of its own essential power some prenotion of things to come. Now this appears most in sleep, in extasies, and near death; and more rarely in waking apprehensions, or when the body is healthy and strong [6]. But this state of mind is commonly induced or furthered by those abstinences and observances which most withdraw the mind from exercising the duties of the body, so that it may enjoy its own nature, free from external restraints. Divination by influxion is grounded upon this other conceit; that the mind, as a mirror or glass, receives a kind of secondary illumination from the foreknowledge of God and spirits; and this also is furthered by the same state and regimen of the body as the other. For the retiring of the mind within itself gives it the fuller benefit of its own nature, and makes it the more susceptible of divine influxions; save that in divinations by influxion the mind is seized with a kind of fervency and impatience as it were of the present Deity (a state which the ancients noted by the name of divine fury); while in primitive divination it is more in a state of quiet and repose.

Fascination is the power and act of imagination intensive upon the body of another (for of the power of imagination upon the body of the imaginant I have spoken above); wherein the school of Paracelsus and the disciples of pretended natural magic have been so intemperate, that they have exalted the power and apprehension of the imagination to be much one with the power of miracle-working faith [7]. Others, that draw nearer to probability, looking with a clearer eye at the secret workings and impressions of things, the irradiations of the senses, the passage of contagion from body to body, the conveyance of magnetic virtues, have concluded that it is much more probable there should be impressions, conveyances, and communications from spirit to spirit (seeing that the spirit is above all other things both strenuous to act and soft and tender to be acted on); whence have arisen those conceits (now become as it were popular) of the mastering spirit, of men unlucky and ill omened, of the glances of love, envy, and the like. With this is joined the inquiry how to raise and fortify the imagination; for if the imagination fortified have so much power, it is worth while to know how to fortify and exalt it. And here comes in crookedly and dangerously a palliation and defence of a great part of ceremonial magic. For it may be speciously pretended that ceremonies, characters, charms, gesticulations, amulets, and the like, do not derive their power from any tacit or sacramental contract with evil spirits, but serve only to strengthen and exalt the imagination of him who uses them. As likewise in religion the use of images to fix the cogitations and raise the devotions of those who pray before them has grown common. My own judgment however is this: though it be admitted, that imagination has power, and further that ceremonies fortify and strengthen that power; and that they be used sincerely and intentionally for that purpose, and as a physical remedy, without any the least thought of inviting thereby the aid of spirits; they are nevertheless to be held unlawful, as opposing and disputing that divine sentence passed upon man for sin, "In the sweat of thy face shalt thou eat bread" [8]. For magic of this kind proposes to attain those noble fruits which God ordained to be bought at the price of labour by a few easy and slothful observances.

[6] A curious illustration of this remark is mentioned in the geography ascribed to Ibn Haukal. When a prince among the Khazars was made Khakan, he was strangled with a piece of taffeta, and asked, when he could scarcely breathe, how long he had to reign. He answered so many years; and if he reached the term, was then put to death. This was also a Turkish usage, except that it does not seem that they put the prince to death if he lived as long as he had foretold. See Klaproth, *Tableaux Hist. de l'Asie*, p. 273.

On the subject of natural divination see Campanella, *De Sensu Rerum*, iii. 7-11. He says of himself: "Ast ego, cum mali quippiam mihi imminet, inter somnium et vigiliam audire soleo vocem clarè loquentem mihi 'Campanella, Campanella,' et interdum alia addentem, et ego attendo nec intelligo quis sit".

[7] See Paracelsus's tract *De Vi imaginativâ* and many other parts of his writings.

[8] Gen. iii. 19.

There remain two doctrines, which refer principally to the faculties of the inferior or sensible soul,—as that which is most connected with the organs of the body; the one concerning Voluntary Motion, the other concerning Sense and the Sensible. In the first of these, which has in other respects also been very barrenly inquired, one entire part almost is wanting. For the proper office and structure of the nerves and muscles, and of the other parts required for this motion; and what part of the body is at rest, while another moves; and that the imagination is as it were the director and driver of this motion, insomuch that when the image which is the object of the motion is withdrawn the motion itself is immediately interrupted and stopped (as in walking, if you begin to think eagerly and fixedly of something else, you immediately stand still); these, I say, and some other subtleties which are not amiss, have long ago come into observation and inquiry. But how the compressions, dilatations, and agitations of the spirit (which is doubtless the source of motion) can sway, excite, or impel the corporeal and gross mass of the parts, has not as yet been diligently inquired and handled. And no wonder; seeing the sensible soul has been regarded rather as a function than as a substance[9]. But since it is now known that it is itself a corporeal and material substance, it is necessary to inquire by what efforts a spirit so small and tender can put in motion bodies so gross and hard. Of this part therefore, since it is deficient, let inquiry be made.

Concerning Sense and the Sensible there has been much fuller and more diligent inquiry, both in general treatises concerning them and also in particular arts, as perspective and music; how correctly, is nothing to the purpose, seeing they cannot be ranked as deficients. Yet there are two noble and distinguished parts, which I pronounce deficient in this doctrine; the one concerning the Difference of Perception and Sense, the other concerning the Form of Light.

A good explanation of the difference between Perception and Sense should have been prefixed by philosophers to their treatises on Sense and the Sensible, as a matter most fundamental. For we see that all natural bodies have a manifest power of perception, and also a kind of choice in receiving what is agreeable, and avoiding what is hostile and foreign. Nor am I speaking only of the more subtle perceptions, as when the magnet attracts iron, flame leaps towards naphtha, one bubble coming near another unites with it, rays of light start away from a white object, the body of an animal assimilates things that are useful and excerns things that are not so, part of a sponge attracts water (though held too high to touch it) and expels air, and the like. For what need is there of enumerating such things? since no body when placed near another either changes it or is changed by it, unless a reciprocal *perception* precede the operation. A body perceives the passages by which it enters; it perceives the force of another body to which it yields; it perceives the removal of another body which held it fast, when it recovers itself; it perceives the disruption of its continuity, which for a time it resists; in short there is Perception everywhere. And air perceives heat and cold so acutely, that its perception is far more subtle than that of the human touch, which yet is reputed the normal measure of heat and cold. It seems then that in regard to this doctrine men have committed two

[9] In the school philosophy, at least among the Realists, every substantial form (and the soul among the rest) was regarded as a *substance*. This of course implies the possibility of its independent existence, though, as form and matter are correlatives, it is difficult to understand how either can exist apart from the other. This difficulty however seems to have been completely surmounted or set aside; and thus, for instance, St. Thomas Aquinas affirms that angels are immaterial forms (*Sum. Theol.* i. 61). Bacon's remark that the soul had hitherto been looked on rather as a function than a substance refers, I think, to Melancthon's exposition of the Aristotelian doctrine. For Melancthon, whose views of the Peripatetic philosophy had long great influence in the Protestant universities, affirms that, according to the true view of Aristotle's opinion, the soul is not a substance but an ἐντελέχεια or functio. The word ἐντελέχεια he conceives to be only a modification of ἐνδελέχεια, which he proposes to render "habitualis agitatio seu δύναμις quædam ciens actiones." See his *De Animâ*, c. 15.

faults; one, that they have for the most part left it untouched and unhandled (though it be a most noble subject); the other, that they who have happened to turn their minds to it have gone too far, and attributed *sense* to all bodies; so that it were a kind of impiety to pluck off the branch of a tree, lest it should groan, like Polydorus [10]. But they should have examined the difference between perception and sense, not only in sensible as compared with insensible bodies (as plants with animals), one body with another; but also in the sensible body itself they should have observed what is the reason why so many actions are performed without any sense at all; why food is digested and ejected; humours and juices carried up and down; the heart and the pulse beat; the entrails, like so many workshops, perform every one its own work; and yet all these and many other things are done without sense. But men have not seen clearly enough of what nature the action of sense is; and what kind of body, what length of time, or what repetition of impression is required to produce pleasure or pain. In a word, they do not seem at all to understand the difference between simple perception and sense; nor how far perception may take place without sense. Neither is this a dispute about words merely, but about a matter of great importance. Concerning this doctrine then (being of great use and bearing upon very many things) let a better inquiry be set on foot. For ignorance on this point drove some of the ancient philosophers to suppose that a soul was infused into all bodies without distinction; for they could not conceive how there could be motion at discretion without sense, or sense without a soul [11].

That no due investigation has been made concerning the Form of Light (especially as men have taken great pains about perspective) may be considered an astonishing piece of negligence. For neither in perspective nor otherwise has any inquiry been made about Light which is of any value. The radiations of it are handled, not the origins. But it is the placing of perspective among the mathematics that has caused this defect, and others of the kind; for thus a premature departure has been made from Physics. Again the manner in which Light and its causes are handled in Physics is somewhat superstitious, as if it were a thing half way between things divine and things natural; insomuch that some of the Platonists have made it older than matter itself; asserting upon a most vain notion that when space was spread forth it was filled first with light, and afterwards with body; whereas the Holy Scriptures distinctly state that there was a dark mass of heaven and earth before light was created [12]. And where the subject is handled physically and according to sense, it comes at once to questions of radiations; so that there is but little physical inquiry extant on the matter. Now men ought to have sunk their speculations for awhile, and inquired what that is which is common to all lucid bodies; in other words, into the Form of Light. For see what an immense difference of body

10 Virg. *Æn.* iii. 39.

11 There is a remarkable similarity between the view which Bacon here maintains and that which we find in several passages in the writings of Leibnitz. See his *Monadologie*, §§ 14 and 19, or his *Principes de la Nature et de la Grace*, § 4. The distinction between perceptio and sensus corresponds in Leibnitz's language to that between perception and apperception, a distinction on which the classification of the different orders of monads essentially depends. It is not probable that Bacon was acquainted with the most celebrated treatise on the doctrine of universally diffused sensation, namely the *De Sensu Rerum* of Campanella, as it was not published much before the appearance of the *De Augmentis*; but the same doctrine had, as Brucker remarks, been taught, though not in so formal a manner, by Telesius, with whose works Bacon was as we know familiar; and it may in truth be traced in the writings of Giordano Bruno, of Cæsalpinus, and of Gilbert, and probably in those of many of their contemporaries. See, for Leibnitz's remarks as to the origin of this doctrine, his letter to Thomasius, referred to in the note at p. 31.

12 Bacon appears to refer to the visionary opinions of Fludd. See the first part of Fludd's great work referred to in the note at p. 445. The process of creation is illustrated by some curious engravings. There is an account of Fludd's views on this and other subjects in Tennemann's *History of Philosophy*, ix. p. 218.

there is (if they be considered according to their dignity) between the sun and rotten wood, or even the putrified scales of fish? They should also have inquired why some things take fire and throw out light when heated, and others not. Iron, metals, stones, glass, wood, oil, tallow, when they are subjected to fire, either break into flame, or at least become red; but water and air do not acquire any light from the most intense and raging heat, nor cast forth any brightness. And if any one thinks that this is because it is the property of fire to shine, and air and water are entirely hostile to fire, he can never have rowed on the sea on a dark night in hot weather; when he would have seen the drops of water that are struck up by the oars glittering and shining: a thing which happens likewise in the boiling sea-froth, which they call "sea-lungs"[13]. Lastly what connexion with fire and lighted matter have glowworms and fireflies, and the Indian fly, which lights up a whole room; or the eyes of some animals in the dark; or sugar while it is being scraped or broken; or the sweat of a horse, hard-ridden on a hot night; and the like? Nay, so little is this subject understood, that most people think sparks from flint to be but air in friction. And yet since the air does not take fire with heat, and manifestly conceives light, how happens it that owls and cats and some other animals can see by night? It must needs be (since sight cannot pass without light) that the air has some pure and natural light of its own[14], which, though very faint and dull, is nevertheless suited to the visual organs of such animals, and enables them to see. But the reason of this error (as of most others) is that men have not from particular instances elicited the Common Forms of natures; which I have laid down as the proper subject of Metaphysic, which is itself a part of Physic, or of the doctrine concerning nature. Wherefore let inquiry be made of the Form and Origins of Light, and in the meantime let it be set down as deficient. And so much for the doctrine concerning the substance of the soul both rational and sensible, with its faculties; and for the appendices of that doctrine.

[13] See note on *Nov. Org.* ii. 12, p. 312.

[14] A doctrine of Telesius. See note p. 662.

Book V.

CHAPTER I.

Division of the Doctrine concerning the Use and Objects of the Faculties of the Human Soul into Logic *and* Ethic. *Division of Logic into the Arts of* Discovering, *of* Judging, *of* Retaining, *and of* Transmitting.

THE doctrine concerning the Intellect (most excellent King), and the doctrine concerning the Will of man, are as it were twins by birth. For purity of illumination and freedom of will began and fell together [1]; and nowhere in the universal nature of things is there so intimate a sympathy as between truth and goodness. The more should learned men be ashamed, if in knowledge they be as the winged angels, but in their desires as crawling serpents; carrying about with them minds like a mirror indeed, but a mirror polluted and false [2].

I come now to the knowledge which respects the use and objects of the faculties of the human soul. It has two parts, and those well known and by general agreement admitted; namely, Logic and Ethic; only Civil Knowledge, which is commonly ranked as a part of Ethic, I have already emancipated and erected into an entire doctrine by itself,—the doctrine concerning man congregate, or in society; and in this place I treat only of man segregate. Logic discourses of the Understanding and Reason; Ethic of the Will, Appetite, and Affections: the one produces determinations, the other actions. It is true indeed that the imagination performs the office of an agent or messenger or proctor in both provinces, both the judicial and the ministerial. For sense sends all kinds of images over to imagination for reason to judge of; and reason again when it has made its judgment and selection, sends them over to imagination before the decree be put in execution. For voluntary motion is ever preceded and incited by imagination; so that imagination is as a common instrument to both,—both reason and will; saving that this Janus of imagination has two different faces; for the face towards reason has the print of truth, and the face towards action has the print of goodness; which nevertheless are faces

quales decet esse sororum [3].

Neither is the imagination simply and only a messenger; but it is either invested with or usurps no small authority in itself, besides the simple duty of the message. For it was well said by Aristotle, "That the mind has over the body that commandment which the lord has over a bondman; but that reason has over the imagination that commandment which a magistrate has over a free citizen" [4], who may come also to rule in his turn. For we see that in matters of faith and religion our imagination raises itself above our reason; not that divine illumination resides in the imagination; its seat being rather in the very citadel of the mind and understanding; but that the divine grace uses the motions of the imagination as an instrument of illumination, just as it uses the motions of the will as an instrument of virtue; which is the reason why religion ever sought access to the mind by similitudes, types, parables, visions, dreams. And again it is no

[1] Namely at the fall; as St. Thomas Aquinas observes: "Homo peccando liberum arbitrium dicitur perdidisse, non quantum ad libertatem naturalem quæ est a coactione, sed quantum ad libertatem quæ est a culpâ et miseriâ".—*Sum. Theol.* i. 83, 2.

[2] Orig. *menstruati*. For an account of the notion on which this use of the word menstruatus is founded, see Aristotle *De Insomniis*, 2. 8., or Pliny [vii. 13].

[3] Ov. *Met.* ii. 14:—Such as sisters' faces should be. [4] Arist. *Pol.* i. 3.

small dominion which imagination holds in persuasions that are wrought by eloquence; for when by arts of speech men's minds are soothed, inflamed, and carried hither and thither, it is all done by stimulating the imagination till it becomes ungovernable, and not only sets reason at nought, but offers violence to it, partly by blinding, partly by incensing it. Nevertheless, I see no cause to alter the former division; for imagination hardly produces sciences; poesy (which in the beginning was referred to imagination) being to be accounted rather as a pleasure or play of wit than a science. And for the power of the imagination in nature, I have just now assigned it to the doctrine concerning the soul. And its relation to rhetoric I think best to refer to that art itself, which I shall handle hereafter.

That part of human philosophy which regards Logic is less delightful to the taste and palate of most minds, and seems but a net of subtlety and spinosity. For as it is truly said that "knowledge is the food of the mind"[5], so in their choice and appetite for this food most men are of the taste and stomach of the Israelites in the desert, that would fain have returned to the flesh-pots, and were weary of manna; which though it were celestial, yet seemed less nutritive and comfortable. And in like manner those sciences are (for the most part) best liked which have some infusion of flesh and blood; such as civil history, morality, policy, about which men's affections, praises, fortunes, turn and are occupied. But this same "dry light" parches and offends most men's soft and watery natures. But to speak truly of things as they are in worth, rational knowledges are the keys of all other arts. And as the hand is the instrument of instruments, and mind is the form of forms, so these are truly said to be the arts of arts[6]. Neither do they only direct, but likewise confirm and strengthen; even as the habit of shooting not only enables one to take a better aim, but also to draw a stronger bow.

The logical arts are four in number; divided according to the ends at which they aim. For men's labour in rational knowledges is either to invent that which is sought, or to judge that which is invented, or to retain that which is judged, or to deliver over that which is retained. So therefore the Rational Arts must be four; Art of Inquiry or Invention; Art of Examination or Judgment; Art of Custody or Memory; and Art of Elocution or Tradition[7]. Of these I will now speak separately.

CHAPTER II.

Division of the Art of Discovery into discovery of Arts *and discovery of* Arguments: *and that the former of these (which is the most important) is wanting. Division of the Art of Discovery of Arts into* Learned Experience *and the* New Organon. *Description of* Learned Experience.

INVENTION is of two kinds, very different; the one of arts and sciences, and the other of speech and arguments. The former of these I report altogether deficient, which seems to me to be such a deficience, as if in the making of an inventory touching the estate of a deceased person, it should be set down that "there is no ready money". For as money will fetch all other commodities, so by this art all the rest are obtained. And as the West Indies would never have been discovered if the use of the mariner's needle had not been discovered first, though the one be vast regions and the other a small motion; so it cannot be found strange if no further progress has been made in the discovery and advancement of the arts, when the art itself of discovery and invention has as yet been passed over.

[5] Mr. Markby, in his edition of the *Advancement of Learning*, refers to Cicero, *Acad. Qu.* ii. 41.: "Est enim animorum ingeniorumque naturale quoddam quasi pabulum consideratio contemplatioque naturæ".—*J. S.*

[6] That the hand is the instrument of instruments, and the soul the form of forms, is said by Aristotle. See the *De Animâ*, iii. 8.

[7] These divisions are adopted from Peter Ramus; the artes logicæ including what Ramus calls Dialectic and Rhetoric, of which the former is divided into Inventio and Judicium, and the latter into Elocutio and Pronunciatio.

That this part of knowledge is wanting stands plainly confessed. For in the first place, Logic says nothing, no nor takes any thought, about the invention of arts, whether mechanical or what are termed liberal, or about eliciting the works of the one or the axioms of the other; but passes on, merely telling men by the way that for the principles of each art they must consult the professor of it [1]. Celsus, a wise man as well as a physician, (though all men are wont to be large in praise of their own art) acknowledges gravely and ingenuously, speaking of the empirical and dogmatical sects of physicians, that medicines and cures were first found out, and then afterwards the reasons and causes were discovered; and not the causes first found out from the nature of things, and by light from them the medicines and cures discovered [2]. And Plato more than once intimates "that particulars are infinite; and the higher generalities give no sufficient direction; that the pith therefore of all sciences, which makes the artsman differ from the inexpert, is in the middle propositions, which in every particular knowledge are taken from tradition and experience" [3]. Moreover they who have written about the first inventors of things or the origins of sciences have celebrated chance rather than art, and represented brute beasts, quadrupeds, birds, fishes, serpents, as the doctors of sciences, rather than men:

> Dictamnum genitrix Cretæa carpit ab Ida,
> Puberibus caulem foliis, et flore comantem
> Purpureo: non illa feris incognita capris
> Gramina, cum tergo volucres hæsere sagittæ [4].

So that it is no marvel (the manner of antiquity being to consecrate inventors of useful things) that the ancient Egyptians (to whom very many of the arts owe their origin) had so few human idols in their temples, but almost all brute;

> Omnigenumque Deum monstra, et latrator Anubis,
> Contra Neptunum, et Venerem, contraque Minervam, etc.[5].

And if you like better, according to the tradition of the Greeks, to ascribe the first inventions to men; yet you would not say that Prometheus was led by speculation to the discovery of fire, or that when he first struck the flint he expected the spark; but rather that he lighted on it by accident, and (as they say) stole it from Jupiter. So that in the invention of arts it would seem that hitherto men are rather beholden to a wild goat for surgery, to a nightingale for music, to the ibis for clysters [6], to the pot lid that flew open for artillery, and in a word to

[1] See Aristotle, *Prior Analytics*, i. 30.

[2] See *Nov. Org.* i. 73. This is not what Celsus himself confesses in the passage to which Bacon apparently refers, but what he represents the Empirics as urging against the Rationalists.—*J. S.*

[3] See note above p. 360.

[4] Virg. *Æn.* xii. 412:—

> Far off in Cretan Ide a plant there grew
> With downy leaves and flower of purple hue,
> The dittany, whose medicinable power
> The wild goat proves whene'er in evil hour
> The hunter's arrow lodges in his side.

[5] Virg. *Æn.* viii. 698:—

> All kinds and shapes of gods, a monstrous host,
> The dog Anubis foremost, stood arrayed
> 'Gainst Neptune, Venus, Pallas, etc.

[6] See Plutarch, *De Solertiâ Animalium*, or *De Iside*. Compare Pliny. The story of the accidental invention of gunpowder by Schwartz is well known. So too is it said that the Jesuit's bark was discovered by the lions who cured their fevers by drinking the water into which it had fallen. It is obvious that all stories of this kind are more or less mythical. The subject has been systematically discussed by Virey. (*Journal de Pharmacie*, 1818.)

chance, or anything else, rather than to Logic. Neither is that form of invention much other which is well described by Virgil,

> Ut varias usus meditando extunderet artes
> Paulatim [7].

For here no other method of invention is proposed than that which the brute beasts are capable of and frequently use; which is an extreme solicitude about some one thing, and perpetual practising of it, such as the necessity of self-preservation imposes on such animals. For so Cicero says very truly, "that practice constantly applied to one thing often does more than either nature or art." [8] And therefore if it be said of men,

> Labor omnia vincit
> Improbus, et duris urgens in rebus egestas [9];

it is likewise said of brutes,

> Quis expedivit psittaco suum χαῖρε [10]?

Who taught the raven in a drought to throw pebbles into a hollow tree where she espied water, that the water might rise till she could reach it with her beak? Who showed the way to the bees, that sail through such a vast sea of air to fields in flower far removed from their hive, and back again [11]? Who taught the ant to bite the grains of corn that she lays up in her hill, lest they should sprout and so disappoint her hope [12]? And in that line of Virgil, if the word "extundere," which imports the difficulty, and the word "paulatim," which imports the slowness of the thing, be observed, we shall find ourselves where we were, amongst those gods of the Egyptians; for men have hitherto used the faculty of reason but little, and the office of art not at all, for the discovery of inventions.

Secondly, this very thing which I assert is demonstrated (if you observe it

[7] Virg. *Georg.* i. 133:—

> So might long use, with studious thought combined,
> The various arts by slow endeavour find.

[8] Cicero, *Pro Balbo*, c. 20.

[9] Virg. *Georg.* i. 145:—

> Stern labour masters all,
> And want in poverty importunate.

[10] Persius, *Prolog.*:—Who taught the parrot to say how d'ye do?

[11] Much more remarkable than the return of the bees to their hive is the appearance of mathematical knowledge shown in the construction of their cells. In every case of instinct, the impulse in obedience to which the instinctive act is performed is a matter at the nature of which we can only guess; but the case just mentioned has a difficulty of its own. The bees may be supposed to know when they have reached their hive; but how do they perceive that the cell has acquired its just proportions? Several attempts have been made to explain away this especial difficulty; but those which I am acquainted with appear to be quite unsatisfactory. It is worthy of remark that the degree of accuracy with which the cells are constructed has been exaggerated; one writer after another having repeated, on the supposed authority of Maraldi, what Maraldi never said. According to his observations the angles of the terminal rhomb are about 108° and 72°. He does not attempt to determine them more precisely, although he has generally been supposed to do so. It has been recently stated that the mathematical problem which the cells of bees suggest was first correctly solved by Lord Brougham in the notes to his edition of Paley's *Natural Theology*; but this statement is, it need scarcely be said, erroneous. [The problem has been cleared up by Darwin, *Origin of Species*, ch. viii.—Ed.]

[12] This statement is probably taken from Plutarch, *De Solertiâ Animalium*. The supposed grains of corn are no doubt the nymphæ. Huber repeatedly observed ants in the act of tearing the integument in which the young ant was enclosed, in order to facilitate its exit. This practice is, it may be presumed, the origin of the notion mentioned in the text.

carefully) by the form of induction which Logic proposes, as that whereby the principles of sciences may be invented and proved ; which form is utterly vicious and incompetent, and so far from perfecting nature, that contrariwise it perverts and distorts her. For he that shall attentively observe how the mind gathers this excellent dew of knowledge, like to that the poet speaks of,

aërei mellis cœlestia dona [13],

(for the sciences themselves are extracted out of particular instances, partly natural, partly artificial, as the flowers of the field and the garden) shall find that the mind does of herself by nature manage and act an induction much better than logicians describe it ; for to conclude upon a bare enumeration of particulars (as the logicians do) without instance contradictory, is a vicious conclusion; nor does this kind of induction produce more than a probable conjecture. For who can assure himself, when the particulars which he knows or remembers only appear on one side, that there are not others on the contrary side which appear not ? As if Samuel should have rested upon those sons of Jesse who were brought before him in the house, and not sought for David, who was in the field [14]. And this form of induction (to say truth) is so gross and stupid, that it had not been possible for wits so acute and subtle as those that have studied these things to offer it to the world, but that they were hurrying on to their theories and dogmaticals, and were too dainty and lofty to pay due attention to particulars, and especially to dwell any time upon them. For they used examples or particular instances but as serjeants or whifflers to drive back the crowd and make way for their opinions, and never called them into council from the first, for the purpose of legitimate and mature deliberation concerning the truth of things. Certainly it is a thing that may touch a man with a religious wonder to see how the footsteps of seducement are the very same in divine and human truth. For as in the perception of divine truth man cannot induce himself to become as a child ; so in the study of human truth, for grown-up men to be still reading and conning over the first elements of inductions like boys, is accounted poor and contemptible.

Thirdly, even if it be granted that the principles of sciences may, by the induction which is in use, or by sense and experience, be rightly established ; yet it is very certain that the lower axioms cannot (in things natural, which participate of matter) be rightly and safely deduced from them by syllogism. For in the syllogism propositions are reduced to principles through intermediate propositions. Now this form of invention, or of probation may be used in popular sciences, such as ethics, politics, laws, and the like ; yea, and in divinity also, because it has pleased God of his goodness to accommodate himself to the capacity of man ; but in Physics, where the point is not to master an adversary in argument, but to command nature in operation, truth slips wholly out of our hands, because the subtlety of nature is so much greater than the subtlety of words ; so that, syllogism failing, the aid of induction (I mean the true and reformed induction) is wanted everywhere, as well for the more general principles as for intermediate propositions. For syllogisms consist of propositions, and propositions of words ; and words are but the current tokens or marks of popular notions of things ; wherefore if these notions (which are the souls of words) be grossly and variably collected out of particulars, the whole structure falls to pieces [15]. And it is not the laborious examination either of consequences of arguments or of the truth of propositions that can ever correct that error ; being

[13] Virg. *Georg.* iv. 1 :—The heavenly gift of aërial honey.

[14] 1 Sam. xvi.

[15] Compare *Novum Organum*, i. 13. and 14. The formation of abstract conceptions is one of the objects of Bacon's inductive method, as well as the establishment of axioms. See Gen. Pref. p. 23. It is difficult to understand how the subtlety of language and the subtlety of natural operations can be compared. Bacon must be understood to mean that scientific terms and the conceptions which they express are not an adequate representation of the natural phenomena which have led to their formation.

(as the physicians say) in the first digestion ; which is not to be rectified by the subsequent functions. And therefore it was not without great and evident reason that so many philosophers (some of them most eminent) became Sceptics and Academics, and denied any certainty of knowledge or comprehension ; affirming that the knowledge of man extended only to appearances and probabilities. It is true that Socrates, when he disclaimed certainty of knowledge for himself is thought by some to have done it only in irony[16], and to have enhanced his knowledge by dissembling it ; pretending not to know that which it was plain he knew in order that he might be thought to know also that which he knew not. And in the later academy too (which Cicero embraced) that opinion of the incapacity of the mind to comprehend truth was not held very sincerely. For those who excelled in eloquence commonly chose that sect, for the glory of speaking copiously on either side of the question ; whereby they were led astray from the straight road, which they ought to have followed in pursuit of truth, into certain pleasant walks laid out for amusement and recreation. It is certain however that there were some here and there in both academies (both old and new) and much more among the Sceptics, who held this opinion in simplicity and integrity. But their great error was, that they laid the blame upon the perceptions of the sense, and thereby pulled up the sciences by the very roots. Now the senses, though they often deceive us or fail us, may nevertheless, with diligent assistance, suffice for knowledge ; and that by the help not so much of instruments (though these too are of some use) as of those experiments which produce and urge things which are too subtle for the sense to some effect comprehensible by the sense. But they ought rather to have charged the defect upon the mind—as well its contumacy (whereby it refuses to submit itself to the nature of things) as its errors,—and upon false forms of demonstration, and ill-ordered methods of reasoning and concluding upon the perception of the senses. But this I say not to disable the intellect, or to urge the abandonment of the enterprise ; but to stir men to provide the intellect with proper helps for overcoming the difficulties and obscurities of nature. For no steadiness of hand or amount of practice will enable a man to draw a straight line or perfect circle by hand alone, which is easily done by help of a ruler or compass. And this is the very thing which I am preparing and labouring at with all my might,—to make the mind of man by help of art a match for the nature of things ; to discover an art of Indication and Direction, whereby all other arts with their axioms and works may be detected and brought to light. For this I have with good reason set down as wanting.

This Art of Indication (for so I call it) has two parts. For the indication either proceeds from one experiment to another ; or else from experiments to axioms ; which axioms themselves suggest new experiments. The one of these I will term Learned Experience[17], the other Interpretation of Nature, or the New Organon.

[16] Cicero, *Acad. Quæst.* ii. 5. 15.—*J. S.*

[17] With reference to the question how far Bacon thought it possible for observation to be carried on apart from theory (see General Preface, p. 35), it is, I think, important to remark that this notion of an Experientia *Literata*, as an intermediate step between simple experimentation *absque ullâ serie aut methodo* and the Interpretation of Nature, was not an after-thought, but formed part of his original design in the earliest shape in which it is known to us. "This part of Invention (he says in the *Advancement of Learning*) concerning the Invention of Sciences, I purpose (if God give me leave) hereafter to propound : having digested it into two parts : whereof the one I term *Experientia literata*, and the other *Interpretatio naturæ ; the former being but a degree or rudiment of the latter*." Now if he meant by "Experientia literata" the same thing which he describes here, or anything like it,—which I see no reason to doubt—he must have seen even then the impossibility of making a collection of facts sufficient for the purposes of Interpretation without the help of some principle of arrangement, some "series et methodus", some "sagacitas" in seeking and selecting ; which necessarily implied some amount of *theory*. Such theory was indeed to be provisional only, and subject at all times to revision. It was not to be allowed as an axiom. But it does not appear that he would have put any other restriction upon the exercise of human sagacity in this way. The process might have been carried therefore to an indefinite length, and the further the better.

But the former (as I have hinted elsewhere [18]) must hardly be esteemed an art or a part of philosophy, but rather a kind of sagacity; whence likewise (borrowing the name from the fable) I sometimes call it the Hunt of Pan. Nevertheless as a man may proceed on his path in three ways: he may grope his way for himself in the dark; he way be led by the hand of another, without himself seeing anything; or lastly, he may get a light, and so direct his steps; in like manner when a man tries all kinds of experiments without order or method, this is but groping in the dark; but when he uses some direction and order in experimenting, it is as if he were led by the hand; and this is what I mean by Learned Experience For the light itself, which was the third way, is to be sought from the Interpretation of Nature, or the New Organon.

Learned Experience, or the Hunt of Pan, treats of the methods of experimenting; and (since I have set it down as wanting, and the thing itself is not altogether obvious) I will here, according to my plan and custom, give some shadow of it. The method of experimenting proceeds principally either by the Variation, or the Production, or the Translation, or the Inversion, or the Compulsion, or the Application, or the Conjunction, or finally the Chances, of experiment. None of these however extend so far as to the invention of any axiom. For all transition from experiments to axioms, or from axioms to experiments, belongs to that other part, relating to the New Organon.

Variation of experiment takes place first in the Matter; that is, when in things already known an experiment has scarcely been tried except in a certain kind of matter, but now is tried in other things of a like kind; as the manufacture of paper has been only tried in linen, not in silks (except perhaps among the Chinese) nor yet in hair stuffs, from which what are called chamblets are made; nor in wools, cotton [19], and skins; though these three last seem to be more heterogeneous, so that they may be more useful if mixed together than separate. Grafting again is common in fruit trees, but has been seldom tried on wild trees; though it is said that the elm when grafted on the elm produces a wonderful foliage. Grafting in flowers is likewise very rare, though now it is sometimes done in musk-roses, which are successfully inoculated with the common roses. Variation in the part of a thing I likewise set down among variations in matter. For we see that a sucker grafted on the trunk of a tree grows better than if planted in the ground. Why then should not the seed of an onion, inserted into the head of another onion when green, grow better than if sown by itself in the ground? Here the root is substituted for the trunk, so that this may be regarded as a kind of grafting in the root. Variation of experiment takes place secondly in the efficient. The rays of the sun are so intensified in heat by burning-glasses, that they can set on fire any combustible matter; can the rays of the moon by the

And though it may be true that no amount of diligence and sagacity could ever have made a collection of facts complete *enough* to lead to the discovery of Forms by the method of the *Novum Organum*, it seems impossible to fix a point beyond which, through successive reductions of particular phenomena and groups of phenomena under laws more and more general, further progress could not have been made towards the highest law which includes them all. And such progress men have in fact been making ever since Bacon's time; the whole of our experimental philosophy being what he, I think, would have described as *Experientia literata*, and allowed as legitimate and successful—so far as it goes. Whether, if he could see the results which it has produced during the last two hundred years, he would still believe in the possibility of arriving ultimately at what he would have called "the *Interpretation of Nature*", may be doubted; but that if this "hunt of Pan" were conducted as skilfully and assiduously by the whole body of inquirers through the entire field of nature as it has been by particular inquirers in particular fields, we should be able to approach much nearer to such a consummation than anybody now imagines—this I cannot doubt that he would still believe.—*J. S.*

[18] *Nov. Org.* i. 100.

[19] Cotton paper was known long before that made from rags. It seems probable that the art of making paper came to the west of Europe from Constantinople, and that our word quire, of which the equivalent in Low Latin is manus, is a token of its Greek origin, and means properly a handful of paper.

same process be actuated to any degree of heat however mild ? that we may see whether all heavenly bodies have the power of heating. So again, radiant heats are increased by glasses ; can the same effect be produced on opaque heats (as of stones and metals, before they are red-hot) or has light something to do with it [20] ? So again amber and jet when rubbed attract straws ; will they do the same when warmed by fire ? Variation of experiment takes place thirdly in Quantity ; which must be treated with great care, as it is surrounded by many errors. For men believe that if the quantity be increased or multiplied, the power and virtue is increased or multiplied proportionately. And this they postulate and suppose as if it had a kind of mathematical certainty; which is utterly false. A leaden ball of a pound weight dropped from a tower reaches the ground in (say) ten seconds : will a ball of two pounds weight (in which the force of natural motion, as they call it, ought to be doubled) reach the ground in five seconds ? No, but it will take almost the same time in falling, and will not be accelerated in proportion to the increase of quantity [21]. Again, suppose one drachm of sulphur mixed with half a pound of steel will melt it and make it liquid ; will therefore one ounce of sulphur mixed with four pounds of steel be able to melt it ? This does not follow ; for it is certain that the obstinacy of matter in the patient is more increased by quantity than the active power of the agent. Besides, both overmuch and overlittle equally prejudice the effect. For in the smelting and refining of metals it is a common error to suppose that in order to advance the smelting either the heat of the fire or the quantity of the added ingredient should be increased ; whereas, if these surpass the due proportion, they retard the operation ; because by their power and acrimony they turn much of the pure metal into fumes, and carry it off, so as both to cause a loss of metal and to make the remaining mass more hard and intractable. Men should therefore consider the story of the woman in Æsop, who expected that with a double measure of barley her hen would lay two eggs a day ; whereas the hen grew fat and laid none. As a rule, then, it will not be safe to rely on any experiment in nature, unless it has been tried both in greater and lesser quantities. And so much for Variation of Experiment.

Production of experiment is of two kinds; repetition and extension ; that is, when the experiment is either repeated, or urged to some effect more subtle. As an instance of repetition : spirit of wine is made from wine by a simple distillation, and is much more pungent and stronger than wine itself ; will then spirit of wine, if it be itself distilled and clarified, proportionately exceed itself in strength ? But repetition also is not free from fallacy. For in the second exaltation the difference is not so great as in the first ; and besides, by the repetition of an experiment, after the operation has reached a certain standing point or *acme*, nature oftentimes is so far from advancing that she rather inclines to relapse. Judgment therefore is to be exercised in this matter. Again, quicksilver put into linen or into the middle of molten lead when it is beginning to cool again,

[20] The researches which Bacon here suggests, in which obscure radiant heat is dealt with in the same manner as luminous heat, have been recently carried on with great success, and have led to many interesting results. The question as to the nature of the essential or formal connexion between heat and light remains however as yet unanswered, though it may be hoped that it will shortly be satisfactorily solved.

Telesius, of whom more than of any one else Bacon was a follower, maintained that heat and light were " contubernales naturæ", and that where one was present the other must be present too. Bacon, with a more subtle insight into nature, proposed to trace the analogy which might exist between them in cases where, sensibly at least, the dogma of Telesius seemed unfounded.

[21] Long before the publication of the *De Augmentis*, the theory of the acceleration of falling bodies, which of course includes the fact that all bodies fall from rest with equal velocities (the resistance of the air being set aside), had been made known by Galileo, The experiments which he made about the year 1590 to show the absurdity of the received opinion that the velocity of falling increases as the mass of the falling body, led to his leaving Pisa, where he had made them, and where he had in consequence been involved in disputes with the adherents of the Peripatetic philosophy.

becomes solid, and is no longer fluid; will therefore this same quicksilver, by many repetitions of the operation, become fixed and malleable? As an example of extension; if water be put into a glass with a long neb and hung up, and then the neb be dipped into a mixture of wine and water underneath, it will separate the water from the wine, the wine gradually rising to the top, the water sinking to the bottom [22]; will it likewise happen that as wine and water (being different bodies) are by this device separated, so the finer parts of wine (being of the same body) may be separated from the more gross; so that there shall take place a kind of distillation by gravity, and a substance shall be found on the top much like spirits of wine, but perhaps more delicate? Again, a magnet attracts a solid piece of iron; will a piece of a magnet dipped in a dissolution of iron, attract the iron itself and so get a coating of iron? Again, the magnetic needle turns to the pole; does it in so doing follow the same course as the heavenly bodies? As if one should turn the needle the wrong way, that is point it to the south, and hold it there for a while, and then let it go; would it, in returning to the north, go round by the west rather than by the east? Again, gold imbibes quicksilver when contiguous to it; does it receive this quicksilver into itself without extending its bulk, and so become a body heavier than gold itself? Again, men help the memory by putting images of persons in places; could the same thing be done without the places, by connecting actions or habits with persons? And so much for the Production of Experiment.

Translation of experiment is of three kinds: either from nature or chance into an art; or from one art or practice into another; or from a part of one art into a different part of the same. Of translation from nature or chance into an art there are innumerable examples; for almost all mechanical arts have sprung from small beginnings presented by nature or chance. It was received as an adage "that one cluster of grapes ripens faster by the side of another;" [23] which has grown into a common saying, as applied to the mutual services and offices of friendship. But our cyder makers have an excellent way of imitating the operation. For they take care not to bruise or squeeze the apples till they have lain together for awhile in heaps, and so ripened by mutual contact; that the too great acidity of the drink may be corrected. Again, the artificial imitation of rainbows, with drops thickly sprinkled, is translated by an easy passage from natural rainbows formed by a dripping cloud. Again, the method of distilling may have been drawn either from above, that is, from showers and dew; or from that homely experiment of drops adhering to the lids of pans of boiling water. Nor would a man have ventured to imitate thunder and lightning, if it had not been suggested by the pot lid of the monkish chemist suddenly flying up with great force and a loud report. The more plentiful the examples however, the fewer need be adduced. But if men were at leisure to inquire after useful things, they ought to observe attentively and minutely and systematically all natural works and operations, and be ever eagerly considering which of them may be transferred to the arts. For nature is the mirror of art. Nor are the experiments

[22] This experiment is more minutely described in the *Sylva Sylvarum*, i. 14. The water in the inverted glass or phial is maintained by the pressure of the atmosphere at a higher level than that of the wine and water into which the neck of the vessel containing it is inserted, but as the density of the water is greater than that of the diluted wine, it is in a position of unstable equilibrium. But for friction etc. the equilibrium could not practically exist at all; and after a little while it ceases to do so, the water gradually subsiding to the bottom and forcing the wine and water or some part of it into the vessel, which originally contained only water. The water for a considerable time passes without mixing through the wine and water; but of course there is no separation between the wine and the portion of water with which it was originally mixed, and the experiment succeeds just as well with pure as with diluted wine.

[23] This proverb Bacon doubtless took from Erasmus's collection. The *Promus* contains nearly 200 Latin proverbs (and this among the number) all of which are given by Erasmus. In more than one instance errors of Erasmus's are copied in Bacon's extract, so that there can be no doubt as to the source from which he derived them. See for the proverb in the text, Erasm. iii. 2. 49.

fewer in number which may be transferred from one art to another, or from one practice to another; although the thing is not so common. For nature meets everybody everywhere; but particular arts are only known to their own artists. Spectacles have been invented to assist weak sight; might not some instrument be devised, which being applied to the ear would assist those dull of hearing? Dead bodies are preserved by honey and embalming; could not something of the same kind be transferred to medicine for the benefit of live bodies? The practice of sealing upon wax, cements, and lead is of old date; but it led to impressions on paper, or the art of printing. In cookery, salt preserves meat, and that better in winter than summer; might not this be profitably transferred to baths, to regulate their temperament, when necessary, by impression or extraction? So in the late experiment of artificial freezing, salt is discovered to have great powers of condensing [24], may not this be transferred to the condensations of metals? seeing it is already known that strong waters composed of certain salts precipitate small sands of gold from certain metals not so dense as gold itself [25]? So again, painting revives the memory of a thing by the image of it; has not this been transferred into the art which they call the art of memory? Of these things it may be said generally, that the best chance of bringing down as from heaven a shower of inventions at once useful and new, is to bring within the knowledge of one man, or of a few who may sharpen one another by conference, the experiments of a number of mechanical arts; that by this translation (as I call it) of experiments the arts may mutually cherish and as it were kindle one another by mixture of rays. For though the rational method of inquiry by the Organon promises far greater things in the end, yet this sagacity proceeding by Learned Experience will in the meantime present mankind with a number of inventions which lie near at hand, and scatter them like the donatives that used to be thrown among the people [26]. There remains the translation from one part of an art to a different part; which differs little from the translation from one art into another. But as some arts are so extensive that the translation of experiments may take place within them, I have thought it right to annex this kind also; especially as in some arts it is of great importance. For it would tend greatly to the advancement of the art of medicine if the experiments of that part which relates to the cure of diseases were transferred to the parts that are concerned with the preservation of health and the prolongation of life. For if an excellent opiate is able to assuage the raging fever of the spirits in a dangerous disease, it need not be doubted that something of a similar nature, made familiar to the system by well proportioned doses, may likewise in some measure check and retard that continually advancing and creeping fever which is the effect of age. And so much for the Translation of Experiment.

Inversion of Experiment takes place when trial is made of the contrary of that which has been by the experiment proved. For instance, heat is increased by burning-glasses; is cold also [27]? Again, heat spreads round, but with a tendency

[24] Bacon refers to the experiments exhibited by Drebbel in 1620. One of them was of a boat that would go under water. See Nelli's *Life of Galileo*. I have not been able to see the Chronicle of Alkmaar to which Nelli refers. It is said that in presence of James I. Drebbel produced an intolerable degree of cold in Westminster Hall.

[25] The experiment here referred to, which, as Professor Cumming has suggested to me, may not improbably have been an alchemist's trick, is not sufficiently described to make it possible to ascertain its nature. It appears probable, however, that it was based on a reduction of a solution of perchloride of gold in an excess of acid by some other metal. Of all metallic salts the perchloride of gold appears to be one of the most easy to decompose. Its reduction by a metal is employed as a gilding process.

[26] Compare Suetonius in Calig. c. 18.

[27] With Bacon, as with the Peripaticians, cold is not the negation of heat; it is something positive—the opposite of heat, and not merely its absence. Prevost's experiment, in which two concave mirrors are placed opposite to one another with a piece of ice in the focus of the one and a thermometer in that of the other, shows that the effect apparently due to the radiation of cold may be made more intense in the manner which Bacon suggests: the real explanation of the phenomenon of course depends upon the "theory of exchanges".

upwards; does cold spread round with a tendency downwards? For example: take an iron rod and heat it at one end; then raise it, with the heated part downwards and the hand above; it will burn the hand at once: hold it with the heated part upwards and the hand below, it will be much longer in doing so [28]. But how if the whole bar be heated, and one end touched with snow or with a sponge dipped in cold water? will the cold travel downwards, if the snow or sponge be applied to the top, faster than upwards if applied to the bottom? Again, the rays of the sun are reflected from white but collected on black; are shadows likewise lost on black and collected on white? As we see in a dark room, where the light is only let in by a little chink, that the images of things outside are received on a white paper, but not on a black. Again, the megrims are relieved by opening a vein in the forehead; is a pain in the forehead relieved by scarifying the skull? And so much for the Inversion of Experiment.

Compulsion of Experiment is when an experiment is urged and extended to the annihilation or deprivation of the power; for in the other hunts the prey is only caught, but in this it is killed. Here is an example of compulsion. The magnet attracts iron; urge the iron or the magnet till it can attract no longer; for instance, if the loadstone be burnt, or steeped in aqua fortis, will it lose its power entirely or for a time? On the other hand, if iron or steel be reduced to oxide of iron, or to what is termed prepared steel, or if it be dissolved in aqua fortis, will it still be attracted by the loadstone? Again, the magnet attracts iron through all mediums we know of; as gold, silver, and glass; find some medium, if it be possible, which will intercept the power; try quicksilver; try oil, gums, ignited coal, and other things, hitherto untried. Again, glasses have lately been invented which magnify minute visible objects in a wonderful manner; urge the use of them to objects either so small as to be beyond their power or so large as to confound it. Thus, can they clearly detect in urine things otherwise imperceptible? Can they discover specks or flaws in jewels which appear every way clear and bright? Can they make the motes in the sunbeams (which were objected most unjustly to Democritus as if they were his atoms and principles of things [29]) appear like great bodies? or a thick powder of vermilion and white lead appear so distinct that the red and the white grains shall be seen separately? Again, can they magnify larger figures (say a face, or an eye) as much as they can a flea or a mite? Can they make a piece of cambric, or any of these finer and more open linen textures, appear full of holes, like a net? But on the compulsions of experiments I dwell the less, because they commonly fall outside the limits of learned experience, and are rather referred to causes, and axioms, and the New Organon. For wherever a case is established of negation, privation, or exclusion, there is some light given towards the invention of Forms. And so much for the Compulsion of Experiment.

Application of Experiment is nothing but the ingenious translation of it to some other useful experiment. For instance; all bodies have their own dimensions and gravities; gold has more weight, but less dimension than silver; water than wine. From this is derived a useful experiment; for by taking the bulk and the weight you may know how much silver has been mixed with gold, or how much water with wine; which was the εὕρηκα of Archimedes [30]. Again, flesh

[28] It is obvious that the difference arises simply from the circumstance that the air close to the hot end of the rod rises in the one case to that at which the hand is applied, and in the other case does not do so. In other words, in the first form of the experiment, the effect of conduction is increased by that of convection, and in the second is not.

[29] Democritus maintained the absolute invisibility of his atoms. See Sextus Empiricus, *Adv. Logic.* i. 135; ii. 6, and elsewhere.

[30] The εὕρηκα of Archimedes related to the discovery of a method of determining the specific gravity of a body which could not be made "implere mensuram". If he had had a crown of pure gold of the same size and form as the suspected one, he need only have weighed the one against the other; and if the latter were lighter, the question as to its being alloyed would have been settled. Or if he had been at liberty to melt down a portion of the crown and to run it into a mould in which a piece of pure gold had previously been moulded, he might then have weighed them and determined which was the

putrefies sooner in some cellars than in others; it would therefore be useful to apply this experiment to the discovery of airs more or less healthy to live in; those namely, in which flesh is longest in putrefying. The same method may be applied to discover healthy and unhealthy seasons of the year. But there are innumerable instances of this kind. Let men only watch, and keep their eyes continually turned to the nature of things on one side, and to the uses of man on the other. And so much for the Application of Experiment.

Coupling of Experiment is the link or chain of applications; when things which would be ineffectual singly are effectual in conjunction. For example; you wish to have roses or fruit come late. It will be effected, if you pluck off the earlier buds; it will be effected likewise if you uncover the roots and expose them to the air until the middle of spring; but much more if the two be coupled together. Again, ice and nitre have both of them great power of refrigeration; much more when mixed. This is indeed clear of itself. And yet there may often be a deception in it (as in all things else where axioms are wanting), if the things so coupled be such as operate in different and contrary ways. And so much for Coupling of Experiment.

There remain the Chances of Experiment. This form of experimenting is merely irrational and as it were mad, when you have a mind to try something, not because reason or some other experiment leads you to it, but simply because such a thing has never been attempted before. Yet I know not but in this very process (of which we are now treating) some great thing may be involved; the leaving (I say) of no stone in nature unturned. For the *magnalia* of nature generally lie out of the common roads and beaten paths, so that the very absurdity of the thing may sometimes prove of service. But if reason go along with it; that is, if it be evident that an experiment of this nature has never been tried, and yet there is great reason why it should be tried; then it is one of the best ways, and plainly shakes out the folds of nature. For instance, when fire works upon a natural body, one of two things has hitherto always happened;—either that something flies out (as flame and smoke in common combustion), or at least that there is a local separation of the parts, and to some distance; as in distillation, where the dregs settle at the bottom, and the vapours, after they have had their play, are gathered into receptacles. But of what I may call *close distillation* no man has yet made trial. Yet it seems probable that the force of heat, if it can perform its exploits of alteration within the enclosure of the body, where there is neither loss of the body nor yet means of escape, will succeed at last in handcuffing this Proteus of matter, and driving it to many transformations; only the heat must be so regulated and varied, that there be no fracture of the vessels. For this operation is like that of the womb, where the heat works, and yet no part of the body is either emitted or separated [31]. In the womb indeed alimentation is conjoined; but as far as conversion is concerned it seems to be the same thing. Such then are the chances of Experiment.

Meanwhile I give this advice as to experiments of this nature; that no one should be disheartened or confounded if the experiments which he tries do not answer his expectation. For though a successful experiment be more agreeable, yet an unsuccessful one is oftentimes no less instructive. And it must ever be kept in mind (as I am continually urging) that experiments of Light are even more to be sought after than experiments of Fruit. And so much for learned Experience, which (as I have already said) is rather a sagacity and a kind of hunting by scent, than a science. Of the New Organon I say nothing, nor shall

heaviest. But the problem he had to solve was quite different from this, and required the application of the principles of hydrostatics. Yet both here and in the *Historia Densi et Rari* Bacon refers to the discovery of Archimedes without distinguishing between his own inartificial method of determining specific gravities (which consisted in filling a measure with different substances and then weighing it) and that of Archimedes. Bacon's results are wonderfully accurate (with one remarkable exception), considering the manner in which they were obtained.

[31] This notion is taken from Telesius. See his *De rerum naturâ*, vi. 23.

I give any taste of it here; as I purpose by the divine favour to compose a complete work on that subject,—being the most important thing of all [32].

CHAPTER III.

Division of the art of discovery of Arguments into Promptuary *and* Topics. *Division of Topics into* General *and* Particular. *Example of a Particular Topic in an Inquiry concerning Heavy and Light.*

THE invention of arguments is not properly an invention; for to invent is to discover that we know not, not to recover or resummon that which we already know. Now the use and office of this invention is no other than out of the mass of knowledge which is collected and laid up in the mind to draw forth readily that which may be pertinent to the matter or question which is under consideration. For to him who has little or no knowledge on the subject proposed, places of invention are of no service; and on the other hand, he who is ready provided with matter applicable to the point in question will, even without art and places of invention (although perhaps not so expeditiously and easily), discover and produce arguments. So (as I have said) this kind of invention is not properly an invention, but a remembrance or suggestion with an application. Nevertheless, as the name has come into use, let it be called invention; for the hunting of any wild animal may be called a finding of it, as well in an enclosed park as in a forest at large. But not to be nice about words, let it be clearly understood, that the scope and end of this invention is readiness and present use of our knowledge, rather than addition or amplification thereof.

Provision for discourse may be procured in two ways. The place where a thing is to be looked for may be marked, and as it were indexed; and this is that which I call *Topics;* or arguments concerning such matters as commonly fall out and come under discussion may be composed beforehand and laid up for use: and this I will name the *Promptuary*. This last however scarcely deserves to be spoken of as a part of knowledge, consisting rather of diligence than of any artificial erudition. And herein Aristotle wittily, but hurtfully, derides the sophists of his time, saying, "they did as if one that professed the art of shoemaking, should not teach how to make a shoe, but only exhibit a number of shoes of all fashions and sizes" [1]. But here a man might reply, that if a shoemaker should have no shoes in his shop, but only work as he is bespoken, he should be still a poor man, and have few customers. Far otherwise says our Saviour, speaking of divine knowledge: "Every scribe that is instructed in the kingdom of heaven is like a householder, that bringeth forth old and new store [2]". We see likewise that the ancient rhetoricians gave it in precept to pleaders, that they should have by them a variety of commonplaces, ready prepared, and handled and illustrated on both sides; arguments (for example) for the sense of the law against the words of the law; and the contrary: for inferences against testimony, and the contrary. And Cicero himself, taught by long experience, directly asserts that a diligent orator may have by him premeditated and carefully handled beforehand everything which he shall have occasion to speak of; so that in the pleading of any particular cause, he shall not have to introduce anything

[32] It has been inferred from this passage that this part of the *De Augmentis* was written before the publication of the *Novum Organum*. But it must be remembered that the *Novum Organum*, which was published in 1620, was not an *opus integrum*. Writing to Fulgenzio *after* the publication of the *De Augmentis*, Bacon says, "Debuerat sequi Novum Organum; interposui tamen scripta mea moralia et politica, *quia magis erant in promptu*. Hæc sunt, etc. . . . Tum demum sequetur Organum Novum, cui secunda pars adhuc adjicienda est, quam animo jam complexus et metitus sum." Afterwards he seems to have come to the conclusion that a sample of Natural History was more urgently wanted, and therefore postponed the completion of the *Novum Organum* until he had finished the *Sylva Sylvarum*, which, according to Dr. Rawley, was his last work; and it does not appear that any portion of the second part was ever written.—*J. S.*

[1] Arist. *De Repreh. Sophist.* ii. 9. [2] St. Matt. xiii. 52.

new or on the sudden, except names and some special circumstances [3]. But such was the diligence and exactness of Demosthenes, that seeing what great force the entrance and access into a cause has to make a good impression on the minds of the audience, he thought it worth while to compose and have ready by him a number of prefaces for orations and speeches. All which authorities and precedents may fairly overweigh Aristotle's opinion, who would have us change a rich wardrobe for a pair of shears. Therefore that part of knowledge concerning provision or preparatory store was not to be omitted, though here I have said enough respecting it. For as it is common to both, logic as well as rhetoric, I have chosen in treating of logic only to mention it by the way, referring the fuller discussion of it to rhetoric.

The other part of invention (namely Topics) I will divide into general and particular. General has been sufficiently handled in logic, so that there is no need to dwell on the explanation of it. Only it may be observed by the way, that this kind of Topic is of use not only in argumentations, where we are disputing with another, but also in meditations, where we are considering and resolving anything with ourselves; neither does it serve only to prompt and suggest what we should affirm and assert, but also what we should inquire or ask. For a faculty of wise interrogating is half a knowledge. For Plato says well, "whosoever seeks a thing, knows that which he seeks for in a general notion; else how shall he know it when he has found it" [4]? And therefore the fuller and more certain our anticipation is, the more direct and compendious is our search. The same places therefore which will help us to shake out the folds of the intellect within us, and to draw forth the knowledge stored therein, will also help us to gain knowledge from without; so that if a man of learning and experience were before us, we should know how to question him wisely and to the purpose; and in like manner how to select and peruse with advantage those authors, books, and parts of books, which may best instruct us concerning that which we seek.

But Particular Topics contribute much more to those purposes whereof I speak, and are to be accounted most useful. Of these there is indeed some slight mention in some writers, but they have not been fully handled, according to the dignity of the subject. But leaving the humour which has reigned too long in the schools,—which is to pursue with infinite subtlety the things which are near at hand, and never to go near those which lie a little further off,—I for my part receive particular Topics (that is places of invention and inquiry appropriated to particular subjects and sciences) as things of prime use. They are a kind of mixtures of logic with the proper matter of each science. For he must be a trifler and a man of narrow mind who thinks that the perfect art of invention of knowledge can be devised and propounded all at once; and that then it needs only to be set at work. Let men be assured that the solid and true arts of invention grow and increase as inventions themselves increase; so that when a man first enters into the pursuit of any knowledge, he may have some useful precepts of invention; but when he has made further advances in that knowledge, he may and ought to devise new precepts of invention, to lead him the better to that which lies beyond. It is indeed like journeying in a champagne country; for when we have gone some part of our way, we are not only nearer to our journey's end, but we can likewise see better that part of the way which remains. In the same manner in sciences every step forward on the journey gives a nearer view of that which is to come. But I have thought right to annex an example of this kind of Topic, seeing I set it down among the Deficients.

A Particular Topic, or Articles of Inquiry concerning Heavy and Light.

1. Inquire what bodies are susceptible of the motion of gravity, what of levity, and if there be any of an intermediate and indifferent nature.

2. After the simple inquiry concerning heavy and light proceed to comparative inquiry; as what heavy bodies weigh more, what less, in the same dimensions. Likewise of light bodies, which rise quicker, which slower.

[3] Cic. *De Oratore*, ii. 32–34. [4] Bacon probably refers to the *Meno*, ii. p. 80.

3. Inquire what effect the quantity of a body has in the motion of gravity. At first sight indeed this may appear superfluous; for the proportions of motion ought to follow the proportions of quantity; but the case is otherwise. For although in the scales the quantity of a body makes up the gravity (the force of the body being there collected, by the recoil or resistance of the scales or beam), yet where there is but little resistance (as in the fall of bodies through air) the velocity of the fall is little hastened by the quantity of the body; for a ball of twenty pounds weight falls to the ground in nearly the same time as a ball of one pound.

4. Inquire whether the quantity of a body can be so increased as entirely to lose the motion of gravity; as in the earth, which is pendulous, but falls not. Can there then be other substances so massive as to support themselves? For the motion towards the centre of the earth is a fiction; and every great mass abhors all local motion, unless it be overcome by another stronger appetite.

5. Inquire what power and operation the resistance of an intervening or opposing body may have to control the motion of gravity. For a descending body either cuts and penetrates through an opposing body, or is stopped by it. If it pass through, penetration takes place either with slight resistance, as in air, or with a stronger, as in water. If it be stopped, it is either by an unequal resistance, where there is a superiority of weight, as if wood be placed on the top of wax; or by an equal resistance, as if water be placed on the top of water, or wood upon wood of the same kind; which is what the schoolmen (upon no solid apprehension) term the non-gravitation of a body in its own place. And all these things vary the motion of gravity. For heavy things move one way in the scales, and another in falling; one way (which may seem strange) when the scales are hanging in the air, another when they are sunk in water; one way again in falling through water, another in floating or being carried upon it.

6. Inquire what power and operation the figure of a descending body has in directing the motion of gravity; as if a figure be broad and thin, cubic, oblong, round, pyramidal; also when bodies turn, and when they keep the same position in which they were let fall.

7. Inquire what power and operation the continuation and progression of the descent or fall has in increasing the velocity and impetus, and in what proportion and to what extent that velocity will increase. For the ancients upon slight consideration imagined that this motion, being natural, was continually increasing and strengthening.

8. Inquire what power and operation the distance or nearness of the falling body to the earth has, in making it fall quicker or slower, or not at all (if it be beyond the orb of the earth's activity, according to Gilbert's opinion); and also what is the effect of the plunging of a descending body further into the depths of the earth, or of the location thereof nearer the surface. For this also varies the motion, as is perceived by miners.

9. Inquire what power and operation the difference of the bodies has, through which the motion of gravity is diffused and communicated; and whether it is communicated as well through soft and porous bodies, as through hard and solid ones; as if the beam of a pair of scales on one side of the tongue be made of wood, on the other of silver (though both be reduced to the same weight), will it produce any variation in the scales? Likewise will metal laid on wood, or on a blown bladder, weigh the same as it does on the bottom of the scale?

10. Inquire what power and operation the distance of the body from the fulcrum has in the communication of the motion of gravity; that is, in the sooner or later perception of the weight or pressure: as in scales, if one arm of the beam be longer than the other (though both are of the same weight), does this of itself incline the scale? or in the syphon, where the longer limb will certainly draw the water, though the shorter (being made more capacious) contain a greater weight of it [5].

[5] The theory of the lever, to which the first part of this inquiry relates, was as well understood in Bacon's time as it is now; that of the siphon, inasmuch as it depends on the idea of atmospherical pressure, was then unknown, and could not be established until

11. Inquire what power the mixing or coupling of a light body with a heavy one has in lessening the gravity of a body; as in the weight of animals alive and dead.

12. Inquire of the secret ascents and descents of the lighter and heavier parts in one entire body; whence fine separations often take place; as in the separation of wine and water, the rising of cream, and the like.

13. Inquire what is the line and direction of the motion of gravity; how far it follows the centre or mass of the earth, how far the centre of the body itself, that is the strife and pressure of its parts. For these centres, though convenient for demonstrations, are of no effect in nature.

14. Inquire touching the motion of gravity as compared with other motions; what motions it overcomes, and what overcome it. As in violent motion (as it is called) the motion of gravity is overpowered for a time; and as when a little magnet lifts a piece of iron much heavier than itself, the motion of gravity yields to the motion of sympathy.

15. Inquire touching the motion of air; whether it rise upwards, or is as it were indifferent. And this is hard to discover, except by some subtle experiments. For the springing up of air at the bottom of water is rather caused by the force of the water than the motion of the air; seeing that the same thing happens also with wood. But air mingled with air gives no evidence, seeing that air in air appears no less light, than water in water appears heavy; but in a bubble, where there is a thin film drawn round it, it is stationary for a time.

16. Inquire what is the limit of lightness; for men do not mean (I suppose) that as the centre of the earth is the centre of gravity, so the extreme convexity of the heaven is the term of lightness; is it that as heavy bodies seem to fall till they rest and reach the immovable, so light bodies rise till they begin to rotate, and attain as it were motion without limit?

17. Inquire why vapours and exhalations rise as high as what is called the middle region of the air; seeing they consist of a somewhat heavy matter, and the rays of the sun at intervals (that is, at night) cease their operation.

18. Inquire of the rule which governs the upward motion of flame; which is the more mysterious as flame expires every instant, except perhaps it be in the midst of larger flames; for flames separated from their continuity last but a little while.

19. Inquire of the upward motion of the activity of heat; as when the heat of red-hot iron spreads faster upwards than downwards.

Such then is an instance of a Particular Topic. In the meantime I again repeat my former advice; namely, that men ought so to vary their particular topics, as, after any great advance has been made in the inquiry, to set out another and again another topic, if they desire to climb the heights of the sciences. But so much importance do I attribute to Particular Topics, that I design to construct a special work concerning them in the more important and obscure subjects of nature. For we can command our questions, though we cannot command the nature of things. And so much for Invention.

CHAPTER IV.

Division of the art of Judging into judgment by Induction *and judgment by* Syllogism. *The first whereof is referred to the New Organon. First division of Judgment by Syllogism into* Reduction Direct *and* Reduction Inverse. *Second division of the same into* Analytic *and doctrine concerning* Detection of Fallacies. *Division of the doctrine concerning the detection of fallacies into detection of* Sophistical *fallacies, fallacies of* Interpretation, *and fallacies of* false appearances or Idols. *Division of Idols into Idols of the* Tribe, *Idols of the* Cave, *and Idols of the* Market-place. *Appendix to the Art of Judging; viz. concerning the* Analogy of Demonstrations according to the nature of the subject.

Let us now pass on to Judgment, or the art of judging, which handles the nature of proofs and demonstrations. In this art (as indeed it is commonly received)

this idea was introduced by Torricelli. The experiment which bears his name, and which was in effect the construction of a mercurial barometer, corresponds in the history of physics to the invention of the telescope in that of astronomy.

the conclusion is made either by induction or by syllogism. For enthymems and examples are but abridgments of these two. With regard however to judgment by induction there is nothing to detain us; for here the same action of the mind which discovers the thing in question judges it; and the operation is not performed by help of any middle term, but directly, almost in the same manner as by the sense. For the sense in its primary objects at once apprehends the appearance of the object, and consents to the truth thereof. In the syllogism it is otherwise; for there the proof is not immediate, but by mean. And therefore the invention of the mean is one thing, and the judgment of the consequence is another; for the mind ranges first, and rests afterwards. But the vicious form of induction I entirely disclaim; and as for the legitimate form, I refer it to the New Organon. Enough here therefore of Judgment by Induction.

For the other judgment by Syllogism, what need to speak; seeing it has been beaten over and over by the subtlest labours of men's wits and reduced to many niceties? And no wonder, for it is a thing most agreeable to the mind of man. For the mind of man is strangely eager to be relieved from suspense, and to have something fixed and immovable, upon which in its wanderings and disquisitions it may securely rest. And assuredly as Aristotle endeavours to prove that in all motion there is some point quiescent; and as he very elegantly interprets the ancient fable of Atlas, who stood fixed and supported the heaven on his shoulders, to be meant of the poles or axletree of heaven, whereupon the conversion is accomplished [1]; so do men earnestly desire to have within them an Atlas or axletree of the thoughts, by which the fluctuations and dizziness of the understanding may be to some extent controlled; fearing belike that their heaven should fall. And hence it is that they have been in too great a hurry to establish some principles of knowledge, round which all the variety of disputations might turn, without peril of falling and overthrow; not knowing that he who makes too great haste to grasp at certainties shall end in doubts, while he who seasonably restrains his judgment shall end in certainties.

So then this art of judgment by Syllogism is but the reduction of propositions to principles in a middle term; the principles being understood as agreed upon and exempted from argument; and the invention of the middle terms left to the free exercise of wit and inquiry. Now this reduction is of two kinds: direct and inverse; direct, when the proposition is reduced to the principle; which they term proof *ostensive;* inverse, when the contradictory of the proposition is reduced to the contradictory of the principle; which they call proof *per incommodum*, or by showing that it involves an absurdity. But the number of series or middle terms is greater or less as the proposition stands more or less removed from the principle.

This being premised, I will divide the art of judgment (according to the usual manner) into Analytic, and the doctrine concerning *Elenches*, or detection of fallacies; whereof the one proceeds by way of direction, the other by way of caution. Analytic sets down true forms of consequences in argument; from which if there be any variation or deflexion, the conclusion is detected to be faulty; and this contains in itself a kind of detection, or refutation: for the straight (as they say) indicates what is not straight as well as what is. And yet it is safest to employ Elenches, as monitors, for the better detection of fallacies by which the judgment would otherwise be ensnared. In Analytic however I find no deficiency; but it is rather overladen with superfluities than in need of additions.

The doctrine of detection of fallacies I divide into three parts; detection of *sophistical* fallacies, of fallacies of *interpretation*, and of *false appearances* or *Idols*. The detection of sophistical fallacies is especially useful. For although the grosser kind of fallacies is well compared by Seneca [2] to the feats of jugglers, in which though we know not how the thing is done, yet we know well it is not as it seems to be; yet the more subtle sophisms not only put a man beside his answer, but many times seriously confound his judgment.

This part concerning the detection of sophistical fallacies is excellently handled

1 Arist. *De Mot. Anim.* 2 and 3.

2 Seneca, *Epist.* 45.

by Aristotle in the way of precepts, but still more excellently by Plato in the way of examples: and that not only in the persons of the ancient sophists (Gorgias, Hippias, Protagoras, Euthydemus, and the rest), but even in Socrates himself, who professing to affirm nothing, but to infirm that which was affirmed by another, has most wittily expressed all the forms of fallacy, objection, and redargution. In this part therefore I have no deficiency to report. In the meantime I may observe, that although I have said that the honest and principal use of this doctrine is for redargution of sophisms; yet it is manifest that the degenerate and corrupt use is for raising, by means of these very sophisms, captions and contradictions. And this passes for a great faculty, and no doubt is of very great advantage; though the difference be good which was made between an orator and a sophist, that the one is as the greyhound, which has his advantage in the race, the other as the hare, which has her advantage in the turn.

Next come fallacies of Interpretation; for so (borrowing the name rather than the sense from Aristotle) I will term them. Let me call to mind then what I said above (in speaking of Primitive or Summary Philosophy) touching the Transcendental or Adventitious Conditions or Adjuncts of Essences. These are Greater, Less, Much, Little, Before, After, Identity, Diversity, Potential, Actual, Habit, Privation, Whole, Parts, Active, Passive, Motion, Rest, Entity, Nonentity, and the like. And first let the different ways which I mentioned of viewing these things be remembered and noted; namely that they may be inquired either physically or logically. Now the physical handling of them I referred to Primitive or Summary Philosophy. There remains then the logical. And this is the very thing which at present I mean by the doctrine of the detection of fallacies of Interpretation. Certainly it is a sound and good part of learning. For common and general notions enter necessarily into every discussion; so that unless great care be taken to distinguish them well at the outset, all the light of disputations will be strangely clouded with darkness by them, and the matter end in disputes about words. For equivocations and false acceptations of words (especially of this sort) are the sophisms of sophisms; and therefore I have thought it better that the treatment of them should be made a part by itself, than that it should be either included in Summary Philosophy or Metaphysic, or placed partly under Analytic; as has been done by Aristotle confusedly enough. The name I have given it is taken from the use; because its true use is simply redargution and caution with regard to the use of words. Moreover that part concerning the Predicaments, if rightly managed, relating to cautions against confounding and transposing the terms of definitions and divisions, I hold to be of principal use, and wish it to be referred to this place. And so much for the Detection of Fallacies of Interpretation.

As for the detection of False Appearances or Idols, Idols are the deepest fallacies of the human mind. For they do not deceive in particulars, as the others do, by clouding and snaring the judgment; but by a corrupt and ill-ordered predisposition of mind, which as it were perverts and infects all the anticipations of the intellect. For the mind of man (dimmed and clouded as it is by the covering of the body), far from being a smooth, clear, and equal glass (wherein the beams of things reflect according to their true incidence), is rather like an enchanted glass, full of superstition and imposture. Now idols are imposed upon the mind, either by the nature of man in general; or by the individual nature of each man; or by words, or nature communicative. The first part of these I call Idols of the *Tribe*, the second the Idols of the *Cave*, the third the Idols of the *Market-place*. There is also a fourth kind which I call the Idols of the *Theatre*, superinduced by corrupt theories or systems of philosophy, and false laws of demonstration. But this kind may be rejected and got rid of: so I will leave it for the present. The others absolutely take possession of the mind, and cannot be wholly removed. In these therefore Analytic is not to be looked for; but the doctrine of Elenches is with regard to the idols themselves a primary doctrine. Nor (to say truth) can the doctrine concerning Idols be reduced to an art; all that can be done is to use a kind of thoughtful prudence to guard against them. The full and subtle handling of these however I reserve for the New Organon, making here only a few general observations touching them.

As an example of the Idols of the Tribe, take this. The nature of the human mind is more affected by affirmatives and actives than by negatives and privatives; whereas by right it should be indifferently disposed towards both. But now a few times hitting or presence produces a much stronger impression on the mind than many times failing or absence: a thing which is as the root of all vain superstition and credulity. And therefore it was well answered by one who when the table was shown to him hanging in a temple of such as had paid their vows upon escape from shipwreck, and he was pressed to say whether he did not now acknowledge the power of Neptune, " Yea," asked he in return, " but where are they painted that were drowned after paying their vows "[3]? And so it is in similar superstitions, as astrology, dreams, omens, and the like. Here is another instance. The spirit of man (being of an equal and uniform substance) pre-supposes and feigns in nature a greater equality and uniformity than really is. Hence the fancy of the mathematicians that the heavenly bodies move in perfect circles, rejecting spiral lines. Hence also it happens, that whereas there are many things in nature unique and full of dissimilarity, yet the cogitation of man still invents for them relatives, parallels, and conjugates. Hence sprang the introduction of an element of fire, to keep square with earth, water, and air. Hence the chemists have marshalled the universe in phalanx; conceiving, upon a most groundless fancy, that in those four elements of theirs (heaven, air, water, and earth,) each species in one has parallel and corresponding species in the others[4]. The third example is of kin to the last; Man is as it were the common measure and mirror of nature. For it is not credible (if all particulars be gone through and noted) what a troop of fictions and idols the reduction of the operations of nature to the similitude of human actions has brought into natural philosophy; I mean, the fancy that nature acts as man does. Neither are these much better than the heresy of the Anthropomorphites, bred in the cells of gross and solitary monks; or the opinion of Epicurus answering to the same in heathenism, who supposed the gods to be of human shape. And therefore Velleius the Epicurean needed not to have asked, " Why God should have adorned the heaven with stars and lights, like an ædile "[5]? For if that great workmaster had acted as an ædile, he would have cast the stars into some pleasant and beautiful order, like the frets in the roofs of palaces; whereas one can scarce find a posture in square or triangle or straight line amongst such an infinite number. So differing a harmony is there between the spirit of man and the spirit of nature.

With regard to the Idols of the Cave, they arise from each man's peculiar nature both of mind and body; and also from education and custom, and the accidents which befall particular men. For it is a most beautiful emblem, that of Plato's cave: for (not to enter into the exquisite subtlety of the allegory) if a child were kept in a dark grot or cave under the earth until maturity of age, and then came suddenly abroad, and beheld this array of the heavens and of nature, no doubt many strange and absurd imaginations would arise in his mind. Now we, although our persons live in the view of heaven, yet our spirits are included in the caves of our own bodies; so that they must needs be filled with infinite errors and false appearances, if they come forth but seldom and for brief periods from their cave, and do not continually live in the contemplation of nature, as in the open air. And with this emblem of Plato's concerning the cave the saying of Heraclitus agrees well, " that men seek the sciences in their own proper worlds, and not in the greater world "[6].

[3] See *Nov. Org.* 46. i.

[4] See note on *Nov. Org.* i. 45.—*J. S.*

[5] Cicero *De Nat. Deor.* i. c. 9. Compare the following extract from Galileo's letter to Gallanzone Gallanzoni:—" Uno dei nostri più celebri architetti se avesse avuto a compartire nella gran volta del cielo la moltitudine di tante stelle fisse, credo io che dis tribuite le avrebbe con bei partimenti di quadrati, esagoni, ed ottangoli; interzando le maggiori tra le mezzane e le piccole, con sue intere correspondenze, parendogli in questo modo di valersi di belle proporzione: ma all incontro Iddio, quasi che colla mano del caso le abbia disseminate, pare a noi che senza regola simmetria o eleganza alcuno le abbia colassù sparpagliate".

[6] Plato, *Republ.* vi. For the reference to Heraclitus, see note to *Nov. Org.* i. 42.

But the Idols of the Market-place are most troublesome; which have crept into the understanding through the tacit agreement of men concerning the imposition of words and names. Now words are generally framed and applied according to the conception of the vulgar, and draw lines of separation according to such differences as the vulgar can follow: and where a more acute intellect or a more diligent observation tries to introduce a better distinction, words rebel. And that which is the remedy for this evil (namely definitions) is in most cases unable to cure it; for definitions themselves consist of words, and words beget words. And although we think we govern our words, and it is easy to say "a man should speak as the vulgar, and think as the wise"[7]; and though technical terms (only used by the learned) may seem to answer the purpose; and the setting down of those definitions I spoke of at the entrance of arts (after the prudent course of the mathematicians) may avail to correct the perverted acceptation of terms; yet all is not enough, but the juggleries and charms of words will in many ways seduce and forcibly disturb the judgment, and (after the manner of the Tartar bowmen) shoot back at the understanding from which they proceeded. This evil stands in need therefore of a deeper remedy, and a new one. But here I only glance at these things by the way; in the meantime pronouncing this doctrine (which I call the Great Elenches, or the doctrine concerning the Idols of the Human Mind, native and adventitious) to be wanting. The regular handling of it I refer to the New Organon.

There remains an Appendix to the Art of Judging, of great excellency; which I also set down as deficient: for though Aristotle has noticed the thing, he has nowhere followed out the manner of it. It treats of the application of the differing kinds of proofs to the differing kinds of matters or subjects; and may be called the doctrine of the judgment of judgments. For Aristotle rightly observes, "That we ought not to require either demonstrations from orators or persuasions from mathematicians"[8]. And therefore if there be an error in the kind of proof employed, the judgment itself cannot be truly made. Now whereas there are four kinds of demonstrations,—either by immediate consent and common notions, or by induction, or by syllogism, or by that which Aristotle rightly calls *demonstration in circle*[9], —(that is, not from things higher in the order of nature, but as it were from the same level):—so there are certain subjects and matters in science wherein each of these demonstrations respectively does well, and certain others from which they are respectively excluded. For rigour and curiosity in requiring the more severe kinds of proof in some things, and still more facility and remissness in contenting ourselves with the weaker kinds in others, are to be numbered among the chief causes of detriment and hindrance to knowledge. And so much for the Art of Judging.

CHAPTER V.

Division of the Art of Retaining into the doctrine concerning Helps of Memory, *and the doctrine concerning* Memory itself. *Division of the doctrine concerning Memory itself into* Prenotion *and* Emblem.

The art of retaining or keeping knowledge I will divide into two parts; namely, the doctrine concerning Helps of Memory, and the doctrine concerning Memory itself. The great help to the memory is *writing;* and it must be taken as a rule that memory without this aid is unequal to matters of much length and accuracy; and that its unwritten evidence ought by no means to be allowed. This is particularly the case in inductive philosophy and the interpretation of nature; for a man might as well attempt to go through the calculations of an Ephemeris in his head without the aid of writing, as to master the interpretation of nature by the natural and naked force of thought and memory, without the help of tables duly arranged. But not to speak of the interpretation of nature, which is a new doc-

[7] "Loquendum est ut plures, sententiendum ut pauci."—Niphus's *Comm. on Aristot. de Gen. et Corr.* lib. i. fol. 29. G.

[8] Arist. *Metaph.* ii. 3.

[9] Arist. *Post. Analyt.* ii. 13.

trine, there can hardly be anything more useful even for the old and popular sciences, than a sound help for the memory; that is a good and learned Digest of Common-Places. I am aware indeed that the transferring of the things we read and learn into common-place books is thought by some to be detrimental to learning, as retarding the course of the reader and inviting the memory to take holiday. Nevertheless, as it is but a counterfeit thing in knowledge to be forward and pregnant, except a man be also deep and full, I hold diligence and labour in the entry of common-places to be a matter of great use and support in studying; as that which supplies matter to invention, and contracts the sight of the judgment to a point. But yet it is true that of the methods and frameworks of common-places which I have hitherto seen, there is none of any worth; all of them carrying in their titles merely the face of a school and not of a world; and using vulgar and pedantical divisions, not such as pierce to the pith and heart of things.

For the Memory itself, the inquiry seems hitherto to have been pursued weakly and languidly enough. An art there is indeed extant of it; but it is clear to me that there might be both better precepts for strengthening and enlarging the memory than that art contains, and a better practice of the art itself than that which is received. Not but (if any one chooses to abuse this art for purposes of ostentation) feats can be performed by it that are marvellous and prodigious; but nevertheless it is a barren thing (as now applied) for human uses. At the same time the fault I find with it is not that it destroys and overburdens the natural memory (which is the common objection), but that it is not well contrived for providing assistance to the memory in serious business and affairs. And for my own part (owing perhaps to the life of business I have led) I am ever disposed to make small account of things which make parade of art but are of no use. For the being able to repeat at once and in the same order a great number of names or words upon a single hearing, or to make a number of verses extempore on any subject, or to make a satirical simile of everything that happens, or to turn any serious matter into a jest, or to carry off anything with a contradiction or cavil, or the like, (whereof in the faculties of the mind there is great store, and such as by device and practice may be exalted to an extreme degree of wonder) —all such things I esteem no more than I do the tricks and antics of clowns and rope-dancers. For they are almost the same things; the one an abuse of the powers of the body, the other of the mind; matters perhaps of strangeness, but of no worthiness [1].

The Art of Memory is built upon two intentions; Prenotion and Emblem. By Prenotion I mean a kind of cutting off of infinity of search. For when a man desires to recall anything into his memory, if he have no prenotion or perception of that he seeks, he seeks and strives and beats about hither and thither as if in infinite space. But if he have some certain prenotion, this infinity is at once cut off, and the memory ranges in a narrower compass; like the hunting of a deer within an enclosure [2]. And therefore order also manifestly assists the memory; for we have a prenotion that what we are seeking must be something which agrees with order. So again verse is more easily learned by heart than prose; for if we stick at any word, we have a prenotion that it must be such a word as fits the verse. And this prenotion is the principal part of artificial memory. For in artificial memory we have the *places* digested and prepared beforehand; the *images* we make extempore according to the occasion. But then we have a pre-

[1] Of the art of memory Agrippa remarks: "Solent enim in gymnasiis plerunque hujus artis professione nebulones quidam scholaribus imponere ac rei novitate pecuniolam ab incautis emungere: turpe et impudentis est multarum rerum lectionem instar mercimoniorum ante fores explicare, cum interim vacua domus sit."—*De Incert. et Vanit. Scient.* c. 10.

The illustration at the end of this passage may have suggested that which Bacon employs in speaking of the method of Raymond Lully, vide infrà, p. 533.

In Selden's *Table Talk* he is made to affirm that, whatever may be said of great memories, no man will trust his memory when writing what is to be given to the world. [See *Table Talk*, under title "Minister Divine."]

[2] Compare *Nov. Org.* ii. 26.

notion that the image must be one which has some conformity with the place; and this reminds the memory, and in some measure paves the way to the thing we seek. Emblem, on the other hand, reduces intellectual conceptions to sensible images; for an object of sense always strikes the memory more forcibly and is more easily impressed upon it than an object of the intellect; insomuch that even brutes have their memory excited by sensible impressions, never by intellectual ones. And therefore you will more easily remember the image of a hunter pursuing a hare, of an apothecary arranging his boxes, of a pedant making a speech, of a boy repeating verses from memory, of a player acting on the stage, than the mere notions of invention, disposition, elocution, memory, and action. Other things there are (as I said just now) which relate to the help of memory, but the art as it now is consists of the two above stated. But to follow out the particular defects of arts would be from my purpose. So much therefore for the Art of Retaining or Keeping Knowledge. And now we have arrived in due course at the fourth division of Logic, which treats of the Transmission and Delivery of our knowledge to others.

Book VI.

CHAPTER I.

Division of the art of Transmitting into the doctrine concerning the Organ of Discourse, *the doctrine concerning the* Method of Discourse, *and the doctrine concerning the* Illustration of Discourse. *Division of the doctrine concerning the organ of discourse into the doctrine concerning the* Notations of Things, *concerning* Speech, *and concerning* Writing ; *whereof the two first constitute* Grammar, *and are divisions of it. Division of the doctrine concerning the notations of things into* Hieroglyphics *and* Real Characters. *Second division of Grammar into* Literary *and* Philosophic. *Reference of* Poesy in respect of metre *to the doctrine concerning* Speech. *Reference of the doctrine concerning* Ciphers *to the doctrine concerning* Writing.

It is permitted to every man (excellent King) to make merry with himself and his own matters. Who knows then but this work of mine is copied from a certain old book found in the most famous library of St. Victor, of which Master Francis Rabelais made a catalogue[1]? For there is a book there entitled "The Ant-hill of Arts". And certainly I have raised up here a little heap of dust, and stored under it a great many grains of sciences and arts ; into which the ants may creep and rest for a while, and then prepare themselves for fresh labours. Now the wisest of kings refers sluggards to the ants ; and for my part I hold all men for sluggards who care only to use what they have got, without preparing for new seed-times and new harvests of knowledge.

Let us now proceed to the art of Transmitting, or of producing and expressing to others those things which have been invented, judged, and laid up in the memory ; which I will call by a general name the Art of Transmission. This art includes all the arts which relate to words and discourse. For although reason be as it were the soul of discourse, yet in the handling of them reason and discourse should be kept separate, no less than soul and body. The art of transmission I will divide into three parts ; the doctrine concerning the Organ of Discourse, the doctrine concerning the Method of Discourse, and the doctrine concerning the Illustration or adornment of Discourse.

The doctrine concerning the Organ of Discourse, which is also called Grammar, has two parts ; one relating to Speech, the other to Writing : for Aristotle says rightly that "words are the images of thoughts and letters are the images of words"[2]. Both these I assign to Grammar. But to go a little higher up, before I come to Grammar and the parts thereof just mentioned, I must speak concerning the Organ of Transmission in general. For it seems that the art of transmission has some other children besides Words and Letters. This then may be laid down as a rule ; that whatever can be divided into differences sufficiently numerous to explain the variety of notions (provided those differences be perceptible to the sense) may be made a vehicle to convey the thoughts of one man to another. For we see that nations which understand not one another's language carry on their commerce well enough by means of gestures. And in the practice of some who had been deaf and dumb from their birth and were otherwise clever, I have seen wonderful dialogues carried on between them and their friends who had learned to understand their gestures. Moreover it is now well known that in China and the provinces of the furthest East there are in use at this day certain *real characters*, not nominal ; characters, I mean, which represent neither

[1] *Pantag.* ii. 7. The humour of making catalogues of imaginary books probably began with Rabelais.

[2] Arist. De *Interpret.* i. 1.

letters nor words, but things and notions [3]; insomuch that a number of nations whose languages are altogether different, but who agree in the use of such characters (which are more widely received among them), communicate with each other in writing; to such an extent indeed that any book written in characters of this kind can be read off by each nation in their own language [4].

The Notes of Things then which carry a signification without the help or intervention of words, are of two kinds: one *ex congruo*, where the note has some congruity with the notion, the other *ad placitum*, where it is adopted and agreed upon at pleasure. Of the former kind are Hieroglyphics and Gestures; of the latter the Real Characters above mentioned. The use of Hieroglyphics is very old, and held in a kind of reverence, especially among the Egyptians, a very ancient nation. So that they seem to have been a kind of earlier born writing, and older than the very elements of letters, except perhaps among the Hebrews. Gestures are as transitory Hieroglyphics. For as uttered words fly away, but written words stand, so Hieroglyphics expressed in gestures pass, but expressed in pictures remain. For when Periander, being consulted with how to preserve a tyranny, bade the messenger follow him, and went into his garden and topped the highest flowers, hinting at the cutting off of the nobility [5], he made use of a Hieroglyphic

[3] In Acosta's *History of the New World* [book vi. c. 5], which is a very interesting book, the writer, in giving an account of the way in which the Mexicans used hieroglyphical characters, makes a digression on the writing of the Chinese, in a manner which indicates that at that time their mode of writing was not generally known.

[4] This assertion was made by the early missionaries, and has been constantly repeated since. Within certain limits it is true; just as an Italian and an Englishman may read or write Latin equally well, though they pronounce it differently. But the structure of the spoken languages, or rather dialects, to which written Chinese can correspond must be identical. It is difficult to attach a precise meaning to such statements as Remusat's " Les signes de leur écriture, pris en général, n'expriment pas des prononciations, mais des idées ". Every character has in truth, he immediately afterwards remarks, its sound; and a Chinese book can of course be read aloud in Chinese. Moreover the great majority of Chinese characters carry with them an indication of their pronunciation. They consist of two elements, one being a simpler character of the same sound, although generally speaking of totally different meaning, the other referring more or less precisely to the meaning. Thus the character for a particular kind of tree will contain, besides the phonetic element, the character for tree or wood in general; so too will very frequently that for a thing made of wood. These elements have been termed Phoneticæ and Classificæ. But most of the latter admit of being used in different combinations as Phoneticæ. They correspond precisely with the kind of hieroglyphics which Bunsen calls determinants, and are for the most part the same as the radicals (as they are called) used in arranging words in the Chinese dictionaries. The class of characters of which I have been speaking, is the fourth of the six classes into which Chinese characters are commonly divided. They are called Hiai-Ching, *id est* joined to sound, or Hing-Ching, *id est* representing the sound; and it is said that out of twenty-four thousand characters it was found that twenty-two thousand are of this kind. See Callery, *Systema phoneticum Scripturæ Sinicæ*, i. 9. He refers for his authority to a Chinese encyclopædia.

The view taken of the nature of these characters in Marshman's *Clavis Sinica*, is, as Remusat has pointed out, wholly wrong. It is much to be wished that a person sufficiently acquainted with the subject would investigate the analogy which exists between the Chinese and Egyptian modes of writing; not, of course, with any notion of establishing a historical connexion (as was once attempted) between the two nations. It is exceedingly remarkable, that as early as the fourth dynasty the Egyptians seem to have had a complete and even copious system of purely alphabetic characters, though, as Lepsius has shown, the majority of their alphabetic characters are of later date. I must apologise for the length of this note on a subject not very closely connected with the text.

[5] Compare this with Solyman's lesson to his vizir on the art of sieges. " Come close to me," said the Sultan, " but on your head be it if you tread on the carpet on which I sit." The vizir reflected for a while, then gradually rolling up the carpet, advanced close to his instructor. " All is said," resumed Solyman; " you know now how strong places are to be taken." The lesson was given, it is said, in relation to the siege of Rhodes in 1521.

just as much as if he had drawn it on paper. In the meantime it is plain that Hieroglyphics and Gestures have always some similitude to the thing signified, and are a kind of emblems. Whence I have called them "notes of things by congruity". Real characters on the other hand have nothing emblematic in them, but are merely surds, no less than the elements of letters themselves, and are only framed *ad placitum*, and silently agreed on by custom. It is evident however that a vast multitude of them is wanted for writing; for there ought to be as many of them as there are radical words. This portion therefore of the doctrine of the Organ of Discourse, which relates to the Notes of Things, I set down as wanting. And although it may seem to be of no great use, since words and writing by letters are by far the most convenient organs of transmission; yet I thought good to make some mention of it here, as a thing not unworthy of consideration. For we are handling here the currency (so to speak) of things intellectual, and it is not amiss to know that as moneys may be made of other material besides gold and silver, so other Notes of Things may be coined besides words and letters.

Now therefore I pass on to Grammar, which is as it were the harbinger of other sciences; an office not indeed very noble, yet very necessary; especially as sciences in our age are principally drawn from the learned languages, and are not learned in our mother tongue. Nor must it be esteemed of little dignity, seeing that it serves for an antidote against the curse of the confusion of tongues. For man still strives to renew and reintegrate himself in those benedictions of which by his fault he has been deprived. And as he arms and defends himself against the first general curse of the barrenness of the earth, and of eating bread in the sweat of his face, by the invention of all other arts; so against this second curse of the confusion of tongues he calls in the aid of Grammar; whereof the use in a mother tongue is small; in a foreign tongue more; but most in such foreign tongues as have ceased to be vulgar tongues, and are only extant in books.

Grammar likewise is of two sorts; the one being Literary, the other Philosophical. The one is used simply for languages, that they may be learned more quickly or spoken more correctly and purely; the other ministers in a certain degree to philosophy. And here I am reminded that Cæsar wrote some books on "Analogy"; and a doubt occurs to me, whether they handled this kind of philosophical grammar of which I speak. I suspect however that they did not contain anything very subtle or lofty; but only laid down precepts for a chaste and perfect style, not vitiated or polluted either by a bad habit of speech, or by any particular affectation; in which style himself excelled[6]. Taking the hint however from this, I have thought of a kind of grammar which should diligently inquire, not the analogy of words with one another, but the analogy between words and things, or reason; not going so far however as that interpretation which belongs to Logic. Certainly words are the footsteps of reason, and the footsteps tell something about the body. I will therefore give some sketch of what I mean. But I must first say that I by no means approve of that curious inquiry, which nevertheless so great a man as Plato did not despise[7]; namely concerning the imposition and original etymology of names; on the supposition that they were not arbitrarily fixed at first, but derived and educed by reason and according to significance; a subject elegant indeed, and pliant as wax to be shaped and turned, and (as seeming to explore the recesses of antiquity) not without a kind of reverence,—but yet sparingly true and bearing no fruit. But the noblest species of grammar, as I think, would be this: if some one well seen in a great number of tongues, learned as well as vulgar, would handle the various properties of *languages*; showing in what points each excelled, in what it failed. For so not

[6] Aulus Gellius quotes from the *Analogia* of Cæsar, a precept to avoid an unusual word "veluti scopulum", *Noctes Att.* i. 10. Bacon refers to the *Analogia* in several other places. Vide suprà, p. [476. Observe that he there speaks of it as a *grammatical philosophy* in which Cæsar was endeavouring to bring words, which are the images of things, into congruity with the things themselves. Whence it would seem that he had changed his opinion as to the character of the book; for this would be the very *analogia inter verba et res* from which here he distinguishes it.]

[7] See particularly the *Cratylus*.

only may languages be enriched by mutual exchanges, but the several beauties of each may be combined (as in the Venus of Apelles [8]) into a most beautiful image and excellent model of speech itself, for the right expressing of the meanings of the mind. And at the same time there will be obtained in this way signs of no slight value but well worthy of observation (which a man would hardly think perhaps) concerning the dispositions and manners of peoples and nations, drawn from their languages. I like well that remark of Cicero's that the Greeks had no word to express the Latin *ineptus* ; " because," says he, " that vice was so familiar among the Greeks that they did not perceive it in themselves " [9] ; a censure worthy of the Roman gravity. And how came it that the Greeks used such liberty in composition of words, the Romans on the contrary were so strict and sparing in it ? One may plainly collect from this fact that the Greeks were fitter for arts, the Romans for business : for the distinctions of arts are hardly expressed without composition of words ; whereas for the transaction of business simpler words are wanted. Then again the Hebrews have such a dislike to these compositions that they had rather abuse a metaphor than introduce a compound word : and the words they use are so few and so little mixed, that one may plainly perceive from their very language that they were a Nazarite nation, separated from the rest of the nations. And is it not a fact worthy of observation (though it may be a little shock to the spirits of us moderns) that the ancient languages were full of declensions, cases, conjugations, tenses, and the like, while the modern are nearly stripped of them, and perform most of their work lazily by prepositions and verbs auxiliary ? Surely a man may easily conjecture (how well so ever we think of ourselves) that the wits of the early ages were much acuter and subtler than our own [10]. There are numberless observations of this kind, enough to fill a good volume. And therefore it is not amiss to distinguish Philosophic Grammar from Grammar Simple and Literary, and to set it down as wanting.

To Grammar also I refer all accidents of words, of what kind soever ; such as Sound, Measure, Accent. The primary formation of simple letters indeed (that is, by what percussion of the tongue, by what opening of the mouth, by what meeting of the lips, by what effort of the throat, the sound of each letter is produced) does not belong to Grammar, but is part of the doctrine concerning Sounds, and to be handled under Sense and the Sensible. The sound which I speak of as belonging to Grammar relates only to sweetnesses and harshnesses. Of these some are common to all nations ; for there is no language that does not in some degree shun the hiatus caused by vowels coming together, and the harshnesses caused by consonants coming together. There are others again which are respective, being found pleasing to the ears of some nations and displeasing to others. The Greek language abounds in diphthongs ; the Latin is much more sparing of them. The Spanish dislikes thin letters, and changes them immediately into those of a middle tone [11]. Languages derived from the Goths delight

[8] Not the Venus of Apelles, but the Helen of Zeuxis. [9] Cicero *De Orat.* ii. 4.

[10] On this very interesting question, which Bacon was probably the first to propose, Grimm has some good remarks in his essay on the origin of language, in the *Berlin Transactions* for 1852. He shows that of the two classes of languages here contrasted each has its own merits, observing that mere fulness of grammatical forms is not to be recognised as necessarily an advantage ; else we should be obliged to rate Finnish, in which the noun has thirteen cases, above Sanscrit, in which it has eight, and Greek, in which it has only five. It may be remarked in illustration of this that although there are in Sanscrit past tenses corresponding to the Greek aorists and perfects, yet the accuracy of logical discrimination which appropriates the latter to the completed action belongs to Greek only; so too of the appropriation of the imperfect to express an uncompleted action. See Bopp, *Comparative Grammar*, § 513.

[11] This is somewhat overstated. The Spanish generally retains the Latin tenuis at the beginning of words and often in the middle. The tendency to the flattening Bacon mentions is most marked in the case of *p* and *b*. See Diez, *Grammatik der Romanischen Sprachen*, i. 252, for a general table of consonantal changes in the Roman tongues. A remarkable peculiarity in Spanish is the substitution of *h* (now dropped in pronunciation, for the Latin *f* at the beginning of words. It is not however universal, and belongs

in aspirates[12]. Many things of this kind might be mentioned; but these are perhaps more than enough.

The Measure of words has produced a vast body of art; namely Poesy, considered with reference not to the matter of it (of which I have spoken above) but to the style and form of words: that is to say, metre or verse; wherein the art we have is a very small thing, but the examples are large and innumerable. Neither should that art (which the grammarians call Prosody) be confined to the teaching of the kinds and measures of verse. Precepts should be added as to the kinds of verse which best suit each matter or subject. The ancients used hexameter for histories and eulogies; elegiac for complaints; iambic for invectives; lyric for odes and hymns. Nor have modern poets been wanting in this wisdom, so far as their own languages are concerned. The fault has been, that some of them, out of too much zeal for antiquity, have tried to train the modern languages into the ancient measures (hexameter, elegiac, sapphic, etc.[13]); measures incompatible with the structure of the languages themselves, and no less offensive to the ear. In these things the judgment of the sense is to be preferred to the precepts of art,—as the poet says,

> Cœnæ fercula nostræ
> Mallem convivis quam placuisse cocis[14].

And it is not art, but abuse of art, when instead of perfecting nature it perverts her. But for poesy (whether we speak of stories or metre) it is (as I said before) like a luxuriant plant, that comes of the lust of the earth, without any formal seed. Wherefore it spreads everywhere and is scattered far and wide,—so that it would be vain to take thought about the defects of it. With this therefore we need not trouble ourselves. And with regard to Accents of words, it is too small a matter to speak of; unless perhaps it be thought worth remarking, that while the accentuation of *words* has been exquisitely observed, the accentuation of *sentences* has not been observed at all. And yet it is common to all mankind almost to drop the voice at the end of a period, to raise it in asking a question, and other things of the kind not a few. And so much for the part of Grammar which relates to Speech.

As for Writing, it is performed either by the common alphabet (which is used by everybody) or by a secret and private one, agreed upon by particular persons: which they call *ciphers*. And with regard to the common orthography itself, a controversy and question has been raised among us,—namely, whether words ought to be written as they are pronounced, or in the usual way. But this apparently reformed style of writing (viz. in which the spelling should agree with the pronunciation) belongs to the class of unprofitable subtleties. For the pro-

to a comparatively late period of the language, no trace of it being found, according to Diez, in the poem of the *Cid*.

[12] Bacon no doubt refers to High and Low German. The Gothic itself—commonly called Mœso-Gothic, but which might perhaps be as fitly called Italian-Gothic, as the existing remains of it belong probably to Italy in the time of Theodoric and his successors —is much less charged with aspirates than the tongues which claim descent from it. The last editor of Ulphilas, after pointing out the prevalence of liquids and tenues, observes rather fancifully: "Our ancestors were not a mountain people; they must have dwelt on plains under a moist, mild climate", The analogy of Gothic with Sanscrit is very striking. Bopp remarks: "When I read the venerable Ulphilas, I feel as if I were reading Sanscrit".

[13] This affectation prevailed about the same time in France and Italy, and a little later in England. Jodelle was the first person, according to Pasquier, who produced a French hexameter and pentameter.

Augustus von Schlegel, in his *Indische Bibliothek* has an interesting essay on this subject, especially with respect to the Greek hexameter. He endeavours to determine the modifications necessary in order that it may be really naturalised in modern languages.

[14] Mart. ix. 83.:—

> The dinner is for eating, and my wish is
> That guests and not that cooks should like the dishes.

nunciation itself is continually changing; it does not remain fixed; and the derivations of words, especially from foreign tongues, are thereby completely obscured. And as the spelling of words according to the fashion is no check at all upon the fashion of pronunciation, but leaves it free, to what purpose is this innovation [15]?

[15] Every living language is continually changing; and the orthography gradually follows changes of pronunciation. But to make the pronunciation of the present moment the standard of orthography is to set aside as far as possible the historical element in the development of the language, and thus greatly to diminish its value as a record of the progress of human thought, not to mention the effect which such a system would have in making works composed before the era of the last reformation unintelligible.

[I cannot help thinking that Bacon would have pronounced a less confident judgment on this question, if it had occurred to him that a system of notation might be contrived which should not only represent the pronunciation of the particular time, but *accompany all changes of pronunciation which time might introduce*; so that the written word should be *at all times* a true description of the spoken word. For this purpose nothing more is required than an alphabet containing as many distinct *characters* as there are distinguishable elementary *sounds* in the language, so that the same sound may always be represented by the same character or combination of characters, and no combination of characters may be used to represent more than one combination of sounds. Against a reform of orthography founded upon such a reconstruction of the alphabet, it appears to me that none of the objections either in the text or in the note can be justly urged. With regard to the history of the past, everything would remain as it is. A dictionary containing the old and new spelling of every word in the language would effectually preserve its etymological history (so far as our present orthography does preserve it) up to the present time. For the future pronunciation would still be free to change, and orthography would still follow; but the changes of pronunciation would be less rapid and capricious, and the corresponding changes of orthography would be not gradual but immediate. Pronunciation would change, not according to fashion or accident, but according to the laws of nature; and each change would be registered as it came in the printed records of the language. All this would surely be a great advantage, whether we regard language as a medium of communication, for which it serves best when it is most uniform and constant, or as a record of the progress of human thought, for which it serves the better in proportion as capricious and accidental changes are excluded and natural changes marked and registered.

Bacon was probably thinking of some particular scheme proposed in his own day, in which the existing alphabet was to be used. Many such partial schemes of orthographical reform have been attempted from time to time, all of which may be justly condemned as "useless subtilties," not because the thing aimed at—*ut scilicet scriptio pronunciationi consona sit*—would be useless if accomplished, but because, without such a reconstruction of the alphabet as should enable us to assign to each distinct sound a distinct character, the thing *cannot* be accomplished. With an alphabet of only twenty-six letters, it is impossible to make the spelling of English represent the pronunciation, because there are more than twenty-six distinct sounds used in English speech. It has recently been shown, however, that with an alphabet of only forty letters, every sound used in speaking good English may be represented accurately enough for all practical purposes; and a few more would probably include all the sounds of all the classical languages in Europe.

Two or three alphabets of this kind have been suggested within the last hundred years. There was one proposed by Benjamin Franklin, another by Dr. William Young, another by Sir John Herschell. But the first serious attempt to bring such an alphabet into general use, and fairly to meet and overcome all the practical as well as all the theoretical difficulties, was made by Mr. Alexander Ellis and Mr. Isaac Pitman in 1848. And there can be no doubt that by means of their alphabet every English word now in use may be so written that the spelling shall contain a sufficient direction for the pronunciation. Nor is there any reason to apprehend that it would ever be necessary to remodel it, since, however the fashion of pronunciation may change, it is not likely that any new elementary sounds will be developed; and therefore, though we might have to spell some of our words differently, we should still be able to spell them out of the same alphabet.

Let us proceed then to Ciphers[16]. Of these there are many kinds: simple ciphers; ciphers mixed with non-significant characters; ciphers containing two different letters in one character; wheel-ciphers; key-ciphers; word-ciphers; and the like. But the virtues required in them are three; that they be easy and not laborious to write; that they be safe, and impossible to be deciphered; and lastly that they be, if possible, such as not to raise suspicion. For if letters fall into the hands of those who have power either over the writers or over those to whom they are addressed, although the cipher itself may be safe and impossible to decipher, yet the matter comes under examination and question; unless the cipher be such as either to raise no suspicion or to elude inquiry. Now for this elusion of inquiry, there is a new and useful contrivance for it, which as I have it by me, why should I set it down among the desiderata, instead of propounding the thing itself? It is this: let a man have two alphabets, one of true letters, the other of non-significants; and let him infold in them two letters at once; one carrying the secret, the other such a letter as the writer would have been likely to send, and yet without anything dangerous. Then if any one be strictly examined as to the cipher, let him offer the alphabet of non-significants for the true letters, and the alphabet of true letters for non-significants. Thus the examiner will fall upon the exterior letter; which finding probable, he will not suspect anything of another letter within. But for avoiding suspicion altogether, I will add another contrivance, which I devised myself when I was at Paris in my early youth, and which I still think worthy of preservation. For it has the perfection of a cipher, which is to make anything signify anything; subject however to this condition, that the infolding writing shall contain at least five times as many letters as the writing infolded: no other condition or restriction whatever is required. The way to do it is this: First let all the letters of the alphabet be resolved into transpositions of two letters only. For the transposition of two letters through five places will yield thirty-two differences; much more twenty-four, which is the number of letters in our alphabet[17]. Here is an example of such an alphabet.

Example of an Alphabet in two letters[18].

A	*B*	*C*	*D*	*E*	*F*	*G*	*H*
Aaaaa.	*aaaab.*	*aaaba.*	*aaabb.*	*aabaa.*	*aabab.*	*aabba.*	*aabbb.*
I	*K*	*L*	*M*	*N*	*O*	*P*	*Q*
abaaa.	*abaab.*	*ababa.*	*ababb.*	*abbaa.*	*abbab.*	*abbba.*	*abbbb.*
R	*S*	*T*	*V*	*W*	*X*	*Y*	*Z.*
baaaa.	*baaab.*	*baaba.*	*baabb.*	*babaa.*	*babab.*	*babba.*	*babbb.*

As for the fear that, if such a reformation were adopted, works composed previously would become unintelligible, it has been ascertained by many experiments that children who have learned to read books printed phonetically in the new alphabet easily teach themselves to read books printed in the ordinary way; and therefore, even if the new system should become universal for all new books, no one would have any difficulty in mastering the old ones.—*J. S.*]

[16] See, for an account of these ciphers, the appendix at the end of the treatise. Bacon's biliteral cipher seems, as I have there pointed out, to be connected with one which had been given by Porta, which also depends on the principle of which the Electric Telegraph is now a familiar illustration, that any number of things may be denoted by combinations of two signs, as in the binary scale of numeration.

[17] There is a simpler way of attaining the same end, viz., by using two sets of characters, the differences being, as in Bacon's method, intended to be imperceptible, and making the length of the intervals at which those of one set recur significant of the letters of the "interius scriptum." This is a system mentioned by writers on the subject; whether ever actually used, I do not know.

[18] For this and the following examples, a special character is used in the original edition, resembling handwriting, and apparently cut in wood for the occasion. But as it is only in the *Alphabetum Biforme* and the *Exempla Accommodationis* that anything

Nor is it a slight thing which is thus by the way effected. For hence we see how thoughts may be communicated at any distance of place by means of any objects perceptible either to the eye or ear, provided only that those objects are capable of two differences; as by bells, trumpets, torches, gunshots, and the like. But to proceed with our business: when you prepare to write, you must reduce the interior epistle to this biliteral alphabet. Let the interior epistle be

Fly.

Example of reduction.

F L Y.

aabab. ababa. babba.

Have by you at the same time another alphabet in two *forms;* I mean one in which each of the letters of the common alphabet, both capital and small, is exhibited in two different forms,—any forms that you find convenient.

Example of an Alphabet in two forms [19].

a	*b*	*a*	*b*	*a*	*b*	*a*	*b*	*a*	*b*	*a*	*b*
A	*A*	*a*	*a*	*B*	*B*	*b*	*b*	*C*	*C*	*c*	*c*
D	*D*	*d*	*d*	*E*	*E*	*e*	*e*	*F*	*F*	*f*	*f*
G	*G*	*g*	*g*	*H*	*H*	*h*	*h*	*I*	*J*	*i*	*i*
K	*K*	*k*	*k*	*L*	*L*	*l*	*l*	*M*	*M*	*m*	*m*
N	*N*	*n*	*n*	*O*	*O*	*o*	*o*	*P*	*P*	*p*	*p*
Q	*Q*	*q*	*q*	*R*	*R*	*r*	*r*	*S*	*S*	*s*	*s*
T	*T*	*t*	*t*	*U*	*U*	*u*	*u*	*v*	*v*		
W	*W*	*w*	*w*	*X*	*X*	*x*	*x*	*Y*	*Y*	*y*	*y*
				Z	*Z*	*z*	*z*				

Then take your interior epistle, reduced to the biliteral shape, and adapt to it letter by letter your exterior epistle in the biform character; and then write it out. Let the exterior epistle be,

Do not go till I come.

Example of Adaptation.

F L Y.

aa bab. ab aba.b a bba.

Do not go till I come.

I add another larger example of the same cipher,—of the writing of anything by anything.

The interior epistle; for which I have selected the Spartan despatch, formerly sent in the *Scytale.*

All is lost. Mindarus is killed. The soldiers want food. We can neither get hence, nor stay longer here.

The exterior epistle, taken from Cicero's first letter, and containing the Spartan despatch within it.

depends upon the shape of the letters, I have printed all the rest in the common italic type.—*J. S.*

[19] This biform alphabet is set out somewhat differently in the original edition. The characters are cut to represent handwriting, the distinctions being made by loops or flourishes; and the (*a*) or (*b*) is repeated in every case. By keeping the columns distinct, I have avoided the necessity of this repetition; and I have obtained the requisite distinction between the two sets of characters by using types belonging to two different founts. The particular forms of the letters are of course immaterial, so long as those which stand for *a* can be clearly distinguished from those which stand for *b;* and the table, as I have arranged it, will be found easier of reference.—*J. S.*

In all duty or rather piety towards you I satisfy everybody except myself. ***Myself*** *I never satisfy. For so great are the services which you have rendered me, that seeing you did not rest in your endeavours on my behalf till the thing was done, I feel as if life had lost all its sweetness, because I cannot do as much in this cause of yours. The occasions are these: Ammonius, the king's ambassador, openly besieges us with money: the business is carried on through the same creditors who were employed in it when you were here, etc.*

The doctrine of Ciphers carries along with it another doctrine, which is its relative. This is the doctrine of deciphering, or of detecting ciphers, though one be quite ignorant of the alphabet used or the private understanding between the parties: a thing requiring both labour and ingenuity, and dedicated, as the other likewise is, to the secrets of princes. By skilful precaution indeed it may be made useless; though as things are it is of very great use. For if good and safe ciphers were introduced, there are very many of them which altogether elude and exclude the decipherer, and yet are sufficiently convenient and ready to read and write. But such is the rawness and unskilfulness of secretaries and clerks in the courts of kings, that the greatest matters are commonly trusted to weak and futile ciphers.

It may be suspected perhaps that in this enumeration and *census*, as I may call it, of arts, my object is to swell the ranks of the sciences thus drawn up on parade, that the numbers of them may raise admiration; whereas in so short a treatise, though the numbers may perhaps be displayed, the force and value of them can hardly be explained. But I am true to my design, and in framing this globe of knowledge I do not choose to omit even the smaller and more remote islands. And though my handling of these things be cursory, it is not (as I think) superficial; but out of a large mass of matter I pick out with a fine point the kernels and marrows of them. Of this however I leave those to judge who are most skilful in such arts. For whereas most of those who desire to be thought multiscient are given to parade the terms and externals of arts, thereby making themselves the admiration of those who do not understand those arts and the scorn of those who do; I hope that my labours will have the contrary fate, and arrest the judgment of those who are most skilful in the several arts, and be less cared for by the rest. As for those arts which may appear to be of a lower order, if any one thinks that I make somewhat too much of them, let him look round, and he will see that men who are great and famous in their own countries, when they come up to the metropolis and seat of empire are almost lost in the crowd, and of no mark [20]; and in like manner it is not strange that these lighter arts when placed by the side of the principal and superior ones appear of less dignity; although to such as have spent their chief study upon them they seem great and illustrious things. And so much for the Organ of discourse.

CHAPTER II.

The doctrine of the Method of Discourse *is made a substantive and principal part of the art of transmitting; and is named* Wisdom of Transmission. *Different kinds of Method are enumerated, with a note of their advantages and disadvantages.*

Let us now come to the doctrine concerning the Method of discourse. This has been commonly handled as a part of Logic: and it also finds a place in Rhetoric, under the name of *Disposition*. But the placing of it in the train of other arts has led to the passing over of many things relating to it which it is useful to know. I have therefore thought fit to make the doctrine concerning Method a substantive and principal doctrine, under the general name of *Wisdom of Transmission*. The kinds of method being various, I will begin by enumerating rather than distributing them. And first, for the "one and only method," with its

[20] Being then, as King James used to say, like ships at sea, and when at home like ships in a creek; a comparison which may possibly have been suggested by this passage, which occurs in the *Advancement* as well as here.

distribution of everything into two members, it is needless to speak of it[1]; for it was a kind of cloud that overshadowed knowledge for awhile and blew over: a thing no doubt both very weak in itself and very injurious to the sciences. For while these men press matters by the laws of their method, and when a thing does not aptly fall into those dichotomies, either pass it by or force it out of its natural shape, the effect of their proceeding is this,—the kernels and grains of the sciences leap out, and they are left with nothing in their grasp but the dry and barren husks[2]. And therefore this kind of method produces empty abridgments, and destroys the solid substance of knowledge.

Let the first difference of Method then be this: it is either *Magistral* or *Initiative.* Observe however that in using the word "initiative," I do not mean that the business of the latter is to transmit the beginnings only of sciences, of the former to transmit the entire doctrine. On the contrary I call that doctrine *initiative* (borrowing the term from the sacred ceremonies) which discloses and lays bare the very mysteries of the sciences. The magistral method teaches; the initiative intimates. The magistral requires that what is told should be believed; the initiative that it should be examined. The one transmits knowledge to the crowd of learners; the other to the sons, as it were, of science. The end of the one is the use of knowledges, as they now are; of the other the continuation and further progression of them. Of these methods the latter seems to be like a road abandoned and stopped up; for as knowledges have hitherto been delivered, there is a kind of contract of error between the deliverer and the receiver; for he who delivers knowledge desires to deliver it in such form as may be best believed, and not as may be most conveniently examined; and he who receives knowledge desires present satisfaction, without waiting for due inquiry; and so rather not to doubt, than not to err; glory making the deliverer careful not to lay open his weakness, and sloth making the receiver unwilling to try his strength. But knowledge that is delivered to others as a thread to be spun on ought to be insinuated (if it were possible) in the same method wherein it was originally invented. And this indeed is possible in knowledge gained by induction; but in this same anticipated and premature knowledge (which is in use) a man cannot easily say how he came to the knowledge which he has obtained. Yet certainly it is possible for a man in a greater or less degree to revisit his own knowledge, and trace over again the footsteps both of his cognition and consent; and by that means to transplant it into another mind just as it grew in his own. For it is in knowledges as it is in plants; if you mean to use the plant, it is no matter what you do with the root; but if you mean to remove it to grow, then it is safer to use roots than slips. So the method of transmitting knowledge which is now in use presents trunks as it were of sciences (and fair ones too), but without the roots; good for the carpenter but useless for the planter. But if you will have sciences grow, you need not much care about the body of the tree; only look well to this, that the roots be taken up uninjured, and with a little earth adhering to them. Of which kind of transmission the method of the mathematicians has, in that subject, some shadow, but generally I do not see it either put in use or inquired of. Therefore I note it as deficient, and term it the *Handing on of the Lamp*, or Method of Delivery to Posterity[3].

[1] The allusion is to the method of Peter Ramus, which he made to apply to every kind of science, and which depends, as Bacon says, on a dichotomising arrangement. See for Ramus's tabular statements of the contents of the seven liberal arts, the *Professio Regia P. Rami.* (Basil, 1576; but there is probably an earlier edition.)

[2] Ampère's Essay on the Philosophy of Science, though the work of a very able man is certainly open to this reproach. His classification attempts to introduce uniformity where uniformity is impossible. The objections to a dichotomising method are pointed out by Aristotle, who shows that the last of the classes which we obtain by it can have only a negative character. Professor Owen, in his Lectures on the Invertebrata, remarks that no class thus constituted has been found satisfactory. Such a one for instance is that denoted by Dr. Prichard's word *Allophyl* for tribes not of Indo-Germanic origin. See Trendelenburg, *Elementa Logices*, p. 129.

[3] This illustrates the circumstance that several of Bacon's minor works are addressed as to a son or sons; by whom we are to understand those who are qualified to be disciples

Another diversity of Method there is, which in intention has an affinity with the former, but is in reality almost contrary. For both methods agree in aiming to separate the vulgar among the auditors from the select; but then they are opposed in this, that the former makes use of a way of delivery more open than the common, the latter (of which I am now going to speak) of one more secret. Let the one then be distinguished as the *Exoteric* method, the other as the *Acroamatic*; a distinction observed by the ancients principally in the publication of books, but which I transfer to the method of delivery. Indeed this acroamatic or enigmatical method was itself used among the ancients, and employed with judgment and discretion. But in later times it has been disgraced by many, who have made it as a false and deceitful light to put forward their counterfeit merchandise. The intention of it however seems to be by obscurity of delivery to exclude the vulgar (that is the profane vulgar) from the secrets of knowledges, and to admit those only who have either received the interpretation of the enigmas through the hands of the teachers, or have wits of such sharpness and discernment as can pierce the veil.

Next comes another diversity of Method, of great consequence to science; which is the delivery of knowledge in *aphorisms*, or in *methods*. For it is specially to be noted, that it has become the fashion to make, out of a few axioms and observations upon any subject, a kind of complete and formal art, filling it up with some discourses, illustrating it with examples, and digesting it into method. But that other delivery by aphorisms has many excellent virtues whereto the methodical delivery does not attain. First it tries the writer, whether he be light and superficial in his knowledge, or solid. For aphorisms, not to be ridiculous, must be made out of the pith and heart of sciences. For illustration and excursion are cut off; variety of examples is cut off; deduction and connexion are cut off: descriptions of practice are cut off; so there is nothing left to make the aphorisms of but some good quantity of observation. And therefore a man will not be equal to the writing in aphorisms, nor indeed will he think of doing so, unless he feel that he is amply and solidly furnished for the work. But in methods,

—— Tantum series juncturaque pollet,
Tantum de medio sumptis accedit honoris [4],

that those things many times carry a show of I know not what excellent art, which if they were taken to pieces, separated, and stripped, would shrink to little or nothing. Secondly, methodical delivery is fit to win consent or belief, but of little use to give directions for practice; for it carries a kind of demonstration in circle, one part illuminating another, and therefore more satisfies the understanding; but as actions in common life are dispersed, and not arranged in order, dispersed directions do best for them. Lastly, aphorisms, representing only portions and as it were fragments of knowledge, invite others to contribute and add something in their turn; whereas methodical delivery, carrying the show of a total, makes men careless, as if they were already at the end.

Next comes another diversity of Method, which is likewise of great weight; namely the delivery of knowledge by *assertions with proofs*, or by *questions with determinations*; the latter kind whereof, if it be immoderately followed, is as pre-

In the *Redargutio Philosophiarum*, the speaker addresses his audience as "filii;" and we find a corresponding phrase in the *New Atlantis*.

[I understand by *filios* in this passage not so much those who are qualified to be disciples, as those who will carry on the work. The *traditio lampadis* refers to the Greek torch-races, in which there were relays of runners, and each as he was spent handed the torch to a fresh man. The *methodus ad filios* is the method which, having in view the continual progression of knowledge, hands over its unfinished work to another generation, to be taken up and carried forward. See preface to the *Novum Organum*, note B at the end.—*J. S.*]

[4] Hor. *Ep. ad Pisones*, 242:—
The order and the joining give such graces,
Mean matters take such honour from their places.

judicial to the advancement of learning as it is detrimental to the fortunes and progress of an army to go about to besiege every little fort or hold. For if the field be kept, and the sum of the enterprise pursued, those smaller things will come in of themselves; although it is true that to leave a great and fortified town in the rear would not be always safe. In like manner in the transmission of knowledge confutations should be refrained from; and only employed to remove strong pre-occupations and prejudgments, and not to excite and provoke the lighter kind of doubts.

Next comes another diversity of Method, namely that the *method used should be according to the subject-matter which is handled.* For there is one method of delivery in the mathematics (which are the most abstracted and simple of knowledges), another in politics (which are the most immersed and compounded). And (as I have already said) uniformity of method is not compatible with multiformity of matter. Wherefore as I approved of particular Topics for invention, so to a certain extent I allow likewise of Particular Methods for transmission. Next comes another diversity of Method, which in the delivery of knowledge is to be used with discretion. This is regulated according to the informations and anticipations already infused and impressed on the minds of the learners concerning the knowledge which is to be delivered. For that knowledge which comes altogether new and strange to men's minds is to be delivered in another form than that which is akin and familiar to opinions already taken in and received. And therefore Aristotle, when he thinks to tax Democritus, does in truth commend him, where he says, "If we shall indeed dispute, and not follow after similitudes"[5], etc.; thus making it a charge against Democritus, that he was too fond of comparisons. For those whose conceits are already seated in popular opinions, need but to dispute and prove; whereas those whose conceits are beyond popular opinions, have a double labour; first to make them understood, and then to prove them; so that they are obliged to have recourse to similitudes and metaphors to convey their meaning. We see therefore in the infancy of learning, and in rude times, when these conceits which are now old and trivial were new and unheard of, that the world was full of parables and similitudes. For else would men either have passed over without due mark or attention, or else rejected as paradoxical, that which was laid before them. For it is a rule in the art of transmission, that all knowledge which is not agreeable to anticipations or presuppositions must seek assistance from similitudes and comparisons[6].

And so much for the diversities of Method, which have not hitherto been pointed out by others. For as for those other methods,—Analytic, Systatic, Diæretic, also Cryptic, Homeric[7], and the like,—they are rightly invented and distributed, and I see no reason why I should dwell upon them.

Such then are the kinds of Method. Its parts are two; the one relating to the

[5] Arist. *Nic. Eth.* vi. 3. It is difficult to know why Bacon supposed Aristotle to allude to Democritus, as there is no reason to doubt the correctness of the received opinion that the allusion is to Plato's illustration of the nature of knowledge, which will be found at p. 197 of the *Theætetus.* On different occasions Aristotle blames those who in philosophical questions employ similitudes or comparisons; but it does not appear that in any such passage he refers to Democritus.

Mr. Munro, to whom I am indebted for the substance of this note, has pointed out to me the passage in Sextus Empiricus, *Adversus Logicos,* in which the opinion held by Democritus and others of the Physicists that "like is known of like" is mentioned. If any commentator has asserted that such a view of the nature of knowledge is condemned by Aristotle as would make it dependent upon this notion of ὁμοιότης, and that this notion was held by Democritus, we should get a probable explanation of the error into which Bacon seems to have fallen; but the simplest explanation is that he put the name of Democritus for that of Plato by mere inadvertence.

It may be remarked that Democritus might be charged not only with propounding a materialistic view of the nature of knowledge, but also with employing illustrations in support of it derived from material objects.

[6] Compare Plato, *Politic.* 277.

[7] See, for most of these terms, the *Rhetoric* of Ramus.

disposition of the whole work or argument of a book; the other to the limitation of propositions. For there belongs to architecture not only the frame of the whole building, but also the formation and shape of the several beams and columns thereof; and Method is as it were the architecture of the sciences. And herein Ramus merited better in reviving those excellent rules of propositions (that they should be true, universally, primarily, and essentially [8]), than he did in introducing his uniform method and dichotomies; and yet it comes ever to pass, I know not how, that in human affairs (according to the common fiction of the poets) "the most precious things have the most pernicious keepers." Certainly the attempt of Ramus to amend propositions drove him upon those epitomes and shallows of knowledge. For he must have a lucky and a happy genius to guide him who shall attempt to make the axioms of sciences convertible, and shall not withal make them circular, or returning into themselves. Nevertheless I must confess that the intention of Ramus in this was excellent.

There still remain two limitations of propositions, besides that for making them convertible; the one regarding their extension, the other their production. Certainly sciences, if a man rightly observe it, have, besides profundity, two other dimensions, namely latitude and longitude. The profundity relates to their truth and reality; for it is they which give solidity. As to the other two, the latitude may be accounted and computed from one science to another; the longitude from the highest proposition to the lowest in the same science. The one contains the true bounds and limits of sciences, that the propositions thereof may be handled properly, not promiscuously, and repetition, excursion, and all confusion may be avoided; the other prescribes the rule how far and to what degree of particularity the propositions of a science should be deduced. For certainly something must be left to exercise and practice; since we should avoid the error of Antoninus Pius and not be "splitters of cummin seeds" in the sciences, nor multiply divisions to extreme minuteness. Therefore it is plainly worth inquiry how we are to guide ourselves in this matter. For we see that too remote generalities (unless they be deduced) give little information, and do but offer knowledge to the scorn of practical men; being of no more avail for practice, than an Ortelius's universal map is to direct the way between London and York. Certainly the best sort of rules are not unfitly compared to mirrors of steel, where you may see the images of things, but not before they are polished; so rules and precepts will help if they be laboured and polished by practice, but not otherwise. But if these rules could be made clear and crystalline from the first, it were best; because there would then be less need of continual labour and practice. And so much for the science of method, which I have called the Wisdom of Transmission. And yet I must not omit to mention, that some persons, more ostentatious than learned, have laboured about a kind of method not worthy to be called a legitimate method, being rather a method of imposture, which nevertheless would no doubt be very acceptable to certain meddling wits. The object of it is to sprinkle little drops of a science about, in such a manner that any sciolist may make some show and ostentation of learning. Such was the Art of Lullius: such the Typocosmy traced out by some; being nothing but a mass and heap of the terms of all arts, to the end that they who are ready with the terms may be thought to understand the arts themselves. Such collections are like a fripper's or broker's shop, that has ends of everything, but nothing of worth [9].

[8] Καθόλου πρῶτον, κατὰ παντὸς, καθ' αὑτό, etc. These rules are in reality Ramus's own, though he professed to find them in Aristotle. They were however suggested to him by the fourth chapter of the first book of the *Posterior Analytics*. See the preface to *Valerius Terminus*.

[9] The fundamental idea of Lully's art, and of all similar methods, may be thus stated:—The propositions which in the aggregate make up the sum of human knowledge consist of combinations of a certain number of conceptions. If then we had a complete list of these conceptions so arranged as that all their admissible combinations could be obtained by a mechanical process, such a list would be virtually equivalent to a complete encyclopædia. Even an incomplete list would give a certain portion, greater or less according to circumstances, of all the knowledge which relates to the conceptions

CHAPTER III.

Of the foundations and office of Rhetoric. Three appendices of Rhetoric, which relate only to the Promptuary ; Colours of Good and Evil, *both Simple and Comparative ;* Antitheses of Things ; Lesser Forms of Speeches.

I NOW come to the doctrine concerning the Illustration of Discourse. This is that which is called Rhetoric, or Oratory ; a science certainly both excellent in itself, and excellently well laboured. Truly valued indeed, eloquence is doubtless in-

which enter into it. It is obvious that such a method can give no criterion of the truth of the propositions which it evolves ; but it may be so managed as that every proposition shall be intelligible. To take a very simple instance : I confine myself to a table consisting of three columns, the first column to consist of names of quadrupeds, as horse, stag, mouse etc. ; the second of adjectives, such as large, small, rare, etc. ; the third of names of classes of animals, as ruminant, rodent, and the like. With a few more such columns Lully would have said that the natural history of quadrupeds could be completely made out. Take any word from the first column, any word from the second, any word from the third, and connect them by the logical copula ; and if you are fortunate, you obtain a result as reasonable as this—" a mouse is a small rodent ". But of course it might have appeared that a horse was a ruminant.

Notwithstanding this obvious and incurable defect, different arrangements and modifications of the art were proposed by many writers, some of whom probably believed that it contained a key to all knowledge, while others believed that it would be at least useful as a means of arranging and suggesting to the mind all that could be said truly or falsely on a given subject. It appears to have suggested to Leibnitz one of his early tracts, that on the art of combination, and thus to have led him to his notion of reducing reasoning to a calculus. Analogous to Lully's art is a puerility which has recently been revived, namely, mechanical verse-making. It seems also to have suggested to Trithemius his method of secret writing, the fundamental idea of which may be explained by saying that if there were six and twenty animals in the first column of my table, the same number of adjectives in the second, and of classes in the third, each column might represent a complete alphabet, and the proposition " a mouse is a small rodent " would stand for a word of three letters. With more columns longer words might be spelt, etc., etc. It is obvious that in this case the truth or falsehood of the propositions used would be of little or no moment.

Lully's art was, it is said, revealed to him by an angel, after he had taken the resolution of giving up the world and of devoting himself to studies for which his previous way of life had unfitted him. Cornelius Agrippa, who had himself written an exposition of it, thus condemns it in the *De Vanit. et Incert. Scient.* c. 9. : " Hoc autem admonere vos oportet, hanc artem ad pompam ingenii et doctrinæ ostentationem potius quam ad comparandam eruditionem valere, ac longe plus habere audaciæ quam efficaciæ ". Though much cannot be said in favour of his method, yet Lully himself is one of the most remarkable persons of the middle ages. The story of his renouncing the world in consequence of the intense revulsion of feeling produced by the sudden extinction of a passionate love is well known ; whether authentic or not, it is a striking illustration of the solemn words of Peter Damiani : " Quid ergo sit caro doceat ipsa caro ". Lully says of himself : " I was married, I had begotten children, I was tolerably rich, I was wanton and worldly. All this with a willing mind did I forsake, that I might further God's glory and the public good, and exalt the holy faith ; I learnt Arabic ; many times went I forth to preach to the Saracens ; for the faith's sake I was made prisoner and kept in bonds and beaten ; forty and five years have I laboured to stir up the rulers of the Church and Christian princes to take heed to the public good ; now am I old, now am I poor, yet in the same mind still, by God's help, will so continue to my life's end." Accordingly he went again to Africa, and, preaching the Gospel, was on the feast of St. Peter and St. Paul stoned and left half-dead. Some Genoese merchants put him on board their ship and there he died, and was buried in his native island of Majorca in 1315. See Antonio, *Bibl. Hisp. Vet.* vol. ii. p. 123. See, with respect to Lully in general, and particularly as to the charge of heterodoxy made against him, Perroquet, *Apologie de la Vie et des Ecritz du bien heureux Raymond Lully.*

The foolish story, still occasionally repeated, of Raymond Lully having made gold for

ferior to wisdom. For what a distance there is between them is shown in the words spoken by God to Moses, when he declined the office assigned him on the ground that he was no speaker; "There is Aaron, he shall be thy speaker, thou shalt be to him as God"[1]. Yet in profit and in popular estimation wisdom yields to eloquence; for so Solomon says; "The wise in heart shall be called prudent, but he that is sweet of speech shall compass greater things"[2]; plainly signifying that wisdom will help a man to a name or admiration, but that it is eloquence which prevails most in action and common life. But as to the labouring of this art, the emulation of Aristotle with the rhetoricians of his time, and the eager and vehement zeal of Cicero doing his utmost to ennoble it, coupled with his long experience, has made them in their works on rhetoric exceed themselves. Again, those most brilliant examples of the art which we have in the orations of Demosthenes and Cicero, added to the perfection and skill of the precepts, have doubled the progression in it. And therefore the deficiencies which I shall note will rather be in some collections which may as handmaids attend the art, than in the rules and use of the art itself. For when in treating of Logic I made mention of a certain Promptuary or Preparatory Store, I promised to produce fuller examples of it in Rhetoric.

Notwithstanding, to open and stir the earth a little, according to my custom, about the roots of this science; Rhetoric is subservient to the imagination, as Logic is to the understanding; and the duty and office of Rhetoric, if it be deeply looked into, is no other than to apply and recommend the dictates of reason to imagination, in order to excite the appetite and will. For we see that the government of reason is assailed and disordered in three ways; either by the illaqueation of sophisms, which pertains to Logic; or by juggleries of words, which pertain to Rhetoric; or by the violence of the Passions, which pertains to Ethics. For as in negotiations with others, men are usually wrought either by cunning or by importunity, or by vehemency; so likewise in this negotiation within ourselves, we are either undermined by fallacies of arguments, or solicited and importuned by assiduity of impressions and observations, or agitated and transported by violence of passions. And yet the nature of man is not so unfortunately built, as that those arts and faculties should have power to disturb reason, and no power to strengthen or establish it; on the contrary they are of much more use that way. For the end of logic is to teach a form of argument to secure reason, and not to entrap it; the end likewise of moral philosophy is to procure the affections to fight on the side of reason, and not to invade it; the end of rhetoric is to fill the imagination with observations and images, to second reason, and not to oppress it. For abuses of arts only come in indirectly, as things to guard against, not as things to practise.

And therefore it was great injustice in Plato (though springing out of a just hatred of the rhetoricians of his time) to place rhetoric among arts voluptuary; resembling it to cookery, which did as much to spoil wholesome meats, as by variety and delicacy of sauces to make unwholesome meats more palatable[3]. But God forbid that speech should not be much more conversant in adorning that which is good, than in colouring that which is evil; for this is a thing in use everywhere; there being no man but speaks more honestly than he thinks or acts. And it was excellently noted by Thucydides as a censure passed upon Cleon, that because he used always to hold on the bad side, therefore he was ever inveighing against eloquence and grace of speech; as well knowing that no man can speak fair of courses sordid and base; while it is easy to do it of courses just and honourable[4]. For Plato said elegantly (though it has now grown into a commonplace) "that virtue, if she could be seen, would move great love and affection"[5]; and

Edward the Third, is sufficiently refuted by the date of his death, which occurred, according to authority which there is no reason to doubt, while Edward the Third was a child, and nearly thirty years before the coinage of the nobles said to have been made of Lully's gold. Camden is, I am afraid, responsible for the currency of the story, which in Selden's *Table Talk* seems to be transferred from Lully to Ripley.

[1] Exod. iv. 16. [2] Prov. xvi. 21. [3] Cf. Plato, *Gorgias*, p. 462, etc.
[4] Cf. Plato, *Phædrus*, p. 250. [5] Cf. Thucyd. iii. 42.

it is the business of rhetoric to make pictures of virtue and goodness, so that they may be seen. For since they cannot be showed to the sense in corporeal shape, the next degree is to show them to the imagination in as lively representation as possible, by ornament of words. For the method of the Stoics, who thought to thrust virtue upon men by concise and sharp maxims and conclusions, which have little sympathy with the imagination and will of man, has been justly ridiculed by Cicero[6].

Again, if the affections themselves were brought to order, and pliant and obedient to reason, it is true there would be no great use of persuasions and insinuations to give access to the mind, but naked and simple propositions and proofs would be enough But the affections do on the contrary make such secessions and raise such mutinies and seditions (according to the saying,

> —— Video meliora proboque,
> Deteriora sequor)[7]

that reason would become captive and servile, if eloquence of persausions did not win the imagination from the affections' part, and contract a confederacy between the reason and imagination against them. For it must be observed that the affections themselves carry ever an appetite to apparent good, and have this in common with reason; but the difference is that affection beholds principally the good which is present; reason looks beyond and beholds likewise the future and sum of all. And therefore the present filling the imagination more, reason is commonly vanquished and overcome. But after eloquence and force of persuasion have made things future and remote appear as present, then upon the revolt of imagination to reason, reason prevails.

Let us conclude therefore that rhetoric can be no more blamed for knowing how to colour the worse side, than logic for teaching how to make fine sophisms. For who does not know that the principle of contraries is the same, though the use be opposite? It appears also that logic differs from rhetoric not only (as is commonly said) in that the one is like the fist, and the other like the open hand (that is, the one close, the other at large[8]); but much more in this, that logic handles reason in truth and nature, and rhetoric handles it as it is planted in the opinions of the vulgar. And therefore Aristotle wisely places rhetoric between logic on the one side, and moral and civil knowledge on the other, as participating of both[9]. For the proofs and demonstrations of logic are the same to all men; but the proofs and persuasions of rhetoric ought to differ according to the auditors; so that like a musician accommodating his skill to different ears, a man should be

> Orpheus in silvis, inter delphinas Arion[10];

which application and variety of speech, in perfection of idea, ought to extend so far, that if a man should speak of the same thing to several persons, he should nevertheless use different words to each of them; though this politic and familiar part of eloquence in private discourse it is certain that the greatest orators commonly want; while in observing their well graced forms of speech, they lose that volubility of application, and those characters of style, which it would be better to use in addressing different individuals. And therefore it will not be amiss to recommend this of which I now speak to fresh inquiry, and calling it by the name of *The Wisdom of Private Discourse* to set it down among the deficients; being a thing which the more it is considered the more it will be valued. But whether it be placed in rhetoric or in policy, is a matter of little moment.

6 Cicero, *De Fin.* iv. 18 and 19.

7 Ovid. *Metam.* vii. 20:—

> The better course I know and well approve;
> The worse I follow.

8 Cf. Cicero, *De Fin.* ii. 17.

9 Arist. *Rhet.* i. 2.

10 Virg. *Ecl.* viii. 56:—

> Orpheus by land the trees about him bringing,
> By sea, Arion borne to the dolphins singing.

Let us now descend to the deficiencies in this art, which (as I said before) are rather as appendices than parts of the art itself, and all belong to the Promptuary. First therefore I do not find the wisdom and diligence of Aristotle well pursued and supplied. For he began to make a collection of *the popular signs or colours of apparent good and evil*, both simple and comparative; which are really the sophisms of rhetoric. Now these are of excellent use, especially for business and the wisdom of private discourse. But the labours of Aristotle[11] regarding these colours are in three points defective; one, that he recounts a few only out of many; another, that he does not add the answers to them; and the third, that he seems to have conceived but a part of the use of them. For their use is not more for probation than for affecting and moving. For there are many forms which, though they mean the same, yet affect differently; as the difference is great in the piercing of that which is sharp and that which is flat, though the strength of the percussion be the same. Certainly there is no man who will not be more affected by hearing it said, " Your enemies will be glad of this,"

Hoc Ithacus velit, et magno mercentur Atridæ[12],

than by hearing it said only, " This will be evil for you ". Therefore these points and stings of words are by no means to be neglected. But as I set this down as deficient, I will according to my custom support it by examples; for precepts would not give a sufficient illustration of the thing.

Examples *of the Colours of Good and Evil, both Simple and Comparative.*

SOPHISM.

1. *What men praise and honour is good; what they dispraise and condemn is evil.*

ANSWER.

This Sophism deceives in four ways; by reason of ignorance, of bad faith, of party spirit and factions, of the natural disposition of those who praise and blame. By reason of ignorance; for what is popular judgment worth as a test of good and evil? Better was Phocion's inference, who when the people applauded him more than usual, asked whether he had done wrong[13]. By reason of bad faith, because in praising and blaming, men are commonly thinking of their own business, and not speaking what they think.

Laudet venales, qui vult extrudere, merces[14].

And again; " It is naught, it is naught (says the buyer); but when he is gone his way, he will vaunt "[15]. By reason of factions; for any man may see that men are wont to exalt those of their own party with immoderate praises, and depress below their desert those of the contrary. By reason of natural disposition; for some men are by nature formed and composed for servile adulation, while others on the contrary are crabbed and captious; so that in praising and blaming they do but gratify their own dispositions, with little regard to truth.

SOPHISM.

2. *What is praised even by enemies, is a great good; but what is reproved even by friends, is a great evil.*

This Sophism appears to rest on the ground that that which we speak unwillingly and against our wish and inclination may be supposed to be wrung from us by the force of truth.

ANSWER.

This Sophism deceives by reason of the cunning as well of enemies as of friends. For enemies sometimes bestow praise, not against their will, nor as being com-

11 Arist. *Rhetoric*, i. 6 and 7.

12 Virg. *Æn.* ii. 104:—

This would Ulysses wish, and Atreus' sons
Give much to hear of.

13 Plutarch. in Phocion, c. 8.

14 Hor. *Ep.* ii. 2. 11.:—The merchant praises what he wants to sell.

15 Proverbs, xx. 14.

pelled thereto by the force of truth, but choosing such points for praise as may breed envy and dangers to the subjects of it. And hence there was a prevailing superstition amongst the Greeks, that when a man was praised by another with a malicious purpose to injure him, a pimple would grow upon his nose. It deceives likewise, because enemies sometimes bestow praises merely by way of preface, that they may afterwards calumniate more freely and maliciously. On the other hand, this Sophism deceives also by reason of the cunning of friends. For they too are wont sometimes to acknowledge and proclaim the faults of their friends, not because truth compels them, but choosing such faults as may do them least injury; as if in other respects they were excellent men. It deceives again, because friends also use reprehensions (as I have said that enemies bestow praises) by way of prefaces, whereby they may presently be the more large in commendation.

SOPHISM.

3. *That which it is good to be deprived of, is in itself an evil; that which it is bad to be deprived of, is in itself a good.*

ANSWER.

This Sophism deceives in two ways, by reason either of the comparative degrees of good and evil, or of the succession of good to good, or evil to evil. By reason of comparison: if it was for the good of mankind to be deprived of acorns as food, it does not follow that that food was bad; acorns were good, but corn is better [16]. Nor if it was bad for the Syracusans to be deprived of the elder Dionysius, does it follow that he was good, but that he was not so bad as Dionysius the younger. By reason of succession:—for when a good thing is taken away it is not always succeeded by a bad thing, but sometimes by a greater good; as when the flower falls and the fruit succeeds. Neither when a bad thing is taken away is it always succeeded by a good thing, but sometimes by a worse. For by the removal of his enemy Clodius, Milo lost the "seedbed of his glory" [17].

SOPHISM.

4. *That which approaches to good or evil, is itself good or evil; but that which is remote from good is evil, that from evil, good.*

It is commonly found that things which agree in nature are placed together, and that things of a contrary nature are placed apart; for everything delights to associate with itself that which is agreeable, and to repel that which is disagreeable.

ANSWER.

But this Sophism deceives in three ways; by reason, 1st of destitution, 2ndly of obscuration, and 3rdly of protection. By reason of destitution; for it happens that those things which are most abundant and excellent in their own kind attract everything as far as may be to themselves, spoiling and as it were starving all things in their neighbourhood. Thus you will never find flourishing underwood near great trees. And rightly was it said "that the servants of a rich man are the greatest slaves". So also the lower order of courtiers were pleasantly compared to the vigils of festivals, that are next the feast days, but are themselves devoted to fasting. By reason of obscuration; for all things that are excellent in their own kind have this,—that though they do not impoverish and starve the things next to them, yet they obscure and overshadow them; as astronomers remark of the sun, that it is good in aspect, but evil in conjunction and approximation. By reason of protection; for it is not only for consort and similarity of nature that things unite and collect together; but evil also (especially in civil matters) betakes itself to good for concealment and protection. And hence male-

[16] The allusion is to the following lines:—

"Prima Ceres ferro mortales vertere terram
Instituit, cum jam glandes atque arbuta sacræ
Deficerent silvæ, et victum Dodona negaret."—Virg. *Georg.* i. 147.

[17] Cicero, *Pro. Mil.* 36.

factors seek the protection of sanctuaries, and vice itself resorts to the shadow of virtue;

Sæpe latet vitium proximitate boni[18].

So on the other hand good draws near to evil, not for company, but to convert and reform it. And therefore physicians attend more on the sick than the healthy; and it was objected to our Saviour that he conversed with publicans and sinners.

SOPHISM.

5. *That to which the other parties or sects agree in giving the second place (each putting itself first) seems to be the best; for it seems that in taking the first place they are moved by zeal and partiality, but in bestowing the second by truth and merit.*

So Cicero argues that the sect of the Academics, which maintained the impossibility of comprehending truth, was the best of the philosophies. "For (said he) ask a Stoic which is the best philosophy, and he will prefer his own to the rest; then ask him which is the next best, and he will acknowledge the Academic. So again the Epicurean (who will hardly deign to look at a Stoic), after he has placed his own philosophy at the head, will place the Academic next"[19]. In like manner when a place is vacant, if the prince were to ask each candidate whom he would most recommend next to himself, it is probable that their second votes would meet in the most able and deserving man.

ANSWER.

This Sophism deceives by reason of envy. For next to themselves and their own party, men generally incline to those who are weakest and least formidable, and have given them least trouble; in despite of those who have most insulted or inconvenienced them.

SOPHISM.

6. *That which is better in perfection, is better altogether.*

To this belong the common forms; "Let us not wander in generalities," "Let us compare particular with particular," etc.

ANSWER.

This Sophism appears forcible enough, and rather logical than rhetorical; but still it is sometimes deceptive. First, because there are not a few things which are very much exposed to danger, yet if they escape prove excellent; so that in kind they are inferior, as being oftener imperilled and lost, but individually they are more noble. Of this kind is a blossom in March, whereof the French proverb says: "A March blossom, and a Paris child, if one of them survive, it is worth ten others"[20]. So that generally the blossom of May is superior to the blossom of March; but yet individually the best blossom of March is preferred to the best of May. It deceives secondly, by reason of the nature of things being more equal in some kinds or species, and more unequal in others; as it has been remarked that in general the hotter climates produce the sharper wits; but then the best wits of the colder climates surpass the sharpest of the hotter. So again in many armies if the matter were tried by duel between two champions, the victory would go on the one side, if by the whole army, on the other. For excellencies and superiorities are casual; whereas kinds are governed by nature or discipline. In kind again, metal is more precious than stone; but yet a diamond is more precious than gold.

SOPHISM.

7. *That which keeps the matter open, is good; that which leaves no opening for retreat, is bad. For not to be able to retreat is to be in a way powerless; and power is a good.*

[18] Ovid, *De Art. Amand.* ii. 262:—Vice often lurks 'neath Virtue's shade.

[19] Cf. the fragment of the *Academ. ad Varr.* preserved by St. Augustine.

[20] Bourgeon de Mars, enfant de Paris,
Si un eschape, il en vaut dix.

Hence Æsop derived the fable of the two frogs, who in a great drought, when water was everywhere failing, consulted together what was to be done. The first said, "Let us leap down into a deep well, since it is not likely that the water will fail there." But the other rejoined, "Yes, but if it chance that the water fail there also, how shall we be able to get up again?" And the ground of this Sophism is, that human actions are so uncertain and subject to such risks, that that appears the best course which has the most passages out of it. To this belong those forms which are in use,—"You will tie your hands and engage yourself," "You will not be free to take what fortune may offer," etc.

ANSWER.

This Sophism deceives, first because in human actions fortune insists that some resolution shall be taken. For, as it was prettily said by some one, "not to resolve is itself to resolve"; so that many times suspension of resolution involves us in more necessities than a resolution would. And it seems to be the same disease of mind which is found in misers, only transferred from the desire of keeping money to the desire of keeping freedom of will and power. For as the miser will enjoy nothing, because he will not diminish his store, so this kind of sceptic will execute nothing, because he will still keep all in his own hands. It deceives secondly, because necessity, and the casting of the die (as they call it), is a spur to the courage; as one says, "Being a match for them in the rest, your necessity makes you superior"[21].

SOPHISM.

8. *The evil which a man brings on himself by his own fault is greater; that which is brought on him by external causes, is less.*

The reason of this is that the sting of conscience doubles adversity, while on the other hand the being conscious that a man is clear and free from fault affords great consolation in calamity. And therefore the poets most exaggerate those sufferings, as coming near to despair, where a man accuses and torments himself;

Seque unum clamat, causamque caputque malorum[22].

On the other hand the calamities of worthy persons are lightened and tempered by the consciousness of innocence and merit. Besides when the evil is inflicted by others, a man has something that he may freely complain of, whereby his griefs evaporate and do not suffocate the heart. For in things which come from human injury, we are wont to feel indignation, or to meditate revenge, or to implore, or if not to implore yet to expect, providential retribution; and even if the blow come from fortune, yet is there left a kind of expostulation with the fates themselves;

Atque Deos, atque astra vocat crudelia mater[23].

Whereas if the evil be derived from a man's own fault, the stings of pain strike inward, and more wound and lacerate the heart.

ANSWER.

This Sophism deceives, first by reason of hope, the great antidote of evils. For amendment of a fault is often in our power, but amendment of fortune is not. Hence Demosthenes more than once addressed his countrymen in words like these: "That which, having regard to the time passed, is the worst point and circumstance of all the rest, that as to the time to come is the best. What is that? Even this; that it is your own sloth, irresolution, and misgovernment that have brought your affairs into this ill condition. For had you ordered your means and forces to the best and done your parts every way to the full, and notwithstanding your matters had gone backwards as they do, there had been no hope

[21] Livy, iv. 28.

[22] Virg. *Æn.* xii. 600:—And on herself cries out, as cause of all. Bacon alters the original, which runs:—

Se causam clamat crimenque caputque malorum.

[23] Virg. *Eclog.* v. 23:—And she upbraids the gods and cruel stars.

left of recovery or reparation. But since it has been brought about chiefly by your own errors, you may fairly trust that by amending them you will recover your former condition "[24]. So Epictetus discoursing on the degrees of mental tranquillity, puts those lowest who accuse others next those who accuse themselves, and highest of all those who accuse neither others nor themselves[25]. It deceives secondly, by reason of the innate pride of men's minds, which makes them unwilling to acknowledge their own errors. This to avoid, they exercise far more patience in bearing those ills which they have brought on themselves by their own fault. For as we see that when a fault is committed and it is not yet known who is to blame, men are exceeding angry and make much ado about it; but if afterwards it come out that it was done by a son or a wife or a favourite, all is at once hushed and no more noise made; so it is when anything happens for which we must needs take the blame upon ourselves; as we see it very often in women, that if they have done anything against the wishes of their parents and friends, and it turn out ill, whatever misfortune follows they will keep it to themselves and set a good face upon it[26].

SOPHISM.

9. *From something to nothing appears a greater step than from more to less; and again from nothing to something appears a greater step than from less to more.*

It is a rule in mathematics that there is no proportion between nothing and something; and therefore the degrees of nullity and quiddity appear greater than the degrees of increase and decrease. Thus the loss of an eye is harder for a man with only one eye than for a man with two. In like manner if a man has several children, it is more grief to him to lose the last surviving son than all the rest. Hence also the Sibyl, when she had burned her two first books, doubled the price of the third; for the loss of this would have been a degree of privation, and not of diminution.

ANSWER.

This Sophism deceives, first in respect of those things whereof the use consists in a sufficiency or competency, that is in a determinate quantity. For if a man were bound by penalty to pay a certain sum of money on a stated day, it would be worse for him to be one pound short, than (supposing that that one could not be got) to be short by ten pounds more. So in the wasting of fortunes, the degree of debt which makes the first inroad on the capital seems worse than the last which reduces to beggary. To this belong the common forms; "Sparing comes too late when all is gone"[27]; "as good never a whit as never the better," etc. It deceives secondly, in respect of that principle of nature, that the decay of one thing is the generation of another[28]; so that the degree of extreme privation is sometimes of less disadvantage, because it gives a handle and stimulus to some new course. Hence also Demosthenes often complains to his countrymen; "That the terms which they accepted from Philip, not being profitable nor honourable, were nothing else than aliments of their sloth and indolence; which they would be much better without; because then their industry might be better excited to seek other remedies"[29]. I knew a physician that when delicate women com-

[24] Cf. Demosth. *Philipp.* i. and iii. [25] Epict. *Enchririd.* c. 5.

[26] Bacon makes the same remark in the Essay on Marriage.

[27] Cf. Erasm. *Adag.* ii. 2. 64; and Hesiod. *Op. et Dies*, 339.

[28] Arist. *De Gen. et Corr.* i. 4.

[29] Olynth. iii. 33. Wats refers to the first Philippic, towards the end of which there is a passage not unlike that in the text; but the phrase "alimenta socordiæ," which Bacon has quoted in several parts of his works, is not to be found there. He derived it from H. Wolf's translation of a passage in the third Olynthiac, c. 33., where the Greek is simply ἔστι ταῦτα τὰ τὴν ἑκάστου ῥᾳθυμίαν ὑμῶν ἐπαυξάνοντα, which Wolf renders by "alimenta sunt vestrûm omnium socordiæ." There is no reference to Philip's conduct in the immediate context, the "alimenta socordiæ" being in reality matters of internal arrangement. It seems as if Bacon read the oration in Wolf's version, and adopted the phrase "alimenta socordiæ" (the point of which belongs to the translator and not to

plained that they were ill and yet could not endure to take any medicine, would say to them, not less wittily than sharply, " Your only way is to be worse, for then you will be glad of any medicine." Moreover this degree of privation or extreme want may be useful not only to stimulate energy, but also to enforce patience.

With regard to the second part of this Sophism, it rests on the same foundation as the former (that is on the degrees of nullity and quiddity). Hence the making of a beginning of anything is thought so great a matter :—

Dimidium fact, qui bene cœpit, habet, etc.[30].

Hence also the superstition of astrologers, who make a judgment of the disposition and fortune of a man from the point or moment of his nativity or conception.

ANSWER.

This Sophism deceives first because in some cases the first beginnings of things are no more than that what Epicurus in his philosophy calls *tentamenta*[31], that is imperfect offers and essays, which are nothing unless they be repeated or proceeded with. Therefore in this case the second degree seems more worthy and more powerful than the first, as the wheel-horse in a cart does more work than the leader. Again, it is not a bad saying " that it is the second word which makes the fray." For perhaps the first would have passed. And so the one made a beginning of the mischief, but the other prevented it from coming to an end. It deceives secondly, by reason of the dignity of perseverance ; which lies in the progress, not in the first attempt. For chance or nature may give the first impulse, but only a settled affection and judgment can give constancy. It deceives thirdly, in those things whereof the nature and ordinary course goes against the beginning made ; so that the first start is ever being frustrated unless the force be kept up ; according to the common forms ; " Not to advance, is to retreat " ; " He who is not gaining, is losing " ; as in running up hill, and rowing against stream. But on the other hand, if the motion be down hill, or the rowing be down stream, then the degree of inception is of far greater importance. Besides, this colour extends not only to the degree of inception, which proceeds from power to act, compared with the degree from act to increase ; but also to the degree from impotency to power, compared with the degree from power to act. For the degree from impotency to power seems greater than from power to act.

SOPHISM.

10. *That which has relation to truth is greater than that which has relation to opinion ; and the proof that a thing has relation to opinion is this : it is that which a man would not do if he thought it would not be known.*

So the Epicureans say of the Stoics' Felicity placed in virtue, that it is like the felicity of a player, who if he were left of his auditory and their applause, would straight be out of heart and countenance. And therefore in derision they call virtue a theatrical good. But it is otherwise in riches, of which it is said,

Populus me sibilat ; at mihi plaudo[32].

And likewise of pleasure,

—— Grata sub imo
Gaudia corde premens, vultu simulante pudorem[33].

Demosthenes) without comparing it with the original. [I think, however, that the idea of " alimenta " is really involved in the word ἐπαυξάνοντα, when taken with the context and that no other word could have given the meaning so well. . . .—*J. S.*]

[30] Hor. *Ep.* i. 2. 40 :—Well begun is half done. [31] Cf. Lucretius, v. 835.

[32] Hor. *Sat.* i. 1. 66 :—The people hiss me, but I applaud myself.

[33] Her face said fie, for shame ; but sweet delight
Possessed her heart in secret.

[A quotation from the Latin translation of Theocritus (*Id.* xxvii.) by Hessus (Paris, 1546).

ANSWER.

The fallacy of this Sophism is somewhat more subtle ; though the answer to the example alleged is easy. For virtue is not chosen for the sake of popularity ; since it is a precept that a man should above all things reverence himself [34]. So that a good man will be the same in solitude as on the stage ; though perhaps his virtue may be somewhat strengthened by praise, as heat is increased by reflexion. This however denies the supposition and does not refute the fallacy. Now the answer is this. Allow that virtue (especially such as is attended with labours and conflicts) would not be chosen, except for the sake of the glory and fame accompanying it ; yet it does not therefore follow that the motive and appetite to virtue is not principally for its own sake ; for fame may only be the impulsive cause, or *sine qua non*, and not the efficient or constituent cause. For instance ; if there were two horses, and one of them without the spur could do well, but the other with the spur could do much better, the latter should in my judgment bear off the prize and be accounted the better horse. And to say "Tush, the life of this horse is in the spur," would not move any man of sound judgment ; for since the ordinary instrument of horsemanship is the spur, and that it is no manner of burden or impediment to the rider, the horse that is quickened with the spur is not therefore to be less valued ; nor again is the other that does wonderfully well without the spur to be reckoned on that account the better, but only the finer and daintier. So glory and honour are the spurs of virtue ; and though virtue would somewhat languish without them, yet as they are always at hand to attend virtue, even when not invited, there is no reason why virtue may not be sought for its own sake as well. And thus the proposition that "a thing which is chosen for opinion's sake and not for truth may be known by this—it is what a man would not do if he thought it would not be known," is rightly answered.

SOPHISM.

11. *That which is gained by our own merit and industry is a greater good ; that which is derived from the kindness of others or from the indulgence of fortune a lesser good.*

The reasons of this are,—first, because there is better hope of the future ; for in the favours of others or the good winds of fortune there is little certainty ; but our own virtue and industry are ever with us ; so that after we have obtained some good in this way we have always the same instruments ready to use again ; yea, and by habit and success made more effective. Secondly, because for what we get by the favour of other men we are other men's debtors ; whereas what we obtain of ourselves carries no obligation with it. Nay, even when divine mercy has bestowed any favour on us, it demands a kind of retribution to the goodness of God, which is distressing to depraved and wicked men ; whereas in the former kind, that comes to pass which the prophet speaks of, "Men rejoice and exult, they sacrifice unto their net, and burn incense unto their drag [35]". Thirdly, because what proceeds not by our own merit, carries with it no praise or reputation ; for felicity begets a kind of admiration, but not praise. As Cicero said to Cæsar ; "We have enough to admire, we are looking for something, to praise [36]". Fourthly, because the things obtained by our own industry are generally achieved by labour and exertion, which have some sweetness in themselves ; as Solomon says, "Meat taken in hunting is sweet" [37].

[34] Pythag. *Aur. Vers.* v. 12. :—πάντων δὲ μάλιστ' αἰσχύνεο σαυτόν.

[35] Habakkuk, i. 15, 16.

[36] Cicero, *Pro Marcello*, c. 9. The quotation is inaccurate, though the meaning is preserved.

[37] In the *Colours of Good and Evil*, of which this tract is only an expansion, this sentence is given in Latin as here, but without any reference to Solomon. There are one or two of Solomon's proverbs to the same purpose, but none I think in these words. It was probably suggested to Bacon by something in Solomon, and turned into its present shape by himself. In after years, remembering where the thought came from, he may easily have forgotten that the expression was his own.—*J. S.*

ANSWER.

To these there are four opposing Sophisms, which incline to the contrary side, and may respectively serve as refutations to the former. The first is that felicity seems to be a kind of sign and character of the divine favour ; which both creates confidence and alacrity in ourselves, and wins obedience and respect from others. And this felicity extends to casual things, to which virtue hardly aspires ; as when Cæsar to encourage the pilot said, " You carry Cæsar and his fortune " [38] ; whereas if he had said, " You carry Cæsar and his virtue ", it would have been but cold comfort against the dangers of a storm. The second is that the deeds of virtue and industry are imitable and open to others ; whereas felicity is inimitable, and a kind of prerogative of the individual man. Hence we generally see that natural things are preferred to artificial, because they admit not of imitation ; for whatever is imitable is potentially common. The third is that things which come of felicity appear free gifts, bought without toil ; but things gained by our own virtue seem as paid for. Therefore Plutarch said elegantly, in comparing the actions of Timoleon, a man eminently fortunate, with those of his contemporaries Epaminondas and Agesilaus, " That they were like the verses of Homer, which, as they excel in other respects, so they seem to flow naturally, and as it were at the inspiration of genius [39]." The fourth is that which happens contrary to hope and expectation comes more gratefully and with greater pleasure to men's minds ; but this cannot be the case with things effected by our own care and exertion.

SOPHISM.

12. *That which consists of many divisible parts is greater than that which consists of few parts and is more one ; for all things when viewed part by part appear greater ; whence likewise plurality of parts makes a show of magnitude ; but it has a greater effect if the parts be without order ; for it produces a resemblance to infinity and prevents comprehension.*

The fallacy here is very palpable, even at first sight ; for it is not the plurality of parts alone, but the majority of them, which make the total greater. But yet this Sophism often carries away the imagination ; yea, and deceives the sense. For to the sight it appears a shorter distance on a dead level, where nothing intervenes to break the view, than when there are trees and buildings or some other mark to divide and measure the space. So again when a great monied man has divided and distributed his chests and bags, he seems to himself richer than he was. So likewise in amplifications, the effect is increased if the whole be divided into many parts and each be handled separately. And if this be done without order and promiscuously, it fills the imagination still more ; for confusion gives an impression of multitude ; inasmuch as things set forth and laid out in order, both appear more limited in themselves, and make it evident that nothing has been omitted ; whereas things that are presented confusedly are not only thought to be numerous in themselves, but leave room for suspicion that there are many more behind.

ANSWER.

This Sophism deceives, first when a man is prepossessed with an opinion that a thing is greater than it really is. For then the distribution thereof will destroy that false opinion, and show it in its true dimensions, without amplification. And therefore if a man be in sickness or pain, the hours will seem longer without a clock or an hour-glass than with it.. For if the weariness and pain of disease makes time appear longer than it really is, then the computation of time corrects the error, and makes it appear shorter than had been conceived by the false opinion. So again in the dead plain, the contrary to that which I said just now sometimes happens. For though at first the eye represents the distance to the sense as shorter, because it is undivided ; yet if that give an impression of a much shorter distance than it is afterwards found to be, the disappointment of that false opinion will make it appear longer than it really is. Therefore if a man have an over great opinion of anything and you wish to make it still greater, you must

[38] Plutarch, *De Fortunâ Roman.* p. 319. [39] Plutarch. in Timol. c. 36.

beware of distributions, but extol it in the whole. The Sophism deceives secondly, when the distribution is distracted and scattered, and does not meet or strike the eye at one glance. Thus if flowers in a garden be divided into many beds, they will give the appearance of a greater number than if they were all growing in one bed, provided that all the beds can be seen at once; for otherwise the union will have more effect than the scattered distribution. So again men's revenues seem greater when their farms and properties lie near and contiguous; for if they lie scattered they do not so easily come under view. The Sophism deceives, thirdly, by reason of the superiority of unity to multitude. For all piecing together of things is a sure sign of poverty in the pieces; where it comes to that,

Et quæ non prosunt singula, multa juvant[40].

Therefore Mary's was the better part,—"Martha, Martha, thou art busy about many things, one thing sufficeth[41]". Hence the fable in Æsop of the fox and the cat. For the fox boasted how many tricks and shifts he had to escape the hounds; but the cat said she had only one help to rely on; which was the poor faculty of climbing a tree; yet this was a far better protection than all the fox's tricks; whence the proverb, "The fox knows many tricks, but the cat one good one[42]". And in the moral signification of this fable we see the same thing. For the support of a powerful and faithful friend is a surer protection than all manner of plots and tricks.

These then shall suffice for an example. I have by me indeed a great many more Sophisms of the same kind, which I collected in my youth; but without their illustrations and answers, which I have not now the leisure to perfect; and to set forth the naked colours without their illustrations (especially as those above given appear in full dress) does not seem suitable. Be it observed in the meantime that this matter, whatever may be thought of it, seems to me of no small value; as that which participates of Primary Philosophy, of Politics, and of Rhetoric. And so much for the Popular Signs or Colours of Apparent Good and Evil, both simple and comparative.

The second Collection, which belongs to the *Promptuary* or Preparatory Store, is that to which Cicero alludes (as I said above in treating of Logic), where he recommends the orator to have commonplaces ready at hand, in which the question is argued and handled on either side: such as "for the letter of the law", "for the intention of the law," etc. But I extend this precept to other cases; applying it not only to the judicial kind of oratory, but also to the deliberative and demonstrative. I would have in short all topics which there is frequent occasion to handle (whether they relate to proofs and refutations, or to persuasions and dissuasions, or to praise and blame) studied and prepared beforehand; and not only so, but the case exaggerated both ways with the utmost force of the wit, and urged unfairly, as it were, and quite beyond the truth. And the best way of making such a collection, with a view to use as well as brevity, would be to contract those commonplaces into certain acute and concise sentences[43]; to be as skeins or bottoms of thread which may be unwinded at large when they are wanted. Some such piece of diligence I find in Seneca[44], but in hypotheses or cases. A few instances of the thing, having a great many by me, I think fit to propound by way of example. I call them *Antitheses of Things*[45].

40 Ovid. *Rem. Amor.* 420:—Things of no good separate, are useful together.

41 St. Luke, x. 41, 42.

42 Cf. Erasmus, *Adag.* i. 5. 18.

43 The habit of reducing arguments into this form accounts probably for the difficulty of verifying many of Bacon's quotations. The form fittest for the *promptuaria* was the form easiest to remember and most convenient to use.—*J. S.*

44 The Seneca here referred to is M. Annæus Seneca, the rhetorician, who is supposed to have been the uncle of L. Annæus Seneca, the preceptor of Nero.

45 Of these Antitheta many are Bacon's own, and are to be found in other parts of his writings; others are doubtless quotations, of which I shall mention some, though many more might probably be easily pointed out. [A great many of them will be found in the *Essays*.—*J. S.*]

Examples of *Antitheses.*

I. Nobility.

For.

They whose virtue is in the stock cannot be bad even if they would.

Nobility is the laurel with which Time crowns men.

We reverence antiquity even in dead monuments; how much more in living ones?

If you regard not nobility of birth, where will be the difference between the offspring of men and brutes?

Nobility withdraws virtue from envy, and makes it gracious.

Against.

Seldom comes nobility from virtue; seldomer virtue from nobility.

Noblemen have to thank their ancestors for pardon oftener than for advancement.

New men are commonly so diligent, that noblemen by their side look like statues.

Noblemen look behind them too often in the course; the mark of a bad runner.

II. Beauty.

For.

Deformed persons commonly take revenge on nature.

Virtue is nothing but inward beauty; beauty nothing but outward virtue.

Deformed persons seek to rescue themselves from scorn—by malice.

Beauty makes virtues shine, and vices blush.

Against.

Virtue is like a rich stone, best plain set.

As a fair garment on a deformed body, such is beauty in a bad man.

They that are beautiful and they that are affected by beauty are commonly alike light.

III. Youth.

For.

First thoughts and young men's counsels have more ot divineness.

Old men are wiser for themselves, not so wise for others and for the commonwealth.

Old age, if it could be seen, deforms the mind more than the body.

Old men are afraid of everything except the Gods.

Against.

Youth is the seedbed of repentance.

There is implanted in youth contempt for the authority of age; so every man must grow wise at his own cost.

The counsels to which Time is not called, Time will not ratify.

In old men the Loves are changed into the Graces.

IV. Health.

For.

The care of health humiliates the mind and makes it the beggar of the body.

A healthy body is the soul's host, a sick body her gaoler.

Nothing forwards the conclusion of business so much as good health; weak health on the contrary takes too many holidays.

Against.

Often to recover health, is often to renew youth.

Ill health is a good excuse for many things; which we are glad to use even when well.

Good health makes too close an alliance between the soul and the body.

Great empires have been governed from bed, great armies commanded from the litter.

V. Wife and Children.

For.

Love of his country begins in a man's own house.

Against.

He that has wife and children has given hostages to fortune.

A wife and children are a kind of discipline of humanity; whereas unmarried men are harsh and severe.

To be without wife or children is good for a man only when he wants to run away.

He who begets not children, sacrifices to death.

They that are fortunate in other things are commonly unfortunate in their children; lest men should come too near the condition of Gods.

Man generates and has children; God creates and produces works.

The eternity of brutes is in offspring; of men, in fame, good deserts, and institutions.

Domestic considerations commonly overthrow public ones.

Some persons have wished for Priam's fortune, who survived all his children [46].

VI. RICHES.

For.

They despise riches who despair of them.

It is envy of riches that has made virtue a goddess.

While philosophers are disputing whether virtue or pleasure be the proper aim of life, do you provide yourself with the instruments of both.

Virtue is turned by riches into a common good.

Other goods have but a provincial command; only riches have a general one.

Against.

Of great riches you may have either the keeping, or the giving away, or the fame; but no use.

Do you not see what feigned prices are set upon little stones and such rarities, only that there may be some use of great riches?

Many men while they thought to buy everything with their riches, have been first sold themselves.

I cannot call riches better than the baggage of virtue; for they are both necessary to virtue and cumbersome.

Riches are a good handmaid but the worst mistress.

VII. HONOURS.

For.

Honours are the suffrages not of tyrants (as they are said to be), but of divine providence.

Honours make both virtues and vices conspicuous; therefore they are a spur to the one and a bridle to the other.

No man can tell how far his virtue will go unless honours give him a fair field.

Virtue, like all things else, moves violently to her place, calmly in her place; now the place of virtue is honour.

Against.

While we seek honours we lose liberty.

Honours commonly give men power over those things wherein the best condition is not to will, the next best not to can.

The rising to honours is laborious, the standing slippery, the descent headlong.

Great persons had need to borrow the opinions of the vulgar, to think themselves happy.

VIII. EMPIRE.

For.

The enjoyment of happiness is a great good; but the power of imparting it to others is a still greater.

Kings are not as men, but as the stars; for they have great influence both on individuals and on the times themselves.

To resist the vice-gerent of God is not treason, but a kind of theomachy.

Against.

How wretched to have nothing to desire, and everything to fear!

Kings are like the heavenly bodies, which have much veneration but no rest.

None of human condition is admitted to the banquets of the Gods unless it be in derision.

[46] The allusion is to Tiberius. See Sueton. in Tiber. c. 62.

IX. Praise, Reputation.

For.

Praise is the reflexion of virtue.

Praise is the honour that comes by free votes.

Honours are conferred by many forms of government; but praise comes everywhere of liberty.

The voice of the people has something divine; else how could so many agree in one thing ?

Marvel not if the vulgar speak truer than the great, for they speak safer.

Against.

Fame is a worse judge than messenger.

What has a good man to do with the slaver of the common people ?

Fame is like a river, it bears up the light and lets the solid sink.

The lowest virtues are praised by the common people, the middle are admired; but of the highest they have no sense of perception.

Praise is won by ostentation more than by merit, and follows the vain and windy more than the sound and real.

X. Nature.

For.

Custom advances in an arithmetical ratio, nature in a geometrical.

As common laws are to customs in states, such is nature to custom in individuals.

Custom against nature is a kind of tyranny, and is soon and upon slight occasions overthrown.

Against.

We think according to our nature, speak as we have been taught, but act as we have been accustomed.

Nature is a schoolmaster, custom a magistrate.

XI. Fortune.

For.

Overt and apparent virtues bring forth praise; secret and hidden virtues bring forth fortune.

Virtues of duty bring forth praise; virtues of ability bring forth fortune.

Fortune is like the Milky Way; a cluster of obscure virtues without a name.

Fortune is to be honoured if it be but for her daughters, Confidence and Authority.

Against.

The folly of one man is the fortune of another.

The best that can be said of fortune is that, as she uses no choice in her favours, so she does not care to uphold them.

Great men, to decline the envy of their own virtues, turn worshippers of fortune.

XII. Life.

For.

It is absurd to prefer the accidents of life to life itself.

A long course is better than a short one for everything, even for virtue.

Without a good space of life a man can neither finish, nor learn, nor repent.

Against.

Philosophers in making such preparations against death make death itself appear more fearful.

Men fear death, as children fear to go into the dark, because they know not what is there.

There is no human passion so weak but if it be a little roused it masters the fear of death.

A man might wish to die, though he were neither brave nor miserable nor wise, merely from weariness of being alive[47].

[47] Seneca, *Ep.* 77.

XIII. SUPERSTITION.

For.

They that err from zeal, though we cannot approve them, yet we must love them.

Mediocrities belong to matters moral; extremities to matters divine.

The religious man is called superstitious. I had rather believe the most monstrous fables that are to be found in any religion, than that this world was made without a deity.

Against.

As the likeness of an ape to a man makes him all the more ugly, so does the likeness of superstition to religion.

Look how hateful affectation is in human affairs, so hateful is superstition in divine.

Better have no opinion of God at all than an injurious one.

It was not the Epicureans but the Stoics that troubled the ancient states.

There is no such thing as a mere atheist in opinion; but great hypocrites are the true atheists, who are ever handling holy things without reverencing them.

XIV. PRIDE.

For.

Pride is unsociable to vices among other things; and as poison by poison, so not a few vices are expelled by pride.

The good-natured man is subject to other men's vices as well as his own; the proud man to his own only.

Let pride go a step higher, and from contempt of others rise to contempt of self, and it becomes philosophy.

Against.

Pride is the ivy that winds about all virtues and all good things.

Other vices do but thwart virtues; only pride infects them.

Pride lacks the best condition of vice—concealment.

The proud man while he despises others neglects himself.

XV. INGRATITUDE.

For.

The crime of ingratitude is nothing more than a clear insight into the cause of a benefit conferred.

In our desire to show gratitude to certain persons we sacrifice both the justice we owe to others and the liberty we owe to ourselves.

Before we are called on to be grateful for a benefit, let us be sure as to the value of it.

Against.

The crime of ingratitude is not restrained by punishments, but given over to the Furies.

The bonds of benefits are stricter than the bonds of duties; wherefore he that is ungrateful is unjust and every way bad.

This is the condition of humanity no man is born in so public a fortune but he must obey the private calls both of gratitude and revenge.

XVI. ENVY.

For.

It is natural for a man to hate that which reproaches to him his own fortunes.

Envy in commonwealths is a wholesome kind of ostracism.

Against.

Envy keeps no holidays.

Nothing but death can reconcile envy to virtue.

Envy puts virtues to laborious tasks, as Juno did Hercules.

XVII. UNCHASTITY.

For.

It is owing to jealousy that chastity has been made a virtue.

Against.

Unchastity was the worst of Circe's transformations.

A man must be of a very sad disposition to think love a serious matter.

Why make a virtue of that which is either a matter of diet, or a show of cleanliness, or the child of pride ?

Loves are like wildfowl ; there is no property in them, but the right passes with the possession.

He that is unchaste is without all reverence for himself, which is the bridle of all vices.

All who like Paris prefer beauty, quit like Paris wisdom and power.

It was no vulgar truth that Alexander lighted on, when he said that sleep and lust were earnests of death.

XVIII. Cruelty.

For.

None of the virtues has so many crimes to answer for as clemency.

Cruelty, if it proceeds from revenge, is justice, if from danger, prudence.

He that has mercy on his enemy has no mercy on himself.

Bloodlettings are not oftener necessary in medicine than executions in states.

Against.

To delight in blood, one must be either a wild beast or a Fury.

To a good man cruelty always seems fabulous, and some tragical fiction.

XIX. Vain-Glory.

For.

He that would procure praise for himself must procure the benefit of other men.

He who is so sober that he cares for nothing that is not his own business, I fear he thinks the good of the public to be no business of his.

Dispositions that have in them some vanity are readier to undertake the care of the commonwealth.

Against.

Vain-glorious persons are ever factious, liars, inconstant, extreme.

Thraso is Gnatho's prey [48].

It is a shame for the suitor to woo the waiting-woman, and praise is the waiting-woman to virtue.

XX. Justice.

For.

Kingdoms and governments are but accessories to justice ; for there would be no need of them if justice could be carried on without.

It is owing to justice that man is a god to man, and not a wolf.

Justice though it cannot extirpate vices, yet prevents them from doing hurt.

Against.

If to be just be not to do that to another which you would not have another do to you, then is mercy justice.

If everyone has a right to his own, surely humanity has a right to pardon.

What tell you me of equal measure, when to the wise man all things are equal ?

Consider the condition of accused persons among the Romans, and conclude that justice is not for the good of the common wealth.

The ordinary justice of governments is but as a philosopher in the court—it merely conduces to the reverence of those who govern.

XXI. Fortitude.

For.

Nothing is to be feared except fear itself.

Against.

A noble virtue, to be willing to die yourself in order to kill another !

[48] The allusion is to the *Eunuchus* of Terence.

There is nothing either solid in pleasure, or secure in virtue, where fear intrudes.

He that looks steadily at dangers that he may meet them, sees also how he may avoid them.

Other virtues free us from the domination of Vice, Fortitude only from the domination of fortune.

A noble virtue, which a man may acquire by getting drunk!

He that is prodigal of his own life is dangerous to other men's.

Fortitude is the virtue of the iron age.

XXII. Temperance.

For.

The power of abstinence is not much other than the power of endurance.

Uniformity, concord, and measured motion, are attributes of heaven and characters of eternity.

Temperance is like wholesome cold; it collects and braces the powers of the mind.

Exquisite and restless senses need narcotics; so do passions.

Against.

I like not these negative virtues; for they show innocence and not merit.

The mind grows languid that has no excesses.

I like those virtues which induce excellence of action, not dullness of passion.

If you will have the motions of the mind all consonant, you must have them few—for it is a poor man that can count his stock.

To abstain from the use of a thing that you may not feel the want of it, to shun the want that you may not fear the loss of it, are precautions of pusillanimity and cowardice.

XXIII. Constancy.

For.

Constancy is the foundation on which virtues rest.

Wretched is the man who knows not what himself may become.

Human judgment is too weak to be true to the nature of things, let it then at least be true to itself.

Even vices derive a grace from constancy.

If inconstancy of mind be added to the inconstancy of fortune, in what darkness do we live?

Fortune is like Proteus; if you persevere she returns to her shape.

Against.

Constancy is like a surly porter; it drives much useful intelligence from the door.

It is fit that constancy should bear adversity well, for it commonly brings it on.

The shortest folly is the best.

XXIV. Magnanimity.

For.

If the mind do but choose generous ends to aim at, it shall have not only the virtues but the deities to help.

Virtues induced by habit or by precepts are ordinary; those imposed by a virtuous end are heroical.

Against.

Magnanimity is a poetical virtue.

XXV. Knowledge, Contemplation.

For.

That pleasure is indeed according to nature, of which there is no satiety.

What prospect so sweet as to look down upon the errors of other men?

How good a thing to have the motion of the mind concentric with the universe.

All depraved affections are but false estimations; and goodness and truth are the same thing.

Against.

Contemplation is a specious idleness.

Good thoughts are little better than good dreams.

Providence takes care of the world; do thou take care of thy country.

A politic man uses his very thoughts for seed.

XXVI. Learning.

For.

If books were written about small matters, there would be scarce any use of experience.

In reading a man converses with the wise, in action generally with fools.

Sciences which are of no use in themselves are not to be deemed useless, if they sharpen the wit and put the thoughts in order.

Against.

In colleges men learn to believe.

What art ever taught the seasonable use of art?

To be wise by rule and to be wise by experience are contrary proceedings; he that accustoms himself to the one unfits himself for the other.

Art is often put to a foolish use, that it may not be of no use at all.

Almost all scholars have this—when anything is presented to them, they will find in it that which they know, not learn from it that which they know not.

XXVII. Promptitude.

For.

Wisdom that comes not quick comes not in season.

He that quickly errs quickly amends his error.

He that is wise in deliberation and not upon the moment does no great matters.

Against.

The wisdom that is ready at hand does not lie deep.

Wisdom is like a garment, it must be light if it be for speed.

He whose counsels are not ripened by deliberation, his wisdom will not ripen with age.

Things speedily devised speedily fall out of favour.

XXVIII. Silence in matters of Secrecy.

For.

The silent man hears everything, for everything can be safely communicated.

He that is apt to tell what he knows, is apt to tell also what he knows not.

Mysteries are due to secrecy.

Against.

The best way of keeping the mind secret is to vary the manners.

Silence is the virtue of a confessor.

The silent man has nothing told him, because he gives nothing but silence in exchange.

To be close is next to being unknown.

XXIX. Facility.

For.

I love the man who yields to others' feelings, and yet keeps his judgment

Against.

Facility is a foolish privation of judgment.

free.

To be pliant is to be most like gold.

Favours received from a man of facile disposition pass for debts; denials for injuries.

He that obtains a favour from a man of facile disposition thanks himself for it.

The facile man is oppressed with all difficulties, for he involves himself in all.

The facile man seldom gets out of it without a blush.

XXX. Popularity.

For.

Wise men are commonly pleased with the same things; but to meet the various inclinations of fools is the part of wisdom.

To court the people is to be courted by the people.

Men that are themselves great find no single person to respect, but only the people.

Against.

He who agrees very well with fools may himself be suspected.

He that pleases the mob is apt to raise a mob.

Nothing that is moderate is liked by the common people.

The lowest of all flatteries is the flattery of the common people.

XXXI. Loquacity.

For.

He that is silent betrays want of confidence either in others or in himself.

All kinds of constraint are unhappy, that of silence is the most miserable of all.

Silence is the virtue of a fool, And therefore it was well said to a man that would not speak, "If you are wise you are a fool; if you are a fool, you are wise"[49].

Silence, like night, is convenient for treacheries.

Thoughts are wholesomest when they are like running waters.

Silence is a kind of solitude.

He that is silent lays himself out for opinion.

Silence neither casts off bad thoughts nor distributes good.

Against.

Silence gives to words both grace and authority.

Silence is the sleep which nourishes wisdom.

Silence is the fermentation of thought.

Silence is the style of wisdom.

Silence aspires after truth.

XXXII. Dissimulation.

For.

Dissimulation is a compendious wisdom.

We are not bound to say the same thing, but to aim at the same end[50].

Against.

If we cannot think according to the truth of things, let us at least speak according as we think.

When arts of policy are beyond a

[49] This sarcasm is ascribed by Diogenes Laertius and Plutarch to Theophrastus, the author of the *Characters* (which form the foundation of those of La Bruyère) and of many other works. It has also been ascribed to Simonides. Bacon seems to have taken it from Plutarch.

[50] *Non idem dicere, sed idem spectare, debemus:* a sentence in which I suspect that there is either some misprint or some inaccuracy of expression.—*J. S.*

Nakedness is uncomely in the mind as well is in the body.

Dissimulation is both a grace and a guard.

Dissimulation is the fence of counsels.

There are some for whom it is good to be deceived.

He that does everything without dissimulation is not the less a deceiver; for most people either do not understand him or do not believe him.

Want of dissimulation is nothing but want of power over the mind.

man's capacity, dissimulation must serve him for wisdom.

He that dissembles deprives himself of a principal instrument of action, namely trust and belief.

Dissimulation invites dissimulation.

He that dissembles is not free.

XXXIII. Boldness.

For.

He that shows diffidence invites reproof.

What action is to an orator boldness is to a politician,—the first requisite, the second, and the third.

I love a confessing modesty, hate an accusing one [51].

Confidence of manners brings minds the sooner together.

I like a reserved countenance and an open speech.

Against.

Boldness is the pioneer of folly.

Impudence is of no use except for imposture.

Confidence is the mistress of fools, and the sport of wise men.

Boldness is dullness of the sense joined with malice of the will.

XXXIV. Ceremonies, Puntos, Affectation.

For.

A decorous govèrnment of the countenance and carriage is the true seasoning of virtue.

We comply with the vulgar in our words, why not in habit and gesture?

He that does not preserve decorum in trifles and daily habits may be a great man; but be sure of this, such a man is not wise at all hours.

Virtue and wisdom without forms are like foreign languages; for they are not intelligible to the common people.

He that knows not the sense of the common people by an inward congruity, if he know it not by outward observation either, is of all men the most foolish.

Forms of behaviour are the translation of virtue into vernacular.

Against.

What more uncomely than to mak life a piece of acting?

From ingenuousness comes grac from artifice hatred.

Better painted cheeks and curle hair than painted and curled manner

He that applies his mind to suc small observations, is not capable (great thoughts.

Affectation is the shining putr faction of ingenuousness [52].

XXXV. Jests.

For.

A jest is the orator's altar.

He that throws into everything a

Against.

Who does not despise these hunter after deformities and prettinesses

[51] *Amo confitentem verecundiam, accusantem odi.* I do not understand this sentence —*J. S.* [52] The same image occurs in Ralegh's *Lye*:—

"Go tell the Court it glows
And shines like rotten wood."

dash of modest pleasantry keeps his mind the more at liberty.

To pass easily from jest to earnest and from earnest to jest is a thing more politic than men suppose.

A jest is many times the vehicle of a truth which would not otherwise have been brought in.

It is a dishonest trick to wash away with a jest the real importance of things.

Consider jests when the laugh is over.

These wits hardly penetrate below the surface of things, where jests ever lie.

Where a jest has any weight in serious matters, it is a childish levity.

XXXVI. Love.

For.

See you not that all men seek themselves ? But it is only the lover that finds himself.

There is nothing which better regulates the mind than the authority of some powerful passion.

If you are wise, seek something to desire ; for to him who has not some special object of pursuit all things are distasteful and wearisome [53].

Why should not one be content with one ?

Against.

The stage is much beholden to love, life not at all.

Nothing has so many names as love ; for it is a thing either so foolish that it does not know itself, or so foul that it hides itself with paint.

I hate those men of one thought.

Love is a very narrow contemplation.

XXXVII. Friendship.

For.

Friendship does the same things as fortitude, but more sweetly.

Friendship is a sweet seasoning to all other blessings.

It is the worst solitude, to have no true friendships.

It is a retribution worthy of bad faith to be deprived of friendships.

Against.

He that contracts close friendships imposes upon himself new necessities.

It is the part of a weak mind to go shares in fortune.

XXXVIII. Flattery.

For.

Flattery proceeds more from manners than malice.

To suggest what a man should be, under colour of praising what he is, was ever a form due in civility to the great.

Against.

Flattery is the style of slaves.

Flattery is the refuse of vices.

The flatterer is like the fowler that deceives birds by imitating their cry.

The unseemliness of flattery is matter of comedy, its mischief of tragedy.

Nothing so hard to cure as the ear.

XXXIX. Revenge.

For.

Revenge is a kind of wild justice.

He who requites violence with violence, sins against the law but not against the man.

The fear of private revenge is a useful thing; for laws too often sleep.

Against.

He that did the first wrong made a beginning of mischief, he that returned it made no end.

The more natural revenge is, the more need to restrain it.

He that is ready to return an injury was behindhand more in time perhaps than in will.

[53] Ovid. *Amores*, i. 9, 46. The line occurs in Bacon's *Promus*.—*J. S.*

XL. Innovation.

For.

Every medicine is an innovation.

He that will not have new remedies will have new evils.

Time is the greatest innovator, why then should we not imitate time ?

Ancient precedents are unfit, modern ones corrupt and interested.

Leave it to the unskilful and the contentious to act by precedent.

As those who first bring honour into their family are commonly worthier than their descendants, so are the first precedents commonly better than the imitations of them.

A froward retention of custom is as turbulent a thing as an innovation.

Seeing that things alter of themselves to the worse, if counsel shall not alter them to the better, what shall be the end ?

The slaves of custom are the sport of time.

Against.

Things new born are ill-shapen.

The only author I like is time.

There is no novelty that does not some hurt, for it unsettles what is.

Things settled by custom, though they be not good, yet at least they fit one with another.

What innovator imitates time, who so insinuates his innovations that they are not perceived ?

That which comes unlooked for gets the less thanks from him whom it helps, and gives the more annoyance to him whom it hurts.

XLI. Delay.

For.

Fortune sells many things to him that is in a hurry, which she gives to him that waits.

While we hasten to take hold of the beginnings of things, we grasp shadows.

While things are wavering, watch ; when they have taken their direction, act.

Commit the beginnings of actions to Argus, the end to Briareus.

Against.

Opportunity offers the handle of the bottle first, and afterwards the belly.

Opportunity is like the Sibyl ; she raises the price as she diminishes the offer.

Speed is Pluto's helmet.

Things that are done betimes are done with judgment ; things that are put off too late, by circuit [54].

XLII. Preparation.

For.

He that attempts a great matter with small means, does but provide himself with an occasion of hoping.

With small preparations you may purchase wisdom, but not fortune.

Against.

The time to cease preparing is the instant you can begin acting.

Let no man hope that he can bind fortune by preparation.

To interchange preparation and action is politic, to part them is vain and unfortunate.

Great preparation wastes both time and matter.

XLIII. Meeting the First Move.

For.

More dangers have deceived men than forced them.

It is less trouble to apply the remedy

Against.

He that arms himself to meet danger teaches it to come on, and in remedying fixes it.

[54] *Per ambitum :* meaning, I suppose (if the reading be correct), that at first you can choose the best way, but at last you must take the way that offers.—*J. S.*

to a danger than to keep watch upon the approach of it.

A danger is no more light, if it once seem light.

The very remedies of dangers carry little dangers in them.

It is better to have to deal with a few dangers in their maturity, than with the menaces of every one.

XLIV. VIOLENT COUNSELS.

For.

For those who embrace this mild kind of wisdom an increase of the evil is salutary.

Necessity, which gives violent counsels, also executes them.

Against.

Every violent remedy is pregnant with some new evil.

The only violent counsellors are anger and fear.

XLV. SUSPICION.

For.

Distrust is the sinews of wisdom, but suspicion is a medicine for the joints.

His faith is justly suspected whose faith suspicion shakes.

Suspicion loosens a frail faith, but braces a strong one.

Against.

Suspicion discharges faith.

The distemper of suspicions is a kind of civil madness.

XLVI. THE WORDS OF THE LAW.

For.

The interpretation which departs from the letter is not interpretation but divination.

When the letter is departed from, the judge becomes the law-giver.

Against.

The sense according to which each word is to be interpreted must be gathered from all the words together.

The worst tyranny is the torturing of the law.

XLVII. FOR WITNESSES AGAINST ARGUMENTS.

For.

He who relies on arguments decides according to the merits of the pleader, not of the cause.

He who believes arguments more than witnesses, ought to give more credit to the wit than the senses.

Arguments might be trusted, if men never acted absurdly.

Arguments, when opposed to testimony, may make a fact seem strange, but cannot make it seem not a fact.

Against.

If witnesses are to be believed in spite of arguments, it is enough if the judge be not deaf.

Arguments are the antidote against the poison of testimony.

It is safest to believe those proofs which seldomest lie.

These Antitheses (which I have here set down) are perhaps of no great value; but as I had long ago prepared and collected them, I was loth to let the fruit of my youthful industry perish—the rather because (if they be carefully examined) they are *seeds* only, not *flowers*. In one respect indeed they savour altogether of youth, there being plenty of them in the moral and demonstrative kind, but in the deliberative and judicial very few.

The third Collection, which belongs to the Promptuary, or Preparatory Store, and is likewise deficient, is that of what I call *Lesser Forms* [55]. I mean those parts of speech which answer to the vestibules, back doors, ante-chambers,

[55] The *Promus* contains some of these formulæ.

withdrawing-chambers, passages etc., of a house; and may serve indiscriminately for all subjects. Such are prefaces, conclusions, digressions, transitions, intimations of what is coming, excusations, and a number of the kind. For as in buildings it is a great matter both for pleasure and use that the fronts, doors, windows, approaches, passages, and the like be conveniently arranged, so also in a speech these accessory and interstitial passages (if they be handsomely and skilfully fashioned and placed) add a great deal both of ornament and effect to the entire structure. Of these Forms I will subjoin one or two examples, without dwelling longer upon them. For though they be matters of no small use, yet as I have nothing of my own to add in this part, but merely transcribe the naked forms out of Demosthenes or Cicero or some other chosen author, they are not of that importance that I should spend time upon them.

Examples of *Lesser Forms.*

A CONCLUSION IN A DELIBERATIVE.

So may we redeem the fault passed and at the same time prevent the inconveniences to come.

COROLLARY OF AN ACCURATE DISTRIBUTION.

That all may know that I have no wish either to evade anything by silence or to obscure it by speech [56].

A TRANSITION WITH A HINT.

Let us pass these things, and yet not without marking and turning back to look at them as we go by [57].

A FORM TO PREOCCUPY THE MIND AGAINST AN OPINION PREVIOUSLY FORMED.

I will make you understand in all this business how much is truth, how much error, and how much envy [58]. These few may be enough by way of example; and with these I conclude the Appendices to Rhetoric, which belong to the Promptuary.

CHAPTER IV.

Two General Appendices of the Art of Transmission; Critical and Pedagogical.

THERE remain two appendices touching the transmission of knowledge in general; the one Critical, the other Pedagogical. For as the principal part of transmission of knowledge consists in the writing of books, so the relative part thereof turns on the reading of books. Now reading is either directed by teachers, or attained by each man's own endeavours; and to this these two knowledges which I have mentioned appertain.

To the Critical part belongs, first, the true correction and amended edition of approved authors; whereby both themselves receive justice and their students light. Yet in this the rash diligence of some has done no little harm. For many critics, when they meet a passage which they do not understand, immediately suppose that there is a fault in the copy. As in that passage of Tacitus, where he relates that when a certain colony asserted before the senate the right of asylum, their arguments were not very favourably listened to by the emperor and the senate; whereupon the ambassadors, fearing for the success of their cause, gave a good sum of money to Titus Vinius to support them—by which means they prevailed. "Then" (says Tacitus) "the dignity and antiquity of the colony had its weight" [1]; meaning that the arguments which appeared light before gained

[56] Cic. *Pro. Cluent.* c. i. The quotation is inaccurate.

[57] Cic. *Pro. Sext.* c. 5.

[58] Cic. *Pro. Cluent.* c. 4.

[1] Cf. Tacitus, *Hist.* i. 66. The case is incorrectly stated. That Bacon had but an imperfect recollection of the passage, is plain from his substituting the name of Titus Vinius for that of Fabius Valens, and from his mentioning the senate, as if the trans-

fresh weight by the money. But a critic, and he not one of the worst, here erased the word *tum*, and substituted *tantum*. And this bad habit of critics has brought it to pass that (as some one has wisely remarked) "the most corrected copies are often the least correct". Moreover, to speak truly, unless critics be learned in the sciences which the books they edit treat of, their diligence is not without its danger.

Secondly, there belongs to the Critical part the interpretation and explication of authors,—commentaries, scholia, annotations, collections of beauties, and the like. In labours of this kind however some of the critics have been visited with that very bad disease, of leaping over many of the obscurer places, while they linger and expatiate to tediousness on those which are clear enough; as if the object were not so much to illustrate the author as to display on every possible opportunity the extensive learning and various reading of the critic himself. It were especially to be desired (though this is a matter which belongs rather to the art of transmission in the main, than to the appendices thereof) that every writer who handles arguments of the obscurer and more important kind, should himself subjoin his own explanations; that so the text may not be interrupted by digressions and expositions, and the notes may not be at variance with the writer's meaning. Something of the kind I suspect in Theon's Commentary on Euclid [2].

There belongs thirdly to the Critical part (and from this indeed it derives its name) the insertion of some brief judgment concerning the authors edited, and comparison of them with such other writers on the same subjects; that students may by such censure be both advised what books to read and better prepared when they come to read them. This last office is indeed, so to speak, the critic's chair; which has certainly in our age been ennobled by some great men,—men in my judgment above the stature of critics.

As for the Pedagogical part, the shortest rule would be, "Consult the schools of the Jesuits"; for nothing better has been put in practice. Nevertheless I will as usual give a few hints, gleaning an ear here and there. I am clearly in favour of a collegiate education for boys and young men; not in private houses, nor merely under schoolmasters. For in colleges there is a greater emulation of the youths amongst themselves; there is also the sight and countenance of grave men, which tends to modesty, and forms their young minds from the very first after that model; and in short there are very many advantages in a collegiate education. For the order and manner of teaching, I would say first of all,—avoid abridgments and a certain precocity of learning, which makes the mind over bold, and causes great proficiency rather in show than in fact. Also let some encouragement be given to the free exercise of the pupils' minds and tastes; I mean, if any of them, besides performing the prescribed exercises, shall steal time withal for other pursuits to which he is more inclined, let him not be checked. Observe moreover (what perhaps has not hitherto been remarked) that there are two ways of training and exercising and preparing the mind, which proceed in opposite directions. The one begins with the easier tasks, and so leads on gradually to the more difficult; the other begins by enforcing and pressing the more difficult, that when they are mastered the easier ones may be performed with pleasure. For it is one method to begin swimming with bladders, which keep you up; and another to begin dancing with heavy shoes, which weigh you down. Nor is it easy to tell how much a judicious intermixture of these methods helps to advance the faculties of the mind and body. Again, the application

action had taken place at Rome. It was by a donative to the soldiery that the colony of Vienna was saved, not (directly at least) by a bribe to their leader; though Tacitus adds that it was believed that he also had been bought over,—"ipsum Valentem magnâ pecuniâ emptum".—*Hist.* i. 66.

[2] It seems probable that this remark, showing a kind of reading with which Bacon does not seem to have been familiar (vide suprà p. 476), was derived from his friend Sir Henry Savile. We find Theon's services in relation to Euclid's *Elements* depreciatingly spoken of in Saville's *Prælectiones tresdecim in Principium Elementorum Euclidis* (1621), pp. 12, 13.

and choice of studies according to the nature of the mind to be taught, is a matter of wonderful use and judgment; the due and careful observation whereof is due from the masters to the parents, that they may be able to advise them as to the course of life they should choose for their sons. And herein it should be carefully observed, that as a man will advance far fastest in those pursuits to which he is naturally inclined, so with respect to those for which he is by defect of nature most unsuited there are found in studies properly chosen a cure and remedy for his defects. For example, if one be bird-witted, that is, easily distracted and unable to keep his attention as long as he should, Mathematics provides a remedy; for in them if the mind be caught away but a moment, the demonstration has to be commenced anew. Exercises, again, it is obvious, play the principal part in instruction. But few have observed that there ought to be not only a wise choice and course of exercises, but a wise intermission of them also; for it is well observed by Cicero, "that men in their exercises for the most part exercise their faults as well as their faculties"[3], so that an ill habit is sometimes acquired along with the good. It is safer therefore to intermit exercises from time to time and return to them after a while, than continually to pursue and press them. But enough of this. Certainly these are matters not very grand or imposing at first sight, yet of singular fruit and efficacy. For as the good or ill thriving of plants depends chiefly upon the good or ill treatment they received when they were young and tender; and as the immense increase of the Roman empire is by some deservedly attributed to the virtue and wisdom of the first six kings, who were in truth as the tutors and guardians of it in its infancy[4]; so surely the culture and ordering of youthful or tender years has a power which though latent and not perceptible to everybody, neither length of time nor assiduity and earnestness of labour in mature age can afterwards countervail. It will not be amiss to observe also, that even mean faculties, when they fall into great men or great matters, sometimes work great and important effects. Of this I will adduce a memorable example; the rather, because the Jesuits appear not to despise this kind of discipline; therein judging (as I think) well. It is a thing indeed, if practised professionally, of low repute; but if it be made a part of discipline, it is of excellent use. I mean stage-playing: an art which strengthens the memory, regulates the tone and effect of the voice and pronunciation, teaches a decent carriage of the countenance and gesture, gives not a little assurance, and accustoms young men to bear being looked at. The example which I shall give, taken from Tacitus, is that of one Vibulenus, formerly an actor, then a soldier in the Pannonian legions. This man had at the death of Augustus raised a mutiny, whereupon Blæsus, the lieutenant, committed some of the mutineers to prison. The soldiers however broke in and let them out; whereupon Vibulenus getting up to speak, began thus; "These poor innocent wretches you have restored to light and life; but who shall restore life to my brother, or my brother to me? whom, being sent hither in message from the legions of Germany, to treat of the common cause, this man has murdered last night by some of his swordsmen, whom he keeps and arms for the execution of soldiers. Answer, Blæsus, where have you thrown his body? Enemies themselves deny not burial. When with kisses and tears I shall have satiated my grief, command me also to be slain beside him; only let these my fellows, seeing we are put to death for no crime, but because we consulted for the good of the legions, have leave to bury us"[5]. With which words he excited such excessive jealousy and alarm, that, had it not shortly afterwards appeared that nothing of the sort had happened, nay, that he had never had a brother, the soldiers would hardly have kept their hands off the prefect; but the fact was that he played the whole thing as if it had been a piece on the stage.

And now I am come to the end of my treatise concerning Rational Knowledges; wherein if I have sometimes made the divisions other than those that are received, yet let it not be thought that I disallow all those divisions which I do not use. For there is a double necessity imposed upon me of altering the divisions. First, because to reduce into one class things next in nature, and to gather into one

[3] Cic. *De Orator.* i. 33. [4] Macchiavelli, *Discorsi*, i. 19. [5] Tacit. *Ann.* i. 16–22.

bundle things wanted for use, are operations differing in the very end and intention. For as a secretary of a king or state, when he arranges his papers in his study or general cabinet, puts those things together, no doubt, which are of like nature,—treatises by themselves in one place, instructions by themselves in another, foreign letters, domestic letters, and the like, each apart by themselves, —but when on the contrary he arranges them in his boxes or particular cabinet, he puts those together which, though of different kinds, he thinks he will have occasion to use together; so in this general cabinet of knowledge it was necessary for me to make the divisions according to the nature of the things themselves, whereas if I had been to handle any particular knowledge I should have adopted the divisions fittest for use and practice. Secondly, because the introduction of the *Desiderata*, and the incorporation of them with the rest, involved as a consequence an alteration in the distribution of the existing sciences. For suppose (by way of demonstration) that the arts which we now have are as 15, and that the same with the desiderata added are as 20; I say that the factors of the number 15 are not the same with the factors of the number 20. For the factors of 15 are 3 and 5; the factors of 20 are 2, 4, 5, and 10. It is plain therefore that these things could not be otherwise. And so much for the Logical Sciences.

Book VII.

CHAPTER I.

The Division of Moral Knowledge into the Exemplar or Platform of Good, and the Georgics or Culture of the Mind. The Division of the Platform of Good, into Simple and Comparative Good. The Division of Simple Good into Individual Good, and Good of Communion.

We come now, most excellent king, to moral knowledge, which respects and considers the will of man. The will is governed by right reason, seduced by apparent good, having for its spurs the passions, for its ministers the organs and voluntary motions; wherefore Solomon says, "Above all things keep thy heart with all diligence, for out of it are the issues of life"[1]. In the handling of this science, the writers seem to me to have done as if a man who, professing to teach the art of writing, had exhibited only fair copies of letters, single and joined, without giving any direction for the carriage of the pen and framing of the characters. So have these writers set forth good and fair copies, and accurate draughts and portraitures of good, virtue, duty, and felicity, as the true objects for the will and desires of man to aim at. But though the marks themselves be excellent and well placed, how a man may best take his aim at them; that is, by what method and course of education the mind may be trained and put in order for the attainment of them, they pass over altogether, or slightly and unprofitably. We may discourse as much as we please that the moral virtues are in the mind of man by habit, and not by nature, and we may make a formal distinction that generous spirits are won by doctrines and persuasions, and the vulgar sort by reward and punishment; or we may give it in precept that the mind like a crooked stick must be straightened by bending it the contrary way[2] and the like scattered glances and touches; but they would be very far from supplying the place of that which we require.

The reason of this neglect I suppose to be that hidden rock whereupon both this and so many other barks of knowledge have struck and foundered; which is, that men have despised to be conversant in ordinary and common matters which are neither subtle enough for disputation, nor illustrious enough for ornament. It is hard to compute the extent of the evil thus introduced; namely, how from innate pride and vainglory men have chosen those subjects of discourse, and those methods of handling them, which rather display their own genius than benefit the reader. Seneca says well, "Eloquence is injurious to those whom it inspires with a fondness for itself, and not for the subject[3]"; for writings should be such as should make men in love with the lesson, and not with the teacher. They therefore are on the right path, who can say the same of their counsels as Demosthenes did of his, and conclude with this sentence, "If you do what I advise you will not only praise the orator at the time, but in no long time yourselves also, by reason of the better condition of your affairs"[4]. For myself, most excellent king, I may truly say that both in this present work, and in those I intend to publish hereafter, I often advisedly and deliberately throw aside the dignity of my name and wit (if such thing be) in my endeavour to advance human interests; and being one that should properly perhaps be an architect in philosophy and the sciences, I turn common labourer, hodman, anything that is wanted; taking upon myself the burden and execution of many

[1] Prov. iv. 23. [2] Arist. *Nic. Eth.* ii. 9.

[3] Seneca, *Epist.* 52. Seneca is speaking of the auditors of popular lecturers on philosophy. The only kind of applause which he would allow the lecturer to affect or the audience to bestow, is that of young men so stirred by the matter that they cannot refrain.—*J. S.* [4] Demosth. *Olynth.* ii.

things which must needs be done, and which others through an inborn pride shrink from and decline. But to return to the subject: moral philosophers have chosen for themselves a certain glittering and lustrous mass of matter, wherein they may principally glorify themselves for the point of their wit, or the power of their eloquence; but those which are of most use for practice, seeing that they cannot be so clothed with rhetorical ornaments, they have for the most part passed over.

Neither needed men of so excellent parts to have despaired of a fortune which the poet Virgil promised to himself, and indeed obtained; who got as much glory of eloquence, wit, and learning in the expressing of the observations of husbandry, as of the heroical acts of Æneas:

> Nec sum animi dubius, verbis ea vincere magnum
> Quam sit, et angustis his addere rebus honorem [5].

And surely, if the purpose be in good earnest, not to write at leisure that which men may read at leisure, but really to instruct and suborn action and active life, these Georgics of the Mind are no less worthy to be had in honour than the heroical descriptions of virtue, goodness, and felicity, whereon so much labour has been spent.

Wherefore I will divide moral knowledge into two principal parts; the one the *Exemplar* or *Platform of Good*," the other " the *Regiment* or *Culture of the Mind*," which I also call the *Georgics of the Mind*; the one describing the nature of good, the other prescribing rules how to accommodate the will of man thereunto.

The doctrine touching the platform or nature of good, considers good either Simple or Comparative; either the kinds of good, or the degrees of good; in the latter whereof those infinite disputations and speculations touching the supreme degree thereof, which they termed "Felicity," "Beatitude," or the "Highest Good" (which were as the heathen Divinity), are by the Christian faith removed and discharged. And as Aristotle says, "That young men may be happy, but only by hope "[6], so we, instructed by the Christian faith, must all acknowledge our minority, and content ourselves with that felicity which rests in hope.

Freed therefore happily, and delivered from this doctrine of the heathen heaven, whereby they certainly imagined a higher elevation of man's nature than it is really capable of (for we see in what height of style Seneca writes, "It is true greatness to have the frailty of a man and the security of a god"[7]), we may with more sobriety and truth receive the rest of what they have delivered concerning the doctrine of the Exemplar; wherein, for the nature of good Positive or Simple, they have painted it excellently and to the life, as in a picture, diligently representing the forms of virtues and duties, their situations and their postures, kinds, relations, parts, subjects, provinces, actions, administrations, and the like; nay further, they have commended and insinuated them into man's nature and spirit with great quickness of argument and beauty of persuasions; yea, and fortified and entrenched them, as much as discourse can do, against corrupt and popular opinions. Again, for the nature of Comparative Good, they have also excellently well handled it, in their triplicity of good [8]; in the comparison between a contemplative and active life; in the distinction between virtue with reluctation, and virtue settled and secured; in their encounters between honesty and profit; in their balancing of virtue with virtue, as to which outweighs the other, and the like; so that I find that this part is excellently laboured, and that the ancients have done their work admirably therein, yet so as the pious and earnest diligence of divines, which has been

[5] Virg. *Georg.* iii. 289:—

> How hard the task, alas, full well I know,
> With charms of words to grace a theme so low.

[6] Arist. *Nic. Eth.* i. 10.

[7] Seneca, *Epist.* 53.

[8] Namely the good which relates respectively to mind, body and estate. See Aristot. *Nicom. Eth.* i. 8. 2.

employed in weighing and determining duties, moral virtues, cases of conscience, the bounds of sin, and the like, has left the philosophers far behind.

Notwithstanding (to return to the philosophers), if before they had come to the popular and received notions of virtue and vice, pleasure and pain, and the rest, they had stayed a little longer upon the inquiry concerning the roots of good and evil, and the strings of those roots; they had given in my opinion a great light to those questions which followed; and especially if they had consulted with the nature of things, as well as moral axioms, they had made their doctrines less prolix, and more profound; which being by them in part omitted, and in part handled with much confusion, I will briefly resume; and endeavour to open and cleanse the fountains of morality, before I come to the knowledge of the culture of the mind, which I set down as deficient. For this will in my opinion reinforce the doctrine of the exemplar with new strength.

There is formed and imprinted in everything an appetite toward two natures of good; the one as everything is a total or substantive in itself, the other as it is a part or member of a greater body; whereof the latter is in degree the greater and the worthier, because it tends to the conservation of a more general form. The former of these may be termed "Individual or Self-good," the latter the "Good of Communion". Iron in particular sympathy moves to the loadstone, but yet if it exceed a certain quantity it forsakes its affection to the loadstone, and like a good patriot moves to the earth, which is the region and country of its connaturals; so again, compact and massy bodies move to the earth, the great collection of dense bodies; and yet rather than suffer a divulsion in nature and create a vacuum, they will move upwards from the centre of the earth, forsaking their duty to the earth in regard to their duty to the world. Thus it is ever the case, that the conservation of the more general form controls and keeps in order the lesser appetites and inclinations. This prerogative of the communion of good is much more engraven upon man, if he be not degenerate; according to that memorable speech of Pompey, when being in commission of purveyance for a famine at Rome, and being dissuaded with great vehemency and instance by his friends about him that he should not hazard himself to sea in an extremity of weather, he said only to them, "It is needful that I go, not that I live"[9], so that the love of life, which is the predominant feeling in the individual, did not with him outweigh affection and fidelity to the commonwealth. But why do I dwell on this point? for never in any age has there been any philosophy, sect, religion, law, or other discipline, which did so highly exalt the good which is communicative, and depress the good which is private and particular, as the Holy Christian Faith; well declaring that it was the same God who gave the Christian law to men, that gave also those laws of Nature to inanimate creatures; whence we read that some of the elected saints of God have wished, rather than that their brethren should not obtain salvation, that they themselves should be anathematized and erased out of the book of life, in an ecstasy of charity and infinite feeling of communion[10].

This being set down and strongly planted, judges and determines of some the most important controversies in moral philosophy. For first it decides the question touching the preferment of the contemplative or active life, and decides it against Aristotle. For all the reasons which he brings for the contemplative respect private good, and the pleasure or dignity of a man's self; in which

9 Plut. in Pomp. c. 50.

10 St. Paul, Romans, ix. 3; and Exod. xxxii. 32. Bacon here touches on what theologians call the conditional sacrifice of salvation—a matter frequently referred to in the unhappy controversy between Bossuet and Fénelon. The 33rd of the Articles of Issy, which they both signed, sanctions the notion of this conditional sacrifice. It appears, however, that the article in question was one of the four added at Fénelon's suggestion to Bossuet's original draft, and that the latter did not consent without reluctance to its introduction. Fénelon's own views on the subject are developed in his *Instruction Pastorale*, etc., sec. 10, and elsewhere. St. Chrysostom, according to a passage quoted by Fénelon, disapproved greatly of those who held that St. Paul speaks merely of temporal death.

respects no question the contemplative life has the pre-eminence, being not much unlike that comparison which Pythagoras made for the gracing and magnifying of philosophy and contemplation; who, being asked by Hiero what he was, answered, "that if Hiero were ever at the Olympian games, he knew the manner, that some came to try their fortune for the prizes; and some came as merchants to utter their commodities; and some came to make good cheer, and meet their friends; and some came to look on; and that he was one of them that came to look on" [11]. But men must know that in this theatre of man's life it is reserved only for God and Angels to be lookers on [12]; neither could the like question ever have been raised in the Church (notwithstanding it has been in the mouths of many, "Right dear in the sight of the Lord is the death of his Saints" [13], by which text they used to exalt that civil death of theirs, and the orders and rules of the life monastic); were it not true withal that the monastical life is not simply contemplative, but engaged also in the performance of certain ecclesiastical duties, such as continual prayer, and votive sacrifices offered to God, and the leisurely writing of theological books for advancing the knowledge of the divine law; as Moses did, when he abode so long in the Mount. And so we see, that Enoch, the seventh from Adam, who seems to have been the first contemplative (for he is said to have walked with God [14]), yet also endowed the Church with a book of prophecy, which St. Jude cites [15]. But for mere contemplation which should be finished in itself without casting beams of heat and light upon society, assuredly divinity knows it not. It decides also the question so earnestly argued between the schools of Zeno and Socrates on the one hand, who placed felicity in virtue simple or attended, which is ever chiefly concerned with the duties of life; and on the other hand, the numerous other sects, as the Cyrenaics and Epicureans, who placed it in pleasure, and made virtue (as it is used in some comedies, wherein the mistress and the maid change habits) to be but as a servant, without which pleasure cannot be properly served and attended; and the reformed school of the Epicureans, which pronounced felicity to be nothing else than the tranquillity and serenity of a mind free from perturbation (as if they would have deposed Jupiter again, and restored Saturn with the Golden Age, when there was neither summer nor winter, spring nor autumn, but all after one air and season); and lastly, that exploded school of Pyrrho and Herillus, who placed felicity in the removal from the mind of all doubts and scruples, admitting no fixed and consistent nature of good and evil, but esteeming actions good or evil according as they proceed from the mind acting clearly and regularly, or with reluctance and aversion; which opinion was revived in the heresy of the Anabaptists, who measured all things according to the notions or instincts of the spirit, and the constancy or wavering of belief. Now all the points above enumerated manifestly regard private repose and contentment, and not the good of society.

It censures also the philosophy of Epictetus, who presupposes that felicity must be placed in those things which are in our power, lest we be subject to fortune and disturbance [16]; as if it were not a thing much more happy to fail in

[11] Iamblichus in Vitâ, and Cic. *Tusc. Quæst.* v. 3. "Hiero" is a mistake for Leo (tyrant of Phliuns). The story of the interview between him and Pythagoras is told by Cicero, *Tusc. Quæst.* v. 3. Compare Iamblichus's Life of Pythagoras, in which, though the same sentiment is ascribed to him, it is not put in a dramatic form.

[12] Compare St. Augustine, speaking of St. Paul, *De Civ. Dei*, xiv. 9.

[13] Psalm cxvi. 15. [14] Gen. v. 24. [15] Jude, 14.

[16] The moral philosophy of the Stoics is misunderstood when it is said that they placed happiness in that which is in the wise man's power, in order that he may be happy. They set out from the inquiry, "What is the end and purpose, the summum bonum, of man's life?" in which is involved the assumption that it *has* an end and purpose, and that this is in its own nature attainable. And this assumption may be developed into an answer to the inquiry in which it is involved. For as the wise man, who is the representative of humanity in its best estate, must be capable of attaining the true end of his being, they concluded that whatever might in virtue of outward circumstances be to him unattainable, must be, with reference to that end, a thing indifferent; or, in other words, that

good and virtuous ends for the public, than to obtain all that we can wish to ourselves in our private fortune; as Gonsalvo, addressing his soldiers and pointing to Naples, nobly protested, "He had rather die one foot forwards, than secure a long life by one foot of retreat[17]". Whereunto agrees the wisdom of that heavenly leader, who has affirmed "that a good conscience is a continual feast[18]", showing plainly that the conscience of good intentions howsoever failing in success imparts a joy truer, surer, and more agreeable to nature, than all the provisions which a man can make either for the satisfying of his desires or for the repose of his mind.

It censures likewise that abuse of philosophy which grew general about the times of Epictetus, in converting it into an occupation or profession, as if the business of philosophy had been not to resist and extinguish perturbations, but to fly and avoid the causes and occasions of them, and to shape a particular kind and course of life to that end; introducing such a health of mind as was that health of body cultivated by Herodicus, of whom Aristotle tells us that he did nothing all his life long but attend his health, and accordingly abstained from an infinite variety of things, depriving himself as it were of the use of his body in the meantime[19]. Whereas, if men refer themselves to duties of society, as that state of body is most to be desired which is best able to endure and overcome all changes and extremities; so likewise that mind is to be esteemed truly and properly healthy which can go through the greatest temptations and perturbations: so that Diogenes's opinion seems excellent, who commended that strength of mind which enabled a man not to abstain but to sustain, and which could refrain its impetuosity even in the steepest precipices, and give it the property of a well broken horse, that of stopping and turning most quickly and suddenly[20].

Lastly, it censures also the tenderness and want of compliance in some of the most ancient and reverend philosophers, who retired too easily from civil business that they might avoid indignities and perturbations, and live (as they thought) more pure and saint-like; whereas the resolution of men truly moral ought to be such as the same Gonsalvo required in a soldier, "whose honour," he said, "should be of a stouter web, and not so fine as that everything should catch in it, and rend it".

CHAPTER II.

The Division of Individual or Self-good into Active and Passive Good.—The Division of Passive Good into Conservative and Perfective Good.—The Division of the Good of Communion, into General and Respective Duties.

To resume then, and pursue first private and self good, we will divide it into *Good Active and Good Passive*; for this difference of good, not unlike to that which, amongst the Romans, was expressed in the familiar or household terms of "Promus" and "Condus", is formed also in all things, and is best disclosed in the two several appetites in creatures; the one, to preserve or continue themselves: and the other, to multiply and propagate themselves; whereof the latter, which is active and as it were the promus, seems to be the stronger and more worthy; and the former, which is passive and as it were the condus, seems

the summum bonum must be looked for in that which is in his own power. That felicity in this sense is always in the wise man's power is thus not an arbitrary assertion, but results from the principle that life is not merely a purposeless dream.

[17] "Desiderare piuttosto di avere al presente la sua sepoltura un palmo di terreno più avanti, che col ritirarsi indietro poche braccie allungare la vita cento anni."—Guicciard. vi. 2.

Fernandez Gonsalvo of Cordova, commonly called the Great Captain, and certainly one of the most successful soldiers of the age in which he lived, was employed by the King of Spain in his Italian wars. He died at [Granada] in [December, 1515]. See, for the testimony to his merits of apparently an unwilling witness, Brantôme's *Vies des Grands Capitaines*, and for a panegyrical biography, Paulus Jovius.

[18] Prov. xv. 15.

[19] Rhet. i. 5. 10. See also Plato's *Republic*, b. iii.

[20] [The reference may be to Diogenes Laërt. in *Diogen.* § 4.]

to be inferior. For in the universe, the heavenly nature is mostly the agent, the earthly nature the patient; in the pleasures of living creatures, that of generation is greater than that of food; in divine doctrine, "It is more blessed to give than to receive"[1]; and in common life there is no man's spirit so soft and effeminate but esteems the effecting of somewhat that he has fixed in his desire more than any pleasure or sensuality. And this pre-eminence of the active good is infinitely raised by the consideration that the condition of man is mortal, and exposed to the blows of fortune; for if we might have a certainty and perpetuity in our pleasures, the certainty and continuance of them would advance their price. But when we see it is but thus with us, "We count it much to postpone death for awhile"[2]; "Boast not thyself of the morrow; Thou knowest not what a day may bring forth"[3]; it is no wonder that we earnestly pursue such things as are secured and exempted from the injuries of time, which are only our deeds and our works; as it is said, "Their works follow them"[4]. There is also another important pre-eminence of the active good, produced and upheld by that affection which is inseparable from human nature; the love of novelty and variety; which in the pleasures of the sense (which is the principal part of passive good) is very confined, and can have no great latitude[5]. "Only think how often you do the same thing over and over. Food, Sleep, Play, come round in a perpetual circle; a man might wish to die, not only from fortitude or misery or wisdom, but merely from disgust and weariness of life." But in enterprises, pursuits and purposes of life there is much variety; whereof men are sensible with pleasure in their inceptions, progressions, rests, recoils, reintegrations, approaches, and attainings to their ends; so as it was well said, "Life without a purpose is unsettled and languid"[6]. And this befalls as well the wise as the foolish; as Solomon says, "A heady man seeks to satisfy his desire, and intermeddles with everything"[7]. And we see that the greatest kings who might have at command everything which can gratify the sense, have yet sometimes affected mean and frivolous pursuits (as was the passion of Nero for the harp, of Commodus for gladiatorial combats, of Antoninus for chariot-driving, and the like); which nevertheless they esteemed more of than of the whole abundance of sensual pleasures; so much pleasanter is it to be doing than to be enjoying.

But here it must be more carefully observed, that this active individual good has no identity with the good of society, though in some case it has an incidence into it: for although it many times produces and brings forth acts of beneficence (which is a virtue of communion), yet there is this difference, that these acts are mostly done not with a view to the benefit and happiness of others, but to a man's own power and greatness; as plainly appears when this kind of active good strikes on a subject contrary to the good of society. For that gigantean state of mind, which possesses the troublers of the world (such as was Lucius Sylla, and infinite others in smaller model, who are bent on having all men happy or unhappy as they are their friends or enemies[8], and would shape the world according to their own humours, which is the true Theomachy), this I say aspires to the active good of the individual (apparent good at least), though it recedes farthest of all from the good of society.

But Passive good is subdivided into *Conservative and Perfective*. For there is impressed on all things a triple desire or appetite, in respect of self or individual good; one of preserving, another of perfecting, and a third of multiplying and spreading themselves: whereof the last is that which we have just handled by the name of "Active good", so that there remain only the two other goods which we have mentioned; whereof that of perfecting is the highest; for to preserve a thing in its existing state is the less, to raise the same to a higher nature is the greater. For in all things there are some nobler natures to the dignity and

[1] Luke, xiv. 12-14; Acts, xx. 35.
[2] Seneca, *Nat. Quæst.* ii. 59.
[3] Prov. xxvii. 1.
[4] Rev. xiv. 13.
[5] Seneca, *Ep.* 77.
[6] Seneca, *Ep.* 95.
[7] Cf. Prov. xviii. 1.
[8] The epitaph which Plutarch says Sylla made for himself was probably in Bacon's mind. It boasted that no man had surpassed him in doing good to his friends or evil to his enemies. See Plut. in *Sylla.*

excellence whereof inferior natures aspire as to their sources and origins. So it was not unfitly said of men "that they have a fiery vigour and a heavenly origin"[9], for the assumption or approach of man to the Divine or Angelical nature is the perfection of his form; the false and preposterous imitation of which perfective good is the very plague and stormy whirlwind of human life, which carries off and destroys everything; while men upon the instinct of an advancement formal and essential are carried by a blind ambition to seek an advancement merely local. For as those who are sick, and find no remedy, tumble up and down and change place, as if by a remove local they could obtain a remove internal, and get away from themselves and from the disease that is within them; so is it in ambition, when men possessed by a false idea of exalting their nature obtain nothing else but an eminence and exaltation of place.

The good of conservation consists in the reception and fruition of that which is agreeable to our natures; which, though it seems to be the most pure and natural of pleasures, is yet the softest and the lowest. And this also receives a difference, which has in part been weakly judged, in part not examined; for the good of fruition, or (as it is commonly termed) pleasure, is placed either in the sincerity of the fruition, or in the vigour of it; the one of which is the result of equality; the other of variety and vicissitude; the one having less mixture of evil, the other a stronger and more lively impression of good. Which of these is the greater good, is a question controverted, but whether man's nature may not be capable of both is a question not inquired. The former question being debated in a dispute between Socrates and a sophist, Socrates placing felicity in an equal and constant peace of mind, and the sophist in much desiring and much enjoying, they fell from arguments to ill words; the sophist saying that "Socrates's felicity was the felicity of a block or stone[10]", and Socrates saying, "that the sophist's felicity was the felicity of one that had the itch, who did nothing but itch and scratch". And both these opinions do not want their supports; for the opinion of Socrates is much upheld by the general consent even of the Epicureans, who did not deny that virtue bears a great part in felicity; and if so, certain it is, that virtue has more use in clearing perturbations than in compassing desires. But the sophist's opinion is somewhat favoured by the assertion we last spoke of, "that good of advancement is greater than good of simple preservation", because every obtaining a desire has a show of advancing nature towards perfection; which though it be not really the case, yet motion even in a circle has a show of progression.

But the second question (as to whether a man's nature may not be capable of tranquillity of mind and vigour of fruition both), decided in the true way, makes the former superfluous. For do we not often see some minds so constituted, as to take the greatest delight in enjoying pleasures when present, and yet nevertheless little annoyed at the loss and leaving of them? So that the philosophical progression, "Enjoy not, that you may not desire; desire not, that you may not fear," is the precaution of cowardice and pusillanimity. And indeed most of the doctrines of the philosophers seem to me to be more fearful and cautionary than the nature of things requires: thus they increase the fear of death in offering to cure it; for when they would have a man's whole life to be but a discipline or preparation to die[11], they must needs make men think that it is a terrible enemy, against whom there is no end of preparing. Better says the poet (for a heathen):—

> Fortem posce animum mortis terrore carentem
> Qui finem vitæ extremum inter munera ponat
> Naturæ[12].

[9] Virg. *Æn.* vi. 7. 30:—

> Igneus est ollis vigor et cœlestis origo.

[10] Plato, *Gorgias*, p. 494.

[11] Said by Socrates in the *Phaedo*. Contrast Spinoza, *Ethica*, iv. 67.

[12] Juv. x. 357:—

> Give me a soul which can grim death defy,
> And count it Nature's privilege to die.

Bacon substitutes *finem* for the *spatium* of the original.

So have philosophers sought in all things to make men's minds too uniform and harmonical, not breaking them to contrary motions and extremes; the reason whereof I suppose to be, because they themselves were men dedicated to a private life, free from business and from the necessity of applying themselves to other duties. But men should rather imitate the wisdom of jewellers, who, if there be a grain or a cloud or an ice in a jewel, which may be ground forth without taking too much of the stone, they remove it: otherwise they will not meddle with it. And in like manner men ought so to procure serenity, as they destroy not magnanimity. And so much for Individual good.

Having, therefore, discussed self-good (which we also term "*Private*," "*Particular*," and "*Individual*" good), let us resume the good of communion, which respects and beholds society, which we may term *Duty* because the term of duty is more proper to a mind well framed and disposed towards others, as the term of virtue is applied to a mind well formed and composed in itself. This part may seem at first glance to pertain to science civil and politic, but not if it be well observed; for it concerns the regimen and government of every man over himself, and not over others. And as in architecture it is one thing to direct the framing the posts, beams, and other parts of the building, and another thing to join and fasten them; and as in mechanics it is one thing to direct how to frame an instrument or engine, and another to set it on work and employ it; so the doctrine of the conjugation of men in the state or society, differs from that which teaches them to conform and be well-disposed to the advantages thereof.

This part of duty is likewise subdivided into two parts; whereof the one treats of "the common duty of every man" as a member of a state; the other treats of "the respective or special duties of every man, in his profession, vocation, rank, and character." The first of these is extant and well laboured, as has been said; the second likewise I may report as handled dispersedly, though not digested into an entire body of science; not that I object to this manner of dispersed writing, which on the contrary in this kind of argument I acknowledge to be best. For who is there with such clearness or confidence that he can take upon him to write skilfully and accurately of the proper and relative duty of every several vocation and place? But treatises on matters of this kind which do not savour of experience, but are only drawn from a general scholastic knowledge of the subject, are for the most part empty and unprofitable. For although sometimes a looker-on may see more than a player, and there be a proverb more arrogant than sound, concerning the censure of the people on the actions of their superiors, "That the vale best discovers the hill"; yet it were much to be wished that only men of most practice and experience should meddle with such arguments; for the writing of speculative men on active matter for the most part seems to men of experience, as Phormio's arguments of the wars seemed to Hannibal, to be but dreams and dotage[13]. Only there is one vice which accompanies those who write on their own arts and professions, that they can not refrain from adorning and magnifying in excess those little Sparta's of theirs[14].

In which kind it were inexcusable not to mention (honoris causâ) your Majesty's excellent book touching the duties of a king[15], a work richly compounded of many known and secret treasures of divinity, morality, and policy, with great aspersion of all other arts, and being in my opinion one of the most sound and healthful writings that I have read; not distempered in the heat of invention, nor chilled in the coldness of negligence; not subject to fits of dizziness, and so falling into confusion and disorder; not distracted by digressions, so as to embrace in a discursive narrative things impertinent to the purpose[16]; not savouring of perfumes and paintings, as those do, who attend more to the pleasure of the reader than the nature of the argument; above all, being a book

[13] Cic. *de Orat.* lib. ii. 18.

[14] Cf. Erasm. *Adag.* ii. 5. 1.

[15] The proper title of this work is *Basilicon Doron.* It contains three books. The first is, "Of a king's Christian duetie towards God;" the second, "Of a king's duetie in his office;" and the last, "Of a king's behaviour in things indifferent."

[16] Compare the corresponding passage in the *Advancement*:—"not sick of dizziness as those are who leese themselves in their order; nor of convulsions, as those which cramp in matters impertinent".—*J. S.*

as good in spirit as in body, since it is both agreeable to truth, and apt for action. And it is moreover quite free from that vice which I have noted above (which, if it were tolerated in any, certainly it would be so in a king, writing of the authority of a king), seeing it does not exalt invidiously or above measure the height and summit of kingly power; for your Majesty has represented, not a king of Assyria, or Persia, in the glitter of outward pride and glory; but truly a Moses or a David, that is, shepherds of their people. Neither can I ever forget the observation so truly worthy of a king, which your Majesty delivered [17], in the same sacred spirit of government, in deciding a great cause of judicature; which was, "That kings ruled by the laws of their kingdoms, as God did by the laws of Nature, and ought as rarely to put in use their supreme prerogative, as God does his power of working miracles". And yet, notwithstanding, in your other book of a free monarchy [18] it well appears that you no less perceive and understand the plenitude of the power of a king, and the ultimities (as the schoolmen say) of regal rights, than the circle and bounds of his office and duty. Thus have I presumed to allege this excellent writing of your Majesty, as a prime or eminent example of treatises concerning special and respective duties; wherein I should have said as much if it had been written by any king a thousand years since. Nor am I moved with that rule of manners which is usually laid down, "that one should not praise in presence"; provided that the praise be not beyond the truth, and bestowed unseasonably, or without occasion. Surely Cicero, in that brilliant oration for Marcellus, was but exhibiting an excellent picture of Cæsar's praises, though he was speaking before his face. And the like did Pliny the younger to Trajan [19].

But to return to our purpose. There belongs further to the handling of this part, touching the respective duties of vocations and professions, a relative or opposite doctrine touching the frauds, cautions, impostures, and vices of every profession; for corruptions and vices are opposed to duties and virtues. And it is true that these are not altogether passed over, but there are many treatises and writings in which they are touched upon at least in passing; but how? rather in a satire, and cynically after the manner of Lucian, than seriously and wisely. For men have rather sought by wit to traduce much that is good or useful in professions, and expose it to ridicule, than to discover and sever that which is vicious and corrupt. But Solomon says well, "A scorner seeks wisdom, and finds it not, but knowledge offers itself unto him that is desirous thereof [20]"; for he who comes to seek after knowledge with a mind to scorn and censure will be sure to find matter enough for his humour, but very little for his instruction. But the serious handling of this argument with integrity and sincerity ought, as it appears to me, to be reckoned among the best fortifications for honesty and virtue. For as the fable goes of the basilisk, that if he see you first, you die for it, but if you see him first, he dies; so is it with deceits, impostures, and evil arts, which, if they be first espied, they lose their life, but if they prevent, they endanger; so that we are much beholden to Machiavelli and other writers of that class, who openly and unfeignedly declare or describe what men do, and not what they ought to do. For it is not possible to join the wisdom of the serpent with the innocence of the dove, except men be perfectly acquainted with the nature of evil itself; for without this, virtue is open and unfenced; nay, a virtuous and honest man can do no good upon those that are wicked, to correct and reclaim them, without first exploring all the depths and recesses of their malice [21]. For

[17] Probably in the case of Sir Francis Goodwin in 1604, when the question was whether it belonged to the House of Commons or the Court of Chancery to judge of the validity of an election.—*J. S.*

[18] This second work of James's is, "The Trew Law of Free Monarchies, or the reciprock and mutual duetie betwixt a free King and his naturall Subjects," free being nearly equivalent to absolute. This work was at first published anonymously, but is included in the edition of King James's works which appeared in 1616.

[19] In his *Panegyrica.*

[20] Prov. xiv. 6.

[21] Compare Charron *De la Sagesse*, liv. ii. c. 10. :—" Il faut temperer et marier l'inno-

men of corrupted minds presuppose that honesty grows out of an ignorance or simplicity of manners, and believing of preachers, schoolmasters, books, moral precepts, common discourses, and opinions; so as, except they plainly perceive that you know as much of their corrupt opinions and depraved principles as they do themselves, they despise all honesty of manners and counsel; according to the excellent proverb of Solomon, "The fool receives not the word of the wise, unless thou speakest the very things that are in his heart [22]". But this part, touching respective cautions and vices, we set down as deficient, and will call it by the name of "*Serious Satire*", or the *Treatise of the Inner Nature of Things.*

Unto this part, touching respective duty, do also appertain the mutual duties between husband and wife, parent and child, master and servant; so likewise the laws of friendship and gratitude, the civil bonds of companies, colleges, neighbourhood, and the like; but it must ever be kept in mind, that they are here handled, not as they are parts of civil society (for that is referred to policy), but as to the framing and predisposing of the minds of particular persons towards the preservation of those bonds of society.

The knowledge concerning good respecting Society (as well as that which respects Individual good) handles it not simply alone, but comparatively; whereunto belongs the weighing of duties between person and person, case and case, particular and public, present and future; as we see in the stern and severe proceeding of Lucius Brutus against his own sons, how it was generally extolled to the sky; and yet what did another say of it? "It was an unhappy deed, whatever posterity might say of it [23]".

And we see the same in that supper to which Marcus Brutus, Caius Cassius, and others were invited. When to make trial of their opinions touching the intended murder of Cæsar, the question was cunningly raised, "whether the killing of a tyrant were lawful," they were divided in opinion; some holding that it was clearly lawful, for servitude was the extreme of evils; others, not so, for tyranny was better than a civil war; while a third set affirmed, according to the doctrine of Epicurus, that it was unfit for wise men to endanger themselves in the cause of fools [24]. But there are a number of like cases of comparative duties; amongst which, that is most frequent where the question is, whether injustice may be committed in order to save one's country, or for some great future advantage of that kind; touching which, Jason of Thessaly used to say, "Some things must be done unjustly, that many may be done justly [25]". But the reply is good; "Present justice is in your power, for that which is to come you have no security." Men must pursue things which are good and just at present, leaving the future to the Divine Providence. And so much for the knowledge touching the exemplar and description of good.

CHAPTER III.

The Division of the Doctrine concerning the Culture of the Mind, into the Doctrine concerning the Characters of the Mind, the Affections, and the Remedies or Cures.—An Appendix of this same Doctrine, touching the Congruity between the Good of the Mind and the Good of the Body.

Now therefore that I have spoken of the fruit of life (understanding it in a philosophical sense), it remains to speak of the husbandry which belongs thereto; without which the former part seems to be no better than a fair image or statue, which is beautiful to contemplate, but is without life and motion; whereunto

cence colombine en n'offensant personne avec la prudence et astuce serpentine en se tenant sur ses gardes et se preservant des finesses, trahisons, et ambuches d'autrui." The whole chapter is worth comparing with Bacon's remarks on the art of self-advancement.

[22] Prov. xviii. 2. The words are accurately quoted from the Vulgate: the authorised version is wholly dissimilar.

[23] Virg. *Æn.* vi. 823: Infelix, utcumque ferent ea facta minores.

[24] Plutarch in *Brut.*

[25] Plut. *Reip. ger. Præcep.* 817.

Aristotle eloquently subscribes in these words, " It is necessary then to speak of virtue, both what it is, and whence it proceeds, for it were almost useless to know what virtue is, but to be ignorant of the ways and means of acquiring it; therefore we must inquire not only to what kind virtue belongs, but also how it may be obtained; for we wish both to be acquainted with the thing itself, and to gain possession of it; wherein we shall not fully succeed, unless we know both the whence and the how[1] ". In such express words and with such iteration does he inculcate this part, although he does not himself pursue it. This likewise it is which Cicero bestows on Cato the younger as no ordinary praise; that he had applied himself to philosophy, " not for the sake of disputing as most do, but for the sake of living according to its rules[2] ". And although through the negligence of our times, wherein few men take any care touching the cultivation and disposition of the mind, and the framing of their life to any fixed rule, (as Seneca[3] excellently says, " Everyone takes thought about the parts of life, no one about the whole "): this part may seem superfluous, yet I will not on that account pass it by untouched, but rather conclude with that aphorism of Hippocrates, " That they who are sick and feel no pain are sick in their mind[4] "; they need medicine not only to assuage the disease, but to awake the sense. And if it be objected that the cure of men's minds belongs to sacred divinity, it is most true; but yet moral philosophy may be admitted into the train of theology, as a wise servant and faithful handmaid to be ready at her beck to minister to her service and requirements. For as the Psalm says, " That the eyes of the handmaid look perpetually to the hands of her mistress[5] ", and yet no doubt many things are left to the care and discretion of the handmaid; so ought moral philosophy to give a constant attention to the doctrines of divinity, and be obedient to them, and yet so as it may yield of itself within its own limits many sound and profitable directions.

This part therefore, when I recall the excellency thereof, I cannot but find exceeding strange that it is not yet reduced to written inquiry. Wherefore seeing I set it down among the deficients, I will according to my custom sketch out some of the heads and points thereof.

First therefore in this, as in all things which are practical, we ought to cast up our account what is in our power and what not; for the one may be dealt with by way of alteration, but the other by way of application only. The husbandman cannot command either the nature of the soil or the seasons of the weather; no more can the physician either the natural temper and constitution of the patient, or the variety of accidents. Now in the culture of the mind and the cure for its diseases three things are to be considered; the different characters of dispositions, the affections, and the remedies; just as in the treatment of the body three things are observed; the complexion or constitution of the sick man, the disease, and the cure; but of these three, only the last is in our power, the two former are not. Yet the inquiry into things beyond our power ought to be as careful as into those within it; for the exact and distinct knowledge thereof is the groundwork of the doctrine of remedies, that they may be more conveniently and successfully applied; and we cannot fit a garment, except we first take measure of the body.

So then the first article of this knowledge is concerned with *the different characters of natures and dispositions.* And we are not here speaking of the common inclinations either to virtues and vices, or to disorders and passions, but of those which are more profound and radical. And in truth I cannot sometimes but wonder that this part of knowledge should for the most part be omitted both in Morality and Policy, considering it might shed such a ray of light on both sciences. In the traditions of astrology men's natures and dispositions are not unaptly distinguished according to the predominances of the planets; for some are naturally formed for contemplation, others for business, others for war, others for advancement of fortune, others for love, others for the arts, others for a varied kind of life; so among the poets (heroic, satiric, tragic, comic) are everywhere

[1] Pro Muræna, c. 30.
[2] Magn. Mor. lib. i. 1.
[3] Sen. Ep. 71.
[4] Aph. ii. 6.
[5] Psalm cxxiii. 2.

interspersed representations of characters, though generally exaggerated and surpassing the truth. And this argument touching the different characters of dispositions, is one of those subjects in which the common discourse of men (as sometimes though very rarely happens) is wiser than books. But far the best provision and material for this treatise is to be gained from the wiser sort of historians, not only from the commemorations which they commonly add on recording the deaths of illustrious persons, but much more from the entire body of history as often as such a person enters upon the stage; for a character so worked into the narrative gives a better idea of the man, than any formal criticism and review can; such is that of Africanus and Cato the Elder in Livy, of Tiberius, and Claudius, and Nero in Tacitus, of Septimius Severus in Herodian, of Louis XI., King of France, in Philip de Comines, of Ferdinand of Spain, the Cæsar Maximilian, and the Popes Leo and Clement in Francesco Guicciardini. For these writers, having the images of those persons whom they have selected to describe constantly before their eyes, hardly ever make mention of any of their actions without inserting something concerning their nature. So some of the relations which I have met with touching the conclaves of the popes, present good characters of the Cardinals [6]; as the letters of ambassadors do likewise of the councillors of princes. Wherefore out of these materials (which are surely rich and abundant) let a full and careful treatise be constructed. Not however that I would have these characters presented in ethics (as we find them in history or poetry or even in common discourse), in the shape of complete individual portraits, but rather the several features and simple lineaments of which they are composed, and by the various combinations and arrangements of which all characters whatever are made up, showing how many, and of what nature these are, and how connected and subordinate one to another; that so we may have a scientific and accurate dissection of minds and characters, and the secret dispositions of particular men may be revealed; and that from the knowledge thereof better rules may be framed for the treatment of the mind.

And not only should the characters of dispositions which are impressed by nature be received into this treatise, but those also which are imposed on the mind by sex, by age, by region, by health and sickness, by beauty and deformity, and the like; and again, those which are caused by fortune, as sovereignty, nobility, obscure birth, riches, want, magistracy, privateness, prosperity, adversity, and the like. For we see that Plautus makes it a wonder to see an old man beneficent, "His beneficence is that of a young man [7]". St. Paul advising that severity of discipline should be used towards the Cretans ("Reproach them severely"), accuses the disposition of their country; citing the poet's censure, "the Cretans are always liars, evil beasts, slow bellies [8]". Sallust notes that it is usual with kings to desire contradictories, "the desires of kings, as they are violent, so are they generally changeable and often contrary to themselves [9]." Tacitus observes that honours and fortune more often alter men's dispositions to the worse than to the better; "Vespasian alone was changed for the better [10]". Pindar makes the observation that great and sudden good fortune for the most part defeats and enervates men's minds. "There be, that are not able to digest great prosperity [11]". The Psalm shows it is more easy to keep a measure in the enjoying of fortune, than in the increase thereof, "If riches increase, set not your heart upon them [12]". These observations and the like I deny not but are touched a little by Aristotle in his Rhetoric, and here and there in some other men's writings, but they have never been incorporated into moral philosophy, to which they principally appertain; no less than the knowledge of the diversity of grounds and moulds does to agriculture, and the knowledge

[6] For an account of the writings here referred to, which were generally composed by the "Conclavisti," but sometimes by one of the Cardinals, see Ranke's work "*Die Römischen Päpste*, sect. 5. of the Appendix. Among the *Litteræ Legatorum*, those of the Venetians are especially valuable. They are, properly speaking, reports made to the Senate on the ambassador's return.

[7] Mil. Glor. iii. 1. 40.

[8] Ep. Tit. i. 12. The poet referred to is Epimenides.

[9] In Jugurth. c. 113.

[10] Tac. *Hist.* i. 50.

[11] Cf. Pind. *Olymp.* i. 88.

[12] Psalm lxii. 10.

of the diversity of complexions and constitutions does to medicine. It should be done however now, except we mean to follow the indiscretion of empirics, who minister the same medicines to all patients of every constitution.

Next in order is the *knowledge touching the affections and perturbations*, which are, as I have said, the diseases of the mind. For as the ancient politicians in popular states were wont to compare the people to the sea, and the orators to the winds; because as the sea would of itself be calm and quiet, if the winds did not move and trouble it; so the people would be peaceable and tractable if the seditious orators did not set them in working and agitation [13]; so it may be fitly said, that the mind in its own nature would be temperate and staid; if the affections, as winds, did not put it into tumult and perturbation. And here again I find it strange, that Aristotle should have written divers volumes of ethics, and never handled the affections, as a principal portion thereof; yet in his Rhetoric, where they are considered but collaterally and in a second degree (as they may be moved and excited by speech), he finds a place for them, and handles them acutely and well, for the quantity thereof. For it is not his disputations about pleasure and pain that can satisfy this inquiry: no more than he who should generally handle the nature of light can be said to handle the nature of particular colours; for pleasure and pain are to the particular affections, as light is to particular colours. Better pains, I suppose, had the Stoics taken in this argument, as far as I can gather by that which remains of them; but yet I conceive it was rather in subtlety of definitions than in any full and ample description. So likewise I find some particular writings of an elegant nature, touching some of the affections, as of anger, of tenderness of countenance, and some few others [14]. But to speak the real truth, the poets and writers of history are the best doctors of this knowledge, where we may find painted forth with great life and dissected, how affections are kindled and excited, and how pacified and restrained, and how again contained from act and further degree; how they disclose themselves, though repressed and concealed; how they work; how they vary; how they are enwrapped one within another; how they fight and encounter one with another; and many other particularities of this kind; amongst which this last is of special use in moral and civil matters; how, I say, to set affection against affection, and to use the aid of one to master another; like hunters and fowlers who use to hunt beast with beast, and catch bird with bird, which otherwise perhaps without their aid man of himself could not so easily contrive; upon which foundation is erected that excellent and general use in civil government of reward and punishment, whereon commonwealths lean; seeing those predominant affections of fear and hope suppress and bridle all the rest. For as in the government of states it is sometimes necessary to bridle one faction with another, so it is in the internal government of the mind.

I now come to those points which are within our own command, and have operation on the mind to affect and influence the will and appetite, and so have great power in altering manners; wherein philosophers ought carefully and actively to have inquired of the strength and energy of custom, exercise, habit, education, imitation, emulation, company, friendship, praise, reproof, exhortation, fame, laws, books, studies, and the like. For these are the things that rule in morals; these the agents by which the mind is affected and disposed; and the ingredients of which are compounded the medicines to preserve or recover the health of the mind, as far as it can be done by human remedies; of which number I will select some one or two, upon which to insist, as patterns of the rest. I will therefore make a few observations on *Custom and Habit*.

The opinion of Aristotle seems to me to savour of negligence and narrowness of contemplation, when he asserts that custom has no power over those actions which are natural; using for example, " that if a stone be thrown up a thousand times, it will not learn to ascend of itself; and that by often seeing or hearing we

[13] Cicero *Pro Cluent.* c. 49.

[14] Bacon was probably thinking of Plutarch's tract On Shamefacedness, which is I think the only one on this subject which has come down to us from antiquity. On anger there are two special treatises; Plutarch's and Seneca's.

do not learn to see or hear the better "[15]. For though this principle be true in some things, wherein nature is peremptory (the reasons whereof we have not now leisure to discuss), yet it is otherwise in things wherein nature admits, within certain limits, intension and remission. For he might see that a tight glove will come on more easily with use; that a wand by use and continuance will be bent contrary to its natural growth, and after a while will continue in the same position; that by use of the voice it becomes stronger and louder; that by custom we can better bear heat and cold, and the like; which two latter examples have a nearer resemblance to the subject, than those instances which he alleges. But however it be, the more true it is that virtues and vices consist in habit, he ought so much the more to have taught the rules for acquiring or removing that habit; for there may be many precepts for the wise ordering of the exercises of the mind, as well as of the body; whereof I will recite a few.

The first shall be, that we beware we take not at the first either a greater or a smaller task than the case requires. For if too great a burden be imposed, in a diffident nature you discourage; in a confident nature you breed an opinion, whereby a man promises to himself more than he is able to perform, which produces sloth; and in both these natures the trial will fail to satisfy the expectation, a thing which ever discourages and confounds the mind. But if the tasks be too weak, progress will be much retarded.

The second precept shall be, that to practise any faculty by which a habit may be acquired, two several times should be observed; the one, when the mind is best disposed, the other when it is worst disposed; that by the one, you may gain a great step, by the other, you may through strenuous exertion work out the knots and obstacles of the mind, and so make the middle times the more easy and pleasant.

The third precept shall be that which Aristotle mentions by the way. "To bear ever with all our strength, so it be without vice, towards the contrary extreme of that whereunto we are by nature inclined [16]"; as when we row against the stream, or straighten a wand by bending it contrary to its natural crookedness.

The fourth precept depends on that axiom, which is most true; that the mind is brought to anything with more sweetness and happiness, if that whereunto you pretend be not first in the intention, but be obtained as it were by the way while you are attending to something else; because of the natural hatred of the mind against necessity and constraint. Many other useful precepts there are, touching the regulation of custom; for custom wisely and skilfully conducted proves indeed, according to the saying, a second nature; but governed unskilfully and by chance it will be but an ape of nature, imitating nothing to the life, but bringing forth only that which is lame and counterfeit.

So, if we should handle books and studies and what influence and operation they should have upon manners, are there not divers precepts and directions of great profit appertaining thereunto? Did not one of the fathers[17], in great indignation, call poesy "the wine of demons," because it engenders temptations, desires, and vain opinions? Is not the opinion of Aristotle very wise and worthy to be regarded, "that young men are not fit auditors of moral philosophy"[18], because the boiling heat of their affections is not yet settled, nor tempered with time and experience? And to say the truth, does it not hereof come that those excellent books and discourses of the ancient writers (whereby they have per-

[15] Nic. Eth. ii. 1. [16] Nic. Eth. ii. 9.

[17] St. Augustine. Cf. Agrippa *de Incert.* c. 4.

[18] Nic. Eth. i. 1. Aristotle, however, speaks not of moral but of political philosophy. It is interesting to observe that the error of the text, which occurs also in the *Advancement of Learning*, has been followed by Shakespeare in *Troilus and Cressida*:—

"Not much
Unlike young men, whom Aristotle thought
Unfit to hear *moral* philosophy."

See Hector's speech in the second scene of the second act.

suaded unto virtue most effectually by representing her in state and majesty, and popular opinions against virtue as clad in parasites' cloaks, fit to be scorned and derided) are of so little effect towards honesty of life and amendment of evil manners, because they are not read and revolved by men in their mature and settled years, but confined almost to boys and beginners. But is it not true also that much less are young men fit auditors of matters of policy, till they have been thoroughly seasoned in religion, morality, and duty, lest their judgments be corrupted and made apt to think that there are no true and real differences of things; but all things are to be measured by utility and fortune; as the poet says:—

Prosperum et felix scelus virtus vocatur [19];

and again,

Ille crucem pretium sceleris tulit, hic diadema [20];

which the poets speak satirically and in indignation, but some books of policy speak seriously and positively. For so it pleases Machiavelli [21] to say "That if Cæsar had been overthrown, he would have been more odious than ever was Catiline;" as if there had been no difference but in fortune alone between a very fury of lust and blood, and the most excellent spirit (his ambition reserved) of the unconverted world. And how necessary it is for men to be fully imbued with pious and moral knowledge before they take any part in politics we see from this; that they who are brought up from their infancy in the courts of kings and affairs of state scarce ever attain to a deep and sincere honesty of manners; how much less chance have they then, if to this be added the like discipline in books? Again, is there not a caution likewise to be given of the doctrines of moralities themselves, at least some kinds of them, lest they make men too precise, arrogant, and incompatible? as Cicero says of Marcus Cato, "The divine and noble qualities we see in him, be sure are his own; the defects which we sometimes find, proceed not from his nature, but from his instructors [22]". Many other axioms there are touching those properties which studies and books infuse into men's minds; for the saying is true, "that studies pass into manners [23]", as may likewise be said of all those other points, of company, fame, laws, and the rest, which I a little before recited.

But there is a kind of culture of the mind, which seems yet more accurate and elaborate than the rest, and is built upon this ground; that the minds of all men are at some times in a state more perfect, and at other times in a state more depraved. The purpose therefore and intention of this practice is to cherish the good hours of the mind, and to obliterate and take forth the evil out of the calendar. The fixing of the good has been practised by two means; vows or constant resolutions of the mind, and observances or exercises, which are not to be regarded so much in themselves, as because they keep the mind in continual duty and obedience. The obliteration of the evil can likewise be practised by two means; some kind of redemption or expiration of that which is past, and an inception or new account of life for the time to come. But this part seems clearly to belong to religion, and justly so; for all true and sincere moral philosophy, as was said before, is but a handmaid to religion.

Wherefore I will conclude this part of the culture of the mind with that remedy, which is of all other means the most compendious and summary; and again the most noble and effectual to the reducing of the mind into virtue, and placing it in the state nearest to perfection; which is, *the electing and propounding unto a man's self good and virtuous ends of his life and actions; such as may be in a reason-*

[19] Senec. *Herc. Fur.* 251.:—

Successful guilt will borrow virtue's name.

[20] Juv. xiii. 105.:—

Success is all; and for the self-same thing,
One dies a felon, the other lives a king.

[21] Machiavelli, *Discorsi*, i. 10. [22] Cic. *Pro Murænâ*, c. 29. [23] Ovid. *Epist.* xv. 83.

able sort within his compass to attain. For if these two things be supposed, that a man set before him honest and good ends, and again that his mind be resolute and constant to pursue and obtain them, it will follow that his mind shall address and mould itself to all virtues at once. And this indeed is like the work of Nature; whereas the other courses I have mentioned are like the work of the hand. For as when a carver makes an image, he shapes only that part whereon he works, and not the rest (as if he be upon the face, that part which shall be the body is but a rude and unshaped stone still, till such time as he comes to it); but contrariwise, when Nature makes a flower or living creature, she forms and produces rudiments of all the parts at one time; so in obtaining virtue by habit, while we practise temperance, we do not advance much in fortitude, nor the like; but when we dedicate and apply ourselves entirely to good and honest ends, what virtue soever the pursuit and passage towards those ends suggests and enjoins, we shall find ourselves invested with a precedent disposition and propensity to conform thereto. And this is the state of mind excellently described by Aristotle, and distinguished by him as having a character not of virtue but of divinity; his words are these: "To brutality we may not unaptly oppose that heroic or divine virtue which is above humanity [24]"; and a little after, "For as beasts are incapable of virtue or vice, so likewise is the Deity; for this latter state is something higher than virtue, as the former is somewhat other than vice". Again, Pliny the younger using the license of heathen grandiloquence sets forth the virtue of Trajan, not as an imitation, but rather as a pattern of the divine, where he says, "That men needed not to make any other prayers to the gods, but that they would show themselves as good and kind lords to them, as Trajan had been [25]". But these be heathen and profane passages, which grasp at shadows greater than the substance; but the true religion and holy Christian faith lays hold of the reality itself, by imprinting upon men's souls, Charity, which is excellently called "the bond of Perfection [26]", because it comprehends and fastens all virtues together. And it is elegantly said by Menander [27] of sensual love (which is but a false imitation of divine love), "That love is a better teacher for human life than a left-handed sophist," whereby he means that comeliness of manner is better taught by love than by a clumsy preceptor or sophist, whom he calls left-handed; because with all his laborious rules and precepts he cannot form a man so dexterously, nor with that facility to prize and govern himself in all things, as love can do. So certainly if a man's mind be truly inflamed with charity, it raises him to greater perfection than all the doctrines of morality can do; which is but a sophist in comparison of the other. Nay further, as Xenophon truly observed, "that all other affections though they raise the mind yet they distort and disorder it by their ecstasies and excesses, but only love at the same time exalts and composes it [28]"; so all the other qualities which we admire in man, though they advance nature, are yet subject to excess; whereas Charity alone admits of no excess. The Angels aspiring to be like God in power, transgressed and fell: "I will ascend, and be like unto the most High [29]". Man aspiring to be like God in knowledge, transgressed and fell: "Ye shall be as gods, knowing good and evil [30]"; but by aspiring to a similitude of God in goodness or love, neither angel or man ever transgressed or shall transgress; for unto that imitation we are called, "Love your enemies, bless them which hate you, and pray for them that despitefully use you and persecute you, that ye may be children of your Father who is in heaven, who makes his sun to rise on the evil and the good, and sends his rain on the just and the unjust [31]". So in the first platform of the divine nature itself, the heathen religion speaks thus, "Optimus Maximus," but the sacred Scriptures thus, "His mercy is over all His works [32]".

Here then I conclude this part of moral knowledge concerning the Georgics of the mind, wherein if any man, from viewing the parts thereof which I have enumerated, judge that my labour is but to collect into an art or science that which has been omitted by other writers as matter of common sense and ex-

[24] Nic. Eth. vii. 1. [25] Pliny, *Paneg.* 1. c. 74.
[26] Coloss. iii. 14. [27] Anaxandrides, not Menander. [28] Xenoph. *Sympos.*
[29] Isaiah, xiv. 14. [30] Gen. iii. 5. [31] St. Matt. v. 44. [32] Psalm cxlv. 9.

perience, and sufficiently clear and self-evident, he is welcome to his opinion; but in the mean while let him remember that I am in pursuit, as I said at first, not of beauty but of utility and truth: and let him withal call to mind the ancient parable of the two gates of sleep:—

> Sunt geminæ Somni portæ, quarum altera fertur
> Cornea, qua veris facilis datur exitus umbris;
> Altera candenti perfecta nitens elephanto,
> Sed falsa ad cœlum mittunt insomnia Manes[33].

Great no doubt is the magnificence of the ivory gate, but the true dreams pass through the gate of horn.

To these observations concerning moral philosophy may be added, *That there seems to be a relation or conformity between the good of the mind and the good of the body*. For as I said that the good of the body consisted of health, beauty, strength and pleasure; so the good of the mind considered according to the precepts of moral knowledge tends to this; to make the mind sound and without perturbation; beautiful and graced with decency; and strong and agile for all the duties of life; lastly, not stupid, but retaining a lively sense of pleasure and comfort in an honest way. These three as in the body so in the mind seldom all meet together. For it is easy to observe that many have strength of wit and courage, who are yet disordered by perturbations and have little beauty and decency in their manners; some again have an elegance and fineness of carriage, who have neither honesty of will nor strength for action; and some again have honest and reformed minds who can neither become themselves nor manage business: while others, though perhaps endowed with all these three, yet from a Stoical severity and insensibility have no pleasure in the virtuous actions which they practise. But though it happen that of these four two or three of them sometimes meet, yet the meeting of them all is, as I have said, very rare. I have now handled that general part of human philosophy which contemplates man as he consists of body and spirit, but segregate and apart from society.

[33] Virg. *Æn.* vi. 894:—

> Two gates the entrance of Sleep's house adorn:
> Of ivory one, the other simple horn;
> Through horn a crowd of real visions streams,
> Through ivory portals pass delusive dreams.

Book VIII.

CHAPTER I.

The Division of Civil Knowledge into the Doctrine concerning Conversation, Negotiation, and Empire or State Government.

THERE is an old story, most excellent king, that many philosophers being met together in the presence of the ambassador of a foreign prince, each endeavouring to give a sample of his wisdom, that the ambassador might be able to make a report of the wonderful wisdom of Greece; one of them remained silent and propounded nothing; insomuch that the ambassador turning to him, said, "What have you to say for me to report?" To whom he answered, "Tell your king that you have found a man in Greece, who knew how to hold his tongue [1]". And in truth, in this synopsis of the arts I have forgotten to mention the art of silence, which (since it is commonly deficient) I will now teach by my own example. For since the course of the argument has now brought me down to that point, that I should presently handle the art of government; and since I am writing to so great a king who is such a master in that art, wherein he has been trained from his cradle; and since I cannot altogether forget what position I have held under your majesty; I thought that I should better approve myself by silence on such a matter before your majesty, than by speech. Cicero indeed makes mention not only of an art, but of a kind of eloquence in silence; for in one of his letters to Atticus, after relating a conversation between himself and another person on both sides of a subject, he writes, "Here I borrowed part of your eloquence, for I held my tongue [2]". Pindar again (whose peculiar gift it was to surprise men's minds with some striking expression, as with a magic rod), utters some such saying as this, "Silence sometimes says more than speech [3]". Wherefore in this part I have determined to be silent, or to be very brief, which is next thing to silence; but before I come to the arts of government, I must first make some observations touching the other parts of civil knowledge.

Civil knowledge is conversant about a subject, which of all others is most immersed in matter, and with most difficulty reduced to axioms. Nevertheless there are some circumstances to relieve this difficulty; for first, as Cato the Censor used to say of the Romans, "that they were like sheep, for that a man might better drive a flock of them, than one of them; for in a flock, if you could but get some few of them to go right, the rest would follow [4]"; so in that respect the duty of moral philosophy is more difficult than that of policy. Secondly, moral philosophy propounds to itself to imbue and endow the mind with internal goodness; but civil knowledge requires only an external goodness, for that suffices for society. And therefore it often comes to pass that there be evil times in good governments; as in the sacred history we find it said more than once in speaking of good and pious kings, "Howbeit the people had not yet directed their heart aright to the Lord God of their Fathers [5]"; wherefore in this respect also the office of moral philosophy is more difficult. Thirdly, states as great engines are moved slowly and not without great efforts, whence they are not so soon put out of frame; for as in Egypt the seven good years sustained the seven bad, so in states the good government of previous years prevents the errors of succeeding times from causing immediate ruin; but the resolutions and morality of parti-

[1] This story is told of Zeno. See Plut. *de Garrulitate,* and Diog. Laert. vii. 24.
[2] Cic. *Ep. ad Attic.* xii. 42. The person in question was his nephew, Q. Cicero.
[3] Pind. *Nem.* v. 32. [4] Plut. in *Cato,* c. 8. [5] 2 Chron. xx. 33.

cular persons are more suddenly subverted. And this makes moral knowledge more difficult, but civil knowledge more easy.

Civil knowledge has three parts, according to the three summary actions of Society; *the knowledge of conversation, the knowledge of negotiation*, and *the knowledge of empire or government*. For there are three kinds of good which men seek in society, comfort against solitude, assistance in business, and protection against injuries; and they are three wisdoms of divers natures, which are often separate; wisdom of behaviour, wisdom of business, and wisdom of state.

The wisdom of conversation ought certainly not to be overmuch affected, but much less despised; for a wise management thereof has not only a grace and honour in itself, but an important influence in business and government. For as action in an orator, though an external quality, is held of such account as even to be preferred to those other parts which appear more important and internal; so in a man of business conversation and the management thereof, though employed on external objects, finds, if not the highest, yet at all events an eminent place. For look what an effect is produced by the countenance and the carriage of it. Well says the poet,

Nec vultu destrue verba tuo[6].

For a man may destroy and betray the force of his words by his countenance; nay, and the effect of his deeds also, if we believe Cicero; who in recommending to his brother affability towards the provincials, said that it did not so much consist in affording them easy access, as in receiving them with a courteous and open countenance. "It is nothing to have your door open, if your countenance be shut[7]". So we see Atticus before the first interview between Cæsar and Cicero, the war still depending, carefully and seriously advised Cicero touching the composing and ordering of his countenance and gesture[8]. And if the government of the face and countenance alone be of such effect; much more is that of the speech and other carriage appertaining to conversation. Indeed all grace and dignity of behaviour may be summed up in the even balancing of our own dignity and that of others, as has been well expressed by Livy, (though not meant for this purpose) in that description which he gives of personal character. "Lest I should appear (says he) either arrogant or servile, whereof the one were to forget the liberty of others, the other to forget my own[9]". On the other side, if behaviour and outward carriage be intended too much it may pass into a deformed and spurious affectation. "And then, what is more uncomely than to bring the manners of the stage into the business of life?" And even if it proceed not to that faulty extreme, yet too much time is consumed in these frivolous matters, and the mind is employed more than is right in the care of them. And therefore as in the universities preceptors use to advise young students from too much company-keeping, by saying, "Friends are the thieves of time"; so certainly the constant attention of the mind to the discretion of the behaviour is a great thief of more serious meditation. Again such as are accomplished in urbanity, and seem as formed by nature for that alone, generally find satisfaction enough therein, and seldom aspire to higher and more solid virtue; whereas those who are conscious of a defect in this point seek comeliness by reputation; for where reputation is, there almost everything is becoming; but where that is not, it must be supplied by manners and behaviour. Again

[6] Ovid, *De Art. Am.* ii. 312:—

Let not your looks your words betray.

[7] No such remark occurs in the letter of advice which Marcus Cicero wrote to his brother Quintus, when the latter was about to take possession of his province. But in Quintus's tract *De Petitione Consulatsû* in which he gives his brother advice as to his conduct in canvassing for the consulship, we find the antithesis quoted in the text, though somewhat differently worded. But of course the passage in which it occurs has no reference to any class of "provinciales."

[8] See Cicero, *Ep. ad Att.* ix. 12; and compare the eighteenth letter of the same book, in which the interview with Cæsar is described.

[9] Livy, xxiii. 12.

there is no greater nor more common impediment of action than an overcurious observance of external decency, and the attendant of decency, which is an anxious watching of Time and Season. For as Solomon well observes, "He that regards the winds does not sow, and he that regards the clouds does not reap [10]"; a man must make his opportunity as oft as find it. To conclude, this behaviour is as the garment of the mind, and ought to have the conditions of a garment. For first, it ought to be made in fashion; secondly it should not be too curious or costly; thirdly, it ought to be so framed as to best set forth any virtue of the mind, and supply and hide any deformity; lastly, and above all, it ought not be too strait, so as to confine the mind and interfere with its freedom in business and action. But this part of civil knowledge touching conversation has been elegantly handled, and therefore I cannot report it for deficient.

CHAPTER II.

The Division of the Doctrine concerning Negotiation into the Doctrine concerning Scattered Occasions and the Doctrine concerning Advancement in Life.—Example of the Doctrine concerning Scattered Occasions from some of the Proverbs of Solomon.—Precepts concerning Advancement in Life.

THE Doctrine concerning Negotiation is divided into *the Doctrine concerning Scattered Occasions*, and *the Doctrine concerning Advancement in Life*; whereof the one comprises all variety of business, and is as it were the secretary for the whole department of life; the other merely selects and suggests such things as relate to the improvement of a man's own fortune, and may serve each man for a private notebook or register of his own affairs. But before I descend to the species, I will make some preparatory remarks touching the doctrine concerning negotiation in general. The science of negotiation has not hitherto been handled in proportion to the importance of the subject, to the great derogation of learning and the professors thereof. For from this root springs chiefly that evil, with which the learned have been branded; "*That there is no great concurrence between learning and practical wisdom*". For if it be rightly observed, of the three wisdoms which we have set down to pertain to civil life, the wisdom of behaviour is by learned men for the most part despised, as a thing servile, and moreover an enemy to meditation. For wisdom of government, it is true that as often as learned men are called to the helm, they acquit themselves well, but that happens to few. But for the wisdom of business (of which I am now speaking), wherein man's life is most conversant, there are no books at all written of it, except some few civil advertisements collected in one or two little volumes, which have no proportion to the magnitude of the subject. For if books were written of this as of the other, I doubt not but learned men with but little experience would far excel men of long experience without learning, and outshoot them (as they say) in their own bow.

Neither is there any reason to fear that the matter of this knowledge should be so variable that it falls not under precept; for it is much less infinite than that science of government, which notwithstanding we see is excellently laboured and reduced. Of this kind of wisdom it seems some of the ancient Romans in the best times were professors: for Cicero reports that a little before his age senators who had most name and opinion for wisdom and practice in affairs (as Coruncanius, Curius, Lælius, and many others) used to walk at certain hours in the Forum, where they might give audience to their fellow-citizens, who would consult with them not merely on subjects of law but on all sorts of business; as on the marriage of a daughter, the education of a son, the purchase of a farm, a contract, accusation, defence, and every other occasion incident to man's life [1]. Whence it appears that there is a wisdom of counsel and advice even in private causes, arising out of a universal insight and experience of the affairs of the world; which is used indeed upon particular causes, but is gathered by general observation of causes of like nature. For so we see in that book which Cicero wrote to his brother, "on Canvassing for the Consulship" [2] (being the only book of particular business that I know written by the ancients), although it especially concerned an action then

[10] Eccles. xi. 4. [1] Cicero, *De Orat.* iii. 33. [2] De Petit. Consulatûs.

on foot, yet it contained many political axioms which prescribe not only a temporary but a perpetual direction in the case of popular elections. But of this kind there is nothing any way comparable to those Aphorisms composed by Solomon the King, of whom the Scriptures testify, "that his heart was as the sands of the sea [3]"; for as the sands of the sea encompass all the coasts of the earth, so did his wisdom embrace all things human as well as divine. But in these Aphorisms, besides those of a theological character, there are not a few excellent civil precepts and cautions, springing from the inmost recesses of wisdom and extending to much variety of occasions. Wherefore seeing I set down this knowledge of scattered occasions (which is the first part of the knowledge touching negotiation) among the deficients, I will stay awhile upon it according to my custom, and offer to consideration an example of the same, taken from the Aphorisms or Proverbs of Solomon. Nor in my opinion can I be with reason blamed for seeking a politic meaning in one of the Sacred writers; for if those commentaries were extant which this same Solomon wrote touching the nature of things, (wherein he treated of every vegetable, from the moss upon the wall to the cedar of Lebanon, and likewise of all animals) [4], it would surely be lawful to interpret them in a natural sense; and therefore we may here use the same liberty in matters political.

An Example of a portion of the doctrine concerning scattered occasions, *taken from some of the Proverbs of Solomon.*

PROVERB.

(1.) A soft answer turneth away wrath [5].

Explanation.

If the anger of a prince or a superior be kindled against you, and it is your turn to speak, Solomon gives two directions; first, "that an answer be made," and secondly, "that it be soft"; the former contains three precepts. First, to beware of a sullen and obstinate silence, which either takes the fault entirely on yourself, as if you had no answer to make; or secretly impugns your superior of injustice, insinuating that his ears are not open to a defence, though a just one. Secondly, to beware of postponing the matter and demanding another time for defence; which either conveys the same impression as the preceding one, that your superior is carried away by too great an excitement of temper; or plainly intimates that having no answer ready you are meditating a false and artificial defence. Wherefore it will ever be the best course to bring forward something in excuse directly as the occasion arises. Thirdly, that an answer at all events be made; not a simple confession or submission, but with a mixture of defence and excuse; for a different course is unsafe, except with very generous and noble characters, which are extremely rare. It follows in the other precept, that the answer be soft, not harsh or rough.

PROVERB.

(2.) A wise servant shall have rule over a foolish son, and shall have part of the inheritance among the brethren [6].

Explanation.

In all disordered and discordant families there is ever some servant or humble friend of great influence, who acts as arbiter and settles their disputes, and to whom on that account both the whole family and the master himself are subject. Such a man, if he is pursuing his own interests, foments and aggravates the family feuds; but if he is truly faithful and upright he deserves a great reward; even to be counted as one of the brothers, or at least to have the direction of the inheritance in trust.

PROVERB.

(3.) If a wise man contend with a fool, whether he rage or laugh, he shall find no rest [7].

[3] 1 Kings, iv. 29. [4] Kings, iv. 33. [5] Prov. xv. 1.
[6] Prov. xvii. 2. [7] Prov. xxix. 9.

Explanation.

We are often advised to avoid an unequal contest, meaning that we should not contend with those that are too strong for us. But the advice here given by Solomon is no less useful, "Not to contend with one that is unworthy"; for herein the chances are altogether unequal; seeing it is no victory to conquer, and a great disgrace to be conquered. And it makes no difference in this kind of contest, whether we take it in jest, or in scorn and contempt; for, whichever way we turn, we must lose in dignity, and can no ways quit ourselves well of it. But the worst of all is if, as Solomon says, our adversary has somewhat of the fool in him, that is, if he be bold and presumptuous.

PROVERB.

(4.) Lend not thine ear to all words that are spoken, lest thou hear thy servant curse thee [8].

Explanation.

It is scarcely credible what confusion is created in life by a useless curiosity about the things which concern us; that is, when we set to work to inquire into those secrets which when discovered produce uneasiness of mind, but are of no use to forward our designs. For first there ensues vexation and disquiet of mind, seeing all things human are full of treachery and ingratitude. And therefore if we could obtain a magic glass wherein we might view all the enmities and all the hostile designs that are at work against us, it were better for us to throw it down at once and break it to pieces; for these matters are but as the rustling of leaves, and have short duration. Secondly, this curiosity overcharges the mind with suspicions, a thing which ruins counsels, and renders them inconstant and perplexed. Thirdly, it often renders permanent those very evils which would otherwise blow over; for it is a dangerous thing to alarm the consciences of men; who, if they imagine themselves undetected, may come to a better mind; but if they perceive that they are discovered, they repel mischief with mischief. Rightly therefore was it considered great wisdom in Pompey that he burned all Sertorius's papers unperused either by himself or others [9].

PROVERB.

(5.) Poverty comes as one that travelleth, and want as an armed man [10].

Explanation.

It is elegantly described in this proverb, how the shipwreck of fortunes comes upon prodigals and those that are careless of their estates; for debt and diminution of capital come on at first slowly and step by step like a traveller, and are scarce perceived; but soon after want rushes in like an armed man, so strong and powerful as no longer to be resisted; for it was rightly said by the ancients, "that necessity was of all things the strongest [11]". Wherefore we must meet the traveller on his way, but against the armed man we must fortify.

PROVERB.

(6.) He that instructs a scorner gets to himself shame, and he that rebukes the wicked gets himself a blot [12].

Explanation.

This agrees with the command of our Saviour, "not to cast pearls before swine [13]", but a difference is made between the actions of instruction and reproof; and also between the persons of the scorner and the wicked; and lastly, there is a difference in the return; for in the former case the labour is but lost, in the latter it is repaid with a stain and blot. For when a man informs and instructs a scorner, in the first place he loses his time; and secondly, the attempt is laughed at by others as a vain thing and labour misapplied; and lastly, the scorner himself despises the knowledge which he has received. But there is greater danger in the reproval of the wicked; for not only does a wicked man lend no ear to ad-

[8] Eccles. vii. 21. [9] Plut. in Pomp. c. 20.; and in Sertor. c. 27. [10] Prov. vi. 11.
[11] Cf. Eurip. *Helena*, 513. [12] Prov. ix. 7. [13] St. Matt. vii. 6.

vice, but turns again on his reprover, whom being now made odious to him he either directly assails with abuse, or afterwards traduces to others.

PROVERB.

(7.) A wise son makes a glad father, but a foolish son is the heaviness of his mother [14].

Explanation.

Here are distinguished the domestic comforts and tribulations of a father and mother respectively, touching their children. For a wise and prudent son is of most comfort to the father, who knows the value of virtue better than the mother, and accordingly has more joy in the virtuous inclination of his son; he may feel a satisfaction also in the course he has pursued, whereby he has brought up his son so well and implanted sound morality in him by precept and example. But the mother has most sorrow and discomfort at the ill fortune of her son, both because the affection of a mother is more gentle and tender, and because she is conscious perhaps that she has spoiled and corrupted him by her indulgence.

PROVERB.

(8.) The memory of the just is blessed, but the name of the wicked shall rot [15].

Explanation.

A distinction is here made between the character of the good and the bad after death. For when the envy which carped at the reputation of the good in their lifetime is quenched, their name forthwith shoots up and flourishes, and their praises daily increase; but for the wicked, though their reputation through the favour of their friends and partizans last for a time, yet soon it turns into contempt, and in the end their fleeting glory changes into infamy and as it were a foul and noxious odour.

PROVERB.

(9.) He that troubles his own house shall inherit the wind [16].

Explanation.

A very useful admonition, touching domestic discords and disturbances. For many from the separation of their wives, the disinheriting of their children, the frequent changes in their family, promise to themselves great things; as if they would thence obtain peace of mind and a better management of their affairs; but commonly their hopes vanish into the winds. For both those changes generally turn out ill, and such disturbers of their family often experience trouble and ingratitude from those whom to the neglect of others they select and adopt; nay further, they thus draw upon themselves ill reports and dishonourable rumours; for it is well said by Cicero, "Every man's reputation proceeds from those of his own household [17]". And both these evils Solomon elegantly expresses by "the inheritance of the winds;" for both the disappointment of expectation and the raising of rumours are not unaptly compared to winds.

PROVERB.

(10.) Better is the end of a speech than the beginning thereof [18].

Explanation.

This proverb reproves a very common error, not only of those who make an especial study of words, but even of the more wise and prudent; which is, that men are more careful of the entrances and commencement of their speeches than of the end, and study more diligently the prefaces and inducements than the conclusions and issues; whereas for the former, they ought not indeed to neglect them, but the latter as being of far greater importance they ought to have ready prepared and arranged at hand; considering within themselves and endeavouring as much as possible to anticipate what shall be the end of their speech, and how

[14] Prov. x. 1. [15] Prov. x. 7. [16] Prov. xi. 29. [17] De Petit. Cons. 5.
[18] Eccles. vii. 8. *Vulgate.* The English version has, "Better is the end of a *thing*," etc.

their business may be advanced and ripened thereby. Nor is this all; for it is not only proper to study perorations and conclusions of such speeches as relate to the business itself, but also to be prepared with some discourse which may be conveniently and gracefully thrown in at the close, although foreign to the matter in hand. Indeed I knew two great and wise councillors on whom the weight of business principally rested, with whom it was a constant care and especial art, whenever they conferred with their princes on matters of state, not to end their discourse with matters relating to the business itself, but always by way of divertisement to draw it away to some jest or some agreeable news, and so end by washing off (as the proverb has it) their salt water discourses with fresh [19]. Nor was this the least valuable of their arts.

PROVERB.

(11.) As dead flies do cause the best ointment to stink, so does a little folly him that is in reputation for wisdom and honour [20].

Explanation.

It is a very hard and unhappy condition (as the Proverb well remarks) of men pre-eminent for virtue, that their errors, be they ever so trifling, are never excused. But as in the clearest diamond every little cloud or speck catches and displeases the eye, which in a less perfect stone would hardly be discerned; so in men of remarkable virtue the slightest faults are seen, talked of, and severely censured, which in ordinary men would either be entirely unobserved, or readily excused. Hence a little folly in a very wise man, a small offence in a very good man, a slight impropriety in a man of polite and elegant manners detracts greatly from their character and reputation; and therefore it would be no bad policy for eminent men to mingle some harmless absurdities with their actions; so that they may retain some liberty for themselves, and make small defects less distinguishable.

PROVERB.

(12.) Scornful men bring a city to destruction, but wise men turn away wrath [21].

Explanation.

It may seem strange that Solomon in his description of men formed as it were by nature for the ruin and destruction of states, should have selected the character not of a proud and insolent, not of a tyrannical and cruel, not of a rash and violent, not of a wicked and impious, not of an unjust and oppressive, not of a seditious and turbulent, not of an incontinent and sensual, not finally of a foolish and incapable person, but the character of a scorner. And this selection is worthy of the wisdom of a king who well knew how states were overthrown or preserved; for there is hardly a greater danger to kingdoms and states than that councillors or senators and those who stand at the helm should be of a scornful disposition. For such men ever undervalue dangers, that they may appear bold councillors, and insult those who make a just estimate of them, as cowards. They sneer at seasonable delays and careful discussions in consultation and deliberation, as mere matter of oratory, full of weariness, and contributing nothing to the completion of business. As for reputation, with a view to which the counsels of princes ought to be specially framed, they despise it as the breath of the people, that will quickly be blown away. They make no more account of the power and authority of laws, than of cobwebs which ought not to be in the way of more important business. Counsels and precautions looking far into the future they despise as dreams and melancholy apprehensions. They scorn with gibes and jests men of real wisdom and experience, of great minds, and deep judgment. In short, they weaken all the foundations of civil government; a thing the more to be attended to, because the mischief is wrought, not openly, but by secret engines and intrigues; and the matter is not yet regarded by men with as much apprehension as it deserves.

PROVERB.

(13.) A prince who readily hearkens to lies, has all his servants wicked [22].

[19] Erasm. *Adag.* iii. 3. 26. This proverbial phrase Erasmus found in the *Phædrus* of Plato, and in Athenæus. [20] Eccles. x. 1. [21] Prov. xxix. 8. [22] Prov. xxix. 12.

Explanation.

When the prince is one who lends an easy and credulous ear without discernment to whisperers and informers, there breathes as it were from the king himself a pestilent air, which corrupts and infects all his servants. Some probe the fears and jealousies of the prince, and increase them with false tales; others excite in him passions of envy, especially against the most virtuous objects; others seek to wash away their own vileness and evil consciences by accusing others; others make way for the honours and wishes of their friends by traducing and calumniating their opponents; while others get up stage plots and a number of the like fables against their enemies. These are the machinations of servants who are of a more dishonest nature. But those also who are naturally of greater honesty and principle, when they find no safeguard in their innocence (the prince not being able to distinguish truth from falsehood), throw off their honesty, and catching the court breezes allow themselves to be carried where they blow. "For," as Tacitus says of Claudius, "there is no safety with that prince, who has nothing in his mind, but what others put into it [23]". And Comines well remarks, "It is better to be the servant of a prince whose suspicions have no end, than of one whose credulity has no measure [24]".

PROVERB.

(14.) A righteous man regardeth the life of his beast, but the tender mercies of the wicked are cruel [25].

Explanation.

There is implanted in man by nature a noble and excellent spirit of compassion, that extends itself even to the brutes which by the divine ordinance are subject to his command. This compassion therefore has a certain analogy with that of a prince towards his subject. Moreover it is most true, that the nobler a spirit is, the more objects of compassion it has. For narrow and degenerate spirits think that these things concern them not; but the spirit which forms a nobler portion of the universe has a feeling of communion with them. Whence we see that under the old law there were many commandments, not so much purely ceremonial as institutions of mercy; as was that of not eating the flesh with the blood thereof, and the like. The Essenes and Pythagoreans even abstained altogether from eating flesh: and the same superstition still prevails among some of the inhabitants of the Mogul Empire. Nay, the Turks, though by race and habits a cruel and bloody people, yet are wont to give alms to brute creatures, and cannot endure to see them ill used or tortured [26]. But lest these things which we have mentioned should seem to countenance every kind of mercy, Solomon wisely adds, "That the mercies of the wicked are cruel." Such is the sparing to use the sword of justice upon wicked and guilty men; which kind of mercy is more cruel than cruelty itself; for cruelty is only practised upon individual persons, but this mercy to crime by granting impunity arms and lets loose upon the innocent the whole army of villains.

PROVERB.

(15.) A fool utters all his mind, but a wise man reserves something for the future [27].

Explanation.

This proverb seems to be espceially aimed not at the levity of foolish men, who with equal readiness let out what should be uttered, and what should be concealed; not at that plain speaking, with which they inveigh without discrimination and judgment against everybody and everything; not at that talkativeness whereby they weary others *usque ad nauseam;* but at another fault which is less observed, namely, a method of discourse of all others most unwise and impolitic; I mean, when a man in private conversation so frames his discourse as to produce whatsoever he has to say pertinent to the matter in hand all at once and in a breath, without any stop or pause. Now this is a great impediment to business.

[23] Annal. xii. 3. [24] Histoire de Comines, i. 16. [25] Prov. xii. 10.
[26] See Busbequius, Ep. 3.—*J. S.* [27] Prov. xxix. 11.

For in the first place a speech that is broken and let fall part by part makes far more impression than a continuous one; because in the latter the matters touched are not distinctly and severally apprehended and weighed; and they have not time enough to settle; but one reason drives out another before it has taken firm hold. Secondly, no one is endowed with such powerful and persuasive eloquence as with the first stroke of his discourse to make his listeners dumb and speechless, but the other party will always have some answer to make, and will perhaps raise objections; and then it falls out, that the arguments which should have been reserved for refutation or reply, having been used and tasted beforehand, lose their strength and grace. Thirdly, if a man does not use all his arguments at once, but delivers them in parts, throwing in one after the other, he will detect by the countenance and answer of his opponent how each is taken, and what effect it produces, and he may thence take warning what to suppress and what to select in that which is to follow.

PROVERB.

(16.) If the spirit of the ruler rise up against thee, leave not thy place; for management pacifies great offences [28].

Explanation.

This proverb directs a man how to behave when he has incurred the wrath and indignation of his prince, and contains two precepts; first, that he resign not his place; and secondly, that he carefully and prudently apply himself to the remedy, as he would in the case of a serious disease. For generally, when men perceive the anger of princes stirred against them, partly through impatience of disgrace, partly that they may not by their presence irritate the wound, and partly that their princes may see their sorrow and contrition, they withdraw from their offices and appointments, nay sometimes they resign their places and dignities into his hands. But Solomon censures this remedy as injurious, and with good reason. For in the first place it makes the disgrace too public, whereby enemies and enviers become the bolder to attack, and friends the more timid to assist. Secondly, it thus happens that the anger of the prince, which if it had not been divulged might have died away of its own accord, is more deeply rooted, and having as it were commenced by displacing the person proceeds to his overthrow. Lastly, this resignation savours somewhat of a malcontent spirit and one offended with the times which aggravates anger with suspicion. The precepts for the remedy are these; first, let a man take care above all things neither by dullness on the one hand nor high spirit on the other to let it appear that he is insensible to the indignation of the prince, or not properly affected by it: that is, let him compose his countenance not to a sullen gloom but to a grave and modest sadness; and in all his duties and actions let him exhibit less cheerfulness and pleasure. It will be also advantageous for him to engage the assistance and mediation of some friend with the prince, who should take occasion at fit times to insinuate how deeply the offender is grieved. Secondly, let him carefully avoid all, even the slightest occasions, which may lead to the reopening of the subject of offence, or draw upon him fresh indignation or open rebuke, for whatever cause, from the prince. Thirdly, let him diligently seek for every occasion of making his services acceptable to the prince, that he may both show an anxious wish to redeem his past fault, and that the prince may perceive of how good a servant he will be deprived if he loses him. Fourthly, let him either contrive to transfer the fault to others, or insinuate that it was committed with no bad intention, or even let him point out the malice of those who complained of him to the king or exaggerated the matter more than it deserved. Lastly, let him be watchful in everything, and intent on the remedy.

PROVERB.

(17.) He that is first in his own cause is just, then comes the other side, and searches him [29].

[28] Eccles. x. 4.

[29] Prov. xviii. 17.

Explanation.

In every cause the first information, if it have dwelt for a little in the judge's mind, takes deep root, and colours and takes possession of it; insomuch that it will hardly be washed out, unless either some clear falsehood be detected in the matter of the information, or some deceit in the statement thereof. A bare and simple defence, though it be just and of greater weight, will hardly counterbalance the prejudice of the first information, or restore to an equilibrium the scales of justice which have once inclined. Wherefore as it is safest for the judge to know nothing of the merits of the case, till both parties are heard together, so it is the best course for the defendant, if he finds the judge prejudiced, to apply himself, as far as the case allows, to detect some fraud or deceit employed by the opposite party to abuse the judge.

PROVERB.

(18.) He that delicately brings up his servant from a child, shall afterwards find him insolent [30].

Explanation.

According to the advice of Solomon, princes and masters ought to keep a measure in conferring grace and favour on their servants. In this three points are to be observed; first, that the promotion be by steps, and not by jumps; secondly, that they be accustomed to an occasional disappointment; and thirdly, as Machiavelli well advises, that they should have ever before their eyes some ulterior object of ambition [31]. Otherwise princes will be requited by their servants with disrespect and contumacy instead of duty and gratitude; for sudden promotion begets insolence; continual obtaining of desires begets impatience of refusal; and if there be nothing further to aspire to, there will be an absence of alacrity and industry.

PROVERB.

(19.) Seest thou a man swift of despatch? he shall stand before kings, and shall not be reckoned among the mean [32].

Explanation.

Of all the qualities which kings especially look to and require in the choice of their servants, that of despatch and energy in the transactions of business is the most acceptable. Men of deep wisdom are objects of jealousy to kings, as being too close observers, and being able to use their abilities as an engine to turn and wind their masters against their will and knowledge. Popular men are disliked as standing in the light of kings and drawing the eyes of the people upon themselves. Men of great spirit and courage are often accounted turbulent and overdaring. Men of honour and integrity are reputed unmanageable and not pliant enough to all their masters' commands. Lastly, there is no other virtue which does not present some shadow of offence to the minds of kings. Expedition in the execution of their commands is the only one which contains nothing that is not acceptable. Moreover the minds of kings are hasty and impatient of delay; for they imagine that they have power to do what they will; all they want is, that it be done quickly; whence of all things despatch is most pleasing to them.

PROVERB.

(20.) I considered all the living which walk under the sun, with the second child who shall rise in his stead [33].

Explanation.

This proverb remarks upon the vanity of men, who are wont to crowd around the appointed heirs of princes. The root hereof is in that madness, deeply im-

[30] Prov. xxix. 21. The English version has, "Shall have him become his son at the length."

[31] We find Machiavelli's opinion, as to what the conduct of princes towards their ministers ought to be, in the twenty-second chapter of *Il Principe*. It hardly appears to justify the reference here made to him; but I have not met with any passage in his writings which contains precisely the remark in the text.

[32] Prov. xxii. 29.

[33] Eccles. iv. 15, *Vulgate*. The English version differs considerably.

planted by nature in human minds, of being too fond of their own hopes. For there is scarcely any one but takes more delight in what he hopes for than in what he has. Novelty also is very pleasing to man, and is eagerly sought after. Now in a prince's heir hope and novelty are combined. And this proverb implies the same as that which was said of old, first by Pompey to Sylla, and afterwards by Tiberius respecting Macro. "That there be more who worship the rising than the setting sun[34]." And yet princes are not much disturbed at this, nor do they care much for it, as neither Sylla nor Tiberius did; but they rather scorn the fickleness of mankind, and do not care to strive with dreams; and hope, as was said, is but the dream of a waking man[35].

PROVERB.

(21.) There was a little city, and few men within it; and there came a great king against it, and built great bulwarks round against it, and besieged it. Now there was found in it a poor wise man, and he by his wisdom delivered the city, yet no man remembered that same poor man[36].

Explanation.

This proverb notes the corrupt and ungrateful nature of mankind, who in distress and adversity have commonly recourse to the wise and active men, whom they formerly held in contempt; but as soon as the storm has passed over, they are found ungrateful to their preservers. Machiavelli might well make it a question, "Which was more ungrateful to their benefactors, a prince or a people?[37]" but meanwhile he implies that both are guilty of ingratitude. But the ingratitude of the prince or the people is not the only cause of this; there is added the envy of nobles, who are secretly displeased with the issue though fortunate and prosperous, because it did not originate in themselves; whence they both depreciate the merit of the work, and depress the author.

PROVERB.

(22.) The way of the slothful is as an hedge of thorns[38].

Explanation.

This proverb very elegantly expresses the fact, that sloth is in the end laborious. Diligence and careful preparation remove the obstacles against which the foot would otherwise stumble, and smooth the path before it is entered; but he who is sluggish and defers everything to the last moment of execution, must needs walk every step as it were amidst briars and thorns, which catch and stop him. This likewise may be noted in the management of a family; wherein if care and forethought be used, everything goes on smoothly and of itself, without noise and discord; but if they be wanting, on any important emergency everything has to be done at once, the servants are in confusion, and the house in an uproar.

PROVERB.

(23.) To have respect of persons in judgment is not good; for, for a piece of bread will that man forsake the truth[39].

Explanation.

This proverb most wisely marks that in a judge facility of disposition is more pernicious than bribery; for it is not every one that offers a bribe, but there is scarcely a case wherein something may not be found to bias the mind of the judge, if he be a respecter of persons. One man will be respected because he is popular; another because he has a shrewd tongue; another because he is rich; another because he is agreeable; another because he is recommended by a friend. In fine, where respect of persons prevails, there will be unequal measures everywhere, and for the most trifling reason, as it were for a morsel of bread, judgment will be perverted.

34 Tac. Ann. vi. 46., and Plut. in Pomp. c. 14.

35 Cf. Quintil. vi. 2. 30.

36 Eccl. ix. 14, 15.

37 Machiavelli, *Discorsi*, i. 29.

38 Prov. xv. 19.

39 Prov. xxviii. 21.

PROVERB.

(24.) A poor man that oppresses the poor, is like a sweeping rain, which causes famine [40].

Explanation.

This proverb was anciently figured and represented under the fable of the full and hungry horseleech; for the oppression of a poor and hungry man is far more severe than that of a rich and full one, inasmuch as the former practises all the arts of exactions, and searches every corner for money. The same used also to be likened to a sponge, which when dry sucks in strongly, but not so when wet. And it contains a useful warning for princes and peoples; for princes, that they commit not offices or the government of provinces to needy persons and such as are in debt; for peoples, that they allow not their rulers to be too much in want of money.

PROVERB.

(25.) A righteous man falling down before the wicked is as a troubled fountain and a corrupt spring [41].

Explanation.

This proverb teaches that an unjust and scandalous judgment in any conspicuous and weighty cause is above all things to be avoided in a state; especially where it involves, not the acquittal of the guilty, but the condemnation of the innocent. For particular injuries passing unpunished do indeed trouble and pollute the waters of justice, but it is only in the streamlets; whereas unjust judgments, such as we spoke of, infect and corrupt the very fountain-heads. For when the judgment seat takes the part of injustice, there succeeds a state of general robbery, and men turn wolves to each other, according to the adage [42].

PROVERB.

(26.) Make no friendship with an angry man, and walk not with a furious man [43].

Explanation.

The more religiously the laws of friendship are to be observed and honoured among good men, the more care should be taken to make a prudent selection of friends at the first. Now the disposition and manners of our friends, so far as they affect ourselves only, should by all means be borne with; but when they compel us to alter our bearing and deportment towards other men, the condition of the friendship becomes very hard and unfair. Wherefore, as Solomon advises, it is of the first importance for the peace and security of life to have no dealings with passionate men, or such as easily engage in disputes and quarrels; for they will perpetually involve us in strife and faction, so that we shall be compelled either to break off our friendship, or disregard our own safety.

PROVERB.

(27.) He that covers a transgression seeks friendship, but he that repeats a matter separates very friends [44].

Explanation.

There are two ways of making peace and reconciling differences; the one begins with an amnesty, the other with a recital of injuries, combined with apologies and excuses. Now, I remember that it was the opinion of a very wise man and a great politician, that "he who negotiates a peace, without recapitulating the grounds of difference, rather deludes the minds of the parties by representing the sweetness of concord, than reconciles them by equitable adjustment". But Solomon, a wiser man than he, is of a contrary opinion, approving of amnesty and forbidding recapitulation of the past. For in it are these disadvantages; it is as the chafing of a sore; it creates the risk of a new quarrel, (for the parties will never agree as to the proportions of injuries on either side);

[40] Prov. xxviii. 3. [41] Prov. xxv. 26. [42] Erasm. *Adag.* i. 1. 70.
[43] Prov. xxii. 24. [44] Prov. xvii. 9.

and lastly, it brings it to a matter of apologies; whereas either party would rather be thought to have forgiven an injury than to have accepted an excuse.

PROVERB.

(28.) In every good work there is abundance; but where there are many words there is commonly penury [45].

Explanation.

Herein Solomon makes a distinction between the fruit of the labour of the hand and that of the tongue; from the one proceeds abundance, from the other penury. For it generally happens that they who talk much, boast much, and make many promises, are needy persons, who make no profit of the things whereof they discourse. For the most part also they are no ways industrious and active in point of work, but merely feed and fill themselves with words, as with wind. Surely, as the poet says, "He that is silent is sure [46]"; —he who knows that he is succeeding in what he is about, is satisfied and holds his tongue; whereas he who feels that he has got hold of nothing but wind, betakes himself to talking and boasting.

PROVERB.

(29.) Open rebuke is better than secret love [47].

Explanation.

This proverb rebukes the mistaken kindness of friends, who do not use the privilege of friendship freely and boldly to admonish their friends, as well of their errors as their dangers. "What can I do," says a man of this character, "or what steps can I take? I love him as much as any one, and if any misfortune were to befall him I would gladly substitute myself in his place; but I know his disposition well; if I deal freely with him I shall offend him, or at all events put him out of humour, and do no good by it; and I should sooner estrange him from his friendship for me, than from those things which he has fixed his heart upon." Now a friend of this sort Solomon reprehends as weak and useless, affirming that more advantage may be gained from an open enemy than from such a man; for a man may chance to hear by way of reproach from an enemy, what the friend is too good-natured to utter.

PROVERB.

(30.) A wise man looketh well to his ways, but a fool turneth to deceit [48].

Explanation.

There are two kinds of wisdom; the one true and sound, the other degenerate and false, which Solomon does not hesitate to term folly. He who applies himself to the former takes heed of his own ways, foreseeing dangers, preparing remedies, employing the assistance of the good, guarding himself against the wicked, cautious in entering upon a work, not unprepared for a retreat, watchful to seize opportunities, strenuous to remove impediments, and attending to many other things which concern the government of his own actions and proceedings. But the other kind is entirely made up of deceits and cunning tricks, laying all its hopes in the circumventing of others, and moulding them to its pleasure; which kind the proverb denounces as being not only dishonest, but also foolish. For in the first place, it is not among the things which are in our own power, nor does it even depend on any certain rule; but fresh stratagems are daily to be contrived, as the old ones are used up and worn out. Secondly, a man who has once earned a character for deceit and trickery, entirely loses one of the principal instruments of business, which is credit; whence he will find everything turn out otherwise than he expects. Lastly, these very arts, however fair and specious they may appear, generally fail: as Tacitus has well remarked, "Bold and crafty counsels are fair in promise, hard in execution, and unfortunate in issue [49]".

[45] Prov. xiv. 23. [46] Ovid. *Rem. Amor.* 697. [47] Prov. xxvii. 5.
[48] Prov. xiv. 8. 15. [49] The words occur in Livy, xxxv. 32, and not in Tacitus.

PROVERB.

(31.) Be not righteous overmuch, neither make thyself over wise; why shouldest thou destroy thyself before thy time [50].

Explanation.

"There are seasons," says Tacitus [51], "wherein great virtues are the surest causes of ruin." And upon men eminent for virtue and justice it comes sometimes suddenly, sometimes long foreseen. But if they have also the gift of wisdom, that is, if they are cautious and watchful for their own safety, they gain this advantage; that their ruin comes upon them all at once and entirely through dark and secret plots, whereby envy is avoided, and destruction assails them unawares. But with regard to that *overmuch* which the proverb speaks of (as these are not the words of a Periander, but of Solomon, who, though he often takes notice of what is bad in human life, never enjoins it), we must not understand it of virtue itself (in which there can be no overmuch), but of the vain and invidious affectation and show thereof. Something of the same kind is suggested by Tacitus in speaking of Lepidus; placing it in the light of a miracle that he never uttered a servile opinion, and yet lived safely in such dangerous times: "The thought occurs to me," says he, "whether these things are controlled by fate, or whether it is in our power to steer an intermediate course between slavish obedience and abrupt contumacy, free alike from danger and from indignity [52]"

PROVERB.

(32.) Give opportunity to a wise man, and he will increase his wisdom [53].

Explanation.

Here distinction is made, between the wisdom which is grown and ripened into a true habit, and that which floats only in the conceit of the brain, or is boasted in talk and has no deep root. The former, upon occasion presented for its exercise, is instantly excited, made alert, and enlarged, so that it appears greater than it was; but the latter, which before the occasion was eager and active, when the emergency occurs, becomes amazed and confused; so that even he who considered himself possessed of it, begins to doubt whether the notions he had formed of it were not mere dreams and idle speculations.

PROVERB.

(33.) He who praises his friend with a loud voice, rising early in the morning, it shall be counted a curse to him [54].

Explanation.

Praises, when moderate and seasonable, and expressed on fit occasion, contribute greatly both to the reputation and fortune of men; but when immoderate, noisy, and unseasonably lavished, they do no good; nay rather (if we believe the Proverb), they do great harm. For in the first place they openly betray themselves as either springing from excessive partiality, or got up and affected for the purpose of gratifying the object of them by false encomiums, rather than of honouring him with his just attributes. Secondly, sparing and moderate praises generally invite the audience to add something to them; whereas lavish and immoderate praises provoke them to take off and detract. Thirdly (which is the principal point), he that is over-praised becomes an object of envy; for all excessive praises seem to point to the reproach of others who are no less deserving.

PROVERB.

(34.) As the face is reflected in the water, so is the heart of man manifest to the wise [55].

Explanation.

Here is distinguished between the mind of a wise man, and that of others; the former being compared to water or a glass which represents the forms and

50 Eccles. vii. 16.
51 Tac. Hist. i. 2.
52 Tac. Ann. iv. 20.
53 Prov. ix. 9.
54 Prov. xxvii. 14.
55 Prov. xxvii. 19.

images of things; the other to the earth, or an unpolished stone, which gives no reflection. And this comparison of the mind of a wise man to a glass is the more proper; because in a glass he can see his own image together with the images of others, which the eye itself without a glass cannot do. But if the mind of a wise man is sufficiently large to observe and distinguish an infinite variety of dispositions and characters, it only remains to take care that the application be as various as the representation. "A wise man will know how to adapt himself to all sorts of characters[56]."

Thus have I stayed perhaps somewhat longer on these Proverbs of Solomon than is agreeable to the proportion of an example, being led on by the dignity of the subject, and the renown of the author. Neither was this in use only with the Hebrews, but it is generally to be found in the wisdom of the ancients, that as men found out any observation which they thought good for life, they would gather it and express it in some short proverb, parable, or fable. Fables, as has been said elsewhere, were formerly substitutes and supplements of examples, but now that the times abound with history, the aim is more true and active when the mark is alive. And therefore the form of writing, which of all others is fittest for such variable argument as that of negotiation and scattered occasions, is that which Machiavelli most wisely and aptly chose for government; namely, Observations or Discourses upon Histories and Examples. For knowledge drawn freshly and in our view out of particulars knows best the way back to particulars again; and it contributes much more to practice, when the discourse or discussion attends on the example, than when the example attends upon the discourse. And this is not only a point of order, but of substance also. For when the example is laid down as the ground of the discourse, it is set down with all the attendant circumstances, which may sometimes correct the discourse thereupon made, and sometimes supply it, as a very pattern for imitation and practice; whereas examples alleged for the sake of the discourse are cited succinctly and without particularity, and like slaves only wait upon the demands of the discourse.

But it is worth while to observe this difference; that as Histories of Times are the best ground for such discourse upon governments as Machiavelli handles; so Histories of Lives are the most proper for discourse on business, because they include all kinds of occasions and transactions, both great and small. Nay, we may find a ground for discourse on business fitter than them both, which is discourse upon such Letters as are wise and weighty, like those of Cicero to Atticus, and others; for letters have a closer and more lively representation of business, than either annals or lives. Thus have we spoken both of the matter and form of this first part of the knowledge of negotiation touching scattered occasions, which we note to be deficient.

But yet there is another part of this knowledge, which differs as much from that whereof we have spoken, as general wisdom differs from wisdom for oneself; the one moving as it were from the centre to the circumference, the other from the circumference to the centre. For there is a wisdom of imparting counsel to others, and there is a wisdom of foresight for one's own fortunes; and these sometimes meet, but oftener sever. For many are wise in their own ways, who yet are weak for government or counsel, like ants, which are wise creatures for themselves, but very hurtful for the garden. This wisdom for oneself the Romans, though excellent guardians of their country, took much knowledge of; "For," says the comic poet, "a wise man fashions his fortune for himself[57]". And it grew into an adage amongst them, "Every man is the maker of his own fortune". And Livy attributes it to the elder Cato, "Such was his vigour of mind and understanding, that wherever he had been born he would have made his fortune[58]".

This kind of wisdom, if it be too much declared and professed, has always been regarded as not only impolitic, but unlucky and ill-omened; as was observed in Timotheus the Athenian, who having done many great services to the

[56] Ovid, *De Arte Amat.* i. 760:—Qui sapit innumeris moribus aptus erit.
[57] Plaut. *Trinummus*, ii. 2. 84. [58] Livy, xxxix. 40.

state in his government, and giving the customary account thereof to the people, concluded every particular with this clause, "And in this fortune had no part [59]". But it happened that he never prospered in anything he took in hand afterwards; for this is too high and too arrogant, savouring of that which Ezekiel says of Pharaoh, "Thou sayest, my river is mine own, and I have made it for myself [60]"; or of that which Habakkuk says, "They exult and offer sacrifices to their net [61]"; or of that which the poet expresses of Mezentius the despiser of the gods :—

> Dextra mihi Deus, et telum quod missile libro
> Nunc adsint [62].

Lastly, Julius Cæsar never, as far as I recollect, betrayed the weakness of his secret thoughts, except in a similar kind of speech. For when the augur brought him word that the entrails were not favourable, he murmured in a low voice, "They will be more favourable when I choose [63]"; which speech did not long precede the misfortune of his death. For this excess of confidence was ever as unlucky as unhallowed; and therefore great and truly wise men have thought it right to ascribe their successes to their fortune, and not to their skill or energy. Thus Sylla surnamed himself "the Fortunate [64]", not "the great"; and Cæsar (better in this instance than the last) said to the master of the ship, "You carry Cæsar and his fortune [65]".

Nevertheless, proverbs such as these, "Every man is the architect of his own fortune"; "A wise man shall rule over the stars;" "No path is impervious to virtue [66];" and the like; if taken and used as spurs to industry, and not as stirrups to insolence, rather to beget in men resolution and strength of judgment than arrogance or outward declaration, have ever been rightly held round and good, and are doubtless imprinted in the greatest minds, so as sometimes they can scarce contain such opinions within; as we see in Augustus Cæsar, (who, compared with his uncle, was rather unlike than inferior, though decidedly a man of more moderation), how on his death-bed he desired his friends around him to give him a "Plaudite" when he expired, as if he were conscious to himself that he had well played his part in life [67]. This part of knowledge I report also as deficient; not but that it is used and practised even more than is fit, but it has not been handled in books. And therefore according to my custom, I will as before set down some heads or passages of it, and call it the *Architect of Fortune, or the Knowledge of Advancement in Life.*

Wherein at first sight I shall appear to handle a new and unwonted argument, in teaching men how to raise and make their fortune; a doctrine indeed, wherein every man perchance will be ready to yield himself a disciple, till he has experience of the difficulty thereof. For the things necessary for the acquisition of fortune are neither fewer or less difficult nor lighter than those to obtain virtue; and it is as hard and severe a thing to be a true politician, as to be truly moral. But the handling hereof concerns learning greatly, both in honour and substance; in honour principally, that pragmatical men may not imagine that learning is like a lark, which can mount and sing and please itself and nothing else; but may know that it rather partakes of the nature of a hawk, which can soar aloft, and can also descend and strike upon its prey at pleasure. Again, it tends to the perfection of learning, because it is the perfect law of the inquiry of truth, "that nothing be in the globe of matter which has not its parallel in the globe of crystal or the understanding"; that is, that there be nothing in practice, whereof there is no theory and doctrine. Not however that learning admires or esteems this architecture of fortune otherwise than as an inferior work. For no man's fortune can be an end worthy of the gift of being that

[59] Plut. in Sylla, c. 6. [60] Ezek. xxix. 3. [61] Habak. i. 16.

[62] Virg. *Æn.* x. 773 [Cf. vii. 648; x. 880] :—

> My own right hand and sword assist my stroke,
> These gods alone Mezentius will invoke.

[63] Sueton, in Julio, c. 77. [64] Plut. in Sylla, c. 6. [65] Plut. *de Roman. Fort.* p. 319.

[66] Ovid, *Met.* xiv. 113. [67] Sueton. in Aug. c. 99.

has been given him by God; and often the worthiest men abandon their fortunes willingly, that they may have leisure for higher pursuits. But nevertheless, fortune as an instrument of virtue and merit deserves its own speculation and doctrine.

To this doctrine are attached certain precepts, some summary, and some scattered or various; whereof the former relate to the just knowledge of ourselves and others. Let the first precept then (on which the knowledge of others turns) be set down as this: that we obtain (as far as we can) that window which Momus required [68]; who seeing in the frame of man's heart such angles and recesses found fault that there was not a window to look into its mysterious and tortuous windings. This window we shall obtain by carefully procuring good information of the particular persons with whom we have to deal; their natures, their desires and ends, their customs and fashions, their helps and advantages, with their principal means of support and influence; so again their weaknesses and disadvantages, where they lie most open and obnoxious; their friends, factions, patrons and clients; their enemies, enviers, and competitors; their moods and times;

(Sola viri molles aditus et tempora noris) [69]:

lastly, their principles, fashions, prescribed rules, and the like; and this not only of particular persons, but also of the particular actions which are on foot from time to time, and as it were under the anvil; how they are directed and succeed, by whom promoted or opposed, what is their weight and importance, what consequences they involve, and the like. For the knowledge of present actions is not only material in itself, but without it also the knowledge of persons will be very treacherous and erroneous; for men change with actions, and whilst they are involved and engaged in them they are one, and when they return to their nature they are another. These informations of particulars touching persons and actions, are as the minor propositions in every active syllogism; for no truth or excellence of observations or axioms (whence the major political propositions are drawn) can suffice to ground a conclusion, if there be error in the minor proposition. For the possibility of this knowledge Solomon is our surety; who says, "Counsel in the heart of man is like deep water, but a man of understanding will draw it out [70]". And although the knowledge itself falls not under precept, because it is of individuals, yet the instructions for obtaining it may be laid down with advantage.

Knowledge of men may be derived and obtained in six ways; by their countenances and expressions, their words, their actions, their dispositions, their ends, and lastly, by the reports of others. With regard to the countenance, be not influenced by the old adage, "Trust not to a man's face [71]"; for though this may not be wrongly said of the general outward carriage of the face and action, yet there are some more subtle motions and labours of the eyes, mouth, countenance, and gesture, by which (as Q. Cicero elegantly expresses it), the "door of the mind [72]" is unlocked and opened. Who more close than Tiberius Cæsar? Yet Tacitus, in noting the different character and manner of speaking which he employed in commending the exploits of Germanicus and Drusus to the Senate, says, that his praises of Germanicus were set forth "in words which appeared rather studied for outward effect, than as if he really felt them"; but of his praises of Drusus, he says, "that he said less, but spoke more earnestly and

[68] See Lucian in Hermotim. 20. But as Bacon, in the Essay on Building, alludes to a part of the story which Lucian does not tell, it is probable that his knowledge of it was derived from some other source. The most obvious one is the Æsopic fable; but there Momus's wish is not quite the same as in the text. In the fable he complains not that there are no shutters, *δυρίδες*, in the breast, but that *αἱ φρένες* are inside it, and not on the surface.

[69] Virg. *Æn.* iv. 423:—

His times of access you alone can find,
And know the soft approaches to his mind.

[70] Prov. xx. 5. [71] Juv. ii. 8.:—Fronti nulla fides. [72] De Pet. Cons. § 11.

sincerely [73]". Again, Tacitus in speaking of this same Tiberius, and remarking on some speech, as being somewhat less ambiguous, says, "At other times he appeared to have a difficulty with his words, but he spoke more freely when he took anybody's part [74]"; so that it is hard to find any man so skilled and perfect in the art of dissimulation, or any countenance so controlled or commanded (as he calls it) as to sever from a feigned and dissembling tale all these marks, and prevent the style from being either more careless, or more adorned, or more tedious and wandering, or more dry and hard, than usual.

As for words, though they be (as physicians say of waters) full of trickery and deceit, yet they are excellently detected in two ways; namely, either when they are spoken on the sudden, or in passion. Thus we see Tiberius, being suddenly incensed at some stinging words of Agrippina, and thrown a little off his guard, advanced a step out of his natural dissimulation. "These words (says Tacitus) drew from him a voice seldom heard from that dark bosom, and taking her up sharply, he reminded her in a Greek verse that she was thus offended, because she did not reign [75]". And therefore the poet elegantly calls passions "tortures", which urge men to confess their secrets:

Vino tortus et irâ [76].

And experience shows that there are very few men so true to their own secrets, and so settled in their purpose, but that sometimes through anger, sometimes through bravado, sometimes through affection for their friends, sometimes through a weakness of mind unable any longer to bear the burden of its thoughts, and sometimes through some other affection, they open and communicate their secret thoughts and feelings; especially if they be put to it with a counter-dissimulation, according to the Spanish proverb, "Tell a lie, and find a truth".

Neither should deeds, though the most assured pledges which the human mind can give, be entirely trusted, without a judicious and careful consideration of their magnitude and nature. For the saying is most true, "that fraud begins by winning credit in small things, that it may deceive with greater advantage [77]"; and the Italian thinks himself upon the point of being bought and sold, if he is better used than he was wont to be, without manifest cause [78]. For small favours do but quiet and lull to sleep men's caution and industry, whence they are rightly called by Demosthenes, "sops to feed sloth [79]". Again, the treacherous and ambiguous character of some deeds, even such as are taken for favours, may be seen from that deception which Mucianus practised on Antonius Primus; when after the hollow and unfaithful reconciliation which was made between them, he advanced many of the friends of Antonius to great offices: "At the same time he bestows on his friends tribuneships and governments [80]"; wherein under pretence of strengthening Antonius, he entirely disarmed and isolated him by winning his friends.

But the surest key to unlock the minds of men is by searching and thoroughly understanding either their natures and characters, or their intentions and ends; wherein the weaker and more simple sort are best interpreted by their natures, but the wiser and more reserved by their ends. For it was both pleasantly and wisely said, though I think very untruly, by a nuncio of the Pope, on his return from an embassy to a certain nation, where he had served as legate; who, when

[73] Tacit. *Ann.* i. 52. In the *Advancement of Learning*, this passage stands thus:—"None more close than Tiberius, and yet Tacitus saith of Gallus, *Etenim vultu offensionem conjectaverat.* So again, noting," etc. The passage referred to is in *Annals*, i. 12, and was probably omitted by an oversight, for it is quite in point.—*J. S.*

[74] Tac. *Ann.* iv. 31.

[75] Ib. iv. 52.

[76] Hor. *Ep.* i. 18. 38:—

Tortured with wine and wrath.

[77] Livy, xxviii. 42.

[78] Bacon alludes to the Italian proverb:—

Chi mi fa più caresse chè non suole
O m'a ingannato, o ingannar mi vuole.

[79] Cf. Dem. *Olynth.* iii. 33, and 1 *Philipp.* towards the end.

[80] Tac. *Hist.* iv. 39.

his opinion was asked touching the appointment of his successor, gave as his advice, "in no case to send one who was remarkably wise, but one rather of moderate abilities; because (said he) no very wise man would ever imagine what they in that country were likely to do". And certainly it is a very frequent error, especially among wise men, to measure others by the standard of their own genius, and to shoot over the mark, by supposing that men have deeper ends in view, and more subtle schemes than ever entered into their minds; as is elegantly expressed by the Italian proverb, which remarks, "There is always less money, less wisdom, and less faith, than men imagine[81]". Wherefore in men of a meaner capacity, because they do many foolish things, we must form our opinion rather from the propensities of their natures, than from their designs and ends.

Princes also (though for a very different reason) are best interpreted by their natures, but private persons by their ends. For princes, being at the summit of human desires, have for the most part no particular ends whereto they earnestly and constantly aspire, by their position and distance from which a measure and scale of the rest of their actions might be taken; and this is one of the principal causes why their hearts are (as the Scriptures declare) inscrutable[82]. But every private person is like a traveller striving earnestly to arrive at the end of his journey where he may rest; whereby it is not difficult to conjecture what he will do, and what he will not do. For if it be a means to his end he will probably do it; but if opposed to his end, he will probably not do it. Nor is it enough to inform ourselves only of the variety of men's natures and ends simply; but we should also examine them comparatively, and find what it is that predominates and directs the rest. Thus, we see, when Tigellinus saw himself outstripped by Petronius Turpilianus in providing pleasures and catering to Nero's humours, "he wrought" (says Tacitus) "on Nero's fears[83]", whereby he displaced his rival.

As for the knowing of men at second hand from the reports of others, a few words will suffice. Men's weaknesses and faults are best known from their enemies, their virtues and abilities from their friends, their customs and times from their servants, their opinions and thoughts from their familiar friends, with whom they discourse most. General fame is light, and the judgments of superiors are not much to be trusted; for to them men are more masked. "The truest character comes from a man's own household[84]."

But to all this inquiry the most compendious way rests in three things; the first is to have a general acquaintance with those who have a varied and extensive knowledge both of persons and things; but especially to endeavour to have at least some particular friends who, according to the diversity of business and the diversity of persons, can give perfect and solid intelligence in every several kind. The second is to keep a discreet temper and mediocrity both in liberty of speech and in secrecy; in most cases using liberty, but secrecy when the occasion requires it. For liberty of speech invites and provokes a similar liberty in others, and so brings much to a man's knowledge; but secrecy induces trust, so that men like to deposit their secrets there, as in their own bosom. The last is the gradual reducing of a man's self to such a watchful and ready habit of mind, that in every conference and action he may both carry on the matter in hand, and also observe other incidents. For as Epictetus lays down that a philosopher in every particular action should say to himself, "I both wish to do this, and also to keep to my rule[85]": so a political man in everything should inwardly resolve, "I will both do this, and learn something more for future use." Wherefore those who are so intent and absorbed in the matter which they have in hand, that they have not even a thought to spare for anything that may turn up by the way (which Montaigne confesses to have been his weakness[86]), are indeed the best servants of kings and commonwealths, but fail in advancing their own fortunes. Meanwhile special care should be taken to restrain too great an

[81] Quoted also in the *Advancement of Learning*. [82] Pro. xxv. 3.
[83] Tac. Ann. xiv. 57. [84] Q. Cicero, De Pet. Cons. 5. [85] Epict. *Enchir.* c. 9.
[86] Cf. Essay *De l'Utilité et de l'Honnêteté*.

energy and zeal of mind, lest by much knowledge we be drawn on to much meddling, than which nothing is more unfortunate and rash. So that this variety of knowledge of persons and things, which I recommend to be gained, returns in the end only to this, that we make a more judicious choice of the actions we undertake, and of the persons whose assistance we use; whereby we may manage and conduct everything with more safety and dexterity.

Next to the knowledge of others comes the knowledge of self. And here we must use even greater care in gaining good and accurate information touching ourselves, than touching others; since the oracle "know thyself" is not only a rule of universal wisdom, but has a special place in politics. For St. James says well, "That he who looks at his face in a glass, yet suddenly forgets what manner of man he was [87]"; so that there is need of very frequent inspection. And this holds good likewise in politics, though the glasses are different; for the divine glass in which we ought to behold ourselves is the Word of God, but the political glass is nothing else than the state of the world or times wherein we live.

Men ought therefore to take an accurate and impartial survey of their own abilities, virtues, and helps; and again, of their wants, inabilities, and impediments; making the account in such a manner that the former are always estimated rather more, and the latter rather less than they really are. From this examination they should frame the following considerations.

First, to consider how their natural and moral constitution sort with the general state of the times; which if they find agreeable and consonant, then in all things to give themselves more scope and liberty, and indulge their dispositions; but if there be anything differing and discordant, then in the whole course of their life to be more close, retired, and reserved. And this we see in Tiberius, who being conscious that his tastes did not well suit with the age, never attended the public games, and during the twelve last years of his life never even went into the Senate; whereas Augustus lived ever in men's eyes, which Tacitus observes: "Tiberius's habits (says he) were different [88]". Pericles also acted on the same principle.

Secondly, to consider how their nature sorts with the professions and courses of life which are in use and repute, and whereof they have to make election; so that if their profession is not already determined, they may make choice of that which is most fit and agreeable to their disposition; but if they have already entered on a path of life for which they are not naturally suited, that they may leave it the first opportunity, and adopt a fresh profession. And this we see was done by Valentine Borgia [89], who was brought up by his father to the priesthood, but afterwards quitted it in obedience to his own inclination, and betook himself to a military life; although equally unworthy of the office of prince and priest, seeing that he dishonoured both.

Thirdly, to consider how they sort with their equals and rivals, whom they are like to have as competitors in their fortune; and to take that course of life wherein there is the greatest scarcity of distinguished men, and they themselves are likely to be most eminent. As Julius Cæsar did, who at first was an orator and pleader, and devoted himself entirely to a civil life; but when he saw how Cicero, Hortensius, and Catulus excelled in eloquence, and that there was no man of any great reputation in military matters but Pompey, he forsook the course he had begun, and bidding a long farewell to a civil greatness, transferred his designs to the arts of a soldier and a general; whereby he mounted to the highest power of the state [90].

Fourthly, to consider their own nature and disposition in the choice of their friends and dependences. For different natures require different kinds of friends: to some is suited such as are solemn and silent; to others such as are bold and arrogant, and so on. And it is worthy of mark what kind of men the friends of Julius Cæsar were (namely, Antony, Hirtius, Pansa, Oppius, Balbus, Dola-

87 St. James, i. 23, 24. 88 Tacit. *Ann.* i. 54.

89 Better known as Cæsar Borgia, son of Pope Alexander the Sixth. After his change of profession, for an account of which see Guicciardini, vi. 3, he was made Duke of the Valentinois, and is therefore spoken of by Italian writers as "il duca Valentino". Bacon has here used this title as a prænomen. 90 See Plutarch, *Cæsar*, c. 3.

bella, Pollio, and the rest), who used to swear, "that they were ready to die, so Cæsar might live[91]", displaying an infinite affection for Cæsar, but arrogance and contempt towards every one else; men active in the execution of business, but of no great character or reputation.

Fifthly, to take especial heed how they guide themselves by examples, and not vainly to endeavour to frame themselves upon other men's models; as if what is open to others must needs be open to them, not at all reflecting how far the nature and character of their models may differ from their own. And it was this error into which Pompey evidently fell, who, as Cicero has recorded, was so often wont to say, "Sylla could do this, why should not I[92]?" Wherein he was much deceived, the nature and proceedings of himself and Sylla being as far removed as the heaven from the earth; the one being fierce, violent, and in everything pressing on to the end; the other solemn, respectful of the laws, and regulating everything with a view to his dignity and character, which made him far less strong and effectual in accomplishing his designs. There are likewise other precepts of this nature, but these will be enough for an example of the rest.

But it is not enough for a man only to know himself; for he should consider also of the best way to set himself forth to advantage; to disclose and reveal himself; and lastly, to turn and shape himself according to occasion. Now for the first we see nothing more usual than for the worse man to make the better external show. It is therefore no unimportant attribute of prudence in a man to be able to set forth to advantage before others, with grace and skill, his virtues, fortunes, and merits (which may be done without arrogance or breeding disgust); and again, to cover artificially his weaknesses, defects, misfortunes, and disgraces; dwelling upon the former and turning them to the light, sliding from the latter or explaining them away by apt interpretations, and the like. Tacitus says of Mucianus, the wisest and most active politician of his time, "That he had a certain art of setting forth to advantage every thing he said or did[93]". And it requires indeed some art, lest it become wearisome and contemptible; but yet it is true that ostentation, though carried to the first degree of vanity, is rather a vice in morals than in policy. For as it is said of calumny "calumniate boldly, for some of it will stick," so it may be said of ostentation (except it be in a ridiculous degree of deformity), "boldly sound your own praises, and some of them will stick". It will stick with the more ignorant and the populace, though men of wisdom may smile at it; and the reputation won with many will amply countervail the disdain of a few. But if this self-display whereof I am speaking be carried with decency and judgment, as with a natural, candid, and ingenuous bearing; or if it be employed in times of danger, as by military persons in the time of war, or at times when others are most envied; or if what a man says in his own praises appears to drop carelessly and unintentionally, without being dwelt upon too long or too seriously; or if a man at the same time that he praises does not refrain from ridiculing and finding fault with himself; or if he do it not spontaneously, but appears provoked and challenged to it by the reproaches and insolence of others, it adds greatly to his reputation. And surely no small number of those who are of a solid nature, and who from the want of this ventosity cannot spread all sail in pursuit of their own honour, suffer some prejudice and lose dignity by their moderation.

But for this enhancement of virtue, though some persons of weaker judgment and perhaps too scrupulous morality may disapprove of it, yet no one will deny that we ought at least to take care that virtue be not undervalued and unduly debased through neglect. This depreciation in the price of virtue may be effected in three ways: first, by a man offering and obtruding himself and his services in any business when he is unasked and uncalled for; wherein men think he is rewarded, if he be not rejected. Secondly, by doing too much at the commencement of an action, and by performing all at once what ought to be done by degrees; which in matters well managed procures a premature favour at first, but in the end induces satiety. Thirdly, by feeling too soon and easily the fruit of

[91] The phrase in Balbus's letter to Cicero *Ep. ad Att.* ix. 8.
[92] Cic. *Ep. ad Att.* ix. 10. [93] Tac. *Hist.* ii. 80.

virtue in commendation, applause, honour, and favour, and being content therewith; on which there is a prudent warning, "Take care lest you appear unaccustomed to great things if you are thus delighted by a small thing, as if it were great [94]".

But a diligent covering of defects is of no less importance than a prudent and skilful display of virtues. Defects may be principally concealed in three ways, and as it were under three coverts; namely, caution, colour, and confidence. Caution is, when men discreetly avoid those things to which they are not equal; whereas contrariwise bold and unquiet spirits thrust themselves without reflection into matters of which they have no experience, and so publish and proclaim all their defects. Colour is, when men warily and skilfully make and prepare a way for themselves, for a favourable and convenient construction of their faults or wants; as proceeding from a better cause, or intended for some other purpose, than is commonly imagined. For as to the concealment of vice, it is well said by the poet, that "vice often hides itself in the neighbourhood of virtue [95]". And therefore, whatsoever want a man has, he must take care to borrow the mask and colour of the neighbouring virtue that shadows it; as if he be dull, he must affect gravity; if a coward, mildness; and so on. It will be of advantage also for a man to frame and spread abroad some probable reason why he shrunk from doing his best, that the want of power may be imputed to want of will. As to confidence, it is indeed an impudent, but yet the surest and most effectual remedy; namely, for a man to profess to depreciate and despise whatsoever he cannot obtain; after the principle of prudent merchants, whose business and custom it is to raise the price of their own commodities, and to beat down the price of others. But there is a confidence which surpasses this other in impudence; and this is, for a man to brazen out his own defects, by putting them forward and displaying them to view; as if he believed himself especially eminent in those things wherein he is deficient. And the more easily to impose on others, he should appear to have least opinion of himself in those things wherein he is really the best: just as we see it is the practice of poets, who when they recite their verses, and you except to any, will immediately say "that that line cost them more labour than any of the rest"; and presently they will bring forward some other verse, which they know well enough to be the best in the number and the least open to objection, and seeming to suspect it themselves they will ask your opinion of it. But above all, if a man means to make a good figure and maintain his just position in the world, I consider it of the greatest importance to him, not to show himself disarmed and exposed to scorn and injury by too much goodness and sweetness of nature; but rather in everything to exhibit from time to time some sparks of a free and noble spirit and one that carries with it no less of the sting than of the honey. This kind of fortified carriage, with a spirit ready and prepared to defend itself against insults, is sometimes accidentally forced upon men by something inherent in their person or fortune; as in the case of persons deformed, illegitimate, or disgraced. Whence men of this nature, if ability be not wanting, commonly turn out fortunate.

With regard to the disclosing of a man's self, it is a very different thing from the self-display of which I have been speaking. For it relates not to a man's virtues or faults, but to his particular actions in life; wherein there is nothing more politic for a man than to preserve a sound and wise mediocrity in declaring or concealing his meaning in particular actions. For although depth of secrecy and concealment of designs, and that manner of action which effects everything by dark arts and methods (or *menées sourdes* as the French call them), be both useful and admirable; yet frequently, as is said, dissimulation breeds errors which ensnare the dissembler himself. Whence we see that the greatest and most noted politicians have not hesitated to declare freely and undisguisedly the objects which they had in view. So Lucius Sylla made open profession "that he wished all men happy or unhappy, as they stood his friends or enemies". So Cæsar, when he first went into Gaul, did not scruple to profess, "that he had

[94] Rhetor. ad Heren. iv. 4.

[95] Ovid, *Art. Amat.* ii. 662: Sæpe latet vitium proximitate boni.

rather be first in a village than second in Rome[96]". And again, as soon as he had begun the war, he by no means played the part of a dissembler, if we may judge by what Cicero says of him, "The other (meaning Cæsar) does not refuse, but rather demands to be called a tyrant, as he really is[97]". So we see in a letter of Cicero to Atticus, how little of a dissembler Augustus Cæsar was; for on his very entrance into public life, when he was still the darling of the senate, yet in his harangues to the people he would use this form of oath; "As I hope to attain to the honours of my parent[98]"; which was nothing less than the tyranny. It is true indeed that to lessen the envy of it, he would at the same time stretch forth his hand towards a statue of Julius Cæsar which was erected in the place; whereat men laughed, and applauded, and wondered, and said to one another, "What is this? What sort of young man is this?" and yet thought a man could mean no mischief who spoke his feelings so openly and ingenuously. Now all these, whom I have mentioned, were prosperous; whereas Pompey, who tended to the same ends, but in a more dark and dissembling manner (as Tacitus says of him, "A more reserved, but not a better character[99]"; wherein Sallust concurs, "Of honest tongue and shameless mind[100]"), made it his design, and attempted by innumerable intrigues to keep his own ambition and desires quite secret, and in the mean time to drive the state into such anarchy and confusion that it should be forced of necessity to cast itself into his arms, and the sovereign power might thus be thrust upon him, apparently against his will and inclination. But when he had brought it, as he thought, to this point, when he was chosen sole consul (as no one had ever been before), yet was he no way nearer to his ends; because they who certainly would have assisted him did not understand what he wanted; so that in the end he was fain to go on the common and beaten track of procuring arms and raising an army under colour of opposing Cæsar; so tedious, uncertain, and mostly unfortunate are those designs which are concealed beneath a deep dissimulation. And this appears to have been the feeling of Tacitus, when he constitutes the artifices of dissimulation as a wisdom of an inferior form to the arts of true policy, attributing the former to Tiberius, but the latter to Augustus; for speaking of Livia, he says, "That she was equally suited to the arts of her husband, and the dissimulation of her son[101]".

With regard to turning and shaping the mind, we must strive with all possible endeavour to render the mind obedient to occasions and opportunities, and to be no ways obstinate and refractory towards them. For nothing hinders men's actions or fortunes so much as this, "to remain the same, when the same is unbecoming[102]"; that is, for men to be as they were, and follow their own nature, when occasions change: whence Livy, in introducing Cato the Elder, as a most skilful architect of his fortune, adds well of him, "That he had a wit that could turn[103]". This also is the reason why grave and solemn wits, which know not how to change, have generally more dignity than good fortune. But this viscous and knotty temper which is so averse to change is nature in some; in others it is the result of habit (which is a second nature), and an opinion (which easily steals into men's minds), namely, that men can hardly make themselves believe that they ought to change that course which they have found by experience to be prosperous and successful. For Machiavelli notes wisely, how Fabius Maximus would have retained to the last his old habit of temporising and protracting the war, when the nature of the war was altered and required more vigorous measures[104]. In others again the fault arises from weakness of judgment, that they do not discern in time when things or actions have reached a period, but come in too late, when the occasion has passed by; as Demosthenes says, when in reproving the Athenians he compares them to country fellows, who, in playing in the fencing school, when they have received a blow, always remove their shield to that ward, and not before[105]. In others again it is a dislike to lose

[96] Plutarch's *Apophthegms*.
[97] Cic. *Ep. ad Attic.* x, 4.
[98] Cic. *ad Attic.* xvi. 15.
[99] Tac. *Hist.* ii. 38.
[100] Cf. Sueton. *de Claris Grammaticis*, c. 15.
[101] Tac. *Ann.* v. 1.
[102] Cf. Cic. *de Clar. Orat.* c. 95: Idem manebat neque idem decebat.
[103] Livy. xxxix. 40.
[104] Machiavelli, *Discorsi*, iii. 9.
[105] Demosth. 1 *Philipp.* 46.

their labours in the path which they have once entered, and an unwillingness to sound the retreat, with a confidence that by perseverance they will overcome the occasion. But from whatsoever root this stubbornness and restiveness of mind proceeds, it is a thing most prejudicial to man's actions and fortunes ; and nothing is more politic than to make the wheels of the mind concentric and voluble with the wheels of fortune. And so much for the two summary precepts of this Architecture of Fortune ; whereof the scattered precepts are numerous, but I will select a few for example's sake.

The first precept is that the carpenter of fortune should make a good use and a right application of his rule ; that is, that he should accustom his mind to judge of the proportion and value of all things, as they conduce more or less to his fortune and ends, and that he do this substantially, not superficially. It is a thing strange, but true, that the logical part (if I may so term it) of many men's minds is good, but the mathematical part erroneous ; that is, they can judge well enough of the consequences, but most unskilfully of the values of things ; whence it happens that some take delight in private and secret converse with princes, others in popular fame and applause, supposing them to be things of great value ; whereas in many cases they are full both of envy and peril. Others again measure things according to the labour and difficulty bestowed upon them, and think if they be only moving they must needs advance and proceed ; as Cæsar said ironically of Cato of Utica, when he describes how laborious and assiduous and indefatigable he was to no great purpose, " All these things he did with much earnestness [106] ". Hence too it comes that men often deceive themselves, in thinking that if they procure the assistance of any man of worth and reputation, they are certain to succeed ; whereas it is not the greatest but the fittest instruments that finish the work both quickest and best. Now for the true information of the mathematical part of the mind, it is worth while to know and have a description of what should be set down first for the raising and advancing of a man's own fortune, what second, and so on. First I set down the amendment of the mind ; for the removing of impediments and working out the knots of the mind will sooner open the passage to fortune, than the obtaining of fortune will remove the impediments of the mind. In the second place I set down wealth and means, which many perhaps would have placed first, because of their great use in everything; but that opinion I may condemn, for the reason which Machiavelli gave in a case not much unlike. For whereas there was an old proverb, " that money is the sinews of war [107] ", yet he maintained on the contrary that the true sinews of war are nothing else than the sinews of a valiant and military people. And so in like manner it may be truly affirmed, that it is not money that is the sinews of fortune, but it is rather the sinews of the mind, wit, courage, audacity, resolution, temper, industry, and the like. In the third place, I set down character and reputation, the rather because they have certain tides and seasons, which if they be not taken in due time are difficult to be recovered, it being extremely hard to restore a falling reputation. And lastly, I place honour, which is more easily won by any of the other three, much more by all combined, than if you begin with honour, and then proceed to the rest. But as it is of no little consequence to preserve order in matter, so it is of no less consequence to preserve order in time, the confusion whereof is one of the commonest errors ; while men fly to their ends, when they should only be attending to their beginnings ; and carelessly passing over the things which lie before them they rush at once to the highest and greatest of all ; whereas it is a good precept, " Attend to present business [108] ".

A second precept is to beware of being carried by an excess of magnanimity and confidence to things beyond our strength, and not to row against the stream. It is excellent counsel regarding men's fortunes, " Be ruled by the Fates and the Gods [109] " ; for we ought to look round and observe where things lie open to us and where they are closed and obstructed, where they are difficult and where

[106] Cæs. *Bell. Civil.* i. 30. [107] Machiavelli, *Discorsi,* ii. 10 ; and cf. Cicero, Philip. v.
[108] Virg. *Eclog.* ix. 66 : Quod nunc instat agamus.
[109] Lucan, viii. 486 : Fatis accede Deisque.

easy, that we may not waste our strength on things to which convenient access is forbidden. For in this way we shall avoid repulse, not occupy ourselves too much about one matter, earn a character for moderation, offend fewer persons, and get the credit of continual success ; whilst things which would perhaps have happened of themselves will be attributed to our industry.

The third precept seems to be somewhat repugnant to the former two, though not so if rightly understood. The nature of it is this, that we should not always wait for occasions, but sometimes challenge and induce them ; and it is that to which Demosthenes alludes in high terms, " In the same manner as it is a received principle that the general should lead the army, so should wise men lead affairs, causing things to be done which they think good, and not themselves waiting upon events [110] ". For if we diligently observe, we shall find two different kinds of sufficiency in performing actions and managing business. Some can make an apt use of occasions, but plot or invent nothing of themselves ; others are wholly bent on their own plots, but cannot take advantage of accidental opportunities ; either of which abilities without the other is very lame and imperfect.

A fourth precept is to undertake nothing which of necessity takes up a great quantity of time, but to have this sound ever ringing in our ears, " time is flying, time which cannot be retrieved [111] ". And this is the reason why those who have devoted themselves to laborious professions and the like, as lawyers, orators, learned divines, and writers of books, are not so clever in founding and promoting their own fortunes ; because their time is so much occupied with other things that they cannot investigate particulars, wait occasions, and devise and meditate on plots to advance their fortunes. Moreover, in the courts of princes and in commonwealths you will find that the ablest persons both to improve their own fortunes and to assail the fortunes of others are those who have no public duty to perform, but are ever occupied in this study of advancement in life.

A fifth precept is to a certain degree to imitate nature, which does nothing in vain ; no very difficult task, if a man will skilfully mix and interlace his several kinds of business. For in every particular action a man ought so to direct and prepare his mind, and should have one intention so underlying and subordinate to another, that if he cannot obtain his wishes in the best degree, he may yet be satisfied if he succeed in a second, or even a third ; and if he cannot obtain them at all in that particular, then he may turn the labour spent in it to some other end besides the one intended ; and if he cannot reap any fruit of it for the present, he may yet make it as a seed of somewhat in time to come ; and if he can derive no substance from it either now or hereafter, he may try at all events to win some good opinion by it, or the like ; by always exacting an account of himself, by which it may appear that each action and scheme has borne him some fruit more or less, and never allowing himself to stand amazed and confused, or to despond immediately that he fails to hit his chief mark. For nothing is more impolitic than to be entirely bent on one action. He that is so loses an infinite number of occasions, which indirectly fall out by the way, and are perhaps more proper and propitious for the future use than for the present matter ; wherefore men must be perfect in that rule, " These things ought ye to do, and not to leave the others undone [112] ".

A sixth precept is not to engage oneself too peremptorily in anything, though at first sight it seem not liable to accident ; but ever to have either a window open to fly out at, or a secret way to retire by.

A seventh precept is that ancient precept of Bias, not construed to any point of perfidiousness, but only to caution and moderation, " Love as if you were sometime to hate, and hate as if you were sometime to love [113] " ; for it utterly

[110] Demosth. *Philip.* i. 45.

[111] Virg. *Georg.* iii. 284 :—

Sed fugit interea, fugit irreparabile tempus.

[112] St. Matth. xxiii. 23 ; St. Luke, xi. 42.

[113] La Bruyère's remarks on this precept are, I think, worth transcribing —" Vivre avec nos ennemis comme s'ils devoient un jour être nos amis, et vivre avec nos amis comme s'ils pouvoient devenir nos ennemis, n'est ni selon la nature de la haine, ni selon les règles

betrays and destroys all utility, for men to embark themselves too far in unfortunate friendships, troublesome and turbulent quarrels, or foolish and childish jealousies and emulations.

These will suffice for an example of the doctrine of advancement in life. I would however have it frequently remembered, that I am far from meaning that these sketches of things which I note as deficients should be set down as complete treatises, but only as shreds or fragments to serve as samples of the whole piece. Nor again am I so foolish as to assert that fortunes are not gained without all this contrivance which I have mentioned. For I well know they come tumbling into some men's laps; and that others only obtain them by simple diligence and attention (using only a little caution), without any great or laborious art. But as Cicero, in his portrait of a perfect orator, does not mean that every pleader should be or can be such; and again, as in the description of a prince or courtier by such as have handled those subjects, the model is always framed according to the perfection of the art, and not according to common practice; so likewise have I done in the description of a politic man, I mean politic for his own fortune.

But it must be remembered all this while, that the precepts which I have selected and set down on this subject are of that kind which may be called *Good Arts*. As for *Evil Arts*, if a man would propose to himself that principle of Machiavelli, "that virtue itself a man should not trouble himself to attain, but only the appearance thereof to the world, because the credit and reputation of virtue is a help, but the use of it is an impediment"; or again, that other principle of his "that a politic man should have for the basis of his policy the assumption that men cannot fitly or safely be wrought upon otherwise than by fear; and should therefore endeavour to have every man, as far as he can contrive it, dependent and surrounded by straits and perils [114]"; so that his politician would appear to be what the Italians call "A sower of thorns:" or that principle embodied in the verse quoted by Cicero, "Let friends fall, provided our enemies perish with them [115]"; as the Triumvirs did, who with the lives of their friends purchased the destruction of their enemies: or if he would be an imitator of L. Catiline, to set on fire and trouble states, that he may the better fish in muddy waters and make way for his own fortune; "For," said he, "if a fire be lighted in my fortunes, I will quench it, not with water, but with destruction [116]": or if he would make his own that saying of Lysander, "that children are to be deceived with comfits, and men with oaths [117]"; with the like depraved and pernicious doctrines, whereof (as in all other things) there are a greater number than of the wise and good: if any one, I say, takes pleasure in such kind of corrupt wisdom, I

de l'amitié : ce n'est point une maxime morale, mais politique. On ne doit pas se faire des ennemis de ceux qui mieux connus pourroient avoir rang entre nos amis. On doit faire choix d'amis si sûrs et d'une si exacte probité que venant à cesser de l'être, ils ne veuillent pas abuser de notre confiance, ni se faire craindre comme nos ennemis."—*Les Caractères*, c. 4.

[La Bruyère's rule would, I think, be perfect, if it were possible to make a certain judgment of each man's character beforehand. The defect of it is, that, taking no account of the necessary uncertainty of all such judgments, it fails to give any practical direction in the real affairs of life. Put it thus:—"Treat no man as your enemy unless you are *sure* that he can never deserve to be your friend, make no man your friend unless you are *sure* that he will never become your enemy;"—and your practical direction becomes much the same as that of Bias, The question which in morals is really disputable is whether a man should encourage himself to doubt other men, or not to doubt; and this, being a question of more or less, cannot be determined except in reference to particular cases. No man will say generally either that you cannot doubt too much, or that you cannot doubt too little. Perhaps the best general direction that can be given is to lean *against* your natural inclination whichever way it goes. If you are naturally inclined to distrust appearances, trust them more; if to trust, trust them less.—*J. S.*]

114 Cf. Machiavelli, *Principle*. c. 17, 18.

115 Cic. *pro Deiotaro*, c. 9.

116 Cic. *pro Murœn*. c. 25; Sallust, *Cat.* c. 31.

117 Plut. in Lysand. c. 8. The saying seems, however, not to be Lysander's. He apparently adopted it from Polycrates of Samos.

will not certainly deny that (with these dispensations from all the laws of charity and virtue, and an entire devotion to the pressing of his fortune,) he may advance it quicker and more compendiously. But it is in life as it is in ways, the shorter way is commonly the foulest and muddiest, and surely the fairer way is not much about.

But men ought to be so far removed from devoting themselves to wicked arts of this nature, that rather (if they are only in their own power, and can bear and sustain themselves without being carried away by a whirlwind or tempest of ambition) they ought to set before their eyes not only that general map of the world, "that all things are vanity and vexation of spirit [118]", but also that more particular chart, namely, "that being without well-being is a curse, and the greater being the greater curse"; and "that all virtue is most rewarded, and all wickedness most punished in itself;" as the poet excellently says:—

> Quæ vobis, quæ digna, viri, pro laudibus istis
> Præmia posse rear solvi? pulcherrima primum
> Dii moresque dabunt vestri [119].

And so on the other hand, it is no less truly said of the wicked, "His own manners will be his punishment [120]". Secondly, men in projecting their schemes and diffusing their thoughts abroad on every side, in order to forecast and advance their fortunes, ought in the midst of these flights of the mind to look up to the Eternal Providence and Divine Judgment, which often overthrows and brings to nought the machinations and evil designs of the wicked, however deeply laid; according to that Scripture, "He hath conceived mischief, and shall bring forth a vain thing [121]". Moreover, although men should refrain themselves from injury and evil arts, yet this incessant, restless, and as it were sabbathless pursuit of fortune leaves not the tribute which we owe to God of our time; whom we see demands and separates for himself a tenth part of our substance, but a seventh of our time. For what advantage is it to have a face erected towards heaven, with a spirit perpetually grovelling upon earth, eating dust like the serpent? As the heathen also observed, "the particle of the Divine Spirit cleaves to the ground [122]". But if here any man flatter himself, that he will employ his fortune well, though he should obtain it ill; as was said concerning Augustus Cæsar, and Septimius Severus, "that either they should never have been born, or else they should never have died [123]", they did so much mischief in their rise to greatness, and so much good when they were established; yet, let him bear in mind that such compensations of evil with good are to be approved after the evil is done, but that such counsels are to be condemned. Lastly, it will not be amiss for men, in this eager and excited chase of fortune, to cool themselves a little with that conceit which is elegantly expressed by the Emperor Charles the Fifth in his instructions to his son, "That fortune has somewhat of the nature of a woman, who, if she is too much wooed, is commonly the further off [124]". But this last is only a remedy for those whose tastes are corrupted by a disorder of the mind. Let men rather build upon that foundation, which is as a corner stone of both Divinity and Philosophy, wherein they nearly agree as to that which ought to be

118 Eccles. ii. 11.

119 Virg. *Æn.* ix. 252:—

> Ye brave young men, what equal gifts can we,
> In recompense of such desert, decree?
> The greatest, sure, and best you can receive,
> The gods and your own conscious worth will give.

120 Cic. *Ep. ad Att.* ix. 12.

121 Psalm vii. 14, or Job, xv. 35.

122 Hor. *Sat.* ii. 2. 79: Atque affigit humo divinæ particulam auræ.

123 Aurelius Victor, Epit. c. 1. Lampridius de Severo.

124 It was on being obliged to raise the siege of Metz that Charles V. remarked that Fortune was like a woman,—that, after having favoured him in his youth, she turned against him when he was no longer young. There are, I believe, several papers of instructions addressed by him to Philip II. In one or two which I have seen the remark mentioned in the text does not appear to occur.

sought first. For Divinity says, "Seek ye first the kingdom of God, and all these things shall be added unto you[125]"; and philosophy says something like it, "Seek ye first the good things of the mind, and the rest will either be supplied or their loss will not be felt". And although the human foundation is sometimes built upon the sand, as we see in Marcus Brutus, when he brake forth into that speech at his death,—

Te colui, Virtus, ut rem; ast tu nomen inane es[126]:

yet the same foundation, laid by the hand of heaven, is ever laid upon the rock. Here then I conclude the doctrine concerning advancement in life, and with it the general doctrine concerning negotiation.

CHAPTER III.

The Divisions of the Doctrine concerning Empire or Government are omitted;—An Introduction only is made to two Deficients; namely, the Doctrine concerning the Extension of the Bounds of Empire, and the Doctrine concerning Universal Justice, or the Fountains of Law.

I NOW come to the Art of Empire or Civil Government, which includes Œconomics, as a state includes a family. On this subject, as I before said, I have imposed silence on myself, though perhaps I might not be entirely unqualified to handle such topics with some skill and profit, as being one who has had the benefit of long experience, and who by your Majesty's most gracious favour, without any merit of his own, has risen through so many gradations of office and honour to the highest dignity in the realm and borne the same for four whole years; and, what is more, being one who has been accustomed for eighteen successive years to the commands and conversation of your Majesty (whereby a very stock might be turned into a statesman), and who also, besides other arts, has spent much time in the study of laws and histories. All which I report to posterity, not through any vain boasting, but because I think that it is of no little importance to the dignity of literature, that a man naturally fitted rather for literature than for anything else, and borne by some destiny against the inclination of his genius into the business of active life, should have risen to such high and honourable civil appointments under so wise a king. But if my leisure time shall hereafter produce anything concerning political knowledge, the work will perchance be either abortive or posthumous. In the mean time, now that all the sciences are ranged as it were in their proper seats, lest so eminent a seat should be left entirely vacant, I have determined to mark as deficient only two parts of Civil Knowledge, which do not belong to the secrets of Empire, but have a wider and more common nature; and according to my custom I will set down examples thereof.

The Arts of Government contain three political duties; first, "the preservation," secondly, "the happiness and prosperity," and thirdly, "the extension," of empire. Of these the two former have in a great measure been excellently handled, but nothing has been said on the last. I will therefore set it down among the deficients, and according to custom, propose an example thereof, calling it "*the Statesman in Armour,*" or the "*Doctrine concerning the Extension of the Bounds of Empire*".

125 St. Matth. vi. 33.

126 Virtue I worshipped, and as real, sought,
But found her empty, and a thing of nought—

which, according to Dio Cassius, xlvii., was the dying exclamation of Brutus. From the way in which the lines are introduced by Dio Cassius, they appear to be a fragment of a speech of Hercules in some lost tragedy. The first line and the first portion of the second (which, in effect, is all that is here translated) occur not only in Dio Cassius, but also in Plutarch, *De Superstitione*, where, however, no reference is made to Brutus. Most editions of Dio Cassius are accompanied by a Latin translation. In the earlier ones of those which I have seen, the words in question are given in prose, and in the later in iambic verse.

Example of a Summary Treatise touching the Extension of Empire.[1]

THE speech of Themistocles if applied to himself was certainly haughty and arrogant, but if generally applied to others it seems to contain both a wise observation and a severe censure. Desired at a feast to touch a lute, he said, " he could not fiddle, but yet he could make a small town a great city [2]". Now these words, transferred to a political meaning, excellently describe and distinguish two very different abilities in those that deal in business of state. For if a true survey be taken of the councillors, senators, and other public statesmen who have ever been, there will be found some, though very few, who can make a small city or kingdom great, and yet cannot fiddle ; and on the other hand, there will be found many very cunning on the lute or lyre (that is, in the follies of courts), who yet are so far from having the power to make a small state great, that they appear rather to be naturally gifted to bring a great and flourishing state to ruin and decay. And certainly those degenerate arts and shifts, whereby many councillors and ministers often gain both favour with their masters and estimation with the people, deserve no other name than a certain knack of fiddling ; being things rather pleasing for the time, and graceful to themselves only, than advantageous to the weal and advancement of the state, whereof they are ministers. There will no doubt be found other councillors and ministers, of no mean character, equal to their business, and able to govern the state well, so as to preserve it from manifest precipices and inconveniences, who nevertheless are far removed from the ability to raise and amplify an empire in power, means, and fortune.

But be the workmen what they may be, let us consider the work itself ; that is, what is the true greatness of kingdoms and states, and how it can be obtained It is a subject indeed fit for princes to have ever in their hands and carefully to consider ; to the end that neither by over-measuring their forces they may engage in vain enterprises beyond their power ; nor on the other hand by undervaluing them they may demean themselves to timid and pusillanimous counsels.

The greatness of an empire as regards its size and territory falls under measure ; as regard its revenue under computation. The number of the population may be taken by a census ; the number and greatness of cities and towns by maps and surveys. But yet there is nothing among civil affairs more subject to error than the forming a true and right valuation of the power and forces of an empire. The kingdom of heaven is likened not to an acorn or any larger nut, but to a grain of mustard seed [3] ; which is the smallest of all seeds, but yet has within itself a certain property and spirit hastily to get up and spread. So there are some kingdoms and states very great in extent of territory, and yet not apt to enlarge or command ; and some that have but a small dimension of stem, and yet are apt to be the foundations of great monarchies.

Walled towns, stored arsenals and armouries, goodly races of horse, chariots of war, elephants, ordnance, artillery, and the like ; all this is but a sheep in a lion's skin, except the breed and disposition of the people be stout and warlike. Nay, number itself in armies is not much advantage where the people are of weak courage ; for, as Virgil says, it never troubles the wolf how many the sheep be [4]. The army of the Persians in the plains of Arbela was such a vast sea of people, that it somewhat alarmed the commanders in Alexander's army ; who came to him therefore and wished him to set upon them by night ; but he answered, He would not pilfer the victory : and the defeat was easy [5]. When Tigranes the Armenian, being encamped upon a hill with four hundred thousand men, discovered the army of the Romans being not above fourteen thousand marching towards him, he made himself merry with it, and said, " Yonder men are too many for an embassage, and too few for a fight [6]". But before the sun set, he found them enough to give him the chase with infinite slaughter. Many are the examples of the great

[1] This Latin treatise on the Extension of Empire is nearly an exact translation of the Essay on the True Greatness of Kingdoms. I have therefore retained the original English with very few alterations. [2] Plut. in Them. 2.

[3] St. Matth. xiii. 31. [4] Virg. *Ecl.* vii. 52. [5] Plut. in Alex. c. 31.

[6] Plutarch in Lucull. 27. and Appian, *Bell. Mithrid.* c. 85.

odds between number and courage; so that it may be set down as a sure and tried rule, that the principal point of greatness in any state is that the people itself be by race and disposition warlike. Neither is money the sinews of war, as it is trivially said, where the sinews of men's arms in base and effeminate people are failing. For Solon said well to Crœsus, when in ostentation he showed him his gold, "Sir, if any other come that has better iron than you, he will be master of all this gold [7]". Therefore let any prince or state think soberly of his forces, except his militia of natives be of good and valiant soldiers. And let princes, on the other side, who have subjects of martial disposition, know their own strength, unless they be otherwise wanting unto themselves. As for mercenary forces (which is the usual help in this case), all examples show that whatsoever state or prince rests upon them, he may spread his feathers for a time, but he will mew them soon after.

The blessing of Judah and Issachar will never meet; that the same people or nation should be both the lion's whelp, and the ass between burdens [8]. Neither will it be that a people over-laid with taxes should ever become valiant and martial. It is true that taxes levied by consent of the state do abate men's courage less; as it has been seen notably in the excises of the Low Countries [9]; and, in some degree, in the subsidies of England. For you must note, that we speak now of the heart and not of the purse. So that although the same tribute and tax, laid by consent or by imposing, be all one to the purse, yet it works differently upon the courage. So that you may conclude that no people over-charged with tribute is fit for empire.

Let states that aim at greatness take heed that their nobility and gentlemen do not multiply too fast; for that makes the common subject grow to be a peasant and base swain, driven out of heart, and in effect but the gentlemen's labourer. Even as you may see in coppice woods; if you leave your staddles too thick, you will never have clean under-wood, but shrubs and bushes. So in countries, if the gentlemen be too many, the commons will be base; and you will bring it to that, that scarce one man in an hundred will be fit for an helmet; especially as to the infantry, which is the nerve of an army: and so there will be great population and little strength. This which I speak of has been nowhere better seen than by comparing of England and France; whereof England, though far less in territory and population, has been nevertheless an overmatch; and for this reason, that the yeomen and lower classes of England make good soldiers, which the peasants of France do not. And herein the device of King Henry the Seventh (whereof I have spoken largely in the history of his life) was profound and admirable: in making farms and houses of husbandry of a standard; that is, maintained with such a proportion of land attached inseparably to them, as may allow a subject to live in convenient plenty and no servile condition; and to keep the plough in the hands of the owners, or at least the tenants, and not mere hirelings. And thus indeed you shall attain to Virgil's character which he gives to ancient Italy:

Terra potens armis, atque ubere glebæ [10].

[7] Cf. Lucian's *Charon*.

[8] Gen. xlix. 9. 14.

[9] The excise, or accise (from acciisse) was originally in the Low Countries a municipal tax; it seems to have arisen from a privilege granted by Charles V. in 1536 to certain towns, of imposing duties on wine, beer, and woollen and silken stuffs. See *Histoire Générale des Provinces-Unies*, i. 236. That the inhabitants of these countries were from an early time jealous of the administration of public money appears from the following passage from *Meteranus*: "Status Belgici, Italico et Gallico gravatai bello, novennalem exactionem Regi consentiunt: harum autem pecuniarum administrationem et præsidiorum atque turmarum publice merentium satisfactionem suo officio reservant: quæ res multis sibi id arrogantibus non parum displicuit: hinc Regis subditorumque mutua alienatio et offensio orta, cum Regi esset persuasum, hoc modo suæ majestati summopere derogari".—*Hist. Belg. Meter.* in anno 1554.

[10] Virg. *Æn.* i. 531:—

Hesperia called, a land divinely blessed,
Of strength in arms and fruitful soil possessed.

Neither is that state, (which, for anything I know, is almost peculiar to England, and hardly to be found anywhere else, except it be perhaps in Poland,) to be passed over; I mean the state of free servants and attendants upon noblemen and gentlemen, who are no ways inferior to the yeomanry as infantry. And therefore out of all question, the splendour and magnificence and great retinues and hospitality of noblemen and gentlemen received into custom conduce much unto martial greatness; whereas, contrariwise, the close and reserved living of noblemen and gentlemen causes a penury of military forces.

By all means it is to be looked to, that the trunk of Nebuchadnezzar's tree of monarchy [11] be great enough to bear the branches and the boughs; that is, that the natural subjects of the crown or state bear a sufficient proportion to the stranger subjects that they govern. Therefore all states that are liberal of naturalization towards strangers are fit for empire. For to think that an handful of people can, with the greatest courage and policy in the world, embrace too large extent of dominion, it may hold for a time, but it will fail suddenly. The Spartans were a difficult and jealous people in point of naturalization; whereby, while they kept their compass, they stood firm; but when they did spread, and their boughs were become too great for their stem, they became a windfall upon the sudden. Never was any state in this point so open to receive strangers into their body as were the Romans; therefore it sorted with them accordingly, for they grew to the greatest monarchy. Their manner was to grant naturalization, which they called the right of citizenship, and to grant it in the highest degree, that is, not only the right of commerce, the right of marriage, the right of inheritance; but also, the right of voting, and the right of bearing office; and this not to single persons alone, but likewise to whole families; yea, to cities, and sometimes to nations. Add to this their custom of plantation of colonies, whereby the Roman plant was removed into the soil of other nations: and putting both constitutions together, you will say that it was not the Romans that spread upon the world, but it was the world that spread upon the Romans: and that was the surest way of greatness. I have marvelled sometimes at Spain, how they clasp and contain so large dominions with so few natural Spaniards; but sure the whole compass of Spain is a very great body of a tree, far above Rome and Sparta at the first. And besides, though they have not had that usage to naturalize liberally yet they have that which is next to it; that is, to employ, almost indifferently, all nations in their militia of ordinary soldiers; yea, and sometimes in their highest commands [12]. Nay, it seems at this instant they are sensible of this want of natives and desire to remedy it; as appears by the pragmatical sanction published in this year [13].

[11] Daniel, c. iv.

[12] *E.g.* Bourbon, Prosper, Colonna, Pescara, Egmont, Castaldo, Parma, Piccolomini, Spinola. Of these, however, one or two might almost be called Spaniards; and it must be remembered that the dominions of Charles V. and his successors extended beyond the natural limits of the Spanish monarchy.

[13] In 1618, the Cortes, among other projects of reformation, petitioned the king not to grant any licences for monastic foundations.

The excessive multiplication of religious houses had attracted the attention of the government long before; and the opinions of a number of ecclesiastics were taken on the subject, in 1603, but nothing further seems to have been done. Subsequently, however, to the representation of the Cortes, the state of the kingdoms belonging to the crown of Castile was referred by the king to the council of Castile; and their report, which is given at full length in Davila's *Life of Philip the Third* (see chap. 86), is known as the *Gran Consulta de 1619*. The distress and depopulation of the parts of Spain to which it refers are stated in very strong language, the causes assigned being mainly excessive and oppressive taxation, the increase of luxury, and the non-residence of the rich on their estates. To relieve the revenue, the revocation of royal grants, when any fair reason could be found for doing so, is recommended. Sumptuary laws are also proposed, and some regulations tending to the relief of the agricultural class. The king is also advised to be cautious in granting licenses to religious houses. Ortiz states expressly that no measures were taken to carry out the recommendation of the council during the reign

It is certain that sedentary and within-door arts, and delicate manufactures (that require rather the finger than the arm), have in their nature a contrariety to a military disposition. And generally all warlike people are a little idle, and love danger better than work; neither must they be too much broken of it, if they shall be preserved in vigour. Therefore it was great advantage in the ancient states of Sparta, Athens, Rome, and others, that they had the use of slaves who commonly dispatched those manufactures. But that is abolished in greatest part by the Christian law. That which comes nearest to it is to leave those arts chiefly to strangers, who for that purpose are to be invited or at least easily received, and to contain the principal bulk of the vulgar natives within those three kinds; tillers of the ground, free servants, and handicraftsmen of strong and manly arts, as smiths, masons, carpenters, and the like; not reckoning professed soldiers.

But above all, for empire and greatness, it is of most importance that a nation profess arms as their principal honour, study, and occupation. For the things which we have formerly spoken of are but qualifications for the use of arms; and what is qualification without intention and act? Romulus after his death (as they report or feign), sent an injunction to the Romans, that above all they should attend to arms, and then they should prove the greatest empire of the world [14]. The fabric of the state of Sparta was wholly and carefully (though not wisely) framed and composed to that scope and end to make the people warriors. The Persians and Macedonians had it for a flash. The Britons, Gauls, Germans, Goths, Saxons, Normans, and others, had it for a time. The Turks have it at this day, (being not a little stimulated thereto by their law,) though in great declination. Of Chris-

of Philip the Third; a statement which seems to be fully confirmed by the silence of so copious and seemingly so painstaking an annalist as Gonzalez Davila. The assertion to be found in some French and English books, that the king made a decree in virtue of which those who introduced agricultural improvements on their estates were ennobled, is in itself exceedingly improbable, and has perhaps no other foundation than the imagination of some French economist who may have been misled by the circumstance that in the Cortes of 1618 something was done with respect to proofs of nobility. I speak, however, without having seen Navarrete's *Conservacion della Monarquia*. Soon after the accession of Philip the Fourth a royal decree or *Pragmatica* was published which attempted to carry out some of the recommendations of the council, and which gave certain privileges to persons who married, and further immunities to those who had six children. For some account of its provisions, see Cespedes' *History of the first Six Years of Philip the Fourth* (published at Lisbon in 1631, and reprinted in Spain in 1634), book 3, cc. 17, 18. Cespedes does not precisely fix the date of the decree, but it was plainly issued some time in the summer of 1622, and is no doubt that to which Bacon refers. The date assigned by Desormeaux, namely, the 10th of February 1624, is manifestly wrong; the sumptuary part of the enactment was suspended on the occasion of the visit of Prince Charles in 1623. See Mead's Letters to Stuteville, in *Ellis's Letters*.

It is a historical commonplace to assert that the depopulation of Spain was caused by the expulsion of the Moriscos, but this alone could not have produced so permanent an effect. The energies of the country were exhausted by excessive and unequal taxation; and the increase of the number of religious houses, especially of those belonging to the Mendicant Orders, aggravated the evil. Ranke has justly remarked that Spain must always have been a thinly peopled country; and he might have added, a country in which there seems always to have been a tendency to become depopulated. Thus in a passage of the *Siete Partidas*, quoted in the *Gran Consulta*, it is said to be part of the duty of the king to see that the population of places does not fall off. Even the word *despoblado* suggests a different idea from that which is expressed by weald or wilderness. It may be well to remark that there seems no reason to doubt that the population of Spain is much greater now than it was in the sixteenth century, although for a considerable time there must have been a decrease. Cassmany, in an interesting essay on the subject, has shown how much exaggeration there is in the statements made by Spanish writers of the sixteenth and seventeenth centuries, as to the population and manufacturing industry of the country in earlier times. According to him the population reached its minimum about 1700.

[14] Livy, i. 16.

tian Europe they that still have it are in effect only the Spaniards. But it is so plain that every man profits most in that to which he most attends, that it needs not to be stood upon. It is enough to point at it; that no nation, which does not directly profess arms and devote themselves to the practice thereof, may look to have any special greatness fall into their mouths. And on the other side it is a most certain oracle of time, that those states that continue long in that profession (as the Romans and Turks principally have done), do wonders in extension of empire: and those that have professed arms but for an age have notwithstanding commonly attained that greatness in that age which maintained them long after, when their profession and exercise of arms has grown to decay.

Incident to this point is for a state to have those laws or customs which may reach forth unto them just occasions or at least pretexts for making war. For there is that justice imprinted in the nature of men, that they enter not upon wars (whereof so many calamities do ensue), but upon some weighty, at the least specious, grounds and quarrels. The Turk has at hand for the cause of war the propagation of his law or sect; a quarrel that he may always command. The Romans, though they esteemed the extending the limits of their empire to be great honour to their generals when it was done, yet they never rested upon that alone to begin a war. First therefore, let nations that pretend to greatness have this, that they be quickly sensible of wrongs, either upon borderers, merchants, or public ministers; and that they sit not too long upon a provocation. Secondly, let them be prompt and ready to give aids and succours to their confederates and allies, as it ever was with the Romans, insomuch, as if the confederate had leagues defensive with divers other states, and upon invasion offered did implore their aids severally, yet the Romans would ever be the foremost and leave it to none other to have the honour. As for the wars which were anciently made on the behalf of a kind of party or tacit conformity of estate, I do not see how they may be well justified; as when the Romans made a war for the liberty of Greece; or when the Lacedemonians and Athenians made wars, to set up or pull down democracies and oligarchies; or when wars were made by commonwealths and princes, under the pretence of justice or protection, to deliver the subjects of others from tyranny and oppression, and the like. Let it suffice for the present, that no estate expect to be great that is not awake upon any just occasion of arming.

No body can be healthful without exercise, neither natural body nor politic; and certainly to a kingdom or estate a just and honourable war is the true exercise. A civil war indeed is like the heat of a fever; but a foreign war is like the heat of exercise, and serves most of all to keep the body in health. For in a slothful peace both courage will effeminate and manners corrupt. But howsoever it be for happiness, without all question for greatness, it makes to be still for the most part in arms; and the strength of a veteran army (though it be doubtless a costly business), always on foot, is that which commonly gives the law, or at least the reputation amongst all neighbour states, as may be well seen in Spain; which has had, in one part or other, a veteran army almost continually, now by the space of six-score years [15].

To be master of the sea, is an abridgment of a monarchy [16], Cicero writing to Atticus of Pompey's preparation against Cæsar, says, "Pompey's counsel is plainly that of Themistocles, for he thinks that whoever is master of the sea is master of the empire[17]". And without doubt Pompey had tired out and reduced Cæsar, if upon vain confidence he had not left that way. We see the great effects of battles by sea from many instances. The battle of Actium decided the empire of the world. The battle of Lepanto arrested the greatness of the Turk [18]. There be certainly many examples where sea-fights have put an end to the war; but

[15] Commencing, that is, with the wars in Italy which arose out of the invasion of that country by Charles VIII.

[16] [*Orig.* "Maris dominium monarchiæ quædam *epitome* est." The sense is obscure. See the same passage in essays, p. 773.—ED.]

[17] Cic. *Ep. ad Att.* x. 8.

[18] Orig. *Pugna ad Insulas Cursolares.* The Insulæ Cursolares or Kurzolari islands are the ancient Echinades. The naval engagement generally, though perhaps incorrectly, called the Battle of Lepanto, took place off these islands in 1571. The Turkish fleet was defeated with great loss. It was on this occasion that Cervantes lost his hand.

this is when princes or states have risked their whole fortune upon the battles. But thus much is certain, that he that commands the sea is at great liberty, and may take as much and as little of the war as he will; whereas those that be strongest by land are many times nevertheless in great straits. Surely at this day with us of Europe the advantage of strength at sea (which is one of the principal dowries of this kingdom of Great Britain) is great; both because most of the kingdoms of Europe are not merely inland, but girt with the sea most part of their compass; and because the wealth and treasures of both Indies seem in great part but an accessory to the command of the sea.

The wars of latter ages seem to be made in the dark, in respect of the glory and honour which reflected upon men from the wars in ancient time. There be now for martial encouragement some degrees and orders of chivalry, which nevertheless are conferred promiscuously upon soldiers and no soldiers; and some remembrance perhaps upon the escutcheon, and some hospitals, for maimed soldiers, and such like things. But in ancient times, the trophies erected upon the place of the victory, the funeral laudatives and monuments for those that died in the wars, the crowns and garlands personal, the style of emperor, which the great kings of the world afterwards borrowed, the triumphs of the generals upon their return, the great donatives and largesses to the soldiers upon the disbanding of the armies, these, I say, and such like dazzling honours, were things able to inflame all men's courage and excite even the coldest breast; but above all, that of the triumph among the Romans was not a pageant or gaudery but one of the wisest and noblest institutions that ever was. For it contained three things, honour to the general, riches to the treasury out of the spoils, and donatives to the army. But that honour perhaps were not fit for monarchies, except it be in the person of the monarch himself or his sons; as it came to pass in the times of the Roman emperors, who did appropriate the actual triumphs to themselves and their sons for such wars as they achieved in person, and left only for wars achieved by subjects some triumphal garments and ensigns to the general.

To conclude: no man can, by taking thought, as the Scripture saith, "add one cubit to his stature[19]" in this little model of a man's body; but in the great frame of kingdoms and commonwealths it is in the power of princes or states to add amplitude and greatness to their kingdoms. For by wisely introducing such ordinances, constitutions, and customs, as we have now touched, and others like them, they may sow greatness to their posterity and succession. But these counsels are commonly not observed, but left to take their chance.

Such then are the thoughts that now occur to me touching the extension of empire. But what avails this consideration, seeing that the Roman is supposed to have been the last of earthly monarchies? Yet because the extension of empire was set down as the last of the three political duties, I could not have passed it by altogether without deviating from my proposed course. There remains now the other of the two deficients which I mentioned; namely, the treatise of Universal Justice, or the Fountains of Equity.

All who have written concerning laws have written either as philosophers or lawyers. The philosophers lay down many precepts fair in argument, but not applicable to use: the lawyers, being subject and addicted to the positive rules either of the laws of their own country or else of the Roman or Pontifical, have no freedom of opinion, but as it were talk in bonds. But surely the consideration of this properly belongs to statesmen, who best understand the condition of civil society, welfare of the people, natural equity, customs of nations, and different forms of government; and who may therefore determine laws by the rules and principles both of natural equity and policy. Wherefore let it be my present object to go to the fountains of justice and public expediency, and endeavour with reference to the several provinces of law to exhibit a character and idea of justice, in general comparison with which the laws of particular states and kingdoms may be tested and amended. I will now therefore according to my custom set forth an example thereof in one of its heads.

[19] St. Matth. vi. 27; St. Luke, xii. 25.

Example of a Treatise on Universal Justice or the Fountains of Equity, by Aphorisms : one Title of it.

PREFACE.

APHORISM I.

IN Civil Society, either law or force prevails. But there is a kind of force which pretends law, and a kind of law which savours of force rather than equity. Whence there are three fountains of injustice ; namely, mere force, a malicious ensnarement under colour of law, and harshness of the law itself.

APHORISM 2.

The ground on which private right rests is this. He who commits an injury, receives either pleasure or profit from the act, but incurs danger from the precedent. For others do not share in the particular pleasure or profit, but look upon the precedent as concerning themselves. And hence they readily agree to protect themselves by laws, that the course of injury may not come round to them in turn. But if through the state of the times, and a communion of guilt, it happen that those whom a law protects are not so numerous or so powerful as those whom it endangers, a party is made to overthrow the law ; and this is often the case [20].

[20] The doctrine of this aphorism resembles that of Hobbes, inasmuch as there is no recognition of the principle that moral ideas lie at the root of civil rights. All the evidence of which the nature of the subject admits tends to show that society has always been held together, not by fear, but by notions more or less perfectly developed of the distinction between right and wrong ; and to assert that in the absence of any such notions selfish fear could serve as the " firmamentum juris privati," is at best to assert that which never has been proved and never can be.

Of course it is not meant to deny that fear is the principle by means of which the moral force of society becomes efficient in the repression of crime.

[That a notion of the distinction between right and wrong in general lies at the bottom of all our notions of individual rights and wrongs ; that when we think of one man as doing an *injury* to another, we think of him as doing something not only in its effect hurtful, but in its nature unjust ; I do not think Bacon would have denied. That in the absence of any such notion the interest which all men have in protection from injury would lead them to concur in the measures necessary to secure protection to each, he would not, I think, have affirmed. But such questions did not enter into the practical problem with which he had to deal ; which was this : Given our common notions of right and wrong, *jus* and *injuria*, with all their constituent elements, what is the principle by which they are made to bear upon the protection of individuals ? To this he answers : It is the interest which each individual has in being himself protected. That the personal interest would be insufficient without the sanction of the " moral idea " to stimulate and support it, is probably true ; for we see that actions the most dangerous to society, if committed by madmen, and therefore not objects of moral disapprobation, are exempted from punishment ; the necessity of self-defence requiring only that measures be taken to prevent the recurrence of them, and the sense of justice refusing to sanction any further severity. But that the " moral idea," unassisted by the sense of personal interest, could be still less relied upon as a " firmamentum privati juris," seems to me still more certain ; for we see that the penalties exacted or denounced by the laws, though proportioned with tolerable accuracy to the danger of the offence, bear no proportion at all to the moral disapprobation of which it is the object. Actions which are morally wrong in the highest degree, if they be such as every man may protect himself against, are not punished at all. Actions which the moral sense scarcely condemns, if such that the general permission of them would entail a general insecurity of property, are punished with great severity. And the truth seems to be, that to make an action seem a fit object of punishment, there must be *something* morally offensive in it, but that the nature and amount of punishment varies according to the interest of society in preventing it, and the difficulty of effecting that end. Men are not content with less severity than they think necessary for their protection, nor do they feel justified in using more.—*J. S.*]

APHORISM 3.

Private right depends upon the protection of public right. For the law protects the people, and magistrates protect the laws; but the authority of the magistrates depends on the sovereign power of the government, the structure of the constitution, and the fundamental laws. Wherefore, if this part of the constitution be sound and healthy, the laws will be of good effect, but if not, there will be little security in them.

APHORISM 4.

It is not however the only object of public law, to be attached as the guardian of private right, to protect it from violation and prevent injuries; but it extends also to religion, arms, discipline, ornaments, wealth, and in a word, to everything that regards the well-being of a state.

APHORISM 5.

The end and scope which laws should have in view, and to which they should direct their decrees and sanctions, is no other than the happiness of the citizens. And this will be effected, if the people be rightly trained in piety and religion, sound in morality, protected by arms against foreign enemies, guarded by the shield of the laws against civil discords and private injuries, obedient to the government and the magistrates, and rich and flourishing in forces and wealth. And for all these objects laws are the sinews and instruments.

APHORISM 6.

This end the best laws attain, but many pass wide it. For there is a strange and extreme difference in laws; some being excellent, some moderately good, and others entirely vicious. I will therefore set down, according to the best of my judgment, what may be called certain "laws of laws," whereby we may derive information as to the good or ill set down and determined in every law.

APHORISM 7.

But before I proceed to the actual body of particular laws, I will take a brief survey of the virtues and dignities of laws in general. That law may be set down as good, which is certain in meaning, just in precept, convenient in execution, agreeable to the form of government, and productive of virtue in those that live under it.

TITLE I.

Of the Primary Dignity of Laws, that they be certain.

APHORISM 8.

Certainty is so essential to law, that law cannot even be just without it. "For if the trumpet give an uncertain sound, who shall prepare himself to the battle[21]?" So if the law give an uncertain sound, who shall prepare to obey it? It ought therefore to warn before it strikes. It is well said also, "That that is the best law which leaves least to the discretion of the judge[22]"; and this comes from the certainty of it.

APHORISM 9.

Uncertainty of laws is of two kinds; the one, where no law is prescribed; the other, where the law is ambiguous and obscure. We must therefore speak first of cases omitted by the law, that in these also we may find some rule of certainty.

Of Cases omitted by the Law.

APHORISM 10.

The narrow compass of human wisdom cannot take in all the cases which time may discover; whence new and omitted cases often present themselves. For

[21] 1 Corinth. xiv. 8.

[22] Arist. *Rhet.* i. 1.

these, the remedy or supplement is threefold ; namely, by reference to similar cases, by employment of examples which have not yet grown into law, and by jurisdictions empowered to decide according to the arbitration of a good man and sound discretion, whether they be Prætorian or Censorian Courts.

Of reference to Similar Cases, and the Extensions of Laws.

APHORISM 11.

In omitted cases, the rule of law is to be drawn from cases similar to them, but with caution and judgment ; wherein the following rules are to be observed : Let reason be esteemed prolific, and custom barren. Custom must not make cases. Whatever therefore is received contrary to the reason of a law, or even where its reason is obscure, must not be drawn into consequence [23].

APHORISM 12.

Great public good draws omitted cases to itself. Wherefore when any law notoriously and to an extraordinary degree respects and procures the good of the people, let its interpretation be wide and comprehensive.

APHORISM 13.

It is harsh to torture laws, in order that laws may torture men. We would not therefore that penal, much less capital laws be extended to new offences. If however the offence be old and taken cognizance of by the laws, but the prosecution thereof fall upon a new case, unprovided for by the laws, we ought by all means to depart from the decrees of law rather than leave offences unpunished.

APHORISM 14.

In statutes which directly repeal the common law (especially in matters of frequent occurrence and long standing), we approve not the proceeding by similarity to cases omitted. For when the state has long been without the entire law, and that too in expressed cases, there is little danger in allowing the cases omitted to wait for a remedy from a new statute.

APHORISM 15.

Statutes which have a manifest relation to the time when they were made, and spring out of a temporary emergency of state, when the state of the times is altered, should have all their due, if they retain their authority in the cases proper to them ; for it would be preposterous to wrest them to omitted cases.

APHORISM 16.

Consequence does not draw consequence, but the extension should stop within the next cases ; otherwise there will be a gradual lapse into dissimilar cases, and sharpness of wit will have greater power than authority of law.

APHORISM 17.

When laws and statutes are concise in style, extend freely ; when they enumerate particular cases, more cautiously. For as exception corroborates the application of law in cases not excepted, so enumeration invalidates it in cases not enumerated.

APHORISM 18.

An explanatory statute stops the streams of the statute which it explains, and neither of them admits of extension afterwards. For the judge must not make a super-extension, when the law has once begun an extension.

APHORISM 19.

Formality of words and acts admits not of an extension to similar cases. For formality loses its character when it passes from custom to discretion ; and the introduction of new things destroys the majesty of the old.

[23] Paulus, *Digest.* 141, Ff. De Div. Reg. Jur.

APHORISM 20.

The extension of the law to posthumous cases, which had no existence at the time of the passing of the law, is easy. For where a case could not be expressed, as having no existence, a case omitted is taken for a case expressed, if there be the same reason for it.

Enough then on the extensions of laws in cases omitted. I will now speak of the employment of examples.

On Examples, and their Use [24].

APHORISM 21.

I now come to speak of examples, from which justice is to be derived when the law is deficient. Of custom, which is a kind of law, and of examples which by frequent use have passed into custom as a tacit law, I will speak in their place. But here I will speak of such examples as happen seldom and at distant intervals, and have not yet acquired the force of law; to show when, and with what caution, the rule of justice may be sought from them where the law is deficient.

APHORISM 22.

Examples are to be sought from good and moderate times, not from such as are tyrannical, factious, or dissolute. For those belonging to such times are spurious in their origin, and rather injurious than instructive.

APHORISM 23.

Of examples the latest are to be accounted the safest. For why should not that which has been lately done without any subsequent inconvenience be done again? But yet they have less authority; and if it happen that a reform be needed, modern examples savour more of their own age than of right reason.

APHORISM 24.

Ancient examples are to be received cautiously, and with proper selection. For the lapse of time makes many alterations, so that what in respect of time appears ancient is, by reason of the confusion which it makes and its inconformity to the present state of things, really new. Wherefore the best examples are those of the middle time, or else such a time as is most in conformity with the present age; and this is sometimes to be found in a more remote age rather than in that immediately preceding.

APHORISM 25.

Keep within, or rather on this side, of the limits of the example, and on no account go beyond them. For where there is no rule of law, everything should be

[24] It is to be observed, that the principle on which the English courts have proceeded —namely, that a decision on a point not previously decided on is to be accepted merely as a declaration of an already existing law virtually contained in the unwritten corpus juris entitled the Common Law—has had the effect of giving nearly equal weight to all cases decided by a competent tribunal. On the other hand, we find in the history of French jurisprudence that great uncertainty has existed as to the degree of authority to which a "res judicata" was entitled; the principle that "res judicata pro veritate accipitur" extending only to the parties between whom the actual decision was had. Thus it is related that De Thou was in the habit of saying, when it was mentioned that in a case similar to the one before him a decree had been given in favour of the plaintiff or defendant, "C'est bon pour lui"; implying that it was not of authority in any other case. The Parliament of Paris was for a long time in the habit of distinguishing the decisions to the principle of which it intended to give force of law from other decisions by a more solemn form of delivering judgment; thereby in effect claiming what our courts have never claimed, namely, a power of making new law. A collection has been published of these quasi-legislative decisions, with the title of "Arrêts rendus en robe rouge". It is evident that the practice of the Parliament of Paris, which was probably followed by other of the French Parliaments, escapes from some of the inconveniences of the English theory.

looked on with suspicion; and therefore, as in obscure cases, be very careful how you proceed.

APHORISM 26.

Beware of fragments, and epitomes of examples, and look carefully into the whole of the examples with all the process thereof. For if it be unreasonable to judge of part of a law, without examining the whole [25]; much more ought this to have weight in examples, the use whereof is doubtful, if they do not exactly correspond.

APHORISM 27.

It is of great importance through what hands examples have passed, and by whom they have been sanctioned. For if they have only passed among clerks and secretaries, in the ordinary course of the court, without the manifest knowledge of the higher officers; or among the teacher of all errors, the people; they are to be condemned and held of little account. But if they have passed under the eyes of senators, judges, or the principal courts, in such a manner that they must needs have been strengthened by at least the tacit approval of the judges, they are entitled to more authority.

APHORISM 28.

Examples, which even though they have been little used have been published, yet having been well debated and ventilated in discourse and discussion, deserve more authority; but those which have lain as it were buried in desks and archives and have openly passed into oblivion, deserve less. For examples like waters are most wholesome in a running stream.

APHORISM 29.

Examples which have reference to laws should not be sought from historians, but from public acts and the more careful traditions. For it is a misfortune even of the best historians, that they do not dwell sufficiently upon laws and judicial acts; or if by chance they use some diligence therein, yet they differ greatly from the authentic reporters.

APHORISM 30.

An example, which the same or the succeeding age has upon the recurrence of the case rejected, should not be readily re-admitted. For the fact that it was once adopted does not tell so much in its favour, as the subsequent abandonment tells against it.

APHORISM 31.

Examples are to be used for advice, not for rules and orders. Wherefore let them be so employed as to turn the authority of the past to the use of the present.

Enough then of instruction from examples where the law is deficient. I must now speak of the Courts Prætorian and Censorian.

On Courts Prætorian and Censorian [26]

APHORISM 32.

Let there be courts and jurisdictions to determine, by the judgment and discre-

[25] Celsus, *Digest*, i. 3, 24.

[26] M. Bouillet remarks that every one who has commented on this tract of Bacon's has condemned the institution of these Courts. M. Dupin is evidently much perplexed by them. "Hic mera utopia proponitur" is the commencement of his note on the thirty second aphorism. Doubtless it is odd that in inquiring how the law may be made certain Bacon should have introduced two Courts, of which the distinguishing character is the absence of any kind of certainty. But to every one who is acquainted with the history of English law, it is manifest that Bacon's intention was to give an idealised description of the Court of Star-Chamber, and of the equity jurisdiction of the Court of Chancery. Of the two institutions which he thus indirectly praises it is not necessary to say much. The Court of Star-Chamber, though of use in particular cases, was un-

tion of a conscientious man, when the rule of the law is deficient. For the law (as has been before said) cannot provide for all cases, but is adapted to meet such as

questionably on the whole an instrument of injustice and oppression; while, on the other hand, if equity had continued to be as indefinite as the jurisdiction of the "curiæ prætoriæ," it would soon have become a more intolerable evil than any which it could have been applied to relieve.

[The apparent inconsistency of introducing these discretionary tribunals into a scheme specially designed to make the operation of the law *certain*, admits in my opinion of a satisfactory explanation. The uncertainty of the law is injurious in two ways. On the one hand, it may lead me to expect that if I observe certain prescribed conditions, my liberty will not be interfered with; and when I think I have observed them, it may, by some arbitrary or unexpected interpretation, take me up and send me to prison. On the other hand, it may lead me to expect protection against particular kinds of injury or (failing protection) redress; and, from some defect in its provisions, it may fail to prevent the injury or to afford the redress. The first kind of uncertainty resides in the interpretation, the second in the framing, of the law; and against *both* it is necessary, as far as may be, to provide. The *perfect* remedy is a code of laws so framed as to provide expressly for every possible case, coupled with a rule of interpretation which leaves no discretion whatever to the judge. But this is for Utopia. No lawgiver can perfectly foresee either the conditions of cases or the effect of words. Laws will therefore pass occasionally, which, if strictly construed, will punish the man whom they were intended to protect, and protect the man whom they were intended to punish. To correct such errors, a discretion must be allowed somewhere in the administration of the law; and the question is, where? According to Bacon's scheme, the necessary discretion is to be confided, not to the ordinary tribunals, but to others specially constituted for the purpose, and acting under restrictions and regulations specially framed to prevent them from abusing it; lest in correcting one kind of uncertainty, uncertainties of another kind be introduced. What these restrictions and regulations should be, the rest of the section is occupied in explaining.

Now, to supply the defects of the law by the exercise of this kind of discretion was the proper function of the Star-Chamber and the Court of Chancery; and I see no occasion to seek further for Bacon's motive in introducing "an idealised description" of those Courts,—or, I should rather say, a description of two Courts constituted as, in a perfect administrative system, the Star-Chamber and the Court of Chancery ought to be.

With regard to the character of the actual Star-Chamber, we are not to forget that Bacon was not the only eminent jurist who approved of it. Sir Edward Coke, in the fourth book of his *Institutes*, which was written in his old age, when he was regarded as the great champion of the people against the Crown, speaks of it in terms as favourable as ever Bacon did. "It is the most honourable Court" (he says)—"our parliament excepted—that is in the Christian world, both in respect of the Judges of the Court, and of their honourable proceeding according to their just jurisdiction, and the ancient and just orders of the Court." And I cannot help thinking that modern constitutional writers have judged of it too hastily from the accidental and exceptional circumstances which led to its abolition. It was an instrument of *government*. When the government was oppressive and unjust, it was an instrument of oppression and injustice. So, also, at many periods of our history have the Courts of Common Law been. But if we would know whether a Court constituted like the Star-Chamber had any necessary tendency to become an instrument of oppression, we must consider it in connexion with the rest of the constitution. Was it in any special manner under the command of the Crown? Certainly not: it was under the command of the Crown so far only and so long only as the whole powers of government were under the command of the Crown. So far and so long as the King could appoint his own ministers and maintain them and carry on the government with them in spite of the House of Commons, so far and so long he could exercise an effectual control over the proceedings of a Court constituted like the Star-Chamber; no farther and no longer. The body of the Court was composed of the chief officers of the government; less than eight did not make a quorum; their proceedings were public; each member gave his own sentence with the reasons; the majority decided; the decree was solemnly recorded. As soon as the theory of a responsible ministry was recognised, and the impossibility of carrying on the government without money

generally occur. And time, according to the ancient saying, is the wisest of all things [27], and daily creates and invents new cases.

APHORISM 33.

Fresh cases happen both in criminal causes, which require punishment, and in civil causes, which require relief. The courts which take cognizance of the former I call Censorian, those which respect the latter, Prætorian.

APHORISM 34.

Let the Censorian Courts have power and jurisdiction, not only to punish new offences, but also to increase the punishments appointed by law for old ones, where the cases are heinous and enormous, provided they are not capital. For an enormous crime has somewhat of the nature of a new one.

APHORISM 35.

In like manner let the Prætorian Courts have power both to abate the rigor of the law and to supply its defects. For if relief is due to a person whom the law has neglected, much more is it due to one whom it has wounded.

APHORISM 36.

Let these Prætorian and Censorian Courts entirely confine themselves to monstrous and extraordinary cases, and not encroach upon the ordinary jurisdictions, lest they rather tend to supplant than to supply the law.

APHORISM 37.

Let these jurisdictions reside only in the supreme courts, and not be shared by the lower; for the power of supplying, extending, and moderating laws, differs little from that of making them.

APHORISM 38.

Let not these courts be entrusted to the charge of one man, but let them consist of many. And let not the decrees go forth in silence, but let the judges give the reasons of their decision, and that openly and in full court; so that what is free in point of power may yet be restrained by regard to character and reputation.

APHORISM 39.

Let there be no authority to shed blood; nor let sentence be pronounced in any court upon capital cases, except according to a known and certain law. God himself denounced death before he inflicted it. Nor should a man be deprived of his life, who did not first know that he was risking it.

APHORISM 40.

In the Censorian Courts, let there be opportunity for three verdicts; that the judges may not be obliged to acquit or condemn, but be at liberty to declare the fact "not proven". And besides the penalty, let there be power also to inflict a note or mark; such I mean as shall not extend to actual punishment, but may end either in admonition only, or in a light disgrace; punishing the offender as it were with a blush.

voted by the House of Commons gave the people an effective check upon the Crown, they would have had a check equally effective upon the proceedings of a court of justice so constituted. Any abuse of its authority would have led to a change of ministry, and to the transfer of that authority to other hands.

With regard to the Court of Chancery, it is less easy to say how it would have worked had its jurisdiction been exercised according to the conditions here prescribed for the Curiæ Prætoriæ; one of which is, that it was not to be confided to a single man. "Curiæ illæ" (i.e. Curiæ Censoriæ *et Prætoriæ*, see Aph. 36) "uni viro ne committantur, sed ex pluribus constent." And in speculating upon the evil which it might have become with powers so indefinite, we must not forget how great an evil it has actually become, in consequence of the rules by which its discretion has been defined and limited. The nearest approach to *certainty* attained by the existing system appears to be the certainty of damage to both parties.—*J. S.*]

[27] Xen. *Hellen.* iii. 3, 2.

APHORISM 41.

In Censorian Courts, let the commencements and middle acts of all great crimes and offences be punished, even though the end be not consummated [28]. And let this be even the principal use of these courts; for it is as well the part of severity to punish the commencements of crimes, as of mercy to prevent their completion, by punishing the intermediate acts.

APHORISM 42.

Especial care must be taken, in Prætorian Courts, not to afford relief in such cases as the law has not so much omitted, as depised for their unimportance, or for their odious nature judged unworthy of redress.

APHORISM 43.

It is of the greatest importance to the certainty of laws (of which I am now treating), that Prætorian Courts be not allowed to swell and overflow, so as, under colour of mitigating the rigour of the law, to break its strength and relax its sinews, by drawing everything to be a matter of discretion.

APHORISM 44.

Let not the Prætorian Courts have authority, under any pretext of equity, to decree against an express statute. For in that case the judge would pass into the legislator, and everything would be at discretion.

APHORISM 45.

Some hold that the jurisdiction which decrees according to equity and conscience, and that which proceeds according to strict justice, ought to be deputed to the same courts; but others would have them kept separate. I am clearly for keeping them separate. For if there be a mixture of jurisdictions, the distinction of cases will not be retained, but discretion will in the end supersede the law.

APHORISM 46.

The Prætor's Table at Rome, wherein he set down and published the rules by which he meant to judge, was not established without good reason. And after this example, judges in the Prætorian Courts ought, as far as possible, to propose certain rules for themselves, and set them up where they can be seen by the people. For as that is the best law which leaves the least to the discretion of the judge, so he is the best judge who leaves the least to himself.

But I will treat more fully of these courts when I come to speak of judgments; for here I have only noticed of them in passing, in what way they remedy and supply the omissions of law.

Of the Retrospective Aspect of Laws.

APHORISM 47.

There is likewise another kind of supplement to omitted cases; namely, when one law follows and amends another, and draws the omitted cases along with it. And this is done by those laws and statutes which are called retrospective. But laws of this kind must be used seldom, and with great caution; for we approve not of a Janus in laws.

APHORISM 48.

He who evades and narrows the words or meaning of a law by fraud and cavil deserves to be himself ensnared by a subsequent law. And therefore in cases of fraud and captious evasion it is just that laws should be retrospective, and be of assistance one to the other; that a man who plots to deceive and upset the present laws may at least feel apprehensions from future ones.

[28] Of the Star-Chamber Bacon has said in his *History of Henry VII.*, that it took cognizance of "forces, frauds, crimes various, of stellionate, and the incusations or middle acts towards crimes, capital or heinous, not actually committed or perpetrated".

APHORISM 49.

Laws which strengthen and confirm the real intentions of acts and instruments against the defects of forms and usages very properly include past actions. For the principal inconvenience of a retrospective law is that it creates disturbance; but confirmatory laws of this sort tend rather to peace and the settlement of past transactions. We must however take care not to call in question matters already adjudged.

APHORISM 50.

It must be observed that not those only are to be considered retrospective laws which invalidate acts passed; but those likewise which prohibit and restrain future acts as necessarily connected with the past. Thus a law which should prohibit certain artisans from henceforth selling their wares seems only to bear upon the future, yet it operates on the past; for such persons have not now the power to seek their living in another way.

APHORISM 51.

Every declaratory law, though it does not mention the past, yet by the very force of the declaration must needs apply to past transactions. For the interpretation does not date from the time of the declaration, but is made as it were contemporary with the law itself. And therefore enact no declaratory laws, except in cases where they may be justly retrospective.

And here I end that part which treats of Uncertainty of laws where no law exists. I must now speak of that other part, where some law is extant, but confused and obscure.

Of Obscurity of Laws.

APHORISM 52.

Obscurity of laws arises from four sources; either from an excessive accumulation of laws, especially if they be mixed with such as are obsolete; or from an ambiguity or want of clearness and distinctness in the drawing of them; or from negligent and ill-ordered methods of interpreting law; or lastly, from a contradiction and inconsistency of judgments.

Of Excessive Accumulation of Laws.

APHORISM 53.

The prophet says, "He shall rain snares upon them[29]". But there are no worse snares than legal snares, especially in penal laws; if being infinite in number, and useless through the lapse of time, instead of being as a lantern to the feet they are as nets in the path.

APHORISM 54.

There are two ways in use of making a new statute. The one confirms and strengthens former statutes on the same subject, and then makes a few additions and alterations. The other repeals and cancels all former enactments, and substitutes an entirely new and uniform law. The last method is the best. For by the former the enactments become confused and complicated, and though indeed the immediate object is effected, yet the body of laws is in the meantime corrupted. But in the latter, though greater care is required in deliberating about the law itself, and former acts must be carefully searched and canvassed before it pass; yet it is the best course for securing harmony in times to come.

APHORISM 55.

The Athenians had a custom of appointing six men to examine every year the contradictory titles of their laws (which they called *Antinomies*) and to report to the people such as could not be reconciled, that a definite resolution might be passed concerning them. After their example let the legislators of every state

[29] Psalm xi. 6.

every three or five years, or as often as it appears good, review their Antinomies. And let these be first examined and drawn up by commissioners appointed for the purpose, and then laid before the Parliament, that the matter may be settled and established by vote.

APHORISM 56.

But let there not be too great an eagerness and anxiety to reconcile or salve (as they term it) these contradictory titles by fine and far-fetched distinctions. For this is a web of the wit; which, whatever appearance of modesty and reverence it may bear, must yet be regarded as prejudicial, seeing that it makes the whole body of laws ill-assorted and incoherent. It is therefore far better to let the worse titles drop, and leave the best to stand alone.

APHORISM 57.

It should also be a part of the office of the Commissioners to propose that obsolete laws and such as have fallen into disuse should be repealed, as well as antinomies. For since an express statute is not regularly abolished by disuse, it comes to pass that through the contempt of obsolete laws the authority of the rest is somewhat impaired. And from this ensues a torment like that of Mezentius, whereby the living laws are stifled in the embraces of the dead. And above all things a gangrene in our laws is to be avoided.

APHORISM 58.

But in the meantime let the Prætorian Courts have power to decree against laws and statutes which are obsolete, and have not lately passed. For though it has been well said, "that no one should be wiser than the laws[30]," yet this must be understood of waking and not of sleeping laws. Not so however with more recent statutes, which are found to be injurious to public justice. The power of giving relief in this case should be left not to the judge, but to kings, councils, and the supreme authorities of the state, who should be empowered to suspend the execution of them by Acts or Proclamations, till the re-assembling of Parliament or of that body which has the power of repealing them; lest in the meantime the welfare of the community be endangered[31].

30 Bacon refers perhaps to D'Argentré's maxim, "Stulta videtur sapientia quæ lege vult sapientior videri". In the passage from which these words are taken, he is condemning the presumption of judges who depart from the text on the pretence of equity. D'Argentré died in 1590. Cf. Arist. *Rhet.* i. 15, 12; and Thucyd. iii. 37.

31 Here, as in the description of the Curiæ Censoriæ and Prætoriæ, reference is made to what actually existed in England in Bacon's time. In the concluding part of this aphorism he sanctions the doctrine that an act of Parliament may provisionally at least be suspended or set aside by an Order in Council. This doctrine was undoubtedly commonly maintained in Bacon's time, but it was nevertheless even then protested against.

[When the rights of the people were not sufficiently secured against the powers of the Crown, and therefore to weaken those powers was a patriotic object, such doctrines were naturally protested against. For when the Crown could successfully and safely abuse the powers it had, the evil could only be remedied or mitigated by taking them away. And it was doubtless by restricting its authority in matters like this that the people were in fact enabled to win the game, and exact sufficient securities for themselves. But we must remember that throughout this treatise Bacon assumes the existence of a government otherwise well constituted. And I am much inclined to think that these securities being once attained, and the House of Commons having in fact a veto upon all the proceedings of the Crown, such an authority might be intrusted to the government both safely and beneficially. Bacon was not considering what powers could be exercised *constitutionally*, i.e. according to law and precedent, by the *English* government, but generally what powers it was good for a people that the governing authority should have. —*J. S.*]

Of New Digests of Laws[32].

APHORISM 59.

But if the laws by accumulation have grown so voluminous, or become so confused, that it is expedient to remodel them entirely and reduce them to a sound and manageable body, let it by all means be done ; and let it be considered a heroic work ; and let the authors thereof be justly and deservedly reckoned among legislators and reformers of law.

APHORISM 60.

This kind of expurgation and new digest of laws is effected by five processes. First, let obsolete laws, which Justinian calls old fables[33], be omitted. Secondly, let the most approved antinomies be received, and the rest abolished. Thirdly, let *Homoionomies*, or laws which are of the same import, and nothing else but reiterations of the same thing, be erased, and let the one which is the most perfect among them be retained in place of all the rest. Fourthly, let such laws as determine nothing, but only propose questions, and leave them undecided, be dismissed in like manner. Lastly, let those laws which are found to be wordy and too prolix be more compressed and abridged.

APHORISM 61.

It will be very useful in a new digest of laws to digest and arrange separately on the one side all the laws received as Common Law, the existence whereof is as it were from time immemorial ; and on the other side the statutes which have from time to time been superadded. For in many points, in passing judgment, the interpretation and administration of the Common Law are not the same as the Statute Law. And this was the plan followed by Trebonianus in the Digest and Code[34].

APHORISM 62.

But in this regeneration and reconstruction of the laws, by all means retain the words and text of the old laws and law-books, though it be necessary to extract them by scraps and fragments : and afterwards connect them together in proper order. For although this might perhaps be done more conveniently, and, if you look to right reason, more correctly also by a new text than by patching up the old ; yet in laws we ought not so much to look to style and drawing up as to authority, and its patron, antiquity. Otherwise the work would appear rather a matter of scholarship and method than a body of commanding laws.

32 This section, and especially the 64th Aphorism, is spoken of with great commendation by perhaps the highest authority on such subjects. See Savigny " On the Vocation of our Time to Legislation," 3d edition, p. 20.

33 Institut. Procem. § 3. The great bulk of Justinian's *Institutiones* are merely a reproduction of those of Gaius.

34 The Digest consists of Excerpta from the works of a great number of jurists, so arranged as to form a connected view of the whole of the Roman law. The Codex is a collection of imperial ordinances, most of which relate to particular cases, but are nevertheless of general authority, while others are in form as well as in effect legislative enactments.

The Digest cannot be regarded as a Corpus of customary law : we find in every portion of it continual references to every source of law,—to leges, plebiscita, edicta, senatûs consulta, and imperial rescripts and constitutions, as well as to jus civile, in the narrow sense in which the phrase is equivalent to immemorial custom. It is scarcely necessary to mention that Tribonianus was Justinian's chief instrument in the compilation of the *Digest*, *Codex*, and the *Institutes*. The first of these three works is the greatest in extent and importance. It was drawn up by a commission of seventeen persons, of which Tribonianus was the head, as he was likewise of the smaller commissions by which the other two were compiled. By the *Codex* I mean the *Codex Repetitæ Prælectionis* : Tribonianus was not at the head of the commission by which the original *Codex* was drawn up, and it has been conjectured that his dissatisfaction at this circumstance occasioned the revision.

APHORISM 63.

It will be expedient in this new digest of laws that the old volumes do not altogether perish and pass into oblivion; but that they be preserved at least in libraries, though the ordinary and promiscuous use of them be prohibited. For in important cases it will not be amiss to examine and consider the successive changes which have taken place in past laws. And surely it is a reverent thing to intermingle antiquity with things present. But this new body of laws ought to be regularly confirmed by the legislative power of the state; lest under pretence of digesting old laws, new laws be secretly imposed.

APHORISM 64.

It were desirable that this instauration of the laws should be undertaken in such times as are superior in learning and experience to those more ancient times whose works and acts they revise. But the reverse of this happened in the work of Justinian. For it is an unfortunate circumstance, when by the taste and judgment of a less wise and less learned generation the works of the ancients are mutilated and reconstructed. But that is often necessary which is not best.

So much then for obscurity of laws arising from an excessive and confused accumulation thereof. I now come to speak of the ambiguous and obscure drawing up of them.

Of the Confused and Obscure Drawing up of Laws.

APHORISM 65.

Obscure drawing up of laws arises either from their loquacity and verbosity or on the other hand from an excess of conciseness, or from the preamble of the law being at variance with the body.

APHORISM 66.

I must now speak of the obscurity of laws which arises from their being ill drawn up. The loquacity and prolixity used in the drawing up of laws I do not approve. For it does not at all secure its intention and purpose; but rather the reverse. For while it tries to enumerate and express every particular case in apposite and appropriate words, expecting greater certainty thereby; it does in fact raise a number of questions about words; so that, by reason of the noise and strife of words, the interpretation which proceeds according to the meaning of the law (which is the juster and sounder kind of interpretation) is rendered more difficult.

APHORISM 67.

Not that I therefore approve of a too concise and affected brevity, as being the style of majesty and command, especially in these times; lest by chance the law should become like a Lesbian rule[35]. We must therefore aim at a mean, and look out for a well-defined generality of words; which though it does not attempt to express all the cases comprehended, yet excludes with sufficient clearness the cases not comprehended.

APHORISM 68.

In ordinary laws and proclamations of state however, in which lawyers are not generally consulted, but every man trusts to his own judgment, everything should be more fully explained, and pointed out, as it were with the finger, to the capacity of the people.

[35] "Lesbia regula dicitur quoties præpostere, non ad rationem factum, sed ratio ad factum accommodatur."—*Erasm. Adag.* i. 93.

Bacon's meaning is, that if the law be too concisely stated it may be bent by the interpretations which its excessive brevity will render necessary, so as to operate in a way which the legislator did not contemplate. This will more clearly appear to be his meaning from the passage in the *Nicomachean Ethics*, v. c. 10, to which Erasmus refers. In building with irregularly shaped stones, flexible rules might be found of use, and it would appear that the Lesbians were in the habit of employing them.

APHORISM 69.

Nor should I at all approve of the preambles of laws, which were formerly deemed impertinent, and which represent laws disputing and not commanding, if we could endure the ancient manners[33]. But as times now are, these preambles are necessarily used in most cases, not so much to explain the law, as to persuade Parliament to pass it, and also to satisfy the people. But avoid preambles as much as possible, and let the law commence with the enactment.

APHORISM 70.

Though the intention and purport of a law may sometimes be well gathered from the prefaces and preambles, yet the latitude or extension thereof should by no means be sought from thence. For the preamble often selects a few of the most plausible and specious points by way of example, even when the law contains many things besides. Or on the other hand, the law sometimes makes many restrictions and limitations, the reasons whereof need not be inserted in the preamble. Wherefore the extent and latitude of a law is to be taken from the body thereof; for the preamble often either exceeds or falls short of it.

APHORISM 71.

There is one very faulty method of drawing up laws. And this is, when the case at which the law aims is fully set forth in the preamble; and then from the force of the word "such" or some like relative, the body of the law is reflected back upon the preamble, which is thereby inserted and incorporated into the law, and renders it both more obscure and less safe. For the same care is not usually employed in weighing and examining the words of the preamble which is bestowed on the body of the law itself.

But this part of the uncertainty of laws, arising from their being ill drawn up, I will treat of more fully, when I come afterwards to the interpretation of laws. And so much for the obscure drawing up of laws; I must now speak of the methods of expounding law.

Of the Methods of Expounding Law, and Removing Ambiguities.

APHORISM 72.

There are five methods of expounding law, and removing ambiguities: namely, by reports of judgments; by authentic writers; by auxiliary books; by prelections; or by the answers and decrees of learned men. All these if properly instituted will be of great service against the obscurity of laws.

Of the Reporting of Judgments.

APHORISM 73.

Above all things, let the Judgments delivered in the Supreme and Principal Courts on important cases, especially if they be doubtful and contain some difficulty or novelty, be diligently and accurately taken down. For judgments are the anchors of laws, as laws are of the state.

APHORISM 74.

Let this be the method of taking down judgments and committing them to writing. Record the cases precisely, the judgments themselves word for word; add the reasons which the judges allege for their judgments; do not mix up the authority of cases brought forward as examples with the principal case; and omit the perorations of counsel, unless they contain something very remarkable.

APHORISM 75.

Let the reporters be taken from the most learned counsel, and receive a libera salary from the state. But let not the judges themselves meddle with the reports; lest from being too fond of their opinions, and relying on their own authority, they exceed the province of a reporter.

[33] "Jubeat," says Seneca, speaking of law, "non disputet. Nihil videtur mihi ineptius quam lex cum prologo."—*Ep.* 95.

APHORISM 76.

Let these judgments be digested in chronological order, and not by method and titles. For such writings are a kind of history or narrative of the laws. And not only the acts themselves, but the times also when they passed, give light to a wise judge.

Of Authentic Writers.

APHORISM 77.

Let the body of law be composed only of the laws that constitute the Common Law, of the constitutional laws or statutes, and of reported judgments. Besides these, let no others be deemed authentic, or at least let them be sparingly accepted.

APHORISM 78.

Nothing contributes so much to the certainty of laws (whereof I am now treating), as to keep the authentic writings within moderate bounds, and to get rid of the enormous multitude of authors and doctors of laws. For by them the meaning of laws is distracted, the judge is perplexed, the proceedings are made endless, and the advocate himself, as he cannot peruse and master so many books, takes refuge in abridgment. Perhaps some one good commentary, and a few classic authors, or rather some few selections from some few of them, may be received as authentic. Let the rest however be kept for use in libraries, that the judges or counsel may inspect them if necessary; but let them not be allowed to be pleaded in court, or to pass into authorities.

Of Auxiliary Books.

APHORISM 79.

The science and practice of the law should not be deprived of auxiliary books, but rather well furnished with them. These are of six kinds; namely, Institutes: On Terms of Law: On Rules of Law: Antiquities of Laws: Summaries, and Forms of Pleading.

APHORISM 80.

Students and novices are to be educated and trained by Institutes to take in more readily and profoundly the higher parts of the law. Let these Institutes be arranged in a clear and perspicuous order. Let them run through the whole private law, not omitting some things, and dwelling too long on others; but giving a slight sketch of all; so that when the student comes to peruse the body of law he may find nothing entirely new, or of which he has not had a slight notion beforehand. But touch not the public law in these institutes, but let that be drawn from the fountains themselves.

APHORISM 81.

Construct a commentary on legal terms; but let it not enter too curiously or laboriously into an explanation of their full sense. For the object is not so much to look for exact definitions of the words, as for explanations to make the way easier in reading law books. And let not this treatise be digested in the order of the alphabet, but leave that to an index; and let the words which relate to the same thing be arranged together, that the one may serve to explain the other.

APHORISM 82.

A good and careful treatise on the different rules of law conduces as much as anything to the certainty thereof; and it deserves to be entrusted to the ablest and wisest lawyers. For I am not content with the works of this kind which are now extant. The collection should consist not only of the common and well known rules, but of others likewise more subtle and abstruse, which may be gathered from the harmony of laws and decided cases; such as are sometimes found in the best tables of contents; and are in fact the general dictates of reason, which run through the different matters of law, and act as its ballast.

APHORISM 83.

But let not every decree or position of law be taken for a rule; as is commonly done, ignorantly enough. For if this were admitted there would be as many rules as laws; for law is nothing else than a commanding rule. But let those be considered rules which are inherent in the very form of justice; and whereby it comes that for the most part nearly the same rules are found in the civil laws of different states; except perhaps that they may sometimes vary with reference to the forms of constitutions.

APHORISM 84.

After a rule has been stated in a concise and solid form of words, let examples, and such decisions of cases as are most clear, be added for the explanation; distinctions and exceptions for the limitation; and kindred cases for the amplification of the rule.

APHORISM 85.

It is a sound precept not to take the law from the rules, but to make the rule from the existing law [37]. For the proof is not to be sought from the words of the rule, as if it were the text of law. The rule, like the magnetic needle, points at the law, but does not settle it.

APHORISM 86.

Besides the body of law, it will be of service likewise to take a survey of the antiquities thereof; of which though the authority has perished, yet the reverence still remains. And by antiquities of laws, I understand those writings on laws and judgments, whether published or unpublished, which preceded the body of law; for these should not be lost. Wherefore select the most useful of them, (for many will be found to be frivolous and unprofitable) and collect them into one volume; that old fables, as Trebonianus calls them, may not be mixed up with the actual laws.

APHORISM 87.

It is of great importance to practice, that the whole law should be arranged in order under titles and heads; to which reference may be made at once, when occasion shall require, as to a store-house provided for present wants. Summaries of this kind both reduce to order what is dispersed in the law, and abridge what is diffuse and prolix. But we must take care that while they make men ready in practice, they do not make them idlers in the science itself; for their business is to facilitate the recollection of the law, not to teach it. But these summaries are by all means to be constructed with great care, accuracy, and judgment, lest they cheat the laws.

APHORISM 88.

Collect the different forms of pleading of every sort. For this is both a help to practice; and besides, these forms disclose the oracles and mysteries of laws. For many things lie concealed in the laws, which in these forms of pleading are more fully and clearly revealed; the one being as the fist, the other as the open hand.

Of Answers and Opinions.

APHORISM 89.

Some means should be devised for solving and clearing away the particular doubts which from time to time arise. For it is hard that they who desire to secure themselves from error should not be able to find a guide; but that their actions must themselves run the risk, there being no means of knowing the law before the thing is done.

APHORISM 90.

I do not approve that the answers of learned men [38], whether advocates or doctors of law, given to those who ask their advice on a point of law, should have

[37] Paulus, Digest, *De diversis regulis antiqui juris*, i. 1.

[38] Orig. *Responsa Prudentum.* By the Roman Jurists the Responsa prudentum are

such authority that the judge should not be allowed to depart from their opinions. Let the laws be taken from sworn judges.

APHORISM 91.

I do not approve that men should make trial of judgments under feigned persons and causes, for the purpose of ascertaining what the rule of law will be [39]. For this lowers the majesty of the laws and is a kind of prevarication. Besides, it is unseemly for judicial proceedings to borrow anything from the stage.

APHORISM 92.

Therefore, let judgments, as well as answers and opinions, proceed from the judges alone; the former in questions on pending suits, the latter on difficult points of law. And let not these opinions, whether on public or private matters, be demanded from the judges themselves, (for that were to turn the judge into an advocate); but from the king or state. Let the king or state refer them to the judges. Let the judges thus authorised hear the pleadings of the advocates, whether selected by the parties themselves, or (if necessary) appointed by the judges themselves, and the arguments on both sides; and after deliberating on the case let them deliver and declare the law. Let these opinions be recorded and published among judgments, and be of equal authority with them [40].

reckoned among the Fontes Juris, but there are few points in the history of Roman law on which it is more difficult to form a satisfactory opinion. We have no satisfactory information either as to the form in which these Responsa were given, or as to the degree of authority with which they were invested. The common opinion is, that they received absolute force of law in virtue of an ordinance of Augustus, and that more precise regulations with respect to cases in which a diversity of opinion existed were made by Hadrian. The connexion between them and the law of citations of Honorius and Valentinian is also a matter of much obscurity. See Böcking's *Pandekten*, i. p. 36. Walter, *Gesch. d. R. Rechts*, § 409 and 421. Hug, *Gesch. d. R. Rechts*, § 313 and 385.

39 Lord Ellenborough refused to try a case in which a bet had been made on a point of law. He asked, it is said, to see the record, and threw it down "with much indignation". Tradition adds that he threw it at the head of the plaintiff's attorney. Until lately, when it was found necessary in proceedings in equity to have the decision of a jury on a question of fact, recourse was had to the machinery of a feigned issue; that is, an action was brought on an imaginary wager as to the truth or falsehood of an agreed-upon statement of facts. Possibly in Bacon's time a similar course may have been adopted in order to obtain the opinion of the judges on points of law. In modern times the practice has been in accordance with what he a little further on recommends; the point of law being referred to the judges directly, who, after hearing counsel, certify their opinion of it to the Chancellor.

40 Bacon refers to the practice of extra-judicial consultations as it existed in his own time. It does not, I believe, appear that it was ever the practice for private persons to obtain through the intervention of the Privy Council authoritative decisions on legal questions, but it is well known that the Court occasionally obtained "præjudicia" from the judges on points in which it was itself interested. The effect of this practice in promoting judicial servility is well seen in the case of ship-money; the extra-judicial decision of the judges in favour of its legality being unanimous, whereas when the case came on in the exchequer chamber, it was affirmed to be legal by a bare majority of seven against five.

[I cannot think that Bacon alludes to extra-judicial consultations of this kind; which were conducted in a different way from those he recommends, and resorted to for a different purpose. The object of the Government in asking the judges' opinions on the case privately before commencing a prosecution, was to ascertain that the case was a good one, and so avoid the scandal and disrepute which then attended the failure of a Crown prosecution. The object of the proceeding which Bacon here advocates, is to provide a means of settling any disputed point of law, without either waiting for a real cause in which it may be involved, or getting up a fictitious one; and the manner of it is to be public and formal. The case is to be regularly argued and the judgment formally recorded.—*J. S.*]

Of Prelections.

APHORISM 93.

Let the lectures and exercises of those who study and labour at the law be so ordered and instituted, as rather to set legal questions and controversies at rest, than to raise and excite them. For at present they are nothing but schools and institutions for multiplying altercations and controversies on points of law, as if for the display of wit. And this evil is also an old one. For it was likewise the pride of antiquity, as by sects and factions, to keep alive a number of questions of law, rather than to settle them [41]. Let this however be provided against.

Of Inconsistency of Judgments.

APHORISM 94.

Inconsistency of judgments arises either from an immature and hasty decision, or from the rivalry of Courts, or from a bad and ignorant reporting of judgments, or from too great facility being given for their reversal. Care therefore should be taken that judgments proceed after mature deliberation; that courts preserve mutual respect for one another; that judgments be faithfully and wisely reported; and that the way to a repeal of judgments be narrow, rocky, and as it were paved with flint stones.

APHORISM 95.

If judgment be given on any case in a principal court, and a similar case occur in any other, do not proceed to pass judgment till a consultation has been held in some general assembly of the judges. For if it be that previous decisions must be rescinded, at least let them be interred with honour.

APHORISM 96.

That Courts should fence and dispute about jurisdiction is natural to humanity; the rather because of a foolish doctrine, that it is the part of a good and active judge to extend the jurisdiction of his Court; which stimulates the disease and applies a spur where a bit is needed. But that through this spirit of contention courts should freely rescind each other's judgments (judgments having nothing to do with the question of jurisdiction) is an intolerable evil, that should by all means be put down by kings or senates or governments. For it is a most pernicious example, that Courts, whose business it is to keep the subjects at peace, should be at war with one another.

APHORISM 97.

Let not the way to a repeal of judgments by appeals, writs of error, new trials, and the like, be much too easy and open. Some hold that a suit should be withdrawn to a higher court, as quite a new cause, the previous judgment being completely laid aside and suspended. Others are of opinion that the judgment itself should remain in full force, whilst only its execution should be deferred. I do not like either of these ways; unless the courts wherein judgment has been delivered be of a low and inferior character; but I would rather let both the judgment stand, and the execution proceed, the defendant only giving security for costs and damages if the judgments be reversed.

[41] Our knowledge of the history of the two sects or schools of jurists which existed during what is called the middle period of Roman jurisprudence is still imperfect, though less so than before the discovery of the Institutes of Gaius. It appears probable that the importance of the differences of opinion between them has been exaggerated, and that the sects themselves had died out before the time of Justinian. The two schools respectively regarded Ateius Capito and Anstitius Labeo as their head or founder; but the followers of the former were called Sabinians or Cassians; the other school being that of the Proculeians; all these names being derived from those of certain eminent followers of the two jurists just mentioned. Gaius, the author of the *Institutes*, belonged to the former school, which is said to have been distinguished from the other by a closer ad herence to the letter of the law. Probably the best writer on the subject is Dirksen.

This Title then touching Certainty of Laws shall stand as a model of the rest of the Digest which I have in mind.

But here I have concluded Civil Knowledge (as far as I have thought right to handle it), and together with it Human Philosophy, and, with Human Philosophy, Philosophy in General. At length therefore having arrived at some pause, and looking back into those things which I have passed through, this treatise of mine seems to me not unlike those sounds and preludes which musicians make while they are tuning their instruments, and which produce indeed a harsh and unpleasing sound to the ear, but tend to make the music sweeter afterwards. And thus have I intended to employ myself in tuning the harp of the muses and reducing it to perfect harmony, that hereafter the strings may be touched by a better hand or a better quill. And surely, when I set before me the condition of these times, in which learning seems to have now made her third visitation to men; and when at the same time I attentively behold with what helps and assistances she is provided; as the vivacity and sublimity of the many wits of this age; the noble monuments of ancient writers, which shine like so many lights before us; the art of printing, which brings books within reach of men of all fortunes; the opened bosom of the ocean, and the world travelled over in every part, whereby multitudes of experiments unknown to the ancients have been disclosed, and an immense mass added to Natural History; the leisure time which the greatest wits in the kingdoms and states of Europe everywhere have at their disposal, not being so much employed in civil business as were the Greeks in respect of their popular governments, and the Romans in respect of the greatness of their monarchy; the peace which Britain, Spain, Italy, France too at last, and many other countries now enjoy; the consumption and exhaustion of all that can be thought or said on religious questions, which have so long diverted many men's minds from the study of other arts; the excellence and perfection of your Majesty's learning, which calls whole flocks of wits around you, as birds round a phœnix; and lastly, the inseparable property of time, ever more and more to disclose Truth; I cannot, I say, when I reflect on these things but be raised to this hope, that this third period will far surpass the Greek and Roman in learning; if only men will wisely and honestly know their own strength and their own weakness; and take from one another the light of invention and not the fire of contradiction; and esteem the inquisition of truth as a noble enterprise, and not a pleasure or an ornament; and employ wealth and magnificence on things of worth and excellence, not on things vulgar and of popular estimation. As for my labours, if any man shall please himself or others in the reprehension of them, they shall make at all events that ancient and patient request, " Strike, but hear [42] ". Let men reprehend them as much as they please, if only they observe and weigh what is said. For the appeal is lawful, though perhaps it may not be necessary, from the first cogitations of men to their second, and from the present age to posterity. Now let us come to that learning which the two former periods have not been so blessed as to know, namely, *Sacred and Inspired Divinity*, the most noble Sabbath and port of all men's labours and peregrinations.

whose work was published in 1825. The distinction between the character of the doctrines of the two schools is not very strongly marked.

[42] Plut. in Themist. c. 11.

Book IX.

CHAPTER I.

The Divisions of Inspired Divinity are omitted—Introduction only is made to three Deficients; namely, the Doctrine concerning the Legitimate Use of the Human Reason in Divine Subjects; the Doctrine concerning the Degrees of Unity in the Kingdom of God; and the Emanations of the Scriptures.

SEEING now, most excellent king, that my little bark, such as it is, has sailed round the whole circumference of the old and new world of sciences (with what success and fortune it is for posterity to decide), what remains but that having at length finished my course I should pay my vows? But there still remains Sacred or Inspired Divinity; whereof however if I proceed to treat I shall step out of the bark of human reason, and enter into the ship of the church; which is only able by the Divine compass rightly to direct its course. Neither will the stars of philosophy, which have hitherto so nobly shone upon us, any longer supply their light. So that on this subject also it will be as well to keep silence. I will accordingly omit the proper divisions thereof, contributing however a few remarks upon it, according to my slender ability, by way of paying my vows. And I am the more inclined to do this, because in the body of Theology I find no region or district entirely desert and uncultivated; such has been the diligence of man in sowing wheat or tares.

I will propose therefore three Appendices of Theology, which treat, not of the matter concerning which theology gives or shall give information, but only of the manner in which the information is imparted. I will not however, as in other like cases, either introduce examples or give precepts. That I will leave to theologians; for these, as I have said, are only in the place of vows.

The prerogative of God comprehends the whole man, extending to the reason as well as to the will; that man may deny himself entirely, and draw near unto God. Wherefore as we are bound to obey the divine law though we find a reluctation in our will, so are we to believe His word though we find a reluctation in our reason. For if we believe only that which is agreeable to our sense, we give consent to the matter and not to the author, which is no more than we would do to a suspected witness. But that faith which was accounted to Abraham for righteousness was of such a nature that Sarah laughed at it, who therein was an image of natural reason. The more discordant therefore and incredible the Divine mystery is, the more honour is shown to God in believing it, and the nobler is the victory of faith. Nay, even sinners, the more they are oppressed in their conscience, trusting nevertheless to be saved through the mercy of God, the more do they honour Him; for all despair is a kind of reproach towards God. Howbeit, if we will truly consider it, it is more worthy to believe, than to know as we now know. For in knowledge man's mind suffers from sense, which is the reflection of things material, but in faith the spirit suffers from spirit, which is a worthier agent. Otherwise it is in the state of man glorified, for then faith shall cease, and we shall know even as we are known.

Wherefore we conclude that Sacred Theology ought to be derived from the word and oracles of God, and not from the light of nature, or the dictates of reason. For it is written, "The heavens declare the glory of God[1]", but it is nowhere written, "The heavens declare the will of God"; but of that it is said, "To the law and to the testimony; if men do not according to this word," etc.[2]. And this holds not only in those great mysteries which concern the Deity, the

[1] Psalm xix. 1. [2] Isaiah, viii. 20.

Creation, and the Redemption ; but it pertains likewise to a more perfect interpretation of the moral law, " Love your enemies " ; " do good to them that hate you", and so on ; " that ye may be the children of your father who is in heaven, that sendeth rain upon the just and the unjust [3]". To which words this applause may well be applied, " that they do not sound human [4] " ; since it is a voice beyond the light of nature. Again, we see the heathen poets, especially when they discourse of the passions, often expostulate with laws and moral doctrines (which yet are far more easy and indulgent than the divine laws), as if they were contradictory and malignant to the liberty of nature ; " What nature grants the envious laws deny [5]". So said Dendamis the Indian to Alexander's messengers, " That he had heard somewhat of the name of Pythagoras and some other wise men of Greece, and that he held them for excellent men ; but that they had a fault, which was that they had too great reverence and veneration for a kind of phantom, which they called law and manners [6]". Wherefore it must be confessed that a great part of the moral law is higher than the light of nature can aspire to. Nevertheless what is said, that man has by the light and law of nature some notions of virtue and vice, justice and injustice, good and evil, is most true. For we must observe that the light of nature is used in two several senses ; the one, as far as it springs from sense, induction, reason, argument, according to the laws of heaven and earth ; the other, as far as it flashes upon the spirit of man by an inward instinct, according to the law of conscience ; which is a spark and relic of his primitive and original purity. And in this latter sense chiefly does the soul partake of some light to behold and discern the perfection of the moral law, a light however not altogether clear, but such as suffices rather to reprove the vice in some measure, than to give full information of the duty. So then religion, whether considered with regard to morals or mysteries, depends on revelation from God.

The use notwithstanding of reason in spiritual things is manifold and very general. For it is not for nothing that the Apostle called religion, "Our reasonable service of God [7]". If we review the types and ceremonies of the old law we see that they were full of reason and signification, differing widely from the ceremonies of idolatry and magic, which were like surds and non-significants, mostly without meaning, and not even suggestive of anything. But especially the Christian faith, as in all things, so in this is pre-eminent ; holding the golden mean touching the use of reason and discussion (the child of reason) between the law of the heathen and the law of Mahomet, which embrace the two extremes. For the religion of the heathen had no constant belief or confession ; and the religion of Mahomet on the other side interdicts argument altogether ; so that the one has the very face of vague and manifold error, the other of crafty and cautious imposture ; whereas the holy Christian faith both admits and rejects the use of reason and disputation, but according to just limitations.

The use of human reason in matters of religion is of two sorts ; the former in the explanation of the mystery, the latter in the inferences derived from it. With regard to the explanation of the mysteries, we see that God vouchsafes to descend to the weakness of our apprehension, by so expressing his mysteries that they may be most sensible to us ; and by grafting his revelations upon the notions and conceptions of our reason ; and by applying his inspirations to open

[3] St. Matt. v. 44, 45.

[4] Virg. *Æn.* i. 328 : Nec vox hominum sonat.

[5] Ovid, *Metam.* x. 330 :—

Et quod natura remittit,
Invida jura negant.

[6] Cf. Plut. in Alex. c. 65 ; and Strabo, i. xv. The name of the person of whom this story is told by Plutarch is Dandamis, but wherever Bacon has mentioned it, he spells it as in the text. Dandamis is also mentioned by Arrian, who, however, does not relate this anecdote. We find the same story in Strabo ; but the name of the Indian is with him not Dandamis, but Mandanis. In the *Temporis Partus Masculus*, Bacon speaks of these remarks of Dandamis as one of the exceptions to his general assertion of the worthlessness of the speculations of the philosophers of antiquity.

[7] Romans, xii. 1.

our understanding, as the form of the key to the ward of the lock. But here we ought by no means to be wanting to ourselves ; for as God uses the help of our reason to illuminate us, so should we likewise turn it every way, that we may be more capable of receiving and understanding His mysteries ; provided only that the mind be enlarged, according to its capacity, to the grandeur of the mysteries, and not the mysteries contracted to the narrowness of the mind.

But with regard to inferences, we should know that there is allowed us a use of reason and argument (in regard to mysteries) secondary and respective, though not original and absolute. For after the articles and principles of religion have been set in their true place, so as to be completely exempted from the examination of reason, it is then permitted us to derive and deduce inferences from them according to their analogy. In nature indeed this holds not. For both the principles themselves are examinable, though not by a syllogism, yet by induction ; and besides, these same principles have no discordance with reason, so that the first and middle propositions are derived from the same fountain. It is otherwise in religion, where the first propositions are not only self-existent and self-supporting ; but likewise unamenable to that reason which deduces consequent propositions. Nor yet does this hold in religion alone, but also in other sciences both of a greater and smaller nature ; namely, wherein the primary propositions are arbitrary and not positive ; for in these also there can be no use of absolute reason. For we see in games, as chess or the like, that the first rules and laws are merely positive, and at will ; and that they must be received as they are, and not disputed ; but how to play a skilful and winning game is scientific and rational. So in human laws there are many maxims, as they call them, which are mere Placets of Law, depending on authority rather than upon reason, and therefore not to be disputed. But what is most just, not absolutely but relatively (that is, according to these maxims), that is a matter of reason, and opens a wide field for disputation. Such therefore is that secondary reason which has place in Divinity, which is grounded upon the Placets of God.

But as the use of the human reason in things divine is of two kinds, so likewise in the use there are two kinds of excess ; the one when it inquires too curiously into the manner of the mystery ; the other when the same authority is attached to inferences as to principles. For he may appear to be the disciple of Nicodemus who persists in asking, " How can a man be born when he is old [8] ? " And he can be nowise considered the disciple of Paul who does not sometimes insert in his doctrines, " I, not the Lord " ; or again, "According to my counsel [9] "; which style is generally suited to inferences. Wherefore it appears to me that it would be of especial use and benefit, if a temperate and careful treatise were instituted, which, as a kind of divine logic, should lay down proper precepts touching the use of human reason in theology. For it would act as an opiate, not only to lull to sleep the vanity of curious speculations, wherewith sometimes the schools labour, but also in some degree to assuage the fury of controversies, wherewith the church is troubled. Such a treatise I reckon among the things deficient ; and call it *Sophron*, or *The Legitimate Use of Human Reason in Divine Subjects.*

It is of extreme importance to the peace of the Church, that the Christian covenant ordained by our Saviour be properly and clearly explained in those two heads, which appear somewhat discordant ; whereof the one lays down, " He that is not with us is against us ; " and the other, " He that is not against us is with us [10]." From this it is evident that there are some articles, wherein if a man dissent he is placed beyond the pale of the covenant ; but that there are others in which he may dissent, and yet remain within it. For the bonds of the Christian Communion are set down, "one Lord, one Faith, one Baptism, etc. [11] ",

[8] St. John, iii. 4. 10. [9] 1 Corinth. vii. 10.

[10] The two passages Bacon refers to are St. Luke, xi. 23. (or St. Matth. xii. 30.), and St. Luke, ix. 50. But the former he has not quoted accurately. The words of our version are, " He that is not with me is against me " ; while the passage in the ninth chapter is, "He that is not against us is for us".

[11] Ephes. iv. 5. Compare *St. August. Ep. ad Casulan. de jejuniis priscorum.* He

not one Ceremony, one Opinion. So we see the coat of our Saviour was without seam, but the garment of the church was of divers colours. The chaff should be separated from the corn in the ear, but the tares should not be pulled up from the corn in the field. Moses, when he saw the Egyptian fighting with the Israelite, did not say, "Why strive ye ?" but drew his sword and slew the Egyptian [12]. But when he saw the two Israelites fighting, though it were not possible for both to be in the right, yet he addresses them thus, "Ye are brethren, why strive ye ?"

And therefore on these considerations it appears a thing of great use and importance, well to define what and of what latitude those points are, which disincorporate men from the Church of God, and excommunicate them from the communion of the faithful. And if any one think that this has already been done, let him think again and again, and say whether it has been done with sincerity and moderation. Meanwhile if a man talks of peace, he is very like to get the answer of Jehu to the message, ("Is it peace, Jehu?") "What hast thou to do with peace ? turn thee behind me [13]" ; for it is not peace but party that most men care for. Nevertheless I have thought right to set down among the deficients a treatise *on the degrees of Unity in the kingdom of God*, being as a wholesome and profitable undertaking.

Since the Holy Scriptures are the principal sources of information in theology, we must especially look to their interpretation. And I am not now speaking of the authority of interpreting them, which rests in the consent of the church ; but of the manner thereof ; which is of two sorts ; methodical and free. For this divine water, which excels so much that of Jacob's well, is drawn forth and employed much in the same manner as natural water is out of wells and fountains. For it is either first forced up into cisterns, whence it may be conveniently fetched and derived by pipes for use ; or else it is poured into buckets and vessels to be used as it is wanted. The former method has in the end produced to us the scholastical divinity ; whereby divinity has been reduced into an art, as into a cistern, and the streams of doctrines and positions have been derived and conveyed from thence to water every part. But in the free way of interpreting Scripture, there occur two excesses. The one presupposes such perfection in Scripture, that all philosophy likewise should be derived from its sources ; as if all other philosophy were something profane and heathen. This distemper has principally grown up in the school of Paracelsus and some others ; but the beginnings thereof came from the Rabbis and Cabalists [14]. But these men do not gain their object ; and instead of giving honour to the Scriptures as they suppose they rather embase and pollute them. For to seek the materiate heaven and earth in the word of God, (whereof it is said, "Heaven and earth shall pass away, but my word shall not pass away [15]"), is rashly to seek for temporary things amongst eternal ; and as to seek divinity in philosophy is to seek the living among the dead, so to seek philosophy in divinity is to seek the dead among the living. The other method of interpretation which I set down as an excess, appears at the first glance sober and modest, yet in reality it both dishonours the Scriptures themselves, and is very injurious to the Church. This is (in a word), when the divinely-inspired Scriptures are explained in the same way as human writings. But we ought to remember that there are two things which are known to God the author of the Scriptures, but unknown to man ; namely, the secrets of the heart, and the successions of time. And therefore as the dictates of Scripture are written to the hearts of men, and comprehend the vicissitudes of all ages ; with an eternal and certain foreknowledge of all heresies, contradictions, and differing and changing estates of the Church, as well in general as of the individual

has elsewhere said, "Desuper texta tunica, quid significat nisi unitatem ?" See his *Exp. in Evan. Joan.* in c. 3. and other passages. Compare St. Jerome, *Pro Libris adversus Jovin. Apolog.*, where the many-coloured coat of Joseph is expressly mentioned, as well as the passage in the Psalms to which St. Augustine refers.

[12] Exod. ii. 12.

[13] 2 Kings, ix. 19.

[14] In support of this statement see Tennemann's History of Philosophy.

[15] St. Mark, xiii. 31.

elect, they are not to be interpreted only according to the latitude and obvious sense of the place; or with respect to the occasion whereon the words were uttered; or in precise context with the words before or after; or in contemplation of the principal scope of the passage; but we must consider them to have in themselves, not only totally or collectively, but distributively also in clauses and words, infinite springs and streams of doctrines, to water every part of the Church and the souls of the faithful. For it has been well observed that the answers of our Saviour to many of the questions which were propounded to Him do not appear to the point, but as it were impertinent thereto. The reason whereof is twofold; the one, that knowing the thoughts of his questioners not as we men do by their words, but immediately and of himself, he answered their thoughts and not their words; the other, that He did not speak only to the persons then present, but to us also now living, and to men of every age and nation to whom the Gospel was to be preached. And this also holds good in other passages of Scripture.

Having made then these prefatory remarks, I come to that treatise which I pronounce deficient. There are found indeed among theological writings too many books of controversies, a great mass of that theology which I have termed Positive, commonplaces, special tracts, cases of conscience, sermons and homilies, and many prolix commentaries upon the Scriptures. But what we want is a concise, sound, and judicious collection of annotations and observations on particular texts of Scripture; neither dilated into commonplaces, nor chasing after controversies, nor reduced into method of art, but entirely unconnected and natural. It is indeed a thing sometimes found in the more learned sermons, which for the most part do not last; but not yet introduced into books, which may be handed down to posterity. But certainly, as wines which flow gently from the first treading of the grape are sweeter than those which are squeezed out by the wine-press; because these last have some taste of the stones and skin of the grape; so those doctrines are very sweet and healthy, which flow from a gentle pressure of the Scriptures, and are not wrested to controversies or commonplaces. Such a treatise then I will denominate *the Emanations of the Scriptures.*

Now therefore have I made as it were a small globe of the intellectual world, as faithfully as I could; with a note and description of those parts which I find either not constantly occupied, or not well cultivated by the labour and industry of man. Wherein, if I have in any point receded from the opinion of the ancients, let it be understood that I have done so not from a desire of innovation or mere change, but of change for the better. For I could not be true and constant to myself or the argument I handle, if I had not determined to add as much as I could to the inventions of others; being however no less willing that my own inventions should be surpassed by posterity. But how fair I am in this matter may appear from this; that I have propounded my opinions everywhere naked and unarmed, without seeking to prejudice the liberty of men's judgments by disputes and confutations. For in anything which is well set down, I am in good hope that if the first reading move a scruple or objection, the second reading will of itself make an answer. And in those things wherein it has been my lot to err, I am sure I have not prejudiced the truth by litigious arguments; which commonly have this effect, that they add authority to error, and diminish the authority of that which is well invented; for question is an honour to falsehood, but it is a repulse to honour. Meanwhile I am reminded of the sarcastic reply of Themistocles to the ambassador, who coming from a small town used great words, "Friend, (said he) your words require a city[16]". And certainly it may be objected to me with truth, that my words require an age; a whole age perhaps to prove them, and many ages to perfect them. But yet as even the greatest things are owing to their beginnings, it will be enough for me to have sown a seed for posterity and the Immortal God; whose Majesty I humbly implore through His Son and our Saviour, that He will vouchsafe favourably to accept these and the like offerings of the human intellect, seasoned with religion as with salt, and sacrificed to His Glory.

[16] Lysander, not Themistocles. Cf. Plut. *Lac. Apopthegmata.*

APPENDIX

[The following Notes on some old treatises on the art of writing in cipher are referred to by Mr. Ellis, at p. 527, note 16.—*J. S.*]

THE earliest writer, I believe, on ciphers, except Trithemius whom he quotes, is John Baptist Porta, whose work *De occultis literarum notis* was reprinted in Strasburg in 1606. The first edition was published when Porta was a young man. The species of ciphers which Bacon mentions are described in this work. What he calls the *ciphra simplex* is doubtless that in which each letter is replaced by another in accordance with a secret alphabet. (Porta, ii. c. 5.) The manner of modifying this by introducing non-significants and by other contrivances is described in the following chapter. The *wheel cipher* is described in chapters 7, 8, 9. It is that in which the ordinary alphabet and a secret one are written respectively on the rim of two concentric disks, so that each letter of the first corresponds in each position of the second (which is movable) to a letter of the secret alphabet. Thus in each position of the movable disk we have a distinct cipher, and in using the instrument this disk is made to turn through a given angle after each letter has been written. The *ciphra clavis* is described by Porta, book ii. 15, 16. It is a cipher of position; that is, one in which the difficulty is obtained not by replacing the ordinary alphabet by a new one, but by deranging the order in which the letters of a sentence or paragraph succeed each other. This is done according to a certain form of words or series of numbers which constitute the key. The *cipher of words* was given by Trithemius and in another form by Porta, ii. 19 (and in a different shape, v. 16). It is a cipher which is meant to escape suspicion. Each letter of the alphabet corresponds to a variety of words arranged in columns. Any word of the first column followed by any of the second, and that followed by any of the third, etc., will make, with the help of a non-significant word occasionally introduced, a perfectly complete sense; and by the time the last alphabet has been used a letter on some indifferent subject has been written. Only sixty alphabets are given by Porta, and therefore the secret communication can consist only of sixty letters. It is worth remarking that when Porta wrote it was usual to put the sign of the cross at the head of an ordinary epistle. The first of his alphabets corresponds not to a series of words but to two and twenty different modifications of the figure of a cross, and his second alphabet similarly corresponds to two and twenty different modifications of the introductory flourish. His sixtieth alphabet is of the same kind. We see here perhaps whence Bacon derived his idea of giving significance to seemingly accidental modifications of the characters of ordinary writing.

The idea of a *biliteral alphabet*, which Bacon seems to claim as his own, is employed, though in a different manner, by Porta. His method is in effect this. He reduces the alphabet to sixteen letters, and then takes the eight different arrangements *aaa*, *aba*, etc., to represent them; each arrangement representing two letters indifferently: the ambiguity arising from hence he seems to disregard. In this manner he reduces any given word or sentence to a succession of *a*'s and *b*'s. At this point his method, of which he has given several modifications, departs wholly from Bacon's. Let us suppose the biliteral series to commence with *aababb*. A word of two syllables and beginning with A indicates that two *a*'s commence the series; any monosyllable will serve to show that one *b* follows, another that it is succeeded by one *a*, and then any dissyllable will stand for *bb*. Thus *Amo te mi fili* or *Amat qui non sapit* will represent the biliteral arrangement *aababb*, and so on on a larger scale. Porta's method is therefore not, like Bacon's, a method *scribendi omnia per omnia*, but only *omnia per multa*. Still the analogy of the two methods is to be remarked; both aim at concealing that there is any but the obvious meaning, and both depend essentially on representing all letters by combinations of two only. See the *De oc. Lit. Signis*. v. c. 3.

The *Polygraphia* of Trithemius (dedicated to Maximilian in 1508[1]) consists of six books. The first four contain extensive tables constituting four different *ciphræ verborum*; the first and second of which are significant, and relate, the former to the second person of the Trinity, and the latter to the Blessed Virgin. The fifth and sixth books are of less importance. *Trithemius*, written in the cipher of the second book, becomes "Charitatem pudicissimæ Virginis Mariæ productricis coexistentis verbi, robustissimi commilitonis mei dilectissimi devotissime benedicamus; vivificatrix omnium," etc.

Traicté des Chiffres, ou secrètes manières d'escrire, par Blaise de Vigenère, Bourbonnois. (Paris, 1587.)

This work is described by the author as what he had saved of his work "*Du Secrétaire*," written in Italy in 1567 and 68. The two first books were stolen at Turin in 1569. The third is the foundation of the present work. (v. f. 285. verso.) He says he had revealed nothing of its contents.

The two authors whom he chiefly mentions are Trithemius and Porta; that is, modern authors; for there is a great deal said of the Cabala. The key ciphers of which Porta speaks he ascribes to a certain Belasio, who employed it as early as 1549: Porta's book not being published until 1563, "auquel il a inseré ce chiffre sans faire mention dont il le tenoit." Porta's book, he goes on to say, was not *en vente* until 1568. The invention was ascribed to Belasio by the grand vicar of St. Peter at Rome, who had great skill in deciphering (f. 35. rect. and 37. verso).

At f. 199. Vigenère gives an account of ciphers in which letters are represented by combinations of other letters,—which Porta had already done, but which he varies in a number of ways.

f. 200. A table where the twenty-three letters of the alphabet, and four other characters, are represented by combinations of *abc*. D (e. gr.)=*aaa*, S=*bac* etc.)

f. 201. A smaller table where an alphabet of twenty-one letters is similarly represented.

f. 202. An alphabet of twenty letters represented by binary combinations of five letters, *a*=ED, etc.

f. 202. goes on to what Bacon speaks of, a cipher within a cipher. You write in a common cipher with an alphabet of eighteen letters; the cipher being such that the five vowels are used as nulls; then by the last cipher these five vowels are made significant, and give the hidden sense. He seems to speak of this as his own.

After mentioning a cipher described by Cardan, he goes on, f. 205, to Porta's ciphers by transposition, etc.

At f. 240. he shows how characters may be multiplied by different ways of writing them; which Porta had not done.

f. 241. An alphabet and &, each character written in four ways.

f. 241, verso, An application of these variations.

f. 242. He remarks that a great variety of uses may be made of this idea, and gives some.

f. 244. He goes on "De ce même retranchement et de la variété de figure, part une autre invention encore d'un chiffre carré à double entente, le plus exquis de tous ceux qui ayent esté decouvers jusqu'à icy," &c. You write with twelve letters only, as in the subjoined table, in which however I have not followed his ways of diversifying.

[1] The edition of 1600 is that I use.

	P	*C*	*T*	*E*	*I*	*L*	*M*	*N*	*A*	*R*	*S*	*V*
E, *P*	a_1	a_2	a_3	b_1	b_2	b_3	c_1	c_2	c_3	d_1	d_2	d_3
V, *C*	e_1	e_2	e_3	f_1	f_2	f_3	g_1	g_2	g_3	h_1	h_2	h_3
T, *L*	i_1	i_2	i_3	k_1	k_2	k_3	l_1	l_2	l_3	m_1	m_2	m_3
I, *M*	n_1	n_2	n_3	o_1	o_2	o_3	p_1	p_2	p_3	q_1	q_2	q_3
A, *N*	r_1	r_2	r_3	s_1	s_2	s_3	t_1	t_2	t_3	u_1	u_2	u_3
R, *S*	x_1	x_2	x_3	y_1	y_2	y_3	z_1	z_2	z_3	$\&_1$	$\&_2$	$\&_3$

In this table, Z, for instance, represents 1st M, and 2nd R or S ; to distinguish whether R or S, he has recourse to a supplementary contrivance by nulls.

f. 242. v. He refers to table at 200., and says the three letters *abc*, (which there represent I) may be replaced by a single character ; for this table represents in another column letters by dots. Thus T is .. .. ; D . . . ; or if we will we may put *o*'s for dots ; so that D=*o oo* and T=*oo ooo o* ; and the spaces may be filled up by a slightly varied *o*. Thus D=*ooooo*, T=*oooooooo*, and thus the whole cipher will apparently consist of *o*'s.

The transition from this to Bacon's cipher is so easy that the credit given to him must be reduced.

ON PRINCIPLES AND ORIGINS ACCORDING TO THE FABLES OF CUPID AND COELUM, ETC.

[TRANSLATION OF *DE PRINCIPIIS ATQUE ORIGINIBUS, ETC.*]

PREFACE.

BY ROBERT LESLIE ELLIS.

THE following tract is one of those which were published by Gruter. It seems to be of later date than many of the others, as it contains several phrases and turns of expression which occur also in the *Novum Organum*.

Bacon's design was to give a philosophical exposition of two myths; namely that of the primeval Eros or Cupid, and that of Uranos or Cœlum. Only the first however is discussed in the fragment which we now have, and even that is left incomplete.

The philosophy of Democritus appeared to Bacon to be nearly in accordance with the hidden meaning of these fables; but we are not well able to judge of his reasons for thinking so, as the only system spoken of in detail is that of Telesius.

Touching the origin of Eros, Bacon remarks that no mention is made anywhere of his progenitors. In this he is supported by the authority of Plato, or rather by that of one of the interlocutors in the *Symposium*, who affirms that no one, whether poet or not, has spoken of the parents of Eros; but that Hesiod in the order of his theogony places Gaia and Eros next after primeval Chaos [1]. It seems in truth probable that the fables which make Eros the son of Zeus and Aphrodite are of later origin. From the *Symposium* Bacon may also have derived the recognition of an elder and a younger Eros, of whom the former was allied to the heavenly Aphrodite, and the latter to Aphrodite Pandemos [2]. But it is more probable that his account of the distinction between them comes from some later writer.

Hesiod, to whom the first speaker in the *Symposium* refers, though he places Eros and Gaia next to Chaos, says nothing of Eros as the progenitor of the universe. His existence is recognised, but nothing is said of his offspring. In this the theogony of Hesiod differs essentially from that which is contained in the Orphic poems, and shows I think signs of greater antiquity. To recognise as a deity an abstract feeling of love or desire, is in itself to recede in some measure from the simplicity of the old world: we find no such recognition in Homer; and the transition from him to Hesiod is doubtless a transition from an earlier way of thinking to a later. But even in Hesiod Eros is not the producing principle of the universe, nor is his share in its production explained. On the other hand in the Orphic poems, Phanes, whom we are entitled to identify with Eros, is the progenitor of gods and men, the light and life of the universe. He comes forth from Chaos, uniting in his own essence the poles of the mysterious antithesis on which all organic production depends. From him all other things derive their existence. There seems clearly more of a philosopheme in this than in the simpler statements of Hesiod.

[1] Sympos. p. 178.; and see Valcknaer's Diatribe, to whom Stallbaum refers. On the other hand Pausanias mentions as an early myth that Eros was the son of Ilithyia. See Pausan. Bœot. ix. 27.

[2] Sympos. p. 180, and see also p. 195.

The identification of Eros with Phanes or Ericapeus rests on a passage in the *Argonautics* in which it is said that he was called Phanes by the men of later time because he was manifested before all other beings; πρῶτος γὰρ ἐφάνθη[3]. It is confirmed by the authority of Proclus.

Phanes, in the common form of the Orphic theogony, comes out of the egg into which Chaos had formed itself[4]. But I am not aware that any one except Aristophanes makes Night lay the egg from which Eros afterwards emerges[5]; and it seems that this is only a playful modification of the common myth, not unsuitable to the chorus of birds by whom it is introduced[6]. It does not appear necessary to suppose, as Cudworth seemingly does, that Aristophanes had in some unexplained way become acquainted with a peculiar form of "the old atheistic cabala[7]".

The most remarkable passage in which Eros (not Phanes) is spoken of as the producer of all things, is in the *Argonautics*:—

πρῶτα μὲν ἀρχαίου χάεος μεγαλήφατον ὕμνον,
ὡς ἐπάμειψε φύσεις, ὥς τ' οὐραεὸς ἐς πέρας ἦλθεν,
γῆς τ' εὐρυστέρνου γένεσιν, πυθμένας τε θαλάσσης,
πρεσβύτατόν τε καὶ αὐτοτελῆ πολύμητιν Ἔρωτα,
ὅσσά τ' ἔφυσεν ἅπαντα, τὰ δ' ἔκριθεν ἄλλου ἀπ' ἄλλο[8].

Nothing is said here, or elsewhere I believe, of his having mingled with Uranos in the engendering of the universe; and I am inclined to think that when Bacon says, "Ipse cum Cœlo mistus, et deos et res universos progenuit," we ought to substitute Chao for Cœlo[9]. For the passage in Aristophanes goes on to say that in wide Tartarus Eros and Chaos mingled in love and produced first the race of birds and then gods and men.

Of Phanes nothing of this kind is mentioned, except his intercourse with Night[10]; so that Bacon's statement does not seem to be in any way justified.

It would be endless to cite passages in which the attributes of Eros are described, nor is it necessary to do so.

The form in which Bacon connects the myth of the primeval Eros with philosophy is far less artificial and unreal than most of the interpretations which he has given in the Wisdom of the Ancients. Chaos represents uninformed matter; Eros matter actually existing, and possessed of the law or principle by which it is energised; the first principle, in short, which is the cause of all phenomena. The parents of Eros are unknown; that is to say, it is in vain to seek to carry our inquiries beyond the fact of the existence of matter possessed of such and such primitive qualities. On what do those primary qualities ultimately depend? On the "lex summa essentiæ atque naturæ . . . vis scilicet primis particulis a Deo indita, ex cujus multiplicatione omnis rerum varietas emergat et confletur". Whether this highest law can ever be discovered is by Bacon left here as elsewhere doubtful; but he does not forbid men to seek for it. But what he utterly condemns is the attempt to make philosophy rise above the theory of matter. We must ever remember that Eros has no progenitors, "ne forte intellectus ad inania deflectat"—that we turn not aside to transcendental fancies; for in these the mind can make no real progress, and "dum ad ulteriora tendit ad proximiora recidit". We must of necessity take as the starting point of our

3 Orph. *Argon.* 14. In the preceding line, Eros is made, according to Gesner's reading, the son of Night. But for υἷα there is another reading, πατέρα.

4 See Lobeck, *Aglaoph.* i. 474.

5 Aves, 650.

6 This seems to be confirmed by the half ludicrous epithet ὑπηνέμιον.

7 See Cudworth, *Intellect. Syst.*

8 Argonaut. 423. In the third line πυθμένας is admitted to be corrupt. I would venture to suggest πολιᾶς, making σαλάσσης the genitive case after γένεσιν.

9 This conjecture is confirmed by the corresponding passage in the *De Sap. Vet.*, where for *cum cœlo mistus* we have *ex chao.—J. S.*

10 Lobeck, i. 501. It is to this intercourse that the line quoted by Proclus refers:—
Αὐτὸς ἑῆς γὰρ παιδὸς ἀφείλετο κούριον ἄνθος.

philosophy, matter possessed of its primitive qualities; and this principle is in accordance with the wisdom of those by whom the myth of Eros was constructed. And certainly, Bacon goes on to say, "that despoiled and merely passive matter is a figment of the human mind"; a statement which refers to the Aristotelian doctrine in which the primitive ὕλη is not conceived of as a thing actually existing, but as that which first receives existence through the εἶδος, wherewith it is united. Of this doctrine Bacon asserts that it is altogether trifling: "For that which primarily exists must no less exist than that which thence derives its existence"; that is to say, matter must in itself exist actually and not potentially. And the same conclusion follows from the Scriptures, "wherein it is not said that God created hyle, but that he created heaven and earth".

This application of Scripture certainly does not deserve the indignation which De Maistre, perhaps in honest ignorance, has poured out upon it[11]. "He asserts the eternity of matter," is De Maistre's commentary on the passage in which it occurs. Beyond doubt he denies that hyle was created, but he also denies that it exists; treating it as the mere figment of the Aristotelian philosophy.

But although De Maistre's remark is only a fair specimen of his whole work, in which ignorance and passion are so mixed together that it is hard to say how much is to be ascribed to the one and how much to the other, yet it cannot be denied that Bacon does not appear to have understood Aristotle. So far from putting at the origin of things that which is potential, and educing the actual from it, Aristotle asserts that any system which does this is untenable; and it is curious that he refers particularly to the theogonists, οἱ ἐκ νυκτὸς γεννῶντες, who engender realities out of night[12]. For night and chaos may not unfitly be taken to represent uninformed matter[13]. The doctrine of Aristotle being in this as in other matters followed by the schoolmen, it was a question with them how the words "and the earth was without form", which come immediately after the declaration that in the beginning God created the heaven and the earth, ought to be understood. For to create the earth is to give it actual existence; how then can it be without form? To this the most satisfactory answer was that the words without form do not imply the absence of substantial form, failing which the earth could have no actual existence, but simply mean that as yet the earth was unadorned and in disorder; a solution in which we see how far they were from supposing that according to Aristotle the first created thing ought to be uninformed matter. They insist on the contrary that the Scripture cannot mean that any created thing can be mere matter: "non enim datur ens actu sine actu".

Aristotle, as I have said, condemns the theogonists in whose system Night is a producing principle,—a remark in which he may refer either to Hesiod or to the Orphic writers, but which probably relates to the former only. In the reason of this condemnation Bacon agrees with him, and yet takes into the myth which he proposes to explain, Aristophanes's fancy that the egg from which Eros came forth was laid by Night. His reason for doing so is that this part of the fable appears to him to relate not to essence but to cognition, that is to the method whereby we may arrive at a knowledge of Eros, or of the fundamental properties of matter. For conclusions obtained by means of affirmatives are, so to speak, brought forth by Light: whereas those which are obtained by negatives and exclusions are the offspring of Night and Darkness. Therefore the egg is laid by Night, seeing that the knowledge of Eros, though it is assuredly attainable, can yet only be attained by exclusions and negatives; that is, to express the same opinion in the language of the *Novum Organum*, the knowledge of Forms necessarily depends on the *Exclusiva*. That this method of exclusions must of necessity be ultimately successful is intimated by the myth itself; for the incubation of the primeval egg is not eternal. In due time the egg is hatched and Eros is made manifest. If it be asked what analogy there is between darkness and the method of exclusions, Bacon's answer is satisfactory,—that darkness is as ignorance, and that in employing the method of exclusions we are all along ignorant of that

11 Examen de la Philosophie de Bacon, ii. p. 143.

12 Arist. *Metaph.* xii. 6.

13 See Brandis's Schol. in Aristot. p. 803, and for the remarks of Alexander Aphrodisiensis, Lobeck, *Aglaoph.* i. 488.

which at any stage of the process still remains unexcluded. It may again be asked why the method of exclusions is the only one whereby Eros may be disclosed,—a question to which Bacon suggests an answer by saying that Democritus did excellently well in teaching that atoms are devoid of all sensible qualities. Bacon's opinion seems therefore to be, that any method but a negative one would necessarily fail, because that which is sought bears no analogy to any of the sensible objects by which we are surrounded. The parable, he says, maintains throughout the principles of heterogeneity and exclusion: meaning by heterogeneity a strongly marked antithesis between the fundamental qualities of matter and the sensible qualities of which we are directly cognisant. In accordance with this he censures Democritus for departing from this principle in giving his atoms the downward motion of gravity and the impulsive motion (*motus plagæ*) which belong to ordinary bodies. Not only are atoms and bodies different as touching their qualities, but also in their motions.

In these views, which however do not show either that the method of exclusions is the only one which can succeed or that it will always do so, there is much which deserves attention. They show that Bacon had obtained a deep insight into the principles of the atomic theory. The earlier developments of this theory have always been encumbered by its being thought necessary, in order to explain phenomena, to ascribe to the atoms properties which in reality belong only to the bodies which they compose; that is, by its being thought necessary to break through Bacon's principle of heterogeneity. Thus the atoms have been supposed of definite sizes and figures, thereby resembling other and larger bodies, and to be perfectly hard and unyielding. When freed from these subsidiary hypotheses, the atomic theory becomes a theory of forces only, and of whatever ulterior developments it may be capable, these can only be introduced when it has assumed this form. The speculations of Boscovich do not mark the farthest point to which the atomic theory may be carried, but they were nevertheless an essential step in advance, and altogether in accordance with what Bacon has here said, though in an obscure and somewhat abrupt manner. "We do well," remarks Leibnitz, "to think highly of Verulam, for his hard sayings have a deep meaning in them:" a judgment which may not improbably have had a particular reference to the views now spoken of. For Leibnitz's own monadism is in effect only an abstract atomic theory [14]: more abstract doubtless than any thing which Bacon had conceived of, but yet a system which might have been derived from that of Democritus by insisting on and developing Bacon's principle of heterogeneity. And again, in a different point of view, it seems not unlikely that Leibnitz perceived an analogy between his own doctrine and that of Bacon. In the earlier part of his philosophical life, Leibnitz was disposed to agree with the opinion common among the reformers of philosophy, that what Aristotle had said of matter, of form and of mutation, was to be explained by means of magnitude, figure, and motion. This opinion he ascribes to all the reformers of the seventeenth century, mentioning by name Bacon and several others [15]. Thirty years afterwards, in giving some account of the history of his opinions, he says that he came to perceive, "que la seule considération d'une masse étendue ne suffisoit pas, et qu'il falloit employer encore la notion de la force, qui est très-intelligible, quoiqu'elle soit du ressort de la Métaphysique [16]". In introducing this notion of force, he conceived that he was rehabilitating the Aristotelian or scholastic philosophy, seeing "que les formes des Anciens ou Entelechies ne sont autre chose que les forces [17]". These primitive forces [18] being the constituent forms of substances, he supposed them, with one exception (founded on dogmatic grounds), to have been created at the beginning of the world. The "lex a Deo lata" at the creation "reliquit aliquod sui expressum in rebus vestigium", namely an efficacy, or form, or force by virtue of which and in accordance with the divine precept all phenomena had been engendered [19].

[14] The monad, Leibnitz himself remarks, is a metaphysical point, or formal atom.
[15] Epist. ad Thomas. p. 48 of Erdmann's edition of Leibnitz's Phil. Works.
[16] *Système nouveau*, p. 124, Erdmann.
[17] Lettre à Bouvet, p. 146, Erdmann.
[18] *Forces primitives*, v. *Syst. Nouv.*
[19] See his *De ipsâ Naturâ*, p. 156.

If we compare these expressions, which contain the fundamental idea of Leibnitz's philosophy, with those which have already been quoted from the following tract, we shall I think perceive more than an accidental analogy between them. Leibnitz speaks of the primitive forces impressed by the divine word on created things, "ex quâ series phenomenorum ad primi jussûs præscriptum consequeretur,"—and Bacon of the "lex summa essentiæ et naturæ, vis scilicet primis particulis a Deo indita, ex cujus multiplicatione omnis rerum varietas emergat et confletur". It does not seem improbable that Leibnitz, who in the letter to Thomasius classes Bacon, so far as relates to the present subject, with Gassendi and Descartes, came afterwards to find in Bacon's language hints of the deeper view which he had himself been led to adopt, and which constitutes the point of separation between his system and the Cartesian. This supposition would at least be in accordance with the emphatic manner in which he has contrasted the physical theories of Descartes and Bacon, taking the former as a type of acuteness and the latter of profundity, and asserting that compared with Bacon, Descartes seems to creep along the ground [20].

It may not be out of place here to remark that there are other traces of Bacon's influence on Leibnitz. In Erdmann's edition of his philosophical works, we find several fragmentary papers which Leibnitz wrote under the name of Gulielmus Pacidius. The title of one of these is "Gulielmi Pacidii Plus Ultra, sive initia et specimina scientiæ generalis de instauratione et augmentis scientiarum ac de perficiendâ mente rerumque inventione ad publicam fœlicitatem". Plus Ultra was the motto to Bacon's device of a ship sailing through the Pillars of Hercules, and the remainder of the title is both in tone and language clearly Baconian. The work itself was to have concluded with an exhortation "ad viros dignitate doctrinâque egregios de humanâ fœlicitate exiguo tempore, si velimus modo, in immensum augendâ [21]".

Another of these fragments contains some account of himself, or rather of Wilhelmus Pacidius, in which he mentions it as one of the happy incidents of his youth, that when he had perceived the defects of the scholastic philosophy the writings of several of the reformers came into his hands—among which he gives the first place to the "consilia magni viri Francisci Baconi Angliæ Cancellarii de augmentis Scientiarum [22]".

To return to the fable of Cupid. After interpreting the statement that all things come from Eros to mean that all phenomena must be referred to the fundamental and originally inherent properties of matter as the first ground of their production, Bacon goes on to say that next to the error of these who make formless matter an original principle, is the error of ascribing secondary qualities to primitive matter. This he expresses by saying that though Eros is endued with personality, he is nevertheless naked, "ita personatus [23] ut sit tamen nudus". Those who have committed the error of clothing him have either merely covered him with a veil, or have dressed him up in a tunic, or lastly have wrapped him round with a cloak.

These three errors are respectively the erors of those who have sought to explain everything by the transformations of one element, as air or fire,—of those who assume a plurality of elements,—and of those who assume an infinity of first principles (the homœomeria of Anaxagoras), each possessed of specific properties.

Contrasted with these errors is the doctrine that there is one first material principle, "idque fixum et invariabile," and that all phenomena are to be explained "per hujusmodi principii . . . magnitudines figuras et positiones,"—a statement which includes along with the old atomic theory every such hypothesis as the Cartesian. By those only who hold this opinion is Eros rightly displayed; they show him as he really is, "nativus et exutus".

In the interval between writing this tract and the *Novum Organum* Bacon's

[20] *Leibnitiana*, vol. vi. p. 303, ed. Genev. 1768.—*J. S.*

[21] Leibnitz, ed. Erd. p. 89.

[22] Leibnitz, ed. Erd. p. 91.

[23] The meaning of *personatus* appears from the phrase Bacon previously uses: "Cupidinis est persona quædam".

opinions seem to have undergone some change, as he has there condemned the atomists for asserting the existence of " materia non fluxa " ; an obscure phrase, but which appears irreconcilable with the expression which I have just quoted— " fixum et invariabile ".

However this may be, Bacon next proceeds to enumerate the different forms of doctrine into which the doctrine of a single element has been subdivided. The first principle or primitive matter has been asserted to be water, or air, or fire. Something is then said of the opinions of Thales, of Anaximenes, and of Heraclitus, and they are collectively commended for having given Eros but a single garment, that is, for having ascribed to primitive matter only a single form, substantially homogeneous with any of the forms of secondary existences.

The Anaxagorean doctrine of an infinity of elements is then set aside as belonging to the interpretation of the fable of Cœlum, and thus Bacon comes to the doctrine of two opposing principles, with which the remainder of the tract is taken up. Parmenides, he observes, among the ancients, and Telesius in modern times, have made fire and earth, or heaven and earth, the two first principles.

In connecting together Telesius and Parmenides Bacon overlooked an essential point of difference. For the system of Telesius is merely physical, it deals only with phenomena, and seeks for no higher grounds of truth than the evidence of the senses. Parmenides, on the other hand, recognized the antithesis of τὸ ὄν and τὸ φαινόμενον, of that which exists and that which is apparent. His doctrine is ontological rather than physical, and he does not admit that phenomena have any connexion with real or essential truth. He seeks for a deeper insight into things than any which a mere " Welt-anschauung," a mere contemplation of the universe, could be made to furnish. The hypothesis which he framed to explain the phenomena by which we are surrounded, is with him a hypothesis merely; and though, like Telesius's, this hypothesis refers every phenomenon to the antagonism of heat and cold, yet it has a character of its own, inasmuch as in a way not distinctly conceivable it also serves to represent the metaphysical antithesis of τὸ ὄν and τὸ μὴ ὄν.

It is however to be remembered that with the ontological aspect of the philosophy of Parmenides Bacon has here no concern.

The fundamental notion of Telesius's system was doubtless suggested both to him and to Parmenides [24], by certain obvious phenomena, and especially by the growth, decay, and reproduction of plants and animals. But it is essentially derived from the delight which the mind takes in every form of antithetic dualism, and especially in the idea of the reciprocal action of opposing forces. It comes from the same source as the love and strife of Empedocles, and as the good and evil principles of the Persian theology.

By the help of this notion, namely that heat and cold are the constituent principles of the universe, Telesius attempts to give general explanations of all phenomena, leaving it to others to study them in detail. The largeness of his plan and the grave eloquence with which it is set forth won for him some celebrity, notwithstanding the extreme obscurity of his style and the vagueness of his whole doctrine.

The academy of Cosenza (it was at Cosenza that Telesius was born) adopted his views, and both there and elsewhere men were for some time to be found who called themselves Telesiani. Spiriti, in his *Scrittori Cosentini*, gives a list of the disciples of Telesius ; it contains however no name of much note, except that of Campanella, and the fame of Campanella rests much more on his moral and political speculations than on his defence of Telesius. Giordano Bruno and Patricius cannot be called disciples of Telesius, though the writings of both bear traces of his influence [25]. Among real students of nature it was not to be ex-

[24] The same notion is ascribed also to Hippo of Rhegium, and to others of the Greek philosophers. See Ps, Orig. [Hippolytus, i. 14] *Philos.* (16), as to Hippo

[25] The influence of Telesius on Bruno is not, I think, mentioned by historians of philosophy, yet there is no doubt of its existence. In the following passage the fundamental principle of Telesius is plainly assumed, mingled with ideas derived from Copernicus. " Così vien distinto l'universo in fuoco et acqua, che sono soggetti di doi primi principii

pected that so indefinite a system as that of Telesius could find much acceptance, and accordingly it is but seldom mentioned by scientific writers. Grassi, in the *Libra Astronomica*[26], seems to reproach Galileo with having taken some notion about comets from Cardan and Telesius; remarking that their philosophy was sterile and unfruitful, and that they had left to posterity "libros non liberos". To this Galileo answers that as for what Cardan and Telesius might have said on the matter he had never read it, and it would seem as if he means to disclaim all knowledge of their writings. Though he protests against the argumentum ex consensu which Grassi brings against them, yet it is plain that he does so only to confute his opponent, and not because he thought them worthy of a greater fame than they had received. Even among the large class of men who are content to acquiesce in general views and are not careful to inquire whether these views are accurate or ill-defined, Telesius's popularity could not last long. For he had left nothing for his followers to do. All that could be said in favour of his fundamental idea he had said himself, and any attempt to develop it further could only show how insecure a foundation it was built on. His works are however not undeserving of attention, even apart from the influence which they had on the opinions of Bacon. They show much of the peculiar character of mind which distinguishes southern from northern Italy, and which is yet more conspicuous in the writings of Campanella and of Vico: grave and melancholy earnestness,—a fondness for symbol and metaphor, and for wide-reaching but dreamy theories.

The first two books of his principal work, the *De Rerum Naturâ*, were published at Rome in 1565. The complete work was not published until 1586, only two years before his death[27]. In 1590 a number of tracts, some of which had appeared in his lifetime, were published by Antonius Persius, one of his chief disciples, with a dedication to Patricius, which seems to claim him as at least half an adherent to the Telesian philosophy[28]. For some account of Telesius's minor works I may refer to Spiriti's *Scrittori Cosentini*, or to what Salsi has said of them in Ginguené's *Histoire Littéraire de l'Italie*[29].

Of Lotter's work, *De Vita et Scriptis B. Telesii*, Leipsic, 1733, I much regret that I only know what is said of it in the *Acta Eruditorum* for that year. It appears to contain much information not easily to be found elsewhere.

The view which Bacon gives of the doctrines of Telesius seems to have been much used and trusted by the historians of philosophy[30],—a natural result of the involved and obscure style in which they were originally propounded. Whether it is altogether an accurate representation of these doctrines may at least be doubted: it seems as if Bacon, in some matters of detail, mingles with what he finds in Telesius some further developments of his own. Perhaps he is in some measure influenced by his jural habits of thought, and tries in all fairness and equity to put a favourable construction on that on which he sits in judgment[31].

formali et attivi, freddo et caldo. Que' corpi che spirano il caldo, son le sole, che per se stesso son lucenti et caldi; que' corpi che spirano il freddo son le terre".—*Cena di Cenere*, p. 174 of Wagner's edition.

[26] Published in 1618, with the pseudonym of Lotario Sarsi. It is incorporated in the new edition of Galileo's works, iv. p. 61.

[27] It was reprinted in 1588, along with the *Contemplationes* of Mocenicus and the *Quæstiones Peripateticæ* of Cæsalpinus. The volume containing these three works is entitled "Tractationum Philosophicarum tomus unus", and is apparently not easily met with. It is this edition that I have been in the habit of using.

[28] This dedication is prefixed to the tract "De Mari".

[29] The account of Telesius in Ginguené was written by Salsi. See Ginguené, vii. p. 500.

[30] See what Brucker says of Morhof and Sosellus, *Hist. Crit. Phil.* iv. 453.

[31] Bacon's own language suggests this impression. "Nos enim," he declares, "in omnium inventis summâ cum fide et tanquam faventes versamur." And that he does not conceive himself bound to minute accuracy in reproducing the opinions of the philosophers of whom he speaks, appears from several expressions: "Hujusmodi quædam de diversitate calorum a Telesio dicuntur;" "Hæc, aut iis meliora, cogitabant illi," etc.

However this may be, I have certainly found it difficult to support all his statements by quotations from his author, and in some cases have noticed at least apparent discrepancies.

The tract ends abruptly with the discussion of the system of Telesius. A similar discussion of the atomic theory would have been of far greater interest, for Bacon's own opinions are much more closely connected with those of Democritus than with Telesius's, from whom he derived only isolated doctrines. The most important of these doctrines is that of the duality of the soul, of which and of its relation to the orthodox opinion I have elsewhere had occasion to speak [32].

32 See General Preface, p. 29.—*J. S.*

ON PRINCIPLES AND ORIGINS ACCORDING TO THE FABLES OF CUPID AND CŒLUM: ETC.

THE stories told by the ancients concerning Cupid, or Love, cannot all apply to the same person; and indeed they themselves make mention of two Cupids, very widely differing from one another; one being said to be the oldest, the other the youngest of the gods. It is of the elder that I am now going to speak. They say then that this Love was the most ancient of all the gods, and therefore of all things else, except Chaos, which they hold to be coeval with him. He is without any parent of his own; but himself united with Chaos[1] begat the gods and all things. By some however it is reported that he came of an egg[2] that was laid by Nox. Various attributes are assigned to him; as that he is always an infant, blind, naked, winged, and an archer. But his principal and peculiar power is exercised in uniting bodies; the keys likewise of the air, earth and sea were entrusted to him. Another younger Cupid, the son of Venus, is also spoken of, to whom the attributes of the elder are transferred, and many added of his own.

This fable, with the following one respecting Cœlum, seems to set forth in the small compass of a parable a doctrine concerning the principles of things and the origins of the world, not differing in much from the philosophy which Democritus held, excepting that it appears to be somewhat more severe, sober, and pure. For the speculations of that philosopher, acute and diligent as he was, could not rest nor keep within bounds, nor put a sufficient check and control over themselves. And even the opinions which are veiled in the parable, though somewhat more correct, are yet no better than such as proceed from the intellect left to itself and not resting constantly on experience and advancing step by step; a fault to which I suppose the primitive ages were likewise subject. It must be understood however, in the first place, that the things here brought forward are drawn and concluded from the authority of human reason alone, according to the belief of the sense, whose expiring and failing oracles are deservedly rejected since a better and more certain light has been shed upon us from divine revelation. This Chaos then, which was contemporary with Cupid, signified the rude mass or congregation of matter. But matter itself, and the force and nature thereof, the principles of things in short, were shadowed in Cupid himself. He is introduced without a parent, that is to say, without a cause; for the cause is as the parent of the effect; and it is a familiar and almost continual figure of speech to denote cause and effect as parent and child. Now of this primary matter and the proper

[1] *Cœlo* in the original. For the grounds of the correction, see Preface, p. 640.—*J. S.*

[2] Kellgren, *De Ovo mundano* (Helsingfors, 1849), has collected the passages on the egg cosmogony in the Institutes of Menu, the Puranas, and certain Commentaries. He remarks that, so far as he is aware, no trace of the mythus occurs in the Vedas. It follows that he did not perceive any reference to it in the 129th hymn of the 10th book of the Rig Veda, with which he was certainly acquainted, as he has quoted a portion of Colebrook's translation of it. In this translation it is difficult to recognise even the germ of the mythus, but in that which has since been given by Max Müller it seems more easy to do so. It would be interesting to ascertain how far the mythus was developed at the time at which the older portions of the Rig Veda were composed. The subject may be said to have a natural interest at Helsingfors, as the egg cosmogony exists among the Finns. For the hymn referred to see Colebrook's *Miscellaneous Essays*, i. p. 34, and Müller's Addenda to Bunsen's *Hippolytus*, p. 140.

virtue and action thereof there can be no cause in nature (for we always except God), for nothing was before it. Therefore there was no efficient cause of it, nor anything more original in nature; consequently neither genus nor form. Wherefore whatsoever this matter and its power and operation be, it is a thing positive and inexplicable, and must be taken absolutely as it is found, and not to be judged by any previous conception. For if the manner could be known, yet it cannot be known by cause, seeing that next to God it is the cause of causes, itself only without a cause. For there is a true and certain limit of causes in nature; and it is as unskilful and superficial a part to require or imagine a cause when we come to the ultimate force and positive law of nature, as not to look for a cause in things subordinate. And hence Cupid is represented by the ancient sages in the parable as without a parent, that is to say, without a cause,—an observation of no small significance; nay, I know not whether it be not the greatest thing of all. For nothing has corrupted philosophy so much as this seeking after the parents of Cupid; that is, that philosophers have not taken the principles of things as they are found in nature, and accepted them as a positive doctrine, resting on the faith of experience; but they have rather deduced them from the laws of disputation, the petty conclusions of logic and mathematics, common motions, and such wanderings of the mind beyond the limits of nature. Therefore a philosopher should be continually reminding himself that Cupid has no parents, lest his understanding turn aside to unrealities; because the human mind runs off in these universal conceptions, abuses both itself and the nature of things, and, struggling towards that which is far off, falls back on that which is close at hand. For since the mind, by reason of its narrowness, is commonly most moved by things of familiar occurrence and which may enter and strike it directly and at once, it comes to pass that when it has advanced to those things which are most universal in experience, and yet cannot be content to rest in them that then, as if striving after things still more original, it turns to those by which itself has been most affected or ensnared, and fancies these to be more causative and demonstrative than those universals themselves.

It has been said then that the primitive essence, force and desire of things has no cause. How it proceeded, having no cause, is now to be considered. Now the manner is itself also very obscure; and of this we are warned by the parable, where Cupid is elegantly feigned to come of an egg which was laid by Nox. Certainly the divine philosopher declares that "God hath made everything beautiful in its season, also he hath given the world to their disputes; yet so that man cannot find out the work that God worketh from the beginning to the end[3]". For the summary law of being and nature, which penetrates and runs through the vicissitudes of things (the same which is described in the phrase, "the work which God worketh from the beginning to the end"), that is, the force implanted by God in these first particles, from the multiplication whereof all the variety of things proceeds and is made up, is a thing which the thoughts of man may offer at but can hardly take in. Now that point concerning the egg of Nox bears a most apt reference to the demonstrations by which this Cupid is brought to light. For things concluded by affirmatives may be considered as the offspring of light; whereas those concluded by negatives and exclusions are extorted and educed as it were out of darkness and night. Now this Cupid is truly an egg hatched by Nox; for all the knowledge of him which is to be had proceeds by exclusions and negatives: and proof made by exclusion is a kind of ignorance, and as it were night, with regard to the thing included. Whence Democritus excellently affirmed that atoms or seeds, and the virtue thereof, were unlike anything that could fall under the senses; but distinguished them as being of a perfectly dark and hidden nature; saying of themselves, "that they resembled neither fire nor anything else that could be felt or touched[4]"; and of their

[3] Eccles. iii. 11.

[4] Lucret. i. 688:—

> Neque sunt igni simulata, neque ulli
> Præterea rei quæ corpora mittere possit
> Sensibus, et nostros adjectu tangere tactus.

virtue, "that in the generation of things the first beginnings must needs have a dark and hidden nature, lest something should rise up to resist and oppose them[5]". Atoms therefore are neither like sparks of fire, nor drops of water, nor bubbles or air, nor grains of dust, nor particles of spirit or ether. Neither is their power and form heavy or light, hot or cold, dense or rare, hard or soft, such as those qualities appear in greater bodies; since these and others of the kind are results of composition and combination. And in like manner the natural motion of the atom is not that motion of descent which is called natural, nor the one contrary to it (that of percussion), nor the motion of expansion and contraction, nor the motion of impulse and connection, nor the motion or rotation of the celestial bodies, nor any of the other motions of large bodies simply. Notwithstanding in the body of the atom are the elements of all bodies, and in the motion and virtue of the atom are the beginnings of all motions and virtues. But yet on this point, namely, the motion of the atom compared with the motion of larger bodies, the philosophy of the parable seems to differ from that of Democritus. For Democritus is found to be not only at variance with the parable, but inconsistent and almost in contradiction with himself in that which he says further on this point. For he should have attributed to the atom a heterogeneous motion, as well as a heterogeneous body and a heterogeneous virtue; whereas, out of the motions of the larger bodies, he has selected two motions; namely the descent of heavy things and the ascent of light (which latter he explained as the effect of force or percussion of the heavier driving the less heavy upwards), and ascribed them as primitive motions to the atom. The parable on the contrary preserves the heterogeneity and exclusion throughout, both in substance and motion. But it further intimates, that there is some end and limit to these exclusions; for Nox does not sit for ever. And certainly it is the prerogative of God alone, that when his nature is inquired of by the sense, exclusions shall not end in affirmations. But here the case is different; and the result is, that after due exclusions and negations something is affirmed and determined, and an egg laid, as it were, after a proper course of incubation; and not only that Nox lays her egg, but that from this egg is hatched the person of Cupid: that is to say, not only is some notion of the thing educed and extracted out of ignorance, but a distinct and definite notion. With regard then to the kind of demonstrations which are possible concerning primary matter, this is what I conceive to be most in accordance with the meaning of the parable. Let us now proceed to Cupid himself, that is, primary matter, together with its properties, which are surrounded by so dark a night: and see what light the parable throws upon this. Now I am well aware that opinions of this kind sound harsh and almost incredible to the senses and thoughts of men. As we see it has been tried and proved in this very philosophy of Democritus respecting atoms[6], which because it penetrated somewhat more sharply and deeply into nature and was further removed from common ideas, was treated as childish by the vulgar; and was moreover by the disputes of other philosophies more adapted to their capacity blown about and almost extinguished. And yet this man was much admired in his day[7], and was called Pentathlus from the variety of his knowledge, and by consent of all was esteemed the greatest physical philosopher, so that he obtained also the surname of Magus. Nor could either the battles and contests of Aristotle (who after the Ottoman fashion felt insecure about his own kingdom of philosophy till he had slain his

[5] Id. i. 779:—

> At primordia gignundis in rebus oportet
> Naturam clandestinam cæcamque adhibere,
> Emineat ne quid, quod contra pugnet et obstet.

[6] Lucret. ii. 82.

> Cuncta necesse est
> Aut gravitate sua ferri primordia rerum,
> Aut ictu forte alterius.

But Democritus himself did not ascribe gravity to the atom, and in this as in some other points Bacon was misled by assuming that Lucretius always represents the opinions of Democritus. See Stobæus, *Eclog. Phys.* i. 15.

[7] Diog. Laërt. ix. 37.

brethren; and who was likewise anxious, as appears from his own words, that posterity should doubt about nothing), or the majesty and solemnity of Plato, so far prevail—the one by violence, the other by reverence—as to obliterate entirely this philosophy of Democritus. But while that of Plato and Aristotle was noised and celebrated in the schools amid the din and pomp of professors, this of Democritus was held in great honour with the wiser sort, and those who embraced more closely the more silent and arduous kinds of speculation. Certainly in the times of Roman learning that of Democritus was not only extant but well accepted; for Cicero mentions him everywhere in terms of the highest praise; and the well-known lines of the poet, who appears to have spoken (as poets commonly do) according to the judgment of his own time, were written not long after; wherein he is quoted as an instance to prove that great men may be born in heavy climates[8]. Therefore it was not Aristotle or Plato, but Genseric and Attila and the barbarians, who destroyed this philosophy. For at that time, when all human learning had suffered shipwreck, these planks of Aristotelian and Platonic philosophy, as being of a lighter and more inflated substance, were preserved and came down to us, while the more solid parts sank and almost passed into oblivion. But to me the philosophy of Democritus seems worthy to be rescued from neglect; especially as in most things it agrees with the authority of the earliest ages. First therefore Cupid is described as a person; and to him are attributed infancy, wings, arrows, and other things of which I will afterwards speak separately. But in the mean time I make this assumption; that the ancients set down the first matter (such as may be the beginning of things) as having form and qualities, not as abstract, potential and unshapen. And certainly that despoiled and passive matter seems altogether a fiction of the human mind, arising from this, that to the human mind those things most seem to exist, which itself imbibes most readily, and by which it is most affected. It follows therefore that forms (as they call them) seem to exist more than either matter or action; because the former is hidden, the latter variable; the former does not strike so strongly, the latter does not rest so constantly. These images on the other hand are thought to be both manifest and constant; so that the first and common matter seems to be as an accessory and support; and action, of whatever kind, to be merely an emanation from the form; and altogether the first place is assigned to forms. And hence appears to have come the reign of forms and ideas in essences; with the addition (that is to say) of a kind of fantastical matter. All which was increased, moreover, by superstition (intemperance following error); and abstract ideas and their dignities were also introduced, with so much confidence and majesty, that the dreamers almost overpowered the wakers. These things however have now for the most part vanished; though an individual in our age has tried, with more boldness (as it appears to me) than success, to prop them up in their decline and resuscitate them[9]. But how contrary to reason it is to lay down abstract matter as a principle is easily seen, if prejudices be not in the way. For the actual existence of separate forms has been asserted by some[10], of separate

[8] Juv. x. 48:—

> Cujus prudentia monstrat,
> Magnos posse viros, et magna exempla daturos,
> Vervecum in patria crassoque sub aere nasci.

[9] The allusion is apparently to Patricius, whose *Nova Philosophia* was published in 1593; a work long since so rare that Sorellus (apud Brucker, iv. 28) says that a small library might be purchased for the price of this single book. See for an account of it Brucker, ubi modo.

[10] Angels are regarded by the schoolmen as forms not immersed in matter. Thus St. Thomas says, "Angeli sunt formæ immateriales".—*Sum. Theol.* i. q. 61. Even the soul of man is spoken of as a form "non penitus materiæ immersa"; a way of speaking probably employed for two reasons,—to save the possibility of the soul's separate existence, and to obviate the difficulty of the Scotists, that an unextended, or intense, form like the soul cannot give extension or corporeity. From this difficulty Duns Scotus deduced the existence of a "forma corporeitatis" distinct from the soul; a doctrine not to be confounded with that of Avicenna, who, from the impossibility of conceiving

matter by no one; not even by those who have taken it for a principle; and to constitute entities from things imaginary seems hard and perverse, and not consonant with the inquiry concerning principles. For the inquiry is not how we may most conveniently embrace and distinguish the nature of entities in our thoughts, but what are really the first and most simple entities from which the rest are derived. Now, the first entity must exist no less really than the things derived from it; and in a certain way more. For it is self-subsisting, and other things subsist by it. But the things which are said about this abstract matter are not much better than if a man were to assert that the world and all things are made of categories and such like logical notions, as principles. For it makes little difference whether you say that the world is made of matter, form, and privation, or of substance and contrary qualities. But almost all the ancients, as Empedocles, Anaxagoras, Anaximenes, Heraclitus, and Democritus, though in other respects they differed about the first matter, agreed in this, that they set down matter as active, as having some form, as dispensing that form, and as having the principle of motion in itself. Nor can any one think otherwise, unless he plainly deserts experience. Therefore all these submitted their minds to the nature of things. Whereas Plato made over the world to thoughts; and Aristotle made over thoughts to words; men's studies even then tending to dispute and discourse, and forsaking the stricter inquiry of truth. Hence such opinions are rather to be condemned in the whole, than confuted separately in the parts; for they are the opinions of those who wish to talk much, and know little. And this abstract matter is the matter of disputation, not of the universe. But one who philosophises rightly, and in order, should dissect nature and not abstract her (but they who will not dissect are obliged to abstract); and must by all means consider the first matter as united to the first form, and likewise to the first principle of motion, as it is found. For the abstraction of motion also has begotten an infinite number of fancies about souls, lives, and the like; as if these were not satisfied by matter and form, but depended on principles of their own. But these three are by no means to be separated, only distinguished; and matter (whatever it is) must be held to be so adorned, furnished, and formed, that all virtue, essence, action, and natural motion, may be the consequence and emanation thereof. Nor need we fear that the result will be general torpor, or that the variety of things which we see cannot be explained; as I will show hereafter. Now that the first matter has some form is demonstrated in the fable by making Cupid a person: yet so that matter as a whole, or the mass of matter, was once without form; for Chaos is without form; Cupid is a person. And this agrees well with Holy Writ; for it is not written that God in the beginning created matter, but that he created the heaven and the earth.

There is subjoined likewise some description of the state of things as it was before the work of the six days, wherein distinct mention is made of earth and water, which are the names of forms; but yet in the whole the mass was still unformed [11]. But though Cupid is represented in the allegory as a person, he is yet naked. Therefore, next to those who make matter abstract they are most in error (though on the contrary side) who make it clothed. I have slightly touched on this in what has been already said of the demonstrations applicable to the first matter, and of the heterogeneous nature of matter itself. But this part on which I am now entering is the proper place for treating of them. We must see therefore among those who have grounded the principles of things in formed matter, who they are who have attributed a native and naked form to matter, and who one appareled and clothed. Now, in all there are four different opinions on this. The first is that of those who assert that there is some one principle of things, but make the diversity of beings to consist in the inconstant and dispensable nature of that same principle. The second is that of those who make the principle of things one in substance, and that fixed and invariable; but deduce the diversity of beings from the different magnitudes, configurations, and positions of that same principle. The third is that of those who set up many principles of things, and lay

unextended matter, was led to assert the existence of a form of corporeity primitively inherent in all matter.

[11] Compare St. Thomas, *Summa*, i. 66. 1.

the diversity of beings to the tempering and mixing of them. The fourth is that of those who constitute infinite or at least numerous principles, but make them specific and formed; and these have no need of any device to account for the multiplicity of things; for they diversify nature at the very outset [12]. Of these sects the second alone appears to me to represent Cupid as he is—native and naked. For the first introduces him as separated by a veil, the third as wearing a tunic, and the fourth as cloaked and almost masked. But on each of these I will speak a few words, for the better explanation of the allegory. First therefore among those who have asserted one principle of things, I have found no one who would affirm that principle to be the earth. For the quiet, sluggish, and inactive nature of the earth which submits patiently to the heaven, fire, and other things, prevented such an assertion from entering into any one's mind [13]. Nevertheless the wisdom of the ancients made Earth to come next to Chaos, and to be first the parent, then the bride of Cœlum, from which marriage all things were born [14]. But it is not to be therefore understood that the ancients ever constituted the earth the principle of essence; but only the principle, or rather origin, of configuration or system. I refer this point therefore to the following allegory respecting Cœlum, when I shall inquire about origins; which inquiry is posterior to that of principles.

Thales asserted Water to be the principle of things [15]. For he saw that matter was principally dispensed in moisture, and moisture in water; and it seemed proper to make that the principle of things, in which the virtues and powers of beings, and especially the elements of their generations and restorations, were chiefly found. He saw that the breeding of animals is in moisture; that the seeds and kernels of plants (as long as they are productive and fresh), are likewise soft and tender; that metals also melt and become fluid, and are as it were concrete juices of the earth, or rather a kind of mineral waters; that the earth itself is fertilised and revived by showers or irrigation, and that earth and mud seem nothing else than the lees and sediment of water; that air most plainly is but the exhalation and expansion of water; nay, that even fire itself cannot be lighted, nor kept in and fed, except with moisture and by means of moisture. He saw, too, that the fatness which belongs to moisture, and which is the support and life of flame and fire, seems a kind of ripeness and concoction of the water. Again, that the body and bulk of water is distributed throughout the universe, as the common support of everything; that the earth is encircled by the ocean; that there is a vast supply of fresh water within the earth, whence come springs and rivers, which like the veins of a body carry off water over the surface and through the bowels of the earth. That there are also immense masses and collections of vapours and waters in the upper regions;—another universe of waters, as it were, for the repair and refreshment of those below, and indeed of the ocean itself. He also supposed that even the celestial fires fed on these vapours and waters, inasmuch as they could neither subsist without aliment nor be nourished by anything else; also that the configuration of water as seen in its particles (I mean drops), is the same as the configuration of the universe, namely, round and spherical; moreover that the undulation of water is seen and observed likewise in air and flame; and lastly, that the motion of water is easy, neither sluggish nor too rapid, and that the generation of fish and water animals is very numerous. But Anaximenes selected Air to be the one principle of things. For if mass is to be

[12] In enumerating these four sects, Bacon alludes successively to the Ionian physicists; to the atomists; to Parmenides, Telesius, Empedocles, and many others; and lastly to Anaxagoras.

[13] This remark Bacon may have derived from Aristotle, *Metaph.* i. 7. However, Hippo of Rhegium, or rather Hippo the atheist, who is probably the same person, made earth the principle of all things, at least according to the scholiast on Hesiod's *Theogony*. (See Heinsius's Hesiod, p. 237.) Others, however, give a different account of Hippo's opinions, and it is possible that the scholiast's story was suggested to him merely by what Aristotle says of him in the third chapter of the same book.

[14] As I have remarked in the preface, reference is here made to Hesiod.

[15] Plutarch, *De Plac. Philosoph.* i. 3.

regarded in constituting the principles of things, air seems to occupy by far the greatest space in the universe. For unless a separate vacuum be allowed, or the superstition concerning the heterogeneity between celestial and sublunary bodies be received, it would seem that the whole extent of space between the globe of the earth and the bodies of the heaven,—all of it that is not either star or meteor,—is filled with an airy substance. Now the terrestrial globe is but as a point compared with the heaven that surrounds it. But in the ether itself how small a portion is studded with stars! In the spheres next the earth each star is seen singly; and in that furthest from it, though the number of them is immense, yet they occupy a small space in comparison to the distances between them; so that all things seem to float as it were in a vast sea of air. Nor is it a small portion of air and spirit which resides in waters and in cavities of the earth; whence waters receive their fluidity, and sometimes also spread and swell; and the earth, besides its porosity, has its tremblings and shakings, which are evident signs of confined air and wind. And if a kind of middle nature be suited to principle, as being susceptible of so great a variety, this seems to be found in perfection in the air. For air is as the common link of things, not only because it is everywhere present, and comes in and occupies vacuities, but much rather because it seems to have a middle and indifferent nature. For it is a body which receives and conveys light, opacity, the tints of all colours, and obscurations of shade; which likewise distinguishes with the greatest accuracy the different impressions and notes of musical and (what is greater) articulate sounds; which admits without confusion the differences of smells, not only the general ones of sweet, foul, heavy, brisk, and the like, but also the peculiar and specific, as the smell of a rose or violet; which is indifferently disposed towards the great and powerful qualities of heat, cold, moisture and dryness; in which watery vapours, fat exhalations, spirits of salts and fumes of metals hang and float; lastly, in which the radiations of the heavenly bodies, and the closer agreements and disagreements of things, secretly communicate and dispute; so that air is like a second chaos, in which the seeds of so many things act, wander, endeavour, and experiment. Finally you consult the generative and vivifying power in things as that which may lead us up to principles, and make them manifest; in these likewise air seems to play the principal part; so that the names of air and spirit and life or breath are sometimes confounded. And rightly; since respiration is as it were the inseparable companion of the more advanced states of life, (that is, excepting the first rudiments of life in embryos and eggs); insomuch that fish are suffocated when the surface of the water is hard frozen. Even fire itself, unless it be animated by the surrounding air, dies out, and seems to be nothing else than air rubbed together, excited and kindled; as water, on the other hand, seems to be but a congelation and contraction of air. The earth also perpetually exhales air, and has no need to pass through water into the form of air. Heraclitus, on the other hand, with more acuteness but less credibility, made Fire to be the principle of things [16]. For he did not look for a middle nature, which is usually the most vague and corruptible, to constitute the principles of things; but for a consummate and perfect one, such as might be the end and period of corruption and alteration. Now he saw that the greatest variety and confusion was found in solid and consistent bodies. For such bodies may be organic, and like machines, which from their very configuration admit innumerable variations; such as are the bodies of plants and animals. And even those which are not organic, are yet on minute inspection found to be very dissimilar. For how great is the dissimilarity between those parts of animals which are called similar? the brain, the crystalline humour, the white of the eye, bone, membrane, cartilage, nerve, vein, flesh, fat, marrow, blood, seed, spirit, chyle, and the like? and likewise in the parts of vegetables, the root, bark, stem, leaf, flower, seed, and the like? Fossils certainly are not organic, but yet they exhibit both a great mixture in one species and a very plentiful variety compared one with another. Wherefore this broad, ample, and extensive basis of the diversity of beings, wherein so large an array of things displays itself and comes into action, seems to consist in the nature

[16] Plutarch, l. c.

of solidity and consistence. But the bodies of liquids are plainly without the power of organic structure. For through the whole world of visible nature there is found no animal or plant in a body simply fluid; and therefore this infinite variety is precluded and cut off from the nature of liquidity. And yet the liquid nature has its variety, and that in no small degree, as is manifested in the great diversity of melted bodies, juices, distilled liquors, and the like. Whereas in airy and pneumatic bodies this variety is much more limited, and a sort of promiscuous resemblance of things takes its place. Certainly that virtue of colours and tastes, whereby liquids are sometimes distinguished, absolutely ceases; that of odours indeed and some other things remains, yet only transitory, confused, and separable; so that as a general rule, the nearer bodies approach to the nature of fire, the more do they lose of variety. And after they have assumed the nature of fire, and that in a rectified and pure state, they throw off every organ, every property, and every dissimilarity; and nature seems as it were to gather to a point in the vertex of the pyramid, and to have reached the limit of her proper action. Therefore this kindling or catching fire Heraclitus called peace; because it composed nature and made her one; but generation he called war, because it multiplied and made her many [17]. And that this process (by which things flowed and ebbed, like the tide, from variety to unity, and from unity to variety) might be some way explained, he maintained that fire was condensed and rarefied, yet so that its rarefaction towards a fiery nature was the direct and progressive action of nature; while its condensation was a kind of retrograde action or failing of the same. Both of these he considered to take place by fate, and (in the sum of things) at certain periods; so that this revolving world would some time or other be set on fire, and afterwards renewed again, and that this series and succession of conflagration and generation would go on for ever. Only the inflammation and the extinction were according to him (if one studies diligently the scanty account which has come down to us of the man and his opinions) to take place in a different order. For as to the process of inflammation, he no way differed from the common opinions; that the progress of rarefaction and extenuation went on from earth to water, from water to air, from air to fire. But the way back was not by the same stages; the order being directly inverted [18]. For he affirmed that fire by its extinction produces earth, as its dregs and soot; that these then conceive and collect moisture, whence proceeds an overflow of water, which again emits and exhales air; so that the change from fire to earth is sudden, not gradual.

Such then, or better than these, were the opinions of those who laid down one principle of things; regarding nature simply, not contentiously. And they deserve commendation for giving Cupid but one garment, which is the next degree to nakedness; and that garment too (as I have said) like a veil, and of no thicker texture. Now by the garment of Cupid I mean some form attributed to the primary matter, that may be said to be substantially homogeneous with the form of any of the secondary essences. But the assertions made by them with respect to water, air, and fire, which rest on no very firm grounds, it will not be difficult to confute; nor does there seem to be any reason for discussing them severally, so I will only touch on them in general. First therefore, in the inquiry of principles these ancient philosophers do not appear to have adopted a very perfect system; but what they did was only to seek out among apparent and manifest bodies that which seemed most excellent, and set down that as the principle of all things; by right, as it were, of its excellence; not as being truly and really so. For they thought that such a nature was the only one of which it could be said that it is what it seems; other things they thought were this same nature, though not according to appearance; so that they seem to have spoken either figuratively, or under the influence of fascination; the stronger impression carrying the rest with it. But a true philosopher should look at all things alike, and lay down those as the principles of things, which agree as well with the

17 Diog. Laërt. ix. 8.

18 Plutarch, l. c. Diogenes Laërtius, however, does not support the statement of the text.

smallest, rarest, and most neglected of essences, as with the greatest and most numerous and vigorous. For though we men most admire the beings which are most universal, yet the bosom of nature is opened to all. If on the other hand they hold this principle of theirs not by excellence, but simply, they seem certainly to have fallen on a harsh figure of speech ; for it brings it plainly to a matter of equivocation, what they assert not being predicated either of natural fire or natural air or natural water, but of some imaginary and ideal fire, air, etc., which retains the name, but does not answer the definition. They seem further to be driven to the same difficulties in which the assertors of abstract matter are involved ; for as those introduce a potential and imaginary matter altogether, so do these likewise in part. Moreover they make matter formed and actual with respect to one thing (namely, that principle of theirs), but potential with respect to everything else. Nor does there seem to be any advantage in this kind of single principle, more than in that of abstract matter ; except that it has something to offer to the human understanding, whereon the thoughts of men may better fix themselves and rest, and by which the notion of the principle itself becomes somewhat fuller, than that of all others things more abstruse and hard. But the fact is that at that time the *Predicaments* had not commenced their reign, whereby this principle of an abstract nature might have sheltered itself under the credit and protection of the predicament of substance ; and therefore no one durst feign a matter quite imaginary, but asserted a principle according to sense ; some true entity ; the manner only of dispensation (for in that they used greater license) being imaginary [10]. For they do not discover, nay, do not even speculate, by what appetite or spur, or by what reason, way, or inducement, that principle of theirs degenerates from, and again recovers its nature. But seeing there are such armies of contraries in the world, as of dense and rare, hot and cold, light and darkness, animate and inanimate, and many others, which oppose, deprive, and destroy one another in turn ; to suppose that all these emanate from some one source of a material substance, and yet not to show any manner in which the thing can be, seems but a confused speculation, and an abandonment of inquiry. For if the thing itself were ascertained by the sense, you must receive it, though the manner thereof be hidden ; and again, if by force of reason any convenient and credible manner could be discovered, you must perhaps give up appearances ; but you should by no means be required to assent to those things whereof neither the being is manifest by the sense, nor the explanation probable by the reason. Besides, if there were but one principle of things, it ought to have a visible mark, and as it were a superiority and predominance in all things ; nor should anything of importance be found diametrically opposite to that principle. Likewise it should hold a middle position, so as to be more conveniently available for everything, and diffuse itself around. But there is nothing of this found in the principles of those philosophers. For the earth, which is cut off and excluded from the honour of a principle, seems to receive and cherish natures opposed to those three principial natures ; for to the mobility and lucidity of fire it opposes rest and opacity; to the tenuity and softness of air, in like manner, it opposes density and hardness ; and to the moistness and sequacity of water, dryness, rigidity, and asperity ; besides, the earth itself occupies the central place, the rest being turned out. And further, if there were only one principle of things, it ought to have a nature indifferently disposed towards the generation and the dissolution of them. For it is as much the condition of a principle that things should resolve themselves into it, as that things should be produced from it. But this is not the case ; for of these bodies, air and fire seem unsuited to supply matter for the generation of things, though ready to receive their dissolution ; whereas water, on the other hand, is favourable and good for generation, but more unfit and averse to dissolution or restoration ; as would be easily seen if for some time there were to be no rain. Moreover, putrefaction itself in no way reduces things to crude and pure water. But by far the greatest error is that they set up for a principle that which is corruptible and mortal ; for they do no less when they introduce such a principle

[10] Cf. Aristot. *Met.* i. 3.

as forsakes and lays aside its nature in compositions. "For when a thing shifts and changes, that which it was dies [20]."

But I shall have to make further use of this reason presently, now that our discourse has come down in order to the third sect, which asserted many principles of things; a sect which seems to have more strength on its side, and certainly has more prejudice. Therefore I will not examine their opinions in general or in common, but one by one.

Amongst those who have maintained that there are many principles I shall set aside those who hold them to be infinite; for the point concerning infinity belongs to the allegory respecting Cœlum. But among the ancients Parmenides maintained two principles of things, fire and earth, or heaven and earth [21]. For he asserted that the sun and stars were real fire, pure and limpid [22], not degenerate as fire is with us, which is only as Vulcan thrown down from heaven, and lamed by the fall. And these opinions of Parmenides Telesius has in our age revived; a man strong and well armed with the reasonings of the Peripatetics (if they were worth anything), which likewise he has turned against themselves; but embarrassed in his affirmations, and better at pulling down than at building up. Of the discoveries of Parmenides himself the account is very scanty and shadowy; yet the foundations of a similar opinion seem plainly laid in the book written by Plutarch on the "Primal Cold"; which appears to have been derived from some ancient treatise, at that time extant but now lost. For it contains not a few things both more acute and more sound than the speculations of the author himself commonly are, and by these Telesius appears to have been prompted and incited to take them up studiously and follow them out strenuously in his "Commentaries on the Nature of Things". Now the opinions of this sect are as follows [23]: That the first forms, and the first active entities, and therefore the first substances are heat and cold; yet that these have no body, but a passive and potential matter, which supplies a corporeal bulk, and is equally susceptible of both natures; itself without any action at all. That light is a sprouting of heat, but of heat dissipated, which being multiplied by collection becomes robust and sensible. That darkness in like manner is the destitution and confusion of the radiating nature from cold. That density and rarity are but the textures and, as it were, the webs of heat and cold; heat and cold being the producers and operatives thereof; cold condensing and thickening the work, heat separating and extending it. That from such textures a disposition towards motion, either apt or reverse, is impressed upon bodies; that is, prompt and apt upon rare bodies, sluggish and averse upon dense. Therefore that heat by tenuity excites and creates motion; cold by density checks and quiets it. Hence that there are four co-essential and conjugate natures, and those of two kinds, preserving the respective order I have mentioned (for the source is heat and cold, the rest are emanations); but yet always concomitant and inseparable. These are heat, lucidity, rarity, mobility; and again, their four opposites cold, opacity, density, immobility. That the seats and stages of the first conjugation are placed in the heaven, stars, and especially in the sun; of the second, in the earth. For that the heaven from its perfect and entire heat and the extreme extension of matter is most hot, lucid, rarefied, and movable; whereas the earth, on the contrary, from its entire and unrefracted cold, and the extreme contraction of matter, is most cold, dark, and dense, completely immovable, and exceedingly averse to motion. That the summits of the heaven preserve their nature entire and

20 Lucret. iii. 518:—

Nam quodcunque suis mutatum finibus exit,
Continuo hoc mors est illius, quod fuit ante.

21 This opinion, or something analogous to it, was held by many of the older physicists. (See Karsten's Parmenides, p. 230.) Beside those whom Karsten mentions we know that Hippo Rheginus is said to have made fire and earth, or heat and cold, his first principles. (See Pseudo-Origen. [Hippolytus, i. 14] *Philosoph.* c. 16.)

22 Stobæus, *Eclog. Phys.* i. 23.

23 [In his notes to the original, Mr. Ellis supplies parallel passages from the *De natura rerum* of Telesius to the statements in Bacon's text.]

inviolate, admitting some diversity among themselves, but completely removed from the violence and insult of a contrary body; that there is a like constancy in the depths or innermost parts of the earth; and that it is only the extremities, where contrary bodies approach and meet together, which struggle and suffer, and are assailed by one another. That the heaven therefore, in its whole bulk and substance, is hot, and quite free from every contrary nature, but that heat is unequal; some parts being more, others less hot. For that in the body of the stars heat is more intense, in the space between them less so; moreover that of the stars themselves, some are more burning than others, and have a livelier and more radiant fire; yet so that the contrary nature of cold, or any gradation thereof, never penetrates there; for that it admits a difference of nature, but not a contrariety. That you must not however judge of the heat or fire of celestial bodies, which is entire and native, from common fire. For that our fire is out of its place, trembling, surrounded with contrary bodies, needy, dependent for its preservation on the fuel given it, and fugitive; whereas in heaven it is in its true position, apart from the violence of any contrary body, constant, kept up by itself and similar things, and performing its proper operations free and undisturbed. Also that the heaven is bright everywhere, but with differences of more or less. For that, seeing there are among the known and numbered stars some which are only visible in a clear sky, and in the milky way there are knots of small stars which show a kind of whiteness all together, but are not severally distinguishable as bright bodies; no one can doubt but that there are many stars invisible to us; and indeed that the whole heaven is endued with light, though not with a splendour so strong and far-darting, nor with rays so thick and close, as can travel so great a distance and come to our view. So again, that the whole heaven consists of a fine and rare substance, nothing in that substance being crowded or packed closer than it likes, but yet that in some parts matter is more extended, in others less. Lastly, that the motion of the heaven is found to be that which suits the most movable body: namely, the motion of conversion or rotation. For circular motion is interminable and for its own sake. Motion in a straight line is to an end, and for the sake of something, and as it were to obtain rest. Therefore that the whole heaven moves with a circular motion, and no part of it is free from that motion; but yet as in the heat, light, and rarity of heaven there is inequality, the same likewise is observed in its motion; an inequality the more conspicuous because it more invites and bears numan observation, and may even be reduced to calculation. Now, orbicular motion may differ both in velocity and direction; in velocity it may be quicker or slower; in direction, it may be in a perfect circle, or it may have something of a spiral course, and not return exactly to the same spot; for a spiral line is made up of a straight line and a circle. Therefore that these very inequalities have place in the heaven—variety of velocity, and deviation from the point of return, or spirality. For both the fixed stars and the planets are unequal in their velocity; and the planets evidently deviate from tropic to tropic; and the higher the heavenly bodies are, the greater is their velocity, and the more spiral their course. For if phenomena be taken simply and as they are seen, and there be set down one natural and simple daily motion in the heavenly bodies, and the mathematical propriety of reducing motions to perfect circles be rejected, and spiral lines be admitted, and those contrarieties of motions following the order from east to west (which they call the motion of primitive mobility) and again from west to east (which they call the proper motion of the planets) be reduced to one; difference of time in the return being accounted for by differences of speed, and difference of position with regard to the zodiac by spiral lines, it is plain that what I have said must come to pass—that the moon, for instance, which is the lowest of the planets, must proceed both slowest of all, and in the rarest and most open spirals. Such then appears to be the opinion of this sect respecting the nature of that portion of the heaven which (by reason of its distance from a contrary) is firm and perpetual. But whether Telesius kept to the old limits, and imagined that such was the nature of everything above the moon, together with the moon itself, or whether he held that the hostile force could ascend higher, he does not clearly lay down. But of the earth (which is the stage and seat of the contrary nature) he asserts likewise, that the greatest part is inviolate and undisturbed, and that the heavenly bodies do not penetrate thither. But of what kind it is, he says, need not be inquired. It is enough to consider it endowed with these four natures—coldness, opacity, density, and rest, and those absolute and in no degree impaired. Now the part of the earth towards the surface, being like a crust or rind, he assigns to the generation of things; and supposes all entities any way known to us, even the heaviest, hardest, and those which lie the deepest, as metals, stones, the sea, to consist of earth in some degree changed and wrought by the heat of the heaven, and which has already conceived some heat, radiation, tenuity, and mobility, and partakes in short of an intermediate nature between the sun and pure earth. It follows therefore that this pure earth must be depressed below the lowest depths of the sea, the deepest mines, and all generated bodies; and that beween this pure earth and the moon, or perhaps higher, there must be situated a middle nature composed by the temperaments and refractions of heaven and earth. Having thus sufficiently fortified the interior of both kingdoms he gets up an invasion and war. For he supposes that in the regions lying between the furthest parts of the heaven and the innermost of the earth, there is found all tumult, conflict, and perturbation, as we see in empires whose borders are ravaged by incursions and violence, while the interior provinces enjoy secure peace; that such natures therefore, with their concretions, have the appetite and faculty of constantly generating, multiplying, and spreading themselves in all directions, of occupying the whole mass of matter, of mutually assailing and invading one another, of turning one another out from their proper seats and settling themselves therein; and moreover of perceiving and apprehending the force and actions of another nature as well as their own, and by means of such perception of shifting and adjusting themselves; and that from this contest every variety of entity, action, and virtue is derived. Yet he seems in some places, though hesitatingly and cursorily, to assign to matter some quality of its own; as first, that it is neither increased nor diminished by forms and active entities, but consists of a universal sum; secondly, that to it is referred the motion of gravity or descent; and he also adds something about the blackness of matter. But this is set down plainly, that heat and cold, in the same power and quantity, remit or increase their strength accordingly as the matter in which they exist is opened out or folded up; since they fill the measure of the matter, not their own. But Telesius proceeds to devise and explain the manner in which, by means of this strife and contest, so fruitful and manifold a generation of beings may be induced and turned out. He begins by securing the earth, as being the inferior principle; and shows the reason why it has not been long ago destroyed and absorbed by the sun, nor ever can be [24]. The first and principal point which he alleges is the immense distance of the earth from the fixed stars, and its very great distance from the sun—a distance tolerably well measured. The second point is the declination of the sun's rays from the perpendicular with respect to the different parts of the earth; that is, that over the greatest part of the earth the sun is never vertical, nor his rays perpendicular; so that he never affects the whole globe of the earth with any remarkable force of heat. The third point is the obliquity of the sun's motion in passing through the zodiac with respect to the same parts of the earth; whence the heat, whatever be its force, is not continually

[24] The tenth chapter of Telesius's first book is teleological. "Summa Dei bonitas . . . ens nullum . . . perdi velit." For the preservation of the universe and the balance of heat and cold, the earth is put in the middle point of the heavens. The heavens and the earth are both spherical—the former according to the free and uniform motion of the different orbs, and the latter that half of it may always be exposed to the sun's influence. If the earth were larger and not in the centre of the universe, the power of cold would predominate and destroy the lower part of heaven. For the security of the earth, the density and heat of the heavens are not uniform, and both sun and stars are at a great distance; and the oblique and unequal motion of the sun prevents his remaining too long over any part of the earth's surface. All this agrees tolerably well with Bacon's account of it, but to his fifth reason I do not find anything corresponding in the text.

redoubled, but returns after long intervals. The fourth point is the velocity of the sun in its diurnal motion, performing as it does so large a circuit in so short a time, whereby the heat stays the less, and is not stationary for an instant. The fifth point is the continuation of bodies between the sun and the earth, whereby the heat of the sun does not come through a vacuum with its force entire, but by passing through so many resisting bodies, with each of which it has to struggle and dispute, is immensely weakened and enfeebled; and so much the more because the further it goes and the weaker it becomes the more stubborn are the bodies it meets and most of all when it arrives at the surface of the earth, where there seems not only resistance, but a direct repulsion. But the process of mutation laid down by Telesius is as follows. The war (he holds) is absolutely inexpiable and internecine. These contrary natures do not agree in any one point, nor do they meet in a third, excepting in Hyle. Therefore the one nature desires, strives, and contends absolutely to destroy the other, and to impress matter with itself only and its own image; so that the sun's work (as he says clearly and often) is plainly to turn the earth into sun, and *vice versâ* the earth's work is to turn the sun into earth. This however does not prevent everything being done in certain order, definite times, and just measure; and every action in its due course beginning, working, flourishing, languishing, and ceasing; but this is not caused by any laws of alliance or concord, but entirely by a want of power; for all more or less in virtue and action proceeds not from the regulation of the intensive power (which desires something entire) but from the stroke and curb of the opposite nature. The diversity, multiplicity, and likewise the perplexity of operation must certainly proceed from one of three things; namely, the force of heat, the disposition of matter, or the manner of working; which three are nevertheless united together by a mutual bond and are causes one of the other. Heat itself differs in power, quantity, continuance, mean, and succession; succession again has its own manifold variations in approaching and withdrawing, or in intension and remission; in sudden or graduated accession; in return or repetition at longer or shorter intervals; and such like alterations. Heats therefore are far the most varied in their force and nature, according as they are made purer or less pure, with reference to the first fountain thereof, namely, the sun. Neither does all heat cherish heat, but when two heats differ many degrees from one another, either kills and destroys the other no less than cold; each having its proper actions, and thwarting and opposing the actions of the other; so that Telesius makes lesser heats to be as traitors and deserters towards great ones, and as conspiring with cold. Therefore the feeble heat which creeps in water destroys the lively heat which vibrates in fire; and in like manner the preternatural heat of putrid humours in the human body suffocates and extinguishes the natural heat. But that quantity of heat makes a great difference, is too manifest to need explanation. For one or two burning coals are not so hot as a whole heap; but the effect of quantity is most remarkably shown in the multiplication of the sun's heat, by the reflection of the rays; for the number of rays is doubled by simple and multiplied by various reflection. But to quantity of heat there should be added also union; which is likewise best shown in the oblique and perpendicular direction of rays, since the nearer the direct and reflected rays coincide, and the acuter the angle which they make with each other, the stronger is the force of heat thrown out. Moreover the sun himself when he is present among the larger and stronger fires of the fixed stars, Regulus, the Dog Star, and Spica, sends out stronger heats. But continuance of heat is most plainly an operation of the greatest importance; as all natural virtues respect and observe their times, some time being required to put their strength in action, and a good deal to give it full vigour. Therefore continuance of heat converts an equal heat into a progressive and unequal one, because both the preceding and the subsequent heat are united together; and this is clearly shown in the heats of autumn, inasmuch as they are felt to be more burning than the heats of summer, and in the heats of summer afternoons, inasmuch as they are felt to be more burning than those of noonday. So also the weakness of heat in the colder countries is sometimes compensated by the continuance and length of the days in summer. But the power and efficacy of the medium in conveying heat is wonderful. For hence the temperature of the seasons is exceedingly varied, so that with an unspeakable changeableness it is sometimes found to be chilly in summer and sunny in winter; the sun meanwhile keeping his course and distance constantly and regularly. Crops of corn likewise and grapes ripen sooner with a south wind and a cloudy sky. And every disposition and excretion of the heaven in the various revolutions of years; sometimes pestilent and diseased, sometimes healthy and favourable, derives its cause and origin from this; namely, from the variation of the intermediate air, which gathers a different disposition from the very change and alteration of the seasons, perhaps in a long series. But the succession of heat, and the order in which one follows another, as the reasons of it are manifold, so its virtue is supreme. For the sun could not have generated so numerous and prolific an offspring, did not the configuration of the sun's body as it moves, with respect to the earth and the parts of the earth, partake of very much inequality and variation. For the sun moves both in a circle and rapidly and obliquely, and changes himself, so as to be both absent and present, nearer and further off, more perpendicular and more oblique, returning slowly and quickly, and never for a single moment is the heat emanating from the sun constant, and nowhere (unless it be in the tropics) does it return at a short interval; so that such variation of the generator excellently agrees with such variety of the thing generated. Whereto may be added the extreme diversity of the nature of the medium or conductor. The other things also, which have been said of the inequality and degrees of a single heat, may be referred to the changes and varieties of succession in different heats. Therefore not without reason did Aristotle attribute the generation and corruption of things to the oblique course of the sun, and set down that as the efficient cause thereof; had he not from his love of laying down the law and of acting as the arbiter of nature, and of distinguishing and arranging things according to his own pleasure, spoiled a sound conception. For he should have assigned generation and corruption (which is never merely privative, but is still pregnant with the generation of something else) to the inequality of the sun's heat as a whole; that is, to his advance and retreat both together; not generation to the advance and corruption to the retreat separately; a thing which he did stupidly and almost according to the vulgar judgment. And if any one is surprised that generation of things is attributed to the sun; seeing the sun is asserted and supposed to be fire, and fire generates nothing; it is a weak objection. For that notion of the heterogeneity of the heats of the sun and of fire is plainly a dream. For there are infinite operations in which the action of the sun and of fire agree; as in the ripening of fruits, the preservation in cold climates of tender plants accustomed to warm skies, the hatching of eggs, the clarifying of urine (for I put the heat of the sun and of animals together), the reviving of small animals stiffened with cold, the raising of dews and vapours, and the like. Nevertheless our fire is a bad actor, and cannot well imitate or come near to the actions of the sun, for the sun's heat has three properties, which common fire can by scarce any device represent. First by reason of its distance it is less in degree and gentler; this however is a property which may in some measure be matched; for such a measure of heat is rather unknown than unprocurable. Secondly by flowing and shooting through so many and such different mediums, it borrows and obtains a certain dissimilar and generative force. But above all, it is so regular in the inequality with which it increases and diminishes, advances, and retreats, never succeeding by starts or precipitately. Which two latter properties are almost inimitable by fire, though the matter may be advanced by a perspicacious and well-considered industry. Such then are the opinions of Telesius respecting the diversity of heats.

But cold, that is, the contrary principle, and the distribution thereof, he scarce mentions; unless he thought that in treating of the disposition of matter (to which I now proceed in the second place) he had sufficiently provided for it. Yet this he should not have done; seeing that he held cold to be by no means the privation of heat, but a decidedly active principle; a rival as it were and competitor with heat. And what he says concerning the disposition of matter goes to show how matter suffers and is worked upon and converted by heat, without any mention or thought about cold. Of cold however (for I wish to deal quite fairly with every man's opinions, and to give them the benefit of a favourable construction) he might have said something of this kind:—that the immovable and fixed seat of cold answers excellently to the movable and changeable structure of heat; as the anvil to the hammer. For if both principles had been subject to variety and alteration they would doubtless have produced hourly and momentary beings. Likewise that the immense regions of heat (namely, the heavens) are somewhat balanced by the compact nature of the earth and surrounding objects; since it is not space that is regarded, but the quantity of matter in space. But for the nature of cold and its virtues and proportions it is fit they should be passed over in silence, or with few words; seeing no certain and well-approved information can be had concerning it from experience. For we have common fire, as a kind of substitute for the sun, to manifest the nature of heat. But for the cold of the earth there is no substitute which is at man's command and available for experiment. For those chills and rigours of cold which in winter time and in the coldest countries are exhaled into the air from the globe and circumference of the earth are merely tepid airs and baths, compared to the nature of the primal cold shut up in the bowels of the earth; insomuch that that cold whereof men have perception and command is much the same as if they had no other heat than that of the summer sun in hot countries; which as compared with the fire of a burning furnace may be regarded as coolness. However not to dwell longer on supposititious suggestions, let us next see what Telesius says of the disposition of matter on which heat acts; and which has such power as to promote, impede, and change the very action of heat. It falls under four heads. The first difference is derived from the pre-existence or non-pre-existence of heat. The second, from the abundance or paucity of matter. The third, from the degree in which it is worked. The fourth, from the closeness or openness of the body worked upon. With regard to the first, Telesius supposes that in all known beings there exists some latent heat, though imperceptible to the touch, which unites itself to the new or supervenient heat; and which is itself moreover excited and inflamed by this same foreign heat to perform its own actions in its proper sphere: of this he says it is a notable argument, that there is no being,—neither metal, nor stone, nor water, nor air,—which does not grow warm at the touch, or even at the approach, of fire or a hot body; which would probably not be so, were there not some pre-existing and latent heat to prepare the way for this new and manifest heat. Also that the more or less in this respect, that is the greater or less readiness to catch fire, which is found in beings, corresponds with the measure of pre-existing heat. For air warms with a little heat, and such as in the body of water would not be perceptible by the sense. Water likewise warms sooner than stone, or metal, or glass. For though it is true that some of these, as metal and stone, seem to warm sooner than water, that is only on the surface and not in the inner part of the body; for consistent bodies have less free communication in their parts than liquids. Therefore the exterior of metal is warmed sooner than the exterior of water, but the whole body not so soon. The second difference is laid in the collection and bulk of matter. For if this be closed, the strength and heat is more limited, and by union more increased and intensified; on the other hand, if it be loose, the strength is more dispersed, and by dispersion more diminished and weakened. Therefore the heat of ignited metals is stronger than that of boiling water, even than of flame itself, except that flame, by reason of its tenuity, finds easier entrance. For the flame of coals or wood, unless it be excited by blowing, so that the motion may help to drive it in and make it penetrate, is not very furious; nay, some flame (as the flame of spirit of wine, for instance, especially in a small quantity and scattered), has so gentle a heat that the hand may almost bear it. The third difference, which is taken from the degree in which matter is wrought upon, is manifold; for he mentions some seven degrees of this working: of which the first is pliancy, or that disposition of matter which makes a body yield a little to any great violence, or bear compression, and especially extension; in a word, flexible or ductile. The second is softness, when there is no need of any great violence, but the body yields upon the slightest impulse and at a touch, without any apparent resistance. The third is viscosity or tenacity, which is a kind of beginning of fluidity. For a viscous body seems at the touch and embrace of another body to begin to flow and continue itself, and not to be terminated in itself; though it does not flow spontaneously and of its own accord; for a fluid follows itself, a viscous body rather something else. The fourth is fluidity itself, where a body partaking of an inner spirit is glad to be in motion, and follows itself, and is not easily defined or fixed. The fifth is vapour, when the body is attenuated into something intangible, which likewise gives, flows, undulates, and trembles with greater agility and mobility. The sixth is exhalation, which is a kind of vapour more concocted and ripened, and prepared for the reception of a fiery nature. The seventh is air itself; which Telesius contends is actually endowed with a native heat of its own, and that not small or weak; because even in the coldest regions the air is never congealed or frozen. Likewise that we have an evident proof that the air in its own nature is hot, in this: that all air enclosed, separated from the universal body of air, and left to itself, manifestly contracts warmth, as appears in wool and fibrous bodies. Again, in close and confined places, the air, when breathed, feels somewhat suffocating; which comes from heat. And that the reason of this is that air, when confined, begins to exercise its nature, whereas the open air out of doors is refrigerated by the cold which the globe of the earth perpetually emits and discharges. Moreover our common air has some slender portion of the qualities of the heavenly bodies; since it contains some light in itself: as is shown by the sight of animals, who can see at night and in dark places[25]. Such then, according to Telesius, is the order of the disposition of matter; in intermediate bodies, that is; for the extremes, namely hard and rigid bodies on the one side, fire itself on the other, as being the limits of those lying between, are not counted. But besides these simple gradations he finds a great diversity in the disposition of matter, by reason of similarity and dissimilarity of body; since the various portions of matter, which are compounded and united together in one body, may either be referred equally to some one of the above-mentioned gradations, or unequally to different ones. For thence arises by far the greatest difference in the operation of heat. Therefore the fourth difference necessarily depends on the nature and also the position of the body on which heat acts, whether it be close or porous and open. For when heat works on an open and exposed nature, it works in succession and part by part, attenuating and at the same time drawing out and separating. But when it works in a confined and compact nature, it operates in the whole and in the mass, without losing any heat, but the old and the new heat plainly uniting and conspiring together; whence it comes that it effects more powerful, more profound, and more exquisite alterations and preparations; of these however I shall presently speak more when I come to the manner of preparation. Meanwhile Telesius labours hard and strangely perplexes himself to explain the manner of the divorce and separation of his primary connatural qualities, heat, light, tenuity, and mobility, and the four opposed to them, according as they take place in bodies; for there are found some bodies hot or excellently prepared for heat, which are also dense, quiet, and dark; others rare, movable, bright or white, yet cold; and in like manner with regard to the rest: there being some one quality existing in things, with which the rest do not agree; and again, others partake of two of these natures, without the other two, with a great variety of permutations and assortments. In which part Telesius does not acquit himself very happily, but behaves like his opponents; who, having formed their

[25] That certain animals can see at night is with Telesius a proof that the apparently obscure parts of the heavens—the highest and lowest,—give out a perceptible amount of light, not that the air is itself luminous,—unless the "infima cœli portio" be understood to mean our atmosphere. (See De Rer. Nat. i. 3.) It is remarkable that Bacon omits Telesius's chief argument in favour of the opinion that the air is generated by and contains heat, namely that it partakes in some measure of the circular motion which the heavens derive from the pure and effectual heat by which they are constituted. The natural motion of the air is made manifest according to Telesius by the sound heard when a shell is put to the ear.

opinion before they made the trial, when they come to particulars, abuse both their own wit and the facts of nature, and miserably mangle and torture both; and yet they proceed confidently and (if you believe themselves) victoriously, and by one means or another still find enough to say for themselves. In the end, however, he gives up in despair, and falls to wishes, intimating that though both the power and quantity of heat and the disposition of matter may be grossly and in sum distinguished and determined, yet their exact and accurate proportions, and their distinct and as it were measured methods are placed beyond the reach of human inquiry; and yet so that (if of two impossibles one can be said to be easier than another) the diversity of the disposition of matter may be better discerned than the strength and gradations of heat; and nevertheless that in these very things (if the fates allow) is to be found the summit and culmination both of the knowledge and the power of man. But having plainly professed despair, he yet does not cease from vows and prayers. For his words are: "Further, what heat and how much,—that is, what strength and what amount of it,—can turn what earth and what entities into what,—is a question not to be asked; being a thing impossible (as it seems to me) for man to know. For how is it possible to divide, as it were, into degrees either the force of heat or heat itself, or to have a distinct perception of the amount and quantity of matter into which it is infused, and to assign to a certain and determinate force and quantity of heat a certain quality and disposition and certain actions of matter; or on the other hand, to a certain quantity and certain actions of matter a certain and determinate quantity of heat? Would that they who enjoy leisure and a clearer intellect, and who have the means of searching the nature of things in perfect tranquillity, may find this out; that men may not only understand all things, but likewise be masters of all[26]!" wherein he shows himself somewhat honester than his adversaries usually are, who set down as absolutely unattainable by art everything which the arts that they themselves have made do not attain; so that no art can be found guilty, being itself both party and judge. There remains the method which was mentioned in the third place, that of working upon; which Telesius disposes of by three dogmas. The first is, what I before remarked by the way, that we know of no concordance (as in the doctrine of the Peripatetics), whereby things are cherished and conspire as by agreement. For all generation, and therefore all effect in the natural body, is accomplished by victory and predominance of one or other, and not league or compact between the two. And this is no new thing, as Aristotle likewise remarked it in the doctrine of Empedocles[27]; namely, that Empedocles, although he had set down strife and friendship as the efficient principles of things, yet in his explanations of causes commonly makes use of hostility, as if forgetful of the other. The second is that heat, by its own action, always turns being into moisture, and that neither dryness has any agreement with heat, nor moisture with cold. For to attenuate is the same as to moisten; and what is rarest is also moistest; by moisture understanding that which yields, divides into parts, and restores itself again most easily, and is defined and fixed with difficulty. All which qualities exist more in flame than in air, which is made by the Peripatetics to be most moist. Therefore heat perpetually attracts, feeds upon, extends, supplies, and generates moisture; and on the other hand cold drives all things into dryness, concretion and hardness: and here he holds Aristotle to be both dull in observation and inconsistent with himself, and imperious and wilful as regards experience, because he unites heat with dryness. For that heat sometimes dries beings, happens by accident; that is to say, in a body dissimilar, and made up of parts some grosser and some finer, heat attracts and (by attenuation) gives an outlet to the finer part, while the grosser part is thereby forced together and more constrained; which grosser part nevertheless, if a stronger heat be applied, itself becomes fluid, as is shown in bricks. For, in the first place, a moderate heat forces the clay to become brick, the finer part having evaporated; but a stronger heat melts this brick substance into glass. Now these two dogmas may be regarded as confutations of errors; the third plainly affirms, and not only that, but also clearly dis-

[26] This quotation is inaccurate.

[27] Arist. *Meteor.* iii. 4.

tinguishes the manner of working and preparation. This is twofold, either by rejection or conversion; either of which methods is carried out into acts according to the force of heat and disposition of matter. Yet in this there seem to be as it were two rules; one, that when heat and cold meet together in great quantities and in regular armies, there follows an ejection. For the beings are dislodged like armies, and driven from their place. But when a smaller quantity is engaged then there follows conversion; for the beings are destroyed and rather change their nature than their place. Of this there is a remarkable and noble instance in the upper regions of the air, which, though they are situated nearer to the heat of heaven, are yet found to be colder than the confines of the earth. For in those places where a nearer approach is made to the seat of primitive heat, the heat, collecting itself at once, drives out and thrusts down the entire force of cold that had risen up, and prevents its approach. And it may be, in like manner, that in the depths of the earth the heats are more intense than on the surface; for that as the seat of primitive cold is approached, the cold, exciting itself, drives back and puts to flight the heat with great impetuosity, and converts it into itself. The other rule is, that in an open place there follows ejection; in a confined, conversion. Now this is wonderfully shown in close vessels, where the emission of the rarefied body (which we commonly call spirit) being prevented and driven back, there follows deep and radical alterations and fermentations in bodies. But this in like manner happens when a body, from the compactness of its parts, is itself like a close vessel. Such then are the opinions of Telesius, and perhaps also of Parmenides, concerning the principles of things, except that Telesius has added something of his own respecting Hyle, being led astray by the Peripatetic notions.

Now what Telesius says would have been probable, if man were removed from the world, and with him the mechanical arts which vex matter, and the fabric of the world were regarded simply. For this philosophy of his seems a kind of pastoral philosophy, which contemplates the world placidly and at its ease. Of the system of the world he discourses well enough, but of principles most unskilfully. Moreover in his system itself there is a great mistake; namely, that he frames such a system as may apparently be eternal, without supposing a chaos, or any changes of the great configuration of things. For whatever philosophy it be, whether the Telesian or the Peripatetic, or any other, that professes a system so furnished, balanced, and guarded, that it may seem not to have come from chaos, it is a philosophy of little value, and conceived in the narrowness of the human breast. For by one who philosophises according to the sense alone, the eternity of matter is asserted, the eternity of the world (such as we now see it) is denied; and this was the conclusion both of the primitive wisdom, and of him who comes nearest to it, Democritus. The same thing is testified by Sacred Writ; the principal difference being, that the latter represents matter also as proceeding from God; the former, as self-existing. For there seem to be three things with regard to this subject which we know by faith. First, that matter was created from nothing. Secondly, that the development of a system was by the word of Omnipotence; and not that matter developed itself out of chaos into the present configuration. Thirdly, that this configuration (before the fall) was the best of which matter (as it had been created) was susceptible. These however were doctrines to which those philosophies could not rise. Creation out of nothing they cannot endure; the existing configuration of the world they suppose to have grown out of many indirect and circuitous processes, and many attempts and efforts of matter: and as for its being the best possible, they do not trouble themselves about that, seeing they maintain it to be perishable and variable. In these points therefore we must rest upon faith and the firmaments of faith. But whether it would have been possible for this created matter, in a long course of ages, by the force which was given to it, to have gathered and shaped itself into that perfect configuration (as it did at once without any rounding about at the word of command), is a question perhaps not to be asked. For the anticipation of time is as much a miracle, and belongs to the same omnipotence as the formation of being. Now the Divine nature seems to have chosen to manifest itself by both these emanations of omnipotence, by operating omnipotently, first on being and matter in the creation of something out of nothing; secondly on motion and time in anticipating the order of nature and accelerating the process of being. But these things belong to the allegory of Cœlum, where I will discuss more fully what I now briefly glance at. Let us proceed then to the principles of Telesius. And would that this were but agreed on for once by all, that beings are not to be made out of things which have no being; nor principles out of what are not principles; and that a manifest contradiction is not to be admitted. Now an abstract principle is not a being; and again, a mortal being is not a principle; so that a necessity plainly inevitable drives men's thoughts (if they would be consistent) to the atom; which is a true being, having matter, form, dimension, place, resistance, appetite, motion and emanations; which likewise, amid the destruction of all natural bodies, remains unshaken and eternal. For seeing the corruptions of the greater bodies are so many and various, it must needs be that that which remains as the centre immutable should be either something potential or infinitely small. But it is not potential; for the original potentiality cannot be like other potentialities, which are one thing actually and another potentially. But it must necessarily be something entirely abstract, since it refuses all act and contains all power. It remains therefore that this immutable thing must be infinitely small; unless indeed it be asserted that there are no principles at all, but that one thing is as a principle to another; that the law and order of change are things constant and eternal, but essence itself inconstant and mutable. And it would be better to affirm directly something of this kind than, from a desire to maintain some eternal principle, to fall into the greater inconvenience of making that principle imaginary. For the former method seems to have some issue; namely that things change in a circle; whereas this would have none at all, which regards as beings things that are merely notional and instruments of the mind. And yet that this is no way the case, shall be shown hereafter. Telesius however chose *Hyle*, which, though the offspring of a later age, he transferred into the philosophy of Parmenides. But he institutes a strange and altogether unequal contest between his active principles, unequal both in numbers and the method of fighting. For as to numbers, the earth with him is single, while the heaven has a great army; the earth likewise is almost like a point, whereas the space and region of the heaven are immense. Nor can this inconvenience be removed by the assertion that the earth and its connaturals are of a matter most compact, whereas the heavens and the ethereal bodies are of a matter most spread out. For though this certainly makes a considerable difference, yet it by no means equalises the forces, not by a wide interval. But the strength of Telesius's doctrine depends principally upon the possibility of assigning, as it were, an equal portion of Hyle (equal in quantity, not in bulk) to each of the active principles; so that things may last, and a system be constituted and established. For whoever, agreeing with Telesius in other respects, shall admit the superabundance of Hyle in one principle as compared with the other, especially in so great an excess, will find himself in a difficulty, and will not be able to make it out. Therefore in the dialogue of Plutarch respecting the face in the moon's orb, this consideration is wisely proposed, that it is not probable that in the dispersion of matter nature enclosed every compact body in the globe of the earth alone, when there were so many globes of stars revolving. But Gilbert has indulged this thought to such excess, as to assert that not only the earth and moon, but many other solid and opaque globes are scattered amid the shining globes throughout the expanse of heaven[28]. Nay, the Peripatetics themselves, when they had set down the heavenly bodies as eternal in their own state, and sublunary bodies as eternal by succession and renovation, were not confident of being able to maintain that doctrine without assigning as it were equal portions of matter to the elements. For this is what they are thinking of in that dream of theirs about the tenfold proportion of the ambient to the interior element. Nor do I adduce these things because none of them please me, but to show that it is an inconceivable thing and a thought altogether ill-measured, to

[28] Gilbert, *Nov. Phys.* i. 10.

set down the earth as the contrary active principle to heaven, which Telesius did. And the supposition becomes much harder, if, besides the difference in quantity between heaven and earth, a man shall consider the difference in virtue and act. For the conditions of battle are entirely destroyed if the weapons on one side take effect, and on the other do not reach their distance, but fall short. Now, it is certain that the sun's force reaches the earth; but who will undertake to say that the earth's force reaches the sun? For of all the virtues which nature produces, that of light and shade is emitted furthest, and spreads round in the widest circle. But the shade of the earth stops on this side of the sun, whereas the light of the sun, if the earth were transparent, would strike quite through the globe of the earth. Heat and cold again (of which we are now speaking) are never found to carry their virtue so far as light and shade. Therefore if the shade of the earth does not reach the sun, much less is it probable that the cold of the earth reaches thither. If it be the case that the sun and heat act on certain intermediate bodies to which the virtue of the contrary principle does not ascend, and where it does not in any way interfere with their action, it must needs be that they (the sun, I say, and heat) first occupy all bodies near them, and then take in those also which are further off, till it would end in the conflagration of Heraclitus, the solar and celestial nature gradually descending, and approaching nearer to the earth and its confines. Nor does it well agree with the supposition, that this power of imposing and multiplying its nature and converting other things into itself, which Telesius attributes to principles, does not operate on similar things equally or more than on contraries; in which case the heaven should now be of a white heat, and the stars united with one another. But to come closer, it seems there are four demonstrations to be proposed, by any one of which, much more by all together, Telesius's philosophy respecting principles may be pulled to pieces and destroyed. Of these the first is, that there are found in nature certain actions and effects, even among the most powerful and universal, which can in no way be referred to heat and cold. The second is, that there are found some natures, of which heat and cold are the effects and consequences; and that not by the excitation of pre-existing heat, or the application of an adventitious heat, but in which heat and cold, in their original essence, are implanted and generated. Therefore the condition of a principle fails here in both ways; as there is both something that does not proceed from them, and they themselves proceed from something. The third is, that even those things which derive their origin from heat and cold (which certainly are very many) yet proceed from them as from their efficient and instrument, not as from their proper and intimate cause. The last is, that this coordination of four connatural bodies is altogether disordered and confused. I will speak therefore on each of these points separately. And to some it may perhaps seem scarce worth while to take such pains in refuting the philosophy of Telesius, a philosophy not much spoken of or received. But I do not stand upon such points of dignity. For of Telesius himself I have a good opinion, and acknowledge him as a lover of truth, useful to the sciences, the reformer of certain opinions, and the first of the moderns; at the same time it is not as Telesius that I have to do with him, but as the restorer of the philosophy of Parmenides, to whom much respect is due. But my principal reason for being more full in this part is that, in dealing with him who comes first, I take occasion to discuss many questions which may be transferred to the refutation of other sects, of which I shall have to treat hereafter; that I may not be obliged to say the same things many times over. For errors, though different, have their fibres strangely entangled and intertwined; yet so that they may often be mowed down by one refutation as by a sweep of a scythe. But, as I was going to say, we must see what virtues and actions there are in nature, which can by no consent of things or force of wit be attributed to heat and cold. First therefore let us assume what Telesius grants, that the sum of matter is eternal and without increase or diminution. This property, by which matter preserves and supports itself, he dismisses as passive, and as belonging rather to quantity than to form and action; as if there were no need to ascribe it to heat and cold, which are set down as the sources only of active forms and virtues; for that matter is not destitute simply, but only destitute of all active virtue. Now in these assertions there is a great mental error,—an error truly wonderful, were it not that consent and common and inveterate opinion take away the wonder. For there is scarce any error comparable to that of taking this virtue implanted in matter (by which it saves itself from destruction, insomuch that not the smallest portion of matter can either be overpowered by the whole mass of the world, or destroyed by the force and power of all agents together, or any way so annihilated and reduced to order, but that it both occupies some space, and maintains a resistance with impenetrable dimensions, and itself attempts something in its turn, and never deserts itself) not to be an active virtue; whereas, on the contrary, it is of all virtues far the most powerful, and plainly insuperable, and as it were mere fate and necessity. And yet Telesius does not even attempt to refer this virtue to heat and cold. And rightly so; for it is a thing which neither conflagration, nor torpor and congelation, can add anything to or detract anything from or have any power over while itself meantime is active both in the sun and at the centre of the earth, and everywhere else. But his mistake appears to have lain here—that while he acknowledges a certain and definite mass of matter, he is blind to the virtue by which that matter keeps itself undiminished in quantity, and (buried in the deepest darkness of the Peripatetics) ranks this as an accessory; whereas it is the very principal,—vibrating one body, removing another, solid and adamantine in itself, and the fountain whence emanate the decrees of possible and impossible with inviolable authority. The common school philosophy likewise childishly attempts to grasp it in a set of words; thinking it enough to set it down as a rule that there cannot be two bodies in the same place; but the virtue and the process thereof it never contemplates with its eyes open, nor dissects to the quick; little knowing how much depends on it, and what a light may thence rise to the sciences. But (to return to the present business) this virtue, however great it be, falls beyond the principles of Telesius. I must now pass on to that virtue which is as the converse of the former, namely, that which maintains the connexion of matter. For as matter refuses to be overpowered by matter, so does matter refuse to be separated from matter. Notwithstanding there is great doubt whether this law of nature be as peremptory as the other. For Telesius maintained, and so did Democritus, the existence of a collective vacuum without any limit, in order that individual beings may lay aside and sometimes even forsake the one contiguous to them, with difficulty (as they say) and against their will,—that is, when subdued and forced by some greater violence; and this he tries to prove by certain experiments, especially adducing those which are everywhere cited for the contradiction and refutation of a vacuum, and as it were making extracts from them, and amplifying them so as to allow beings to be under some slight necessity of holding to that which is contiguous, but so that if they be more strongly pressed, they will admit a vacuum; as we see in water-clocks, in which if the hole through which the water runs is too small, they will want an air-hole to enable the water to descend; but if the hole be larger, even though there be no air-hole, the water, pressing with a heavier weight on the hole, flows downwards, not caring for the vacuum above. In like manner, in bellows, if you shut them and then stop the mouth so that there is no passage for the air to enter, and then raise and expand them,—if the leather be thin and weak it bursts; if it be thick and not liable to burst it holds; and so in other things. But these experiments are neither exactly proved nor do they altogether satisfy the inquiry or settle the question; and though by them Telesius thinks that he is applying himself to things and inventions, and endeavours to distinguish more accurately what has been observed confusedly by others, yet he is no way equal to the work, nor does he unravel the matter to the end, but falls off in the middle,—a habit common both to him and the Peripatetics; who are very owls in looking at experiments; and that not so much from weakness of vision, as because it is clouded by opinions, as by cataracts, and from impatience of full and fixed consideration. But this question (one of the most difficult) as to how far a vacuum is allowed, and at what distances seeds may attract or repel each other, and what there is in this matter peremptory and invariable, I refer to the place where I shall treat of a vacuum. For it is not of much importance to the present question whether Nature utterly abhors a vacuum, or whether beings (as Telesius thinks it more correct to say) delight in mutual contact. For I make it plain that this, whether it be abhorrence of vacuum or desire of contact, no way depends on heat and cold; nor is it ascribed thereto by Telesius himself, nor can it be ascribed to them upon any evidence in the nature of things; seeing matter when moved from its place cannot but draw other matter to it, whether it be hot or cold, wet or dry, hard or soft, friendly or unfriendly; insomuch that a hot body will sooner attract the coldest body to its side, than suffer itself to be deserted and separated from all. For the bond of matter is stronger than the enmity of heat and cold; nor does the sequacity of matter care for the diversity of special forms. Therefore this virtue of connexion does not at all depend upon those principles of heat and cold. Next come two virtues opposed to each other, by which this kingdom of principles has been transferred (as may be thought) to heat and cold, but on a claim of right not well made out; I mean those virtues by which beings open and rarefy, dilate and expand themselves, so as to occupy a greater space and spread themselves over a larger sphere; or contrariwise close and condense, confine and contract themselves, so as to cover less space and shrink into a smaller sphere. We must show therefore how far this virtue has its origin from heat and cold, and how far it keeps separate and unmixed with them. Now it is most true, as Telesius affirms, that density and rarity are as it were the proper work of heat and cold; for they have far the most to do in making bodies occupy a larger or less space; but yet these things are understood confusedly. For bodies seem sometimes to migrate and transfer themselves from one natural dimension to another, and that freely and as it were willingly, and with a change of form; sometimes they seem only to be forced away from their natural dimension, and their old form still remaining, to return to their usual dimension again. Now that virtue of progression into a new space is almost governed by heat and cold. But it is not so with that other virtue of restitution; since water expands itself into vapour and air, oil likewise and fat things into exhalation of flame, by the power of heat; nor (if the transmigration be perfect) do they care to return; nay the air itself also swells and is extended by heat. Whereas if the migration be only half effected, then after the heat is withdrawn it easily returns to itself; so that even in the virtue of restitution heat and cold have something to do. But things which are extended and drawn asunder not by means of heat, but by some violence, as soon as the violence ceases return most eagerly (even without any accession of cold or diminution of heat) to their former dimensions; as we see in the sucking of the glass egg, and the raising of the bellows. But this is still more evident in solid and gross bodies. For if a piece of cloth or a harp-string be stretched, on the removal of the force they rebound with great velocity; and it is the same with compression. For air compressed and imprisoned by any violence bursts out with a great force; and indeed all that mechanical motion caused by the striking of one hard body by another, commonly termed violent motion, by which solid bodies are sent flying through the air and water, is nothing but an endeavour of the parts of the discharged body to free themselves from compression; and yet here there are no apparent traces of heat and cold. Nor can any such fine argument be made upon this doctrine of Telesius as to say, that to every natural dimension there is assigned a quantity of heat and cold, in a certain proportion; therefore it may be that although no heat and cold are added, yet if the dimensions of the material body be extended or contracted it will come to the same thing; because more or less of matter is put in the space than is proportionate to the heat and cold. Such things, though not absurd in words, are yet the suggestions of men who are always seeking some device by which they may maintain their first thought, and do not follow out the inquiry in nature and fact. For if heat and cold be added to such extended or compressed bodies, and that in a greater measure than is proportionate to the nature of the body itself (let the stretched cloth for instance be warmed by the fire), yet it will by no means restore the balance, nor extinguish the force of restitution. I have therefore now made it plain that this virtue of dimension does not depend in any notable proportion on heat or cold; although it is this very virtue which has given most authority to these principles. Next come two virtues, which are in everybody's mouth, and are spread far and wide, namely those by which bodies are carried towards the greater masses and collections of their connaturals ; in the observation whereof, as in the rest, men either trifle or go quite wrong. For the common philosophy of the school holds it enough to distinguish natural from violent motion ; and to assert that heavy bodies by a natural motion are borne downwards and light bodies upwards. But such speculations are of little help to philosophy. For these words, *nature*, *art*, and *violence*, are but compendious phrases and trifles. They ought not only to refer this motion to nature, but likewise to seek in this very motion for the particular and proper affection and appetite of the natural body. For there are a great many other natural motions arising from very different passions of things. Therefore the thing is to be propounded according to its differences. Nay, those very motions which they call violent may be said to be more according to nature than that which they call natural ; if that be more according to nature which is stronger, or even which is more according to the system of the universe. For this motion of ascent and descent is not very imperious, nor even universal ; but provincial as it were, and confined to certain regions ; and it is moreover obedient and subject to other motions. And as for saying that heavy things move downwards and light upwards, it is the same as saying that heavy things are heavy and light light. For that which is predicated is assumed in the subject by the very force of the term. But if by heavy they mean dense and by light rare, they do advance somewhat ; yet so as to arrive at an adjunct and concomitant rather than a cause. Those on the other hand who explain the appetites of heavy and light things by contending that the one are borne to the centre of the earth, and the other to the circumference and compass of the heaven, as to their proper places, certainly assert something, and likewise point towards a cause ; but altogether wrongly. For place has no forces, nor is body acted on except by body ; and all swift motion of a body, which seems as if it were seeking a place for itself, is really in pursuit not of location or position simply, but with reference to some other body.

*   *   *   *   *   *

A DESCRIPTION OF THE INTELLECTUAL GLOBE.

[TRANSLATION OF THE *DESCRIPTIO GLOBI INTELLECTUALIS*.]

PREFACE.

BY ROBERT LESLIE ELLIS.

THIS tract, published by Gruter in 1653, must have been written about 1612. This follows from what is said of the new star in Cygnus [1], which was first observed in 1600. It is therefore intermediate in date between the *Advancement of Learning* and the *De Augmentis*; and though on a larger scale than either, it is to be referred to the same division of Bacon's writings. The design of all three is the same, namely, a survey of the existing state of knowledge. The commendation of learning which forms the first book of the other two works being in this one omitted, it commences with the tripartite division of knowledge which Bacon founded on the corresponding division of the faculties of man—memory, imagination, and reason. History, which corresponds to memory, is here as in the *De Augmentis* primarily divided into natural and civil, whereas in the *Advancement* the primary division of history is quadripartite, literary and ecclesiastical history being made co-ordinate with civil history, instead of being as here subordinated to it.

The divisions of natural history are then stated, and are the same as in the *De Augmentis*; and the remainder of the tract relates to one of these divisions, namely the history of things celestial, or in other words to astronomy. The problems which it should consider, and the manner in which they ought to be solved, are treated of at some length; but even with respect to astronomy much which it is proposed to do is left undone, the whole tract being merely a fragment.

Bacon has nowhere else spoken so largely of astronomy; the reason of which apparently is, that he was writing just after Galileo's discoveries had been made known in the *Sydereus Nuncius*, published in 1611; a circumstance which makes the *Descriptio Globi Intellectualis* one of the most interesting of his minor writings. The oracles of his mind were in this case evoked by the contemplation, not of old errors, but of new truths.

The *Thema Cœli*, which contains a provisional statement of his own astronomical opinions, is immediately connected with the astronomical part of the *Descriptio Globi Intellectualis*. They are clearly of the same date, and form in reality but one work.

In the *De Augmentis* Bacon has expressed the same general views on the subject of astronomy as in these tracts; and they are in truth views which it was natural for a man not well versed in the phenomena of the science to entertain and to promulgate. What had been done by the old astronomers seemed to him full of useless subtleties and merely mathematical conceptions; men therefore were to be exhorted to cast all these aside, and to study the phenomena of the heavens independently of arbitrary hypotheses. Let us first obtain an accurate knowledge of the phenomena, and then begin to search out their real causes. Orbs, eccentrics, and epicycles must not stand between the astronomer and the facts with which he has to deal. In this language, which had been held by others, there is

[1] Stella nova in pectore Cygni . . . jam per duodecim annos integros duravit.

something not wholly untrue ; yet the counsel which it contains would, if it could have been followed, have put an end to the progress of astronomical science. Let us obtain an accurate knowledge of the phenomena—this no doubt is necessary, but then how is it to be done ? To say that instead of trying to resolve the motion of the planets into a combination of elementary circular motions, we ought to be content to save the appearance by means of spirals, is to no purpose unless we are prepared to give an accurate definition of the kind of spiral we mean. Failing this, a statement that the apparent path of a planet is a spiral or irregular line along which it moves with varying velocity, is much too vague to be of any scientific value whatever ; and if we seek to give precision to this statement, we find ourselves led back again into the region of mathematical conceptions, or, if the phrase be preferred, of mathematical hypotheses. The distinction between what is real and what is only apparent lies at the root of all astronomy ; and it is in vain to seek for a physical cause of that which has only a phenomenal existence, as for instance of the stations and regressions of the planets. Thus in two points of view, astronomy must of necessity employ mathematical hypotheses, firstly in order to the distinct conception of the phenomena, and secondly in order to be able to state the problems which a higher science is afterwards to solve. If the hypotheses employed are inappropriate, as in the systems of Ptolemy or Tycho Brahe, they may nevertheless have done good service in making it possible to conceive the phenomena, and moreover may serve to suggest the truer views by which they are to be replaced. Almost any hypothesis is better than none: "citius enim," as Bacon has elsewhere said, "emergit veritas ex errore quam ex confusione". The wrong hypotheses doubtless lead to premature speculation touching physical causes ; but this is a mischief which in course of time tends to correct itself, as we see in the Ptolemaic system, of which the overthrow was in good measure due to the cumbrous machinery of solid orbs, which had been constructed to explain the motions mechanically. It came to be seen that even if this system could save the phenomena, it was unable to give a basis on which a just explana tion of their causes could be founded.

I have said that almost any hypothesis is better than none. But the truth is that as soon as men begin to speculate at all an hypothesis of some kind or other is a matter of necessity. On merely historical grounds and apart from any consideration of the relation between facts and ideas, questions might be propounded to a writer who was trying to describe the phenomena of the heavens without introducing any portion of theory, to which he would not find it easy to give clear answers. Thus we know that one of the philosophers of antiquity affirmed that the sun is new every day ;—are you prepared, we might ask, to set aside the authority of Heraclitus, and to maintain your theory in opposition to his ? If you affirm that the sun which set last night is the same as that which rose this morning, you are no longer a describer of phenomena, but, like those whom you condemn, a dealer in hypotheses.

However this difficulty is got over, you will at any rate not venture to confound Hesperus and the morning star. It is true that one of the great teachers of Greece long since asserted that they are the same ; but the speculative fancies of Pythagoras must be rejected not less than those of Ptolemy or Regiomontanus.

We find that Bacon, both in the *De Augmentis* and in the following tract, speaks of the constructions of astronomy as purely hypothetical. In this he agrees with many other writers. It was a common opinion that these constructions had no foundation in reality, but were merely employed as the basis of mathematical calculations. They served to represent the phenomena, and that was all. This view, which has not been without influence on the history of astronomy, inasmuch as it made the transition from one hypothesis to another more easy than it would have been if either had been stated as of absolute truth, connected itself with a circumstance not unfrequently overlooked. The struggle between the peripatetic philosophers and the followers of Copernicus has caused an earlier struggle of the same kind to be forgotten. The Ptolemaic system is in reality not much more in accordance with the philosophy of Aristotle than the Copernican ; and therefore, while the authority of Aristotle was unshaken, it could only be accepted, if accepted at all, as a means of representing the phenomena. The motions of the several orbs of heaven must, if our astronomy is to accord with Aristotle, be absolutely simple and concentric. On these conditions only can the incorruptibility of the heavens be secured. Consequently eccentrics and epicycles must be altogether rejected; and as the Ptolemaic system necessarily employs them, it follows that this system is only of value as a convenient way of expressing the result of observation. Such was the view of those who, while they adopted Aristotle's principles, were aware that the astronomical system with which he was satisfied, and of which he has given an account in the twelfth book of the *Metaphysics*, was wholly inadequate as a representation of the phenomena. But his more strenuous adherents went further, and followed Averroes in speaking with much contempt of Ptolemy and of his system; an excess of zeal which Melancthon, in the spirit of conciliation which belongs to his gentle nature, has quietly condemned [2].

Out of this antinomy, if the word may be so used, sprang several attempts to replace the Ptolemaic system by a construction which should be in accordance both with the phenomena and with Aristotle. Of these the best known is the Homocentrica of Fracastorius. As the name implies, all the orbs have on this hypothesis the same centre, and of these homocentric orbs he employs seventy-seven. But a fatal objection to this and all similar attempts is that they can give no explanation of changes in apparent distance. Fracastorius tries to set aside this objection by asserting that although the distance of some of the heavenly bodies from the earth may seem to vary, yet it never does so in reality, the apparent variation being caused by the varying medium through which they are seen.

Though this explanation is wholly unsatisfactory, the wish to get rid of eccentrics and epicycles was sufficiently strong to win for Fracastorius a much more favourable reception than his complex and imperfect hypothesis deserved. He was spoken of as a man who had succeeded in overcoming the divorce which had so long separated astronomy from philosophy [3].

Of the similar attempt made by D'Amico I know no more than what is mentioned by Spiriti in his *Scrittori Cozentini*.

The Ptolemaic system being thus treated as a mere hypothesis by the followers of Aristotle, for of course the astronomers who accepted Purbach's theory of solid orbs must have regarded it as a reality, it was natural that Bacon should have thought that what we now call physical astronomy, that is the causal explanation of the phenomena, ought to be studied independently of this system. Whatever it had accomplished might be as well done without it. Spirals and dragons would be found sufficient to represent the phenomena, if the perverse love of simplicity which had led the mathematicians to confine themselves to circles and combinations of circles was once got rid of. Galileo's view of this matter is however undoubtedly the true one, "Le linee irregolari son quelle che, non avendo determinazion veruna sono infinite e casuali, e perciò indefinibili, ne di esse si può in conseguenza dimostrar proprietà alcuna, ne in somma saperne nulla; sicchè il voler dire, il tale accidente accade mercè di una linea irregolare, è il medesimo che dire io non so perchè ei si accagia [4]".

Bacon was not the first who proposed to sweep away from astronomy the mathematical constructions by which it seemed to be encumbered. We find in Lucretius nearly the same views as those of Bacon. The astronomers, Bacon often says, insist on explaining the retardation of the inferior orbs by giving them a proper motion of their own, opposite to that which they derive from the starry heaven: surely it would be simpler to say that all the orbs move in the same direction with unequal velocities; the inequality depending on their remoteness from the prime mover.

[2] See his *Initia Physicæ*.

[3] See Flaminius. [Carmin. lib. ii. f. 30. Ed. Lutet. per Nicol. Divitem.] It is remarkable that Delambre declares that he cannot see why Fracastorius should have thought his own system better than the old one. The reason is perfectly obvious if we consider the matter in connection with the history of philosophy.

[4] Saggiatore, ii. p. 187.

Compare with this the following lines of Lucretius :—

> " Quanto quæque magis sint terram sidera propter,
> Tanto posse minus cum cœli turbine ferri :
> Evanescere enim rapidas illius, et acreis
> Imminui subter, viries ; ideoque relinqui
> Paullatim solem cum posterioribu' signis,
> Inferior multum quum sit quam fervida signa :
> Et magis hoc lunam ; " etc[5].

But it was probably not from Lucretius that Bacon derived this way of considering the matter. For Telesius, whom Bacon esteemed " the best of the novelists," and whose pastoral philosophy, as he has not unhappily called it, was contented with vague speculations as to the causes of phenomena without any accurate knowledge of their details, had suggested to his followers that it was nowise necessary to resolve the motion of the sun into the motion of the starry heaven and the motion of his own orb, and that on the contrary this composition of motions is unintelligible. You may see, he affirms, with your own eyes the way in which the sun, moving with one motion only, advances continually from east to west, and alternately towards the north and south ; all that is necessary is to admit that the poles on which he revolves are not constantly at the same distance from the poles of heaven, but on the contrary are always receding from or advancing towards them[6].

Amongst those who called themselves Telesians the view here suggested received a fuller development ; they adopted the doctrine of Alpetragius, a Latin translation of whose *Theorica Planetarum* was published at Venice in 1531. Alpetragius professes that he found the complication of the Ptolemaic system intolerable, and that the foundation of his own is much simpler. " Apparet sensu quod quilibet planeta revolvitur singulo die super circulis æquidistantibus ab æquinoctiali ; attamen diuturnitate temporis et revolutione planetæ multis revolutionibus ex periodis diurnis, videtur ille planeta moveri a puncto in quo visus est primum æquinoctialis et respectu motus similis ei postponi in longitudine et declinare a suo primo loco in latitudine", of which the reason is that it does not really revolve in circles parallel to the equator, "sed est revolutio girativa dicta laulabina ex declinatione planetæ a loco suo in latitudine[7]". Of this the reason is twofold: the planet's orb moves more slowly than the prime mover in consequence of its essential inferiority, an inferiority which increases in the case of different planets with their nearness to the earth ; and its poles revolve on two small circles parallel to the equator. Alpetragius goes on to apply these hypotheses to each of the planets. It is needless to point out of how little value his speculations necessarily are. Such as they are, however, the Telesians, as we learn from Tassoni[8], were content to accept them. Of the astronomical writings of the Telesians I have not been able to find any account. None of those who are mentioned by Spiriti appear to have published anything on the subject. However this may be, the authority of Tassoni is sufficient to show that the school of Telesius rejected the Ptolemaic system and especially the notion that the planets etc. have a proper motion from west to east ; and that their views are therefore in accordance with those which Bacon propounds in the *Thema Cœli*, so far at least as relates to the general conception of the planetary motions.

Patricius, on whom the influence of Telesius is manifest, and who furnished Bacon with many of the facts contained in the following treatises, also rejected, and more contemptuously than Telesius, the common astronomical hypotheses. The planetary motions, their stations and regressions, are, he says, explained by astronomers by the help of epicycles and eccentrics ; but we ascribe them to the natures and spirits of the planets, and in a higher degree to their souls and minds. Of this idle talk Gilbert remarks that it destroys the study of astronomy. " Quid autem," he observes, " tum postea spectabit otiosus incassum philosophus, opinione suâ satiatus, cœlum sine usu sine motuum prævidentiâ : ita nullius usus erit

[5] Lucret. v. 622.

[6] Telesius, *De Rer. Nat.* iv. 25.

[7] Alpetragius, fo. 14. v.

[8] *Pensieri diversi*, ii. 4. (Venice, 1636.)

illa scientia[9]". But Patricius's opinions on astronomy could clearly not be of much value, seeing that he was sufficiently ignorant to blame astronomers for not taking into account the distance of the place where their observations are made, from the centre of the earth; and speaks of this omission as "a most evident fallacy": a remark which proves that he had either never heard of the correction for parallax, or having heard of it was unable to understand its nature.

From him, however, Bacon derived some of the most remarkable statements in the *Descriptio Globi Intellectualis*; particularly the incredible account of the mutations which Venus underwent in 1578. That, setting aside Patricius's loose way of speaking, the real phenomenon was simply that Venus was visible before sunset, is probably the safest explanation of the whole story; of which I have found no mention elsewhere. Thus much however is certain, that there could have been no such peculiarity in her appearance as to suggest to well-informed persons the notion that she had undergone any real change, since in the controversy whether there were any evidence of corruption or generation in the heavens a fact like this could not have been passed over.

Of the discoveries announced by Galileo in the *Sydereus Nuncius* Bacon does not speak at much length, though it is difficult not to believe that he was led to say so much of astronomical theories by the interest which these discoveries must have excited when they were first made known. The discovery of Jupiter's satellites, the resolution into stars of the nebula Præsepe, and the irregularities in the moon's surface, are all mentioned in the following tract; but, as I have said, somewhat briefly[10].

It is remarkable that neither in the following tracts nor in his subsequent writings has Bacon mentioned the discoveries of Kepler. The treatise *De Stellâ Martis* was published in 1609, and became known in England at least as early as 1610. Harriot, it appears from Professor Rigaud's account of his papers, was then in correspondence with him, and repeated his calculations. That Bacon was acquainted with his writings we can hardly believe; they bear so directly on the questions which he has discussed that he could scarcely have neglected to notice them, had he known them even by report. In the very first page of Kepler's great work we find a quotation from Peter Ramus, declaring that he would resign his professorship in favour of any one who should produce an astronomy without hypotheses. To this Kepler subjoins an apostrophe to Ramus, telling him that it is well that death had relieved him of the necessity of redeeming his pledge, and vindicating Copernicus from the charge of having explained the phenomena of astronomy by unreal hypotheses. The same subject is resumed in the preface, and elsewhere throughout the book. Again, in another point of view, it makes Bacon's complaints that astronomers cling superstitiously to perfect circles appear somewhat out of date, to find that before the time at which he wrote the man who professedly both by his genius and his official position stood at the head of the astronomers of Europe and, so to speak, represented them, had succeeded in saving the phenomena more accurately than had been done before, by means of ellipses. A great change had just taken place; two most remarkable laws, the foundations of modern physical astronomy, had just been propounded, namely the law of elliptic motion, and that of the equable description of areas; and the whole state of the question with respect to the truth or falsehood of the Copernican system was thus changed. In truth this system was inextricably connected not only with Kepler's results, but with his method. In his dedication to the Emperor he says, "Locum (that is, the place of Mars) indagine cinxi, curribus

[9] *Physiol. Nov.* ii. 9.

[10] The interest which these discoveries excited must have been very great. Sir William Lower writes to Harriot, "I gave your letter a double welcome, both because it came from you and contained news of that strange nature . . . Methinks my diligent Galileus hath done more in his threefold discovery than Magellane in opening the straits to the South Sea, or the Dutchmen that were eaten by bears in Nova Zembla". The news had just reached him. His date is "the longest day of 1610." It had taken rather more than three months to travel from Italy to Wales.—*Professor Rigaud's Supplement*, etc., p. 26.

magnæ Matris Telluris in gyrum circumactis". He means by this that he used observations of Mars made when in the same point of his orbit, the earth being at the time of the different observations in different points of hers. The same idea of the connection of the Copernican hypothesis with Kepler's method, is expressed in one of the complimentary stanzas prefixed to the book :—

Cœlos Keplerus terrarum oppugnat alumnus :
De scalis noli quærere : terra volat.

In one of Kepler's letters to David Fabricius, nothing can be more decided than his rejection of the notion that all motions of the heavenly bodies are in perfect circles. "Quod ais non dubium quin omnes motus fiant per circulum perfectum, si de compositis (id est de realibus) loqueris, falsum : fiunt enim Copernico, ut dixi, per orbitam ad latera circuli excedentem, Ptolemæo et Braheo insuper per spiras. Sin autem loqueris de componentibus, de fictis igitur hoc est de nullis loqueris. Nihil enim in cœlo circumit præter ipsum corpus planetæ, nullus orbis, nullus epicyclus : quod Braheanæ Astronomiæ initiatus ignorare non potes." And it is interesting to observe how clearly he distinguishes between the real motions and the component elements into which they may be resolved.

Until the language of modern analysis had enabled us to express the nature and properties of curves merely quantatively, without reference to genesis or construction, it was difficult to attain to a clear way of thinking as to the relation which astronomical hypotheses bear to reality. In order to define the motion which actually takes place, it was necessary to refer to simpler motions which have only an abstract or ideal existence. But then it was asked, how can the result be real if the elements are not so ? In this point of view the unpicturesqueness of symbolical language, though it has led to other inconveniences, has delivered us from a great deal of confused thinking. If Poinsot's illustration of the motion of a rigid body by means of a central ellipsoid rolling on a fixed plane, had been proposed at the beginning of the seventeenth century, most people would have said that the hypothesis was absurd, though it might correspond to the phenomena.

To take the matter more generally, it must be remembered that positive truth or falsehood belongs only to the region of the actual and individuated. To say that two and three make five is not to deny that four and one do so too, although if I assert that of five houses, first three were built and then two added, I contradict that four were built at first and that only one is of later date. Not merely in the relation between cinematical or formal and physical astronomy, but generally, the question whether an hypothesis be true or false does not arise unless it is presented as a causal explanation. Thus when Berosus taught that one half of the moon is luminous, and that her phases arise from this half being always turned towards the sun in virtue of their mutual sympathy, both being bright, the explanation is unexceptionable, except so far as relates to the efficient cause. One half of the moon always is bright ; and always is turned to the sun ; and this Berosus saw as clearly as we do. It is in this way that false hypotheses are transformed into true ones ; not by the transformation of anything false into truth, but by the severance of the causal or real element from that which is neither true nor false, namely the abstract conception. But the interest of the subject has led me to dwell on it at too much length.

It is curious to observe that in the interval between the composition of the following tracts and that of the *De Augmentis* Bacon's leaning against the Copernican system became more decided, though in the same interval the system had received an accession of strength, of which doubtless he was not aware, in the discovery of Kepler's third law [11]. This law, connecting as it does the planets with

[11] This discovery was made, as Kepler has informed us, on the 15th of May, 1618. In Professor Rigaud's account of Harriot's papers, published in 1833, it is mentioned that Harriot, who was apparently the first person to determine the periods of Jupiter's satellites, committed an error of calculation, in consequence of which that of the first satellite is given at about half its real length, but that Harriot, even before the publication in 1614 of Marius's *Mundus Jovialis*, seems to have suspected his error. The Pro-

the sun by an uniform relation which is fulfilled also by the earth, is in some respects the most remarkable of the three, and points the most directly to the sun as the great centre of our system. No doubt neither this law, nor all three together, amounts to a positive demonstration: it has sometimes been forgotten that after all they are but approximations to the truth; but of all approximations these laws are the most remarkable, and it would be very difficult to doubt, even without the knowledge we now possess, that they are grounded on a physical basis. This basis is their correspondence with a causal or physical approximation. They would be absolutely true if the lesser bodies of the solar system did not attract one another, and if all were attracted by the sun as if he and they were physical points. It would be possible to crowd together a number of epicycles whereby the orbit of the earth would be better represented than on the elliptic hypothesis; but such a system would have no physical significance. No doubt too, all the laws might be true and yet the earth at rest; but we could not adopt such an opinion without doing violence to all our ideas of symmetry and harmony, —ideas which influence our judgments of natural things more than we are aware of. Such a doctrine would be felt "primam violare fidem". We may well believe that had Bacon been acquainted with the discoveries of Kepler, he would not only have been impressed by their astronomical importance, but have felt the full force of the lesson which they convey. He would have felt that they constituted a sufficient reason for transferring the allegiance which had been paid to Mother Earth to a nobler object more justly entitled to the homage which she had so long received. We now know that neither Earth nor Sun is the true Hestia of the old Philosopheme. We know too, that in all the orbs of heaven that we can see or dream of, there can be nothing fully entitled to the appellation,—nothing wholly fixed, or wholly unperturbed. Happy for us if we feel also that there is a Sun of suns whose absolute existence transcends our conceptions of space and time[12].

fessor enquires why he did not try his result by means of Kepler's third law, as we know that he was a student of the work in which this law is stated; forgetting that only the first two laws were given in the *De Stellâ Martis*, and that in the interval referred to, between 1610 and 1614, Harriot could no more have known of Kepler's third law than of Newton's *Principia*. But it is really curious that Kepler does not seem to have applied his law to the satellites. The application is said to have been first made by Vendelinus. See Narrien, *Hist. of Astronomy*, p. 398.

[12] Deus, sine qualitate bonus, sine quantitate magnus, sine indigentia creator, sine situ præsens, sine habitu omnia continens, sine loco ubique totus, sine tempore sempiternus, sine ulla mutatione mutabilia faciens, nihilque patiens.—St. Augustine, *De Trin.*

A DESCRIPTION OF THE INTELLECTUAL GLOBE.

CHAPTER I.

Division of all Human Learning into History. Poesy, and Philosophy, ***according to the three faculties of the mind,*** Memory, Imegination, and Reason : ***and that the same division holds good likewise in Theology ; the vessel (that is, the human understanding) being the same, though the matter and the manner of conveyance be different.***

I ADOPT that division of human learning which corresponds to the three faculties of the understanding. Its parts therefore are three ; History, Poesy, and Philosophy. History is referred to the Memory ; poesy to the Imagination ; philosophy to the reason. And by poesy here I mean nothing else than feigned history. History is properly concerned with individuals ; the impressions whereof are the first and most ancient guests of the human mind, and are as the primary material of knowledge. With these individuals and this material the human mind perpetually exercises itself, and sometimes sports. For as all knowledge is the exercise and work of the mind, so poesy may be regarded as its sport. In philosophy the mind is bound to things ; in poesy it is released from that bond, and wanders forth, and feigns what it pleases. That this is so any one may see, who seeks ever so simply and without subtlety into the origins of intellectual impressions. For the images of individuals are received by the sense and fixed in the memory. They pass into the memory whole, just as they present themselves. Then the mind recalls and reviews them, and (which is its proper office) compounds and divides the parts of which they consist. For the several individuals have something in common one with another, and again something different and manifold. Now this composition and division is either according to the pleasure of the mind, or according to the nature of things as it exists in fact. If it be according to the pleasure of the mind, and these parts are arbitrarily transposed into the likeness of some individual, it is the work of imagination ; which, not being bound by any law and necessity of nature or matter, may join things which are never found together in nature and separate things which in nature are never found apart ; being nevertheless confined therein to these primary parts of individuals. For of things that have been in no part objects of the sense, there can be no imagination, not even a dream. If on the other hand these same parts of individuals are compounded and divided according to the evidence of things, and as they really show themselves in nature, or at least appear to each man's comprehension to show themselves, this is the office of reason ; and all business of this kind is assigned to reason. And hence it is evident that from these three fountains flow these three emanations, History, Poesy, and Philosophy ; and that there cannot be other or more than these. For under philosophy I include all arts and sciences, and in a word whatever has been from the occurrence of individual objects collected and digested by the mind into general notions. Nor do I think that there is need of any other division than this for Theology. For the informations of revelation and of sense differ no doubt both in matter and in the manner of entrance and conveyance ; but yet the human spirit is one and the same ; and it is but as if different liquors were poured through different funnels into one and the same vessel. Therefore I say that Theology itself likewise consists either of sacred history, or of divine precepts and doctrines, as a kind of perennial philosophy. And that part which seems to fall outside this division (that is, prophecy) is itself a species of history, with the prerogative of divinity wherein times are joined together, that the narrative may precede the fact; and the manner of delivery, both of prophecies by means of visions and of divine doctrine by parables, partakes of poesy.

CHAPTER II.

Division of History into Natural and Civil; Ecclesiastical and Literary History being included under Civil. Division of Natural History into History of Generations, Pretergenerations, and Arts, according to the three states of Nature, namely, Nature Free, *Nature* Erring, *and Nature* Constrained.

History is either Natural or Civil. Natural history relates the deeds and actions of nature; civil history those of men. Matter of Divinity shows itself no doubt in both, but more in civil; so much so indeed as to constitute a distinct species in history, which we call Sacred or Ecclesiastical. This therefore I attribute to Civil; but first I will speak of Natural. Natural history does not treat of particular objects separately. Not that I was wrong in saying that history deals with individuals, circumscribed by place and time. For properly it is so. But since there is in natural objects a promiscuous resemblance one to another, insomuch that if you know one you know all, it would be a superfluous and endless labour to speak of them severally. And therefore we see that where there is no such promiscuous resemblance, natural history does take in individuals; such I mean of which there is not a body, or nation as it may be called. For of the sun, moon, earth, and the like, which are unique in their species, it is very right that separate histories should be written; nor less of such things as notably deviate from their species, and are prodigies; since in their case a description and knowledge of the species itself is neither sufficient nor competent. These two kinds of individuals therefore natural history does not reject; but for the most part (as has been said) it is concerned with species. But I will make the division of natural history according to the force and condition of nature itself; which is found in three states, and subject as it were to three kinds of regimen. For nature is either free, and allowed to go her own way and develop herself in her ordinary course; that is when she works by herself, without being any way obstructed or wrought upon; as in the heavens, in animals, in plants, and in the whole array of nature;—or again she is forced and driven quite out of her course by the perversities and insubordination of wayward and rebellious matter, and by the violence of impediments; as in monsters and heteroclites of nature;—or lastly, she is constrained, moulded, translated, and made as it were new by art and the hand of man; as in things artificial. For in things artificial nature seems as it were made, whereby a new array of bodies presents itself, and a kind of second world. Natural history therefore treats either of the *liberty* of nature or her *errors* or her *bonds*. And if any one dislike that arts should be called the bonds of nature, thinking they should rather be counted as her deliverers and champions, because in some cases they enable her to fulfil her own intention by reducing obstacles to order; for my part I do not care about these refinements and elegancies of speech; all I mean is, that nature, like Proteus, is forced by art to do that which without art would not be done; call it which you will,—force and bonds, or help and perfection. I will therefore divide natural history into history of generations, history of preter-generations, and history of arts; which I also call mechanical and experimental history. And I am the rather induced to set down the history of arts as a species of natural history, because it is the fashion to talk as if art were something different from nature, so that things artificial should be separated from things natural, as differing totally in kind; whence it comes that most writers of natural history think it enough to make a history of animals or plants or minerals, without mentioning the experiments of mechanical arts (which are far the most important for philosophy); and not only that, but another and more subtle error finds its way into men's minds; that of looking upon art merely as a kind of supplement to nature; which has power enough to finish what nature has begun or correct her when going aside, but no power to make radical changes, and shake her in the foundations; an opinion which has brought a great deal of despair into human concerns. Whereas men ought on the contrary to have a settled conviction, that things artificial differ from things natural, not in form or essence, but only in the efficient; that man has in truth no power over nature, except that of motion—the power, I say of putting natural bodies together or separating them—and that the rest is done by nature working within. Whenever therefore there is a possibility of moving natural bodies towards one another or away from one another, man and art can do everything; when there is no such possibility, they can do nothing. On the other hand, provided this motion to or from, which is required to produce any effect, be duly given, it matters not whether it be done by art and human means, or by nature unaided by man; nor is the one more powerful than the other. As for instance when a man makes the appearance of a rainbow on a wall by the sprinkling of water, nature does the work for him, just as much as when the same effect is produced in the air by a dripping cloud; and on the other hand when gold is found pure in sands, nature does the work for herself just as much as if it were refined by the furnace and human appliance. Sometimes again the ministering office is by the law of the universe deputed to other animals; for honey, which is made by the industry of the bee, is no less artificial than sugar, which is made by man; and in manna (which is a thing of like kind) nature asks no help, but does all herself. Therefore as nature is one and the same, and her power extends through all things, nor does she ever forsake herself, these three things should by all means be set down as alike subordinate only to nature; namely, the course of nature; the wandering of nature; and art, or nature with man to help. And therefore in natural history all these things should be included in one continuous series of narratives; as indeed Pliny has in great part done; who conceived an idea of natural history suitable to its dignity, but handled it in a manner most unworthy of the conception. Let this then be the first division of natural history.

CHAPTER III.

Division of Natural History according to its use and end; and that by far the noblest end of Natural History is to lay a foundation for Philosophy; and that such a history (a history framed with a view to that end) is wanting.

NATURAL history, which in subject (as I said) is threefold, is in use twofold. For it is used either for the sake of the knowledge of the things themselves which are committed to it, or as the primary material of philosophy. Now the noblest end of natural history is this; to be the stuff and matter of true and lawful induction; and to draw from the sense enough to inform the intellect. For that other kind which aims either to please by the agreeableness of the narrative, or to help by the use of experiments, and is pursued for the sake of such pleasure or such profit, is an inferior thing, and in its very kind of less value, than that which is qualified to be a proper preparative for the building up of philosophy. For this is that natural history which constitutes a solid and eternal basis of true and active philosophy; this it is which gives the first spark to the pure and real light of nature; and whose genius being neglected and not propitiated, has caused us to be visited most unhappily by that host of spectres and kingdom of shadows which we see flitting about among the philosophies, afflicting them with utter barrenness in respect of works. Now I affirm and bear witness that a natural history properly adapted to this end is not extant, but is wanting, and should be set down among the deficients. And let no man be so dazzled either by the great names of ancient writers or the great volumes of modern, as to think this complaint of mine unjust. I know well that a natural history is extant, large in bulk, pleasing in variety, curious often in diligence; and yet strip it of fables, antiquities, quotations and opinions of authors, empty disputes and controversies, philology and ornaments (which are more fitted for table-talk and the *noctes* of learned men than for the institution of philosophy), and it will shrink into small compass; so that it would seem as if people were engaged in getting up a treasure-house of eloquence, rather than a sound and faithful narrative of facts. Besides, it is not of much use to recount or to know the exact varieties of flowers, as of the iris or tulip, no, nor of shells or dogs or hawks. For these and the like are but sports and wanton freaks of nature, and almost approach to the nature of individuals. And though they involve an exquisite knowledge of the particular objects, the information which they afford to the sciences is slight and almost useless. And yet these are the things which our ordinary natural history takes pride in. And while it descends to matters which do not belong to it, and indulges to excess in matters superfluous, on the other hand its great and solid parts are either entirely omitted or carelessly and lightly treated. And indeed in the whole course of inquiry pursued and the whole mass of matter gathered, it appears to be in no way adapted or qualified for the end which I have mentioned, namely the building up of philosophy. This will be best shown in the particular branches of it, and by comparing the history of which I am now going to set forth a description, with that which we have.

CHAPTER IV.

Beginning of a treatise showing of what nature the required history should be ; namely the Natural History which is to serve as a foundation of Philosophy. *For the clearer explanation of this, a division of the* History of Generations *is first subjoined. This is digested into five parts. The first the* History of the Heavens ; *the second, the* History of Meteors ; *the third, the* History of Earth *and* Sea ; *the fourth, the* History of Collegia Majora, *or* Elements *or* Masses ; *the fifth, the* History of Collegia Minora, *or* Species. *The history of* Primary Virtues *is postponed, till the explanation of this first division, of* Generations, Preter-generations, *and* Arts, *is concluded.*

ALTHOUGH I consider myself bound not to leave the completion of this history which I pronounce deficient to others, but to take it upon myself ; because the more it may seem a thing open to every man's industry, the greater fear there is that they will go astray from my design ; and I have therefore marked it out as the third part of my instauration ; yet that I may still keep true to my plan of giving either explanations or specimens of those things which are wanting, and likewise that in case of my death there may be something saved, I think fit now in this place to set down my opinion and advice in this matter. Of the History of Generations or Nature at large I set down five parts. These are the History of Ether ; the History of Meteors and of the Regions of the Air, as they are called ; for the sublunar region down to the surface of the earth, and the bodies situated upon it, I assign to the history of meteors ; Comets likewise of all kinds (however the truth may be) yet for the sake of order I include among meteors. Third comes the History of the Earth and Sea, which together make up one globe. And so far the nature of things is distributed according to places and positions. The two remaining parts distinguish the substances of things or rather masses. For connatural bodies are congregated into greater and lesser masses ; which I commonly term greater and lesser Colleges, and which are related to one another in the polity of the world as tribes or families. Therefore fourth in order is placed the History of Elements or the Greater Colleges ; fifth and last, the History of Species or the Lesser Colleges. For I mean by Elements not the commencements of things, but only the greater masses of connatural bodies. Now this greatness of mass is owing to the texture of the matter of which they are composed being easy, simple, obvious, and prepared ; whereas species are sparingly supplied by nature, because the texture of matter is complex, and in most cases organic. As for those virtues which may be regarded as cardinal and universal in nature, as Dense, Rare, Light, Heavy, Hot, Cold, Consistent, Fluid, Similar, Dissimilar, Specific, Organic, and the like, together with the motions contributing to them, as Resistance, Connexion, Contraction, Expansion, and the rest (the history of which I would by all means have collected and constructed, even before we come to the work of the intellect), I will treat of the history of these and of the manner of constructing it, when I have completed the explanation of this triple division, of Generations, Preter-generations, and Arts. For I have not included it in that threefold division, because it is not properly a history, but as it were a middle term between history and philosophy. But now I will speak of the History of the Celestial Bodies, and give precepts concerning them, and then of the rest.

CHAPTER V.

The history of Celestial Bodies *is resumed ; showing both what it should be in kind, and that the legitimate ordering of such a history turns on three kinds of precepts ; namely, the End, the Matter, and the Manner of Construction.*

I WOULD have the History of Celestial Bodies simple, and without any infusion of dogmas ; all theoretical doctrine being as it were suspended : a history embracing only the phenomena themselves (now almost incorporated with the dogmas) pure and separate ; a history in short, setting forth a simple narrative of the facts, just as if nothing had been settled by the arts of astronomy and astrology, and only experiments and observations had been accurately collected and described with perspicuity. In which kind of history there is nothing extant which satisfies me. Something of the kind indeed Pliny has touched on cursorily and loosely ; but that would be the best history of the celestial bodies which might be extracted and worked out from Ptolemæus and Copernicus and the more learned writers on astronomy, taking the experiments detached from the art, and adding the observations of more modern writers. It may seem strange that I should wish to recall to their primitive rudeness and the simplicity of naked observations things so laboriously produced, advanced, and amended. But the truth is that, without meaning to throw away the benefit of former inventions, I am attempting a far greater work : for it is not merely calculations or predictions that I aim at, but philosophy : such a philosophy I mean as may inform the human understanding, not only of the motion of the heavenly bodies and the period of that motion, but likewise of their substance, various qualities, powers, and influences, according to natural and certain reasons, free from the superstition and frivolity of traditions ; and again such as may discover and explain in the motion itself, not what is accordant with the phenomena, but what is found in nature herself, and is actually and really true. Now it is easy to see, that both they who think the earth revolves, and they who hold the *primum mobile* and the old construction, are about equally and indifferently supported by the phenomena. Nay, and the author of the new construction in our own day, who made the sun the centre of the *secundum mobile*, as the earth of the *primum mobile*, whereby the planets in their proper revolutions would seem to wheel in dance round the sun (as some of the ancients suspected to be the case with Venus and Mercury), if he had thought the matter fairly out, might probably have brought it to a very good conclusion [1]. Nor have I any doubt but that other similar constructions might by wit and severe thought be invented. Neither indeed do they who propose these theories mean to say that the things they allege are actually true, but only that they are convenient hypotheses for calculations and the construction of tables. But my plan has a different aim ; for I seek not for ingenious adjustments, which may be many, but for the truth of the thing, which is simple. And to this a history of phenomena kept pure and simple will open the way, while one tinctured with dogma will obstruct it. I may say also, that as I hope for the discovery of the truth regarding the heavenly bodies from a history made and compiled according to my principle, by itself alone ; so I rest that hope much more upon observation of the common passions and desires of matter in both globes. For those supposed divorces between ethereal and sublunary things seem to me but figments, superstitions mixed with rashness ; seeing it is most certain that very many effects, as of expansion, contraction, impression, cession, collection into masses, attraction, repulsion, assimilation, union, and the like, have place not only here with us, but also in the heights of the heaven and the depths of the earth. Nor have we any more faithful interpreters to consult in order that the human understanding may penetrate the depths of the earth, which are never seen at all, and the heights

[1] The reference is to Tycho Brahe, and by *nonnulli ex antiquis* Bacon probably meant Martianus Capella and Vitruvius.

of heaven which are for the most part seen untruly. Most excellently therefore did the ancients represent Proteus, him of the many shapes, to be likewise a prophet triply great; as knowing the future, the past, and the secrets of the present. For he who knows the universal passions of matter and thereby knows what is possible to be, cannot help knowing likewise what has been, what is, and what will be, according to the sums of things. Therefore the best hope and security for the study of celestial bodies I place in physical reasons; meaning by physical reasons not such as are commonly supposed, but only the doctrine concerning those appetites of matter which no diversity of regions or places can distract or dissever. Not that on this account (to return to my design) I would have any diligence spared in descriptions and observations of the celestial phenomena themselves. For the fuller our supply of such appearances, the readier and surer will everything be. But before I speak more of this, I have to congratulate both the industry of mechanics, and the zeal and energy of certain learned men, that now of late by the help of optical instruments, as by skiffs and barks, they have opened a new commerce with the phenomena of the heavens; an undertaking which I regard as being both in the end and in the endeavour a thing noble and worthy of the human race; the rather because these men are as much to be praised for their honesty as for their boldness; seeing that they have ingenuously and perspicuously explained the manner in which each point of their proceeding in each case has been made out. All that is wanted further is constancy and great severity of judgment, to change the instruments, to increase the number of witnesses, to try each particular experiment many times and many ways; lastly, to suggest to themselves and open to others every objection that can be made, not despising even the minutest scruple; lest it fare with them as with Democritus in the matter of the sweet figs, when it turned out that the old woman was wiser than the philosopher, and that a vast and wonderful speculation was built upon a trifling and ridiculous mistake. But now having made these general remarks by way of preface, let us go on to a description of the history of celestial bodies more at large, to show what and what kind of things are to be sought concerning them. First, therefore, I will set down the questions in nature, at least some of them, and those the chief; to these I will add the uses which may probably be derived to man from the study of celestial bodies; both of these as being the mark at which the history aims; that they who undertake to compose a history of the heavens may know what we are about, and may keep these questions, together with these operations and effects, in mind and view; and so proceed to form such a history as shall be adapted to the solution of the said questions, and the procuring of such fruits and benefits to the human race. Now the questions I mean are of that kind which inquire of the fact in nature, not of causes. For this is the proper business of history. Next, I will show distinctly in what the history of celestial bodies consists, and what are its parts; what things are to be understood or inquired, what experiments to be collected and procured, what observations to be employed and sifted; propounding as it were certain Inductive Topics, or Articles of Interrogation concerning the heavens. Lastly, I will give some precepts, not only concerning that which should be sought, but also how the matters under inquiry are to be examined and how presented and put in writing; that the diligence of the first inquiry may not be lost in passing it on, nor (what is worse) the beginning of the work, on which the subsequent progress depends, prove weak and fallacious. In short I will explain both what should be inquired with regard to the heavenly bodies, and with what view, and in what manner.

GHAPTER VI.

That philosophical questions concerning the Celestial Bodies, *even such as are contrary to opinion, and somewhat harsh, should be received. Five questions are propounded concerning the system itself; namely,* is there a system? *if there be,* what is the centre of it, what the depth, what the connexion, and what the position of the parts?

Most men no doubt will think that I am digging up the remains of old questions long since laid up and buried, and in a manner raising their ghosts, and mixing fresh questions with them. But since the philosophy of which we are hitherto in possession concerning the heavens has no soundness; and since it is my constant determination to refer everything to a new trial by legitimate induction; and since if any questions are passed over, there will be so much less pains and diligence bestowed on the history, because it will perhaps seem superfluous to inquire of things concerning which no question has been raised; I hold it necessary to take in hand all questions which the nature of things anywhere presents. Nay, the less certain I am concerning the questions which are to be determined by my method, the less difficulty do I make in entertaining them. For I see an end of the matter. The first question therefore is, *whether there be a system?* that is, whether the world or universe compose altogether one globe, with a centre; or whether the particular globes of earth and stars be scattered dispersedly each on its own roots, without any system or common centre? Certainly the school of Democritus and Epicurus boasted that their founders had overthrown the walls of the world; yet this did not absolutely follow from their words. For when Democritus had set down matter or seeds as infinite in quantity and finite in attributes and power, as moving about, and never located in any position from all eternity, he was driven by the very force of this opinion to constitute multiform worlds, subject to birth and death, some well ordered, others badly put together, even essays of worlds and vacant spaces between. But yet though this were admitted, there was no reason why that part of matter which is assigned to this particular world which is visible to us, should not have the shape of a globe. For each one of those worlds must have received some shape; and although there can be no middle point in infinity, yet in the parts of infinity a round figure may exist, no less in a world than in a ball. Now Democritus was a good dissector of the world, but in the integral parts of the world inferior even to the ordinary philosophers. But the opinion of which I am now speaking, which destroyed and confounded system, was that of Heraclides Ponticus, Ecphantus, and Nicetas of Syracuse, and most of all Philolaus, and likewise, in our own day, of Gilbert, and all those (except Copernicus) who believed that the earth was a planet and movable, and as it were one of the stars[1]. And the effect of this opinion is that the several planets and stars, together with innumerable other stars which elude our sight by reason of their distance, and others again which are invisible to us from their nature being not lucent but opaque, having each of them obtained their own globes and primary forms, are scattered and suspended through that immense expanse which we behold above us, whether it be of vacuum or some thin and almost indifferent body, like so many islands in an immense sea, and revolve not round any common centre, but each separately round its own; some simply, others with some progressive motion of the

[1] All the persons here mentioned affirmed that the earth moved, but their opinions are not accurately represented. Thus Ecphantus and Heraclides denied that the earth changes its place. According to them it moves, but οὐ μήν γε μεταβατικῶς (Plutarch, *De Placit. Philos.* iii. 13): and with respect to Ecphantus we are expressly told by the pseudo-Origen, *Philos.* c. 15, that he affirmed τὴν γῆν μέσον κόσμου κινεῖσθαι περὶ τὸ αὑτῆς κέντρον, ὡς πρὸς ἀνατολήν; so far was he from rejecting the notion of a κόσμος or system. Philolaus undoubtedly admitted the motion of the earth through space, and so probably did Niaetas, or rather Hicetas; but neither of them rejected the notion of a system. For Philolaus, see Boeckh's *Philolaus* and the second dissertation *De Platonico Systemate.* The Philolaic system (although Martin appears to doubt it) was probably the same as that of the Pythagoreans in general. According to it, neither the earth nor the sun is at rest, but both, with the planets, revolve about a central fire, the light from which is reflected to us from the sun. It never reaches us directly, because between us and it revolves the Antichthon, which is either a separate planet, or simply the other side of the earth, for the point does not seem quite settled. [See Berry, *Short Hist. of Astron.*, p. 25, for a solution.—*Ed.*] The passage in the text is apparently taken from Gilbert, *De Magnete*, vi. 3. Heraclides, though he did not believe in the earth's moving through space, yet affirmed, as did also the Pythagoreans, that each of the heavenly bodies constitutes a κόσμος in itself. See Stobæus, *Ec. Phys.* i. 25. On the other hand, Philolaus and Ecphantus distinctly asserted the unity of the universe. See Stobæus, i. 16, 23.

centre. Now the harshest thing in this opinion is, that they take away quiet or immobility from nature [2]. But it seems that as there are bodies in the universe which revolve, that is, which move with an infinite and perpetual motion, so on the other hand there should be some body which is at rest; between which comes a middle nature, of such as move in a straight line; seeing that motion in a straight line suits the parts of globes, and things banished from their native countries, which move towards the globes of their connaturals, that being united with them they may themselves also either revolve or rest. But this question (namely, *whether there be a system*) will be answered by that which shall be determined concerning the motion of the earth, that is, *whether the earth stands still or revolves*, and the substance of the stars, *whether they are solid or flamy*, and the ether or interstellar spaces in the heaven, *whether they consist of body or vacuum*. For if the earth be stationary and the heavens revolve in a diurnal motion, there is doubtless a system; but if the earth revolve, it does not necessarily follow that there is no system; because there may be some other centre of the system; the sun, for instance, or something else. Again, if the globe of the earth be the only one dense and solid, it would seem that the matter of the universe is collected and condensed to that centre; but if it be found that the moon or some of the planets consist likewise of dense and solid matter, it would seem that dense bodies collect not to any one centre, but dispersedly, and as it were fortuitously. Lastly, if it be asserted that there is a collective vacuum in the interstellar spaces it would seem that each globe has round it an emanation of rarer substance, and beyond that a vacuum [3]. But if these spaces be filled with body, it would seem that there is a union of dense things in the middle, and a repulsion of rarer things to the circumference. Now it is of great importance to science to know the conjugations of questions; because in some cases there is history or inductive matter by which they may be settled, in others not so. But granting that there is a system, we come next to the second question, *what is the centre of that system?* For if any one of the globes is to occupy the position of centre, there are two especially, which offer themselves as having the nature of a middle or centre; namely, the earth and the sun. In favour of the earth, we have the evidence of our sight, and an inveterate opinion; and most of all this, that as dense bodies are contracted into a narrow compass, and rare bodies are widely diffused (and the area of every circle is contracted to the centre), it seems to follow almost of necessity that the narrow space about the middle of the world be set down as the proper and peculiar place for dense bodies. In favour of the sun, on the other hand, we have this consideration, that that body which has the chief office in the system should occupy that place from which it may best act on the whole system and communicate its influence. And since the sun is that which seems most to vivify the world by imparting heat and light, it appears to be altogether right and in order that it should be placed in the middle of the world. Besides, the sun manifestly has Venus and Mercury as his satellites [4], and in the opinion

[2] Yet Bacon would have found, by referring to Cicero, that Nicetas at least denied that any part of the universe except the earth is in motion.

[3] Compare Gilbert, *Physiol.* ii. 27.

[4] It is difficult to see why Bacon should speak of this as manifest; the theory that Mercury and Venus are satellites of the sun constitutes a distinct system, often called the Egyptian. See with respect to it Martin, *Etudes*, etc., vol. ii. p. 129. According to Gassendi, Copernicus was much struck by the passage of Martianus Capella in which this system is mentioned. Apelt has remarked that the Copernican system includes two distinct elements: the first the reference of the motion of the planets to the sun as a common centre; the second the doctrine of the motion of the earth. The first was common to Copernicus with Tycho Brahe; the second was his own exclusively. Tycho's system, as Apelt well observes, is the natural transition from Ptolemy's to the Copernican, and must of necessity have been arrived at as soon as the true distances between the sun and the planets were introduced into the Ptolemaic hypothesis. Thus Tycho's system is a step backwards, although it saved the phenomena as well as that of Copernicus; but, as Apelt goes on to remark, Tycho was an observer, and Copernicus a philosopher who sought not merely for an astronomical hypothesis, but for a new idea of the universe.

of Tycho the other planets also; whence it is plain that the sun can sustain the nature of a centre, and perform its office in some things, and so has the better title to be constituted the centre of the universe; as was asserted by Copernicus. Nevertheless, in the system of Copernicus there are found many and great inconveniences; for both the loading of the earth with a triple motion is very incommodious[5], and the separation of the sun from the company of the planets, with which it has so many passions in common, is likewise a difficulty, and the introduction of so much immobility into nature, by representing the sun and stars as immovable, especially being of all bodies the highest and most radiant, and making the moon revolve about the earth in an epicycle, and some other assumptions of his, are the speculations of one who cares not what fictions he introduces into nature, provided his calculations answer. But if it be granted that the earth moves, it would seem more natural to suppose that there is no system at all, but scattered globes, according to the opinion of those I have already mentioned, than to constitute a system in which the sun is the centre. And this the consent of ages and of antiquity has rather embraced and approved. For the opinion concerning the motion of the earth is not new, but revived from the ancients, as I said; whereas the opinion that the sun is the centre of the world and immovable is altogether new (except one verse, wrongly translated[6]),

Copernicus says of himself, that he had set the sun, the great light of the universe, in the midst of the temple of nature, and as on a kingly throne. No man less deserved to be spoken of as a merely calculating astronomer. Bacon's difficulty, that in the Copernican system the moon revolves about the earth, had been felt by others. Galileo, at the end of the *Sydereus Nuncius*, points out the analogy of this hypothesis with what he had discovered to be the case with respect to Jupiter and his satellites, remarking that it removed the difficulty in question.

[5] Copernicus conceived the earth's motion round the sun to be as if the earth were rigidly attached to the line which joins them. Thus the motion round the sun results from the composition of two simpler motions, namely that of the earth's centre and the change of the direction of its axis. The second of these components is eliminated from the hypothesis by supposing that the earth, besides the motion round the sun and about its own axis, has a third motion, namely a change in the direction of its axis equal and opposite to that which results from the motion round the sun. Galileo showed, by an illustrative experiment, that this kind of motion was in reality only an unnecessary complication; and Gilbert also makes the same remark. See the *Saggiatore*, ii. 304, and the *Physiol. Nova*. In Germany the same thing was remarked by Rothman; but I am not aware whether he or Gilbert was the first person to introduce the simplification, which is indeed obvious. Nevertheless the notion of a triple motion long adhered to the Copernican hypothesis. See *Paradise Lost*, viii. 130. Of course the earth's axis really has a third motion which gives rise to the phenomena of precession and mutation; but this is exceedingly slow. In justice to Copernicus it should be added, that though his notion of an annual third motion was unnecessary, yet he employed it, and in a correct manner, to explain precession. Boeckh's notion that the movement of the fixed stars in the theory of Philolaus was introduced for the same purpose, does not seem to be well made out. No doubt, as the earth revolved every day round Hestia, the fixed stars might have been allowed to remain at rest; but we have a remarkable example of a similar pleonasm in the astronomical theory of Cardinal Cusanus. See the fragment of Cusanus first published by Clemens in 1843: it is given *in extenso* in Apelt *On the Reformation of Astronomy*, p. 23.

[6] Bacon alludes to Job, ix. 6. On this verse, "Qui commovet terram de loco suo et columnæ ejus concutiuntur," Didacus à Stunicâ, in his Commentary on Job, published in 1584, founded an argument in favour of the Copernican hypothesis, alleging that no text could be found in which the earth's motion is as distinctly denied as it is here asserted:—"Nullus dabitur scripturæ sacrosanctæ locus qui tam aperte dicat terram non moveri quam hic moveri dicit".—*Stunica on Job*, p. 41. (I quote from the edition of 1591). This argument of Stunica's seems to have attracted some attention. Galileo mentions it in his letter to the Grand Duchess Christina, which was written about 1615. See the new edition of his works (Florence, 1843), ii. p. 52. The passage of Stunica's Commentary in which it occurs is inserted in Salisbury's *Mathematical Collections and*

and was first introduced by Copernicus. Then comes the third question, concerning the depth of the system; not with a view to find its exact measure, but to ascertain *whether the starry heaven be like one region, or orb, as it is commonly called; or whether of the fixed stars, as they call them, some are higher than others, with an immeasurable depth between?* For it cannot be that they are of equal height, if the words be taken exactly; since the stars are certainly not situated as in a plain, so as to have a superficial dimension only, like spots or bubbles, but they are entire globes, great and deep; and being of such different magnitudes, it must needs be that some protrude more than others either upwards or downwards, nor is it possible for them to be united in one surface, either above or below. And if this be the case in the parts of stars, it would plainly be rash to assert that there are not some stars higher than others in their whole body. But though this be true, it may nevertheless be maintained that the width of that region which they call the sphere or starry heaven, though great, is definite; and that within this those prominences and degrees of altitude are in a manner limited; for we see from the apogees and perigees of the planets that every one of their heavens through which they ascend and descend has an observable width. But the question proposed relates only to this,—whether some stars are above others, like planet above planet, and as it were in different orbs. And this question is in like manner related to that other concerning the motion or fixedness of the earth. For if the stars move in a diurnal motion round the earth, since they all move with the same velocity, and as it were with one spirit (and since it is very evident in planets that as they vary in height and lowness of position, so they vary in quickness and slowness of motion), it is probable that the stars, being equal in velocity, are situated likewise in one region of the ether, the width or profundity of which, although it be great, yet is not so great as to make a difference in the velocity or quickness of motion; but so that throughout the whole of that region everything being united together by a kind of bond of connaturality revolves equally, or at least with so little difference that at this distance it is not visible to the sight. But if the earth moves, the stars may either be stationary, as Copernicus thought, or, as is far more probable, and has been suggested by Gilbert, they may revolve each round its own centre in its own place, without any motion of its centre, as the earth itself does; if only you separate that diurnal motion of the earth from those two supposititious motions which Copernicus superadded. But either way, there is no reason why there

Translations (1661), which contains, beside a translation of the *Dialogi dei Sistemi*, translations of certain tracts on the religious question involved in the Copernican controversy. —I am not sure, though *versiculus* is an odd expression for anything except a verse in the Bible, that M. Bouillet is not right in thinking that the reference is to what is said of Philolaus by Plutarch, *De Placit. Phil.*, which (as Apelt has remarked) Copernicus has always mistranslated, confounding the central fire, the seat of the gods, with the sun: See Apelt's *Reformation of Astronomy*, 1852, p. 128.

Apelt altogether agrees with Bacon as to the complete originality of Copernicus, and, apparently forgetting what is said of Aristarchus by Archimedes and others, says that the idea of the annual movement of the earth sprang out of the mind of Copernicus, as Minerva from the head of Jupiter. But yet, as Humboldt has remarked, he may have been acquainted with the doctrine of Aristarchus. See *Cosmos*, vol. ii. p. 349. Bacon was not, or he would not have said that the immobility of the sun was, "excepting one versicle," a wholly new doctrine.

A third hypothesis as to Bacon's meaning is that he refers to some passage in which the sun is spoken of as τὸ μέσον, as Boeckh has pointed out in his *Commendatio Academica altera de Platonico Systemate*, etc. The sun is sometimes called μέση on musical grounds and sometimes spoken of as μέσον πάντων, simply because it occupies a middle place among the planets. Such a passage occurs in the *Placit. Phil.*; and it is perhaps to this that Bacon refers. Compare Martin, *Études sur le Timée*, vol. ii. pp. 103. and 128. I have not seen Gruppe's recent work on the Cosmical System of the Ancients. The notion that Plato was the first proposer of the Copernican system seems altogether unfounded. According to Apelt, Gruppe relies on a passage in the seventh book of the *De Legibus*.

should not be stars above stars till they go beyond our sight. The fourth question is *concerning the connexion of the system.* Now of the nature and essence of the body or thing which is regarded as pure ether, and occupies the space between the stars, I will inquire afterwards. At present I will only speak of the coherence of the system. This may be in three ways. For there is either vacuum, or contiguity, or continuity; therefore we must first inquire, *whether there be a collective vacuum in the interstellar spaces?* a thing which Gilbert distinctly affirmed [7], and which likewise some of those among the ancients who thought that the globes were dispersed without any system, seem to intimate; especially those who asserted that the bodies of the stars are compact. The opinion is this: that all the globes, as well the stars as the earth, consist of solid and dense matter: that these are immediately surrounded by a kind of bodies which are to a certain extent connatural with the globe itself, but yet more imperfect, languid, and attenuated; and are in fact nothing else than the effluvia and emanations of the globes themselves; such as vapours are, and exhalations, and indeed the air itself, when compared with the earth: that these emanations do not extend for any great distance round each globe; and that the remaining space (which is far the most extensive) is empty. Which opinion is countenanced by the fact that the bodies of the stars are seen at such an immense distance. For if all that space were filled, especially with bodies which are doubtless very unequal in density and rarity, the refraction of rays would be so great that they would not reach our sight; whereas if far the greatest part of that space be a vacuum, it is natural to suppose that they traverse it more easily. And indeed this question will in great part depend on the question which I shall next bring forward concerning the substance of the stars, *whether it be dense, or rare and open.* For if their substance be solid, it will seem as if nature were only busy and anxious about the globes and their immediate neighbourhood; and that she leaves and passes by, as it were, the intermediate spaces. Therefore it would not be improbable that the globes are denser about the centre, more open towards the circumference, in the surrounding atmosphere and effluvia almost exhausted, and so terminated at last in vacuum. On the other hand, if the nature of the stars be rare and flamy, it will appear that the nature of rarity is not merely the diminution of density, but powerful and primary of itself, no less than the nature of solidity; and that it abounds both in the stars themselves, and in the ether, and in the air, so that there is no need of a collective vacuum. This question concerning a vacuum in the interstellar spaces will depend likewise on that question which relates to the principles of nature; *Does nature admit a vacuum?* Not however on this absolutely, without proper distinction. For it is one thing to deny a vacuum absolutely, another to deny a collective vacuum. For the reasons which may be advanced in favour of a vacuum interspersed, whereby bodies are relaxed and opened, are far stronger than those on which the assertion of a collective vacuum, that is, a vacuum extending over great spaces, is supported. And it was not Hero alone, a man of wit and a mechanician, who saw this [8], but Leucippus likewise and Democritus, the founders of the opinion concerning the vacuum, which Aristotle endeavours by certain fine reasons to attack and destroy; which two philosophers, certainly most acute and famous men, in admitting an interspersed vacuum, do in fact deny a collective one. For in the opinion of Democritus vacuity is bounded and circumscribed, so that beyond certain limits distraction or divulsion of bodies is no more possible than compulsion or compaction [9]. For although in those works of Democritus which have come down to us this is never expressly declared, yet he seems to imply as much when he asserts that bodies as well as spaces are infinite: using as his argument, that otherwise (that is, if space were infinite and bodies finite) bodies would never cohere. Therefore by reason of matter and space being equally infinite, vacuity is necessarily confined within certain bounds, which seems to have been his real opinion rightly understood; that is, that there is a certain limit to the expansion of bodies by reason of the vacuum with which they are

[7] Gilbert, *Physiol. Nova*, i. 22.

[8] See Hero, *Spiritalia*, proëm.

[9] Cf. Lucretius, i. 983.

coupled; and that there is no solitary vacuum, not enclosed in a body. But if there be no vacuum amounting to a solution of continuity in the system, yet as there is found so great a diversity of bodies in the parts and regions of the system that they seem to belong as it were to different nations and countries, there arises a second question, which relates to the connexion of the system, this is, *whether the pure ether be one perpetual and continuous fluid, or consist of many contiguous to one another?* Now it is not for me to refine about words, but by a contiguous body I understand a body which lies on another without mixing with it. I do not mean however a series of hard rigid floors, like the stories of a house, such as the vulgar astronomers imagine, but such a succession, as fluids admit of, as when water floats on quicksilver, oil on water, air on oil. For no one can doubt but that in that immense tract of pure ether there are wonderful differences as to density and rarity and many other things; but upon either supposition (that is, whether you assume continuity or contiguity) this may be the case. For it is certain that even in the sea the water at the top and the water at the bottom are not of the same consistency and taste; while in the air, there is a very great difference between the air contiguous to the earth and the upper air, and yet the fluidity is one and entire and uninterrupted. The question therefore is brought to this, *whether the differences in the tract of pure air insinuate themselves gradually and with a continuous flow; or whether they take place at certain distinguishable limits, where the bodies which cannot mix are joined to each other; as with us air lies on water.* For indeed to a simple observer the whole of that pure and clear body in which the globes of the earth and stars float and hang as in an immense sea, and which is infinitely greater both in quantity and the space which it occupies than the globes between which it is interposed, seems to be a thing undivided and completely united. But to a deeper searcher of nature it will plainly appear that nature is accustomed to proceed for some distance by degrees, and then suddenly by jumps, and to take these processes in turn. Otherwise, if a man examine it well, no structure of things or organic figure could be formed, if the proceeding were always by imperceptible degrees. Therefore this gradual progress may do for the spaces between the worlds, but not for the world, the construction of which requires that things very dissimilar be separated one from the other, and yet brought into approximation. Thus air succeeds to and touches earth and water, a body very different from them, and yet placed in immediate proximity; not first mud, then vapour or mist, and then pure air; but air at once, without any thing between. But in air and ether (for I put the two together) the most remarkable and radical division of all may be derived from a greater or less susceptibility of the starry nature. Between the globe of the earth then and the summits of heaven there seem to be generally three regions especially remarkable; namely, the tract of the air, the tract of the planetary heaven, and the tract of the starry heaven. Now in the lowest of these tracts, the starry nature is not consistent; in the middle it is consistent, but gathers into separate globes; in the highest it diffuses itself among a great number of globes, till at the summits thereof it seems to pass as it were into the perfect empyrean. But in the meantime I must not forget what I said just now, that nature is accustomed to adopt the gradual and the sudden process by turns, so that the confines of the first region communicate with the second, and the second with the third. For both in the higher air, when the air has begun to be cleared from the emanations of the earth and to be more rarefied by the emanations of the heavens, flame tries and endeavours to be consistent; as we see in the lower comets, which are of a middle nature between the starry nature in consistence and in evanescence; and again in the neighbourhood of the sun (it may be) where the heaven seems to become starry, and to begin to pass into the nature of the starry heaven. For it may be that those spots which have been discovered in the sun, certainly by faithful and diligent observation, are a kind of rudiments of starry matter; whereas in the heaven of Jupiter absolute and perfect stars are discernible, though too small to be seen without the aid of telescopes; and again in the summits of the starry heaven it seems from the innumerable sparklings of the ether between the numbered stars (for which other causes bald enough are usually given) that the starry nature is more diffused and continuous[10]. Of these things however I will speak further in the questions which I shall presently propose about the substance of the stars and the interstellar heaven. For the things which I have just said relate only to connexion of system. There remains the fifth question, *concerning the collocation of the parts of the system, or the order of the heavens.* And whether it be assumed that there is no system, but that the globes are scattered, or that there is a system, of which the sun is the centre ; or even though astronomers look for some new system ; yet there still remains the inquiry, *which planet is nearer to another planet, or further off ;* and in like manner *which planet is more or less elongated from the earth or from the sun.* Now if the ancient system be received, there seems to be no reason why we should insist much upon a new inquiry concerning the four superior heavens, namely, the heavens of the fixed stars, of Saturn, of Jupiter, and of Mars. For with respect to their position and order the consent of ages is agreed, and there is no contrary phenomenon ; the calculations of their motions also (whence is derived the chief proof of the heights of the heavens) are agreeable, and present no difficulty. But with regard to the Sun, Venus, Mercury, and the Moon, according to the old system, the ancients were in doubt [11] ; and among the moderns also there is a question with regard to Venus and Mercury which of them is superior. For in favour of Venus being superior, there is the reason that she moves somewhat slower ; and in favour of Mercury, that he is fixed at a less distance from the sun, whence one might assert that he ought to be placed next to the sun. But with regard to the moon, no one has ever doubted that she is placed nearest to the earth, though there are various opinions about her approximation to the sun. Nor should any one who is seriously considering the subject let another kind of question escape him, pertaining to the constitution of the system ; that is, *whether one planet sometimes goes above another and sometimes again comes below ;* a thing which seems to be proved with regard to Venus by some tolerably diligent demonstrations, that she is found sometimes above and sometimes below the sun. It is a very fit inquiry also, *whether the apogee of the lower planet does not cut the perigee of the higher and enter its boundaries.* There remains the last question, *concerning the position of the parts of the system ;* that is, *whether there be many different centres in the system, and as it were many dances ;* especially as not only the earth is set down as the centre of the *primum mobile,* and the sun (according to Tycho) of the *secundum mobile ;* but Jupiter likewise is supposed by Galileo to be the centre of those smaller and recently discovered wanderers. Such then are these five questions, which seem fit to be proposed concerning the system itself, namely, *is there a system ? what is the centre of it ? what the depth ? what the connexion ? and what the order of the position of the parts ?*

As for the extremities of the heaven and the empyrean, I do not draw up any propositions or questions concerning them. For there is no history of these things nor any phenomenon extant. And therefore what can be known about them

[10] See the *Sydereus Nuncius* of Galileo, which had just appeared when this tract was written ; and compare the following passage in the letter of Sir Wm. Lower to Harriot, written when he had first heard of Galileo's discoveries. "We Traventane philosophers were a consideringe of Kepler's reasons by which he indeavers to overthrow Nolanus' and Gilbert's opinions concerninge the immensitie of the spheare of the starres . . Said I (havinge heard you say often as much) what is [if ?] in that huge space betweene the starres and Saturne, ther remaine ever fixed infinite numbers which by reason of their lesser magnitudes doe flie our sighte . . . what if about ♄ ♃ ♂ etc., ther move other planets also which appear not. Just as I was saying this comes your letter, which when I had redd, Loe, quod I, what I spoke probablie, experience hath made good ". The name "Traventane" is taken from his house Traventi. It probably alludes to the title Cosentine philosophers, affected by the disciples of Telesius. Bruno affected to talk of the Nolan philosophy.

[11] It was doubted whether the orbs of Venus and Mercury are superior to the sun's or inferior to it. The former was the older hypothesis and is preferred by Ptolemy ; who however remarks that some had dissented from it. See the *Megal. Syntaxis*, iv. 1 Bacon's information is apparently derived from Patricius, *Pancosmia*, 13.

can only be known by consequence, and not at all by induction. For such inquiry however there will come a fit time, and a plan and method. But with regard to the immateriate heavens and spaces, we must rest entirely upon religion, and leave them to it. For as for what the Platonists and of late Patricius [12] (by way of giving their philosophy a diviner character) have alleged, not without superstition, arrogance, and some disorder of mind, and in a word, with too much presumption and no fruit, like the images and dreams of Valentinus [13]; I regard all such things as idle fancies. For an apotheosis of Folly, like that of the Emperor Claudius, is a thing not to be endured; and most mischievous it is, and a very pest and destruction of the understanding, for vanity to be made an object of veneration.

CHAPTER VII.

Then follow questions concerning the substance of heavenly bodies; namely, what is the substance of heavenly bodies generally as compared with sublunary bodies; what is the substance of the interstellar ether as compared with the body of a star; what is the substance of the stars as compared with one another, with our fire, and in their own nature; what is the substance of the Milky Way, and the black spots in the antarctic hemisphere? *Then is proposed the first question,* Is there a heterogeneity between celestial and sublunary bodies, and of what nature may it be?

HAVING finished the questions concerning the system, we must proceed to those concerning the substance of the heavenly bodies. For the inquiry concerning the substance of the heavenly bodies, and the causes of their motion, belongs principally to philosophy; the inquiry concerning the motion itself and the accidents thereof, to astronomy; the inquiry concerning their influence and power, to both. Now it ought to have been so arranged between astronomy and philosophy, that astronomy should prefer those hypotheses which are most convenient for compendious calculations; philosophy those which come nearest to the truth of nature. And further, that while the hypotheses adopted by astronomy for convenience should by no means prejudice the truth of the thing, the judgments of philosophy in their turn should be such as are perfectly reconcileable with the phenomena of astronomy. But now it comes to pass, contrariwise, that the fictions of astronomy have been introduced into philosophy and corrupted it; while the speculations of philosophers about the celestial bodies please none but themselves, and almost forsake astronomy, looking at the celestial regions in general, but not at all addressing themselves to particular phenomena and their causes. Therefore since both sciences (as now practised) are slight and superficial, we must plant our footing deeper; and treat these two, which by reason of the narrowness of men's views and the practice of professors have been for so many ages separated, as one and the same thing, and making up together one body of science. The first question proposed therefore is, *whether the substance of the heavenly bodies is different in kind from the substance of those below?* For Aristotle's temerity and cavilling has begotten for us a fantastic heaven, composed of a fifth essence, free from change, and free likewise from heat [1]. Now to say nothing at present about the four elements, which this fifth essence sup-

[12] Patricius, from whom much of the latter part of the present tract is taken, was born at Cherso in 1529, and died in 1597. He wrote a treatise on philosophy—*Nova de Universis Philosophia*—[which was published in 1591]. It is an attempt, of no great value, to conciliate Plato and Aristotle. In the last book, entitled *Pancosmia,* there is some interesting information touching theories of the tides.

[13] Valentinus is the alchemist Basil Valentine. He is said to have been a Benedictine of the congregation of St. Peter's, at Erfurdt, and to have lived in the beginning of the fifteenth century. But it seems that the writings which bear his name are spurious. See Sprengel, *Hist. Med.* iii. p. 267, and Morhof, *Polyhistor,* i. p. 84, who mentions that Piaccius, in the *Pseudon. Catalog.*, is disposed to deny the existence of any such person, and does not believe that his name could be found either in the provincial catalogue of Benedictines at Erfurdt, or in the general one at Rome.

[1] Arist. *De Cœlo,* ii. 7.

poses, it was certainly an act of great boldness to destroy altogether the relationship between the elementary, as they call them, and the celestial bodies ; seeing two of the elements, namely air and fire, agree so well with the stars and ether ; only that it was his way to abuse his wit, and make difficulties for himself, and prefer those things which were more obscure. Yet there is no doubt that the regions above and below the moon, together with the bodies contained in the same space, differ in many important points ; but then again there is as little doubt that the bodies of both regions have many common inclinations, passions and motions ; so that, with due regard to the unity of nature, we should rather distinguish these than separate them. But as for that point of heterogeneity, that the heavenly bodies should be supposed eternal, the inferior corruptible ; the opinion seems to fail both ways, for neither does such eternity as they feign belong to the heaven, nor such mutability to the earth. For with respect to the earth, if the matter be truly considered, judgment is not to be made from the things which are visible to us, since among the bodies seen by man's eye there is none that has been disinterred or cast up from a depth of above three miles at the most, which is as nothing compared with the extent of the whole terrestrial globe. Therefore there is no reason for thinking that the interior of the earth is not endowed with the same eternity as the heaven itself. For if the earth underwent changes in its inmost depths, it could not be but that the consequences of those changes would produce, even in this region where we tread, greater accidents than we see take place. For of the changes visible to us here towards the surface of the earth, there appears almost always some manifest cause sent from above, due to the state of the atmosphere, to rains, heats, and the like ; so that the earth itself, of its own proper force, does not seem to cause any considerable change. And if it be granted (which certainly is probable) that the earth itself also, as well as the heavenly bodies, acts upon the regions of the air, either by exhaling cold, or by emitting winds, or the like ; yet all that variety may be referred to the parts of the earth close at hand, in which no man in his senses would deny that very many changes and alterations take place. It must certainly be confessed that of all terrestrial phenomena, those which penetrate deepest into the earth are earthquakes and things of that sort, as eruptions of water, vomitings of flames, yawnings and rents of the earth, and the like ; yet even these seem to rise from no great distance, seeing most of them occupy only a small space in the surface of the earth. For the wider the space an earthquake or anything of that kind extends on the surface of the earth, the deeper must we suppose its roots and sources to penetrate into the interior ; and the narrower the less deep. And if it be said that there are sometimes earthquakes which shake vast and extensive districts of country, so no doubt it is. But these certainly happen seldom, and are to be numbered among the greater accidents ; and may be compared therefore with the higher comets, which are also uncommon. For I am not attempting to prove simply that the earth is eternal, but only (as I said at first) that between heaven and earth, as regards constancy and change, there is not much difference. Neither is it worth while to reason of eternity from the principles of motion ; for as circular motion may be without limits, so may rest ; and the consistency of dense bodies in the place and great congregation of their connaturals is not less susceptible of eternity than the rotation of rare bodies ; seeing that the parts of both when separated from the rest move in a straight line. That the interior of the earth is not more subject to corruption than the heaven itself, may be inferred also from this, that waste commonly takes place where there are means of supply. Now as rains and things falling from above, which renew the surface of the earth, cannot penetrate far into the interior, which nevertheless remains undiminished in bulk and quantity, it must be that nothing is lost, since there is nothing to take its place. Lastly, the mutability which is discovered in the exterior of the earth seems itself to be by accident. For that small incrustation which seems to extend a few miles downwards (within which those noble workshops and fabrics of plants and minerals are enclosed) would scarce receive any variety, much less such beautiful and elaborate contrivances, unless that part of the earth were acted upon and perpetually stimulated by the heavenly bodies. And if any one think that the heat and active power of the sun and heavenly bodies can strike through the thickness of the whole earth, he may be regarded as superstitious and fanatical; seeing it is very evident by how small an obstacle they may be repelled and restrained. So much then for the constancy of the earth; we must now inquire concerning the mutability of the heavens.

First then we are not to infer that changes in the heavens do not take place because they are not visible to us. For the sight is disabled both by distance of place, and by excess or deficiency of light, and by the fineness or smallness of the body; and if a man were to look from the moon he would not be able to see the changes which take place here with us on the surface of the earth, such as inundations, earthquakes, buildings, structures, and the like; which would not show so big as little straws at so great a distance. Nor from the fact that the interstellar heaven is transparent, and in clear nights the stars are seen the same in number and appearance, can a man conclude that the whole body of ether is clear, pure, and immutable. For we know that the air below admits innumerable varieties of heat, cold, odours, and all kinds of mixture with the finer vapours, and does not thereby lose its transparency; in like manner therefore we must not trust to the face or appearance of the heaven. For if those great masses of clouds which sometimes obscure the heaven, and by reason of their proximity to our view take away from us the light of the sun and stars, were hung in the higher parts of the heaven, they would no way alter the face of a clear sky; since they would neither be visible themselves by reason of the distance, nor would they at all eclipse the stars, by reason of the smallness of their bodies, in respect to the magnitude of the stars. Nay the body of the moon itself, except in the part which the light strikes, does not change the appearance of the sky; so that, if that light were absent, so great a body as that would be altogether imperceptible to us. On the other hand it is quite plain from the masses of bodies which by their bulk and magnitude can overcome the distance of space, and by the luminous nature and brilliancy of their matter can affect our sight, that wonderful changes and unusual appearances do happen in the heaven. For this is shown in the higher comets, those I mean which have appeared in the figure of a star without a tail, and are not only proved from the doctrine of parallax [2] to be situated above the moon, but have likewise had a certain and constant position relative to the fixed stars, and kept their places, and not been wanderers; such as our age has witnessed more than once, first in Cassiopea [3], and again not so long ago in Ophiuchus. And as for the notion that this constancy visible in comets proceeds from their following some star (which was the opinion of Aristotle, who affirmed that there was the same relation between a comet and a single star as between the Milky Way and the collection of stars, an assertion false both ways), this has long ago been exploded, not without a censure on the wit of Aristotle, who ventured to invent such theories on slight grounds [4]. Neither does that change in the celestial regions with regard to new stars hold with regard to those stars only which seem to be of an evanescent nature, but likewise in those which remain. For in the case of the new star of Hipparchus, mention is made by the ancients of the appearance of it [5]; but no mention of the disappearance. There appeared also of late a new star in the breast of Cygnus, which has now lasted for twelve whole years [6], having already exceeded the age (as it is held) of a comet, without as yet any diminution or preparation for flight.

[2] Galileo (in the opening of his first lecture on the new star in 1604) "showed *from the absence of parallax* that the new star could not be, as the vulgar hypothesis represented, a mere meteor engendered in our atmosphere and nearer the earth than the moon, but must be situated among the most remote heavenly bodies".—*Life of Galileo, L. U. K.* p. 16.—*J. S.*

[3] A new star was observed in Cassiopeia by Cornelius Gemma and Tycho Brahe in 1572; it disappeared in 1574. The star in Ophiuchus was observed by Kepler in 1604, and disappeared about the end of 1605. Compare with the argument in the text, Galileo, *Dialogi dei Sistemi.*

[4] Cf. Arist. *Meteorol.* i. 8.

[5] Cf. Pliny, ii. 23.

[6] This star, which is of variable magnitude, was first observed by Jansen in 1600, so that the *Descriptio Globi Intellectualis* must have been written in 1612.

Nor again can it be affirmed as a fact without exception that the old stars suffer no change at all, but only those that have appeared more recently; in which it is no wonder that a change should take place, seeing their very generation and origin is not immemorial. For setting aside the fable of the Arcadians about the first appearance of the moon, which they assert to be younger than themselves [7], there are not wanting examples within trustworthy memory, when the sun on three several occasions, without eclipse or interposition of clouds, the air being clear and serene, appeared for many days with an altered visage; yet not affected in the same manner each time, but once faint, and twice of a reddish brown. For such phenomena happened in the year 790 for seventeen days, and in the times of Justinian for half a year [8], and after the death of Julius Cæsar for several days. Of the Julian darkness there remains that notable testimony of Virgil:—

> Ille etiam extincto miseratus Cæsare Romam,
> Cum caput obscurâ nitidum ferrugine texit,
> Impiaque æternam timuerunt secula noctem [9].

The narrative of Varro, a man most learned in antiquity respecting the star Venus, which is found in Augustine [10],—namely, that in the time of King Ogyges she changed colour, size, and shape,—might have been of doubtful credit, had not a like event recurred in our age, in the year 1578, and attracted much notice. For then also through a whole year a remarkable alteration took place in the star Venus, which appeared of unusual magnitude and brilliancy, and redder than Mars himself, and changed her shape several times, becoming sometimes triangular, sometimes quadrangular, and even sometimes round, as if her very mass and substance were affected [11]. Again, that old star in the hip of Cani-

[7] Cf. Ovid. *Fasti*, i. 469.

[8] These phenomena are mentioned in juxtaposition by Patricius, *Pancosmia*, p. 111, from whom Bacon probably derived his knowledge of them. For the darkness in 790 Patricius quotes Paul the Deacon, and for that in Justinian's time Peter Messias or Mexia, who was almost a contemporary of his own. The original authority for it is Procopius, *Bell. Vandal.* ii. 14. It is to this darkening of the sun that Bacon refers in the phrase "semel luce exili". Compare a list of seventeen examples of obscuration of the sun's light in the third volume of Humboldt's *Cosmos*. He does not mention that which took place in 790; and the obscuration in the time of Justinian is said on the authority of Abul-Faragius to have lasted fourteen months. Humboldt compares it to that which took place in 1783.

[9] Georg. i. 469:—

> Then did the sun in pity dim his light,
> And drew a dusk veil o'er his visage bright,
> And shook the impious times with dread of endless night.

[10] St. August. *De Civit. Dei*, xxi. 8.

[11] Patricius was Bacon's authority for this story. After mentioning what Augustine repeats from Varro, he goes on thus: "Quæ res ævo etiam nostro accidit anno M.D. LXX. VIII. Romæque visum id est die xvi. Novembris. In Germaniâ vero die Decembris XXVI. Perque totum eum annum, sub vesperam, sole nondum merso visa est magnitudine insolitâ, figurâ vero modo triangulâ, modo quadrangulâ, modo rotundâ, et splendore maximo, et rubedine majore quam sit Martis rubedo. Cursum tamen non mutavit".—Patricius, *Pancosmia*, p. 107. This is given as evidence against the Aristotelian doctrine of the immutability of the heavens; and that it is not mentioned by Galileo and the other writers who so constantly refer to the new stars in Cassiopeia and Serpentarius for similar evidence seems to show that the story has no other foundation than that Venus was then visible before sunset. The story would, if true, have been a better proof of a change in the superlunary heavens than the new stars, seeing that it could not be said that Venus was a merely sublunary meteor. So wonderful a fact ought not to have been quoted on the authority of a loose and somewhat rhetorical writer like Patricius. [We must not forget however that this is an unfinished work, not published, nor prepared, nor perhaps intended, for publication by the writer.—*J.S.*] [And see p. 700.—ED.]

It is possible that Patricius's story may be connected with the phenomenon observed

cula, which Aristotle says that he himself saw with somewhat of a tail, and that tail, especially when cursorily looked at, vibrating, seems now to be changed and to have lost its tail; since nothing of the kind can now in our time be detected[12]. Besides, many changes of heavenly bodies, especially in the smaller stars, may easily from neglect of observation pass unnoticed, and be lost to us. That these things are due to vapours and the disposition of the medium will occur at once to any sciolist; but changes which are found to attend the body of any star constantly, equably, and for a long continuance, and to revolve along with it, must be regarded as being in the star itself, or at least in the ether near it, not in the lower regions of the air; which is likewise confirmed by the fact, that such changes take place seldom, and at long intervals; whereas those which are caused in the air by the interposition of vapours take place more frequently. And if any man concludes from the order of the heaven and the equability of the motion itself that the heaven is immutable; taking this certainty of revolutions and restitutions for a sure token of eternity, inasmuch as constancy of motion can hardly belong to a corruptible substance; he should look about him a little more attentively, and observe that this return of things by turns and as it were in circle at fixed times, is found even with us here below in some things; most of all in the tide of the ocean; while those smaller differences which may take place in the heavens both in the revolutions and restitutions escape our sight and reckoning. No more again can the circular motion of the heaven be taken as a proof of eternity; on the ground that circular motion has no limit[13], and eternal motion belongs to eternal substance. For the lower comets that are situated below the moon revolve likewise, and that of their own force; unless you had rather believe the fiction of their being attached to a star. And assuredly if we argue of the eternity of the heavenly bodies from their circular motion, we must apply the argument to the whole heaven, and not to parts of it; for we know that the air, sea, earth, though eternal in their masses, are perishable in their parts. But it may rather be said, contrariwise, that this argument from the motion of rotation does not tell in favour of the eternity of the heaven; because this motion itself is not perfect in the heaven, and does not restore itself exactly in a pure and perfect circle, but with deviations, curves, and spirals. If again a man retort upon me that which I said concerning the earth (namely, that the changes which take place in it happen by accident, because the earth is acted on by the heaven), and assert that the case of the heaven is different, seeing that the heaven cannot in any way be acted upon in its turn by the earth, inasmuch as all emanation from the earth stops on this side of the heaven, and therefore it is probable that the heaven, being set apart beyond the reach of any hostile force, is susceptible of eternity, not being disturbed or shaken by an opposite nature; his objection is not to be despised. For I have no respect for the simple notion of Thales, who thought that the celestial fires fed on the clarified vapours of the earth and ocean, and were thence nourished and repaired[14]; (whereas these vapours fall back again in almost the same quantity as they rose, and are far from being enough to refresh both the earth and the celestial globes, nor can they at all mount so high); but yet admitting that these materiate emanations of the earth stop far below the heaven, nevertheless if the earth be, as Parmenides and Telesius supposed, the original source of cold, it is not easy to say for certain to what height this opposite and rival power to the heaven may insinuate itself by series and succession; especially as rare bodies

in China in 1578, and which is thus mentioned in Biot's extracts from the annals of the Ming dynasty. "1578. 22 Fevrier (période Wanli 6me année 1me lune jour Woutchin) il parut une étoile grande comme le soleil."

Humboldt observes that it is extraordinary that no mention was made in the 16th century by European astronomers of this phenomenon. It seems that Bacon has mistaken Patricins's expression "totum eum annum"; which appears to mean, not that the phenomenon lasted a year, but that it was visible to the end of the year in which it appeared. See *Connaissances des Temps* for 1846.

[12] Arist. *Meteorol.* i. 6.

[13] Arist. *De Cœlo*, i. 9.

[14] Plutarch, *De Placit. Philosoph.* i. 3.

imbibe the nature and impression of heat and cold, and transmit it to a great distance. Grant however that the heaven is not acted upon by the earth, why may not celestial bodies be affected and changed one by another,—the sun by the stars, the stars by the sun, the planets by both, and all by the ambient ether, especially at the borders of their globes ? Then again the opinion of the eternity of the heaven derives much apparent strength from the very machinery and construction of the heaven, about which astronomers have taken such pains. For great provision seems to be made thereby to exempt the celestial bodies from all change besides simple rotation, and leave them in other respects at rest and without perturbation. Therefore they have supposed the bodies of the stars to be fixed in their orbs, as if they were nailed ; while to each of their declinations, elevations, depressions, and sinuous movements they have assigned so many perfect circles of suitable width ; carefully turning and smoothing both the concave and the convex parts of those circles, so as to leave no prominence or roughness, but that one may fit into another, and, being by reason of the polish at once exactly contiguous and free to slide easily, may move quietly and happily; which immortal contrivance removes all violence and perturbation, the inseparable forerunners of corruption. For certainly if such great bodies as the globes of stars are, do pass through ether, and yet do not always travel through the same parts of it, but through parts and tracts very different, sometimes invading the higher regions, sometimes descending to the lower, sometimes turning to the south, sometimes to the north, there is danger no doubt of very many impressions, concussions, reciprocations, and fluctuations in the heaven, and that hence may ensue condensations and rarefactions of bodies, which may procure and prepare the way to generations and alterations. But since it will clearly appear from physical reasons, and withal from the phenomena themselves, that this last is really the fact ; and those figments of astronomers of which I spoke are, as any man of sound judgment will see, mere mockeries of nature, without any reality in them ; it is but reasonable that the opinion of the eternity of the heaven, connected as it is with them, should undergo the same judgment. And if objection be here made on religious grounds, I answer that it is only heathen arrogance that attributes this eternity to the heaven alone ; sacred writ assigns eternity to earth and heaven alike [15]. For we read not only that "the Sun and Moon are eternal and faithful witnesses in the heaven", but also that "Generations come and go, but the Earth remaineth for ever" [16]. And for the transitory and perishable nature of both, we find it concluded in one oracle, "Heaven and earth shall pass away, but the Word of the Lord shall not pass away" [17]. Again, if it be still urged that for all this it must be admitted that there are innumerable changes in the surface of the earth and the parts next to it, whereas it is not so in the heaven ; I reply that in the first place I do not maintain them to be in all respects alike; and yet that if we take what are called the upper and middle regions of the air for the surface or inner covering of the heaven, as we take that space with us in which animals, plants, and minerals are contained, for the surface or outer covering of the earth, various and multiform generations are found there likewise. It would seem therefore that all tumult, conflict, and disorder take place only in the confines of heaven and earth ; just as it is in civil matters, in which it is commonly found that the border land of two kingdoms is troubled by continual inroads and violence, while the interior provinces of both countries are in the enjoyment of long peace, and are not disturbed except by

[15] Lansberg makes a curious remark as to the difficulties which may arise from a literal interpretation of Scripture. "You may so interpret it," he says, "as to make it interfere not only with astronomy but with geometry ; as when it is said that one of the ewers in the temple was ten cubits across and thirty cubits round". Campanella, in his *Apologia pro Galilæo*, tells a story of one Ulysses Albergettus, who wishing to show that the moon shines by her own light, quoted the text "Luna non dabit lumen *suum*"—"faciens vim in ly suum".—*Ly*, it may be well to remark, is used by the schoolmen as τὸ in Greek ; probably because transcribers were often ignorant of Greek, and copying by eye changed the form of what they did not understand.

[16] Ecclesiastes, i. 4. [17] St. Matth. xxiv. 35.

the more serious wars, which happen rarely. As for that other point of heterogeneity in the celestial bodies (as asserted by Aristotle [18],) that they are not hot in themselves (for otherwise the conflagration of Heraclitus might ensue), but only the cause of heat by accident, through the friction and diverberation of the air; I know not what a man can mean who abandons experience in this way, and that too against the consent of the ancients. But it is nothing new in him to snatch some one thing from experience, and straightway proceed to trample on nature, joining pusillanimity with audacity. Of this however I shall speak presently upon the question, *whether the stars are real fires;* and more fully and accurately in my precepts concerning the history of Virtues, where I shall treat of the origins and cradles of Heat and Cold, a subject hitherto unknown and untouched by men. Let the question then of the heterogeneity of the celestial bodies be propounded in this manner; for though the case calls perhaps for judgment against the opinion of Aristotle without adjournment, yet my plan of proceeding does not allow of it.

Another question is, *what is contained in the interstellar spaces?* For they are either empty, as Gilbert thought; or filled with a body which is to the stars what air is to flame,—a supposition which comes familiarly to the sense; or filled with a body homogeneous with the stars themselves, lucid and almost empyreal, but in a less degree, that is with a light not so refulgent and flashing,—which seems to be the meaning of the received opinion [19] that a star is the denser part of its sphere. Nor is there any reason why a lucid body should not be a transparent medium for the transmission of a stronger light. For Telesius has acutely remarked that even common air contains some light, using as an argument that there are some animals which see by night, their sight being (it would seem) adapted to receive and cherish this feeble light [20]: for that it is not credible that the action of light can take place without any light, or merely by the internal light itself of the visual spirit. But we see that flame itself is a transparent medium for the transmission even of the species of an opaque body, as is shown in the wick of a candle; much more of the species of an intenser light. Of flames likewise some are more pellucid than others. And this is caused either by the nature of the lighted body or the quantity. For the flame of tallow or wax is more luminous and (so to speak) more fiery; whereas the flame of spirit of wine is more opaque, and as it were airy, especially if it be in a small quantity, so that the flame does not thicken itself. Of this I have myself made trial. For I took a wax candle and set it upright in a socket (making use of a metal one for the purpose, that the body of the candle might be protected against the flame by which it was to be surrounded); and having placed the socket in a porringer where there was a little spirit of wine, I lighted first the candle, and then the spirit of wine; when it was easy to see the flame of the candle coruscating and white, through the middle of the flame of the spirit of wine, which was weak and inclining to transparency. And in like manner lucid beams are often seen along the heaven, emitting a manifest light, and wonderfully illuminating the darkness of the night; through the borders of which nevertheless the stars are visible. This inequality however between the stars and interstellar ether is not well defined by rarity and density; as if the star were denser, the ether rarer. For in general here with us flame is a body more subtle than air,—more expansive, I say, and having less matter in proportion to the space it occupies; and it is probable that this is the case also in the heavens. But the error is more harsh, if they mean that the star is a portion of the sphere fixed as with a nail, and the ether that which carries the star [21]. For

[18] Aristot. *De Cœlo*, ii. [19] Aristot. *De Cœlo*, ii. 7. [20] Telesius, *De Rer. Nat.* i. 3.

[21] The phrase fixed stars, Sidera infixa cœlo, was originally connected with the notion of the stars being fastened to the vault of heaven. The substitution, as Humboldt has remarked, of *fixa* for *infixa* or *affixa*, indicates the transition to our notion of fixed stars, which relates only to their relative immobility. See *Cosmos*, vol. iii., chapter on Fixed Stars. There is a curious passage in Acosta's *History of the Indies* on this subject. He conceives that both the Milky Way and what are commonly called the Coal Bags belong to the substance of the heaven itself, and prove by their motion that the heavens turn as well as the stars [i. 2].

this is a fiction, like that series of orbs ranged one above another which is described. For either the body of star passes through the body of the ether in its course, or else the ether itself revolves at the same time with an equal motion. For if the motion be not equal, in that case also must the star pass through the ether. And as for that structure of contiguous circles, whereby the concave part of the outer admits the convex of the inner, and yet by reason of the smoothness of both the one does not obstruct the other in its rotations, though they are unequal,—it is not a reality; the body of ether being uninterrupted and continuous, as that of the air is; although, there being so great difference between the two as regards rarity and other things, their regions are for convenience of explanation very properly distinguished. Let this question therefore be admitted, as I have thus explained it. Next comes another question, and that likewise not a simple one; concerning the substance of the stars themselves. For it is asked first, *whether there be other globes or masses of solid and compact matter besides the earth itself?* For it is a speculation soberly proposed in a book concerning the face in the moon's orb, that it is not probable that in the dispersion of matter nature included all compact body in the globe of the earth alone, when there is so great an array of globes composed of rare and expansive matter [22]. But Gilbert carried the same idea so immoderately far (wherein however he had some of the ancients as precursors, or rather guides), as to assert that not only the earth and moon, but likewise many other globes, solid and opaque, are scattered among the shining globes throughout the expanse of heaven [23]. Neither did his opinion stop here, but he thought likewise that those globes which are shining in appearance, namely, the sun and the brightest stars, consisted of a kind of solid matter, though more splendid and equal; confusing primitive light with luminous matter, which is regarded as its image (for he thought that even our sea throws out light of its own for a proportionate distance); but he acknowledged no conglobation, except in solid matter; of which matter he held those rare and fine bodies that surround it to be a kind of effluvia, and as it were defections; and beyond them a vacuum. Now that the moon is composed of solid matter is a thought which might occur to the most diligent and sober investigator of nature. For it reflects light, it does not transmit light, it is without any proper light of its own, and it is full of inequality; which are all properties of solid bodies. For we see that the ether itself and the air, which are rare bodies, receive the sun's light but do not reflect it; which the moon does. The sun's rays are so vigorous that they can penetrate and pass through very thick clouds, which are of a watery matter; but they cannot pass through the moon. The moon itself in some eclipses gives some degree of light, though obscure; but in new moons and the quarters no light at all is visible except in the part which is touched by the sun's rays. Moreover, though it be true that impure and feculent flames (of which kind of substance Empedocles [24] thought the moon consisted) are unequal, yet the inequalities have no fixed places, but are commonly movable; whereas the spots in the moon are supposed to be constant. Besides, it is now ascertained by telescopes that these spots also have their own inequalities, so that the moon is found to be clearly of manifold configuration, and that selenography or map of the moon, which Gilbert conceived [25], seems now by the industry of Galileo and others to be nearly attained. But if it may be that the moon is made of a certain solid matter, as being kindred to the earth, or the dregs of heaven (and such things are talked of), we must next inquire whether it be the only one of this kind. For Mercury too is sometimes

[22] Plutarch, *De Facie in Orbe Lunæ*, p. 924.

[23] Gilbert, *Physiol. Nov.* ii. 10. Thales is said to have been the first person who asserted that the moon is illuminated by the sun. Ocellus, and perhaps Heraclides, said that she consists of earth surrounded by a mist. Diogenes Apolloniates, probably following Anaxagoras, affirmed that along with the visible stars revolve in the heavens ἀφανεῖς λίθοι, which occasionally fall to the earth. Stobæus, *Eclog. Phys.* i. 25.

[24] Stobæus, *Eclog. Phys.* i. 27. Heeren remarks that Stobæus is the only author by whom this opinion is mentioned.

[25] See his *Physiologia Nova*, ii. 14, and the map of the moon by which it is illustrated.

found in conjunction with the sun, like a spot or little eclipse. But those dusky spots which are observed in the antarctic hemisphere, and which are fixed, like the milky way, suggest a greater doubt concerning the existence of opaque globes in the higher parts of the heaven [26]. For that they are caused by the heaven in those places being rare and as it were perforated, is not probable; because such a diminution and as it were privation of a visible object could not affect our sight at so great a distance; since the rest of the body of ether is itself invisible, and can only be distinguished by comparison with the bodies of stars. It would perhaps be more probable to attribute these blacknesses to defect of light, because the stars are fewer in that part of the heaven, as on the other hand in the neighbourhood of the milky way they are more crowded; so that the one place would seem to be continuously luminous, the other interspersed with shadows. For the celestial fires appear to be more joined together in the antarctic hemisphere than in ours; there being larger stars there, but not so many, and greater spaces between. But the report itself concerning those spots is not much to be relied on; at least there has not been enough diligence used in the observation to justify us as yet in drawing any consequences therefrom. A fact which touches the present inquiry nearer is, that there may possibly be other opaque bodies scattered through the ether, which are not seen at all. For the moon herself when new, though the horn and thin rim of the outer circle, as far as the sun's rays touch, strike the sight, is not visible at all in the middle of the disc: that part is not distinguishable in appearance from the rest of the ether; and those wandering stars discovered (if the report may be trusted) about Jupiter by Galileo are lost to our sight in that sea of ether, like so many small and invisible islands; and in like manner also those stars whereof the collection makes the Milky Way, if they were placed each apart, and not assembled in a crowd, would escape our sight altogether; as likewise many others, that in clear nights, especially in winter, sparkle; besides, those nebulous stars or openings in *Præsepe* [27] are now resolved by telescopes into a number of distinct stars; nay, and it seems that in the very purest fountain of light (I mean the sun), there is some reason, on the evidence of these same telescopes, to suspect the existence of spots, opacity, and inequalities. But if there were no other evidence, the very gradation of light among the celestial stars, descending as it does from the most brilliant to those which are obscure and misty, is enough to prove that there may likewise be globes which are completely opaque. For there seems less difference between a nebulous and opaque star than between the brightest star and a nebulous one. But our sight is plainly deceived and circumscribed; for whatever is dispersed in the heaven, and has not great magnitude and likewise a strong and vivid light, is concealed from us, and does not alter the face of the heaven. And let not any unskilful person be astonished if it be made a question whether globes of compact matter can remain pendulous. For both the earth itself floats pendulous in the middle of the surrounding air, which is an exceedingly soft thing; and great

[26] See for this Patricius, fol. 90, and Acosta's *History of the Indies*, bk. i. ch. 2.

[27] The nebula Præsepe in Cancer, and the one in the head of Orion, were the two first nebulæ ever resolved into distinot stars. Galileo gave figures of them as they appeared through his telescope in the *Sydereus Nuncius*. What Bacon goes on to say of spots in the sun is particularly interesting. Galileo did not publish on the subject before 1613; so that Bacon's information was probably not derived from Galileo, though it is believed that Galileo's first observations were made in November 1610. The earliest account which is known to have been printed of these spots is that of Fabricius, whose father's interesting correspondence with Kepler has recently been published. His tract *De Maculis in Sole observatis* was published at Wittenberg, 1611. It seems difficult to decide the question of priority of observation between him and Galileo. Harriot observed the spots in December 1610, but did not apparently know what to make of the appearance, and does not designate the phenomena by the specific name of spots until December 1611, before which time their existence had been fully ascertained by others. He drew a picture however of what he had seen on the first occasion, of which a facsimile has been published by Professor Rigaud, to whom I am indebted for most of the substance of this note. See his Supplement to Bradley's Works, pp. 32, 35, 37.

masses of watery clouds and stores of hail hang in the regions of the air, whence they are rather forced down than fall of themselves, before they begin to feel the neighbourhood of the earth. Excellently therefore did Gilbert remark, that heavy bodies when removed to a great distance from the earth gradually lose their motion downwards ; inasmuch as that motion rises from no other appetite of bodies than that of uniting and collecting themselves to the earth (which is the mass of bodies of the same nature with them), and is confined within the orb of its own virtue [28]. For as for what is said of motion to the earth's centre, it would indeed be a potent kind of Nothing that should draw such great things to it ; nor is body acted on except by body. Therefore let this question concerning solid and opaque globes, though new and harsh to vulgar opinions, be admitted ; and let there be joined with it the old though still unsettled question, *which of the stars emit a primitive light, and from themselves, and which a light derived from the sun ?* whereof the one seems to be consubstantial with the sun, the other with the moon. And in short, all inquiry concerning the different substance of the stars as compared one with the other, which appears to be multifarious, some stars looking fiery, others lead-coloured, others white, others brilliant, others manifestly and constantly nebulous, I mean to be referred to this seventh question. Another question is, *are the stars true fires ?* a question however which requires some care to understand it rightly. For it is one thing to say, that *the stars are true fires ;* and another thing to say that *the stars (admitting them to be true fires) exert all the powers and produce the same effects which common fire does.* Nor does this require us to suppose some notional or imaginary fire, retaining the name of fire without its properties. For our fire also, if it were placed in the ether in such a quantity as the stars are, would perform different operations to those which it does here with us ; seeing things acquire very different virtues, both from quantity and from relative position or location. For the greater masses I mean, connatural bodies which are collected in such quantity as to bear a due proportion to the sum of the universe, assume cosmical virtues, which are not to be found in the portions of them. Thus the ocean, which is the largest collection of waters, ebbs and flows ; whereas pools and lakes do not. In like manner the whole earth hangs suspended ; a piece of earth falls. And the relative position of a thing is of great importance in all respects both in the larger and smaller parts, by reason of the contiguity and neighbourhood of friendly or unfriendly bodies. But there must also be a far greater diversity of actions between the fire of the stars and our own, because it varies not only in quantity and relative position, but also to some extent in substance. For the fire of the stars is pure, perfect, and native ; whereas our fire is degenerate, like Vulcan thrown from heaven and halting with the fall. For if a man observe it, fire as we have it here is out of its place, trembling, surrounded by contraries, needy, depending for sustenance upon fuel, and fugitive. Whereas in heaven fire exists in its true place, removed from the assault of any contrary body, constant, sustained by itself and things like itself, and performing its proper operations freely and without molestation. And therefore Patricius had no need, in order to preserve the pyramidal form of flame, as it is found with us, to fancy that the upper part of a star, which is turned towards the ether, may be pyramidal, though the lower, which is visible to us, be globular [29]. For that pyramid of flame comes by accident, from the air closing in and crushing it ; since the flame, which is fuller in the region of its aliment, is by the hostility of the air insensibly contracted and moulded into the form of a pyramid. Hence flame is broad at the base and pointed at the apex, smoke on the other hand is pointed at the bottom and broad at the apex, and like a pyramid inverted ; because the air receives smoke, but quenches flame. It is natural therefore that flame should with us be pyramidal, and in the heaven globular. In like manner also flame with us is a momentary body, in ether permanent and durable. And yet even with us flame might last and subsist in its own form, if it were not destroyed by the things about it ; which is most manifest in the larger flames. For all that part of a flame which is situated in the midst and surrounded by flame on all sides, perishes not, but remains the

[28] Gilbert, *Physiol. Nova*, i. 21.

[29] Patricius, *Pancosmia*, xv.

same in quantity, unextinguished, and rising rapidly upwards; whereas at the sides it is troubled, and it is there that extinction commences. The manner whereof (I mean the permanency of the inner flame in a globular figure, and the vanishing and pyramidal form of the outer flame) may be experimentally demonstrated by using flames of two colours. Then again in point of fierceness there may be a great deal of variation between the celestial flame and ours. For the celestial flame unfolds itself freely and calmly, as being at home, whereas our flame, as being a stranger, is pent in and violent and furious. All fire likewise when close-packed and imprisoned becomes fiercer. For the rays of celestial flame themselves when they reach the denser and more obstinate bodies, lay aside their gentleness, and become more scorching. Aristotle ought not therefore to have feared the conflagration of Heraclitus for his world, although he had determined the stars to be real fires. This question then may be received according to this explanation. Next comes another question; *whether the stars are nourished*, and likewise, *whether they are increased, diminished, generated, and extinguished*. There was one of the ancients indeed who with a plebeian kind of observation thought that the stars are nourished as fire is, and that they feed on the waters and ocean and moisture of the earth, and are repaired by vapours and exhalations. But this opinion does not seem worthy to supply matter for a question. For such vapours are both exhausted long before they reach the heights of the stars, nor is there enough of them to repair the waters and the earth with rains and dews, and withal to refresh so many and great celestial globes; especially as it is evident that the earth and ocean have continued now for many ages without decrease of moisture; whereby it seems that no more is drawn out than comes back again. Nor again does the principle of aliment apply to the stars as it does to our fire. For the principle is that wherever anything perishes and departs there likewise something is replaced and assimilated; which kind of assimilation belongs to the region of confusions, and comes of being surrounded by contrary or dissimilar bodies; whereas in the similar and inner mass of the stars nothing of the kind happens, no more than in the bowels of the earth, which themselves also receive no nourishment, but preserve their substance in its identity, not by assimilation. With regard however to the outer borders of the sidereal bodies, the question is rightly asked, *whether these remain of one and the some tenor, or whether they prey on the surrounding ether, and likewise infect it?* In this sense therefore a question may be put concerning the aliments of the stars. And to this is rightly joined a question as to the augmentations and diminutions of stars in their whole; though the phenomena are very few which can give occasion to this doubt. For in the first place there is no example of the thing, nor anything resembling it among the things found with us, to countenance such a question; seeing that our globe of earth and water does not seem to be liable to any evident or notable augmentation or diminution on the whole, but to preserve its mass and quantity. But the stars (it will be said) appear to our eyes sometimes of a greater, sometimes of a smaller body. True; but that greatness and smallness of a star is due either to distance and vicinity, as in the apogees and perigees of planets, or to the constitution of the medium. Now that which is caused by the constitution of the medium is easily distinguished, because it changes the appearance, not of some one particular star, but of all alike; as we see in winter nights, in hard frost, when the stars appear increased in magnitude, because vapours both rise more sparingly and are harder strained, and the whole body of the air is somewhat condensed, and inclines to the aqueous or crystalline, which shows forms more large. And if there chance to be any particular interposition of vapours between our sight and one particular star, which magnifies its apparent size (as is frequently and manifestly the case with the sun and moon, and may happen with the rest), neither can this appearance deceive; because this change of magnitude does not last, nor does it follow the star or move with the body of it, but the star is soon freed from it and recovers its usual appearance. Nevertheless although these things be so, yet since both formerly in ancient times and likewise in our own age—when it was a great sight and much talked of—a great change took place in the star of Venus both as to magnitude and colour and even shape; and since a change which perpetually and constantly follows one particular star, and is seen to revolve along with it, must necessarily be set down as being in the star and not in the medium; and since through neglect of observation many things that are conspicuous in the heavens are passed by and lost to us; I think that this part of the ninth question is rightly admitted. The other part of the question is of the same kind; *whether stars are in long revolutions of ages created and dissipated?* There is a greater number of phenomena indeed to challenge this question than that about their augmentations and diminutions; but yet only of one kind. For as to the old stars, neither have we in all the memory of ages any record of the first birth of any of them (except the stories which the Arcadians of old told about the moon), nor is one of them missing. Of those however which have been regarded as comets, yet having the form and motion of stars, and being exactly like new stars [30], we have witnessed both appearances (of which we have likewise heard from the ancients) and dissappearances; when they looked to some persons as if consumed, to some as if taken up (that is, as if having come down to us in their perigees, they returned again to the higher regions), to others as if rarefying and dissolving into ether. But all this question concerning new stars I refer to that place where I shall speak of comets. There remains another question, namely concerning the Milky Way; *is the Milky Way a collection of small stars, or a continuous body, and part of the ether, of a middle nature between the ethereal and the starry?* For that opinion concerning exhalations has itself long ago exhaled, not without censure of the wit of Aristotle, who ventured to invent such a matter [31], ascribing to a thing so constant and fixed a nature transitory and variable. And this question moreover, as I put it, seems on the point of being settled, if we believe the report of Galileo, who has resolved this confused appearance of light into stars numbered and placed. For the fact that the milky way does not hide from view those stars which are found within it, certainly does not settle the question, nor incline the balance either way. Only perhaps it proves by way of negation that the milky way is not situated below the starry heaven. For if it were, and if withal that continuous body of the milky way had any depth, our view would probably be intercepted. But if be situated at the same altitude as the stars which are seen through it, why may not stars be scattered in the milky way itself, as well as in the rest of the ether? This question therefore I admit likewise. And these six questions pertain to the substance of the heavenly bodies; namely, what is the substance of the heaven in kind, what that of the interstellar ether, what that of the milky way, and what that of the stars themselves, compared either with one another, or with our fire, or with their own body. As to the number, magnitude, configuration, and distance of the stars, besides the phenomena themselves and historical questions, of which I shall speak afterwards, the philosophical problems are mostly simple. With regard to the number there follows this other question; *is the number of the stars that which appears, and which has been observed and set down by the diligence of Hipparchus, and included in his model of the celestial globe?* For not only is that a poor reason that is given for the countless multitude of hidden stars not distinctly visible, which is usually seen in clear nights, especially during the winter; namely that these appearances are not smaller stars, but only

30 This mode of speaking of the new stars confirms Professor Rigaud's explanation of a curious phrase in one of Sir William Lower's letters to Harriot. "His elliptical Iter planetarum, methinkes, shewes a way to the solving of the unknown walks of comets" (he is speaking of Kepler); "for as his ellipsis in the earth's motion is more a circle, and in Mars is more longe, and in some of the other planets may be longer againe, so in thos commets that appeare fixed the ellipsis may be neere a right line". The Professor remarks that he may possibly allude to phenomena like the new star of 1572. It is this letter of Sir William Lower's, the first part of which Baron Zach ascribed to the Earl of Northumberland, an error which is repeated by Apelt in his *Reformation of Astronomy*. See Rigaud's Supplement to Bradley's Works, pp. 43, 49.

The idea that the new star of 1572 moved alternately towards and from the earth in a right line, was proposed by John Dee. See Narrien's *Hist. of Astronomy*, p. 384.

31 Aristot. *Meteor*. i. 8.

radiations and flashings and as it were darts cast from the known stars; but the census now made by Galileo of the celestial population contains additional heads, not only in that cluster denominated the Milky Way, but likewise among the very stations and ranks of the planets. And stars become invisible, either by reason of smallness of body or by reason of opacity (for I do not much approve of the term "tenuity," seeing that pure flame is a body of extreme tenuity), or by reason of elongation and distance. As for the question respecting the increase of the number of the stars by the generation of new ones, I refer it as before to the place where I shall speak of comets. Now with regard to the magnitude of the stars, the apparent magnitude belongs to phenomena, but the true magnitude to philosophical inquiry, within the limits of that twelfth problem; *what is the true magnitude of each star, either measured, or at least compared?* for it is easier to discover and prove that the globe of the moon is smaller than the globe of the earth, than that the globe of the moon is so many miles in circumference. We must therefore find exact magnitudes, if we can; and if they cannot be had we must make use of comparative. Now true magnitudes are taken and concluded either by eclipses and shadows; or by extensions as well of light as of other virtues which each body shoots out and diffuses to a greater or less distance in proportion to its magnitude; or lastly by the symmetry of the universe, which by a kind of necessity governs and defines the portions of connatural bodies. We are not however to be bound by the statements of astronomers regarding the true magnitudes of stars; statements made (though it may seem a matter of great accuracy and subtlety) loosely and carelessly enough; but we must seek proofs (if there be any) more trustworthy and genuine. Now the magnitude and the distance of the stars mutually indicate each other from optical calculations; which themselves however require sifting. This question then concerning the true magnitude of the stars is the twelfth in number. Next comes another concerning their figure; *whether the stars are globes;* that is collections of matter in a solid round figure? To appearance there seem to be three figures of heavenly bodies; globular and beamy like the sun, globular and angular like the stars (the beams and angles referring only to sight, the globular form only to substance); globular simply, like the moon. For there is no star to be seen which is oblong or triangular or square, or of any other figure. And it seems natural that the greater masses of things should for their preservation and more perfect union collect into globes. The fourteenth question relates to distance; *what is the true distance of any star in the depth of heaven?* For the distances of the planets both from one another and from the fixed stars, laterally, or in the superficial compass of the heaven, are governed by their motions. But as I said before concerning the magnitude of the stars, that if an exact and measured magnitude is not to be had, we must take a comparative magnitude; so I say with respect to their distances; namely that if the distance (say from the earth to Saturn or Jupiter) cannot be exactly taken, yet let us make it certain that Saturn is higher than Jupiter. For neither is the interior system of the heaven, I mean the order of the planets in point of attitude, entirely without controversy; nor were the doctrines now prevalent believed in former times. And even now the question whether Mercury or Venus be the higher, is still pending. Now distances are discovered either from parallaxes, or eclipses, or calculations of motions, or differences in apparent magnitude. And other aids are to be provided for the determination of this, which may be devised by human industry. The thicknesses or depths of the spheres also have relation to distances.

THEORY OF THE EARTH.

[TRANSLATION OF THE *THEMA COELI*.]

SEEING then that there are such difficulties on all sides, we must be content if something be asserted that is not harsh. I will myself therefore construct a Theory of the Universe, according to the measure of the history as yet known to us; keeping my judgment however in all points free, for the time when history, and by means of history my inductive philosophy, shall have been further advanced. Wherein I will first propound some things respecting the matter of the heavenly bodies, whereby their motion and construction may be better understood; and then I will bring forward my thoughts and views concerning the motion itself, which is now the principal question. It seems then that nature has in the distribution of matter separated fine bodies from gross; and assigned the globe of the earth to the gross, and the whole space from the surface of the earth and waters to the very extremities of the heaven, to the fine or pneumatic, as the two primary classes of things, in proportions not equal indeed, but suitable. And this is the natural and proper collocation of things, nor is it confounded either by water hanging in the clouds or wind pent within the earth. Now this distinction of fine or pneumatic and gross or tangible, is quite primordial, and the one which is most employed in the system of the universe. And it is derived from that condition of things which is of all the simplest, namely the quantity and paucity of matter in proportion to bulk. The pneumatic bodies which are found here with us (I speak of such as exist simple and perfect, not compound and imperfectly mixed), are those two, Air and Flame. And these are to be regarded as bodies altogether heterogeneous; not as is commonly imagined, that flame is only air on fire. To these correspond, in the upper world, the ethereal and the starry nature; as in the lower, water and oil; and lower still, mercury and sulphur; and generally, crude bodies and fat bodies, or in other words, bodies which abhor and bodies which conceive flame (salts being of a compound nature, consisting at once of crude and inflammable parts [1]). Now for these two great families of things, the Airy and the Flamy, we have to inquire upon what conditions they have taken possession of by far the greater part of the universe, and what office they have in the system. In the air next the earth, flame only lives for a moment and at once perishes. But when the air begins to be cleared of the exhalations of the earth and well rarefied, the nature of flame makes divers trials and experiments to attain consistency therein, and sometimes acquires a certain duration, not by succession as with us, but in identity; as happens for a time in some of the lower comets, which are of a kind of middle nature between successive and consistent flame; it does not however become fixed or constant till we come to the body of the Moon. There flame ceases to be extinguishable, and in some way or other supports itself; but yet such flame is weak and without vigour, having little radiation, and being neither vivid in its own nature, nor much excited by the contrary nature. Neither is it pure and entire, but spotted and crossed by the substance of ether (such as it exists there), which mixes with it. Even in the region of Mercury flame is not very happily placed, seeing that by uniting together it makes but a little planet; and that with a great perturbation, variety, and

[1] Salt is mentioned here, because Mercury, Sulphur, and Salt are according to Paracelsus the three constituent principles of all substances. Bacon however, as we see in the *Historia Sulphuris, Mercurii et Salis*, of which only the aditus or preface has been preserved, refuses to recognise salt as a co-ordinate principle with the other two.

fluctuation of motions, like *ignis fatuus*, labouring and struggling, and not bearing to be separated from the protection of the sun except for a little distance. When we come to the region of Venus, the flamy nature begins to grow stronger and brighter, and to collect itself into a globe of considerable size; yet one which itself also waits on the sun and cannot bear to be far away from him. In the region of the Sun, flame is as it were on its throne, midway between the flames of the planets, stronger, likewise and more vibrating than the flames of the fixed stars, by reason of the greater reaction, and exceeding intensity of union. In the region of Mars flame appears even robust; acknowledging the vicinity of the sun by its redness, but now independent, and bearing to be separated from the sun by the whole diameter of the heavens. In the region of Jupiter, flame, gradually ceasing to be contentious, seems calmer and whiter, not so much from its own nature (as the star Venus is, being more fiery), but from the surrounding nature being less irritated and exasperated; in which region it is probable, according to the discovery of Galileo, that the heaven begins to be set with stars, though stars invisible from their smallness [2]. But in the region of Saturn, the flamy nature appears again to grow somewhat feeble and dull, as being both further removed from the support of the sun, and exhausted by the proximity of the starry heaven. Last of all, the flamy and sidereal nature, victorious over the ethereal, produces the starry heaven, which is compounded of the ethereal and sidereal nature (as the globe of the earth is compounded of land and water variously diffused, yet with the ethereal substance so converted, wrought, and assimilated, as to be completely patient and obedient to the sidereal. Thus we have between the earth and the summits of heaven three general regions, and as it were three stages, in respect of the flamy nature; the region of the extinction of flame, the region of its union, and the region of its dispersion. Now to argue of contiguity and continuity in the case of soft bodies and fluids would be vulgar. But it must be understood, that it is the way of nature to proceed a certain distance by gradations, and then suddenly by jumps; and to alternate this process; otherwise there could be no structural fabric, if all changes proceeded by insensible gradations. For how great a leap it is (in respect of expansion of matter) from earth and water to air, even the grossest and most nebulous! And yet these bodies so different in nature are in place and surface joined together, without any medium or interval. Nor is it a less leap (in respect of substantial nature) from the region of the air to the region of the moon: an immense leap again from the heaven of the moon to the starry heaven. Therefore if continuity and contiguity be understood with reference not to the manner of connexion, but to the diversity of the bodies connected, these three regions which I have mentioned may be regarded as being in their boundaries only contiguous. But now we must examine clearly and perspicuously what and what kind of points this theory of mine on the substances of the system affirms, and what and what kind it denies; that it may the more easily be either maintained or overthrown. It denies the common theory, that *flame is air on fire*; affirming that these two bodies, air and flame, are completely heterogeneous, like water and oil, sulphur and mercury. It denies Gilbert's doctrine of *a collective vacuum between the scattered globes*; affirming that space is filled with either an airy or a flamy nature. It denies that *the moon is either a watery or a dense or a solid body*; affirming that it is of a flamy nature, though slow and weak, as being the first rudiment and last sediment of celestial flame; flame admitting (as regards density), no less than air and liquids, of innumerable degrees. It affirms that *flame, in its true place and left to itself, is fixed and constant*, no less than air and water; and that it is not a thing momentary, and preserved in its mass only by succession through renovation and aliment, as it is here with us. It affirms that *flame has a nature apt to unite and gather into globes*, like the nature of earth; not like that of air and water, which are collected in the circles and interstices of globes, but never into entire globes. It affirms that *the same flamy nature in its own place* (*that is, the starry heaven*) *is scattered about in infinite clusters*, yet in such sort that the dualism of

[2] This reference to Jupiter's satellites shows that the *Thema Cœli* was written after the publication of the *Sydereus Nuncius*.

ether and star is still maintained, and flame does not continue into the perfect empyrean. It affirms that *the stars are real flames*, but that the actions of flame in heavenly bodies are in no way to be applied to the actions of our flame, most of which operate only by accident. It affirms that *the interstellar ether and the stars bear to each other the relations of air and flame, but sublimed and rectified.* Regarding the Substance then of the System of the Universe, such are the thoughts which occur to me. I must now speak of the Motions of the Heavenly Bodies with reference to which I have brought these things forward. It seems reasonable to suppose that rest is not excluded from nature, as regards any whole (for I am not now talking of particles). This (discarding logical and mathematical subtleties) appears most clearly from the fact, that the speed and velocities of the celestial motions relax themselves gradually, as if about to end in something immovable; and that even the celestial bodies have a share of rest in respect of the poles; and that if immobility be excluded, the sytem is dissolved and dispersed. Now if there be any collection and mass of the immovable nature, we need not look further to show that this mass is the globe of the earth. For close and strict compaction of matter induces a disposition towards motion torpid and averse; as on the other hand free explication of it induces a disposition prompt and apt. Nor was it ill done by Telesius (who revived the philosophy and discussions of Parmenides in his book on the original source of cold) to introduce into nature, not indeed coessentiality and conjugation (which he would have), but yet affinity and conspiration; making heat, light, tenuity, and mobility to be allied on one part; cold, darkness, density, and immobility on the opposite; and placing the seat of the first set in the heaven, of the second in the earth. But if *rest and immobility* be admitted, it seems that *motion without limit and perfect mobility* should likewise be admitted, especially in opposite natures. Now this motion is the motion of rotation, such as is generally found in the celestial bodies. For motion in a circle has no limit, and seems to proceed from an appetite of the body which moves merely for the sake of moving and following itself and seeking its own embraces, and exciting and enjoying its own nature, and performing its own operation; whereas contrariwise motion in a straight line seems like a journey to an end, as seeking both to reach the limit where it may cease and rest, and to attain some object and then discontinue its motion. We must see therefore how this motion of rotation, which is the true and perennial motion, and commonly considered peculiar to the heavenly bodies, acquits itself, and by what control it spurs and bridles itself, and generally how it is affected; in the explanation of which things I shall not stand upon that piece of mathematical elegance, the reduction of motions to perfect circles, either eccentric or concentric, or that high speech, that the earth in comparison to heaven is a point and not a quantity, or many other fictitious inventions of astronomers; but remit them to calculations and tables. But first I will make a division of the motions of the heavenly bodies. *Some are cosmical, others mutual.* Those I call *cosmical*, which celestial bodies assume by consent, not only of the heavens, but likewise of the universe; those *mutual*, in which one celestial body depends on another. And this is a true and necessary division. The earth then being stationary (for that I now think the truer opinion [3]), it is manifest that the heaven revolves in a diurnal motion, the measure whereof is the space of twenty-four hours or thereabouts, the direction from east to west, the axis of revolution certain points (which they call poles) north and south. For the heavens do not travel on movable poles, nor are there any other points than those I have mentioned. And this motion appears to be truly cosmical, and therefore one and the same; except in so far as it admits both diminutions and deviations; according to which diminutions and deviations this motion strikes through the whole universe of things movable, and penetrates from the starry

[3] Bacon, in his later writings, rejected more decidedly than in this passage the doctrine of the earth's motion. Thus in the *Nov. Org.* ii. 46, it is said that Galileo's theory of the tides is founded on a "concessum non concessibile," namely, that the earth moves; and in the third book of the *De Augmentis*, Bacon, in speaking of the cumbrous machinery of the Ptolemaic system, remarks, "harum suppositionum absurditas in motum terræ diurnum (quod nobis constat falsissimum esse) homines impegit".

heaven to the bowels and depths of the earth; not forcing them along with violence or vexation, but by a perpetual consent. And this motion is in the starry heaven perfect and entire, as well in just measure of time, as in exact restitution of place. But the lower down we come, the more imperfect is this motion, in respect of slowness, and in respect likewise of deviation from circular motion. And first I must speak of the slowness separately. I say then that the diurnal motion of Saturn is too slow to allow of its completing the circle or coming back to the same place within twenty-four hours; but that the starry heaven moves faster, and outstrips Saturn each day by a distance which multiplied by the number of days in thirty years makes up the whole circuit of the heaven. So also with regard to the other planets, according to the diversity of their several periods; so that the diurnal motion of the starry heaven (speaking of the period only, without reference to the magnitude of the circle) is about one hour quicker than the diurnal motion of the moon. For if the moon completed its course in twenty-four days, it would be quicker by an hour exactly. Therefore that motion of opposition and resistance from west to east which they talk of, and which is attributed to the planets as peculiar to them, is not a real motion, but only in appearance, owing to the starry heaven advancing faster to the west, and so leaving the planets behind towards the east. Upon which supposition, it is manifest that the velocity of this cosmical motion decreases in regular order as it descends, so that the nearer every planet approaches the earth the slower it moves; whereas the received opinion disturbs and inverts the order; and by attributing a peculiar motion to the planets falls into the absurdity of supposing that the nearer the planets approach the earth (which is the seat of immobility) the quicker they move; a thing which astronomers idly and unsuccessfully endeavour to account for by supposing a remission of the violence of the *primum mobile*. And if it seem strange that in so great a space as lies between the starry heaven and the moon this motion diminishes so little; namely less than one hour, which is a twenty-fourth part of the diurnal motion; it is to be remembered that the nearer a planet is to the earth the smaller is the circle of its revolution; so that if we add the decrease in the magnitude of the circle to the decrease in the time of revolution, we shall see that the motion is diminished very considerably. Thus far I have spoken of velocity separately; as if the planets (placed, for instance, under the equinoctial, or any of the parallels) were only outrun by the starry heaven and by one another, but yet in the same circle. For this would be simple leaving behind without obliquity of motion. But it is manifest that the planets not only move with unequal velocity, but do not return to the same point of the circle, deflecting to the north and south; the limits of which deflexion are the tropics; and to this deflexion it is that we owe the Oblique Circle and the Difference of Polarity; just as we owe to the inequality of velocity the motion of resistance. But the nature of things does not stand in need of this device, more than of the other; seeing that by adopting spiral lines (the supposition which comes nearest to the sense and the fact) the thing is accomplished, and those phenomena are saved. And (which is the chief point) these spirals are nothing else than defections from perfect circular motion, whereof the planets are impatient. For in proportion as substances degenerate in purity and freedom of development, so do their motions degenerate. Now it happens, that as in point of velocity the higher planets move faster, and the lower less fast; so also the higher planets make spirals more closely coincident and coming nearer to circles the lower make spirals more disjoined and further apart. For continually as they descend they recede more and more both from that height of velocity and that perfection of circular motion, in regular order. Yet in this the planets agree (as being bodies that retain much of a common nature, though otherwise differing) that they have the same limits of deflexion [4]. For neither does Saturn come back within the tropics, nor the moon go forth beyond the tropics (and yet with regard to the wandering of Venus there are certain traditions and observations not to be overlooked); but all the planets, whether the higher or lower, as

[4] It appears from this that Bacon was not aware of the obliquity to the ecliptic of all the planetary orbits.

soon as they reach the tropics, turn back and retrace their course, disliking the smaller spiral in which they would have to move if they approached nearer the poles; and shrinking from that loss of motion, as from the destruction of their nature. For however in the starry heaven both the stars near the poles and the stars about the equinoctial maintain their ranks and stations, one being kept in order by another, with a perfect and equable constancy; yet the planets seem to be of such a mixed nature as not willingly to endure either a shorter circle or a larger. These views then concerning the celestial motions appear to me a little better than the carrying by force, the repugnance of motions, the different polarity of the zodiac, the inverted order of velocity, and the like; which have no manner of agreement with the nature of things, however they keep peace, such as it is, with the calculations. Neither were the better astronomers blind to these things; but being intent on their art, and foolishly attached to perfect circles, and catching at subtleties, and too servile to philosophy, they scorned to follow nature. But this imperious disposition of philosophers towards nature is worse even than the simplicity and credulity of the vulgar; if a man disdains a plain thing because it is plain. And yet a vast evil it is and of very wide extent, that the human wit, not being able to match nature, must needs put itself above nature. But now we must inquire whether that single and simple motion, in a circle and spiral, from east to west, on certain poles south and north, ends and terminates with the heaven, or extends likewise to things below. For it will not be open to us to invent here in the regions next us such things as they suppose in the heavens. If therefore in these regions also this motion be found, it will appear that in heaven likewise it is, under the conditions of a common or cosmical nature, such as we experience it. First then it is plain that it is not confined within the limits of heaven. But the demonstrations and evidences on this point I have fully treated in my "anticipation" concerning the ebb and flow of the sea; to which men are therefore referred; and taking this for settled and concluded, I will proceed to the other motions of celestial bodies. These I have said are not cosmical, but mutual, or having relation one to another. There are four kinds of motions visible in heavenly bodies besides that which I have called cosmical, which is the diurnal motion by spirals within the tropics. For the stars either rise higher and again sink lower, so as to be further off and nearer the earth; or they turn and wind from side to side of the zodiac, running out more to the south or more to the north, and forming what they call dragons; or they vary in velocity and likewise in direction of motion (for I put these two together), proceeding sometimes quicker, sometimes slower, sometimes in progression, sometimes in regression, sometimes likewise stopping and remaining stationary; or they are attached and circumscribed at a greater or less distance from the sun. The causes and natures of these motions I will only give in general and by heads; for this the plan of my work here demands. But to pave and open the way for this, I must say without reserve what I think with regard to certain philosophical doctrines, as well as astronomical hypotheses, and likewise with regard to the observations of astronomers in various ages, upon which they build their art; all which appear to me full of error and confusion. There are some axioms then, or rather opinions, which being received by philosophers, transferred into astronomy, and unhappily believed, have corrupted the art. Of these my rejection and judgment will be simple: for I have no time to spend in confutations. The first is, that all things above the moon inclusive are incorruptible, and not subject to new generations or changes of any kind. Of this I have spoken elsewhere, as being a superstition and a vanity. But it is the fountain from which springs that vast evil, that upon every anomaly astronomers frame new and (as they think) corrected theories, and often apply to things that are as it were fortuitous, causes eternal and invariable. The second is, that the heaven (as consisting of a fifth essence, and of no elementary substance) admits not of those turbulent actions of compression, relaxation, repulsion, submission, and the like, that seem to be produced by a certain hardness and softness of bodies, which are regarded as elementary qualities. But this assertion is an insolent and licentious repudiation of fact and sense. For wherever a natural body is placed, there also is resistance, and that in proportion to the body. And wherever there are natural bodies and local motion, there is either repulsion, or yielding, or division ; for these things above mentioned, namely, compression, relaxation, repulsion, yielding with many others, are universal passions of matters every where, and yet from this fountain has flowed that multiplication of circles complicated at pleasure, which they will nevertheless have to be so adapted to each other, and to move and turn with such smoothness and slipperiness one within the other, that there is no obstruction at all, no fluctuation ; all which are plainly fanciful, and trample upon the nature of things. The third is, that all natural bodies have their own proper motions ; and if any be found to have more than one, that all the rest come from elsewhere, and from some separate moving body. Than which nothing falser can be devised, seeing all bodies by the manifold consent of things are endued likewise with many motions, some ruling, some obeying, and some also lying dormant unless exerted ; and proper motions of things there are none, except exact measures and modes of common motions. Hence again has come forth a separate *primum mobile*, and heavens above heavens, and a continuous chain of new structures, to meet the demands of such different motions. The fourth is, that all celestial motions are performed in perfect circles ; a thing very cumbrous, which has produced for us those prodigies of eccentrics and epicycles ; whereas if they had consulted nature, they would have found that while motion orderly and uniform is in a perfect circle, motion orderly but multiform, such as is found in many heavenly bodies, is in other lines ; and deservedly does Gilbert laugh at this, saying that it is not probable nature would have formed wheels of one or two miles for instance in circuit, to carry a ball the size of a palm [5]. For it seems that the body of a planet is no bigger, as compared with those circles which they invent for it to move in. The fifth is, that the stars are parts of their own orb fixed as it were by a nail. But this is very evidently a conceit of those who deal with mathematics, not with nature, and fixing all their attention on the motion of bodies entirely forget their substances. For that fixation is a particular affection of compact and consistent things, which keep firm hold by reason of the pressure of their parts. But it is quite inconceivable, if it be transferred to soft or liquid bodies. The sixth is, that a star is the denser part of its own orb ; whereas the stars are neither parts, nor denser [6]. For they are not homogeneous with the air, differing only in degree, but they are quite heterogeneous and differ in substance ; which substance also is in respect of density rarer and more open than the ethereal. There are likewise many other opinions equally vain ; but these will suffice for the present business. So much then for the doctrines of philosophy concerning celestial bodies. As for the hypotheses of astronomers, it is useless to refute them, because they are not themselves asserted as true, and they may be various and contrary one to the other, yet so as equally to save and adjust the phenomena. Let it then be arranged, if you will, between philosophy and astronomy, as by a convenient and legitimate compact, that astronomy shall prefer those hypotheses which are most suitable for compendious calculation, philosophy those which approach nearest the truth of nature ; and that the hypotheses of astronomy shall not prejudice the truth of the thing, while the decisions of philosophy shall be such as are explicable on the phenomena of astronomy. And so much for hypotheses. But with respect to astronomical observations, which are assiduously accumulated, and are continually dropping like waters from the heaven, I would by all means have men beware, lest Æsop's pretty fable of the fly that sate on the pole of a chariot at the Olympic races and said, "What a dust do I raise," be verified in them. For so it is that some small observation, and that disturbed sometimes by the instrument, sometimes by the eye, sometimes by the calculation, and which may be owing to some real change in the heaven, raises new heavens and new spheres and circles. Nor do I say this because I would have any relaxation of industry in observations and history, which I say should be sharpened and strengthened in all ways, but only that prudence and a perfect and settled maturity of judgment may be employed in rejecting or altering hypotheses. Having therefore now opened the way, I will make a few general observations on

[5] Gilbert *Physiol. Nov.* ii. 11.

[6] Cf. Arist. *De Cœlo*, ii. 7.

the motions themselves. I have said that there are four kinds of greater motions in the heavens. *Motion in the depth of heaven*, upward or downward; *motion through the latitude of the zodiac*, deviating to south and north; *motion in the direction of the zodiac*, quick, slow, progressive, retrograde, and stationary; and *motion of elongation* from the sun. And let no one object that this second motion of latitude, or the dragons, might have been referred to that great cosmical motion, being an alternate inclination towards north and south; inasmuch as these spirals move in like manner from tropic to tropic; only that the cosmical motion is spiral simply, whereas the other is likewise sinuous and with much smaller intervals. For this has not escaped me. But the fact is, that the constant and perpetual motion of the sun in the ecliptic without latitude and dragons, which sun nevertheless has a common motion with the other planets in respect of spirals between the tropics, forbids me to agree with this opinion. We must therefore seek other sources both of this and of the three other motions. Such are the ideas with regard to the celestial motions which seem to me to have least inconvenience. Let us see then what they deny and what they affirm. They *deny* that the earth revolves. They *deny* that there are two motions in the heavenly bodies, one being from west to east; and *affirm* a difference in speed, one out-stripping and leaving the other behind. They *deny* an oblique circle with a different position of its poles; and *affirm* spirals. They *deny* separate *primum mobile*, and carriage by force; and *affirm* a cosmical consent as the common bond of the system. They *affirm* that the diurnal motion is found not in the heaven only [7], but also in the air, water, and even the exterior of the earth, in respect of its verticity. They *affirm* that this cosmical motion of flowing and rolling in fluids, becomes verticity and direction in solids, until it passes into pure immobility. They *deny* that the stars are fixed like knots in a board. They *deny* that eccentrics, epicycles, and such structures are real. They *affirm* that the magnetic motion, or that which brings bodies together, is active in the stars, whereby fire evokes and raises fire. They *affirm* that in the planetary heavens the bodies of the planets move and revolve with greater velocity than the rest of the heaven in which they are situated, which does indeed revolve, but more slowly. They *affirm* that from this inequality come the fluctuations, waves, and reciprocations of the planetary ether, and from them a variety of motions. They *affirm* a necessity in the planets of revolving faster and slower, according as they are situated high or low in the heaven, and that by consent of the universe. But at the same time they *affirm* a dislike in the planets of preternatural velocity as well of the greater as of the lesser circle. They *affirm* a tendency to follow the sun, by reason of neediness of nature, in the weaker fires of Venus and Mercury; the rather, because Galileo has discovered certain small wandering stars attendant upon Jupiter. These then are the things I see, standing as I do on the threshold of natural history and philosophy; and it may be that the deeper any man has gone into natural history the more he will approve them. Nevertheless I repeat once more that I do not mean to bind myself to these; for in them as in other things I am certain of my way, but not certain of my position. Meanwhile, I have introduced them by way of interlude, lest it be thought that it is from vacillation of judgment or inability to affirm that I prefer negative questions. I will preserve therefore, even as the heavenly bodies themselves do (since it is of them I am discoursing), a variable constancy.

[7] *Motum diurnum inveniri non in cœlo, sed et in aere, aquis, etiam extimis terræ, quoad verticitatem.* So the sentence stands in the original. But it seems that *tantum* or some equivalent word has dropped out.—*J. S.*

THE NEW ATLANTIS

PREFACE.

BY JAMES SPEDDING.

THE *New Atlantis* seems to have been written in 1624, and, though not finished, to have been intended for publication as it stands. It was published accordingly by Dr. Rawley in 1627, at the end of the volume containing the *Sylva Sylvarum;* for which place Bacon had himself designed it, the subjects of the two being so near akin; the one representing his idea of what should be the end of the work which in the other he supposed himself to be beginning. For the story of Solomon's House is nothing more than a vision of the practical results which he anticipated from the study of natural history diligently and systematically carried on through successive generations.

In this part of it, the work may probably be considered as complete. Of the state of Solomon's House he has told us all that he was as yet qualified to tell. His own attempts to "interpret nature" suggested the apparatus which was necessary for success: he had but to furnish Solomon's House with the instruments and preparations which he had himself felt the want of. The difficulties which had baffled his single efforts to provide that apparatus for himself suggested the constitution and regulations of a society formed to overcome them: he had but to furnish Solomon's House with the helps in head and hand which he had himself wished for. His own intellectual aspirations suggested the result: he had but to set down as known all that he himself most longed to know. But here he was obliged to stop. He could not describe the *process* of a perfect philosophical investigation; because it must of course have proceeded by the method of the *Novum Organum,* which was not yet expounded. Nor could he give a particular example of the result of such investigation, in the shape of a Form or an Axiom; for that presupposed the completion, not only of the *Novum Organum,* but (at least in some one subject) of the Natural History also; and no portion of the Natural History complete enough for the purpose was as yet producible. Here therefore he stopped; and it would almost seem that the nature of the difficulty which stood in his way had reminded him of the course he ought to take; for just at this point (as we learn from Dr. Rawley) he did in fact leave his fable and return to his work. He had begun it with the intention of exhibiting a model political constitution, as well as a model college of natural philosophy; but "his desire of collecting the natural history diverted him, which he preferred many degrees before it". And in this, according to his own view of the matter, he was no doubt right; for though there are few people now who would not gladly give all the *Sylva Sylvarum,* had there been ten times as much of it, in exchange for an account of the laws, institutions, and administrative arrangements of Bensalem, it was not so with Bacon; who being deeper read in the phenomena of the human heart than in those of the material world, probably thought the perfect knowledge of nature an easier thing than the perfect government of men,—easier and not so far off; and therefore preferred to work where there was fairest hope of fruit.

To us, who can no longer hope for the fruits which Bacon expected, the *New Atlantis* is chiefly interesting as a record of his own feelings. Perhaps there is no single work of his which has so much of himself in it. The description of Solomon's House is the description of the vision in which he lived,—the vision not of an ideal world released from the natural conditions to which ours is subject, but of our own world as it might be made if we did our duty by it; of a state of things which he believed would one day be actually seen upon this earth such as it is by men such as we are; and the coming of which he believed that his own labours were sensibly hastening. The account of the manners and customs of the people of Bensalem is an account of his own taste in humanity; for a man's ideal, though not necessarily a description of what he is, is almost always an indication of what he would be; and in the sober piety, the serious cheerfulness, the tender and gracious courtesy, the open-handed hospitality, the fidelity in public and chastity in private life, the grave and graceful manners, the order, decency, and earnest industry, which prevail among these people we recognise an image of himself made perfect,— of that condition of the human soul which he loved in others, and aspired towards in himself. Even the dresses, the household arrangements, the order of their feasts and solemnities, their very gestures of welcome and salutation, have an interest and significance independent of the fiction, as so many records of Bacon's personal taste in such matters. Nor ought the stories which the Governor of the House of Strangers tells about the state of navigation and population in the early post-diluvian ages, to be regarded merely as romances invented to vary and enrich the narrative, but rather as belonging to a class of serious speculations to which Bacon's mind was prone. As in his visions of the future, embodied in the achievements of Solomon's House, there is nothing which he did not conceive to be really practicable by the means which he supposes to be used; so in his speculations concerning the past, embodied in the traditions of Bensalem, I doubt whether there be any (setting aside, of course, the particular history of the fabulous island) which he did not believe to be historically probable. Whether it were that the progress of the human race in knowledge and art seemed to him too small to be accounted for otherwise than by supposing occasional tempests of destruction, in which all that had been gathered was swept away, or that the vicissitudes which had actually taken place during the short periods of which we know something had suggested to him the probability of similar accidents during those long tracts of time of which we know nothing,—or merely that the imagination is prone by nature to people darkness with shadows,—certain it is that the tendency was strong in Bacon to credit the past with wonders; to suppose that the world had brought forth greater things than it remembered, had seen periods of high civilisation buried in oblivion, great powers and peoples swept away and extinguished. In the year 1607, he avowed before the House of Commons a belief that in some forgotten period of her history (possibly during the Heptarchy) England had been far better peopled than she was then. In 1609, when he published the *De Sapientiâ Veterum*, he inclined to believe that an age of higher intellectual development than any the world then knew of had flourished and passed out of memory long before Homer and Hesiod wrote; and this upon the clearest and most deliberate review of all the obvious objections; and more decidedly than he had done four years before when he published the *Advancement of Learning*. And I have little doubt that when he wrote the *New Atlantis* he thought it not improbable that the state of navigation in the world 3000 years before was really such as the Governor of the House of Strangers describes; that some such naval expeditions as those of Coya and Tyrambel may really have taken place; and that the early civilisation of the Great Atlantis may really have been drowned by a deluge and left to begin its career again from a state of mere barbarism.

Among the few works of fiction which Bacon attempted, the *New Atlantis* is much the most considerable; which gives an additional interest to it, and makes one the more regret that it was not finished according to the original design. Had it proceeded to the end in a manner worthy of the beginning, it would have stood, as a work of art, among the most perfect compositions of its kind.

The notes to this piece, which are not marked with Mr. Ellis's initials, are mine.

J. S.

TO THE READER.

THIS fable my Lord devised, to the end that he might exhibit therein a model or description of a college instituted for the interpreting of nature and the producing of great and marvellous works for the benefit of men, under the name of Salomon's House, or the College of the Six Days' Works. And even so far his Lordship hath proceeded, as to finish that part. Certainly the model is more vast and high than can possibly be imitated in all things; notwithstanding most things therein are within men's power to effect. His Lordship thought also in this present fable to have composed a frame of Laws, or of the best state or mould of a commonwealth; but foreseeing it would be a long work, his desire of collecting the Natural History [1] diverted him, which he preferred many degrees before it.

This work of the *New Atlantis* (as much as concerneth the English edition) his Lordship designed for this place [2]; in regard it hath so near affinity (in one part of it) with the preceding Natural History.

W. RAWLEY.

NEW ATLANTIS.

WE sailed from Peru, (where we had continued by the space of one whole year,) for China and Japan, by the South Sea [3]; taking with us victuals for twelve months; and had good winds from the east, though soft and weak, for five months' space and more. But then the wind came about, and settled in the west for many days so as we could make little or no way, and were sometimes in purpose to turn back. But then again there arose strong and great winds from the south, with a point east; which carried us up (for all that we could do) towards the north: by which time our victuals failed us, though we had made good spare of them. So that finding ourselves in the midst of the greatest wilderness of waters in the world, without victuals, we gave ourselves for lost men, and prepared for death. Yet we did lift up our hearts and voices to God above, who *showeth his wonders in the deep*; beseeching him of his mercy, that as in the beginning he discovered [4] the face of the deep, and brought forth dry land, so he would now discover land to us, that we might [5] not perish. And it came to pass that the next day about evening, we saw within a kenning before us, towards the north, as it were thick clouds, which did put us in some hope of land; knowing how that part of the South Sea was utterly unknown; and might have islands or continents that hitherto were not come to light. Wherefore we bent our course thither, where we saw the appearance of land, all that night; and in the dawning of the next day, we might plainly discern that it was a land; flat to our sight, and full of boscage; which made it shew the more dark. And after an hour and a half's sailing, we entered into a good haven, being the port of a fair city; not great indeed, but well built, and that gave a pleasant view from the sea: and we thinking every minute long till we were on land, came close to the shore, and offered to land. But straightways we saw divers of the people, with bastons in their hands, as it were forbidding us to land; yet without any cries or fierceness, but only as warning us off by signs that they made. Whereupon being not a little discomforted, we were advising with ourselves what we should do. During which time there made forth to us a small boat, with about eight persons in it; whereof one of them had in his hand a tipstaff of a yellow cane, tipped at both

[1] In the Latin translation Rawley adds, *aliarumque Instaurationis partium contexendarum;* alluding probably to the *De Augmentis*, the only portion of the Instauration not belonging to the Natural History, which he seems to have been employed upon afterwards.

[2] It was published at the end of the volume containing the *Sylva Sylvarum.*

[3] The words "by the South Sea" are omitted in the translation.

[4] So in the original. If *discovered* be the right word, it must mean *removed the covering of* the face of the deep. But I think there must be some mistake. The Latin version has *quemadmodum in principio congregationes aquarum mandavit et Aridam apparere fecit.* The allusion is, no doubt, to Genes. i. 9: "Let the waters under the heaven be gathered together unto one place, and let the dry land appear".

[5] *mought* in the original; a form of the word frequently, though not uniformly, adopted by Bacon. I have always substituted *might.*

ends with blue, who came aboard our ship, without any show of distrust at all. And when he saw one of our number present himself somewhat afore the rest, he drew forth a little scroll of parchment, (somewhat yellower than our parchment, and shining like the leaves of writing tables, but otherwise soft and flexible,) and delivered it to our foremost man. In which scroll were written in ancient Hebrew and in ancient Greek, and in good Latin of the School, and in Spanish, these words ; " Land ye not, none of you ; and provide to be gone from this coast within sixteen days, except you have further time given you. Meanwhile, if you want fresh water, or victual, or help for your sick, or that your ship needeth repair, write down your wants, and you shall have that which belongeth to mercy." This scroll was signed with a stamp of cherubins' wings, not spread but hanging downwards, and by them a cross. This being delivered, the officer returned, and left only a servant with us to receive our answer. Consulting hereupon amongst ourselves, we were much perplexed. The denial of landing and hasty warning us away troubled us much ; on the other side, to find that the people had languages and were so full of humanity, did comfort us not a little. And above all, the sign of the cross to that instrument was to us a great rejoicing, and as it were a certain presage of good. Our answer was in the Spanish tongue ; " That for our ship, it was well ; for we had rather met with calms and contrary winds than any tempests. For our sick, they were many, and in very ill case ; so that if they were not permitted to land, they ran danger of their lives." Our other wants we set down in particular ; adding, " that we had some little store of merchandise, which if it pleased them to deal for, it might supply our wants without being chargeable unto them". We offered some reward in pistolets unto the servant, and a piece of crimson velvet to be presented to the officer ; but the servant took them not, nor would scarce look upon them ; and so left us, and went back in another little boat which was sent for him.

About three hours after we had dispatched our answer, there came towards us a person (as it seemed) of place. He had on him a gown with wide sleeves, of a kind of water chamolet, of an excellent azure colour, far more glossy than ours ; his under apparel was green ; and so was his hat, being in the form of a turban, daintily made, and not so huge as the Turkish turbans ; and the locks of his hair came down below the brims of it. A reverend man was he to behold. He came in a boat, gilt in some part of it, with four persons more only in that boat ; and was followed by another boat, wherein were some twenty. When he was come within a flight-shot [6] of our ship, signs were made to us that we should send forth some to meet him upon the water ; which we presently did in our ship-boat, sending the principal man amongst us save one, and four of our number with him. When we were come within six yards of their boat, they called to us to stay, and not to approach farther ; which we did. And thereupon the man whom I before described stood up, and with a loud voice in Spanish, asked, " Are ye Christians ? " We answered, " We were " ; fearing the less, because of the cross we had seen in the subscription. At which answer the said person lifted up his right hand towards heaven, and drew it softly to his mouth, (which is the gesture they use when they thank God) and then said : " If ye will swear (all of you) by the merits of the Saviour that ye are no pirates, nor have shed blood lawfully nor unlawfully within forty days past, you may have licence to come on land." We said, " We were all ready to take that oath". Whereupon one of those that were with him, being (as it seemed) a notary, made an entry of this act. Which done, another of the attendants of the great person, which was with him in the same boat, after his lord had spoken a little to him, said aloud ; " My lord would have you know, that it is not of pride or greatness that he cometh not aboard your ship but for that in your answer you declare that you have many sick amongst you, he was warned by the Conservator of Health of the city that he should keep a distance ". We bowed ourselves towards him, and answered, " We were his humble servants ; and accounted for great honour and singular humanity towards us

[6] *spiculi jactum.* When archers try which can shoot furthest, they call it flight-shooting. The distance would be between 200 and 300 yards. Old Double, according to Justice Shallow, would have " carried you a forehand shaft a fourteen and fourteen and half " ; that is, 284 or 294 yards. See *Hen. IV*. Part II. act 3. sc. 2.

that which was already done; but hoped well that the nature of the sickness of our men was not infectious". So he returned; and a while after came the notary to us aboard our ship; holding in his hand a fruit of that country, like an orange, but of colour between orange-tawney and scarlet, which cast a most excellent odour. He used it (as it seemeth) for a preservative against infection. He gave us our oath: "By the name of Jesus and his merits": and after told us that the next day by six of the clock in the morning we should be sent to, and brought to the Strangers' House, (so he called it,) where we should be accommodated of things both for our whole and for our sick. So he left us; and when we [7] offered him some pistolets, he smiling said, "He must not be twice paid for one labour:" meaning (as I take it) that he had salary sufficient of the state for his service. For (as I after learned) they call an officer that taketh rewards, *twice paid.*

The next morning early, there came to us the same officer that came to us at first with his cane, and told us, "He came to conduct us to the Strangers' House; and that he had prevented the hour, because we might have the whole day before us for our business. "For," said he, "if you will follow my advice, there shall first go with me some few of you, and see the place, and how it may be made convenient for you; and then you may send for your sick, and the rest of your number which ye will bring on land." We thanked him, and said, "That this care which he took of desolate strangers God would reward". And so six of us went on land with him: and when we were on land, he went before us, and turned to us, and said, "He was but our servant, and our guide". He led us through three fair streets; and all the way we went there were gathered some people on both sides standing in a row; but in so civil a fashion, as if it had been not to wonder at us but to welcome us; and divers of them, as we passed by them, put their arms a little abroad; which is their gesture when they bid any welcome. The Strangers' House is a fair and spacious house, built of brick, of somewhat a bluer colour than our brick; and with handsome windows, some of glass, some of a kind of cambric oiled. He brought us first into a fair parlour above stairs, and then asked us, "What number of persons we were? And how many sick?" We answered, "We were in all (sick and whole) one and fifty persons, whereof our sick were seventeen". He desired us to have patience a little, and to stay till he came back to us; which was about an hour after; and then he led us to see the chambers which were provided for us, being in number nineteen: they having cast it (as it seemeth) that four of those chambers, which were better than the rest, might receive four of the principal men of our company, and lodge them alone by themselves; and the other fifteen chambers were to lodge us two and two together. The chambers were handsome and cheerful chambers, and furnished civilly. Then he led us to a long gallery, like a dorture [8], where he showed us all along the one side (for the other side was but wall and window) seventeen cells, very neat ones, having partitions of cedar wood. Which gallery and cells, being in all forty, (many more than we needed,) were instituted as an infirmary for sick persons. And he told us withal, that as any of our sick waxed well, he might be removed from his cell to a chamber; for which purpose there were set forth ten spare chambers, besides the number we spake of before. This done, he brought us back to the parlour, and lifting up his cane a little, (as they do when they give any charge or command [9]), said to us, "Ye are to know that the custom of the land requireth, that after this day and to-morrow, (which we give you for removing of your people from your ship), you are to keep within doors for three days. But let it not trouble you, nor do not think yourselves restrained, but rather left to your rest and ease. You shall want nothing, and there are six of our people appointed to attend you, for any business you may have abroad." We gave him thanks with all affection and respect, and said, "God surely is manifested in this land". We offered him also twenty pistolets; but he smiled, and only said; "What? twice paid!" And so he left us. Soon after our dinner was served in; which was right good viands, both for bread and meat [10]; better than any colle-

[7] So ed. 1635. Ed. 1629 has *he*.

[8] Dormitory. The Latin translation has, *qualia solent esse dormitoria monachorum.*

[9] *i.e.*, any charge which they have received from superior authority.

[10] The translation has both for meat and *drink: tam respectu ciborum quam potus.*

giate diet that I have known in Europe. We had also drink of three sorts, all wholesome and good; wine of the grape; a drink of grain, such as is with us our ale, but more clear; and a kind of cider made of a fruit of that country; a wonderful pleasing and refreshing drink. Besides, there were brought in to us great store of those scarlet oranges for our sick; which (they said) were an assured remedy for sickness taken at sea. There was given us also a box of small grey or whitish pills, which they wished our sick should take, one of the pills every night, before sleep; which (they said) would hasten their recovery. The next day, after that our trouble of carriage and removing of our men and goods out of our ship was somewhat settled and quiet, I thought good to call our company together, and when they were assembled said unto them; "My dear friends, let us know ourselves, and how it standeth with us. We are men cast on land, as Jonas was out of the whale's belly, when we were as buried in the deep: and now we are on land, we are but between death and life; for we are beyond both the old world and the new; and whether ever we shall see Europe, God only knoweth. It is a kind of miracle hath brought us hither: and it must be little less that shall bring us hence. Therefore in regard of our deliverance past, and our danger present and to come, let us look up to God, and every man reform his own ways. Besides we are come here amongst a Christian people, full of piety and humanity: let us not bring that confusion of face upon ourselves, as to show our vices or unworthiness before them. Yet there is more. For they have by commandment (though in form of courtesy) cloistered us within these walls for three days: who knoweth whether it be not to take some taste of our manners and conditions? and if they find them bad, to banish us straightways; if good, to give us further time. For these men that they have given us for attendance may withal have an eye upon us. Therefore for God's love, and as we love the weal of our souls and bodies, let us so behave ourselves as we may be at peace with God, and may find grace in the eyes of this people." Our company with one voice thanked me for my good admonition, and promised me to live soberly and civilly, and without giving any the least occasion of offence. So we spent our three days joyfully and without care, in expectation what would be done with us when they were expired. During which time, we had every hour joy of the amendment of our sick; who thought themselves cast into some divine pool of healing, they mended so kindly and so fast.

The morrow after our three days were past, there came to us a new man that we had not seen before, clothed in blue as the former was, save that his turban was white, with a small red cross on the top. He had also a tippet of fine linen. At his coming in, he did bend to us a little, and put his arms abroad. We of our parts saluted him in a very lowly and submissive manner; as looking that from him we should receive sentence of life or death. He desired to speak with some few of us: whereupon six of us only stayed, and the rest avoided the room. He said, "I am by office governor of this House of Strangers, and by vocation I am a Christian priest; and therefore am come to you to offer you my service, both as strangers and chiefly as Christians. Some things I may tell you, which I think you will not be unwilling to hear. The state hath given you licence to stay on land for the space of six weeks: and let it not trouble you if your occasions ask further time, for the law in this point is not precise; and I do not doubt but myself shall be able to obtain for you such further time as may be convenient. Ye shall also understand, that the Strangers' House is at this time rich, and much aforehand; for it hath laid up revenue these thirty-seven years; for so long it is since any stranger arrived in this part: and therefore take ye no care; the state will defray you all the time you stay; neither shall you stay one day the less for that. As for any merchandise ye have brought, ye shall be well used, and have your return either in merchandise or in gold and silver: for to us it is all one. And if you have any other request to make, hide it not. For ye shall find we will not make your countenance to fall by the answer ye shall receive. Only this I must tell you, that none of you must go above a *karan*" (that is with them a mile and an half) "from the walls of the city, without especial leave." We answered, after we had looked awhile one upon another admiring this gracious and parent-like usage; "That we could not tell what to say: for we wanted words to express our thanks ; and his noble free offers left us nothing to ask. It seemed to us that we had before us a picture of our salvation in heaven ; for we that were awhile since in the jaws of death, were now brought into a place where we found nothing but consolations. For the commandment laid upon us, we would not fail to obey it, though it was impossible but our hearts should be inflamed to tread further upon this happy and holy ground." We added ; " That our tongues should first cleave to the roofs of our mouths, ere we should forget either his reverend person or this whole nation in our prayers." We also most humbly besought him to accept of us as his true servants, by as just a right as ever men on earth were bounden ; laying and presenting both our persons and all we had at his feet. He said : " He was a priest, and looked for a priest's reward : which was our brotherly love and the good of our souls and bodies ". So he went from us, not without tears of tenderness in his eyes ; and left us also confused with joy and kindness, saying amongst ourselves, " That we were come into a land of angels, which did appear to us daily and prevent us with comforts, which we thought not of, much less expected ".

The next day, about ten of the clock, the governor came to us again, and after salutations said familiarly, " That he was come to visit us," and called for a chair, and sat him down : and we, being some ten of us, (the rest were of the meaner sort, or else gone abroad,) sat down with him. And when we were set, he began thus : " We of this island of Bensalem," (for so they call it in their language,) " have this : that by means of our solitary situation, and of the laws of secrecy which we have for our travellers, and our rare admission of strangers, we know well most part of the habitable world, and are ourselves unknown. Therefore because he that knoweth least is fittest to ask questions, it is more reason, for the entertainment of the time, that ye ask me questions, than that I ask you." We answered ; " That we humbly thanked him that he would give us leave so to do : and that we conceived by the taste we had already, that there was no worldly thing on earth more worthy to be known than the state of that happy land. But above all," (we said), " since that we were met from the several ends of the world, and hoped assuredly that we should meet one day in the kingdom of heaven, (for that we were both parts Christians,) we desired to know (in respect that land was so remote, and so divided by vast and unknown seas, from the land where our Saviour walked on earth,) who was the apostle of that nation, and how it was converted to the faith ? " It appeared in his face that he took great contentment in this our question : he said, " Ye knit my heart to you, by asking this question in the first place ; for it showeth that you *first seek the kingdom of heaven ;* and I shall gladly and briefly satisfy your demand.

" About twenty years after the ascension of our Saviour, it came to pass that there was seen by the people of Renfusa, (a city upon the eastern coast of our island,) within night, (the night was cloudy and calm,) as it might be some mile into the sea, a great pillar of light ; not sharp, but in form of a column or cylinder, rising from the sea a great way up towards heaven : and on the top of it was seen a large cross of light, more bright and resplendent than the body of the pillar. Upon which so strange a spectacle, the people of the city gathered apace together upon the sands, to wonder ; and so after put themselves into a number of small boats, to go nearer to this marvellous sight. But when the boats were come within about sixty yards of the pillar, they found themselves all bound, and could go no further ; yet so as they might move to go about, but might not approach nearer : so as the boats stood all as in a theatre, beholding this light as an heavenly sign [11]. It so fell out, that there was in one of the boats one of the wise men of the society of Salomon's House ; which house or college (my good brethren) is the very eye of this kingdom ; who having awhile attentively and devoutly viewed and contemplated this pillar and cross, fell down upon his face ; and then raised himself upon his knees, and lifting up his hands to heaven, made his prayers in this manner :

" ' Lord God of heaven and earth, thou hast vouchsafed of thy grace to those of our order, to know thy works of creation, and the secrets of them ; and to dis-

[11] *tanquam* scenam *cœlestem* in the translation.

cern (as far as appertaineth to the generations of men) between divine miracles, works of nature, works of art, and impostures and illusions of all sorts [12]. I do here acknowledge and testify before this people, that the thing which we now see before our eyes is thy Finger and a true Miracle; and forasmuch as we learn in our books that thou never workest miracles but to a divine and excellent end, (for the laws of nature are thine own laws and thou exceedest them not but upon great cause,) we most humbly beseech thee to prosper this great sign, and to give us the interpretation and use of it in mercy; which thou dost in some part secretly promise by sending it unto us.'

"When he had made his prayer, he presently found the boat he was in movable and unbound; whereas all the rest remained still fast; and taking that for an assurance of leave to approach, he caused the boat to be softly and with silence rowed towards the pillar. But ere he came near it the pillar and cross of light brake up, and cast itself abroad, as it were, into a firmament of many stars; which also vanished soon after, and there was nothing left to be seen but a small ark or chest of cedar, dry, and not wet at all with water, though it swam. And in the fore-end of it, which was towards him, grew a small green branch of palm; and when the wise man had taken it with all reverence into his boat, it opened of itself, and there were found in it a Book and a Letter; both written in fine parchment, and wrapped in sindons of linen. The Book contained all the canonical books of the Old and New Testament, according as you have them, (for we know well what the Churches with you receive); and the Apocalypse itself [13], and some other books of the New Testament which were not at that time written, were nevertheless in the book. And for the letter, it was in these words:

"'I Bartholomew, a servant of the Highest, and Apostle of Jesus Christ, was warned by an angel that appeared to me in a vision of glory, that I should commit this ark to the floods of the sea. Therefore I do testify and declare unto that people where God shall ordain this ark to come to land, that in the same day is come unto them salvation and peace and goodwill, from the Father, and from the Lord Jesus.'

"There was also in both these writings, as well the Book as the Letter, wrought a great miracle, conform to that of the Apostles in the orginal Gift of Tongues. For there being at that time in this land Hebrews, Persians, and Indians, besides the natives, every one read upon the Book and Letter, as if they had been written in his own language. And thus was this land saved from infidelity (as the remain of the old world was from water) by an ark, through the apostolical and miraculous evangelism of St. Bartholomew." And here he paused, and a messenger came, and called him from us. So this was all that passed in that conference.

The next day, the same governor came again to us immediately after dinner, and excused himself, saying, "That the day before he was called from us somewhat abruptly, but now he would make us amends, and spend time with us, if we held his company and conference agreeable". We answered, "That we held it so agreeable and pleasing to us, as we forgot both dangers past and fears to come, for the time we heard him speak; and that we thought an hour spent with him, was worth years of our former life". He bowed himself a little to us, and after we were set again, he said: "Well, the questions are on your part". One of our number said, after a little pause; "That there was a matter we were no less desirous to know, than fearful to ask, lest we might presume too far. But encouraged by his rare humanity towards us, (that could scarce think ourselves strangers, being his vowed and professed servants), we would take the hardiness to propound it: humbly beseeching him, if he thought it not fit to be answered, that he would pardon it, though he rejected it." We said: "We well observed those his words, which he formerly spake, that this happy island where we now stood was known to few, and yet knew most of the nations of the world; which

[12] *illusiones dæmonum cum imposturis omnimodis.*

[13] The original has a semicolon after "itself", which would seem to connect this clause with the last. But the translation (*Apocalypsis ipsa*) shows that it was meant to be the beginning of a new sentence.

we found to be true, considering they had the languages of Europe, and knew much of our state and business; and yet we in Europe (notwithstanding all the remote discoveries and navigations of this last age,) never heard any of the least inkling or glimpse of this island. This we found wonderful strange; for that all nations have inter-knowledge [14] one of another either by voyage into foreign parts, or by strangers that come to them: and though the traveller into a foreign country doth commonly know more by the eye, than he that stayeth at home can by relation of the traveller; yet both ways suffice to make a mutual knowledge, in some degree, on both parts. But for this island, we never heard tell of any ship of theirs that had been seen to arrive upon any shore of Europe; no, nor of either the East or West Indies; nor yet of any ship of any other part of the world that had made return from them. And yet the marvel rested not in this. For the situation of it (as his lordship said) in the secret conclave of such a vast sea might cause it. But then that they should have knowledge of the languages, books, affairs, of those that lie such a distance from them, it was a thing we could not tell what to make of; for that it seemed to us a condition and propriety of divine powers and beings, to be hidden and unseen to others, and yet to have others open and as in a light to them." At this speech the governor gave a gracious smile, and said; "That we did well to ask pardon for this question we now asked; for that it imported as if we thought this land a land of magicians, that sent forth spirits of the air into all parts, to bring them news and intelligence of other countries". It was answered by us all, in all possible humbleness, but yet with a countenance taking knowledge that we knew that he spake it but merrily, "That we were apt enough to think there was somewhat supernatural in this island; but yet rather as angelical than magical. But to let his lordship know truly what it was that made us tender and doubtful to ask this question, it was not any such conceit, but because we remembered he had given a touch in his former speech, that this land had laws of secrecy touching strangers." To this he said; "You remember it aright; and therefore in that I shall say to you I must reserve some particulars, which it is not lawful for me to reveal; but there will be enough left to give you satisfaction.

"You shall understand (that which perhaps you will scarce think credible that about three thousand years ago, or somewhat more, the navigation of the world, (specially for remote voyages,) was greater than at this day. Do not think with yourselves that I know not how much it is increased with you within these six-score years; I know it well, and yet I say greater then than now; whether it was, that the example of the ark, that saved the remnant of men from the universal deluge, gave men confidence to adventure upon the waters; or what it was; but such is the truth. The Phœnicians, and especially the Tyrians, had great fleets. So had the Carthaginians, their colony, which is yet further west. Toward the east, the shipping of Egypt and of Palestina was likewise great. China also, and the great Atlantis (that you call America), which have now but junks and canoes [15], abounded then in tall ships. This island (as appeareth by faithful registers of those times) had then fifteen hundred strong ships, of great content. Of all this there is with you sparing memory, or none; but we have large knowledge thereof.

"At that time, this land was known and frequented by the ships and vessels of all the nations before named. And (as it cometh to pass) they had many times men of other countries, that were no sailors, that came with them; as Persians, Chaldeans, Arabians; so as almost all nations of might and fame resorted hither; of whom we have some stirps and little tribes with us at this day. And for our own ships, they went sundry voyages, as well to your Straits, which you call the Pillars of Hercules [16], as to other parts in the Atlantic and Mediterrane Seas; as

[14] *enterknowledge* in the original. [15] *Canoa's* in the original.

[16] Hercules is called by Edrisi Dhoulcarnain. He says he lived in the time of Abraham, and has been confounded with Iscander Dhoulcarnain, or Alexander the two-horned. That the limits beyond which it is impossible to pass were set up by Dhoulcarnain gives the obvious explanation of the passage in Chaucer's *Troilus and Cressida*:—

"I am tyl God me bettre mynde sende,
At Dulcarnon, right at my wytte's end"—

"qui interpretes mire torsit".—*R. L. E.*

to Paguin [17] (which is the same with Cambaline [18]) and Quinzy [19], upon the Oriental Seas, as far as to the borders of the East Tartary.

"At the same time, and an age after, or more, the inhabitants of the great Atlantis did flourish [20]. For though the narration and description which is made by a great man with you, that the descendants of Neptune planted there; and of the magnificent temple, palace, city, and hill; and the manifold streams of goodly navigable rivers, (which, as so many chains, environed the same site and temple); and the several degrees of ascent whereby men did climb up to the same, as if it had been a *scala cœli*, be all poetical and fabulous: yet so much is true, that the said country of Atlantis, as well that of Peru, then called Coya, as that of Mexico, then named Tyrambel, were mighty and proud kingdoms in arms, shipping, and riches: so mighty, as at one time (or at least within the space of ten years) they both made two great expeditions; they of Tyrambel through the Atlantic to the Mediterrane Sea; and they of Coya through the South Sea upon this our island. And for the former of these, which was into Europe, the same author amongst you (as it seemeth) had some relation from the Egyptian priest whom he citeth. For assuredly such a thing there was. But whether it were the ancient Athenians that had the glory of the repulse and resistance of those forces, I can say nothing: but certain it is, there never came back either ship or man from that voyage. Neither had the other voyage of those of Coya upon us had better fortune, if they had not met with enemies of greater clemency. For the king of this island (by name Altabin) a wise man and a great warrior, knowing well both his own strength and that of his enemies, handled the matter so, as he cut off their land-forces from their ships; and entoiled both their navy and their camp with a greater power than theirs, both by sea and land; and compelled them to render themselves without striking stroke; and after they were at his mercy, contenting himself only with their oath that they should no more bear arms against him, dismissed them all in safety. But the Divine Revenge overtook not long after those proud enterprises. For within less than the space of one hundred years, the great Atlantis was utterly lost and destroyed: not by a great earthquake, as your man saith, (for that whole tract is little subject to earthquakes,) but by a particular deluge or inundation; those countries having, at this day, far greater rivers and far higher mountains to pour down waters, than any part of the old world. But it is true that the same inundation was not deep; not past forty foot, in most places, from the ground: so that although it destroyed man and beast generally, yet some few wild inhabitants of the wood [21] escaped. Birds also were saved by flying to the high trees and woods. For as for men, although they had buildings in many places higher than the depth of the water, yet that inundation, though it were shallow, had a long continuance; whereby they of the vale that were not drowned, perished for want of food and other things necessary. So as marvel you not at the thin population of America, nor at the rudeness and ignorance of the people; for you must account your inhabitants of America as a young people; younger a thousand years, at the least, than the rest of the world; for that there was so much time between the universal flood and their particular inundation. For the poor remnant of human seed which remained in their mountains peopled

[17] Peking. It seems as if Bacon supposed that Peking was a sea-port.—*R. L. E.* [The translation adds *civitatem in Chinâ antiquissimam*].

[18] Cambalu is the reading of the common text of Marco Polo. The word is properly Khambalik. It is the Tartar name for Peking.—*R. L. E.* [It is *Cambalu* in the translation; and in the English Bacon probably wrote *Cambalue.*—*J. S.*]

[19] The Quinsai of Marco Polo, now Hangchowfoo.—*R. L. E.*

[20] See Plato, *Critias*, p. 113, and *Timæus*, p. 25. Everything relating to the story of Atlantis has been collected by Humboldt, *Examen critique de l'Histoire de la Géographie*, etc., i. p. 167. Compare Martin, *Études sur le Timée*; and see Gesenius, *Monumenta Phœnicia*, for an account of a spurious Phœnician inscription, purporting to give the history of the destruction of Atlantis. It may be a question whether there be not some affinity between Atlantis and Homer's Phæacia.—*R. L. E.*

[21] The translation says, of the *mountains*: *silvestres habitatores quidam montium.*

the country again slowly, by little and little ; and being simple and savage people, (not like Noah and his sons, which was the chief family of the earth,) they were not able to leave letters, arts, and civility to their posterity ; and having likewise in their mountainous habitations been used (in respect of the extreme cold of those regions) to clothe themselves with the skins of tigers, bears, and great hairy goats, that they have in those parts ; when after they came down into the valley, and found the intolerable heats which are there, and knew no means of lighter apparel, they were forced to begin the custom of going naked, which continueth at this day. Only they take great pride and delight in the feathers of birds, and this also they took from those their ancestors of the mountains, who were invited unto it by the infinite flights of birds that came up to the high grounds, while the waters stood below. So you see, by this main accident of time, we lost our traffic with the Americans, with whom of all others, in regard they lay nearest to us, we had most commerce. As for the other parts of the world, it is most manifest that in the ages following (whether it were in respect of wars, or by a natural revolution of time,) navigation did everywhere greatly decay ; and specially far voyages (the rather by the use of galleys, and such vessels as could hardly brook the ocean), were altogether left and omitted. So then, that part of intercourse [22] which could be from other nations to sail to us, you see how it hath long since ceased ; except it were by some rare accident, as this of yours. But now of the cessation of that other part of intercourse, which might be by our sailing to other nations, I must yield you some other cause. For I cannot say (if I shall say truly,) but our shipping, for number, strength, mariners, pilots, and all things that appertain to navigation, is as great as ever ; and therefore why we should sit at home, I shall now give you an account by itself : and it will draw nearer to give you satisfaction to your principal question.

" There reigned in this island, about nineteen hundred years ago, a King, whose memory of all others we most adore ; not superstitiously, but as a divine instrument, though a mortal man ; his name was Solamona : and we esteem him as the law-giver of our nation. This king had a *large heart*, inscrutable for good ; and was wholly bent to make his kingdom and people happy. He therefore, taking into consideration how sufficient and substantive this land was to maintain itself without any aid at all of the foreigner ; being five thousand six hundred miles in circuit, and of rare fertility of soil in the greatest part thereof ; and finding also the shipping of this country might be plentifully set on work, both by fishing and by transportations from port to port, and likewise by sailing unto some small islands that are not far from us, and are under the crown and laws of this state ; and recalling into his memory the happy and flourishing estate wherein this land then was, so as it might be a thousand ways altered to the worse, but scarce any one way to the better ; thought nothing wanted to his noble and heroical intentions, but only (as far as human foresight might reach) to give perpetuity to that which was in his time so happily established. Therefore amongst his other fundamental laws of this kingdom, he did ordain the interdicts and prohibitions which we have touching entrance of strangers ; which at that time (though it was after the calamity of America) was frequent ; doubting novelties, and commixture of manners. It is true, the like law against the admission of strangers without licence is an ancient law in the kingdom of China, and yet continued in use. But therein it is a poor thing ; and hath made them a curious, ignorant, fearful, foolish nation. But our lawgiver made his law of another temper. For first, he hath preserved all points of humanity, in taking order and making provision for the relief of strangers distressed ; whereof you have tasted." At which speech (as reason was) we all rose up, and bowed ourselves. He went on. " That king also, still desiring to join humanity and policy together ; and thinking it against humanity to detain strangers here against their wills, and against policy that they should return and discover their knowledge of this state, he took this course : he did ordain that of the strangers that should be permitted to land, as many (at all times) might depart as would ; but as many as would stay should have very good conditions and means to live from the state. Wherein he saw so far, that now in so

[22] *entercourse* in orig.

many ages since the prohibition, we have memory not of one ship that ever returned; and but of thirteen persons only, at several times, that chose to return in our bottoms. What those few that returned may have reported abroad I know not. But you must think, whatsoever they have said could be taken where they came but for a dream. Now for our travelling from hence into parts abroad, our Lawgiver thought fit altogether to restrain it. So is it not in China. For the Chineses sail where they will or can; which sheweth that their law of keeping out strangers is a law of pusillanimity and fear. But this restraint of ours hath one only exception, which is admirable; preserving the good which cometh by communicating with strangers, and avoiding the hurt; and I will now open it to you. And here I shall seem a little to digress, but you will by and by find it pertinent. Ye shall understand (my dear friends) that amongst the excellent acts of that king, one above all hath the pre-eminence. It was the erection and institution of an Order or Society which we call *Salomon's House;* the noblest foundation (as we think) that ever was upon the earth; and the lanthorn of this kingdom. It is dedicated to the study of the Works and Creatures of God. Some think it beareth the founder's name a little corrupted, as if it should be Solamona's House. But the records write it as it is spoken. So as I take it to be denominate of the King of the Hebrews, which is famous with you, and no stranger to us. For we have some parts of his works which with you are lost; namely, that Natural History which he wrote, of all plants, from the *cedar of Libanus* to the *moss that groweth out of the wall*, and of all *things that have life and motion*. This maketh me think that our king, finding himself to symbolize in many things with that king of the Hebrews (which lived many years before him), honoured him with the title of this foundation [23]. And I am the rather induced to be of this opinion, for that I find in ancient records this Order or Society is sometimes called Salomon's House, and sometimes the College of the Six Days Works; whereby I am satisfied that our excellent king had learned from the Hebrews that God had created the world and all that therein is within six days; and therefore he instituting that House for the finding out of the true nature of all things, (whereby God might have the more glory in the workmanship of them, and men the more fruit in the use of them), did give it also that second name. But now to come to our present purpose. When the king had forbidden to all his people navigation into any part that was not under his crown, he made nevertheless this ordinance; That every twelve years there should be set forth out of this kingdom two ships, appointed to several voyages; That in either of these ships there should be a mission of three of the Fellows or Brethren of Salomon's House; whose errand was only to give us knowledge of the affairs and state of those countries to which they were designed, and especially of the sciences, arts, manufactures, and inventions of all the world; and withal to bring unto us books, instruments, and patterns in every kind; That the ships, after they had landed the brethren, should return; and that the brethren should stay abroad till the new mission. These ships are not otherwise fraught, than with store of victuals, and good quantity of treasure to remain with the brethren, for the buying of such things and rewarding of such persons as they should think fit. Now for me to tell you how the vulgar sort of mariners are contained from being discovered at land; and how they that must be put on shore for any time, colour themselves under the names of other nations; and to what places these voyages have been designed; and what places of *rendez-vous* are appointed for the new missions; and the like circumstances of the

[23] Bacon in speaking of this king who symbolizes with Solomon seems to allude to James I.—*R. L. E.* [If the *New Atlantis* had been written in the earlier part of James's reign, Bacon might have been suspected perhaps of some such allusion. He might have hoped to encourage James to justify the parallel by going and doing likewise. But since James had now reigned above twenty years without doing or attempting to do anything for the furtherance of Natural Philosophy; without showing any interest in it or any taste or capacity for it: I cannot understand what the allusion can be or where the resemblance. Nor does it seem necessary to suppose anything of the kind in order to explain why a model-king for wisdom and knowledge should be likened to Solomon.—*J. S.*

practique ; I may not do it : neither is it much to your desire. But thus you see we maintain a trade, not for gold, silver, or jewels ; nor for silks ; nor for spices ; nor any other commodity of matter ; but only for God's first creature, which was *Light :* to have *light* (I say) of the growth of all parts of the world [24]." And when he had said this, he was silent ; and so were we all. For indeed we were all astonished to hear so strange things so probably told. And he, perceiving that we were willing to say somewhat but had it not ready, in great courtesy took us off, and descended to ask us questions of our voyage and fortunes ; and in the end concluded, that we might do well to think with ourselves what time of stay we would demand of the state ; and bade us not to scant ourselves ; for he would procure such time as we desired. Whereupon we all rose up, and presented ourselves to kiss the skirt of his tippet ; but he would not suffer us ; and so took his leave. But when it came once amongst our people that the state used to offer conditions to strangers that would stay, we had work enough to get any of our men to look to our ship, and to keep them from going presently to the governor to crave conditions. But with much ado we refrained them, till we might agree what course to take.

We took ourselves now for free men, seeing there was no danger of our utter perdition ; and lived most joyfully, going abroad and seeing what was to be seen in the city and places adjacent within our tedder ; and obtaining acquaintance with many of the city, not of the meanest quality ; at whose hands we found such humanity, and such a freedom and desire to take strangers as it were into their bosom, as was enough to make us forget all that was dear to us in our own countries : and continually we met with many things right worthy of observation and relation ; as indeed, if there be a mirror in the world worthy to hold men's eyes, it is that country. One day there were two of our company bidden to a Feast of the Family, as they call it. A most natural, pious, and reverend custom it is, shewing that nation to be compounded of all goodness. This is the manner of it. It is granted to any man that shall live to see thirty persons descended of his body alive together, and all above three years old, to make this feast ; which is done at the cost of the state. The Father of the Family, whom they call the *Tirsan*, two days before the feast, taketh to him three of such friends as he liketh to choose ; and is assisted also by the governor of the city or place where the feast is celebrated ; and all the persons of the family, of both sexes, are summoned to attend him. These two days the Tirsan sitteth in consultation concerning the good estate of the family. There, if there be any discord or suits between any of the family, they are compounded and appeased. There, if any of the family be distressed or decayed, order is taken for their relief and competent means to live. There, if any be subject to vice, or take ill courses, they are reproved and censured. So likewise direction is given touching marriages, and the courses of life which any of them should take, with divers other the like orders and advices. The governor assisteth, to the end to put in execution by his public authority the decrees and orders of the Tirsan, if they should be disobeyed ; though that seldom needeth ; such reverence and obedience they give to the order of nature. The Tirsan doth also then ever choose one man from amongst his sons, to live in house with him : who is called ever after the Son of the Vine. The reason will hereafter appear. On the feast-day, the Father or Tirsan cometh forth after divine service into a large room where the feast is celebrated ; which room hath an half-pace [25] at the upper end. Against the wall, in the middle of the half-pace, is a chair placed for him, with a table and carpet before it. Over the chair is a state [26], made round or oval, and it is of ivy ; an ivy somewhat whiter than ours, like the leaf of a silver asp, but more shining ; for it is green all winter. And the state is curiously wrought with silver and silk of divers colours building or binding in the ivy ; and is ever of the work of some of the daughters of the family ; and veiled over at the top with a fine net of silk and silver. But the substance of it is true ivy ; whereof, after it is taken down, the friends of the family are desirous to have

[24] *i.e.*, in whatever parts of the world it is to be found.

[25] Half-pace or daïs, the part raised by a low step above the rest of the floor.—*R. L. E.*

[26] *i.e.* a canopy, *conopeum.*

some leaf or sprig to keep. The Tirsan cometh forth with all his generation or lineage,[27] the males before him, and the females following him; and if there be a mother from whose body the whole lineage[27] is descended, there is a traverse placed in a loft above on the right hand of the chair, with a privy door, and a carved window of glass, leaded with gold and blue; where she sitteth, but is not seen. When the Tirsan is come forth, he sitteth down in the chair; and all the lineage place themselves against the wall, both at his back and upon the return of the half-pace, in order of their years without difference of sex; and stand upon their feet. When he is set; the room being always full of company, but well kept and without disorder; after some pause there cometh in from the lower end of the room a *Taratan* (which is as much as an herald) and on either side of him two young lads; whereof one carrieth a scroll of their shining yellow parchment; and the other a cluster of grapes of gold, with a long foot or stalk. The herald and children are clothed with mantles of sea-water-green sattin; but the herald's mantle is streamed with gold, and hath a train. Then the herald with three curtesies, or rather inclinations, cometh up as far as the half-pace; and there first taketh into his hand the scroll. This scroll is the King's Charter, containing gift of revenew, and many privileges, exemptions, and points of honour, granted to the Father of the Family; and is ever styled and directed, *To such an one our well-beloved friend and creditor:* which is a title proper only to this case. For they say the king is debtor to no man, but for propagation of his subjects. The seal set to the king's charter is the king's image, imbossed or moulded in gold; and though such charters be expedited of course, and as of right, yet they are varied by discretion, according to the number and dignity of the family. This charter the herald readeth aloud; and while it is read, the father or Tirsan standeth up, supported by two of his sons, such as he chooseth. Then the herald mounteth the half-pace, and delivereth the charter into his hand: and with that there is an acclamation by all that are present in their language, which is thus much: *Happy are the people of Bensalem.* Then the herald taketh into his hand from the other child the cluster of grapes, which is of gold, both the stalk and the grapes. But the grapes are daintily enamelled; and if the males of the family be the greater number, the grapes are enamelled purple, with a little sun set on the top; if the females, then they are enamelled into a greenish yellow, with a crescent on the top. The grapes are in number as many as there are descendants of the family. This golden cluster the herald delivereth also to the Tirsan; who presently delivereth it over to that son that he had formerly chosen to be in house with him: who beareth it before his father as an ensign of honour when he goeth in public, ever after; and is thereupon called the Son of the Vine. After this ceremony ended, the father or Tirsan retireth; and after some time cometh forth again to dinner, where he sitteth alone under the state, as before; and none of his descendants sit with him, of what degree or dignity soever, except he hap to be of Salomon's House. He is served only by his own children, such as are male; who perform unto him all service of the table upon the knee; and the women only stand about him, leaning against the wall. The room below the half-pace hath tables on the sides for the guests that are bidden; who are served with great and comely order; and towards the end of dinner (which in the greatest feasts with them lasteth never above an hour and an half) there is an hymn sung, varied according to the invention of him that composeth it, (for they have excellent poesy), but the subject of it is (always) the praises of Adam and Noah and Abraham; whereof the former two peopled the world, and the last was the Father of the Faithful: concluding ever with a thanksgiving for the nativity of our Saviour, in whose birth the births of all are only blessed. Dinner being done, the Tirsan retireth again; and having withdrawn himself alone into a place where he maketh some private prayers, he cometh forth the third time, to give the blessing; with all his descendants, who stand about him as at the first. Then he calleth them forth by one and by one, by

[27] *linage* in the original; which seems to be the proper form of the word. The *e* may have been introduced originally as a direction for the lengthening of the first syllable, and then the resemblance of the word to such words as *lineal* may have suggested the modern pronunciation.

name, as he pleaseth, though seldom the order of age be inverted. The person that is called (the table being before removed) kneeleth down before the chair, and the father layeth his hand upon his head, or her head, and giveth the blessing in these words: *Son of Bensalem, (or Daughter of Bensalem,) thy father saith it; the man by whom thou hast breath and life speaketh the word; The blessing of the everlasting Father, the Prince of Peace, and the Holy Dove be upon thee, and make the days of thy pilgrimage good and many.* This he saith to every of them; and that done, if there be any of his sons of eminent merit and virtue, (so they be not above two,) he calleth for them again; and saith, laying his arm over their shoulders, they standing; *Sons, it is well ye are born, give God the praise, and persevere to the end.* And withal delivereth to either of them a jewel, made in the figure of an ear of wheat, which they ever after wear in the front of their turban or hat. This done, they fall to music and dances, and other recreations, after their manner, for the rest of the day. This is the full order of that feast.

By that time six or seven days were spent, I was fallen into strait acquaintance with a merchant of that city, whose name was Joabin. He was a Jew, and circumcised: for they have some few stirps of Jews yet remaining among them, whom they leave to their own religion. Which they may the better do, because they are of a far differing disposition from the Jews in other parts. For whereas they hate the name of Christ, and have a secret inbred rancour against the people amongst whom they live: these (contrariwise) give unto our Saviour many high attributes, and love the nation of Bensalem extremely. Surely this man of whom I speak would ever acknowledge that Christ was born of a virgin, and that he was more than a man; and he would tell how God made him ruler of the Seraphims which guard his throne; and they call him also the *Milken Way*, and the *Eliah* of the *Messiah;* and many other high names; which though they be inferior to his divine Majesty, yet they are far from the language of other Jews. And for the country of Bensalem, this man would make no end of commending it: being desirous, by tradition among the Jews there, to have it believed that the people thereof were of the generations of Abraham, by another son, whom they call Nachoran; and that Moses by a secret cabala ordained the laws of Bensalem which they now use; and that when the Messiah should come, and sit in his throne at Hierusalem, the king of Bensalem should sit at his feet, whereas other kings should keep a great distance. But yet setting aside these Jewish dreams, the man was a wise man, and learned, and of great policy, and excellently seen in the laws and customs of that nation. Amongst other discourses, one day I told him I was much affected with the relation I had from some of the company, of their custom in holding the Feast of the Family; for that (methought) I had never heard of a solemnity wherein nature did so much preside. And because propagation of familes proceedeth from the nuptial copulation, I desired to know of him what laws and customs they had concerning marriage; and whether they kept marriage well; and whether they were tied to one wife? For that where population is so much affected, and such as with them it seemed to be, there is commonly permission of plurality of wives. To this he said, " You have reason for to commend that excellent institution of the Feast of the Family. And indeed we have experience, that those families that are partakers of the blessing of that feast do flourish and prosper ever after in an extraordinary manner. But hear me now, and I will tell you what I know. You shall understand that there is not under the heavens so chaste a nation as this of Bensalem; nor so free from all pollution or foulness. It is the virgin of the world. I remember I have read in one of your European books, of an holy hermit amongst you that desired to see the Spirit of Fornication; and there appeared to him a little foul ugly Æthiop [28]. But if he had desired to see the Spirit of Chastity of Bensalem, it would have appeared to him in the likeness of a fair beautiful Cherubin. For there is nothing amongst mortal men more fair and admirable, than the chaste minds of this people. Know therefore, that with them there are no stews, no dissolute houses, no courtesans, nor any thing of that kind. Nay they wonder (with detestation) at you in Europe, which permit such things. They say ye have put marriage out

[28] The Klein Meister of La Motte Fouqué's *Sintram.*—*R. L. E.*

of office: for marriage is ordained a remedy for unlawful concupiscence; and natural concupiscence seemeth as a spur to marriage. But when men have at hand a remedy more agreeable to their corrupt will, marriage is almost expulsed. And therefore there are with you seen infinite men that marry not, but chuse rather a libertine and impure single life, than to be yoked in marriage; and many that do marry, marry late, when the prime and strength of their years is past. And when they do marry, what is marriage to them but a very bargain; wherein is sought alliance, or portion, or reputation, with some desire (almost indifferent) of issue; and not the faithful nuptial union of man and wife, that was first instituted. Neither is it possible that those that have cast away so basely so much of their strength, should greatly esteem children, (being of the same matter [29]), as chaste men do. So likewise during marriage, is the case much amended, as it ought to be if those things were tolerated only for necessity? No, but they remain still as a very affront to marriage. The haunting of those dissolute places, or resort to courtesans, are no more punished in married men than in bachelors. And the depraved custom of change, and the delight in meretricious embracements, (where sin is turned into art), [30] maketh marriage a dull thing, and a kind of imposition or tax. They hear you defend these things, as done to avoid greater evils; as advoutries, deflouring of virgins, unnatural lust, and the like. But they say this is a preposterous wisdom; and they call it *Lot's offer*, who to save his guests from abusing, offered his daughters: nay they say farther that there is little gained in this; for that the same vices and appetites do still remain and abound; unlawful lust being like a furnace, that if you stop the flames altogether, it will quench; but if you give it any vent, it will rage. As for masculine love, they have no touch of it; and yet there are not so faithful and inviolate friendships in the world again as are there; and to speak generally, (as I said before,) I have not read of any such chastity in any people as theirs. And their usual saying is, *That whosoever is unchaste cannot reverence himself*; and they say, *That the reverence of a man's self is, next religion, the chiefest bridle of all vices*". And when he had said this, the good Jew paused a little; whereupon I, far more willing to hear him speak on than to speak myself, yet thinking it decent that upon his pause of speech I should not be altogether silent, said only this; "That I would say to him, as the widow of Sarepta said to Elias; that he was come to bring to memory our sins; and that I confess the righteousness of Bensalem was greater than the righteousness of Europe". At which speech he bowed his head, and went on in this manner: "They have also many wise and excellent laws touching marriage. They allow no polygamy. They have ordained that none do intermarry or contract, until a month be passed from their first interview. Marriage without consent of parents they do not make void, but they mulct it in the inheritors: for the children of such marriages are not admitted to inherit above a third part of their parents' inheritance. I have read in a book of one of your men, of a Feigned Commonwealth, where the married couple are permitted, before they contract, to see one another naked [31]. This they dislike; for they think it a scorn to give a refusal after so familiar knowledge: but because of many hidden defects in men and women's bodies [32], they have a more civil way; for they have near every town a couple of pools, (which they call *Adam and Eve's pools*,) where it is permitted to one of the friends of the man, and another of the friends of the woman, to see them severally bathe naked."

And as we were thus in conference, there came one that seemed to be a messenger, in a rich huke [33], that spake with the Jew: whereupon he turned to me and said; "You will pardon me, for I am commanded away in haste". The

[29] *liberi* (*pars nostri altera*).

[30] Non v'era giunto ancor Sardanapalo
A mostrar cio ch' in camera si puote.
Dante, *Paradiso*, xiv. —*R. L. E.*

[31] See More's Utopia, book ii.—*R. L. E.*

[32] The translation adds *qui matrimonium postea infelix reddere possint.*

[33] *indutus tunicâ pictâ et inauratâ.*

next morning he came to me again, joyful as it seemed, and said, "There is word come to the governor of the city, that one of the Fathers of Salomon's House will be here this day seven-night: we have seen none of them this dozen years. His coming is in state; but the cause of his coming is secret. I will provide you and your fellows of a good standing to see his entry". I thanked him, and told him, "I was most glad of the news". The day being come, he made his entry. He was a man of middle stature and age, comely of person, and had an aspect as if he pitied men. He was clothed in a robe of fine black cloth, with wide sleeves and a cape. His under garment was of excellent white linen down to the foot, girt with a girdle of the same; and a sindon or tippet of the same about his neck. He had gloves that were curious, and set with stone; and shoes of peach-coloured velvet. His neck was bare to the shoulders. His hat was like a helmet, or Spanish Montera; and his locks curled below it decently: they were of colour brown. His beard was cut round, and of the same colour with his hair, somewhat lighter [34]. He was carried in a rich chariot without wheels, litter-wise; with two horses at either end, richly trapped in blue velvet embroidered; and two footmen on each side in the like attire. The chariot was all of cedar, gilt, and adorned with crystal; save that the fore-end had pannels of sapphires, set in borders of gold, and the hinder-end the like of emeralds [35] of the Peru colour. There was also a sun of gold, radiant, upon the top, in the midst [36]; and on the top before, a small cherub of gold, with wings displayed. The chariot was covered with cloth of gold tissued upon blue. He had before him fifty attendants, young men all, in white sattin loose coats to the mid-leg; and stockings of white silk; and shoes of blue velvet; and hats of blue velvet; with fine plumes of divers colours, set round like hat-bands. Next before the chariot went two men, bare-headed, in linen garments down to the foot, girt, and shoes of blue velvet; who carried the one a crosier, the other a pastoral staff like a sheep-hook; neither of them of metal, but the crosier of balm-wood, the pastoral staff of cedar. Horsemen he had none, neither before nor behind his chariot: as it seemeth, to avoid all tumult and trouble. Behind his chariot went all the officers and principals of the Companies of the City. He sat alone, upon cushions of a kind of excellent plush, blue: and under his foot curious carpets of silk of divers colours, like the Persian, but far finer. He held up his bare hand as he went, as blessing the people, but in silence. The street was wonderfully well kept: so that there was never any army had their men stand in better battle-array, than the people stood. The windows likewise were not crowded, but every one stood in them as if they had been placed. When the shew was passed, the Jew said to me; "I shall not be able to attend you as I would, in regard of some charge the city hath laid upon me, for the entertaining of this great person". Three days after, the Jew came to me again, and said; "Ye are happy men; for the Father of Salomon's House taketh knowledge of your being here, and commanded me to tell you that he will admit all your company to his presence, and have private conference with one of you that ye shall choose: and for this hath appointed the next day after to-morrow. And because he meaneth to give you his blessing, he hath appointed it in the forenoon." We came at our day and hour, and I was chosen by my fellows for the private access. We found him in a fair chamber, richly hanged, and carpeted under foot, without any degrees to the state. He was set upon a low throne richly adorned, and a rich cloth of state over his head, of blue sattin embroidered. He was alone, save that he had two pages of honour, on either hand one, finely attired in white. His under-garments were the like that we saw him wear in the chariot; but instead of his gown, he had on him a mantle with a cape, of the same fine black, fastened about him. When we came in, as we were taught, we bowed low at our first entrance; and when we were come near his chair, he stood up, holding forth his hand ungloved, and in posture of blessing; and we every one of us stooped down, and kissed the hem of his tippet. That done, the rest departed, and I

[34] The words "somewhat lighter" are omitted in the translation.

[35] *emerauds* in orig.

[36] *Etiam in medio verticis cathedræ, sol erat, ex auro radians.* The English in the original has a comma after "gold," and no stop after "radiant"; a misprint probably.

remained. Then he warned the pages forth of the room, and caused me to sit down beside him, and spake to me thus in the Spanish tongue :

"God bless thee, my son ; I will give thee the greatest jewel I have. For I will impart unto thee, for the love of God and men, a relation of the true state of Salomon's House. Son, to make you know the true state of Salomon's House, I will keep this order. First, I will set forth unto you the end of our foundation. Secondly, the preparations and instruments we have for our works. Thirdly, the several employments and functions whereto our fellows are assigned. And fourthly, the ordinances and rites which we observe.

" The End of our Foundation is the knowledge of Causes, and secret motions of things ; and the enlarging of the bounds of Human Empire, to the effecting of all things possible.

" The Preparations and Instruments are these. We have large and deep caves of several depths : the deepest are sunk six hundred fathom ; and some of them are digged and made under great hills and mountains : so that if you reckon together the depth of the hill and the depth of the cave, they are (some of them) above three miles deep. For we find that the depth of a hill, and the depth of a cave from the flat, is the same thing ; both remote alike from the sun and heaven's beams, and from the open air. These caves we call the Lower Region. And we use them for all coagulations, indurations, refrigerations, and conservations of bodies. We use them likewise for the imitation of natural mines ; and the producing also of new artificial metals, by compositions and materials which we use and lay there for many years. We use them also sometimes, (which may seem strange,) for curing of some diseases, and for prolongation of life in some hermits that choose to live there, well accommodated of all things necessary ; and indeed live very long ; by whom also we learn many things.

" We have burials in several earths, where we put divers cements, as the Chineses do their porcellain. But we have them in greater variety, and some of them more fine. We have also great variety of composts, and soils, for the making of the earth fruitful.

" We have high towers ; the highest about half a mile in height ; and some of them likewise set upon high mountains ; so that the vantage of the hill with the tower is in the highest of them three miles at least. And these places we call the Upper Region : accounting the air between the high places and the low, as a Middle Region. We use these towers, according to their several heights and situations, for insolation, refrigeration, conservation ; and for the view of divers meteors ; as winds, rain, snow, hail, and some of the fiery meteors also. And upon them, in some places, are dwellings of hermits, whom we visit sometimes, and instruct what to observe.

" We have great lakes both salt and fresh, whereof we have use for the fish and fowl. We use them also for burials of some natural bodies : for we find a difference in things buried in earth or in air below the earth, and things buried in water. We have also pools, of which some do strain fresh water out of salt ; and others by art do turn fresh water into salt. We have also some rocks in the midst of the sea, and some bays upon the shore for some works wherein is required the air and vapour of the sea. We have likewise violent streams and cataracts, which serve us for many motions : and likewise engines for multiplying and enforcing of winds, to set also on going divers motions.

" We have also a number of artificial wells and fountains, made in imitation of the natural sources and baths ; as tincted upon vitriol, sulphur, steel, lead, brass, nitre, and other minerals. And again we have little wells for infusions of many things, where the waters take the virtue quicker and better than in vessels or basons. And amongst them we have a water which we call Water of Paradise, being, by that we do to it, made very sovereign for health, and prolongation of life.

" We have also great and spacious houses, where we imitate and demonstrate [37]

[37] *i.e.* exhibit : *in quibus imitamenta et repræsentationes meteororum exhibemus.*

meteors; as snow, hail, rain, some artificial rains of bodies and not of water, thunders, lightnings [38]; also generations of bodies in air; as frogs, flies, and divers others.

"We have also certain chambers, which we call Chambers of Health, where we qualify the air as we think good and proper for the cure of divers diseases, and preservation of health [39].

"We have also fair and large baths, of several mixtures, for the cure of diseases, and the restoring of man's body from arefaction; and others for the confirming of it in strength of sinews, vital parts, and the very juice and substance of the body.

"We have also large and various orchards and gardens, wherein we do not so much respect beauty, as variety of ground and soil, proper for divers trees and herbs; and some very spacious, where trees and berries are set whereof we make divers kinds of drinks, besides the vineyards. In these we practise likewise all conclusions of grafting and inoculating, as well of wild-trees as fruit-trees, which produceth many effects. And we make (by art) in the same orchards and gardens, trees and flowers to come earlier or later than their seasons; and to come up and bear more speedily than by their natural course they do. We make them also by art greater much than their nature; and their fruit greater and sweeter and of differing taste, smell, colour, and figure, from their nature. And many of them we so order, as they become of medicinal use.

"We have also means to make divers plants rise by mixtures of earths without seeds; and likewise to make divers new plants, differing from the vulgar; and to make one tree or plant turn into another.

"We have also parks and inclosures of all sorts of beasts and birds, which we use not only for view or rareness, but likewise for dissections and trials; that thereby we may take light what may be wrought upon the body of man. Wherein we find many strange effects; as continuing life in them, though divers parts, which you account vital, be perished and taken forth; resuscitating of some that seem dead in appearance; and the like. We try also all opisons and other medicines upon them, as well of chirurgery as physic [40]. By art likewise, we make them greater or taller than their kind is; and contrariwise dwarf them, and stay their growth: we make them more fruitful and bearing than their kind is; and contrariwise barren and not generative. Also we make them differ in colour, shape, activity, many ways. We find means to make commixtures and copulations of different kinds; which have produced many new kinds, and them not barren, as the general opinion is. We make a number of kinds of serpents, worms, flies, fishes, of putrefaction; whereof some are advanced (in effect) to be perfect creatures, like beasts or birds; and have sexes, and do propagate. Neither do we this by chance, but we know beforehand of what matter and commixture what kind of those creatures will arise [41].

"We have also particular pools, where we make trials upon fishes, as we have said before of beasts and birds.

"We have also places for breed and generation of those kinds of worms and flies which are of special use; such as are with you your silk-worms and bees.

"I will not hold you long with recounting of our brew-houses, bake-houses, and kitchens, where are made divers drinks, breads, and meats, rare and of special effects. Wines we have of grapes; and drinks of other juice of fruits, of grains,

[38] The translation adds *coruscationum*.

[39] This experiment has been tried, especially by Dr. Beddoes of Clifton, but without any marked result. Some relief has been obtained in cases of phthisis by inhaling oxygenated air.—*R. L. E.*

[40] The translation adds *ut corpori humano melius caveamus*.

[41] This passage is quoted with great approbation by Geoffroi St. Hilaire at the end of a memoir on the results of artificial incubation read before the Academy of Sciences in 1826, and published in the *Annales du Muséum* for that year. It may be said that he was the first by whom the scientific importance of monstrosities was fully appreciated, and in answer to the objections which were made to the study of Teratology on the ground of its inutility, he invokes the authority of Bacon.—*R. L. E.*

and of roots : and of mixtures with honey, sugar, manna, and fruits dried and decocted. Also of the tears or woundings of trees, and of the pulp of canes. And these drinks are of several ages, some to the age or last of forty years. We have drinks also brewed with several herbs, and roots, and spices ; yea with several fleshes, and white meats ; whereof some of the drinks are such, as they are in effect meat and drink both [42] : so that divers, especially in age, do desire to live with them, with little or no meat or bread. And above all, we strive to have drinks of extreme thin parts, to insinuate into the body, and yet without all biting, sharpness, or fretting ; insomuch as some of them put upon the back of your hand will, with a little stay, pass through to the palm, and yet taste mild to the mouth. We have also waters which we ripen in that fashion, as they become nourishing ; so that they are indeed excellent drink ; and many will use no other. Breads we have of several grains, roots, and kernels ; yea and some of flesh and fish dried ; with divers kinds of leavenings and seasonings : so that some do extremely move appetites ; some do nourish so, as divers do live of them, without any other meat ; who live very long. So for meats, we have some of them so beaten and made tender and mortified, yet without all corrupting, as a weak heat of the stomach will turn them into good chylus, as well as a strong heat would meat otherwise prepared. We have some meats also and breads and drinks, which taken by men enable them to fast long after ; and some other, that used make the very flesh of men's bodies sensibly more hard and tough, and their strength far greater than otherwise it would be.

" We have dispensatories, or shops of medicines. Wherein you may easily think, if we have such variety of plants and living creatures more than you have in Europe, (for we know what you have,) the simples, drugs, and ingredients of medicines, must likewise be in so much the greater variety. We have them likewise of divers ages, and long fermentations. And for their preparations, we have not only all manner of exquisite distillations and separations, and especially by gentle heats and percolations through divers strainers, yea and substances ; but also exact forms of composition, whereby they incorporate almost, as they were natural simples.

" We have also divers mechanical arts, which you have not ; and stuffs made by them ; as papers, linen, silks, tissues ; dainty works of feathers of wonderful lustre ; excellent dyes, and many others ; and shops likewise, as well for such as are not brought into vulgar use amongst us as for those that are. For you must know that of the things before recited, many of them are grown into use throughout the kingdom ; but yet if they did flow from our invention, we have of them also for patterns and principals.

" We have also furnaces of great diversities, and that keep great diversity of heats ; fierce and quick ; strong and constant ; soft and mild ; blown, quiet ; dry, moist ; and the like. But above all, we have heats in imitation of the sun's and heavenly bodies' heats, that pass divers inequalities and (as it were) orbs, progresses, and returns, whereby we produce admirable effects. Besides, we have heats of dungs, and of bellies and maws of living creatures, and of their bloods and bodies ; and of hays and herbs laid up moist ; of lime unquenched ; and such like. Instruments also which generate heat only by motion [43]. And farther, places for strong insolations ; and again, places under the earth, which by nature or art yield heat. These divers heats we use, as the nature of the operation which we intend requireth.

" We have also perspective-houses, where we make demonstrations of all lights and radiations ; and of all colours ; and out of things uncoloured and transparent, we can represent unto you all several colours ; not in rain-bows, as it is in gems and prisms, but of themselves single. We represent also all multiplications of

[42] Chocolate, which however was well known in Bacon's time, seems to fulfil this description. It long since gave rise to a doubt whether drinking it amounted to breaking fast. See the treatise of the Jesuit Hurtado, " Utrum potio chocolatica frangat jejunium Ecclesiæ".—*R. L. E.*

[43] Bacon seems to refer to the result of his investigation into the form of heat, namely that heat is a kind of motion.—*R. L. E.*

light, which we carry to great distance, and make so sharp as to discern small points and lines; also all colorations of light: all delusions and deceits of the sight, in figures, magnitudes, motions, colours: all demonstrations of shadows. We find also divers means, yet unknown to you, of producing of light orginally from divers bodies. We procure means of seeing objects afar off; as in the heaven and remote places; and represent things near as afar off, and things afar off as near; making feigned distances. We have also helps for the sight, far above spectacles and glasses in use. We have also glasses and means to see small and minute bodies perfectly and distinctly; as the shapes and colours of small flies and worms, grains and flaws in gems, which cannot otherwise be seen; observations in urine [44] and blood, not otherwise to be seen [45]. We make artificial rainbows, halos, and circles about light. We represent also all manner of reflexions, refractions, and multiplications of visual beams of objects.

"We have also precious stones of all kinds, many of them of great beauty, and to you unknown; crystals likewise; and glasses of divers kinds; and amongst them some of metals vitrificated, and other materials besides those of which you make glass. Also a number of fossils, and imperfect minerals, which you have not. Likewise loadstones of prodigious virtue; and other rare stones, both natural and artificial.

"We have also sound-houses, where we practise and demonstrate all sounds, and their generation. We have harmonies which you have not, of quarter-sounds, and lesser slides of sounds. Divers instruments of music likewise to you unknown, some sweeter than any you have; together with bells and rings that are dainty and sweet. We represent small sounds as great and deep; likewise great sounds extenuate and sharp; we make divers tremblings and warblings of sounds, which in their original are entire. We represent and imitate all articulate sounds and letters, and the voices and notes of beasts and birds. We have certain helps which set to the ear do further the hearing greatly. We have also divers strange and artificial echos, reflecting the voice many times, and as it were tossing it: and some that give back the voice louder than it came; some shriller, and some deeper; yea, some rendering the voice differing in the letters or articulate sound from that they receive. We have also means to convey sounds in trunks and pipes, in strange lines and distances [46].

"We have also perfume-houses; wherewith we join also practices of taste. We multiply smells, which may seem strange. We imitate smells, making all smells to breathe out of other mixtures than those that give them [47]. We make divers imitations of taste likewise, so that they will deceive any man's taste. And in this house we contain also a confiture-house; where we make all sweet-meats, dry and moist [48], and divers pleasant wines, milks, broths, and sallets, far in greater variety than you have.

[44] It has been proposed to facilitate the examination of diabetic urine by an apparatus in which the amount of sugar present in it is to be measured by its effect on the plane of polarisation of polarised light transmitted through it.—*R. L. E.*

[45] Nothing that has been accomplished with the microscope would have interested Bacon more than the discoveries of Schleiden and Schwann, because nothing has brought us so near the latens processus by which the tissues of organic life are formed. It is remarkable that when Schleiden had as he conceived destroyed the analogy between the developments of vegetable and animal life, by showing that all vegetable tissues are developed by cells, Schwann should have re-established it more clearly than before by showing that this is true of all animal tissues also.—*R. L. E.*

[46] [*ad magnam distantiam, et in lineis tortuosis.*] This is now done very effectively by means of gutta percha tubing.—*R. L. E.*

[47] This power of imitating smells is one of the recent achievements of chemistry. From fusil oil, a product of the distillation of spirits from potatoes, itself exceedingly offensive, may be got oil of apples, oil of pears, oil of grapes, and oil of cognac. The oil of pine-apples and that of bitter almonds enable confectioners to imitate perfectly the scent and flavour of pine-apples and bitter almonds respectively, and both, like the perfumes already mentioned, are got from very offensive substances.—*R. L. E.*

[48] The translation adds *imò et condimus ea cum rebus aliis dulcibus, gratissimis, præter saccharum et mel.*

"We have also engine-houses, where are prepared engines and instruments for all sorts of motions. There we imitate and practise to make swifter motions than any you have, either out of your muskets or any engine that you have; and to make them and multiply them more easily, and with small force, by wheels and other means: and to make them stronger, and more violent than yours are; exceeding your greatest cannons and basilisks. We represent also ordnance and instruments of war, and engines of all kinds: and likewise new mixtures and compositions of gun-powder, wildfires burning in water, and unquenchable. Also fire-works of all variety both for pleasure and use. We imitate also flights of birds; we have some degrees of flying in the air; we have ships and boats for going under water [49], and brooking of seas; also swimming-girdles and supporters. We have divers curious clocks, and other like motions of return, and some perpetual motions. We imitate also motions of living creatures, by images of men, beasts, birds, fishes, and serpents. We have also a great number of other various [50] motions, strange for equality, fineness, and subtilty.

"We have also a mathematical-house, where are represented all instruments, as well of geometry as astronomy, exquisitely made.

"We have also houses of deceits of the senses; where we represent all manner of feats of juggling, false apparitions, impostures, and illusions; and their fallacies. And surely you will easily believe that we that have so many things truly natural which induce admiration, could in a world of particulars deceive the senses, if we would disguise those things and labour to make them seem more miraculous. But we do hate all impostures and lies; insomuch as we have severely forbidden it to all our fellows, under pain of ignominy and fines, that they do not shew any natural work or thing, adorned or swelling; but only pure as it is, and without all affectation of strangeness.

"These are (my son) the riches of Salomon's House.

"For the several employments and offices of our fellows; we have twelve that sail into foreign countries, under the names of other nations, (for our own we conceal;) who bring us the books, and abstracts, and patterns of experiments of all other parts. These we call Merchants of Light.

"We have three that collect the experiments which are in all books. These we call Depredators.

"We have three that collect the experiments of all mechanical arts; and also of liberal sciences; and also of practices which are not brought into arts. These we call Mystery-men [51].

"We have three that try new experiments, such as themselves think good. These we call Pioners or Miners.

"We have three that draw the experiments of the former four into titles and tables, to give the better light for the drawing of observations and axioms out of them. These we call Compilers [52].

"We have three that bend themselves, looking into the experiments of their fellows, and cast about how to draw out of them things of use and practice for man's life, and knowledge as well for works as for plain demonstration of causes, means of natural divinations, and the easy and clear discovery of the virtues and parts of bodies. These we call Dowry-men or Benefactors [53].

"Then after divers meetings and consults of our whole number, to consider of

49 A boat for going under water was one of Drebbel's inventions exhibited in 1620. Bacon in the *De Augmentis* refers to another, namely Drebbel's method of producing cold.—*R. L. E.*

50 The word "various," which seems to be redundant, is omitted in the translation.

51 In the translation they are called *Venatores*, hunters; a name, however, which does not seem to distinguish their peculiar office so accurately as "mystery-men," that is, men whose business was to inquire after mysteries, *i.e.* crafts.

52 These represent the formation of the tables comparentiæ, absentiæ in proximo, and graduum. See *Novum Organum*, ii. § 11-13.—*R. L. E.*

For "compilers," the translation has *divisores*, distributors.

53 These represent the Vindemiatio prima. See *Nov. Org.* ii. § 20.—*R. L. E.*

the former labours and collections, we have three that take care, out of them, to direct new experiments, of a higher light, more penetrating into nature than the former. These we call Lamps.

" We have three others that do execute the experiments so directed, and report them. These we call Inoculators.

" Lastly, we have three that raise the former discoveries by experiments into greater observations, axioms, and aphorisms [54]. These we call Interpreters of Nature.

" We have also, as you must think, novices and apprentices, that the succession of the former employed men do not fail ; besides a great number of servants and attendants, men and women. And this we do also : we have consultations, which of the inventions and experiments which we have discovered shall be published, and which not : and take all an oath of secrecy, for the concealing of those which we think fit to keep secret : though some of those we do reveal sometimes to the state, and some not.

" For our ordinances and rites : we have two very long and fair galleries : in one of these we place patterns and samples of all manner of the more rare and excellent inventions : in the other we place the statua's of all principal inventors. There we have the statua of your Columbus, that discovered the West Indies : also the inventor of ships : your monk that was the inventor of ordnance and of gunpowder : the inventor of music : the inventor of letters : the inventor of printing : the inventor of observations of astronomy : the inventor of works in metal : the inventor of glass : the inventor of silk of the worm : the inventor of wine : the inventor of corn and bread : the inventor of sugars : and all these by more certain tradition than you have. Then have we divers inventors of our own, of excellent works ; which since you have not seen, it were too long to make descriptions of them ; and besides, in the right understanding of those descriptions you might easily err. For upon every invention of value, we erect a statua to the inventor, and give him a liberal and honourable reward. These statua's are some of brass ; some of marble and touch-stone ; some of cedar and other special woods gilt and adorned : some of iron ; some of silver ; some of gold.

" We have certain hymns and services, which we say daily, of laud and thanks to God for his marvellous works : and forms of prayers, imploring his aid and blessing for the illumination of our labours, and the turning of them into good and holy uses.

" Lastly, we have circuits or visits of divers principal cities of the kingdom ; where, as it cometh to pass, we do publish such new profitable inventions as we think good. And we do also declare natural divinations of diseases, plagues, swarms of hurtful creatures, scarcity, tempests, earthquakes, great inundations, comets, temperature of the year, and divers other things ; and we give counsel thereupon what the people shall do for the prevention and remedy of them ".

And when he had said this, he stood up ; and I, as I had been taught, kneeled down ; and he laid his right hand upon my head, and said,, " God bless thee, my son, and God bless this relation which I have made. I give thee leave to publish it for the good of other nations ; for we here are in God's bosom, a land unknown ". And so he left me ; having assigned a value of about two thousand ducats, for a bounty to me and my fellows. For they give great largesses where they come upon all occasions.

[THE REST WAS NOT PERFECTED.]

[54] The translation adds that this was only done after consultation with the whole body. *Quod faciunt non nisi consultatione et colloquis prius habitis cum sociis universis.*

ESSAYS OR COUNSELS CIVIL AND MORAL

PREFACE.

BY JAMES SPEDDING.

AMONG the innumerable editions of Bacon's Essays that have been published, there are only four which, as authorities for the text, have any original or independent value; namely those published by Bacon himself in 1597, in 1612, and in 1625; and the Latin version published by Dr. Rawley in 1638. The rest are merely reprints of one or other of these.

The edition of 1597 contained ten essays, together with the *Meditationes Sacræ*, and the *Colours of Good and Evil*. That of 1612, a small volume in 8vo., contained essays only; but the number was increased to thirty-eight, of which twenty-nine were quite new, and all the rest more or less corrected and enlarged. That of 1625, a 4to. and one of the latest of Bacon's publications, contained fifty-eight essays, of which twenty were new, and most of the rest altered and enlarged.

The gradual growth of this volume, containing as it does the earliest and the latest fruits of Bacon's observation in that field in which its value has been most approved by universal and undiminished popularity, is a matter of considerable interest; and as the successive changes are not such as could be represented by a general description or conveniently specified in foot-notes, I have thought it best to reprint the two first editions entire, and add them in an appendix[1]. Considering also that, although it has been thought expedient throughout the text of this edition of Bacon's works to modernize the spelling, it may nevertheless be convenient to the reader to have a specimen of the orthography of Bacon's time, I have taken this opportunity in giving one; and preserved the original spelling throughout both these reprints.

I have also been able to supply from a manuscript in the British Museum evidence of another stage in the growth of this volume, intermediate between the editions of 1597 and 1612; of which manuscript, in connexion with the reprint of the latter, a complete account will be given.

The text of the Essays is taken of course from the edition of 1625; a correct representation of which is nearly all that a modern reader requires. The only points in which the audience to which they now address themselves stands in a different position towards them from that to which they were originally addressed, appear to be,—first, knowledge of Latin, which is probably a less general accomplishment among the readers of books now than it was then; and secondly, familiarity with the ordinary language of that day, in which some expressions have worn out of use with time, and some have acquired new meanings. To meet these changes, I have in the first place translated the Latin quotations, in the same manner and upon the same principle which I have explained at length in my preface to the Advancement of Learning (pp. 40–41.); and in the second place, I have added an explanatory note wherever I have observed any expression which a modern reader is likely to misunderstand or not to understand. But I have not attempted to develope allusions, or to canvass historical statements, or to point out inaccuracies of quotation, where the difference does not affect the argument,—still less to entertain the reader with discourses of my own; conceiving that the worth of writings of this kind depends in great part

[1] [Not given in the present edition.]

upon the rejection of superfluities, and that an annotator who is too diligent in producing all that he can find to say about his text runs a great risk of merely encumbering the reader with the very matter from which it was the author's labour to disembarrass him. I have even had my doubts whether in writings which remain as fresh as these, the very insertion of references to passages quoted be not an unwelcome interruption and an unwarrantable liberty. When a modern writer introduces, for ornament or illustration or impression, a line from Virgil or Milton, he never thinks of adding a reference to the book and verse; and I suppose that Mr. Singer would not look upon an asterisk and a foot-note, with *Hor. Carm.* I. 12.45 as any improvement to the elegant motto which occupies the blank page fronting the title of his very elegant edition of these Essays. Bacon's philosophical works stand in many respects in a different position. Their value is in great part historical and antiquarian. They no longer speak to us as to contemporaries. To understand their just import, we must be carried back to the time, and it is of importance to know what books were then in estimation and what authors were familiarly appealed to, and carried weight as vouchers. The Essays, on the contrary, have for us precisely the same sort of interest which they had for the generation to which they were immediately addressed; they "come home to men's business and bosoms" just in the same way; they appeal to the same kind of experience; the allusions and citations are still familiar, and produce the same kinds of impression on the imagination. So that I do not see why the reason which induced Bacon to cite an ancient saying, a tradition of the poets, an observation of one of the fathers, or a sentence from some classical writer, without specifying the volume and page where he found it, should not still, be held a reason for leaving them to produce the effect which he intended, unincumbered with a piece of information which I suppose he thought superfluous or inconvenient.

The Latin translation of the Essays, published by Dr. Rawley in 1638 among the *Opera Moralia et Civilia*, under the weightier [2] title of *Sermones Fideles sive Interiora Rerum*, has (as I said) an original and independent value. Whether any of them were actually translated by Bacon himself, or how far he superintended the work, it seems impossible to know. Mr. Singer indeed represents them, on the authority of the title [3], as having been put into Latin by Bacon himself "*præterquam in paucis*": but the words which he quotes occur in the title not of the *Sermones Fideles*, but of the whole volume, which contains four other works; the *Sermones Fideles* forming less than a fourth of the whole: so that for any thing these words imply they may themselves have been among the things excepted [4]. As it is certain however that Bacon himself regarded the Latin version as that in which they were to live, we may be sure that he took care to have it properly done: only as it was not published till twelve years after his death, we cannot be sure that it was all finished before he died. Several hands are said to have been employed in the work, and in the absence of all specific information, it is not improbable that there are parts of it which he did not live to see completed. Taken with this caution however, the Latin translation must

[2] Deinde sequetur libellus ille quem vestra lingua *Saggi Morali* appellastis. Verum illi libro nomen gravius impono: scilicet ut inscribatur *Sermones Fideles, sive Interiora Rerum.—Bacon's Letter to Fulgentio.*

[3] "In the year 1638, Dr. Rawley, who had been Bacon's chaplain, published a folio volume, containing, amongst other works in Latin, a translation of the Essays, under the title of 'Sermones Fideles, ab ipso Honoratissimo Auctore, præterquam in paucis, Latinitate donati."—Pref. p. xvi.

[4] Francisci Baconi . . . operum moralium et civilium tomus.

Qui continet { *Historiam Regni Henrici Septimi Regis Angliæ.*
Sermones Fideles, sive Interiora Rerum.
Tractatum de Sapientia Veterum.
Dialogum de Bello Sacro.
Et Novam Atlantidem. }

Ab ipso Honoratissimo Auctore, præterquam in paucis, Latinitate donatus.

be accepted as a work of authority, and in one respect of superior authority to the original, because of later date.

I am not aware that any such value belongs to any of the translations into modern languages. An Italian translation of the Essays and the De Sapientia Veterum, published in London in 1618, with a dedicatory letter from Tobie Matthew to Cosmo de' Medici, may be presumed to have been made with Bacon's sanction; both because Matthew was so intimate a friend, and because it includes one essay which had not then been published [5], as well as a large extract from the letter to Prince Henry which Bacon had intended to prefix to the edition of 1612, but was prevented by his death. But there is no reason to suppose that Bacon had anything more to do with it. It is true that Andrea Cioli, who by Cosmo's direction brought out a new and revised edition of this volume at Florence in 1619, seems at first sight to speak of the translation as if it were Bacon's own composition—(ma non hò già voluto alterare alcuna di quelle parole, che forse nella lingua nostra non appariscono interamente proprie del senso, à che sono state in detta Opera destinate, *per non torre all' Autore la gloria, che merita di havere cosi ben saputo esprimere i suoi Concetti* in Idioma altretanto diverso dal suo, quanto è lontana da questa nostra la sua Regione)—but the supposition is hardly reconcilable with the words of Matthew's dedicatory letter (non può mancar la scusa à chi s' è ingegnato tradur li concetti di questo Autore, etc.); and in the absence of all other evidence is too improbable to be believed. Nor do Cioli's words necessarily imply more than that the translator was an Englishman. That the translation was not the work of an Italian,—and therefore not (according to Mr. Singer's conjecture) by Father Fulgentio,—they afford evidence which may be considered conclusive.

THE EPISTLE DEDICATORY.

To the Right Honourable my very good Lo. the DUKE of BUCKINGHAM his Grace Lo. High Admiral of England.

EXCELLENT LO.

SALOMON says, *A good name is as a precious ointment;* and I assure myself, such will your Grace's name be with posterity. For your fortune and merit both have been eminent. And you have planted things that are like to last. I do now publish my Essays; which, of all my other works, have been most current; for that, as it seems, they come home to men's business and bosoms. I have enlarged them both in number and weight; so that they are indeed a new work. I thought it therefore agreeable to my affection and obligation to your Grace, to prefix your name before them, both in English and in Latin. For I do conceive that the Latin volume of them (being in the universal language) may last as long as books last. My Instauration I dedicated to the King; my History of Henry the Seventh (which I have now also translated into Latin), and my portions of Natural History, to the Prince; and these I dedicate to your Grace; being of the best fruits that by the good encrease which God gives to my pen and labours I could yield. God lead your Grace by the hand.

Your Grace's most obliged and
faithful seruant,
FR. ST. ALBAN.

[5] Mr. Singer says two: but one of those he quotes—the Essay "Of Honour and Reputation"—will be found in the edition of 1597.

I. Of Truth.

What is Truth? said jesting Pilate; and would not stay for an answer. Certainly there be that delight in giddiness, and count it a bondage to fix a belief; affecting free-will in thinking, as well as in acting. And though the sects of philosophers of that kind be gone, yet there remain certain discoursing wits which are of the same veins, though there be not so much blood in them as was in those of the ancients. But it is not only the difficulty and labour which men take in finding out of truth; nor again that when it is found it imposeth upon men's thoughts; that doth bring lies in favour; but a natural though corrupt love of the lie itself. One of the later school of the Grecians examineth the matter, and is at a stand to think what should be in it, that men should love lies; where neither they make for pleasure, as with poets, nor for advantage, as with the merchant; but for the lie's sake. But I cannot tell; this same truth is a naked and open day-light that doth not shew the masks and mummeries and triumphs of the world, half so stately and daintily as candle-lights. Truth may perhaps come to the price of a pearl, that sheweth best by day; but it will not rise to the price of a diamond or carbuncle, that sheweth best in varied lights. A mixture of a lie doth ever add pleasure. Doth any man doubt, that if there were taken out of men's minds vain opinions, flattering hopes, false valuations, imaginations as one would, and the like, but it would leave the minds of a number of men poor shrunken things, full of melancholy and indisposition, and unpleasing to themselves? One of the Fathers, in great severity, called poesy *vinum dæmonum* [devil's-wine], because it filleth the imagination; and yet it is but with the shadow of a lie. But it is not the lie that passeth through the mind, but the lie that sinketh in and settleth in it, that doth the hurt; such as we spake of before. But howsoever these things are thus in men's depraved judgments and affections, yet truth, which only doth judge itself, teacheth that the inquiry of truth, which is the love-making or wooing of it, the knowledge of truth, which is the presence of it, and the belief of truth, which is the enjoying of it, is the sovereign good of human nature. The first creature of God, in the works of the days, was the light of the sense; the last was the light of reason; and his sabbath work ever since is the illumination of his Spirit. First he breathed light upon the face of the matter or chaos; then he breathed light into the face of man; and still he breatheth and inspireth light into the face of his chosen. The poet that beautified the sect that was otherwise inferior to the rest [6], saith yet excellently well: *It is a pleasure to stand upon the shore, and to see ships tossed upon the sea; a pleasure to stand in the window of a castle, and to see a battle and the adventures thereof below: but no pleasure is comparable to the standing upon the vantage ground of Truth* (a hill not to be commanded, and where the air is always clear and serene), *and to see the errors, and wanderings, and mists, and tempests, in the vale below;* so always that this prospect be with pity, and not with swelling or pride. Certainly, it is heaven upon earth, to have a man's mind move in charity, rest in providence, and turn upon the poles of truth.

To pass from theological and philosophical truth, to the truth of civil business; it will be acknowledged even by those that practise it not, that clear and round dealing is the honour of man's nature; and that mixture of falsehood is like allay in coin of gold and silver, which may make the metal work the better, but it embaseth it. For these winding and crooked courses are the goings of the serpent; which goeth basely upon the belly, and not upon the feet. There is no vice that doth so cover a man with shame as to be found false and perfidious. And therefore Montaigne saith prettily, when he inquired the reason, why the word of the lie should be such a disgrace and such an odious charge? Saith he, *If it be well weighed, to say that a man lieth, is as much to say, as that he is brave towards God and a coward towards men* [7]. For a lie faces God, and shrinks from man. Surely the wickedness of falsehood and breach of faith cannot possibly be so highly expressed, as in that it shall be the last peal to call the judgments of God upon the generations of men; it being foretold, that when Christ cometh, *he shall not find faith upon the earth.*

[6] Lucretius. B. II: beginning. [7] Essais, II. 18. Cf. Plutarch, Lysand. c. 8.

II. Of Death.

Men fear Death, as children fear to go in the dark; and as that natural fear in children is increased with tales, so is the other. Certainly, the contemplation of death, as the wages of sin and passage to another world, is holy and religious; but the fear of it, as a tribute due unto nature, is weak. Yet in religious meditations there is sometimes mixture of vanity and of superstition. You shall read in some of the friars' books of mortification, that a man should think with himself what the pain is if he have but his finger's end pressed or tortured, and thereby imagine what the pains of death are, when the whole body is corrupted and dissolved; when many times death passeth with less pain than the torture of a limb: for the most vital parts are not the quickest of sense. And by him that spake only as a philosopher and natural man, it was well said, *Pompa mortis magis terret, quam mors ipsa*[8]: [it is the accompaniments of death that are frightful rather than death itself]. Groans and convulsions, and a discoloured face, and friends weeping, and blacks, and obsequies, and the like, shew death terrible. It is worthy the observing, that there is no passion in the mind of man so weak, but it mates and masters the fear of death; and therefore death is no such terrible enemy when a man hath so many attendants about him that can win the combat of him. Revenge triumphs over death; Love slights it; Honour aspireth to it; Grief flieth to it[9]; Fear pre-occupateth it; nay we read, after Otho the emperor had slain himself, Pity (which is the tenderest of affections) provoked many to die, out of mere compassion to their sovereign, and as the truest sort of followers. Nay, Seneca adds niceness and satiety: *Cogita quamdiu eadem feceris; mori velle, non tantum fortis, aut miser, sed etiam fastidiosus potest.* A man would die, though he were neither valiant nor miserable, only upon a weariness to do the same thing so oft over and over. It is no less worthy to observe, how little alteration in good spirits the approaches of death make; for they appear to be the same men till the last instant. Augustus Cæsar died in a compliment: *Livia, conjugii nostri memor, vive et vale* [farewell, Livia; and forget not the days of our marriage]. Tiberius in dissimulation; as Tacitus saith of him, *Jam Tiberium vires et corpus, non dissimulatio, deserebant:* [his powers of body were gone, but his power of dissimulation still remained]. Vespasian in a jest; sitting upon the stool, *Ut puto Deus fio* [I think I am becoming a god]. Galba with a sentence; *Feri, si ex re sit populi Romani* [strike, if it be for the good of Rome]; holding forth his neck. Septimius Severus in despatch; *Adeste si quid mihi restat agendum* [make haste, if there is anything more for me to do]. And the like. Certainly the Stoics bestowed too much cost upon death, and by their great preparations made it appear more fearful. Better saith he, *qui finem vitæ extremum inter munera ponat naturæ* [who accounts the close of life as one of the benefits of nature]. It is as natural to die as to be born; and to a little infant, perhaps, the one is as painful as the other. He that dies in an earnest pursuit, is like one that is wounded in hot blood; who, for the time, scarce feels the hurt; and therefore a mind fixed and bent upon somewhat that is good doth avert the dolours of death. But above all, believe it, the sweetest canticle is, *Nunc dimittis*; when a man hath obtained worthy ends and expectations. Death hath this also; that it openeth the gate to good fame, and extinguisheth envy. *Extinctus amabitur idem* [the same man that was envied while he lived, shall be loved when he is gone].

[8] Seneca, Ep. 24. Tolle istam pompam sub qua lates et stultos territas: mors es, quem nuper servus meus, quem ancilla contempsit. See the rest of the passage, and my note on Rawley's Life of Bacon, *supra*, p. 8.

[9] The translation adds, *metus ignominiæ eligit:* a sentence which is also found in the edition of 1612,—"Delivery from ignominy chooseth it"; omitted here probably by accident.

III. Of Unity in Religion.

Religion being the chief band of human society, it is a happy thing when itself is well contained within the true band of Unity. The quarrels and divisions about religion were evils unknown to the heathen. The reason was, because the religion of the heathen consisted rather in rites and ceremonies, than in any constant belief. For you may imagine what kind of faith theirs was, when the chief doctors and fathers of their church were the poets. But the true God hath this attribute, that he is a *jealous God ;* and therefore his worship and religion will endure no mixture nor partner. We shall therefore speak a few words concerning the Unity of the Church; what are the Fruits thereof; what the Bounds; and what the Means.

The Fruits of Unity (next unto the well pleasing of God, which is all in all) are two; the one towards those that are without the church, the other towards those that are within. For the former; it is certain that heresies and schisms are of all others the greatest scandals; yea, more than corruption of manners. For as in the natural body a wound or solution of continuity is worse than a corrupt humour; so in the spiritual. So that nothing doth so much keep men out of the church, and drive men out of the church, as breach of unity. And therefore, whensoever it cometh to that pass, that one saith *Ecce in deserto*, another saith *Ecce in penetralibus :* that is, when some men seek Christ in the conventicles of heretics, and others in an outward face of a church, that voice had need continually to sound in man's ears, *Nolite exire,—Go not out.* The Doctor of the Gentiles (the propriety of whose vocation drew him to have a special care of those without) saith, *If an heathen come in, and hear you speak with several tongues, will he not say that you are mad ?* And certainly it is little better, when atheists and profane persons do hear of so many discordant and contrary opinions in religion; it doth avert them from the church, and maketh them *to sit down in the chair of the scorners.* It is but a light thing to be vouched in so serious a matter, but yet it expresseth well the deformity. There is a master of scoffing, that in his catalogue of books of a feigned library sets down this title of a book, *The morris-dance of Heretics.* For indeed every sect of them hath a diverse posture or cringe by themselves, which cannot but move derision in worldlings and depraved politics, who are apt to contemn holy things.

As for the fruit towards those that are within; it is peace; which containeth infinite blessings. It establisheth faith. It kindleth charity. The outward peace of the church distilleth into peace of conscience. And it turneth the labours of writing and reading of controversies into treatises[10] of mortification and devotion.

Concerning the Bounds of Unity; the true placing of them importeth excedingly. There appear to be two extremes. For to certain zelants all speech of pacification is odious. *Is it peace, Jehu ? What hast thou to do with peace ? turn thee behind me.* Peace is not the matter, but following and party. Contrariwise, certain Laodiceans and lukewarm persons think they may accommodate points of religion by middle ways, and taking part of both, and witty reconcilements; as if they would make an arbitrement between God and man. Both these extremes are to be avoided; which will be done, if the league of Christians penned by our Saviour himself were in the two cross clauses thereof soundly and plainly expounded: *He that is not with us is against us ;* and again, *He that is not against us is with us ;* that is, if the points fundamental and of substance in religion were truly discerned and distinguished from points not merely of faith, but of opinion, order, or good intention. This is a thing may seem to many a matter trivial, and done already. But if it were done less partially, it would be embraced more generally.

Of this I may give only this advice, according to my small model. Men ought to take heed of rending God's church by two kinds of controversies. The one is, when the matter of the point controverted is too small and light, not worth the heat and strife about it, kindled only by contradiction. For as it is noted by one of the fathers, *Christ's coat indeed had no seam, but the church's vesture was of divers colours ;* whereupon he saith, *In veste varietas sit, scissura non sit* [let there be variety in the garment, but let there be no division]: they be two things, Unity and Uniformity. The other is, when the matter of the point controverted

[10] *treaties*, in the original.

is great, but it is driven to an overgreat subtilty and obscurity; so that it becometh a thing rather ingenious than substantial. A man that is of judgment and understanding shall sometimes hear ignorant men differ, and know well within himself that those which so differ mean one thing, and yet they themselves would never agree. And if it come so to pass in that distance of judgment which is between man and man, shall we not[11] think that God above, that knows the heart, doth not discern that frail men in some of their contradictions intend the same thing; and accepteth of both? The nature of such controversies is excellently expressed by St. Paul in the warning and precept that he giveth concerning the same, *Devita profanas vocum novitates, et oppositiones falsi nominis scientiæ* [Avoid profane novelties of terms, and oppositions of science falsely so called]. Men create oppositions which are not; and put them into new terms so fixed, as whereas the meaning ought to govern the term, the term in effect governeth the meaning. There be also two false peaces or unities: the one, when the peace is grounded but upon an implicit ignorance; for all colours will agree in the dark: the other, when it is pieced up upon a direct admission of contraries in fundamental points. For truth and falsehood, in such things, are like the iron and clay in the toes of Nabuchadnezzar's image; they may cleave, but they will not incorporate.

Concerning the Means of procuring Unity; men must beware, that in the procuring or muniting of religious unity they do not dissolve and deface the laws of charity and of human society. There be two swords amongst Christians, the spiritual and temporal; and both have their due office and place in the maintenance of religion. But we may not take up the third sword, which is Mahomet's sword, or like unto it; that is, to propagate religion by wars or by sanguinary persecutions to force consciences; except it be in cases of overt scandal, blasphemy, or intermixture of practice against the state; much less to nourish seditions; to authorise conspiracies and rebellions; to put the sword into the people's hands; and the like; tending to the subversion of all government, which is the ordinance of God. For this is but to dash the first table against the second; and so to consider men as Christians, as we forget that they are men. Lucretius the poet, when he beheld the act of Agamemnon, that could endure the sacrificing of his own daughter, exclaimed:

> Tantum Relligio potuit suadere malorum:

[to such ill actions Religion could persuade a man]. What would he have said, if he had known of the massacre in France, or the powder treason of England? He would have been seven times more Epicure and atheist than he was. For as the temporal sword is to be drawn with great circumspection in cases of religion; so it is a thing monstrous to put it into the hands of the common people. Let that be left unto the Anabaptists, and other furies. It was great blasphemy when the devil said, *I will ascend and be like the Highest;* but it is greater blasphemy to personate God, and bring him in saying, *I will descend, and be like the prince of darkness:* and what is it better, to make the cause of religion to descend to the cruel and execrable actions of murthering princes, butchery of people, and subversion of states and governments? Surely this is to bring down the Holy Ghost, instead of the likeness of a dove, in the shape of a vulture or raven; and set out of the bark of a Christian church a flag of a bark of pirates and Assassins. Therefore it is most necessary that the church by doctrine and decree, princes by their sword, and all learnings, both Christian and moral, as by their Mercury rod, do damn and send to hell for ever those facts and opinions tending to the support of the same; as hath been already in good part done. Surely in counsels concerning religion, that counsel of the apostle would be prefixed, *Ira hominis non implet justitiam Dei* [The wrath of man worketh not the righteousness of God]. And it was a notable observation of a wise Father, and no less ingenuously confessed, *that those which held and persuaded pressure of consciences, were commonly interested therein themselves for their own ends.*

[11] So in the original. One of the *nots* should obviously be struck out.

IV. Of Revenge.

Revenge is a kind of wild justice; which the more man's nature runs to, the more ought law to weed it out. For as for the first wrong, it doth but offend the law; but the revenge of that wrong putteth the law out of office. Certainly, in taking revenge, a man is but even with his enemy; but in passing it over, he is superior; for it is a prince's part to pardon. And Salomon I am sure, saith, *It is the glory of a man to pass by an offence.* That which is past is gone, and irrevocable; and wise men have enough to do with things present and to come; therefore they do but trifle with themselves, that labour in past matters. There is no man doth a wrong for the wrong's sake; but thereby to purchase himself profit, or pleasure, or honour, or the like. Therefore why should I be angry with a man for loving himself better than me? And if any man should do wrong merely out of ill-nature, why, yet is it but like the thorn or briar, which prick and scratch, because they can do no other. The most tolerable sort of revenge is for those wrongs which there is no law to remedy; but then let a man take heed the revenge be such as there is no law to punish; else a man's enemy is still beforehand, and it is two for one. Some, when they take revenge, are desirous the party should know whence it cometh. This the more generous. For the delight seemeth to be not so much in doing the hurt as in making the party repent. But base and crafty cowards are like the arrow that flieth in the dark. Cosmus, duke of Florence, had a desperate saying against perfidious or neglecting friends, as if those wrongs were unpardonable; *You shall read* (saith he) *that we are commanded to forgive our enemies; but you never read that we are commanded to forgive our friends.* But yet the spirit of Job was in a better tune: *Shall we* (saith he) *take good at God's hands, and not be content to take evil also?* And so of friends in a proportion. This is certain, that a man that studieth revenge keeps his own wounds green, which otherwise would heal and do well. Public revenges are for the most part fortunate; as that for the death of Cæsar; for the death of Pertinax; for the death of Henry the Third [12] of France; and many more. But in private revenges it is not so. Nay rather, vindictive persons live the life of witches; who, as they are mischievous, so end they infortunate.

V. Of Adversity.

It was a high speech of Seneca (after the manner of the Stoics), *that the good things which belong to prosperity are to be wished; but the good things that belong to adversity are to be admired. Bona rerum secundarum optabilia; adversarum mirabilia.* Certainly if miracles be the command over nature, they appear most in adversity. It is yet a higher speech of his than the other (much too high for a heathen), *It is true greatness to have in one the frailty of a man, and the security of a God. Vere magnum habere fragilitatem hominis, securitatem Dei.* This would have done better in poesy, where transcendences are more allowed. And the poets indeed have been busy with it; for it is in effect the thing which is figured in that strange fiction of the ancient poets, which seemeth not to be without mystery; nay, and to have some approach to the state of a Christian; that *Hercules, when he went to unbind Prometheus* (by whom human nature is represented), *sailed the length of the great ocean in an earthen pot or pitcher;* lively describing Christian resolution, that saileth in the frail bark of the flesh thorough the waves of the world. But to speak in a mean. The virtue of Prosperity is temperance, the virtue of Adversity is fortitude; which in morals is the more heroical virtue. Prosperity is the blessing of the Old Testament; Adversity is the blessing of the New; which carrieth the greater benediction, and the clearer revelation of God's favour. Yet even in the Old Testament, if you listen to David's harp, you shall hear as many hearse-like airs as carols; and

[12] So the original. The Latin translation has *Henrici Quarti, magni illius Galliæ regis.* It is probable therefore that we should read here *fourth* instead of *third.* But the observation is true to a certain extent with regard to both.

the pencil of the Holy Ghost hath laboured more in describing the afflictions of Job than the felicities of Salomon. Prosperity is not without many fears and distastes; and Adversity is not without comforts and hopes. We see in needle-works and embroideries, it is more pleasing to have a lively work upon a sad and solemn ground, than to have a dark and melancholy work upon a lightsome ground: judge therefore of the pleasure of the heart by the pleasure of the eye. Certainly virtue is like precious odours, most fragrant when they are incensed or crushed: for Prosperity doth best discover vice, but Adversity doth best discover virtue.

VI. Of Simulation and Dissimulation.

Dissimulation is but a faint kind of policy or wisdom, for it asketh a strong wit and a strong heart to know when to tell truth, and to do it. Therefore it is the weaker sort of politics that are the great dissemblers.

Tacitus saith, *Livia sorted well with the arts of her husband and dissimulation of her son;* attributing arts or policy to Augustus, and dissimulation to Tiberius. And again, when Mucianus encourageth Vespasian to take arms against Vitellius, he saith, *We rise not against the piercing judgment of Augustus, nor the extreme caution or closeness of Tiberius.* These properties of arts or policy and dissimulation or closeness, are indeed habits and faculties several, and to be distinguished. For if a man have that penetration of judgment as he can discern what things are to be laid open, and what to be secreted, and what to be shewed at half lights, and to whom and when, (which indeed are arts of state and arts of life, as Tacitus well calleth them), to him a habit of dissimulation is a hindrance and a poorness. But if a man cannot obtain to that judgment, then it is left to him generally to be close, and a dissembler. For where a man cannot choose or vary in particulars, there it is good to take the safest and wariest way in general; like the going softly, by one that cannot well see. Certainly the ablest men that ever were have had all an openness and frankness of dealing; and a name of certainty and veracity; but then they were like horses well managed; for they could tell passing well when to stop or turn; and at such times when they thought the case indeed required dissimulation, if then they used it, it came to pass that the former opinion spread abroad of their good faith and clearness of dealing made them almost invisible.

There be three degrees of this hiding and veiling of a man's self. The first, Closeness, Reservation, and Secrecy; when a man leaveth himself without observation, or without hold to be taken, what he is. The second, Dissimulation, in the negative; when a man lets fall signs and arguments, that he is not that he is. And the third, Simulation, in the affirmative; when a man industriously and expressly feigns and pretends to be that he is not.

For the first of these, Secrecy; it is indeed the virtue of a confessor. And assuredly the secret man heareth many confessions. For who will open himself to a blab or babbler? But if a man be thought secret, it inviteth discovery; as the more close air sucketh in the more open; and as in confession the revealing is not for worldly use, but for the ease of a man's heart, so secret men come to the knowledge of many things in that kind; while men rather discharge their minds than impart their minds. In few words, mysteries are due to secrecy. Besides (to say truth) nakedness is uncomely, as well in mind as body; and it addeth no small reverence to men's manners and actions, if they be not altogether open. As for talkers and futile persons, they are commonly vain and credulous withal. For he that talketh what he knoweth, will also talk what he knoweth not. Therefore set it down, *that an habit of secrecy is both politic and moral.* And in this part, it is good that a man's face give his tongue leave to speak. For the discovery of a man's self by the tracts of his countenance is a great weakness and betraying; by how much it is many times more marked and believed than a man's words.

For the second, which is Dissimulation; it followeth many times upon secrecy by a necessity; so that he that will be secret must be a dissembler in some degree. For men are too cunning to suffer a man to keep an indifferent carriage between both, and to be secret, without swaying the balance on either side. They will so beset a man with questions, and draw him on, and pick it out of him, that, without an absurd silence, he must shew an inclination one way; or if he do not, they will gather as much by his silence as by his speech. As for equivocations, or oraculous speeches, they cannot hold out long. So that no man can be secret except he give himself a little scope of dissimulation; which is, as it were, but the skirts or train of secrecy.

But for the third degree, which is Simulation and false profession; that I hold more culpable, and less politic; except it be in great and rare matters. And therefore a general custom of simulation (which is this last degree) is a vice, rising either of a natural falseness or fearfulness, or of a mind that hath some main faults, which because a man must needs disguise, it maketh him practise simulation in other things, lest his hand should be out of use.

The great [13] advantages of simulation and dissimulation are three. First, to lay asleep opposition, and to surprise. For where a man's intentions are published, it is an alarum to call up all that are against them. The second is, to reserve to a man's self a fair retreat. For if a man engage himself by a manifest declaration, he must go through or take a fall. The third is, the better to discover the mind of another. For to him that opens himself men will hardly shew themselves adverse; but will (fair) [14] let him go on, and turn their freedom of speech to freedom of thought. And therefore it is a good shrewd proverb of the Spaniard, *Tell a lie and find a troth.* As if there were no way of discovery but by simulation. There be also three disadvantages, to set it even. The first, that simulation and dissimulation commonly carry with them a shew of fearfulness which in any business doth spoil the feathers of round flying up to the mark. The second, that it puzzleth and perplexeth the conceits of many that perhaps would otherwise co-operate with him; and makes a man walk almost alone to his own ends. The third and greatest is, that it depriveth a man of one of the most principal instruments for action; which is trust and belief. The best composition and temperature is to have openness in fame and opinion; secrecy in habit; dissimulation in seasonable use; and a power to feign, if there be no remedy.

VII. Parents and Children.

The joys of parents are secret; and so are their griefs and fears. They cannot utter the one; nor they will not utter the other. Children sweeten labours; but they make misfortunes more bitter. They increase the cares of life; but they mitigate the remembrance of death. The perpetuity by generation is common to beasts; but memory, merit, and noble works are proper to men. And surely a man shall see the noblest works and foundations have proceeded from childless men; which have sought to express the images of their minds, where those of their bodies have failed. So the care of posterity is most in them that have no posterity. They that are the first raisers of their houses are most indulgent towards their children; beholding them as the continuance not only of their kind but of their work; and so both children and creatures.

The difference in affection of parents towards their several children is many times unequal; and sometimes unworthy; especially in the mother; as Salomon saith, *A wise son rejoiceth the father, but an ungracious son shames the mother.* A man shall see, where there is a house full of children, one or two of the eldest respected, and the youngest made wantons; but in the midst some that are as it were forgotten, who many times nevertheless prove the best. The illiberality of parents in allowance towards their children is an harmful error; makes them base; acquaints them with shifts; makes them sort with mean company; and makes them surfeit more when they come to plenty. And therefore the proof is best, when men keep their authority towards their children, but not their purse. Men

[13] So in original, and in ed. 1639. *Great* is omitted in the translation, and in some modern editions, including Mr. Singer's.

[14] So in the original, and also in edition 1639. The translation has: *Etenim ei qui consilia sua profert, non facile quis se adversarium profiteatur, verum assentabitur potius.* I do not remember to have met with this use of *fair* anywhere else; but it is intelligible enough, and may, I should think, be right.

have a foolish manner (both parents and schoolmasters and servants) in creating and breeding an emulation between brothers during childhood, which many times sorteth to discord when they are men, and disturbeth families. The Italians make little difference between children and nephews or near kinsfolks; but so they be of the lump, they care not though they pass not through their own body. And, to say truth, in nature it is much a like matter; insomuch that we see a nephew sometimes resembleth an uncle or a kinsman more than his own parent; as the blood happens. Let parents choose betimes the vocations and courses they mean their children should take; for then they are most flexible; and let them not too much apply themselves to the disposition of their children, as thinking they will take best to that which they have most mind to. It is true, that if the affection or aptness of the children be extraordinary, then it is good not to cross it; but generally the precept is good, *optimum elige, suave et facile illud faciet consuetudo* [choose the best—custom will make it pleasant and easy]. Younger brothers are commonly fortunate, but seldom or never where the elder are disinherited.

VIII. Of Marriage and Single Life.

He that hath wife and children hath given hostages to fortune; for they are impediments to great enterprises, either of virtue or mischief. Certainly the best works, and of greatest merit for the public, have proceeded from the unmarried or childless men; which both in affection and means have married and endowed the public. Yet it were great reason that those that have children should have greatest care of future times; unto which they know they must transmit their dearest pledges. Some there are, who though they lead a single life, yet their thoughts do end with themselves and account future times impertinences. Nay, there are some other that account wife and children but a bill of charges. Nay, more, there are some foolish rich covetous men, that take a pride in having no children, because they may be thought so much the richer. For perhaps they have heard some talk, *Such an one is a great rich man*, and another except to it *Yea, but he hath a great charge of children;* as if it were an abatement to his riches. But the most ordinary cause of a single life is liberty, especially in certain self-pleasing and humorous minds, which are so sensible of every restraint, as they will go near to think their girdles and garters to be bonds and shackles. Unmarried men are best friends, best masters, best servants; but not always best subjects; for they are light to run away; and almost all fugitives are of that condition. A single life doth well with churchmen; for charity will hardly water the ground where it must first fill a pool. It is indifferent for judges and magistrates; for if they be facile and corrupt, you shall have a servant five times worse than a wife. For soldiers, I find the generals commonly in their hortatives put men in mind of their wives and children; and I think the despising of marriage amongst the Turks maketh the vulgar soldier more base. Certainly wife and children are a kind of discipline of humanity; and single men, though they may be many times more charitable, because their means are less exhaust, yet, on the other side, they are more cruel and hardhearted (good to make severe inquisitors), because their tenderness is not so oft called upon. Grave natures, led by custom, and therefore constant, are commonly loving husbands; as was said of Ulysses, *vetulam suam prætulit immortalitati* [he preferred his old wife to immortality]. Chaste women are often proud and froward, as presuming upon the merit of their chastity. It is one of the best bonds both of chastity and obedience in the wife, if she think her husband wise; which she will never do if she find him jealous. Wives are young men's mistresses; companions for middle age; and old men's nurses. So as a man may have a quarrel to marry when he will. But yet he was reputed one of the wise men that made answer to the question, when a man should marry?—*A young man not yet, an elder man not at all.* It is often seen that bad husbands have very good wives; whether it be that it raiseth the price of their husband's kindness when it comes; or that the wives take a pride in their patience. But this never fails, if the bad husbands were of their own choosing, against their friends' consent; for then they will be sure to make good their own folly.

IX. Of Envy.

There be none of the affections which have been noted to fascinate or bewitch, but love and envy. They both have vehement wishes; they frame themselves readily into imaginations and suggestions; and they come easily into the eye, especially upon the presence of the objects; which are the points that conduce to fascination, if any such thing there be. We see likewise the scripture calleth envy an *evil eye*; and the astrologers call the evil influences of the stars *evil aspects*; so that still there seemeth to be acknowledged, in the act of envy, an ejaculation or irradiation of the eye. Nay, some have been so curious as to note that the times when the strokes of percussion of an envious eye doth most hurt, are when the party envied is beheld in glory or triumph; for that sets an edge upon envy: and besides, at such times the spirits of the person envied do come forth most into the outward parts, and so meet the blow.

But leaving these curiosities, (though not unworthy to be thought on in fit place), we will handle, what persons are apt to envy others; what persons are most subject to be envied themselves; and what is the difference between public and private envy.

A man that hath no virtue in himself, ever envieth virtue in others. For men's minds will either feed upon their own good or upon other's evil; and who wanteth the one will prey upon the other; and whoso is out of hope to attain to another's virtue, will seek to come at even hand by depressing another's fortune.

A man that is busy and inquisitive is commonly envious. For to know much of other men's matters cannot be because all that ado may concern his own estate; therefore it must needs be that he taketh a kind of play-pleasure in looking upon the fortunes of others. Neither can he that mindeth but his own business find much matter for envy. For envy is a gadding passion, and walketh the streets, and doth not keep home; *Non est curiosus, quin idem sit malevolus* [There is no curious man but has some malevolence to quicken his curiosity].

Men of noble birth are noted to be envious towards new men when they rise. For the distance is altered; and it is like a deceit of the eye, that when others come on they think themselves go back.

Deformed persons, and eunuchs, and old men, and bastards, are envious. For he that cannot possibly mend his own case will do what he can to impair another's; except these defects light upon a very brave and heroical nature, which thinketh to make his natural wants part of his honour; in that it should be said, that an eunuch, or a lame man, did such great matters; affecting the honour of a miracle; as it was in Narses the eunuch, and Agesilaus and Tamberlanes, that were lame men.

The same is the case of men that rise after calamities and misfortunes. For they are as men fallen out with the times; and think other men's harms a redemption of their own sufferings.

They that desire to excel in too many matters, out of levity and vain glory, are ever envious. For they cannot want work [15]; it being impossible but many in some one of those things should surpass them. Which was the character of Adrian the Emperor; that mortally envied poets and painters and artificers, in works wherein he had a vein to excel.

Lastly, near kinsfolks, and fellows in office, and those that have been bred together, are more apt to envy their equals when they are raised. For it doth upbraid unto them their own fortunes, and pointeth at them, and cometh oftener into their remembrance, and incurreth likewise more into the note of others; and envy ever redoubleth from speech and fame. Cain's envy was the more vile and malignant towards his brother Abel, because when his sacrifice was better accepted there was nobody to look on. Thus much for those that are apt to envy.

Concerning those that are more or less subject to envy: First, persons of eminent virtue, when they are advanced, are less envied. For their fortune seemeth but due unto them; and no man envieth the payment of a debt, but rewards and liberality rather. Again, envy is ever joined with the comparing of a man's self; and where there is no comparison, no envy; and therefore kings are not envied

[15] *i.e.* Matter for envy to work upon: *ubique enim occurrunt objecta invidiæ.*

but by kings. Nevertheless it is to be noted that unworthy persons are most envied at their first coming in, and afterwards overcome it better; whereas contrariwise, persons of worth and merit are most envied when their fortune continueth long. For by that time, though their virtue be the same, yet it hath not the same lustre; for fresh men grow up that darken it.

Persons of noble blood are less envied in their rising. For it seemeth but right done to their birth. Besides, there seemeth not much added to their fortune; and envy is as the sunbeams, that beat hotter upon a bank or steep rising ground, than upon a flat. And for the same reason those that are advanced by degrees are less envied than those that are advanced suddenly and *per saltum.*

Those that have joined with their honour great travels, cares, or perils, are less subject to envy. For men think that they earn their honours hardly, and pity them sometimes; and pity ever healeth envy. Wherefore you shall observe that the more deep and sober sort of politic persons, in their greatness, are ever bemoaning themselves, what a life they lead; chanting a *quanta patimur.* Not that they feel it so, but only to abate the edge of envy. But this is to be understood of business that is laid upon men, and not such as they call unto themselves. For nothing increaseth envy more than an unnecessary and ambitious engrossing of business. And nothing doth extinguish envy more than for a great person to preserve all other inferior officers in their full rights and pre-eminences of their places. For by that means there be so many screens between him and envy.

Above all, those are most subject to envy, which carry the greatness of their fortunes in an insolent and proud manner; being never well but while they are shewing how great they are, either by outward pomp, or by triumphing over all opposition or competition; whereas wise men will rather do sacrifice to envy, in suffering themselves sometimes of purpose to be crossed and overborne in things that do not much concern them. Notwithstanding so much is true, that the carriage of greatness in a plain and open manner (so it be without arrogancy and vain glory) doth draw less envy than if it be in a more crafty and cunning fashion. For in that course a man doth but disavow fortune; and seemeth to be conscious of his own want in worth; and doth but teach others to envy him.

Lastly, to conclude this part; as we said in the beginning that the act of envy had somewhat in it of witchcraft, so there is no other cure of envy but the cure of witchcraft; and that is, to remove the *lot* (as they call it) and to lay it upon another. For which purpose, the wiser sort of great persons bring in ever upon the stage somebody upon whom to derive[16] the envy that would come upon themselves; sometimes upon ministers and servants; sometimes upon colleagues and associates; and the like; and for that turn there are never wanting some persons of violent and undertaking natures, who so they may have power and business will take it at any cost.

Now, to speak of public envy. There is yet some good in public envy, whereas in private there is none. For public envy is as an ostracism, that eclipseth men when they grow too great. And therefore it is a bridle also to great ones to keep them within bounds.

This envy, being in the Latin word *invidia,* goeth in the modern languages by the name of *discontentment*; of which we shall speak in handling Sedition. It is a disease in a state like to infection. For as infection spreadeth upon that which is sound, and tainteth it; so when envy is gotten once into a state, it traduceth even the best actions thereof, and turneth them into an ill odour. And therefore there is little won by intermingling of plausible actions. For that doth argue but a weakness and fear of envy, which hurteth so much the more; as it is likewise usual in infections; which if you fear them, you call them upon you.

This public envy seemeth to beat chiefly upon principal officers or ministers, rather than upon kings and estates themselves. But this is a sure rule, that if the envy upon the minister be great, when the cause of it in him is small; or if the envy be general in a manner upon all the ministers of an estate; then the envy (though hidden) is truly upon the state itself. And so much of public envy or discontentment, and the difference thereof from private envy, which was handled in the first place.

[16] Turn from its course.

We will add this in general, touching the affection of envy; that of all other affections it is the most importune and continual. For of other affections there is occasion given but now and then; and therefore it was well said, *Invidia festos dies non adit* [Envy keeps no holidays]: for it is ever working upon some or other. And it is also noted that love and envy do make a man pine, which other affections do not, because they are not so continual. It is also the vilest affection and the most depraved; for which cause it is the proper attribute of the devil, who is called *The envious man, that soweth tares amongst the wheat by night;* as it always cometh to pass, that envy worketh subtilly, and in the dark; and to the prejudice of good things, such as is the wheat.

X. Of Love.

The stage is more beholding to Love, than the life of man. For as to the stage, love is ever matter of comedies, and now and then of tragedies; but in life it doth much mischief; sometimes like a syren, sometimes like a fury. You may observe, that amongst all the great and worthy persons (whereof the memory remaineth, either ancient or recent,) there is not one that hath been transported to the mad degree of love; which shews that great spirits and great business do keep out this weak passion. You must except nevertheless Marcus Antonius, the half partner of the empire of Rome, and Appius Claudius, the decemvir and lawgiver; whereof the former was indeed a voluptuous man, and inordinate; but the latter was an austere and wise man: and therefore it seems (though rarely) that love can find entrance not only into an open heart, but also into a heart well fortified, if watch be not well kept. It is a poor saying of Epicurus, *satis magnum alter alteri theatrum sumus* [Each is to other a theatre large enough]; as if man, made for the contemplation of heaven and all noble objects, should do nothing but kneel before a little idol, and make himself a subject, though not of the mouth (as beasts are), yet of the eye; which was given him for higher purposes. It is a strange thing to note the excess of this passion, and how it braves the nature and value of things, by this; that the speaking in a perpetual hyperbole is comely in nothing but in love. Neither is it merely in the phrase, for whereas it hath been well said that the arch-flatterer, with whom all the petty flatterers have intelligence, is a man's self; certainly the lover is more. For there was never proud man thought so absurdly well of himself as the lover doth of the person loved; and therefore it was well said, *That it is impossible to love and to be wise*[17]. Neither doth this weakness appear to others only, and not to the party loved; but to the loved most of all, except the love be reciproque. For it is a true rule, that love is ever rewarded either with the reciproque or with an inward and secret contempt. By how much the more men ought to beware of this passion, which loseth not only other things, but itself. As for the other losses, the poet's relation doth well figure them; That he that preferred Helena quitted the gifts of Juno and Pallas. For whosoever esteemeth too much of amorous affection quitteth both riches and wisdom. This passion hath his floods in the very times of weakness; which are great prosperity and great adversity; though this latter hath been less observed: both which times kindle love, and make it more fervent, and therefore shew it to be the child of folly. They do best, who if they cannot but admit love, yet make it keep quarter; and sever it wholly from their serious affairs and actions of life; for if it check once with business, it troubleth men's fortunes, and maketh men that they can no ways be true to their own ends. I know not how, but martial men are given to love; I think it is but as they are given to wine; for perils commonly ask to be paid in pleasures. There is in man's nature a secret inclination and motion towards love of others, which if it be not spent upon some one or a few, doth naturally spread itself towards many, and maketh men become humane and charitable; as it is seen sometime in friars. Nuptial love maketh mankind; friendly love perfecteth it; but wanton love corrupteth and embaseth it.

[17] [See note above, p. 425.]

XI. Of Great Place.

Men in great place are thrice servants; servants of the sovereign or state; servants of fame; and servants of business. So as they have no freedom; neither in their persons, nor in their actions, nor in their times. It is a strange desire, to seek power and to lose liberty: or to seek power over others and to lose power over a man's self. The rising unto place is laborious; and by pains men come to greater pains; and it is sometimes base; and by indignities men come to dignities. The standing is slippery, and the regress is either a downfall, or at least an eclipse, which is a melancholy thing. *Cum non sis qui fueris, non esse cur velis vivere* [Where a man feels that he is no longer what he was, he loses all his interest in life]. Nay, retire men cannot when they would, neither will they when it were reason; but are impatient of privateness, even in age and sickness, which require the shadow; like old townsmen, that will be still sitting at their street door, though thereby they offer age to scorn. Certainly great persons had need to borrow other men's opinions, to think themselves happy; for if they judge by their own feeling, they cannot find it: but if they think with themselves what other men think of them, and that other men would fain be as they are, then they are happy as it were by report; when perhaps they find the contrary within. For they are the first that find their own griefs, though they be the last that find their own faults. Certainly men in great fortunes are strangers to themselves, and while they are in the puzzle of business they have no time to tend their health either of body or mind. *Illi mors gravis incubat, qui notus nimis omnibus, ignotus moritur sibi* [It is a sad fate for a man to die too well known to every-body else, and still unknown to himself]. In place there is licence to do good and evil; whereof the latter is a curse: for in evil the best condition is not to will; the second not to can. But power to do good is the true and lawful end of aspiring. For good thoughts (though God accept them) yet towards men are little better than good dreams, except they be put in act; and that cannot be without power and place, as the vantage and commanding ground. Merit and good works is the end of man's motion; and conscience of the same is the accomplishment of man's rest. For if a man can be partaker of God's theatre, he shall likewise be partaker of God's rest. *Et conversus Deus, ut aspiceret opera quæ fecerunt manus suæ, vidit quod omnia essent bona nimis* [And God turned to look upon the works which his hands had made and saw that all were very good]; and then the sabbath. In the discharge of thy place set before thee the best examples; for imitation is a globe of precepts. And after a time set before thee thine own example; and examine thyself strictly whether thou didst not best at first. Neglect not also the examples of those that have carried themselves ill in the same place; not to set off thyself by taxing their memory, but to direct thyself what to avoid. Reform therefore, without bravery or scandal of former times and persons; but yet set it down to thyself as well to create good precedents as to follow them. Reduce things to the first institution, and observe wherein and how they have degenerate; but yet ask counsel of both times; of the ancient time, what is best; and of the latter time, what is fittest. Seek to make thy course regular, that men may know beforehand what they may expect; but be not too positive and peremptory; and express theyself well when thou digressest from thy rule. Preserve the right of thy place; but stir not questions of jurisdiction: and rather assume thy right in silence and *de facto*, than voice it with claims and challenges. Preserve likewise the rights of inferior places; and think it more honour to direct in chief than to be busy in all. Embrace and invite helps and advices touching the execution of thy place; and do not drive away such as bring thee information, as meddlers; but accept of them in good part. The vices of authority are chiefly four; delays, corruption, roughness, and facility. For delays; give easy access; keep times appointed; go through with that which is in hand, and interlace not business but of necessity. For corruption; do not only bind thine own hands or thy servants' hands from taking, but bind the hands of suitors also from offering. For integrity used doth the one; but integrity professed, and with a manifest detestation of bribery, doth the other. And avoid not only the fault, but the suspicion. Whosoever is found variable, and changeth manifestly without manifest cause, giveth suspicion of corruption. Therefore always when thou changest thine opinion or course, profess it plainly, and declare it, together with the reasons that move thee to change; and do not think to steal it. A servant or a favourite, if he be inward, and no other apparent cause of esteem, is commonly thought but a by-way to close corruption. For roughness; it is a needless cause of discontent: severity breedeth fear, but roughness breedeth hate. Even reproofs from authority ought to be grave, and not taunting. As for facility; it is worse than bribery. For bribes come but now and then; but if importunity or idle respects lead a man, he shall never be without. As Salomon saith, *To respect persons is not good; for such a man will transgress for a piece of bread*. It is most true that was anciently spoken, *A place sheweth the man*. And it sheweth some to the better, and some to the worse. *Omnium consensu capax imperii, nisi imperasset*, [a man whom every body would have thought fit for empire if he had not been emperor], saith Tacitus of Galba; but of Vespasian he saith, *Solus imperantium, Vespasianus mutatus in melius* [He was the only emperor whom the possession of power changed for the better]; though the one was meant of sufficiency, the other of manners and affection. It is an assured sign of a worthy and generous spirit, whom honour amends. For honour is, or should be, the place of virtue; and as in nature things move violently to their place and calmly in their place, so virtue in ambition is violent, in authority settled and calm. All rising to great place is by a winding stair; and if there be factions, it is good to side a man's self whilst he is in the rising, and to balance himself when he is placed. Use the memory of thy predecessor fairly and tenderly; for if thou doest not, it is a debt will sure be paid when thou art gone; if thou have colleagues, respect them and rather call them when they look not for it, than exclude them when they have reason to look to be called. Be not too sensible or too remembering of thy place in conversation and private answers to suitors; but let it rather be said, *When he sits in place he is another man*.

XII. Of Boldness.

It is a trivial grammar-school text, but yet worthy a wise man's consideration. Question was asked of Demosthenes, *what was the chief part of an orator?* he answered, *action*: what next? *action*: what next again? *action*. He said it that knew it best, and had by nature himself no advantage in that he commended. A strange thing, that that part of an orator which is but superficial, and rather the virtue of a player, should be placed so high, above those other noble parts of invention, elocution, and the rest; nay almost alone, as if it were all in all. But the reason is plain. There is in human nature generally more of the fool than of the wise; and therefore those facilities by which the foolish part of men's minds is taken are most potent. Wonderful like is the case of Boldness, in civil business; what first? Boldness: what second and third? Boldness. And yet boldness is a child of ignorance and baseness, far inferior to other parts. But nevertheless it doth fascinate and bind hand and foot those that are either shallow in judgment or weak in courage, which are the greatest part; yea and prevaileth with wise men at weak times. Therefore we see it hath done wonders in popular states; but with senates and princes less; and more ever upon the first entrance of bold persons into action than soon after; for boldness is an ill keeper of promise. Surely as there are mountebanks for the natural body, so are there mountebanks for the politic body; men that undertake great cures, and perhaps have been lucky in two or three experiments, but want the grounds of science, and therefore cannot hold out. Nay you shall see a bold fellow many times do Mahomet's miracle. Mahomet made the people believe that he would call an hill to him, and from the top of it offer up his prayers for the observers of his law. The people assembled; Mahomet called the hill to come to him, again and again; and when the hill stood still, he was never a whit abashed, but said, *If the hill will not come to Mahomet, Mahomet will go to the hill*. So these men, when they have promised great matters and failed most shamefully, yet (if they have the perfection of boldness) they will but slight it over, and make a turn, and no more ado. Certainly to men of great judgment, bold persons are a sport to behold; nay and to the vulgar also, boldness has somewhat of the ridiculous. For if absurdity be the subject of laughter, doubt you not but great boldness is seldom without some absurdity. Especially it is a sport to see, when a bold fellow is out of countenance; for that puts his face into a most shrunken and wooden posture; as needs it must; for in bashfulness the spirits do a little go and come; but with bold men, upon like occasion, they stand at a stay; like a stale at chess, where it is no mate, but yet the game cannot stir. But this last were fitter for a satire than for a serious observation. This is well to be weighed; that boldness is ever blind; for it seeth not dangers and inconveniences. Therefore it is ill in counsel, good in execution; so that the right use of bold persons is, that they never command in chief, but be seconds, and under the direction of others. For in counsel it is good to see dangers; and in execution not to see them, except they be very great.

XIII. Of Goodness and Goodness of Nature.

I take Goodness in this sense, the affecting of the weal of men, which is that the Grecians call *Philanthropia*; and the word *humanity* (as it is used) is a little too light to express it. Goodness I call the habit, and Goodness of Nature the inclination. This of all virtues and dignities of the mind is the greatest; being the character of the Deity; and without it man is a busy, mischievous, wretched thing; no better than a kind of vermin. Goodness answers to the theological virtue Charity, and admits no excess, but error. The desire of power in excess caused the angels to fall; the desire of knowledge in excess caused man to fall: but in charity there is no excess; neither can angel or man come in danger by it. The inclination to goodness is imprinted deeply in the nature of man; insomuch that if it issue not towards men, it will take unto other living creatures; as it is seen in the Turks, a cruel people, who nevertheless are kind to beasts, and give alms to dogs and birds; insomuch as Busbechius reporteth, a Christian boy in Constantinople had like to have been stoned for gagging in a waggishness a long-billed fowl [18]. Errors indeed in this virtue of goodness or charity may be committed. The Italians have an ungracious proverb, *Tanto buon che val niente; So good, that he is good for nothing.* And one of the doctors of Italy [19], Nicholas Machiavel, had the confidence to put in writing, almost in plain terms, *That the Christian faith had given up good men in prey to those that are tyrannical and unjust.* Which he spake, because indeed there was never law, or sect, or opinion did so much magnify goodness, as the Christian religion doth. Therefore, to avoid the scandal and the danger both, it is good to take knowledge of the errors of an habit so excellent. Seek the good of other men, but be not in bondage to their faces or fancies; for that is but facility or softness; which taketh an honest mind prisoner. Neither give thou Æsop's cock a gem, who would be better pleased and happier if he had a barley-corn. The example of God teacheth the lesson truly; *He sendeth his rain, and maketh his sun to shine, upon the just and unjust;* but he doth not rain wealth, nor shine honour and virtues, upon men equally. Common benefits are to be communicate with all; but peculiar benefits with choice. And beware how in making the portraiture thou breakest the pattern. For divinity maketh the love of ourselves the pattern; the love of our neighbours but the portraiture. *Sell all thou hast, and give it to the poor, and*

[18] The Latin translation has, more correctly, *adeo ut (referente Busbequio) aurifex quidam Venetus, Byzantii agens, vix furorem populi effugerit, quod avis cujusdam rostri oblongi fauces inserto baculo diduxisset.* The bird was a goat-sucker, which the goldsmith ("homo alioqui ridiculus") fastened over his door with wings spread and jaws distended. The story will be found in Busbequius's letter from Constantinople, p. 179 of ed. 1633.

[19] These words are omitted in the translation; no doubt as likely to give offence at Rome. The Italian translation has "quel empio Nicolo Macciavello".

follow we: but sell not all thou hast, except thou come and follow me; that is, except thou have a vocation wherein thou mayest do as much good with little means as with great; for otherwise in feeding the streams thou driest the fountain. Neither is there only a habit of goodness, directed by right reason; but there is in some men, even in nature, a disposition towards it; as on the other side there is a natural malignity. For there be that in their nature do not affect the good of others. The lighter sort of malignity turneth but to a crossness, or forwardness, or aptness to oppose, or difficilness, or the like; but the deeper sort to envy and mere mischief. Such men in other men's calamities are as it were in season, and are ever on the loading part: not so good as the dogs that licked Lazarus' sores; but like flies that are still buzzing upon any thing that is raw; *misanthropi*, that make it their practice to bring men to the bough, and yet have never a tree for the purpose in their gardens [20], as Timon had. Such dispositions are the very errours of human nature; and yet they are the fittest timber to make great politiques of; like to knee timber, that is good for ships, that are ordained to be tossed; but not for building houses, that shall stand firm. The parts and signs of goodness are many. If a man be gracious and courteous to strangers it shews he is a citizen of the world, and that his heart is no island cut off from other lands, but a continent that joins to them. If he be compassionate towards the afflictions of others, it shews that his heart is like the noble tree that is wounded itself when it gives the balm. If he easily pardons and remits offences, it shews that his mind is planted above injuries; so that he cannot be shot. If he be thankful for small benefits, it shews that he weighs men's minds, and not their trash. But above all, if he have St. Paul's perfection, that he would wish to be an *anathema* from Christ for the salvation of his brethren, it shews much of a divine nature, and a kind of conformity with Christ himself.

XIV. Of Nobility.

We will speak of Nobility first as a portion of an estate; then as a condition of particular persons. A monarchy where there is no nobility at all, is ever a pure and absolute tyranny; as that of the Turks. For nobility attempers sovereignty, and draws the eyes of the people somewhat aside from the line royal. But for democracies, they need it not; and they are commonly more quiet and less subject to sedition, than where there are stirps of nobles. For men's eyes are upon the business, and not upon the persons; or if upon the persons, it is for the business sake, as fittest, and not for flags and pedigree. We see the Switzers last well, notwithstanding their diversity of religion and of cantons. For utility is their bond, and not respects. The united provinces of the Low Countries in their government excel; for where there is an equality, the consultations are more indifferent, and the payments and tributes more cheerful. A great and potent nobility addeth majesty to a monarch, but diminisheth power; and putteth life and spirit into the people, but presseth their fortune. It is well when nobles are not too great for sovereignty nor for justice; and yet maintained in that height, as the insolency of inferiors may be broken upon them before it come on too fast upon the majesty of kings. A numerous nobility causeth poverty and inconvenience in a state; for it is a surcharge of expense; and besides, it being of necessity that many of the nobility fall in time to be weak in fortune, it maketh a kind of disproportion between honour and means.

As for nobility in particular persons; it is a reverend thing to see an ancient castle or building not in decay; or to see a fair timber tree sound and perfect. How much more to behold an ancient noble family, which hath stood against the waves and weathers of time. For new nobility is but the act of power, but ancient nobility is the act of time. Those that are first raised to nobility are commonly more virtuous, but less innocent, than their descendants; for there is rarely any rising but by a commixture of good and evil arts. But it is reason

[20] That is, I suppose, without openly professing it. The Italian translation introduces the word *palesemente*: " et con tutto ciò non hanno palesemente nei loro giardini à tal proposito l'albero di Timone ".

the memory of their virtues remain to their posterity, and their faults die with themselves. Nobility of birth commonly abateth industry; and he that is not industrious, envieth him that is. Besides, noble persons cannot go much higher; and he that standeth at a stay when others rise, can hardly avoid motions of envy. On the other side, nobility extinguisheth the passive envy from others towards them; because they are in possession of honour. Certainly, kings that have able men of their nobility shall find ease in employing them, and a better slide into their business; for people naturally bend to them, as born in some sort to command.

XV. Of Seditions and Troubles.

Shepherds of people had need know the calendars of tempests in state; which are commonly greatest when things grow to equality; as natural tempests are greatest about the *Equinoctia*. And as there are certain hollow blasts of wind and secret swellings of seas before a tempest, so are there in states:

> ——— Ille etiam cæcos instare tumultus
> Sæpe monet, fraudesque et operta tumescere bella.
> [Of troubles imminent and treasons dark
> Thence warning comes, and wars in secret gathering.]

Libels and licentious discourses against the state, when they are frequent and open; and in like sort, false news often running up and down to the disadvantage of the state, and hastily embraced; are amongst the signs of troubles. Virgil giving the pedigree of Fame, saith *she was sister to the Giants:*

> Illam Terra parens, irâ irritata Deorum,
> Extremam (ut perhibent) Cœo Enceladoque sororem
> Progenuit.

As if fames were the relics of seditions past; but they are no less indeed the preludes of seditions to come. Howsoever he noteth it right, that seditious tumults and seditious fames differ no more but as brother and sister, masculine and feminine; especially if it come to that, that the best actions of a state, and the most plausible, and which ought to give greatest contentment, are taken in ill sense, and traduced: for that shews the envy great, as Tacitus saith, *conflata magna invidia, seu bene seu male gesta premunt* [when dislike prevails against the government, good actions and bad offend alike]. Neither doth it follow, that because these fames are a sign of troubles, that [21] the suppressing of them with too much severity should be a remedy of troubles. For the despising of them many times checks them best; and the going about to stop them doth but make a wonder long-lived. Also that kind of obedience which Tacitus speaketh of, is to be held suspected: *Erant in officio, sed tamen qui mallent mandata imperantium interpretari quam exequi* [ready to serve, and yet more disposed to construe commands than execute them]; disputing, excusing, cavilling upon mandates and directions, is a kind of shaking off the yoke, and assay of disobedience; especially if in those disputings they which are for the direction speak fearfully and tenderly, and those that are against it audaciously.

Also, as Machiavel [22] noteth well, when princes, that ought to be common parents, make themselves as a party, and lean to a side, it is as a boat that is overthrown by uneven weight on the one side; as was well seen in the time of Henry the Third of France; for first himself entered league for the extirpation of the Protestants; and presently after the same league was turned upon himself. For when the authority of princes is made but an accessary to a cause, and that there be other bands that tie faster than the band of sovereignty, kings begin to be put almost out of possession.

Also, when discords and quarrels and factions are carried openly and audaciously, it is a sign the reverence of government is lost. For the motions of the greatest persons in a government ought to be as the motions of the planets

[21] So in original. One of the *thats* should of course be omitted.

[22] The Italian translation omits the name of Machiavel, and says only *un scrittore*.

under *primum mobile ;* (according to the old opinion), which is, that every of them is carried swiftly by the highest motion, and softly in their own motion. And therefore, when great ones in their own particular motion move violently, and, as Tacitus expresseth it well, *liberius quam ut imperantium meminissent* [unrestrained by reverence for the government], it is a sign the orbs are out of frame. For reverence is that wherewith princes are girt from God ; who threateneth[23] the dissolving thereof ; *Solvam cingula regum* [I will unbind the girdles of kings].

So when any of the four pillars of government are mainly shaken or weakened (which are Religion, Justice, Counsel, and Treasure), men had need to pray for fair weather. But let us pass from this part of predictions (concerning which, nevertheless, more light may be taken from that which followeth) ; and let us speak first of the Materials of seditions ; then of the Motives of them ; and thirdly of the Remedies.

Concerning the Materials of seditions. It is a thing well to be considered; for the surest way to prevent seditions (if the times do bear it) is to take away the matter of them. For if there be fuel prepared, it is hard to tell whence the spark shall come that shall set it on fire. The matter of seditions is of two kinds, much poverty and much discontentment. It is certain, so many overthrown estates, so many votes for troubles. Lucan noteth well the state of Rome before the civil war:

> Hinc usura vorax, rapidumque in tempore fœnus,
> Hinc concussa fides, et multis utile bellum

[estates eaten up by usurious rates of interest, credit shaken, war a gain to many].

This same *multis utile bellum* is an assured and infallible sign of a state disposed to seditions and troubles. And if this poverty and broken estate in the better sort be joined with a want and necessity in the mean people, the danger is imminent and great. For the rebellious of the belly are the worst. As for discontentments, they are in the politic body like to humours in the natural, which are apt to gather a preternatural heat and to inflame. And let no prince measure the danger of them by this, whether they be just or unjust : for that were to imagine people to be too reasonable ; who do often spurn at their own good ; nor yet by this, whether the griefs whereupon they rise be in fact great or small : for they are the most dangerous discontentments where the fear is greater than the feeling : *Dolendi modus, timendi non item* [Suffering has its limit, but fears are endless]. Besides, in great oppressions, the same things that provoke the patience, do withal mate the courage ; but in fears it is not so. Neither let any prince or state be secure concerning discontentments, because they have been often, or have been long, and yet no peril hath ensued : for as it is true that every vapour or fume doth not turn into a storm ; so it is nevertheless true that storms, though they blow over divers times, yet may fall at last ; and, as the Spanish proverb noteth well, *The cord breaketh at the last by the weakest pull.*

The Causes and Motives of seditions are, innovation in religion ; taxes ; alteration of laws and customs ; breaking of privileges ; general oppression ; advancement of unworthy persons ; strangers ; dearths ; disbanded soldiers ; factions grown desperate ; and whatsoever, in offending people, joineth and knitteth them in a common cause.

For the Remedies : there may be some general preservatives, whereof we will speak : as for the just cure, it must answer to the particular disease ; and so be left to counsel rather than rule.

The first remedy or prevention is to remove by all means possible that material cause of sedition whereof we spake ; which is, want and poverty in the estate. To which purpose serveth, the opening and well-balancing of trade ; the cherishing of manufactures ; the banishing of idleness ; the repressing of waste and excess by sumptuary laws ; the improvement and husbanding of the soil ; the

[23] That is, holds it out as a threat. A manuscript copy of this Essay in an earlier form has, " who threateneth the dissolving thereof as one of his greatest judgments".

regulating of prices of things vendible ; the moderating of taxes and tributes, and the like. Generally, it is to be foreseen that the population of a kingdom (especially if it be not mown down by wars) do not exceed the stock of the kingdom which should maintain them. Neither is the population to be reckoned only by number ; for a smaller number that spend more and earn less, do wear out an estate sooner than a greater number that live lower and gather more. Therefore the multiplying of nobility and other degrees of quality in an over proportion to the common people, doth speedily bring a state to necessity ; and so doth likewise an overgrown clergy ; for they bring nothing to the stock ; and in like manner when more are bred scholars than preferments can take off.

It is likewise to be remembered, that forasmuch as the increase of any estate must be upon the foreigner (for whatsoever is somewhere gotten is somewhere lost), there be but three things which one nation selleth unto another ; the commodity as nature yieldeth it ; the manufacture ; and the vecture, or carriage. So that if these three wheels go, wealth will flow as in a spring tide. And it cometh many times to pass, that *materiam superabit opus ;* that the work and carriage is more worth than the material, and enricheth a state more ; as is notably seen in the Low-Countrymen, who have the best mines above ground in the world.

Above all things, good policy is to be used that the treasure and monies in a state be not gathered into few hands. For otherwise a state may have a great stock, and yet starve. And money is like muck, not good except it be spread. This is done chiefly by suppressing, or at the least keeping a strait hand upon the devouring trades of usury, ingrossing, great pasturages, and the like.

For removing discontentments, or at least the danger of them ; there is in every state (as we know) two portions of subjects ; the nobless and the commonalty. When one of these is discontent, the danger is not great ; for common people are of slow motion, if they be not excited by the greater sort ; and the greater sort are of small strength, except the multitude be apt and ready to move of themselves. Then is the danger, when the greater sort do but wait for the troubling of the waters amongst the meaner, that then they may declare themselves. The poets feign, that the rest of the gods would have bound Jupiter, which he hearing of, by the counsel of Pallas, sent for Briareus, with his hundred hands, to come in to his aid. An emblem, no doubt, to show how safe it is for monarchs to make sure of the good will of common people.

To give moderate liberty for griefs and discontentments to evaporate (so it be without too great insolency or bravery), is a safe way. For he that turneth the humours back, and maketh the wound bleed inwards, endangereth malign ulcers and pernicious imposthumations.

The part of Epimetheus mought well become Prometheus, in the case of discontentments ; for there is not a better provision against them. Epimetheus, when griefs and evils flew abroad, at last shut the lid, and kept hope in the bottom of the vessel. Certainly, the politic and artificial nourishing and entertaining of hopes, and carrying men from hopes to hopes, is one of the best antidotes against the poison of discontentments. And it is a certain sign of a wise government and proceeding, when it can hold men's hearts by hopes, when it cannot by satisfaction ; and when it can handle things in such manner, as no evil shall appear so peremptory but that it hath some outlet of hope : which is the less hard to do, because both particular persons and factions are apt enough to flatter themselves, or at least to brave that they believe not.

Also the foresight and prevention, that there be no likely or fit head whereunto discontented persons may resort, and under whom they may join, is a known, but an excellent point of caution. I understand a fit head to be one that hath greatness and reputation ; that hath confidence with the discontented party, and upon whom they turn their eyes ; and that is thought discontented in his own particular : which kind of persons are either to be won and reconciled to the state, and that in a fast and true manner ; or to be fronted with some other of the same party, that may oppose them, and so divide the reputation. Generally, the dividing and breaking of all factions and combinations that are adverse to the state, and setting them at distance, or at least distrust, amongst themselves, is not one of the worst remedies. For it is a desperate case, if those that hold with the proceeding of the state be full of discord and faction, and those that are against it be entire and united.

I have noted that some witty and sharp speeches which have fallen from princes have given fire to seditions. Cæsar did himself infinite hurt in that speech, *Sylla nescivit literas, non potuit dictare* [Sylla was no scholar, he could not dictate] : for it did utterly cut off that hope which men had entertained, that he would at one time or other give over his dictatorship. Galba undid himself by that speech, *legi a se militem, non emi* [that he did not buy his soldiers, but levied them] : for it put the soldiers out of hope of the donative. Probus likewise, by that speech, *si vixero, non opus erit amplius Romano imperio militibus* [if I live, the Roman empire shall have no more need of soldiers] : a speech of great despair for the soldiers. And many the like. Surely princes had need, in tender matters and ticklish times, to beware what they say ; especially in these short speeches, which fly abroad like darts, and are thought to be shot out of their secret intentions. For as for large discourses, they are flat things, and not so much noted.

Lastly, let princes, against all events, not be without some great person, one or rather more, of military valour, near unto them, for the repressing of seditions in their beginnings. For without that, there useth to be more trepidation in court upon the first breaking out of troubles than were fit. And the state runneth the danger of that which Tacitus saith ; *Atque is habitus animorum fuit, ut pessimum facinus auderent pauci, plures vellent, omnes paterentur* [A few were in a humour to attempt mischief, more to desire, all to allow it]. But let such military persons be assured, and well reputed of, rather than factious and popular ; holding also good correspondence with the other great men in the state; or else the remedy is worse than the disease.

XVI. Of Atheism.

I HAD rather believe all the fables in the Legend [24], and the Talmud, and the Alcoran [25], than that this universal frame is without a mind. And therefore God never wrought miracle to convince atheism, because his ordinary works convince it. It is true, that a little philosophy inclineth man's mind to atheism ; but depth in philosophy bringeth men's minds about to religion. For while the mind of man looketh upon second causes scattered, it may sometimes rest in them, and go no further ; but when it beholdeth the chain of them, confederate and linked together, it must needs fly to Providence and Deity. Nay, even that school which is most accused of atheism doth most demonstrate religion ; that is, the school of Leucippus and Democritus and Epicurus. For it is a thousand times more credible, that four mutable elements, and one immutable fifth essence, duly and eternally placed, need no God, than that an army of infinite small portions or seeds unplaced, should have produced this order and beauty without a divine marshal. The scripture saith, *The fool hath said in his heart, there is no God ;* it is not said, *The fool hath thought in his heart ;* so as he rather saith it by rote to himself, as that he would have, than that he can throughly believe it, or be persuaded of it. For none deny there is a God, but those for whom it maketh that there were no God. It appeareth in nothing more, that atheism is rather in the lip than in the heart of man, than by this ; that atheists will ever be talking of that their opinion, as if they fainted in it within themselves, and would be glad to be strengthened by the consent of others. Nay more, you shall have atheists strive to get disciples, as it fareth with other sects. And, which is most of all, you shall have of them that will suffer for atheism and not recant ; whereas if they did truly think that there were no such thing as God, why should they trouble themselves ? Epicurus is charged that he did but dissemble for his credit's sake, when he affirmed there were blessed natures, but such as enjoyed themselves without having respect to the government of

[24] [*I.e.* the " Golden Legend ".]

[25] In the edition of 1612, it stood, " all the fables in the Legend and the Alcoran ". The Italian translation omits the Legend, and has only " tutte le favole dell' Alcorano ".

the world. Wherein they say he did temporize; though in secret he thought there was no God. But certainly he is traduced; for his words are noble and divine: *Non Deos vulgi negare profanum; sed vulgi opiniones Diis applicare profanum* [There is no profanity in refusing to believe in the Gods of the vulgar: the profanity is in believing of the Gods what the vulgar believe of them]. Plato could have said no more. And although he had the confidence to deny the administration, he had not the power to deny the nature. The Indians of the west have names for their particular gods, though they have no name for God: as if the heathens should have had the names Jupiter, Apollo, Mars, etc. but not the word *Deus*; which shews that even those barbarous people have the notion, though they have not the latitude and extent of it. So that against atheists the very savages take part with the very subtlest philosophers. The contemplative atheist is rare: a Diagoras, a Bion, a Lucian perhaps, and some others; and yet they seem to be more than they are; for that all that impugn a received religion or superstition are by the adverse part branded with the name of atheists. But the great atheists indeed are hypocrites; which are ever handling holy things, but without feeling; so as they must needs be cauterized in the end. The causes of atheism are, divisions in religion, if they be many; for any one main division addeth zeal to both sides; but many divisions introduce atheism. Another is, scandal of priests; when it is come to that which St. Bernard saith, *Non est jam dicere, ut populus sic sacerdos; quia nec sic populus ut sacerdos* [One cannot now say, the priest is as the people, for the truth is that the people are not so bad as the priest]. A third is, custom of profane scoffing in holy matters; which doth by little and little deface the reverence of religion. And lastly, learned times, specially with peace and prosperity; for troubles and adversities do more bow men's minds to religion. They that deny a God destroy man's nobility; for certainly man is of kin to the beasts by his body; and, if he be not of kin to God by his spirit, he is a base and ignoble creature. It destroys likewise magnanimity, and the raising of human nature; for take an example of a dog, and mark what a generosity and courage he will put on when he finds himself maintained by a man; who to him is instead of a God, or *melior natura*; which courage is manifestly such as that creature, without that confidence of a better nature than his own, could never attain. So man, when he resteth and assureth himself upon divine protection and favour, gathereth a force and faith which human nature in itself could not obtain. Therefore, as atheism is in all respects hateful, so in this, that it depriveth human nature of the means to exalt itself above human frailty. As it is in particular persons, so it is in nations. Never was there such a state for magnanimity as Rome. Of this state hear what Cicero saith: *Quam volumus licet, patres conscripti, nos amemus, tamen nec numero Hispanos, nec robore Gallos, nec calliditate Pœnos, nec artibus Græcos, nec denique hoc ipso hujus gentis et terræ domestico nativoque sensu Italos ipsos et Latinos; sed pietate, ac religione, atque hac una sapientia, quod Deorum immortalium numine omnia regi gubernarique perspeximus, omnes gentes nationesque superavimus* [Pride ourselves as we may upon our country, yet are we not in number superior to the Spaniards, nor in strength to the Gauls, nor in cunning to the Carthaginians, nor to the Greeks in arts, nor to the Italians and Latins themselves in the homely and native sense which belongs to this nation and land; it is in piety only and religion, and the wisdom of regarding the providence of the Immortal Gods as that which rules and governs all things, that we have surpassed all nations and peoples].

XVII. Of Superstition [26].

It were better to have no opinion of God at all, than such an opinion as is unworthy of him. For the one is unbelief, the other is contumely: and certainly superstition is the reproach of the Deity. Plutarch saith well to that purpose: *Surely* (saith he) *I had rather a great deal men should say there was no such man at all as Plutarch, than that they should say that there was one Plutarch that would eat his children as soon as they were born*; as the poets speak of Saturn. And as

[26] This Essay is omitted in the Italian translation.

the contumely is greater towards God, so the danger is greater towards men. Atheism leaves a man to sense, to philosophy, to natural piety, to laws, to reputation; all which may be guides to an outward moral virtue, though religion were not; but superstition dismounts all these, and erecteth an absolute monarchy in the minds of men. Therefore atheism did never perturb states; for it makes men wary of themselves, as looking no further: and we see the times inclined to atheism (as the time of Augustus Cæsar) were civil times. But superstition hath been the confusion of many states, and bringeth in a new *primum mobile*, that ravisheth all the spheres of government. The master of superstition is the people; and in all superstition wise men follow fools; and arguments are fitted to practice, in a reversed order [27]. It was gravely said by some of the prelates in the council of Trent, where the doctrine of the schoolmen bare great sway, *that the schoolmen were like astronomers, which did feign eccentrics and epicycles, and such engines of orbs, to save the phænomena; though they knew there were no such things*; and in like manner, that the schoolmen had framed a number of subtle and intricate axioms and theorems, to save the practice of the church. The causes of superstition are, pleasing and sensual rites and ceremonies; excess of outward and pharisaical holiness; over-great reverence of traditions, which cannot but load the church; the stratagems of prelates for their own ambition and lucre; the favouring too much of good intentions, which openeth the gate to conceits and novelties; the taking an aim at divine matters by human, which cannot but breed mixture of imaginations: and, lastly, barbarous times, especially joined with calamities and disasters. Superstition, without a veil, is a deformed thing; for as it addeth deformity to an ape to be so like a man, so the similitude of superstition to religion makes it the more deformed. And as wholesome meat corrupteth to little worms, so good forms and orders corrupt into a number of petty observances. There is a superstition in avoiding superstition, when men think to do best if they go furthest from the superstition formerly received; therefore care would be had that (as it fareth in ill purgings) the good be not taken away with the bad; which commonly is done when the people is the reformer.

XVIII. Of Travel.

Travel, in the younger sort, is a part of education; in the elder, a part of experience. He that travelleth into a country before he hath some entrance into the language, goeth to school, and not to travel. That young men travel under some tutor, or grave servant, I allow well; so that he be such a one that hath the language, and hath been in the country before; whereby he may be able to tell them what things are worthy to be seen in the country where they go; what acquaintances they are to seek; what exercises or discipline the place yieldeth. For else young men shall go hooded, and look abroad little. It is a strange thing, that in sea voyages, where there is nothing to be seen but sky and sea, men should make diaries; but in land-travel, wherein so much is to be observed, for the most part they omit it; as if chance were fitter to be registered than observation. Let diaries therefore be brought in use. The things to be seen and observed are, the courts of princes, specially when they give audience to ambassadors; the courts of justice, while they sit and hear causes; and so of consistories ecclesiastic; the churches and monasteries, with the monuments which are therein extant; the walls and fortifications of cities and towns, and so the havens and harbours; antiquities and ruins; libraries; colleges, disputations, and lectures, where any are; shipping and navies; houses and gardens of state and pleasure, near great cities; armories; arsenals; magazines; exchanges; burses; warehouses; exercises of horsemanship, fencing, training of soldiers, and the like; comedies, such whereunto the better sorts of persons do resort; treasuries of jewels and robes; cabinets and rarities; and, to conclude, whatsoever is memorable in the places where they go. After all which, the tutors or servants ought to make diligent inquiry. As for triumphs, masks, feasts, weddings,

[27] That is, reason is governed by practice, instead of practice by reason. *Argumenta practicæ succumbunt, ordine perverso.*

funerals, capital executions, and such shows, men need not to be put in mind of them; yet are they not to be neglected. If you will have a young man to put his travel into a little room, and in short time to gather much, this you must do. First, as was said, he must have some entrance into the language before he goeth. Then he must have such a servant or tutor as knoweth the country, as was likewise said. Let him carry with him also some card or book describing the country where he travelleth; which will be a good key to his inquiry. Let him keep also a diary. Let him not stay long in one city or town; more or less as the place deserveth, but not long; nay, when he stayeth in one city or town, let him change his lodging from one end and part of the town to another; which is a great adamant of acquaintance. Let him sequester himself from the company of his countrymen, and diet in such places where there is good company of the nation where he travelleth. Let him upon his removes from one place to another, procure recommendation to some person of quality residing in the place whither he removeth; that he may use his favour in those things he desireth to see or know. Thus he may abridge his travel with much profit. As for the acquaintance which is to be sought in travel; that which is most of all profitable, is acquaintance with the secretaries and employed men of ambassadors: for so in travelling in one country he shall suck the experience of many. Let him also see and visit eminent persons in all kinds, which are of great name abroad; that he may be able to tell how the life agreeth with the fame. For quarrels, they are with care and discretion to be avoided. They are commonly for mistresses, healths, place, and words. And let a man beware how he keepeth company with choleric and quarrelsome persons; for they will engage him into their own quarrels. When a traveller returneth home, let him not leave the countries where he hath travelled altogether behind him; but maintain a correspondence by letters with those of his acquaintance which are of most worth. And let his travel appear rather in his discourse than in his apparel or gesture; and in his discourse let him be rather advised in his answers, than forward to tell stories; and let it appear that he doth not change his country manners for those of foreign parts; but only prick in some flowers of that he hath learned abroad into the customs of his own country.

XIX. Of Empire.

It is a miserable state of mind to have few things to desire, and many things to fear; and yet that commonly is the case of kings; who, being at the highest, want matter of desire, which makes their minds more languishing; and have many representations of perils and shadows, which makes their minds the less clear. And this is one reason also of that effect which the Scripture speaketh of, *That the king's heart is inscrutable*. For multitude of jealousies, and lack of some predominant desire that should marshal and put in order all the rest, maketh any man's heart hard to find or sound. Hence it comes likewise, that princes many times make themselves desires, and set their hearts upon toys; sometimes upon a building; sometimes upon erecting of an order; sometimes upon the advancing of a person; sometimes upon obtaining excellency in some art or feat of the hand; as Nero for playing on the harp, Domitian for certainty of the hand with the arrow, Commodus for playing at fence, Caracalla for driving chariots, and the like. This seemeth incredible unto those that know not the principle *that the mind of man is more cheered and refreshed by profiting in small things, than by standing at a stay in great*. We see also that kings that have been fortunate conquerors in their first years, it being not possible for them to go forward infinitely, but that they must have some check or arrest in their fortunes, turn in their latter years to be superstitious and melancholy; as did Alexander the Great; Dioclesian; and in our memory, Charles the Fifth; and others: for he that is used to go forward, and findeth a stop, falleth out of his own favour, and is not the thing he was.

To speak now of the true temper of empire; it is a thing rare and hard to keep; for both temper and distemper consist of contraries. But it is one thing to mingle contraries, another to interchange them. The answer of Apollonius to Vespasian is full of excellent instruction. Vespasian asked him, *what was Nero's overthrow?* He answered, *Nero could touch and tune the harp well; but in government sometimes he used to wind the pins too high, sometimes to let them down too low.* And certain it is that nothing destroyeth authority so much as the unequal and untimely interchange of power pressed too far, and relaxed too much.

This is true, that the wisdom of all these latter times in princes' affairs is rather fine deliveries and shiftings of dangers and mischiefs when they are near, than solid and grounded courses to keep them aloof. But this is but to try masteries with fortune. And let men beware how they neglect and suffer matter of trouble to be prepared; for no man can forbid the spark, nor tell whence it may come. The difficulties in princes' business are many and great; but the greatest difficulty is often in their own mind. For it is common with princes (saith Tacitus[28]) to will contradictories, *Sunt plerumque regum voluntates vehementes, et inter se contrariæ* [Their desires are commonly vehement and incompatible one with another]. For it is the solecism of power, to think to command the end, and yet not to endure the mean.

Kings have to deal with their neighbours, their wives, their children, their prelates or clergy, their nobles, their second nobles or gentlemen, their merchants, their commons, and their men of war; and from all these arise dangers, if care and circumspection be not used.

First for their neighbours; there can no general rule be given (the occasions are so variable), save one, which ever holdeth; which is, that princes do keep due sentinel, that none of their neighbours do overgrow so (by increase of territory, by embracing of trade, by approaches, or the like), as they become more able to annoy them than they were. And this is generally the work of standing counsels to foresee and to hinder it. During that triumvirate of kings, King Henry the Eighth of England, Francis the First King of France, and Charles the Fifth Emperor, there was such a watch kept, that none of the three could win a palm of ground, but the other two would straightways balance it, either by confederation, or, if need were, by a war; and would not in any wise take up peace at interest. And the like was done by that league (which Guicciardine saith was the security of Italy) made between Ferdinando King of Naples, Lorenzius Medices, and Ludovicus Sforza, potentates, the one of Florence, the other of Milan. Neither is the opinion of some of the schoolmen to be received, *that a war cannot justly be made but upon a precedent injury or provocation.* For there is no question but a just fear of an imminent danger, though there be no blow given, is a lawful cause of a war.

For their wives; there are cruel examples of them. Livia is infamed for the poisoning of her husband; Roxalana, Solyman's wife, was the destruction of that renowned prince Sultan Mustapha, and otherwise troubled his house and succession; Edward the Second of England his queen had the principal hand in the deposing and murder of her husband. This kind of danger is then to be feared chiefly, when the wives have plots for the raising of their own children; or else that they be advoutresses.

For their children; the tragedies likewise of dangers from them have been many. And generally, the entering of fathers into suspicion of their children hath been ever unfortunate. The destruction of Mustapha (that we named before) was so fatal to Solyman's line, as the succession of the Turks from Solyman until this day is suspected to be untrue, and of strange blood; for that Selymus the Second was thought to be suppositious. The destruction of Crispus, a young prince of rare towardness, by Constantinus the Great, his father, was in like manner fatal to his house; for both Constantinus and Constance, his sons, died violent deaths; and Constantius, his other son, did little better; who died indeed of sickness, but after that Julianus had taken arms against him. The destruction of Demetrius, son to Philip the Second of Macedon, turned upon the father, who died of repentance. And many like examples there are; but few or none where the fathers had good by such distrust; except it were where the sons were up in open arms against them; as was Selymus the First against Bajazet; and the three sons of Henry the Second, King of England.

[28] Not Tacitus, but Sallust. *Bell. Jug.* 113

For their prelates; when they are proud and great, there is also danger from them; as it was in the times of Anselmus and Thomas Becket, Archbishops of Canterbury; who with their crosiers did almost try it with the king's sword; and yet they had to deal with stout and haughty kings; William Rufus, Henry the First, and Henry the Second. The danger is not from that state, but where it hath a dependance of foreign authority; or where the churchmen come in and are elected, not by the collation of the king, or particular patrons, but by the people.

For their nobles; to keep them at a distance, it is not amiss; but to depress them may make a king more absolute, but less safe; and less able to perform any thing that he desires. I have noted it in my History of King Henry the Seventh of England, who depressed his nobility; whereupon it came to pass that his times were full of difficulties and troubles; for the nobility, though they continued loyal unto him, yet did they not co-operate with him in his business. So that in effect he was fain to do all things himself.

For their second-nobles; there is not much danger from them, being a body dispersed. They may sometimes discourse high, but that doth little hurt; besides, they are a counterpoise to the higher nobility, that they grow not too potent; and, lastly, being the most immediate in authority with the common people, they do best temper popular commotions.

For their merchants; they are *vena porta* [29]; and if they flourish not, a kingdom may have good limbs, but will have empty veins, and nourish little. Taxes and imposts upon them do seldom good to the king's revenue; for that that he wins in the hundred he leeseth in the shire; the particular rates being increased, but the total bulk of trading rather decreased.

For their commons; there is little danger from them, except it be where they have great and potent heads; or where you meddle with the point of religion, or their customs, or means of life.

For their men of war; it is a dangerous state where they live and remain in a body, and are used to donatives; whereof we see examples in the janizaries, and pretorian bands of Rome; but trainings of men, and arming them in several places, and under several commanders, and without donatives, are things of defence, and no danger.

Princes are like to heavenly bodies, which cause good or evil times; and which have much veneration, but no rest. All precepts concerning kings are in effect comprehended in those two remembrances; *memento quod es homo;* and *memento quod es Deus*, or *vice Dei* [Remember that you are a man; and remember that you are a God, or God's lieutenant]: the one bridleth their power, and the other their will.

XX. Of Counsel.

The greatest trust between man and man is the trust of giving counsel. For in other confidences men commit the parts of life; their lands, their goods, their

[29] Upon this phrase, which recurs two or three times in Bacon (see for instance the History of Henry VII.: "being a king that loved wealth and treasure, he could not endure to have trade sick, nor any obstruction to continue *in the gate-vein, which disperseth that blood*)," I am indebted to Mr. Ellis for the following characteristic note. "The metaphor," he writes "is historically curious; for no one would have used it since the discovery of the circulation of the blood and of the lacteals. But in Bacon's time it was supposed that the chyle was taken up by the veins which converge to the *vena porta*. The latter immediately divides into branches, and ultimately into four ramifications, which are distributed throughout the substance of the liver, so that it has been compared to the trunk of a tree giving off roots at one extremity and branches at the other. Bacon's meaning therefore is, that commerce concentrates the resources of a country in order to their redistribution. The *heart*, which receives blood from all parts of the body and brings it into contact with the external air, and then redistributes it everywhere, would I think have taken the place of the *vena porta*, after Harvey's discovery had become known; especially as the latter is a mere conduit, and not a source of motion."

child [30], their credit, some particular affair ; but to such as they make their counsellors, they commit the whole : by how much the more they are obliged to all faith and integrity. The wisest princes need not think it any diminution to their greatness, or derogation to their sufficiency, to rely upon counsel. God himself is not without, but hath made it one of the great names of his blessed Son ; *The Counsellor*. Salomon hath pronounced that *in counsel is stability*. Things will have their first or second agitation : if they be not tossed upon the arguments of counsel, they will be tossed upon the waves of fortune ; and be full of inconstancy, doing and undoing, like the reeling of a drunken man. Salomon's son found the force of counsel, as his father saw the necessity of it. For the beloved kingdom of God was first rent and broken by ill counsel ; upon which counsel there are set for our instruction the two marks whereby bad counsel is for ever best discerned ; that it was young counsel, for the persons ; and violent counsel, for the matter.

The ancient times do set forth in figure both the incorporation and inseparable conjunction of counsel with kings, and the wise and politic use of counsel by kings: the one, in that they say Jupiter did marry Metis, which signifieth counsel ; whereby they intend that Sovereignty is married to Counsel : the other in that which followeth, which was thus : They say, after Jupiter was married to Metis, she conceived by him and was with child, but Jupiter suffered her not to stay till she brought forth, but eat her up ; whereby he became himself with child, and was delivered of Pallas armed, out of his head. Which monstrous fable containeth a secret of empire ; how kings are to make use of their counsel of state. That first they ought to refer matters unto them, which is the first begetting or impregnation ; but when they are elaborate, moulded, and shaped in the womb of their counsel, and grow ripe and ready to be brought forth, that then they suffer not their counsel to go through with the resolution and direction, as if it depended on them ; but take the matter back into their own hands, and make it appear to the world that the decrees and final directions (which, because they come forth with prudence and power, are resembled to Pallas armed) proceeded from themselves ; and not only from their authority, but (the more to add reputation to themselves) from their head and device.

Let us now speak of the inconveniences of counsel, and of the remedies. The inconveniences that have been noted in calling and using counsel, are three. First, the revealing of affairs, whereby they become less secret. Secondly, the weakening of the authority of princes, as if they were less of themselves. Thirdly, the danger of being unfaithfully counselled, and more for the good of them that counsel than of him that is counselled. For which inconveniences, the doctrine of Italy [31], and practice of France, in some kings' times, hath introduced *cabinet* counsels ; a remedy worse than the disease [32].

As to secrecy ; princes are not bound to communicate all matters with all counsellors ; but may extract and select. Neither is it necessary that he that consulteth what he should do, should declare what he will do. But let princes beware that the unsecreting of their affairs comes not from themselves. And as for cabinet counsels, it may be their motto, *plenus rimarum sum* [they are full of leaks] : one futile person that maketh it his glory to tell, will do more hurt than many that know it their duty to conceal. It is true there be some affairs which require extreme secrecy, which will hardly go beyond one or two persons besides the king : neither are those counsels unprosperous ; for, besides the

[30] So edd. 1612 and 1625. Ed. 1639 has *children*.

[31] *doctrina quorundam ex Italis*. The Italian translation has *l'uso d'Italia e di Francia*.

[32] The sentence ends here in both the printed editions. But in a manuscript copy, with alterations in Bacon's own hand, and which appears to have been written a little earlier than 1612, the following clause is added : "which hath turned Metis the wife to Metis the mistress ; that is counsels of state, to which princes are married, to counsels of favoured persons, recommended chiefly by flattery and affection". *Cabinet Counsels* therefore (translated *concilia interiora quæ vulgo vocantur* Cabinetti) are not to be understood in the modern sense. What we call the Cabinht answers exactly to what Bacon calls a Counsel of State.

secrecy, they commonly go on constantly in one spirit of direction, without distraction. But then it must be a prudent king, such as is able to grind with a hand-mill; and those inward counsellors had need also be wise men, and especially true and trusty to the king's ends; as it was with King Henry the Seventh of England, who in his greatest business imparted himself to none, except it were to Morton and Fox.

For weakening of authority; the fable[33] showeth the remedy. Nay, the majesty of kings is rather exalted than diminished when they are in the chair of counsel; neither was there ever prince bereaved of his dependances by his counsel; except where there hath been either an over-greatness in one counsellor or an over-strict combination in divers; which are things soon found and holpen.

For the last inconvenience, that men will counsel with an eye to themselves; certainly, *non inveniet fidem super terram* [he will not find faith on the earth], is meant of the nature of times, and not of all particular persons. There be that are in nature faithful, and sincere, and plain, and direct; not crafty and involved; let princes, above all, draw to themselves such natures. Besides, counsellors are not commonly so united, but that one counsellor keepeth sentinel over another; so that if any do counsel out of faction or private ends, it commonly comes to the king's ear. But the best remedy is, if princes know their counsellors as well as their counsellors know them:

Principis est virtus maxima nosse suos.

And on the other side, counsellors should not be too speculative into their sovereign's person. The true composition of a counsellor is rather to be skilful in their master's business, than in his nature; for then he is like to advise him and not feed his humour. It is of singular use to princes if they take the opinions of their counsel both separately and together. For private opinion is more free; but opinion before others is more reverent. In private, men are more bold in their own humours; and in consort, men are more obnoxious to others' humours; therefore it is good to take both; and of the inferior sort rather in private, to preserve freedom; of the greater rather in consort, to preserve respect. It is in vain for princes to take counsel concerning matters, if they take no counsel likewise concerning persons; for all matters are as dead images; and the life of the execution of affairs resteth in the good choice of persons. Neither is it enough to consult concerning persons *secundum genera*, as in an idea, or mathematical description, what the kind and character of the person should be; for the greatest errors are committed, and the most judgment is shown, in the choice of individuals. It was truly said, *optimi consiliarii mortui* [the best counsellors are the dead]: books will speak plain when counsellors blanch. Therefore it is good to be conversant in them, specially the books of such as themselves have been actors upon the stage.

The counsels at this day in most places are but familiar meetings, where matters are rather talked on than debated. And they run too swift to the order or act of counsel. It were better that in causes of weight, the matter were propounded one day and not spoken to till the next day; *in nocte consilium* [night is the season for counsel]. So was it done in the Commission of Union between England and Scotland; which was a grave and orderly assembly. I commend set days for petitions; for both it gives the suitors more certainty for their attendance, and it frees the meetings for matters of estate, that they may *hoc agere*. In choice of committees for ripening business for the counsel, it is better to choose indifferent persons, than to make an indifferency by putting in those that are strong on both sides. I commend also standing commissions; as for trade, for treasure, for war, for suits, for some provinces; for where there be divers particular counsels and but one counsel of estate (as it is in Spain), they are, in effect, no more than standing commissions: save that they have greater authority. Let such as are to inform counsels out of their particular professions, (as lawyers, seamen, mintmen, and the like), be first heard before committees; and then, as occasion serves, before the counsel. And let them not come in

[33] That is, the fable of Jupiter and Metis.

multitudes, or in a tribunitious manner; for that is to clamour counsels, not to inform them. A long table and a square table, or seats about the walls, seem things of form, but are things of substance; for at a long table a few at the upper end, in effect, sway all the business; but in the other form there is more use of the counsellors' opinions that sit lower. A king, when he presides in counsel, let him beware how he opens his own inclination too much in that which he propoundeth; for else counsellors will but take the wind of him, and instead of giving free counsel, sing him a song of *placebo*.

XXI. Of Delays.

Fortune is like the market; where many times, if you can stay a little, the price will fall. And again, it is sometimes like Sibylla's offer; which at first offereth the commodity at full, then consumeth part and part, and still holdeth up the price. For occasion (as it is in the common verse) *turneth a bald noddle, after she hath presented her locks in front, and no hold taken;* or at least turneth the handle of the bottle first to be received, and after the belly, which is hard to clasp. There is surely no greater wisdom than well to time the beginnings and onsets of things. Dangers are no more light, if they once seem light; and more dangers have deceived men than forced them. Nay, it were better to meet some dangers half way, though they come nothing near, than to keep too long a watch upon their approaches; for if a man watch too long, it is odds he will fall asleep. On the other side, to be deceived with too long shadows (as some have been when the moon was low and shone on their enemies' back), and so to shoot off before the time; or to teach dangers to come on, by over early buckling towards them; is another extreme. The ripeness or unripeness of the occasion (as we said) must ever be well weighed; and generally it is good to commit the beginnings of all great actions to Argos with his hundred eyes, and the ends to Briareus with his hundred hands; first to watch, and then to speed. For the helmet of Pluto, which maketh the politic man go invisible, is secrecy in the counsel and celerity in the execution. For when things are once come to the execution, there is no secrecy comparable to celerity; like the motion of a bullet in the air, which flieth so swift as it outruns the eye.

XXII. Of Cunning.

We take Cunning for a sinister or crooked wisdom. And certainly there is a great difference between a cunning man and a wise man; not only in point of honesty, but in point of ability. There be that can pack the cards, and yet cannot play well; so there are some that are good in canvasses and factions that are otherwise weak men. Again, it is one thing to understand persons and another thing to understand matters; for many are perfect in men's humours that are not greatly capable of the real part of business, which is the constitution of one that hath studied men more than books. Such men are fitter for practice than for counsel; and they are good but in their own alley: turn them to new men, and they have lost their aim; so as the old rule to know a fool from a wise man, *Mitte ambos nudos ad ignotos, et videbis* [Send them both naked to those they know not], doth scarce hold for them. And because these cunning men are like haberdashers of small wares, it is not amiss to set forth their shop.

It is a point of cunning, to wait upon him with whom you speak, with your eye; as the Jesuits give it in precept: for there be many wise men that have secret hearts and transparent countenances. Yet this would be done with a demure abasing of your eye sometimes, as the Jesuits also do use.

Another is, that when you have any thing to obtain of present despatch, you entertain and amuse the party with whom you deal with some other discourse, that he be not too much awake to make objections. I knew a counsellor and secretary, that never came to Queen Elizabeth of England with bills to sign, but he would always first put her into some discourse of estate, that she mought the less mind the bills.

The like surprise may be made by moving things when the party is in haste, and cannot stay to consider advisedly of that is moved.

If a man would cross a business that he doubts some other would handsomely and effectually move, let him pretend to wish it well, and move it himself in such sort as may foil it.

The breaking off in the midst of that one was about to say, as if he took himself up, breeds a greater appetite in him with whom you confer to know more.

And because it works better when any thing seemeth to be gotten from you by question, than if you offer it of yourself, you may lay a bait for a question by showing another visage and countenance than you are wont; to the end to give ocasion for the party to ask what the matter is of the change? As Nehemias did; *And I had not before that time been sad before the king.*

In things that are tender and unpleasing, it is good to break the ice by some whose words are of less weight, and to reserve the more weighty voice to come in as by chance, so that he may be asked the question upon the other's speech; as Narcissus did, in relating to Claudius the marriage of Messalina and Silius.

In things that a man would not be seen in himself, it is a point of cunning to borrow the name of the world; as to say, *The world says*, or *There is a speech abroad.*

I knew one that, when he wrote a letter, he would put that which was most material in the postscript, as if it had been a bye-matter.

I knew another that, when he came to have speech, he would pass over that that he intended most; and go forth, and come back again, and speak of it as of a thing that he had almost forgot.

Some procure themselves to be surprised at such times as it is like the party that they work upon will suddenly come upon them; and to be found with a letter in their hand, or doing somewhat which they are not accustomed; to the end they may be apposed of those things which of themselves they are desirous to utter.

It is a point of cunning, to let fall those words in a man's own name, which he would have another man learn and use, and thereupon take advantage. I knew two that were competitors for the secretary's place in queen Elizabeth's time, and yet kept good quarter between themselves; and would confer one with another upon the business; and the one of them said, That to be a secretary *in the declination of a monarchy* was a ticklish thing, and that he did not affect it: the other straight caught up those words, and discoursed with divers of his friends, that he had no reason to desire to be secretary in the declination of a monarchy. The first man took hold of it, and found means it was told the Queen; who hearing of *a declination of a monarchy*, took it so ill, as she would never after hear of the other's suit.

There is a cunning, which we in England call *The turning of the cat in the pan* which is, when that which a man says to another, he lays it as if another had said it to him. And to say truth, it is not easy, when such a matter passed between two, to make it appear from which of them it first moved and began.

It is a way that some men have, to glance and dart at others by justifying themselves by negatives; as to say *This I do not;* as Tigellinus did, towards Burrhus; *Se non diversas spes, sed incolumitatem imperatoris simpliciter spectare* [That he had not several hopes to rest on, but looked simply to the safety of the Emperor].

Some have in readiness so many tales and stories, as there is nothing they would insinuate, but they can wrap it into a tale; which serveth both to keep themselves more in guard, and to make others carry it with more pleasure.

It is a good point of cunning, for a man to shape the answer he would have in his own words and propositions; for it makes the other party stick the less.

It is strange how long some men will lie in wait to speak somewhat they desire to say; and how far about they will fetch; and how many other matters they will beat over, to come near it. It is a thing of great patience, but yet of much use.

A sudden, bold, and unexpected question doth many times surprise a man, and lay him open. Like to him that, having changed his name, and walking in Paul's, another suddenly came behind him and called him by his true name, whereat straightways he looked back.

But these small wares and petty points of cunning are infinite; and it were a good deed to make a list of them, for that nothing doth more hurt in a state than that cunning men pass for wise.

But certainly some there are that know the resorts and falls of business, that cannot sink into the main of it; like a house that hath convenient stairs and entries, but never a fair room. Therefore you shall see them find out pretty looses in the conclusion, but are no ways able to examine or debate matters. And yet commonly they take advantage of their inability, and would be thought wits of direction. Some build rather upon the abusing of others, and (as we now say) *putting tricks upon them*, than upon soundness of their own proceedings. But Salomon saith, *Prudens advertit ad gressus suos: stultus divertit ad dolos* [The wise man taketh heed to his steps: the fool turneth aside to deceits].

XXIII. Of Wisdom for a Man's Self.

An ant is a wise creature for itself, but it is a shrewd thing in an orchard or garden. And certainly men that are great lovers of themselves waste the public. Divide with reason between self-love and society; and be so true to thyself, as thou be not false to others; specially to thy king and country. It is a poor centre of a man's actions, *himself*. It is right earth. For that only stands fast upon his own centre; whereas all things that have affinity with the heavens move upon the centre of another, which they benefit. The referring of all to a man's self is more tolerable in a sovereign prince; because themselves are not only themselves, but their good and evil is at the peril of the public fortune. But it is a desperate evil in a servant to a prince, or a citizen in a republic. For whatsoever affairs pass such a man's hands, he crooketh them to his own ends; which must needs be often eccentric to the ends of his master or state. Therefore let princes, or states, choose such servants as have not this mark; except they mean their service should be made but the accessary. That which maketh the effect more pernicious is that all proportion is lost. It were disproportion enough for the servant's good to be preferred before the master's; but yet it is a greater extreme, when a little good of the servant shall carry things against a great good of the master's. And yet that is the case of bad officers, treasurers, ambassadors, generals, and other false and corrupt servants; which set a bias upon their bowl, of their own petty ends and envies, to the overthrow of their master's great and important affairs. And for the most part, the good such servants receive is after the model of their own fortune; but the hurt they sell for that good is after the model of their master's fortune. And certainly it is the nature of extreme self-lovers, as they will set an house on fire, and it were but to roast their eggs; and yet these men many times hold credit with their masters, because their study is but to please them and profit themselves; and for either respect they will abandon the good of their affairs.

Wisdom for a man's self is, in many branches thereof, a depraved thing. It is the wisdom of rats, that will be sure to leave a house somewhat before it falls. It is the wisdom of the fox, that thrusts out the badger, who digged and made room for him. It is the wisdom of crocodiles, that shed tears when they would devour. But that which is specially to be noted is, that those which (as Cicero says of Pompey) are *sui amantes, sine rivali* [lovers of themselves without rival], are many times unfortunate. And whereas they have all their times sacrificed to themselves, they become in the end themselves sacrifices to the inconstancy of fortune; whose wings they thought by their self-wisdom to have pinioned.

XXIV. Of Innovations.

As the births of living creatures at first are ill-shapen, so are all Innovations which are the births of time. Yet notwithstanding, as those that first bring honour into their family are commonly more worthy than most that succeed, so the first precedent (if it be good) is seldom attained by imitation. For Ill, to man's nature as it stands perverted, hath a natural motion, strongest in continuance; but Good, as a forced motion, strongest at first. Surely every medicine is an innovation; and he that will not apply new remedies must expect new evils; for time is the greatest innovator; and if time of course alter things to the worse, and wisdom and counsel shall not alter them to the better, what shall be the end? It is true, that what is settled by custom, though it be not good, yet at least it is fit; and those things which have long gone together, are as it were confederate within themselves; whereas new things piece not so well; but though they help by their utility, yet they trouble by their inconformity. Besides, they are like strangers; more admired and less favoured. All this is true, if time stood still; which contrariwise moveth so round, that a froward retention of custom is as turbulent a thing as an innovation; and they that reverence too much old times, are but a scorn to the new. It were good therefore that men in their innovations would follow the example of time itself; which indeed innovateth greatly, but quietly, and by degrees scarce to be perceived. For otherwise, whatsoever is new is unlooked for; and ever it mends some, and pairs other; and he that is holpen takes it for a fortune, and thanks the time; and he that is hurt, for a wrong, and imputeth it to the author. It is good also not to try experiments in states, except the necessity be urgent, or the utility evident; and well to beware that it be the reformation that draweth on the change, and not the desire of change that pretendeth the reformation. And lastly, that the novelty, though it be not rejected, yet be held for a suspect; and, as the Scripture saith, *that we make a stand upon the ancient way, and then look about us, and discover what is the straight and right way, and so to walk in it.*

XXV. Of Dispatch.

Affected dispatch is one of the most dangerous things to business that can be. It is like that which the physicians call *predigestion*, or hasty digestion; which is sure to fill the body full of crudities and secret seeds of diseases. Therefore measure not dispatch by the times of sitting, but by the advancement of the business. And as in races it is not the large stride or high lift that makes the speed; so in business, the keeping close to the matter, and not taking of it too much at once, procureth dispatch. It is the care of some only to come off speedily for the time; or to contrive some false periods of business, because they may seem men of dispatch. But it is one thing to abbreviate by contracting, another by cutting off. And business so handled at several sittings or meetings goeth commonly backward and forward in an unsteady manner. I knew a wise man that had it for a by-word, when he saw men hasten to a conclusion, *Stay a little, that we may make an end the sooner.*

On the other side, true dispatch is a rich thing. For time is the measure of business, as money is of wares; and business is bought at a dear hand where there is small dispatch. The Spartans and Spaniards have been noted to be of small dispatch; *Mi venga la muerte de Spagna; Let my death come from Spain;* for then it will be sure to be long in coming.

Give good hearing to those that give the first information in business; and rather direct them in the beginning, than interrupt them in the continuance of their speeches; for he that is put out of his own order will go forward and backward, and be more tedious while he waits upon his memory, than he could have been if he had gone on in his own course. But sometimes it is seen that the moderator is more troublesome than the actor.

Iterations are commonly loss of time. But there is no such gain of time as to iterate often the state of the question; for it chaseth away many a frivolous speech as it is coming forth. Long and curious speeches are as fit for dispatch as a robe or mantle with a long train is for race. Prefaces and passages, and excusations, and other speeches of reference to the person, are great wastes of time; and though they seem to proceed of modesty, they are bravery. Yet beware of being too material[34] when there is an impediment or obstruction in men's wills; for pre-occupation of mind ever requireth preface of speech; like a fomentation to make the unguent enter.

Above all things, order, and distribution, and singling out of parts, is the life

[34] That is, of keeping too close to the matter. *Cave ne in rem ipsam ab initio descendas.*

of dispatch; so as the distribution be not too subtle: for he that doth not divide will never enter well into business; and he that divideth too much will never come out of it clearly. To choose time is to save time; and an unseasonable motion is but beating the air. There be three parts of business; the preparation, the debate or examination, and the perfection. Whereof, if you look for dispatch, let the middle only be the work of many, and the first and last the work of few. The proceeding upon somewhat conceived in writing doth for the most part facilitate dispatch: for though it should be wholly rejected, yet that negative is more pregnant of direction than an indefinite; as ashes are more generative than dust.

XXVI. Of Seeming Wise.

It hath been an opinion, that the French are wiser than they seem, and the Spaniards seem wiser than they are. But howsoever it be between nations, certainly it is so between man and man. For as the Apostle saith of godliness, *Having a shew of godliness, but denying the power thereof;* so certainly there are in point of wisdom and sufficiency, that do nothing or little very solemnly: *magno conatu nugas.* It is a ridiculous thing and fit for a satire to persons of judgment, to see what shifts these formalists have, and what prospectives to make *superficies* to seem body that hath depth and bulk. Some are so close and reserved, as they will not shew their wares but by a dark light; and seem always to keep back somewhat; and when they know within themselves they speak of that they do not well know, would nevertheless seem to others to know of that which they may not well speak. Some help themselves with countenance and gesture, and are wise by signs; as Cicero saith of Piso, that when he answered him, he fetched one of his brows up to his forehead, and bent the other down to his chin; *Respondes, altero ad frontem sublato, altero ad mentum depresso supercilio, crudelitatem tibi non placere.* Some think to bear it by speaking a great word, and being peremptory; and go on, and take by admittance that which they cannot make good. Some, whatsoever is beyond their reach, will seem to despise or make light of it as impertinent or curious; and so would have their ignorance seem judgment. Some are never without a difference, and commonly by amusing men with a subtilty, blanch the matter; of whom A. Gellius saith, *Hominem delirum, qui verborum minutiis rerum frangit pondera* [a trifler, that with verbal points and niceties breaks up the mass of matter]. Of which kind also, Plato in his Protagoras bringeth in Prodicus in scorn, and maketh him make a speech that consisteth of distinctions from the beginning to the end. Generally, such men in all deliberations find ease to be of the negative side, and affect a credit to object and foretell difficulties; for when propositions are denied there is an end of them; but if they be allowed, it requireth a new work; which false point of wisdom is the bane of business. To conclude, there is no decaying merchant, or inward beggar, hath so many tricks to uphold the credit of their wealth, as these empty persons have to maintain the credit of their sufficiency. Seeming wise men may make shift to get opinion; but let no man choose them for employment for certainly you were better take for business a man somewhat absurd than over-formal.

XXVII. Of Friendship.

It had been hard for him that spake it to have put more truth and untruth together in few words, than in that speech, *Whosoever is delighted in solitude is either a wild beast or a god.* For it is most true that a natural and secret hatred and aversation towards society in any man, hath somewhat of the savage beast; but it is most untrue that it should have any character at all of the divine nature except it proceed, not out of a pleasure in solitude, but out of a love and desire to sequester a man's self for a higher conversation: such as is found to have been falsely and feignedly in some of the heathen; as Epimenides the Candian, Numa the Roman, Empedocles the Sicilian, and Apollonius of Tyana; and truly and really in divers of the ancient hermits and holy fathers of the church. But little do men perceive what solitude is, and how far it extendeth. For a crowd is not company; and faces are but a gallery of pictures; and talk but a tinkling cymbal, where there is no love. The Latin adage meeteth with it a little: *Magna civitas, magna solitudo* [a great town is a great solitude]; because in a great town friends are scattered; so that there is not that fellowship, for the most part, which is in less neighbourhoods. But we may go further, and affirm most truly that it is a mere and miserable solitude to want true friends; without which the world is but a wilderness; and even in this sense also of solitude, whosoever in the frame of his nature and affections is unfit for friendship, he taketh it of the beast, and not from humanity.

A principal fruit of friendship is the ease and discharge of the fulness and swellings of the heart, which passions of all kinds do cause and induce. We know diseases of stoppings and suffocations are the most dangerous in the body; and it is not much otherwise in the mind; you may take sarza to open the liver, steel to open the spleen, flower [35] of sulphur for the lungs, castoreum for the brain; but no receipt openeth the heart, but a true friend; to whom you may impart griefs, joys, fears, hopes, suspicions, counsels, and whatsoever lieth upon the heart to oppress it, in a kind of civil shrift or confession.

It is a strange thing to observe how high a rate great kings and monarchs do set upon this fruit of friendship whereof we speak: so great, as they purchase it many times at the hazard of their own safety and greatness. For princes, in regard of the distance of their fortune from that of their subjects and servants, cannot gather this fruit, except (to make themselves capable thereof) they raise some persons to be as it were companions and almost equals to themselves, which many times sorteth to inconvenience. The modern languages give unto such persons the name of favourites, or privadoes; as if it were matter of grace or conversation. But the Roman name attaineth the true use and cause thereof, naming them *participes curarum;* for it is that which tieth the knot. And we see plainly that this hath been done, not by weak and passionate princes only, but by the wisest and most politic that ever reigned; who have oftentimes joined to themselves some of their servants; whom both themselves have called friends, and allowed others likewise to call them in the same manner; using the word which is received between private men.

L. Sylla, when he commanded Rome, raised Pompey (after surnamed the Great) to that height, that Pompey vaunted himself for Sylla's over-match. For when he had carried the consulship for a friend of his [36], against the pursuit of Sylla, and that Sylla did a little resent thereat, and began to speak great, Pompey turned upon him again, and in effect bade him be quiet; *for that more men adored the sun rising than the sun setting.* With Julius Cæsar, Decimus Brutus had obtained that interest, as he set him down in his testament for heir in remainder after his nephew. And this was the man that had power with him to draw him forth to his death. For when Cæsar would have discharged the senate, in regard of some ill presages, and specially a dream of Calpurnia; this man lifted him gently by the arm out of his chair, telling him he hoped he would not dismiss the senate till his wife had dreamt a better dream. And it seemeth his favour was so great, as Antonius, in a letter which is recited *verbatim* in one of Cicero's Philippics, calleth him *venefica, witch;* as if he had enchanted Cæsar. Augustus raised Agrippa (though of mean birth) to that height, as when he consulted with Mæcenas about the marriage of his daughter Julia, Mæcenas took the liberty to tell him, *that he must either marry his daughter to Agrippa, or take away his life: there was no third way, he had made him so great.* With Tiberius Cæsar, Sejanus had ascended to that height, as they two were termed and reckoned as a pair of friends. Tiberius in a letter to him saith, *hæc pro amicitiâ nostrâ non occultavi* [these things, as our friendship required, I have not concealed from you]; and the whole senate dedicated an altar to Friendship as to a goddess, in respect of the great dearness of friendship between them two. The like or more was between Septimius Severus and Plautianus [37]. For he

[35] So Ed. 1639. The original edition has *flowers.*

[36] Lepidus. See Plutarch in Pompey. But the occasion on which Pompey made the remark in question was Sylla's opposition to his triumph.

[37] *Plantianus* in the original, and also in Ed. 1639, and in the Latin translation, in all the places.

forced his eldest son to marry the daughter of Plautianus; and would often maintain Plautianus in doing affronts to his son; and did write also in a letter to the senate, by these words: *I love the man so well, as I wish he may over-live me.* Now if these princes had been as a Trajan or a Marcus Aurelius, a man might have thought that this had proceeded of an abundant goodness of nature; but being men so wise, of such strength and severity of mind, and so extreme lovers of themselves, as all of these were, it proveth most plainly that they found their own felicity (though as great as ever happened to mortal men) but as an half piece, except they mought have a friend to make it entire; and yet, which is more, they were princes that had wives, sons, nephews; and yet all these could not supply the comfort of friendship.

It is not to be forgotten what Comineus observeth of his first master, Duke Charles the Hardy; namely, that he would communicate his secrets with none; and least of all, those secrets which troubled him most. Whereupon he goeth on and saith that towards his latter time *that closeness did impair and a little perish his understanding.* Surely Comineus mought have made the same judgment also, if it had pleased him, of his second master Lewis the Eleventh, whose closeness was indeed his tormentor. The parable of Pythagoras is dark, but true: *Cor ne edito; Eat not the heart.* Certainly, if a man would give it a hard phrase, those that want friends to open themselves unto are cannibals of their own hearts. But one thing is most admirable (wherewith I will conclude this first fruit of friendship), which is, that this communicating of a man's self to his friend works two contrary effects; for it redoubleth joys, and cutteth griefs in halfs. For there is no man that imparteth his joys to his friend but he joyeth the more: and no man that imparteth his griefs to his friend, but he grieveth the less. So that it is in truth of operation upon a man's mind, of like virtue as the alchymists use to attribute to their stone for man's body; that it worketh all contrary effects, but still to the good and benefit of nature. But yet without praying in aid of alchymists, there is a manifest image of this in the ordinary course of nature. For in bodies, union strengtheneth and cherisheth any natural action; and on the other side weakeneth and dulleth any violent impression; and even so it is of minds.

The second fruit of friendship is healthful and sovereign for the understanding, as the first is for the affections. For friendship maketh indeed a fair day in the affections, from storm and tempests; but it maketh daylight in the understanding, out of darkness and confusion of thoughts. Neither is this to be understood only of faithful counsel, which a man receiveth from his friend; but before you come to that, certain it is that whosoever hath his mind fraught with many thoughts, his wits and understanding do clarify and break up, in the communicating and discoursing with another; he tosseth his thoughts more easily; he marshalleth them more orderly; he seeth how they look when they are turned into words: finally, he waxeth wiser than himself; and that more by an hour's discourse than by a day's meditation. It was well said by Themistocles to the king of Persia, *That speech was like cloth of Arras, opened and put abroad; whereby the imagery doth appear in figure; whereas in thoughts they lie but as in packs.* Neither is this second fruit of friendship, in opening the understanding, restrained only to such friends as are able to give a man counsel; (they indeed are best); but even without that, a man learneth of himself, and bringeth his own thoughts to light, and whetteth his wits as against a stone, which itself cuts not. In a word, a man were better relate himself to a statua or picture, than to suffer his thoughts to pass in smother.

Add now, to make this second fruit of friendship complete, that other point which lieth more open and falleth within vulgar observation; which is faithful counsel from a friend. Heraclitus saith well in one of his enigmas, *Dry light is ever the best.* And certain it is, that the light that a man receiveth by counsel from another, is drier and purer than that which cometh from his own understanding and judgment; which is ever infused and drenched in his affections and customs. So as there is as much difference between the counsel that a friend giveth, and that a man giveth himself, as there is between the counsel of a friend and of a flatterer. For there is no such flatterer as is a man's self; and there is no such remedy against flattery of a man's self as the liberty of a friend. Counsel is of two sorts; the one concerning manners, the other concerning business. For the first, the best preservative to keep the mind in health is the faithful admonition of a friend. The calling of a man's self to a strict account is a medicine, sometime, too piercing and corrosive. Reading good books of morality is a little flat and dead. Observing our faults in others is sometimes improper for our case. But the best receipt (best, I say, to work, and best to take) is the admonition of a friend. It is a strange thing to behold what gross errors and extreme absurdities many (especially of the greater sort) do commit, for want of a friend to tell them of them; to the great damage both of their fame and fortune: for, as St. James saith, they are as men *that look sometimes into a glass, and presently forget their own shape and favour*. As for business, a man may think, if he will, that two eyes see no more than one; or that a gamester seeth always more than a looker-on; or that a man in anger is as wise as he that hath said over the four and twenty letters; or that a musket may be shot off as well upon the arm as upon a rest; and such other fond and high imaginations, to think himself all in all. But when all is done, the help of good counsel is that which setteth business straight. And if any man think that he will take counsel, but it shall be by pieces; asking counsel in one business of one man, and in another business of another man; it is well (that is to say, better perhaps than if he asked none at all); but he runneth two dangers; one, that he shall not be faithfully counselled; for it is a rare thing, except it be from a perfect and entire friend, to have counsel given, but such as shall be bowed and crooked to some ends which he hath that giveth it. The other, that he shall have counsel given, hurtful and unsafe (though with good meaning), and mixed partly of mischief and partly of remedy; even as if you would call a physician that is thought good for the cure of the disease you complain of, but is unacquainted with your body; and therefore may put you in way for a present cure, but overthroweth your health in some other kind; and so cure the disease and kill the patient. But a friend that is wholly acquainted with a man's estate will beware, by furthering any present business, how he dasheth upon other inconvenience. And therefore rest not upon scattered counsels; they will rather distract and mislead, than settle and direct.

After these two noble fruits of friendship (peace in the affections, and support of the judgment), followeth the last fruit; which is like the pomegranate, full of many kernels; I mean aid and bearing a part in all actions and occasions. Here the best way to represent to life the manifold use of friendship, is to cast and see how many things there are which a man cannot do himself; and then it will appear that it was a sparing speech of the ancients, to say *that a friend is another himself*; for that a friend is far more than himself. Men have their time, and die many times in desire of some things which they principally take to heart; the bestowing of a child, the finishing of a work, or the like. If a man have a true friend he may rest almost secure that the care of those things will continue after him. So that a man hath, as it were, two lives in his desires. A man hath a body, and that body is confined to a place; but where friendship is, all offices of life are as it were granted to him and his deputy. For he may exercise them by his friend. How many things are there which a man cannot, with any face or comeliness, say or do himself? A man can scarce allege his own merits with modesty, much less extol them; a man cannot sometimes brook to supplicate or beg; and a number of the like. But all these things are graceful in a friend's mouth, which are blushing in a man's own. So again, a man's person hath many proper relations which he cannot put off. A man cannot speak to his son but as a father; to his wife but as a husband; to his enemy but upon terms; whereas a friend may speak as the case requires, and not as it sorteth with the person. But to enumerate these things were endless; I have given the rule, where a man cannot fitly play his own part; if he have not a friend, he may quit the stage.

XXVIII. Of Expense.

Riches are for spending, and spending for honour and good actions. Therefore extraordinary expense must be limited by the worth of the occasion; for voluntary undoing may be as well for a man's country as for the kingdom of heaven. But ordinary expense ought to be limited by a man's estate; and governed with such regard, as it be within his compass; and not subject to deceit and abuse of servants; and ordered to the best shew, that the bills may be less than the estimation abroad. Certainly, if a man will keep but of even hand, his ordinary expenses ought to be but to the half of his receipts; and if he think to wax rich, but to the third part. It is no baseness for the greatest to descend and look into their own estate. Some forbear it, not upon negligence alone, but doubting to bring themselves into melancholy, in respect they shall find it broken. But wounds cannot be cured without searching. He that cannot look into his own estate at all, had need both choose well those whom he employeth, and change them often; for new are more timorous and less subtle. He that can look into his estate but seldom, it behoveth him to turn all to certainties. A man had need, if he be plentiful in some kind of expense, to be as saving again in some other. As if he be plentiful in diet, to be saving in apparel; if he be plentiful in the hall, to be saving in the stable; and the like. For he that is plentiful in expenses of all kinds will hardly be preserved from decay. In clearing of a man's estate, he may as well hurt himself in being too sudden, as in letting it run on too long. For hasty selling is commonly as disadvantageable as interest. Besides, he that clears at once will relapse; for finding himself out of straits, he will revert to his customs: but he that cleareth by degrees induceth a habit of frugality, and gaineth as well upon his mind as upon his estate. Certainly, who hath a state to repair, may not despise small things; and commonly it is less dishonourable to abridge petty charges than to stoop to petty gettings. A man ought warily to begin charges which once begun will continue: but in matters that return not he may be more magnificent.

XXIX. Of the True Greatness of Kingdoms and Estates.

The speech of Themistocles the Athenian, which was haughty and arrogant in taking so much to himself, had been a grave and wise observation and censure, applied at large to others. Desired at a feast to touch a lute, he said, *He could not fiddle, but yet he could make a small town a great city.* These words (holpen a little with a metaphor) may express two differing abilities in those that deal in business of estate. For if a true survey be taken of counsellors and statesmen, there may be found (though rarely) those which can make a small state great, and yet cannot fiddle: as on the other side, there will be found a great many that can fiddle very cunningly, but yet are so far from being able to make a small state great, as their gift lieth the other way; to bring a great and flourishing estate to ruin and decay. And, certainly those degenerate arts and shifts, whereby many counsellors and governors gain both favour with their masters and estimation with the vulgar, deserve no better name than fiddling; being things rather pleasing for the time, and graceful to themselves only, than tending to the weal and advancement of the state which they serve. There are also (no doubt) counsellors and governors which may be held sufficient (*negotiis pares*), able to manage affairs, and to keep them from precipices and manifest inconveniences; which nevertheless are far from the ability to raise and amplify an estate in power, means, and fortune. But be the workmen what they may be, let us speak of the work; that is, the true Greatness of Kingdoms and Estates, and the means thereof. An argument fit for great and mighty princes to have in their hand; to the end that neither by over-measuring their forces, they leese themselves in vain enterprises; nor on the other side, by undervaluing them, they descend to fearful and pusillanimous counsels.

The greatness of an estate in bulk and territory, doth fall under measure; and the greatness of finances and revenew doth fall under computation. The population may appear by musters; and the number and greatness of cities and towns by cards and maps. But yet there is not any thing amongst civil affairs more subject to error, than the right valuation and true judgment concerning the power and forces of an estate. The kingdom of heaven is compared, not to any great kernel or nut, but to a grain of mustard-seed; which is one of the least grains, but hath in it a property and spirit hastily to get up and spread. So are there states great in territory, and yet not apt to enlarge or command; and some that have but a small dimension of stem, and yet apt to be the foundations of great monarchies.

Walled towns, stored arsenals and armories, goodly races of horse, chariots of war, elephants, ordnance, artillery, and the like; all this is but a sheep in a lion's skin, except the breed and disposition of the people be stout and warlike. Nay, number (itself) in armies importeth not much, where the people is of weak courage; for (as Virgil saith) *It never troubles a wolf how many the sheep be*. The army of the Persians in the plains of Arbela was such a vast sea of people, as it did somewhat astonish the commanders in Alexander's army; who came to him, therefore, and wished him to set upon them by night; but he answered, *He would not pilfer the victory*. And the defeat was easy. When Tigranes, the Armenian, being encamped upon a hill with four hundred thousand men, discovered the army of the Romans, being not above fourteen thousand, marching towards him, he made himself merry with it and said, *Yonder men are too many for an ambassage, and too few for a fight*. But, before the sun set, he found them enow to give him the chase with infinite slaughter. Many are the examples of the great odds between number and courage: so that a man may truly make a judgment, that the principal point of greatness in any state is to have a race of military men. Neither is money the sinews of war (as it is trivially said), where the sinews of men's arms, in base and effeminate people, are failing. For Solon said well to Crœsus (when in ostentation he showed him his gold), *Sir, if any other come that hath better iron than you, he will be master of all this gold*. Therefore let any prince or state think soberly of his forces, except his militia of natives be of good and valiant soldiers. And let princes, on the other side, that have subjects of martial disposition, know their own strength; unless they be otherwise wanting unto themselves. As for mercenary forces (which is the help in this case), all examples show that whatsoever estate or prince doth rest upon them, *he may spread his feathers for a time, but he will mew them soon after*.

The blessing of Judah and Issachar will never meet; *that the same people or nation should be both the lion's whelp and the ass between burthens*; neither will it be, that a people overlaid with taxes should ever become valiant and martial. It is true that taxes levied by consent of the estate do abate men's courage less: as it hath been seen notably in the excises of the Low Countries; and, in some degree, in the subsidies of England. For you must note that we speak now of the heart, and not of the purse. So that although the same tribute and tax, laid by consent or by imposing, be all one to the purse, yet it works diversely upon the courage. So that you may conclude, *that no people over-charged with tribute is fit for empire*.

Let states that aim at greatness, take heed how their nobility and gentlemen do multiply too fast. For that maketh the common subject grow to be a peasant and base swain, driven out of heart, and in effect but the gentleman's labourer. Even as you may see in coppice woods; if you leave your staddles too thick, you shall never have clean underwood, but shrubs and bushes. So in countries, if the gentlemen be too many, the commons will be base; and you will bring it to that, that not the hundred poll will be fit for an helmet; especially as to the infantry, which is the nerve of an army; and so there will be great population and little strength. This which I speak of hath been no where better seen than by comparing of England and France; whereof England, though far less in territory and population, hath been (nevertheless) an over-match; in regard the middle people of England make good soldiers, which the peasants of France do not. And herein the device of king Henry the Seventh (whereof I have spoken largely in the history of his life) was profound and admirable; in making farms and houses of husbandry of a standard; that is, maintained with such a proportion of land unto them, as may breed a subject to live in convenient plenty, and no servile condition; and to keep the plough in the hands of the owners, and not mere hirelings. And thus indeed you shall attain to Virgil's character which he gives to ancient Italy :

Terra potens armis atque ubere glebæ :

[A land powerful in arms and in productiveness of soil]. Neither is that state (which, for any thing I know, is almost peculiar to England, and hardly to be found any where else, except it be perhaps in Poland) to be passed over ; I mean the state of free servants and attendants upon noblemen and gentlemen ; which are no ways inferior unto the yeomanry for arms. And therefore out of all question, the splendour and magnificence and great retinues and hospitality of noblemen and gentlemen, received into custom, doth much conduce unto martial greatness. Whereas, contrariwise, the close and reserved living of noblemen and gentlemen causeth a penury of military forces.

By all means it is to be procured, that the trunk of Nebuchadnezzar's tree of monarchy be great enough to bear the branches and the boughs ; that is, that the natural subjects of the crown or state bear a sufficient proportion to the stranger subjects that they govern. Therefore all states that are liberal of naturalisation towards strangers are fit for empire. For to think that an handful of people can, with the greatest courage and policy in the world, embrace too large extent of dominion, it may hold for a time, but it will fail suddenly. The Spartans were a nice people in point of naturalisation ; whereby, while they kept their compass, they stood firm ; but when they did spread, and their boughs were becomen too great for their stem, they became a windfall upon the sudden. Never any state was in this point so open to receive strangers into their body as were the Romans. Therefore it sorted with them accordingly ; for they grew to the greatest monarchy. Their manner was to grant naturalisation (which they called *jus civitatis*), and to grant it in the highest degree ; that is, not only *jus commercii, jus connubii, jus hæreditatis ;* but also *jus suffragii*, and *jus honorum.* And this not to singular persons alone, but likewise to whole families ; yea to cities, and sometimes to nations. Add to this their custom of plantation of colonies ; whereby the Roman plant was removed into the soil of other nations. And putting both constitutions together, you will say that it was not the Romans that spread upon the world, but it was the world that spread upon the Romans ; and that was the sure way of greatness. I have marvelled sometimes at Spain, how they clasp and contain so large dominions with so few natural Spaniards ; but sure the whole compass of Spain is a very great body of a tree ; far above Rome and Sparta at the first. And besides, though they have not had that usage to naturalise liberally, yet they have that which is next to it ; that is, to employ almost indifferently all nations in their militia of ordinary soldiers ; yea and sometimes in their highest commands. Nay it seemeth at this instant they are sensible of this want of natives ; as by the Pragmatical Sanction, now published [38], appeareth.

It is certain, that sedentary and within-door arts, and delicate manufactures (that require rather the finger than the arm), have in their nature a contrariety to a military disposition. And generally, all warlike people are a little idle, and love danger better than travail. Neither must they be too much broken of it, if they shall be preserved in vigour. Therefore it was great advantage in the ancient states of Sparta, Athens, Rome, and others, that they had the use of slaves, which commonly did rid those manufactures. But that is abolished, in greatest part, by the Christian law. That which cometh nearest to it, is to leave those arts chiefly to strangers (which for that purpose are the more easily to be received), and to contain the principal bulk of the vulgar natives within those three kinds,— tillers of the ground ; free servants ; and handicraftsmen of strong and manly arts, as smiths, masons, carpenters, etc. : not reckoning professed soldiers.

But above all, for empire and greatness, it importeth most, that a nation do profess arms as their principal honour, study, and occupation. For the things

[38] *hoc anno promulgata.* A royal decree, or *pragmatica*, was published in the summer of 1622, which gave certain privileges to persons who married, and further immunities to those who had six children. See Mr. Ellis's note, pp. 609–610.

which we formerly have spoken of are but habilitations towards arms; and what is habilitation without intention and act? Romulus, after his death (as they report or feign), sent a present to the Romans, that above all they should intend arms; and then they should prove the greatest empire of the world. The fabric of the state of Sparta was wholly (though not wisely) framed and composed to that scope and end. The Persians and Macedonians had it for a flash. The Gauls, Germans, Goths, Saxons, Normans, and others, had it for a time. The Turks have it at this day, though in great declination. Of Christian Europe, they that have it are, in effect, only the Spaniards. But it is so plain *that every man profiteth in that he most intendeth*, that it needeth not to be stood upon. It is enough to point at it; that no nation which doth not directly profess arms, may look to have greatness fall into their mouths. And on the other side, it is a most certain oracle of time, that those states that continue long in that profession (as the Romans and Turks principally have done) do wonders. And those that have professed arms but for an age, have notwithstanding commonly attained that greatness in that age which maintained them long after, when their profession and exercise of arms hath grown to decay.

Incident to this point is, for a state to have those laws or customs which may reach forth unto them just occasions (as may be pretended) of war. For there is that justice imprinted in the nature of men, that they enter not upon wars (whereof so many calamities do ensue) but upon some, at the least specious, grounds and quarrels. The Turk hath at hand, for cause of war, the propagation of his law or sect; a quarrel that he may always command. The Romans, though they esteemed the extending the limits of their empire to be great honour to their generals when it was done, yet they never rested upon that alone to begin a war. First therefore, let nations that pretend to greatness have this; that they be sensible of wrongs, either upon borderers, merchants, or politic ministers; and that they sit not too long upon a provocation. Secondly, let them be prest and ready to give aids and succours to their confederates; as it ever was with the Romans; insomuch as, if the confederates had leagues defensive with divers other states, and, upon invasion offered, did implore their aids severally, yet the Romans would ever be the foremost, and leave it to none other to have the honour. As for the wars which were anciently made on the behalf of a kind of party, or tacit conformity of estate, I do not see how they may be well justified: as when the Romans made a war for the liberty of Græcia; or when the Lacedæmonians and Athenians made wars to set up or pull down democracies and oligarchies; or when wars were made by foreigners, under the pretence of justice or protection, to deliver the subjects of others from tyranny and oppression; and the like. Let it suffice, that no estate expect to be great, that is not awake upon any just occasion of arming.

No body can be healthful without exercise, neither natural body nor politic; and certainly to a kingdom or estate, a just and honourable war is the true exercise. A civil war indeed is like the heat of a fever; but a foreign war is like the heat of exercise, and serveth to keep the body in health; for in a slothful peace, both courages will effeminate and manners corrupt. But howsoever it be for happiness, without all question, for greatness it maketh, to be still for the most part in arms; and the strength of a veteran army (though it be a chargeable business) always on foot, is that which commonly giveth the law, or at least the reputation, amongst all neighbour states; as may well be seen in Spain, which hath had, in one part or other, a veteran army almost continually, now by the space of six score years.

To be master of the sea is an abridgment of a monarchy. Cicero, writing to Atticus of Pompey his preparation against Cæsar, saith, *Consilium Pompeii plane Themistocleum est; putat enim, qui mari potitur, eum rerum potiri* [Pompey is going upon the policy of Themistocles; thinking that he who commands the sea commands all]. And, without doubt, Pompey had tired out Cæsar, if upon vain confidence he had not left that way. We see the great effects of battles by sea. The battle of Actium decided the empire of the world. The battle of Lepanto arrested the greatness of the Turk. There be many examples where sea-fights have been final to the war; but this is when princes or states have set up their rest upon the battles. But thus much is certain, that he that commands the sea is at great liberty, and may take as much and as little of the war as he will. Whereas those that be strongest by land are many times nevertheless in great straits. Surely, at this day, with us of Europe, the vantage of strength at sea (which is one of the principal dowries of this kingdom of Great Britain) is great; both because most of the kingdoms of Europe are not merely inland, but girt with the sea most part of their compass; and because the wealth of both Indies seems in great part but an accessary to the command of the seas.

The wars of latter ages seem to be made in the dark, in respect of the glory and honour which reflected upon men from the wars in ancient time. There be now for martial encouragement, some degrees and orders of chivalry; which nevertheless are conferred promiscuously upon soldiers and no soldiers; and some remembrance perhaps upon the scutcheon; and some hospitals for maimed soldiers; and such like things. But in ancient times, the trophies erected upon the place of the victory; the funeral laudatives and monuments for those that died in the wars; the crowns and garlands personal; the style of Emperor, which the great kings of the world after borrowed; the triumphs of the generals upon their return; the great donatives and largesses upon the disbanding of the armies; were things able to inflame all men's courages. But above all, that of the Triumph, amongst the Romans, was not pageants or gaudery, but one of the wisest and noblest institutions that ever was. For it contained three things; honour to the general; riches to the treasury out of the spoils; and donatives to the army. But that honour perhaps were not fit for monarchies; except it be in the person of the monarch himself, or his sons, as it came to pass in the times of the Roman emperors, who did impropriate the actual triumphs to themselves and their sons, for such wars as they did achieve in person; and left only, for wars achieved by subjects, some triumphal garments and ensigns to the general.

To conclude: no man can *by care taking* (as the Scripture saith) *add a cubit to his stature*, in this little model of a man's body; but in the great frame of kingdoms and commonwealths, it is in the power of princes or estates to add amplitude and greatness to their kingdoms; for by introducing such ordinances, constitutions, and customs, as we have now touched, they may sow greatness to their posterity and succession. But these things are commonly not observed, but left to take their chance.

XXX. Of Regiment of Health.

There is a wisdom in this beyond the rules of physic: a man's own observation, what he finds good of, and what he finds hurt of, is the best physic to preserve health. But it is a safer conclusion to say, *This agreeth not well with me, therefore I will not continue it;* than this, *I find no offence of this, therefore I may use it.* For strength of nature in youth passeth over many excesses, which are owing a man till his age. Discern of the coming on of years, and think not to do the same things still; for age will not be defied. Beware of sudden change in any great point of diet, and if necessity enforce it, fit the rest to it. For it is a secret both in nature and state, that it is safer to change many things than one. Examine thy customs of diet, sleep, exercise, apparel, and the like; and try, in any thing thou shalt judge hurtful, to discontinue it by little and little; but so, as if thou dost find any inconvenience by the change, thou come back to it again: for it is hard to distinguish that which is generally held good and wholesome, from that which is good particularly, and fit for thine own body. To be free-minded and cheerfully disposed at hours of meat and of sleep and of exercise, is one of the best precepts of long lasting. As for the passions and studies of the mind; avoid envy; anxious fears; anger fretting inwards; subtle and knotty inquisitions; joys and exhilarations in excess; sadness not communicated. Entertain hopes; mirth rather than joy; variety of delights, rather than surfeit of them; wonder and admiration, and therefore novelties; studies that fill the mind with splendid and illustrious objects, as histories, fables, and contemplations of nature. If you fly physic in health altogether, it will be too strange for your body when you shall need it. If you make it too familiar, it will work no extraordinary effect when sickness cometh. I commend rather some diet for certain seasons, than frequent use of physic, except it be grown into a custom. For those diets alter the body more, and trouble it less. Despise no new accident in your body, but ask opinion of it. In sickness, respect health principally ; and in health, action. For those that put their bodies to endure in health, may in most sicknesses which are not very sharp be cured only with diet and tendering. Celsus could never have spoken it as a physician, had he not been a wise man withal, when he giveth it for one of the greatest precepts of health and lasting, that a man do vary and interchange contraries, but with an inclination to the more benign extreme : use fasting and full eating, but rather full eating ; watching and sleep, but rather sleep ; sitting and exercise, but rather exercise ; and the like. So shall nature be cherished, and yet taught masteries. Physicians are some of them so pleasing and conformable to the humour of the patient, as they press not the true cure of the disease ; and some other are so regular in proceeding according to art for the disease, as they respect not sufficiently the condition of the patient. Take one of a middle temper ; or if it may not be found in one man, combine two of either sort ; and forget not to call as well the best acquainted with your body, as the best reputed of for his faculty.

XXXI. Of Suspicion.

Suspicions amongst thoughts are like bats amongst birds, they ever fly by twilight. Certainly they are to be repressed, or at the least well guarded : for they cloud the mind ; they leese friends ; and they check with business, whereby business cannot go on currently and constantly. They dispose kings to tyranny, husbands to jealousy, wise men to irresolution and melancholy. They are defects, not in the heart, but in the brain ; for they take place in the stoutest natures : as in the example of Henry the Seventh of England. There was not a more suspicious man, nor a more stout. And in such a composition they do small hurt. For commonly they are not admitted, but with examination, whether they be likely or no. But in fearful natures they gain ground too fast. There is nothing makes a man suspect much, more than to know little ; and therefore men should remedy suspicion by procuring to know more, and not to keep their suspicions in smother. What would men have ? Do they think those they employ and deal with are saints ? Do they not think they will have their own ends, and be truer to themselves than to them ? Therefore there is no better way to moderate suspicions, than to account upon such suspicions as true and yet to bridle them as false. For so far a man ought to make use of suspicions, as to provide, as if that should be true that he suspects, yet it may do him no hurt. Suspicions that the mind of itself gathers are but buzzes ; but suspicions that are artificially nourished, and put into men's heads by the tales and whisperings of others, have stings. Certainly, the best mean to clear the way in this same wood of suspicions, is frankly to communicate them with the party that he suspects ; for thereby he shall be sure to know more of the truth of them than he did before ; and withal shall make that party more circumspect not to give further cause of suspicion. But this would not be done to men of base natures ; for they, if they find themselves once suspected, will never be true. The Italian says, *Sospetto licentia fede ;* as if suspicion did give a passport to faith ; but it ought rather to kindle it to discharge itself.

XXXII. Of Discourse.

Some in their discourse desire rather commendation of wit, in being able to hold all arguments, than of judgment, in discerning what is true ; as if it were a praise to know what might be said, and not what should be thought. Some have certain common places and themes wherein they are good, and want variety ; which kind of poverty is for the most part tedious, and when it is once perceived, ridiculous. The honourablest part of talk is to give the occasion ; and again to moderate and pass to somewhat else ; for then a man leads the dance. It is good, in discourse and speech of conversation, to vary and intermingle speech of the present occasion with arguments, tales with reasons, asking of questions with telling of opinions, and jest with earnest : for it is a dull thing to tire, and, as we say now, to jade, any thing too far. As for jest, there be certain things which ought to be privileged from it ; namely, religion, matters of state, great persons, any man's present business of importance, and any case that deserveth pity. Yet there be some that think their wits have been asleep, except they dart out somewhat that is piquant, and to the quick. That is a vein which would be bridled ;

Parce, puer, stimulis, et fortius utere loris.

And generally, men ought to find the difference between saltness and bitterness Certainly, he that hath a satirical vein, as he maketh others afraid of his wit, so he had need be afraid of other's memory. He that questioneth much, shall learn much, and content much ; but especially if he apply his questions to the skill of the persons whom he asketh ; for he shall give them occasion to please themselves in speaking, and himself shall continually gather knowledge. But let his questions not be troublesome ; for that is fit for a poser [39]. And let him be sure to leave other men their turns to speak. Nay, if there be any that would reign and take up all the time, let him find means to take them off, and to bring others on ; as musicians use to do with those that dance too long galliards. If you dissemble sometimes your knowledge of that you are thought to know, you shall be thought another time to know that you know not. Speech of a man's self ought to be seldom, and well chosen. I knew one was wont to say in scorn, *He must needs be a wise man, he speaks so much of himself :* and there is but one case wherein a man may commend himself with good grace ; and that is in commending virtue in another ; especially if it be such a virtue whereunto himself pretendeth. Speech of touch towards others should be sparingly used ; for discourse ought to be as a field, without coming home to any man. I knew two noblemen, of the West part of England, whereof the one was given to scoff, but kept ever royal cheer in his house ; the other would ask of those that had been at the other's table, *Tell truly, was there never a flout or dry blow given ?* to which the guest would answer, *Such and such a thing passed.* The lord would say, *I thought he would mar a good dinner.* Discretion of speech is more than eloquence ; and to speak agreeably to him with whom we deal, is more than to speak in good words or in good order. A good continued speech, without a good speech of interlocution, shews slowness ; and a good reply or second speech, without a good settled speech, sheweth shallowness and weakness. As we see in beasts, that those that are weakest in the course, are yet nimblest in the turn ; as it is betwixt the greyhound and the hare. To use too many circumstances ere one come to the matter, is wearisome ; to use none at all, is blunt.

XXXIII. Of Plantations [40].

Plantations are amongst ancient, primitive, and heroical works. When the world was young it begat more children ; but now it is old it begets fewer : for I may justly account new plantations to be the children of former kingdoms. I like a plantation in a pure soil ; that is, where people are not displanted to the end to plant in others. For else it is rather an extirpation than a plantation. Planting of countries is like planting of woods ; for you must make account to leese almost twenty years profit, and expect your recompense in the end. For the principal thing that hath been the destruction of most plantation, hath been the base and hasty drawing of profit in the first years. It is true, speedy profit is not to be neglected, as far as may stand with the good of the plantation, but no further. It is a shameful and unblessed thing to take the scum of people, and wicked condemned men, to be the people with whom you plant ; and not only so, but it spoileth the plantation ; for they will ever live like rogues, and not fall to work, but be lazy, and do mischief, and spend victuals, and be quickly weary, and then certify over to their country to the discredit of the plantation. The people wherewith you plant ought to be gardeners, ploughmen, labourers, smiths, carpenters, joiners, fishermen, fowlers, with some few apothecaries, surgeons, cooks, and

[39] That is, an examimer. *Id enim examinatori convenit.*

[40] *De Plantationibus populorum et coloniis.* This Essay seems to have been carefully translated ; and revised in the translation, probably by Bacon himself.

bakers[41]. In a country of plantation, first look about what kind of victual the country yields of itself to hand ; as chesnuts, wallnuts, pine-apples, olives, dates, plums, cherries, wild honey, and the like ; and make use of them. Then consider what victual or esculent things there are, which grow speedily, and within the year ; as parsnips, carrots, turnips, onions, radish[42], artichokes of Hierusalem, maize, and the like. For wheat[43], barley, and oats, they ask too much labour ; but with pease and beans you may begin, both because they ask less labour, and because they serve for meat as well as for bread. And of rice likewise cometh a great increase, and it is a kind of meat. Above all, there ought to be brought store of biscuit, oat-meal, flour, meal, and the like, in the beginning, till bread may be had. For beasts, or birds, take chiefly such as are least subject to diseases, and multiply fastest ; as swine, goats, cocks, hens, turkeys, geese, house-doves[44], and the like. The victual in plantations ought to be expended almost as in a besieged town ; that is, with certain allowance. And let the main part of the ground employed to gardens or corn, be to a common stock ; and to be laid in, and stored up, and then delivered out in proportion ; besides some spots of ground that any particular person will manure for his own private. Consider likewise what commodities the soil where the plantation is doth naturally yield, that they may some way help to defray the charge of the plantation (so it be not, as was said, to the untimely prejudice of the main business), as it hath fared with tobacco in Virginia[45]. Wood commonly aboundeth but too much[46] ; and therefore timber is fit to be one. If there be iron ore[47], and streams whereupon to set the mills, iron is a brave commodity where wood aboundeth. Making of bay-salt, if the climate be proper for it, would be put in experience. Growing silk likewise, if any be, is a likely commodity. Pitch and tar, where store of fires and pines are, will not fail. So drugs and sweet woods, where they are, cannot but yield great profit. Soap-ashes likewise, and other things that may be thought of. But moil not too much under ground ; for the hope of mines is very uncertain, and useth to make the planters lazy in other things. For government, let it be in the hands of one, assisted with some counsel ; and let them have commission to exercise martial laws, with some limitation. And above all, let men make that profit of being in the wilderness, as they have God always, and his service, before their eyes. Let not the government of the plantation depend upon too many counsellors and undertakers in the country that planteth, but upon a temperate number ; and let those be rather noblemen and gentlemen, than merchants ; for they look ever to the present gain. Let there be freedoms from customs, till the plantation be of strength ; and not only freedom from custom, but freedom to carry their commodities where they may make their best of them, except there be some special cause of caution. Cram not in people, by sending too fast company after company ; but rather hearken how they waste, and send supplies proportionably ; but so as the number may live well in the plantation, and not by surcharge be in penury. It hath been a great endangering to the health of some plantations, that they have built along the sea and rivers, in marish and unwholesome grounds. Therefore, though you begin there, to avoid carriage and other like discommodities, yet build still rather upwards from the streams than along. It concerneth likewise the health of the plantation that they have good store of salt with them, that they may use it in their victuals when it shall be necessary. If you plant where savages are, do not only entertain them with trifles and gingles ; but use them justly and graciously, with sufficient guard nevertheless ; and do

[41] The translation adds, *cervisiarii, et hujusmodi.*

[42] The translation adds, *melones, pepones, cucumeres.*

[43] The translation adds *siliquam.*

[44] The translation adds, rabbits : *cuniculi.*

[45] *ut exportatio eorum in loca ubi maxime in pretio sunt sumptus levet ; ut usuvenit in Nicotiano apud Virginiam ; modo non sit*, etc. I have inserted the marks of parenthesis, which are not in the original ; the construction being ambiguous without them.

[46] The words "but too much", are omitted in the translation.

[47] Spelt *ure* in the original ; as the same word is in one place in the manuscript of the History of Henry VII. The translation has *vena ferri.*

not win their favour by helping them to invade their enemies, but for their defence it is not amiss; and send oft of them over to the country that plants, that they may see a better condition than their own, and commend it when they return. When the plantation grows to strength, then it is time to plant with women as well as with men; that the plantation may spread into generations and not be ever pieced from without. It is the sinfullest thing in the world, to forsake or destitute a plantation once in forwardness; for besides the dishonour, it is the guiltiness of blood of many commiserable persons.

XXXIV. Of Riches.

I cannot call Riches better than the baggage of virtue. The Roman word is better, *impedimenta*. For as the baggage is to an army, so is riches to virtue. It cannot be spared nor left behind, but it hindereth the march; yea and the care of it sometimes loseth or disturbeth the victory. Of great riches there is no real use, except it be in the distribution; the rest is but conceit. So saith Salomon, *Where much is, there are many to consume it; and what hath the owner but the sight of it with his eyes?* The personal fruition in any man cannot reach to feel great riches: there is a custody of them; or a power of dole and donative of them; or a fame of them; but no solid use to the owner. Do you not see what feigned prices are set upon little stones and rarities? and what works of ostentation are undertaken, because there might seem to be some use of great riches? But then you will say, they may be of use to buy men out of dangers or troubles. As Salomon saith, *Riches are as a strong hold, in the imagination of the rich man.* But this is excellently expressed, that it is in imagination, and not always in fact. For certainly great riches have sold more men than they have bought out. Seek not proud riches, but such as thou mayest get justly, use soberly, distribute cheerfully, and leave contentedly. Yet have no abstract nor friarly contempt of them. But distinguish, as Cicero saith well of Rabirius Posthumus, *In studio rei amplificandæ apparebat, non avaritiæ prædam, sed instrumentum bonitati quæri;* [In seeking to increase his estate it was apparent that he sought not a prey for avarice to feed on but an instrument for goodness to work with]. Hearken also to Salomon, and beware of hasty gathering of riches; *Qui festinat ad divitias, non erit insons* [He that maketh haste to be rich shall not be innocent]. The poets feign, that when Plutus (which is Riches) is sent from Jupiter, he limps and goes slowly; but when he is sent from Pluto, he runs and is swift of foot. Meaning that riches gotten by good means and just labour pace slowly; but when they come by the death of others (as by the course of inheritance, testaments, and the like), they come tumbling upon a man. But it mought be applied likewise to Pluto, taking him for the devil. For when riches come from the devil (as by fraud and oppression and unjust means), they come upon speed. The ways to enrich are many, and most of them foul. Parsimony is one of the best, and yet is not innocent, for it withholdeth men from works of liberality and charity. The improvement of the ground is the most natural obtaining of riches; for it is our great mother's blessing, the earth's; but it is slow. And yet where men of great wealth do stoop to husbandry, it multiplieth riches exceedingly. I knew a nobleman in England, that had the greatest audits of any man in my time; a great grazier, a great sheep-master, a great timber man, a great collier, a great corn-master, a great lead-man, and so of iron, and a number of the like points of husbandry. So as the earth seemed a sea to him, in respect of the perpetual importation. It was truly observed by one, that himself came very hardly to a little riches, and very easily to great riches. For when a man's stock is come to that, that he can expect the prime of markets, and overcome those bargains which for their greatness are few men's money, and be partner in the industries of younger men, he cannot but increase mainly. The gains of ordinary trades and vocations are honest; and furthered by two things chiefly, by diligence, and by a good name for good and fair dealing. But the gains of bargains are of a more doubtful nature; when men shall wait upon others' necessity, broke by servants and intruments to draw them on, put off others cunningly that would be better chapmen, and the like practices, which are crafty and naught. As for the chopping of bargains, when a man buys not to hold but to sell over again, that commonly grindeth double, both upon the seller and upon the buyer. Sharings do greatly enrich, if the hands be well chosen that are trusted. Usury is the certainest means of gain, though one of the worst; as that whereby a man doth eat his bread *in sudore vultus alieni* [in the sweat of another man's face] ; and besides, doth plough upon Sundays. But yet certain though it be, it hath flaws ; for that the scriveners and brokers do value unsound men to serve their own turn. The fortune in being the first in an invention or in a privilege, doth cause sometimes a wonderful overgrowth in riches ; as it was with the first sugar man in the Canaries. Therefore if a man can play the true logician, to have as well judgment as invention, he may do great matters ; especially if the times be fit. He that resteth upon gains certain, shall hardly grow to great riches ; and he that puts all upon adventures, doth oftentimes break and come to poverty : it is good therefore to guard adventures with certainties, that may uphold losses. Monopolies, and coemption of wares for resale, where they are not restrained, are great means to enrich ; especially if the party have intelligence what things are like to come into request, and so store himself beforehand. Riches gotten by service, though it be of the best rise, yet when they are gotten by flattery, feeding humours, and other servile conditions, they may be placed amongst the worst. As for fishing for testaments and executorships (as Tacitus saith of Seneca, *testamenta et orbos tamquam indagine capi*), it is yet worse ; by how much men submit themselves to meaner persons than in service. Believe not much them that seem to despise riches ; for they despise them that despair of them ; and none worse when they come to them. Be not penny-wise ; riches have wings, and sometimes they fly away of themselves, sometimes they must be set flying to bring in more. Men leave their riches either to their kindred, or to the public ; and moderate portions prosper best in both. A great state left to an heir, is as a lure to all the birds of prey round about to seize on him, if he be not the better stablished in years and judgment. Likewise glorious gifts and foundations are like *sacrifices without salt* ; and but the painted sepulchres of alms, which soon will putrefy and corrupt inwardly. Therefore measure not thine advancements by quantity, but frame them by measure : and defer not charities till death ; for, certainly, if a man weigh it rightly, he that doth so is rather liberal of another man's than of his own.

XXXV. Of Prophecies [48].

I MEAN not to speak of divine prophecies ; nor of heathen oracles ; nor of natural predictions ; but only of prophecies that have been of certain memory, and from hidden causes. Saith the Pythonissa to Saul, *To-morrow thou and thy son shall be with me.* Homer hath these verses :

At domus Æneæ cunctis dominabitur oris,
Et nati natorum, et qui nascentur ab illis

[The house of Æneas shall reign in all lands, and his children's children, and their generations]. A prophecy, as it seems, of the Roman empire. Seneca the tragedian hath these verses :

——— Venient annis
Sæcula seris, quibus Oceanus
Vincula rerum laxet, et ingens
Pateat Tellus, Tiphysque novos
Detegat orbes ; nec sit terris
Ultima Thule

[There shall come a time when the bands of ocean shall be loosened, and the vast earth shall be laid open ; another Tiphys shall disclose new worlds, and lands shall be seen beyond Thule] : a prophecy of the discovery of America. The daughter of Polycrates dreamed that Jupiter bathed her father, and Apollo anointed him ; and it came to pass that he was crucified in an open place, where

48 There is no Latin translation of this Essay.

the sun made his body run with sweat, and the rain washed it. Philip of Macedon dreamed he sealed up his wife's belly; whereby he did expound it, that his wife should be barren; but Aristander the soothsayer told him his wife was with child, because men do not use to seal vessels that are empty. A phantasm that appeared to M. Brutus in his tent, said to him, *Philippis iterum me videbis* [Thou shalt see me again at Philippi]. Tiberius said to Galba, *Tu quoque, Galba, degustabis imperium* [Thou likewise shall taste of empire]. In Vespasian's time, there went a prophecy in the East, that those that should come forth of Judea should reign over the world: which though it may be was meant of our Saviour, yet Tacitus expounds it of Vespasian. Domitian dreamed the night before he was slain, that a golden head was growing out of the nape of his neck: and indeed the succession that followed him, for many years, made golden times. Henry the Sixth of England said of Henry the Seventh, when he was a lad, and gave him water, *This is the lad that shall enjoy the crown for which we strive.* When I was in France, I heard from one Dr. Pena, that the Queen mother, who was given to curious arts, caused the King her husband's nativity to be calculated, under a false name; and the astrologer gave a judgment, that he should be killed in a duel; at which the Queen laughed, thinking her husband to be above challenges and duels: but he was slain upon a course at tilt, the splinters of the staff of Montgomery going in at his beaver. The trivial prophecy, which I heard when I was a child, and queen Elizabeth was in the flower of her years, was,

> When hempe is sponne
> England's done:

whereby it was generally conceived, that after the princes had reigned which had the principial letters of that word *hempe* (which were Henry, Edward, Mary, Philip, and Elizabeth) England should come to utter confusion; which thanks be to God, is verified only in the change of the name; for that the King's style is now no more of England, but of Britain. There was also another prophecy, before the year of eighty-eight, which I do not well understand.

> There shall be seen upon a day,
> Between the Baugh and the May,
> The black fleet of Norway.
> When that that is come and gone,
> England build houses of lime and stone,
> For after wars shall you have none.

It was generally conceived to be meant of the Spanish fleet that came in eighty-eight: for that the king of Spain's surname, as they say, is Norway. The prediction of Regiomontanus,

> Octogesimus octavus mirabilis annus,

was thought likewise accomplished in the sending of that great fleet, being the greatest in strength, though not in number, of all that ever swam upon the sea. As for Cleon's dream, I think it was a jest. It was, that he was devoured of a long dragon; and it was expounded of a maker of sausages, that troubled him exceedingly. There are numbers of the like kind; especially if you include dreams and predictions of astrology. But I have set down these few only of certain credit, for example. My judgment is, that they ought all to be despised; and ought to serve but for winter talk by the fireside. Though when I say *despised*, I mean it as for belief; for otherwise, the spreading or publishing of them is in no sort to be despised. For they have done much mischief; and I see many severe laws made to suppress them. That that hath given them grace, and some credit, consisteth in three things. First that men mark when they hit, and never mark when they miss; as they do generally also of dreams. The second is, that probable conjectures, or obscure traditions, many times turn themselves into prophecies; while the nature of man, which coveteth divination, thinks it no peril to foretell that which indeed they do but collect. As that of Seneca's verse. For so much was then subject to demonstration, that the globe of the earth had great parts beyond the Atlantic, which mought be probably conceived not to be all sea: and adding thereto the tradition in Plato's Timæus and his Atlanticus[49], it mought encourage one to turn it to a prediction. The third and last (which is the great one) is, that almost all of them, being infinite in number, have been impostures, and by idle and crafty brains merely contrived and feigned after the event past.

XXXVI. Of Ambition.

Ambition is like choler; which is an humour that maketh men active, earnest, full of alacrity, and stirring, if it be not stopped. But if it be stopped, and cannot have his way, it becometh adust, and thereby malign and venomous. So ambitious men, if they find the way open for their rising, and still get forward, they are rather busy than dangerous; but if they be checked in their desires, they become secretly discontent, and look upon men and matters with an evil eye, and are best pleased when things go backward; which is the worst property in a servant of a prince or state. Therefore it is good for princes, if they use ambitious men, to handle it so as they be still progressive and not retrogade; which because it cannot be without inconvenience, it is good not to use such natures at all. For if they rise not with their service, they will take order to make their service fall with them. But since we have said it were good not to use men of ambitious natures, except it be upon necessity, it is fit we speak in what cases they are of necessity. Good commanders in the wars must be taken, be they never so ambitious; for the use of their service dispenseth with the rest; and to take a soldier without ambition is to pull off his spurs. There is also great use of ambitious men in being screens to princes in matters of danger and envy; for no man will take that part, except he be like a seeled dove, that mounts and mounts because he cannot see about him. There is use also of ambitious men in pulling down the greatness of any subject that overtops; as Tiberius used Macro in the pulling down of Sejanus. Since therefore they must be used in such cases, there resteth to speak how they are to be bridled, that they may be less dangerous. There is less danger of them if they be of mean birth, than if they be noble; and if they be rather harsh of nature, than gracious and popular: and if they be rather new raised, than grown cunning and fortified in their greatness. It is counted by some a weakness in princes to have favourites; but it is of all others the best remedy against ambitious great-ones. For when the way of pleasuring and displeasuring lieth by the favourite, it is impossible any other should be over-great. Another means to curb them is to balance them by others as proud as they. But then there must be some middle counsellors, to keep things steady; for without that ballast the ship will roll too much. At the least, a prince may animate and inure some meaner persons, to be as it were scourges to ambitious men. As for the having of them obnoxious to ruin; if they be of fearful natures, it may do well; but if they be stout and daring, it may precipitate their designs, and prove dangerous. As for the pulling of them down, if the affairs require it, and that it may not be done with safety suddenly, the only way is, the interchange continually of favours and disgraces; whereby they may not know what to expect, and be as it were in a wood. Of ambitions, it is less harmful, the ambition to prevail in great things, than that other to appear in everything; for that breeds confusion, and mars business. But yet it is less danger to have an ambitious man stirring in business, than great in dependances. He that seeketh to be eminent amongst able men hath a great task; but that is ever good for the public. But he that plots to be the only figure amongst ciphers is the decay of a whole age. Honour hath three things in it; the vantage ground to do good; the approach to kings and principal persons; and the raising of a man's own fortunes. He that hath the best of these intentions when he aspireth, is an honest man; and that prince that can discern of these intentions in another that aspireth, is a wise

[49] That is the Critias.

prince. Generally, let princes and states choose such ministers as are more sensible of duty than of rising; and such as love business rather upon conscience than upon bravery; and let them discern a busy nature from a willing mind.

XXXVII. Of Masques and Triumphs[50].

These things are but toys, to come amongst such serious observations. But yet, since princes will have such things, it is better they should be graced with elegancy than daubed with cost. Dancing to song, is a thing of great state and pleasure. I understand it, that the song be in quire, placed aloft, and accompanied with some broken music; and the ditty fitted to the device. Acting in song, especially in dialogues, hath an extreme good grace; I say acting, not dancing (for that is a mean and vulgar thing); and the voices of the dialogue would be strong and manly (a base and a tenor; no treble); and the ditty high and tragical; not nice or dainty. Several quires, placed one over against another, and taking the voice by catches, anthem-wise, give great pleasure. Turning dances into figure is a childish curiosity. And generally let it be noted, that those things which I here set down are such as do naturally take the sense, and not respect petty wonderments. It is true, the alterations of scenes, so it be quietly and without noise, are things of great beauty and pleasure; for they feed and relieve the eye, before it be full of the same object. Let the scenes abound with light, specially coloured and varied; and let the masquers, or any other, that are to come down from the scene, have some motions upon the scene itself before their coming down; for it draws the eye strangely, and makes it with great pleasure to desire to see that it cannot perfectly discern. Let the songs be loud and cheerful, and not chirpings or pulings. Let the music likewise be sharp and loud, and well placed. The colours that shew best by candle-light, are white, carnation, and a kind of sea-water-green; and oes, or spangs, as they are of no great cost, so they are of most glory. As for rich embroidery, it is lost and not discerned. Let the suits of the masquers be graceful, and such as become the person when the vizards are off; not after examples of known attires; Turks, soldiers, mariners, and the like. Let antimasques not be long; they have been commonly of fools, satyrs, baboons, wild-men, antics, beasts, sprites, witches, Ethiops, pigmies, turquets, nymphs, rustics, Cupids, statua's moving, and the like. As for angels, it is not comical enough to put them in anti-masques; and any thing that is hideous, as devils, giants, is on the other side as unfit. But chiefly, let the music of them be recreative, and with some strange changes. Some sweet odours suddenly coming forth, without any drops falling, are, in such a company as there is steam and heat, things of great pleasure and refreshment. Double masques, one of men, another of ladies, addeth state and variety. But all is nothing except the room be kept clear and neat.

For justs and tourneys and barriers; the glories of them are chiefly in the chariots wherein the challengers make their entry; especially if they be drawn with strange beasts: as lions, bears, camels, and the like; or in the devices of their entrance; or in the bravery of their liveries; or in the goodly furniture of their horses and armour. But enough of these toys.

XXXVIII. Of Nature in Men.

Nature is often hidden; sometimes overcome; seldom extinguished. Force maketh nature more violent in the return; doctrine and discourse maketh nature less importune; but custom only doth alter and subdue nature. He that seeketh victory over his nature, let him not set himself too great nor too small tasks; for the first will make him dejected by often failings; and the second will make him a small proceeder, though by often prevailings. And at the first let him practise with helps, as swimmers do with bladders or rushes; but after a time let him practise with disadvantages, as dancers do with thick shoes. For it breeds great perfection, if the practice be harder than the use.

[50] This Essay is not translated.

Where nature is mighty, and therefore the victory hard, the degrees had need be, first to stay and arrest nature in time; like to him that would say over the four and twenty letters when he was angry; then to go less in quantity; as if one should, in forbearing wine, come from drinking healths to a draught at a meal; and lastly, to discontinue altogether. But if a man have the fortitude and resolution to enfranchise himself at once, that is the best:

Optimus ille animi vindex lædentia pectus
Vincula qui rupit, dedoluitque semel.

[Wouldst thou be free? The chains that gall thy breast
With one strong effort burst, and be at rest].

Neither is the ancient rule amiss, to bend nature as a wand to a contrary extreme, whereby to set it right; understanding it, where the contrary extreme is no vice. Let not a man force a habit upon himself with a perpetual continuance, but with some intermission. For both the pause reinforceth the new onset; and if a man that is not perfect be ever in practice, he shall as well practise his errors as his abilities, and induce one habit of both; and there is no means to help this but by seasonable intermissions. But let not a man trust his victory over his nature too far; for nature will lay [51] buried a great time, and yet revive upon the occasion or temptation. Like as it was with Æsop's damsel, turned from a cat to a woman, who sat very demurely at the board's end, till a mouse ran before her. Therefore let a man either avoid the occasion altogether; or put himself often to it, that he may be little moved with it. A man's nature is best perceived in privateness, for there is no affectation; in passion, for that putteth a man out of his precepts; and in a new case or experiment, for there custom leaveth him. They are happy men whose natures sort with their vocations; otherwise they may say, *multum incola fuit anima mea* [my soul hath been a stranger and a sojourner]; when they converse in those things they do not affect [52]. In studies, whatsoever a man commandeth upon himself, let him set hours for it; but whatsoever is agreeable to his nature, let him take no care for any set times; for his thoughts will fly to it of themselves; so as the spaces of other business or studies will suffice. A man's nature runs either to herbs or weeds; therefore let him seasonably water the one, and destroy the other.

XXXIX. Of Custom and Education.

Men's thoughts are much according to their inclination; their discourse and speeches according to their learning and infused opinions; but their deeds are after as they have been accustomed. And therefore as Machiavel well noteth (though in an ill-favoured instance), there is no trusting to the force of nature nor to the bravery of words, except it be corroborate by custom. His instance is, that for the achieving of a desperate conspiracy, a man should not rest upon the fierceness of any man's nature, or his resolute undertakings; but take such an one as hath had his hands formerly in blood. But Machiavel knew not of a friar Clement, nor a Ravillac, nor a Jaureguy, nor a Baltazar Gerard [53]; yet his rule holdeth still, that nature, nor the engagement of words, are not so forcible as custom. Only superstition is now so well advanced, that men of the first blood [54] are as firm as butchers by occupation; and votary resolution is made equipollent to custom even in matter of blood. In other things the predomi-

[51] So in original, and also in Ed. 1639. I have not thought it right to substitute *lie*, as has been usually done; because it may be that the form of the word was not settled in Bacon's time; and the correction of obsolete forms tends to conceal the history of the language.

[52] This clause is omitted in the translation.

[53] The translation adds: *aut Guidone Faulxio.*

[54] The translation has *primæ classis sicarii* (murderers of the first class): which seems to me to miss the meaning of the English. "Men of the first blood" must mean here, *men whose hands have not been in blood before.*

nancy of custom is every where visible; insomuch as a man would wonder to hear men profess, protest, engage, give great words, and then do just as they have done before; as if they were dead images, and engines moved only by the wheels of custom. We see also the reign or tyranny of custom, what it is. The Indians (I mean the sect of their wise men) lay themselves quietly upon a stack of wood, and so sacrifice themselves by fire. Nay the wives strive to be burned with the corpses of their husbands. The lads of Sparta, of ancient time, were wont to be scourged upon the altar of Diana, without so much as queching [55]. I remember, in the beginning of Queen Elizabeth's time of England, an Irish rebel condemned, put up a petition to the Deputy that he might be hanged in a with, and not in an halter; because it had been so used with former rebels. There be monks in Russia, for penance, that will sit a whole night in a vessel of water, till they be engaged with hard ice. Many examples may be put of the force of custom both upon mind and body. Therefore, since custom is the principal magistrate of man's life, let men by all means endeavour to obtain good customs. Certainly custom is most perfect when it beginneth in young years: this we call education; which is, in effect, but an early custom. So we see, in languages the tongue is more pliant to all expressions and sounds, the joints are more supple to all feats of activity and motions, in youth than afterwards. For it is true that late learners cannot so well take the ply; except it be in some minds that have not suffered themselves to fix, but have kept themselves open and prepared to receive continual amendment, which is exceeding rare. But if the force of custom simple and separate be great, the force of custom copulate and conjoined and collegiate is far greater. For there example teacheth, company comforteth, emulation quickeneth, glory raiseth: so as in such places the force of custom is in his exaltation. Certainly the great multiplication of virtues upon human nature resteth upon societies well ordained and disciplined. For commonwealths and good governments do nourish virtue grown, but do not much mend the seeds. But the misery is, that the most effectual means are now applied to the ends least to be desired.

XL. Of Fortune.

It cannot be denied, but outward accidents conduce much to fortune; favour, opportunity, death of others, occasion fitting virtue. But chiefly, the mould of a man's fortune is in his own hands. *Faber quisque fortunæ suæ*, saith the poet [56]. And the most frequent of external causes is, that the folly of one man is the fortune of another. For no man prospers so suddenly as by others' errors. *Serpens nisi serpentem comederit non fit draco* [A serpent must have eaten another serpent, before he can become a dragon]. Overt and apparent virtues bring forth praise; but there be secret and hidden virtues that bring forth fortune; certain deliveries of a man's self, which have no name. The Spanish name *desemboltura*, partly expresseth them; when there be not stonds nor restiveness in a man's nature; but that the wheels of his mind keep way with the wheels of his fortune. For so Livy (after he had described Cato Major in these words, *In illo viro tantum robur corporis et animi fuit, ut quocunque loco natus esset, fortunam sibi facturus videretur*) [Such was his strength of body and mind, that wherever he had been born he could have made himself a fortune]; falleth upon that, that he had *versatile ingenium* [a wit that could turn well]. Therefore if a man look sharply and attentively, he shall see Fortune: for though she be blind, yet she is not invisible. The way of fortune is like the milken way in the sky; which is a meeting or knot of a number of small stars; not seen asunder, but giving light together. So are there a number of little and scarce discerned virtues, or rather faculties and customs, that make men fortunate. The Italians note some of them, such as a man would little think. When they speak of one that cannot do amiss, they will throw in into his other conditions, that he hath *Poco di matto*. And certainly there be not two more fortunate

[55] *Quech*, according to Dr. Whately, means to *move* or *stir*.
[56] *inquit Comicus*. The poet is Plautus, *Trinum*. ii. 2. 34.

properties, than to have a little of the fool, and not too much of the honest. Therefore extreme lovers of their country or masters were never fortunate, neither can they be. For when a man placeth his thoughts without himself, he goeth not his own way. An hasty fortune maketh an enterpriser and remover (the French hath it better, *entreprenant*, or *remuant*) ; but the exercised fortune maketh the able man. Fortune is to be honoured and respected, and it be but for her daughters, Confidence and Reputation. For those two felicity breedeth ; the first within a man's self, the latter in others towards him [57]. All wise men, to decline the envy of their own virtues, use to ascribe them to Providence and Fortune ; for so they may the better assume them : and, besides, it is greatness in a man to be the care of the higher powers. So Cæsar said to the pilot in the tempest, *Cæsarem portas, et fortunam ejus* [You carry Cæsar and his fortune]. So Sylla chose the name of *Felix*, and not of *Magnus*. And it hath been noted, that those who ascribe openly too much to their own wisdom and policy, end infortunate. It is written that Timotheus the Athenian, after he had, in the account he gave to the state of his government, often interlaced this speech, *and in this Fortune had no part*, never prospered in any thing he undertook afterwards. Certainly there be, whose fortunes are like Homer's verses, that have a slide and easiness more than the verses of other poets ; as Plutarch saith of Timoleon's fortune, in respect of that of Agesilaus or Epaminondas. And that this should be, no doubt it is much in a man's self.

XLI. Of Usury.

Many have made witty invectives against Usury. They say that it is a pity the devil should have God's part, which is the tithe. That the usurer is the greatest sabbath breaker, because his plough goeth every Sunday. That the usurer is the drone that Virgil speaketh of :

Ignavum fucos pecus a præsepibus arcent.

That the usurer breaketh the first law that was made for mankind after the fall, which was, *in sudore vultûs tui comedes panem tuum* ; not, *in sudore vultûs alieni* [in the sweat of thy face shalt thou eat bread—not in the sweat of another's face]. That usurers should have orange-tawny bonnets, because they do judaize. That it is against nature for money to beget money ; and the like. I say this only, that usury is a *concessum propter duritiem cordis* [a thing allowed by reason of the hardness of men's hearts] : for since there must be borrowing and lending, and men are so hard of heart as they will not lend freely, usury must be permitted. Some others have made suspicious and cunning propositions of banks, discovery of men's estates, and other inventions. But few have spoken of usury usefully. It is good to set before us the incommodities and commodities of usury, that the good may be either weighed out or culled out ; and warily to provide, that while we make forth to that which is better, we meet not with that which is worse.

The discommodities of usury are, First, that it makes fewer merchants. For were it not for this lazy trade of usury, money would not lie still, but would in great part be employed upon merchandizing ; which is the *vena porta* [58] of wealth in a state. The second, that it makes poor merchants. For as a farmer cannot husband his ground so well if he sit at a great rent ; so the merchant cannot drive his trade so well, if he sit at great usury. The third is incident to the other two ; and that is the decay of customs of kings or states, which ebb or flow with merchandizing. The fourth, that it bringeth the treasure of a realm or state into a few hands. For the usurer being at certainties, and others at uncertainties, at the end of the game [59] most of the money will be in the box ; and ever a state flourisheth when wealth is more equally spread. The fifth, that it beats down the price of land ; for the employment of money is chiefly either merchandizing or purchasing ; and usury waylays both. The sixth, that it

[57] The translation adds, *Eæque vicissim pariunt animos et auctoritatem.*

[58] See p. 759, note.

[59] So Ed. 1639. The original has *gaine ;* the translation, *in fine ludi.*

doth dull and damp all industries, improvements, and new inventions, wherein money would be stirring, if it were not for this slug. The last, that it is the canker and ruin of many men's estates; which in process of time breeds a public poverty.

On the other side the commodities of usury are, first, that howsoever usury in some respect hindereth merchandizing, yet in some other it advanceth it; for it is certain that the greatest part of trade is driven by young merchants, upon borrowing at interest; so as if the usurer either call in or keep back his money there will ensue presently a great stand of trade. The second is, that were it not for this easy borrowing upon interest, men's necessities would draw upon them a most sudden undoing; in that they would be forced to sell their means (be it lands or goods) far under foot; and so, whereas usury doth but gnaw upon them, bad markets would swallow them quite up. As for mortgaging or pawning, it will little mend the matter, for either men will not take pawns without use; or if they do, they will look precisely for the forfeiture. I remember a cruel monied man in the country, that would say, The devil take this usury, it keeps us from forfeitures of mortgages and bonds. The third and last is, that it is a vanity to conceive that there would be ordinary borrowing without profit; and it is impossible to conceive the number of inconveniences that will ensue, if borrowing be cramped. Therefore to speak of the abolishing of usury is idle. All states have ever had it, in one kind or rate, or other. So as that opinion must be sent to Utopia.

To speak now of the reformation and reiglement of usury; how the discommodities of it may be best avoided, and the commodities retained. It appears by the balance of commodities and discommodities of usury two things are to be reconciled. The one, that the tooth of usury be grinded, that it bite not too much; the other, that there be left open a means to invite monied men to lend to the merchants, for the continuing and quickening of trade. This cannot be done, except you introduce two several sorts of usury, a less and a greater. For if you reduce usury to one low rate, it will ease the common borrower, but the merchant will be to seek for money. And it is to be noted, that the trade of merchandize, being the most lucrative, may bear usury at a good rate: other contracts not so.

To serve both intentions, the way would be briefly thus. That there be two rates of usury; the one free, and general for all; the other under licence only, to certain persons and in certain places of merchandizing. First therefore let usury in general be reduced to five in the hundred; and let that rate be proclaimed to be free and current; and let the state shut itself out to take any penalty for the same. This will preserve borrowing from any general stop or dryness. This will ease infinite borrowers in the country. This will, in good part, raise the price of land, because land purchased at sixteen years' purchase will yield six in the hundred, and somewhat more; whereas this rate of interest yields but five. This by like reason will encourage and edge industrious and profitable improvements; because many will rather venture in that kind than take five in the hundred, especially having been used to greater profit. Secondly let there be certain persons licensed to lend to known merchants upon usury at a higher rate; and let it be with the cautions following. Let the rate be, even with the merchant himself, somewhat more easy than that he used formerly to pay; for by that means all borrowers shall have some ease by this reformation, be he merchant, or whosoever. Let it be no bank or common stock, but every man be master of his own money. Not that I altogether mislike banks, but they will hardly be brooked, in regard of certain suspicions[60]. Let the state be answered some small matter for the licence, and the rest left to the lender; for if the abatement be but small, it will no whit discourage the lender. For he, for example, that took before ten or nine in the hundred, will sooner descend to eight in the hundred, than give over his trade of usury, and go from certain gains to gains of hazard. Let these licensed lenders be in number indefinite, but

60 These two sentences are omitted in the translation.

restrained to certain principal cities and towns of merchandizing; for then they will be hardly able to colour other men's monies in the country: so as the licence of nine will not suck away the current rate of five [61]; for no man will lend his monies far off, nor put them into unknown hands.

If it be objected that this doth in a sort authorize usury, which before was in some places but permissive; the answer is, that it is better to mitigate usury by declaration, than to suffer it to rage by connivance [62].

XLII. Of Youth and Age.

A man that is young in years may be old in hours, if he have lost no time. But that happeneth rarely. Generally, youth is like the first cogitations, not so wise as the second. For there is a youth in thoughts, as well as in ages. And yet the invention of young men is more lively than that of old; and imaginations stream into their minds better, and as it were more divinely. Natures that have much heat and great and violent desires and perturbations, are not ripe for action till they have passed the meridian of their years; as it was with Julius Cæsar, and Septimius Severus. Of the latter of whom it is said, *Juventutem egit èrroribus, imo furoribus, plenam* [He passed a youth full of errors, yea of madnesses]. And yet he was the ablest emperor, almost, of all the list. But reposed natures may do well in youth. As it is seen in Augustus Cæsar, Cosmus Duke of Florence, Gaston de Fois, and others. On the other side, heat and vivacity in age is an excellent composition for business. Young men are fitter to invent than to judge; fitter for execution than for counsel; and fitter for new projects than for settled business. For the experience of age in things that fall within the compass of it, directeth them; but in new things, abuseth them. The errors of young men are the ruin of business; but the errors of aged men amount but to this, that more might have been done, or sooner. Young men, in the conduct and manage of actions, embrace more than they can hold; stir more than they can quiet; fly to the end, without consideration of the means and degrees; pursue some few principles which they have chanced upon absurdly; care not to innovate, which draws unknown inconveniences [63]; use extreme remedies at first; and that which doubleth all errors, will not acknowledge or retract them; like an unready horse, that will neither stop nor turn. Men of age object too much, consult too long, adventure too little, repent too soon, and seldom drive business home to the full period, but content themselves with a mediocrity of success. Certainly it is good to compound employments of both; for that will be good for the present, because the virtues of either age may correct the defects of both; and good for succession, that young men may be learners, while men in age are actors; and, lastly, good for extern accidents, because authority followeth old men, and favour and popularity youth. But for the moral part, perhaps youth will have the pre-eminence, as age hath for the politic. A certain rabbin, upon the text, *Your young men shall see visions, and your old men shall dream dreams,* inferreth that young men are admitted nearer to God than old, because vision is a clearer revelation than a dream. And certainly, the more a man drinketh of the world, the more it intoxicateth: and age doth profit rather in the powers of understanding, than in the virtues of the will and affections. There be some have an over-early ripeness in their years, which fadeth betimes. These are, first, such as have brittle wits, the edge whereof is soon turned; such as was Hermogenes the rhetorician, whose books are exceeding subtle; who afterwards waxed stupid. A second sort is of those that have some natural dispositions which have better grace in youth than in age; such as is a fluent and luxuriant speech; which becomes youth well, but not age: so Tully saith of Hortensius, *Idem manebat, neque idem decebat* [He continued the same, when

[61] To "colour another man's money" is to pass it for one's own. See Whateley's edition of Bacon's Essays, p. 382.

[62] The last paragraph is omitted in the translation.

[63] This clause is omitted in the translation.

the same was not becoming]. The third is of such as take too high a strain at the first, and are magnanimous more than tract of years can uphold. As was Scipio Africanus, of whom Livy saith in effect, *Ultima primis cedebant* [His last actions were not equal to his first].

XLIII. Of Beauty.

Virtue is like a rich stone, best plain set; and surely virtue is best in a body that is comely, though not of delicate features; and that hath rather dignity of presence, than beauty of aspect. Neither is it almost seen, that very beautiful persons are otherwise of great virtue; as if nature were rather busy not to err, than in labour to produce excellency. And therefore they prove accomplished, but not of great spirit; and study rather behaviour than virtue. But this holds not always: for Augustus Cæsar, Titus Vespasianus, Philip le Bel of France, Edward the Fourth of England, Alcibiades of Athens, Ismael the Sophy of Persia, were all high and great spirits; and yet the most beautiful men of their times. In beauty, that of favour is more than that of colour; and that of decent and gracious motion more than that of favour. That is the best part of beauty, which a picture cannot express; no nor the first sight of life. There is no excellent beauty that hath not some strangeness in the proportion. A man cannot tell whether Apelles or Albert Durer were the more trifler; whereof the one would make a personage by geometrical proportions; the other, by taking the best parts out of divers faces, to make one excellent. Such personages, I think, would please nobody but the painter that made them. Not but I think a painter may make a better face than ever was; but he must do it by a kind of felicity [64], (as a musician that maketh an excellent air in music), and not by rule. A man shall see faces, that if you examine them part by part, you shall find never a good; and yet altogether do well. If it be true that the principal part of beauty is in decent motion, certainly it is no marvel though persons in years seem many times more amiable [65], *pulchrorum autumnus pulcher* [beautiful persons have a beautiful Autumn]; for no youth can be comely but by pardon, and considering the youth as to make up the comeliness. Beauty is as summer fruits, which are easy to corrupt, and cannot last; and for the most part it makes a dissolute youth, and an age a little out of countenance; but yet certainly again, if it light well, it maketh virtue shine, and vices blush.

XLIV. Of Deformity.

Deformed persons are commonly even with nature; for as nature hath done ill by them, so do they by nature; being for the most part (as the Scripture saith) *void of natural affection;* and so they have their revenge of nature [66]. Certainly there is a consent between the body and the mind; and where nature erreth in the one, she ventureth in the other. *Ubi peccat in uno, periclitatur in altero.* But because there is in man an election touching the frame of his mind, and a necessity in the frame of his body, the stars of natural inclination are sometimes obscured by the sun of discipline and virtue. Therefore it is good to consider of deformity, not as a sign, which is more deceivable; but as a cause, which seldom faileth of the effect. Whosoever hath any thing fixed in his person that doth induce contempt, hath also a perpetual spur in himself to rescue and deliver himself from scorn. Therefore all deformed persons are extreme bold. First, as in their own defence, as being exposed to scorn; but in process

[64] Keats seems to have felt that this is true also with regard to his own art:—

"When I behold upon the night's starred face
Huge cloudy symbols of a high romance,
And think that I may never live to trace
Their shadows, *with the magic hand of chance.*"

Life, Letters, etc., of John Keats, vol. ii. p. 293.

[65] The translation adds: *Secundum illud Euripidis.*

[66] This clause is omitted in the translation.

of time by a general habit. Also it stirreth in them industry, and especially of this kind to watch and observe the weakness of others, that they may have somewhat to repay. Again, in their superiors, it quencheth jealousy towards them, as persons that they think they may at pleasure despise: and it layeth their competitors and emulators asleep; as never believing they should be in possibility of advancement, till they see them in possession. So that upon the matter, in a great wit, deformity is an advantage to rising. Kings in ancient times (and at this present in some countries) were wont to put great trust in eunuchs, because they that are envious towards all are more obnoxious and officious towards one. But yet their trust towards them hath rather been as to good spials and good whisperers than good magistrates and officers. And much like is the reason of deformed persons. Still the ground is, they will, if they be of spirit, seek to free themselves from scorn; which must be either by virtue or malice; and therefore let it not be marvelled if sometimes they prove excellent persons; as was Agesilaus, Zanger the son of Solyman, Æsop, Gasca President of Peru; and Socrates may go likewise amongst them; with others.

XLV. Of Building.

Houses are built to live in, and not to look on; therefore let use be preferred before uniformity, except where both may be had. Leave the goodly fabrics of houses, for beauty only, to the enchanted palaces of the poets; who build them with small cost. He that builds a fair house upon an ill seat, committeth himself to prison. Neither do I reckon it an ill seat only where the air is unwholesome; but likewise where the air is unequal; as you shall see many fine seats set upon a knap of ground, environed with higher hills round about it, whereby the heat of the sun is pent in, and the wind gathereth as in troughs; so as you shall have, and that suddenly, as great diversity of heat and cold as if you dwelt in several places. Neither is it ill air only that maketh an ill seat, but ill ways, ill markets: and, if you will consult with Momus [67], ill neighbours. I speak not of many more; want of water; want of wood, shade, and shelter; want of fruitfulness and mixture [68] of grounds of several natures; want of prospect; want of level grounds; want of places at some near distance for sports of hunting, hawking, and races; too near the sea, too remote; having the commodity of navigable rivers [69], or the discommodity of their overflowing; too far off from great cities, which may hinder business, or too near them, which lurcheth all provisions, and maketh every thing dear; where a man hath a great living laid together, and where he is scanted: all which, as it is impossible perhaps to find together, so it is good to know them, and think of them, that a man may take as many as he can; and if he have several dwellings, that he sort them so, that what he wanteth in the one he may find in the other. Lucullus answered Pompey well; who, when he saw his stately galleries, and rooms so large and lightsome, in one of his houses, said, *Surely an excellent place for summer, but how do you in winter?* Lucullus answered, *Why, do you not think me as wise as some fowl are, that ever change their abode towards the winter?*

To pass from the seat to the house itself; we will do as Cicero doth in the orator's art; who writes books *De Oratore*, and a book he entitles *Orator*; whereof the former delivers the precepts of the art, and the latter the perfection. We will therefore describe a princely palace, making a brief model thereof. For it is strange to see, now in Europe, such huge buildings as the Vatican and Escurial and some others be, and yet scarce a very fair room in them.

First therefore, I say you cannot have a perfect palace, except you have two several sides; a side for the banquet, as is spoken of in the book of Hester, and

[67] For an explanation of this allusion to Momus, about which there has been some controversy of late, I am indebted to Mr. Ellis. "In one of Æsop's fables," he writes, "Minerva makes a house; and Momus says it should have been on wheels, to get away from bad neighbours." [68] That is, *want of* mixture.

[69] So in the original, and also in Ed. 1639. It seems as if *not* had dropped out; or as if *the* should be *no*. The translation has *commoditas nulla fluviorum navigabilium*.

a side for the household; the one for feasts and triumphs, and the other for dwelling. I understand both these sides to be not only returns, but parts of the front; and to be uniform without, though severally partitioned within; and to be on both sides of a great and stately tower in the midst of the front, that, as it were, joineth them together on either hand. I would have on the side of the banquet, in front, one only goodly room above stairs of some forty foot high [70]; and under it a room for a dressing or preparing place at times of triumphs. On the other side, which is the household side, I wish it divided at the first into a hall and a chapel, (with a partition between); both of good state and bigness; and those not to go all the length, but to have at the further end a winter and a summer parlour, both fair. And under these rooms, a fair and large cellar sunk under ground; and likewise some privy kitchens, with butteries and pantries, and the like. As for the tower, I would have it two stories, of eighteen foot high a piece, above the two wings; and a goodly leads upon the top, railed with statua's interposed; and the same tower to be divided into rooms, as shall be thought fit [71]. The stairs likewise to the upper rooms, let them be upon a fair open newel, and finely railed in with images of wood cast into a brass colour; and a very fair landing-place at the top. But this to be, if you do not point any of the lower rooms for a dining place of servants. For otherwise you shall have the servants' dinner after your own: for the steam of it will come up as in a tunnel. And so much for the front. Only I understand the height of the first stairs to be sixteen foot, which is the height of the lower room.

Beyond this front is there to be a fair court, but three sides of it, of a far lower building than the front. And in all the four corners of that court fair staircases, cast into turrets, on the outside, and not within the row of buildings themselves. But those towers are not to be of the height of the front, but rather proportionable to the lower building. Let the court not be paved, for that striketh up a great heat in summer, and much cold in winter. But only some side alleys, with a cross and the quarters to graze, being kept shorn, but not too near shorn. The row of return on the banquet side, let it be all stately galleries: in which galleries let there be three, or five, fine cupolas in the length of it, placed at equal distance; and fine coloured windows of several works. On the household side, chambers of presence and ordinary entertainments, with some bedchambers; and let all three sides be a double house, without thorough lights on the sides, that you may have rooms from the sun both for forenoon and afternoon. Cast it also, that you may have rooms both for summer and winter; shady for summer, and warm for winter. You shall have sometimes fair houses so full of glass, that one cannot tell where to become to be out of the sun or cold. For in bowed windows, I hold them of good use (in cities, indeed, upright do better, in respect of the uniformity towards the street); for they be pretty retiring places for conference; and besides they keep both the wind and sun off; for that which would strike almost thorough the room doth scarce pass the window. But let them be but few, four in the court, on the sides only.

Beyond this court let there be an inward court, of the same square and height; which is to be environed with the garden on all sides; and in the inside, cloistered on all sides, upon decent and beautiful arches, as high as the first story. On the under story, towards the garden, let it be turned to a grotta, or place of shade, or estivation. And only have opening and windows towards the garden; and be level upon the floor, no whit sunken under ground, to avoid all dampishness. And let there be a fountain, or some fair work of statua's in the midst of this court; and to be paved as the other court was. These buildings to be for privy lodgings on both sides; and the end for privy galleries. Whereof you must foresee that one of them be for an infirmary, if the prince or any special person should be sick, with chambers, bed-chamber, antecamera, and recamera, joining to it. This upon the second story. Upon the ground story, a fair gallery, open, upon pillars; and upon the third story likewise, an open gallery, upon pillars, to take the prospect and freshness of the garden. At both corners of the further

[70] The translation raises it to fifty feet. *Eamque supra gradus ad quinguaginta pedes ad minus altam.*

[71] This clause is omitted in the translation.

side, by way of return, let there be two delicate or rich cabinets, daintily paved, richly hanged, glazed with crystalline glass, and a rich cupola in the midst; and all other elegancy that may be thought upon. In the upper gallery too, I wish that there may be, if the place will yield it, some fountains running in divers places from the wall, with some fine avoidances [72]. And thus much for the model of the palace [73]; save that you must have before you come to the front, three courts. A green court plain, with a wall about it; a second court of the same, but more garnished, with little turrets, or rather embellishments, upon the wall; and a third court, to make a square with the front, but not to be built, nor yet enclosed with a naked wall, but enclosed with tarrasses, leaded aloft, and fairly garnished, on the three sides; and cloistered on the inside, with pillars, and not with arches below. As for offices, let them stand at distance, with some low galleries, to pass from them to the palace itself.

XLVI. Of Gardens.

God Almighty first planted a Garden. And indeed it is the purest of human pleasures. It is the greatest refreshment to the spirits of man; without which buildings and palaces are but gross handyworks: and a man shall ever see that when ages grow to civility and elegancy, men come to build stately sooner than to garden finely; as if gardening were the greater perfection. I do hold it, in the royal ordering of gardens, there ought to be gardens for all the months in the year; in which severally things of beauty may be then in season [74]. For December, and January, and the latter part of November, you must take such things as are green all winter [75]: holly; ivy; bays; juniper; cypress-trees; yew; pine-apple-trees [76]; fir-trees; rosemary; lavender; periwinkle, the white, the purple, and the blue; germander; flags; orange-trees; lemon-trees; and myrtles, if they be stoved; and sweet marjoram, warm set. There followeth for the latter part of January and February, the mezereon-tree, which then blossoms; crocus vernus, both the yellow and the grey; primroses; anemones; the early tulippa; hyacinthus orientalis; chamaïris; fritellaria. For March, there come violets, specially the single blue, which are the earliest; the yellow daffodil; the daisy; the almond-tree in blossom; the peach-tree in blossom; the cornelian-tree in blossom; sweet-briar. In April follow, the double white violet; the wall-flower; the stock-gilliflower; the cowslip; flower-de-lices, and lilies of all natures [77]; rosemary-flowers; the tulippa; the double-

[72] *qui per secretos tubos iterum transeant.* The following sentence is inserted here in the translation: *Interior autem pars in solario superiore, versus aream, formetur in porticus et ambulacra, bene munita et obducta, ad usum convalescentium.*

[73] The translation adds: *nam de balneis et piscinis non loquor.*

[74] *in quibus separatim plantæ quæ illo mense florent et vigent producantur.* The scene in the "Winter's Tale," where Perdita presents the guests with flowers suited to their ages, has some expressions which, if this Essay had been contained in the earlier edition, would have made me suspect that Shakespeare had been reading it. As I am not aware that the resemblance has been observed, I will quote the passages to which I allude in connexion with those which remind me of them.

[75] Reverend Sirs,
For you there's Rosemary and Rue; these keep
Seeming and savour all the winter long.
Grace and Remembrance be to you both,
And welcome to our shearing.
Pol. Shepherdess,
(A fair one are you) well you fit our ages
With flowers of winter.

[76] In place of "pine-apple-trees," the translation has *buxus, pinus, abies.*

[77] Now, my fair'st friend,
I would I had some flowers o' the Spring, that might
Become your time of day . . .

piony; the pale daffodil; the French honeysuckle; the cherry-tree in blossom; the dammasin and plum-trees in blossom; the white thorn in leaf; the lilac-tree. In May and June come pinks of all sorts, specially the blush-pink; roses of all kinds, except the musk, which comes later; honey-suckles; strawberries; bugloss; columbine; the French marigold; flos Africanus [78]; cherry-tree in fruit; ribes; figs in fruit; rasps; vine flowers; lavender in flowers; the sweet satyrian, with the white flower; herba muscaria; lilium convallium; the apple-tree in blossom [79]. In July come gilliflowers of all varieties [80]; musk roses; the lime-tree in blossom; early pears and plums in fruit; genitings, quadlins. In August come plums of all sorts in fruit; pears; apricocks; berberries; filberds; musk-melons; monks-hoods, of all colours. In September come grapes; apples; poppies of all colours; peaches; melocotones; nectarines; cornelians; wardens; quinces. In October and the beginning of November come services; medlars; bullaces; roses cut or removed to come late; holly-oaks; and such like. These particulars are for the climate of London; but my meaning is perceived, that you may have *ver perpetuum*, as the place affords.

And because the breath of flowers is far sweeter in the air (where it comes and goes like the warbling of music) than in the hand, therefore nothing is more fit for that delight, than to know what be the flowers and plants that do best perfume the air. Roses, damask and red, are fast flowers of their smell; so that you may walk by a whole row of them, and find nothing of their sweetness; yea though it be in a morning's dew. Bays likewise yield no smell as they grow. Rosemary little; nor sweet marjoram. That which above all others yields the sweetest smell in the air, is the violet, specially the white double violet, which comes twice a year; about the middle of April, and about Bartholomew-tide. Next to that is the musk-rose. Then the strawberry-leaves dying, with a most excellent cordial smell [81]. Then the flower of the vines; it is a little dust, like the dust of a bent, which grows upon the cluster in the first coming forth. Then sweet-briar. Then wall-flowers, which are very delightful to be set under a palour or lower chamber window. Then pinks and gilliflowers [82], specially the

Daffodils,
That come before the swallow dares, and take
The winds of March with beauty: Violets (dim
But sweeter than the lids of Juno's eyes,
Or Cytherea's breath): pale Prime-roses,
That die unmarried, ere they can behold
Bright Phœbus in his strength . . .
bold Oxlips, and
The Crown Imperial: Lilies of all kinds,
(The Flower-de-Luce being one).

[78] *Flos Africanus, simplex et multiplex.* The "French Marigold" is omitted in the translation.

[79] The translation adds; *flos cyaneus*: [the corn-cockle].

[80]
Sir, the year growing ancient,
Not yet on Summer's death, nor on the birth
Of trembling Winter, the fairest flowers o' the season
Are our Carnations and streaked Gilly-vors
(Which some call Nature's bastards) . . .
Here's flowers for you:
Hot Lavender, Mints, Savory, Marjoram,
The Mary-gold, that goes to bed wi' the Sun,
And with him rises, weeping; These are flowers
Of middle Summer, and I think they are given
To men of middle age.

[81] So Ed. 1639. The original has "which a most excellent cordialsmell". Possibly it should be *which yield*. The translation has *quæ halitum emittunt plane cardiacum.*

[82] The British Museum copy (see note at the end) omits *and gilliflowers*. The translation has *tum cariophyllatæ tam minores quam majores.*

matted pink and clove gilliflower. Then the flowers of the lime-tree. Then the honeysuckles, so they be somewhat afar off[83]. Of bean-flowers I speak not, because they are field flowers. But those which perfume the air most delightfully, not passed by as the rest, but being trodden upon and crushed, are three; that is, burnet, wild-thyme, and watermints. Therefore you are to set whole alleys of them, to have the pleasure when you walk or tread.

For gardens (speaking of those which are indeed prince-like, as we have done of buildings), the contents ought not well to be under thirty acres of ground; and to be divided into three parts; a green in the entrance; a heath or desert in the going forth; and the main garden in the midst; besides alleys on both sides. And I like well that four acres of ground be assigned to the green; six to the heath; four and four to either side; and twelve to the main garden. The green hath two pleasures: the one, because nothing is more pleasant to the eye than green grass kept finely shorn; the other, because it will give you a fair alley in the midst, by which you may go in front upon a stately hedge, which is to enclose the garden. But because the alley will be long, and, in great heat of the year or day, you ought not to buy the shade in the garden by going in the sun thorough the green, therefore you are, of either side the green, to plant a covert alley, upon carpenter's work, about twelve foot in height, by which you may go in shade into the garden. As for the making of knots or figures with divers coloured earths, that they may lie under the windows of the house on that side which the garden stands[84]; they be but toys: you may see as good sights many times in tarts. The garden is best to be square, encompassed on all the four sides with a stately arched hedge. The arches to be upon pillars of carpenter's work, of some ten foot high, and six foot broad; and the spaces between of the same dimension with the breadth of the arch. Over the arches let there be an entire hedge of some four foot high, framed also upon carpenter's work; and upon the upper hedge, over every arch, a little turret, with a belly, enough to receive a cage of birds: and over every space between the arches some other little figure, with broad plates of round coloured glass gilt, for the sun to play upon. But this hedge I intend to be raised upon a bank, not steep, but gently slope, of some six foot, set all with flowers. Also I understand, that this square of the garden should not be the whole breadth of the ground, but to leave on either side ground enough for diversity of side alleys; unto which the two covert alleys of the green may deliver you. But there must be no alleys with hedges at either end of this great enclosure; not at the hither end, for letting your prospect upon this fair hedge from the green; nor at the further end, for letting your prospect from the hedge through the arches upon the heath.

For the ordering of the ground within the great hedge, I leave it to variety of device; advising nevertheless that whatsoever form you cast it into, first[85], it be not too busy, or full of work. Wherein I, for my part, do not like images cut out in juniper or other garden stuff; they be for children. Little low hedges, round, like welts, with some pretty pyramides, I like well; and in some places, fair columns upon frames of carpenter's work. I would also have the alleys spacious and fair. You may have closer alleys upon the side grounds, but none in the main garden. I wish also, in the very middle, a fair mount, with three ascents, and alleys, enough for four to walk abreast; which I would have to be perfect circles, without any bulwarks or embossments; and the whole mount to be thirty foot high; and some fine banqueting house; with some chimneys neatly cast, and without too much glass.

83 The translation adds *tum flores lavendulæ*.

84 This clause is omitted in the translation.

85 My copy of Ed. 1625 has a comma after *first* and no comma after *into*. The copy in the British Museum has a comma after *into*, and no comma after *first*. So also Ed. 1639. The translation has *quæcunque ea tandem sit, nimis curiosa et operosa ne sit.* I suspect that the direction was to add the second comma and leave the first, and that it was misunderstood, or imperfectly executed; an accident which may easily happen, and would account for the occasional introduction of a change which could not have been intended.

For fountains, they are a great beauty and refreshment; but pools mar all and make the garden unwholesome, and full of flies and frogs. Fountains intend to be of two natures: the one that sprinkleth or spouteth water; the other a fair receipt of water, of some thirty or forty foot square, but without fish, or slime, or mud. For the first, the ornaments of images gilt, or of marble, which are in use, do well: but the main matter is so to convey the water, as it never stay, either in the bowls or in the cistern; that the water be never by rest discoloured, green or red or the like; or gather any mossiness or putrefaction. Besides that, it is to be cleansed every day by the hand. Also some steps up to it, and some fine pavement about it, doth well. As for the other kind of fountain, which we may call a bathing pool, it may admit much curiosity[86] and beauty; wherewith we will not trouble ourselves; as, that the bottom be finely paved, and with images; the sides likewise; and withal embellished with coloured glass, and such things of lustre; encompassed also with fine rails of low statua's. But the main point is the same which we mentioned in the former kind of fountain; which is, that the water be in perpetual motion, fed by a water higher than the pool, and delivered into it by fair spouts, and then discharged away under ground, by some equality of bores, that it stay little. And for fine devices, of arching water without spilling, and making it rise in several forms (of feathers, drinking glasses, canopies, and the like), they be pretty things to look on, but nothing to health and sweetness.

For the heath, which was the third part of our plot, I wish it to be framed, as much as may be, to a natural wildness. Trees I would have none in it, but some thickets made only of sweet-briar and honeysuckle, and some wild vine amongst; and the ground set with violets, strawberries, and primroses. For these are sweet, and proper in the shade. And these to be in the heath, here and there, not in any order. I like also little heaps, in the nature of mole-hills (such as are in wild heaths), to be set, some with wild thyme; some with pinks; some with germander, that gives a good flower to the eye; some with periwinkle; some with violets; some with strawberries; some with cowslips; some with daisies; some with red roses; some with lilium convallium; some with sweet-williams red; some with bear's-foot: and the like low flowers, being withal sweet and sightly. Part of which heaps are to be with standards of little bushes pricked upon their top, and part without. The standards to be roses; juniper; holly; berberries (but here and there, because of the smell of their blossom[87]); red currants; gooseberry; rosemary; bays; sweet-briar, and such like. But these standards to be kept with cutting, that they grow not out of course.

For the side grounds, you are to fill them with variety of alleys, private, to give a full shade, some of them, wheresoever the sun be. You are to frame some of them likewise for shelter, that when the wind blows sharp, you may walk as in a gallery. And those alleys must be likewise hedged at both ends, to keep out the wind; and these closer alleys must be ever finely gravelled, and no grass, because of going wet. In many of these alleys likewise, you are to set fruit-trees of all sorts; as well upon the walls as in ranges. And this would be generally observed, that the borders wherein you plant your fruit-trees be fair and large, and low, and not steep; and set with fine flowers, but thin and sparingly, lest they deceive the trees. At the end of both the side grounds, I would have

[86] The copy in the British Museum has a semicolon after *curiosity*: my copy has a comma. And as it has certainly been a change in the type, and not a variety in the impression or an alteration made by the hand, I am inclined to think that the Museum copy was a proof in which corrections were afterwards made.

[87] *sed hæc rarior, propter odoris gravitatem dum floret.* The British Museum copy has a semicolon after *blossom* and no stop after *berberries* (or *beare-berries* as it is spelt): my copy has a semicolon after *beare-berries* and no stop after blossom. It is difficult to say which has been the alteration: for in the original setting of the type room for a semicolon does not seem to have been left in either place. Here (as before) I suspect the intention of the corrector was to insert the first without removing the second. The parenthesis certainly refers to the berberry; the blossom of which has an offensive smell, when too near.

a mount of some pretty height, leaving the wall of the enclosure breast high, to look abroad into the fields.

For the main garden, I do not deny but there should be some fair alleys ranged on both sides, with fruit trees; and some pretty tufts of fruit trees, and arbours with seats, set in some decent order; but these to be by no means set too thick; but to leave the main garden so as it be not close, but the air open and free. For as for shade, I would have you rest upon the alleys of the side grounds, there to walk, if you be disposed, in the heat of the year or day; but to make account that the main garden is for the more temperate parts of the year; and in the heat of summer, for the morning and the evening, or overcast days.

For aviaries, I like them not, except they be of that largeness as they may be turfed, and have living plants and bushes set in them: that the birds may have more scope, and natural nestling, and that no foulness appear in the floor of the aviary. So I have made a platform of a princely garden, partly by precept, partly by drawing, not a model, but some general lines of it; and in this I have spared for no cost. But it is nothing for great princes, that for the most part taking advice with workmen, with no less cost set their things together; and sometimes add statua's, and such things for state and magnificence, but nothing to the true pleasure of a garden.

XLVII. Of Negociating.

It is generally better to deal by speech than by letter; and by the mediation of a third than by a man's self. Letters are good, when a man would draw an answer by letter back again; or when it may serve for a man's justification afterwards to produce his own letter; or where it may be danger to be interrupted, or heard by pieces. To deal in person is good, when a man's face breedeth regard, as commonly with inferiors; or in tender cases, where a man's eye upon the countenance of him with whom he speaketh may give him a direction how far to go; and generally, where a man will reserve to himself liberty either to disavow or to expound. In choice of instruments, it is better to choose men of a plainer sort, that are like to do that that is committed to them, and to report back again faithfully the success, than those that are cunning to contrive out of other men's business somewhat to grace themselves, and will help the matter in report for satisfaction sake. Use also such persons as affect the business wherein they are employed; for that quickeneth much; and such as are fit for the matter; as bold men for expostulation, fair-spoken men for persuasion, crafty men for inquiry and observation, froward and absurd men for business that doth not well bear out itself. Use also such as have been lucky and prevailed before in things wherein you have employed them; for that breeds confidence, and they will strive to maintain their prescription. It is better to sound a person with whom one deals afar off, than to fall upon the point at first; except you mean to surprise him by some short question. It is better dealing with men in appetite, than with those that are where they would be. If a man deal with another upon conditions, the start or first performance is all; which a man cannot reasonably demand, except either the nature of the thing be such, which must go before; or else a man can persuade the other party that he shall still need him in some other thing; or else that he be counted the honester man. All practice is to discover, or to work. Men discover themselves in trust, in passion, at unawares, and of necessity, when they would have somewhat done and cannot find an apt pretext. If you would work any man, you must either know his nature and fashions, and so lead him; or his ends, and so persuade him; or his weakness and disadvantages, and so awe him; or those that have interest in him, and so govern him. In dealing with cunning persons, we must ever consider their ends, to interpret their speeches; and it is good to say little to them, and that which they least look for. In all negociations of difficulty, a man may not look to sow and reap at once; but must prepare business, and so ripen it by degrees.

XLVIII. Of Followers and Friends.

Costly followers are not to be liked; lest while a man maketh his train longer, he make his wings shorter. I reckon to be costly, not them alone which charge the purse, but which are wearisome and importune in suits. Ordinary followers ought to challenge no higher conditions than countenance, recommendation, and protection from wrongs. Factious followers are worse to be liked, which follow not upon affection to him with whom they range themselves, but upon discontentment conceived against some other; whereupon commonly ensueth that ill intelligence that we many times see between great personages. Likewise glorious followers, who make themselves as trumpets of the commendation of those they follow, are full of inconvenience; for they taint business through want of secrecy; and they export honour from a man, and make him a return in envy. There is a kind of followers likewise which are dangerous, being indeed espials; which inquire the secrets of the house, and bear tales of them to others. Yet such men, many times, are in great favour; for they are officious, and commonly exchange tales. The following by certain estates of men, answerable to that which a great person himself professeth (as of soldiers to him that hath been employed in the wars, and the like), hath ever been a thing civil, and well taken even in monarchies; so it be without too much pomp or popularity. But the most honourable kind of following is to be followed as one that apprehendeth to advance virtue and desert in all sorts of persons. And yet, where there is no eminent odds in sufficiency, it is better to take with the more passable, than with the more able. And besides, to speak truth, in base times active men are of more use than virtuous. It is true that in government it is good to use men of one rank equally: for to countenance some extraordinarily, is to make them insolent, and the rest discontent; because they may claim a due. But contrariwise, in favour, to use men with much difference and election is good; for it maketh the persons preferred more thankful, and the rest more officious: because all is of favour. It is good discretion not to make too much of any man at the first; because one cannot hold out that proportion. To be governed (as we call it) by one, is not safe; for it shews softness, and gives a freedom to scandal and disreputation; for those that would not censure or speak ill of a man immediately, will talk more boldly of those that are so great with them, and thereby wound their honour. Yet to be distracted with many is worse; for it makes men to be of the last impression [88], and full of change. To take advice of some few friends is ever honourable; *for lookers-on many times see more than gamesters; and the vale best discovereth the hill.* There is little friendship in the world, and least of all between equals, which was wont to be magnified. That that is, is between superior and inferior, whose fortunes may comprehend the one the other.

XLIX. Of Suitors.

Many ill matters and projects are undertaken; and private suits do putrefy the public good. Many good matters are undertaken with bad minds; I mean not only corrupt minds, but crafty minds, that intend not performance. Some embrace suits, which never mean to deal effectually in them; but if they see there may be life in the matter by some other means, they will be content to win a thank, or take a second reward, or at least to make use in the meantime of the suitor's hopes. Some take hold of suits only for an occasion to cross some other; or to make an information whereof they could not otherwise have apt pretext; without care what become of the suit when that turn is served; or, generally, to make other men's business a kind of entertainment to bring in their own. Nay some undertake suits, with a full purpose to let them fall; to the end to gratify the adverse party or competitor. Surely there is in some sort a right in every suit; either a right in equity, if it be a suit of controversy; or a right of desert, if it be a suit of petition. If affection lead a man to favour the wrong side in justice, let him rather use his countenance to compound the matter

[88] *postremæ (ut nunc loquuntur) editionis.* Whence it would appear that the metaphor is from the printing-press.

than to carry it. If affection lead a man to favour the less worthy in desert, let him do it without depraving or disabling the better deserver. In suits which a man doth not well understand, it is good to refer them to some friend of trust and judgment, that may report whether he may deal in them with honour: but let him choose well his referendaries, for else he may be led by the nose. Suitors are so distasted with delays and abuses, that plain dealing in denying to deal in suits at first, and reporting the success barely, and in challenging no more thanks than one hath deserved, is grown not only honourable, but also gracious. In suits of favour, the first coming ought to take little place: so far forth consideration may be had of his trust, that if intelligence of the matter could not otherwise have been had but by him, advantage be not taken of the note, but the party left to his other means; and in some sort recompensed for his discovery. To be ignorant of the value of a suit is simplicity; as well as to be ignorant of the right thereof is want of conscience. Secrecy in suits is a great mean of obtaining; for voicing them to be in forwardness may discourage some kind of suitors, but doth quicken and awake others. But timing of the suit is the principal. Timing, I say, not only in respect of the person that should grant it, but in respect of those which are like to cross it. Let a man, in the choice of his mean, rather choose the fittest mean than the greatest mean; and rather them that deal in certain things, than those that are general. The reparation of a denial is sometimes equal to the first grant; if a man shew himself neither dejected nor discontented. *Iniquum petas ut æquum feras* [Ask more than is reasonable, that you may get no less], is a good rule, where a man hath strength of favour: but otherwise a man were better rise in his suit; for he that would have ventured at first to have lost the suitor, will not in the conclusion lose both the suitor and his own former favour. Nothing is thought so easy a request to a great person, as his letter; and yet, if it be not in a good cause, it is so much out of his reputation. There are no worse instruments than these general contrivers of suits; for they are but a kind of poison and infection to public proceedings.

L. Of Studies.

Studies serve for delight, for ornament, and for ability. Their chief use for delight, is in privateness and retiring; for ornament, is in discourse; and for ability, is in the judgment and disposition of business. For expert men can execute, and perhaps judge of particulars, one by one; but the general counsels, and the plots and marshalling of affairs, come best from those that are learned. To spend too much time in studies is sloth; to use them too much for ornament, is affectation; to make judgment wholly by their rules, is the humour of a scholar. They perfect nature, and are perfected by experience; for natural abilities are like natural plants, that need proyning[89] by study; and studies themselves do give forth directions too much at large, except they be bounded in by experience. Crafty men contemn studies, simple men admire them, and wise men use them; for they teach not their own use; but that is a wisdom without them, and above them, won by observation. Read not to contradict and confute; nor to believe and take for granted; nor to find talk and discourse; but to weigh and consider. Some books are to be tasted, others to be swallowed, and some few to be chewed and digested; that is, some books are to be read only in parts; others to be read, but not curiously; and some few to be read wholly, and with diligence and attention. Some books also may be read by deputy, and extracts made of them by others; but that would be only in the less important arguments, and the meaner sort of books; else distilled books are like common distilled waters, flashy things. Reading maketh a full man; conference a ready man; and writing an exact man. And therefore, if a man

[89] So in the original. Compare *Sylva Sylvarum*, § 432: "the lower boughs only maintained, and the higher continually *proined* off:" and again § 823: "many birds do *proine* their feathers:" from which I suppose that it is not a misprint, but another form of the word.

write little, he had need have a great memory; if he confer little, he had need have a present wit: and if he read little, he had need have much cunning, to seem to know that he doth not. Histories make men wise; poets witty; the mathematics subtle; natural philosophy deep; moral grave; logic and rhetoric able to contend. *Abeunt studia in mores* [The studies pass into the manners]. Nay there is no stond or impediment in the wit, but may be wrought out by fit studies: like as diseases of the body may have appropriate exercises. Bowling is good for the stone and reins; shooting for the lungs and breast; gentle walking for the stomach; riding for the head; and the like. So if a man's wit be wandering, let him study the mathematics; for in demonstrations, if his wit be called away never so little, he must begin again. If his wit be not apt to distinguish or find differences, let him study the schoolmen; for they are *cymini sectores* [splitters of hairs]. If he be not apt to beat over matters, and to call up one thing to prove and illustrate another, let him study the lawyers' cases. So every defect of the mind may have a special receipt.

LI. Of Faction.

Many have an opinion not wise, that for a prince to govern his estate, or for a great person to govern his proceedings, according to the respect of factions, is a principal part of policy; whereas contrariwise, the chiefest wisdom is either in ordering those things which are general, and wherein men of several factions do nevertheless agree; or in dealing with correspondence to particular persons, one by one. But I say not that the consideration of factions is to be neglected. Mean men, in their rising, must adhere; but great men, that have strength in themselves, were better to maintain themselves indifferent and neutral. Yet even in beginners, to adhere so moderately, as he be a man of the one faction which is most passable with the other, commonly giveth best way. The lower and weaker faction is the firmer in conjunction; and it is often seen that a few that are stiff do tire out a greater number that are more moderate. When one of the factions is extinguished, the remaining subdivideth; as the faction between Lucullus and the rest of the nobles of the senate (which they called *Optimates*) held out awhile against the faction of Pompey and Cæsar; but when the senate's authority was pulled down, Cæsar and Pompey soon after brake. The faction or party of Antonius and Octavianus Cæsar against Brutus and Cassius, held out likewise for a time; but when Brutus and Cassius were overthrown, then soon after Antonius and Octavianus brake and subdivided. These examples are of wars, but the same holdeth in private factions. And therefore those that are seconds in factions do many times, when the faction subdivideth, prove principals; but many times also they prove cyphers and cashiered; for many a man's strength is in opposition; and when that faileth he groweth out of use. It is commonly seen that men once placed take in with the contrary faction to that by which they enter: thinking belike that they have the first sure, and now are ready for a new purchase. The traitor in faction lightly goeth away with it; for when matters have stuck long in balancing, the winning of some one man casteth them, and he getteth all the thanks. The even carriage between two factions proceedeth not always of moderation, but of a trueness to a man's self, with end to make use of both. Certainly in Italy they hold it a little suspect in popes, when they have often in their mouth *Padre commune:* and take it to be a sign of one that meaneth to refer all to the greatness of his own house. Kings had need beware how they side themselves, and make themselves as of a faction or party; for leagues within the state are ever pernicious to monarchies: for they raise an obligation paramount to obligation of sovereignty, and make the king *tanquam unus ex nobis* [like one of themselves]; as was to be seen in the League of France. When factions are carried too high and too violently, it is a sign of weakness in princes; and much to the prejudice both of their authority and business. The motions of factions under kings ought to be like the motions (as the astronomers speak) of the inferior orbs, which may have their proper motions, but yet still are quietly carried by the higher motion of *primum mobile.*

LII. Of Ceremonies and Respects.

He that is only real, had need have exceeding great parts of virtue; as the stone had need to be rich that is set without foil. But if a man mark it well, it is in praise and commendation of men as it is in gettings and gains: for the proverb is true, *That light gains make heavy purses*; for light gains come thick, whereas great come but now and then. So it is true that small matters win great commendation, because they are continually in use and in note: whereas the occasion of any great virtue cometh but on festivals. Therefore it doth much add to a man's reputation, and is (as queen Isabella said) *like perpetual letters commendatory*, to have good forms. To attain them it almost sufficeth not to despise them; for so shall a man observe them in others; and let him trust himself with the rest. For if he labour too much to express them, he shall lose their grace; which is to be natural and unaffected. Some men's behaviour is like a verse, wherein every syllable is measured; how can a man comprehend great matters, that breaketh his mind too much to small observations? Not to use ceremonies at all, is to teach others not to use them again; and so diminisheth respect to himself; especially they be not to be omitted to strangers and formal natures; but the dwelling upon them, and exalting them above the moon, is not only tedious, but doth diminish the faith and credit of him that speaks. And certainly there is a kind of conveying of effectual and imprinting passages amongst compliments, which is of singular use, if a man can hit upon it. Amongst a man's peers a man shall be sure of familiarity; and therefore it is good a little to keep state. Amongst a man's inferiors one shall be sure of reverence; and therefore it is good a little to be familiar. He that is too much in anything, so that he giveth another occasion of satiety, maketh himself cheap. To apply one's self to others is good; so it be with demonstration that a man doth it upon regard, and not upon facility. It is a good precept generally in seconding another, yet to add somewhat of one's own: as if you will grant his opinion, let it be with some distinction; if you will follow his motion, let it be with condition; if you allow his counsel, let it be with alleging further reason. Men had need beware how they be too perfect in compliments; for be they never so sufficient otherwise, their enviers will be sure to give them that attribute to the disadvantage of their greater virtues. It is loss also in business to be too full of respects, or to be curious in observing times and opportunities. Salomon saith, *He that considereth the wind shall not sow, and he that looketh to the clouds shall not reap*. A wise man will make more opportunities than he finds. Men's behaviour should be like their apparel, not too strait or point device, but free for exercise or motion.

LIII. Of Praise.

Praise is the reflexion of virtue. But it is as the glass or body which giveth the reflexion. If it be from the common people, it is commonly false and naught; and rather followeth vain persons than virtuous. For the common people understand not many excellent virtues. The lowest virtues draw praise from them; the middle virtues work in them astonishment or admiration; but of the highest virtues they have no sense of perceiving at all. But shews, and *species virtutibus similes*, serve best with them. Certainly fame is like a river, that beareth up things light and swoln, and drowns things weighty and solid. But if persons of quality and judgment concur, then it is (as the Scripture saith) *Nomen bonum instar unguenti fragrantis* [a good name like unto a sweet ointment]. It filleth all round about, and will not easily away. For the odours of ointments are more durable than those of flowers. There be so many false points of praise, that a man may justly hold it a suspect. Some praises proceed merely of flattery; and if he be an ordinary flatterer, he will have certain common attributes, which may serve every man; if he be a cunning flatterer he will follow the arch-flatterer, which is a man's self; and wherein a man thinketh best of himself, therein the flatterer will uphold him most: but if he be an impudent flatterer, look wherein a man is conscious to himself that he is most defective, and is most out of countenance in himself, that will the flatterer entitle him to perforce, *spretâ conscientiâ*. Some praises come of good wishes and respects, which is a form due in civility to kings and great persons, *laudando præcipere ;* when by telling men what they are, they represent to them what they should be. Some men are praised maliciously to their hurt, thereby to stir envy and jealousy towards them ; *pessimum genus inimicorum laudantium ;* [the worst kind of enemies are they that praise] ; insomuch as it was a proverb amongst the Grecians, that *he that was praised to his hurt, should have a push rise upon his nose ;* as we say, *that a blister will rise upon one's tongue that tells a lie.* Certainly moderate praise, used with opportunity, and not vulgar, is that which doth the good. Salomon saith, *He that praiseth his friend aloud, rising early, it shall be to him no better than a curse.* Too much magnifying of man or matter doth irritate contradiction, and procure envy and scorn. To praise a man's self cannot be decent, except it be in rare cases ; but to praise a man's office or profession, he may do it with good grace, and with a kind of magnanimity. The Cardinals of Rome, which are theologues, and friars, and schoolmen, have a phrase of notable contempt and scorn towards civil business : for they call all temporal business of wars, embassages, judicature, and other employments, *shirrerie,* which is *under-sheriffries ;* as if they were but matters for under-sheriffs and catch-poles : though many times those under-sheriffries do more good than their high speculations. St. Paul, when he boasts of himself, he doth oft interlace, *I speak like a fool ;* but speaking of his calling, he saith, *magnificabo apostolatum meum* [I will magnify my mission].

LIV. Of Vain-Glory.

It was prettily devised of Æsop ; *the fly sat upon the axle-tree of the chariot wheel, and said, What a dust do I raise !* So are there some vain persons, that whatsoever goeth alone or moveth upon greater means, if they have never so little hand in it, they think it is they that carry it. They that are glorious must needs be factious ; for all bravery stands upon comparisons. They must needs be violent, to make good their own vaunts. Neither can they be secret, and therefore not effectual ; but according to the French proverb, *Beaucoup de bruit, peu de fruit ; Much bruit, little fruit.* Yet certainly there is use of this quality in civil affairs. Where there is an opinion and fame to be created either of virtue or greatness, these men are good trumpeters. Again, as Titus Livius noteth in the case of Antiochus and the Ætolians, *There are sometimes great effects of cross lies ;* as if a man that negotiates between two princes, to draw them to join in a war against the third, doth extol the forces of either of them above measure, the one to the other ; and sometimes he that deals between man and man, raiseth his own credit with both, by pretending greater interest than he hath in either. And in these and the like kinds, it often falls out that somewhat is produced of nothing ; for lies are sufficient to breed opinion, and opinion brings on substance. In militar [90] commanders and soldiers, vain-glory is an essential point ; for as iron sharpens iron, so by glory one courage sharpeneth another. In cases of great enterprise upon charge and adventure, a composition of glorious natures doth put life into business ; and those that are of solid and sober natures have more of the ballast than of the sail. In fame of learning, the flight will be slow without some feathers of ostentation. *Qui de contemnendâ gloriâ libros scribunt, nomen suum inscribunt* [They that write books on the worthlessness of glory, take care to put their names on the title page]. Socrates, Aristotle, Galen, were men full of ostentation. Certainly vain-glory helpeth to perpetuate a man's memory ; and virtue was never so beholding to human nature, as it received his due at the second hand. Neither had the fame of Cicero, Seneca, Plinius Secundus, borne her age so well, if it had not been joined with some vanity in themselves ; like unto varnish, that makes ceilings not only shine but last. But all this while, when I speak of vain-glory, I mean not of that property that Tacitus doth attribute to Mucianus ; *Omnium, quæ dixerat feceratque, arte quâdam ostentator* [A man that had a kind of art of setting forth to advantage

[90] So in the original. It is the form of the word which Bacon generally uses.

all that he had said or done] : for that proceeds not of vanity, but of natural magnanimity and discretion : and in some persons is not only comely, but gracious. For excusations, cessions, modesty itself well governed, are but arts of ostentation. And amongst those arts there is none better than that which Plinius Secundus speaketh of, which is to be liberal of praise and commendation to others, in that wherein a man's self hath any perfection. For saith Pliny very wittily, *In commending another you do yourself right ; for he that you commend is either superior to you in that you commend, or inferior. If he be inferior, if he be to be commended, you much more ; if he be superior, if he be not to be commended, you much less.* Glorious men are the scorn of wise men, the admiration of fools, the idols of parasites, and the slaves of their own vaunts.

LV. Of Honour and Reputation.

The winning of Honour [91] is but the revealing of a man's virtue and worth without disadvantage. For some in their actions do woo and affect honour and reputation ; which sort of men are commonly much talked of, but inwardly little admired. And some, contrariwise, darken their virtue in the shew of it ; so as they be undervalued in opinion. If a man perform that which hath not been attempted before ; or attempted and given over ; or hath been achieved, but not with so good circumstance, he shall purchase more honour than by effecting a matter of greater difficulty or virtue, wherein he is but a follower. If a man so temper his actions, as in some one of them he doth content every faction or combination of people, the music will be the fuller. A man is an ill husband of his honour, that entereth into any action, the failing wherein may disgrace him more than the carrying of it through can honour him. Honour that is gained and broken upon another hath the quickest reflexion, like diamonds cut with fascets. And therefore let a man contend to excel any competitors of his in honour, in outshooting them, if he can, in their own bow. Discreet followers and servants help much to reputation. *Omnis fama a domesticis emanat.* Envy, which is the canker of honour, is best extinguished by declaring a man's self in his ends rather to seek merit than fame ; and by attributing a man's successes rather to divine Providence and felicity, than to his own virtue or policy. The true marshalling of degrees of sovereign honour are these. In the first place are *conditores imperiorum*, founders of states and commonwealths ; such as were Romulus, Cyrus, Cæsar, Ottoman, Ismael. In the second place are *legislatores*, lawgivers ; which are also called *second founders*, or *perpetui principes*, because they govern by their ordinances after they are gone ; such were Lycurgus, Solon, Justinian, Eadgar, Alphonsus of Castile, the Wise, that made the *Siete partidas*. In the third place are *liberatores*, or *salvatores*, such as compound the long miseries of civil wars, or deliver their countries from servitude of strangers or tyrants ; as Augustus Cæsar, Vespasianus, Aurelianus, Theodoricus, King Henry the Seventh of England, King Henry the Fourth of France. In the fourth place are *propagatores* or *propugnatores imperii ;* such as in honourable wars enlarge their territories, or make noble defence against invaders. And in the last place are *patres patriæ* [fathers of their country] ; which reign justly, and make the times good wherein they live. Both which last kinds need no examples, they are in such numbers. Degrees of honour in subjects are, first, *participes curarum*, those upon whom princes do discharge the greatest weight of their affairs ; their *right hands*, as we call them. The next are *duces belli*, great leaders ; such as are princes' lieutenants, and do them notable services in the wars. The third are *gratiosi*, favourites ; such as exceed not this scantling, to be solace to the sovereign, and harmless to the people. And the fourth, *negotiis pares ;* such as have great places under princes, and execute their places with sufficiency. There is an honour, likewise, which may be ranked amongst the greatest, which happeneth rarely ; that is, of such as sacrifice themselves to death or danger for the good of their country ; as was M. Regulus, and the two Decii.

[91] *Honoris et existimationis vera et jure optimo acquisitio ea est, ut quis*, etc. Harl. MS. 5106 (referred to above, p. 760) has "The true winning of honour :" which is probably the true reading.

LVI. Of Judicature.

Judges ought to remember that their office is *jus dicere*, and not *jus dare* ; to interpret law, and not to make law, or give law. Else will it be like the authority claimed by the church of Rome, which under pretext of exposition of Scripture doth not stick to add and alter ; and to pronounce that which they do not find ; and by shew of antiquity to introduce novelty. Judges ought to be more learned than witty, more reverend than plausible, and more advised than confident. Above all things, integrity is their portion and proper virtue. *Cursed* (saith the law) *is he that removeth the landmark*. The mislayer of a mere-stone is to blame. But it is the unjust judge that is the capital remover of landmarks, when he defineth amiss of lands and property. One foul sentence doth more hurt than many foul examples. For these do but corrupt the stream, the other corrupteth the fountain. So saith Salomon, *Fons turbatus, et vena corrupta, est justus cadens in causâ suâ coram adversario* [A righteous man falling down before the wicked is as a troubled fountain or a corrupt spring]. The office of judges may have reference unto the parties that sue, unto the advocates that plead, unto the clerks and ministers of justice underneath them, and to the sovereign or state above them.

First, for the causes or parties that sue. *There be* (saith the Scripture) *that turn judgment into wormwood* ; and surely there be also that turn it into vinegar ; for injustice maketh it bitter, and delays make it sour. The principal duty of a judge is to suppress force and fraud ; whereof force is the more pernicious when it is open, and fraud when it is close and disguised. Add thereto contentious suits which ought to be spewed out, as the surfeit of courts. A judge ought to prepare his way to a just sentence, as God useth to prepare his way, by raising valleys, and taking down hills : so when there appeareth on either side an high hand, violent prosecution, cunning advantages taken, combination, power, great counsel, then is the virtue of a judge seen, to make inequality equal ; that he may plant his judgment as upon an even ground. *Qui fortiter emungit, elicit sanguinem* [Violent blowing makes the nose bleed] ; and where the wine-press is hard wrought, it yields a harsh wine, that tastes of the grape-stone. Judges must beware of hard constructions and strained inferences ; for there is no worse torture than the torture of laws. Specially in case of laws penal, they ought to have care that that which was meant for terror be not turned into rigour ; and that they bring not upon the people that shower whereof the Scripture speaketh, *Pluet super eos laqueos* ; for penal laws pressed are a *shower of snares* upon the people. Therefore let penal laws, if they have been sleepers of long, or if they be grown unfit for the present time, be by wise judges confined in the execution : *Judicis officium est, ut res, ita tempora rerum, etc.* [A judge must have regard to the time as well as to the matter]. In causes of life and death, judges ought (as far as the law permitteth) in justice to remember mercy ; and to cast a severe eye upon the example, but a merciful eye upon the person.

Secondly, for the advocates and counsel that plead. Patience and gravity of hearing is an essential part of justice ; and an overspeaking judge is no well-tuned cymbal. It is no grace to a judge first to find that which he might have heard in due time from the bar ; or to show quickness of conceit in cutting off evidence or counsel too short ; or to prevent information by questions, though pertinent. The parts of a judge in hearing are four : to direct the evidence ; to moderate length, repetition, or impertinency of speech ; to recapitulate, select, and collate the material points of that which hath been said ; and to give the rule or sentence. Whatsoever is above these is too much ; and proceedeth either of glory and willingness to speak, or of impatience to hear, or of shortness of memory, or of want of a staid and equal attention. It is a strange thing to see that the boldness of advocates should prevail with judges ; whereas they should imitate God, in whose seat they sit ; who *represseth the presumptuous, and giveth grace to the modest*. But it is more strange, that judges should have noted favourites ; which cannot but cause multiplication of fees, and suspicion of bye-ways. There is due from the judge to the advocate some commendation and gracing, where causes are well handled and fair pleaded ; especially towards the side which obtaineth not ; for that upholds in the client the reputation of his counsel, and beats down in him the conceit of his cause. There is likewise due to the public a civil reprehension of advocates, where there appeareth cunning counsel, gross neglect, slight information, indiscreet pressing, or an over-bold defence. And let not the counsel at the bar chop with the judge, nor wind himself into the handling of the cause anew after the judge hath declared his sentence; but on the other side, let not the judge meet the cause half way, nor give occasion for the party to say his counsel or proofs were not heard.

Thirdly, for that that concerns clerks and ministers. The place of justice is an hallowed place; and therefore not only the bench, but the foot-pace and precincts and purprise thereof, ought to be preserved without scandal and corruption. For certainly *Grapes* (as the Scripture saith) *will not be gathered of thorns or thistles*; neither can justice yield her fruit with sweetness amongst the briars and brambles of catching and polling clerks and ministers. The attendance of courts is subject to four bad instruments. First, certain persons that are sowers of suits; which make the court swell, and the country pine. The second sort is of those that engage courts in quarrels of jurisdiction, and are not truly *amici curiæ*, but *parasiti curiæ*, in puffing a court up beyond her bounds, for their own scraps and advantage. The third sort is of those that may be accounted the left hands of courts; persons that are full of nimble and sinister tricks and shifts, whereby they pervert the plain and direct courses of courts, and bring justice into oblique lines and labyrinths. And the fourth is the poller and exacter of fees; which justifies the common resemblance of the courts of justice to the bush whereunto while the sheep flies for defence in weather, he is sure to lose part of his fleece. On the other side, an ancient clerk, skilful in precedents, wary in proceeding, and understanding in the business of the court, is an excellent finger of a court; and doth many times point the way to the judge himself.

Fourthly, for that which may concern the sovereign and estate. Judges ought above all to remember the conclusion of the Roman Twelve Tables; *Salus populi suprema lex* [The supreme law of all is the weal of the people]; and to know that laws, except they be in order to that end, are but things captious, and oracles not well inspired. Therefore it is an happy thing in a state when kings and states do often consult with judges; and again when judges do often consult with the king and state: the one, when there is matter of law, intervenient in business of state; the other, when there is some consideration of state intervenient in matter of law. For many times the things deduced to judgment may be *meum* and *tuum*, when the reason and consequence thereof may trench to point of estate: I call matter of estate, not only the parts of sovereignty, but whatsoever introduceth any great alteration or dangerous precedent; or concerneth manifestly any great portion of people. And let no man weakly conceive that just laws and true policy have any antipathy; for they are like the spirits and sinews, that one moves with the other. Let judges also remember, that Salomon's throne was supported by lions on both sides: let them be lions, but yet lions under the throne; being circumspect that they do not check or oppose any points of sovereignty. Let not judges also be so ignorant of their own right, as to think there is not left to them, as a principal part of their office, a wise use and application of laws. For they may remember what the apostle saith of a greater law than theirs; *Nos scimus quia lex bona est, modo quis ea utatur legitime* [We know that the law is good, if a man use it lawfully].

LVII. Of Anger.

To seek to extinguish Anger utterly is but a bravery of the Stoics. We have better oracles: *Be angry, but sin not. Let not the sun go down upon your anger.* Anger must be limited and confined both in race and in time. We will first speak how the natural inclination and habit to be angry may be attempered and calmed. Secondly, how the particular motions of anger may be repressed, or at least refrained from doing mischief. Thirdly, how to raise anger or appease anger in another.

For the first; there is no other way but to meditate and ruminate well upon the effects of anger, how it troubles man's life. And the best time to do this, is to look back upon anger when the fit is thoroughly over. Seneca saith well, *That anger is like ruin, which breaks itself upon that it falls.* The Scripture exhorteth us *To possess our souls in patience.* Whosoever is out of patience, is out of possession of his soul. Men must not turn bees

. . . animasque in vulnere ponunt

[that put their lives in the sting].

Anger is certainly a kind of baseness; as it appears well in the weakness of those subjects in whom it reigns; children, women, old folks, sick folks. Only men must beware that they carry their anger rather with scorn than with fear; so that they may seem rather to be above the injury than below it; which is a thing easily done, if a man will give law to himself in it.

For the second point; the causes and motives of anger are chiefly three. First, to be too sensible of hurt; for no man is angry that feels not himself hurt; and therefore tender and delicate persons must needs be oft angry; they have so many things to trouble them, which more robust natures have little sense of. The next is, the apprehension and construction of the injury offered to be, in the circumstances thereof, full of contempt: for contempt is that which putteth an edge upon anger, as much or more than the hurt itself. And therefore when men are ingenious in picking out circumstances of contempt, they do kindle their anger much. Lastly, opinion of the touch of a man's reputation doth multiply and sharpen anger. Wherein the remedy is, that a man should have, as Consalvo was wont to say, *telam honoris crassiorem* [an honour of a stouter web]. But in all refrainings of anger, it is the best remedy to win time; and to make a man's self believe, that the opportunity of his revenge is not yet come, but that he foresees a time for it; and so to still himself in the mean time, and reserve it.

To contain anger from mischief, though it take hold of a man, there be two things whereof you must have special caution. The one, of extreme bitterness of words, especially if they be aculeate and proper; for *communia maledicta* are nothing so much; and again, that in anger a man reveal no secrets; for that makes him not fit for society. The other, that you do not peremptorily break off, in any business, in a fit of anger; but howsoever you shew bitterness, do not act anything that is not revocable.

For raising and appeasing anger in another; it is done chiefly by choosing of times, when men are forwardest and worst disposed, to incense them. Again, by gathering (as was touched before) all that you can find out to aggravate the contempt. And the two remedies are by the contraries. The former to take good times, when first to relate to a man an angry business; for the first impression is much; and the other is, to sever, as much as may be, the construction of the injury from the point of contempt; imputing it to misunderstanding, fear, passion, or what you will.

LVIII. Of Vicissitude of Things.

Salomon saith, *There is no new thing upon the earth.* So that as Plato had an imagination, *That all knowledge was but remembrance;* so Salomon giveth his sentence, *That all novelty is but oblivion.* Whereby you may see that the river of Lethe runneth as well above ground as below. There is an abstruse astrologer that saith, *if it were not for two things that are constant* (*the one is, that the fixed stars ever stand at like distance one from another, and never come nearer together, nor go further asunder; the other that the diurnal motion perpetually keepeth time*), *no individual would last one moment.* Certain it is, that the matter is in a perpetual flux, and never at a stay. The great winding-sheets, that bury all things in oblivion, are two; deluges and earthquakes. As for conflagrations and great droughts, they do not[92] merely dispeople and destroy. Phaëton's car went but a day. And the three years' drought in the time of Elias was but particular, and

92 [This "not" appears from the context to be a misprint, though it is followed in the Latin translation. That, from its diffuseness, appears in many places not to be by Bacon. The edition of 1838 rightly drops the "not." The Bohn edition has "not merely dispeople, but destroy," which is no improvement.—Ed.]

left people alive. As for the great burnings by lightnings, which are often in the West Indies, they are but narrow. But in the other two destructions, by deluge and earthquake, it is further to be noted, that the remnant of people which hap to be reserved, are commonly ignorant and mountainous people, that can give no account of the time past; so that the oblivion is all one as if none had been left. If you consider well of the people of the West Indies, it is very probable that they are a newer or a younger people than the people of the old world. And it is much more likely that the destruction that hath heretofore been there, was not by earthquakes (as the Ægyptian priest told Solon concerning the island of Atlantis, *that it was swallowed by an earthquake*), but rather that it was desolated by a particular deluge. For earthquakes are seldom in those parts. But on the other side, they have such pouring rivers, as the rivers of Asia and Africk and Europe are but brooks to them. Their Andes likewise, or mountains, are far higher than those with us; whereby it seems that the remnants of generation of men were in such a particular deluge saved. As for the observation that Machiavel hath, that the jealousy of sects doth much extinguish the memory of things; traducing Gregory the Great, that he did what in him lay to extinguish all heathen antiquities; I do not find that those zeals do any great effects, nor last long; as it appeared in the succession of Sabinian, who did revive the former antiquities.

The vicissitude or mutations in the Superior Globe are no fit matter for this present argument. It may be, Plato's great year, if the world should last so long, would have some effect; not in renewing the state of like individuals (for that is the fume of those that conceive the celestial bodies have more accurate influences upon these things below than indeed they have), but in gross. Comets, out of question, have likewise power and effect over the gross and mass of things; but they are rather gazed upon, and waited upon in their journey, than wisely observed in their effects; specially in their respective effects; that is, what kind of comet, for magnitude, colour, version of the beams, placing in the region of heaven, or lasting, produceth what kind of effects.

There is a toy which I have heard, and I would not have it given over, but waited upon a little. They say it is observed in the Low Countries (I know not in what part) that every five and thirty years the same kind and suit of years and weathers comes about again; as great frosts, great wet, great droughts, warm winters, summers with little heat, and the like; and they call it the *Prime*. It is a thing I do the rather mention, because, computing backwards, I have found some concurrence.

But to leave these points of nature, and to come to men. The greatest vicissitude of things amongst men, is the vicissitude of sects and religions. For those orbs rule in men's minds most. The true religion is *built upon the rock*; the rest are tossed upon the waves of time. To speak therefore of the causes of new sects; and to give some counsel concerning them, as far as the weakness of human judgment can give stay to so great revolutions.

When the religion formerly received is rent by discords; and when the holiness of the professors of religion is decayed and full of scandal; and withal the times be stupid, ignorant, and barbarous; you may doubt the springing up of a new sect; if then also there should arise any extravagant and strange spirit to make himself author thereof. All which points held when Mahomet published his law. If a new sect have not two properties, fear it not; for it will not spread. The one is, the supplanting or the opposing of authority established; for nothing is more popular than that. The other is, the giving licence to pleasures and a voluptuous life. For as for speculative heresies (such as were in ancient times the Arians, and now the Arminians), though they work mightily upon men's wits, yet they do not produce any great alterations in states; except it be by the help of civil occasions. There be three manner of plantations of new sects. By the power of signs and miracles; by the eloquence and wisdom of speech and persuasion; and by the sword. For martyrdoms, I reckon them amongst miracles; because they seem to exceed the strength of human nature: and I may do the like of superlative and admirable holiness of life. Surely there is no better way to stop the rising of new sects and schisms, than to reform abuses; to compound the smaller differences; to proceed mildly, and not with sanguinary persecutions; and rather to take off the principal authors by winning and advancing them, than to enrage them by violence and bitterness.

The changes and vicissitude in wars are many ; but chiefly in three things ; in the seats or stages of the war ; in the weapons ; and in the manner of the conduct. Wars, in ancient time, seemed more to move from east to west ; for the Persians, Assyrians, Arabians, Tartars (which were the invaders), were all eastern people. It is true, the Gauls were western ; but we read but of two incursions of theirs : the one to Gallo-Græcia, the other to Rome. But East and West have no certain points of heaven ; and no more have the wars, either from the east or west, any certainty of observation. But North and South are fixed ; and it hath seldom or never been seen that the far southern people have invaded the northern, but contrariwise. Whereby it is manifest that the northern tract of the world is in nature the more martial region : be it in respect of the stars of that hemisphere ; or of the great continents that are upon the north, whereas the south part, for aught that is known, is almost all sea ; or (which is most apparent) of the cold of the northern parts, which is that which, without aid of discipline, doth make the bodies hardest, and the courages warmest.

Upon the breaking and shivering of a great state and empire, you may be sure to have wars. For great empires, while they stand, do enervate and destroy the forces of the natives which they have subdued, resting upon their own protecting forces ; and then when they fail also, all goes to ruin, and they become a prey. So was it in the decay of the Roman empire ; and likewise in the empire of Almaigne, after Charles the Great, every bird taking a feather ; and were not unlike to befal to Spain, if it should break. The great accessions and unions of kingdoms do likewise stir up wars : for when a state grows to an over-power, it is like a great flood, that will be sure to overflow. As it hath been seen in the states of Rome, Turkey, Spain, and others. Look when the world hath fewest barbarous peoples, but such as commonly will not marry or generate, except they know means to live (as it is almost every where at this day, except Tartary), there is no danger of inundations of people : but when there be great shoals of people, which go on to populate, without foreseeing means of life and sustentation, it is of necessity that once in an age or two they discharge a portion of their people upon other nations ; which the ancient northern people were wont to do by lot ; casting lots what part should stay at home, and what should seek their fortunes. When a warlike state grows soft and effeminate, they may be sure of a war. For commonly such states are grown rich in the time of their degenerating ; and so the prey inviteth, and their decay in valour encourageth a war.

As for the weapons, it hardly falleth under rule and observation : yet we see even they have returns and vicissitudes. For certain it is, that ordnance was known in the city of the Oxidrakes in India ; and was that which the Macedonians called thunder and lightning, and magic. And it is well known that the use of ordnance hath been in China above two thousand years. The conditions of weapons, and their improvement, are, First, the fetching afar off ; for that outruns the danger ; as it is seen in ordnance and muskets. Secondly, the strength of the percussion ; wherein likewise ordnance do exceed all arietations and ancient inventions. The third is, the commodious use of them ; as that they may serve in all weathers ; that the carriage may be light and manageable ; and the like.

For the conduct of the war : at the first, men rested extremely upon number : they did put the wars likewise upon main force and valour ; pointing days for pitched fields, and so trying it out upon an even match : and they were more ignorant in ranging and arraying their battles. After they grew to rest upon number rather competent than vast ; they grew to [93] advantages of place, cunning diversions, and the like : and they grew more skilful in the ordering of their battles.

In the youth of a state, arms do flourish ; in the middle age of a state, learning ; and then both of them together for a time ; in the declining age of a state, mechanical arts and merchandise. Learning hath his infancy, when it is but begin-

[93] So in original. A word appears to have dropped out, such as *seek*, or something equivalent. The translation has *captabant*.

ning and almost childish: then his youth, when it is luxuriant and juvenile: then his strength of years, when it is solid and reduced: and lastly, his old age, when it waxeth dry and exhaust. But it is not good to look too long upon these turning wheels of vicissitude, lest we become giddy. As for the philology of them, that is but a circle of tales, and therefore not fit for this writing.

[NOTE.—In speaking of the original edition, I have referred to a copy in my own possession; from which the title is copied. I have since found that there is a copy in the British Museum bearing the same date, but not in all respects the same. In the title-page, instead of *newly enlarged*, it has *newly written*. It professes to be "printed by John Haviland, for Hanna Barret," omitting the name of Richard Whittaker, and the words which follow. In the text, it is difficult even on a careful examination to detect any differences whatever. But upon referring to the passages in which I had noticed an error, or a doubt, or a variety of reading, I find that in three of them it differs from my copy. In p. 760 it has *children* not *child*: in p. 767 *flower* not *flowers*: in p. 785 *game* not *gaine*. One or two other variations which occur in the later essays I have noticed in their places. Of these copies, one must certainly have been a proof in which corrections were afterwards made. And the fact that all the later editions have "newly enlarged" in the title page, instead of "newly written," favours the supposition that mine is the corrected copy. That in some cases (as for instance in pages 767 and 785) the reading of the other copy is unquestionably the right one, may possibly be explained by accidents of the press. The last letter in *flowers* may have failed to take the ink; the *m* in *game* may have been injured, and being mistaken for an imperfect *in* may have been replaced by a perfect *in*.—*J.S.*]

APPENDIX TO THE ESSAYS.

I.

A FRAGMENT OF AN ESSAY ON FAME [94]

THE poets make Fame a monster. They describe her in part finely and elegantly; and in part gravely and sententiously. They say, look how many feathers she hath, so many eyes she hath underneath; so many tongues; so many voices; she pricks up so many ears.

This is a flourish. There follow excellent parables; as that she gathereth strength in going: that she goeth upon the ground, and yet hideth her head in the clouds: that in the day-time she sitteth in a watch tower, and flieth most by night: that she mingleth things done with things not done: and that she is a terror to great cities. But that which passeth all the rest is; they do recount that the Earth, mother of the Giants that made war against Jupiter and were by him destroyed, thereupon in an anger brought forth Fame; for certain it is that rebels, figured by the giants, and seditious fames and libels, are but brothers and sisters; masculine and feminine. But now, if a man can tame this monster, and bring her to feed at the hand, and govern her, and with her fly other ravening fowl and kill them, it is somewhat worth. But we are infected with the stile of the poets. To speak now in a sad and a serious manner. There is not in all the politics a place less handled, and more worthy to be handled, than this of fame. We will therefore speak of these points. What are false fames; and what are true fames; and how they may be best discerned; how fames may be sown and raised; how they may be spread and multiplied; and how they may be checked and laid dead. And other things concerning the nature of fame. Fame is of that force, as there is scarcely any great action wherein it hath not a great part; especially in the war. Mucianus undid Vitellius, by a fame that he scattered,

[94] This fragment was first published by Dr. Rawley, in the *Resuscitatio* (1657), p. 281 Though unfinished, therefore, it may be regarded as a genuine and undoubted work of Bacon's, as far as it goes. Two other Essays, which have been ascribed to Bacon upon very doubtful authority (and at least one of them in my opinion very improbably), will be printed by themselves at the end of this Appendix.

that Vitellius had in purpose to remove the legions of Syria into Germany, and the legions of Germany into Syria; whereupon the legions of Syria were infinitely inflamed. Julius Cæsar took Pompey unprovided, and laid asleep his industry and preparations, by a fame that he cunningly gave out, how Cæsar's own soldiers loved him not; and being wearied with wars, and laden with the spoils of Gaul, would forsake him as soon as he came into Italy. Livia settled all things for the succession of her son Tiberius, by continual giving out that her husband Augustus was upon recovery and amendment. And it is an usual thing with the Bashaws to conceal the death of the great Turk from the Janizaries and men of war, to save the sacking of Constantinople and other towns, as their manner is. Themistocles made Xerxes King of Persia post apace out of Græcia, by giving out that the Grecians had a purpose to break his bridge of ships which he had made athwart Hellespont. There be a thousand such like examples, and the more they are, the less they need to be repeated; because a man meeteth with them every where. Therefore let all wise governors have as great a watch and care over fames, as they have of the actions and designs themselves.

The rest was not finished.

II.

ESSAYS ATTRIBUTED TO BACON WITHOUT AUTHORITY.

[At the end of the *Resuscitatio* (published in 1657) Dr. Rawley gives what he entitles "A perfect list of his Lordship's true works both in English and Latin"; which he concludes with these words: "as for other pamphlets, whereof there are several put forth under his Lordship's name, they are not to be owned for his".

Any work therefore (not contained in this list) which had appeared before 1657 in any publication which Dr. Rawley knew of, and had been there ascribed to Bacon, must be regarded as distinctly denied by him to be Bacon's.

Now in December 1642, in which year several of Bacon's smaller political pieces were published in separate pamphlets without any editor's name or any account of the source from which they were taken, there appeared among others a 4to of eight pages with the following title: *An Essay of a King, with an explanation what manner of persons those should be that are to execute the power or ordinance of the King's Prerogative. Written by the Right Honourable Francis, Lord Verulam Viscount Saint Alban. December 2. London, Printed for Richard Best*, 1642.

In 1648 appeared a 4to volume of 103 pages, entitled *The Remaines of the Right Honorable Francis, Lord Verulam, Viscount of St. Albanes, sometimes Lord Chancellour of England; being Essayes and severall letters to severall great Personages, and other pieces of various high concernment not heretofore published. A table whereof for the reader's more ease is adjoyned.—London, printed by B. Alsop for Laurence Chapman and are to be sold at his shop neer the Savoy in the Strand*, 1648.

Most of the pieces in the volume are genuine, and were afterwards published by Rawley from the originals. And it is probably to this collection that he alludes, when he alleges as a reason for publishing some things which Bacon himself did not design for publication, that "through the loose keeping of his Lordship's papers whilst he lived, divers surreptitious copies have been taken, which have since employed the press with sundry corrupt and mangled editions; Whereby nothing hath been more difficult than to find the Lord Saint Alban in the Lord Saint Alban; and which have presented (some of them) rather a fardle of nonsense, than any true expression of his Lordship's happy vein"; and that therefore he "thought himself in a sort tied to vindicate those injuries and wrongs done to his Lordship's pen; and at once, by setting forth the true and genuine writings themselves, to prevent the like invasions for the time to come". But whatever the publications may have been to which he alluded, it is hardly conceivable that the existence of this volume was unknown to him; and we must therefore regard all those pieces which it contains, and which are not directly or by implication contained in his own "perfect list", as included in his general repudiation. It does not indeed follow that none of them are genuine; because Rawley may have been mistaken; but that every such piece was *in his opinion* spurious, can hardly be disputed: and he had such very good means of judging, that his opinion is not to be set aside except upon very strong evidence.

Now the two first pieces in the "Remains" are the contents of the pamphlet of which I have quoted the title. Standing where they do, they could not have been overlooked: yet neither of them is to be found in any of the publications cited in Rawley's "perfect list". The inevitable inference is, that Rawley did not believe them to be the work of Bacon; and certainly in this case there is no evidence internal or external which can justify us in overruling his judgment. The *Essay of a King* does indeed contain several sentences which are much in Bacon's manner, and which might have been written by him. But the total composition does not read like his; and even if the external evidences had been equally balanced (which is by no means the case; for the fact that *somebody* thought it was Bacon's cannot be taken as a counterpoise to the fact that *Rawley* thought it was not), I should myself have been inclined, upon consideration of the internal evidence alone, to reject it.

The other piece is still less like Bacon's work. Mr. Heath*, finding it printed among his writings, and knowing nothing of its history, was at once led to doubt its genuineness, from a consideration of the matter and opinions as well as the style.

Passing over a little piece entitled *Short Notes for Civil Conversation* (the claims of which to a place among Bacon's writings have other evidence to support them), we come next to a very remarkable composition—*An Essay on Death.* This stands fourth in the volume, and being also too conspicuous to have been overlooked must be regarded as disclaimed by Dr. Rawley. I do not know whether it had been printed before. It is an eloquent and touching composition, very peculiar in style, and marked with a "humorous sadness" which reminds me of nobody so much as Sir Thomas Browne. Sir Thomas Browne was born in 1605, and therefore there is nothing in the date to preclude the supposition that he was the author of it. How far his never having claimed it is to be taken as an objection, or what other difficulties the supposition may involve, I am not well enough acquainted with his biography to judge. But whoever may have written it, I am fully convinced that Bacon did not. Nothing is less probable than that he would have written so grave a thing on so grave a subject merely as an exercise in imitating another man's style; and the style is so unlike his own, that if we suppose him the author of it we must suppose no less. And the only reason we have for imputing it to him is, that within twenty-four years after his death there was *somebody* or other who thought it was his; against which must be set the fact that Rawley thought it was not.—*J.S.*]

AN ESSAY OF A KING,

Written by Sir Francis Bacon.

1. A KING is a mortal God on Earth, unto whom the living God hath lent his own name as a great honour: But withal told him he should die like a man, lest he should be proud and flatter himself, that God hath with his name imparted unto him his nature also.

2. Of all kinds of men, God is least beholding unto them, for he doth most for them, and they do ordinarily least for him.

3. A King that would not feele his Crown too heavy for him, must weare it every day, but if he think it too light, he knoweth not of what mettall it is made of.

* [Who edited the legal writings in the Ellis-Spedding edition of Bacon's Works —ED.]

4. He must make Religion the Rule of government, and not the Scale[1]; for he that casteth in Religion onely to make the scales even, his own weight is contained in these Characters, *Tekel uphrasin*, he is found too light, his Kingdom shall be taken from him.

5. And that King that holds not Religion the best reason of state, is void of all piety and justice, the Supporters of a King.

6. He must be able to give Counsell himself, but not to relye thereupon; for though happy events justifie their Counsells, yet it is better that the evill event of good advice be rather imputed to a Subject then a Sovereigne.

7. He is the Fountain of Honour, which should not run with a wast pipe, lest the Courtiers sell the waters, and then (as papists say of their holy Wels) to lose the vertue.

8. He is the life of the Law, not onely as he is *lex loquens* himself, but because he animateth the dead letter, making it active towards all his Subjects *præmio et pœna*.

9. A wise King must doe lesse in altering his Laws, than he may; for new government is ever dangerous, it being true in the body politique, as in the corporall, that *omnis subita mutatio est periculosa*, and though it be for the better, yet it is not without a fearfull apprehension; For he that changeth the fundamentall Laws of a Kingdome, thinketh there is no good title to a Crown but by conquest.

10. A King that setteth to sale Seats of Justice, oppresseth the People; for he teacheth his Judges to sell justice, and *pretio parata pretio venditur Justitia*.

11. Bounty and Magnificence are vertues *vere regiæ*, but a prodigall King is neerer a Tyrant then a percimonious: for store at home draweth his contemplations abroad: but want supplieth itself of what is next, and many times the next way, and herein he must be wise, and know what he may justly doe.

12. That King which is not feared, is not loved, and he that is well seen in his craft, must as well study to be feared as loved, yet not loved for feare, but feared for love.

13. Therefore as hee must alwayes resemble him whose great name he beareth, and that in manifesting the sweet influence of his mercy on the severe stroke of his Justice sometimes, so in this not to suffer a man of death to live, for besides that the Land doth mourn, the restraint of Justice towards sin doth more retard the affection of love, than the extent of mercy doth inflame it; and sure where love is bestowed[2], feare is quite lost.

14. His greatest Enemies are his Flatterers, for though they ever speak on his side, yet their words still make against him.[3]

15. The love which a King oweth to the weal-publike, should not be restrained to any one particular, yet that his more speciall favour do reflect upon some worthy ones, is somwhat necessary, because there are so few of that capacity.

16. Hee must have a speciall care of five things, if hee would not have his Crown to be put upon him.[4]

First, that *simulata sanctitas* be not in the Church, for that is *duplex iniquitas*.

Secondly, that *inutilis æquitas* sit not in the Chancery, for that is *inepta misericordia*.

Thirdly, that *utilis iniquitas* keep not the Exchequer, for that is *crudele latrocinium*.

Fourthly, that *fidelis temeritas* be not his Generall, for that will bring but *seram pœnitentiam*.

Fiftly, that *infidelis prudentia* be not his Secretary, for that he is *Anguis sub viridi herba*.

To conclude, as hee is of the greatest power, so hee is subject to the greatest cares, made the servant of his people, or else he were without a calling at all.

He then that honoureth him not, is next an Atheist, wanting the feare of God in his heart.

[1] not to Ballance the Scale. *Remains*.

[2] So in the original, and in the *Remains* also.

[3] So in the *Remains*. The original has "against them".

[4] So in the original. The *Remains* gives "to be put on him *In felix felicitatis*". Modern editions substitute, correctly perhaps, "to be but to him *infelix felicitas*".

An explanation what manner of persons those should be, that are to execute the power or Ordinance of the Kings Prerogative, written by the said Sir Francis Bacon, *late Lord Chancellour, and Lord St.* Albans.

THAT absolute Prerogative according to the Kings pleasure revealed by his Lawes, may be exercised and executed by any Subject, to whom power may be given by the King, in any place of Judgement or Commission, which the King by his Law hath ordained, in which the Judge-subordinate cannot wrong the people, the Law laying downe a measure by which every Judge should governe or execute ; Against which Law if any Judge proceed, he is by the Law questionable and punishable for his transgression.

In this nature are all the Judges and Commissioners of the Land no otherwise then in their Courts, in which the King in person is supposed to sit, who cannot make [5] that trespasse, Felony or treason which the Law hath not made so to be, neither can punish the guilty by other punishment then the Law hath appointed.

This Prerogative or power as it is over all the Subjects, so being known by the Subjects, they are without excuse if they offend ; and suffer no wrong, if they be punished. And by this prerogative the King governeth all sorts of people according unto knowne will.

The absolute prerogative which is in Kings according to their private will and judgement cannot be executed by any Subject, neither is it possible to give such power by Commission, or fit to subject the people to the same. For the King in that he is the substitute of God, immediatly the Father of his people, and head of the Commonwealth, hath [6] by participation with God and his subjects, Discretion, Judgement, and feeling love towards those over whom he raigneth only proper to himselfe, or to his places and person, who seeing he cannot in any others diffuse his wisedome, power, or gifts, which God in respect of his place and charge hath enabled him withall, can neither subordinate any other Judge to governe by that knowledge, which the King can no otherwise then by his knowne will participate unto him. And if any subordinate Judge shall obtaine Commission according to the discretion [7] of such Judge to govern the people, that Judge is bound to think that to be his sound discretion, which [8] the law in which the Kings known will sheweth unto him [9] to be that Justice which hee ought to administer : otherwise he might seeme to esteeme himselfe above the Kings law, who will not governe by him, or to have a power derived from other then from the King, which in the Kingdome will administer Justice contrarie to the justice of the Land. Neither can such a Judge or Commissioner under the name of his high Authoritie shrowde his owne high affection, seeing the Conscience and discretion of every man is particular and private to himselfe ; As the discretion of the Judge cannot be properly or possibly the discretion of the King, or conscience of the King ; And if not his discretion, neither the Judgement that is ruled by another mans only. Therefore it may seeme they rather desire to bee Kings then to rule the people under the King, which will not administer Justice by law, but by their owne wills.

This Administration in a subject is derogative to the Kings Prerogative, for he administreth Justice out of a private direction, being not capable of a generall direction, how to use the Kings pleasure in Causes of particular respect, which if another then the King himselfe can doe, how can it be so, that any man should desire that which is unfit and impossible, but that it must proceed out of some exorbitant affection, the rather seeing such places to be full of trouble, and being altogether unnecessary, no man will seeke to thrust himselfe into it, but for hope of gaine. Then is not any prerogative oppugned but maintained, though it be desired that every subordinate Magistrate may not be made supreame, whereby he may seale up the hearts of the people, take from the King the respect due unto him only, or to judge the people otherwise then the King doth himselfe.

[5] So *Remains*. The original has " worke".
[6] So *Remains*. The original omits " hath".
[7] So *Remains*. The words " to the discretion " are omitted in the original.
[8] So *Remains*. The original has " in which".
[9] So both copies. It should probably be " in which the king's known will is contained".

And although the Prince be not bound to render any accompt to the Law, which in person administreth it selfe[10]: Yet every subordinate Judge must render an accompt to the King by his lawes how hee hath administred Justice in his place where he is set. But if he hath power to rule by private direction, for which there is no law, how can he be questioned by a law, if in his private censure he offendeth.

Therefore it seemeth that in giving such authority the King ordaineth not subordinate Magistrates, but absolute Kings; And what doth the King leave to himselfe, who giveth so much to others as he hath himself? neither is there a greater bond to tie the subject to his Prince in particular then when he shal have recourse unto him in his person or in his power for releif of the wrongs which from private men be offered, or for reformation of the oppressions which any subordinate Magistrate shall impose upon the people: there can be no offence in the Judge, who hath power to execute according to his discretion, when the discretion of any Judge shall be thought fit to be unlimited[11]; And therefore there can be therein no reformation, whereby the King in this useth no prerogative to gaine his Subjects right. Then the subject is bound to suffer helplesse wrong, and the discontent of the people is cast upon the King, the laws being neglected, which with their equitie in all other Causes and Judgements, saving this, interpose themselves and yeeld remedy.

And to conclude, Custome cannot confirme that which is any ways unreasonable of it selfe; Wisedome will not allow that which is many wayes dangerous, and no wayes profitable; Justice will not approve that government, where it cannot be but wrong must be committed. Neither can there be any rule by which to try it, nor meanes for reformation of it.

Therefore whosoever desireth government, must seeke such as he is capable of, not such as seemeth to himselfe most easie to execute; For it appeareth that it is easie to him that knoweth not law nor justice to rule as he listeth, his will never wanting a power to it selfe: but it is safe and blamelesse both for the Judge and People, and honour to the King, that Judges bee appointed who know the Law, and that they bee limited to governe according to the Law.

An Essay on Death,

By the Lord Chancellor Bacon.[12]

I have often thought upon death, and find it the least of all evils. All that which is past is as a dream; and he that hopes or depends upon time coming, dreams waking. So much of our life as we have discovered is already dead; and all those hours which we share, even from the breasts of our mother, until we return to our grand-mother the earth, are part of our dying days; whereof even this is one, and those that succeed are of the same nature; for we die daily; and as others have given place to us, so we must in the end give way to others.

Physicians, in the name of death include all sorrow, anguish, disease, calamity, or whatsoever can fall in the life of man, either grievous or unwelcome: but these things are familiar unto us, and we suffer them every hour; therefore we die daily, and I am older since I affirmed it.

I know many wise men that fear to die; for the change is bitter, and flesh would refuse to prove it: besides, the expectation brings terror, and that exceeds the evil. But I do not believe that any man fears to be dead, but only the stroke of death: and such are my hopes, that if heaven be pleased, and nature renew but my lease for twenty-one years more, without asking longer days, I shall be strong enough to acknowledge without mourning that I was begotten mortal. Virtue walks not in the highway, though she go *per alta*; this is strength and the blood to virtue, to contemn things that be desired, and to neglect that which is feared.

Why should man be in love with his fetters, though of gold? Art thou drowned in security? Then I say thou art perfectly dead. For though thou movest, yet thy soul is buried within thee, and thy good angel either forsakes his guard or sleeps. There is nothing under heaven, saving a true friend, who cannot be

[10] So both copies. It should probably be "himself".

[11] So the original. The *Remains* has "limited". [12] *Remains*, p. 7.

counted within the number of moveables, unto which my heart doth lean. And this dear freedom hath begotten me this peace, that I mourn not for that end which must be, nor spend one wish to have one minute added to the incertain date of my years. It was no mean apprehension of Lucian, who says of Menippus, that in his travels through hell he knew not the kings of the earth from other men, but only by their louder cryings and tears: which was fostered in them through the remorseful memory of the good days they had seen, and the fruitful havings which they so unwillingly left behind them: he that was well seated, looked back at his portion, and was loth to forsake his farm; and others either minding marriages, pleasures, profit, or preferment, desired to be excused from death's banquet: they had made an appointment with earth, looking at the blessings, not the hand that enlarged them, forgetting how unclothedly they came hither, or with what naked ornaments they were arrayed.

But were we servants of the precept given, and observers of the heathens rule *memento mori*, and not become benighted with this seeming felicity, we should enjoy them as men prepared to lose, and not wind up our thoughts upon so perishing a fortune; he that is not slackly strong (as the servants of pleasure), how can he be found unready to quit the veil and false visage of his perfection? The soul having shaven off her flesh, doth then set up for herself, and contemning things that are under, shews what finger hath enforced her; for the souls of idiots are of the same piece with those of statesmen, but now and then nature is at a fault, and this good guest of ours takes soil in an unperfect body, and so is slackened from shewing her wonders; like an excellent musician, which cannot utter himself upon a defective instrument.

But see how I am swarved, and lose my course, touching at the soul, that doth least hold action with death, who hath the surest property in this frail act; his stile is the end of all flesh, and the beginning of incorruption.

This ruler of monuments leads men for the most part out of this world with their heels forward, in token that he is contrary to life; which being obtained, sends men headlong into this wretched theatre, where being arrived, their first language is that of mourning. Nor in my own thoughts can I compare men more fitly to any thing than to the Indian fig-tree, which being ripened to his full height, is said to decline his branches down to the earth; whereof she conceives again, and they become roots in their own stock. So man having derived his being from the earth, first lives the life of a tree, drawing his nourishment as a plant; and made ripe for death he tends downwards, and is sowed again in his mother the earth, where he perisheth not, but expects a quickening.

So we see death exempts a man not from being, but only presents an alteration; yet there are some men, I think, that stand otherwise persuaded. Death finds not a worse friend than an alderman, to whose door I never knew him welcome; but he is an importunate guest, and will not be said nay.

And though they themselves shall affirm that they are not within, yet the answer will not be taken; and that which heightens their fear is, that they know they are in danger to forfeit their flesh, but are not wise of the payment day: which sickly uncertainty is the occasion that (for the most part) they step out of this world unfurnished for their general account, and being all unprovided, desire yet to hold their gravity, preparing their souls to answer in scarlet.

Thus I gather that death is unagreeable to most citizens, because they commonly die intestate: this being a rule, that when their will is made, they think themselves nearer a grave than before: now they, out of the wisdom of thousands, think to scare destiny, from which there is no appeal, by not making a will, or to live longer by protestation of their unwillingness to die. They are for the most part well made in this world (accounting their treasure by legions, as men do devils), their fortune looks toward them, and they are willing to anchor at it, and desire (if it be possible) to put the evil day far off from them, and to adjourn their ungrateful and killing period. No, these are not the men which have bespoken death, or whose looks are assured to entertain a thought of him.

Death arrives gracious only to such as sit in darkness, or lie heavy burdened with grief and irons; to the poor Christian, that sits bound in the galley; to despairful widows, pensive prisoners, and deposed kings: to them whose fortune runs back, and whose spirit mutinies ; unto such death is a redeemer, and the grave a place for retiredness and rest.

These wait upon the shore of death, and waft unto him to draw near, wishing above all others to see his star, that they might be led to his place, wooing the remorseless sisters to wind down the watch of their life, and to break them off before the hour.

But death is a doleful messenger to an usurer, and fate untimely cuts their thread : for it is never mentioned by him but when rumours of war and civil tumults put him in mind thereof.

And when many hands are armed, and the peace of a city in disorder, and the foot of the common soldiers sounds an alarm on his stairs, then perhaps such a one, (broken in thoughts of his moneys abroad, and cursing the monuments of coin which are in his house), can be content to think of death, and (being hasty of perdition) will perhaps hang himself, lest his throat should be cut ; provided that he may do it in his study, surrounded with wealth, to which his eye sends a faint and languishing salute, even upon the turning off ; remembering always, that he have time and liberty, by writing, to depute himself as his own heir. For that is a great peace to his end, and reconciles him wonderfully upon the point.

Herein we all dally with ourselves, and are without proof of necessity.[13] I am not of those that dare promise to pine away myself in vain-glory, and I hold such to be but feat boldness, and them that dare commit it to be vain. Yet for my part, I think nature should do me great wrong, if I should be so long in dying, as I was in being born.[14]

To speak truth, no man knows the lists of his own patience ; nor can divine how able he shall be in his sufferings, till the storm come (the perfectest virtue being tried in action), but I would (out of a care to do the best business well) ever keep a guard, and stand upon keeping faith and a good conscience.

And if wishes might find place, I would die together, and not my mind often, and my body once ; that is, I would prepare for the messengers of death, sickness and affliction, and not wait long, or be tempted by the violence of pain. Herein I do not profess myself a Stoic, to hold grief no evil, but opinion, and a thing indifferent.

But I consent with Cæsar, that the suddenest passage is easiest, and there is nothing more awakens our resolve and readiness to die, than the quieted conscience, strengthened with opinion that we shall be well spoken of upon earth by those that are just, and of the family of virtue ; the opposite whereof is a fury to man, and makes even life unsweet.

Therefore, what is more heavy than evil fame deserved ? Or, likewise, who can see worse days, than he that yet living doth follow at the funerals of his own reputation ? I have laid up many hopes, that I am privileged from that kind of mourning, and could wish the like peace to all those with whom I wage love.

I might say much of the commodities that death can sell a man ; but briefly, death is a friend of ours, and he that is not ready to entertain him, is not at home. Whilst I am, my ambition is not to fore-flow the tide ; I have but so to make my interest of it, as I may account for it ; I would wish nothing but what might better my days, nor desire any greater place than the front of good opinion. I make not love to the continuance of days, but to the goodness of them ; nor wish to die, but refer myself to my hour, which the great dispenser of all things hath appointed me ; yet as I am frail, and suffered for the first fault, were it given me to choose, I should not be earnest to see the evening of my age ; that extremity of itself being a disease, and a mere return into infancy : so that if perpetuity of life might be given me, I should think what the Greek poet said, *Such an age is a mortal evil.* And since I must needs be dead, I require it may not be done before mine enemies, that I be not stript before I be cold ; but before my friends. The night was even now ; but that name is lost ; it is not now late, but early. Mine eyes begin to discharge their watch, and compound with this fleshly weakness for a time of perpetual rest ; and I shall presently be as happy for a few hours, as I had died the first hour I was born.

[13] So the original. Modern editions read "till necessity ;" probably a conjectura correction ; and (I suspect) not the true reading.

[14] *them* in the last sentence, and *yet* in this, are omitted in the original.

OF THE WISDOM OF THE ANCIENTS

[Translation of *De Sapientia Veterum.*]

PREFACE.

BY JAMES SPEDDING.

THE treatise *De Sapientia Veterum* was first published in 1609, in a small duodecimo volume, carefully and beautifully printed in the elegant italic type then in use. It appears to have become speedily popular, and was once or twice reprinted during Bacon's life, and translated both into English and Italian. In 1623, he introduced three of the fables, revised and considerably enlarged, into the *De Augmentis Scientiarum*, as a specimen of one of the *Desiderata*. Two others he had designed for the foundation of an elaborate discussion of the philosophy of Democritus, Parmenides, and Telesius; of which a considerable fragment has been preserved. [The *De Principiis*, above, pp. 647–69.] A year or two before his death he designed to include the whole volume among the *Opera Moralia et Civilia*, of which he was then preparing a collection, and in which it was afterwards published by Dr. Rawley, along with the Latin translations of the History of Henry VII., the Essays, the New Atlantis, and the Dialogue of a Holy War. There can be no doubt therefore that it was a work which he thought well of, and meant to live.

Of the history of it all I know further is that four of the fables,—namely, *Metis sive Consilium*, *Soror Gigantum sive Fama*, *Cœlum sive Origines* and *Proteus sive Materia*,—are found in the same form in the fragment which I have entitled *Cogitationes de Scientiâ Humanâ*, and which I suppose to have been written before 1605.

The object of the work was probably to obtain a more favourable hearing for certain philosophical doctrines of Bacon's own; for it seems certain that the fables themselves could never have suggested the ideas, however a man to whom the ideas had suggested themselves might find or fancy he found them in the fables. But the theory on which his interpretation rests, namely that a period of high intellectual cultivation had existed upon the earth and passed out of memory long before the days of Homer, was, I suppose, seriously entertained by him; nor was it a thing so difficult to believe then as it seems now. When a new continent was first discovered, in which the savage inhabitants were found laden with golden ornaments, it was easy to believe in the rumours of El Dorado; and when the buried fragments of Greek and Roman civilisation were first brought up for the examination of a new age, they might easily suggest to the imagination a world of wonders still unrecovered. But when voyager after voyager returned from America, bringing no confirmation of the first rumours, they ceased to be credible; and now that men have been employed for centuries in diligently collecting and discussing the monuments of antiquity, and yet no further evidence of that period of primeval wisdom has been discovered, the balance of probability turns against the speculation. Comparative philology, coupled with comparative mythology, teaches us to seek for an explanation of the ancient mythes in a new direction; and from these sciences Bacon, though I think he would have accepted them as the best guides in the inquiry, could have no help; for they could hardly be said to exist at all in his time. Regarded therefore as attempts to explain the true historical origin of these fables, his interpretations, however elegant and ingenious, may be set aside, as having lost their serious interest for us. And though they would furnish an editor possessed of the requisite learning, and so minded, with an opportunity of displaying a vast deal of erudition, it would, I think, be wasted in this place. In so far as the question could be settled by the light of common sense with such knowledge as Bacon had, little could be added probably on either side to what he has himself said in his prefatory disquisition. In so far as it depends upon the knowledge which has since been acquired concerning the ancient languages and literature of the East, it should be discussed without reference to Bacon, who had no such knowledge, and would in all probability, if it had been revealed to him, have given up his own conjecture as untenable.

The interest which the book still possesses for us (and it has always been a great favourite with me) is of quite another kind; nor has either change of times or increase of knowledge at all abated its freshness. It is an interest precisely of the same kind with that which in the Essays shows no symptoms of becoming obsolete. The interpretation of each fable is in fact an "essay or counsel," civil, moral, or philosophical; embodying the results of Bacon's own thought and observation upon the nature of men and things, and replete with good sense of the best quality.

The great popularity of this book during the first half of the seventeenth century may have been partly due to the reputation which it then had among scholars as a work of learning and authority; and if so, the decline of its popularity may be accounted for by the abatement of that reputation. Students of Greek naturally neglect it, because it passes no longer for an orthodox exposition of the meaning of the Greek fables. Students of nature and the business of modern life naturally pass it by, not expecting to find under such a title and in a dead language the sort of entertainment they are in search of. But I see no other reason why it should not be as great a favourite with modern readers and be found as amusing and instructive as the Essays are; the matter being of as good quality, and the form not less attractive.

Upon this view of its character, and having a due regard to my own qualifications, I have thought it best to leave points of learning to those who are more competent to handle them (for the most I could do in that way would be to report conclusions which I am not in a condition to verify), and content myself with endeavouring by means of a new translation to bring the book within reach of the less learned. For though three English translations of it have been published, one of which was once very popular, and all are extant and accessible, I do not find any of them much quoted or referred to now, as if they had obtained any real currency among English readers. Whether my attempt will fare better, remains to be seen; but if I have succeeded in putting into the translation so much of the life of the original, that those who are fond of the Essays may read it with something of the same feeling, I shall not regret the pains I have taken in the matter.

With regard to the enigma which these ancient mythes present us with, I have said that the researches of modern science teach us to look for the true solution of it in a direction quite different from that which Bacon took. And without affecting to offer anything that can be called an opinion on the subject for myself, I am fortunately able to illustrate my meaning by an example of a modern solution, derived from one whose information includes probably everything that is known with reference to the question at issue, up to the latest dates. I allude to Professor Max Müller's paper on Comparative Mythology in the *Oxford Essays* of 1856. The difficulty to be explained, as stated by him, is substantially the same as that which Bacon puts forward most prominently among his reasons for concluding that these old fables involved an allegorical meaning. "Let us think," says Professor Müller, "of the times which could bear a Lykurgos and a Solon,—which could found an Areopagos and the Olympic games, and how can we imagine that a few generations before that time, the highest notions of the Godhead among the Greeks were adequately expressed by the story of Uranos maimed by Kronos,—of Kronos eating his children, swallowing a stone, and vomiting out alive his whole progeny? . . . The difficulty is, how at first the human mind was led to such imaginings,—how the names and the tales arose; and unless this question can be answered, our belief in a regular and consistent progress of the human intellect, through all ages and in all countries, must be given up as a false theory[1]." "A fable that is probable," says Bacon, "may be thought to have been composed merely for pleasure, in imitation of history. But when a story is told which could never have entered into any man's head, either to conceive or relate on its own account, we must presume that it had some further reach. What a fiction (for instance) is that of Jupiter and Metis! Jupiter took Metis to wife; as soon as he saw that she was with child, he ate her up: whereupon he grew to be with child himself, and so brought forth out of his head Pallas in armour! Surely I think no man had ever a dream so monstrous, and extravagant, and out of all natural ways of thinking[2]." Both agree likewise in concluding that the original story must have involved another meaning; that the names and incidents must have survived after that meaning had been forgotten; and that they have suffered in the hands of poets a variety of alterations, applications, and corruptions. So far the two speculations go together: but at this point they part, and part in opposite directions. Bacon, having only the Greek language and mythology to interpret the Greek fables by, conceived it possible that a generation of wise men had once flourished upon the earth who taught the mysteries of nature in parables; that they died and their wisdom with them; the parables remaining in memory, merely as tales without meaning. Professor Müller, furnished with materials for a wider induction in the languages and mythologies of all the Eastern nations and races, and finding similar traditions flourishing among them all,—"stories identical in form and in character, whether we find them on Indian, Persian, Greek, Italian, Slavonic, or Teutonic soil,"—and being able likewise to trace the names which figure in many of these stories through their Greek corruptions to their original meaning in the language from which they came,—able, for instance, by help of the *Veda* to identify Daphne with the Dawn (see p. 57)—is led, through a course of reasoning too long for quotation and yet too close for abridgement, to a conclusion much more in accordance with all we know of the progress and vicissitudes of human things; yet one which, if accepted, will be held, I think, to justify me in treating the ideas which Bacon finds in these fables as valuable only for the truth and sense they contain, and not as illustrating antiquity. He traces the origin of these mythes to a time when abstract nouns had not been invented; when men had not learnt to express by single words collective or abstract ideas; when therefore everything was spoken of as a person, with a name and a sex. He conceives that they were in fact merely descriptions of the great phenomena of nature; conveying to those who first uttered them the ideas of morning and evening, summer and winter, dawn, twilight, darkness, etc.; indicating the relations between them by words expressing human relations, human feelings and passions; and thus making every metaphor a story; which, passing into another language in which the original name no longer suggested the original image, lost its metaphorical signification, came to be received and repeated as a story simply, and so grew into what we call a *Mythe*. It would not be difficult to suggest analogies even from our own experience, by which it would be seen that the process is a natural one; but I should do injustice to Professor Müller's argument if I attempted to give an idea of the evidence which he brings to support his view. I have said enough, however, to enable the reader to enter into his exposition of the fable of Endymion, which will sufficiently illustrate his theory; and which, as we have Bacon's exposition to contrast it with, will serve better than anything else to exhibit the difference between the rival methods of interpretation.

"We can best enter," says he, "into the original meaning of a Greek mythe, when some of the persons who act in it have preserved names intelligible in Greek. When we find the names of Eos, Selene, Helios, or Herse, we have words which tell their own story, and we have a ποῦ στῶ for the rest of the mythe. Let us take the beautiful mythe of Selene and Endymion. Endymion is the son of Zeus and Kalyke, but he is also the son of Aethlios, a king of Elis, who is himself called a son of Zeus, and whom Endymion is said to have succeeded as King of Elis. This localises our mythe, and shows, at least, that Elis is its birth place, and that,

[1] *Essay on Comparative Mythology*, pp. 8, 11.
[2] *De Sap. Vet. Præfatio*, p. 816 of this volume.

according to Greek custom, the reigning race of Elis derived its origin from Zeus. The same custom prevailed in India, and gave rise to the two great royal families of ancient India—the so-called Solar and the Lunar races; and Purûravas, of whom more by and by, says of himself,—

The great king of day,
And monarch of the night are my progenitors;
Their grandson I . . .

There may, then, have been a King of Elis, Aethlios, and he may have had a son, Endymion; but what the mythe tells of Endymion could not have happened to the King of Elis. The mythe transfers Endymion into Karia, to Mount Latmos, because it was in the Latmian cave that Selene saw the beautiful sleeper, loved him and lost him. Now about the meaning of Selene, there can be no doubt; but even if tradition had only preserved her other name, Asterodia, we should have had to translate this synonyme, as Moon, as 'Wanderer among the stars.' But who is Endymion? It is one of the many names of the sun, but with special reference to the setting or dying sun. It is derived from ἐν-δύω, a verb which, in classical Greek, is never used for setting, because the simple verb δύω had become the technical term for sunset. Δυσμαὶ ἡλίου, the setting of the Sun, is opposed to ἀνατόλαι, the rising. Now, δύω meant, originally, to dive into; and expressions like ἠέλιος δ' ἄρ' ἔδυ, the sun dived, presupposes an earlier conception of ἔδυ πόντον, he dived into the sea. Thus Thetis addresses her companions, *Il.* xviii. 140.

Ὑμεῖς μὲν νῦν δῦτε αλάσσης εὐρέα κόλπον,

You may now dive into the broad bosom of the sea.

Other dialects, particularly of maritime nations, have the same expression. In Lat. we find 'Cur *mergat* seras æquore flammas'. In Old Norse, 'Sôl gengr i aegi'. Slavonic nations represent the sun as a woman stepping into her bath in the evening, and rising refreshed and purified in the morning; or they speak of the Sea as the mother of the Sun, and of the Sun as sinking into her mother's arms at night. We may suppose, therefore, that in some Greek dialect ἐνδύω was used in the same sense; and that from ἐνδύω, ἔνδυμα was formed to express sunset. From this was formed ἐνδυμίων, like οὐρανίων from οὐρανός, and like most of the names of the Greek months. If ἐνδύμα had become a common name for sunset, the mythe of Endymion could have never arisen. But the original meaning of Endymion being once forgotten, what was told originally of the setting sun was now told of a name, which, in order to have any meaning, had to be changed into a god or a hero. The setting sun *once* slept in the Latmian cave, or cave of night,—Latmos being derived from the same root as Leto, Latona, the night,—but *now* he sleeps on Mount Latmos, in Karia. Endymion, sinking into eternal sleep after a life of but one day, was *once* the setting sun, the son of Zeus, the brilliant Sky, and Kalyke, the covering night (from καλύπτω); or, according to another saying, of Zeus and Protogeneia, the first-born goddess, or the Dawn, who is always represented either as the mother, the sister, or the forsaken wife of the Sun. *Now* he is the son of a King of Elis, probably for no other reason except that it was usual for kings to take names of good omen, connected with the sun, or the moon, or the stars,—in which case a mythe, connected with a solar name, would naturally be transferred to its human namesake. In the ancient poetical and proverbial language of Elis, people said 'Selene loves and watches Endymion', instead of 'it is getting late'; 'Selene embraces Endymion', instead of 'the sun is setting and the moon is rising'; 'Selene kisses Endymion into sleep,' instead of 'it is night'. These expressions remained long after their meaning had ceased to be understood; and as the human mind is generally as anxious for a reason as ready to invent one, a story arose by common consent, and without any personal effort, that Endymion must have been a young lad loved by a young lady, Selene; and if children were anxious to know still more, there would always be a grandmother happy to tell them that this young Endymion was the son of the Protogeneia,—she half meaning and half not meaning by that name the Dawn, who gave birth to the sun; or of Kalyke, the dark and covering night. This name, once touched, would set many chords vibrating; three or four different reasons might be given (as they really were given by ancient poets) why Endymion fell into this everlasting sleep, and if any of these was alluded to by a popular poet, it became a mythological fact, repeated by later poets; so that Endymion grew at last almost into a type, no longer of the setting sun, but of a handsome boy beloved of a chaste maiden, and therefore a most likely name for a young prince. Many mythes have thus been transferred to real persons, by a mere similarity of name, though it must be admitted that there is no historical evidence whatever that there ever was a Prince of Elis, called by the name of Endymion.

"Such is the growth of a legend, originally a mere word, a *μύθος*, probably one of those many words which have but a local currency, and lose their value if they are taken to distant places,—words useless for the daily intercourse of thought,—spurious coins in the hands of the many,—yet not thrown away, but preserved as curiosities and ornaments, and deciphered at last, after many centuries, by the antiquarian [3]."

I give this specimen merely to explain and illustrate the modern theory. For the argument in support of it I must refer to the Essay itself; though even there it suffers much for want of room. But that the process described is possible and natural, may be shewn meanwhile without going out of our own literature or our own times.

> The poetry of earth is never dead:

and even within the last ten years an instance has occurred of the simple language of poetic passion being translated out of poetry into mythology. Alfred Tennyson speaks in *In Memoriam* of returning home in the evening

> Before the crimson-circled star
> Had fallen into her father's grave:

not thinking at all of any traditional pedigree, (no more than when he speaks of

> Sad Hesper, o'er the buried Sun,
> And ready thou to die with him,)

but expressing, by such an image, as the ancient Elian might have resorted to, his sympathy with the pathetic aspect of the dying day. Critics however asked for explanations: what star, whose daughter, what grave? And it turns out curiously enough that all these questions can be answered out of Greek mythology quite satisfactorily. "The planet Venus (says a Belgravian correspondent of *Notes and Queries*, 1851, iii. 506), when she is to the east of the sun, is our *evening star* (and as such used to be termed Hesperus by the ancients). The evening star in a summer twilight is seen surrounded with the glow of sunset, crimson-circled. . . . Venus sinking into the sea, which in setting she would appear to do, falls into the grave of *Uranus*,—her father according to the theory of Hesiod (190). The part cast into the sea from which Aphrodite sprung, is here taken by a becoming licence (which softens the grossness of the old tradition) for the whole; so that the ocean, beneath the horizon of which the evening star sinks, may be well described by the poet as '*her* father's grave'."

I would not indeed have any one remember this explanation when he is reading the poem, for it is fatal to the poetic effect; but the coincidence of the expression with the mythic tradition is curious; and might almost make one think that Tennyson, while merely following the eternal and universal instincts of the human imagination and feeling, had unconsciously reproduced the very image out of which the tradition originally grew.

In Dr. Rawley's list of works composed by Bacon during the last five years of his life, he mentions "his revising of his book *De Sapientia Veterum*". And as he professes to give them in the order in which they were written, and this comes

[3] Oxford Essays, 1856, p. 49.

near the end, I suppose he does not allude merely to the three fables introduced into the second book of the *De Augmentis*, which was published in 1623; but to some further revision of the whole previous to the reprinting of the work among the *Opera Moralia et Civilia*. I have therefore treated that posthumous edition (which varies in a few, though very few, passages from the original of 1609), as the latest authority for the text. But as it is not so carefully printed as the other I have collated the two throughout, and noticed the variations. I have also kept the title-page of the original edition; and I have followed modern editors in making the interpretation of each fable commence a new paragraph.

[NOTE.—Since Spedding wrote, the science of Mythology has been carried further than the stage reached in the early essay of Max Müller, which, moreover, did not take account of sound principles established by K. O. Müller, and other preceding writers. It may suffice here to say that the etymological principle, brought to bear on the phenomena of nature, is quite insufficient to explain the mass of myths, though it is valid in a number of cases. Etymology itself has been greatly perfected through a completer checking of deduction by induction; and mythology has latterly been much advanced by the same discipline. Bacon, failing to apply to it his own most frequently iterated rules of investigation, gives purely *à priori* solutions, which are in all cases wholly removed from scientific truth. Later students, in the usual way of scientific advance, have resorted more and more to analysis, comparison and induction; and it may be that the induction is still far from complete. To the etymological and meteorological keys must be added others, including that of the derivation of many Greek myths from fanciful interpretations of works of art, often originally Eastern astronomical symbols. But the whole inquiry must have regard to the psychology of savages, the bulk of mythology being a derivation from primitive life and thought. To the anthropological or psychological study of myth the best modern guides are Tylor, Mannhardt, Frazer, Spencer, Vignoli, Grant Allen, Lang, and the later works of Max Müller, whose earlier method, further, was ably developed by Sir George Cox.—*Ed.*]

AUTHOR'S DEDICATION.

TO THE MOST ILLUSTRIOUS THE EARL OF SALISBURY LORD HIGH TREASURER OF ENGLAND, AND CHANCELLOR OF THE UNIVERSITY OF CAMBRIDGE.

THINGS dedicated to the University of Cambridge accrue to you as Chancellor; to all that proceeds from me you have a personal title. The question is, whether as these things are yours, so they are worthy of you. Now for that which is least worth in them (the wit of the author), your kindness towards me will let that pass; and there is nothing else in the matter to disgrace you. For if time be regarded,—primæval antiquity is an object of the highest veneration; if the form of exposition,—parable has ever been a kind of arc, in which the most precious portions of the sciences were deposited; if the matter of the work,—it is philosophy, the second grace and ornament of life and the human soul. For be it said, that however philosophy in this our age, falling as it were into a second childhood, be left to young men and almost to boys, yet I hold it to be of all things, next to religion, the most important and most worthy of human nature. Even the art of politics, wherein you are so well approved both by faculty and by merits, and by the judgment of a most wise king, springs from the same fountain, and is a great part thereof. And if any man think these things of mine to be common and vulgar, it is not for me of course to say what I have effected; but my aim has been, passing by things obvious and obsolete and commonplace, to give some help towards the difficulties of life and the secrets of science. To the vulgar apprehension therefore they will be vulgar; but it may be that the deeper intellect will not be left aground by them, but rather (as I hope) carried along. While however I strive to attach some worth to this work, because it is dedicated to you, I am in danger of transgressing the bounds of modesty, seeing it is undertaken by myself. But you will accept it as a pledge of my affection, observance, and devotion to yourself, and will accord it the protection of your name. Seeing therefore that you have so many and so great affairs on your shoulders, I will not take up more of your time, but make an end, wishing you all felicity, and ever remaining yours,

Most bounden to you both by my zeal and your benefits,
FRA. BACON.

TO HIS NURSING-MOTHER THE FAMOUS UNIVERSITY OF CAMBRIDGE.

SINCE without philosophy I care not to live, I must needs hold you in great honour, from whom these defences and solaces of life have come to me. To you on this account I profess to owe both myself and all that is mine; and therefore it is the less strange, if I requite you with what is your own; that with a natural motion it may return to the place whence it came. And, yet I know not how it is, but there are few footprints pointing back towards you, among the infinite number that have gone forth from you. Nor shall I take too much to myself (I think), if by reason of that little acquaintance with affairs which my kind and plan of life has necessarily carried with it, I indulge a hope that the inventions of the learned may receive some accession by these labours of mine. Certainly I am of opinion that speculative studies when transplanted into active life acquire some new grace and vigour, and having more matter to feed them, strike their roots perhaps deeper, or at least grow taller and fuller leaved. Nor do you yourselves (as I think) know how widely your own studies extend, and how many things they concern. Yet it is fit that all should be attributed to you and be counted to your honour, since all increase is due in great part to the beginning. You will not however expect from a man of business anything exquisite; any miracles or prerogatives of leisure; but you will attribute to my great love for you and yours even this,—that among the thorns of business these things have not quite perished, but there is preserved for you so much of your own.

Your most loving pupil,
FRA. BACON.

AUTHOR'S PREFACE.

THE most ancient times (except what is preserved of them in the scriptures) are buried in oblivion and silence: to that silence succeeded the fables of the poets: to those fables the written records which have come down to us. Thus between the hidden depths of antiquity and the days of tradition and evidence that followed there is drawn a veil, as it were, of fables, which come in and occupy the middle region that separates what has perished from what survives.

Now I suppose most people will think I am but entertaining myself with a toy, and using much the same kind of licence in expounding the poets' fables which the poets themselves did in inventing them; and it is true that if I had a mind to vary and relieve my severer studies with some such exercise of pleasure for my own or my readers' recreation, I might very fairly indulge in it. But that is not my meaning. Not but that I know very well what pliant stuff fable is made of, how freely it will follow any way you please to draw it, and how easily with a little dexterity and discourse of wit meanings which it was never meant to bear may be plausibly put upon it. Neither have I forgotten that there has been old abuse of the thing in practice; that many, wishing only to gain the sanction and reverence of antiquity for doctrines and inventions of their own, have tried to twist the fables of the poets into that sense; and that this is neither a modern vanity nor a rare one, but old of standing and frequent in use; that Chrysippus long ago, interpreting the oldest poets after the manner of an interpreter of dreams, made them out to be Stoics; and that the Alchemists more absurdly still have discovered in the pleasant and sportive fictions of the transformation of bodies, allusion to experiments of the furnace. All this I have duly examined and weighed; as well as all the levity and looseness with which people indulge their fancy in the matter of allegories; yet for all this I cannot change my mind. For in the first place to let the follies and licence of a few detract from the honour of parables in general is not to be allowed; being indeed a boldness savouring of profanity; seeing that religion delights in such veils and shadows, and to take them away would be almost to interdict all communion between divinity and humanity. But passing that and speaking of human wisdom only, I do certainly for my own part (I freely and candidly confess) incline to this opinion,—that beneath no small number of the fables of the ancient poets there lay from the very beginning a mystery and an allegory. It may be that my reverence for the primitive time carries me too far, but the truth is that in some of these fables, as well in the very frame and texture of the story as in the propriety of the names by which the persons that figure in it are distinguished, I find a conformity and connection with the thing signified, so close and so evident, that one cannot help believing such a signification to have been designed and meditated from the first, and purposely shadowed out. For who is there so impenetrable and that can so shut his eyes to a plain thing, but when he is told that after the *Giants* were put down, *Fame* sprung up, as their posthumous sister, he will at once see that it is meant of those murmurs of parties and seditious rumours which always circulate for a time after the suppression of a rebellion? Or again who can hear that the *Giant Typhon* cut off and carried away *Jupiter's* sinews, and that *Mercury* stole them from Typhon and gave them back to Jupiter; without at once perceiving that it relates to successful rebellions, by which kings have their sinews both of money and authority cut off;

yet not so but that by fair words and wise edicts the minds of the subjects may be presently reconciled, and as it were stolen back, and so kings recover their strength? Or who can hear that in that memorable expedition of the gods against the giants the braying of *Silenus's ass* had a principal stroke in putting the giants to flight, and not be sure that the incident was invented in allusion to the vast attempts of rebels, dissipated as they commonly are by empty rumours and vain terrors? Then again there is a conformity and significancy in the very names, which must be clear to everybody. Metis, Jupiter's wife, plainly means counsel; Typhon, swelling; Pan, the universe; Nemesis, revenge; and the like. And what if we find here and there a bit of real history underneath, or some things added only for ornament, or times confounded, or part of one fable transferred to another and a new allegory introduced? Such things could not but occur in stories invented (as these were) by men who both lived in different ages and had different ends, some being more modern, some more ancient, some having in their thoughts natural philosophy, others civil affairs; and therefore they need not trouble us.

But there is yet another sign, and one of no small value, that these fables contain a hidden and involved meaning; which is, that some of them are so absurd and stupid upon the face of the narrative taken by itself, that they may be said to give notice from afar and cry out that there is a parable below. For a fable that is probable may be thought to have been composed merely for pleasure, in imitation of history. But when a story is told which could never have entered any man's head either to conceive or relate on its own account, we must presume that it had some further reach. What a fiction (for instance) is that of Jupiter and Metis! Jupiter took Metis to wife: as soon as he saw that she was with child, he ate her up; whereupon he grew to be with child himself; and so brought forth out of his head Pallas in armour! Surely I think no man had ever a dream so monstrous and extravagant, and out of all natural ways of thinking.

But the consideration which has most weight with me is this, that few of these fables were invented, as I take it, by those who recited and made them famous,—Homer, Hesiod, and the rest. For had they been certainly the production of that age and of those authors by whose report they have come down to us, I should not have thought of looking for anything great or lofty from such a source. But it will appear upon an attentive examination that they are delivered not as new inventions then first published, but as stories already received and believed. And since they are told in different ways by writers nearly contemporaneous, it is easy to see that what all the versions have in common came from ancient tradition, while the parts in which they vary are the additions introduced by the several writers for embellishment—a circumstance which gives them in my eyes a much higher value; for so they must be regarded as neither being the inventions nor belonging to the age of the poets themselves, but as sacred relics and light airs breathing out of better times, that were caught from the traditions of more ancient nations and so received into the flutes and trumpets of the Greeks.

Nevertheless, if any one be determined to believe that the allegorical meaning of the fable was in no case original and genuine, but that always the fable was first and the allegory put in after, I will not press that point; but allowing him to enjoy that gravity of judgment (of the dull and leaden order though it be) which he affects, I will attack him, if indeed he be worth the pains, in another manner upon a fresh ground. Parables have been used in two ways, and (which is strange) for contrary purposes. For they serve to disguise and veil the meaning, and they serve also to clear and throw light upon it. To avoid dispute then, let us give up the former of these uses. Let us suppose that these fables were things without any definite purpose, made only for pleasure. Still there remains the latter use. No force of wit can deprive us of that. Nor is there any man of ordinary learning that will object to the reception of it as a thing grave and sober, and free from all vanity; of prime use to the sciences, and sometimes indispensable: I mean the employment of parables as a method of teaching, whereby inventions that are new and abstruse and remote from vulgar opinions may find an easier passage to the understanding. On this account it was that in the old times, when the inventions and conclusions of human reason (even those that are now trite and vulgar) were as yet new and strange, the world was full of all kinds of fables, and enigmas, and parables, and similitudes ; and these were used not as a device for shadowing and concealing the meaning, but as a method of making it understood ; the understandings of men being then rude and impatient of all subtleties that did not address themselves to the sense,—indeed scarcely capable of them. For as hieroglyphics came before letters, so parables came before arguments. And even now if any one wish to let new light on any subject into men's minds, and that without offence or harshness, he must still go the same way and call in the aid of similitudes.

Upon the whole I conclude with this : the wisdom of the primitive ages was either great or lucky ; great, if they knew what they were doing and invented the figure to shadow the meaning ; lucky, if without meaning or intending it they fell upon matter which gives occasion to such worthy contemplations. My own pains, if there be any help in them, I shall think well bestowed either way : I shall be throwing light either upon antiquity or upon nature itself.

That the thing has been attempted by others I am of course aware, but if I may speak what I think freely without mincing it, I must say that the pains which have been hitherto taken that way, though great and laborious, have gone near to deprive the inquiry of all its beauty and worth ; while men of no experience in affairs nor any learning beyond a few commonplaces, have applied the sense of the parables to some generalities and vulgar observations, without attaining their true force, their genuine propriety, or their deeper reach. Here, on the other hand, it will be found (if I mistake not) that though the subjects be old, yet the matter is new ; while leaving behind us the open and level parts we bend our way towards the nobler heights that rise beyond.

OF THE WISDOM OF THE ANCIENTS

I.

CASSANDRA;

OR PLAINNESS OF SPEECH.

THEY say that Cassandra was beloved by Apollo; that she contrived by various artifices to elude his desires, and yet to keep his hopes alive until she had drawn from him the gift of divination: that she had no sooner obtained this, which had all along been her object, than she openly rejected his suit; whereupon he, not being permitted to recal the boon once rashly promised, yet burning with revenge, and not choosing to be the scorn of an artful woman, annexed to it this penalty,—that though she should always foretell true, yet nobody should believe her. Her prophecies therefore had truth, but not credit: and so she found it ever after, even in regard to the destruction of her country; of which she had given many warnings, but could get nobody to listen to her or believe her.

This fable seems to have been devised in reproof of unreasonable and unprofitable liberty in giving advice and admonition. For they that are of a froward and rough disposition, and will not submit to learn of Apollo, the god of harmony, how to observe time and measure in affairs, flats and sharps (so to speak) in discourse, the differences between the learned and the vulgar ear, and the times when to speak and when to be silent; such persons, though they be wise and free, and their counsels sound and wholesome, yet with all their efforts to persuade they scarcely can do any good; on the contrary, they rather hasten the destruction of those upon whom they press their advice; and it is not till the evils they predicted have come to pass that they are celebrated as prophets and men of a far foresight. Of this we have an eminent example in Marcus Cato of Utica, by whom the ruin of his country and the usurpation that followed, by means first of the conjunction and then of the contention between Pompey and Cæsar, was long before foreseen as from a watch-tower, and foretold as by an oracle; yet all the while he did no good, but did harm rather, and brought the calamities of his country faster on; as was wisely observed and elegantly described by Marcus Cicero, when he said in a letter to a friend, *Cato means well: but he does hurt sometimes to the State; for he talks as if he were in the republic of Plato and not in the dregs of Romulus.*

II.

TYPHON;

OR THE REBEL.

THE poets tell us that Juno being angry that Jupiter had brought forth Pallas by himself without her help, implored of all the gods and goddesses that she also might bring forth something without the help of Jupiter: to which when wearied with her violence and importunity they had assented, she smote the earth, which quaking and opening gave birth to Typhon, a huge and hideous monster. He was given to a serpent by way of foster-father to be nursed. As soon as he was grown up he made war upon Jupiter, whom in the conflict he took prisoner; and bearing him on his shoulders to a remote and obscure region, cut out the sinews of his hands and feet, and carrying them away, left him there helpless and mutilated. Then came Mercury, and having stolen the sinews from Typhon gave them back to Jupiter, who finding his strength restored attacked the monster again. And first he struck him with a thunderbolt, which made a wound the blood whereof engendered serpents; then, as he fell back and fled, threw upon him the mountain Ætna and crushed him beneath the weight.

The fable has been composed in allusion to the variable fortune of kings and the rebellions that occur from time to time in monarchies. For kings and their kingdoms are properly, like Jupiter and Juno, man and wife. But it sometimes happens that the king, depraved by the long habit of ruling, turns tyrant and takes all into his own hands; and not caring for the consent of his nobles and senate, brings forth as it were by himself; that is to say, administers the government by his own abitrary and absolute authority. Whereat the people aggrieved endeavour on their part to set up some head of their own. This generally begins with the secret solicitation of nobles and great persons, whose connivancy being obtained, an attempt is then made to stir the people. Thence comes a kind of swelling in the State, which is signified by the infancy of Typhon. And this condition of affairs is fostered and nourished by the innate depravity and malignant disposition of the common people, which is to kings like a serpent full of malice and mischief; till the disaffection spreading and gathering strength breaks out at last into open rebellion; which because of the infinite calamities it inflicts both on kings and peoples is represented under the dreadful image of Typhon, with a hundred heads, denoting divided powers; flaming mouths, for devastations by fire; belts of snakes, for the pestilences which prevail, especially in sieges; iron hands, for slaughters; eagle's talons, for rapine; feathery body, for perpetual rumours, reports, trepidations, and the like. And sometimes these rebellions grow so mighty that the king is forced, as if carried off on the shoulders of the rebels, to abandon the seat and principal cities of his kingdom, and to contract his forces, and betake himself to some remote and obscure province; his sinews both of money and majesty being cut off. And yet if he bears his fortune wisely, he presently by the skill and industry of Mercury recovers those sinews again; that is to say, by affability and wise edicts and gracious speeches he reconciles the minds of his subjects, and awakens in them an alacrity to grant him supplies, and so recovers the vigour of his authority. Nevertheless, having learned prudence and caution, he is commonly unwilling to set all upon the toss of fortune, and therefore avoids a pitched battle, but tries first by some memorable exploit to destroy the reputation of the rebels: in which if he succeed, the rebels feeling themselves shaken and losing their confidence, resort first to broken and empty threats, like serpent's hisses, and then finding their case desperate take to flight. And then is the time, when they are beginning to fall to pieces, for the king with the entire forces and mass of his kingdom, as with the mountain Ætna, to pursue and overwhelm them.

III.

THE CYCLOPES;

OR MINISTERS OF TERROR.

The story is that the Cyclopes were at first on account of their fierceness and brutality driven by Jupiter into Tartarus, and condemned to perpetual imprisonment; but afterwards he was persuaded by the Earth that it would be for his interest to release them and employ them to make thunderbolts for him; which he accordingly did; and they with officious industry laboured assiduously with a terrible din in forging thunderbolts and other instruments of terror. In course of time it happened that Jupiter's wrath was kindled against Æsculapius, son of Apollo, for raising a man from the dead by medicine; but because the deed was pious and famous and no just cause of displeasure, he concealed his anger and secretly set the Cyclopes upon him: who made no difficulty, but presently dispatched him with their thunderbolts; in revenge whereof Apollo (with Jupiter's permission) slew them with his arrows.

This fable seems to relate to the doings of kings; by whom cruel and bloody and exacting ministers are in the first instance punished and put out of office. But afterwards by counsel of the Earth, that is by ignoble and dishonourable counsel, yielding to considerations of utility, they take them into service again, when they have need either of severity of executions or harshness in exactions.

They on their part being by nature cruel and by their former fortune exasperated, and knowing well enough what they are wanted for, apply themselves to this kind of work with wonderful diligence; till for want of caution and from over eagerness to ingratiate themselves, they at one time or another (taking a nod or an ambiguous word of the prince for a warrant) perpetrate some execution that is odious and unpopular. Upon which the prince, not willing to take the envy of it upon himself, and well knowing that he can always have plenty of such instruments, throws them overboard, and leaves them to the course of law and the vengeance of the friends and relatives of their victims, and to popular hatred; and so amid much applause of the people and great acclamations and blessings on the king, they meet at last, though late, the fate they deserve.

IV.
NARCISSUS;
OR SELF-LOVE.

NARCISSUS is said to have been a young man of wonderful beauty, but intolerably proud, fastidious, and disdainful. Pleased with himself and despising all others, he led a solitary life in the woods and hunting-grounds; with a few companions to whom he was all in all; followed also whenever he went by a nymph called Echo. Living thus, he came by chance one day to a clear fountain, and (being in the heat of noon) lay down by it; when beholding in the water his own image, he fell into such a study and then into such a rapturous admiration of himself, that he could not be drawn away from gazing at the shadowy picture, but remained rooted to the spot till sense left him; and at last he was changed into the flower that bears his name; a flower which appears in the early spring; and is sacred to the infernal deities,—Pluto, Proserpine, and the Furies.

In this fable are represented the dispositions, and the fortunes too, of those persons who from consciousness either of beauty or some other gift with which nature unaided by any industry of their own has graced them, fall in love as it were with themselves. For with this state of mind there is commonly joined an indisposition to appear much in public or engage in business; because business would expose them to many neglects and scorns, by which their minds would be dejected and troubled. Therefore they commonly live a solitary, private, and shadowed life; with a small circle of chosen companions, all devoted admirers, who assent like an echo to everything they say, and entertain them with mouth-homage; till being by such habits gradually depraved and puffed up, and besotted at last with self-admiration, they fall into such a sloth and listlessness that they grow utterly stupid, and lose all vigour and alacrity. And it was a beautiful thought to choose the flower of spring as an emblem of characters like this: characters which in the opening of their career flourish and are talked of, but disappoint in maturity the promise of their youth. The fact too that this flower is sacred to the infernal deities contains an allusion to the same thing. For men of this disposition turn out utterly useless and good for nothing whatever; and anything that yields no fruit, but like the way of a ship in the sea passes and leaves no trace, was by the ancients held sacred to the shades and infernal gods.

V.
STYX;
OR TREATIES.

IT is a very common tradition that of the one oath by which the gods bound themselves when they meant to leave no room for repentance; and finds a place in a great many fables. In that case they invoked in witness, not any majesty of heaven or any divine attribute, but Styx; a river in the infernal regions which with many windings encircled the palace of Dis. This form of oath alone, and no other, was held to be sure and inviolable: the penalty of breaking it being one which the deities most dreaded,—namely that the breaker should for a certain period of years be excluded from the banquets of the gods.

The fable seems to have been invented in allusion to treaties and compacts of princes : in respect of which it is but too true that whatever be the solemnity and sanctity of the oath they are confirmed with, yet they are little to be depended on ; insomuch that they are used in fact rather with an eye to reputation and fame and ceremony, than for confidence and security and effect. And even when the ties of relationship (which are as the sacraments of nature) or of mutual good services come in to aid, yet in most cases all are too weak for ambition and interest and the licence of power : the rather because princes can always find plenty of plausible pretexts (not being accountable to any arbiter) wherewith to justify and veil their cupidity and bad faith. There is adopted therefore but one true and proper pledge of faith ; and it is not any celestial divinity. This is Necessity (the great god of the powerful), and peril of state, and communion of interest. Now Necessity is elegantly represented under the figure of Styx ; the fatal river across which no man can return. This is the deity which Iphicrates the Athenian invoked to witness treaties ; and since he was one that spoke out plainly what most men think and keep to themselves, his words are worth quoting. Finding that the Lacedæmonians were devising and propounding various cautions and sanctions and securities and bonds to hold the treaty fast, *There is only one bond and security* (said he, interrupting them) *that can hold between you and us :—you must prove that you have yielded so much into our hands that you cannot hurt us if you would.* And so it is that if the means of hurting be taken away, or if a breach of the treaty would endanger the existence or the integrity of the state and revenue,—then the treaty may be considered to be ratified and sanctioned and confirmed as by the oath of Styx : for then it is upon peril of being interdicted from the banquets of the gods ; which was the ancient expression for the rights and prerogatives of empire, and wealth, and felicity.

VI.

PAN ;

OR NATURE [4].

THE ancients have given under the person of Pan an elaborate description of universal nature. His parentage they leave in doubt. Some call him the son of Mercury ; others assign him an origin altogether different ; saying that he was the offspring of a promiscuous intercourse between Penelope and all her suitors. But in this the name of Penelope has doubtless been foisted by some later author into the original fable. For it is no uncommon thing to find the more ancient narrations transferred to persons and names of later date ; sometimes absurdly and stupidly, as in this instance ; for Pan was one of the oldest gods, and long before the times of Ulysses ; and Penelope was for her matronly chastity held in veneration by antiquity. But there is yet a third account of his birth, which must not be passed over ; for some have called him the son of Jupiter and Hybris, or Insolence.

Whatever was his origin, the Fates are said to have been his sisters.

His person is described by ancient tradition as follows : With horns, and the tops of the horns reaching heaven ; his whole body shaggy and hairy ; his beard especially long. In figure, biform ; human in the upper parts, the other half brute ; ending in the feet of a goat. As emblems of his power he carried in his left hand a pipe compact of seven reeds, in his right a sheep hook or staff crooked at the top ; and he was clothed in a scarf, made of panther's skin. The powers and offices assigned to him are these,—he is the god of hunters, of shepherds, and generally of dwellers in the country : also he presides over mountains ; and is (next to Mercury) the messenger of the gods. He was accounted moreover the captain and commander of the nymphs, who were always dancing and frisking about him : the Satyrs, and their elders, the Sileni, were also of his company. He had the power likewise of exciting sudden terrors,—empty and superstitious ones especially ;—thence called Panics. The actions that are recorded of him are not many ; the principal is that he challenged Cupid to wrestle ; and was

[4] For an enlarged version of this fable see above, pp. 442–447.

beaten by him. He also entangled and caught the giant Typhon in a net; and they say besides, that when Ceres, out of grief and indignation at the rape of Proserpina, had hid herself, and all the gods were earnestly engaged in seeking her out, and had dispersed several ways in search of her, it was Pan's good fortune to light upon and discover her by accident while he was hunting. He had also the presumption to match himself against Apollo in music; and was by Midas's judgment pronounced victor; for which judgment Midas had to wear the ears of an ass, but not so as to be seen. There are no amours reported of Pan, or at least very few: which among a crowd of gods so excessively amorous may seem strange. The only thing imputed to him in this kind is a passion for Echo, who was also accounted his wife; and for one nymph called Syringa, with love of whom he was smitten by Cupid in anger and revenge because of his presumption in challenging him to wrestle. Nor had he any issue (which is again strange, seeing that the gods, especially the males, were remarkably prolific) except one daughter, a little serving woman called Iambe, who used to amuse guests with ridiculous stories, and was supposed by some to be Pan's offspring by his wife Echo.

A noble fable this, if there be any such; and big almost to bursting with the secrets and mysteries of Nature.

Pan, as the very word declares, represents the universal frame of things, or Nature. About his origin there are and can be but two opinions; for Nature is either the offspring of Mercury—that is of the Divine Word (an opinion which the Scriptures establish beyond question, and which was entertained by all the more divine philosophers); or else of the seeds of things mixed and confused together. For they who derive all things from a single principle, either take that principle to be God, or if they hold it to be a material principle, assert it to be though actually one yet potentially many; so that all difference of opinion on this point is reducible to one or other of these two heads,—the world is sprung either from Mercury, or from all the suitors. He sang, says Virgil,

> How through the void of space the seeds of things
> Came first together; seeds of the sea, land, air,
> And the clear fire; how from these elements
> All embryos grew, and the great world itself
> Swelled by degrees and gathered in its globe.

The third account of the generation of Pan, might make one think that the Greeks had heard something, whether through the Egyptians or otherwise, concerning the Hebrew mysteries; for it applies to the state of the world, not at its very birth, but as it was after the fall of Adam, subject to death and corruption. For that state was the offspring of God and Sin,—and so remains. So that all three stories of the birth of Pan (if they be understood with a proper distinction as to facts and times) may be accepted as indeed true. For true it is that this Pan, whom we behold and contemplate and worship only too much, is sprung from the Divine Word, through the medium of confused matter (which is itself God's creature), and with the help of sin and corruption entering in.

To the Nature of things, the Fates or destinies of things are truly represented as sisters. For natural causes are the chain which draws after it the births and durations and deaths of all things; their fallings and risings, their labours and felicities:—in short all the fates that can befall them.

That the world is represented with horns, and that such horns are broad at bottom and narrow at top, has relation to the fact that the whole frame of nature rises to a point like a pyramid. For individuals are infinite: these are collected into species, which are themselves also very numerous; the species are gathered up into genera, and these again into genera of a higher stage; till nature, contracting as it rises, seems to meet at last in one point. Nor need we wonder that Pan's horns touch heaven; since the summits, or universal forms, of nature do in a manner reach up to God; the passage from metaphysic to natural theology being ready and short.

The body of Nature is most elegantly and truly represented as covered with hair; in allusion to the rays which all objects emit; for rays are like the hairs or bristles of nature; and there is scarcely anything which is not more or less radiant. This is very plainly seen in the power of vision, and not less so in all kinds of magnetic virtue, and in every effect which takes place at a distance. For whatever produces an effect at a distance may be truly said to emit rays. But Pan's hair is longest in the beard, because the rays of the celestial bodies operate and penetrate from a greater distance than any other; and we see also that the sun, when the upper part of him is veiled by a cloud and the rays break out below, has the appearance of a face with a beard.

Again, the body of Nature is most truly described as biform; on account of the difference between the bodies of the upper and the lower world. For the upper or heavenly bodies are for their beauty and the equability and constancy of their motion, as well as for the influence they have upon earth and all that belongs to it, fitly represented under the human figure: but the others, by reason of their perturbations and irregular motions, and because they are under the influence of the celestial bodies, may be content with the figure of a brute. The same description of Nature's body may be referred also to the mixture of one species with another. For there is no nature which can be regarded as simple; every one seeming to participate and be compounded of two. Man has something of the brute; the brute has something of the vegetable; the vegetable something of the inanimate body; and so all things are in truth biformed and made up of a higher species and a lower. There is also a very ingenious allegory involved in that attribute of the goat's feet; which has reference to the motion upwards of terrestrial bodies towards the regions of air and sky: for the goat is a climbing animal, and loves to hang from rocks and cling to the sides of precipices: a tendency which is also exhibited in a wonderful manner by substances that belong properly to the lower world—witness clouds and meteors.

The emblems in Pan's hands are of two kinds—one of harmony, the other of empire. The pipe compact of seven reeds evidently indicates that harmony and concent of things, that concord mixed with discord, which results from the motions of the seven planets. Also the sheep-hook is a noble metaphor, alluding to the mixture of straight and crooked in the ways of nature. But the staff is curved chiefly towards the top; because all the works of Divine Providence in the world are wrought by winding and roundabout ways—where one thing seems to be doing, and another is doing really—as in the selling of Joseph into Egypt, and the like. So also in all the wiser kinds of human government, they who sit at the helm can introduce and insinuate what they desire for the good of the people more successfully by pretexts and indirect ways than directly; so that every rod or staff of empire is truly crooked at the top. The scarf or mantle of Pan is very ingeniously feigned to be made of a panther's skin; on account of the spots scattered all over it. For the heavens are spotted with stars, the sea with islands, the earth with flowers; and even particular objects are generally variegated on the surface, which is as it were their mantle or scarf.

Now the office of Pan can in no way be more lively set forth and explained than by calling him god of hunters. For every natural action, every motion and process of nature, is nothing else than a hunt. For the sciences and arts hunt after their works, human counsels hunt after their ends, and all things in nature hunt either after their food, which is like hunting for prey, or after their pleasures, which is like hunting for recreation;—and that too by methods skilful and sagacious.

After the wolf the lion steals; the wolf the kid doth follow;
The kid pursues the cytisus o'er hillock and thro' hollow.

Also Pan is the god of country people in general; because they live more according to nature; whereas in courts and cities nature is corrupted by too much culture; till it is true what the poet said of his mistress,—*the girl herself is the least part of the matter.*

Pan is likewise especially called president of mountains—because it is in mountains and elevated places that the nature of things is most spread abroad, and lies most open to view and study. As for Pan's being, next to Mercury, the messenger of the gods, that is an allegory plainly divine; seeing that next to the Word of God, the image itself of the world is the great proclaimer of the divine wisdom and goodness. So sings the Psalmist: *The heavens declare the glory of God, and the firmament sheweth his handiwork.*

Again, Pan takes delight in the nymphs; that is the souls; for the souls of the living are the delight of the world. And Pan is truly called their commander, since they follow the guidance each of her several nature; leaping and dancing about it with infinite variety, every one in her country's fashion, and with motion that never ceases. And in their company are ever found the Satyrs and the Sileni; that is old age and youth; for all things have their merry and dancing time, and likewise their heavy and tippling time. And yet to one who truly considers them, the pursuits of either age appear perhaps, as they did to Democritus, ridiculous and deformed,—like to a Satyr or Silenus.

In the Panic terrors there is set forth a very wise doctrine; for by the nature of things all living creatures are endued with a certain fear and dread, the office of which is to preserve their life and essence, and to avoid or repel approaching mischief. But the same nature knows not how to keep just measure—but together with salutary fears ever mingles vain and empty ones; insomuch that all things (if one could see into the heart of them) are quite full of Panic terrors; human things most of all; so infinitely tossed and troubled as they are with superstition (which is in truth nothing but a Panic terror), especially in seasons of hardship, anxiety, and adversity.

With regard to the audacity of Pan in challenging Cupid to fight, it refers to this,—that matter is not without a certain inclination and appetite to dissolve the world and fall back into the ancient chaos; but that the overswaying concord of things (which is represented by Cupid or Love) restrains its will and effort in that direction and reduces it to order. And therefore it is well for man and for the world that in that contest Pan was foiled. The same thing is alluded to in that other circumstance of the catching of Typhon in a net: because however it be that vast and strange swellings (for that is the meaning of Typhon) take place occasionally in nature—whether of the sea, or the clouds, or the earth, or any other body—nevertheless all such exuberancies and irregularities are by the nature of things caught and confined in an inextricable net, and bound down as with a chain of adamant.

As for the tale that the discovery of Ceres was reserved for this god, and that while he was hunting, and denied to the rest of the gods though diligently and specially engaged in seeking her; it contains a very true and wise admonition—namely that the discovery of things useful to life and the furniture of life, such as corn, is not to be looked for from the abstract philosophies, as it were the greater gods, no not though they devote their whole powers to that special end—but only from Pan; that is from sagacious experience and the universal knowledge of nature, which will often by a kind of accident, and as it were while engaged in hunting, stumble upon such discoveries.

Then again that match in music and the result of it exhibits a wholesome doctrine, fit to restrain and reduce to sobriety the pride and overweening confidence of human reason and judgment. For it seems there are two kinds of harmony and music; one of divine providence, the other of human reason; and to the human judgment, and the ears as it were of mortals, the government of the world and nature, and the more secret judgments of God, sound somewhat harsh and untunable; and though this be ignorance, such as deserves to be distinguished with the ears of an ass, yet those ears are worn secretly and not in the face of the world—for it is not a thing observed or noted as a deformity by the vulgar.

Lastly, it is not to be wondered at that no amours are attributed to Pan, except his marriage with Echo. For the world enjoys itself and in itself all things that are. Now he that is in love wants something, and where there is abundance of everything want can have no place. The world therefore can have no loves, nor any want (being content with itself) unless it be of discourse. Such is the nymph Echo, or, if it be of the more exact and measured kind, Syringa. And it is excellently provided that of all discourses or voices Echo alone should be chosen for the world's wife. For that is in fact the true philosophy which echoes most faithfully the voice of the world itself, and is written as it were from the world's own dictation ; being indeed nothing else than the image and reflexion of it, which it only repeats and echoes, but adds nothing of its own. That the world has no issue, is another allusion to the sufficiency and perfection of it in itself. Generation goes on among the parts of the world, but how can the whole generate, when no body exists out of itself ? As for that little woman, Pan's putative daughter, it is an addition to the fable with a great deal of wisdom in it ; for by her are represented those vain babbling doctrines about the nature of things, which wander abroad in all times and fill the world—doctrines barren in fact, counterfeit in breed, but by reason of their garrulity sometimes entertaining ; and sometimes again troublesome and annoying.

VII.

PERSEUS ;

OR WAR [5].

PERSEUS was sent, it is said, by Pallas to cut off the head of Medusa, from whom many nations in the westernmost parts of Spain suffered grievous calamities :—a monster so dreadful and horrible that the mere sight of her turned men into stone. She was one of the Gorgons ; and the only one of them that was mortal, the others not being subject to change. By way of equipment for this so noble exploit, Perseus received arms and gifts from three several Gods. Mercury gave him wings for his feet ; Pluto gave him a helmet ; Pallas a shield and a mirror. And yet, though so well provided and equipped, he did not proceed against Medusa directly, but went out of his way to visit the Græœ. These were half-sisters to the Gorgons ; and had been born old women with white hair. They had but one eye and one tooth among them, and these they used to wear by turns ; each putting them on as she went abroad, and putting them off again when she came back. This eye and tooth they now lent to Perseus. Whereupon, judging himself sufficiently equipped for the performance of his undertaking, he went against Medusa with all haste, flying. He found her asleep ; but not daring to face her (in case she should wake) he looked back into Pallas's mirror, and taking aim by the reflexion, cut off her head. From the blood which flowed out of the wound, there suddenly leaped forth a winged Pegasus. The severed head was fixed by Perseus in Pallas's shield ; where it still retained its power of striking stiff, as if thunder or planet stricken, all who looked on it.

The fable seems to have been composed with reference to the art and judicious conduct of war. And first, for the kind of war to be chosen it sets forth (as from the advice of Pallas) three sound and weighty precepts to guide the deliberation.

The first is, not to take any great trouble for the subjugation of the neighbouring nations. For the rule to be followed in the enlarging of a patrimony does not apply to the extension of an empire. In a private property, the vicinity of the estates to each other is of importance ; but in extending an empire, occasion, and facility of carrying the war through, and value of conquest, should be regarded instead of vicinity. We see that the Romans, while they had hardly penetrated westward beyond Liguria, had conquered and included in their empire eastern provinces as far off as Mount Taurus. And therefore Perseus, though he belonged to the east, did not decline a distant expedition to the uttermost parts of the west.

The second is that there be a just and honourable cause of war : for this begets alacrity as well in the soldiers themselves as in the people, from whom the supplies are to come : also it opens the way to alliances, and conciliates friends ; and has a great many advantages. Now there is no cause of war more pious than the overthrow of a tyranny under which the people lies prostrate without spirit or vigour, as if turned to stone by the aspect of Medusa.

Thirdly, it is wisely added that whereas there are three Gorgons (by whom are represented wars), Perseus chose the one that was mortal, that is, he chose such a

[5] For an enlarged version of this fable, see above, pp. 448–450.

war as might be finished and carried through, and did not engage in the pursuit of vast or infinite projects.

The equipment of Perseus is of that kind which is everything in war, and almost ensures success; for he received swiftness from Mercury, secrecy of counsel from Pluto, and providence from Pallas. Nor is the circumstance that those wings of swiftness were for the heels and not for the shoulders without an allegorical meaning, and a very wise one. For it is not in the first attack, so much as in those that follow up and support the first, that swiftness is required; and there is no error more common in war than that of not pressing on the secondary and subsidiary actions with an activity answerable to the vigour of the beginnings. There is also an ingenious distinction implied in the images of the shield and the mirror (for the parable of Pluto's helmet which made men invisible needs no explanation) between the two kinds of foresight. For we must have not only that kind of foresight which acts as a shield, but that other kind likewise which enables us (like Pallas's mirror) to spy into the forces and movements and counsels of the enemy.

But Perseus, however provided with forces and courage, stands yet in need of one thing more before the war be commenced, which is of the highest possible importance,—he must go round to the Græae. These Græae are treasons; which are indeed war's sisters, yet not sisters german, but as it were of less noble birth. For wars are generous; treasons degenerate and base. They are prettily described, in allusion to the perpetual cares and trepidations of traitors, as old and white from their birth. Their power (before they break out into open revolt) lies either in the eye or the tooth; for all factions when alienated from the state, both play the spy and bite. And the eye and tooth are as it were common to them all: the eye because all their information is handed from one to another and circulates through the whole party; the tooth, because they all bite with one mouth and all tell one tale,—so that when you hear one you hear all. Therefore Perseus must make friends of those Græae, that they may lend him their eye and tooth,—the eye for discovery of information, the tooth to sow rumours, raise envy, and stir the minds of the people.

These matters being thus arranged and prepared, we come next to the carriage of the war itself. And here we see that Perseus finds Medusa asleep; for the undertaker of a war almost always, if he is wise, takes his enemy unprepared and in security. And now it is that Pallas's mirror is wanted. For there are many who before the hour of danger can look into the enemy's affairs sharply and attentively; but the chief use of the mirror is in the very instant of peril, that you may examine the manner of it without being confused by the fear of it; which is meant by the looking at it with eyes averted.

The conclusion of the war is followed by two effects: first the birth and springing up of Pegasus, which obviously enough denotes fame, flying abroad and celebrating the victory. Secondly the carrying of Medusa's head upon the shield; for this is incomparably the best kind of safeguard. A single brilliant and memorable exploit, happily conducted and accomplished, paralyses all the enemies' movements, and mates malevolence itself.

VIII.

ENDYMION;

OR THE FAVOURITE.

Tradition says that Endymion, a shepherd, was beloved by the moon. But the intercourse between them was of a strange and singular kind. For while he lay reposing according to his habit in a natural cave under the rocks of Latmos, the moon would come down from heaven and kiss him as he slept, and go up into heaven again. And yet this idleness and sleeping did not hurt his fortunes; for the moon in the mean time so ordered it that his sheep fattened and increased exceedingly; insomuch that no shepherd had finer flocks or fuller.

The fable relates (as I take it) to the dispositions and manners of princes. For princes being full of thoughts and prone to suspicions, do not easily admit to familiar intercourse men that are perspicacious and curious, whose minds are always on the watch and never sleep ; but choose rather such as are of a quiet and complying disposition, and submit to their will without inquiring further, and shew like persons ignorant and unobserving, and as if asleep ; displaying simple obedience rather than fine observation. With men of this kind princes have always been glad to descend from their greatness, as the moon from heaven ; and to lay aside their mask, the continual wearing of which becomes a kind of burden ; and to converse familiarly ; for with such they think they can do so safely. It was a point especially noted in Tiberius Cæsar, a prince extremely difficult to deal with ; with whom those only were in favour who, though they really understood him, yet dissembled their knowledge with a pertinacity which seemed like dulness. The same thing was observable in Louis XI. of France, a most cautious and crafty king. The circumstance of the *cave* also, in which according to the fable Endymion used to lie, is not without its elegance. For those who enjoy this kind of favour with princes have commonly some pleasant places of retirement to invite them to, where they may have the comfort of leisure and relaxation of mind, discharged of the incumbrances which their position lays upon them. And it is true that favourites of this class are commonly prosperous in their private fortunes ; for princes though they may not raise them to honours, yet since their favour springs from true affection and not from considerations of utility, they generally enrich them with their bounty.

IX.

THE SISTER OF THE GIANTS ;

OR FAME.

THE poets tell us that the Giants, being brought forth by Earth, made war upon Jupiter and the gods, and were routed and vanquished with thunderbolts, whereupon Earth, in rage at the wrath of the gods, to revenge her sons brought forth Fame, youngest sister of the giants.

The meaning of the fable appears to be this : by Earth is meant the nature of the common people ; always swelling with malice towards their rulers, and hatching revolutions. This upon occasion given brings forth rebels and seditious persons, who with wicked audacity endeavour the overthrow of princes. And when these are suppressed, the same nature of the common people, still leaning to the worse party and impatient of tranquillity, gives birth to rumours and malignant whispers, and querulous fames, and defamatory libels, and the like ; tending to bring envy upon the authorities of the land : so that seditious fames differ from acts of rebellion, not in race and parentage, but only in sex : the one being feminine and the other masculine.

X.

ACTÆON AND PENTHEUS ;

OR CURIOSITY.

THE curiosity and unhealthy appetite of man for the discovery of secrets, is reproved by the ancients in two examples : one of Actæon, the other of Pentheus. Actæon having unawares and by chance seen Diana naked, was turned into a stag and worried by his own dogs. Pentheus having climbed a tree for the purpose of seeing the secret mysteries of Bacchus, was struck with madness ; and the form of his madness was this : he thought everything was double ; saw two suns, and again two cities of Thebes : insomuch that when he set out towards Thebes, he presently saw another Thebes behind, which made him go back ; and so was kept continually going backwards and forwards without any rest.

> As to distracted Pentheus there appear
> Furies in troops, and in the sky two suns,
> And on the earth two several Thebes at once.

The first of these fables seems to relate to the secrets of princes, the other to the secrets of divinity. For whoever becomes acquainted with a prince's secrets without leave and against his will, is sure to incur his hatred : and then knowing that he is marked and that occasions are sought against him, he lives the life of a stag; a life full of fears and suspicions. Often too it happens that his own servants and domestics, to curry favour with the prince, accuse and overthrow him. For when the displeasure of the prince is manifest, a man shall scarcely have a servant but will betray him; and so he may expect the fate of Actæon.

The calamity of Pentheus is of a different kind. For the punishment assigned to those who with rash audacity, forgetting their mortal condition, aspire by the heights of nature and philosophy, as by climbing a tree, to penetrate the divine mysteries, is perpetual inconstancy, and a judgment vacillating and perplexed. For since the light of nature is one thing and the light of divinity another, they are as men that see two suns; and since the actions of life and the determinations of the will depend upon the intellect, it follows that they are perplexed in will no less than in opinion, and cannot be consistent with themselves: in which sense they in like manner see two Thebes; for by Thebes is meant the ends and aims of our actions; Thebes being Pentheus's home and resting-place. And hence it comes that they know not which way to turn, but being uncertain and fluctuating as to the sum and end of all, they are carried round and round from one thing to another, according to the impulse of the moment.

XI.

ORPHEUS;

OR PHILOSOPHY.

THE story of Orpheus, which though so well known has not yet been in all points perfectly well interpreted, seems meant for a representation of universal Philosophy. For Orpheus himself,—a man admirable and truly divine, who being master of all harmony subdued and drew all things after him by sweet and gentle measures,—may pass by an easy metaphor for philosophy personified. For as the works of wisdom surpass in dignity and power the works of strength, so the labours of Orpheus surpass the labours of Hercules.

Orpheus, moved by affection for his wife who had been snatched from him by an untimely death, resolved to go down to Hell and beg her back again of the Infernal Powers; trusting to his lyre. Nor was he disappointed. For so soothed and charmed were the infernal powers by the sweetness of his singing and playing, that they gave him leave to take her away with him; but upon one condition; she was to follow behind him, and he was not to look back until they had reached the confines of light. From this however in the impatience of love and anxiety he could not refrain. Before he had quite reached the point of safety, he looked back; and so the covenant was broken, and she suddenly fell away from him and was hurried back into Hell. From that time Orpheus betook himself to solitary places, a melancholy man and averse from the sight of women; where by the same sweetness of his song and lyre he drew to him all kinds of wild beasts, in such manner that putting off their several natures, forgetting all their quarrels and ferocity, no longer driven by the stings and furies of lust, no longer caring to satisfy their hunger or to hunt their prey, they all stood about him gently and sociably, as in a theatre, listening only to the concords of his lyre. Nor was that all: for so great was the power of his music that it moved the woods and the very stones to shift themselves and take their stations decently and orderly about him. And all this went on for some time with happy success and great admiration; till at last certain Thracian women, under the stimulation and excitement of Bacchus, came where he was; and first they blew such a hoarse and hideous blast upon a horn that the sound of his music could no longer be heard for the din: whereupon the charm being broken that had been the bond of that order and good fellowship, confusion began again; the beasts returned each to his several nature and preyed one upon the other as before; the stones and woods stayed no longer in their places: while Orpheus himself was torn to pieces by the women in their fury, and his limbs scattered about the fields: at whose death, Helicon (river sacred to the Muses) in grief and indignation buried his waters under the earth, to reappear elsewhere.

The meaning of the fable appears to be this. The singing of Orpheus is of two kinds; one to propitiate the infernal powers, the other to draw the wild beasts and the woods. The former may be best understood as referring to natural philosophy; the latter to philosophy moral and civil. For natural philosophy proposes to itself as its noblest work of all, nothing less than the restitution and renovation of things corruptible, and (what is indeed the same thing in a lower degree) the conservation of bodies in the state in which they are, and the retardation of dissolution and putrefaction. Now certainly if this can be effected at all, it cannot be otherwise than by due and exquisite attempering and adjustment of parts in nature, as by the harmony and perfect modulation of a lyre. And yet being a thing of all others the most difficult, it commonly fails of effect; and fails (it may be) from no cause more than from curious and premature meddling and impatience. Then Philosophy finding that her great work is too much for her, in sorrowful mood, as well becomes her, turns to human affairs; and applying her powers of persuasion and eloquence to insinuate into men's minds the love of virtue and equity and peace, teaches the people to assemble and unite and take upon them the yoke of laws and submit to authority, and forget their ungoverned appetites, in listening and conforming to precepts and discipline; whereupon soon follows the building of houses, the founding of cities, the planting of fields and gardens with trees; insomuch that the stones and the woods are not unfitly said to leave their places and come about her. And this application of Philosophy to civil affairs is properly represented, and according to the true order of things, as subsequent to the diligent trial and final frustration of the experiment of restoring the dead body to life. For true it is that the clearer recognition of the inevitable necessity of death sets men upon seeking immortality by merit and renown. Also it is wisely added in the story that Orpheus was averse from women and from marriage; for the sweets of marriage and the dearness of children commonly draw men away from performing great and lofty services to the commonwealth; being content to be perpetuated in their race and stock, and not in their deeds.

But howsoever the works of wisdom are among human things the most excellent, yet they too have their periods and closes. For so it is that after kingdoms and commonwealths have flourished for a time, there arise perturbations and seditions and wars; amid the uproars of which, first the laws are put to silence, and then men return to the depraved conditions of their nature, and desolation is seen in the fields and cities. And if such troubles last, it is not long before letters also and philosophy are so torn in pieces that no traces of them can be found but a few fragments, scattered here and there like planks from a shipwreck; and then a season of barbarism sets in, the waters of Helicon being sunk under the ground, until, according to the appointed vicissitude of things, they break out and issue forth again, perhaps among other nations, and not in the places where they were before.

XII.

CŒLUM;

OR THE ORIGIN OF THINGS.

It is a tradition of the poets that Cœlum was the most ancient of all the gods: that his parts of generation were cut off by his son Saturn with a scythe; that Saturn himself begot a numerous progeny, but devoured his sons as fast as they were born; that at last Jupiter escaped this fate, and as soon as he grew up overthrew his father Saturn, cast him into Tartarus, and took possession of his kingdom; also that he cut off his genitals with the same scythe with which he, Saturn, had cut off those of Cœlum, and threw them into the sea; and that from them was born Venus. Afterwards they say that the kingdom of Jupiter, when as yet it was scarcely settled, had to stand the brunt of two memorable wars; the first, the war of the Titans, in the subduing of whom the assistance of the Sun (the only one of the Titans that was on Jupiter's side) was conspicuous; the second, the war of the Giants, who were likewise by thunder and the arms of Jupiter defeated; and that when these were put down Jupiter reigned afterwards in security.

This fable seems to be an enigma concerning the origin of things, not much differing from the philosophy afterwards embraced by Democritus: who more openly than any one else asserted the eternity of matter, while he denied the eternity of the world; a point in which he came somewhat nearer to the truth as declared in the divine narrative; for that represents matter without form as existing before the six days' works.

The fable may be explained in this manner. By Cœlum is meant the concave or circumference which encloses all matter. By Saturn is meant matter itself; which, inasmuch as the sum total of matter remains always the same and the absolute quantum of nature suffers neither increase nor diminution, is said to have deprived its parent of all power of generation. Now the agitation and motions of matter produced at first imperfect and ill-compacted structures of things, that would not hold together,—mere attempts at worlds. Afterwards in process of time a fabric was turned out which could keep its form. Of these two divisions of time, the first is meant by the reign of Saturn; who by reason of the frequent dissolutions and short durations of things in his time, was called the devourer of his children: the second by the reign of Jupiter, who put an end to those continual and transitory changes, and thrust them into Tartarus—that is to say the place of perturbation: which place seems to be midway between the lowest parts of heaven and the innermost parts of the earth: in which middle region perturbation and fragility and mortality or corruption have their chief operation. And while that former system of generation lasted which had place under the reign of Saturn, Venus, according to the story, was not yet born. For so long as in the universal frame of matter discord was stronger than concord and prevailed over it, there could be no change except of the whole together; and in this manner did the generation of things proceed before Saturn was castrated. But as soon as this mode of generation ceased, it was immediately succeeded by that other which proceeds by Venus, and belongs to a state in which concord being powerful and predominant, change proceeds part by part only, the total fabric remaining entire and undisturbed. Nevertheless Saturn is represented as thrust out and overthrown only, not as cut off and extinguished; because it was the opinion of Democritus that the world might yet relapse into its ancient confusion and intervals of no government: an event which Lucretius prayed might not happen in his own times.

> Which may all-ruling Fortune keep far hence,
> And reason teach it, not experience.

Again, after the world was established and settled in respect of its mass and moving force, yet it did not from the first remain in quiet. For first there followed notable commotions in the heavenly regions; which however, by the power of the Sun predominating in those regions, were so composed that the world survived and kept its state; afterwards in like manner followed convulsions in the lower regions, by inundations, tempests, winds, earthquakes of more universal character than any we now have; and when these likewise were subdued and dispersed, things settled at last into a more durable state of consent and harmonious operation.

It must be said however of all this, that as there is philosophy in the fable so there is fable in the philosophy. For we know (through faith) that all such speculations are but the oracles of sense which have long since ceased and failed; the world, both matter and fabric, being in truth the work of the Creator.

XIII.

PROTEUS;

OR MATTER.

Proteus, the poets tell us, was herdsman to Neptune. He was an old man and a prophet; a prophet moreover of the very first order, and indeed thrice excellent; for he knew all three,—not the future only, but likewise the past and the present; insomuch that besides his power of divination, he was the messenger and interpreter of all antiquity and all secrets. His dwelling was under an immense cave. There it was his custom every day at noon to count his flock of seals and then go to sleep. And if any one wanted his help in any matter, the only way was first to secure his hands with handcuffs, and then to bind him with chains. Whereupon he on his part, in order to get free, would turn himself into all manner of strange shapes—fire, water, wild beasts, etc., till at last he returned again to his original shape.

The sense of this fable relates, it would seem, to the secrets of nature and the conditions of matter. For under the person of Proteus, Matter—the most ancient of all things, next to God—is meant to be represented. Now matter has its habitation under the vault of heaven, as under a cave. And it may be called the servant of Neptune, inasmuch as all the operation and dispensation of matter is effected principally in liquids. The herd or flock of Proteus seems to be nothing else than the ordinary species of animals, plants, minerals, etc., in which matter may be said to diffuse and use itself up ; insomuch that having once made up and finished those species it seems to sleep and rest, as if its task were done ; without applying itself or attempting or preparing to make any more. And this is what is meant by Proteus counting his herd and then going to sleep. Now this is said to take place not in the morning or in the evening, but at noon : that is to say, when the full and legitimate time has come for completing and bringing forth the species out of matter already duly prepared and predisposed : which is the middle point between the first rudiments of them and their declination. And this we know from the sacred history to have been in fact at the very time of the creation. For then it was that by virtue of the divine word *producat* matter came together at the command of the Creator, not by its own circuitous processes, but all at once ; and brought its work to perfection on the instant, and constituted the species. And here the story is complete, as regards Proteus free and at large with his herd. For the universe with its several species according to their ordinary frame and structure, is merely the face of matter unconstrained and at liberty, with its flock of materiate creatures. Nevertheless if any skilful Servant of Nature shall bring force to bear on matter, and shall vex it and drive it to extremities as if with the purpose of reducing it to nothing, then will matter (since annihilation or true destruction is not possible except by the omnipotence of God) finding itself in these straits, turn and transform itself into strange shapes, passing from one change to another till it has gone through the whole circle and finished the period ; when, if the force be continued, it returns at last to itself. And this constraint and binding will be more easily and expeditiously effected, if matter be laid hold on and secured by the hands ; that is, by its extremities. And whereas it is added in the fable that Proteus was a prophet and knew the three times ; this agrees well with the nature of matter : for if a man knew the conditions, affections, and processes of matter, he would certainly comprehend the sum and general issue (for I do not say that his knowledge would extend to the parts and singularities) of all things past, present, and to come.

XIV.

MEMNON ;

OR THE EARLY-RIPE.

MEMNON, according to the poets, was the son of Aurora. Conspicuous for the beauty of his arms, and great in popular reputation, he came to the Trojan war ; where rushing with breathless haste and headlong courage at the highest mark, he engaged Achilles, the bravest of all the Greeks, in single fight ; and fell by his hand. In pity of his fate Jupiter sent birds to grace his funeral, that kept up a continual cry of grief and lamentation. His statue also, as often as the rays of the rising sun touched it, is said to have uttered a mournful sound.

The fable seems meant to apply to the unfortunate deaths of young men of high promise. For such are as it were the sons of the morning, and it commonly happens that, being puffed up with empty and outward advantages, they venture upon enterprises that are beyond their strength, provoke and challenge to combat the bravest heroes, and falling in the unequal conflict are extinguished But the death of such persons is wont to be followed by infinite commiseration for of all mortal accidents there is none so lamentable, none so powerful to move pity, as this cropping of the flower of virtue before its time : the rather because their life has been too short to give occasion of satiety or of envy, which might otherwise mitigate sorrow at their death and temper compassion. And not only do lamentations and wailings hover like those mourner birds about the funeral pile ; but the same feeling of pity lasts long after : and more especially upon all fresh accidents and new movements and beginnings of great events, as by the touch of sunrise, the regret for them is stirred up again and renewed.

XV.

TITHONUS ;

OR SATIETY.

It is an elegant fable they relate of Tithonus ; that Aurora was in love with him, and, desiring to enjoy his company for ever, begged of Jupiter that he might never die ; but forgot, with a woman's thoughtlessness, to add to her petition that neither might he suffer the infirmities of age. So he was exempted from the condition of dying ; but there came upon him a strange and miserable old age, such as he must needs undergo to whom death is denied, while the burden of years continues to grow heavier and heavier ; so that Jupiter, pitying such a condition, changed him at last into a grasshopper.

The fable seems to be an ingenious picture and description of Pleasure ; which in its beginning, or morning-time, is so agreeable that men are fain to pray that such delights may last and be their own for ever ; forgetting that satiety and loathing of the same will come upon them, like old age, before they are aware. So that at last when men have become incapable of the acts of pleasure and yet retain the desire and appetite, they fall to talking and telling stories about the pleasures of their youth, and find their delight in that : as we see in lewd persons, who are always harping upon indecent stories, and in soldiers that are for ever recounting their deeds, like grasshoppers, whose vigour is only in their voice.

XVI.

JUNO'S SUITOR ;

OR DISHONOUR.

The poets tell us that Jupiter in pursuit of his loves assumed many different shapes,—a bull, an eagle, a swan, a shower of gold ; but that when he courted Juno, he turned himself into the ignoblest shape that could be, a very object of contempt and ridicule ; that of a wretched cuckoo, drenched with rain and tempest amazed, trembling, and half dead.

It is a wise fable, derived from the depths of moral science. The meaning is that men are not to flatter themselves that an exhibition of their virtue and worth will win them estimation and favour with everybody. For that depends upon the nature and character of those to whom they apply themselves. If these be persons of no gifts or ornaments of their own, but only a proud and malignant disposition (the character represented by Juno), then they should know that they must put off everything about them that has the least show of honour or dignity, and that it is mere folly in them to proceed any other way ; nay that it is not enough to descend to the baseness of flattery, unless they put on the outward show and character of abjectness and degeneracy.

XVII.

CUPID ;

OR THE ATOM.

The accounts given by the poets of Cupid, or Love, are not properly applicable to the same person ; yet the discrepancy is such that one may see where the confusion is and where the similitude, and reject the one and receive the other.

They say then that Love was the most ancient of all the gods ; the most ancient therefore of all things whatever, except Chaos, which is said to have been coeval with him ; and Chaos is never distinguished by the ancients with divine honour or the name of a god. This Love is introduced without any parent at all ; only that some say he was an egg of Night. And himself out of Chaos begot all things, the gods included. The attributes which are assigned to him are in number four ; he is always an infant ; he is blind ; he is naked ; he is an archer. There was also another Love, the youngest of all the gods, son of Venus, to whom the attributes of the elder are transferred, and whom in a way they suit.

The fable relates to the cradle and infancy of nature, and pierces deep. This Love I understand to be the appetite or instinct of primal matter ; or to speak more plainly, *the natural motion of the atom ;* which is indeed the original and unique force that constitutes and fashions all things out of matter. Now this is entirely without parent ; that is, without cause. For the cause is as it were parent of the effect ; and of this virtue there can be no cause in nature (God always excepted) : there being nothing before it, therefore no efficient ; nor anything more original in nature ; therefore neither kind nor form. Whatever it be therefore, it is a thing positive and inexplicable. And even if it were possible to know the method and process of it, yet to know it by way of cause is not possible ; it being, next to God, the cause of causes—itself without cause. That the method even of its operation should ever be brought within the range and comprehension of human inquiry, is hardly perhaps to be hoped ; with good reason therefore it is represented as an egg hatched by night. Such certainly is the judgment of the sacred philosopher, when he says, *He hath made all things beautiful according to their seasons ; also he hath submitted the world to man's inquiry, yet so that man connot find out the work which God worketh from the beginning to the end.* For the summary law of nature, that impulse of desire impressed by God upon the primary particles of matter which makes them come together, and which by repetition and multiplication produces all the variety of nature, is a thing which mortal thought may glance at, but can hardly take in.

Now the philosophy of the Greeks, which in investigating the material principles of things is careful and acute, in inquiring the principles of motion, wherein lies all vigour of operation, is negligent and languid ; and on the point now in question seems to be altogether blind and babbling; for that opinion of the Peripatetics which refers the original impulse of matter to privation, is little more than words —a name for the thing rather than a description of it. And those who refer it to God, though they are quite right in that, yet they ascend by a leap and not by steps. For beyond all doubt there is a single and summary law in which nature centres and which is subject and subordinate to God ; the same in fact which in the text just quoted is meant by the words, *The work which God worketh from the beginning to the end.* Democritus considered the matter more deeply ; and having first given the atom some dimension and shape, attributed to it a single desire or primary motion simply and absolutely, and a second by comparison. For he thought that all things move by their proper nature towards the centre of the world ; but that that which has more matter, moving thither faster, strikes aside that which has less, and forces it to go the other way. This however was but a narrow theory, and framed with reference to too few particulars : for it does not appear that either the motion of the heavenly bodies in circle, or the phenomena of contraction and expansion, can be reduced to this principle, or reconciled with it. As for Epicurus's opinion of the declination and fortuitous agitation of the atom, it is a relapse to trifling and ignorance. So it is but too plain that the parentage of this Cupid is wrapped in night.

Let us now consider his attributes. He is described with great elegance as a little child, and a child for ever ; for things compounded are larger and are affected by age ; whereas the primary seeds of things, or atoms, are minute and remain in perpetual infancy.

Most truly also is he represented as naked : for all compounds (to one that considers them rightly) are masked and clothed ; and there is nothing properly naked, except the primary particles of things.

The blindness likewise of Cupid has an allegorical meaning full of wisdom. For it seems that this Cupid, whatever he be, has very little providence ; but directs his course, like a blind man groping, by whatever he finds nearest ; which makes the supreme divine Providence all the more to be admired, as that which contrives out of subjects peculiarly empty and destitute of providence, and as it were blind, to educe by a fatal and necessary law all the order and beauty of the universe.

His last attribute is archery : meaning that this virtue is such as acts at a distance : for all operation at a distance is like shooting an arrow. Now whoever maintains the theory of the atom and the vacuum (even though he suppose the vacuum not to be collected by itself but intermingled through space), necessarily implies the action of the virtue of the atom at a distance ; for without this no motion could be originated, by reason of the vacuum interposed ; but all things would remain fixed and immovable.

As for that younger Cupid, it is with reason that he is reported to be the youngest of the gods ; since until the species were constituted he could have no operation. In the description of him the allegory changes its aim and passes to morals. And yet there remains a certain conformity between him and the elder Cupid. For Venus excites the general appetite of conjunction and procreation ; Cupid, her son, applies the appetite to an individual object. From Venus therefore comes the general disposition, from Cupid the more exact sympathy. Now the general disposition depends upon causes near at hand, the particular sympathy upon principles more deep and fatal, and as if derived from that ancient Cupid, who is the source of all exquisite sympathy.

XVIII.

DIOMEDES ;

OR RELIGIOUS ZEAL.

DIOMEDES, a hero of high renown and a special favourite of Pallas, was incited by her (being of himself apt enough) if he chanced to encounter Venus in the battle, not to spare her. He boldly did as he was bid, and wounded Venus in the hand. This for the time he carried with impunity, and returned to his own country in great fame and reputation : but meeting there with domestic troubles he took refuge abroad in Italy. Here also he had a good enough fortune at first. King Daunus entertained him with hospitality and enriched him with honours and presents, and many statues were raised to him throughout the country. But no sooner did a calamity befall the people among whom he had taken up his abode, than Daunus bethought him that he was entertaining under his roof a man impious and hated by the gods, a fighter against heaven, who had violently assaulted and wounded with the sword a goddess whom it was forbidden even to touch. Whereupon, to free his country from the curse under which it lay, he suddenly (setting aside the bond of hospitality, in respect of the more ancient bond of religion) puts Diomedes to death, and orders his statues to be thrown down and his honours cancelled. Nor was it safe in such a case even to pity so grievous an accident ; but his comrades likewise, when they bewailed the death of their chief and filled the land with lamentations, were changed into a kind of swans,—a bird which at the approach of its own death also utters a sweet and plaintive sound.

The subject of this fable is rare and almost singular ; for there is no other story in which any hero is represented as having wounded a god. This is told of Diomedes only : and in him certainly seems meant to be portrayed the character and fortunes of a man who makes it his declared object to persecute and overthrew by violence and the sword some religious worship or sect, though a vain and light one. For though religious wars were unknown to the ancients (the heathen gods having no touch of jealousy, which is the attribute of the true God), yet so great appears to have been the wisdom of the primitive ages and so wide the range of it, that what they did not know by experience they nevertheless attained in idea by reflexion and imagination.

Now those who make war against any religious sect, though a vain, corrupt, and infamous one (and this is signified in the person of Venus), proceeding not by force of reason and doctrine and by sanctity of life and by weight of examples and authorities to correct and confute, but by fire and sword and sharpness of punishment to cut out and exterminate the same ;—such persons are perhaps set upon the work by Pallas,—that is, by a certain keenness of discernment and severity of judgment which gives them a thorough insight into the fallacies and falsehoods of such errors, joined with hatred of evil and honest zeal ;—and for a time they commonly acquire great glory, and are by the vulgar (who can never like what is moderate) celebrated and almost worshipped as the only champion of truth and religion ; all others appearing lukewarm and timid. And yet this glory and felicity seldom endures to the end ; but almost every kind of violence, unless by an early death it escape the vicissitudes of fortune, is in the end unprosperous. And if it so happen that an alteration takes place in the state, whereby that proscribed and depressed sect gathers strength and raises its head, then are the zealous and contentious courses of these men condemned, their very name hated, and all their honours turned into reproach. The murder of Diomedes by the hands of his host alludes to the fact that difference in matter of religion breeds falsehood and treachery even among the nearest and dearest friends. And where it is said that the very grief and lamentations of his comrades were not tolerated, but visited with punishment, the meaning is that whereas almost every crime is open to pity, insomuch that they who hate the offence may yet in humanity commiserate the person and calamity of the offender, —and it is the extremity of evil to have the offices of compassion interdicted,— yet where religion and piety are in question, the very expression of pity is noted and disliked. On the other hand, the sorrows and lamentations of the comrades of Diomedes, that is of those who are of the same sect and opinion, are commonly very piercing and musical, like the notes of swans, or birds of Diomedes. And this part of the allegory has a further meaning which is striking and noble ; namely that in the case of persons who suffer for religion the words which they speak at their death, like the song of the dying swan, have a wonderful effect and impression upon men's minds, and dwell long after in their memory and feelings.

XIX.

DÆDALUS ;

OR THE MECHANIC.

UNDER the person of Dædalus, a man of the greatest genius but of very bad character, the ancients drew a picture of mechanical skill and industry, together with its unlawful artifices and depraved applications. Dædalus had been banished for murdering a fellow-pupil and rival ; yet found favour in his banishment with kings and states. Many and excellent works, as well in honour of the gods as for the adornment and ennobling of cities and public places, had been built and modelled by him ; but it is for unlawful inventions that his name is most famous. For he it was who supplied the machine which enabled Pasiphae to satisfy her passion for the bull ; so that the unhappy and infamous birth of the monster Minotaurus, which devoured the ingenuous youth, was owing to the wicked industry and pernicious genius of this man. Then to conceal the first mischief he added another, and for the security of this pest devised and constructed the Labyrinth ; a work wicked in its end and destination, but in respect of art and contrivance excellent and admirable. Afterwards again, that his fame might not rest on bad arts only, and that he might be sought to for remedies as well as instruments of evil, he became the author likewise of that ingenious device of the clue, by which the mazes of the labyrinth should be retraced. This Dædalus was persecuted with great severity and diligence and inquisition by Minos ; yet he always found both means of escape and places of refuge. Last of all, he taught his son Icarus how to fly ; who being a novice and ostentatious of his art fell from the sky into the water.

The parable may be interpreted thus. In the entrance is noted that envy which is strongly predominant in great artists and never lets them rest ; for there is no class of men more troubled with envy, and that of the bitterest and most implacable character.

Then is touched the impolitic and improvident nature of the punishment inflicted ; namely banishment. For it is the prerogative of famous workmen to be acceptable all over the world, insomuch that to an excellent artisan exile is scarcely any punishment at all. For whereas other modes and conditions of life cannot easily flourish out of their own country, the admiration of an artisan spreads wider and grows greater among strangers and foreigners; it being the nature of men to hold their own countrymen, in respect of mechanical arts, in less estimation.

The passages which follow concerning the use of mechanical arts are plain enough. Certainly human life is much indebted to them, for very many things which concern both the furniture of religion and the ornament of state and the culture of life in general, are drawn from their store. And yet out of the same fountain come instruments of lust, and also instruments of death. For (not to speak of the arts of procurers) the most exquisite poisons, also guns, and such like engines of destruction, are the fruits of mechanical invention; and well we know how far in cruelty and destructiveness they exceed the Minotaurus himself.

Very beautiful again is that allegory of the labyrinth; under which the general nature of mechanics is represented. For all the more ingenious and exact mechanical inventions may, for their subtlety, their intricate variety, and the apparent likeness of one part to another, which scarcely any judgment can order and discriminate, but only the clue of experiment, be compared to a labyrinth. Nor is the next point less to the purpose; viz. that the same man who devised the mazes of the labyrinth disclosed likewise the use of the clue. For the mechanical arts may be turned either way, and serve as well for the cure as for the hurt and have power for the most part to dissolve their own spell.

Moreover the unlawful contrivances of art, and indeed the arts themselves, are often persecuted by Minos; that is by the laws; which condemn them and forbid people to use them. Nevertheless they are secretly preserved, and find everywhere both hiding-places and entertainment; as was well observed by Tacitus in his times, in a case not much unlike; where speaking of the mathematicians and fortune-tellers, he calls them *a class of men which in our state will always be retained and always prohibited.* And yet these unlawful and curious arts do in tract of time, since for the most part they fail to perform their promises, fall out of estimation, as Icarus from the sky, and come into contempt, and through the very excess of ostentation perish. And certainly if the truth must be told, they are not so easily bridled by law as convicted by their proper vanity.

XX.

ERICTHONIUS;

OR IMPOSTURE.

THE poets tell us that Vulcan wooed Minerva, and in the heat of desire attempted to force her; that in the struggle which followed his seed was scattered on the ground; from which was born Ericthonius, a man well made and handsome in the upper parts of the body, but with thighs and legs like an eel, thin and deformed: and that he, from consciousness of this deformity, first invented chariots, whereby he might shew off the fine part of his body and hide the mean.

This strange and prodigious story seems to bear this meaning: that Art (which is represented under the person of Vulcan, because it makes so much use of fire) when it endeavours by much vexing of bodies to force Nature to its will and conquer and subdue her (for Nature is described under the person of Minerva on account of the wisdom of her works) rarely attains the particular end it aims at; and yet in the course of contriving and endeavouring, as in a struggle, there fall out by the way certain imperfect births and lame works, specious to look at but weak and halting in use: yet impostors parade them to the world with a great deal of false shew in setting forth, and carry them about as in triumph. Such things may often be observed among chemical productions, and among mechanical subtleties and novelties; the rather because men being too intent upon their end to recover themselves from the errors of their way, rather struggle with Nature than woo her embraces with due observance and attention.

XXI.

DEUCALION;

OR RESTORATION.

THE poets relate that when the inhabitants of the old world were utterly extinguished by the universal deluge, and none remained except Deucalion and Pyrrha, these two, being inflamed with a pious and noble desire to restore the human race, consulted the oracle and received answer to the following effect: they should have their wish if they took their mother's bones and cast them behind their backs. This struck them at first with great sorrow and despair, for the face of nature being laid level by the deluge, to seek for a sepulchre would be a task altogether endless. But at last they found that the stones of the earth (the earth being regarded as the mother of all things) were what the oracle meant.

This fable seems to disclose a secret of nature, and to correct an error which is familiar to the human mind. For man in his ignorance concludes that the renewal and restoration of things may be effected by means of their own corruption and remains; as the Phœnix rises out of her own ashes; which is not so; for matters of this kind have already reached the end of their course, and can give no further help towards the first stages of it: so we must go back to more common principles.

XXII.

NEMESIS;

OR THE VICISSITUDE OF THINGS.

NEMESIS, according to the tradition, was a goddess, the object of veneration to all, to the powerful and fortunate of fear also. They say she was the daughter of Night and Ocean. She is represented with wings, and a crown: an ashen spear in her right hand; a phial, with Ethiops in it, in her left; sitting upon a stag.

The parable may be understood thus. The very name Nemesis plainly signifies Revenge or Retribution: for it was the office and function of this goddess to interrupt the felicity of fortunate persons, and let no man be constantly and perpetually happy, but step in like a tribune of the people with her *veto*; and not to chastise insolence only, but to see also that prosperity however innocent and moderately borne had its turn of adversity: as if no one of human race could be admitted to the banquets of the gods, except in derision. And certainly when I have read that chapter of Caius Plinius in which he has collected the misfortunes and miseries of Augustus Cæsar,—him whom I thought of all men the most fortunate, and who had moreover a certain art of using and enjoying his fortune, and in whose mind were no traces of swelling, of lightness, of softness, of confusion, or of melancholy—(insomuch that he had once determined to die voluntarily),—great and powerful must this goddess be, I have thought, when such a victim was brought to her altar.

The parents of this goddess were Ocean and Night; that is, the vicissitude of things, and the dark and secret judgment of God. For the vicissitude of things is aptly represented by the Ocean, by reason of its perpetual flowing and ebbing; and secret providence is rightly set forth under the image of Night. For this Nemesis of the Darkness (the human not agreeing with the divine judgment) was matter of observation even among the heathen.

> Ripheus fell too,
> Than whom a juster and a truer man
> In all his dealings was not found in Troy.
> But the gods judged not so.

Nemesis again is described as winged; because of the sudden and unforeseen revolutions of things. For in all the records of time it has commonly been found that great and wise men have perished by the dangers which they most despised. So was it with M. Cicero; who when warned by Decimus Brutus to beware of Octavius Cæsar's bad faith and evil mind towards him, only answered, *I am duly grateful to you, my dear Brutus, for giving me that information, though it is but folly.*

Nemesis is distinguished also with a crown; in allusion to the envious and malignant nature of the vulgar; for when the fortunate and the powerful fall, the people commonly exult and set a crown upon the head of Nemesis.

The spear in her right hand relates to those whom she actually strikes and transfixes. And if there be any whom she does not make victims of calamity and misfortune, to them she nevertheless exhibits that dark and ominous spectre in her left: for mortals must needs be visited, even when they stand at the summit of felicity, with images of death, diseases, misfortunes, perfidies of friends, plots of enemies, changes of fortune, and the like; even like those Ethiops in the phial. It is true that Virgil, in describing the battle of Actium, adds elegantly concerning Cleopatra:—

> Midmost the Queen with sounding timbrel cheers
> Her armies to the fight; nor dreams the while
> Of those two aspics at her back.

But it was not long before, turn which way she would, whole troops of Ethiops met her eyes.

Lastly, it is wisely added that Nemesis is mounted on a stag: for the stag is a very long lived animal; and it may be that one who is cut off young may give Nemesis the slip; but if his prosperity and greatness endure for any length of time, he is without doubt a subject of Nemesis, and carries her as it were on his back.

XXIII.

ACHELOUS;

OR THE BATTLE.

The ancients relate that when Hercules and Achelous disputed which should marry Deianira, they agreed to decide the question by a fight. Now Achelous began by trying a variety of different shapes, which he was at liberty to do, and presented himself before Hercules at last in the shape of a savage and roaring bull, and so prepared for the combat. Hercules on the other hand retaining his wonted human figure, fell upon him. A close fight followed; the end of which was that Hercules broke off one of the bull's horns: whereupon he, greatly hurt and terrified, to redeem his own horn gave Hercules the horn of Amalthea, or Abundance, in exchange.

The fable alludes to military expeditions. The preparations for war on the part defensive (which is represented by Achelous) is various and multiform. For the form assumed by the invader is one and simple, consisting of an army only, or perhaps a fleet. Whereas a country preparing to receive an enemy on its own ground sets to work in an infinity of ways; fortifies one town, dismantles another, gathers the people from the fields and villages into cities and fortified places; builds a bridge here, breaks down a bridge there; raises, and distributes, forces and provisions; is busy about rivers, harbours, gorges of hills, woods, and numberless other matters; so that it may be said to try a new shape and put on a new aspect every day; and when at last it is fully fortified and prepared, it represents to the life the form and threatening aspect of a fighting bull. The invader meanwhile is anxious for a battle, and aims chiefly at that; fearing to be left without supplies in an enemy's country; and if he win the battle, and so break as it were the enemy's horn, then he brings it to this: that the enemy, losing heart and reputation, must, in order to recover himself and repair his forces, fall back into his more fortified positions, leaving his cities and lands to the conqueror to be laid waste and pillaged; which is indeed like giving him Amalthea's horn.

XXIV.

DIONYSUS;[6]

OR DESIRE.

THEY say that Semele, Jupiter's paramour, made him take an inviolable oath to grant her one wish, whatever it might be, and then prayed that he would come to her in the same shape in which he was used to come to Juno. The consequence was that she was scorched to death in his embrace. The infant in her womb was taken by its father and sewed up in his thigh, until the time of gestation should be accomplished. The burden made him limp, and the infant, because while it was carried in his thigh it caused a pain or pricking, received the name of Dionysus. After he was brought forth he was sent to Proserpina for some years to nurse; but as he grew up his face was so like a woman's, that it seemed doubtful of which sex he was. Moreover he died and was buried for a time, and came to life again not long after. In his early youth he discovered and taught the culture of the vine, and therewithal the composition and use of wine, which had not been known before: whereby becoming famous and illustrious, he subjugated the whole world and advanced to the furthest limits of India. He was borne in a chariot drawn by tigers; about him tripped certain deformed demons called Cobali,—Acratus and others. The Muses also joined his train. He took to wife Ariadne, whom Theseus had abandoned and deserted. His sacred tree was the Ivy. He was accounted likewise the inventor and founder of sacred rites and ceremonies; yet such as were fanatical and full of corruption, and cruel besides. He had power to excite phrensy. At least it was by women excited to phrensy in his orgies that two illustrious persons, Pentheus and Orpheus, are said to have been torn to pieces; the one having climbed a tree to see what they were doing; the other in the act of striking his lyre. Moreover the actions of this god are often confounded with those of Jupiter.

The fable seems to bear upon morals, and indeed there is nothing better to be found in moral philosophy. Under the person of Bacchus is described the nature of Desire, or passion and perturbation. For the mother of all desire, even the most noxious, is nothing else than the appetite and aspiration for apparent good: and the conception of it is always in some unlawful wish, rashly granted before it has been understood and weighed. But as the passion warms, its mother (that is, the nature of good), not able to endure the heat of it, is destroyed and perishes in the flame. Itself while still in embryo remains in the human soul (which is its father and represented by Jupiter), especially in the lower part of the soul, as in the thigh; where it is both nourished and hidden; and where it causes such prickings, pains, and depressions in the mind, that its resolutions and actions labour and limp with it. And even after it has grown strong by indulgence and custom, and breaks forth into acts, it is nevertheless brought up for a time with Proserpina; that is to say, it seeks hiding-places, and keeps itself secret and as it were underground; until casting off all restraints of shame and fear and growing bold, it either assumes the mask of some virtue or sets infamy itself, at defiance. Most true also it is that every passion of the more vehement kind is as it were of doubtful sex, for it has at once the force of the man and the weakness of the woman. It is notably said too that Bacchus came to life again after death. For the passions seem sometimes to be laid asleep and extinguished; but no trust can be placed in them, no not though they be buried; for give them matter and occasion, they rise up again.

It is a wise parable too, that of the invention of the Vine; for every passion is ingenious and sagacious in finding out its own stimulants. And there is nothing we know of so potent and effective as wine, in exciting and inflaming perturbations of every kind; being a kind of common fuel to them all. Very elegantly too is Passion represented as the subjugator of provinces, and the undertaker of an endless course of conquest. For it never rests satisfied with what it has, but goes on and on with infinite insatiable appetite panting after new triumphs.

[6] For a slightly enlarged version of this fable, see pp. 450–452.

Tigers also are kept in its stalls and yoked to its chariot; for as soon as Passion ceases to go on foot and comes to ride in its chariot, as in celebration of its victory and triumph over reason, then is it cruel, savage, and pitiless towards every thing that stands in its way. Again, there is humour in making those ridiculous demons dance about the chariot: for every passion produces motions in the eyes, and indeed in the whole countenance and gesture, which are uncomely, unsettled, skipping and deformed; insomuch that when a man under the influence of any passion, as anger, scorn, love, or the like, seems most grand and imposing in his own eyes, to the lookers-on he appears unseemly and ridiculous. It is true also that the Muses are seen in the train of Passion, there being scarce any passion which has not some branch of learning to flatter it. For herein the majesty of the Muses suffers from the licence and levity of men's wits, turning those that should be the guides of man's life into mere followers in the train of his passions.

And again that part of the allegory is especially noble which represents Bacchus as lavishing his love upon one whom another man had cast off. For most certain it is that passion ever seeks and aspires after that which experience has rejected. And let all men who in the heat of pursuit and indulgence are ready to give any price for the fruition of their passion, know this—that whatever be the object of their pursuit, be it honour or fortune or love or glory or knowledge, or what it will, they are paying court to things cast off,—things which many men in all times have tried, and upon trial rejected with disgust.

Nor is the consecration of the Ivy to Bacchus without its mystery. For this has a double propriety. First because the Ivy flourishes in winter; next because it has the property of creeping and spreading about so many things,—as trees, walls, buildings. For as to the first, every passion flourishes and acquires vigour by being resisted and forbidden, as by a kind of antiperistasis; like the ivy by the cold of winter. As to the second, the master passion spreads itself like ivy about all human actions and resolutions, forcing itself in and mixing itself up with them. Nor is it wonderful that superstitious rites are attributed to Bacchus, since every insane passion grows rank in depraved religions; or if phrensies are supposed to be inflicted by him, seeing that every passion is itself a brief madness, and if it be vehement and obstinate ends in insanity. Again, that circumstance of the tearing of Pentheus and Orpheus has an evident allegorical meaning; since curious inquisition and salutary and free admonition are alike hateful and intolerable to an overpowering passion.

Lastly, the confusion of the persons of Bacchus and Jupiter may be well understood as a parable; inasmuch as deeds of high distinction and desert proceed sometimes from virtue and right reason and magnanimity, and sometimes (however they may be extolled and applauded) only from some lurking passion or hidden lust; and thus the deeds of Bacchus are not easily distinguished from the deeds of Jupiter.

XXV.

ATALANTA;

OR PROFIT.

Atalanta, who was remarkable for swiftness, was matched to run a race with Hippomenes. The conditions were that if Hippomenes won he was to marry Atalanta, if he lost he was to be put to death; and there seemed to be no doubt about the issue, since the matchless excellence of Atalanta in running had been signalized by the death of many competitors. Hippomenes therefore resorted to an artifice. He provided himself with three golden apples, and carried them with him. The race began. Atalanta ran ahead. He seeing himself left behind bethought him of his stratagem, and rolled forward one of the golden apples, so that she might see it,—not straight forwards, but a little on one side, that it might not only delay her but also draw her out of the course. She, with a woman's eagerness, attracted by the beauty of the apple, left the course, ran after it, and stooped to take it up. Hippomenes in the meantime made good way along the course and got before her. She however by force of her natural swiftness made good the loss of time and was again foremost; when Hippomenes a second and a third time interrupted her in the same way, and so at last by craft not speed won the race.

The story carries in it an excellent allegory, relating to the contest of art with Nature. For Art, which is meant by Atalanta, is in itself, if nothing stand in the way, far swifter than Nature and, as one may say, the better runner, and comes sooner to the goal. For this may be seen in almost everything; you see that fruit grows slowly from the kernel, swiftly from the graft; you see clay harden slowly into stones, fast into baked bricks: so also in morals, oblivion and comfort of grief comes by nature in length of time; but philosophy (which may be regarded as the art of living) does it without waiting so long, but forestalls and anticipates the day. But then this prerogative and vigour of art is retarded, to the infinite loss of mankind, by those golden apples. For there is not one of the sciences or arts which follows the true and legitimate course constantly forth till it reach its end; but it perpetually happens that arts stop in their undertakings half way, and forsake the course, and turn aside like Atalanta after profit and commodity,—

Leaving the course the rolling gold to seize.

And therefore it is no wonder if Art cannot outstrip Nature, and according to the agreement and condition of the contest put her to death or destroy her; but on the contrary Art remains subject to Nature, as the wife is subject to the husband.

XXVI.

PROMETHEUS;

OR THE STATE OF MAN.

Tradition says that Man was made by Prometheus, and made of clay; only that Prometheus took particles from different animals and mixed them in. He, desiring to benefit and protect his own work, and to be regarded not as the founder only but also as the amplifier and enlarger of the human race, stole up to heaven with a bundle of fennel-stalks in his hand, kindled them at the chariot of the sun, and so brought fire to the earth and presented it to mankind. For this so great benefit received at his hands, men (it is said) were far from being grateful; so far indeed, that they conspired together and impeached him and his invention before Jupiter. This act of theirs was not so taken as justice may seem to have required. For the accusation proved very acceptable both to Jupiter and the rest of the gods; and so delighted were they, that they not only indulged mankind with the use of fire, but presented them likewise with a new gift, of all others most agreeable and desirable,—perpetual youth. Overjoyed with this, the foolish people put the gift of the gods on the back of an ass. The ass on his way home, being troubled with extreme thirst, came to a fountain; but a serpent that was set to guard it, would not let him drink unless he gave in payment whatever that was that he carried on his back. The poor ass accepted the condition; and so for a mouthful of water the power of renewing youth was transferred from men to serpents. After mankind had lost their prize, Prometheus made up his quarrel with them; but retaining his malice, and being bitterly incensed against Jupiter, he did not scruple to tempt him with deceit, even in the act of sacrifice. Having slain (it is said) two bulls, he stuffed the hide of one of them with the flesh and fat of both, and bringing them to the altar, with an air of devotion and benignity offered Jupiter his choice. Jupiter, detesting his craft and bad faith, but knowing how to requite it, chose the mock bull; then bethinking him of vengeance, and seeing that there was no way to take down the insolence of Prometheus except by chastising the human race (of which work he was extravagantly vain and proud), ordered Vulcan to make a fair and lovely woman. When she was made, each of the gods bestowed upon her his several gift; whence she was called Pandora. Then they placed in her hands an elegant vase, in which were enclosed all mischiefs and calamities; only at the bottom there remained Hope. With her vase in her hand she repaired first of all to Prometheus, to see if he would take and open it, which he, cautious and cunning, declined. Thus rejected she went away to Epimetheus, Prometheus's brother, but of a character entirely different, who opened it without hesitation; but as soon as he saw all the mischiefs rushing out, growing wise when it was too late, he struggled to get the lid on again as fast as possible; but it was all he could do to keep in the last of the party, which was Hope, that lay at the bottom. In the end Jupiter seized Prometheus, and upon many and grave charges,—as that of old he had stolen fire, that he had made a mock of Jupiter's majesty in that deceitful sacrifice, that he had scorned and rejected his gift, together with another not mentioned before, that he had attempted to ravish Minerva,—threw him into chains and condemned him to perpetual tortures. For by Jupiter's command he was dragged to Mount Caucasus, and there bound fast to a column so that he could not stir. And there was an eagle which gnawed and consumed his liver by day; but what was eaten in the day grew again in the night, so that matter was never wanting for the torture to work upon. Yet they say that this punishment had its end at last; for Hercules sailed across the ocean in a cup that was given to him by the Sun, came to Caucasus, shot the eagle with his arrows, and set Prometheus free. In honour of Prometheus there were instituted in some nations games called torch-races, in which the runners carried lighted torches in their hands; and if any went out the bearer stood aside, leaving the victory to those that followed; and the first who reached the goal with his torch still burning received the prize.

This fable carries in it many true and grave speculations both on the surface and underneath. For there are some things in it that have been long ago observed: others have never been touched at all.

Prometheus clearly and expressly signifies Providence: and the one thing singled out by the ancients as the special and peculiar work of Providence was the creation and constitution of Man. For this one reason no doubt was, that the nature of man includes mind and intellect, which is the seat of providence; and since to derive mind and reason from principles brutal and irrational would be harsh and incredible, it follows almost necessarily that the human spirit was endued with providence not without the precedent and intention and warrant of the greater providence. But this was not all. The chief aim of the parable appears to be, that Man, if we look to final causes, may be regarded as the centre of the world; insomuch that if man were taken away from the world, the rest would seem to be all astray, without aim or purpose, to be like a besom without a binding, as the saying is, and to be leading to nothing. For the whole world works together in the service of man; and there is nothing from which he does not derive use and fruit. The revolutions and courses of the stars serve him both for distinction of the seasons and distribution of the quarters of the world. The appearances of the middle sky afford him prognostications of weather. The winds sail his ships and work his mills and engines. Plants and animals of all kinds are made to furnish him either with dwelling and shelter or clothing or food or medicine, or to lighten his labour, or to give him pleasure and comfort; insomuch that all things seem to be going about man's business and not their own. Nor is it without meaning added that in the mass and composition of which man was made, particles taken from the different animals were infused and mixed up with the clay; for it is most true that of all things in the universe man is the most composite, so that he was not without reason called by the ancients the little world. For though the Alchemists, when they maintain that there is to be found in man every mineral, every vegetable, etc., or something corresponding to them, take the word *microcosm* in a sense too gross and literal, and have so spoiled the elegance and distorted the meaning of it, yet that the body of man is of all existing things both the most mixed and the most organic, remains not the less a sober and solid truth. And this is indeed the reason it is capable of such wonderful powers and faculties; for the powers of simple bodies, though they be certain and rapid, yet being less refracted, broken up, and counteracted by mixture, they are few; but abundance and excellence of power resides in mixture and composition. Nevertheless we see that man in the first stage of his existence is a naked and defenceless thing, slow to help himself, and full of want. Therefore Prometheus applied himself with all haste to the invention of fire; which in all human necessities and business is the great minister of relief and help; insomuch that if the soul be the form of forms and the hand the instrument of instruments, fire may rightly be called the help of helps and the mean of means. For through it most operations are effected, through it the arts mechanical and the sciences themselves are furthered in an infinite variety of ways.

Now the description of the manner in which the theft of fire was accomplished is apt and according to the nature of the thing. It was by applying a stalk of fennel to the chariot of the Sun. For fennel is used as a rod to strike with. The meaning therefore clearly is that Fire is produced by violent percussions and collisions of one body with another; whereby the matter they are made of is attenuated and set in motion, and prepared to receive the heat of the celestial bodies, and so by clandestine processes, as by an act of theft, snatches fire as it were from the chariot of the Sun.

There follows a remarkable part of the parable. Men, we are told, instead of gratulation and thanksgiving fell to remonstrance and indignation, and brought an accusation before Jupiter both against Prometheus and against Fire; and this act was moreover by him so well liked, that in consideration of it he accumulated fresh benefits upon mankind. For how should the crime of ingratitude towards their maker, a vice which includes in itself almost all others, deserve approbation and reward? and what could be the drift of such a fiction? But this is not what is meant. The meaning of the allegory is, that the accusation and arraignment by men both of their own nature and of art, proceeds from an excellent condition of mind and issues in good; whereas the contrary is hated by the gods, and unlucky. For they who extravagantly extol human nature as it is and the arts as received; who spend themselves in admiration of what they already possess, and hold up as perfect the sciences which are professed and cultivated; are wanting, first, in reverence to the divine nature, with the perfection of which they almost presume to compare, and next in usefulness towards man; as thinking that they have already reached the summit of things and finished their work, and therefore need seek no further. They on the other hand who arraign and accuse nature and the arts, and abound with complainings are not only more modest (if it be truly considered) in their sentiment, but are also stimulated perpetually to fresh industry and new discoveries. And this makes me marvel all the more at the ignorance and evil genius of mankind, who being overcrowed by the arrogance of a few persons, hold in such honour that philosophy of the Peripatetics, which was but a portion, and no large portion either, of the Greek philosophy, that every attempt to find fault with it has come to be not only useless, but also suspected and almost dangerous. Whereas certainly in my opinion both Empedocles and Democritus, who complain, the first madly enough, but the second very soberly, that all things are hidden away from us, that we know nothing, that we discern nothing, that truth is drowned in deep wells, that the true and the false are strangely joined and twisted together, (for the new Academy carried it a great deal too far), are more to be approved than the school of Aristotle, so confident and dogmatical. Therefore let all men know that the preferring of complaints against nature and the arts is a thing well pleasing to the gods, and draws down new alms and bounties from the divine goodness; and that the accusation of Prometheus, our maker and master though he be, yea sharp and vehement accusation, is a thing more sober and profitable than this overflow of congratulation and thanksgiving: let them know that conceit of plenty is one of the principal causes of want.

Now for the gift which men are said to have received as the reward of their accusation, namely the unfading flower of youth; it seems to show that methods and medicines for the retardation of age and the prolongation of life were by the ancients not despaired of, but reckoned rather among those things which men once had and by sloth and negligence let slip, than among those which were wholly denied or never offered. For they seem to say that by the true use of fire, and by the just and vigorous accusation and conviction of the errors of art, such gifts might have been compassed; and that it was not the divine goodness that was wanting to them therein, but they that were wanting to themselves; in that having received this gift of the gods, they committed the carriage of it to a lazy and slow-paced ass. By this seems to be meant experience; a thing stupid and full of delay, whose slow and tortoise-like pace gave birth to that ancient complaint that *life is short and art is long.* And for my own part I certainly think that those two faculties—the Dogmatical and the Empirical—have not yet been well united and coupled ; but that the bringing down of new gifts from the gods has ever been left either to the abstract philosophies, as to a light bird ; or to sluggish and tardy experience, as to an ass. And yet it must be said in behalf of the ass, that he might perhaps do well enough, but for that accident of thirst by the way. For if a man would put himself fairly under the command of experience, and proceed steadily onward by a certain law and method, and not let any thirst for experiments either of profit or ostentation seize him by the way and make him lay down and unsettle his burthen in order that he may taste them,—such a man I do think would prove a carrier to whom new and augmented measures of divine bounty might be well enough entrusted.

As for the transfer of the gift to serpents, it seems to be an addition merely for ornament ; unless it were inserted in shame of mankind, who with that fire of theirs and with so many arts, cannot acquire for themselves things which nature has of herself bestowed on many other animals.

The sudden reconciliation of men with Prometheus after the frustration of their hope, contains likewise a wise and useful observation. It alludes to the levity and rashness of men in new experiments ; who if an experiment does not at once succeed according to wish, are in far too great a hurry to give up the attempt as a failure, and so tumble back to where they were and take on with the old things again.

Having thus described the state of man in respect of arts and matters intellectual, the parable passes to Religion ; for with the cultivation of the arts came likewise the worship of things divine ; and this was immediately seized on and polluted by hypocrisy. Therefore under the figure of that double sacrifice is elegantly represented the person of the truly religious man and the hypocrite. For in the one there is the fat, which is God's portion, by reason of the flame and sweet savour, whereby is meant affection and zeal burning and rising upward for the glory of God. In him are the bowels of charity ; in him wholesome and useful meat. In the other is found nothing but dry and bare bones, with which the skin is stuffed out till it looks like a fair and noble victim : whereby are signified those external and empty rites and ceremonies with which men overload and inflate the service of religion : things rather got up for ostentation than conducing to piety. Nor is it enough for men to offer such mockeries to God, but they must also lay and father them upon himself, as though he had himself chosen and prescribed them. It is against such a kind of choice that the prophet in God's person remonstrates, when he says, *Is this such a fast as I have* CHOSEN, *that man should afflict his soul for one day and bow his head like a bulrush ?*

After touching the state of Religion, the parable turns to morals and the conditions of human life. Pandora has been generally and rightly understood to mean pleasure and sensual appetite ; which after the introduction of civil arts and culture and luxury, is kindled up as it were by the gift of fire. To Vulcan therefore, who in like manner represents fire, the making of Pleasure is imputed. And from her have flowed forth infinite mischief upon the minds, the bodies, and the fortunes of men, together with repentance when too late ; nor upon individuals only, but upon kingdoms also and commonwealths. For from this same fountain have sprung wars and civil disturbances and tyrannies. But it is worth while to observe how prettily and elegantly the two conditions and as it were pictures or models of human life are set forth in the story, under the persons of Prometheus and Epimetheus. The followers of Epimetheus are the improvident, who take no care for the future but think only of what is pleasant at the time ; and on this account it is true that they suffer many distresses, difficulties, and calamities, and are engaged in a perpetual struggle with them ; and yet in the mean time they indulge their genius, and amuse their minds moreover, as their ignorance allows them to do, with many empty hopes, in which they take delight as in pleasant dreams, and so sweeten the miseries of life. The school of Prometheus on the other hand, that is the wise and fore-thoughtful class of men, do indeed by their caution decline and remove out of their way many evils and misfortunes ; but with that good there is this evil joined, that they stint themselves of many pleasures and of the various agreeableness of life, and cross their genius, and (what is far worse) torment and wear themselves away with cares and solicitude and inward fears. For being bound to the column of Necessity, they are troubled with innumerable thoughts (which because of their flightiness are represented by the eagle), thoughts which prick and gnaw and corrode the liver : and if at intervals, as in the night, they obtain some little relaxation and quiet of mind, yet new fears and anxieties return presently with the morning. Very few therefore are they to whom the benefit of both portions falls,—to retain the advantages of providence and yet free themselves from the evils of solicitude and perturbation. Neither is it possible for any one to attain this double blessing, except by the help of Hercules ; that is, fortitude and constancy of mind, which being prepared for all events and equal to any fortune, foresees without fear, enjoys without fastidiousness, and bears without impatience. It is worth noting too that this virtue was not natural to Prometheus, but adventitious, and came by help from without ; for it is not a thing which any inborn and natural fortitude can attain to ; it comes from beyond the ocean, it is received and brought to us from the Sun ; for it comes of Wisdom, which is as the Sun, and of meditation upon the inconstancy and fluctuations of human life, which is as the navigation of the ocean : two things which Virgil has well coupled together in those lines :—

> Ah, happy, could we but the causes know
> Of all that is ! Then should we know no fears :
> Then should the inexorable Fate no power
> Possess to shake us, nor the jaws of death.

Most elegantly also is it added for the consolation and encouragement of men's minds, that that mighty hero sailed in a cup or pitcher ; lest they should too much mistrust the narrowness and frailty of their own nature, or plead it in their own excuse, as though it were altogether incapable of this kind of fortitude and constancy : the true nature of which was well divined by Seneca, when he said, *It is true greatness to have in one the frailty of man and the security of God.*

But I must now return to a part which, that I might not interrupt the connexion of what precedes, I have purposely passed by. I mean that last crime of Prometheus, the attempt upon the chastity of Minerva. For it was even for this offence,—certainly a very great and grave one,—that he underwent that punishment of the tearing of his entrails. The crime alluded to appears to be no other than that into which men not unfrequently fall when puffed up with arts and much knowledge,—of trying to bring the divine wisdom itself under the dominion of sense and reason : from which attempt inevitably follows laceration of the mind and vexation without end or rest. And therefore men must soberly and modestly distinguish between things divine and human, between the oracles of sense and of faith ; unless they mean to have at once a heretical religion and a fabulous philosophy.

The last point remains,—namely the races with burning torches instituted in honour of Prometheus. This again, like that fire in memory and celebration of which these games were instituted, alludes to arts and sciences, and carries in it a very wise admonition, to this effect,—that the perfection of the sciences is to be looked for not from the swiftness or ability of any one inquirer, but from a succession. For the strongest and swiftest runners are perhaps not the best fitted to keep their torch alight ; since it may be put out by going too fast as well as too slow. It seems however that these races and games of the torch have long been intermitted ; since it is still in their first authors,—Aristotle, Galen, Euclid, Ptolemy,—that we find the several sciences in highest perfection ; and no great matter has been done, nor hardly attempted, by their successors. And well were it to be wished that these games in honour of Prometheus, that is of Human Nature, were again revived ; that the victory may no longer depend upon the unsteady and wavering torch of each single man ; but competition, emulation, and good fortune be brought to aid. Therefore men should be advised to rouse themselves, and try each his own strength and the chance of his own turn, and not to stake the whole venture upon the spirits and brains of a few persons.

Such are the views which I conceive to be shadowed out in this so common and hacknied fable. It is true that there are not a few things beneath which have a wonderful correspondency with the mysteries of the Christian faith. The voyage of Hercules especially, sailing in a pitcher to set Prometheus free, seems to present an image of God the Word hastening in the frail vessel of the flesh to redeem the human race. But I purposely refrain myself from all licence of speculation in this kind, lest peradventure I bring strange fire to the altar of the Lord.

XXVII.

THE FLIGHT OF ICARUS; ALSO SCYLLA AND CHARYBDIS;

OR THE MIDDLE WAY.

MODERATION, or the Middle Way, is in Morals much commended; in Intellectuals less spoken of, though not less useful and good; in Politics only, questionable and to be used with caution and judgment.

The principle of moderation in Morals is represented by the ancients in the path which Icarus was directed to take through the air; the same principle in relation to the intellect, by the passage between Scylla and Charybdis, so famous for its difficulty and danger.

Icarus was instructed by his father to beware, when he came to fly over the sea, of taking either too high or too low a course. For his wings being fixed on with wax, the fear was that if he rose too high the wax would be melted by the sun's heat; if he kept down too near the vapour of the sea, it would lose its tenacity by the moisture. Icarus, in the adventurous spirit of youth, made for the heights, and so fell headlong down.

It is an easy and a familiar parable. The path of virtue goes directly midway between excess on the one hand and defect on the other. Icarus, being in the pride of youthful alacrity, naturally fell a victim to excess. For it is on the side of excess that the young commonly sin, as the old on the side of defect. And yet if he was to perish one way, it must be admitted that of two paths, both bad and mischievous, he chose the better. For sins of defect are justly accounted worse than sins of excess; because in excess there is something of magnanimity,—something, like the flight of a bird, that holds kindred with heaven; whereas defect creeps on the ground like a reptile. Excellently was it said by Heraclitus, *Dry light is the best soul.* For when the moisture and humours of earth get into the soul, it becomes altogether low and degenerate. And yet here too a measure must be kept: the dryness, so justly praised, must be such as to make the light more subtle, but not such as to make it catch fire. But this is what everybody knows.

Now for the passage between Scylla and Charybdis (understood of the conduct of the understanding), certainly it needs both skill and good fortune to navigate it. For if the ship run on Scylla, it is dashed on the rocks; if on Charybdis, it is sucked in by the whirlpool: by which parable (I can but briefly touch it, though it suggests reflexions without end) we are meant to understand that in every knowledge and science, and in the rules and axioms appertaining to them, a mean must be kept between too many distinctions and too much generality,—between the rocks of the one and the whirlpools of the other. For these two are notorious for the shipwreck of wits and arts.

XXVIII.

SPHINX;

OR SCIENCE.

SPHINX, says the story, was a monster combining many shapes in one. She had the face and voice of a virgin, the wings of a bird, the claws of a griffin. She dwelt on the ridge of a mountain near Thebes and infested the roads, lying in ambush for travellers, whom she would suddenly attack and lay hold of; and when she had mastered them, she propounded to them certain dark and perplexed riddles, which she was thought to have obtained from the Muses. And if the wretched captives could not at once solve and interpret the same, as they stood hesitating and confused she cruelly tore them to pieces. Time bringing no abatement of the calamity, the Thebans offered to any man who should expound the Sphinx's riddles (for this was the only way to subdue her) the sovereignty of Thebes as his reward. The greatness of the prize induced Œdipus, a man of wisdom and penetration, but lame from wounds in his feet, to accept the condition and make the trial: who presenting himself full of confidence and alacrity before the Sphinx, and being asked what kind of animal it was which was born four-footed, afterwards became two-footed, then three-footed and at last four-footed again, answered readily that it was man; who at his birth and during his infancy sprawls on all four, hardly attempting to creep; in a little while walks upright on two feet; in later years leans on a walking-stick and so goes as it were on three; and at last in extreme age and decrepitude, his sinews all failing, sinks into a quadruped again, and keeps his bed. This was the right answer and gave him the victory; whereupon he slew the Sphinx; whose body was put on the back of an ass and carried about in triumph; while himself was made according to compact King of Thebes.

The fable is an elegant and a wise one, invented apparently in allusion to Science; especially in its application to practical life. Science, being the wonder of the ignorant and unskilful, may be not absurdly called a monster. In figure and aspect it is represented as many-shaped, in allusion to the immense variety of matter with which it deals. It is said to have the face and voice of a woman, in respect of its beauty and facility of utterance. Wings are added because the sciences and the discoveries of science spread and fly abroad in an instant; the communication of knowledge being like that of one candle with another, which lights up at once. Claws, sharp and hooked, are ascribed to it with great elegance, because the axioms and arguments of science penetrate and hold fast the mind, so that it has no means of evasion or escape; a point which the sacred philosopher also noted: *The words of the wise are as goads, and as nails driven deep in.* Again, all knowledge may be regarded as having its station on the heights of mountains; for it is deservedly esteemed a thing sublime and lofty, which looks down upon ignorance as from an eminence, and has moreover a spacious prospect on every side, such as we find on hill-tops. It is described as infesting the roads, because at every turn in the journey or pilgrimage of human life, matter and occasion for study assails and encounters us. Again, Sphinx proposes to men a variety of hard questions and riddles which she received from the Muses. In these, while they remain with the Muses, there is probably no cruelty; for so long as the object of meditation and inquiry is merely to know, the understanding is not oppressed or straitened by it, but is free to wander and expatiate, and finds in the very uncertainty of conclusion and variety of choice a certain pleasure and delight; but when they pass from the Muses to Sphinx, that is from contemplation to practice, whereby there is necessity for present action, choice, and decision, then they begin to be painful and cruel; and unless they be solved and disposed of, they strangely torment and worry the mind, pulling it first this way and then that, and fairly tearing it to pieces. Moreover the riddles of the Sphinx have always a twofold condition attached to them; distraction and laceration of mind, if you fail to solve them; if you succeed, a kingdom. For he who understands his subject is master of his end; and every workman is king over his work.

Now of the Sphinx's riddles there are in all two kinds; one concerning the nature of things, another concerning the nature of man; and in like manner there are two kinds of kingdom offered as the reward of solving them; one over nature, and the other over man. For the command over things natural,—over bodies, medicine, mechanical powers, and infinite other of the kind—is the one proper and ultimate end of true natural philosophy; however the philosophy of the School, content with what it finds, and swelling with talk, may neglect or spurn the search after realities and works. But the riddle proposed to Œdipus, by the solution of which he became King of Thebes, related to the nature of man;

for whoever has a thorough insight into the nature of man may shape his fortune almost as he will, and is born for empire; as was well declared concerning the arts of the Romans,—

> Be thine the art,
> O Rome, with government to rule the nations,
> And to know whom to spare and whom to abate,
> And settle the condition of the world.

And therefore it fell out happily that Augustus Cæsar, whether on purpose or by chance, used a Sphinx for his seal. For he certainly excelled in the art of politics if ever man did; and succeeded in the course of his life in solving most happily a great many new riddles concerning the nature of man, which if he had not dexterously and readily answered he would many times have been in imminent danger of destruction. The fable adds very prettily that when the Sphinx was subdued, her body was laid on the back of an ass: for there is nothing so subtle and abstruse, but when it is once thoroughly understood and published to the world, even a dull wit can carry it. Nor is that other point to be passed over, that the Sphinx was subdued by a lame man with club feet; for men generally proceed too fast and in too great a hurry to the solution of the Sphinx's riddles; whence it follows that the Sphinx has the better of them, and instead of obtaining the sovereignty by works and effects, they only distract and worry their minds with disputations.

XXIX.

PROSERPINA;

OR SPIRIT.

THEY say that when Pluto upon that memorable partition of the kingdoms received for his portion the infernal regions, he despaired of gaining any of the goddesses above in marriage by addresses and gentle methods, and so was driven to take measures for carrying one of them off by force. Seizing his opportunity therefore, while Proserpina, daughter of Ceres, a fair virgin, was gathering flowers of Narcissus in the Sicilian meadows, he rushed suddenly upon her and carried her off in his chariot to the subterranean regions. Great reverence was paid her there: so much that she was even called the Mistress or Queen of Dis. Meanwhile her mother Ceres, filled with grief and anxiety by the disappearance of her dearly beloved daughter, took a lighted torch in her hand, and wandered with it all round the world in quest of her. Finding the search fruitless, and hearing by chance that she had been carried down to the infernal regions, she wearied Jupiter with tears and lamentations, praying to have her restored; till at last she won a promise from him that if her daughter had not eaten of anything belonging to the under world, then she might bring her back. This condition was unfortunate for the mother; for Proserpina had eaten (it was found) three grains of a pomegranate. But this did not prevent Ceres from renewing her prayers and lamentations; and it was agreed at last that Proserpina should divide the year between the two, and live by turns six months with her husband and the other six with her mother.

Afterwards a very daring attempt to carry away the same Proserpina from the chamber of Dis was made by Theseus and Pirithous. But having sate down to rest by the way on a stone in the infernal regions, they were unable to rise again, and continued sitting there for ever. So Proserpina remained Queen of the under world, where a great and new privilege was granted in honour of her; for whereas they who went down to the under world were not permitted to go back, a singular exception was made in favour of any who should bring a certain golden branch as a present to Proserpina; such present entitling the bearer to go and return. It was a single branch growing by itself in a vast and dark wood; neither had it a stock of its own, but grew like misseltoe upon a tree of different kind; and as soon as it was plucked off, another came in its place.

The fable relates, as I take it, to Nature, and explains the source of that rich and fruitful supply of active power subsisting in the under world, from which all the growths of our upper world spring, and into which they again return and are resolved. By Proserpina the ancients signified that ethereal spirit which, having been separated by violence from the upper globe, is enclosed and imprisoned beneath the earth (which earth is represented by Pluto); as was well expressed in those lines,—

> Whether that the Earth yet fresh, and from the deeps
> Of heaven new-sundered, did some seeds retain,
> Some sparks and motions of its kindred sky.

This spirit is represented as having been ravished, that is suddenly and forcibly carried off, by the Earth; because there is no holding it in if it have time and leisure to escape, and the only way to confine and fix it is by a sudden pounding and breaking up; just as if you would mix air with water, you can only do it by sudden and rapid agitation: for thus it is that we see these bodies united in foam, the air being as it were ravished by the water. It is prettily added that Proserpina was carried off while in the act of gathering flowers of Narcissus in the valleys: for Narcissus takes its name from torpor or stupor; and it is only when beginning to curdle, and as it were to gather torpor, that spirit is in the best state to be caught up and carried off by earthly matter. It is right too that Proserpina should have that honour, which is not conceded to the wife of any other God,—to be called the Mistress or Queen of Dis: for the spirit does in fact govern and manage everything in those regions, without the help of Pluto, who remains stupid and unconscious.

The air meanwhile and the power of the celestial region (which is represented by Ceres) strives with infinite assiduity to win forth and recover this imprisoned spirit again; and that torch which the air carries—the lighted torch in Ceres's hand—means no doubt the Sun, which does the office of a lamp all over the earth, and would do more than anything else for the recovery of Proserpina, were the thing at all possible. But Proserpina remains fixed where she is; the reason and manner whereof is accurately and admirably set forth in those two agreements between Jupiter and Ceres. For with regard to the first, most certain it is that there are two ways of confining and restraining spirit in solid and earthy matter; one by constipation and obstruction, which is simple imprisonment and violence; the other by administering some suitable aliment, which is spontaneous and free. For when the imprisoned spirit begins to feed and nourish itself, it is no longer in a hurry to escape, but becomes settled as in its own land. And this is what is meant by Proserpina's tasting of the pomegranate; which if she had not done, she would have been long since carried off by Ceres as she traversed the globe with her torch in quest of her. For though the spirit which is contained in metals and minerals is prevented from getting out chiefly perhaps by the solidity of the mass, that which is contained in plants and animals dwells in a porous body, from which it could easily escape if it were not by that process of tasting reconciled to remain. As for the second agreement,—that she should stay six months at a time with either party,—it is nothing else but an elegant description of the division of the year; since that spirit which is diffused through the earth does (in regard to the vegetable kingdom) live in the upper world during the summer months, and retires to the under world in the winter months.

Now for that attempt of Theseus and Pirithous to carry Proserpina away, the meaning is that the subtler spirits which in many bodies descend to the earth often fail to draw out and assimilate and carry away with them the subterranean spirit, but contrariwise are themselves curdled and never reascend again, and so go to increase the number of Proserpina's people and the extent of her empire.

As for that golden branch, it may seem difficult for me to withstand the Alchemists, if they attack me from that side; seeing they promise us by that same stone of theirs not only mountains of gold, but also the restitution of natural bodies as it were from the gates of the Infernals. Nevertheless for Alchemy and those that are never weary of their wooing of that stone, as I am sure they have no ground in theory, so I suspect that they have no very good pledge of success in practice. And therefore putting them aside, here is my opinion as to the meaning of that last part of the parable. From many figurative allusions I am satisfied that the ancients regarded the conservation, and to a certain extent the restoration, of natural bodies as a thing not desperate, but rather as abstruse and out of the way. And this is what I take them in the passage before us to mean, by placing this branch in the midst of the innumerable other branches of a vast and thick wood. They represented it as golden; because gold is the emblem of duration; and grafted, because the effect in question is to be looked for as the result of art, not of any medicine or method which is simple or natural.

XXX.

METIS;

OR COUNSEL.

THE ancient poets tell us that Jupiter took Metis, whose name plainly signifies Counsel, to wife; that she conceived by him and was with child; which he perceiving did not wait till she brought forth, but ate her up; whereby he became himself with child; but his delivery was of a strange kind; for out of his head or brain he brought forth Pallas armed.

This monstrous and at first sight very foolish fable contains, as I interpret it, a secret of government. It describes the art whereby kings so deal with the councils of state as not only to keep their authority and majesty untouched, but also to increase and exalt it in the eyes of their people. For kings by a sound and wise arrangement tie themselves to their councils with a bond like that of wedlock, and deliberate with them concerning all their greatest matters, rightly judging that this is no diminution to their majesty. But when the question grows ripe for a decision (which is the bringing forth) they do not allow the council to deal any further in it, lest their acts should seem to be dependent upon the council's will; but at that point, (unless the matter be of such a nature that they wish to put away the envy of it) they take into their own hands whatever has been by the council elaborated and as it were shaped in the womb; so that the decision and execution (which, because it comes forth with power and carries necessity, is elegantly represented under the figure of Pallas armed) may seem to emanate from themselves. Nor is it enough that it be seen to proceed from their free and unconstrained and independent authority and will, but they must have the world think that the decision comes out of their own head, that is out of their proper wisdom and judgment.

XXXI.

THE SIRENS;

OR PLEASURE.

THE fable of the Sirens is truly applied to the pernicious allurements of pleasure; but in a very poor and vulgar sense. For I find the wisdom of the ancients to be like grapes ill-trodden: something is squeezed out, but the best parts are left behind and passed over.

The Sirens were daughters (we are told) of Achelous and of Terpsichore, one of the Muses. Originally they had wings; but being beaten in a contest with the Muses which they had rashly challenged, their wings were plucked off, and turned by the Muses into crowns for themselves, who thenceforward all wore wings on their heads, except only the mother of the Sirens. These Sirens had their dwelling in certain pleasant islands, whence they kept watch for ships; and when they saw any approaching, they began to sing; which made the voyagers first stay to listen, then gradually draw near, and at last land; when they took and killed them. Their song was not all in one strain; but they varied their measures according to the nature of the listener, and took each captive with those which best suited him. So destructive the plague was, that the islands of the Sirens were seen afar off white with the bones of unburied carcasses. For this evil two different remedies were found; one by Ulysses, the other by Orpheus. Ulysses caused the ears of his crew to be stopped with wax; and himself (wishing to make trial of the thing without incurring the danger) to be bound to the mast; at the same time forbidding any one at his peril to loose him even at his own request. Orpheus, not caring to be bound, raised his voice on high, and singing to his lyre the praises of the Gods, drowned the voices of the Sirens, and so passed clear of all danger.

The fable relates to Morals, and contains an elegant though obvious parable. Pleasures spring from the union of abundance and affluence with hilarity and exultation of mind. And formerly they carried men away at once, as if with wings, by the first view of their charms. But doctrine and instruction have succeeded in teaching the mind, if not to refrain altogether, yet to pause and consider consequences; and so have stripped the Pleasures of their wings. And this redounded greatly to the honour of the Muses—for as soon as it appeared by some examples that Philosophy could induce a contempt of Pleasures, it was at once regarded as a sublime thing, which could so lift the soul from earth, and make the cogitations of man (which live in his head) winged and ethereal. Only the mother of the Sirens still goes on foot and has no wings; and by her no doubt are meant those lighter kinds of learning which are invented and applied only for amusement; such as those were which Petronius held in estimation; he who being condemned to die, sought in the very waiting-room of death for matter to amuse him, and when he turned to books among other things for consolation, would read (says Tacitus) none of those which teach constancy of mind, but only light verses. Of this kind is that of Catullus,

> Let's live and love, love, while we may;
> And for all the old men say
> Just one penny let us care;

and that other,—

> Of Rights and Wrongs let old men prate, and learn
> By scrupulous weighing in fine scales of law
> What is allowed to do and what forbid.

For doctrines like these seem to aim at taking the wings away from the Muses' crowns and giving them back to the Sirens. The Sirens are said to live in islands; because Pleasures commonly seek retiring-places aloof from the throngs of men. As for the song of the Sirens, its fatal effect and various artifice, it is everybody's theme, and therefore needs no interpreter. But that circumstance of the bones being seen from a distance like white cliffs, has a finer point: implying that the examples of other men's calamities, however clear and conspicuous, have little effect in deterring men from the corruptions of pleasure.

The parable concerning the remedies remains to be spoken of: a wise and noble parable, though not at all abstruse. For a mischief so fraught with cunning, and violence alike, there are proposed three remedies; two from philosophy, the third from religion. The first method of escape is to resist the beginnings, and sedulously to avoid all occasions which may tempt and solicit the mind. This is the waxing up of the ears, and for minds of ordinary and plebeian cast—such as the crew of Ulysses—is the only remedy. But minds of a loftier order, if they fortify themselves with constancy of resolution, can venture into the midst of pleasures; nay and they take delight in thus putting their virtue to a more exquisite proof; besides gaining thereby a more thorough insight—as lookers-on rather than followers—into the foolishness and madness of pleasures: which is that which Solomon professes concerning himself, when he closes his enumeration of the pleasures with which he abounded in these words: *Likewise my wisdom remained with me.* Heroes of this order may therefore stand unshaken amidst the greatest temptations, and refrain themselves even in the steep-down paths of pleasures; provided only that they follow the example of Ulysses, and forbid the pernicious counsels and flatteries of their own followers, which are of all things most powerful to unsettle and unnerve the mind. But of the three remedies, far the best in every way is that of Orpheus; who by singing and sounding forth the praise of the gods confounded the voices of the Sirens and put them aside: for meditations upon things divine excel the pleasures of the sense, not in power only, but also in sweetness.

APOPHTHEGMS NEW AND OLD.

PREFACE.

BY JAMES SPEDDING.

Bacon's collection of Apophthegms, though a sick man's task, ought not to be regarded as a work merely of amusement; still less as a jest-book. It was meant for a contribution, though a slight one, towards the supply of what he had long considered as a desideratum in literature. In the *Advancement of Learning* he had mentioned Apophthegms with respect, along with Orations and Letters, as one of the appendices to Civil History; regretting the loss of Cæsar's collection; "for as for those which are collected by others (he said) either I have no taste in such matters, or their choice hath not been happy". This was in 1605. In revising and enlarging that treatise in 1623, he had spoken of their use and worth rather more fully. "They serve (he said) not for pleasure only and ornament, but also for action and business; being, as one called them, *mucrones verborum*,—speeches with a point or edge, whereby knots in business are pierced and severed. And as former occasions are continually recurring, that which served once will often serve again, either produced as a man's own or cited as of ancient authority. Nor can there be any doubt of the utility in business of a thing which Cæsar the Dictator thought worthy of his own labour; whose collection I wish had been preserved; for as for any others that we have in this kind, but little judgment has in my opinion been used in the selection[1]." Of this serious use of apophthegms Bacon himself had had long experience, having been all his life a great citer of them; and in the autumn of 1624, when he was recovering from a severe illness, he employed himself in dictating from memory a number that occurred to him as worth setting down.

The fate of this collection has been singular. The original edition [2] (a very small octavo volume dated 1625, but published about the middle of December 1624 [3]) consisted of 280 apophthegms, with a short preface. Of this volume Dr. Rawley, in the first edition of the *Resuscitatio* (1657), makes no mention whatever, either where he enumerates the works composed during the last five years of Bacon's life, or in the "perfect list of his Lordship's true works both in English and Latin" at the end of the volume. And his words, taken strictly, would seem to imply (since he cannot have been ignorant of its existence) that he did not acknowledge it as Bacon's. But I suppose he had either forgotten it, or did not think it important or original enough to be worth mentioning.

In 1658 there came forth a small volume, without any editor's name, under the following title: *Witty Apophthegms delivered at several times and upon several*

[1] "Neque apophthegmata ipsa ad delectationem et ornatum tantum prosunt, sed ad res gerendas etiam et usus civiles. Sunt enim (ut aiebat ille) veluti *secures* aut *mucrones verborum*; qui rerum et negotiorum nodos acumine quodam secant et penetrant; occasiones autem redeunt in orbem, et quod olim erat commodum rursus adhiberi et prodesse potest, sive quis ea tanquam sua proferat, sive tanquam vetera. Neque certe de utilitate ejus rei ad civilia dubitari potest, quam Cæsar Dictator operâ suâ honestavit; cujus liber utinam extaret, cum ea quæ usquam habentur in hoc genere nobis parum cum delectu congesta videantur."—*De Aug. Sci.* ii. 12.

[2] *Apophthegmes new and old. Collected by the Right Honourable Francis Lo. Verulam Viscount St. Alban. London. Printed for Hanna Barret and Richard Whittaker, and are to be sold at the King's Head in Paul's Church-yard.* 1625.

A copy in Gray's Inn Library has the date 1626: but appears to be in all other respects exactly the same.

[3] Chamberlain to Carlton, 18 Dec. 1624. *Court and Times of James I.*, ii. p. 486.

occasions, by King James, King Charles, the Marquess of Worcester, Francis Lord Bacon, and Sir Thomas Moore. Collected and revised. In this volume the apophthegms attributed to Bacon are in all 184; of which 163 are copied verbatim from his own collection of 1625, and follow (with one or two slight exceptions, probably accidental) in the same order. The remaining 21, which are mostly of a very inferior character, are not added but interspersed.

In 1661 appeared a second edition, or rather a reissue, of the *Resuscitatio*, edited as before by Dr. Rawley, and with some additions; among which was a collection of "Apophthegms, new and old". This, though introduced without a word of preface or advertisement from editor or publisher, was so far from being a reprint of the original collection of 1625, that I do not think the editor can have had a copy of it to refer to. Of the original 280 no less than 71 are entirely omitted; 39 new ones are introduced; the order is totally changed, the text considerably altered. The alterations in the text are indeed (though I think not generally for the better) no more than might have been made by Bacon himself in revising the book. A few of the omissions also might be accounted for in the same way; but very many of the omitted ones are among the best in the volume, and such as he could have no motive for suppressing. Still less is it possible to imagine a reason for the change of *order*, which could hardly have been more complete or more capricious if the leaves of the book had been first separated and then shuffled. Whoever will take a copy of the bound volume and endeavour to write directions in it for any such change in the arrangement, will see that it could not have been done without a great deal of time and trouble. And seeing that it was now more than thirty years since that volume appeared, that it had never been reprinted, nor ever much valued and (being so small) might easily be lost, the more probable supposition is that Dr. Rawley had no copy of it, and made up his collection from loose and imperfect manuscripts.

In 1671, three or four years after Dr. Rawley's death, appeared a *third* edition of the *Resuscitatio*, in two parts. The first part contains a collection of Apophthegms, which from the publisher's preface one would expect to find a mere reprint from the second edition. But it is in fact a new collection, made up by incorporating the "Witty Apophthegms" of 1658, of which it contains all but 12, with Dr. Rawley's collection of 1661. By this means the number of apophthegms is increased from 248 to 296; the new ones being not added as a supplement, but interspersed among the old. Of the 71 which formed part of Bacon's original collection but not of Dr. Rawley's, 32 are thus supplied. Eight more might have been supplied from the same source, but were left out perhaps by accident. There remained therefore 39 genuine ones still to be recovered; a fact which may be best explained by supposing that the editor of the third edition of the *Resuscitatio* had not been able, any more than Dr. Rawley when he edited the second, to procure a copy of the original volume.

In 1679, a new volume of remains, under the title of *Baconiana*, was published by Dr. Tenison from original manuscripts; with an introduction containing "an account of all the Lord Bacon's works". In this introduction he tells us (p. 59) that the best edition of the Apophthegms was the first (1625); and censures as spurious, or at least as including spurious matter, the additions contained in the two collections last mentioned of 1658 and 1671; but of Dr. Rawley's collection in 1661 he strangely enough makes no mention whatever. In the body of the work he gives 27 additional apophthegms, found among Bacon's papers, and never before printed.

Next came Blackbourne, in 1730, with an edition of Bacon's works complete in 4 volumes folio. His plan in dealing with the Apophthegms was to reprint, 1st, the whole collection (repetitions omitted) as it stood in the third edition of the *Resuscitatio*; 2ndly, the 27 additional ones in Tenison's *Baconiana* (all but 3; which he omitted, not very judiciously, because they are to be found in the Essays); 3rdly, the remaining 39, contained in the original edition, but omitted in all later copies. Thus we had for the first time a collection which included *all* the genuine apophthegms. But it was defective in this,—that it included likewise all, or all but one or two, of those which Tenison had alluded to in general terms as spurious; and that no attempt was made in it to distinguish those which had Dr. Rawley's sanction from those which had not.

Succeeding editors followed Blackbourne, without either noticing or trying to remedy this defect; until Mr. Montagu took up the task in his edition of 1825, in which he made an attempt, more laudable than successful, to separate the genuine from the spurious. Taking Tenison's remark as his guide, he reprinted the original collection of 1625 exactly as it stood, (or at least meant to do so; for there are more than 130 places in which his copy differs from the original,) and then added the supplementary collection in the *Baconiana.* The rest he concluded to be spurious, and gathered them (or meant to gather them and thought he had done so) into an appendix, under that title. But in this he took no account of the *second* edition of the *Resuscitatio*, which must certainly be considered as having the sanction of Dr. Rawley; and the principle, whatever it was, upon which he proceeded to eliminate the spurious apophthegms was altogether fallacious. Observing that the last apophthegm in the third edition of the *Resuscitatio* was numbered 308, whereas in the original collection there were only 280; and *not* observing that of those 308, 12 were given twice over; he seems to have concluded that the number of the spurious must be 28, and that they might be found by simply going through the later collection, and marking off all those which were not given in the earlier. And the first 25 in his spurious list were probably selected in that way; for they are the first 25 (one only excepted, which *is* given in the original collection, and was probably marked off by mistake) which answer the conditions; and they are set down in the order in which to a person so proceeding they would naturally present themselves. Upon what principle he selected the other three which make up the 28, I cannot guess. One of them he has himself printed a few pages before among the genuine; another he quotes in his preface as one which he can hardly believe *not* to be genuine; and before he came to the third, he must, if he took them as they stand in the book, have passed by 20 others which have precisely the same title to the distinction. But howsoever he went about it, his result is certainly wrong; for among his 28 spurious apophthegms there are several which were undoubtedly sanctioned by Dr. Rawley, besides the two which had been previously printed among the genuine ones by himself; and when all is done, there remain no less than 30 others, silently omitted and entirely unaccounted for.

Such is the latest shape in which this little work appears [4]. The common editions contain *all* the apophthegms; but some that are spurious are printed in them as genuine. Mr. Montagu's edition does *not* contain all: and some that are genuine are printed in it as spurious.

I have now to explain the plan upon which I have myself proceeded in order to set the matter right.

First. Considering that the edition of 1625 was published during Bacon's life with his name on the title-page; that there is no reason for supposing that he revised or altered it afterwards; and that there is some reason for suspecting that the collection published by Dr. Rawley in 1661, far from being a revised edition of the former, was made up, when a copy of the original volume was not procurable, from some imperfect manuscript or from old note-books; I regard the 280 apophthegms printed in 1625 as those which we are most certain that Bacon himself thought worth preserving. I begin therefore by reprinting these from the original edition; and so far I follow Mr. Montagu's example.

Secondly. Considering nevertheless that Bacon may *possibly* have revised this collection, and struck out some and altered others; and that Dr. Rawley may possibly have had by him some portions of that revised copy, or some

[4] This was written before the appearance of Mr. Bohn's volume of the *Moral and Historical Works of Lord Bacon*, edited by Joseph Devey, M.A., which professes to contain the "Apophthegms; omitting those known to be spurious". Of the collection there given however it is not necessary to take any further notice. It is merely a selection from a selection, in which no attempt has really been made to distinguish the spurious from the genuine.

memoranda of those omissions and alterations; I regard the variations as worth preserving [5]. I have therefore compared the two collections, marked with a † all the apophthegms which are not found in the later, and recorded in foot-notes all the more considerable differences of reading that occur in those which are; adding also for convenience of reference the numbers which they bear in the later collection.

Thirdly. Considering that Rawley had access to all Bacon's unpublished papers [6], and had been in constant personal communication with him during his later years; and that Bacon had been in the habit of setting down such things from time to time in note books, and may very likely have made a supplementary collection with a view to publication, I regard all the additional apophthegms which appear in the collection of 1661 as probably genuine, and as resting on authority second only to that which belongs to the original edition. These therefore I reprint from the second edition of the *Resuscitatio*, in the order in which they occur; and for more convenient reference, with the original numbers affixed. And at the same time, because in a common-place book of Dr. Rawley's which is preserved in the Lambeth Library and appears to have been begun soon after Bacon's death I find several of these additional apophthegms set down in a form somewhat different; and because I think it probable that Dr. Rawley, in preparing them for publication, occasionally introduced variations of his own in order to correct the language or clear the meaning; I have thought the original form worth preserving, and have therefore compared the versions and set down the variations in foot-notes.

Fourthly. Considering that many of Bacon's original papers passed through the hands of Dr. Rawley or his son into those of Dr. Tenison, I regard the supplementary collection in the *Baconiana* as also probably genuine, and next in authenticity to the collection of 1661. These therefore I print next; also preserving in foot-notes such various readings as I find in Dr. Rawley's common-place book above mentioned.

Fifthly. In this same common-place book I find other apophthegms and anecdotes, not included in any of the three collections,—Bacon's, Rawley's, or Tenison's; a few of which I have thought worth preserving; some for their independent value, and some for a little light they throw on Bacon's personal character, manners, or habits. These I print next. They have probably as good a right to be considered genuine as any that were not published by Bacon himself; for they are set down in Rawley's own hand.

Sixthly. When all this is done, there remain 16 which rest upon no better authority than that of the unknown editor of the "Witty Apophthegms." These I regard as having no right to appear at all under Bacon's name, and accordingly remit them to a note*, as spurious.

In a note to Bacon's preface, as given in the second editon of the *Resuscitatio*, Dr. Rawley expressly states that the collection was made from memory, "with

[5] The substitution, in almost every case, of "the House of Commons" for "the Lower-House" has a kind of historical significance.

[6] In a catalogue of Bacon's extant MSS. (Add. MSS.Brit. Mus. 629. fo. 271.), not dated, but drawn up by Rawley after Bacon's death, I find the three following entries:—

"Apophthegms cast out of my Lord's book, and not printed.
"Apophthegms of K. James.
"Some few apophthegms not chosen."

There is no allusion to any revision of the printed book. The first of these entries evidently refers to some apophthegms which had been struck out of the MS. before it was published; the last probably to some which had not been included in it. The "apophthegms of K. James" may have been the seven which stand first among the additions introduced by Rawley in his collection of 1661. If the MS. from which the collection of 1625 was printed remained in Dr. Rawley's hands, it would not be mentioned in this catalogue, which relates only to what had not been printed. We may easily suppose therefore that some of the loose sheets were still preserved; and that, when the original volume was not procurable, he made up his collection by incorporating these with the unpublished ones mentioned in the catalogue. [*Not here reprinted.—Ed.]

out turning any book." If I am right in conjecturing that the only *collection* made by Bacon himself was that of 1625, we must understand Dr. Rawley's remark as applying to that; and we must beware of attributing to it any great *historical* authority. It will be found that some of the sayings, especially those of the ancient philosophers, are assigned to the wrong persons. But what is interesting or memorable in them depends in general so little upon the persons who spoke them; and the traditional sayings of famous wits must always be in great part so apocryphal; that I have not thought it worth while to investigate the authorities, or expedient to encumber the text with notes of that kind. The authenticity of the anecdotes relating to persons of more recent times would be better worth investigation; but in these cases Bacon is himself (either as a personal witness or as a preserver of traditions then current) one of the original authorities, whom it would not be easy to correct by a better. In these cases also his memory is less likely to have deceived him [7]. But the whole collection is to be read with this qualification. Dr. Tenison adds that it was one morning's work. But he does not tell us upon what authority; and certainly Dr. Rawley has left no such statement on record. Perhaps he was confounding what Dr. Rawley said of "The beginning of the History of Henry VIII." with what he said about the Apophthegms, and so put the two together. The statement is not to be believed without very good and very express authority.

The use and worth of the collection will be best understood by those who have studied Bacon's own manner of quoting apophthegms, to suggest, illustrate, or enliven serious observations. And it was greater in his time than it is now, not only because they were fresher then and carried more authority in popular estimation, but also because the ingenuities of the understanding were then more affected and in greater request. A similar collection adapted to modern times would be well worth making.

NOTE.—In this edition, where a note is signed R., it means that such is the reading of the *Resuscitatio*, ed. 1661. The numbers within brackets are the numbers by which the several apophthegms are distinguished in that collection. The apophthegms marked † are not contained in it at all.

HIS LORDSHIP'S PREFACE [8].

JULIUS CÆSAR did write a Collection of Apophthegms, as appears in an epistle of Cicero [9]. I need say no more for the worth of a writing of that nature. It is pity his book [10] is lost: for I imagine they were collected with judgment and choice; whereas that of Plutarch and Stobæus, and much more the modern ones, draw much of the dregs. Certainly they are of excellent use. They are *mucrones verborum, pointed speeches* [11]. Cicero prettily calls them *salinas, saltpits*; that you may extract salt out of, and sprinkle it where you will. They serve to be interlaced in continued speech. They serve to be recited upon occasion of themselves. They serve if you take out the kernel of them, and make them your own. I have, for my recreation in my sickness, fanned the old [12]; not omitting any because they are vulgar, (for many vulgar ones are excellent good), nor for the meanness of the person, but because they are dull and flat; and added [13] many new, that otherwise would have died [14].

7 I have however noted two or three cases in which he appears to have relied upon an imperfect recollection of the *Floresta española*; a circumstance which was pointed out to me by Mr. Ellis.

8 So R. There is no heading in the original.

9 So did Macrobius, a Consular man. R.

10 Cæsar's book. R.

11 *The words of the wise are as goads*, saith Solomon. (Added in R.)

12 I have for my recreation, amongst more serious studies, collected some few of them; therein fanning the old. R.

13 adding. R.

14 This collection his Lp. made out of his memory, without turning any book. R. (Note in margin.)

APOPHTHEGMS NEW AND OLD.

† 1. When Queen Elizabeth had advanced Raleigh, she was one day playing on the virginals, and my Lo. of Oxford and another nobleman stood by. It fell out so, that the ledge before the jacks was taken away, so as the jacks were seen: My Lo. of Oxford and the other nobleman smiled, and a little whispered: The Queen marked it, and would needs know *What the matter was?* My Lo. of Oxford answered; *That they smiled to see that when Jacks went up Heads went down.*

2. (16.) Henry the Fourth of France his Queen was great [15] with child. Count Soissons, that had his expectation upon the crown, when it was twice or thrice thought that the Queen was with child before, said to some of his friends, *That it was but with a pillow.* This had some ways come to the King's ear; who kept it till when [16] the Queen waxed great; called [17] the Count Soissons to him, and said, laying his hand upon the Queen's belly, *Come, cousin, it is no pillow* [18]. *Yes Sir,* (answered the Count of Soissons,) [19] *it is a pillow for all France to sleep upon.*

3. (26.) There was a conference in Parliament between the Upper house and the Lower [20], about a Bill of Accountants, which came down from the Lords to the Commons; which bill prayed, that the lands of accountants, whereof they were seized when they entered upon their office, mought be liable to their arrears to the Queen. But the Commons desired that the bill mought not look back to accountants that were already, but extend only to accountants hereafter. But the Lo. Treasurer said, *Why, I pray* [21], *if you had lost your purse by the way, would you look forwards, or would you look back? The Queen hath lost her purse.*

4. (1.) Queen Elizabeth, the morrow of her coronation, went to the chapel; and in the great chamber, Sir John Rainsford, set on by wiser men, (a knight that had the liberty of a buffone,) besought the Queen aloud; *That now this good time when prisoners were delivered, four prisoners amongst the rest mought likewise have their liberty, who were like enough to be kept still in hold.* The Queen asked; *Who they were;* And he said; *Matthew, Mark, Luke, and John, who had long been imprisoned in the Latin tongue; and now he desired they mought go abroad among the people in English.* The Queen answered, with a grave countenance; *It were good (Rainsford) they were spoken with themselves, to know of them whether they would be set at liberty?* [22]

5. (29.) The Lo. Keeper, Sir Nicholas Bacon, was asked his opinion by Queen Elizabeth of one of these Monopoly Licences. And he answered; *Will you have me speak truth, Madam? Licentiâ omnes deteriores sumus: We are all the worse for a licence* [23].

6. (206.) Pace, the bitter Fool, was not suffered to come at the Queen [24], because of his bitter humour. Yet at one time some persuaded the Queen that he

[15] young. R. [16] such time as. R. [17] Then he called. R.

[18] is this a pillow? R. [19] The C. of S. answered, Yes Sir, &c. R.

[20] between the Lords' House and the House of Commons. R.

[21] I pray you. R.

[22] Queen Elizabeth, the morrow of her coronation; (it being the custom to release prisoners at the inauguration of a prince;) went to the Chapel; and in the Great Chamber, one of her courtiers who was well known to her, either out of his own motion, or by the instigation of a wiser man, presented her with a petition; and before a great number of courtiers besought her with a loud voice; *That now this good time there might be four or five principal prisoners more released; those were the four Evangelists and the Apostle Saint Paul, who had been long shut up in an unknown tongue, as it were in prison; so as they could not converse with the common people.* The Queen answered very gravely, *That it was best first to enquire of them, whether they would be released or no.* R.

[23] for licences. R. [24] at Queen Elizabeth. R.

should come to her ; undertaking for him that he should keep compass [25]. So he was brought to her, and the Queen said : *Come on, Pace ; now we shall hear of our faults.* Saith Pace ; *I do not use to talk of that that all the town talks of.*

7. (30.) My Lo. of Essex, at the succour of Rhoan, made twenty-four knights, which at that time was a great matter [26]. Divers of those gentlemen were of weak and small means ; which when Queen Elizabeth heard, she said, *My Lo. mought have done well to have built his alms-house before he made his knights.*

† 8. A great officer in France was in danger to have lost his place ; but his wife by her suit and means making, made his peace ; whereupon a pleasant fellow said, *That he had been crushed, but that he saved himself upon his horns.*

9. (2.) Queen Anne Bullen, at the time when she was led to be beheaded in the Tower, called one of the King's privy chamber to her, and said to him, *Commend me to the King, and tell him he is* [27] *constant in his course of advancing me. From a private gentlewoman he made me a marquisse* [28] *; and from a marquisse* [28] *a queen ; and now he had left* [29] *no higher degree of earthly honour, he hath made me a martyr* [30].

10. (207.) Bishop Latimer said, in a sermon at court ; *That he heard great speech that the King was poor and many ways were propounded to make him rich : For his part he had thought of one way, which was, that they should help the King to some good office, for all his officers were rich.*

11. (122.) Cæsar Borgia, after long division between him and the Lords of Romagna, fell to accord with them. In this accord there was an article, that he should not call them at any time all together in person : The meaning was, that knowing his dangerous nature, if he meant them treason, some one mought be free to revenge the rest [31]. Nevertheless he did with such art and fair usage win their confidence, that he brought them all together to counsel at Sinigalia [32]; where he murthered them all. This act, when it was related unto Pope Alexander his father by a Cardinal, as a thing happy, but very perfidious, the Pope said, *It was they that had broke their covenant first, in coming altogether.*

12. (54.) Pope Julius the third, when he was made Pope, gave his hat unto a youth, a favourite of his, with great scandal. Whereupon at one time a Cardinal that mought be free with him, said modestly to him : *What did your Holiness see in that young man, to make him Cardinal ?* Julius answered, *What did you see in me, to make me Pope ?*

13. (55.) The same Julius, upon like occasion of speech, why he should bear so great affection to the same young man, would say; *That he had found by astrology that it was the youth's destiny to be a great prelate ; which was impossible except himself were Pope ; And therefore that he did raise him, as the driver-on of his own fortune.*

14. (56.) Sir Thomas Moore had only daughters at the first ; and his wife did ever pray for a boy. At last he had a boy ; which after, at man's years, proved simple [33]. Sir Thomas said to his wife ; *Thou prayedst so long for a boy, that he will be a boy as long as he lives.*

15. (58.) Sir Thomas Moore, the day [34] he was beheaded, had a barber sent to him, because his hair was long, which was thought would make him more commiserable [35] with the people. The barber came to him and asked him, *whether he would be pleased to be trimmed ? In good faith, honest fellow,* (said Sir Thomas) *the King and I have a suit for my head, and till the title be cleared I will do no cost upon it.*

16. (59.) Stephen Gardiner, Bishop of Winchester, a great champion of the

[25] within compass. R.
[26] number. R.
[27] hath been ever. R.
[28] marchioness. R.
[29] now that he hath left. R.
[30] he intends to crown my innocency with the glory of martyrdom. R.
[31] he mought [qy mought not ?] have opportunity to oppress them altogether at once. R.
[32] he used such fine art and fair carriage that he won their confidence to meet altogether in counsel at Cinigalia. R.
[33] but simple. R.
[34] on the day that. R.
[35] commiserated. R.

Papists [36], was wont to say of the Protestants, who ground upon the Scripture, *That they were like posts, that bring truth in their letters, and lies in their mouths.*

17. (125.) The Lacedæmonians were besieged by the Athenians in the Fort [37] of Peile; which was won, and some slain and some taken. There was one said to one of them that was taken, by way of scorn, *Were not they brave men that lost their lives at the Fort of Peile?* He answered, *Certainly a Persian arrow is much to be set by, if it can choose out a brave man.*

18. (208.) After the defeat of Cyrus the younger, Falinus was sent by the King to the Grecians, (who had for their part rather victory than otherwise), to command them to yield their arms. Which when it was denied, Falinus said to Clearchus; *Well then, the King lets you know, that if you remove from the place where you are now encamped, it is war: if you stay, it is truce. What shall I say you will do?* Clearchus answered, *It pleaseth us as it pleaseth the King. How is that?* saith Falinus. Saith Clearchus, *If we remove, war: if we stay, truce.* And so would not disclose his purpose.

19. (126.) Clodius was acquit by a corrupt jury, that had palpably taken shares of money. Before they gave up their verdict, they prayed of the Senate a guard, that they might do their consciences freely; for Clodius was a very seditious young nobleman. Whereupon all the world gave him for condemned. But acquitted he was. Catulus, the next day, seeing some of them that had acquitted him together, said to them; *What made you to ask of us a guard? Were you afraid your money should have been taken from you?*

20. (127). At the same judgment, Cicero gave in evidence upon oath: and the jury (which consisted of fifty-seven) passed against his evidence. One day in the Senate, Cicero and Clodius being in altercation, Clodius upbraided him and said; *The jury gave you no credit.* Cicero answered, *Five-and-twenty gave me credit: but there were two-and-thirty that gave you no credit, for they had their money aforehand.*

21. (80.) Many men, especially such as affect gravity, have a manner after other men's speech to shake their heads. Sir Lionel Cranfield would say [38], *That it was as men shake a bottle, to see if there were any wit in their head or no.*

† 22. Sir Thomas Moore (who was a man in all his lifetime that had an excellent vein in jesting) at the very instant of his death, having a pretty long beard, after his head was upon the block, lift it up again, and gently drew his beard aside, and said, *This hath not offended the King.*

23. (60.) Sir Thomas Moore had sent him by a suitor in the chancery two silver flagons. When they were presented by the gentleman's servant, he said to one of his men; *Have him to the cellar, and let him have of my best wine.* And turning to the servant, said, *Tell thy master, friend, if he like it, let him not spare it.*

24. (129.) Diogenes, having seen that the kingdom of Macedon, which before was contemptible and low, began to come aloft, when he died, was asked; *How he would be buried?* He answered, *With my face downward; for within a while the world will be turned upside down, and then I shall lie right.*

25. (130.) Cato the elder was wont to say, *That the Romans were like sheep: A man were better drive a flock of them than one of them.*

26. (201.) Themistocles in his lower fortune was in love with a young gentleman who scorned him. When he grew to his greatness, which was soon after, he sought to him: but Themistocles said; *We are both grown wise, but too late.*

27. Demonax the philosopher, when he died, was asked touching his burial. He answered, *Never take care for burying me, for stink will bury me.* He that asked him, said again: *Why would you have your body left to dogs and ravens to feed upon?* Demonax answered, *Why, what great hurt is it, if having sought to do good, when I lived, to men, my body do some good to beasts, when I am dead.*

28. Jack Roberts was desired by his tailor, when the reckoning grew somewhat high, to have a bill of his hand. Roberts said: *I am content, but you must let no man know it.* When the tailor brought him the bill, he tore it, as in choler,

[36] the Popish religion. R. [37] Port. R. Phyle? or Pylus?
[38] A great officer of this land would say. R.

and said to him; *You use me not well; you promised me nobody should know it, and here you have put in, Be it known unto all men by these presents.*

29. (131.) When Lycurgus was to reform and alter the state of Sparta, in the consultation one advised that it should be reduced to an absolute popular equality. But Lycurgus said to him: *Sir, begin it in your own house.*

† 30. Phocion the Athenian, (a man of great severity, and no ways flexible to the will of the people), one day when he spake to the people, in one part of his speech was applauded: Whereupon he turned to one of his friends and asked, *What have I said amiss?*

† 31. Sir Walter Ralegh was wont to say of the ladies of Queen Elizabeth's privy-chamber and bed-chamber; *That they were like witches; they could do hurt, but they could do no good.*

32. (122.) Bion, that was an atheist, was shewed in a port-city, in a temple of Neptune, many tables or pictures of such as had in tempests made their vows to Neptune, and were saved from shipwrack; and was asked; *How say you now, do you not acknowledge the power of the Gods?* But he said; *Yes, but where are they painted that have been drowned after their vows?*

33. (202.) Bias[39] was sailing, and there fell out a great tempest, and the mariners, that were wicked and dissolute fellows, called upon the Gods; But Bias said to them; *Peace*, let them not know ye are here.

† 34. Bion was wont to say; *That Socrates, of all the lovers of Alcibiades, only held him by the ears.*

† 35. There was a minister deprived for inconformity, who said to some of his friends; *That if they deprived him, it should cost an hundred men's lives.* The party understood it as if, being a turbulent fellow, he would have moved sedition and complained of him. Whereupon being convented and apposed upon that speech, he said; *His meaning was, that if he lost his benefice, he would practise physic; and then he thought he should kill an hundred men in time.*

36. (61.) Michael Angelo, the famous painter, painting in the Pope's chapel the portraiture of hell and damned souls, made one of the damned souls so like a Cardinal that was his enemy, as everybody at first sight knew it: Whereupon the Cardinal complained to Pope Clement, desiring[40] it might be defaced; Who said to him, *Why, you know very well, I have power to deliver a soul out of purgatory, but not out of hell*[41].

† 37. There was a philosopher about Tiberius, that looking into the nature of Caius, said to him; *That he was mire mingled with blood.*

38. (209.) Alcibiades came to Pericles, and stayed a while ere he was admitted. When he came in, Pericles civilly excused it, and said; *I was studying how to give my account.* But Alcibiades said to him; *If you will be ruled by me, study rather how to give no account.*

39. (133.) Cicero was at dinner, where there was an ancient lady that spake of her years, and said, *She was but forty years old.* One that sat by Cicero rounded him in the ear, and said; *She talks of forty years old, and she is far more, out of question.* Cicero answered him again; *I must believe her, for I have heard her say so any time these ten years.*

40. (68.) Pope Adrian the sixth was talking with the Duke of Sesa; *That Pasquil gave great scandal, and that he would have him thrown into the river.* But Sesa answered; *Do it not (holy father) for then he will turn frog; and whereas now he chants but by day, he will then chant both by day and night*[42].

41. (134.) There was a soldier that vaunted before Julius Cæsar of hurts he had received in his face. Julius Cæsar knowing him to be but a coward, told him; *You were best take heed, next time you run away, how you look back.*

†42. There was a Bishop that was somewhat a delicate person, and bathed twice a day. A friend of his said to him; *My lord, why do you bathe twice a day?* The Bishop answered; *Because I cannot conveniently bathe thrice.*

43. (210.) Mendoza that was viceroy of Peru, was wont to say; *That the*

[39] Bion. R. [40] humbly praying. R.

[41] See Melchior (Floresta española, de apoteghmas ó sentencias, etc., 1614), I. 1. 3.

[42] Melch. I. 1. 5.

government of Peru was the best place that the King of Spain gave, save that it was somewhat too near Madrid.

†44. Secretary Bourn's son kept a gentleman's wife in Shropshire, who lived from her husband with him. When he was weary of her, he caused her husband to be dealt with to take her home, and offered him five hundred pounds for reparation. The gentleman went to Sir Henry Sidney, to take his advice upon this offer; telling him; *That his wife promised now a new life; and, to tell him truth, five hundred pounds would come well with him; and besides, that sometimes he wanted a woman in his bed. By my troth,* (said Sir Henry Sidney) *take her home, and take the money; and then whereas other cuckolds wear their horns plain you may wear yours gilt.*

45. (69.) There was a gentleman in Italy that wrote to a great friend of his, upon his advancement [43] to be Cardinal; *That he was very glad of his advancement, for the Cardinal's own sake; but he was sorry that himself had lost so good a friend* [44].

†46. When Rabelais lay on his death-bed, and they gave him the extreme unction, a familiar friend of his came to him afterwards, and asked him; *How he did?* Rabelais answered; *Even going my journey, they have greased my boots already.*

47. (70.) There was a King of Hungary took a Bishop in battle, and kept him prisoner. Whereupon the Pope writ a monitory to him, for that he had broken the privilege of Holy Church, and taken his son. The King sent an embassage to him, and sent withal the armour wherein the Bishop was taken, and this only in writing, *Vide num hæc sit vestis tui* [45].

48. (135). There was a suitor to Vespasian, who to lay his suit fairer, said; *It was for his brother;* whereas indeed it was for a piece of money. Some about Vespasian, to cross him, told the Emperor, *That the party his servant spake for was not his brother; but that it was upon a bargain.* Vespasian sent for the party interested, and asked him; *Whether his mean* [46] *was his brother or no?* He durst not tell untruth to the Emperor, and confessed; *That he was not his brother.* Whereupon the Emperor said, *This do, fetch me the money, and you shall have your suit dispatched.* Which he did. The courtier, which was the mean, solicited Vespasian soon after about his suit. *Why,* (saith Vespasian,) *I gave it last day to a brother of mine.*

49. (211.) When Vespasian passed from Jewry to take upon him the empire, he went by Alexandria, where remained two famous philosophers, Apollonius and Euphrates. The Emperor heard them discourse touching matter of state, in the presence of many. And when he was weary of them, he brake off, and in a secret derision, finding their discourses but speculative, and not to be put in practice, said; *O that I might govern wise men, and wise men govern me.*

50. (212.) Cardinal Ximenes, upon a muster which was taken against the Moors, was spoken to by a servant of his to stand a little out of the smoke of the harquebuss; but he said again; *That that was his incense* [47].

51. (136.) Vespasian asked of Apollonius, *what was the cause of Nero's ruin?* who answered; *Nero could tune the harp well; but in government he did always wind up the strings too high, or let them down too low.*

†52. Mr. Bromley, Solicitor, giving in evidence for a deed which was impeached to be fraudulent, was urged by the counsel on the other side with this presumption; that in two former suits, when title was made, that deed was passed over in silence, and some other conveyance stood upon. Mr. Justice Catyline taking in with that side, asked the Solicitor, *I pray thee, Mr. Solicitor, let me ask you a familiar question; I have two geldings in my stable, and I have divers*

43 whom the Pope had newly advanced. R.

44 a good friend. R. Melchior (I. 2. 1.) gives this as written to Cardinal Ximenes on his being made archbishop of Toledo.

45 *Know now whether this be thy son's coat?* (Added in R.)

46 his mean employed by him. R.

47 Melch. I. 2. 5. where however the occasion is said to have been not the taking a muster against the Moors, but the going to see an altar erected at Madrid, "fuera de la puerta de Moros," and being saluted by the harquebusseers.

times business of importance, and still I send forth one of my geldings, and not the other; would you not think I set him aside for a jade? No, my Lord, (saith Bromley,) *I would think you spared him for your own saddle.*

53. (45.) Alonso Cartilio was informed by his steward of the greatness of his expense, being such as he could not hold out with. The Bishop asked him *Wherein it chiefly rose?* His steward told him; *In the multitude of his servants.* The Bishop bad him make a note of those that were necessary, and those that mought be put off [48]. Which he did. And the Bishop taking occasion to read it before most of his servants, said to his steward; *Well, let these remain because I need them; and these other also because they need me.*

54. (19.) Queen Elizabeth was wont to say, upon the Commission of Sales; *That the commissioners used her like strawberry wives, that laid two or three great strawberries at the mouth of their pot, and all the rest were little ones; so they made her two or three good prices of the first particulars, but fell straightways.*

55. (20). Queen Elizabeth was wont to say of her instructions to great officers; *That they were like to garments, strait at the first putting on, but did by and by wear loose enough.*

56. (46.) Mr. Marbury the preacher would say; *that God was fain to do with wicked men, as men do with with frisking jades in a pasture, that cannot take them up, till they get them at a gate. So wicked men will not be taken up till the hour of death.*

†57. Thales, as he looked upon the stars, fell into the water; Whereupon it was after said; *That if he had looked into the water he might have seen the stars; but looking up to the stars he could not see the water.*

58. (22.) The book of deposing Richard [49] the second, and the coming in of Henry the fourth, supposed to be written by Dr. Hayward, who was committed to the Tower for it, had much incensed queen Elizabeth. And she asked Mr. Bacon, being then of her learned counsel; *Whether there were no treason contained in it?* Mr. Bacon intending to do him a pleasure, and to take off the queen's bitterness with a jest [50], answered; *No, madam, for treason I cannot deliver opinion that there is any, but very much felony.* The Queen, apprehending it gladly, asked, *How, and wherein?* Mr. Bacon answered; *Because he had stolen many of his sentences and conceits out of Cornelius Tacitus.*

59. (199.) Mr. Popham [51], when he was Speaker, and the Lower House [52] had sat long, and done in effect nothing; coming one day to Queen Elizabeth, she said to him; *Now, Mr. Speaker, what hath passed in the Lower House [53]?* He answered, *If it please your Majesty, seven weeks.*

60. (47.) Pope Xystus the fifth, who was a poor [54] man's son, and his father's house ill thatched, so that the sun came in in many places, would sport with his ignobility, and say; He was *nato di casa illustre: son of an illustrious house.*

61. (48.) When the King of Spain conquered Portugal, he gave special charge to his lieutenant that the soldiers should not spoil, lest he should alienate the hearts of the people. The army also suffered much scarcity of victual. Whereupon the Spanish soldiers would afterwards say; *That they had won the King a kingdom, as the kingdom of heaven useth to be won; by fasting and abstaining from that that is another man's.*

62. (108.) Cicero married his daughter to Dolabella, that held Cæsar's party: Pompey had married Julia, that was Cæsar's daughter. After, when Cæsar and Pompey took arms one against the other, and Pompey had passed the seas, and Cæsar possessed Italy, Cicero stayed somewhat long in Italy, but at last sailed over to join with Pompey; who when he came unto him, Pompey said; *You are welcome; but where left you your son-in-law?* Cicero answered; *With your father-in-law.*

63. (213.) Nero was wont to say of his master Seneca; *That his stile was like mortar of sand without lime.*

[48] spared. R. This is told in Melchior I. 3. 2.

[49] King Richard. R.

[50] merry conceit. R.

[51] (afterwards Lord Chief Justice Popham.) R.

[52] House of Commons. R.

[53] Commons' House. R.

[54] very poor. R.

64. (240.) Sir Henry Wotton used to say, *That critics are like brushers of noblemen's clothes.*

65. (23.) Queen Elizabeth, being to resolve upon a great officer, and being by some, that canvassed for others, put in some doubt of that person whom she meant to advance, called for Mr. Bacon, and told him, *She was like one with a lanthorn seeking a man ;* and seemed unsatisfied in the choice she had of men for that place. Mr. Bacon answered her ; *That he had heard that in old time there was usually painted on the church walls the Day of Doom, and God sitting in judgement, and St. Michael by him with a pair of balance*[55] *; and the soul and the good deeds in the one balance, and the faults and the evil deeds in the other ; and the soul's balance went up far too light : Then was our Lady painted with a great pair of beads casting them into the light balance, to make up the weight*[56] *:* so (he said) *place and authority, which were in her hands to give, were like our lady's beads, which though men, through divers imperfections, were too light before, yet when they were cast in, made weight competent.*

66. (128.) Mr. Savill[57] was asked by my lord of Essex his opinion touching poets ; who[58] answered my lord ; *He thought*[59] *them the best writers, next to those that write*[60] *prose.*

†67. Mr. Mason of Trinity college sent his pupil to another of the fellows, to borrow a book of him ; who told him ; *I am loth to lend my books out of my chamber ; but if it please thy tutor to come and read upon it in my chamber, he shall as long as he will.* It was winter ; and some days after, the same fellow sent to Mr. Mason to borrow his bellows ; but Mr. Mason said to his pupil ; *I am loth to lend my bellows out of my chamber ; but if thy tutor would come and blow the fire in my chamber, he shall as long as he will.*

68. (100.) Nero did cut a youth, as if he would have transformed him into a woman[61], and called him *wife.* There was a senator of Rome that said secretly to his friend ; *It was pity Nero's father had not such a wife.*

69. (111.) Galba succeeded Nero, and his age being much despised, there was much licence and confusion in Rome. Whereupon a senator said in full senate; *It were better live where nothing is lawful, than where all things are lawful.*

†70. In Flanders by accident a Flemish tiler fell from the top of a house upon a Spaniard, and killed him, though he escaped himself. The next of the blood prosecuted his death with great violence against the tiler. And when he was offered pecuniary recompence, nothing would serve him but *lex talionis.* Whereupon the judge said to him ; *That if he did urge that kind of sentence, it must be, that he should go up to the top of the house, and thence fall down upon the tiler.*

71. (24.) Queen Elizabeth was dilatory enough in suits, of her own nature ; and the lord treasurer Burleigh, to feed her humour[62], would say to her ; *Madam, you do well to let suitors stay ; for I shall tell you, Bis dat, qui cito dat : If you grant them speedily, they will come again the sooner.*

72. (49.) They feigned[63] a tale of Sixtus Quintus[64], that after his death he went to hell ; and the porter of hell said to him : *You have some reason to offer yourself to this place*[65] *; but yet*[66] *I have order not to receive you : you have a place of your own, purgatory ; you may go thither.* So he went away and sought purgatory a great while, and could find no such place. Upon that he took heart, and went to heaven, and knocked ; and St. Peter asked ; *Who was there ?* He said, *Sixtus Pope.* Whereunto St. Peter said, *Why do you knock ? you have the keys.* Sixtus answered, *It is true ; but it is so long since they were given, as I doubt the wards of the lock be altered.*

73. (50.) Charles King of Swede, a great enemy of the Jesuits, when he took

55 balances. R.
56 and brought down the scale. R.
57 Sir Henry Savill. R.
58 He. R.
59 That he thought. R.
60 writ. R.
61 Nero loved a beautiful youth, whom he used viciously. R.
62 being a wise man, and willing therein to feed her humour. R.
63 So R. The original has "faigne."
64 whom they called *Size-Ace.* R.
65 because you were a wicked man. R.
66 But yet, because you were a Pope. R.

any of their colleges, he would hang the old Jesuits, and put the young to his mines, saying; *That since they wrought so hard above ground, he would try how they could work under ground.*

74. (51.) In Chancery, one time, when the counsel of the parties set forth the boundaries of the land in question, by the plot; And the counsel of one part said, *We lie on this side, my lord*; And the counsel of the other part said, *We lie on this side*: the Lord Chancellor Hatton stood up and said, *If you lie on both sides, whom will you have me to believe.*

75. (109.) Vespasian and Titus his eldest son were both absent from Rome when the empire was cast upon him [67]. Domitian his younger son was at Rome, who took upon him the affairs; and being of a turbulent spirit, made many changes, and displaced divers officers and governors of provinces, sending them successors. So when Vespasian came to Rome, and Domitian came into his presence, Vespasian said to him; *Son, I looked when you would have sent me a successor.*

76. (71.) Sir Amice [68] Pawlet, when he saw too much haste made in any matter, was wont to say, *Stay a while, that we may make an end the sooner.*

77. (31.) The deputies of the reformed religion, after the massacre which was [69] upon St. Bartholomew's day, treated with the King and Queen-Mother, and some other of the counsel, for a peace. Both sides were agreed upon the articles. The question was, upon the security of performance [70]. After some particulars propounded and rejected, the Queen-Mother said; *Why, is not the word of a King sufficient security?* One of the deputies answered; *No, by St. Bartholomew, Madam.*

78. (12.) When the Archduke did raise his siege from Grave, the then secretary came to queen Elizabeth; and the Queen, having intelligence first [71], said to the secretary, *Wot you what? The Archduke is risen from the Grave.* He answered, *What, without the trumpet of the Archangel?* The Queen replied; *Yes, without sound of trumpet.*

† 79. Francis the first used for his pleasure sometimes to go disguised. So walking one day in the company of the Cardinal of Bourbon near Paris, he met a peasant with a new pair of shoes upon his arm. So he called him to him and said; *By our lady, these be good shoes, what did they cost thee?* The peasant said; *Guess.* The King said; *I think some five sols.* Saith the peasant; *You have lyed; but a carolois. What, villain,* (saith the Cardinal of Bourbon) *thou art dead; it is the King.* The peasant replied; *The devil take him, of you and me, that knew so much.*

80. (217.) There was a conspiracy against the Emperor Claudius by Scribonianus, examined in the senate; where Claudius sat in his chair, and one of his freed servants stood at the back of his chair. In the examination, that freed servant, who had much power with Claudius, very saucily had almost all the words: and amongst other things, he asked in scorn one of the examinates, who was likewise freed servant of Scribonianus; *I pray, sir, if Scribonianus had been Emperor what would you have done?* He answered; *I would have stood behind his chair, and held my peace.*

81. (137.) Dionysius the tyrant, after he was deposed, and brought to Corinth, kept a school. Many used to visit him; and amongst others, one, when he came in, opened his mantle and shook his clothes; thinking to give Dionysius a gentle scorn; because it was the manner to do so for them that came in to him while he was tyrant. But Dionysius said to him; *I pray thee do so rather when thou goest out, that we may see thou stealest nothing away.*

82. (241.) Hannibal said of Fabius Maximus and of Marcellus (whereof the former waited upon him, that he could make no progress; and the latter had many sharp fights with him); *that he feared Fabius like a tutor; and Marcellus like an enemy.*

83. (138.) Diogenes, one terrible frosty morning, came into the market-place

[67] Vespasian. R.
[68] Amyas. R.
[69] which was at Paris. R.
[70] for the performance. R.
[71] having first intelligence thereof. R.

and stood naked, quaking, to shew his tolerancy [72]. Many of the people came about him, pitying him. Plato passing by, and knowing he did it to be seen, said to the people, as he went by, *If you pity him indeed, leave him alone.*

84. (72.) Sackford, Master of the Requests [73] to Queen Elizabeth, had divers times moved for audience, and been put off. At last he came to the Queen in a progress, and had on a new pair of boots. When he came in, the Queen [74] said to him, *Fie sloven, thy new boots stink. Madam,* (saith he,) *it is not my new boots that stink, but it is the stale bills that I have kept so long.*

85. (218.) One was saying; *That his great grandfather and grandfather and father died at sea.* Said another that heard him; *And I were as you, I would never come at sea. Why,* (saith he,) *where did your great grandfather and grandfather and father die?* He answered; *Where but in their beds?* Saith the other; *And I were as you, I would never come in bed.*

86. (139.) Aristippus was earnest suitor to Dionysius for somewhat, who would give no ear to his suit. Aristippus fell at his feet. Then Dionysius granted it. One that stood by said afterwards to Aristippus; *You a philosopher, and to be so base as to throw yourself at the tyrant's feet to get a suit?* Aristippus answered; *The fault is not mine, but the fault is in Dionysius, that carries his ears in his feet.*

† 87. There was a young man in Rome, that was very like Augustus Cæsar. Augustus took knowledge of it, and sent for the man, and asked him; *Was your mother never at Rome?* He answered; *No, sir, but my father was.*

† 88. A physician advised his patient, that had sore eyes, that he should abstain from wine. But the patient said, *I think rather, sir, from wine and water* [75] *; for I have often marked it in blear eyes, and I have seen water come forth, but never wine.*

† 89. When Sir Thomas Moore was Lord Chancellor, he did use, at mass, to sit in the chancel; and his lady in a pew. And because the pew stood out of sight, his gentleman-usher ever after service came to the lady's pew, and said; *Madam, my Lord is gone.* So when the Chancellor's place was taken from him, the next time they went to church, Sir Thomas himself came to his lady's pew, and said; *Madam, my Lord is gone.*

90. (73.) At an act of the Commencement, the answerer gave for his question; *That an aristocracy was better than a monarchy.* The replier, who was a dissolute fellow [76], did tax him; *That being a private bred man, he would give a question of state.* The answerer said; *That the replier did much wrong the privilege of scholars; who would be much straitened if they should give questions of nothing but such things wherein they are practised.* And added; *We have heard yourself dispute of virtue, which no man will say you put much in practice.*

91. (219.) There was a dispute, whether great heads or little heads had the better wit? And one said; *It must needs be the little. For* [77] *it is a maxim, Omne majus continet in se minus.*

92. (140.) Solon, when he wept for his son's death, and one said to him; *Weeping will not help;* answered, *Alas, therefore I weep, because weeping will not help.*

93. (141.) Solon being asked; *Whether he had given the Athenians the best laws?* answered; *Yes, the best of those that they would have received.*

94. (142.) One said to Aristippus; *It is a strange thing, why should men rather give to the poor, than to philosophers.* He answered; *Because they think themselves may sooner come to be poor, than to be philosophers.*

95. (145.) Alexander used to say of his two friends, Craterus and Hephæstion: *That Hephæstion loved Alexander, and Craterus loved the King.*

96. (146.) It fell out so, that as Livia went abroad in Rome, there met her naked young men that were sporting in the streets; which Augustus was [78] about severely to punish in them; but Livia spake for them, and said, *It was no more to chaste women than so many statua's.*

72 tolerance. R. 73 A Master of Requests. R. (omitting the name.)
74 The Queen who loved not the smell of new leather. R.
75 So in the original. But I think it should be *from water.* 76 man. R.
77 For that. R. 78 went. R.

97. (75.) Alonso of Arragon was wont to say, in commendation of age, That age appeared to be best in four things: *Old wood best to burn; old wine to drink; old friends to trust; and old authors to read* [79].

98. (76.) It was said of Augustus, and afterwards the like was said of Septimius Severus, both which did infinite mischief in their beginnings, and infinite good towards their ends; *That they should either have never been born or never died.*

99. (74.) Queen Isabell [80] of Spain used to say; *Whosoever hath a good presence and a good fashion, carries letters* [81] *of recommendation.*

100. (143.) Trajan would say of the vain jealousy of princes, that seek to make away those that aspire to their succession; *That there was never King that did put to death his successor.*

101. (144.) When it was represented to Alexander, to the advantage of Antipater, who was a stern and imperious man, that he only of all his lieutenants wore no purple, but kept the Macedonian habit of black, Alexander said; *Yes, but Antipater is all purple within* [82].

102. (77.) Constantine the Great, in a kind of envy, himself being a great builder, as Trajan likewise was, would call Trajan *Wall-flower* [83]; because his name was upon so many walls.

103. (147.) Philip of Macedon was wished to banish one for speaking ill of him. But Philip said [84]; *Better he speak where we are both known, than where we are both unknown.*

† 104. A Grecian captain, advising the confederates that were united against the Lacedæmonians touching their enterprise, gave opinion that they should go directly upon Sparta, saying; *That the state of Sparta was like rivers; strong when they had run a great way, and weak towards their head.*

105. (78.) Alonso of Arragon was wont to say of himself, *That he was a great necromancer, for that he used to ask counsel of the dead:* meaning books [85].

106. (148.) Lucullus entertained Pompey in one of his magnificent houses. Pompey said, *This is a marvellous fair and stately house for the summer: but methinks it should be very cold for winter.* Lucullus answered; *Do you not think me as wise as divers fowl are, to remove with the season?* [86]

107. (149.) Plato entertained some of his friends at a dinner, and had in the chamber a bed or couch, neatly and costly furnished. Diogenes came in, and got upon the bed, and trampled upon it, and said [87]; *I trample upon the pride of Plato.* Plato mildly answered; *But with greater pride.*

† 108. One was examined upon certain scandalous words spoken against the King. He confessed them, and said; *It is true I spake them, and if the wine had not failed I had said much more.*

109. (150.) Pompey being commissioner for sending grain to Rome in time of dearth, when he came to the sea, he found it very tempestuous and dangerous, insomuch as those about him advised him by no means to embark. But Pompey said; *It is of necessity that I go, not that I live.*

† 110. Trajan would say; *That the King's exchequer was like the spleen; for when that did swell, the whole body did pine.*

† 111. Charles the Bald allowed one, whose name was Scottus, to sit at the table with him, for his pleasure. Scottus sat on the other side of the table. One time the King being merry with him, said to him; *What is there between Scot and Sot?* Scottus answered; *The table only.*

112. (79.) Ethelwold, Bishop of Winchester, in a famine, sold all the rich vessels and ornaments of the Church, to relieve the poor with bread; and said, *There was no reason that the dead temple of God should be sumptuously furnished, and the living temples suffer penury.*

† 113. There was a marriage made between a widow of great wealth, and a gentleman of great house that had no estate or means. Jack Roberts said;

[79] Melch. II. 1. 20. [80] Isabella. R. [81] continual letters. R.

[82] [In a note on the *De Augmentis*, B. i., Mr. Ellis points out that in this story Bacon follows Erasmus, who misunderstood it as told in Plutarch.—*Ed.*]

[83] *Parietaria*, wall-flower. R. [84] answered. R.

[85] Of books. R. [86] to change my habitation in the winter season. R.

[87] and trampled it; saying. R.

That marriage was like a black pudding ; the one brought blood, and the other brought suet and oatmeal [88].

114. (151.) Demosthenes was upbraided by Æschines, that his speeches did smell of the lamp. But Demothenes said ; *Indeed there is a great deal of difference between that that you and I do by lamp-light.*

115. (152.) Demades the orator, in his age, was talkative, and would eat hard. Antipater would say of him, *That he was like a sacrifice, that nothing was left of it but the tongue and the paunch.*

116. (242.) When King Edward the Second was amongst his torturers, who hurried him to and fro, that no man should know where he was, they set him down upon a bank ; and one time, the more to disguise his face, shaved him, and washed him with cold water of a ditch by : The King said ; *Well, yet I will have warm water for my beard.* And so shed abundance of tears.

117. (203.) The Turk [89] made an expedition into Persia, and because of the strait jaws of the mountains of Armenia, the basha's consulted which way they should get in. Says a natural fool that stood by [90] ; *Here's much ado how you should* [91] *get in ; but I hear nobody take care how you should get out.*

118. (220.) Sir Thomas Moore, when the counsel of the party pressed him for a longer day [92], said ; *Take Saint Barnaby's day, which is the longest day in the year.* Now Saint Barnaby's day was within few days following.

119. (221.) One of the Fathers saith ; *That there is but this difference between the death of old men and young men ; that old men go to death, and death comes to young men.*

120. (154.) Philo Judæus saith ; *That the sense is like the sun ; For the sun seals up the globe of heaven, and opens the globe of earth ; so the sense doth obscure heavenly things, and reveal earthly things.*

121. (222.) Cassius, after the defeat of Crassus by the Parthians, whose weapons were chiefly arrows, fled to the city of Carras, where he durst not stay any time, doubting to be pursued and besieged. He had with him an astrologer, who said to him ; *Sir, I would not have you go hence, while the moon is in the sign of Scorpio.* Cassius answered, *I am more afraid of that of Sagittarie* [93].

122. (155.) Alexander, after the battle of Granicum, had very great offers made him by Darius. Consulting with his captains concerning them, Parmenio said ; *Sure I would accept of these offers, if I were as Alexander.* Alexander answered ; *So would I, if I were as Parmenio.*

123. (156.) Alexander was wont to say ; *He knew he was mortal* [94] *by two things ; sleep and lust.*

† 124. Augustus Cæsar was invited to supper by one of his old friends that had conversed with him in his less fortunes, and had but ordinary entertainment. Whereupon, at his going, he said ; *I did not know you and I were so familiar* [95].

125. (157.) Augustus Cæsar would say ; *That he wondered that Alexander feared he should want work, having no more* [96] *to conquer ; as if it were not as hard a matter to keep as to conquer.*

126. (158.) Antigonus, when it was told him that the enemy had such vollies of arrows that they did hide the sun, said ; *That falls out well, for it is hot weather, and we shall fight in the shade.*

127. (112.) Augustus Cæsar did write to Livia, who was over-sensible of some ill-words that had been spoken of them both : *Let it not trouble thee, my Livia, if any man speak ill of us ; for we have enough, that no man can do ill unto us.*

128. (113.) Chilon said ; *That kings' friends and favourites were like casting counters ; that sometimes stood for one, sometimes for ten, sometimes for a hundred.*

129. (114.) Theodosius, when he was pressed by a suitor, and denied him,

[88] Melch. IV. 4. 13 where the remark is attributed to a nameless Hidalgo, upon a marriage between a rich labourer's daughter and the son of a poor gentleman.

[89] Turks. R.

[90] One that heard the debate said. R.

[91] Shall. R.

[92] a longer day to perform the decree. R.

[93] *sagittarius.* R.

[94] knew himself to be mortal chiefly. R.

[95] Melch. VI. 8. 14. told of two squires.

[96] no more worlds. R.

the suitor said; *Why, Sir, you promised it.* He answered; *I said it, but I did not promise it, if it be unjust.*

130. (200.) Agathocles, after he had taken Syracusa, the men whereof, during the siege, had in a bravery spoken of him all the villany that mought be, sold the Syracusans for slaves, and said; *Now if you use such words of me, I will tell your masters of you.*

† 131. Dionysius the elder, when he saw his son in many things very inordinate, said to him; *Did you ever know me do such things? His son answered; No, but you had not a tyrant to your father.* The father replied; *No, nor you, if you take these courses, will have a tyrant to your son.*

† 132. Callisthenes the philosopher, that followed Alexander's court, and hated the King, was asked by one; *How one should become the famousest man in the world?* and answered; *By taking away him that is.*

133. (52.) Sir Edward Coke was wont to say, when a great man came to dinner to him, and gave him no knowledge of his coming; *Well, since you sent me no word of your coming, you shall dine with me; but if I had known of your coming* [97], *I would have dined with you.*

134. (115.) The Romans, when they spake to the people, were wont to call [98] them; *Ye Romans.* When commanders in war spake to their army, they called [99] them; *My soldiers.* There was a mutiny in Cæsar's army, and somewhat the soldiers would have had, but they would not declare themselves in it; only they demanded a dimission [100] or discharge, though with no intention it should be granted; but knowing that Cæsar had at that time great need of their service, thought by that means to wrench him to their other desires. Whereupon with one cry they asked dimission [101]. But Cæsar, after silence made, said; *I for my part, ye Romans:* which admitted them [102] to be dismissed. Which voice they had no sooner heard, but they mutined again [103], and would not suffer him to go on [104] until he had called them by the name of *soldiers.* And so with that one word he appeased the sedition.

135. (116.) Cæsar would say of Sylla, for that he did resign his dictatorship; *That he* [105] *was ignorant of letters, he could not dictate.*

136. (117.) Seneca said of Cæsar; *that he did quickly sheath the sword, but never laid it off* [106].

137. (118.) Diogenes begging, as divers philosophers then used, did beg more of a prodigal man, than of the rest that were present: Whereupon one said to him; *See your baseness, that when you find a liberal mind, you will take most of* [107]. *No* (said Diogenes), *but I mean to beg of the rest again.*

138. (223.) Jason the Thessalian was wont to say; *That some things must be done unjustly, that many things may be done justly.*

139. (25.) Sir Nicholas Bacon being Keeper of the Seal [108], when Queen Elizabeth, in progress, came to his house at Redgrave [109], and said to him; *My Lord, what a little house have you gotten?* said [110], *Madam, my house is well, but it is you that have made me too great for my house.*

140. (119.) Themistocles, when an embassador from a mean state did speak great matters, said to him, *Friend, your words would require a city.*

† 141. Agesilaus, when one told him there was one did excellently counterfeit a nightingale, and would have had him hear him, said; *Why I have heard the nightingale herself.*

142. (53.) A great nobleman [111], upon the complaint of a servant of his, laid a citizen by the heels, thinking to bend him to his servant's desire. But the fellow being stubborn, the servant came to his lord, and told him; *Your lordship, I know, hath gone as far as well you may, but it works not; for yonder fellow is more*

[97] known of it in due time. R. [98] stile. R. [99] stiled. R.
[100] but only demanded a mission. R. [101] mission. R.
[102] This title did actually speak them. R. [103] mutinied. R.
[104] to go on with his speech. R. [105] Sylla. R.
[106] did quickly shew the sword, but never leave it off. R. [107] of him. R.
[108] who was Keeper of the Great Seal of England. R. [109] Gorhambury. R.
[110] Answered her. R. [111] William Earl of Pembroke. R.

perverse than before. Said my lord, *Let's forget him a while, and then he will remember himself.*

† 143. One came to a Cardinal in Rome, and told him ; *That he had brought his lordship a dainty white palfrey, but he fell lame by the way.* Saith the Cardinal to him ; *I'll tell thee what thou shalt do ; go to such a Cardinal, and such a Cardinal,* (naming him some half a dozen Cardinals,) *and tell them as much ; and so whereas by thy horse, if he had been sound, thou couldest have pleased but one, with thy lame horse thou mayest please half a dozen.*

144. (120.) Iphicrates the Athenian, in a treaty that he had with the Lacedæmonians for peace, in which question was about security for observing the same [112], said, *The Athenians would not accept of any security, except the Lacedæmonians did yield up unto them those things, whereby it mought be manifest that they could not hurt them if they would.*

† 145. Euripides would say of persons that were beautiful, and yet in some years, *In fair bodies not only the spring is pleasant, but also the autumn.*

146. (81.) After a great fight, there came to the camp of Consalvo, the great captain, a gentleman proudly horsed and armed. Diego de Mendoza asked the great captain ; *Who's this ?* Who answered ; *It is Saint Ermin, who never appears but after a storm.*[113]

† 147. There was a captain sent to an exploit by his general, with forces that were not likely to achieve the enterprize. The captain said to him ; *Sir, appoint but half so many. Why ?* (saith the general). The captain answered ; *Because it is better fewer die than more* [114].

148. (121.) They would say of the Duke of Guise, Henry, that had sold and oppignerated all his patrimony, to suffice the great donatives that he had made ; *That he was the greatest usurer of France. because all his state was in obligations* [115].

† 149. Crœsus said to Cambyses ; *That peace was better than war ; because in peace the sons did bury their fathers, but in war the fathers did bury their sons.*

150. (224.) There was a harbinger who had lodged a gentleman in a very ill room, who expostulated with him somewhat rudely ; but the harbinger carelessly said ; *You will take pleasure in it when you are out of it* [116].

† 151. There was a curst page, that his master whipt naked ; and when he had been whipt, would not put on his clothes ; and when his master bade him, said to him ; *Take them you, for they are the hangman's fees.*

152. (82.) There was one that died greatly in debt. When it was reported in some company, where divers of his creditors were, that he was dead, one began to say ; *In good faith* [117], *then he hath carried five hundred ducats of mine with him into the other world.* And another of them said ; *And two hundred of mine.* And some others spake of several sums of theirs [118]. Whereupon one that was amongst them said; *Well I see* [119] *now that though a man cannot carry any of his own with him into the other world, yet he may carry other men's* [120].

153. (83.) Francis Carvajall, that was the great captain of the rebels of Peru, had often given the chase to Diego Centeno, a principal commander of the Em-

[112] the same peace. R.

[113] the storm. R. Compare Melch. II. 3. 3. : where the story is in one respect better told. Consalvo having just disembarked, three ships were seen approaching ; " Venia delante in uno dellos un cavallero armado que se avia quedado atrás ". A collection of French apophthegms gives it thus : " Le grand Capitaine Gonsalvo voiant venir un sien gentilhomme au devant de lui bien en ordre et richement armé après la journée de Serignolle ; et que les affaires estoient à seurté ; dit à la compagnie : nous ne devons desormais avoir peur de la tourmente. Car Saint Herme nous est apparu."—*Apophthegmata Græca, Latina, Italica, Gallica, Hispanica, collecta a Gerærdo Suningro. Leidensi*, 1609.

[114] Melch. II. 3. 12.

[115] They would say of the Duke of Guise, Henry ; That he was the greatest usurer in France, for that he had turned all his estate into obligations ; meaning that he had sold and oppignorated all his patrimony to give large donatives to other men. R.

[116] Melch. II. 6. 2. ; differently told.

[117] well, if he be gone. R.

[118] And a third spake of great sums of his. R.

[119] perceive. R.

[120] into the next world, yet he may carry that which is another man's. R.

peror's party. He was afterwards taken by the Emperor's lieutenant, Gasca, and committed to the custody of Diego Centeno, who used him with all possible courtesy; insomuch as Carvajall asked him; *I pray, Sir, who are you that use me with this courtesy?* Centeno said; *Do not you know Diego Centeno?* Carvajall answered; *In good faith, Sir,*[121] *I have been so used to see your back, as I knew not your face.*

† 154. Carvajall, when he was drawn to execution, being fourscore and five years old, and laid upon the hurdle, said; *What? young in cradle, old in cradle?*

155. (84.) There is a Spanish adage, [122] *Love without end* [123] *hath no end:* meaning, that if it were begun not upon particular ends it would last.

156. (159.) Cato the elder, being aged, buried his wife, and married a young woman. His son came to him, and said; *Sir, what have I offended you, that you have brought a step-mother into your house?* The old man answered; *Nay, quite contrary, son; thou pleasest me so well, as I would be glad to have more such.*

157. (160.) Crassus the orator had a fish, which the Romans called [124] *Muræna,* that he had made very tame and fond of him. The fish died, and Crassus wept for it. One day falling in contention with Domitius in the senate, Domitius said; *Foolis Crassus, you wept for your Muræna.* Crassus replied; *That's more than you did for both your wives.*

158. (161.) Philip, Alexander's father, gave sentence against a prisoner, what time he was drowsy, and seemed to give small attention. The prisoner, after sentence was pronounced, said; *I appeal,* The King somewhat stirred, said; *To whom do you appeal?* The prisoner answered; *From Philip when he gave no ear, to Philip when he shall give ear.*

159. (204.) The same Philip [125] maintained argument with a musician, in points of his art, somewhat peremptorily. But the musician said to him; *God forbid, Sir, your fortune were so hard that you should know these things better than I* [126].

160. (162.) There was a philosopher that disputed with Adrian the Emperor, and did it but weakly. One of his friends that had been by, afterwards said to him; *Methinks you were not like yourself, last day, in argument with the Emperor; I could have answered better myself. Why,* said the philosopher, *would you have me contend with him that commands thirty legions?*

† 161. Diogenes was asked in a kind of scorn; *What was the matter, that philosophers haunted rich men, and not rich men philosophers?* He answered; *Because the one knew what they wanted, the other did not.*

† 162. Demetrius, King of Macedon, had a petition offered him divers times by an old woman, and still answered; *He had no leisure. Whereupon the woman said aloud; Why then give over to be King.*

163. (225.) The same Demetrius [127] would at times retire himself from business, and give himself wholly to pleasures. One day of those his retirings [128], giving out that he was sick, his father Antigonus came on the sudden to visit him, and met a fair dainty youth coming out of his chamber. When Antigonus came in, Demetrius said; *Sir, the fever left me right now.* Antigonus replied, *I think it was he that I met at the door.*

164. (85.) There was a merchant far in debt that died [129]. His goods and household stuff were set forth to sale. There was one that bought only a pillow, and said [130]; *This pillow sure is good to sleep upon, since he could sleep that owed so many debts* [131].

165. (86.) A lover met his lady in a close chair, she thinking to go [132] unknown. He came and spake to her. She asked him; *How did you know me?* He said; *Because my wounds bleed afresh.* Alluding to the common tradition, that the

[121] Truly, Sir. R. [122] Gondomar would say. R. [123] ends. R.
[124] call. R. [125] Philip King of Macedon. R.
[126] myself. R. [127] Demetrius King of Macedon. R. [128] One of those his retirings.
[129] There was a merchant died, that was very far in debt. R.
[130] A stranger would needs buy a pillow there, saying. R.
[131] The saying is attributed by Macrobius to Augustus Cæsar; and quoted in Erasmus's collection, No. 31. [132] to have gone. R.

wounds of a body slain, in the presence of him that killed him, will bleed afresh [133].

166. (87.) A gentleman brought music to his lady's window, who [134] hated him and had warned him oft away; and when he persisted [135], she threw stones at him. Whereupon a friend of his that was in his company, said to him [136]; *What greater honour can you have to your music, than that stones come about you, as they did to Orpheus?*

167. (226.) Cato Major would say; *That wise men learned more by fools, than fools by wise men.*

168. (227.) When it was said to Anaxagoras; *The Athenians have condemned you to die:* he said again; *And nature them.*

† 169. Demosthenes when he fled from the battle, and that it was reproached to him, said; *That he that flies mought fight again.*

170. (205.) Antalcidas, when an Athenian said to him; *Ye Spartans are unlearned;* said again; *True, for we have learned no evil nor vice of you.*

171. (228.) Alexander, when his father wished him to run for the prize of the race at the Olympian games, (for he was very swift,) said; *He would, if he might run with kings.*

172. (163.) When Alexander passed into Asia, he gave large donatives to his captains, and other principal men of virtue; insomuch as Parmenio asked him; *Sir, what do you keep for yourself?* He answered; *Hope.*

173. (229.) Antigonus used oft to go disguised, and listen at the tents of his soldiers: and at a time heard some that spoke very ill of him. Whereupon he opened the tent a little, and said to them; *If you will speak ill of me, you should go a little further off.*

174. (164.) Vespasian set a tribute upon urine. Titus his son emboldened himself to speak to his father of it: and represented it as a thing indign and sordid. Vespasian said nothing for the time; but a while after, when it was forgotten, sent for a piece of silver out of the tribute money, and called to his son, bidding him smell to it; and asked him; *Whether he found any offence?* Who said, *No. Why lo* [137], (saith Vespasian again), *and yet this comes out of urine.*

† 175. There were two gentlemen, otherwise of equal degree, save that the one was of the ancienter house [138]. The other in courtesy asked his hand to kiss: which he gave him; and he kissed it; but said withal, to right himself, by way of friendship; *Well, I and you, against any two of them:* putting himself first.

176. (165.) Nerva the Emperor succeeded Domitian, who was tyrannical; so as [139] in his time many noble houses were overthrown by false accusations; the instruments whereof were chiefly Marcellus and Regulus. The Emperor [140] one night supped privately with some six or seven: amongst which there was one that was a dangerous man, and began to take the like courses as Marcellus and Regulus had done. The Emperor fell into discourse of the injustice and tyranny of the former time, and by name of the two accusers; and said; *What should we do with them, if we had them now?* One of them that were [141] at supper, and was a free-spoken senator, said; *Marry, they should sup with us.*

[133] that the wounds of a body slain will bleed afresh upon the approach of the murtherer. R. [134] She. R. [135] would not desist. R.

[136] a gentleman said unto him, that was in his company. R. [137] Why so. R.

[138] According to Melchior's version (VI. 6. 4,) *mas anciano*: the older man.

[139] who had been tyrannical; and. R. [140] The Emperor Nerva. R.

[141] was. R. This variation (which is obviously wrong), coupled with others of the same kind, makes me suspect that the text of the edition of 1661 has suffered from a correcting editor. It may be that he had no choice: for the collection may have been made up from a rough imperfect or illegible copy, containing passages which could only be supplied by conjecture. But it strikes me that very few of these different readings are such as Bacon himself would have thought improvements. In this case the history of the change may be easily divined. "One of them that *were* at supper, and *was* a free-spoken senator," struck the editor as an incorrect sentence: *were* and *was* could not both

177. (166.) There was one that found a great mass of money, digged under ground in his grandfather's house. And being somewhat doubtful of the case, signified it to the Emperor that he had found such treasure. The Emperor made a rescript thus; *Use it.* He writ back again, that the sum was greater than his estate or condition could use. The Emperor writ a new rescript thus: *Abuse it.*

178. (198.) A Spaniard was censuring to a French gentleman the want of devotion amongst the French; in that, whereas in Spain, when the Sacrament goes to the sick, any that meets with it turns back and waits upon it to the house whither it goes; but in France they only do reverence, and pass by. But the French gentleman answered him; *There is reason for it; for here with us Christ is secure amongst his friends; but in Spain there be so many Jews and Maranos, that it is not amiss for him to have a convoy.*

179. (88.) Coranus the Spaniard, at a table at dinner, fell into an extolling of his own father, and said; *If he could have wished of God, he would not have chosen amongst men a better father.* Sir Henry Savill said, *What, not Abraham?* Now Coranus was doubted to descend of a race of Jews.

180. (98.) Consalvo would say; *The honour of a soldier ought to be of a good strong web;* meaning, that it should not be so fine and curious, that every little disgrace should[142] catch and stick in it.

181. (243.) One of the Seven was wont to say; *That laws were like cobwebs; where the small flies were caught, and the great brake through.*

† 182. Bias gave in precept; *Love as if you should hereafter hate; and hate as if you should hereafter love.*

183. (169.) Aristippus being reprehended of luxury by one that was not rich, for that he gave six crowns for a small fish, answered; *Why what would you have given?* The other said; *Some twelve pence.* Aristippus said again; *And six crowns is no more with me.*

184. (32.) There was a French gentleman speaking with an English, of the law Salique; that women were excluded to inherit[143] the crown of France. The English said; *Yes, but that was meant of the women themselves, not of such males as claimed by women.* The French gentleman said; *Where do you find that gloss?* The English answered; *I'll tell you, Sir: look on the backside of the record of the law Salique, and there you shall find it indorsed:* meaning[144] there was no such thing at all as the law Salique, but that it was a fiction[145].

185. (33.) There was a friar in earnest dispute[146] about the law Salique, that would needs prove it by Scripture; citing that verse of the Gospel; *Lilia agri non laborant neque nent: which is as much as to say* (saith he) *that*[147] the *flower-de-luces of France cannot descend neither to distaff nor spade: that is, not to a woman, nor to a peasant.*

186. (167.) Julius Cæsar, as he passed by, was by exclamation of some that were suborned called[148] *King*, to try how the people would take it. The people shewed great murmur and distaste at it. Cæsar, finding where the wind stood, slighted it, and said; *I am not King, but Cæsar;* as if they had mistook[149] his name. For *Rex* was a surname amongst the Romans as *King* is with us.

187. (168.) When Crœsus, for his glory, shewed Solon great treasure[150] of gold, Solon said to him; *If another come*[151] *that hath better iron than you, he will be master of all this gold.*

188. (99.) There was a gentleman that came to the tilt all in orange-tawny, and ran very ill. The next day he came[152] all in green, and ran worse. There was

be right; and as "*a* senator" could not be plural, *were* must be replaced by *was*. Unfortunately, in attending to the grammar without attending to the sense, he in effect puts the remark into the mouth of the very person at whom it was aimed. He should have let *were* stand, and put *who* for *and*.

[142] as for every small disgrace to. R.

[143] from inheriting. R. [144] implying. R. [145] is a mere fiction. R.

[146] A friar of France being in an earnest dispute. R.

[147] *The lilies of the field do neither labour nor spin*: applying it thus, that. R.

[148] of some that stood in the way, termed. R. [149] mistaken.

[150] his great treasures. R. [151] if another KING come. [152] came again. R.

one of the lookers-on asked another; *What's the reason that this gentleman changeth his colours?* The other answered; *Sure, because it may be reported that the gentleman in the green ran worse than the gentleman in the orange-tawny.*

189. (230.) Aristippus said; *That those that studied particular sciences, and neglected philosophy, were like Penelope's wooers, that made love to the waiting women* [153].

190. (170.) Plato reproved [154] severely a young man for entering into a dissolute house. The young man said to him; *What,*[155] *for so small a matter?* Plato replied; *But custom is no small matter.*

191. (190.) There was a law made by the Romans against the bribery and extortion of the governors of provinces. Cicero saith, in a speech of his to the people; *That he thought the provinces would petition to the state of Rome to have that law repealed. For* (saith he) *before the governors did bribe and extort as much as was sufficient for themselves; but now they bribe and extort as much as may be enough not only for themselves, but for the judges and jurors and magistrates.*

192. (171.) Archidamus King of Lacedæmon, having received from Philip King of Macedon, after Philip had won the victory of Chæronea upon the Athenians, proud letters, writ back to him; *That if he measured his own shadow, he would find it no longer than it was before his victory.*

193. (172.) Pyrrhus, when his friends congratulated to him his victory over the Romans, under the conduct of Fabricius, but with great slaughter of his own side, said to them again; *Yes, but if we have such another victory, we are undone.*

194. (173.) Cineas was an excellent orator and statesman, and principal friend and counsellor to Pyrrhus; and falling in inward talk with him, and discerning the King's endless ambition [156], Pyrrhus opened himself to him; *That he intended first a war upon Italy* [157], *and hoped to atchieve it.* Cineas asked him; *Sir, what will you do then? Then* (saith he) *we will attempt Sicily* [158]. Cineas said; *Well, Sir, what then? Then* (saith Pyrrhus) *if the Gods favour* [159] *us, we may conquer Africk and Carthage* [160]. *What then, Sir?* saith Cineas. *Nay then* (saith Pyrrhus) *we may take our rest, and sacrifice and feast every day, and make merry with our friends. Alas, Sir,* (said Cineas) *may we not do so now, without all this ado?*

195. (231.) The embassadors of Asia Minor came to Antonius, after he had imposed upon them a double tax, and said plainly to him; *That if he would have two tributes in one year, he must give them two seed-times and two harvests.*

196. (174.) Plato was wont to say of his master Socrates; *That he was like the apothecaries' gally-pots; that had on the outside apes, and owls, and satyrs; but within precious drugs* [161].

† 197. Lamia the courtezan had all power with Demetrius King of Macedon; and by her instigations he did many unjust and cruel acts. Whereupon Lysimachus said; *That it was the first time that ever he knew a whore play in a tragedy.*

† 198. Themistocles would say of himself; *That he was like a plane-tree, that in tempests men fled to him, and in fair weather men were ever cropping his leaves.*

† 199. Themistocles said of speech; *That it was like arras, that spread abroad shews fair images, but contracted is but like packs.*

200. (90.) Brisquet [162], jester to Francis the first of France, did keep a calendar of fools, wherewith he did use to make the King sport; telling him ever the reason why he put every one [163] into his calendar. So when Charles the fifth passed, upon confidence of the noble nature of Francis, thorough France, for the appeasing of the rebellion of Gaunt, Brisquet put him into his calendar. The King asking the cause, he said [164]; *Because you having suffered at the hands of Charles the*

[153] woman. R. [154] reprehended. R. [155] why do you reprehend so sharply. R. [156] when Pyrrhus. R. [157] Sicily R. [158] Italy and Rome. R. [159] succour. R. [160] we may conquer the kingdom of Carthage. R. Compare Erasmus's version of this anecdote (*V. Pyrrh.* 24), from which it seems to be compressed: where the order of the proposed conquests is Rome, Italy, Sicily, Libya, and Carthage, Macedonia and Greece.
[161] See p. 52. [162] Bresquet. R. [163] any one. R.
[164] asked him the cause? He answered. R.

greatest bitterness that ever prince did from other[165], *he would trust his person into your hands.* *Why, Brisquet,* (said the King) *what wilt thou say, if thou seest him pass*[166] *in as great safety as if it were*[167] *thorough the midst of Spain?* Saith Brisquet; *Why then I will put out him, and put in you*[168].

201. (245.) Lewis the eleventh of France, having much abated the greatness and power of the Peers, Nobility, and Court of Parliament, would say; *That he had brought the Crown out of ward.*

202. (57.) Sir Fulke Grevill[169], in Parliament when the Lower House in a great business of the Queen's[170], stood much upon precedents, said unto them; *Why should you stand so much upon precedents? The times hereafter will be good or bad: If good, precedents will do no harm; if bad, power will make a way where it finds none.*

203. (34.) When peace was renewed with the French in England, divers of the great counsellors were presented from the French with jewels. The Lord Henry Howard[171] was omitted. Whereupon the King said to him; *My Lord, how haps it that you have not a jewel as well as the rest?* My Lord answered again, (alluding[172] to the fable in Æsop); *Non sum Gallus, itaque non reperi gemmam.*

204. (232.) An orator of Athens said to Demosthenes; *The Athenians will kill you, if they wax mad.* Demosthenes replied, *And they will kill you, if they be in good sense.*

205. (175.) Alexander sent to Phocion a great present of money. Phocion said to the messenger; *Why doth the King send to me and to none else?* The messenger answered; *Because he takes you to be the only good man in Athens.* Phocion replied; *If he thinks so, pray let him suffer me to be good still*[173].

206. (92.) Cosmus duke of Florence was wont to say of perfidious friends; *That we read that we ought to forgive our enemies; but we do not read that we ought to forgive our friends.*

207. (102.) Æneas Sylvius, that was Pius Secundus[174], was wont to say; *That the former Popes did wisely to set the lawyers on work*[175] *to debate, whether the donation of Constantine the Great to Sylvester*[176] *were good and valid in law or no? the better to skip over the matter in fact, whether there were*[177] *any such thing at all or no?*

208. (176.) At a banquet, where those that were called the Seven Wise Men of Greece were invited by the embassador of a barbarous King, the embassador related, That there was a neighbour King, mightier than his master, picked quarrels with him, by making impossible demands, otherwise threatening war; and now at that present had demanded of him to drink up the sea. Whereunto one of the Wise Men said; *I would have him undertake it.* *Why* (saith the embassador) *how shall he come off?* *Thus* (saith the Wise Man): *Let that King first stop the rivers that run into the sea, which are no part of the bargain, and then your master will perform it.*

209. (177.) At the same banquet, the embassador desired the Seven, and some other wise men that were at the banquet, to deliver every one of them some sentence or parable, that he mought report to his King the wisdom of Græcia. Which they did. Only one was silent. Which the embassador perceiving, said to him; *Sir, let it not displease you, why do not you say somewhat that I may report?* He answered, *Report to your lord, that there are of the Grecians that can hold their peace.*

† 210. One of the Romans said to his friend; *What think you of such an one as was taken with the manner in adultery?* The other answered; *Marry, I think he was slow at dispatch.*

† 211. Lycurgus would say of divers of the heroes of the heathen; *That he*

[165] another, nevertheless. R. [166] pass back. R. [167] he marched. R.
[168] Compare Melch. I. 3. 1., where a different story with a similar point is told of Alonso Carrillo and one of his servants. [169] afterward Lord Brooke. R.
[170] when the House of Commons in a great business stood, etc. R.
[171] being then Earl of Northampton and a Counsellor. R.
[172] answered, according to, etc. R. [173] to be so still. R.
[174] Pope Pius Secundus. R. [175] awork. R.
[176] of St. Peter's patrimony. R. [177] was ever. R.

wondered that men should mourn upon their days for them as mortal men, and yet sacrifice to them as gods.

212. (93.) A Papist being opposed by a Protestant, that they had no Scripture for images, answered; *Yes; for you read that the people laid their sick in the streets, that the shadow of Saint Peter mought come upon them; and that a shadow was an image; and the obscurest of images* [178].

† 213. There is an ecclesiastical writer of the Papists, to prove antiquity of confession in the form that it now is, doth note that in very ancient times, even in the primitive times, amongst other foul slanders spread against the Christians, one was; *That they did adore the genitories of their priests. Which* (he saith) *grew from the posture of the confessant and the priest in confession: which is, that the confessant kneels down, before the priest sitting in a raised chair above him.*

† 214. Epaminondas, when his great friend and colleague in war was suitor to him to pardon an offender, denied him. Afterwards, when a concubine of his made the same suit, he granted it to her; which when Pelopidas seemed to take unkindly, he said; *Such suits are to be granted to whores, but not to personages of worth.*

215. (178.) The Lacedæmonians had in custom to speak very short. Which, being in empire [179], they mought do at pleasure. But after their defeat at Leuctra, in an assembly of the Grecians, they made a long invective against Epaminondas; who stood up, and said no more but this; *I am glad we have taught you to speak long.*

† 216. Fabricius, in conference with Pyrrhus, was tempted to revolt to him; Pyrrhus telling him, that he should be partner of his fortunes, and second person to him. But Fabricius answered, in a scorn, to such a motion; *Sir, that would not be good for yourself: for if the Epirotes once knew me, they will rather desire to be governed by me than by you.*

217. (179.) Fabius Maximus being resolved to draw the war in length, still waited upon Hannibal's progress to curb him; and for that purpose he encamped upon the high grounds. But Terentius his colleague fought with Hannibal, and was in great peril of overthrow. But then Fabius came down [180] the high grounds, and got the day: Whereupon Hannibal said; *That he did ever think that that same cloud that hanged upon the hills, would at one time or other give a tempest.*

218. (246.) There was a cowardly Spanish soldier, that in a defeat the Moors gave, ran away with the foremost. Afterwards, when the army generally fled, this soldier was missing. Whereupon it was said by some, that he was slain. *No, sure* (saith one), *he is alive; for the Moors eat no hare's flesh* [181].

219. (180.) Hanno the Carthaginian was sent commissioner by the state, after the second Carthaginian war, to Rome [182], to supplicate for peace, and in the end obtained it. Yet one of the sharper senators said; *You have often broken with us the peaces whereunto you have been sworn; I pray by what Gods will you swear?* Hanno answered; *By the same Gods that have punished the former perjury so severely.*

† 220. Thales being asked when a man should marry, said; *Young men not yet, old men not at all.*

† 221. Thales said; *That life and death were all one.* One that was present asked him; *Why do not you die then?* Thales said again; *Because they are all one.*

222. (181.) Cæsar after first he had [183] possessed Rome, Pompey being fled, offered to enter the sacred treasury to take the moneys that were there stored. Metellus, tribune of the people, did forbid him. And when Metellus was violent in it, and would not desist, Cæsar turned to him, and said; *Presume no further, or I will lay you dead.* And when Metellus was with those words somewhat astonished, Cæsar added; *Young man, it had been easier for me to do this than to speak it.*

† 223. An Ægyptian priest having conference with Solon, said to him; *You Grecians are ever children; you have no knowledge of antiquity, nor antiquity of knowledge.*

[178] of all images. R.
[179] being an empire. R.
[180] down from. R.
[181] Melch. II. 3. 21.
[182] R. omits "to Rome."
[183] when he had first. R.

224. (14.) The counsel did make remonstrance unto Queen Elizabeth of the continual conspiracies against her life; and namely of a late one: and shewed her a rapier, taken from a conspirator, that had a false chape, being of brown paper, but gilt over, as it could not be known from a chape of metal; which was devised to the end that without drawing the rapier mought give a stab; and upon this occasion advised her [184] that she should go less abroad to take the air, weakly accompanied, as she used. But the Queen answered; *That she had rather be dead, than put in custody.*

225. (194.) Chilón would say, *That gold was tried with the touch-stone, and men with gold.*

226. (101.) Zelim was the first of the Ottomans that did shave his beard, whereas his predecessors wore it long. One of his Basha's asked him; *Why he altered the custom of his predecessors?* He answered; *Because you Basha's shall not lead me by the beard, as you did them.*

† 227. Diogenes was one day in the market-place, with a candle in his hand; and being asked; *What he sought?* he said; *He sought a man.*

† 228. Bias being asked; *How a man should order his life?* answered; *As if a man should live long, or die quickly.*

† 229. Queen Elizabeth was entertained by my Lord Burleigh at Theobalds: and at her going away, my Lord obtained of the Queen to make seven knights. They were gentlemen of the country, of my Lord's friends and neighbours. They were placed in a rank, as the Queen should pass by the hall; and to win antiquity of knighthood, in order, as my Lord favoured; though indeed the more principal gentlemen were placed lowest. The Queen was told of it, and said nothing; but when she went along, she passed them all by, as far as the screen, as if she had forgot it; and when she came to the screen, she seemed to take herself with the manner, and said; *I had almost forgot what I promised.* With that she turned back, and knighted the lowest first, and so upward. Whereupon Mr. Stanhope, of the privy-chamber, a while after told her: *Your Majesty was too fine for my Lord Burleigh.* She answered; *I have but fulfilled the Scripture; The first shall be last, and the last first.*

230. (195.) Simonides being asked of Hiero; *What he thought of God?* asked a seven-night's time to consider of it. And at the seven-night's end he asked a fortnight's time. At the fortnight's end, a month. At which Hiero marvelling, Simonides answered; *That the longer he thought on it* [185]*, the more difficult he found it.*

231. (248.) Anacharsis would say concerning the popular estates of Græcia; *That he wondered how at Athens wise men did propose, and fools did dispose.*

† 232. Solon compared the people unto the sea, and orators to the winds: *For that the sea would be calm and quiet, if the winds did not trouble it.*

233. (197.) Socrates was pronounced by the oracle of Delphos to be the wisest man of Greece; which he would put from himself, ironically [186] saying; *There could be nothing in him* [187] *to verify the oracle, except this; that he was not wise, and knew it; and others were not wise, and knew it not.*

234. (238.) Cato the elder, what time many of the Romans had statua's erected in their honour, was asked by one in a kind of wonder; *Why he had none?* and answered; *He had much rather men should ask and wonder why he had no statua, than why he had a statua.*

† 235. Sir Fulke Grevill had much and private access to Queen Elizabeth, which he used honourably, and did many men good; yet he would say merrily of himself; *That he was like Robin Goodfellow; For when the maids spilt the milkpans, or kept any racket, they would lay it upon Robin; So what tales the ladies about the Queen told her, or other bad offices that they did, they would put it upon him.*

236. (196.) Socrates, when there was shewed him [188] the book of Heraclitus the

[184] and namely, that a man was lately taken who stood ready in a very dangerous and suspicious manner to do the deed; and they shewed her the weapon wherewith he thought to have acted it, and therefore they advised her, etc. R.

[185] thought upon the matter. R.

[186] put from himself in modesty. R.

[187] in himself. R.

[188] unto him. R.

Obscure and was asked his opinion of it, answered; *Those things that I understood were excellent; I imagine, so were those that I understood not; but they require a diver of Delos.*

† 237. Bion asked an envious man that was very sad; *What harm had befallen to him, or what good had befallen to another man?*

† 238. Stilpo the philosopher, when the people flocked about him, and that one said to him; *The people come wondering about you, as if it were to see some strange beast. No,* (saith he) *it is to see a man which Diogenes sought with his lanthorn.*

239. (184.) Antisthenes being asked of one; *What learning was most necessary for man's life?* answered; *To unlearn that which is naught.*

† 240. There was a politic sermon, that had no divinity in it, was preached before the King. The King, as he came forth, said to Bishop Andrews; *Call you this a sermon?* The Bishop [189] answered; *And it please your Majesty, by a charitable construction, it may be a sermon.*

241. (103.) Bishop Andrews was asked at the first coming over of the Bishop [190] of Spalato; *Whether he were a Protestant or no?* He answered; *Truly I know not, but he is a Detestant, of divers opinions of Rome* [191].

242. (162.) Caius Marius was general of the Romans against the Cimbers, who came with such a sea of multitude [192] upon Italy. In the fight, there was a band of the Cadurcians, of a thousand, that did notable service. Whereupon, after the fight, Marius did denizen them all for citizens of Rome, though there was no law to warrant it. One of his friends did represent [193] it unto him, that he had transgressed the law, because that privilege was not to be granted but by the people. Whereto Marius answered; *That for the noise of arms he could not hear the laws.*

243. (105.) Æneas Sylvius would say; *That the christian faith and law, though it had not been confirmed by miracles, yet was worthy to be received for the honesty thereof.*

† 244. Henry Noel would say; *That courtiers were like fasting-days; They were next the holydays, but in themselves they were the most meagre days of the week.*

245. (106.) Mr. Bacon would say; *That it was in business, as it is commonly* [194] *in ways; that the next way is commonly the foulest, and that if a man will go the fairest way, he must go somewhat about.*

246. (215.) Augustus Cæsar, out of great indignation against his two daughters, and Posthumus Agrippa, his grandchild, whereof the first two were infamous, and the last otherwise unworthy, would say; *That they were not his seed, but some imposthumes that had broken from him.*

† 247. Cato said; *The best way to keep good acts in memory, was to refresh them with new.*

248. (183.) Pompey did consummate the war against Sertorius, when Metellus had brought the enemy somewhat low. He did also consummate the war against the fugitives, whom Crassus had before defeated in a great battle. So when Lucullus had had great and glorious victories against Mithridates and Tigranes, yet Pompey, by means his friends made, was sent to put an end to that war. Whereupon Lucullus, taking indignation, as a disgrace offered to himself, said; *That Pompey was a carrion crow, that when others had stricken down bodies, he came to prey upon them* [195].

249. (186.) Diogenes, when mice came about him as he was eating, said; *I see that even Diogenes nourisheth parasites.*

250. (233.) Epictetus used to say; *That one of the vulgar, in any ill that happens to him, blames others; a novice in philosophy blames himself; and a philosopher blames neither the one nor the other.*

251. (187.) Hiero visited by Pythagoras, asked him; *Of what condition he was?* Pythagoras answered; *Sir, I know you have been at the Olympian games. Yes,* saith Hiero. *Thither* (saith Pythagoras) *come some to win the prizes. Some come*

[189] The Lord Bishop. R. [190] Archbishop. R.
[191] *but I think he is a Detestant*: That was, of most of the opinions of Rome. R.
[192] such a sea of people. R. [193] present. R. [194] frequently. R.
[195] then Pompey came and preyed upon them. R.

to sell their merchandize, because it is a kind of mart of all Greece. Some come to meet their friends, and make merry, because of the great confluence of all sorts. Others come only to look on. I am one of them that come to look on. Meaning it of philosophy, and the contemplative life.

252. (107.) Mr. Bettenham[196] used to say; *That riches were like muck; when it lay upon an heap, it gave but a stench and ill odour; but when it was spread upon the ground, then it was cause of much fruit.*

253. (96.) The same Mr. Bettenham said; *That virtuous men were like some herbs and spices, that give not*[197] *their sweet smell, till they be broken and crushed.*

254. (98.) There was a painter became a physician. Whereupon one said to him; *You have done well; for before the faults of your work were seen, but now they are unseen*[198].

255. (189.) One of the philosophers was asked; *What a wise man differed from a fool?* He answered; *Send them both naked to those that know them not, and you shall perceive.*

256. (234.) Cæsar in his book that he made against Cato (which is lost) did write, to shew the force of opinion and reverence of a man that had once obtained a popular reputation; *That there were some that found Cato drunk, and they were ashamed instead of Cato.*

257. (191.) Aristippus, sailing in a tempest, shewed signs of fear. One of the seamen said to him, in an insulting manner; *We that are plebeians are not troubled; you, that are a philosopher, are afraid.* Aristippus answered; *There is not the like wager upon it, for me to perish and you*[199].

258. (192.) There was an orator that defended a cause of Aristippus, and prevailed. Afterwards he asked Aristippus; *Now, in your distress, what did Socrates do you good?* Aristippus answered; *Thus; in making true that good which you said of me*[200].

† 259. Aristippus said; *He took money of his friends, not so much to use it himself, as to teach them how to bestow their money.*

† 260. A strumpet said to Aristippus; *That she was with child by him.* He answered; *You know that no more, than if you went through a hedge of thorns, you could say, This thorn pricked me.*

261. (15.) The Lady Paget, that was very private with Queen Elizabeth, declared herself much against her match[201] with Monsieur. After Monsieur's death, the Queen took extreme grief (at least as she made shew), and kept[202] within her bed-chamber and one antechamber for three weeks space, in token of mourning. At last she came forth into her privy chamber, and admitted her ladies to have access unto her; and amongst the rest my lady Paget presented herself, and came to her with a smiling countenance. The Queen bent her brows, and seemed to be highly displeased, and said to her; *Madam, you are not ignorant of my extreme grief, and do you come to me with a countenance of joy?* My Lady Paget answered; *Alas, and it please your Majesty, it is impossible for me to be absent from you three weeks, but that when I see you I must look cheerfully. No, no,* (said the Queen, not forgetting her former averseness from[203] the match), *you have some other conceit in it; tell me plainly.* My lady answered; *I must obey you. It is this. I was thinking how happy your Majesty was, in that you married not Monsieur; for seeing you take such thought for his death, being but your friend, if it had been your husband, sure it would have cost you your life.*

262. (94.) Sir Edward Dyer, a grave and wise gentleman, did much believe in Kelley the alchymist; that he did indeed the work, and made gold: insomuch as he went himself into Germany, where Kelley then was, to inform himself fully thereof. After his return, he dined with my Lord of Canterbury, where at that time was at the table Dr. Browne, the physician. They fell in talk of Kelley.

[196] Reader of Gray's Inn. R. [197] give not out. R.

[198] Compare Melch. IV. 7. 5. where the remark is represented more gracefully as made by the painter himself.

[199] for you to perish and for me. R.

[200] in making that which you said of me to be true. R.

[201] the match. R. [202] kept in. R. [203] to. R.

Sir Edward Dyer, turning to the Archbishop, said; *I do assure your Grace, that that I shall tell you is truth. I am an eye-witness thereof, and if I had not seen it, I should not have believed it. I saw Master Kelley put of the base metal into the crucible, and after it was set a little upon the fire, and a very small quantity of the medicine put in, and stirred with a stick of wood, it came forth in great proportion perfect gold, to the touch, to the hammer, to the test.* Said the Bishop [204]; *You had need take heed what you say, Sir Edward Dyer, for here is an infidel at the board.* Sir Edward Dyer said again pleasantly; *I would have looked for an infidel sooner in any place than at your Grace's table. What say you, Dr. Browne?* saith the Bishop [205]. Dr. Browne answered, after his blunt and huddling manner, *The gentleman hath spoken enough for me. Why* (saith the Bishop [206]), *what hath he said? Marry,* (saith Dr. Browne) *he said he would not have believed it except he had seen it; and no more will I.*

† 263. Democritus said; *That truth did lie in profound pits, and when it was got, it needed much refining.*

264. (95.) Doctor Johnson said; *That in sickness there were three things that were material; the physician, the disease, and the patient. And if any two of these joined, then they have* [207] *the victory. For, Ne Hercules quidem contra duos. If the physician and the patient join, then down goes the disease; for the patient recovers. If the physician and the disease join, then down goes the patient; that is where the physician mistakes the cure* [208]. *If the patient and the disease join, then down goes the physician; for he is discredited.*

265. (185.) Alexander visited Diogenes in his tub. And when he asked him; *What he would desire of him?* Diogenes answered; *That you would stand a little aside, that the sun may come to me.*

† 266. Diogenes said of a young man that danced daintily, and was much commended; *The better, the worse.*

267. (236.) Diogenes called an ill musician, *Cock. Why?* (saith he.) Diogenes answered; *Because when you crow men use to rise.*

268. (188.) Heraclitus the Obscure said; *The dry light was the best soul.* Meaning, when the faculties intellectual are in vigour, not wet, nor [209], as it were, blooded by the affections.

† 269. There was in Oxford a cowardly fellow that was a very good archer. He was abused grossly by another, and moaned himself to Walter Ralegh, then a scholar, and asked his advice; *What he should do to repair the wrong had been offered him?* Ralegh answered; *Why, challenge him at a match of shooting.*

270. (100.) Whitehead, a grave divine, was much esteemed by Queen Elizabeth, but not preferred, because he was against the government of Bishops. He was of a blunt stoical nature [210]. He came one day to the Queen, and the Queen happened to say to him; *I like thee the better, Whitehead, because thou livest unmarried.* He answered again; *In troth, Madam, I like you the worse for the same cause.*

† 271. There was a nobleman that was lean of visage, but immediately after his marriage he grew pretty plump and fat. One said to him, *Your lordship doth contrary to other married men; for they at the first wax lean and you wax fat.* Sir Walter Ralegh stood by and said; *Why, there is no beast, that if you take him from the common and put him into the several, but he will wax fat.*

† 272. Diogenes seeing one that was a bastard casting stones among the people, bade him *Take heed he hit not his father.*

273. (97.) Dr. Laud [211] said; *That some hypocrites and seeming mortified men, that held down their heads, were like little images that they place in the very bowing of the vaults of churches, that look as if they held up the church, but are but puppets* [212].

[204] My Lord Archbishop said. R.
[205] said the Archbishop. R.
[206] Archbishop. R.
[207] get. R.
[208] If the physician and the disease join, that is a strong disease; and the physician mistaking the cure, then, etc. R.
[209] not drenched, or. R.
[210] This sentence is omitted in R.
[211] The Lord Archbishop Laud. R.
[212] were like the little images in the vaults or roofs of churches, which look and bow down as if they held up the church, when as they bear no weight at all. R.

274. (104.) It was said among some of the grave prelates of the council of Trent, in which the school-divines bore the sway; *That the schoolmen were like the astronomers; who, to save the phenomena, framed to their conceit eccentrics and epicycles, and a wonderful engine of orbs, though no such things were: so they, to save the practice of the church, had devised a number of strange positions.*

† 275. It was also said by many, concerning the canons of that council; *That we are beholding to Aristotle for many articles of our faith.*

276. (35.) The Lo. Henry Howard, being Lord Privy Seal, was asked by the King openly at the table, (where commonly he entertained the King), upon the sudden[213]; *My lord, have you not a desire to see Rome?* My Lord Privy Seal answered, *Yes, indeed, Sir.* The King said, *And why?* My lord answered, *Because, and it please your Majesty, it was once the seat of the greatest monarchy, and the seminary of the bravest men in the world, amongst the heathen: and then again*[214]*, because after it was the see of so many holy Bishops in the primitive church, most of them martyrs.* The King would not give it over, but said; *And for nothing else?* My lord answered; *Yes, and it please your Majesty, for two things especially*[215]*. The one, to see him, who they say hath such a power to forgive other men's sins, to confess his own sins upon his knees before a chaplain or priest; and the other is, to hear Antichrist say his creed.*

277. (235.) There was a nobleman said of a great counsellor; *That he would have made the worst farrier in the world, for he never shod horse but he cloyed him: so he never commended any man to the King for service, or upon occasion of suit, or otherwise, but that he would come in in the end with a* **But**, *and drive in a nail to his disadvantage.*

† 278. There was a lady of the west country, that gave great entertainment at her house to most of the gallant gentlemen thereabout; and amongst others, Sir Walter Ralegh was one. This lady, though otherwise a stately dame, was a notable good housewife; and in the morning betimes she called to one of her maids that looked to the swine, and asked; *Is the piggy served?* Sir Walter Ralegh's chamber was fast by the lady's, so as he heard her. A little before dinner, the lady came down in great state into the great chamber, which was full of gentlemen: And as soon as Sir Walter Ralegh set eye upon her; *Madam,* (saith he) *is the piggy serv'd?* The lady answered, *You know best whether you have had your breakfast.*

279. (237.) There was a gentleman fell very sick, and a friend of his said to him; *Surely, you are in danger; I pray send for a physician.* But the sick man answered; *It is no matter, for if I die, I will die at leisure.*

280. (193.) There was an Epicurean vaunted, that divers of other sects of philosophers did after turn Epicureans, but there was never any Epicurean that turned to any other sect. Whereupon a philosopher that was of another sect, said; *The reason was plain, for that cocks may be made capons, but capons could never be made cocks.*

[213] The same Earl of Northampton, then Lord Privy Seal, was asked by King James openly at the table, where commonly he entertained the King with discourse; the King asked him upon the sudden. R.

[214] secondly. R.

[215] for two things more. R.

APOPHTHEGMS

CONTAINED IN THE SECOND EDITION OF THE RESUSCITATIO (1661), AND NOT IN THE ORIGINAL COLLECTION[1].

3. HIS Majesty James the First, King of Great Britain, having made unto his Parliament an excellent and large declaration, concluded thus : *I have now given you a clear mirror of my mind ; use it therefore like a mirror ; and take heed how you let it fall, or how you soil it with your breath.*

5. His Majesty said to His Parliament at another time, finding there were some causeless jealousies sown amongst them ; *That the King and his people, (whereof the Parliament is the representative body), were as husband and wife ; and therefore that of all other things jealousy was between them most pernicious.*

6. His Majesty, when he thought his counsel mought note in him some variety in businesses, though indeed he remained constant, would say ; *That the sun many times shineth watery ; but it is not the sun which causeth it, but some cloud rising betwixt us and the sun : and when that is scattered, the sun is as it was, and comes to his former brightness.*

7. His Majesty, in his answer to the book of the Cardinal of Evereux, (who had in a grave argument of divinity sprinkled many witty ornaments of poesy and humanity) saith ; *That these flowers were like blue and yellow and red flowers in the corn, which make a pleasant shew to those that look on, but they hurt the corn.*

8. Sir Edward Cook, being vehement against the two Provincial Councils, of Wales and the North, said to the King ; *There was nothing there but a kind of confusion and hotch-potch of justice : one while they were a Star-Chamber ; another while a Kings-bench ; another, a Common-place ; another, a Commission of Oyer and Terminer.* His Majesty answered ; *Why, Sir Edward Cook, they be like houses in progress, where I have not, nor can have, such distinct rooms of state, as I have here at Whitehall, or at Hampton-court.*

9. The Commissioners of the Treasure moved the King, for the relief of his estate, to disafforest some forests of his ; explaining themselves of such forests as lay out of the way, not near any of the King's houses, nor in the course of his progress ; whereof he should never have use nor pleasure. *Why,* (saith the King) *do you think that Salomon had use and pleasure of all his three hundred concubines ?*

10. His Majesty, when the committees of both Houses of Parliament presented unto him the instrument of Union of England and Scotland, was merry with them ; and amongst other pleasant speeches, shewed unto them the laird of Lawreston, a Scotchman, who was the tallest and greatest man that was to be seen ; and said ; *Well, now we are all one, yet none of you will say, but here is one Scotchman greater than any Englishman ;* which was an ambiguous speech ; but it was thought he meant it of himself.

11. His Majesty would say to the lords of his counsel, when they sat upon any great matter, and came from counsel in to him ; *Well, you have sit, but what have you hatched ?*

13. Queen Elizabeth was importuned much by my Lord of Essex, to supply divers great offices that had been long void ; the Queen answered nothing, to the matter ; but rose up on the sudden, and said ; *I am sure my office will not be long void.* And yet at that time there was much speech of troubles and divisions

[1] See Preface, pp. 860, 863.

about the crown, to be after her decease; but they all vanished; and King James came in, in a profound peace.

17. King Henry the fourth of France was so punctual of his word, after it was once passed, that they called him *The King of the Faith* [2].

18. The said King Henry the fourth was moved by his Parliament to a war against the Protestants: he answered; *Yes, I mean it; I will make every one of you captains; you shall have companies assigned you.* The Parliament observing whereunto his speech tended, gave over, and deserted the motion [3].

21. A great officer at court, when my Lord of Essex was first in trouble; and that he and those that dealt for him would talk much of my Lord's friends and of his enemies; answered to one of them; *I will tell you, I know but one friend and one enemy my Lord hath; and that one friend is the Queen, and that one enemy is himself.*

27. The Lord Keeper, Sir Nicholas Bacon, was asked his opinion, by my lord of Leicester, concerning two persons whom the Queen seemed to think well of: *By my troth, my Lord* (said he), *the one is a grave counsellor; the other is a proper young man; and so he will be as long as he lives.*

28. My Lord of Leicester, favourite to queen Elizabeth, was making a large chase about Cornbury-Park; meaning to inclose it with posts and rails; and one day was casting up his charge, what it would come to. Mr. Goldingham, a free spoken man, stood by, and said to my Lord, *Methinks your Lordship goeth not the cheapest way to work. Why, Goldingham?* said My Lord. *Marry, my Lord,* said Goldingham, *count you but upon the posts, for the country will find you railing.*

36. There were fishermen drawing the river at Chelsea: Mr. Bacon came thither by chance in the afternoon, and offered to buy their draught: they were willing. He asked them what they would take? They asked thirty shillings. Mr. Bacon offered them ten. They refused it. *Why then,* saith Mr. Bacon, *I will be only a looker on.* They drew, and catched nothing. Saith Mr. Bacon; *Are not you mad fellows now, that might have had an angel in your purse, to have made merry withal, and to have warmed you thoroughly, and now you must go home with nothing. Ay but* (said the fishermen) *we had hope then to make a better gain of it.* Saith Mr. Bacon; *Well, my masters, then I'll tell you, hope is a good breakfast, but it is a bad supper* [4].

36. A lady walking with Mr. Bacon in Gray's Inn Walks, asked him, *Whose that piece of ground lying next under the walls was?* He answered, *Theirs.* Then she asked him, if those fields beyond the walks were theirs too? He answered. Yes, Madam, these are ours, as you are ours, to look on, and no more [5].

37. His Lordship, when he was newly made Lord Keeper, was in Gray's Inn walks with Sir Walter Raleigh. One came and told him, that the Earl of Exeter was above. He continued upon occasion still walking a good while. At last when he came up, my Lord of Exeter met him, and said; *My Lord, I have made a great venture, to come up so high stairs, being a gouty man.* His Lordship answered; *Pardon me, my lord, I have made the greatest venture of all* [6] *; for I have ventured upon your patience.*

38. When Sir Francis Bacon was made the King's Attorney, Sir Edward Cooke was put up from being Lord Chief Justice of the Common Pleas, to be Lord Chief Justice of the King's Bench; which is a place of greater honour, but of less profit; and withal was made Privy Counsellor. After a few days, the Lord Cooke meeting with the King's Attorney, said unto him; *Mr. Attorney, this is all your doing: It is you that have made this great stir.* Mr. Attorney answered; *Ah my Lord; your Lordship all this while hath grown in breadth; you must needs now grow in height, or else you would be a monster* [7].

39. One day Queen Elizabeth told Mr. Bacon, that my Lord of Essex, after great protestation of penitence and affection, fell in the end but upon the suit of renewing his farm of sweet wines. He answered; *I read that in nature there be two*

[2] Lamb. MS. p. 18. (see above, p. 855.) [3] Id. ibid. (without the last sentence).

[4] See Lamb. MS. p. 1. where the story is set down almost exactly in the same words.

[5] Id. p. 1. (told more compactly). The number 36 is repeated in R.

[6] the greater venture. Lamb. MS. [7] Lamb. MS.

kinds of motions or appetites in sympathy ; the one as of iron to the adamant, for perfection ; the other as of the vine to the stake, for sustentation ; that her Majesty was the one, and his suit the other [8].

40. Mr. Bacon, after he had been vehement in Parliament against depopulation and inclosures ; and that soon after the Queen told him that she had referred the hearing of Mr. Mill's cause to certain counsellors and judges ; and asked him how he liked of it ? answered, *Oh, madam ; my mind is known ; I am against all inclosures, and especially against inclosed justice* [9].

41. When Sir Nicholas Bacon the Lord Keeper lived, every room in Gorhambury was served with a pipe of water from the ponds, distant about a mile off. In the lifetime of Mr. Anthony Bacon, the water ceased. After whose death, his Lordship coming to the inheritance, could not recover the water without infinite charge. When he was Lord Chancellor, he built Verulam House, close by the pond-yard, for a place of privacy when he was called upon to despatch any urgent business. And being asked, why he built that house there ; his Lordship answered, *That since he could not carry the water to his house, he would carry his house to the water* [10].

42. When my Lord President of the Council came first to be Lord Treasurer, he complained to my Lord Chancellor of the troublesomeness of the place ; for that the Exchequer was so empty. The Lord Chancellor answered ; *My Lord, be of good cheer, for now you shall see the bottom of your business at the first* [11].

43. When his Lordship was newly advanced to the Great Seal, Gondomar came to visit him. My Lord said ; *That he was to thank God and the King for that honour ; but yet, so he might be rid of the burthen, he could very willingly forbear the honour ; and that he formerly had a desire, and the same continued with him still, to lead a private life.* Gondomar answered ; *That he would tell him a tale ; of an old rat, that would needs leave the world ; and acquainted the young rats that he would retire into his hole, and spend his days solitarily ; and would enjoy no more comfort : and commanded them upon his high displeasure* [12]*, not to offer to come in unto him. They forbore two or three days ; at last, one that was more hardy than the rest, incited some of his fellows to go in with him, and he would venture to see how his father did ; for he might be dead. They went in, and found the old rat sitting in the midst of a rich Parmesan cheese.* So he applied the fable after his witty manner [13].

44. Rabelais tells a tale of one that was very fortunate in compounding differences. His son undertook the same course [14], but could never compound any. Whereupon he came to his father, and asked him, *what art he had to reconcile differences* [15] *?* He answered, *he had no other but this : to watch when the two parties were much wearied, and their hearts were too great to seek reconcilement at one another's hands ; then to be a means betwixt them, and upon no other terms.* After which the son went home, and prospered in the same undertakings [16].

62. There was an agent here for the Dutch, called Caroon ; and when he used to move the Queen for further succours and more men, my Lord Henry Howard would say ; *That he agreed well with the name of Charon, ferryman of hell ; for he came still for more men to increase* regnum umbrarum.

63. They were wont to call referring to the Masters in Chancery, *committing*. My Lord Keeper Egerton, when he was Master of the Rolls, was wont to ask ; *What the cause had done, that it should be committed ?*

64. They feigned a tale, principally against Doctors' reports in the Chancery ; That Sir Nicholas Bacon, when he came to heaven gate, was opposed, touching an unjust decree which had been made in the Chancery. Sir Nicholas desired to see the order, whereupon the decree was drawn up ; and finding it to begin *Veneris, etc.* *Why,* (saith he) *I was then sitting in the Star-chamber ; this concerns the*

[8] Id. p. 8. [9] Id. p. 8. [10] Id. p. 9. (told more shortly).

[11] Lamb. MS. p. 10. [12] upon his blessing. Lamb. MS. p. 4.

[13] so if he left the world he would retire to some rich place. Lamb. MS.

[14] So Lamb. MS. p. 63. R. has " said course."

[15] what trick he had to make friends. Lamb. MS.

[16] he would even be the means betwixt them. After which time the son prospered in the trade. Lamb. MS.

Master of the Rolls ; let him answer it. Soon after came the Master of the Rolls, Cordal, who died indeed a small time after Sir Nicholas Bacon ; and he was likewise stayed upon it ; and looking into the order, he found, that upon the reading of a certificate of Dr. Gibson, it was ordered, that his report should be decreed. And so he put it upon Dr. Gibson, and there it stuck.

65. Sir Nicholas Bacon, when a certain nimble-witted counsellor at the bar who was forward to speak, did interrupt him often, said unto him ; *There is a great difference betwixt you and me : a pain to me to speak, and a pain to you to hold your peace.*

66. The same Sir Nicholas Bacon, upon bills exhibited to discover where lands lay,—upon proof that they had a certain quantity of land, but could not set it forth, was wont to say ; *And if you cannot find your land in the country, how will you have me find it in the Chancery ?*

67. Mr. Houland, in conference with a young student, arguing a case, happened to say ; *I would ask you but this question.* The student presently interrupted him, to give him an answer. Whereunto Mr. Houland gravely said ; *Nay, though I ask you a question, yet I did not mean you should answer me ; I mean to answer myself.*

91. Archbishop Grindall was wont to say ; *That the physicians here in England were not good at the cure of particular diseases ; but had only the power of the Church, to bind and loose.*

123. Titus Quinctius was in the counsel of the Achaians, what time they deliberated, whether in the war then to follow between the Romans and King Antiochus, they should confederate themselves with the Romans, or with King Antiochus ? In that counsel the Ætolians, who incited the Achaians against the Romans, to disable their forces, gave great words, as if the late victory the Romans had obtained against Philip king of Macedon, had been chiefly by the strength and forces of the Ætolians themselves : And on the other side the embassador of Antiochus did extol the forces of his master ; sounding what an innumerable company he brought in his army ; and gave the nations strange names ; As Elymæans, Caducians, and others. After both their harangues, Titus Quinctius, when he rose up, said ; *It was an easy matter to perceive what it was that had joined Antiochus and the Ætolians together ; that it appeared to be by reciprocal lying of each, touching the other's forces.*

124. Plato was amorous of a young gentleman, whose name was Stella, that studied astronomy, and went oft in the clear nights to look upon the stars. Whereupon Plato *wished himself heaven, that he mought look upon Stella with a thousand eyes.*

153. Themistocles, after he was banished, and had wrought himself into great favour afterwards, so that he was honoured and sumptuously served ; seeing his present glory, said unto one of his friends, *If I had not been undone, I had been undone.*

214. A certain countryman being at an Assizes, and seeing the prisoners holding up their hands at the bar, related to some of his acquaintance ; *That the judges were good fortune-tellers ; for if they did but look upon a man's hand, they could tell whether he should live or die.*

216. A seaman coming before the judges of the Admiralty for admittance into an office of a ship bound for the Indies, was by one of the judges much slighted, as an insufficient person for that office he sought to obtain ; the judge telling him ; *That he believed he could not say the points of his compass.* The seaman answered ; *That he could say them, under favour, better than he could say his Pater-noster.* The judge replied ; *That he would wager twenty shillings with him upon that.* The seaman taking him up, it came to trial : and the seaman began, and said all the points of his compass very exactly : the judge likewise said his *Pater-noster :* and when he had finished it, he required the wager according to agreement ; because the seaman was to say his compass better than he his *Pater-noster*, which he had not performed. *Nay, I pray, Sir, hold* (quoth the seaman), *the wager is not finished : for I have but half done* : and so he immediately said his compass backward very exactly ; which the judge failing of in his *Pater-noster*, the seaman carried away the prize.

239. A certain friend of Sir Thomas Moore's, taking great pains about a book, which he intended to publish, (being well conceited of his own wit, which no man else thought worthy of commendation), brought it to Sir Thomas Moore to peruse it, and pass his judgment upon it; which he did: and finding nothing therein worthy the press, he said to him with a grave countenance; *That if it were in verse, it would be more worthy.* Upon which words, he went immediately and turned it into verse, and then brought it to Sir Thomas again; who looking thereon, said soberly; *Yes, marry, now it is somewhat, for now it is rhyme; whereas before it was neither rhyme nor reason.*

247. A gentleman that was punctual of his word, and loved the same in others, when he heard that two persons had agreed upon a meeting about serious affairs, at a certain time and place; and that the one party failed in the performance, or neglected his hour; would usually say of him, *He is a young man then*[17].

249. His lordship when he had finished this collection of Apophthegms, concluded thus: *Come, now all is well: they say, he is not a wise man that will lose his friend for his wit; but he is less a wise man that will lose his friend for another man's wit*[18].

17 "He broke his promise," said Sir Ralph, "he is a young man then, under twenty years old; and no exception to be taken."—Lamb. MS.

18 "When Sir John Finch and myself had gone over my lord's apophthegms, he said, 'Now it is well: you know it is a common saying that he is an unwise man who will lose his friend for his jest: but he is a more unwise man who will lose his friend for another man's jest.'"—Lamb. MS. p. 10.

APOPHTHEGMS

PUBLISHED BY DR. TENISON IN THE BACONIANA [1].

1. PLUTARCH said well, It is otherwise in a commonwealth of men than of bees. The hive of a city or kingdom is in best condition when there is least of noise or buz in it.

2. The same Plutarch said of men of weak abilities set in great place, That they were like little statues set on great bases, made to appear the less by their advancement.

3. He said again, Good fame is like fire. When you have kindled it, you may easily preserve it; but if once you extinguish it, you will not easily kindle it again; at least, not make it burn as bright as it did.

4. The answer of Apollonius to Vespasian is full of excellent instruction: Vespasian asked him, *What was Nero's overthrow?* He answered, *Nero could touch and tune the harp well; but in government sometimes he used to wind the pins too high, sometimes to let them down too low.* And certain it is, that nothing destroyeth authority so much as the unequal and untimely interchange of power pressed too far, and relaxed too much.

5. Queen Elizabeth seeing Sir Edward —— in her garden, looked out at her window, and asked him in Italian, *What does a man think of when he thinks of nothing?* Sir Edward (who had not had the effect of some of the Queen's grants so soon as he had hoped and desired) paused a little, and then made answer, *Madam, he thinks of a woman's promise.* The Queen shrunk in her head; but was heard to say, *Well, Sir Edward, I must not confute you.* Anger makes dull men witty, but it keeps them poor [2].

6. When any great officer, ecclesiastical or civil, was to be made, the Queen would inquire after the piety, integrity, learning of the man. And when she was satisfied in these qualifications, she would consider of his personage. And upon such an occasion she pleased once to say to me, *Bacon, how can the magistrate maintain his authority when the man is despised* [3] *?*

7. In eighty-eight, when the Queen went from Temple-bar along Fleet-street, the lawyers were ranked on one side, and the companies of the city on the other; said Master Bacon to a lawyer that stood next him, *Do but observe the courtiers; if they bow first to the citizens, they are in debt; if first to us, they are in law* [4].

8. King James was wont to be very earnest with the country gentlemen to go from London to their country houses. And sometimes he would say thus to them; *Gentlemen, at London you are like ships in a sea, which shew like nothing; but in your country villages you are like ships in a river, which look like great things.*

9. Soon after the death of a great officer, who was judged no advancer of the King's matters, the King said to his solicitor Bacon, who was his kinsman; Now

[1] See Preface, pp. 860, 861.

[2] Queen Elizabeth saw Sir Edward Dier in her garden, she looking out at window, and asked him in Italian, *What does a man think of when he thinks of nothing?* Sir Edward Dier, after a little pause, said in Italian, *Madam, of a woman's promise.* The Queen shrunk in her head and shut the window.—Lamb. MS. p. 21.

[3] My Lo. St. Albans hath often told me that Queen Elizabeth when she was to make a bishop or a great officer, besides his learning, piety, and integrity, she would have some respect to the person of the man.—Lamb. MS. p. 34.

[4] Lamb. MS. p. 35.

tell me truly, What say you of your cousin that is gone? Mr. Bacon answered, *Sir, since your Majesty doth charge me, I'll e'en deal plainly with you, and give you such a character of him, as if I were to write his story. I do think he was no fit counsellor to make your affairs better; but yet he was fit to have kept them from growing worse.* The King said, *On my so'l, man, in the first thou speakest like a true man, and in the latter like a kinsman.*

10. King James, as he was a prince of great judgment, so he was a prince of a marvellous pleasant humour; and there now come into my mind two instances of it.

As he was going through Lusen by Greenwich, he asked what town it was? They said Lusen. He asked a good while after, What town is this we are now in? They said, still 'twas Lusen. *On my so'l,* said the King, *I will be King of Lusen* [5].

11. In some other of his progresses, he asked how far it was to a town whose name I have forgotten. They said, *Six miles.* Half an hour after, he asked again One said, *Six miles and a half.* The King alighted out of his coach, and crept under the shoulder of his led horse. And when some asked his Majesty what he meant; *I must stalk* (said he), *for yonder town is shy and flies me* [6].

12. Count Gondomar sent a compliment to my Lord St. Albans, wishing him a good Easter. My Lord thanked the messenger, and said, He could not at present requite the Count better than in returning him the like; *That he wished his Lordship a good Passover* [7].

13. My Lord Chancellor Elsmere, when he had read a petition which he disliked, would say, *What! you would have my hand to this now?* And the party answering, Yes; he would say further; *Well, so you shall. Nay, you shall have both my hands to it.* And so would, with both his hands, tear it in pieces [8].

14. I knew a wise man, that had it for a by-word, when he saw men hasten to a conclusion, *Stay a little, that we may make an end the sooner.*

15. Sir Francis Bacon was wont to say of an angry man who suppressed his passion, That *he thought worse than he spake;* and of an angry man that would chide, That *he spoke worse than he thought* [9].

16. He was wont also to say, *That power in an ill man was like the power of a black witch; she could do hurt, but no good with it.* And he would add, *That the magicians could turn water into blood, but could not turn the blood again to water.*

17. When Mr. Attorney Cook, in the Exchequer, gave high words to Sir Francis Bacon, and stood much upon his higher place; Sir Francis said to him, *Mr. Attorney, the less you speak of your own greatness, the more I shall think of it: and the more, the less* [10].

18. Sir Francis Bacon coming into the Earl of Arundel's garden, where there

[5] King James was going through Lusen by Greenwich. He asked what town it was. They said Lusen. He asked about half an hour after. 'Twas Lusen still. Said the king, *I will be king of Lusen.*—Lamb. MS. p. 84.

[6] He asked how far to a town. They said six miles. Half an hour after he asked again. One said six miles and an half. He lighted from his coach and crept under his horse's shoulder. Some asked him what his M. meant. He said he must stalk, for yonder town fled from him. —Lamb. MS. p. 84.

[7] Lamb. MS. p. 72. Gondomar, I presume, was about to return to Spain. I cannot believe that his message was meant for an insult, as has been supposed; though I can well believe that the popular hatred of Spain and everything Spanish was apt enough to put that construction upon it. But there are no traces of any unkindness between Gondomar and Bacon. These compliments may have been exchanged at Easter-tide in 1622. Easter-day fell on the 21st of April that year, and a new Spanish ambassador arrived a week after.—See *Court and Times of James I.*, ii. 309.

[8] The party would say an it like your Lp. He would answer, *you shall have both my hands to it,* and so would rend it.—Lamb. MS. p. 60.

[9] If one suppresseth his anger he thinks worse than he says; but when he chides, then he says worse than he thinks.—Lamb. MS. p. 24.

[10] When Mr Attorney Cooke gave in the Exchequer high words to Mr Bacon, he replied, *Mr. Attorney,* etc.—Lamb. MS. p. 7.

were a great number of ancient statues of naked men and women, made a stand, and as astonished, cried out, *The resurrection* [11].

19. Sir Francis Bacon (who was always for moderate counsels) when one was speaking of such a reformation of the Church of England as would in effect make it no Church; said thus to him, *Sir, the subject we talk of is the eye of England; and if there be a speck or two in the eye, we endeavour to take them off; but he were a strange oculist who would pull out the eye.*

20. The same Sir Francis Bacon was wont to say, *That those who left useful studies for useless scholastic speculations, were like the Olympic gamesters, who abstained from necessary labours, that they might be fit for such as were not so.*

21. He likewise often used this comparison; *The Empirical philosophers are like to pismires; they only lay up and use their store. The Rationalists are like to spiders; they spin all out of their own bowels. But give me a philosopher, who like the bee, hath a middle faculty, gathering from abroad, but digesting that which is gathered by his own virtue.*

22. The Lord St. Alban, who was not over hasty to raise theories, but proceeded slowly by experiments, was wont to say to some philosophers who would not go his pace, *Gentlemen, Nature is a labyrinth, in which the very haste you move with, will make you lose your way.*

23. The same Lord, when he spoke of the Dutchmen, used to say, *That we could not abandon them for our safety, nor keep them for our profit.* And sometimes he would express the same sense on this manner; *We hold the Belgic lion by the ears* [12].

24. Sir Francis Bacon said upon occasion (meaning it of his old retinew) *That he was all of one piece: his head could not rise but his tail must rise too* [13].

25. The Lord Bacon was wont to commend the advice of the plain old man at Buxton, that sold besoms. A proud lazy young fellow came to him for a besom upon trust; to whom the old man said, *Friend, hast thou no money? borrow of thy back, and borrow of thy belly; they'll ne'er ask thee again, I shall be dunning thee every day* [14].

26. Solon said well to Crœsus, (when in ostentation he shewed him his gold) *Sir, if any other come that has better iron than you, he will be master of all this gold.*

27. Jack Weeks said of a great man (just then dead) who pretended to some religion, but was none of the best livers, *Well, I hope he is in heaven. Every man thinks as he wishes; but if he be in heaven, 'twere pity it were known* [15].

[11] My Lo. St. Albans coming into the Earl of Arundel's garden where there were many statues of naked men and women, made a stand and said, "*The resurrection.*"—Lamb. MS. p. 65.

[12] My Lo. St. Albans was wont to say that it was our greatest unhappiness, that we could not abandon those for our safety who were the greatest enemies to our profit.—Lamb. MS. p. 85.

[13] So Lamb. MS. p. 5. In the Baconiana it is given thus: "The same Lord when a gentleman seemed not much to approve of his liberality to his retinue, said to him, *Sir, I am all of a piece; if the head be lifted up, the inferior parts of the body must too*". It will be observed that Rawley's notes of these apophthegms are in almost every case better than Dr. Tenison's version, by whom they have evidently been dressed for company. In this case I thought the improved version too bad, and made the note and the text change places. That such an alteration could have been sanctioned by Bacon is utterly incredible.

[14] The old man at Buxton that answered him that would have been trusted for brooms: Hast thou no money? borrow of thy back and borrow of thy belly: they'll ne'er ask thee again: I shall be ever asking thee.—Lamb. MS. 5.

[15] Jack Weeks said of the Bishop of London, Montagu; I hope he is in heaven. Every man thinks as he wisheth; but if he be there 'twere pity it were known.—Lamb. MS. p. 55.

SOME ADDITIONAL APOPHTHEGMS

SELECTED FROM A COMMON-PLACE BOOK IN THE HAND-WRITING OF DR. RAWLEY, PRESERVED AT LAMBETH.

MSS. No. 1034[1].

[THE manuscript from which the following apophthegms are selected bears no date or title. But the contents show that it was a common-place book in which Dr. Rawley entered memoranda from time to time; and a few dates occur incidentally; the earliest of which is 8 September 1626, (five months after Bacon's death), and the latest is 25 May 1644. The memoranda are of various kinds, many of them relating to Bacon and his works, many to Dr. Rawley's private affairs. Among them are a number of anecdotes, some very good, but not stated to be derived from Bacon or otherwise connected with him, and therefore not noticed here. It is true that several of the apophthegms printed by Tenison in the *Baconiana* are set down in this manuscript without any hint that Bacon had anything to do with them. It is possible therefore that they too may have been of Dr. Rawley's own selection; who seems to have had a taste for good stories, and seldom spoiled them. But judging by the style, I think it more probable that most of them were copied from Bacon's own notes.]

1. Apophthegms. My Lo.[2]: I was the justest judge that was in England these 50 yeares: But it was the justest censure in Parliament that was these 200 yeares.

2. The same Mr. Bacon[3] went towards Finchley to take the air. There had been growing not long before a pretty shady wood. It was then missing: Said Mr. Bacon, Stay, I've not lost my thoughts in a wood, but methinks I miss a wood here. Saith a country fellow, It is newly cut down. Said Mr. Bacon, Sure he was but a churl that ought it, to cut down a wood of great pleasure and to reap but small profit into his purse. Said the fellow, It was the Bishop of London[4]. Then answered Mr. Bacon, Oh, was it he: he's a learned man: it seems this was an obscure place before, and the Bishop hath expounded the text.

3. A flattering courtier undertook to make a comparison betwixt my Lord St. Alban and Treasurer Cranfield. Said he, My Lord St. Alban had a pretty turning wit, and could speak well: but he wanted that profound judgment and solidity of a statesman that my Lord of Middlesex hath. Said a courtier that stood by: Sir I wonder you will disparage your judgment so much as to offer to make any parallel betwixt these two. I'll tell you what: when these two men shall be recorded in our chronicles to after ages, men will wonder how my Lord St. Alban could fall; and they will wonder how my Lord of Middlesex could rise.

4. There was one would say of one that he thought every man fit for every place[5].

[1] See above, p. 855.

[2] That is, "my Lord St. Alban said of himself". This is the first entry in the book, and is set down in a kind of cipher; the consonants being written in Greek characters, and the six vowels represented by the six numerals; 1=a; 2=e; 3=i; 4=o; 5=u; 6=y.

[3] In the MS. this follows the story of Bacon and the fishermen at Chelsea. Rawley's Collection, No. 36.

[4] Bishop Aylmer, probably; who died in 1594. See Nichols's Progr. Eliz. iii. p. 369.

[5] This sounds to me very like a note of Bacon's; though his name is not mentioned.

5. My Lord Chancellor told the King, that if he bestowed 7000*l.* upon Paul's steeple, he could not lay out his money where it should be more seen.

6. When they sat in commission about re-edifying Paul's steeple, some of the rich aldermen being there, it was motioned to build a new spire upon it. A rich alderman answered; My Lords, you speak of too much cost: Paul's is old: I think a good cap would do well. My Lord Chancellor, who was for the spire, answered: Mr. Alderman, you that are citizens are for the cap; but we that are courtiers are for the hat and feather.

7. [There was] an old woman whom the minister asked, How many commandments there were. She answered, it was above her learning: she was never taught it. Saith the minister, there are ten. Good Lord (said the old woman) a goodly company. He told them her particularly, and then asked her if she had kept them all? Kept them? (said she): alas master, I am a poor woman: I have much ado to keep myself.

8. Sir Harry Mountague came to my Lord Chancellor before he went to the court to Newmarket, and told him; My Lord, I come to do my service to your Lordship: I am even going to Newmarket and I hope to bring the staff [6] with me when I come back. My Lord (said my Lord Chancellor) take heed what you do: I can tell you wood is dearest at Newmarket of any place in England.

9. When the said Lord lost his Treasurer's place, he came to my Lord St. Alban, and told him how they had used him; that though they had taken away the Lord Treasurer's place, yet they had made him Lord President of the Counsel: Why, saith my Lord St. Alban, the King hath made me an example and you a president [7].

10. When Sergeant Heale who is known to be good in giving in evidence, but otherwise unlearned in the law, was made the Queen's sergeant, Mr. Bacon said; The Queen should have a sergeant *de facto et non de jure*.

11. At the King's Bench bar, Sergeant Heale, before he was the Queen's sergeant, contended with Mr. Bacon to be first heard; and said, Why I am your ancient: Mr. Bacon gently answered, Not in this place; for I staid here long, and you are come but right now.

12. There was a tall gentleman and a low gentleman were saying they would go to the Shrive's to dinner; Go, saith the one, and I will be your shadow. Nay, said the other, I will be your shadow. Mr. Bacon standing by said, I'll tell you what you shall do: Go to dinner and supper both; and at dinner when [the shadows are] shorter than the bodies, you shall be the shadow; and at supper you shall be the other's shadow [8].

13. He thought Moses was the greatest sinner that was, for he never knew any break both tables at once but he [9].

14. He said he had feeding swans and breeding swans; but for malice, he thanked God, he neither fed it nor bred it [10].

15. At the Parliament, when King James spied Mr. Gorge, one of my Lord Chancellor's men, who was somewhat fantastical, and stood by there with one rose white and another black; the King called my Lord unto him, and said easily in his ear; My Lord Chancellor, why does your man yonder wear one rose white and another black? My Lord answered; In truth, Sir, I know not, unless it be that his mistress loves a colt with one white foot.

16. Sir Walter Coape and Sir Francis Bacon were competitors for the Mastership of the Wards. Sir Francis Bacon certainly expecting the place had put

[6] The Lord Treasurer's staff.

[7] So *precedent* was usually spelt in those days.

[8] So the MS. It should be "the other shall be your shadow".

But the thing is better told in a common-place book of Bacon's own (Harl. MSS. 7017.). "The two that went to a feast both at dinner and supper, neither known, the one a tall, the other a short man; and said they would be one another's shadows. It was replied, it fell out fit: for at noon the short man might be the long man's shadow and at night the contrary."

[9] This is written in cipher.

[10] This saying is alluded to by Rawley in his Life of Bacon.

most of his men into new cloaks. Afterward when Sir Walter Coape carried the place, one said merrily that Sir Walter was Master of the Wards, and Sir Francis Bacon of the Liveries.

17. My Lord St. Alban said, that wise nature did never put her precious jewels into a garret four stories high; and therefore that exceeding tall men had ever very empty heads [11].

18. My Lord St. Alban invited Sir Ed. Skory to go with him to dinner to a Lord Mayor's feast. My Lord sate still and picked a little upon one dish only. After they returned to York-house, my Lord wished him to stay and sup with him: and told him he should be witness of the large supper he would make: telling him withal: Faith, if I should sup for a wager, I would dine with a Lord Mayor.

19. Sir Robert Hitcham said, He cared not though men laughed at him: he would laugh at them again. My Lord St. Alban answered, If he did so he would be the merriest man in England.

20. My Lord St. Alban would never say of a Bishop *the Lord that spake last*, but *the Prelate that spake last*. King James chid him for it, and said he would have him know that the Bishops were not only *Pares*, as the other Lords were, but *Prælati paribus* [12].

21. He was a wise man [13] that gave the reason why a man doth not confess his faults. It is, *Quia etiam nunc in illis est*.

22. Will you tell any man's mind before you have conferred with him? So doth Aristotle in raising his axioms upon Nature's mind.

23. Old Lord Keeper Sir Nicholas Bacon had his barber rubbing and combing his head. Because it was very hot [14], the window was open to let in a fresh wind. The Lord Keeper fell asleep, and awakened all distempered and in great sweat. Said he to his barber, Why did you let me sleep? Why, my Lord, saith he, I durst not wake your Lordship. Why then, saith my Lord, you have killed me with kindness. So removed into his bed chamber and within a few days died.

24. Four things cause so many rheums in these days, as an old country fellow told my Lord St. Alban. Those were, drinking of beer instead of ale; using glass windows instead of lattice windows; wearing of silk stockings; missing of smoky chimneys.

25. King James and Gondomar were discoursing in Latin. The King spoke somewhat of Tully's Latin. Gondomar spoke very plain stuff. Gondomar laughed. The King asked him, Why he laughed? He answered, Because your Majesty speaks Latin like a scholar, and I speak Latin like a King.

26. Gondomar said, Compliment was too hot for summer, and too cold in winter. He meant it against the French.

27. King Henry the fourth of France having an oration offered him, and the orator beginning "Great Alexander"; said the King, Come let's begone.

28. The beggar, that instructed his son, when he saw he would not be handsome, said, You a beggar! I'll make you a ploughman.

29. Marquis Fiatt's first compliment to my Lord St. Albans was, That he reverenced him as he did the angels, whom he read of in books, but never saw [15].

[11] I have seen this quoted somewhere as Bacon's answer to King James when pressed for his opinion as to the capacity of a French ambassador who was very tall.

[12] This I think must be misreported. It must have been Bacon who defended himself on this ground for preferring "Prelate" to "Peer:" for so Prelate would imply Peer, whereas Peer would not imply Prelate.

[13] Seneca, Ep. 53.

[14] "The 4 of February [21 Eliz. i. e. 1578–9] . . . fell such abundance of snow, etc. . . . It snowed till the eight day and freezed till the tenth. Then followed a thaw, with continual rain a long time after . . . The 20 of February deceased Sir Nicholas Bacon." —*Stowe's Chronicle.*

[15] Bacon being ill and confined to his bed, so that though admitted to his room he could not see him. Compare Rawley's Life of Bacon, Vol. I. p. 16. Tenison (*Baconiana*, p. 101.) makes Fiatt say, "Your Lordship hath been to me *hitherto* like the angels, of

30. My Lord Chancellor Ellesmere's saying of a man newly married; God send him joy, and some sorrow too, as we say in Cheshire. The same my Lord St. Alban said of the Master of the Rolls.

31. My Lord St. Alban said, when Dr. Williams, Dean of Westminster, was made Lord Keeper; I had thought I should have known my successor.

32. My Lord St. Alban having a dog which he loved sick, put him to a woman to keep. The dog died. My Lord met her next day and said, How doth my dog? She answered in a whining tone, and putting her handkerchief to her eye, The dog is well, I hope.

33. The physician that came to my Lord after his recovery, before he was perfectly well, the first time, he told him his pulse was broken-paced; the next time, it tripped; the third day, it jarred a little. My Lord said, he had nothing but good words for his money.

34. Mr. Anthony Bacon chid his man (Prentise) for calling him no sooner. He said, It was very early day. Nay, said Mr. Bacon, the rooks have been up these two hours. He replied, The rooks were but new up: it was some sick rook that could not sleep.

35. [The following is not given in any of these collections, but comes from a letter of Mr. John Chamberlain to Sir Dudley Carleton, 11. Oct. 1617. See *Court and Times of James I.*, ii. p. 38.]

The Queen lately asked the Lord Keeper [Sir F. Bacon], What occasion the Secretary [Sir R. Winwood] had given him to oppose himself so violently against him: who answered prettily, "Madam, I can say no more, but he is proud, and I am proud".

which I have often heard and read, but never saw them *before*" (the words "hitherto" and "before" being his own interpolation, and entirely spoiling the story); and proceeds, "To which piece of courtship he returned such answer as became a man in those circumstances, 'Sir, the charity of others does liken me to an angel, but my own infirmities tell me I am a man;'" of which reply there is no hint in Rawley, either in the common-place book or in the life: an addition, I suspect, by a later hand.

INDEX

Butler & Tanner, The Selwood Printing Works, Frome, and London.

For Product Safety Concerns and Information please contact our EU representative GPSR@taylorandfrancis.com
Taylor & Francis Verlag GmbH, Kaufingerstraße 24, 80331 München, Germany

www.ingramcontent.com/pod-product-compliance
Ingram Content Group UK Ltd.
Pitfield, Milton Keynes, MK11 3LW, UK
UKHW021954180926
475701UK00003B/29
* 9 7 8 0 4 1 5 6 7 9 6 0 2 *